D1258028

CALCULUS

EAGLE MATHEMATICS SERIES

A series of textbooks for an undergraduate program in mathematical analysis

PLANNED AND EDITED BY

Ralph Abraham UNIVERSITY OF CALIFORNIA AT SANTA CRUZ

Ernest Fickas HARVEY MUDD COLLEGE

Jerrold Marsden UNIVERSITY OF CALIFORNIA AT BERKELEY

Kenneth McAloon UNIVERSITÉ DE PARIS

Michael O'Nan RUTGERS UNIVERSITY

Anthony Tromba UNIVERSITY OF CALIFORNIA AT SANTA CRUZ

VOLUMES

1AB *Calculus of Elementary Functions*

1BCD *Calculus*

1BC *Calculus of One Variable*

2A *Linear Algebra*

Volume 1BCD

CALCULUS

Kenneth McAloon

Université de Paris

Anthony Tromba

University of California at Santa Cruz

With the assistance of Jerrold Marsden, Paul King, Harold Abelson, Ellis Cooper, Gerald Grossman, and Lee Rudolph

HARCOURT BRACE JOVANOVICH, INC.

New York Chicago San Francisco Atlanta

ISBN: 0-15-518530-6

Library of Congress Catalog Card Number: 76-186226

Printed in the United States of America

PREFACE

The introductory calculus course has come to assume an unusual number of burdens. The student must master a bewildering variety of problems, and the instructor is expected to communicate the spirit and cogency of modern mathematical rigor. This book, which is the product of a collaboration of professors and undergraduates, is designed to help the instructor to teach the course coherently and to allow the student to develop strong intuitions. All the essential topics of the basic course are covered. We have emphasized geometric understanding, because this is a form of rigorous thinking with which the student is already familiar. The result is a presentation of the calculus which is at times very classical and, other times, somewhat innovative.

The text begins with a discussion of functions and elementary analytic geometry. A special section on "Unsolvable Problems of Analytic Geometry" anticipates the study of the calculus. We introduce limits first as a technique for solving geometric problems involving moving points and then lead the reader to a rigorous definition of limit in terms of open interval neighborhoods. The proofs of basic limit theorems are relatively simple when formulated in terms of neighborhoods, and the abundance of illustrations truly enables the student to "see" the proof. The geometric intuition about limits makes the transition to computing tangents and the derivative very natural and the student can appreciate the many applications of these ideas.

We assume that the reader knows high-school algebra, geometry, and the rudiments of trigonometry. For the sake of simplicity, we have kept set theory to a minimum; about the only sets used are intervals.

Functions are defined as rules and are written using the elegant $f: x \mapsto f(x)$ notation. Occasionally, we use the symbol $\Rightarrow$ to signify implication; we indicate the end of a proof by ∎. From the onset, in addition to the rational functions and various ad hoc examples, we use the trigonometric functions in our discussions (for reference, basic facts about these functions are displayed in §8.1). As a result, there is a much broader range of available examples which, in turn, reinforces the case for the precise formulation of the notions of function, limit, derivative, and so on.

The presentation of the definite integral also begins in an intuitive setting (Chapter 6). Working from a geometric definition based on the concept of area, we carefully set out the basic properties of the integral which are needed to prove the Fundamental Theorem of Calculus. This important result is established and then applied to area and volume problems. To facilitate working with integrals, we include an entire section on functions defined by integrals.

In Chapter 7 we are ready to turn to the task of rigorously defining the definite integral in terms of Riemann sums. There, when the integral is considered as a limit of sums, the basic properties needed to prove the Fundamental Theorem are established precisely. Finally, we use the method of Riemann sums for applications which expressly require it such as arc length, surface area, and work.

Chapter 8 is a thorough study of the transcendental functions and Chapter 9 treats the various techniques of integration. If, for certain courses, it is desirable to teach these topics early, Chapters 8 and 9 can precede Chapter 7.

The next three chapters are, for all practical purposes, independent of one another and are not required for vector calculus. Chapter 10 covers infinite series; the final section, §10.6, deals with sequences and series of functions and is intended for more advanced students. In Chapter 11 we introduce the study of ordinary differential equations and rely extensively on physical examples to explain how such equations arise (the topic of oscillations is treated at some length). Although we use only elementary methods to handle the first-order linear equation and the second-order linear equation with constant coefficients, we do apply the method of power series to second-order linear equations with variable coefficients (thus, §11.6 requires Chapter 10).

Chapter 12, Numerical Methods, is of special interest to courses with students from the computer sciences. The topics we consider here include polynomial interpolation, numerical quadrature (the section on Gaussian quadrature is optional), finite difference interpolation, and differential equations. In addition, there is a formula for numerical differentiation based on Stirling's interpolation formula. This subject matter lends itself remarkably well to a course at this level, demonstrating the wide applicability of elementary calculus in a most effective way.

We begin multivariable calculus in Chapter 13 with a thorough discussion of vectors, curves in space, and the path integral. (For those instructors who wish to introduce the notion of vectors earlier, the two optional sections, §4.3 and §4.4, treat vectors in the plane and plane curves.) The study of functions of several variables entails an increase in geometric emphasis; in order to exploit the available geometry, we first discuss real-valued functions of two variables, their differentiability, partial derivatives, and so on. We then proceed to functions from $\boldsymbol{R}^3$ to $\boldsymbol{R}$, parametrized surfaces, and vector fields; there is an optional section on Taylor's Theorem and extreme values of functions from $\boldsymbol{R}^2$ to $\boldsymbol{R}$.

A brief intuitive discussion introduces the notions of the multiple integral and iterated integral. Hence, the development parallels that of Chapter 6, where the integral was first studied. Our next step is to give a rigorous definition of the integral in terms of Riemann sums and to present the Fubini Theorem. By using an elementary concept of "area zero," we are able to give the discussion a measure-theoretic flavor. This approach, which was already anticipated in Chapter 7, greatly streamlines the work and provides valuable intuition for many subsequent topics, for example, the Change of Variables Theorem. We then apply the multiple integral to problems of surface area and surface integrals; there are also two very instructive (optional) sections devoted to centers of mass and centers of gravity.

Chapter 17, Vector Analysis, is the culmination of the book. It is devoted to a lucid, but rigorous, development of the theorems of Green, Stokes, and Gauss. The statements and proofs of the theorems are carefully formulated to reflect the striking similarity of the results. There are many examples from physics and the final section contains a discussion of the wave equation and Maxwell's equations.

The exercises—perhaps the most important feature of a calculus book—are plentiful and varied. There is an exercise set at the end of each section and review exercises for each chapter. The exercises are tightly integrated to the text; when possible, we have included problems which illustrate the uses of calculus in different scientific areas. We denote more challenging problems with an asterisk. Many exercises contain instructive hints, and selected answers are to be found at the back of the book. There are also tables of values of the transcendental functions and comprehensive listings of derivatives and integrals in the backmatter.

The authors would like to express their especial appreciation to the editor of this volume, Miss Marilyn Davis, whose searching comments and diligence were integral to their work, and also to Mr. Harry Rinehart for his design of the book.

Kenneth McAloon
Anthony Tromba

CONTENTS

1

FUNCTIONS, 1

2

LIMITS, 63

3

DIFFERENTIAL CALCULUS, 101

4

FURTHER RESULTS OF DIFFERENTIAL CALCULUS, 141

5

APPLICATIONS OF DIFFERENTIAL CALCULUS, 215

6

INTEGRAL CALCULUS, 267

7

APPLICATIONS OF THE DEFINITE INTEGRAL, 313

8

TRANSCENDENTAL FUNCTIONS, 371

9

TECHNIQUES OF INTEGRATION, 427

10

INFINITE SERIES, 461

11

ELEMENTARY DIFFERENTIAL EQUATIONS, 511

12

NUMERICAL METHODS, 549

13

VECTORS IN SPACE, 599

14

DIFFERENTIATION, 643

15

MULTIPLE INTEGRATION, 705

16

FURTHER RESULTS OF MULTIPLE INTEGRATION, 751

17

VECTOR ANALYSIS, 779

APPENDIX: MATHEMATICAL TABLES

1

FUNCTIONS

In the modern treatment of calculus, function and limit are central concepts. The idea of function is fundamental in mathematics, and in the applications of mathematics to the physical and social sciences. In this chapter we discuss functions and some additional concepts that will be used throughout the book. We present a brief exposition of analytic geometry—by no means an exhaustive one, but more than sufficient for our needs. In the closing section we introduce some problems that motivate the study in the rest of the book. Limits are considered in Chapter 2.

1 INTERVALS AND FUNCTIONS

Analytic geometry deals with techniques of describing geometric facts in the language of numbers. The simplest example of this is the way we associate a **real number** with each point on a straight line. We may draw a "real number line" as in Figure 1-1. Every point on the line represents a

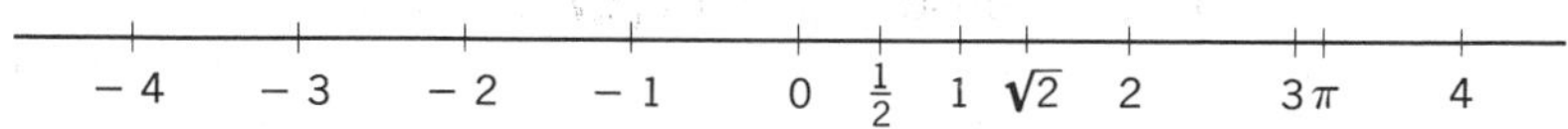

FIGURE 1-1

different real number, and each real number is represented by a different point on the line. We, therefore, can speak of real numbers as points. When we refer to the **real line**, it is this picture that we have in mind.

As the figure indicates, the **integers**, 0, 1, 2, ..., -1, -2, -3, ... correspond to an evenly spaced sequence of points on the real line. The other

rational numbers, that is, numbers that can be expressed as fractions with integers in both the numerator and the denominator, are also points on the line. The rational numbers do not exhaust the points on the real line, however; numbers corresponding to the remaining points are called **irrational numbers**. Examples of irrational numbers are π, $\sqrt{2}$, and $\sqrt[3]{37}$.

In the above paragraph we defined certain collections of real numbers —the integers, the rational numbers, the irrational numbers. Such collections are called **sets**. In general, a **set** is any collection of objects and the members of a set are called its **elements**. For example the collection of real numbers itself is a set and we denote this set by $\boldsymbol{R}$. The number π is a real number and π is an element of $\boldsymbol{R}$. The set of natural numbers is the set whose elements are $1, 2, 3, \ldots$; we denote this set by $\boldsymbol{N}$. The set of natural numbers coincides with the set of positive integers.

To indicate that the element x is a member of the set X we write $x \in X$; when x is not an element of the set X we write $x \notin X$. Thus, for example, we have $6 \in \boldsymbol{N}$, since 6 is a natural number; we also have $6 \in \boldsymbol{R}$, since 6 is a real number. However, $-\sqrt{2}$ is not a natural number, so we have $-\sqrt{2} \notin \boldsymbol{N}$; it is still the case that $-\sqrt{2} \in \boldsymbol{R}$.

Given two real numbers a and b, with $a < b$ (read "a less than b"), we call that portion of the real line between a and b the **open interval** from a to b, and we write $]a, b[$. Neither a nor b is an element of $]a, b[$; only those numbers which lie between them are elements of this open interval. Algebraically, the open interval $]a, b[$ may be described as the set of all real numbers c which satisfy the two inequalities

$$a < c \quad \text{and} \quad c < b$$

The numbers a and b are called the **endpoints** of the open interval $]a, b[$.

If we "tack on" the endpoints of the open interval, that is, if we consider the set of all real numbers c satisfying the inequalities $a \leqslant c$ and $c \leqslant b$ (read "a less than or equal to c" and "c less than or equal to b"), then we call the resulting set of real numbers the **closed interval** from a to b, and write $[a, b]$. Again, the numbers a and b are called the endpoints of the interval. It is important to notice that a closed interval contains both of its endpoints, while an open interval contains neither.

We also define the **half-open intervals** $[a, b[$ and $]a, b]$. The first set is just $]a, b[$ with the extra point a included, and the second is $]a, b[$ with the number b included. A half-open interval contains exactly one of its two endpoints, either a or b.

We have assigned a meaning to the symbols $[a, b]$, $]a, b[$, $]a, b]$, and $[a, b[$ only when a and b are real numbers with $a < b$. We shall also find it convenient to define the following similar symbols:

$]-\infty, a[$ means "the set of all real numbers less than a";
$]-\infty, a]$ means "the set of all real numbers less than or equal to a";
$]a, \infty[$ means "the set of all real numbers greater than a";

$[a, \infty[$ means "the set of all real numbers greater than or equal to a"; and
$]-\infty, \infty[$ means "the set of all real numbers."

Geometrically, $]a, \infty[$ represents that portion of the real number line which lies to the right of a, not including a itself; similar geometric pictures can be drawn for the other "infinite intervals." We assign no meaning to the symbols $-\infty$ and ∞ when occurring alone, nor do we write symbols like $[-\infty, a[$ or $[a, \infty]$. The intervals $]-\infty, a[$, $]a, \infty[$, and $]-\infty, \infty[$ are all **open**; the intervals $]-\infty, a]$ and $[a, \infty[$ are **half open**. The number a is the **only** endpoint of the intervals $[a, \infty[$, $]a, \infty[$, $]-\infty, a]$, and $]-\infty, a[$. The interval $]-\infty, \infty[$ has no endpoints.

Example 1 The interval $]3, \pi[$ contains the real numbers 3.1, 314/1000, and $\pi - 0.001$, but does not contain the real numbers 3, π, $\pi + 0.0001$, or 117. The interval $]0, \infty[$ contains all positive real numbers; the interval $[0, \infty[$ contains all nonnegative real numbers. The interval $]-7, 2]$ has endpoints -7 and 2, and it contains 2 but not -7.

Example 2 If J is the interval $[-1, 1]$, then J contains -1, 0, 0.5, and 1. If K is the interval $]-\infty, -2]$, then K contains any real number less than or equal to -2. For instance, -12 belongs to K.

Example 3 None of the following are meaningful symbols for intervals: $]-3, -5[$, $[6, 0]$, and $]3, \infty]$.

If I is any interval and c is a point in I, then c is an element of I, and we have $\boldsymbol{c \in I}$. If c is not a point in I, then we have $\boldsymbol{c \notin I}$.

Example 4 According to Example 1, $3.1 \in]3, \pi[$, but $3 \notin]3, \pi[$ and $\pi \notin]3, \pi[$.

Example 5 For any real number c, we have $c \in]0, \infty[$ if and only if $c > 0$; and we have $c \in]a, b[$ if and only if $a < c < b$.

Any open interval containing the real number c is called a **neighborhood** of c.

Example 6 The open interval $]0, 10[$ is a neighborhood of 5, 7, and $\sqrt{46}$, but not of 0, -35, or $\sqrt{100}$.

Example 7 The infinite interval $]-\infty, a[$ is a neighborhood of any real number less than a; $]-\infty, \infty[$ is a neighborhood of every real number.

Example 8 The interval $]3, 5]$ is not a neighborhood of 4, although $4 \in]3, 5]$, since $]3, 5]$ is not an open interval.

Example 9 If c is any real number, and if e is a **positive** real number, then $]c - e, c + e[$ is a neighborhood of c.

The **intersection** of two sets A and B, denoted by $A \cap B$, is the set of all elements belonging to both A and B.

Example 10 The intersection of the open intervals $]0, 2[$ and $]1, \pi[$ is the open interval $]1, 2[$ (Figure 1-2).

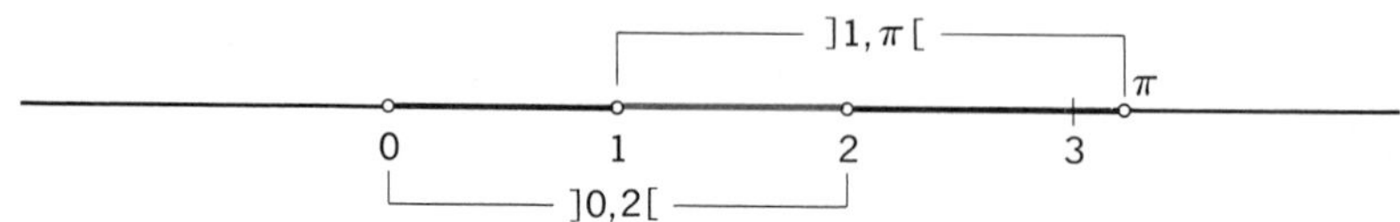

FIGURE 1-2

If two sets A and B have no element in common, we say that A and B are **disjoint**

Example 11 The intervals $]0, \pi[$ and $]\pi, 4[$ are disjoint, as are the intervals $]-1, 0[$ and $]5, 8[$, but the intervals $]0, \pi]$ and $[\pi, 4[$ are not disjoint.

In general, two open intervals are either disjoint or else their intersection is again an open interval (Figure 1-3).

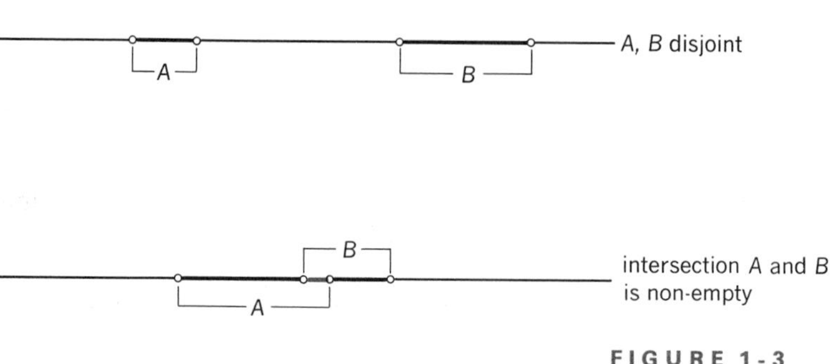

FIGURE 1-3

Example 12 The intervals $]0, 1[$ and $]1, 2[$ are disjoint, but the intersection of $]-\frac{1}{2}, \frac{1}{2}[$ and $]0, 1[$ is the open interval $]0, \frac{1}{2}[$.

If two open intervals I and J are both neighborhoods of the point c, then I and J are not disjoint, since c is an element of both of them. Hence $I \cap J$ is also an open interval containing c; that is, $I \cap J$ is also a neighborhood of c. Thus, the intersection of two neighborhoods of a point c is also a neighborhood of c.

We next discuss a way to relate the elements of one set to the elements of another set. We begin with the definition of **function**.

Definition Given sets X and Y, a rule which assigns to each element of X exactly one element of Y is called a **function** from X to Y. The set X to whose elements the rule is applied is called the **domain** of the function.

Example 13 The rule which assigns to each positive real number the volume of the sphere whose radius is that number is a function whose domain is $]0, \infty[$.

Example 14 The rule which assigns to each physical object its weight is a function whose domain is the set of all physical objects.

Example 15 The rule which assigns to every mirror its index of reflection is a function whose domain is the set of all mirrors.

Example 16 The rule which assigns to the current through a wire the energy dissipated by the wire is a function whose domain is the set of all possible currents which can pass through the wire.

Example 17 The rule which to every nonzero real number assigns its reciprocal is a function whose domain is the set $]0, \infty[$ plus $]-\infty, 0[$.

Example 18 The rule which to every real number assigns its square is a function whose domain is $\boldsymbol{R}$.

Example 19 The rule which to every element of a given set X assigns the element itself is the **identity function** on X, denoted by id_X, and has domain X.

If f is a function from X to Y, we write $\boldsymbol{f: X \rightarrow Y}$. If $x_0 \in X$, the element assigned to x_0 by the function f is called the **value of $\boldsymbol{f}$ at $\boldsymbol{x_0}$** and is denoted $\boldsymbol{f(x_0)}$. We say that f **sends** x_0 to $f(x_0)$.

We emphasize that when f is a function and x_0 is an element of its domain, then $f(x_0)$ is uniquely determined, because a function assigns **precisely one value to each element** of its domain. However, a function might assign the same value to different elements of the domain. Thus, in Example 18, the elements 2 and -2 are both assigned the value $2^2 = (-2)^2 = 4$.

The set of values of a function is called the **range** of the function. A function is **real-valued** if its range contains only real numbers. A function whose domain and range both contain only real numbers is called a **real function**. A function is a **constant function** if its range consists of a single element.

Example 20 The range of the real function which sends x to x^2 is $[0, \infty[$. The range of the constant real function sending every real number to π contains the single element π. The range of the sine function is $[-1, 1]$. The range of the tangent function is all of $\boldsymbol{R}$.

In order to define a function we must specify the domain and give the rule for assigning values to the elements of the domain. In Example 18 the domain is $\boldsymbol{R}$ and the rule is to send x to its square; this rule is denoted by $x \mapsto x^2$. In Example 17 the domain is the set of nonzero real numbers and the rule is $x \mapsto 1/x$. In general, we denote the rule of a function f by $x \mapsto f(x)$; we also often write $f: x \mapsto f(x)$. For example, if f sends x to x^2, we can write $f: x \mapsto x^2$.

For real numbers a rule such as $x \mapsto 1/x$ has meaning only when x is not equal to 0, since $1/0$ is meaningless. Similarly, $x \mapsto x/(x-1)$ has meaning only when x is not equal to 1. Whenever we give a rule for a function by means of an algebraic or trigonometric expression, we shall assume (unless otherwise stated) that the domain of the function is the set of all real numbers for which the expression is meaningful. Thus, when we speak of the function $x \mapsto 1/(x^2-1)$ we assume the domain to be all real numbers except 1 and -1. Strictly speaking therefore the functions $x \mapsto 1$ and $x \mapsto x/x$ are not the same, since 0 is not in the domain of the latter.

Given a function $f: x \mapsto f(x)$ we often substitute y for $f(x)$ and speak of the function as being given by the equation $y = f(x)$. The letters x and y are called **variables** since x may denote any element in the domain of f and y any element in the range. Since the element of the range satisfying the equation is determined as soon as an element of the domain is selected, we sometimes say that y depends on x.

Defining a function simply by giving the equation $y = f(x)$ of its graph is often both elegant and efficient because it accomplishes several tasks simultaneously.

(1) The expression $f(x)$ gives the rule for the function;
(2) "x" is established as notation for a typical element of the domain of f;
(3) "y" is agreed upon as notation for a typical element of the range;
(4) when the letters x and y both appear in subsequent discussion, it is assumed that y equals $f(x)$, and this convention carries over to pairs of fixed numbers such as (x_0, y_0) and (x_1, y_1);
(5) the equation form corresponds to our methods of graphing.

We should emphasize that, as variables, x and y are just standins for elements of the domain and range of the function. Therefore, they could be replaced by any letters (for example, r and s, or u and v) without altering the actual function. Thus, the function given by $y = x^2$ is the same as that defined by $r = s^2$ or $v = u^2$, where "r" and "s," or "u" and "v" are the symbols chosen for variables in the last two equations. Similarly, the rules $x \mapsto x^2$, $s \mapsto s^2$, $u \mapsto u^2$, ... all define the same function.

All the functions we have discussed so far have had a common feature: The same formula was used to evaluate the function at all elements of the domain. This need not always be the case, as the functions in the following examples illustrate.

Example 21 The **signum function, sgn**, has domain $\boldsymbol{R}$ and sends negative real numbers to -1, sends 0 to 0, and sends positive real numbers to $+1$. We express this tripartite rule by the following.

$$x \mapsto \begin{cases} -1, & x < 0 \\ 0, & x = 0 \\ 1, & x > 0 \end{cases}$$

Alternatively, we may also write

$$\operatorname{sgn} x = \begin{cases} -1, & x < 0 \\ 0, & x = 0 \\ 1, & x > 0 \end{cases}$$

Example 22 The **absolute value function** has domain $\boldsymbol{R}$ and its rule is

$$x \mapsto |x| = \begin{cases} -x, & x < 0 \\ x, & x \geqslant 0 \end{cases}$$

Example 23 The following rule,

$$t \mapsto \begin{cases} t, & t \geqslant 0 \\ t-1, & t \leqslant 0 \end{cases}$$

does not define a function because a unique value is not assigned to 0. However, the rule,

$$t \mapsto \begin{cases} t^2-1, & t \geqslant 1 \\ 0, & t \leqslant 1 \end{cases}$$

does define a function: since $1^2-1=0$, the two formulas give the same function value at the overlapping point 1.

EXERCISES

1. The interval $]1,2[$ is an open neighborhood of which of the following numbers: $1\frac{1}{2}$, 7, 2, $\frac{7}{4}$, $1+1/n$, where $n \in \boldsymbol{N}$?
2. Suppose $y \in]a,b[$. Show by a geometric picture and then algebraically (a) $](y+a)/2,(y+b)/2[$ is a neighborhood of y and (b) a subset of $]a,b[$.
3. Suppose $x_0 \in]c,d[$. Show that $]c,2x_0-c[$ and $]2x_0-d,d[$ are neighborhoods of x_0.
4. Prove that the intersection of two neighborhoods of a point x_0 is again a neighborhood of x_0.
5. State the domains of each of the following real-valued functions:

   (a) $x \mapsto x$ (b) $\theta \mapsto \sin\theta$

(c) $\alpha \mapsto \tan \alpha$ (d) $y \mapsto \dfrac{y^3 - 2}{y^2 - 3y + 2}$

(e) $x \mapsto -\sqrt{x}$ (f) $z \mapsto \sqrt[3]{z^2}$

6. Give the ranges of the following functions, all of whose domains are R:

(a) $x \mapsto x$ (b) $\theta \mapsto \cos\theta$

(c) $z \mapsto z^2$ (d) $q \mapsto \dfrac{1}{q^4 + 1}$

(e) $r \mapsto 6$

7. Compute the value of each of the following functions at the points indicated:

(a) $x \mapsto x$ at 0, 2, and π
(b) $\theta \mapsto \sin\theta$ at 0, $\frac{1}{4}\pi$, and $-\pi$
(c) $q \mapsto \tan q$ at 0, $\frac{1}{4}\pi$, and $-\pi$
(d) $y \mapsto \dfrac{y^3 - 2}{y^3 - 3y + 2}$ at 0, 2, and π
(e) $x \mapsto \sqrt{x}$ at 0 and 4
(f) $z \mapsto \sqrt[3]{z}$ at -1, 0, and 81
(g) $r \mapsto 6$ at 0 and π

8. Which of the following rules do not define functions? Why not?

(a) $u \mapsto \begin{cases} 1/u, & u \neq 0 \\ 0, & u = 0 \end{cases}$

(b) $q \mapsto \begin{cases} \dfrac{q - 1}{q^2}, & q \neq 0 \\ \dfrac{q^2 - 1}{q^3 - q^2}, & q < 0 \end{cases}$

(c) $x \mapsto \begin{cases} 4\pi x^2, & x \geqslant 0 \\ -1, & x \leqslant 0 \end{cases}$

(d) $f(x) = \begin{cases} |x|, & x \leqslant 0 \\ -x, & x \geqslant -1 \end{cases}$

(e) $z \mapsto \begin{cases} \sin z, & z \leqslant \frac{1}{2}\pi \\ 1, & \frac{1}{2}\pi \leqslant z \leqslant \pi \\ \cos z, & z \geqslant \pi \end{cases}$

(f) $g(a) = \begin{cases} ha, & 0 \leqslant a \leqslant |h| \\ a, & |h| \leqslant a, \end{cases}$ where h is a fixed constant

9. The **absolute value function**. Prove the following properties of the absolute value function:

(a) $|ab| = |a| \cdot |b|$
(b) $|a|^2 + |b|^2 \leqslant (|a| + |b|)^2$

(c) $\dfrac{|x|}{x} = \begin{cases} 1, & x > 0 \\ -1, & x < 0 \end{cases}$

(d) $||y|| = |y|$
(The absolute value of the absolute value of y equals the absolute value of y.)

(e) The **triangle inequality**: $|a+b| \leqslant |a|+|b|$.
[Hint: Case 1 $a \geqslant 0$, $b \geqslant 0$. Case 2 $a \leqslant 0$, $b \leqslant 0$. Case 3 $a > 0$, $b < 0$; either $|a| \geqslant |b|$ or $|b| \geqslant |a|$.]

(f) $|a-b| \geqslant ||a| - |b||$

10. The **signum function**. Prove the following properties of the signum function:

(a) $\operatorname{sgn} q = \begin{cases} \dfrac{|q|}{q}, & q \neq 0 \\ 0, & q = 0 \end{cases}$

(b) $\operatorname{sgn}|x| = |\operatorname{sgn} x|$

(c) $\operatorname{sgn}\left(\dfrac{1}{x}\right) = \dfrac{1}{\operatorname{sgn} x} = \operatorname{sgn} x \ (x \neq 0)$

(d) $\operatorname{sgn}(\operatorname{sgn} z) = \operatorname{sgn} z$

2 GRAPHS AND LOCI

A **Cartesian coordinate system** for the plane is given by two mutually perpendicular real number lines with a given scale. Suppose we label one of these lines the x axis and the other the y axis. Each point P in the plane is determined by a unique pair (x_P, y_P) of real numbers, called its **coordinates**. We therefore speak of a pair of real numbers as a **point** in the Cartesian plane, just as we speak of a real number being a point on the real number line.

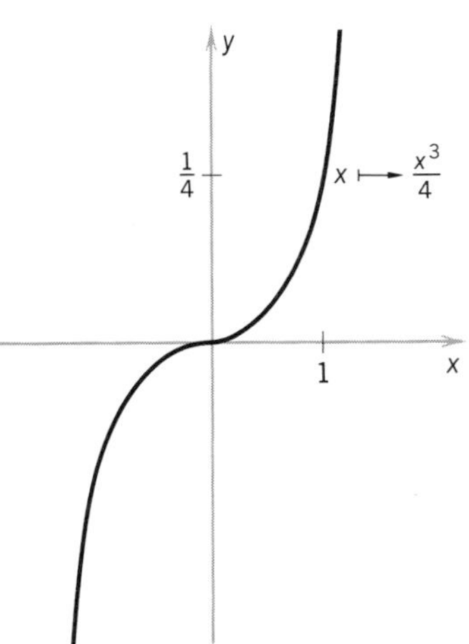

FIGURE 2-1

Given a Cartesian coordinate system for the plane and a real function f, by the **graph** of f we mean the set of all points P such that $f(x_P) = y_P$. The graph of f is also the **graph** of the equation $y = f(x)$.

In Figure 2-1 we have the graph of g: $x \mapsto \frac{1}{4}x^3$ and in Figure 2-2 we have the graph of F: $x \mapsto \frac{1}{3}|x|$. In the following two examples we discuss some features of these graphs.

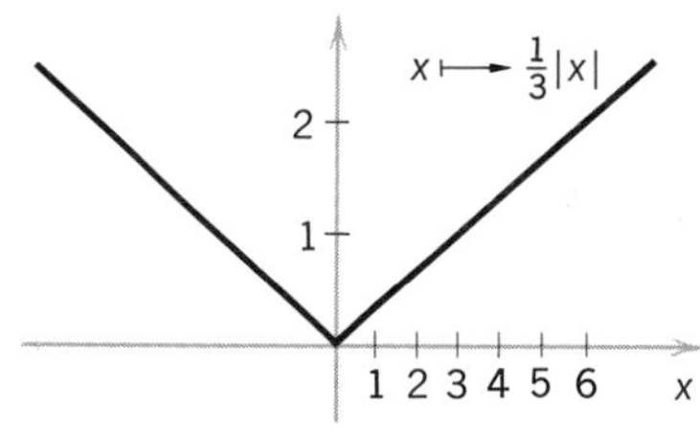

FIGURE 2-2

Example 1 A function g is said to have **odd symmetry** (or to be an **odd function**) if $g(-x) = -g(x)$. Consider g defined by the rule $x \mapsto \frac{1}{4}x^3$ (Figure 2-1). Then $g(-x) = \frac{1}{4}(-x)^3 = -\frac{1}{4}x^3 = -g(x)$, and so g is odd. Since the value of g at any negative real number is -1 times its value at the corresponding positive number, to plot g we need only evaluate g at numbers $x \geqslant 0$. Consider the numbers 1, 2, and 3; we have

$$g(1) = \tfrac{1}{4} = \tfrac{1}{4}, \quad g(2) = 2 = \tfrac{1}{4}2^3, \quad g(3) = \tfrac{1}{4}27 = \tfrac{1}{4}3^3$$

For -1, -2, and -3 we simply write

$$g(-1) = -g(1) = -\tfrac{1}{4}, \quad g(-2) = -g(2) = -2,$$
$$g(-3) = -g(3) = -\tfrac{1}{4}27$$

The value of g at 0 is 0. The points $(0, 0)$, $(-1, -\frac{1}{4})$, $(-2, -2)$, $(-3, -\frac{1}{4}27)$, $(1, \frac{1}{4})$, $(2, 2)$, and $(3, \frac{1}{4}27)$ are plotted, and the freehand curve drawn through them is shown in Figure 2-1.

Example 2 A function F is said to have **even symmetry** (or to be an **even function**) if $F(-x) = F(x)$. For instance F, given by the rule $x \mapsto \frac{1}{3}|x|$, is an even function. Choosing 0, 2, 3, and 6 in the domain of F, we have $F(0) = 0$, $F(2) = \frac{2}{3}$, $F(3) = 1$, and $F(6) = 2$. Applying the condition $F(-x) = F(x)$, we immediately calculate the values of F at -2, -3, and -6 to be $\frac{2}{3}$, 1, and 2, respectively. The graph of F is shown in Figure 2-2.

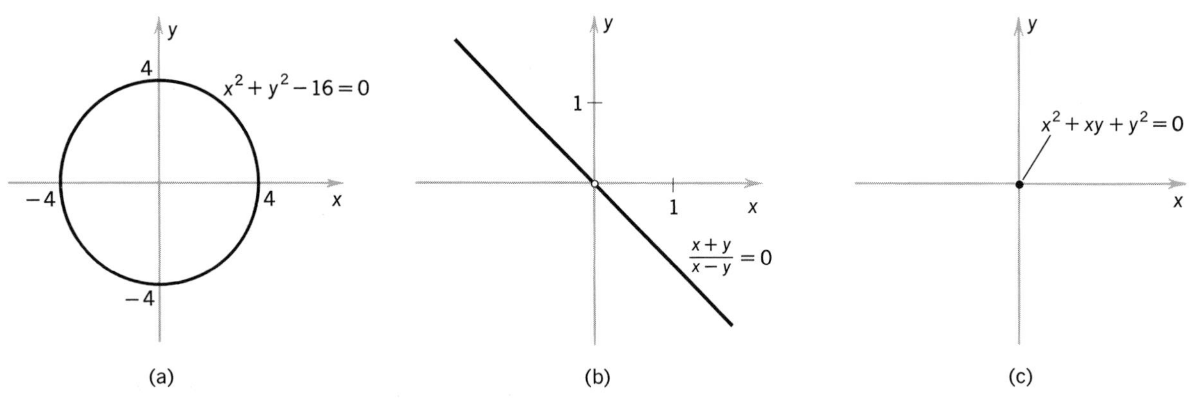

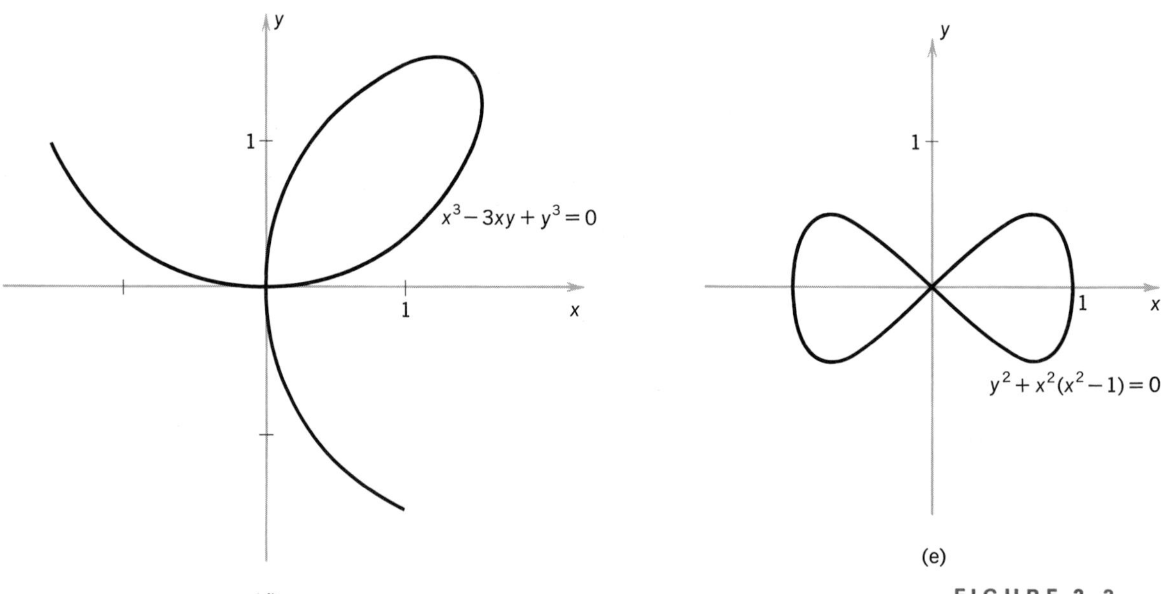

FIGURE 2-3

The odd functions are said to be **symmetric with respect to the origin** and even functions **symmetric with respect to the y axis**.

We now turn to plane curves which are not necessarily the graphs of functions.

A **relation** between two variables x and y is given by an equation of the form $F(x,y)=0$, where $F(x,y)$ is a formula involving x and y. For instance, the equations

$$x^2+y^2-16=0,$$

$$\frac{(x+y)}{(x-y)}=0,$$

$$x^2+xy+y^2=0,$$

$$x^3-3xy+y^3=0,$$

$$\text{and}\quad y^2+x^2(x^2-1)=0$$

all give relations.

If a relation is given by an equation $F(x,y)=0$, then the **locus** or **graph** of the relation is the set of points P which satisfy $F(x_P,y_P)=0$ (Figure 2-3). An open dot in the graph of Figure 2-3 (b) indicates that this point is missing as it does not belong to the domain of the relation. The locus in (c) consists only of $(0,0)$.

We remark that if f is a real function, the equation $y-f(x)=0$ gives a relation whose locus is the graph of f. Thus the graph of $f: x\mapsto 2x+4$ is the locus of the relation $y-2x-4=0$.

The graph of a function intersects any vertical line at most once. On the other hand, the locus of a relation may intersect a vertical line more than once (Figure 2-4).

The process of graphing relations is also simplified by observing symmetries. As with functions we say that a relation $F(x,y)=0$ is **symmetric with respect to the origin** if $F(-x_0,-y_0)=0$ holds whenever we have $F(x_0,y_0)=0$. For example, all the relations in Figure 2-3 are symmetric with respect to the origin except (d). Similarly, we say that a relation $G(x,y)=0$ is **symmetric with respect to the y axis** if $G(x_0,y_0)=0$ implies $G(-x_0,y_0)=0$. In Figure 2-3, (a), (c), and (e) are symmetric with respect to the y axis. A relation $H(x,y)=0$ is said to be **symmetric with respect to the x axis** if $H(x_0,y_0)=0$ implies $H(x_0,-y_0)=0$. In Figure 2-3, (a), (c), and (e) are symmetric with respect to the x axis.

Example 3 Sketch the locus of $9x^2+y^2-9=0$. This relation is symmetric with respect to both the x and y axes, and also with respect to the origin. We can, therefore, restrict our attention to that part of the graph which lies in the first quadrant, and then obtain the rest of the graph by symmetry. Solving our equation for y in terms of x, we find

$$y=\pm\sqrt{9-9x^2}=\pm 3\sqrt{1-x^2}$$

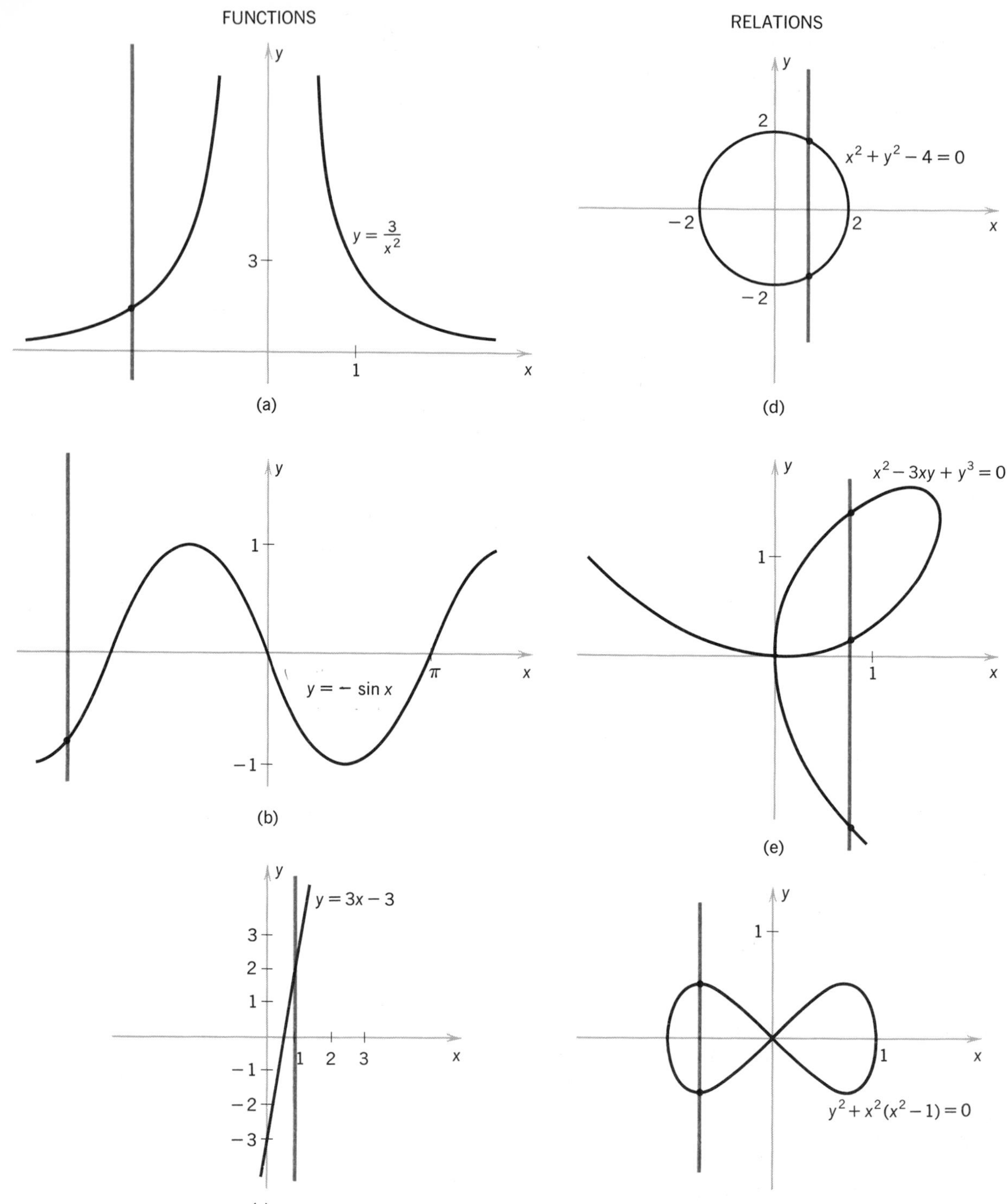
FUNCTIONS
RELATIONS
$y = \frac{3}{x^2}$
(a)
$y = -\sin x$
(b)
$y = 3x - 3$
(c)
$x^2 + y^2 - 4 = 0$
(d)
$x^2 - 3xy + y^3 = 0$
(e)
$y^2 + x^2(x^2 - 1) = 0$
(f)

FIGURE 2-4

For (x, y) in the first quadrant, the relation reduces to

$$y = 3\sqrt{1-x^2}$$

We see that we must have $0 \leqslant x \leqslant 1$ and $0 \leqslant y \leqslant 3$. Plotting points, for example,

$$(0,3),\ (\tfrac{1}{3}, 2\sqrt{2}),\ (\tfrac{1}{2}, \tfrac{3}{2}\sqrt{3}),\ (\tfrac{2}{3}, \sqrt{5}),\ (1,0)$$

we sketch the curve in Figure 2-5. This curve is the locus restricted to the first quadrant. By symmetry, we fill in the curve in the other quadrants to obtain the locus of the relation $9x^2 + y^2 - 9 = 0$ (Figure 2-6).

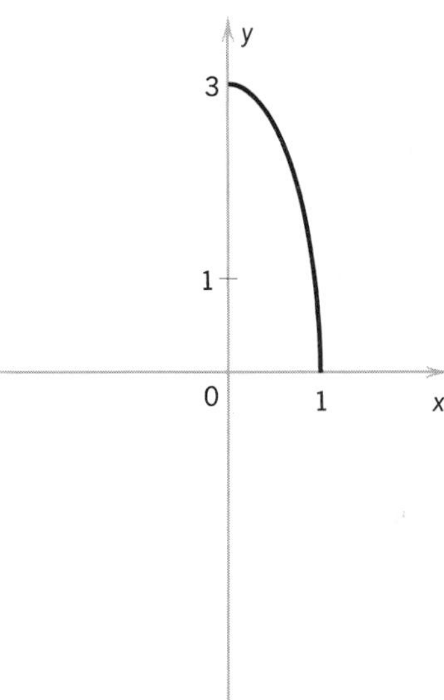

FIGURE 2-5

EXERCISES

1. Sketch the locus of each of the following relations:
   (a) $y^2 - 2 + x = 0$ (b) $9y^2 - 16x^2 = 0$
   (c) $x^2 + y^2 - 64 = 0$ (d) $4x + 6y + 15 = 0$
   (e) $y^2 - \operatorname{sgn} x = 0$ *(f) $x^{2/3} + y^{2/3} - 1 = 0$
   (g) $x - \dfrac{4}{y^3} = 0$
   (h) $y^2(2-x) - x^3 = 0$
   [Hint: Solve for y^2 in terms of x; note symmetry with respect to the x axis.]
   (i) $y^2(1+x) - x^2(3-x) = 0$ (j) $\operatorname{sgn}(x+y) = 1$
   (k) $|x - y| + |x + y| = 3$
2. Sketch the graphs of the following equations and functions:
   (a) $y = x/(1+x^2)$ (b) $y = 4x^2 - 9x$
   (c) $y = -7x^2 - 4$ (d) $y = 7\pi$
   (e) $y = x$ (f) $x \mapsto \dfrac{2}{x}$
   (g) $x \mapsto x^3 - 3x^2 + x$ (h) $t \mapsto t^2$
   (i) $x \mapsto -\sqrt{|x|}$ (j) $\phi \mapsto \sin\phi$
3. Reduce each of the following situations to either relations or functions. Identify your variables. It is not necessary to solve the numerical problems.
   (a) The volume of a rectangular solid is 96 cubic inches, the surface area is 154 square inches, and the girth (twice the sum of the width plus depth) is 16 inches. Find the dimensions of the box.
   (b) The temperature of a gas is directly proportional to pressure and volume and inversely proportional to mass.
   (c) Kinetic energy is one-half the mass times the square of the velocity. Momentum is the product of mass and velocity. Find an equation giving the kinetic energy of an object in terms of its momentum and its mass.
   (d) Two eagles fly away from their nests, which are five miles apart, as shown in Figure 2-7. If they both travel at a constant velocity of 50 mph, show how their separation depends on time.

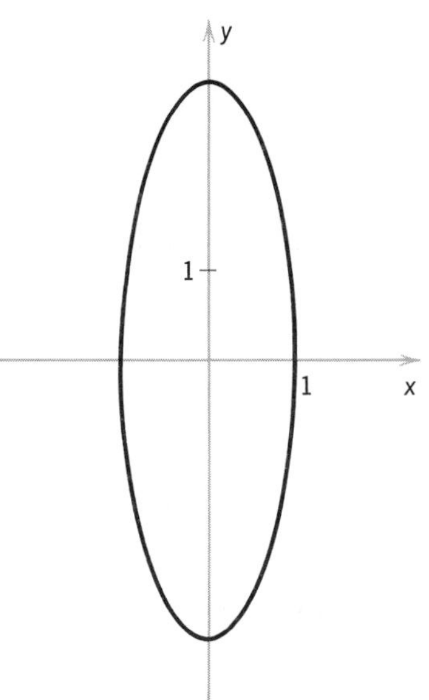

FIGURE 2-6

(e) A page size is needed that will allow for 80 square inches of printed matter in the center with margins of $1\frac{1}{2}$ inches on each side and 2 inches at top and bottom (Figure 2-8). Show how the area of the page depends on the length of one of its sides.

4. Prove that a relation which is symmetric with respect to both axes is also symmetric with respect to the origin.

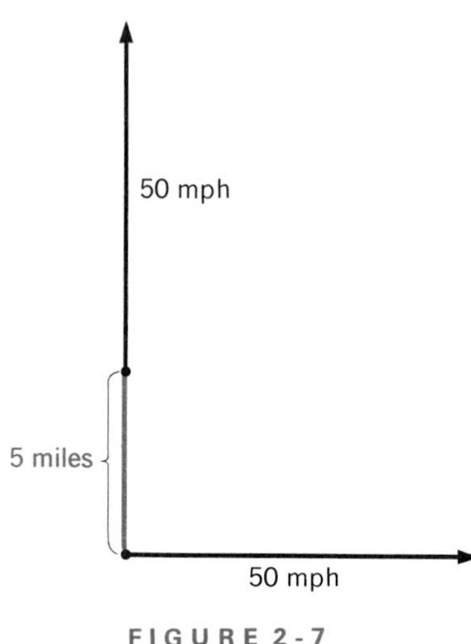

FIGURE 2-7

3 OPERATIONS WITH FUNCTIONS

Often, the rules for two functions can be applied consecutively to form a new rule. Let us consider the functions $f: t \mapsto t^2$ and $g: u \mapsto u^3$. If we apply g to $f(t)$, we have a rule for sending t to t^6:

$$t \overset{f}{\mapsto} t^2 \overset{g}{\mapsto} (t^2)^3 = t^6$$

In general, we can combine a pair of functions g and f in this way provided $f(t)$ is an element of the domain of g whenever t is an element of the domain of f.

Definition If $f: X \to Y$ and $g: Y \to Z$ are functions then $x \mapsto g(f(x))$ is a function which is called the **composite of g on f** (in that order) and denoted by $g \circ f$. One then has $g \circ f: X \to Z$.

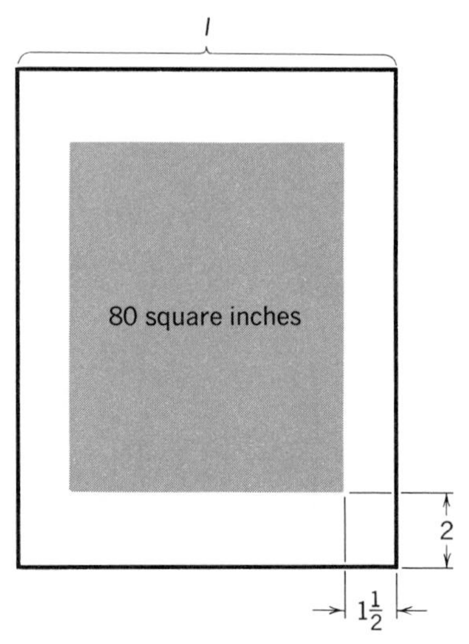

FIGURE 2-8

Example 1 The composite of $g: y \mapsto y^3 - 1$ on $f: x \mapsto x^2$ is $x \mapsto x^2 \mapsto (x^2)^3 - 1$, that is, $g \circ f: x \mapsto x^6 - 1$.

Example 2 The composite of $y \mapsto y^3 - y + 1$ on $x \mapsto \operatorname{sgn} x$ is $x \mapsto \operatorname{sgn} x \mapsto (\operatorname{sgn} x)^3 - \operatorname{sgn} x + 1$. But $(\operatorname{sgn} x)^3 = \operatorname{sgn} x$ for all x; hence $(\operatorname{sgn} x)^3 - \operatorname{sgn} x + 1 = 1$, and the composite is the constant function $x \mapsto 1$.

Example 3 The composite of $y \mapsto g(y)$ on $x \mapsto mx + b$ is $x \mapsto g(mx + b)$, where g is a function whose domain is $\boldsymbol{R}$. This composition is called **affine substitution**, and will be thoroughly discussed in §4. If the rules for the functions are given in the form $y = mx + b$, $z = g(y)$, then the composite can be given in the form $z = g(mx + b)$.

Example 4 The composite of $z = \sqrt{1 - y^2}$ on $y = \sin\theta$ is $z = \sqrt{1 - (\sin\theta)^2} = \sqrt{\cos^2\theta} = |\cos\theta|$.

FORMAL DEVICE

To compose g on f, where $z = g(y)$ and $y = f(x)$, substitute "$f(x)$" for "y" in the equation $z = g(y)$. This yields the composite given in the form $z = g(f(x))$.

Example 5 The composite of $z=y^2+3y+1$ on $y=(x+2)/(x-5)$ is

$$\begin{aligned} z &= \left(\frac{x+2}{x-5}\right)^2 + 3\left(\frac{x+2}{x-5}\right) + 1 \\ &= \frac{(x+2)^2 + 3(x+2)(x-5) + (x-5)^2}{(x-5)^2} \\ &= \frac{x^2 + 4x + 4 + 3x^2 - 9x - 30 + x^2 - 10x + 25}{(x-5)^2} \\ &= \frac{5x^2 - 15x - 1}{(x-5)^2} \end{aligned}$$

Composition can be applied to any functions, real or otherwise, provided the domains and ranges agree properly. For real-valued functions, however, the arithmetic operations are also available. If two functions f and g are real-valued, their function values can be added, subtracted, multiplied, or divided to get new values, and we can associate these values with new functions.

Definition If $f:x \mapsto f(x)$ and $g:x \mapsto g(x)$ are real-valued functions, then the **sum** of f and g, denoted $\boldsymbol{f+g}$, is defined by the rule

$$f+g:x \mapsto f(x)+g(x)$$

The **difference** $\boldsymbol{f-g}$ is

$$f-g:x \mapsto f(x)-g(x)$$

The **product** $\boldsymbol{fg}$ is $fg:x \mapsto f(x)\cdot g(x)$.

The domains of these functions are the intersections of the domains of f and g.

The **quotient** $\boldsymbol{f/g}$ has as its domain the intersection of the domains of f and g excluding the elements x for which $g(x)=0$ and is defined by

$$\frac{f}{g}:x \mapsto \frac{f(x)}{g(x)}$$

Example 6 The sum of $y \mapsto y^2+3y+1$ and $y \mapsto y^3-y+1$ is

$$\begin{aligned} y \mapsto (y^2+3y+1) &+ (y^3-y+1) \\ &= y^3+y^2+2y+2 \\ &= (y+1)(y^2+2) \end{aligned}$$

The product is

$$\begin{aligned} y \mapsto (y^2+3y+1)&(y^3-y+1) \\ &= y^5+3y^4-2y^2+2y+1 \end{aligned}$$

The difference, subtracting the second function from the first, is

$$y \mapsto (y^2+3y+1) - (y^3-y+1)$$
$$= -y^3 + y^2 + 4y$$

The quotient of the first function by the second is

$$y \mapsto \frac{y^2 + 3y + 1}{y^3 - y + 1}$$

As we have already remarked, not all functions have the property that any two different elements of the domain are sent to different values. However, the class of functions having this property is important enough to deserve a special name.

Definition A function f is **one-one** if it has the following property: for any two elements x_0 and x_1 in the domain of f such that $x_0 \neq x_1$, we have $f(x_0) \neq f(x_1)$.

Example 7 The identity function $x \mapsto x$ is clearly one-one. The function $t \mapsto 1/t$ is also one-one. The function $x \mapsto x^2$ is not one-one since it sends x_0 and $-x_0$ to the same value.

If f is a one-one real function, its graph has an important characteristic property: the graph intersects any horizontal line at most once. (Why?) For example, the one-one function $t \mapsto 1/t$ intersects the horizontal line $y = c$ only at the point $(1/c, c)$ if $c \neq 0$. And if $c = 0$, the graph never intersects the line $y = c$. (See Figure 3-1.)

Suppose f is a real function with the property that $x_0 < x_1$ implies $f(x_0) < f(x_1)$ for all elements x_0 and x_1 of its domain: such a function is called **monotone increasing**. The graph of a monotone increasing function is sketched in Figure 3-2; we see that the curve "rises" as x increases.

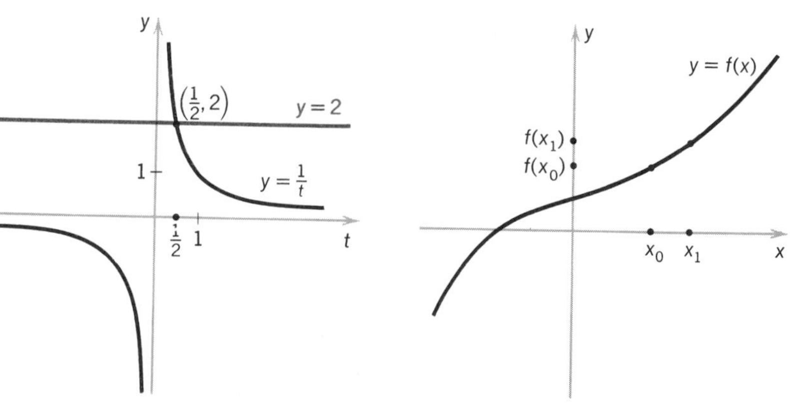

FIGURE 3-1

FIGURE 3-2

A monotone increasing function f is necessarily one-one as the following simple argument shows. If $x_0 \neq x_1$ then either $x_0 < x_1$ or $x_1 < x_0$ and hence either $f(x_0) < f(x_1)$ or $f(x_1) > f(x_0)$; in both cases we have $f(x_0) \neq f(x_1)$, so f is one-one.

Example 8 If $m > 0$, the function $f : x \mapsto mx$, depicted in Figure 3-3, is monotone increasing, as the following calculation shows. Let $x_0 < x_1$; then $x_1 - x_0 > 0$ and $mx_1 - mx_0 = m(x_1 - x_0) > 0$ since $m > 0$. Thus, $f(x_1) - f(x_0) > 0$, or equivalently $f(x_1) > f(x_0)$, which shows f to be monotone increasing.

Similarly, a real function f is **monotone decreasing** (Figure 3-4) if $f(x_0) > f(x_1)$ whenever $x_0 < x_1$ for x_0, x_1 in the domain of f. Monotone decreasing functions are also one-one, as the reader can easily check.

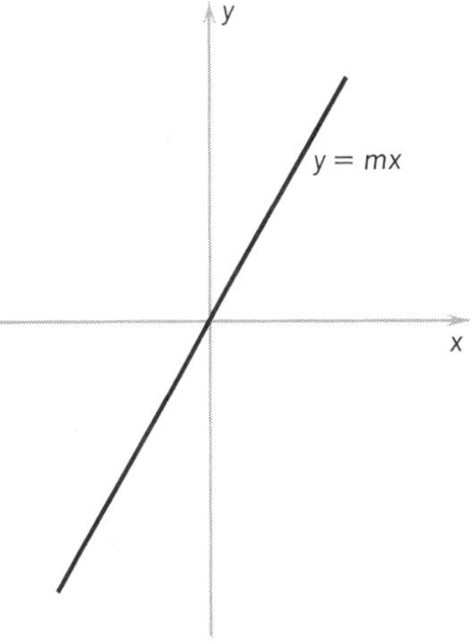

FIGURE 3-3

Example 9 If $m < 0$, the function $x \mapsto mx$ is monotone decreasing.

A one-one real function need not be monotone as the following example shows.

Example 10 Consider the function

$$x \mapsto \begin{cases} x, & x \leqslant 0 \\ 1 - x, & 0 < x < 1 \\ x, & x \geqslant 1 \end{cases}$$

The graph of this function is sketched in Figure 3-5. The function is neither monotone increasing nor decreasing, but it is one-one.

The sine function $x \mapsto \sin x$ is not monotone increasing. However, let us restrict our attention to the interval $[-\frac{1}{2}\pi, \frac{1}{2}\pi]$, as shown in Figure 3-6. For $x_1 < x_2$ in this interval, we always have $\sin x_1 < \sin x_2$. This fact is expressed by saying that the sine function is **monotone increasing on the interval** $[-\frac{1}{2}\pi, \frac{1}{2}\pi]$.

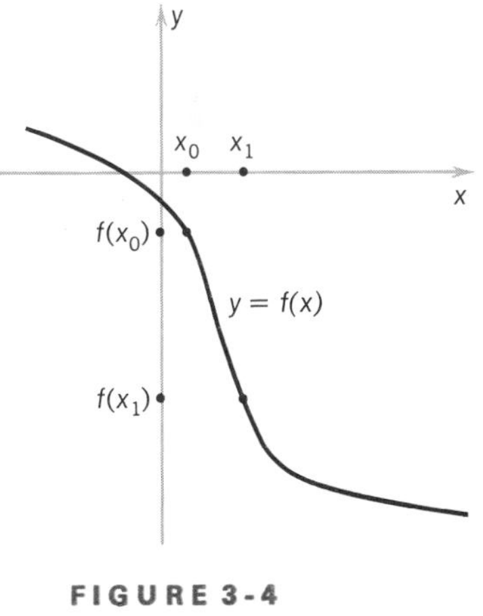

FIGURE 3-4

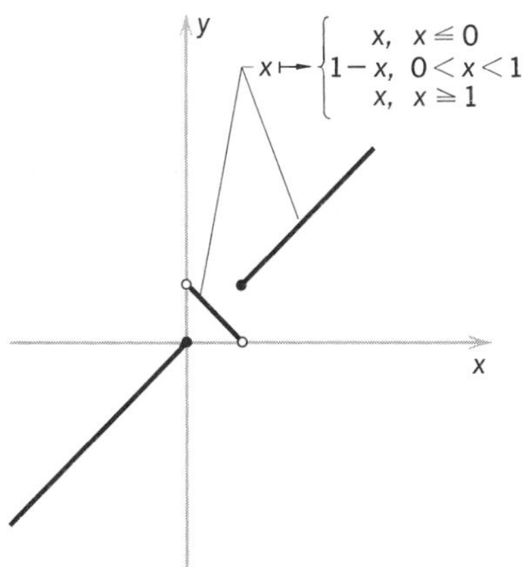

FIGURE 3-5

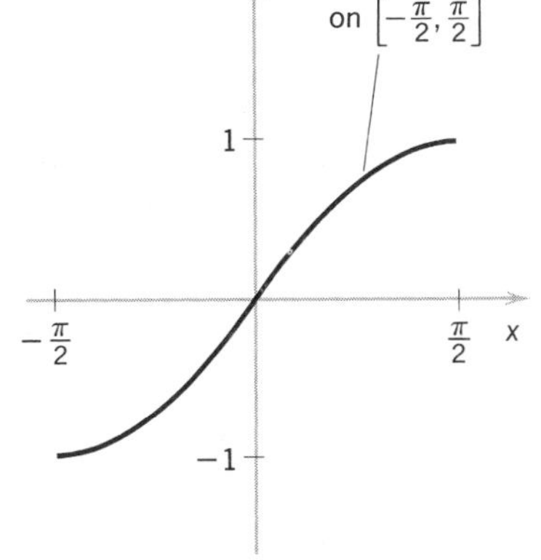

FIGURE 3-6

In general, we say that a function f is **monotone increasing on the interval** I if $f(x_1) < f(x_2)$ whenever x_1, x_2 are elements of I and $x_1 < x_2$. Similarly, we say that f is **monotone decreasing on the interval** I if $f(x_1) > f(x_2)$ whenever x_1, x_2 are in I and $x_1 < x_2$.

Example 11 The cosine function, shown in Figure 3-7, is monotone increasing on $[-\pi, 0]$ and also on $[\pi, 2\pi]$, but it is not monotone increasing on $[-\frac{1}{2}\pi, \frac{1}{2}\pi]$.

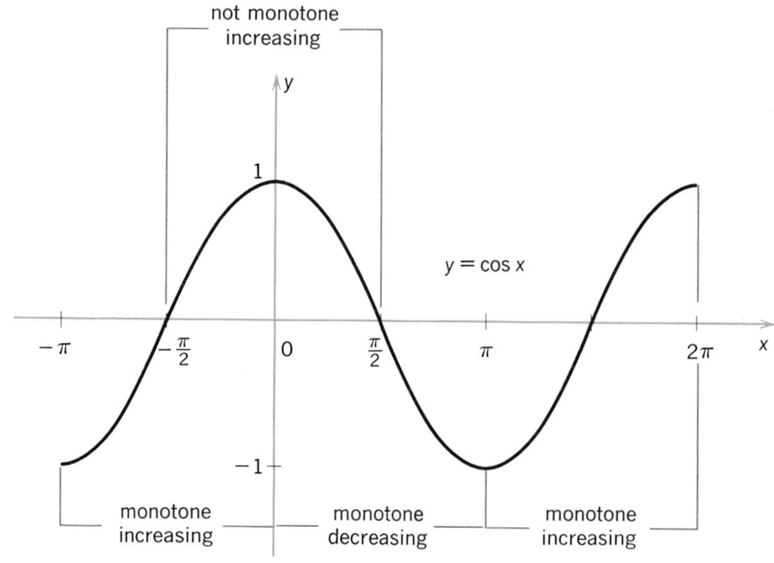

FIGURE 3-7

Example 12 The function of Example 10 is monotone increasing on $]-\infty, 0]$, monotone decreasing on $]0, 1[$, and monotone increasing on $[1, \infty[$.

Suppose a one-one function f is given by the rule $x \mapsto f(x)$ with domain X and range Y. Every element y_0 of Y is equal to $f(x_0)$ for some x_0 in X, since Y is the range of f. Moreover, since f is one-one for each y_0 there is exactly one such x_0. We therefore have a rule for sending elements of Y back to elements of X: y_0 is sent back to the unique element x_0 such that $f(x_0) = y_0$. This rule defines a function which is called the **inverse** of f and is denoted by f^{-1}. Let us make a formal definition.

Definition The **inverse** f^{-1} of a one-one function f is the function whose domain is the range of f and whose range is the domain of f and which satisfies $f^{-1}f(x) = x$ for all x in the domain of f.

Example 13 The inverse of the real function $f: x \mapsto 3x$ is $f^{-1}: y \mapsto \frac{1}{3}y$ since $f^{-1}f(x) = f^{-1}(3x) = \frac{1}{3}(3x) = x$. The inverse of the function

$$f(x) = \begin{cases} \dfrac{x}{3}, & x \leqslant 0 \\ x^2, & x \geqslant 0 \end{cases}$$

is

$$f^{-1}(y) = \begin{cases} 3y, & y \leqslant 0 \\ \sqrt{y}, & y \geqslant 0 \end{cases}$$

FORMAL DEVICE

To find the inverse of a one-one function $y = f(x)$, solve for x in terms of y. The resulting equation $x = g(y)$ gives the rule for the inverse of f; that is, $g = f^{-1}$.

Example 14 Let $y = f(x) = 2x + 1$. If we solve for x in terms of y, we obtain $x = \frac{1}{2}(y-1)$; so $y \mapsto \frac{1}{2}(y-1)$ is the inverse function. To check this, substitute $2x+1$ in this rule, to get

$$\frac{(2x+1)-1}{2} = x$$

Example 15 Find the inverse of the function given by $y = x^2 - 7x + 12$ with domain $]-\infty, 7/2]$. We solve for x in terms of y by applying the quadratic formula to the equation $x^2 - 7x + (12 - y) = 0$, yielding

$$x = \frac{7 \pm \sqrt{49 - 4(12-y)}}{2}$$

$$= \frac{7 \pm \sqrt{1 + 4y}}{2}$$

The negative sign is appropriate because the range of the inverse must be $]-\infty, 7/2]$, the domain of the original function. We also see that the domain of the inverse is $[-\frac{1}{4}, \infty[$, since the expression $1 + 4y$ appearing under the radical sign must be $\geqslant 0$. Thus the inverse function has domain $[-\frac{1}{4}, \infty[$ and is given by

$$y \mapsto \frac{7 - \sqrt{1 + 4y}}{2}$$

To check that $f^{-1}(f(x)) = x$, we substitute $f(x)$ for y in $f^{-1}(y)$

$$\begin{aligned}\frac{7-\sqrt{1+4y}}{2} &= \frac{7-\sqrt{1+4(x^2-7x+12)}}{2} \\ &= \frac{7-\sqrt{4x^2-28x+49}}{2} \\ &= \frac{7-\sqrt{(7-2x)^2}}{2} \\ &= \frac{7-|(7-2x)|}{2} \\ &= x \qquad [\text{since } x \leqslant 7/2]\end{aligned}$$

Equivalently, we verify that $f(f^{-1}(y)) = y$ by

$$\begin{aligned}x^2-7x+12 &= \left(\frac{7-\sqrt{1+4y}}{2}\right)^2 - 7\left(\frac{7-\sqrt{1+4y}}{2}\right) + 12 \\ &= \left(\frac{50+4y}{4} - 7\frac{\sqrt{1+4y}}{2}\right) - \left(\frac{49}{2} - 7\frac{\sqrt{1+4y}}{2}\right) + 12 \\ &= \left(\frac{25}{2}+y\right) - \frac{49}{2} + \frac{24}{2} \\ &= y\end{aligned}$$

Geometrically, the graph of f^{-1} can be obtained by "flipping" the graph of f over the line $y = x$ (Figure 3-8). A further example is shown in Figure 3-9. This topic will be discussed in more detail in §4.8.

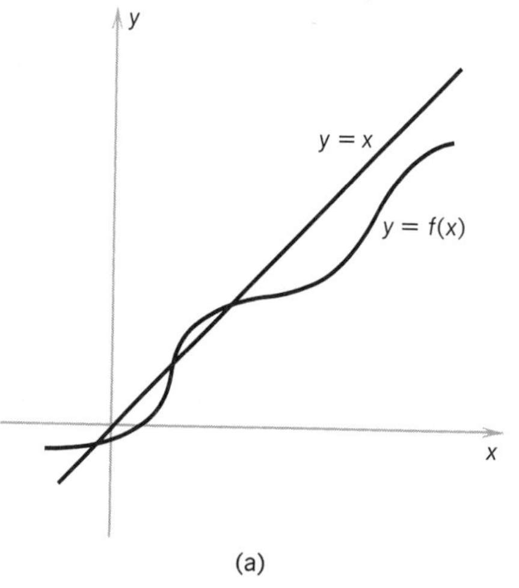

(a)

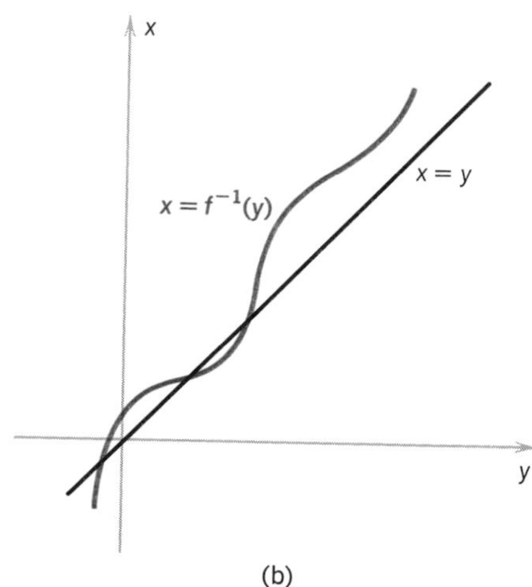

(b)

FIGURE 3-8

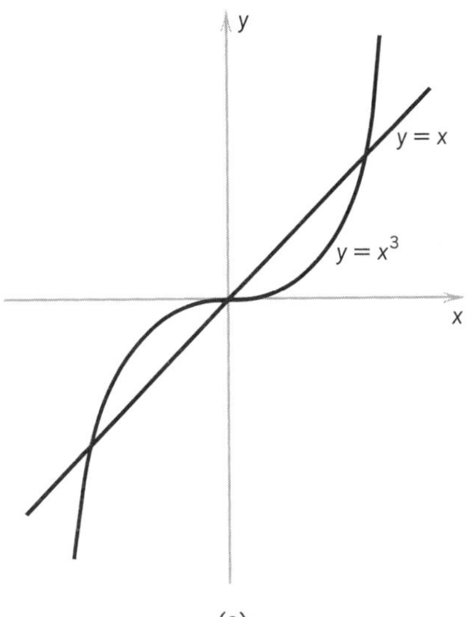

(a)

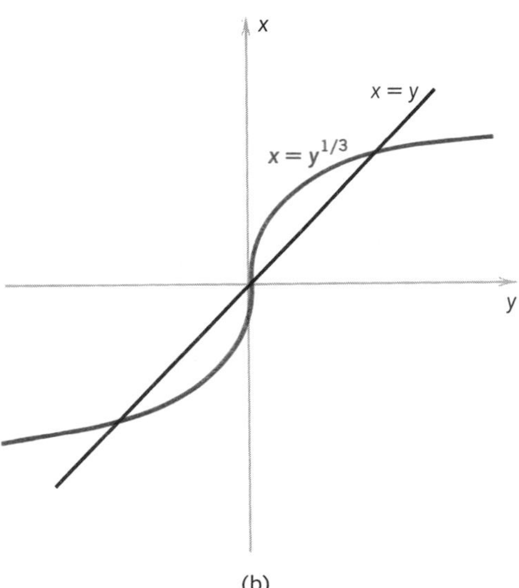

(b)

FIGURE 3-9

EXERCISES

1. Compose the first function on the second function:
   (a) $y \mapsto y^2$; $x \mapsto 4x + 4$
   (b) $y \mapsto \dfrac{4}{y^3}$; $x \mapsto 9x^2 - 4x \pm 36$
   (c) $y \mapsto \dfrac{y^2 - 1}{y^2 - 4}$; $\theta \mapsto 4\sec^2\theta$
   (d) $y \mapsto 3y + 2$; $v \mapsto 7v^2 - 4$
   (e) $z \mapsto \dfrac{4z^2 - 2}{z}$; $q \mapsto \operatorname{sgn} q$
   (f) $y \mapsto \dfrac{\sqrt{2 + y} - 1}{y + 1}$; $\phi \mapsto 2\tan^2\phi$
   *(g) $\alpha \mapsto \sin\alpha$; $t \mapsto \cos^{-1} t$
   (h) $t \mapsto t^3 - 3t^2 + 3t$; $s \mapsto \dfrac{1 + s}{4 + s^2}$
   (i) $a \mapsto \dfrac{a}{1 + a^2}$; $r \mapsto 7\pi$
   (j) $x \mapsto |x|$; $y \mapsto -\sqrt{y}$
   (k) $b \mapsto 4b + 4$; $c \mapsto c^2$
   (l) $x \mapsto 9x^2 - 4x + 36$; $b \mapsto \dfrac{4}{b^2}$
   (m) $e \mapsto -7e^2 - 4$; $k \mapsto 3k + 2$

(n) $r \mapsto \operatorname{sgn} r$; $z \mapsto \dfrac{4z^2 - 2}{z}$

*(o) $q \mapsto \cos^{-1} q$; $\beta \mapsto \sin \beta$

(p) $s \mapsto \dfrac{1 - s}{4 + s^2}$; $t \mapsto t^3 - 3t^2 + 3t$

(q) $x \mapsto 0$; $z \mapsto \dfrac{z}{1 + z^2}$

(r) $y \mapsto -\sqrt{y}$; $q \mapsto |q|$

(s) $y \mapsto y$; $x \mapsto x$

(t) $z \mapsto \dfrac{2}{z}$; $r \mapsto \dfrac{2}{r}$

2. Is $f \circ g$ necessarily the same function as $g \circ f$? [Look at exercises 1(a) and 1(k), for example.] Are they ever the same?

3. Add the following pairs of functions, and simplify as much as possible.

(a) $y \mapsto y^2$; $y \mapsto 4y + 4$

(b) $x \mapsto \dfrac{4}{x^3}$; $x \mapsto 9x^2 - 4x$

(c) $y \mapsto 3y + 2$; $u \mapsto -7u + 4$

(d) $a \mapsto \dfrac{a}{1 + a^2}$; $a \mapsto \dfrac{1}{2}$

(e) $x \mapsto |x|$; $x \mapsto -x$

(f) $n \mapsto \dfrac{7}{m}$; $m \mapsto \dfrac{8}{m^2}$

4. Multiply each pair of functions in exercise 3.

5. In each part of exercise 3, subtract the second function from the first.

6. Divide the first function by the second:

(a) $\theta \mapsto \tan \theta$; $\theta \mapsto \sec \theta$

(b) $x \mapsto \dfrac{4}{x^3}$; $x \mapsto \dfrac{1}{x}$

(c) $t \mapsto |t|$; $t \mapsto -t$

(d) $a \mapsto \dfrac{a}{1 + a^2}$; $a \mapsto \dfrac{1}{2}$

7. In each of the following, state whether the function is one-one or not. If it is, prove your claim; if it is not, give an example where $f(x_0) = f(x_1)$ and $x_0 \neq x_1$. Also indicate whether or not the function is monotone.

(a) $f(x) = |x|$

(b) $y \mapsto \operatorname{sgn} y$

(c) $u \mapsto u + \operatorname{sgn} u$

(d) $g(t) = t - \operatorname{sgn} t$

(e) $v \mapsto v + |v|$

(f) $r \mapsto r^{1/3}$

(g) $f(v) = \tan v$

(h) $x \mapsto x^3$

(i) $s \mapsto -\sqrt{s}\,(s > 0)$

(j) $x \mapsto mx + b(m \neq 0)$

(k) $y \mapsto ay^2 + by + c$

(l) $t \mapsto 60 - 16t^2$

(m) $f(s) = s$

(n) $x \mapsto \begin{cases} x^2, & x \geqslant 0 \\ -x^2, & x \leqslant 0 \end{cases}$

8. Let f and g be monotone increasing functions from $\boldsymbol{R}$ to $\boldsymbol{R}$. Show that $f \circ g$ and $f+g$ are also monotone increasing.
9. Which of the following functions are monotone increasing, monotone decreasing, or neither on the interval specified?
   (a) $\theta \mapsto \tan\theta;\quad [0, \frac{1}{4}\pi]$ (b) $x \mapsto x^2;\quad [-2, 1]$
   (c) $a \mapsto |a|;\quad [12, 30]$ (d) $y \mapsto y^5;\quad [1, 7]$
   (e) $z \mapsto \dfrac{z^3+1}{z};\quad [0, 1]$ (f) $x \mapsto |x|;\quad [-1, 0]$
   (g) $y \mapsto \operatorname{sgn} y;\quad [\pi, 7]$ (h) $y \mapsto y + \operatorname{sgn} y;\quad [a, b]$
   (i) $y \mapsto y - \operatorname{sgn} y;\quad [0, 3]$ (j) $x \mapsto \sqrt{x};\quad [1, 2]$
10. Compute inverses to all the one-one functions of exercise 7. Also find the inverse of the following function:

$$x \mapsto \begin{cases} x, & x \leqslant 0 \\ 1-x, & 0 < x < 1 \\ x, & x \geqslant 1 \end{cases}$$

discussed in Example 10.

4 LINEAR AND AFFINE FUNCTIONS

A function from $\boldsymbol{R}$ to $\boldsymbol{R}$ is an **affine function** if its rule can be written in the form $x \mapsto mx+b$, where m and b are fixed real numbers. When $b=0$, the function is also called a **linear function**. The graph of the affine function $y=mx+b$ is a nonvertical straight line whose y intercept is $(0, b)$. The number m is the **slope** of the line, and the equation $y=mx+b$ is said to be in **slope-intercept form** (Figure 4-1).

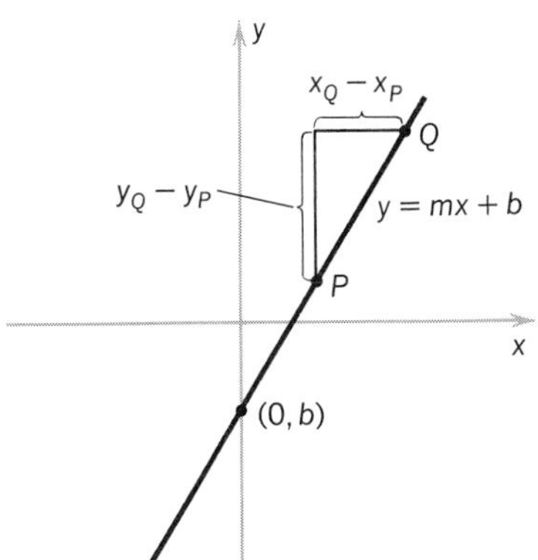

FIGURE 4-1

Let L be the straight line which is the graph of $y=mx+b$. If P and Q are any two distinct points on the line with coordinates (x_P, y_P) and (x_Q, y_Q), then $x_Q - x_P$ does not equal 0 since the line is nonvertical. Moreover, the slope of L is equal to $(y_Q - y_P)/(x_Q - x_P)$: for we know that $y_Q = mx_Q + b$ and $y_P = mx_P + b$, hence $y_Q - mx_Q = y_P - mx_P$ and

$$m = \frac{y_Q - y_P}{x_Q - x_P}$$

Thus, the slope m is the ratio $(y_Q - y_P)/(x_Q - x_P)$ for any pair P and Q of distinct points on the line. We note that since vertical lines do not have slope, a "line with slope" must be a nonvertical straight line.

A line L with slope $m \neq 0$ intersects the x axis at exactly one point; if the equation of L is $y=mx+b$, then the x intercept of L is the point $(-b/m, 0)$. The **angle of inclination** of L is the counterclockwise angle formed by L and the x axis. If α is the angle of inclination of L, then $\tan\alpha = m$, as is evident from Figures 4-1 and 4-2. Since $\tan 0 = 0$, we define the angle of inclination of the x axis to be 0 degrees.

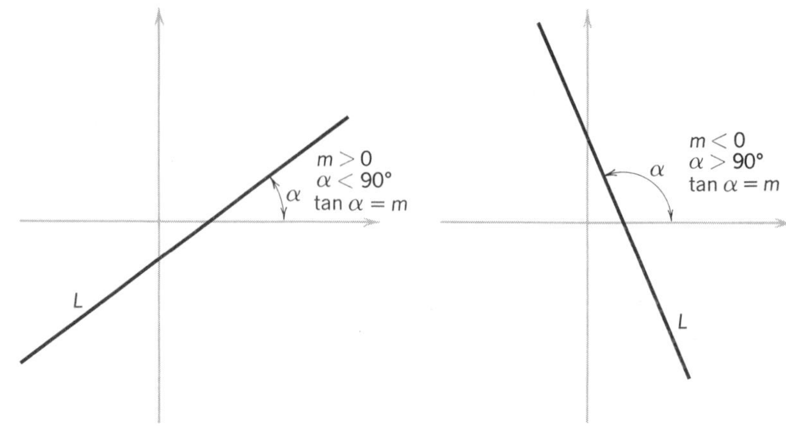

FIGURE 4-2

Given a point P with coordinates (x_P, y_P) and a real number m, we compute the equation of the straight line with slope m which passes through P as follows. Let Q be any point with x coordinate $x_Q \neq x_P$; the line through P and Q has slope $(y_Q - y_P)/(x_Q - x_P)$. We set this ratio equal to m,

$$m = \frac{y_Q - y_P}{x_Q - x_P}$$

and we simplify to obtain

$$y_Q = mx_Q - mx_P + y_P$$

The line of the equation $y = mx - (mx_P - y_P)$ passes through P and has slope m, so this is the equation we seek.

FORMAL DEVICE Point-Slope Formula

Given a point P with coordinates (x_P, y_P), the line passing through P with slope m has the equation

$$\boldsymbol{y - y_P = m(x - x_P)}$$

Example 1 The equation of the line of slope 2, which passes through $(1,0)$, is $y - 0 = 2(x-1)$, or $y = 2x - 2$.

FORMAL DEVICE Point-Point Formula

Given points P and Q with $x_P \neq x_Q$, the line passing through both P and Q has the equation

$$\boldsymbol{\frac{y - y_P}{x - x_P} = \frac{y_Q - y_P}{x_Q - x_P}}$$

Example 2 The line through $(1, 1)$ and $(16, \frac{1}{2})$ has equation

$$\frac{y-1}{x-1} = \frac{16-1}{\frac{1}{2}-1} = \frac{15}{-\frac{1}{2}}$$

or

$$y = -30x + 31$$

Two lines are **parallel** if they do not intersect. Let $y = mx + b$ and $y = mx + b$ $(b \neq b')$ be the equations of two lines with the same slope. These equations have no simultaneous solution; hence the lines do not intersect. Thus two different lines with the same slope are parallel. On the other hand, if two lines have equations $y = mx + b$ and $y = m'x + b'$, and $m \neq m'$ (here b may equal b'), then the equations have the simultaneous solutions

$$x = \frac{b' - b}{m - m'} \quad \text{and} \quad y = \frac{mb' - m'b}{m - m'}$$

so the lines intersect at the point

$$\left(\frac{b' - b}{m - m'}, \frac{mb' - m'b}{m - m'}\right)$$

hence, they are not parallel. We have proved the following theorem.

Theorem 1 Two different nonvertical lines are parallel if and only if they have the same slope (Figure 4-3).

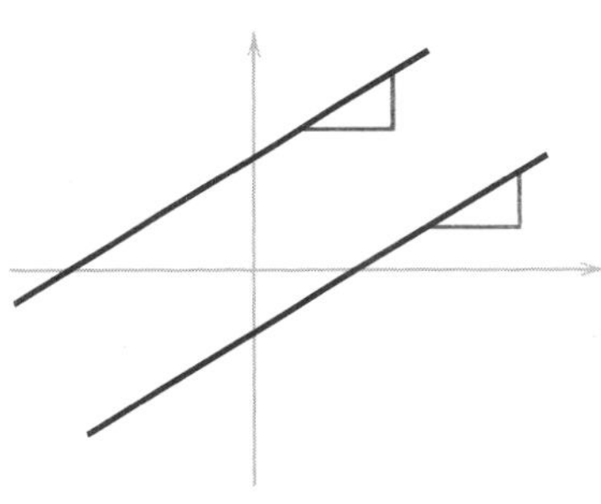

FIGURE 4-3

Two lines are **perpendicular** if they intersect at right angles. The slopes of two parallel lines are equal; the slopes of two perpendicular lines are also related.

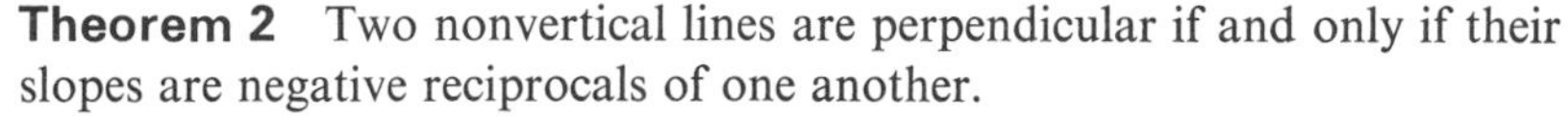

Theorem 2 Two nonvertical lines are perpendicular if and only if their slopes are negative reciprocals of one another.

Proof We wish to show that if $y = mx + b$ and $y = m'x + b'$ are equations of two lines L and L', then L, L' are perpendicular if and only if

$$m' = -\frac{1}{m}$$

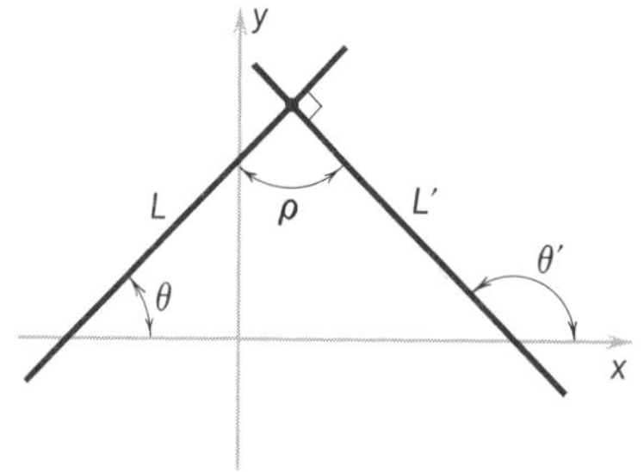

FIGURE 4-4

Now m and m' are the tangents of θ and θ', the angles between the x axis and L and L', respectively (Figure 4-4). Recall from geometry that the exterior angle θ' has measure equal to the sums of the measures of the opposite interior angles θ and ρ; that is, $\theta' = \theta + \rho$. Therefore, $\rho = 90°$ if and only if $\theta' = \theta + 90°$. Taking the tangent of both sides, we get

$$\tan \alpha' = \tan(\theta + 90°)$$

so

$$\tan \theta' = -\cot \theta \qquad (\text{note } \tan(\alpha + 90°) = -\cot \alpha)$$

and

$$\tan\theta' = -\frac{1}{\tan\theta} \qquad \text{(note } \cot\theta = 1/\tan\theta\text{)}$$

But $\tan\theta' = m'$ and $\tan\theta = m$, so this last equation is equivalent to $m' = -(1/m)$. ▨

Example 3 Find the equation of the line through the point $(-7, 2)$ which is parallel to the line L given by $y = 3x + 5$ (Figure 4-5). The line is of the form $y = mx + b$, and since it is parallel to L, we have $m = 3$. Therefore, we have $y = 3x + b$. Since the line is to pass through $(-7, 2)$, we substitute -7 and 2 for x and y to find b:

$$2 = -21 + b$$

$$b = 23$$

Therefore the desired line is given by $y = 3x + 23$.

Example 4 Find the equation of the line through the origin parallel to the line through $(5, -12)$ and $(-4, 6)$ (Figure 4-6). We have

$$m = \frac{y_Q - y_P}{x_Q - x_P} = \frac{6 - (-12)}{-4 - 5} = \frac{18}{-9} = -2$$

so the equation of the line through the origin is $y = -2x$.

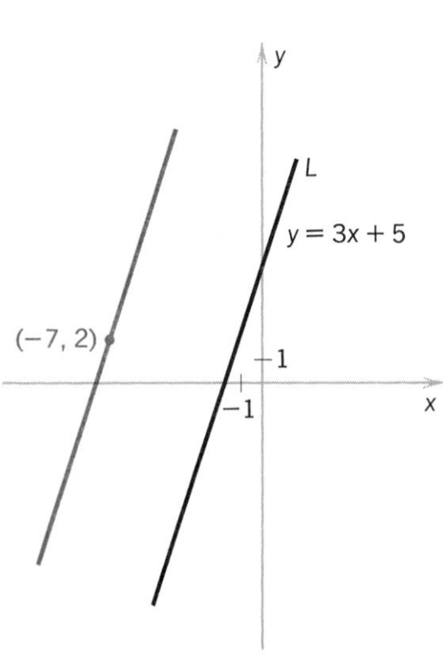

FIGURE 4-5

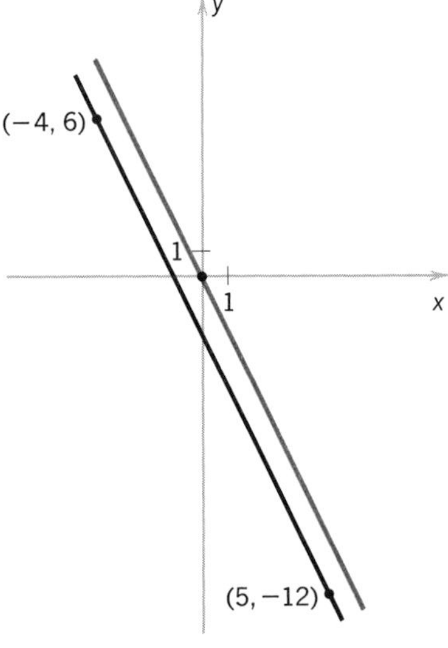

FIGURE 4-6

Example 5 Find the equation of the line perpendicular to $y = 3x - 7$ at the point $(3, 2)$ (Figure 4-7). We use the point-slope form, knowing that $m = -\frac{1}{3}$. We have

$$y - y_P = m(x - x_P)$$
$$y - 2 = -\tfrac{1}{3}(x - 3)$$
$$y = -\tfrac{1}{3}x + 3$$

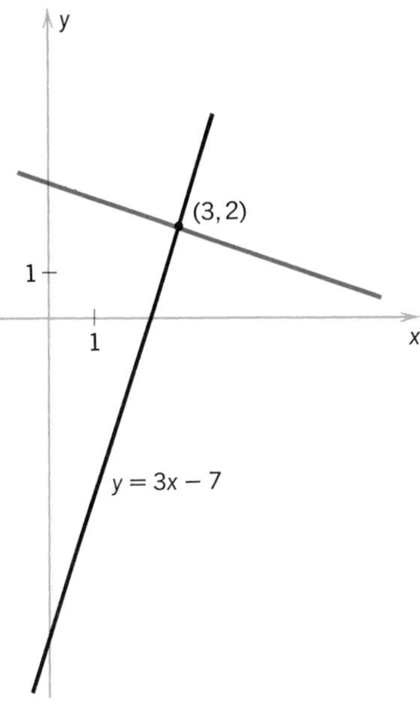

FIGURE 4-7

If P and Q are points in the plane with coordinates (x_P, y_P) and (x_Q, y_Q), respectively, the **distance** $\overline{PQ}$ between them is given by the formula

$$\overline{PQ} = \sqrt{(x_P - x_Q)^2 + (y_P - y_Q)^2}$$

This formula is derived from the Pythagorean theorem, as we see in Figure 4-8.

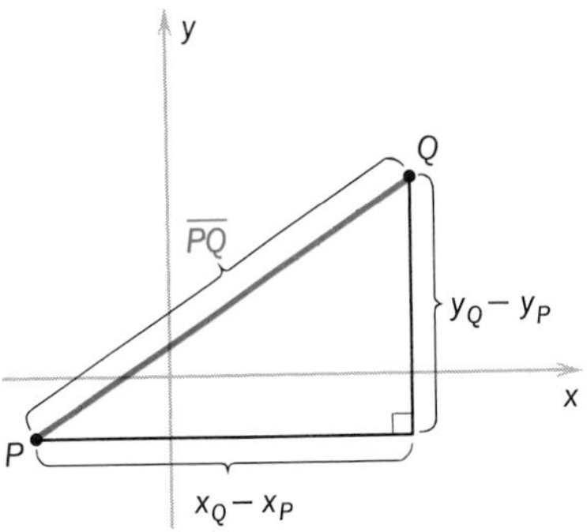

FIGURE 4-8

Example 6 The distance between the points $(3, 2)$ and $(5, -7)$ is

$$\sqrt{(5-3)^2 + (-7-2)^2} = \sqrt{85}$$

The distance d from a point P to a line L is defined to be the length of the perpendicular line segment from P to L (Figure 4-9). When L is given by an equation $y = mx + b$ and P by the coordinates (x_P, y_P), the straight line L' which is perpendicular to L and passes through P has slope $-1/m$, and its equation can be computed from the point-slope formula. The distance from P to L is the distance from P to the point of intersection of L and L'.

Example 7 Find the distance from the point $(7, -3)$ to the straight line L of equation $y = \frac{1}{2}(x - 3)$ (Figure 4-10). The line L has slope $\frac{1}{2}$, so we consider the line L with equation

$$\frac{y - (-3)}{x - 7} = -2 \quad \text{or} \quad y = -2x + 11$$

The coordinates of the point of intersection of L and L' are the solutions to the simultaneous equations

$$y = \tfrac{1}{2}x - \tfrac{3}{2} \qquad (1)$$
$$y = -2x + 11 \qquad (2)$$

Subtracting (2) from (1), we obtain

$$0 = \frac{5x}{2} - \frac{25}{2}$$
$$x = 5$$

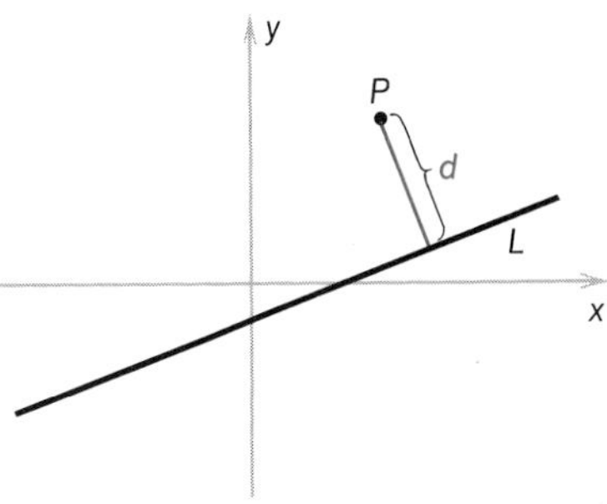

FIGURE 4-9

Substituting $x = 5$ into equation (1), we find $y = 1$. The distance from the point $(7, -3)$ to L is equal to the distance from $(7, -3)$ to $(5, 1)$ which, by the distance formula, is

$$\sqrt{(7-5)^2 + (-3-1)^2} = \sqrt{4 + 16} = \sqrt{20} = 2\sqrt{5}$$

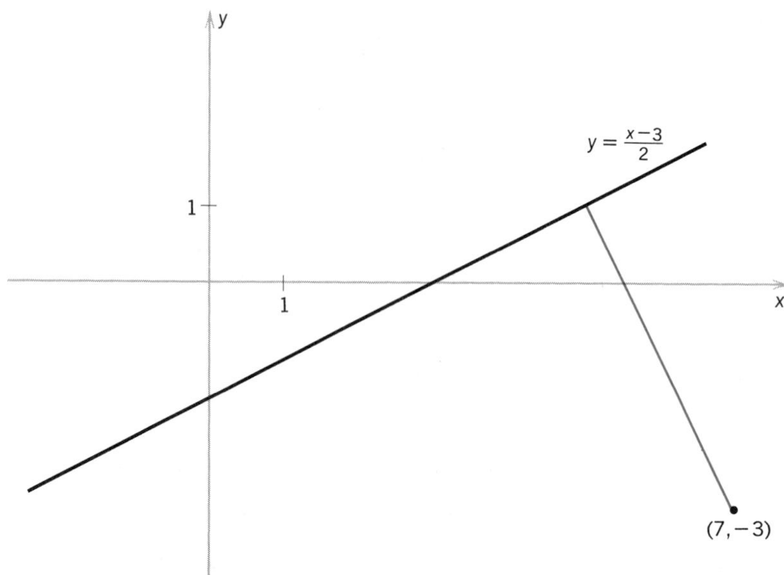

FIGURE 4-10

We now study the effect of composing a function f on any affine function A. Such a function $f \circ A$ is called an **affine substitution**. Let $A(x) = mx + b$ and suppose $f: \mathbf{R} \to \mathbf{R}$ is a real function with domain $\mathbf{R}$. Writing $F = f \circ A$, so that $F(x) = f(mx+b)$, let us compare the graph of F to the graph of f, that is, we shall see what effect the affine substitution has on the graph of f. Since the function A is completely determined by the two constants m and b, it is instructive to examine f "before-and-after" for each of the following cases:

(1) $1 \leqslant m$ and $b = 0$

(2) $0 < m < 1$ and $b = 0$

(3) $m < 0$ and $b = 0$

and

(4) $b \neq 0$

The case when $m = 0$, so that A is the constant function $x \mapsto b$, is left for the reader's examination.

In the first case, $1 \leqslant m$ and $b = 0$. Here $F(x) = f(mx)$, with m not less than 1. If $m = 1$, then the graph of F is identical to the graph of f. Suppose m is greater than 1. If $x > 0$, $-mx < -x$, and the point $(-mx, f(-mx))$ lies to the left of $(-x, F(-x))$ [Figure 4-11]. In words, the value of F at $-x$ is what the value of f is at a point $-mx$, which is farther from the y axis than a point $-x$. Similarly $(mx, f(mx))$ lies to the right of $(x, F(x))$

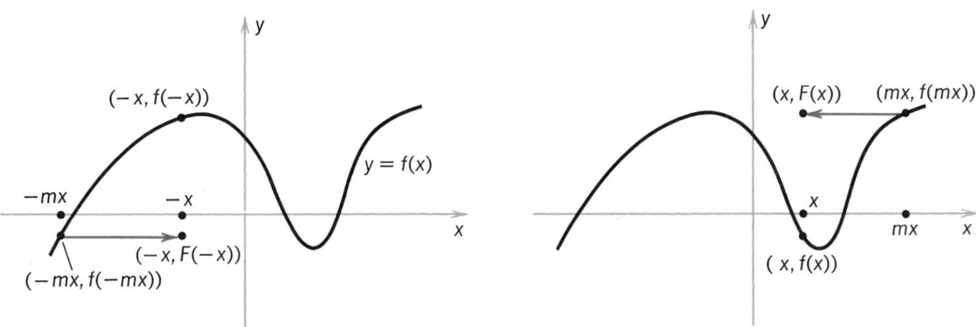

FIGURE 4-11

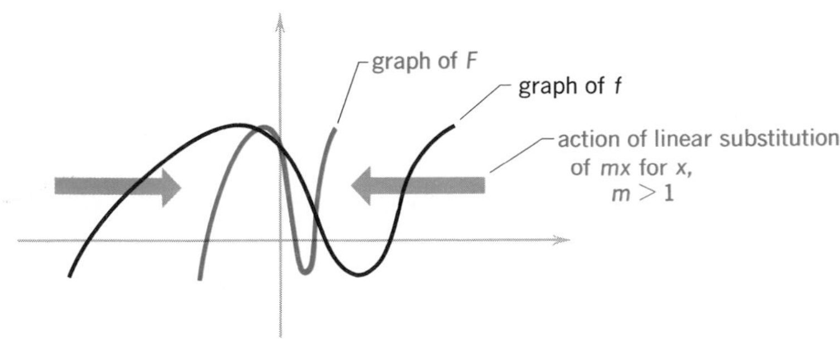

FIGURE 4-12

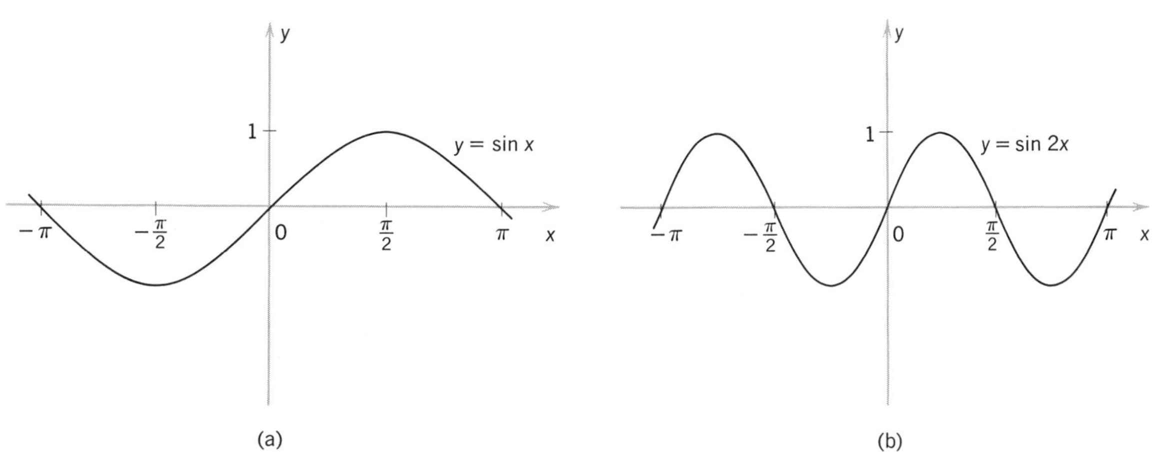

FIGURE 4-13

[Figure 4-11]. Of course, when $x=0$, then $F(x)=f(x)$. Thus for all nonzero x the point $(mx, f(mx))$ is farther from the y axis than the point $(x, F(x))$. We conclude that the effect of the linear substitution of A in f, with $m>1$ and $b=0$, is to "squash" the graph of f symmetrically toward the y axis as shown in Figure 4-12.

Example 8 The linear substitution of $x \mapsto 2x$ in $x \mapsto \sin x$ (Figure 4-13(a)) produces $x \mapsto \sin 2x$ (Figure 4-13(b)).

Our second case is $0<m<1$, $b=0$. Again we have $F(x)=f(mx)$, but this time the point mx is **closer** to the y axis than the point x, because $|mx| = |m||x| = m|x| < |x|$ for $x \neq 0$ (Figure 4-14). Therefore the value of F at x is what the value of f is at a point mx which is closer to the y axis than x, and since $F(0)=f(0)$ as before, we conclude that the graph of F is the result of "stretching" the graph of f symmetrically with respect to the y axis (Figure 4-15).

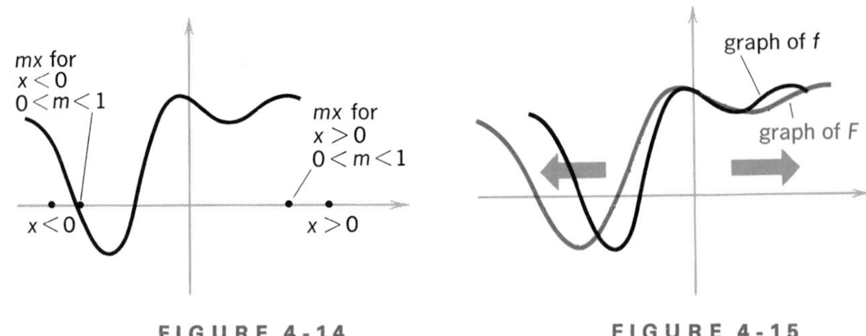

FIGURE 4-14

FIGURE 4-15

Example 9 The linear substitution of $x \mapsto \frac{1}{4}x$ in $x \mapsto \tan x$ (Figure 4-16(a)) produces $x \mapsto \tan \frac{1}{4}x$ (Figure 4-16(b)).

Now we discuss the third case, $m<0$, $b=0$. We have again $F(x)=f(mx)$, but now m is negative. Our analysis proceeds exactly as in the preceding two cases, except that now we deal with $m \leqslant -1$ and with $-1<m<0$, respectively. When $m=-1$ we have $F(x)=f(-x)$, so the value of F at nonzero x is just the value of f at x "flipped" about the y axis to an equal distance on the other side as shown in Figure 4-17. Of course, the value of F at 0 is $f(0)$. We conclude that the graph of F, when $m=-1$ and $b=0$, is obtained from the graph of f by "flipping" the graph of f with respect to the y axis (Figure 4-18).

When $m<-1$, then the point mx on the real line may be located from x first by noting that x goes to $|m|x$, which is farther from the y axis than x, and second by observing $|m|x$ then goes to $-|m|x = mx$, which is

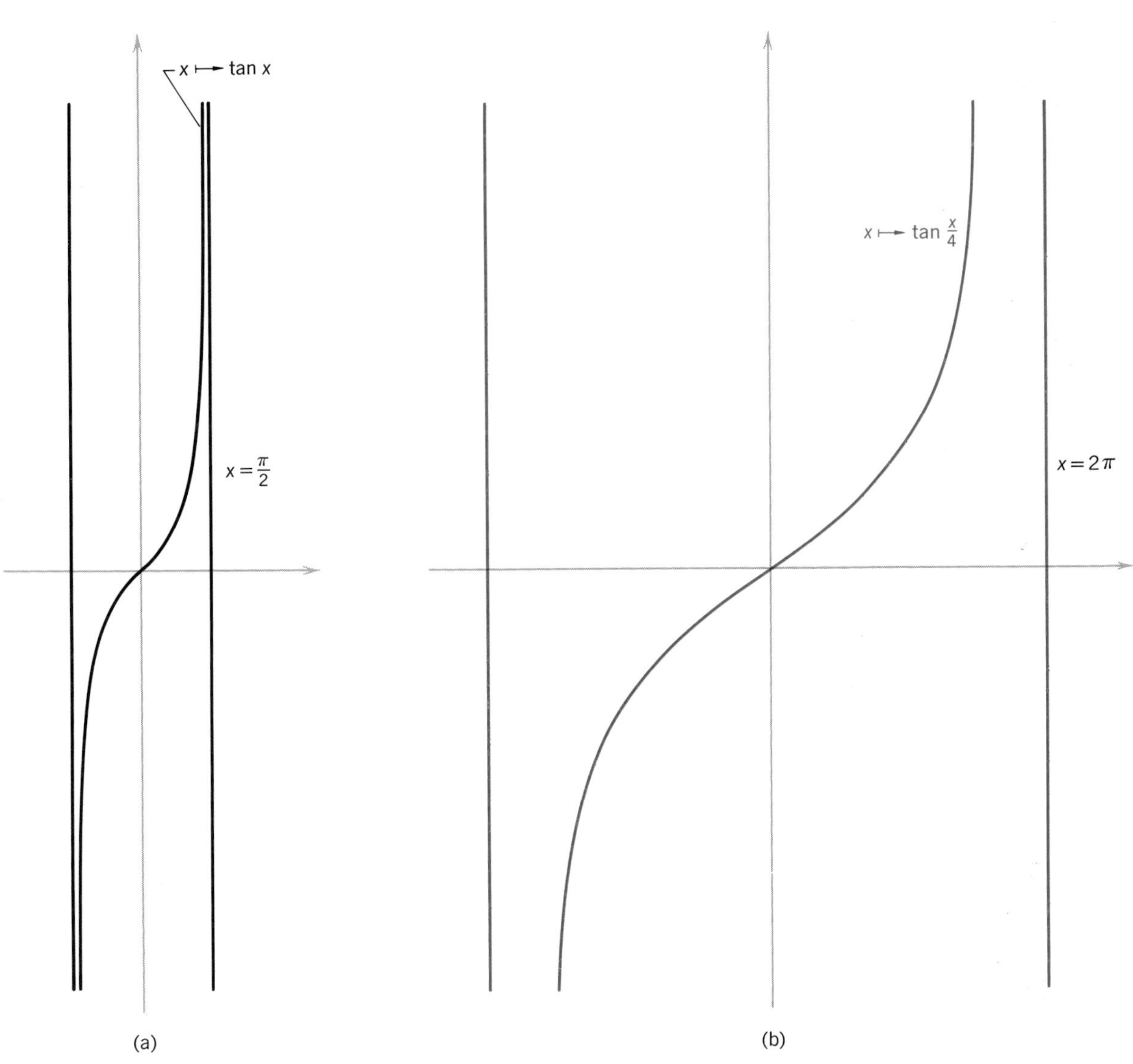

FIGURE 4-16

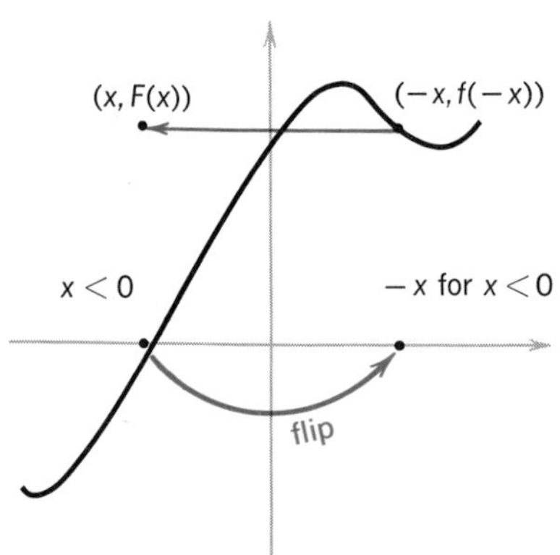

FIGURE 4-17

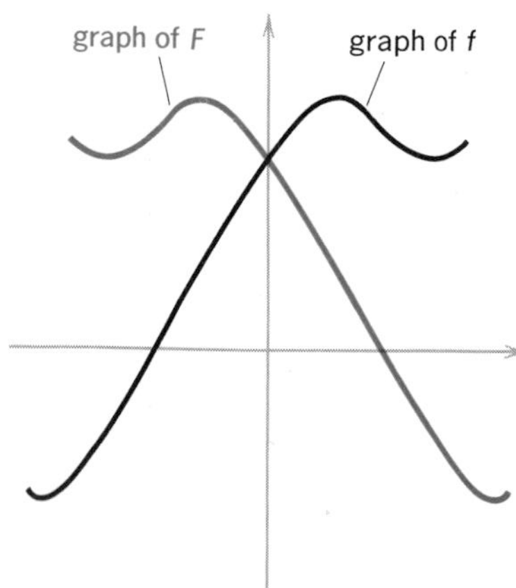

FIGURE 4-18

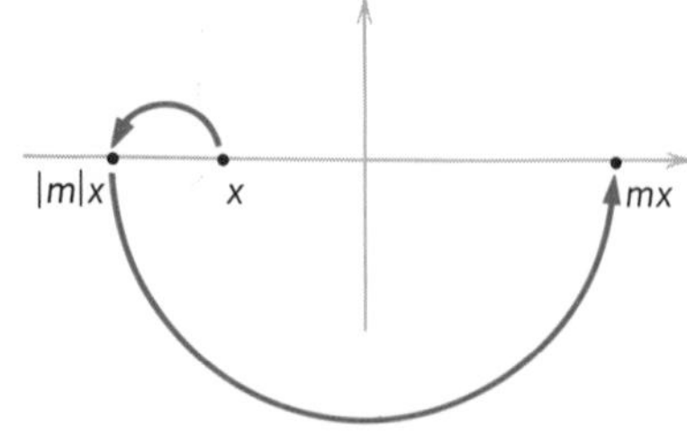

FIGURE 4-19

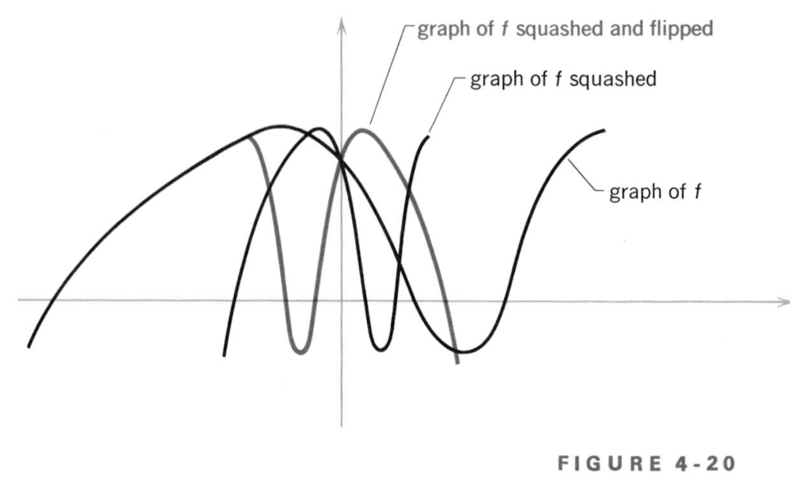

FIGURE 4-20

the "flip side" of $|m|x$ (Figure 4-19). We conclude that the result of the linear substitution of $A(x) = mx$ in f for $m < -1$ is to "squash" the graph of f and then to "flip" the squashed graph with respect to the y axis, as shown in Figure 4-20, to produce a new graph which is the graph of F.

Example 10 Let $A(x) = -\sqrt[3]{5}x$ and $f(x) = x^3$. Draw the graph of $F = f \circ A$. The graph of $f: x \mapsto x^3$ is sketched in Figure 4-21(a). The graph of $-F: x \mapsto 5x^3$ is sketched in Figure 4-21(b) and the graph of $F: x \mapsto -5x^3$ is sketched in Figure 4-21(c).

We leave it as an exercise for the reader to verify that when $-1 < m < 0$, the graph of f is first stretched, then flipped, to yield the graph of F.

Now we examine the fourth case, where m is arbitrary, $b \neq 0$, and $F(x) = f(mx + b)$. First we consider the situation when $m = 1$ and $b < 0$, or $F(x) = f(x + b)$. The effect of adding b to any x is to shift x to the left by the amount $|b|$. Hence the value of F at x is what the value of f is at $x + b$, which is to the left of x (Figure 4-22). Therefore the result of the affine substitution $x \mapsto x + b$, with $b < 0$, in f is to shift the entire graph of f **to the right** by the amount $|b|$ (Figure 4-23). It is not hard to see that if $b > 0$ and $m = 1$, then the effect of the affine substitution will be to shift the graph of f **to the left** by the amount b (Figure 4-23).

Now, if not only $b \neq 0$ but also $0 \neq m \neq 1$, then we can fully characterize the result of the affine substitution of $x \mapsto mx + b$ in f by saying that the graph of $F(x) = f(mx + b)$ is obtained from the graph of f in two steps as follows: first obtain the graph of $G(x) = f(mx)$ from the graph of f by squashing ($|m| > 1$) and flipping (if $m < 0$) or by stretching ($|m| < 1$) and flipping (if $m < 0$); next obtain the graph of $F(x) = G(x + b/m)$ by shifting the graph of G to the right if $b/m < 0$ and to the left if $b/m > 0$.

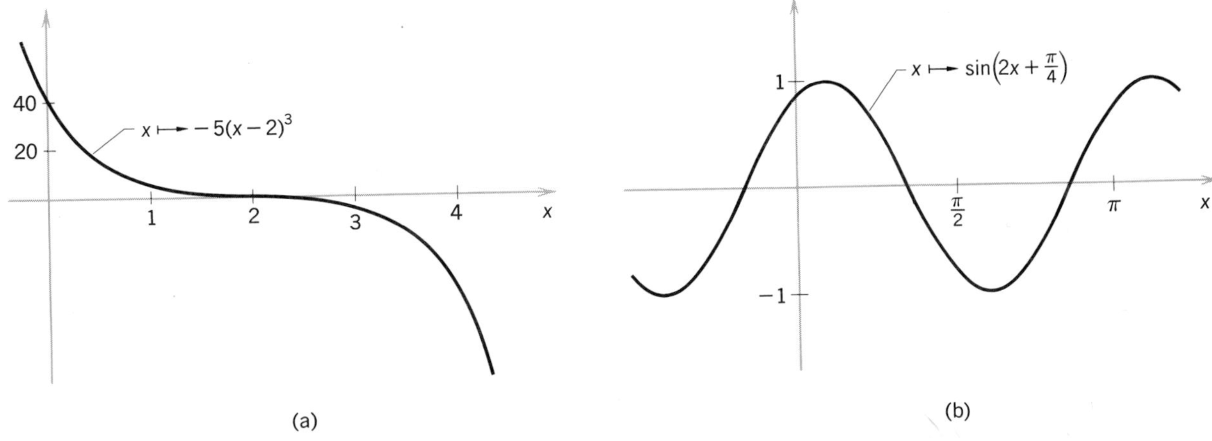

FIGURE 4-24

Example 11 The graph of $x \mapsto -5(x-2)^3$ is sketched in Figure 4-24(a). The graph of $x \mapsto \sin(2x+\frac{1}{4}\pi)$ is drawn in Figure 4-24(b).

EXERCISES

1. Graph the lines:
   (a) $2y+3=0$ (b) $5x-2y+6=0$
   (c) $3x+4y=0$ (d) $5x-2y=0$
2. Find the equation and sketch the graph of the line
   (a) with slope $\frac{2}{3}$ through the point $(4,1)$
   (b) with slope $-\frac{5}{7}$ through the point $(-1,-2)$
   (c) with slope $\frac{5}{2}$ through the point $(1,-3)$
   (d) with slope $-\frac{1}{3}$ through the point $(3,-5)$
   (e) through the points $(4,-1)$ and $(3,2)$
   (f) through the points $(0,-5$ and $(1,5)$
   (g) through the points $(3,-4)$ and $(4,-3)$
   (h) through the points $(5,0)$ and $(-2,-4)$
   (i) through the points $(-3,-2)$ and $(-3,6)$
3. Find the slopes and y intercepts of the lines in exercise 2, parts (e)–(i).
4. Give the equations of the line, through the given point and (i) parallel, (ii) perpendicular to the given line.
   (a) $(3,5)$; $4x-y=5$ (b) $(2,4)$; $2x+3y=6$
   (c) $(-2,-3)$; $3x=4y$ (d) $(0,7)$; $6x+5y=4$
   (e) $(-2,0)$; $5x+6=0$ (f) $(6,-3)$; $5x+2y=5$
   (g) $(-5,0)$; $5x+5y=1$ (h) $(7,-3)$; $3y+4=0$
5. Show that the lines $x+2y+1=0$, $6x-3y=5$, $y=2x-1$, and $4x+8y+7=0$ form a rectangle.

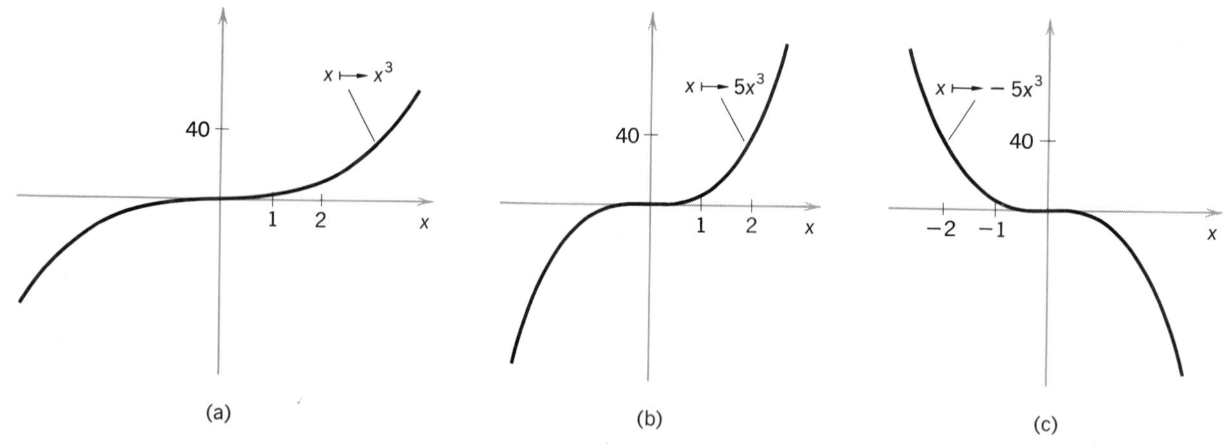

FIGURE 4-21

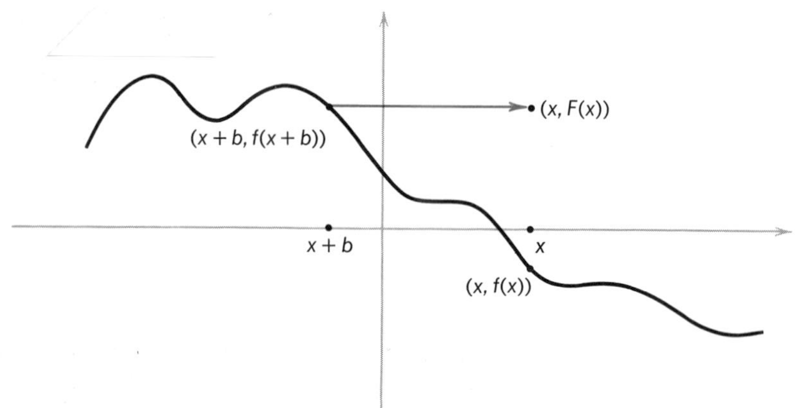

FIGURE 4-22

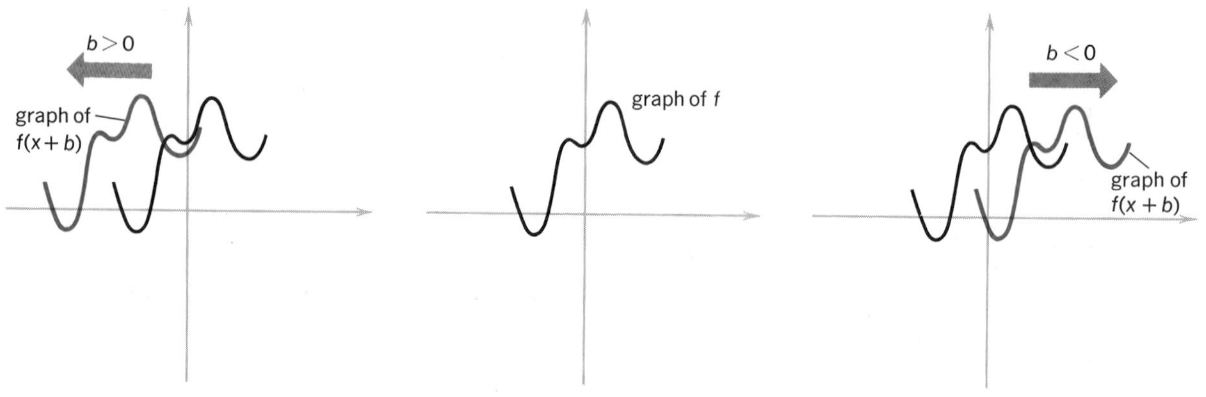

FIGURE 4-23

6. Show that the lines $2x-3y+2=0$, $4x-2y=3$, $4x-6y=1$, and $y=2x+2$ form a parallelogram.
7. Find the distance between each pair of points:
   (a) $(4,2)$ and $(7,8)$
   (b) $(-3,-5)$ and $(4,-1)$
   (c) $(\frac{3}{2},-\frac{7}{4})$ and $(-\frac{3}{4},-\frac{5}{2})$
   (d) $(\frac{5}{6},-\frac{1}{2})$ and $(\frac{1}{4},-\frac{4}{3})$
8. Find the distance between the point and the line:
   (a) $(7,-3)$ and $x-2y=3$
   (b) $(-2,1)$ and $2x+2y=1$
   (c) $(5,2)$ and $3x-4y+5=0$
   (d) $(1,6)$ and $3x+3y=2$
   (e) $(1,4)$ and $x-4y=0$
9. A moving point is equidistant from the point $(a,0)$ and the line $x+y=0$. Find the equation of its locus.
10. The distance of a point from the origin is twice its distance from the line $x+y=a$. Find the equation of its locus.
11. Find the equations of the lines through $(9,-6)$ and at distance 1 from the point $(4,-1)$.
12. Let P and Q be distinct points with coordinates (x_P,y_P) and (x_Q,y_Q), respectively. Show that the point $(\frac{1}{2}(x_P+x_Q),\frac{1}{2}(y_P+y_Q))$ is on the line segment joining P and Q; also show that this point is equidistant from P and Q and is therefore the midpoint of the line segment.
13. Sketch the curve when the given affine substitution is made in the given function. Also sketch the original function.
   (a) $f{:}x \mapsto x^2$; $A{:}x \mapsto x-1$
   (b) $f{:}x \mapsto \sin x$; $A{:}x \mapsto 5x$
   (c) $f{:}x \mapsto \cos x$; $A{:}x \mapsto \frac{1}{3}x$
   (d) $f{:}x \mapsto \sin x$; $A{:}x \mapsto (\frac{1}{2}\pi - x)$
   (e) $f{:}x \mapsto x^3$; $A{:}x \mapsto \frac{1}{20}x$
   (f) $f{:}x \mapsto x^3$; $A{:}x \mapsto -\frac{1}{5}x$

*14. Analyze the problem of composing an affine function on a given function, $A \circ f$. Consider four cases. [Hint: Squash, stretch, and flip with respect to the x-axis.]

*15. Sketch graphs of $\theta \mapsto \sin\theta$, $\theta \mapsto \sin(-2\theta+\frac{1}{6}\pi)$, $\theta \mapsto -2\sin\theta+\frac{1}{2}$, $\theta \mapsto -2\sin(-2\theta+\frac{1}{6}\pi)+\frac{1}{2}$.

16. Give the angle of inclination of each line in exercise 2, either by finding the specific angle, or by identifying it as the arc sine and arc cosine of the appropriate numbers.
17. Find the slope and y intercept of the line $x_0x+y_0y=1$, where x_0 and y_0 are nonzero constants.
18. Find the line through $(4,-1)$ with angle of inclination 30°.
19. The bisector of an angle formed by two intersecting lines is the line each of whose points is equidistant from the sides of the angle. Prove by analytic methods that the angle bisectors of the two pairs of vertical angles, formed by the intersection of lines $y=mx+b$ and $y=m'x+b'$, are perpendicular to each other.

5 CONIC SECTIONS

The term **conic section** is a generic term for four important types of curves: **circles**, **ellipses**, **hyperbolas**, and **parabolas**. The study of these curves and of the equations that define them is an important part of analytic geometry.

We begin with the circle.

Definition A **circle** is the set of points equidistant from a given point called the **center**. The distance between the center and any point on the circle is called the **radius** of the circle.

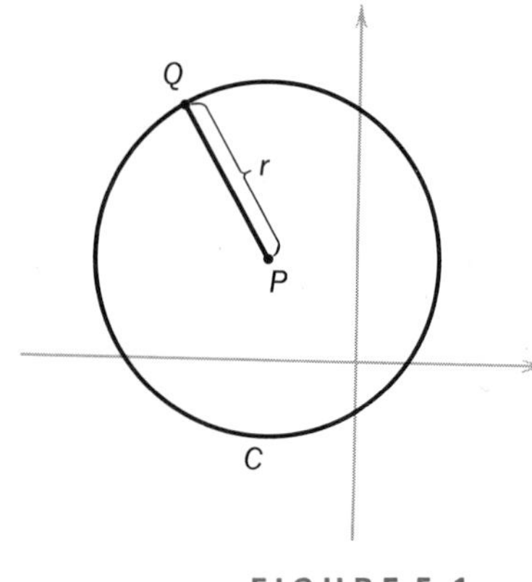

FIGURE 5-1

Given a circle C (Figure 5-1) on a plane with a given rectangular coordinate system, we may ask what relation $F(x,y)=0$ has C as its locus. By definition, the circle C is the set of points Q satisfying $\overline{PQ}=r$, where r is the radius and P is the center of C. Thus, if P has coordinates (x_0, y_0), C is the set of points (x,y) such that

$$\sqrt{(x-x_0)^2+(y-y_0)^2} = r$$

Squaring and then subtracting r^2 from both sides leads to the equation

$$(x-x_0)^2+(y-y_0)^2-r^2 = 0 \tag{1}$$

Equation (1) gives the relation between the variables x and y of which the circle C is the locus. This equation is known as the **standard form** of the circle; from it we can read at once the center and the radius of the circle.

Example 1 Find the relation of the circle with center $(-5,3)$ and radius 2. The equation of the circle is $(x-x_0)^2+(y-y_0)^2-r^2=0$. Since $x_0=-5$, $y_0=3$, and $r=2$, substituting these numbers into the equation of the circle yields $(x+5)^2+(y-3)^2-4=0$, the desired result.

Example 2 Find the equation of the circle with center $(5,-2)$ and passing through $(4,3)$. The radius is the distance between $(5,-2)$ and $(4,3)$, which is

$$r = \sqrt{(5-4)^2+(-2-3)^2} = \sqrt{1+25} = \sqrt{26}$$

Thus the equation of the circle is $(x-5)^2+(y+2)^2-26=0$.

Example 3 Find the equation of the circle passing through the points $(1,4)$, $(-3,6)$, and $(-2,5)$. We can substitute the coordinates of these points into equation (1) to yield three simultaneous equations in the three unknowns x, y, and r:

$$(1-x_0)^2 + (4-y_0)^2 - r^2 = 0 \qquad (2)$$

$$(-3-x_0)^2 + (6-y_0)^2 - r^2 = 0 \qquad (3)$$

$$(-2-x_0)^2 + (5-y_0)^2 - r^2 = 0 \qquad (4)$$

Multiplying out, we get

$$x_0^2 + y_0^2 - 2x_0 - 8y_0 + 17 - r^2 = 0 \qquad (2')$$

$$x_0^2 + y_0^2 + 6x_0 - 12y_0 + 45 - r^2 = 0 \qquad (3')$$

$$x_0^2 + y_0^2 + 4x_0 - 10y_0 + 29 - r^2 = 0 \qquad (4')$$

Subtracting (3′) from (2′) and (4′) from (3′), the second degree terms drop out, and we get, respectively,

$$-8x_0 + 4y_0 - 28 = 0$$

$$2x_0 - 2y_0 + 16 = 0$$

Solving for x_0 and y_0 yields

$$x_0 = 1, \qquad y_0 = 9$$

Substituting these values into equation (2), (3), or (4), we find that $r^2 = 25$, so the desired equation is

$$(x-1)^2 + (y-9)^2 - 25 = 0$$

Any relation $F(x, y) = 0$ which may be manipulated to take the form of equation (1) has a circle as its locus. To recognize equations of that form, let us multiply out the squares in equation (1)

$$x^2 - 2x_0 x + x_0^2 + y^2 - 2y_0 y + y_0^2 - r^2 = 0$$

Collect constant terms together and arrange these terms in the standard form for a second degree polynomial in two variables

$$x^2 + y^2 - 2x_0 x - 2y_0 y + (x_0^2 + y_0^2 - r^2) = 0$$

In particular, if x_0 and y_0 are 0, that is, if the origin is the center of the circle, this becomes the familiar equation

$$x^2 + y^2 - r^2 = 0$$

In general, given an equation of the form

$$Ax^2 + Ay^2 + Dx + Ey + F = 0$$

where $A \neq 0$, we can divide through by A

$$x^2 + y^2 + \frac{D}{A}x + \frac{E}{A}y + \frac{F}{A} = 0$$

Completing the square, we obtain

$$x^2 + \frac{D}{A}x + \left(\frac{D}{2A}\right)^2 + y^2 + \frac{E}{A}y + \left(\frac{E}{2A}\right)^2 + \frac{F}{A} = \left(\frac{D}{2A}\right)^2 + \left(\frac{E}{2A}\right)^2$$

or

$$\left(x+\frac{D}{2A}\right)^2+\left(y+\frac{E}{2A}\right)^2-\frac{D^2+E^2-4AF}{4A^2}=0$$

which is in the form of equation (1) provided that

$$\frac{D^2+E^2-4AF}{4A^2}>0$$

In this case, the center of the circle is at

$$\left(-\frac{D}{2A}, -\frac{E}{2A}\right)$$

and the radius is

$$\sqrt{\frac{D^2+E^2-4AF}{4A^2}}$$

In practice, it is easier to complete the square in each individual problem than to remember the general formulas for the center and the radius.

Example 4 Put the equation $x^2+y^2-10x-6y-2=0$ into the form of equation (1). We complete the square

$$(x^2-10x+25)+(y^2-6y+9)-2 = 25+9$$
$$(x-5)^2+(y-3)^2-36 = 0$$

The **ellipse** is a curve closely related to the circle; it is sort of a "circle with two centers," constructed as follows: Take a string and attach the ends of the string to two points P_1 and P_2. Pulling the string taut (Figure 5-2)

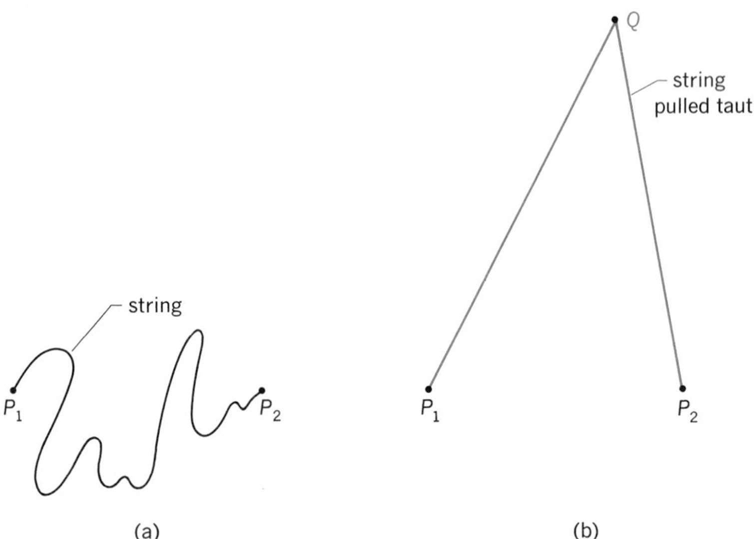

FIGURE 5-2

defines a triangle with vertices at P_1, P_2, and a third point Q. The ellipse is the locus of all such third vertices Q; this locus is shown in Figure 5-3. The points P_1 and P_2 are called the **foci** of the ellipse. (The singular of "foci" is "focus.") In the case $P_1 = P_2$, the construction will yield a circle with diameter equal to the length of the string. We summarize with a definition:

Definition An **ellipse** is the set of points the sum of whose distances from two given points called the **foci** is constant.

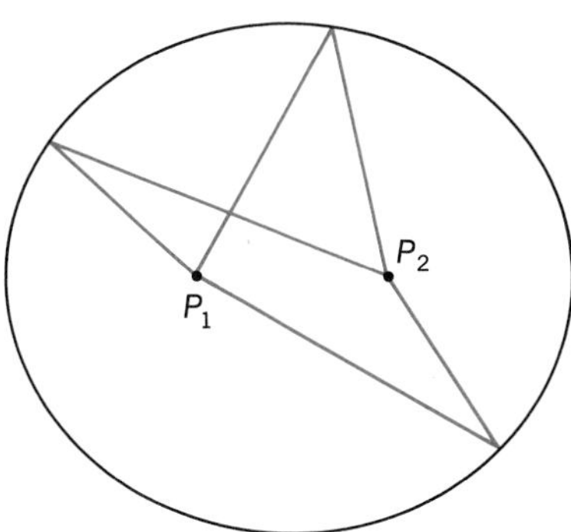

FIGURE 5-3

As we did with the circle, we may ask what relation $F(x,y)=0$ has a given ellipse as its locus. To answer this question we first make a simplifying assumption: we suppose that the foci of the ellipse are located at $(-c,0)$ and $(c,0)$ as in Figure 5-4. Let d denote the constant sum of the distances of a point (x,y) on the ellipse to the foci $(-c,0)$ and $(c,0)$, so that

$$\overline{(-c,0)(x,y)} + \overline{(c,0)(x,y)} = d$$

Using the distance formula, this becomes

$$\sqrt{(x+c)^2 + y^2} + \sqrt{(x-c)^2 + y^2} = d$$

Bringing the second radical to the right-hand side and squaring both sides, we have

$$(x+c)^2 + y^2 = d^2 - 2d\sqrt{(x-c)^2 + y^2} + (x-c)^2 + y^2 \qquad (5)$$

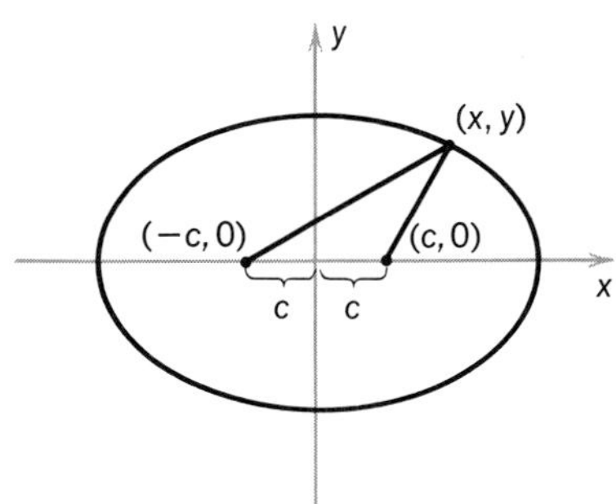

FIGURE 5-4

By multiplying out $(x+c)^2$ and $(x-c)^2$ and simplifying, (5) yields

$$\frac{d}{2} - \frac{2cx}{d} = \sqrt{(x-c)^2 + y^2} \qquad (6)$$

The appearance of $d/2$ and its reciprocal $2/d$ suggests that $d/2$, rather than d, is the important quantity; hence, let $a = d/2$, and (6) becomes

$$a - \frac{cx}{a} = \sqrt{(x-c)^2 + y^2}$$

Squaring both sides to eliminate the radical and simplifying, we obtain

$$x^2\left(\frac{a^2-c^2}{a^2}\right) + y^2 = a^2 - c^2 \qquad (7)$$

Since $d = 2a$ is the sum of two sides of a triangle of which the third side has length $2c$, we have $2a \geqslant 2c$. Hence, $a^2 \geqslant c^2$ and $a^2 - c^2$ is non-negative, and therefore has a real square root which we denote by b. Substituting b^2 for $a^2 - c^2$ in (7) we obtain the equation for the ellipse in standard form

$$\frac{x^2}{a^2} + \frac{y^2}{b^2} = 1 \qquad (8)$$

To discover the geometric significance of the constants a and b, we first choose the arbitrary point (x, y) to be the special point $(0, b)$. Since $(0, b)$ is equidistant from the foci (Why?), a is the length of the hypotenuse of the right triangle with vertices $(0, 0)$, $(0, b)$, and $(c, 0)$ as is shown in Figure 5-5. By the Pythagorean Theorem, we have $b^2 = a^2 - c^2$. The line segment from $(0, b)$ to $(0, -b)$, which is of length $2b$, is called the **minor axis** of the ellipse. If we now let $y = 0$ in equation (8), then $x^2/a^2 = 1$, and $(a, 0)$ and $(-a, 0)$ are therefore the x intercepts of the locus of equation (8). The line segment joining these two points, which is of length $2a$, is called the **major axis** of the ellipse. (Note that $b^2 = a^2 - c^2$ implies $a \geqslant b$.)

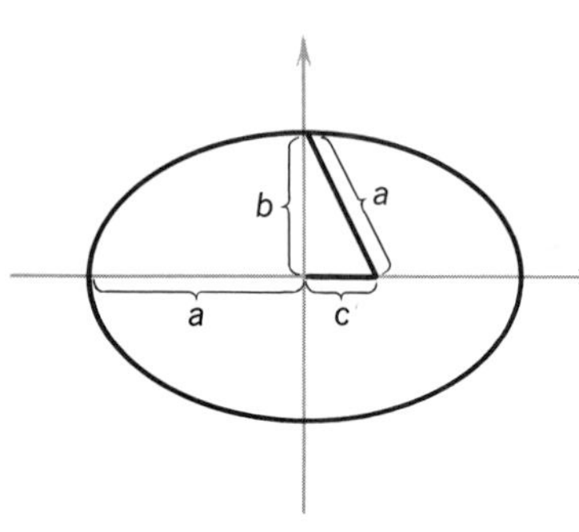

FIGURE 5-5

Example 5 The ellipse

$$\frac{x^2}{169} + \frac{y^2}{144} = 1$$

has a major axis of length 26, and minor axis of length 24. The distance of the foci from the origin is $\sqrt{a^2 - b^2} = \sqrt{169 - 144} = 5$, so the foci are the points $(\pm 5, 0)$.

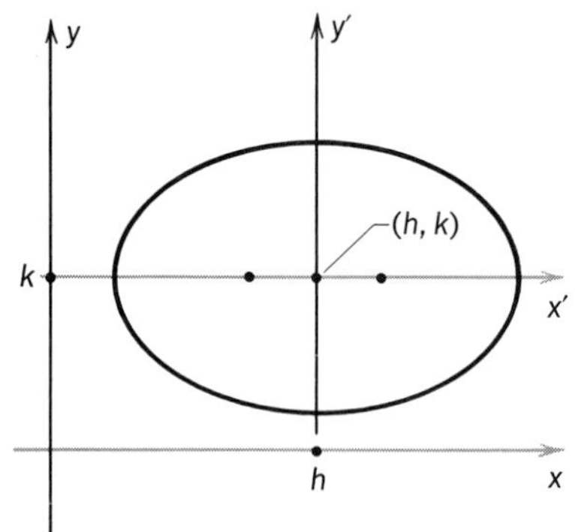

FIGURE 5-6

Example 6 Describe the ellipse whose equation is $4x^2 + 6y^2 = 3$. This equation can be transformed into the form of equation (8) yielding

$$\frac{x^2}{\frac{3}{4}} + \frac{y^2}{\frac{1}{2}} = 1$$

Therefore, the ellipse has major axis of length 3 and minor axis of length $\sqrt{2}/2$ with foci at $(\pm \frac{1}{2}, 0)$.

Suppose we are given an ellipse with its center at the point (h, k) and major axis parallel to the x axis (Figure 5-6). By an analysis similar to that just made we find that the ellipse is the locus of the relation

$$\frac{(x-h)^2}{a^2} + \frac{(y-k)^2}{b^2} = 1 \tag{9}$$

Equation (9) therefore gives the equation of the ellipse with center (h, k) and horizontal major axis. The foci are at $(h \pm \sqrt{a^2 - b^2}, k)$.

An equation of the form

$$Ax^2 + Cy^2 + Dx + Ey + F = 0$$

with A and C having the same sign can be solved by completing the square (as we did for the circle) to get an equation for the ellipse in the form of equation (9).

Example 7 Find the center, foci, and axes of the ellipse $3x^2 + 7y^2 - 12x + 28y + 19 = 0$. Completing the square, we have

$$3(x^2 - 4x + 4) + 7(y^2 + 4y + 4) + 19 = 3(4) + 7(4) = 40$$

or

$$\frac{(x-4)^2}{7}+\frac{(y+4)^2}{3}=1$$

Therefore the ellipse has center $(4, -4)$, major axis $2\sqrt{7}$, minor axis $2\sqrt{3}$ and foci at $(4\pm\sqrt{7-3}, -4)$; that is, at $(6, -4)$ and $(2, -4)$.

Example 8 Find the center, foci, and axes of the ellipse $3x^2+5y^2=10cy$. We complete the square to find

$$3x^2+5(y^2-2cy+c^2)=5c^2$$

$$\frac{x^2}{\frac{5}{3}c^2}+\frac{(y-c)^2}{c^2}=1$$

Therefore, the ellipse has center $(0, c)$, major axis of length $\frac{1}{3}c\sqrt{15}$, minor axis of length c, and foci at $(\pm\sqrt{\frac{5}{3}c^2-c^2}, c)$, that is, at $(\pm\frac{1}{3}c\sqrt{6}, c)$.

The next conic we shall discuss is the parabola.

Definition A **parabola** is the set of all points equidistant from a given line and a given point not on the given line. The given line is called the **directrix** and the given point is called the **focus**.

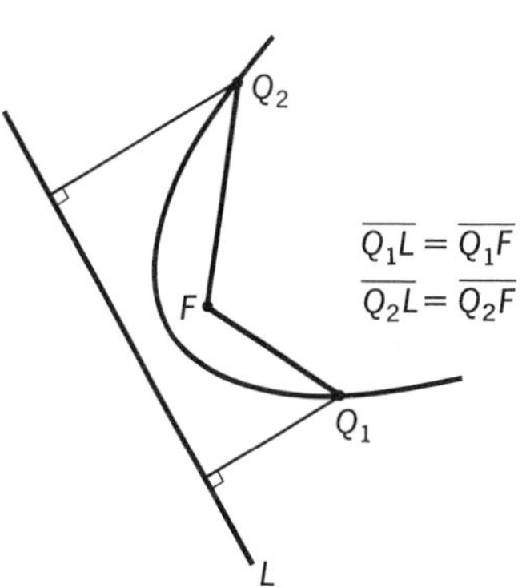

FIGURE 5-7

If we let $\overline{QL}$ denote the distance between the point Q and the line L, then the parabola with focus F and directrix L is the set of all points Q satisfying $\overline{QL}=\overline{QF}$ (Figure 5-7). To find the relation $F(x, y)=0$ of which a parabola is the locus, we suppose the coordinate system is such that the focus F lies at $(0, p)$ on the positive y axis and the directrix L has equation $y=-p$ (Figure 5-8). Let Q be a point on the parabola with coordinates (x, y). Using the distance formula and referring to Figure 5-8, the relation $\overline{QL}=\overline{QF}$ becomes

$$y+p=\sqrt{x^2+(p-y)^2}$$

Squaring both sides, bringing all terms to the left-hand side, and canceling y^2 with $-y^2$ and p^2 with $-p^2$, we obtain

$$\mathbf{4py-x^2=0} \tag{10}$$

Equation (10) is called the "standard form" of the parabola.

The point midway between the focus and the directrix of a parabola is called the **vertex**. The line through the focus and vertex perpendicular to the directrix is the **axis**. Thus the parabola of equation (10) has focus at $(0, p)$, vertex at the origin, directrix line $y=-p$, and axis $x=0$.

Example 9 Find the equation for the parabola with vertex at the origin and focus $(0, 7)$. We have $p=7$, so equation (10) becomes

$$28y-x^2=0$$

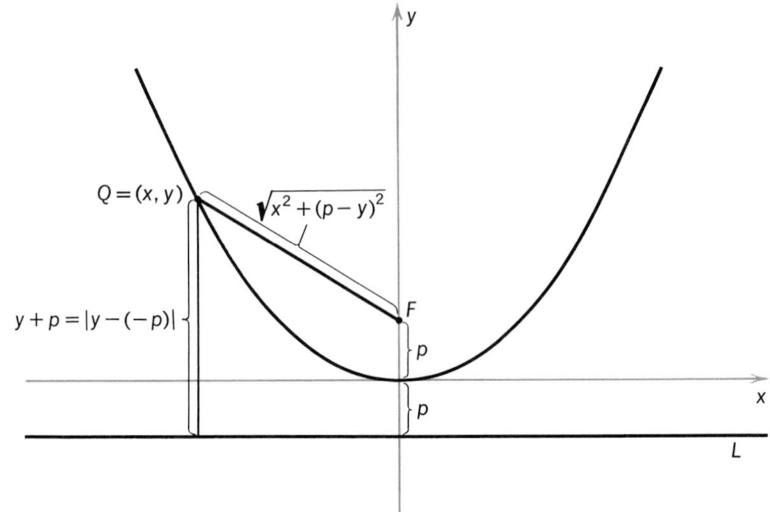

FIGURE 5-8

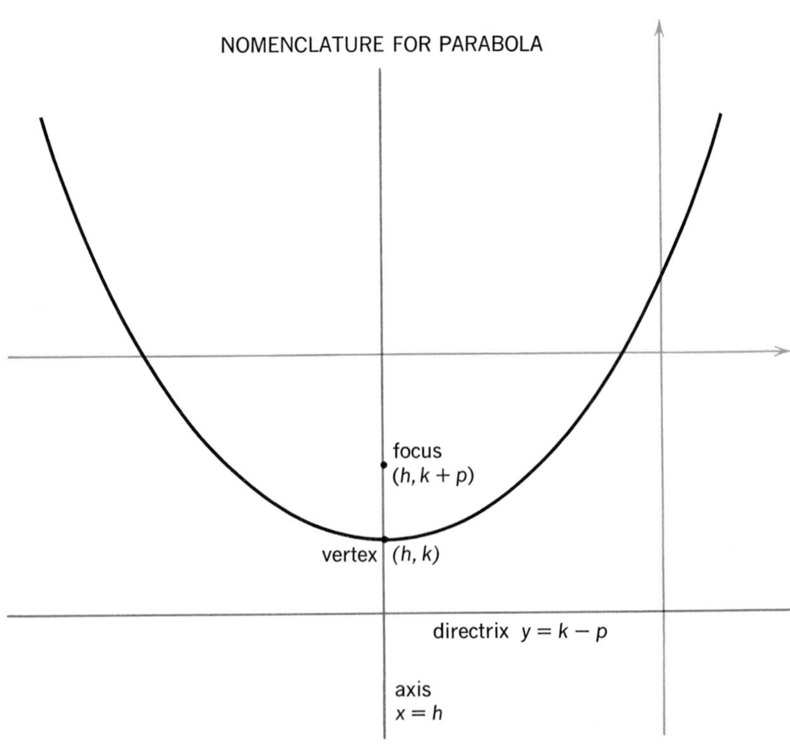

FIGURE 5-9

or

$$y = \frac{x^2}{28}$$

Example 10 Find the equation of the parabola with vertex at the origin and directrix $y=\frac{1}{2}$. We have $p=-\frac{1}{2}$, so equation (10) becomes

$$-2y - x^2 = 0$$

or

$$y = -\frac{x^2}{2}$$

A parabola with a vertical axis and with vertex at (h,k) as shown in Figure 5-9 has equation

$$4p(y-k) - (x-h)^2 = 0 \tag{11}$$

Example 11 Find the vertex and focus of the parabola whose equation is $2x^2+4x+y+6=0$. We complete the square with the x terms

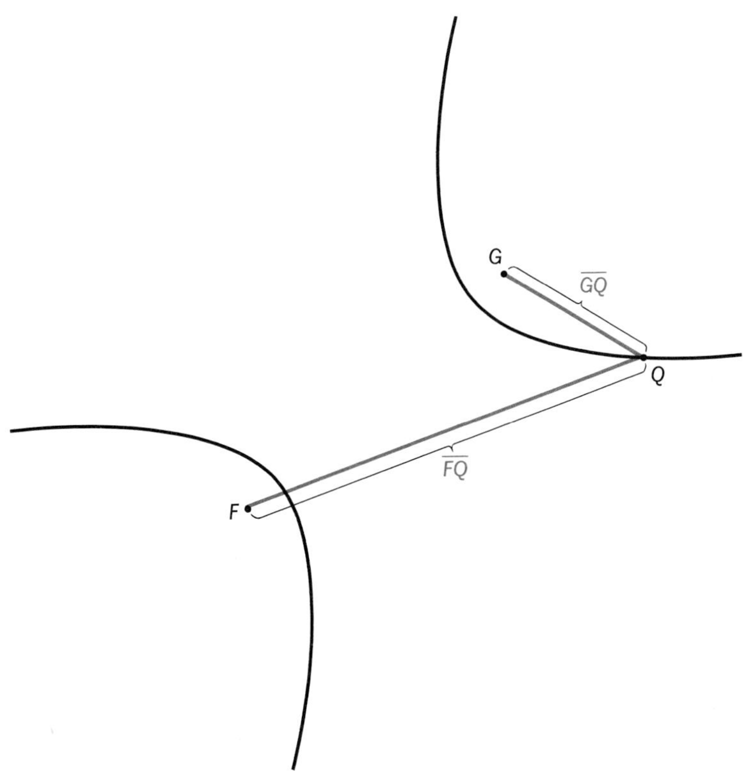

FIGURE 5-10

$$2(x^2+2x+1) + y + 6 = 2$$

$$2(x+1)^2 + (y+4) = 0$$

We rewrite in the form of equation (11) and obtain

$$4(-\tfrac{1}{8})(y+4) - (x+1)^2 = 0$$

Therefore, the vertex is at $(-1, -4)$ and $p = -\frac{1}{8}$, so the focus is at $(-1, -4\frac{1}{8})$.

Finally, we consider the **hyperbola**.

Definition A **hyperbola** is the locus of points, the **difference** of whose distances from two given points is constant. The given points are called the **foci** of the hyperbola.

If F, G denote the foci and Q denotes an arbitrary point on the hyperbola, then the difference between the distances $\overline{FQ}$ and $\overline{GQ}$ is $|\overline{FQ}-\overline{GQ}|$, and the hyperbola itself is the set of points Q such that $|\overline{FQ}-\overline{GQ}| = d$, where d is the constant difference of distances (Figure 5-10).

Let the foci of the hyperbola lie on the x axis at equal distances c from the origin (Figure 5-11). The condition that an arbitrary point (x, y)

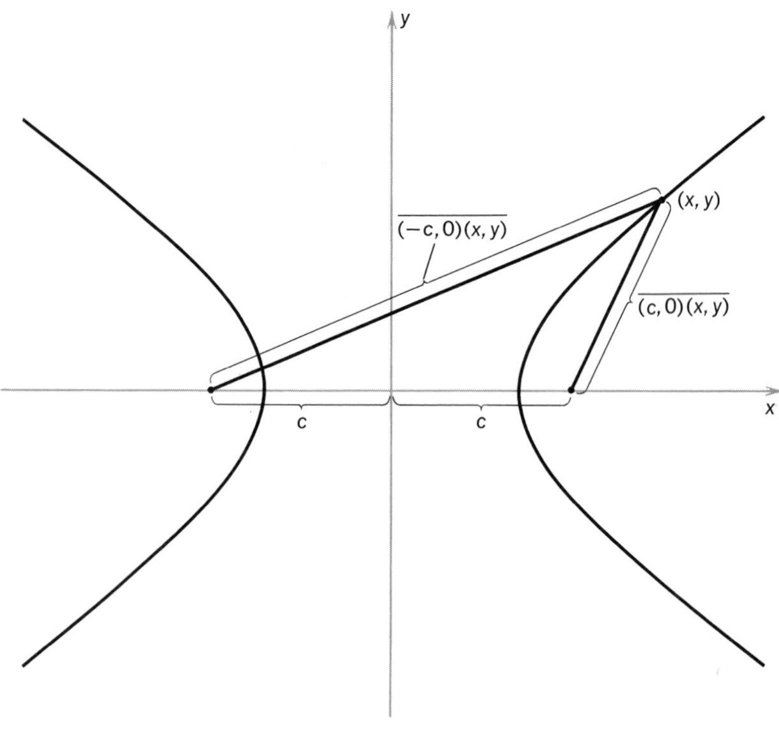

FIGURE 5-11

lie on this hyperbola is that

$$|\overline{(c,0)(x,y)} - \overline{(-c,0)(x,y)}| = d$$

or, by the distance formula and the definition of absolute value

$$d = \sqrt{(x \mp c)^2 + y^2} - \sqrt{(x \pm c)^2 + y^2}$$

according as $\sqrt{(x-c)^2+y^2}$ is greater or less than $\sqrt{(x+c)^2+y^2}$. Adding $\sqrt{(x \pm c)^2+y^2}$ to both sides, squaring, multiplying out squares of sums, and canceling equal quantities occurring with the same sign on both sides, we obtain

$$\boldsymbol{d^2 + 2d\sqrt{(x \pm c)^2 + y^2} = \mp 4cx} \tag{12}$$

Solving this for the radical, squaring, putting all terms on the left, and simplifying, equation (12) becomes

$$\frac{4}{d^2}x^2 + \frac{4}{d^2-4c^2}y^2 - 1 = 0$$

Letting $a = d/2$, we may write the relation whose locus is the hyperbola as

$$\frac{x^2}{a^2} + \frac{y^2}{a^2-c^2} = 1$$

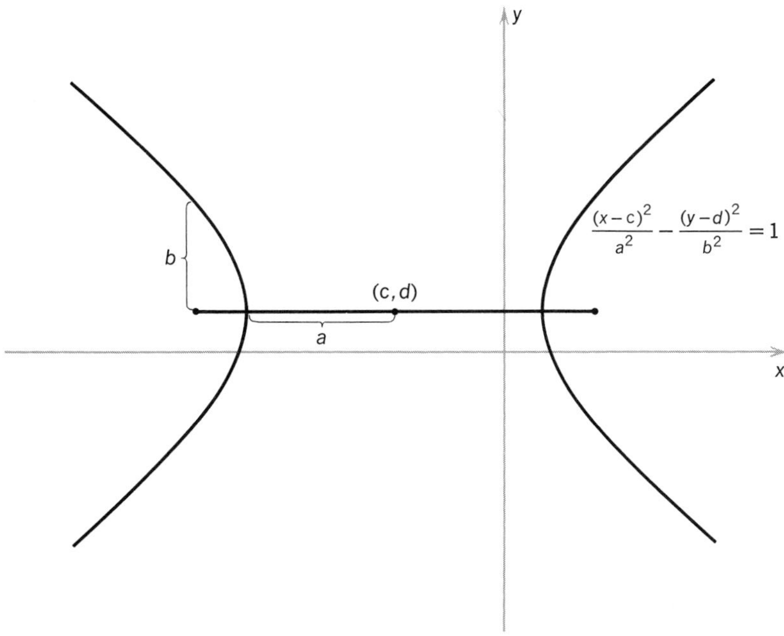

FIGURE 5-12

Note that, since the difference of two sides of a triangle is less than the third side, we have $d \leqslant 2c$, or equivalently, $2a \leqslant 2c$, that is, $a^2 \leqslant c^2$. Therefore, $a^2 - c^2$ is negative, so $c^2 - a^2$ is positive, and has a positive square root which we call b. The "standard form" for the relation of which a hyperbola is the locus is

$$\frac{x^2}{a^2} - \frac{y^2}{b^2} = 1 \tag{13}$$

where a and b are positive real numbers, and the foci are at $(\pm c, 0)$ or $(\pm\sqrt{a^2+b^2}, 0)$.

The midpoint of the line segment connecting the foci is the **center** of the hyperbola. In equation (13), the center is at the origin. If the center is at (h, k) and the foci are on a line parallel to the x axis (Figure 5-12), the equation of the hyperbola is

$$\frac{(x-h)^2}{a^2} - \frac{(y-k)^2}{b^2} = 1 \tag{14}$$

and the foci are at $(h \pm \sqrt{a^2+b^2}, k)$. If the foci are on a line parallel to the y axis (Figure 5-13), we have, interchanging the roles of x and y,

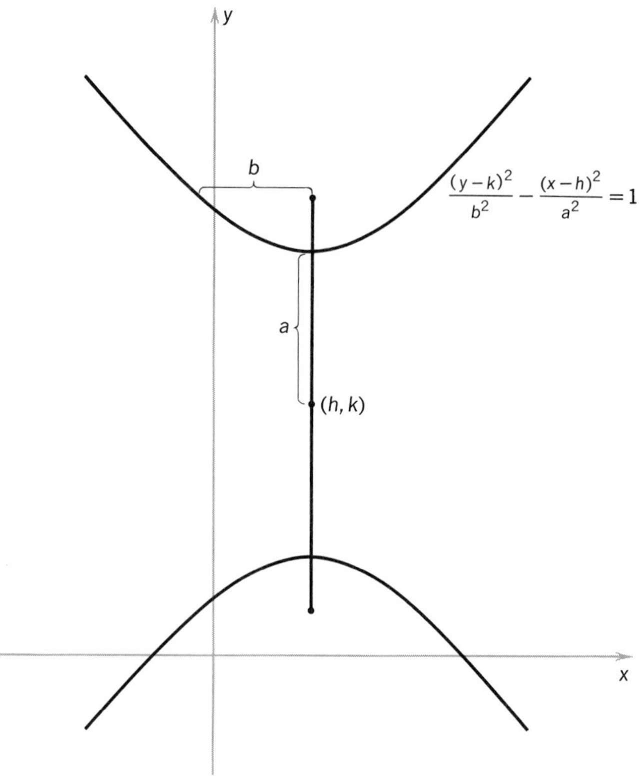

FIGURE 5-13

$$\frac{(y-k)^2}{b^2} - \frac{(x-h)^2}{a^2} = 1 \qquad (15)$$

and the foci are at $(h, k \pm \sqrt{a^2+b^2})$.

Example 12 The hyperbola $\frac{1}{16}x^2 - y^2 = 1$ has foci at $(\pm\sqrt{17}, 0)$ and center at the origin.

Example 13 The hyperbola $3x^2 - 4y^2 + 6x + 6y = 0$ can be put in the form (14) by completing squares

$$3(x^2+2x+1) - 4(y^2 - \tfrac{3}{2}y + \tfrac{9}{16}) = \tfrac{3}{4}$$

or

$$3(x+1)^2 - 4(y-\tfrac{3}{4})^2 = \tfrac{3}{4}$$

or

$$\frac{(x+1)^2}{\frac{1}{4}} - \frac{(y-\frac{3}{4})^2}{\frac{3}{16}} = 1$$

Therefore the center of the hyperbola coordinates has $(-1, \frac{3}{4})$, the foci are at $(-1 \pm \sqrt{\frac{1}{4}+\frac{3}{16}}, \frac{3}{4})$ or $(-1 \pm \frac{1}{4}\sqrt{7}, \frac{3}{4})$.

Example 14 The hyperbola $x^2 - 4y^2 + 4ax - 8ay = 4$ can also be put in standard form by completing the square

$$(x^2+4ax+4a^2) - 4(y^2-2ay+a^2) = 4$$

$$\frac{(x+2a)^2}{4} - (y-a)^2 = 1$$

Therefore the center is at $(-2a, a)$, the foci are at $(-2a \pm \sqrt{4+1}, a)$ or $(-2a \pm \sqrt{5}, a)$, and the vertices are at $(-2a \pm \sqrt{2}, a)$.

The **vertices** of the hyperbola are the points of intersection of the hyperbola with the line segment connecting the foci. This line is called the **major axis** of the hyperbola. The **minor axis** is the line perpendicular to the major axis passing through the center of the hyperbola (Figure 5-14). In Example 12 above, the vertices of the hyperbola are $(\pm 4, 0)$ and in Example 13, $(-1 \pm \frac{1}{2}, \frac{3}{4})$.

EXERCISES

1. Draw the circles with the following equations:
   (a) $x^2 + y^2 + 2x + 8y - 8 = 0$
   (b) $x^2 + y^2 - 10x - 6y - 2 = 0$
   (c) $x^2 + y^2 - 20x = 0$
   (d) $x^2 - 2x + y^2 - 4y = 0$
   (e) $2x^2 + 2y^2 = y + 2$
   (f) $3x^2 + 3y^2 + 2x - 1 = 0$

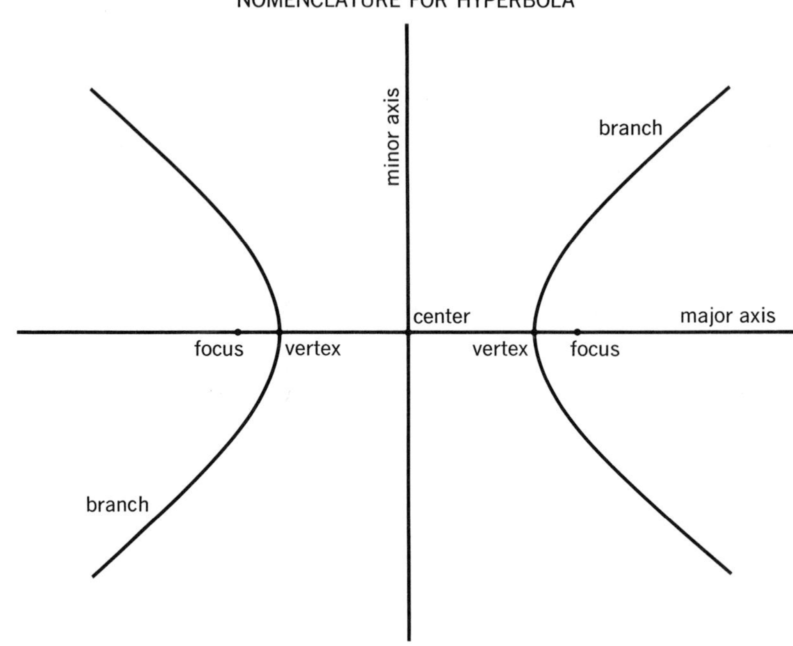

FIGURE 5-14

(g) $4x^2 + 4y^2 + 2x - 6y + 1 = 0$
(h) $5x^2 + 5y^2 - 8x + 6y = 0$
(i) $3x^2 + 3y^2 - 5x - 40 = 0$
(j) $3x^2 + 3y^2 - 4x + 2y + 2 = 0$

2. Find the equation of the circle
   (a) with center $(-5, 3)$ and radius $\sqrt{2}$
   (b) with center $(3, -2)$ and radius $\sqrt{13}$
   (c) with center $(a, 2a)$ and radius $2a$
   (d) with center $(5, -2)$ and passing through $(4, 3)$
   (e) with center $(-1, 2)$ and passing through $(0, 2)$
   (f) with center (h, k) and passing through the origin
   (g) with points $(3, 1)$ and $(5, -3)$ as ends of a diameter
   (h) with points $(2a, 0)$ and $(0, 2b)$ as ends of a diameter
   (i) through the points $(6, 1)$, $(5, -1)$, and $(-2, -3)$
   (j) through the points $(5, 4)$, $(2, 2)$, and $(4, -1)$
   (k) through the points $(a, 0)$, (a, a), and $(0, 2a)$
   (l) with center $(6, 2)$ and through the point $(-1, -3)$
   (m) with radius 10, through $(4, 1)$ and $(6, 3)$
   (n) with radius $\frac{5}{4}\sqrt{2}$, through $(1, 1)$ and $(2, 4)$

3. Find the equation of the line through the origin and tangent to the circle $(x-7)^2+(y+12)^2 = 64$. [Hint: A line tangent to a circle is perpendicular to the radius at the point of tangency.]

4. Put these ellipses in standard form; find the axes and locate the foci; sketch the ellipse.
   (a) $3x^2 + 4y^2 - 12x - 8y - 2 = 0$
   (b) $x^2 + 4y^2 - 12x + 8y + 36 = 0$
   (c) $9x^2 + 10y^2 + 40y + 39 = 0$
   (d) $x^2 + 16y^2 + 2x = 15$
   (e) $3x^2 + 5y^2 = 15$
   (f) $3x^2 + 4y^2 = 12$
   (g) $2x^2 + 4y^2 = 1$
   (h) $144x^2 + 225y^2 = 400$

5. Reduce these parabolas to standard form; find the focus, vertex, and axis of each. Draw each parabola.
   (a) $x^2 + 8y + 8 = 0$
   (b) $x^2 - 55y + 10 = 0$
   (c) $x^2 - 4x - 4y = 0$
   (d) $x^2 + 6x - y - 2 = 0$
   (e) $2x^2 + 4x + y + 6 = 0$
   (f) $x^2 - 20x + y = 20$
   (g) $2x^2 + 9x + 3y + 20 = 0$

6. Find the center, foci, vertices, and axes of each of the following hyperbolas. Draw the hyperbolas.
   (a) $x^2 - 2y^2 + ax = 0$
   (b) $4y^2 = x^2 - 3x + 4$
   (c) $4x^2 - y^2 - 2ay = 0$
   (d) $3x^2 - 4y^2 + 6x + 6y = 0$
   (e) $4x^2 - 4y^2 + 4x - 8y = 1$

7. Suppose the relation $Ax^2 + Cy^2 + Dx + Ey + F = 0$ has a conic section for its locus. Show that
   (a) $AC = 0$ if the locus is a parabola
   (b) $AC > 0$ if the locus is an ellipse
   (c) $AC < 0$ if the locus is a hyperbola

*6 CHANGING COORDINATE SYSTEMS

Before the introduction of a coordinate system, there are no distinguished points, lines, or other special figures in the plane. Relationships of similarity, congruence, tangency, or perpendicularity, etc., between figures are defined independently of possible coordinate systems. Because the choice of a coordinate system is arbitrary and can be changed when the occasion demands, it is natural to investigate how to transcribe the relations expressed in one coordinate system into the analytic "language" of another coordinate system. For now, we consider only rectangular (Cartesian) coordinate systems and how two such systems can differ by a **translation**, and/or by a **rotation**.

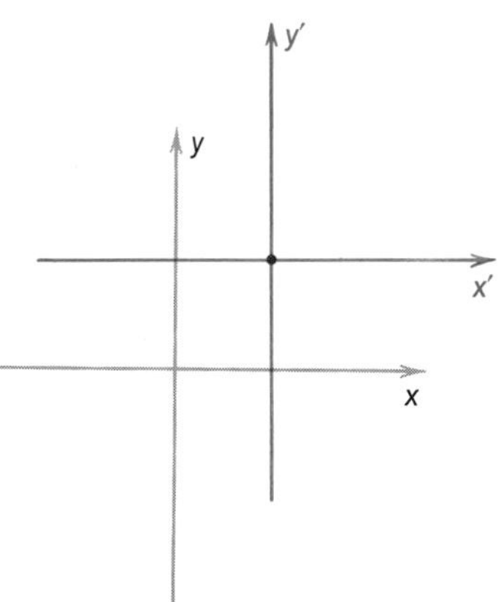

FIGURE 6-1

Suppose that two rectangular coordinate systems, say the xy system and the $x'y'$ system, differ by a translation, that is, we assume that the x axis and the x' axis are parallel, have the same scale, and have their positive rays pointing the same way; similarly we assume that the y axis and the y' axis have the same direction and scale (Figure 6-1). A point P in the plane now has two sets of coordinates, (x_P, y_P) and (x'_P, y'_P).

We shall now derive equations relating x_P to x'_P and y_P to y'_P. Let $\mathbf{O}'$ be the origin of the $x'y'$ system, and let (h, k) be the xy coordinates of $\mathbf{O}'$. Recalling how the Cartesian coordinates of a point are determined, we see (Figure 6-2) that

$$x'_P = x_P - h$$

$$y'_P = y_P - k$$

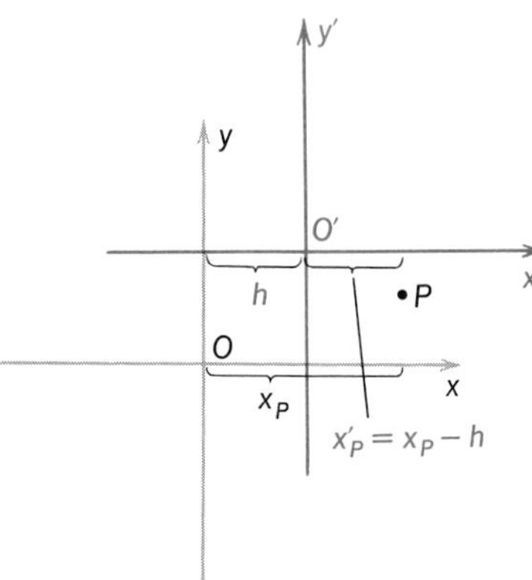

FIGURE 6-2

Given a relation $F(x, y) = 0$, we can substitute $x = x' + h$, $y = y' + k$ to obtain $F(x'+h, y'+k) = 0$, which is a relation between x' and y'. The locus of $F(x, y) = 0$ is the same as the locus of $F(x'+h, y'+k) = 0$.

Example 1 The relation

$$\frac{(x-h)^2}{a^2} + \frac{(y-k)^2}{b^2} - 1 = 0$$

has as its locus an ellipse with center at (h, k) and major axis parallel to the x axis. If we translate according to

$$x' = x - h$$

$$y' = y - k$$

the equation of the ellipse becomes

$$\frac{x'^2}{a^2} + \frac{y'^2}{b^2} - 1 = 0$$

The center of the ellipse is at the origin of the $x'y'$ system and the major axis lies on the x' axis (see Figure 6-3).

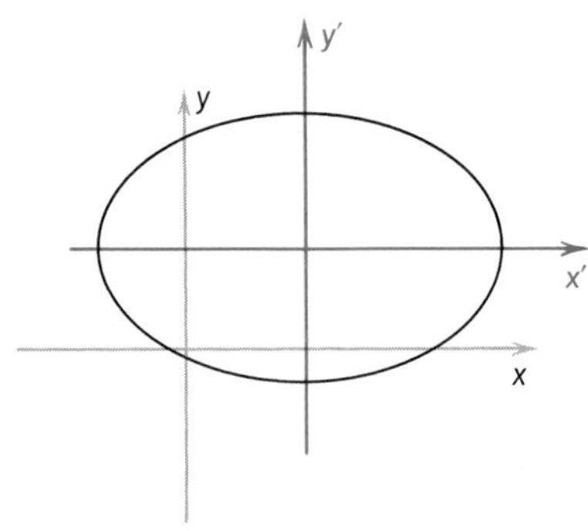

FIGURE 6-3

Example 2 Let a translation be defined by

$$x' = x - b$$

$$y' = y$$

where $b \neq 0$. Suppose $f: \boldsymbol{R} \to \boldsymbol{R}$ is a real function. Consider the equations

(a) $y' - f(x') = 0$

(b) $y - f(x) = 0$

(c) $y - f(x-b) = 0$

(d) $y' - f(x'+b) = 0$

The graphs of (a) and (c) are the same, and the graphs of (b) and (d) are the same.

We turn to the case where two coordinate systems differ by a **rotation**.

By this we mean that the origins of the systems coincide and there is an angle θ such that the x' and y' axes are obtained by a counterclockwise rotation of the x and y axes through the angle θ.

In Figure 6-4, the angle of rotation θ is $\pi/4$, or 45°. We wish to express the $x'y'$ coordinates of an arbitrary point P in terms of the angle of rotation θ and the xy coordinates of P. In Figure 6-4, let α denote the angle between the x' axis and the line segment joining P to the origin, and let r denote the distance

$$\sqrt{x^2+y^2} = \sqrt{x'^2+y'^2}$$

from P to the common origin of the two systems. We can apply the definitions of sine and cosine to write

$$\begin{aligned} x' &= r\cos\alpha \\ y' &= r\sin\alpha \end{aligned} \tag{1}$$

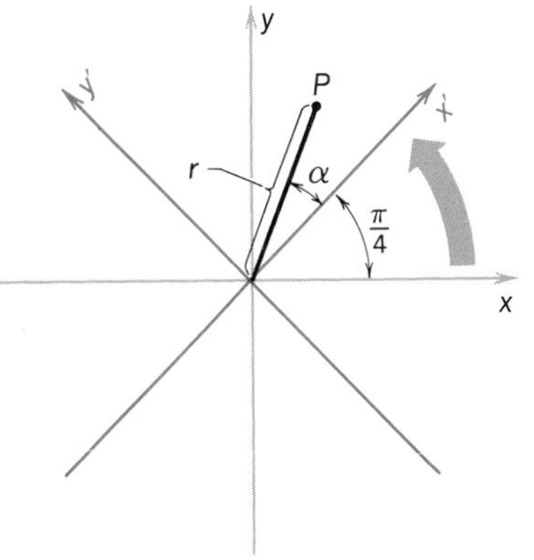

FIGURE 6-4

Similarly, we may also write

$$\begin{aligned} x &= r\cos(\theta+\alpha) \\ y &= r\sin(\theta+\alpha) \end{aligned} \tag{2}$$

Expanding the right-hand side of equations (2) by the formulas for the sine and cosine of sums of angles, we obtain

$$\begin{aligned} x &= r(\cos\theta\cos\alpha - \sin\theta\sin\alpha) \\ y &= r(\sin\theta\cos\alpha + \cos\theta\sin\alpha) \end{aligned} \tag{3}$$

Multiplying through by r and substituting for $r\cos\alpha$ and $r\sin\alpha$ in accord with equations (1), we conclude that

$$\begin{aligned} \boldsymbol{x} &= \boldsymbol{x'\cos\theta - y'\sin\theta} \\ \boldsymbol{y} &= \boldsymbol{x'\sin\theta + y'\cos\theta} \end{aligned} \tag{4}$$

Conversely, to express x' and y' explicitly in terms of x and y we may solve the simultaneous linear equations (4) to yield

$$\begin{aligned} \boldsymbol{x'} &= \boldsymbol{x\cos\theta + y\sin\theta} \\ \boldsymbol{y'} &= \boldsymbol{-x\sin\theta + y\cos\theta} \end{aligned} \tag{5}$$

When we rotate from coordinate system S to S', any relation $F(x,y)=0$ in S is transformed into a relation $G(x',y')=0$ by means of equations (4). The locus of $G(x',y')=0$ is the same as the locus of $F(x,y)=0$.

Example 3 Consider the locus of $x^2-2xy+y^2-x\sqrt{2}-y\sqrt{2}=0$. Let the coordinate system be rotated through 45°. We shall find the equation of this locus in the new coordinate system. Using equations (4) for rotation, we have

$$
\begin{aligned}
x &= x' \cos 45° - y' \sin 45° \\
&= \frac{1}{\sqrt{2}}(x' - y') \\
y &= x' \sin 45° + y' \cos 45° \\
&= \frac{1}{\sqrt{2}}(x' + y')
\end{aligned}
$$

Substituting these values into the original equation, we obtain the relation

$$\tfrac{1}{2}(x'-y')^2 - (x'-y')(x'+y') + \tfrac{1}{2}(x'+y')^2 - (x'-y') - (x'+y') = 0$$

or

$$2y'^2 - 2x' = 0$$

or

$$y'^2 = x'$$

The last equation describes the parabola shown in Figure 6-5.

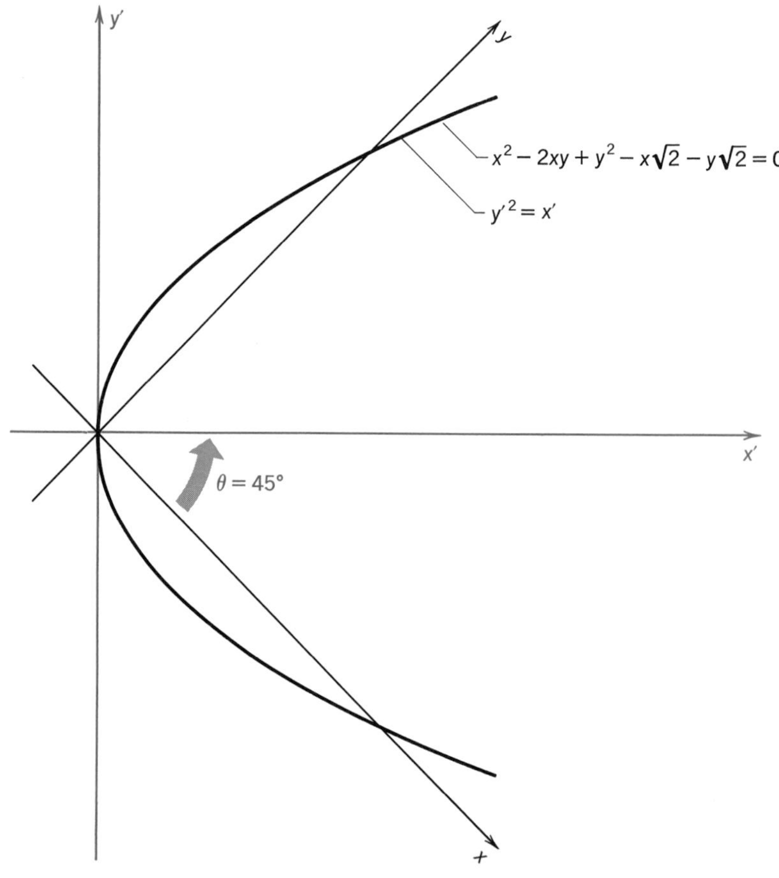

FIGURE 6-5

Example 4 Let us refer back to Example 3. Suppose that the $x'y'$ system is translated 2 units in the x' direction and -3 units in the y' direction to obtain the $x''y''$ coordinate system. Using the equations for translation, we have

$$x' = x'' + 2$$
$$y' = y'' - 3$$

Therefore, with this substitution the equation of the parabola becomes

$$(y''-3)^2 = (x''+2)$$
$$x'' = y''^2 - 6y'' + 7$$

The results of the rotation followed by the translation are shown in Figure 6-6.

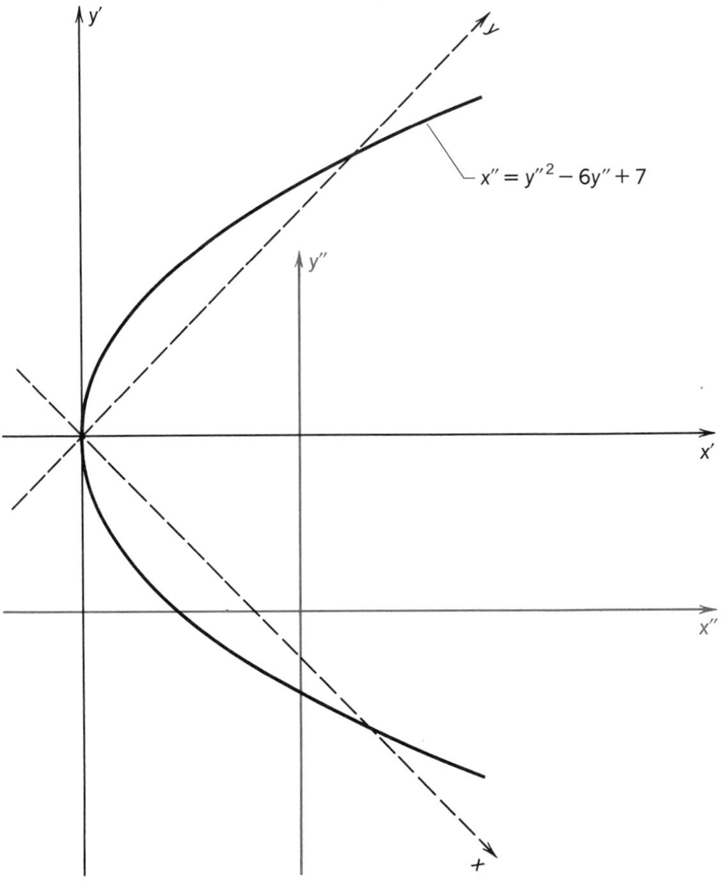

FIGURE 6-6

In §5 we saw that a conic section with an axis of symmetry parallel to a coordinate axis is the locus of a relation of the form

$$Ax^2 + Cy^2 + Dx + Ey + F = 0 \tag{6}$$

If we drop the requirement that the axes of the conic be parallel to the coordinate axes, then the conic is the locus of a relation of the form

$$Ax^2 + Bxy + Cy^2 + Dx + Ey + F = 0 \tag{7}$$

Clearly, (6) is the special case of (7) with $B = 0$.

Example 5 Let $k > 0$. Consider the hyperbola with foci at (k, k) and $(-k, -k)$ with constant difference of distances equal to $2k$. This curve is defined by the equation

$$\sqrt{(x+k)^2 + (y+k)^2} - \sqrt{(x-k)^2 + (y-k)^2} = \pm 2k$$

After elimination of the radical signs, this equation reduces to

$$2xy - k^2 = 0$$

It is a striking theorem of analytic geometry that an equation of the form (7) always has as its locus some conic section (that is, a circle, an ellipse, a parabola, or a hyperbola), or some "degenerate conic" (that is, a line, a pair of lines, or a point), or no locus at all. If we start with an equation

$$Ax^2 + Bxy + Cy^2 + Dx + Ey + F = 0 \tag{8}$$

we can rotate the coordinate system so that upon substitution (8) becomes

$$A'x'^2 + C'y'^2 + D'x' + E'y' + F' = 0 \tag{9}$$

By completing squares, we can analyze (9) and determine its locus. The step from (8) to (9) is the crucial one, and it is accomplished as follows. We substitute

$$x = x' \cos\theta - y' \sin\theta$$

$$y = y' \sin\theta + y' \cos\theta$$

into (8). Multiplying out and collecting terms, we obtain an equation in x' and y'. Then we set the coefficient of $x'y'$ equal to zero and solve for θ (or rather solve for $\sin\theta$ and $\cos\theta$). Having found $\sin\theta$ and $\cos\theta$ such that the $x'y'$ term vanishes, we substitute these values into the equation in x' and y' to obtain an equation of the form (9).

Example 6 Analyze the conic section

$$11x^2 - 24xy + 4y^2 + 6x + 8y - 10 = 0$$

We first substitute for x and y from formula (4)

$$11(x'\cos\theta - y'\sin\theta)^2 - 24(x'\cos\theta - y\sin\theta) + (x'\sin\theta + y'\cos\theta)^2$$
$$+ 4(x'\sin\theta + y'\cos\theta)^2 + 6(x'\cos\theta - y'\sin\theta)$$
$$+ 8(x'\sin\theta + y'\cos\theta) - 10 = 0$$

Multiplying out and collecting terms, we obtain

$$(11\cos^2\theta - 24\cos\theta\sin\theta + 4\sin^2\theta)x'^2$$
$$+ (24\sin^2\theta - 14\cos\theta\sin\theta - 24\cos^2\theta)x'y'$$
$$+ (11\sin^2\theta + 24\cos\theta\sin\theta + 4\cos^2\theta)y'^2 + (6\cos\theta + 8\sin\theta)x'$$
$$+ (8\cos\theta - 6\sin\theta)y' - 10 = 0$$

We set the $x'y'$ coefficient in the above equation equal to zero,

$$24\sin^2\theta - 14\sin\theta\cos\theta - 24\cos^2\theta = 0$$

Solving by the quadratic formula, we get

$$\sin\theta = \cos\theta\left(\frac{7 \pm 25}{24}\right)$$

We choose the + part of the $\pm$ sign to get

$$\sin\theta = \tfrac{4}{3}\cos\theta$$

Combining this with the equation

$$\sin^2\theta + \cos^2\theta = 1,$$

we get $\frac{25}{9}\cos^2\theta = 1$, so $\cos\theta = \frac{3}{5}$ and $\sin\theta = \frac{4}{5}$. Substituting these values into equation (3) and simplifying, we get

$$-5x'^2 + 20y'^2 + 10x' - 10 = 0$$

Completing squares, we get

$$20y'^2 - 5(x'^2 - 2x' + 1) = 5$$

$$\frac{y'^2}{\frac{1}{4}} - \frac{(x'-1)^2}{1} = 1$$

Therefore the conic section is a hyperbola, and in the new coordinate system the center is at $(1, 0)$, the foci are at $(1, \pm\sqrt{5}/2)$, the major axis is $x' = 1$, the minor axis is the x' axis, and the vertices are $(1, \pm\frac{1}{2})$.

Using the values for $\cos\theta$ and $\sin\theta$, we find that in the original coordinate system, the hyperbola has center at $(\frac{3}{5}, \frac{4}{5})$, foci at

$$\left(\frac{3-2\sqrt{5}}{5}, \frac{8+3\sqrt{5}}{10}\right) \quad \text{and} \quad \left(\frac{3+2\sqrt{5}}{5}, \frac{8-3\sqrt{5}}{10}\right),$$

and vertices at $(\frac{1}{5}, \frac{11}{10})$, $(1, \frac{1}{2})$. The equations of the axes are $1 = \frac{3}{5}x + \frac{4}{5}y$ for the major axis, and $0 = \frac{3}{5}y - \frac{4}{5}x$ for the minor axis, or simplifying the major axis is $3x + 4y = 5$ and the minor axis is $3y = 4x$.

EXERCISES

1. Give the equation of the given locus when the coordinate system is translated as directed.
   (a) $9x^2+4y^2=25$; translated by amount -1 in the x direction and 2 in the y direction
   (b) $4x^2-9y^2=36$; 3 units in the x direction, 4 units in the y direction
   (c) $x^2-4y^2-6x-32y=59$; -3 units in the x direction and 4 units in the y direction

2. Make a rotation with the indicated angle and give the new form of the equation.
   (a) $x^2 - \sqrt{3}xy + 2y = 10,\ \theta = \frac{1}{6}\pi$
   (b) $4x^2 + 3\sqrt{3}xy + y^2 = 22,\ \theta = 120°$
   (c) $x^2 - \sqrt{3}xy + 12 = 0,\ \theta = \frac{1}{3}\pi$
   (d) $52x^2 - 72xy + 73y^2 = 100,\ \theta = \sin^{-1}\frac{3}{5}$
   (e) $5x^2 + 24xy - 5y^2 = -325,\ \theta = \tan^{-1}\frac{2}{3}$
   (f) $xy = -16,\ \phi = 135°$

3. With each of the following equations, tell what type of conic section is described and calculate the center, foci, vertices, and axes. Draw the conic section.
   (a) $16x^2 - y^2 - 4x + y - 6 = 0$
   (b) $41x^2 - 84xy + 76y^2 = 208$
   (c) $19x^2 + 6xy + 11y^2 = 20$
   (d) $4xy = 1$
   (e) $xy - x - 2y = 0$
   (f) $3x^2 + 2\sqrt{3}xy + y^2 - 8x + 8\sqrt{3}y + 3 = 0$
   (g) $5x^2 + 4xy + 2y^2 - 6 = 0$
   (h) $2x^2 - \sqrt{3}xy + y^2 - 6 = 0$
   (i) $xy - y = 1$
   (j) $9x^2 - 24xy + 16y^2 - 2x - 14y + 15 = 0$
   (k) $3x^2 - 4xy + 8x - 1 = 0$
   (l) $9x^2 - 6xy + y^2 + 6\sqrt{10}(3x-y) + 50 = 0$
   (m) $17x^2 - 12xy + 8y^2 - 68x + 24y - 12 = 0$
   (n) $2x^2 + y^2 - 2y - 1 = 0$
   (o) $4y^2 - 8y - x + 5 = 0$
   (p) $10000x^2 + y^2 - 10000 = 0$

4. A certain locus has equation $Ax^2+Bxy+Cy^2+Dx+Ey+F=0$, in the xy coordinate system. The coordinate system is rotated through an angle ϕ, and the equation of the locus becomes $A'x'^2+B'x'y'+C'y'^2+D'x'+E'y'+F'=0$, in the new $x'y'$ system. Prove that
   (a) $B' = B(\cos^2\phi-\sin^2\phi) + 2(C-A)\sin\phi\cos\phi$
   (b) $B' = 0$ if $\cot 2\phi = \dfrac{A-C}{B}$
   (c) $B^2 - 4AC = B'^2 - 4A'C'$

5. Suppose that $Ax^2+Bxy+Cy^2+Dx+Ey+F=0$ has a conic section for its locus. Show that
   (a) $B^2 - 4AC > 0$ if the locus is a hyperbola
   (b) $B^2 - 4AC < 0$ if the locus is an ellipse
   (c) $B^2 - AC = 0$ if the locus is a parabola
   The quantity B^2-4AC is called the **discriminant**. [Hint: Use exercise 4(c).]

7 UNSOLVABLE PROBLEMS OF ANALYTIC GEOMETRY

The techniques of analytic geometry prove to be inadequate for certain deeper problems concerning the overall shapes of graphs of functions. The graph of a function is the most important "picture" of how the function "behaves." For example, a graph indicates the extremes of the function's values, what happens for large values of $|x|$, and whether the function undergoes abrupt or smooth transitions between select points of interest.

Succeeding chapters will cover techniques for answering the following specific questions about the overall shape of a graph of a function.

(1) Is the function **discontinuous**? That is do points exist where the function undergoes abrupt transitions of value? The graphs of two **discontinuous** functions are shown in Figure 7-1.

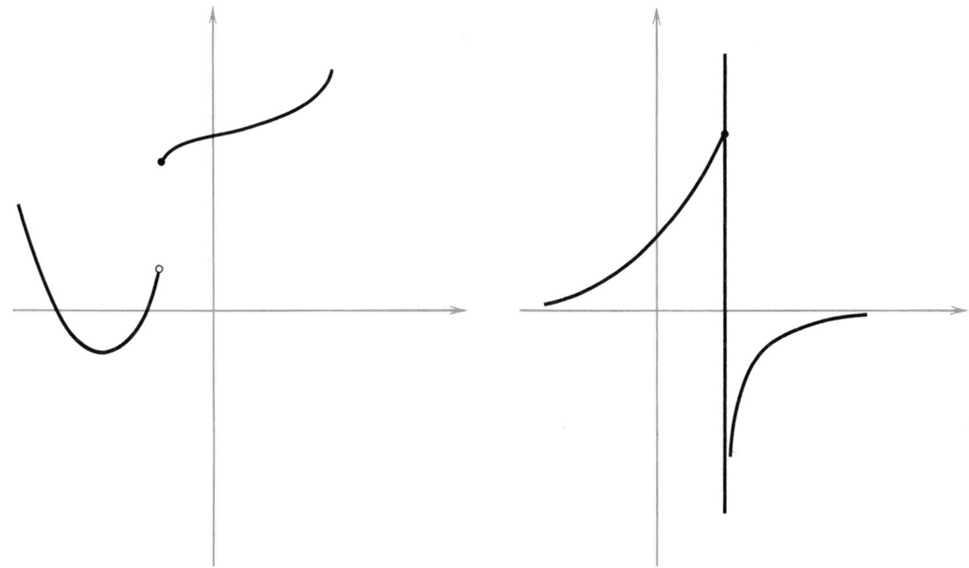

FIGURE 7-1

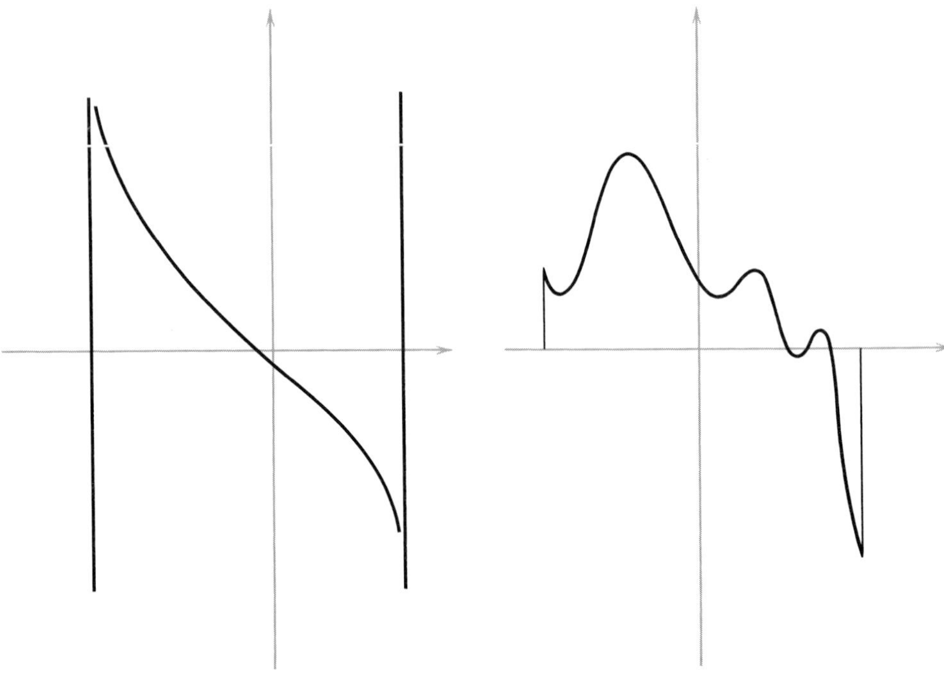

FIGURE 7-2

(2) Does the function have a "highest" and a "lowest" value within its domain of definition (Figure 7-2)?
(3) Is the function increasing or decreasing on given intervals of its domain?

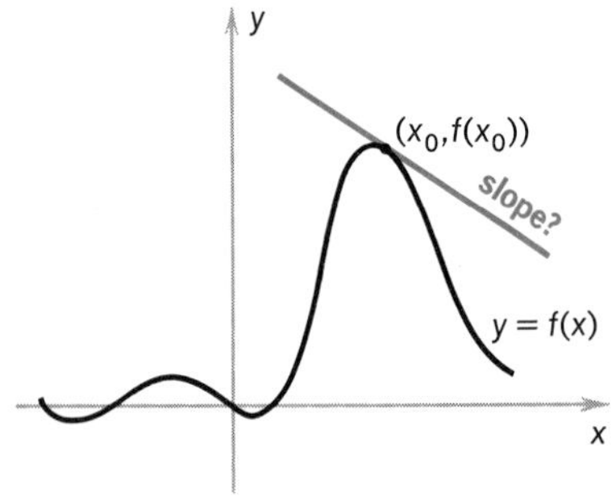

FIGURE 7-3

The theoretical tool for answering such questions about the shapes of graphs, or, in turn, the behavior of functions, is calculus. The two historically significant unsolvable problems of analytic geometry, that are solvable by calculus, are

(4) Given a function f and a point x_0 of its domain, what is the slope of the line tangent to the graph of the function at the point $(x_0, f(x_0))$ (Figure 7-3)?
(5) Given a function and two points of its domain, what is the area of the region between the points and "under" the graph of the function (Figure 7-4)?

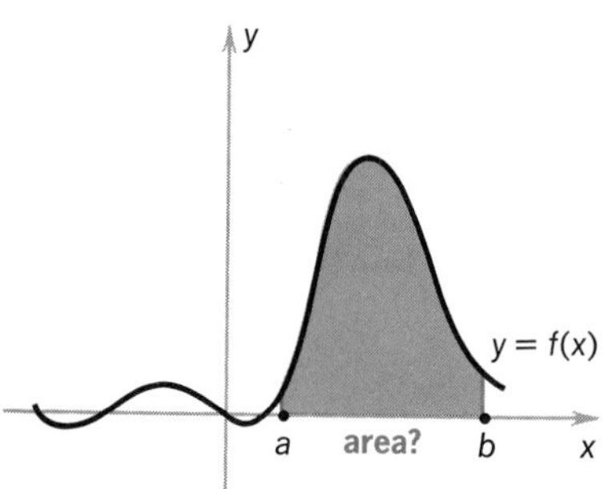

FIGURE 7-4

These problems, traditionally called the Tangent Problem and the Quadrature Problem, were solved by Gottfried Wilhelm Leibniz (1646–1716) and Isaac Newton (1642–1727). The solutions were shown to be intimately related not only to each other, but also to questions (1)–(3) above and many other problems concerning the behavior of functions.

REVIEW EXERCISES FOR CHAPTER 1

1. Define the following terms.

absolute value
affine function
affine substitution
angle of inclination
circle
closed interval
composite function
directrix
discriminant
disjoint
domain
ellipse
endpoints
even function
focus
function
hyperbola
identity function
intersection
inverse function
linear function
major axis
minor axis
monotone decreasing
monotone increasing
neighborhood
odd function
one-one
open interval
parabola
parallel
perpendicular
product of functions
quotient of functions
range
real function
signum function
slope
sum of functions

2. What must be specified in order to define a function?
3. List several ways of specifying the rule for a function.
4. What is the domain of a function given by the quotient of two polynomials?
5. Prove the triangle inequality for the absolute value function.
6. Prove that $x = (\operatorname{sgn} x)\,|x|$, for all $x \in \boldsymbol{R}$.
7. Does an interval always contain some rational numbers? Some natural numbers?
8. Show that the intersection of two closed intervals is not always a closed interval.
9. Discuss the result of composing an affine function A on a real function f; of composing f on A.
10. How does the locus of an arbitrary relation differ from the graph of a function? Give a characteristic property of the graph of a one-one function and tell why this property holds.
11. Suppose a function f is monotone increasing on the interval $[a, b]$. Is $x \mapsto f(x)^2$ also monotone increasing on $[a, b]$?
12. Give an example of a function g which is monotone (increasing or decreasing) with domain an open interval $]a, b[$, such that the range of the function is $]-\infty, +\infty[$.

13. The locus of a relation $F(x, y) = 0$ is said to be symmetric with respect to the point (X, Y) if whenever (x_0, y_0) satisfies $F(x_0, y_0) = 0$, then $F(X - x_0, Y - y_0)$ also equals zero. Show that this definition reduces to the definition of an odd relation if (X, Y) is taken to be the origin.

14. Prove that the cubic $x \mapsto ax^3 + bx^2 + cx + d$ is symmetric with respect the point
$$\left(\frac{-b}{3a}, d + \frac{(3-a)b^3 - 9bc}{27a}\right).$$

15. Without referring back to the text, prove that two nonvertical lines are parallel if and only if they have the same slope.

16. Without referring back to the text, prove that two nonvertical lines are perpendicular if and only if their slopes are negative reciprocals of each other.

17. Prove, using analytic geometry, that the perpendicular bisectors of the sides of a triangle are concurrent, that is, all intersect in a point. [Hint: You may assume, without loss of generality, that the triangle has vertices at the origin and $(a, 0)$, where $a > 0$; let the third vertex be (b, c).]

18. Prove, using analytic geometry, that the angle bisectors of a triangle are concurrent. [Hint: See hint for exercise 17.]

*19. Give a second degree equation whose locus is two straight lines.

*20. Why are conic sections called by that name? (Consult other books.)

In exercises 21–28, two functions f and g are given. In each case, state the domain and range of both functions. Determine which are monotonic, which are one-one, and compute the inverse function of any one-one function. Calculate $g \circ f$, $f + g$, $f - g$, fg, and f/g, giving the domain of each.

21. $f: x \mapsto x^2$ $\qquad g: y \mapsto (y+1)/(y-1)$

22. $f: t \mapsto \sqrt{\dfrac{t}{t+1}}$ $\qquad g: s \mapsto s^2/(s^2 - 1)$

23. $f: x \mapsto 6 + x - x^2$ $\qquad g: z \mapsto z^2 - 2z - 3$

24. $f: t \mapsto t^3 - 3t^2 + 4$ $\qquad g: q \mapsto \operatorname{sgn} q$

25. $f: \theta \mapsto \sin\theta + \cos\theta$ $\qquad g: t \mapsto (t^2 + 2t)/(t^2 + 1)$

26. $f: x \mapsto 43x^3 + 10x^2 - 43x$ $\qquad g: t \mapsto 4$

27. $f: x \mapsto \sin 2x$ $\qquad g: x \mapsto 2\sin x$

28. $f: y \mapsto \tan y$ $\qquad g: \phi \mapsto \sqrt{\phi^2 + 1}$

In exercises 29–38, sketch the locus of the given relation, noting any symmetry.

29. $x^4 + y^4 = 1$

30. $y = x^2 + 1$

31. $x^2 = \dfrac{1 + y^2}{1 - y^2}$

32. $y^2 = x(x-2)$

33. $y = x - 1/x$

34. $x = 1/(y^2 - 1)$

35. $x^2 y - y = 4(x-2)$

36. $y = x^2 + 1/x$

37. $y = \dfrac{x^2 - 4}{x^3}$

38. $y^2 = \dfrac{(x + 1)^3}{x^2 - 2x}$

In exercises 39–42, give the equation of the specified line.

39. The line through $(a, 0)$ and $(0, b)$.
40. The line through the point of concurrence of the angle bisectors of the sides of the triangle with vertices $(-1, 3)$, $(6, 3)$, and $(6, 27)$, and through the point of concurrence of the perpendicular bisectors of the sides of this triangle.
41. The line through $(4, -3)$, the x intercept equaling the cube of the y intercept.
42. The line tangent to the circle of radius 5 centered at $(-4, 3)$ and passing through the point $(9, 12)$.

In exercises 43–49, reduce the second-degree equations to standard form, and sketch the loci.

43. $9x^2 + 25y^2 - 50y = 200$
44. $4x^2 - 25y^2 + 24x + 50y + 22 = 0$
45. $x^2 - 3xy - 3y^2 + 5\sqrt{10}y = 0$
46. $(4x - 3y)^2 = 250x$
47. $xy + 8x - 4y = 36$
48. $3x^2 + 3y^2 + 16x + 8y = 30$
49. $144x^2 + 144y^2 - 216x + 192y = 80$
50. Find the circle passing through the points $(4, -2)$, $(2, 2)$, and $(-5, 1)$.
51. Find the locus of all points whose distance from the point $(-3, 0)$ is 4 more than its distance from $(3, 0)$.

2

LIMITS

The method of limits came to the foreground in the study of calculus in the nineteenth century. Because an understanding of limits is essential to our study of calculus, a thorough analysis of the topic is presented here. Methods of computing limits and theorems about limits are followed by a treatment of the mathematical notion of continuity. Although theorem proofs could be deferred or even omitted from the course, a deeper insight will result from studying them.

1 INTRODUCTION TO LIMITS

Consider the graph of the function $f: x \mapsto x^2+1$ shown in Figure 1-1. Let P be a point moving along the x axis, and let Q be the point directly above P on the graph of f. Thus, Q is moving with P. If P has coordinates $(x, 0)$, then Q has coordinates $(x, (f, x))$. As P tends toward the origin (so x approaches 0), Q slides down the parabola toward $(0, 1)$ and so $f(x)$ approaches 1. We therefore can say that "as x approaches 0, $f(x)$ approaches 1." The notation generally used to state this is

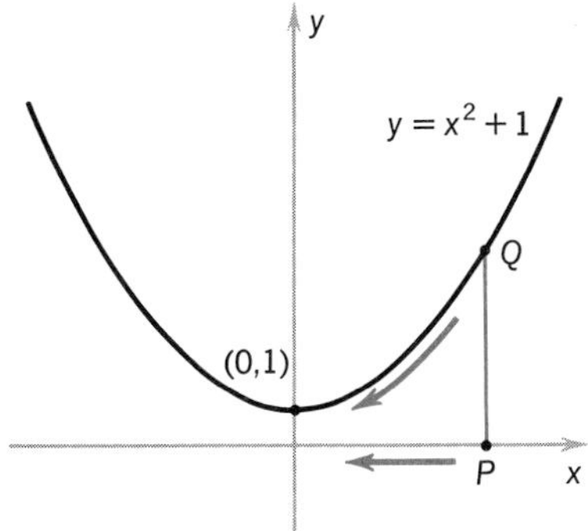

FIGURE 1-1

$$f(x) \to 1 \quad \textit{as} \quad x \to 0$$

Another commonly used notation is

$$\lim_{x \to 0} f(x) = 1$$

which is read as "The limit of $f(x)$ as x approaches 0 is 1."

Now let us consider the graph of $g: x \mapsto (x^3/x)+1$ shown in Figure 1-2. It is the same as the above graph of f except that the point $(0, 1)$ is deleted since $g(0)$ is not defined. Once again, let us consider the point P lying on the x axis and the point Q directly above it on the graph. We again observe that as P moves toward the origin, Q again approaches the point $(0, 1)$. In the case of f, as P reached the origin, Q reached the point $(0, 1)$ on the curve since $f(x)$ is defined when $x=0$. In the case of g, when P reaches the origin Q does not reach a point on the curve since $g(0)$ is not defined. However, the graph of g reveals that the point (0, 1) bears a special relation to g, for we see clearly that as x approaches 0, $g(x)$ approaches 1. Thus we again write that $\lim_{x\to 0} g(x) = 1$ or $g(x)\to 1$ as $x\to 0$.

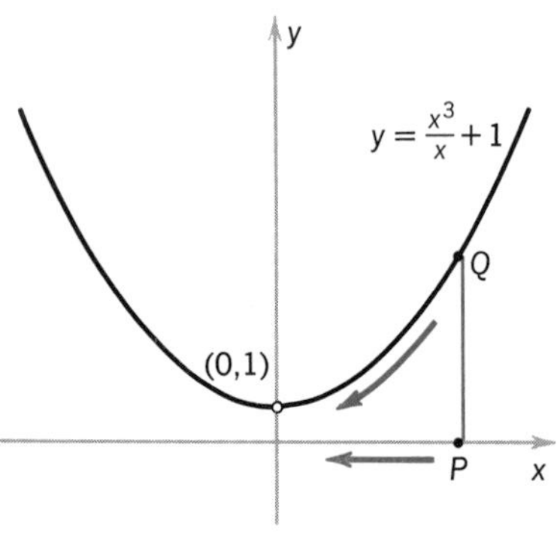

FIGURE 1-2

Often a physical or geometric situation involves a question about the limit of a function. Sometimes (as with the function f above) the value of the function gives the limit directly; sometimes the limit can be guessed by the nature of the problem; sometimes (as with the function g above) the function value is not defined but the limit still exists.

A fundamental fact about limits is the following: If f is any function whose rule can be given by a single algebraic expression $f(x)$, and if $f(x_0)$ is defined, then as x approaches x_0, $f(x)$ approaches $f(x_0)$ and we have

$$\lim_{x\to x_0} f(x) = f(x_0)$$

A more careful formulation of limits and a proof of this fundamental fact will be given below. Let us illustrate it now with some examples.

Example 1 Consider the function $f{:}x\mapsto x^3+1$. It is evident from the graph in Figure 1-3 that as x approaches 0, $f(x)$ approaches $f(0)=1$. We can support this fact algebraically by considering points progressively closer to 0 such as $-\frac{1}{2}, -\frac{1}{4}, -\frac{1}{8}, \ldots$. We tabulate the values

$$x=-\frac{1}{2} \qquad f(x)=\frac{7}{8}=\frac{56}{64}=\frac{448}{512}$$

$$x=-\frac{1}{4} \qquad f(x)=\frac{63}{64}=\frac{504}{512}$$

$$x=-\frac{1}{8} \qquad f(x)=\frac{511}{512}$$

$$x=-\frac{1}{16} \qquad f(x)=\frac{3365}{3366}=\frac{511\frac{7}{8}}{512}$$

We see that $f(x)$ gets progressively closer to 1 as x approaches 0.

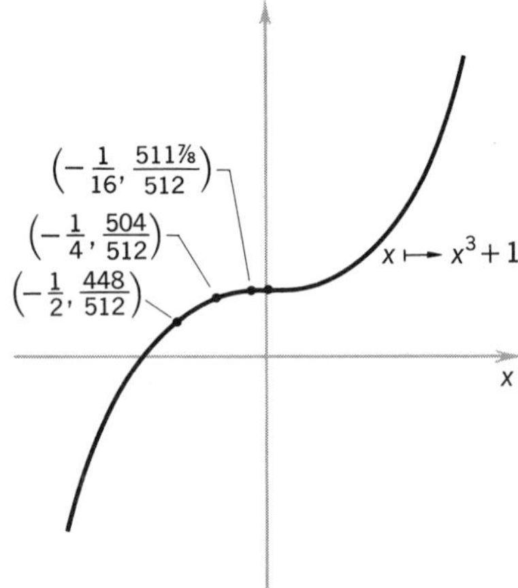

FIGURE 1-3

Example 2 Consider the function $g{:}x\mapsto\sqrt{x}$. We have $g(100)=10$. Let us choose some points approaching 100 and tabulate the function values.

x	$g(x)$
64	8
81	9
90	$3\sqrt{10}\approx 9.48$
96	$4\sqrt{6}\approx 9.80$
99	$3\sqrt{11}\approx 9.96$
99.9	$3\sqrt{11.1}\approx 9.99$

So we see that as x approaches 100, $g(x)$ approaches 10.

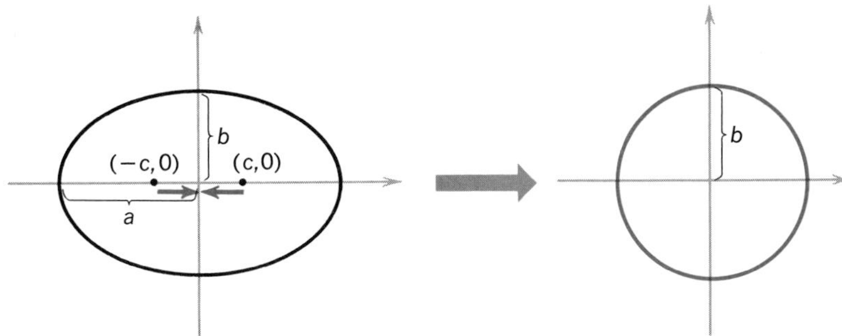

FIGURE 1-4

Example 3 Suppose that the foci $(-c,0)$ and $(c,0)$ of an ellipse (Figure 1-4) come together, while the length b of the semi-minor axis remains constant. What is the limit of the length a of the semi-major axis as the foci approach the origin? Intuitively, we expect the ellipse to become more and more circular; that is, the semi-axes should become like the radii of a circle. Since we are keeping b constant, a should eventually equal b. To solve this problem we must first state it in terms of a function. The lengths of the semi-minor and semi-major axes of an ellipse are related to the distance from the foci to the origin by (see §1.5) $a^2 = b^2 + c^2$ or $a = \sqrt{c^2+b^2}$. The situation is the following: b remains constant and c approaches 0. We want to know what happens to $a = \sqrt{c^2+b^2}$ as c tends toward 0. Hence we want to find $\lim_{c\to 0} f(c)$, where $f(c) = \sqrt{c^2+b^2}$. Since $f(0)$ is defined we find that

$$\lim_{c\to 0} f(c) = \sqrt{0+b^2} = b$$

So a indeed tends to b as c approaches 0, and our intuitive expectation is verified. The graph of $f: c \mapsto \sqrt{c^2+b^2}$ appears in Figure 1-5.

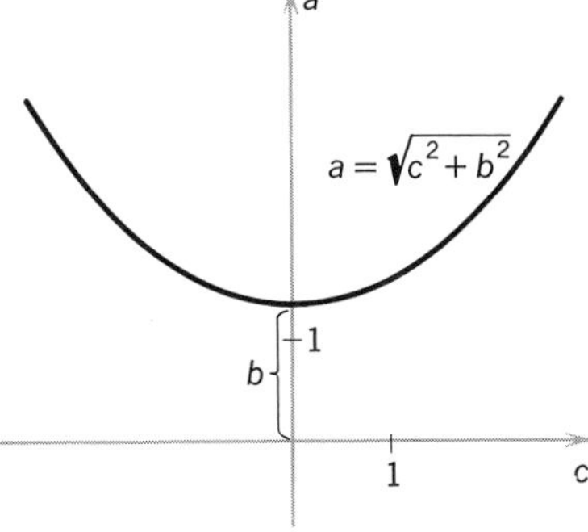

FIGURE 1-5

Limits arise naturally in problems dealing with moving points. The solutions of such problems often involve writing a function f such that the moving point has coordinates $(x, f(x))$. We then want to find what happens to the point $(x, f(x))$ when x approaches some given value x_0. Thus, in effect, we want to find $\lim_{x\to x_0} f(x)$. If $f(x)$ is an algebraic expression and if $f(x_0)$ is meaningful, then we can solve the problem by evaluating $f(x_0)$. However, the expression $f(x_0)$ will often reduce to something meaningless, such as 0/0. When this happens, the method used to evaluate the limit is to simplify the expression $f(x)$ to obtain a form $\tilde{f}(x)$ which is meaningful when $x = x_0$. It is then the case that $\lim_{x\to x_0} f(x) = \tilde{f}(x_0)$.

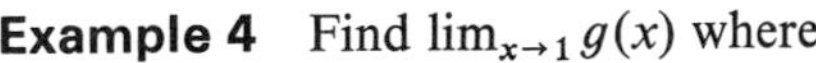

Example 4 Find $\lim_{x\to 1} g(x)$ where

$$g: x \mapsto \frac{x-1}{\sqrt{x}-1}$$

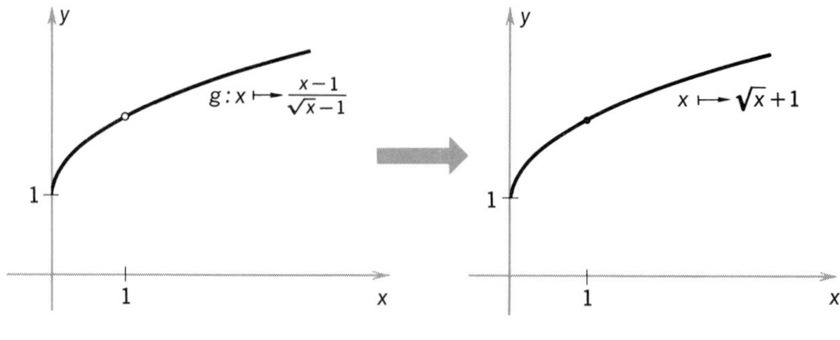

FIGURE 1-6

This function is graphed in Figure 1-6. We see that $g(1)$ is not defined, since division by zero is meaningless. However, if we multiply the numerator and denominator of $g(x)$ by $\sqrt{x}+1$, we find that for all x in the domain of g we have

$$g(x) = \frac{x-1}{\sqrt{x}-1} = \sqrt{x}+1$$

The expression $\tilde{g}(x) = \sqrt{x}+1$ takes the value 2 at $x = 1$; hence, $g(x) \to 2$ as $x \to 1$.

Example 5 Find

$$\lim_{h \to 0} \frac{h^2}{6 - \sqrt{36 - h^2}}$$

(See Figure 1-7.) Let $f: h \mapsto h^2/(6 - \sqrt{36 - h^2})$, so the domain of f is the interval $[6, -6]$ minus the point 0. For all points in the domain of f, we can rationalize the denominator of $f(h)$ to obtain

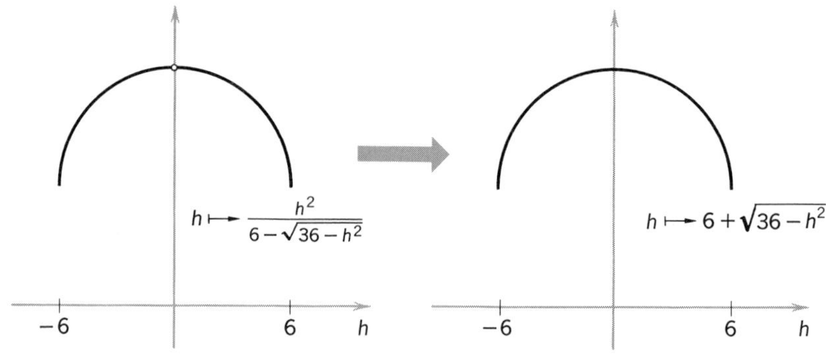

FIGURE 1-7

$$f(h) = \frac{h^2}{6 - \sqrt{36 - h^2}} = \frac{h^2}{6 - \sqrt{36 - h^2}} \cdot \frac{6 + \sqrt{36 - h^2}}{6 + \sqrt{36 - h^2}}$$

$$= \frac{h^2(6 + \sqrt{36 - h^2})}{36 - (36 - h^2)} = \frac{h^2(6 + \sqrt{36 - h^2})}{h^2}$$

$$= 6 + \sqrt{36 - h^2}$$

This last expression also has meaning at $h = 0$, where it takes the value 12; so we have $\lim_{h \to 0} f(h) = 12$.

Example 6 Suppose a point P, moving along the locus of $y = 1/x$ approaches the fixed point $(1, 1)$. For each position of P there is a secant line L through $(1, 1)$ and P (see Figure 1-8). The secant intersects the vertical line $x = b$ (where $b \neq 1$) at some point E, which varies as P moves. What is the limiting position of E as P moves along the curve to the point $(1, 1)$?

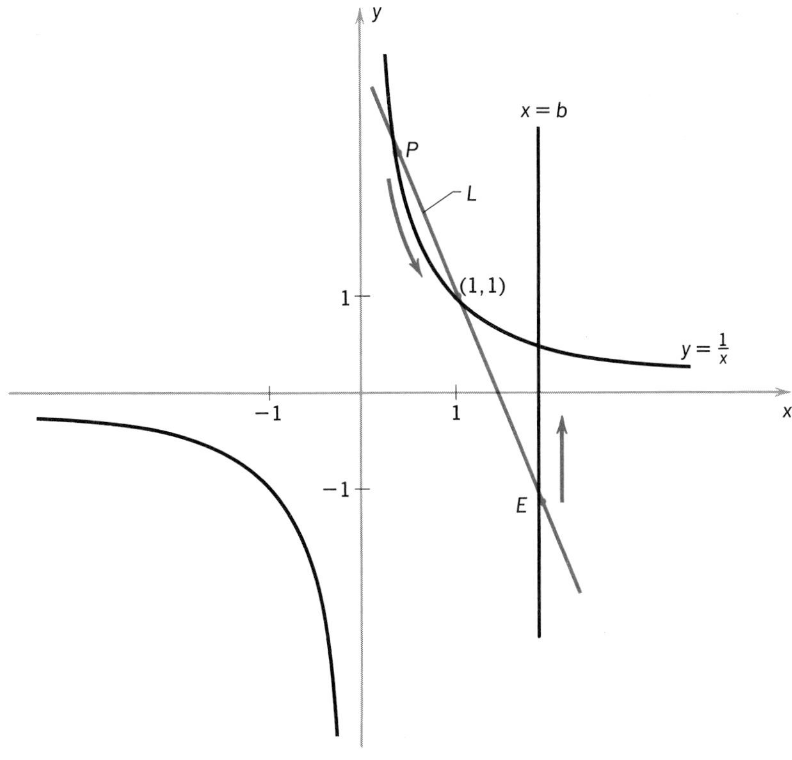

FIGURE 1-8

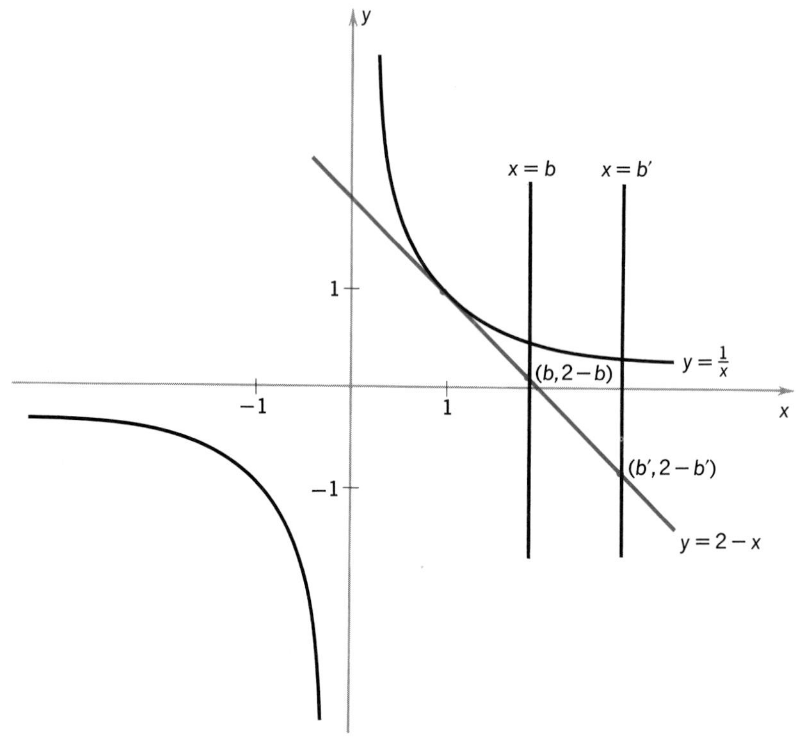

FIGURE 1-9

If the coordinates of P are $(x, 1/x)$, the slope of L is

$$\frac{1/x - 1}{x - 1}$$

and since E lies on L, the y coordinate of E must satisfy the equation

$$\frac{y-1}{b-1} = \frac{(1/x)-1}{x-1}$$

or, equivalently,

$$y = \left[(b-1)\cdot\left(\frac{(1/x)-1}{x-1}\right)\right] + 1$$

To determine the limit position of E as P approaches $(1, 1)$ along the hyperbola, we evaluate the limit as x approaches 1 of the function

$$f{:}x \mapsto \left[(b-1)\cdot\left(\frac{(1/x)-1}{x-1}\right)\right] + 1$$

We simplify the expression for $f(x)$ and obtain $f(x) = [(1-b)/x] + 1$ for all $x \neq 1$. This expression is meaningful for $x = 1$, and by substituting 1 for x, we obtain $\lim_{x\to 1} f(x) = 2-b$. Thus E approaches the position $(b, 2-b)$ as P approaches $(1, 1)$ along the curve.

Now let us consider the same problem with two vertical lines $x = b$ and $x = b'$. If $1 \neq b \neq b' \neq 1$, the points obtained on the lines $x = b$ and $x = b'$ by the above process are $(b, 2-b)$ and $(b', 2-b')$, respectively. The points $(b, 2-b)$, $(b', 2-b')$, and $(1, 1)$ are collinear and all lie on the graph of the straight line $y = 2-x$. This line is tangent to the curve at $(1, 1)$ (see Figure 1-9). The slope of this tangent line is -1. The slope m of the secant line through $(1, 1)$ and a point $(x, 1/x)$, $x \neq 1$, on the curve is given by

$$m(x) = \frac{(1/x) - 1}{x - 1}$$

Since $\lim_{x\to 1} m(x) = -1$, the slope of the tangent line is the limit of slopes of the secant lines as P approaches $(1, 1)$.

Example 7 A circle with center $(3, 0)$ and radius 3 intersects a circle with center $(0, 0)$ and of diminishing radius h (see Figure 1-10). The positive y intercept A with coordinates $(0, h)$ and the first quadrant intersection B of the two circles uniquely determines a line L. What happens to the x intercept C of L as h approaches 0?

Let us develop an equation which expresses the x coordinate of C in terms of h. Since B lies on both circles, its coordinates (x_B, y_B) must satisfy the equation of each circle, so

$$x_B^2 + y_B^2 = h^2 \tag{1}$$

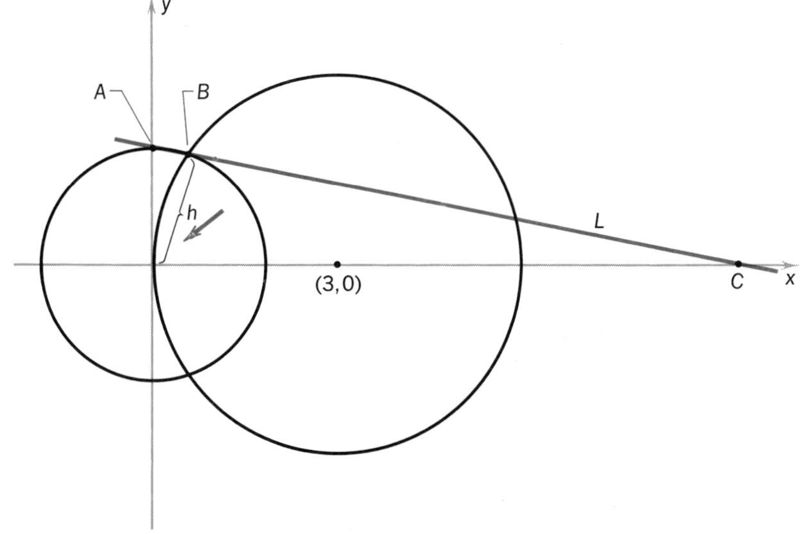

FIGURE 1-10

and

$$(x_B-3)^2+y_B^2=9 \tag{2}$$

Multiplying out (2) and subtracting (1) we obtain

$$6x_B = h^2$$

or solving for x_B,

$$x_B = \frac{h^2}{6} \tag{3}$$

Solving (1) for y_B and substituting (3) into the result, we have, since $y_B > 0$,

$$y_B = h\sqrt{1-\frac{h^2}{36}}$$

The two points (x_A, y_A) and (x_B, y_B) determine L. The y intercept of L is just A itself, that is, the point $(0, h)$, and the slope of L is

$$\frac{y_B-y_A}{x_B-x_A} = \frac{h\sqrt{1-(h^2/36)}-h}{(h^2/6)} = \frac{6\sqrt{1-(h^2/36)}-6}{h}$$

$$= \frac{\sqrt{36-h^2}-6}{h}$$

At the x intercept C of L, the y coordinate is zero, and we have

$$y_C = 0 = \left(\frac{\sqrt{36-h^2}-6}{h}\right)x_C + h \tag{4}$$

Solving for x_C, equation (4) yields

$$x_C = \frac{h^2}{6-\sqrt{36-h^2}}$$

We note that the function

$$f\colon h \mapsto \frac{h^2}{6-\sqrt{36-h^2}}$$

is exactly the function of Example 5, and there we discovered that

$$\lim_{h\to 0} f(h) = 12$$

Hence, as h approaches zero, the x intercept of L approaches the point $(12, 0)$. The graph of f appears in Figure 1-7. The function is not defined when $h > 6$. Moreover, the graph shows that as h goes towards 0, the point C moves progressively to the right along the x axis, approaching but never reaching (except at the limit) the point $(12, 0)$.

EXERCISES

In exercises 1–20, find the indicated limits.

1. $\lim_{y\to 3} \dfrac{1/y - 1/3}{y-3}$

2. $\lim_{x\to 1} \dfrac{x^2+2}{x}$

3. $\lim_{x\to 1} \dfrac{\sqrt{x}-1}{x-1}$

4. $\lim_{h\to 0} \dfrac{(2+h)^2-4}{h}$

5. $\lim_{x\to 3} (x^3 - 5x^2 + 2x - 1)$

6. $\lim_{x\to 2} \dfrac{x^3+17x+4}{2x}$

7. $\lim_{x\to -3} \dfrac{x^2-9}{x+3}$

8. $\lim_{x\to -2} \dfrac{\dfrac{1}{1+x}+1}{x+2}$

9. $\lim_{x\to 2} \dfrac{\dfrac{1}{\sqrt{x}}-\dfrac{1}{2}}{x+2}$

10. $\lim_{t\to 0} t^7 + 21t^2 + 14$

11. $\lim_{z\to 1} \dfrac{z-1}{z^2-1}$

12. $\lim_{h\to 0} \dfrac{(a+h)^3-a^3}{h}$

13. $\lim_{h\to 0} \dfrac{\sqrt{2+h}-\sqrt{2}}{h}$

14. $\lim_{x\to 2} \dfrac{x^2-4}{x-2}$

15. $\lim_{x\to 3} \dfrac{\dfrac{1}{x}-\dfrac{1}{3}}{x-3}$

16. $\lim_{x\to 3} \dfrac{\dfrac{1}{x+2}-\dfrac{1}{5}}{x-3}$

17. $\lim_{h\to 0} \dfrac{\dfrac{1}{5+h}-\dfrac{1}{5}}{h}$

18. $\lim_{h\to 0} \dfrac{(a+h)^2-a^2}{h}$

19. $\lim_{x\to 1} \dfrac{\sqrt{x+3}-2}{x-1}$

20. $\lim_{h\to 0} \dfrac{\sqrt{a+h}-\sqrt{a}}{h}$ $(a>0)$

21. A point P is moving along the parabola $y = x^2$ toward the origin O.
    (a) The secant line through O and P intersects the vertical line $x = b$ $(b \neq 0)$ at a point E. As P approaches the origin, what is the limit position of E?
    (b) The perpendicular bisector of the secant line OP intersects the y axis at a point $(0, C)$. What is the limit value of C, as P approaches the origin?
    (c) The line perpendicular to the secant OP and passing through P intersects the y axis at point $(0, D)$. As P approaches the origin what is the limit value of D? Compare with (b).

22. A point P is moving along the graph of the function $x \mapsto \sqrt{|x|}$ toward the origin O. The perpendicular bisector of the secant line through O and P intersects the y axis at a point $(0, C)$. As P approaches the origin what is the limit value of C? The line perpendicular to OP and passing through P intersects the y axis at the point $(0, D)$. What is the limit value of D as P approaches the origin. Compare with exercises 21(b) and (c).

23. The graph of $f: x \mapsto \sqrt{1-x^2}$ is a semi-circle of radius 1, centered at the origin. Let Q be a fixed point on the semi-circle with $Q \neq (1,0)$ and $Q \neq (-1,0)$. Let P be a point moving toward Q along the semi-circle. The secant line passing through P and Q intersects the vertical line $x=2$ at point E. Find the limit position of E as P approaches Q, and show that this limit position lies on the tangent to the semi-circle at Q.

24. (a) Find the slope of the line joining the pairs of points $(2,4)$ and $(2+h, (2+h)^2), h \neq 0$.
    (b) Find the limit of the slope of the line of part (a) as $h \to 0$.
    (c) Find the limit as $h \to 0$ of the line joining the pairs of points (x, x^2) and $(x+h, (x+h)^2)$.

25. In a circle of radius 4, let $l(h)$ and $L(h)$ be the lengths of chords of distance h and $\frac{1}{2}(4+h)$ from the center, respectively, where $0 < h < 4$.
    (a) Show $l(h) = 2\sqrt{16-h^2}$

$$L(h) = 2\sqrt{16 - \left(\frac{4+h}{2}\right)^2} = \sqrt{(4-h)(12+h)}$$

    (b) Show $\lim_{h \to 4} l(h) = 0$ and $\lim_{h \to 4} L(h) = 0$ but that $\lim_{h \to 4} \dfrac{l(h)}{L(h)} = \sqrt{2}$.

2 NEIGHBORHOODS

In §1 we worked with the concept of limit and applied it to functions whose rules were given by single algebraic expressions. For such functions we can evaluate $\lim_{x \to x_0} f(x)$ in one of two ways: either by computing $f(x_0)$, or, if $f(x_0)$ is not defined, by reducing $f(x)$ to a form meaningful for $x = x_0$ and then evaluating this new expression at $x = x_0$. However, in mathematical analysis we study many functions whose rules are not given by algebraic expressions. Therefore, we must develop a framework for a general theory of limits; in this section we set up the machinery for studying under general conditions what happens to $f(x)$ as x approaches some given point x_0.

For the most part we shall be interested in values of $f(x)$ at points x near x_0. The mathematical translation of "near x_0" is "in a neighborhood of x_0." Thus, neighborhoods will play an important role in the discussion. Furthermore, in studying limits, while we shall want to consider $f(x)$ for points x near x_0, the value $f(x_0)$ itself might not be defined—and even if defined, it might not necessarily be of interest. Therefore, the important point is to consider $f(x)$ for x near x_0 but not equal to x_0. A neighborhood M of x_0 with the point x_0 taken out is called a **deleted neighborhood** of x_0.

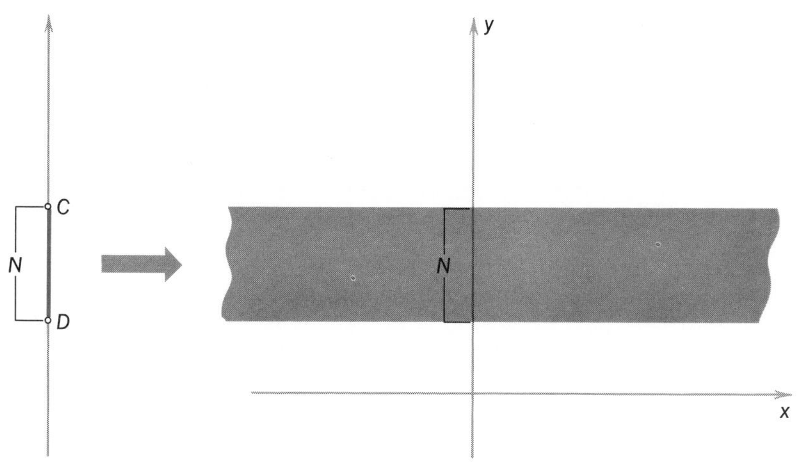

FIGURE 2-1

Thus, $]0,1[$ with the point $\frac{1}{2}$ removed is a deleted neighborhood of $\frac{1}{2}$. If M is a neighborhood of x_0, then we use the symbol $M(x_0)$ to denote the deleted neighborhood of x_0 obtained by removing x_0 from the open interval M. For example, if M is the interval $]-1,2[$ then $M(\frac{1}{4})$ is the interval $]-1,2[$ with the point $\frac{1}{4}$ removed.

If we wish to consider what happens to $f(x)$ as x approaches x_0, we can use intervals to describe the motion of the point $(x,f(x))$. Given an open interval N, we can construct the horizontal strip in the xy plane consisting of all points (x,y) such that $y \in N$. Such a horizontal strip is pictured in Figure 2-1; we see that it is parallel to the x axis. Given N and a function f, we can ask when do points $(x,f(x))$ lie in the strip determined by N as x approaches x_0. An important case occurs when there is a neighborhood M of x_0 such that all the points $(x,f(x))$ lie in the strip so long as $x \in M(x_0)$. In this case, as x tends toward x_0, once x is in $M(x_0)$, the point $(x,f(x))$ must necessarily lie in the strip determined by N. Thus, as x approaches x_0, the point $(x,f(x))$ will eventually remain in the horizontal strip determined by N. Let us illustrate this situation with some pictures.

In Figure 2-2 every point x of M is sent to a point $f(x)$ inside of N. Hence $(x,f(x))$ is eventually in the strip as x approaches x_0.

However, in Figure 2-3, $(x_0,f(x_0))$ does not lie in the strip determined by N, nor is there a neighborhood M of x_0 such that every x of $M(x_0)$ is sent to a point $f(x)$ in N. Hence $(x,f(x))$ is not eventually in the strip determined by N as x tends to x_0.

The point $(x,f(x))$ lies in the strip determined by the interval N exactly when $f(x) \in N$. Since it is long-winded to say "the point $(x,f(x))$ is eventually in the strip determined by N," we use the simpler locution "$f(x)$ is eventually in N." We continue with illustrations.

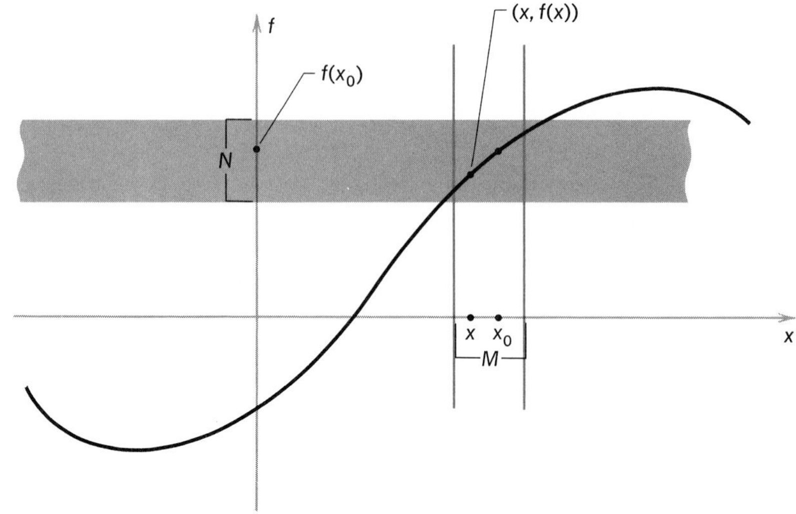

FIGURE 2-2

In Figure 2-4, f is not defined at x_0; nevertheless there is a neighborhood M of x_0 such that all the points $(x, f(x))$ lie in the strip of N for x in M and $x \neq x_0$. Hence even though x_0 is not in the domain of f, still $f(x)$ is eventually in N as x approaches x_0.

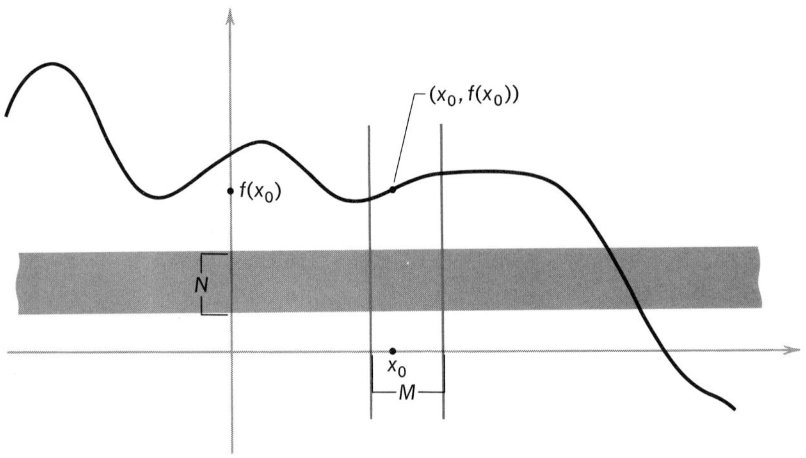

FIGURE 2-3

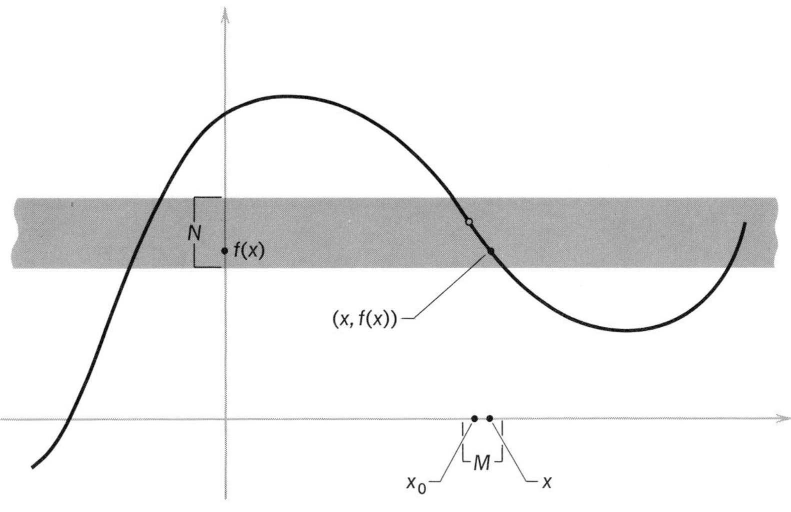

FIGURE 2-4

In Figure 2-5, the function f is not defined at x_0, and every deleted neighborhood of x_0 contains points x such that $f(x) \notin N$. Hence $f(x)$ is **not** eventually in N as x approaches x_0.

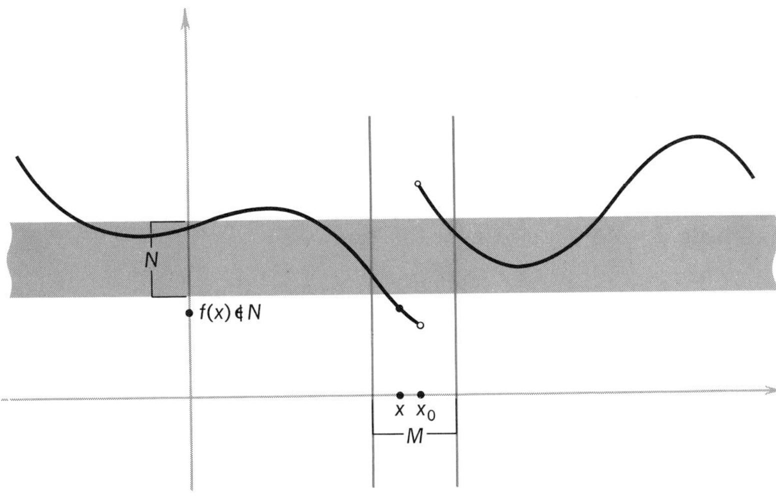

FIGURE 2-5

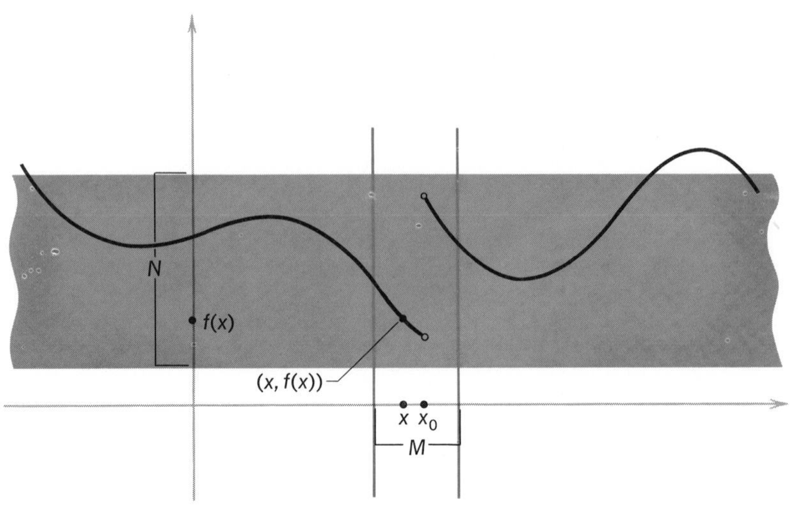

FIGURE 2-6

FIGURE 2-7

In Figure 2-6, the function f is not defined at x_0, but there is a neighborhood M of x_0 (as shown) such that f sends all x in $M(x_0)$ into N. Hence $f(x)$ is eventually in N as x approaches x_0.

Before we treat specific functions and numerical examples, let us make the following formal definition.

Definition Given a function f and an open interval N, we say that **$f(x)$ is eventually in N as x approaches x_0**, if there exists a neighborhood M of x_0 such that $f(x) \in N$ for all $x \in M(x_0)$.

Example 1 The function $h: z \mapsto z + \operatorname{sgn} z$ has all of $\boldsymbol{R}$ for its domain (see Figure 2-7) and $h(0) = 0 + \operatorname{sgn} 0 = 0$. Let N be the neighborhood $]-\frac{1}{2}, \frac{1}{2}[$ of $h(0) = 0$. Is $h(z)$ eventually in N as z approaches 0? The graph in Figure 2-8 makes it clear that the answer is **no**. In fact, if $M =]a, b[$ is any neighborhood of 0 we have $b/2 \in M$ and $h(b/2) \notin N$.

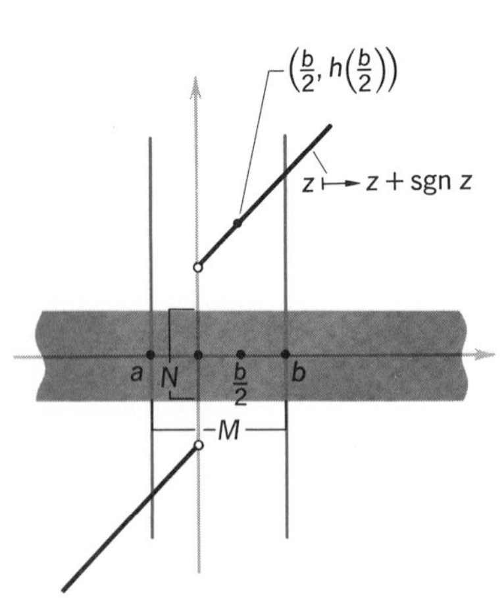

FIGURE 2-8

Example 2 Now consider the function

$$g: t \mapsto \begin{cases} t + \operatorname{sgn} t, & t \leqslant -\frac{1}{2000} \\ 2001t, & -\frac{1}{2000} \leqslant t \leqslant \frac{1}{2000} \\ t + \operatorname{sgn} t, & t \geqslant \frac{1}{2000} \end{cases}$$

which is graphed in Figure 2-9. Let $N =]-\frac{1}{10}, \frac{1}{10}[$, and let $M =]-\frac{1}{20010}, \frac{1}{20010}[$. (See Figure 2-10.) If $t \in M$, we claim that $g(t) \in N$. The statement $t \in M$ is equivalent to $-\frac{1}{20010} < t < \frac{1}{20010}$; multiplying through by 2001 gives $-\frac{1}{10} < 2001t < \frac{1}{10}$. Hence, when t is in $M, g(t) = 2001t$ and $g(t)$ is in N. Therefore $g(t)$ is eventually in N as t tends to 0.

In the example that follows, we study the function $k: s \mapsto 1-\frac{1}{2}s^2$ as s approaches 0. We shall show that for any open interval $]A, B[$ such that $A < 1 < B$, $k(s)$ is eventually in $]A, B[$ as s approaches 0. That is, as s approaches 0, $k(s)$ is eventually in **every** neighborhood $]A, B[$ of 1.

Example 3 Let k be the function $k: s \mapsto 1-\frac{1}{2}s^2$. The graph of k is shown in Figure 2-11. Any neighborhood N of 1 must be of the form $N =]A, B[$, where $A < 1 < B$. We want to compute the endpoints of a neighborhood $M =]a, b[$ such that $k(s) \in N$, whenever $s \in M(0)$. Observe that the graph of k intersects the line $y = A$ in two points; these points are found by solving $A = 1-\frac{1}{2}s^2$ for s and we obtain $s = \pm\sqrt{2-2A}$. Hence let us take $a = -\sqrt{2-2A}$ and $b = \sqrt{2-2A}$. From Figure 2-11 it is clear that $s \in M =]a, b[$ implies $k(s) \in]A, B[$. Let us verify this statement algebraically.

First, that $k(s) < B$ holds is trivial, since $k(s) = 1-\frac{1}{2}s^2$ is always less than 1, and we know that $1 < B$. The statement $s \in M$ is equivalent to saying $|s| < \sqrt{2-2A}$. Squaring this inequality, subtracting 2 from both sides, dividing through by 2, and reversing the direction of $<$ when multiplying by -1, we obtain $1-\frac{1}{2}s^2 > A$. This shows that $k(s) > A$ whenever $s \in]-\sqrt{2-2A}, \sqrt{2-2A}[$.

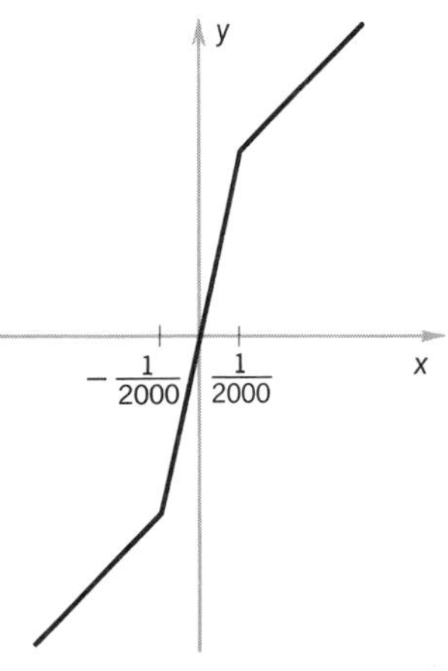

FIGURE 2-9

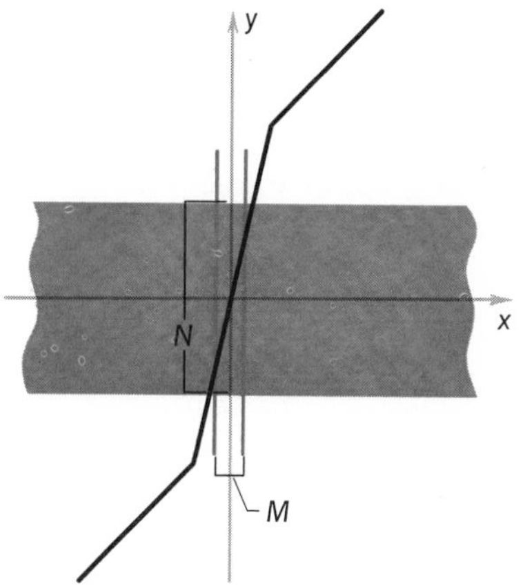

FIGURE 2-10

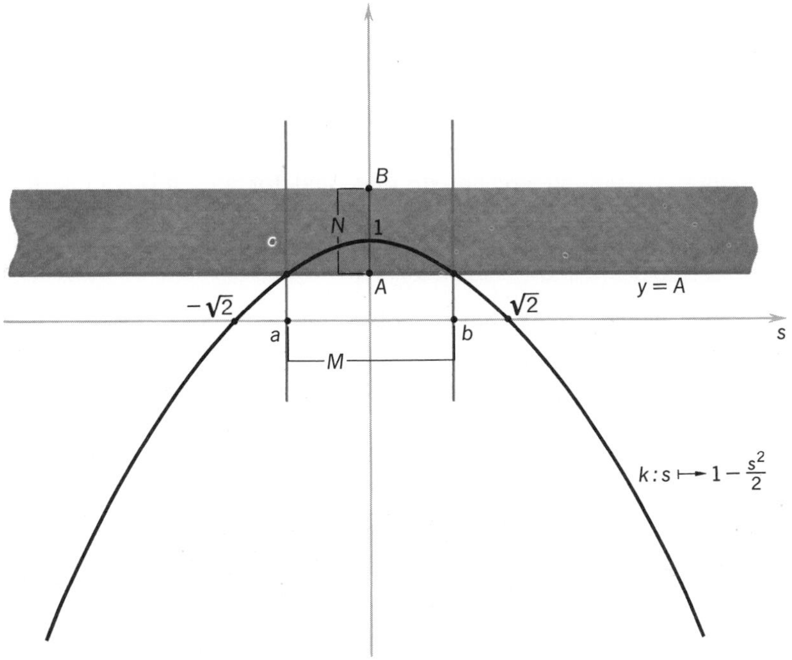

FIGURE 2-11

We conclude that $k(s)$ is eventually in the neighborhood $]A, B[$ as s approaches 0. Since $]A, B[$ could be any neighborhood of 1, $k(s)$ is eventually in every neighborhood of 1 as s approaches 0. This is what we set out to show.

Example 4 Consider a linear function $f{:}x \mapsto mx$, where $m > 0$. Let x_0 be any real number and let $]A, B[$ be any neighborhood of mx_0. It is simple to show that $f(x)$ is eventually in $]A, B[$ as x approaches x_0. We are given that $m > 0$, and since $A < mx_0 < B$, division by m yields $A/m < x_0 < B/m$. Thus $]A/m, B/m[$ is a neighborhood of x_0. If $x \in]A/m, B/m[$, we must have $A < mx < B$. Therefore $f(x)$ is eventually in $]A, B[$ as x approaches x_0.

Recall from §1.1 that two intervals are said to be disjoint if they have no point in common. Two open intervals either are disjoint or intersect to form an open interval. If M and M' are both neighborhoods of some point x_0, then they both contain the point x_0 and their intersection must be an open interval containing x_0. Thus $M \cap M'$ is also a neighborhood of x_0.

Eventuality Lemma If $f(x)$ is eventually in both N_1 and N_2 as x approaches x_0, then N_1 and N_2 are not disjoint and $f(x)$ is eventually in $N_1 \cap N_2$ as x approaches x_0.

Proof By hypothesis, there are neighborhoods M_1 and M_2 of x_0 such that (i) for all $x \in M_1(x_0), f(x) \in N_1$ and (ii) for all $x \in M_2(x_0)$, $f(x) \in N_2$. But $M_1 \cap M_2$ is also a neighborhood of x_0, and it follows that if $x \in M_1 \cap M_2$ and $x \neq x_0$, then we have both $f(x) \in N_1$ and $f(x) \in N_2$. Therefore N_1 and N_2 are not disjoint, and for all $x \in [M_1 \cap M_2](x_0)$, we have $f(x) \in N_1 \cap N_2$. Thus $f(x)$ is eventually in $N_1 \cap N_2$ as x approaches x_0. ▨

The key idea in the above proof is to take the intersection $M_1 \cap M_2$ of two neighborhoods of x_0. Let us look at another application of this technique.

Example 5 Suppose we know that $f(x)$ is eventually in $]0, 1[$ as x approaches x_0 and also that $g(x)$ is eventually in $]0, 1[$ as x approaches x_0. We can then conclude that the product $f(x)g(x)$ is also eventually in $]0, 1[$ as x approaches x_0: Let M_1 be a neighborhood of x_0 such that $x \in M(x_0)$ implies $f(x) \in]0, 1[$ and let M_2 be a neighborhood of x_0 such that $x \in M_2(x_0)$ implies $g(x) \in]0, 1[$. Now take $M = M_1 \cap M_2$. For x in $M(x_0)$ we have both $0 < f(x) < 1$ and $0 < g(x) < 1$; hence, $x \in M(x_0)$ implies that $f(x)g(x)$ is in $]0, 1[$.

EXERCISES

In exercises 1–13 state whether $f(x)$ is eventually in N as x approaches x_0. If the answer is yes, produce a neighborhood M of x_0 such that $x \in M(x_0)$ implies that $f(x) \in N$; if the answer is no, show why not using a sketch or an algebraic argument.

1. $f: x \mapsto 1/x;\ N =]0, 2[;\ x_0 = 1$
2. $f: x \mapsto 1/x;\ N =]0, 2[;\ x_0 = \frac{1}{2}$
3. $f: x \mapsto 1/11;\ N =]0, 11/10[;\ x_0 = 10$
4. $f: x \mapsto x;\ N =]A, B[, A < a < B;\ x_0 = a$
5. (a) $f: x \mapsto 1 - 2x;\ N =]-2\frac{1}{2}, -1\frac{1}{2}[;\ x_0 = 2$
   (b) $f: x \mapsto 1 - 2x;\ N =]-2\frac{1}{20}, -1\frac{19}{20}[; x_0 = 2$
6. $f: x \mapsto \dfrac{\sqrt{x} - 1}{x - 1};\ N =]\frac{1}{4}, \frac{3}{4}[;\ x_0 = 1$
7. (a) $f: x \mapsto |x|;\ N =]-\frac{1}{1000}, +\frac{1}{1000}[; x_0 = 0$
   (b) $f: x \mapsto |x|;\ N =]-\frac{1}{1000}, +\frac{1}{1000}[; x_0 = 1$
   (c) $f: x \mapsto |x|;\ N =]A, B[, A < |1| < B;\ x_0 = 1$
   (d) $f: x \mapsto |x|;\ N =]A, B[;\ A < |a| < B; x_0 = a$
8. $f: x \mapsto c;\ N =]A, B[;\ A < c < B;\ x_0 = a$
9. $f: x \mapsto 20 \operatorname{sgn} x;\ N =]-18, +18[; x_0 = 0$
10. $f: x \mapsto |\operatorname{sgn} x|;\ N =]A, B[, A < 1 < B; x_0 = 0$
11. $f: x \mapsto |1/x|;\ N =]A, B[;\ x_0 = 0$
12. $f: x \mapsto (x-1)^2 + 1;\ N =]A, B[, A < 1 < B; x_0 = 1$
13. (a) $f: x \mapsto 3x;\ N =]2, 4[;\ x_0 = 1$
    (b) $g: x \mapsto 7x;\ N =]6, 8[;\ x_0 = 1$
    (c) $f + g: x \mapsto 10x;\ N =]8, 12[; x_0 = 1$
14. Show that $f: x \mapsto mx + b, m > 0$, is eventually in every neighborhood of $mx_0 + b$ as x approaches x_0. [Hint: Note that if $]A, B[$ is a neighborhood of $mx_0 + b$, then $]A - b, B - b[$ is a neighborhood of mx_0, and use the result of Example 4.]
15. Show that if $]a, b[$ is a neighborhood of 0 and if $]A, B[$ is a neighborhood of 1, and if f is the function $f: s \mapsto 1 - \frac{1}{2}s^2 + ms, m > 0$ then $f(s)$ is eventually in $]A + a, B + b[$ as s approaches 0. [Hint: Refer to Examples 3 and 4 and intersect neighborhoods of 0.]
16. Let f be the function of exercise 15. Show that $f(s)$ is eventually in every neighborhood of 1 as s approaches 0.
17. Show that if $g(x)$ is eventually in $]A, B[$ as x approaches x_0, then $-g(x)/2$ is eventually in $]-B/2, -A/2[$ as $x \to x_0$.
18. Suppose as x approaches x_0, $f(x)$ is eventually in $]A, B[$ and $g(x)$ is eventually in $]C, D[$. Prove that if $A, B, C, D \geqslant 0$, then $f(x)g(x)$ is eventually in $]AC, BD[$ as x approaches x_0.
19. Suppose as x approaches x_0, $f(x)$ is eventually in $]A, B[$. Show that $|f(x)|$ is eventually in $]0, \max(|A|, |B|)[$ as x approaches x_0.

20. (a) Let $]A, B[$ be a neighborhood of x_0. Show that there is a positive number e such that $A \leqslant x_0 - e < x_0 + e \leqslant B$.
    (b) Suppose that $f(x)$ is eventually in $]a, b[$ as x approaches x_0. Show that there is a positive number e such that $x \neq x_0$ and x in $]x_0 - e, x_0 + e[$ implies $f(x) \in]a, b[$.
    (c) Suppose $f(x)$ is eventually in $]a, b[$ as x approaches x_0. Show that there is a positive number e such that $0 < |x - x_0| < e$ implies $f(x) \in]a, b[$.
    (d) Suppose that there is a positive number e such that $0 < |x - x_0| < e$ implies $f(x) \in]a, b[$. Show that $f(x)$ is eventually in $]a, b[$ as x approaches x_0.
    (e) Conclude that $f(x)$ is eventually in $]a, b[$ as x approaches x_0 if and only if there is a positive e such that $0 < |x - x_0| < e$ implies $f(x) \in]a, b[$.
21. Let $\varepsilon > 0$, so that $]y_0 - \varepsilon, y_0 + \varepsilon[$ is a neighborhood of y_0. Show that $f(x)$ is eventually in $]y_0 - \varepsilon, y_0 + \varepsilon[$ as x approaches x_0 if and only if there is a positive number e such that $0 < |x - x_0| < e$ implies $|f(x) - y_0| < \varepsilon$.

3 DEFINITION AND ARITHMETIC OF LIMITS

We begin this section with the formal definition of limit and then employ this definition to compute some important trigonometric limits. Once this is done, we go on to prove that the limit operation commutes with the operations of arithmetic (that is, the limit of a sum is the sum of the limits, etc.).

Definition Let f be a function defined at each point of some deleted neighborhood of a point x_0, and let l be a real number. Then l is a **limit** of f as x approaches x_0 if $f(x)$ is eventually in every neighborhood of l as x approaches x_0. If l is a limit of f as x approaches x_0, then we also say that "$f(x)$ approaches l as x approaches x_0."

In order that we can speak of **the** limit, we prove the following theorem.

Theorem 1 **Uniqueness of the Limit** If $f(x)$ approaches both l and l' as x approaches x_0, then $l = l'$. Thus, f has at most one limit as x approaches x_0.

Proof We suppose that $l \neq l'$ and derive a contradiction. For definiteness, let us also suppose that $l < l'$. Then the interval $]l-1, \frac{1}{2}(l+l')[$ is a neighborhood of l and the interval $]\frac{1}{2}(l+l'), l'+1[$ is a neighborhood of l', and these intervals are disjoint. Since $f(x)$ approaches l as x approaches x_0, $f(x)$ is eventually in $]l-1, \frac{1}{2}(l+l')[$ as x approaches x_0. Also, $f(x)$ is eventually in $]\frac{1}{2}(l+l'), l'+1[$ as x approaches x_0. But then by the Eventuality Lemma (§2.2), $]l-1, \frac{1}{2}(l+l')[$ and $]\frac{1}{2}(l+l'), l'+1[$ are not disjoint; but this is a contradiction. Therefore, l must equal l'. ∎

If l is the (unique) limit of f as x approaches x_0, we write

$$f(x) \to l \text{ as } x \to x_0 \quad \text{or} \quad \lim_{x \to x_0} f(x) = l$$

Combining the definition, Theorem 1, and the notation, we have: **$\lim_{x \to x_0} f(x) = l$ if and only if $f(x)$ is eventually in every neighborhood of l as x approaches x_0.**

Example 1 From the examples and exercises of the last section, we have

$$\lim_{x \to x_0} mx = mx_0, \quad \text{where } m > 0$$

$$\lim_{x \to 0} 1 - \tfrac{1}{2}x^2 = 1$$

$$\lim_{x \to x_0} c = c$$

$$\lim_{x \to x_0} |x| = |x_0|$$

$$\lim_{x \to 1} [(x-1)^2 + 1] = 1$$

In §2.1 we accepted as intuitively clear the fact that if $f(x)$ is an algebraic expression, then

$$\lim_{x \to x_0} f(x) = f(x_0)$$

as long as $f(x_0)$ is defined. This fact can be *proved* using the definition of limit and the various theorems of this section.

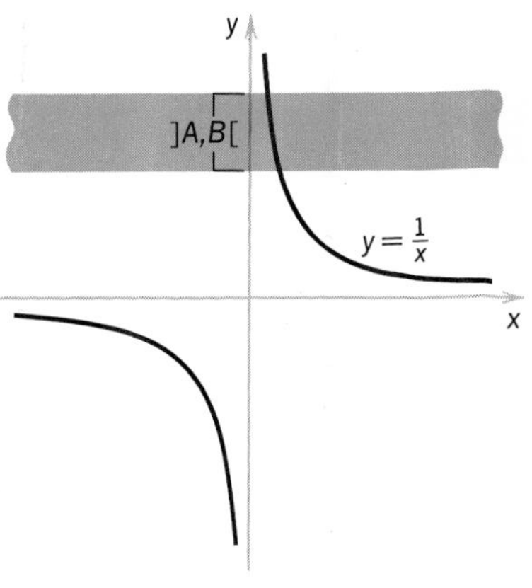

FIGURE 3-1

Now that we have a rigorous definition of limit, we are also in a position to say when a limit does not exist. For example, the signum function $x \mapsto \operatorname{sgn} x$ does not approach a limit as x approaches 0. The $\lim_{x \to 0}(x + \operatorname{sgn} x)$ does not exist either. Also, $\lim_{x \to 0} 1/x$ does not exist; in fact, we cannot find any neighborhood $]A, B[$ with A and B finite such that $1/x$ is eventually in $]A, B[$ as x approaches 0 (see Figure 3-1).

In §2.2 we claimed that a general framework was needed for the definition of limit in order to work with functions other than those whose rules are given by algebraic expressions. In the example that follows we begin our study of limits of trigonometric functions.

Example 2 Show that

$$\lim_{\gamma \to 0} \frac{\sin \gamma}{\gamma} = \lim_{\gamma \to 0} \cos \gamma = 1$$

First consider the case where $0 < \gamma < \frac{1}{2}\pi$. In Figure 3-2, we have

$$\frac{\sin\gamma}{\cos\gamma} = \tan\gamma = 2\,\text{area}(\Delta OAD)$$

$$> \text{area (circular sector } OAB) = \gamma r^2 = \gamma$$

Hence

$$\frac{\sin\gamma}{\gamma} > \cos\gamma \tag{1}$$

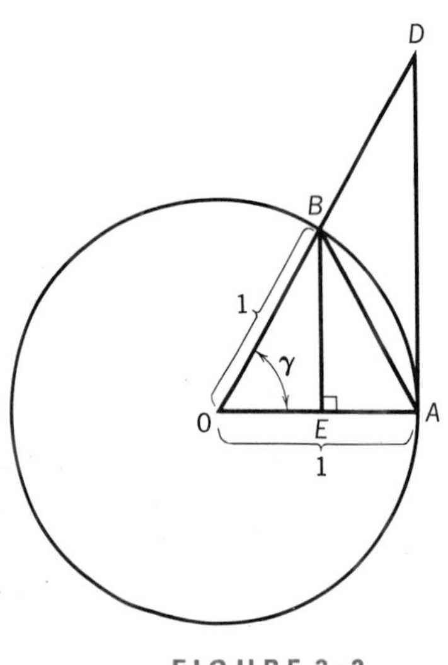

FIGURE 3-2

Moreover, by the law of cosines, we obtain

$$\gamma^2 = (\widehat{AB})^2 > (\overline{AB})^2 = (\overline{OB})^2 + (\overline{OA})^2 - 2\cos\gamma$$
$$= 2 - 2\cos\gamma$$

Thus

$$1 - \tfrac{1}{2}\gamma^2 < \cos\gamma \tag{2}$$

Finally,

$$\sin\gamma = \frac{\overline{BE}}{\overline{OB}} = \overline{BE} < \overline{AB} < \widehat{AB} = \gamma$$

and

$$\frac{\sin\gamma}{\gamma} < 1 \tag{3}$$

We combine equations (1)–(3) to arrive at the inequality

$$1 - \tfrac{1}{2}\gamma^2 < \cos\gamma < \frac{\sin\gamma}{\gamma} < 1 \tag{4}$$

for $0 < \gamma < \frac{1}{2}\pi$. Moreover $\gamma \mapsto \cos\gamma$, $\gamma \mapsto \sin\gamma/\gamma$, $\gamma \mapsto 1 - \gamma^2/2$, and the constant function $\gamma \mapsto 1$ are all even functions (see Example 2, §1.2), so inequality (4) is also valid when $-\frac{1}{2}\pi < \gamma < 0$. Thus throughout the neighborhood $]-\frac{1}{2}\pi, \frac{1}{2}\pi[$ of 0, except at 0 itself, we have inequality (4) holding.

From Example 3 of §2.2, we know that $\lim_{\gamma\to 0} 1 - \frac{1}{2}\gamma^2 = 1$. Therefore, if $]A, B[$ is any neighborhood of 1, there is a neighborhood M of 0 such that $A < 1 - \frac{1}{2}\gamma^2 < B$ whenever $\gamma \in M(0)$. Let $M' = M \cap]-\frac{1}{2}\pi, \frac{1}{2}\pi[$. Then M' is a neighborhood of 0 and for all $\gamma \in M'(0)$, inequality (4) becomes

$$A < 1 - \tfrac{1}{2}\gamma^2 < \cos\gamma < \frac{\sin\gamma}{\gamma} < 1 < B$$

Therefore, $\cos\gamma$ and $\sin\gamma/\gamma$ are eventually in $]A, B[$ as γ approaches 0. Since $]A, B[$ was chosen to be an arbitrary neighborhood of 1, we have shown that $\lim_{\gamma\to 0}\cos\gamma = \lim_{\gamma\to 0}\sin\gamma/\gamma = 1$.

Example 3 Let us now verify that $\lim_{\gamma\to 0}\sin\gamma = 0$. This fact is clear from the graph of the sine function, but it can also be shown by the neighborhood method. When $\frac{1}{2}\pi > \gamma > 0$, we have $0 < \sin\gamma < \gamma$; when $-\frac{1}{2}\pi < \gamma < 0$, we have $y < \sin y < 0$. Thus if $]A, B[$ is any neighborhood of 0 and if $y \in]A, B[\cap]-\frac{1}{2}\pi, \frac{1}{2}\pi[$, we have $A < y < \sin y < 0$ if $y < 0$, and $0 < \sin y < y < B$ if $y > 0$. In both cases $A < \sin y < B$, and so $\sin y$ is eventually in $]A, B[$ as y approaches 0. We conclude that $\lim_{y \to 0} \sin y = 0$.

Example 4 A point P is moving along the sine curve toward the origin. The secant to the curve determined by P and the origin intersects the vertical line $x = 1$ at a point E (Figure 3-3). As P approaches the origin, what is the limiting position of E? If the coordinates of P are $(x, \sin x)$, then by similar triangles those of E are $(1, \sin x/x)$; hence as P approaches the origin, E must approach $(1, \lim_{x \to 0}(\sin x)/x)$, or the point $(1, 1)$.

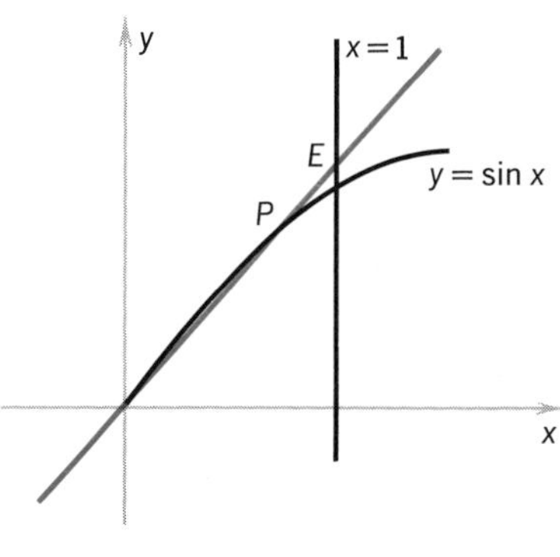

FIGURE 3-3

We now prove a sequence of theorems to show that the usual rules of arithmetic apply to limits; that is taking limits commutes with the arithmetic operations. Most importantly, we demonstrate that if $f(x)$ and $g(x)$ both approach limits as x approaches x_0, then

$$\lim_{x \to x_0} [f(x) + g(x)] = \lim_{x \to x_0} f(x) + \lim_{x \to x_0} g(x) \tag{5}$$

and

$$\lim_{x \to x_0} [f(x) \cdot g(x)] = [\lim_{x \to x_0} f(x)] \cdot [\lim_{x \to x_0} g(x)] \tag{6}$$

Thus we show that "the sum of the limits is the limit of the sum" and "the product of the limits is the limit of the product." The theorems that establish these formulas have quite simple proofs which usually reduce to addition or multiplication of the endpoints of pairs of intervals. The first theorem establishes equation (5) by using little more than the fact that if $]A_1, B_1[$ is a neighborhood of l_1 and if $]A_2, B_2[$ is a neighborhood of l_2, then $]A_1 + A_2, B_1 + B_2[$ is a neighborhood of $l_1 + l_2$.

Theorem 2 Additivity of Limits If $\lim_{x \to x_0} f(x) = l_1$ and $\lim_{x \to x_0} g(x) = l_2$, then

$$\lim_{x \to x_0} (f(x) + g(x)) = l_1 + l_2$$

Proof Let $]A, B[$ be a neighborhood of $l_1 + l_2$. To prove the theorem, we have to show that $f(x) + g(x)$ is eventually in $]A, B[$ as x approaches x_0. Set $\varepsilon = l_1 + l_2 - A$ and $\varepsilon' = B - (l_1 + l_2)$; both ε and ε' are positive since $]A, B[$ is a neighborhood of $l_1 + l_2$. The intervals $]l_1 - \frac{1}{2}\varepsilon, l_1 + \frac{1}{2}\varepsilon'[$ and $]l_2 - \frac{1}{2}\varepsilon, l_2 + \frac{1}{2}\varepsilon'[$ are thus neighborhoods of l_1 and l_2, respectively. Now $\lim_{x \to x_0} f(x) = l_1$, so $f(x)$ is eventually in $]l_1 - \frac{1}{2}\varepsilon, l_1 + \frac{1}{2}\varepsilon'[$ as x approaches x_0; let M_1 be a neighborhood of x_0 such that for all $x \neq x_0$ in M_1, we have $f(x) \in]l_1 - \frac{1}{2}\varepsilon, l_1 + \frac{1}{2}\varepsilon'[$. Similarly, let M_2 be a neighborhood of x_0 such that for all $x \in M_2(x_0)$ we have $g(x) \in]l_2 - \frac{1}{2}\varepsilon, l_2 + \frac{1}{2}\varepsilon'[$. Take $M = M_1 \cap M_2$. If $x \in M(x_0)$ then $l_1 - \frac{1}{2}\varepsilon < f(x) < l_1 + \frac{1}{2}\varepsilon'$ and $l_2 - \frac{1}{2}\varepsilon < g(x) < l_2 + \frac{1}{2}\varepsilon'$.

Adding, we obtain

$$l_1 + l_2 - \varepsilon < f(x) + g(x) < l_1 + l_2 + \varepsilon'$$

or

$$A < f(x) + g(x) < B$$

for all $x \in M(x_0)$. Therefore, $(f(x)+g(x)) \in\ ]A, B[$ for all $x \in M(x_0)$. Since $]A, B[$ is any neighborhood of l_1+l_2 we have proved that $\lim_{x \to x_0}(f(x)+g(x)) = l_1+l_2$. ▨

Example 5 The above theorem combined with Examples 1 and 2 yields

(a) $\lim_{x \to 0}(\cos x + \sin x) = \lim_{x \to 0} \cos x + \lim_{x \to 0} \sin x = 1 + 0 = 1$

(b) $$\begin{aligned} \lim_{x \to 0}(3\cos x + 2\sin x) &= \lim_{x \to 0}(\cos x + \cos x + \cos x + \sin x + \sin x) \\ &= 1+1+1+0+0 = 3 \end{aligned}$$

Theorem 3 **Linearity of Limits** If $\lim_{x \to x_0} f(x) = l$ and if m is any real number, then

$$\lim_{x \to x_0} mf(x) = ml$$

Proof It suffices to prove two cases, $m > 0$ and $m = -1$. (Why?)

The first case is $m > 0$. If $]A, B[$ is a neighborhood of ml, then since m is positive, $]A/m, B/m[$ is a neighborhood of l. Since $\lim_{x \to x_0} f(x) = l$, $f(x)$ is eventually in $]A/m, B/m[$ as x approaches x_0. Therefore there exists a neighborhood M of x_0 such that for all $x \in M(x_0)$ we have $f(x) \in\ ]A/m, B/m[$. Then for all $x \in M(x_0)$, we must have $mf(x) \in\ ]A, B[$; hence, $mf(x)$ is eventually in $]A, B[$ as x approaches x_0. Since $]A, B[$ was arbitrary, we have shown that $\lim_{x \to x_0} mf(x) = ml$.

The second case to consider is $m = -1$. For this case we shall use the fact that $A < -l < B$ is equivalent to $-B < l < -A$ for all real numbers A, B, and l. Thus, given any neighborhood $]A, B[$ of $-l$, the interval $]-B, -A[$ is a neighborhood of l. Since $f(x)$ is eventually in $]-B, -A[$ as x approaches x_0, there exists a neighborhood M of x_0 such that we have $f(x) \in\ ]-B, -A[$ for all $x \in M(x_0)$. But then $-f(x) \in\ ]A, B[$ for all $x \in M(x_0)$. Hence, $-f(x)$ is eventually in $]A, B[$ as x approaches x_0. This establishes that $\lim_{x \to x_0} -f(x) = -l$. ▨

Corollary 1 **Differences of Limits** If $\lim_{x \to x_0} f(x) = l_1$ and $\lim_{x \to x_0} g(x) = l_2$, then $\lim_{x \to x_0}(f(x) - g(x)) = l_1 - l_2$.

Proof The corollary follows at once from the Additivity of Limits Theorem and case 2 of the Linearity of Limits Theorem. ▨

To establish that "the product of the limits is the limit of the product" we shall first prove a lemma. The purpose of this lemma is to show how to "multiply intervals." More precisely, we show that for any neighborhood N of a product $l_1 l_2$ (with $l_1 > 0, l_2 > 0$) there are neighborhoods N_1 of l_1 and N_2 of l_2 such that all products $y_1 y_2$ are in N so long as $y_1 \in N_1$ and $y_2 \in N_2$. All this is quite clear geometrically; a picture appears in Figure 3-4.

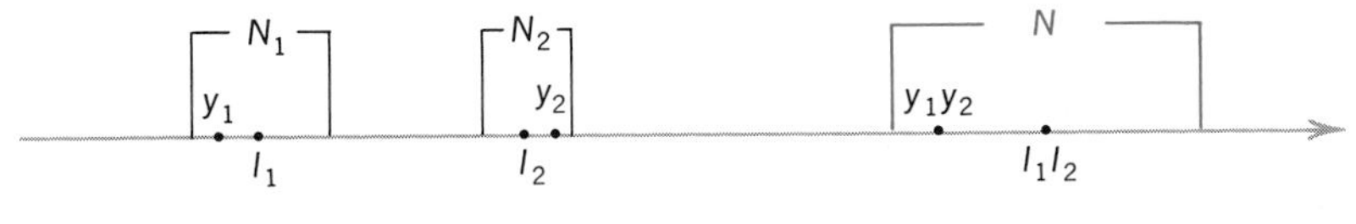

FIGURE 3-4

Lemma Suppose l_1 and l_2 are positive real numbers and that $]a, b[$ is a neighborhood of the product $l_1 l_2$. Then there is a neighborhood $]A, B[$ of l_1 and a neighborhood $]A', B'[$ of l_2 such that $a < AA' < l_1 l_2 < BB' < b$.

Proof Let us find B and B' first. We are given that $l_1 l_2 < b$; hence we have $b/l_1 > l_2$. Setting B' equal to $\frac{1}{2}(b/l_1 + l_2)$ locates B' midway between l_2 and b/l_1, and so $l_2 < B' < b/l_1$. Since l_1 is positive we can multiply this last inequality by l_1 to obtain $l_1 l_2 < B'l_1 < b$; this in turn yields $l_1 < b/B'$. Now we choose $B = \frac{1}{2}(l_1 + b/B')$; thus B is midway between l_1 and b/B'. Thus $l_1 < B < b/B'$, and hence $BB' < b$. Finally, since $l_1 < B$ and $l_2 < B'$, we have $l_1 l_2 < BB'$.

The numbers A and A' are found in similar fashion. Since $a < l_1 l_2$, we have $a/l_2 < l_1$. We want A to be between a/l_2 and l_1, and it is also desirable to ensure $A > 0$; so we set $A = \max(\frac{1}{2}(a/l_2 + l_1), \frac{1}{2}l_1)$. We now have $0 < A$ and $a/l_2 < A < l_1$. Dividing through by A yields $a/Al_2 < 1$ or $a/A < l_2$. We next set $A' = \frac{1}{2}(l_2 + a/A)$, and so we have $a/A < A' < l_2$. This gives $a = A(a/A) < AA'$; and since $l_1 > A$, $l_2 > A'$, we also have $AA' < l_1 l_2$. ∎

Theorem 4 Multiplication of Limits If $\lim f(x) = l_1$ and $\lim g(x) = l_2$, then $\lim_{x \to x_0} (f(x) g(x)) = l_1 l_2$.

Proof We first prove the theorem for the case when both l_1 and l_2 are positive. We then derive the general theorem from this special case.

Suppose $l_1, l_2 > 0$ and let $]a, b[$ be a neighborhood of the product $l_1 l_2$. Now apply the lemma to find A, A', B, and B' with the properties described there.

Since $\lim_{x \to x_0} f(x) = l_1$, $f(x)$ is eventually in $]A, B[$ as x approaches x_0; so let M_1 be a neighborhood of x_0 such that for all $x \in M_1(x_0)$, we have $f(x) \in]A, B[$. Likewise, $g(x)$ is eventually in $]A', B'[$ as x approaches x_0, so let M_2 be a neighborhood of x_0 such that $g(x) \in]A', B'[$ whenever

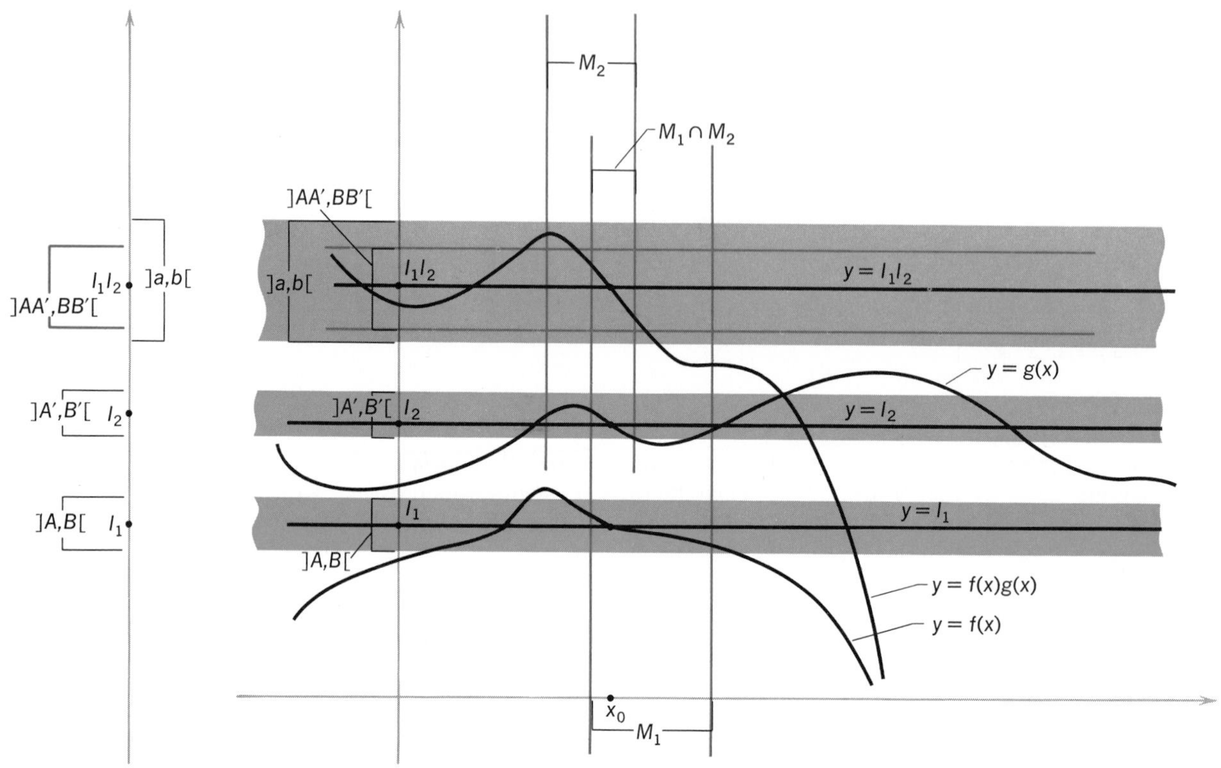

FIGURE 3-5

$x \in M_2(x_0)$. Finally let $M = M_1 \cap M_2$ (Figure 3-5). Then M is a neighborhood of x_0 and for $x \in M$, $x \neq x_0$, we have $A < f(x) < B$ and $A' < g(x) < B'$.

Now $0 < A$ and $0 < A'$, so we multiply these two inequalities to obtain: $a < AA' < f(x)g(x) < BB' < b$ for all $x \in M(x_0)$. We have shown that $f(x)g(x)$ is eventually in $]a,b[$ as x approaches x_0; since $]a,b[$ was an arbitrary neighborhood of $l_1 l_2$, we have proved $\lim_{x\to x_0}(f(x)g(x)) = l_1 l_2$.

We have proved the theorem for the special case $l_1 > 0, l_2 > 0$. We now drop the requirement that l_1 and l_2 be positive. Let C be a number such that $C > |l_1|$ and $C > |l_2|$. By the Additivity of Limits, $\lim_{x\to x_0}(f(x)+C) = l_1 + C$ and $\lim_{x\to x_0}(g(x)+C) = l_2 + C$. Since $l_1 + C$ and $l_2 + C$ are both positive real numbers, by the special case just considered, we have that

$$\lim_{x\to x_0}(f(x)+C)(g(x)+C) = (l_1+C)(l_2+C)$$

We now resort to a simple trick. We note that

$$(f(x)+C)(g(x)+C) = Cf(x) + Cg(x) + C^2 + f(x)g(x)$$

Hence, we have

$$f(x)g(x) = (f(x)+C)(g(x)+C) - Cf(x) - Cg(x) - C^2$$

However, the expression on the right-hand side has a limit as x approaches x_0 as follows from the above special case and our previous limit theorems. We therefore evaluate

$$\begin{aligned}\lim_{x\to x_0} f(x)g(x) &= \lim_{x\to x_0} (f(x)+C)(g(x)+C) - Cf(x) - Cg(x) - C^2 \\ &= (l_1+C)(l_2+C) - Cl_1 - Cl_2 - C^2 \\ &= l_1 l_2\end{aligned}$$

The theorem is now established in the general case. ∎

Example 6 Evaluate $\lim_{x\to 0}(\sin^2 x \cos x)$. We have

$$\begin{aligned}\lim_{x\to 0}(\sin^2 x\cos x) &= \lim_{x\to 0}\sin x \cdot \lim_{x\to 0}\sin x \cdot \lim_{x\to 0}\cos x \\ &= 0\cdot 0\cdot 1 = 0\end{aligned}$$

As an immediate corollary to the above theorem, we have the following.

Corollary 2 If $\lim_{x\to x_0} f(x) = l$, then $\lim_{x\to x_0}(f(x))^n = l^n$.

Example 7 Evaluate $\lim_{x\to -2}(x+3)^{99}$. First we note that $\lim_{x\to -2}(x+3) = 1$. Then, by the above corollary, we find

$$\lim_{x\to -2}(x+3)^{99} = (\lim_{x\to -2}(x+3))^{99} = 1^{99} = 1$$

As a further corollary to the above theorem we have the following.

Corollary 3 Suppose $p(x) = a_0 + a_1 x + \cdots + a_n x^n$ is a polynomial. Then for every real number x_0, we have $\lim_{x\to x_0} p(x) = a_0 + a_1 x_0 + \cdots + a_n x_0^n = p(x_0)$.

We now turn to quotients of limits.

Theorem 5 **Quotients of Limits** If $\lim_{x\to x_0} f(x) = l_1$ and if $\lim_{x\to x_0} g(x) = l_2 \neq 0$, then $\lim_{x\to x_0}(f/g)(x) = l_1/l_2$.

We leave the proof of this theorem to the student. (See exercise 7.)

Example 8 Using the above theorem together with previous results, we find

(a) $$\lim_{\gamma\to 0}\frac{\gamma}{\sin\gamma} = \lim_{\gamma\to 0}\frac{1}{\sin\gamma/\gamma} = \frac{1}{\lim_{\gamma\to 0}(\sin\gamma/\gamma)} = 1$$

(b) $$\lim_{\theta\to 0} 3\tan\theta = \lim_{\theta\to 0}\frac{3\sin\theta}{\cos\theta} = 3\cdot\frac{\lim_{\theta\to 0}\sin\theta}{\lim_{\theta\to 0}\cos\theta} = 3\cdot\frac{0}{1} = 0$$

Example 9 Let us show

$$\lim_{x\to 0}\frac{\cos x-1}{x}=0$$

We have

$$\frac{\cos x-1}{x}=\frac{\cos x-1}{x}\cdot\frac{\cos x+1}{\cos x+1}$$

$$=\frac{\cos^2 x-1}{x(\cos x+1)}=\frac{-\sin^2 x}{x(\cos x+1)}=\frac{\sin x}{x}\cdot\left(\frac{-\sin x}{\cos x+1}\right)$$

So

$$\lim_{x\to 0}\frac{\cos x-1}{x}=\lim_{x\to 0}\left[\frac{\sin x}{x}\cdot\left(\frac{-\sin x}{\cos x+1}\right)\right]$$

$$=\lim_{x\to 0}\frac{\sin x}{x}\cdot -\lim_{x\to 0}\frac{\sin x}{\cos x+1}=1\cdot -\frac{0}{2}=0$$

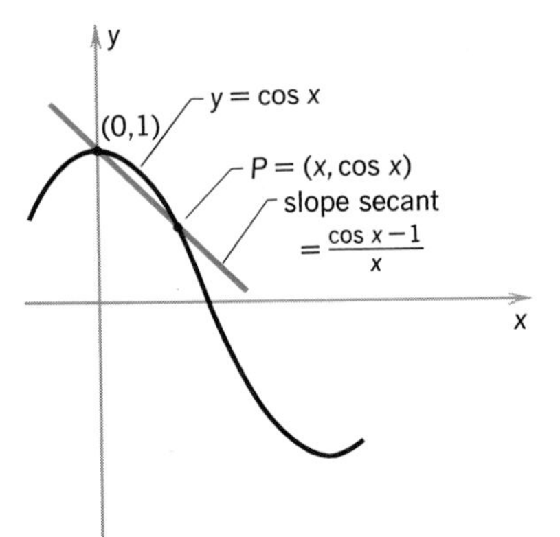

FIGURE 3-6

The limit we have just computed is an important one. To obtain a better intuition, we consider the geometric situation shown in Figure 3-6: Let $P=(x,\cos x)$ be a point on the cosine curve moving towards $(0,1)$. Then the secant through $(0,1)$ and P has slope $(\cos x-1)/x$. As P moves up the curve, the secant line becomes more and more horizontal and its slope $(\cos x-1)/x$ approaches 0.

Further important rules for working with limits are

$$\lim_{x\to x_0}|f(x)|=|\lim_{x\to x_0}f(x)|$$

$$\lim_{x\to x_0}\sqrt[n]{|f(x)|}=\sqrt[n]{|\lim_{x\to x_0}f(x)|}$$

and when n is odd

$$\lim_{x\to x_0}\sqrt[n]{f(x)}=\sqrt[n]{\lim_{x\to x_0}f(x)}$$

all of which are valid whenever $\lim_{x\to x_0}f(x)$ exists. We omit the proofs of these formulas (see exercise 15).

Example 10 Applying the above rules and previous results, we have

(a) $\lim_{x\to 0}\sqrt{|\sin x|}=\sqrt{|\lim_{x\to 0}\sin x|}=\sqrt{0}=0$

(b) $\lim_{x\to 0}\sqrt[3]{\tan x}=\sqrt[3]{\lim_{x\to 0}\tan x}=\sqrt[3]{0}=0$

(c) $\lim_{x\to 2}[(x+2)^{3/2}]=[\lim_{x\to 2}(x+2)^{1/2}]^3$

$$=[\lim_{x\to 2}\sqrt{x+2}]^3=[\sqrt{\lim_{x\to 2}(x+2)}]^3$$

$$=(\sqrt{4})^3=2^3=8$$

EXERCISES

1. Evaluate the following limits:

(a) $\lim_{x \to 3} (x^2 - 3x + 5)$ (b) $\lim_{x \to -2} (2x^3 - 6x^2 + 3x - 2)$

(c) $\lim_{x \to 1} \frac{x + 3}{2x^2 - 6x + 5}$ (d) $\lim_{x \to 4} \frac{2x^2 - 6}{x^3 + 5}$

(e) $\lim_{x \to 1} \frac{x^3 - 1}{x - 1}$ (f) $\lim_{x \to -1} \frac{x^3 + 1}{x + 1}$

(g) $\lim_{x \to -2} \frac{x^2 + x - 2}{x^2 - 4}$ (h) $\lim_{h \to 0} \frac{1}{h}\left(\frac{1}{x+h} - \frac{1}{x}\right)$

(i) $\lim_{x \to -2} \frac{x^2 - 4}{x^2 + 8}$ (j) $\lim_{x \to 0} a\cos^2 x + b\sin^2 x$

(k) $\lim_{x \to 0} \sec x$ (l) $\lim_{\theta \to 0} \theta \cot \theta$

2. Evaluate the following limits:

(a) $\lim_{r \to 1} \sqrt{\frac{2r^2 + 3r - 1}{r^2 + 1}}$ (b) $\lim_{h \to 2} \sqrt{\frac{h^3 - 8}{h^2 - 4}}$

(c) $\lim_{h \to 0} \frac{\sqrt{x + h} - \sqrt{x}}{h} (x > 0)$ (d) $\lim_{y \to 2} \sqrt{\frac{y^2 - 4}{y^2 - 3y + 2}}$

(e) $\lim_{x \to 4} \sqrt{\frac{x^3 - 27}{x^2 + 2x + 1}}$ (f) $\lim_{h \to 0} \frac{1}{h}\left(\frac{1}{\sqrt{1+h}} - 1\right)$

(g) $\lim_{h \to 0} \frac{(1+h)^{3/2} - 1}{h}$

3. Give examples of functions f and g such that
   (a) $\lim_{x \to 0} f(x)$ does not exist, but $\lim_{x \to 0} (fg)(x)$ exists and does not equal zero.
   (b) $\lim_{x \to x_0} g(x) = 0$, $\lim_{x \to x_0} \frac{f(x)}{g(x)}$ exists and does not equal 1.
   (c) $\lim_{x \to x_0} g(x) = 0$, $\lim_{x \to x_0} f(x) = 0$, $\lim_{x \to x_0} \frac{f(x)}{g(x)}$ does not exist.
4. (a) Show that if $\lim_{x \to x_0} f(x) > 0$, then there is a neighborhood of x_0 throughout which $f(x) > 0$ [except possibly for $f(x_0)$].
   (b) Show that if $\lim_{x \to x_0} f(x) < 0$, then there is a neighborhood of x_0 throughout which $f(x) < 0$ [except possibly for $f(x_0)$].
5. Show that if $\lim_{x \to x_0} f(x) > 0$, then $\lim_{x \to x_0} \sqrt{f(x)}$ exists and equals $\sqrt{\lim_{x \to x_0} f(x)}$. [Hint: Use the result of exercise 4(a) above and note that if $0 < a < \sqrt{l}$, then $0 < a^2 < l$.]
6. (a) If $\lim_{x \to x_0} f(x) > 0$, show that $\lim_{x \to x_0} |f(x)| = \lim_{x \to x_0} f(x)$. [Hint: Apply exercise 4.]

(b) Show that if $\lim_{x\to x_0} f(x) = 0$, then $\lim_{x\to x_0} |f(x)| = 0$. [Hint: To show $|f(x)|$ is eventually in $]a, b[$, show $f(x)$ is eventually in $]-b, b[$.]

(c) Conclude that $\lim\limits_{x\to x_0} |f(x)| = |\lim\limits_{x\to x_0} f(x)|$.

(d) Suppose $\lim\limits_{x\to x_0} f(x) = 0$; show that $\lim\limits_{x\to x_0} \sqrt{|f(x)|} = 0$.

(e) Show $\lim\limits_{x\to x_0} \sqrt{|f(x)|} = \sqrt{|\lim\limits_{x\to x_0} f(x)|}$

7. Suppose $\lim\limits_{x\to x_0} g(x) = l \neq 0$.

(a) Show $\lim_{x\to x_0} 1/g(x) = 1/l$. [Hint: Suppose $l > 0$; note that $0 < a < 1/l < b$ if and only if $0 < 1/b < l < 1/a$.]

(b) Suppose $\lim\limits_{x\to x_0} f(x) = l'$; show $\lim\limits_{x\to x_0} f(x)/g(x) = l'/l$.

8. Suppose for all x in some neighborhood K, $|f(x)| < g(x)$.

(a) Show that if g is eventually in $]A, B[$ as x approaches $x_0 \in K$, then f is eventually in $]-B, B[$ as x approaches x_0.

(b) Suppose $\lim_{x\to x_0} g(x) = 0$. Show that $\lim_{x\to x_0} f(x) = 0$. [Hint: To show $f(x)$ is eventually in $]a, b[$, show $g(x)$ is eventually in $]-1, \min(|a|, |b|)]$ and apply part (a).]

9. Let P with coordinates $(x, \sin^2 x)$ be a point on the graph of $y = \sin^2 x$. Suppose that P is different from the origin and that $-\pi < x < \pi$.

(a) The perpendicular bisector of the segment OP intersects the y axis at a point E. As P moves along the graph and approaches 0, what is the limiting position of E?

(b) The line perpendicular to OP and passing through P intersects the y axis at a point D. As P approaches 0 along the graph, what is the limiting position of D?

10. Suppose for all x in an open interval K, $g(x) \leqslant f(x) \leqslant h(x)$. Show that if $x_0 \in K$, and $\lim_{x\to x_0} g(x) = \lim_{x\to x_0} h(x)$, then $\lim_{x\to x_0} f(x) = \lim_{x\to x_0} g(x)$.

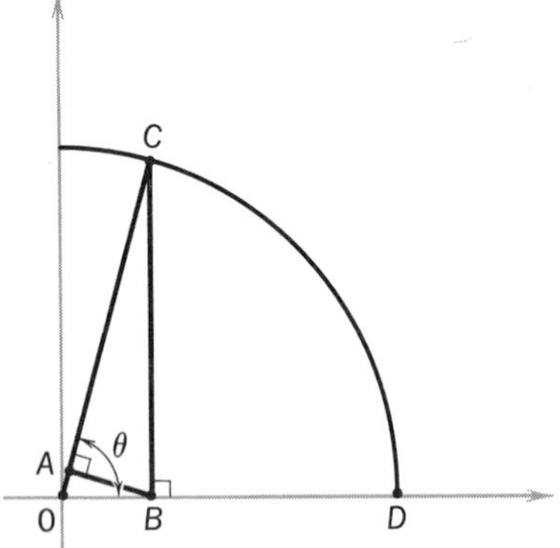

FIGURE 3-7

11. As θ goes to 0, what is the limit of the ratio of the area of the triangle OAB to the area of the triangle OCB? What is the limit of the area of triangle OAB to the area of the sector OCD. (See Figure 3-7.) What is the limiting position of the point A as θ approaches 0?

12. Suppose that $\lim_{x\to x_0} f(x) = k$. Let $\varepsilon > 0$. Show that there is a neighborhood M of x_0 such that $x \in M(x_0)$ implies $k - \varepsilon < k < +\varepsilon$.

13. Suppose that for every $\varepsilon > 0$, $f(x)$ is eventually in $]k - \varepsilon, k + \varepsilon[$ as x approaches x_0. Show that $\lim_{x\to x_0} f(x) = k$.

14. Show that $\lim_{x\to x_0} f(x) = k$ if and only if for every $\varepsilon > 0$, there is an $e > 0$ such that $0 < |x - x_0| < e$ implies $|f(x) - k| < \varepsilon$. [Hint: See §2 exercises 20-21.]

15. Suppose $\lim\limits_{x\to x_0} f(x)$ exists.

(a) Show that $\lim\limits_{x\to x_0} \sqrt[n]{|f(x)|} = \sqrt[n]{|\lim\limits_{x\to x_0} f(x)|}$.

(b) For n odd, show that $\lim\limits_{x\to x_0} \sqrt[n]{f(x)} = \sqrt[n]{\lim\limits_{x\to x_0} f(x)}$.

4 CONTINUOUS FUNCTIONS

The term **continuity** is derived from the study of the physical phenomenon of motion. Our intuitive notion of a moving particle is that it passes from one position to another along a continuous, unbroken course. In §2.1 we considered points P moving along the graph of a function f, and the graph of f was always a smooth, unbroken path for P. We shall now give a formal mathematical definition of what it means for the graph of a function to be continuous and unbroken; **limit** is the essential idea in this definition. Hence, it is more convenient for us to talk about the continuity of a function rather than about the continuity of its graph.

Definition A function f is **continuous at the point x_0** if f is defined on a neighborhood of x_0 and

$$\lim_{x \to x_0} f(x) = f(x_0)$$

Notice that continuity is, as defined, a property of a function **at a point.**

Example 1 The sine and cosine functions are continuous at zero (Figure 4-1). We have $\lim_{x \to 0} \sin x = 0 = \sin 0$, and $\lim_{x \to 0} \cos x = 1 = \cos 0$ and so the condition for continuity is satisfied. (The sine and cosine functions are continuous at every point, as will follow from Theorem 4 of this section.)

Example 2 Consider the function

$$f\colon x \mapsto \begin{cases} \dfrac{\sin x}{x}, & x \neq 0 \\ 1, & x = 0 \end{cases}$$

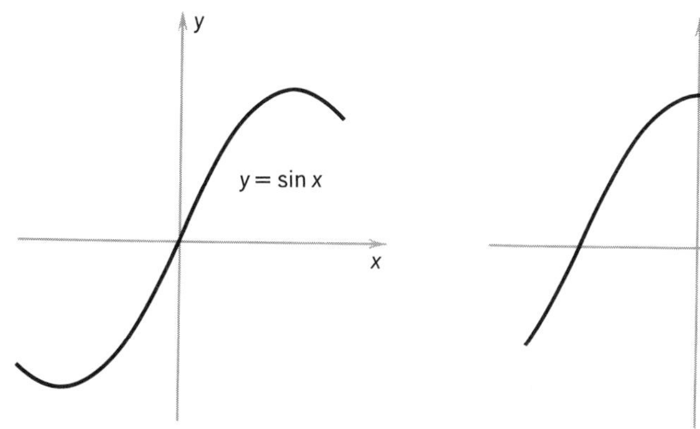

FIGURE 4-1

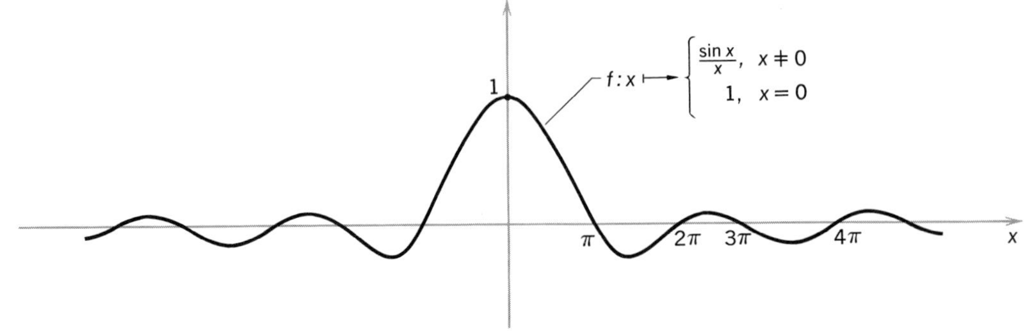

FIGURE 4-2

The function f is continuous at 0, since

$$\lim_{x\to 0} f(x) = \lim_{x\to 0}\frac{\sin x}{x} = 1 = f(0)$$

(See Figure 4-2.)

Example 3 The signum function $z \mapsto \operatorname{sgn} z$ is not continuous at 0, but it is continuous at 1, 2, and every other point $x_0 \neq 0$. The function $z \mapsto z + \operatorname{sgn} z$ is also continuous at every point in its domain except 0. (See Figure 4-3.) Note that the graphs of both these functions have breaks at $z = 0$.

Example 4 A polynomial function $t \mapsto A_0 + A_1 t + A_2 t^2 + \cdots + A_n t^n$ is continuous at every point.

This result follows from the corollary to the theorem on the multiplication of limits in §3.

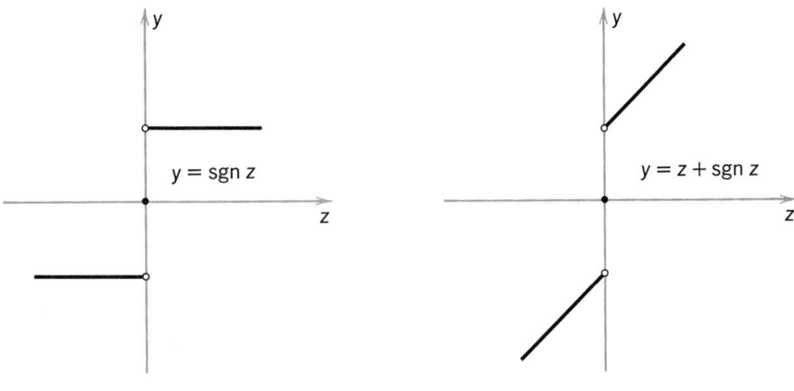

FIGURE 4-3

Theorem 1 Arithmetic of Continuous Functions If the functions f and g are continuous at x_0, then the functions fg, $f+g$, $f-g$, and $|f|$ are also continuous at x_0; if $g(x_0) \neq 0$, the function f/g is continuous at x_0.

Proof Using the arithmetic of limits and the definition of continuity we have

$$\lim_{x \to x_0} (fg)(x) = \lim_{x \to x_0} f(x) \lim_{x \to x_0} g(x) = f(x_0)\, g(x_0) = (fg)(x_0)$$

$$\lim_{x \to x_0} (f+g)(x) = \lim_{x \to x_0} f(x) + \lim_{x \to x_0} g(x) = f(x_0) + g(x_0) = (f+g)(x_0)$$

The proofs for the cases $f-g$, $|f|$, and f/g are analogous. ∎

Example 5 The function $x \mapsto x^2 \sin x + \cos x$ is continuous at 0 since $x \mapsto x^2$, $x \mapsto \sin x$, and $x \mapsto \cos x$ are all continuous at 0.

Theorem 2 Composition of Continuous Functions If the function f is continuous at x_0 and the function g is continuous at the point $f(x_0)$, then the function $g \circ f$ is continuous at x_0; that is,

$$\lim_{x \to x_0} (g \circ f)(x) = (g \circ f)(x_0) = g(\lim_{x \to x_0} f(x)) = \lim_{y \to f(x_0)} g(y).$$

Proof We have to show that $g(f(x))$ is eventually in every neighborhood of $g(f(x_0))$ as x approaches x_0. Hence, let $]A, B[$ be any neighborhood of $g(f(x_0))$. Since g is continuous at $f(x_0)$, there is a neighborhood N of $f(x_0)$, such that $g(y) \in]A, B[$ for all $y \in N$. Since N is a neighborhood of $f(x_0)$, $f(x)$ is eventually in N as x approaches x_0; therefore there is a neighborhood M of x_0 such that $f(x) \in N$ for all $x \in M$ (Figure 4-4). But then for all $x \in M$, $g(f(x)) \in]A, B[$ (since $f(x) \in N$). Therefore, $(g \circ f)(x)$ is eventually in $]A, B[$ as x approaches x_0. ∎

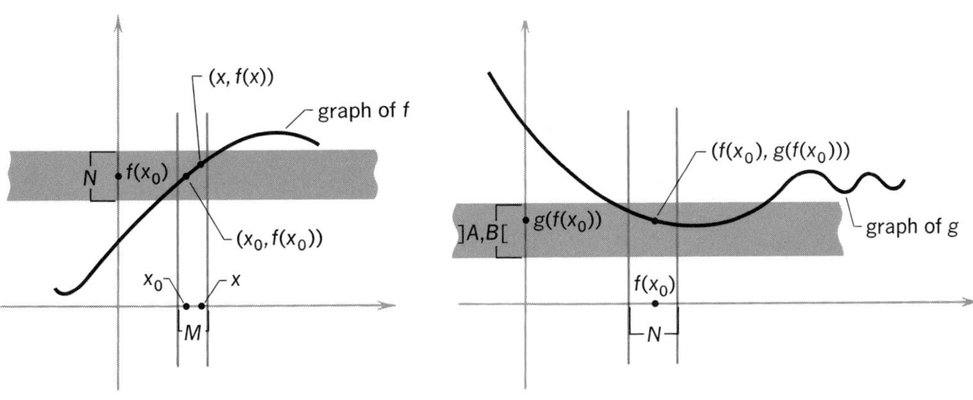

FIGURE 4-4

Example 6 (a) The function $x \mapsto \cos(\sin x)$ is continuous at 0, because the sine function is continuous at 0 and the cosine function is continuous at the point $\sin 0 = 0$. Similarly the functions, $x \mapsto \sqrt{|\sin x|}$ and $x \mapsto \cos^2(x^2)$ are continuous at 0.

(b) If f is continuous at the point $mx_0 + b$, then the composite of f on $x \mapsto mx+b$ is continuous at x_0. (Why?) Therefore, $x \mapsto \cos(mx)$ is continuous at 0 and we have $\lim_{x \to 0} \cos(mx) = 1$. Similarly, we have $\lim_{x \to b} \sin(x-b) = 0$. Furthermore, $\cos x = \sin(\frac{1}{2}\pi - x)$, so the cosine is continuous at $\frac{1}{2}\pi$.

To augment our list of variables, we shall form new variables by prefixing old ones with the symbol Δ (delta). Thus we have the new variables Δx, Δy, Δz, Δu, ... The variables Δx, Δy, ... are used the same way we use $x, y, \ldots$. Thus we can form polynomials such as $\Delta x^2 + \Delta x + 3$ and trigonometric expressions such as $\sin \Delta x + \cos \Delta x$. We can also use them to define functions such as the affine function $\Delta x \mapsto 3 + \Delta x$.

The new variables are also used in working with limits. As examples, we list

(1) $\lim_{\Delta x \to 0} (\Delta x^2 + \Delta x + 3) = 3$

(2) $\lim_{\Delta x \to 0} \cos \Delta x = 1$

(3) $\lim_{\Delta x \to 0} \dfrac{\sin \Delta x}{\Delta x} = 1$

Now let us suppose x_0 is a fixed real number. Then the affine function $\Delta x \mapsto x_0 + \Delta x$ is continuous at all points and, in particular, $\lim_{\Delta x \to 0} (x_0 + \Delta x) = x_0$. Hence, if a function f is continuous at x_0, the composite function $\Delta x \mapsto f(x_0 + \Delta x)$ is continuous at 0, and so

$$\lim_{\Delta x \to 0} f(x_0 + \Delta x) = f(x_0)$$

(See Figure 4-5.)

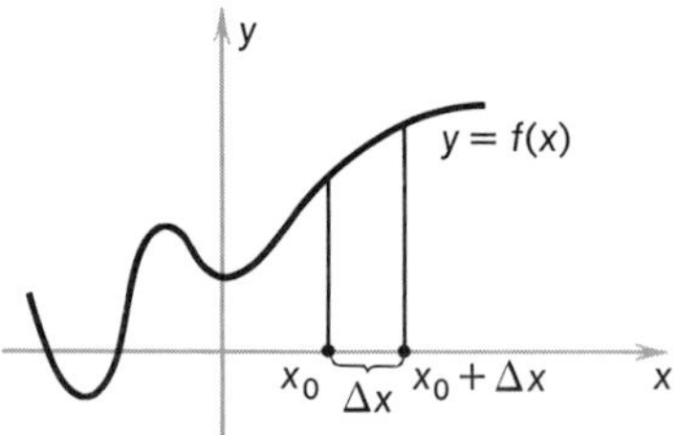

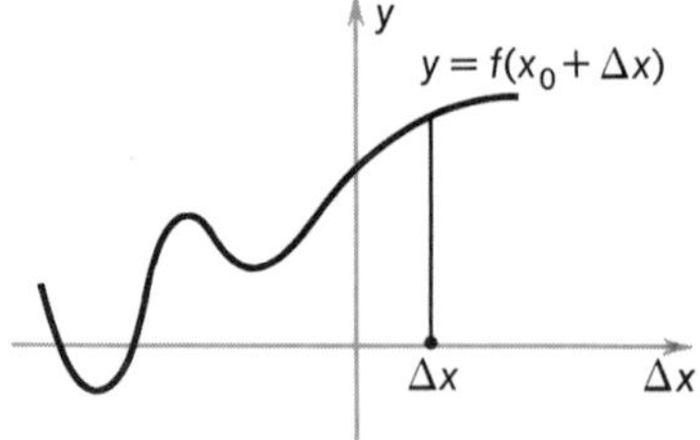

FIGURE 4-5

Example 7 We list several limits as an illustration of the above principle.

(a) $\lim_{\Delta x \to 0} \cos(0+\Delta x) = \lim_{\Delta x \to 0} \cos \Delta x = 1$

(b) $\lim_{\Delta x \to 0} \sin(0+\Delta x) = \lim_{\Delta x \to 0} \sin \Delta x = 0$

(c) $\lim_{\Delta x \to 0} (x_0+\Delta x)^3 = x_0^3$

(d) $\lim_{\Delta u \to 0} (u_0+\Delta u)^2 = u_0^2$

(e) $\lim_{\Delta u \to 0} m(u_0+\Delta u) = mu_0$

Alternatively, assume that $\lim_{\Delta x \to 0} f(x_0+\Delta x) = f(x_0)$. It then follows that the function f is continuous at x_0. (Why?) Combining these two results, we have an important criterion for continuity, whose notational usefulness will become clear later.

The function f is continuous at the point x_0 if and only if

$$\lim_{\Delta x \to 0} f(x_0+\Delta x) = f(x_0)$$

Up to now we have only discussed the **continuity of a function at a point**. We have seen several examples where a function was continuous at some points, but discontinuous at others. A function which is continuous at *every* point in its domain is called a **continuous function**. By way of example, we know that any polynomial function $x \mapsto A_0 + A_1 x + \cdots + A_n x^n$ is continuous at every point, and hence it is a continuous function; thus in particular all the affine functions are continuous.

From Theorem 1 and Theorem 2, we immediately obtain the following theorem.

Theorem 3 If f and g are continuous functions then so are the functions

$$f+g, f-g, fg, |f|, f/g, \text{ and } f \circ g$$

We shall now establish that the trigonometric functions are continuous functions. We first treat the sine and cosine functions. The proof that these functions are continuous is a good illustration of the usefulness of the Δx formulation of continuity.

Theorem 4 The sine and the cosine are continuous functions.

Proof We shall show that $\lim_{\Delta x \to 0} \sin(x_0+\Delta x) = \sin x_0$. Using the double-angle formula for the sine, we have

$$\sin(x_0+\Delta x) = \sin x_0 \cos \Delta x + \cos x_0 \sin \Delta x \tag{1}$$

We know that $\lim_{\Delta x \to 0} \cos \Delta x = 1$ and $\lim_{\Delta x \to 0} \sin \Delta x = 0$, so the limit of the right-hand side of equation (1) exists and we have

$$\lim_{\Delta x\to 0}\sin(x_0+\Delta x) = \lim_{\Delta x\to 0}\sin x_0\cos\Delta x + \lim_{\Delta x\to 0}\cos x_0\sin\Delta x$$
$$= \sin x_0\cdot 1 + \cos x_0\cdot 0 = \sin x_0$$

The proof that $\lim_{\Delta x\to 0}\cos(x_0+\Delta x)=\cos x_0$ is analogous and is left as an exercise. ▨

The continuity of the remaining trigonometric functions can be established using Theorem 3 and some familiar identities. Since $\tan x = \sin x/\cos x$, $\cot x = \cos x/\sin x$, $\sec x = 1/\cos x$, and $\csc x = 1/\sin x$, the continuity of the tangent, cotangent, secant, and cosecant functions follows from the continuity of the sine and cosine. Note that $\tan(\frac{1}{2}\pi)$ is not defined and so the tangent function cannot be continuous at $\frac{1}{2}\pi$. However, the tangent function is continuous at every point in its domain and so it is a continuous function. Similarly, the function $x\mapsto 1/x$ is continuous at all points in its domain. However, the function

$$x\mapsto\begin{cases}\dfrac{1}{x}, & x\neq 0\\ 0, & x=0\end{cases}$$

is not continuous at 0 and so it is not a continuous function.

At the beginning of this section we said that the graph of a continuous function was the mathematical translation of the physical notion of a continuous unbroken path. We have shown that polynomial functions and the sine and cosine functions are continuous, and that the graphs of these functions are indeed continuous and unbroken curves. On the other hand, the graph of the signum function is broken at 0 and this function is not continuous at 0. Likewise, the tangent function is not continuous at $\frac{1}{2}\pi$ and its graph has a break at $\frac{1}{2}\pi$. However, the tangent function is continuous at all points in the open interval $]-\frac{1}{2}\pi,\frac{1}{2}\pi[$ and in the region between the vertical lines $x=-\frac{1}{2}\pi$ and $x=+\frac{1}{2}\pi$, the graph of this function is unbroken.

An important feature of the analysis of continuity in terms of functions and limits is that using this technique we can demonstrate that certain graphs are continuous even though they may be impossible to sketch or to visualize. For example, suppose we are given a continuous function f whose domain is R. The composition of f on an affine function is again a continuous function with domain R. Therefore, the distortion of the graph of f brought about by the affine substitution preserves continuity. Thus we know that the graph of $x\mapsto\sin(10^{100}x)$, a graph impossible to draw, is in fact a continuous curve. We therefore have a good intuition of what this curve is like although we cannot visualize it directly.

We conclude this section with some additional remarks on the limits of composite functions. A close look at the proof of Theorem 2 reveals that a general theorem about limits and composition can be proved. The following theorem differs from Theorem 2 in that here g is not assumed to be continuous at y_0, and f is not assumed to be continuous at x_0.

Theorem 5 Suppose that $\lim_{y\to y_0} g(y)$ exists, $\lim_{x\to x_0} f(x) = y_0$, and $f(x) \neq y_0$ for all $x \neq x_0$. Then $\lim_{x\to x_0} g(f(x))$ exists, and equals $\lim_{y\to y_0} g(y)$.

Proof Since we are only assuming that $\lim_{y\to y_0} g(y)$ exists, $g(y_0)$ might not be defined. However, the hypotheses guarantee that $g(f(x))$ is defined for all x in some deleted neighborhood of x_0. We leave it to the reader to complete the proof of this theorem by following the proof of Theorem 2. ∎

Theorem 5 has many applications. One of them is the following very useful technique for computing limits.

FORMAL DEVICE

To evaluate $\lim_{x\to x_0} g(mx+b)$, set $u = mx+b$ and $u_0 = mx_0+b$. Then

$$\lim_{x\to x_0} g(mx+b) = \lim_{u\to u_0} g(u)$$

Example 8 Evaluate $\lim_{\theta\to 0} (\sin 3\theta)/3\theta$. Set $u = 3\theta$, and $u_0 = 3\cdot 0 = 0$. We then compute

$$\lim_{\theta\to 0} \frac{\sin 3\theta}{3\theta} = \lim_{u\to 0} \frac{\sin u}{u} = 1$$

Therefore $\lim_{\theta\to 0} \sin 3\theta/3\theta = 1$. Let us now evaluate $\lim_{\theta\to 0} \sin 3\theta/\theta$. Since $\sin 3\theta/\theta = 3 \sin 3\theta/3\theta$, we must have $\lim_{\theta\to 0} \sin 3\theta/\theta = 3 \lim_{\theta\to 0} \sin 3\theta/3\theta = 3\cdot 1 = 3$.

Example 9 Evaluate $\lim_{\theta\to 0} \tan 5\theta/\tan\theta$. Recall that

$$\frac{\tan 5\theta}{\tan\theta} = \frac{\sin 5\theta}{\sin\theta}\cdot\frac{\cos\theta}{\cos 5\theta}$$

We know that $\lim_{\theta\to 0} \cos\theta = 1$; substituting $u = 5\theta$ and $u_0 = 0$, we find $\lim_{\theta\to 0} \cos 5\theta = \lim_{u\to 0} \cos u = 1$. Thus $\lim_{\theta\to 0} \cos\theta/\cos 5\theta = 1$. Also,

$$\frac{\sin 5\theta}{\sin\theta} = \frac{\dfrac{\sin 5\theta}{\theta}}{\dfrac{\sin\theta}{\theta}}$$

and we know that $\lim_{\theta\to 0} \sin\theta/\theta = 1$. Again, substituting $u = 5\theta$, $u_0 = 0$, we obtain

$$\lim_{\theta\to 0} \frac{\sin 5\theta}{\theta} = 5 \lim_{u\to 0} \frac{\sin u}{u} = 5$$

Therefore $\lim_{\theta\to 0} \tan 5\theta/\tan\theta = 5$.

EXERCISES

1. Show that
$$\lim_{\theta\to 0}\frac{1-\cos\theta}{\theta} = 0$$
by using the trigonometric identity $\sin^2(\frac{1}{2}\theta) = \frac{1}{2}(1-\cos\theta)$.

2. Evaluate each of the following limits.

(a) $\displaystyle\lim_{\alpha\to\frac{1}{2}\pi}\frac{\cos\alpha}{\frac{1}{2}\pi-\alpha}$

(b) $\displaystyle\lim_{\alpha\to 0}\frac{\sin b\alpha}{b}$

(c) $\displaystyle\lim_{\theta\to 0}\frac{\tan n\theta}{\tan\theta}$

(d) $\displaystyle\lim_{x\to b}\frac{\sin(x-b)}{b-x}$

(e) $\displaystyle\lim_{t\to 0}\frac{15t}{\tan 6t}$

(f) $\displaystyle\lim_{t\to 0}\frac{\cot 4t}{\cot 3t}$

(g) $\displaystyle\lim_{\alpha\to 0}\frac{\alpha\sin\alpha}{1-\cos\alpha}$

(h) $\displaystyle\lim_{\alpha\to 0}\frac{\cos m\alpha}{\cos n\alpha}$

3. Suppose the function f is continuous at x_0. Which of the following statements are necessarily true? Justify your answer.

(a) $\displaystyle\lim_{\Delta x\to 0}(f(x_0+\Delta x)-f(x_0)) = 0$

(b) $\displaystyle\lim_{x\to x_0}(f(x)-f(x_0)) = 0$

(c) $\displaystyle\lim_{\Delta x\to 1} f(x_0+\Delta x) = f(x_0+1)$

(d) $\displaystyle\lim_{\Delta x\to 0} f(x_0-\Delta x) = f(x_0)$

(e) $\displaystyle\lim_{\Delta x\to 0} f(x_0-\Delta x) = -f(x_0)$

(f) $\displaystyle\lim_{t\to 0} f(x_0+\sin t) = f(x_0)$

(g) $\displaystyle\lim_{t\to 0} f(x_0-\tfrac{1}{2}\sin t) = f(x_0)$

(h) $\displaystyle\lim_{\Delta x\to 0} f(x_0+\sin\Delta x) = f(x_0)$

(i) $\displaystyle\lim_{\Delta u\to 0} f(u_0+(\Delta u)^2) = f(u_0)^2$

(j) $\displaystyle\lim_{\Delta u\to 0} f(u_0+(\Delta u)^2) = f(u_0)$

4. Let f be a continuous function. Define functions f^+ and f^- by the rules
$$f^+ : x \mapsto \begin{cases} f(x), & f(x) \geqslant 0 \\ 0, & f(x) \leqslant 0 \end{cases}$$
and
$$f^- : x \mapsto \begin{cases} -f(x), & f(x) \leqslant 0 \\ 0, & f(x) \geqslant 0 \end{cases}$$
Show that f^+ and f^- are continuous. Sketch $\sin^+$ and $\cos^-$.

5. Define a function g by the rule

$$g(x) = \begin{cases} \sin^2 x, & x \leqslant 0 \\ mx + b, & x > 0 \end{cases}$$

For what values of m and b is the function g continuous?

6. Suppose that $\lim_{x \to x_0} g(x) = y_0$ and that f is continuous at y_0. Show $\lim_{x \to x_0} f(g(x)) = f(y_0)$.

7. Complete the proof of Theorem 4.

8. Let the function f be continuous at x_0 and let $\varepsilon > 0$. Show that there is a neighborhood M of x_0 such that $f(x) - \varepsilon < f(x) < f(x_0) + \varepsilon$ for all $x \in M$.

9. Let the function f be continuous at x_0. Show that for every $\varepsilon < 0$ there is an $e > 0$ such that $|x - x_0| < e$ implies $|f(x) - f(x_0)| < \varepsilon$.

10. Suppose we are given points A and B with coordinates $(0, 0)$ and $(0, 1)$, respectively. Let n be a fixed number > 0 and let θ be an angle such that $0 < \theta < \pi/(1+n)$. We can construct a triangle ABC such that AC and AB form the angle θ, and CB and AB form the angle $n\theta$ (Figure 4-6). Let D be the point of intersection of AB with the perpendicular from C to AB. What is the limiting position of D as θ approaches 0?

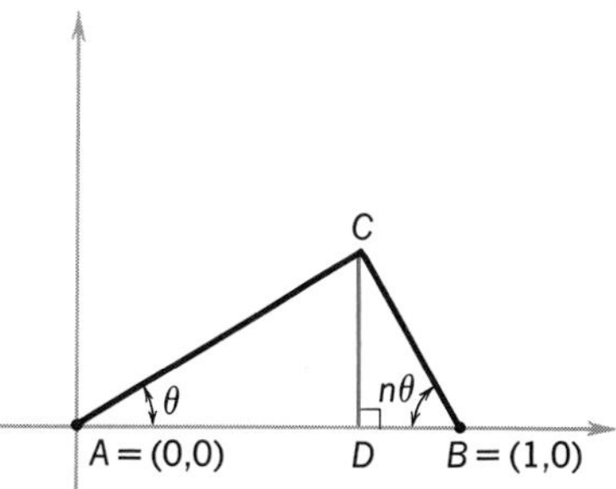

FIGURE 4-6

11. Show how the Formal Device on page 97 follows from Theorem 5.

12. Do exercise 21 of §2.1 for the curve $y = \cos x - 1$.

REVIEW EXERCISES FOR CHAPTER 2

1. Define the following terms.
   continuous at x
   continuous function
   deleted neighborhood
   eventually in
   limit

In exercises 2–12, evaluate the limits.

2. $\lim_{x \to 1} \dfrac{x^2 - x}{x - 1}$

3. $\lim_{x \to 1} \dfrac{2x - 2x^2}{1 - x}$

4. $\lim_{x \to 4} \dfrac{\sqrt{x} - 2}{x - 4}$

5. $\lim_{x \to \sqrt{3}} \dfrac{x^2 - 3}{x - \sqrt{3}}$

6. $\lim_{x \to 4} \dfrac{1/\sqrt{x} - \frac{1}{2}}{x - 4}$

7. $\lim_{x \to 3} \dfrac{1/x - \frac{1}{3}}{x - 3}$

8. $\lim_{t \to 0} \dfrac{(4-t)^3 - 64}{t}$

9. $\lim_{x \to 0} (x^2 - 3x + 4\sqrt{x})$

10. $\lim_{x \to 2} \dfrac{5x\sqrt{x} - 2}{x}$

11. $\lim_{x \to a} \dfrac{x^2 - (a+b)x + ab}{x - a}$

12. $\lim_{x \to 4} \sqrt{2 + \sqrt{x}}$

In exercises 13–15, evaluate $\lim_{x \to a} \dfrac{f(x) - f(a)}{x - a}$ for the given f and a.

13. $f{:}x \mapsto x^3, \qquad a = -1$
14. $f{:}x \mapsto 1/x, \qquad a = 2$
15. $f{:}x \mapsto \sqrt{x} - x, \qquad a = 4$
16. Show that $\lim_{\theta \to 0} \dfrac{\sin \theta}{\theta^2}$ does not exist; show that $\lim_{\theta \to 0} \dfrac{\cos \theta - 1}{\theta^2} = -\dfrac{1}{2}$.
17. A semicircle of radius r has center 0 and base diameter AB as shown in Figure 4-7, with CO perpendicular to AB. A point D is chosen on OC and chord EF is constructed through D, parallel to AB. Line CD is bisected at G, and chord HJ constructed through G parallel to AB. As D moves up along OC toward C, what is the limit of the ratio $\overline{EF}/\overline{HJ}$?

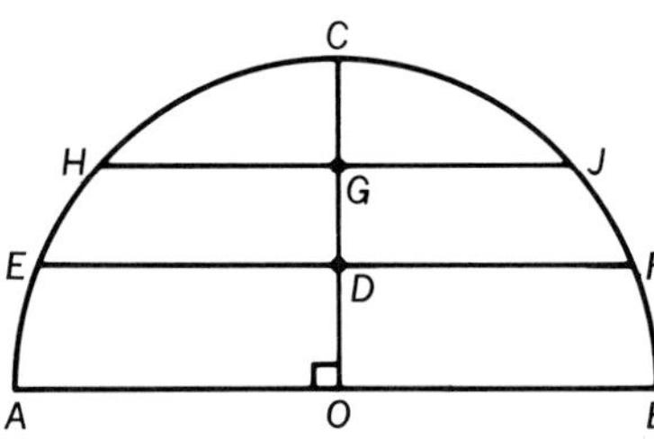

FIGURE 4-7

18. Let $p(x) = Ax^2 + Bx + C$, and let P denote the y intercept of the curve $y = p(x)$. Let Q be a point that is approaching P along the curve. Consider the ratio $\overline{PQ}/|x_Q|$, where x_Q is the x coordinate of Q. This is the ratio of the length of the chord from P to Q to the distance of Q from the y axis. Show that the limit as Q approaches P of this ratio is equal to 1 if and only if $B = O$. Interpret the result geometrically.
19. Show that $\cos x$ is eventually in $]0, 1[$ as x approaches 0. Compare this with the fact that $\lim_{x \to 0} \cos x = 1$.
20. Suppose that $\lim_{x \to x_0} f(x) = l$ and that $f(x)$ is eventually in $]A, l[$ as x approaches x_0 $(A < l)$. Show that $f(x)$ is *not* eventually in $]l, B[$ as x approaches x_0 $(B > l)$.
21. Prove Theorem 5 of §2.4.
22. Let p and q be natural numbers; $q \geqslant 1$. Suppose that $\lim_{x \to x_0} f(x)$ exists.
    (a) Prove that $\lim_{x \to x_0} |f(x)|^{p/q} = |\lim_{x \to x_0} f(x)|^{p/q}$.
    (b) Suppose q is odd. Prove that $\lim_{x \to x_0} (f(x))^{p/q} = (\lim_{x \to x_0} f(x))^{p/q}$.
23. Consider the graph of $y = 1/x$, $x > 0$. Call this curve H. In Example 5 of §2.1, it is stated that the line L of equation $y = 2 - x$ is tangent to H at $(1, 1)$.
    (a) Show that L intersects H only at $(1, 1)$ and otherwise lies below H.
    (b) Find the equation of the line tangent to H at an arbitrary point $(x_0, 1/x_0)$.
24. Consider the curve $y = x^2$. Find the equation of the line tangent to this curve at (x_0, x_0^2).
25. Attempt exercise 21 of §2.1 for the curve $y = x^4$. What goes wrong? Try to explain geometrically.

3

DIFFERENTIAL CALCULUS

In §1.7, we presented five problems, the solutions of which were beyond the scope of analytic geometry. The consideration of continuity in Chapter 2 dealt with the first of these questions; the present chapter is motivated by problems (2)–(4). Our goal is to give precise mathematical meaning to the intuitive idea of a "smooth" curve, that is, one without abrupt changes of direction. Such a curve is characterized by having a unique tangent line at every point, and limits are used to compute the slopes of these tangent lines. In physical problems these slopes will be interpreted as "instantaneous rates of change," such as velocity and acceleration.

1 DEFINITION OF THE DERIVATIVE

The tangent problem for a curve $y=f(x)$ is to determine the slope of the line tangent to the curve at a point $(x,f(x))$. In this chapter and in the following chapters we study this problem in detail. For our study we need the concept of the **derivative**. To help motivate this important definition we begin with an example of a curve whose properties are well known.

Let us consider the graph of the function $f\colon x \mapsto x^2$, which as we know, is a parabola (Figure 1-1). Through each point (x_0, x_0^2) on the graph, there passes a unique tangent line; this tangent line is nonvertical and so it has slope.

This slope, and thus the equation of the tangent line, can be computed using limits as follows. If $(x_0+\Delta x, (x_0+\Delta x)^2)$ is a point on the graph distinct from (x_0, x_0^2), the secant line passing through these two points has slope equal to

$$\frac{(x_0+\Delta x)^2 - x_0^2}{(x_0+\Delta x) - x_0}$$

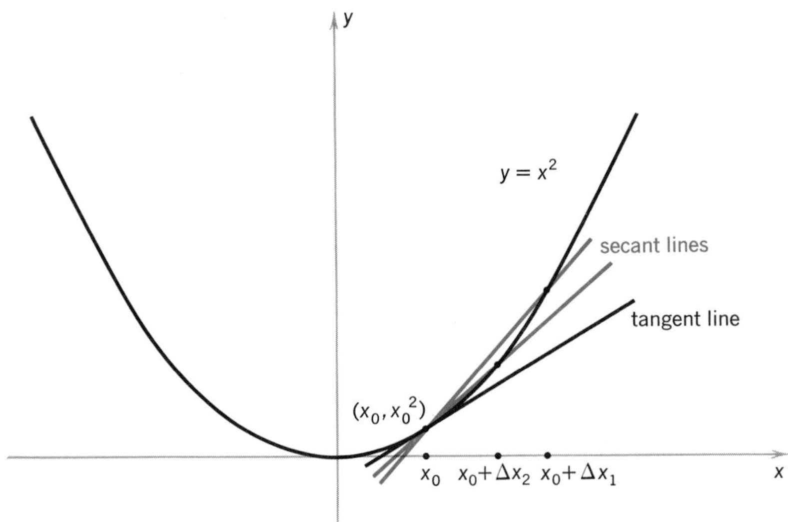

FIGURE 1-1

Simplifying, we find this slope to be

$$x_0^2 + 2x_0\,\Delta x + \Delta x^2 - x_0^2 = 2x_0 + \Delta x$$

Remember that we are interested in the slope of the tangent line at the exact point x_0. As we bring the points x_0 and $x_0+\Delta x$ closer together by making Δx smaller, $f(x_0)$ and $f(x_0+\Delta x)$ also get closer, and the secant line through these points approaches the tangent line to the curve at x_0. Now we take the limit of $2x_0+\Delta x$ as Δx approaches 0 and obtain $\lim_{\Delta x \to 0}(2x_0+\Delta x) = 2x_0$. Hence, we expect that the slope of the tangent to the parabola at (x_0, x_0^2) is $2x_0$. Let us verify this result. The parabola has the property that the line tangent at the point (x_0, x_0^2) passes through this point, and otherwise lies completely below the parabola (Figure 1-2). The equation of the line through (x_0, x_0^2) with slope $2x_0$ is, in point-slope form, $y - x_0^2 = 2x_0(x-x_0)$ which can be rewritten as $y = (2x_0)x - x_0^2$. Therefore, we must check that this line intersects the curve only at (x_0, x_0^2) and that in all other cases the point $(x, 2x_0x - x_0^2)$ lies below the point (x, x^2). This is relatively straightforward if we consider the equations

$$y = x^2 \tag{1}$$

$$y = (2x_0)\,x - x_0^2 \tag{2}$$

Subtracting (2) from (1) yields

$$0 = x^2 - (2x_0)\,x + x_0^2 \tag{3}$$

or factoring,

$$0 = (x - x_0)^2 \tag{4}$$

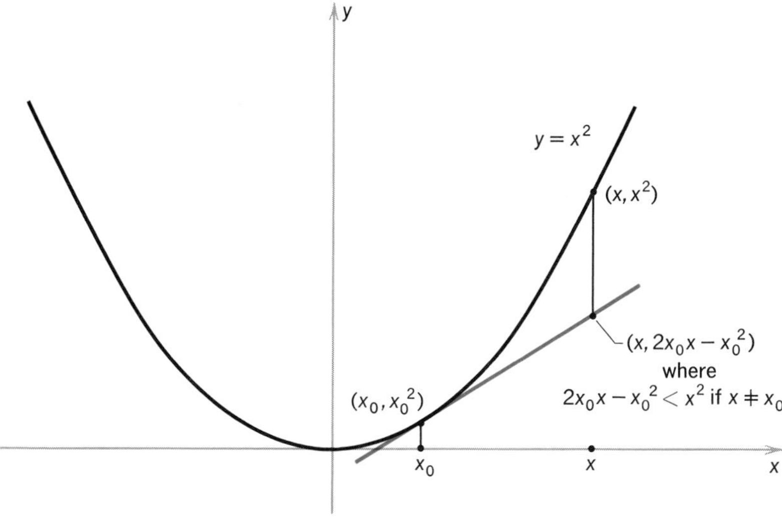

FIGURE 1-2

Hence, the equations have the unique simultaneous root $x = x_0$, $y = x_0^2$. Furthermore, $x_0^2 - 2x_0 x + x^2 = (x - x_0)^2 > 0$ for all $x \neq x_0$; thus for $x \neq x_0$, $x^2 - 2x_0 x + x_0^2 > 0$, or $x^2 > 2x_0 x - x_0^2$. But this implies that the line with equation $y = (2x_0)x - x_0^2$ is indeed the tangent to the parabola at (x_0, x_0^2).

The method of computing tangents by means of limits used in the case of $y = x^2$ above is very general. If f is a function, and if x is in the domain of f, the secant to the graph of f passing through the points $Q = (x, f(x))$ and $P = (x + \Delta x, f(x + \Delta x))$ on the curve, has slope

$$\frac{f(x + \Delta x) - f(x)}{\Delta x}$$

As Δx approaches 0, $x + \Delta x$ approaches x, P approaches Q along the curve, and the slope of QP approaches

$$\lim_{\Delta x \to 0} \frac{f(x + \Delta x) - f(x)}{\Delta x}$$

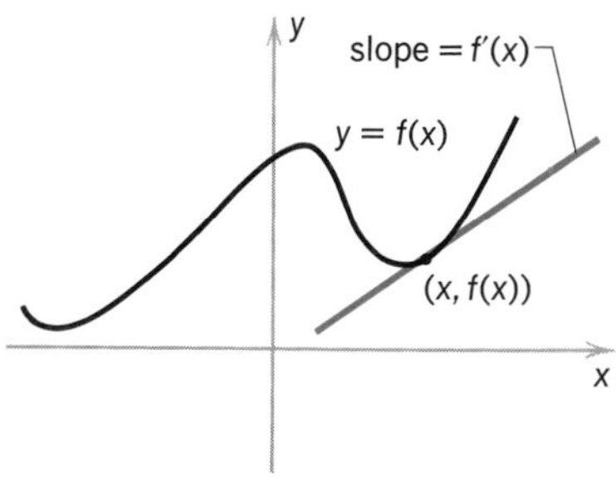

FIGURE 1-3

This limit does not always exist; when it does exist, it is called the **derivative** of f at x and is equal to the slope of the tangent line at $(x, f(x))$ [Figure 1-3].

Definition Let f be defined on a neighborhood of x. The **derivative** of the function f at the point x is the real number

$$\lim_{\Delta x \to 0} \frac{f(x + \Delta x) - f(x)}{\Delta x}$$

provided this limit exists. In that case we say the function is **differentiable** at x. If this limit does not exist, then the derivative of f is not defined at x.

We define f' to be the function given by the rule

$$f' : x \mapsto \lim_{\Delta x \to 0} \frac{f(x+\Delta x) - f(x)}{\Delta x}$$

f' is called the **derivative** of f. Thus the value $f'(x)$ is the derivative of f at x. We emphasize that a point x is in the domain of f' if and only if

(i) f is defined on a neighborhood of x, and

(ii) $\lim\limits_{\Delta x \to 0} \dfrac{f(x+\Delta x) - f(x)}{\Delta x}$ exists.

If f is differentiable at every point in its domain, then f is called a **differentiable function**. Thus, if f is a differentiable function, f and f' have the same domain.

Example 1 Let f be the affine function $x \mapsto 3x-2$. To find f' we compute

$$\frac{f(x+\Delta x) - f(x)}{\Delta x} = \frac{(3(x+\Delta x)-2) - (3x-2)}{\Delta x} = \frac{3\Delta x}{\Delta x} = 3$$

Therefore

$$f'(x) = \lim_{\Delta x \to 0} \frac{f(x+\Delta x) - f(x)}{\Delta x} = \lim_{\Delta x \to 0} 3 = 3$$

so $f'(x) = 3$ for all real numbers x. Thus, f is a differentiable function. The derivative f' has the rule

$$f' : x \mapsto 3$$

As another, more difficult, application of the definition, let us compute the derivatives of the sine and cosine functions. At zero, we have

$$\sin'(0) = \lim_{\Delta x \to 0} \frac{\sin(0+\Delta x) - \sin 0}{\Delta x} = \lim_{\Delta x \to 0} \frac{\sin \Delta x}{\Delta x} = 1$$

In general,

$$\begin{aligned}
\sin'(x) &= \lim_{\Delta x \to 0} \frac{\sin(x+\Delta x) - \sin x}{\Delta x} \\
&= \lim_{\Delta x \to 0} \frac{\sin x \cos \Delta x + \cos x \sin \Delta x - \sin x}{\Delta x} \\
&= \lim_{\Delta x \to 0} \left(\cos x \frac{\sin \Delta x}{\Delta x} - \sin x \frac{1-\cos \Delta x}{\Delta x} \right) \\
&= \cos x \left(\lim_{\Delta x \to 0} \frac{\sin \Delta x}{\Delta x} \right) - \sin x \left(\lim_{\Delta x \to 0} \frac{1-\cos \Delta x}{\Delta x} \right)
\end{aligned}$$

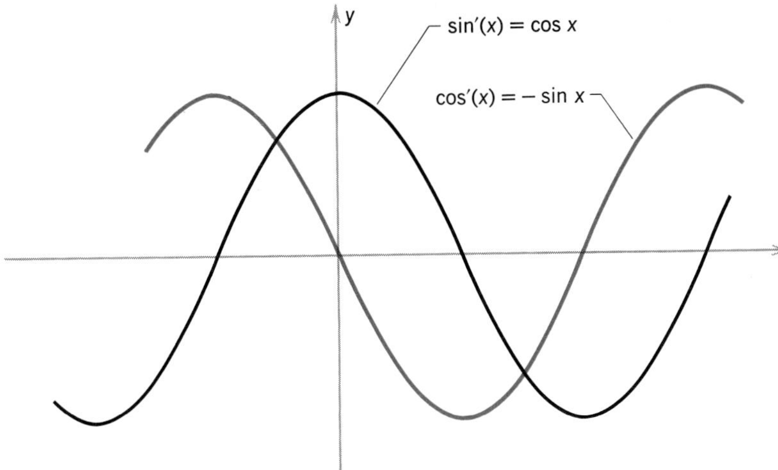

FIGURE 1-4

But $\lim_{\Delta x \to 0} \sin \Delta x/\Delta x = 1$ (Example 2, §2.3), and $\lim_{\Delta x \to 0} (1 - \cos \Delta x)/\Delta x = 0$ (Example 9, §2.3). Therefore, $\sin'(x) = (\cos x) \cdot 1 + (\sin x) \cdot 0 = \cos x$. Thus, the sine function is a differentiable function, and its derivative is the cosine.

Since $\cos x = \sin(x + \frac{1}{2}\pi)$, the graph of the cosine is obtained by shifting the graph of the sine the distance $\frac{1}{2}\pi$ to the left (Figure 1-4). From this geometric consideration, we see that the derivative of the cosine at x must be equal to the derivative of the sine at $(x + \frac{1}{2}\pi)$. But

$$\sin'(x + \tfrac{1}{2}\pi) = \cos(x + \tfrac{1}{2}\pi) = -\sin x$$

so we must have $\cos'(x) = -\sin x$. Let us now establish this formula algebraically by evaluating

$$\lim_{\Delta x \to 0} \frac{\cos(x + \Delta x) - \cos x}{\Delta x}$$

We know that $\cos(x + \Delta x) = \cos x \cos \Delta x - \sin x \sin \Delta x$, therefore,

$$\lim_{\Delta x \to 0} \frac{\cos(x + \Delta x) - \cos x}{\Delta x} = \lim_{\Delta x \to 0} \frac{-\cos x(1 - \cos \Delta x) - \sin x \sin \Delta x}{\Delta x}$$

$$= \lim_{\Delta x \to 0} \left(-\sin x \frac{\sin \Delta x}{\Delta x} \right) - \lim_{\Delta x \to 0} \left(\cos x \frac{(1 - \cos \Delta x)}{\Delta x} \right)$$

$$= -\sin x \cdot 1 - \cos x \cdot 0 = -\sin x$$

We have thus established the formulas

$$\mathbf{\sin'(x) = \cos x}$$

$$\mathbf{\cos'(x) = -\sin x}$$

Example 2 Evaluate $f'(0)$ and $f'(2)$ for the function

$$f(x) = \begin{cases} 1/x, & x \neq 0 \\ 0, & x = 0 \end{cases}$$

(See Figure 1-5.) We have, for $x = 0$,

$$\frac{f(x+\Delta x) - f(x)}{\Delta x} = \frac{\dfrac{1}{0+\Delta x} - 0}{\Delta x} = \frac{1}{(\Delta x)^2}$$

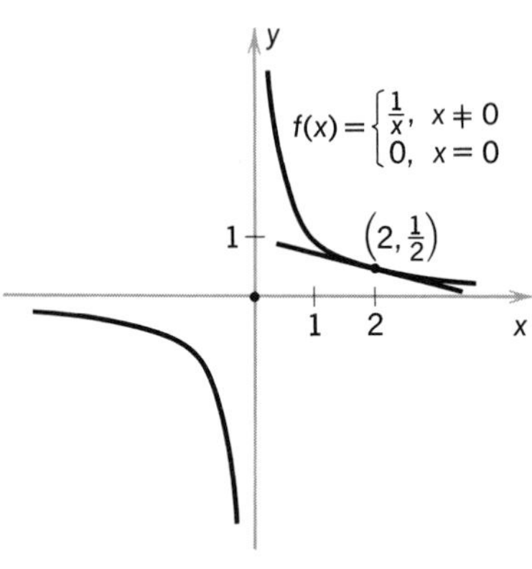

FIGURE 1-5

so

$$f'(0) = \lim_{\Delta x \to 0} \frac{f(x+\Delta x) - f(x)}{\Delta x} = \lim_{\Delta x \to 0} \frac{1}{(\Delta x)^2}$$

which does *not* exist, and so the function f is not differentiable at the point $x = 0$. However, for $x = 2$, we have for Δx in $]-2, \infty[$, $\Delta x \neq 0$,

$$\frac{f(x+\Delta x) - f(x)}{\Delta x} = \frac{\dfrac{1}{x+\Delta x} - \dfrac{1}{x}}{\Delta x}$$

$$= \frac{x - (x+\Delta x)}{(x^2 + x\Delta x)\Delta x} = \frac{-1}{x^2 + x\Delta x}$$

so

$$f'(2) = \lim_{\Delta x \to 0} \frac{-1}{(2)^2 + 2\Delta x} = -\frac{1}{4}$$

In fact, for all $x \neq 0$, we have $f'(x) = -1/x^2$. Notice that f is not continuous at $x = 0$, and that we would find it impossible to construct a tangent line to f at the point $(0, 0)$.

When a function is given in the form $y = f(x)$, the derivative is often denoted by $d(f(x))/dx$ or dy/dx. This is the notation of Leibniz; it is explained in detail in the next chapter (§4.1). Suffice it to say for the moment that this notation is most useful when a function is introduced by a formula (such as $y = x^2$) but no name is explicitly given to the function. That situation often arises in solving physical problems where one works with equations and does not bother to name functions. Thus, if a function is defined by the equation $y = x^2$, its derivative is given by

$$\frac{dy}{dx} = 2x \quad \text{or} \quad \frac{d(x^2)}{dx} = 2x$$

since

$$\frac{(x+\Delta x)^2 - x^2}{\Delta x} = 2x + \Delta x \to 2x$$

as $\Delta x \to 0$.

In this notation, the above formula for the derivatives of the sine and cosine functions can be written

$$\frac{d(\sin x)}{dx} = \cos x \quad \text{and} \quad \frac{d(\cos x)}{dx} = -\sin x$$

The derivative of the function at a specific point x_0 is denoted by

$$\left.\frac{dy}{dx}\right|_{x=x_0} \quad \text{or} \quad \left.\frac{d(f(x))}{dx}\right|_{x=x_0}$$

thus,

$$\left.\frac{d(x^2)}{dx}\right|_{x=2} = 4; \quad \left.\frac{d(\cos x)}{dx}\right|_{x=0} = 0$$

We now establish some fundamental derivative formulas.

Theorem 1 Derivative of an Affine Function The derivative of the affine function $f: x \mapsto mx+b$ is the constant function $f': x \mapsto m$. Thus

$$\frac{d(mx+b)}{dx} = m$$

Proof We have

$$f'(x) = \lim_{\Delta x \to 0} \frac{(m(x+\Delta x) + b) - (mx+b)}{\Delta x}$$

But

$$\frac{m(x+\Delta x) + b - (mx+b)}{\Delta x} = \frac{m\Delta x}{\Delta x} = m$$

for all $x \in \boldsymbol{R}$. Therefore, $f'(x) = \lim_{\Delta x \to 0} m = m$ for all $x \in \boldsymbol{R}$, and $f': x \mapsto m$. ∎

Corollary Derivative of a Constant Function The derivative of a constant function $x \mapsto c$ is $f': x \mapsto 0$. Thus

$$\frac{d(c)}{dx} = 0$$

Theorem 2 Derivative of a Monomial For all $n \in \boldsymbol{N}$, the derivative of the function $f: x \mapsto x^n$ is $f': x \mapsto nx^{n-1}$. Thus

$$\frac{d(x^n)}{dx} = nx^{n-1}$$

Proof We have

$$\frac{f(x+\Delta x)-f(x)}{\Delta x} = \frac{(x+\Delta x)^n - x^n}{\Delta x}$$

$$= \frac{\left[x^n + nx^{n-1}\,\Delta x + \frac{n(n-1)}{2}x^{n-2}(\Delta x)^2 + \cdots + (\Delta x)^n\right] - x^n}{\Delta x}$$

(By the binomial theorem)

$$= nx^{n-1} + \frac{n(n-1)}{2}x^{n-2}(\Delta x) + \cdots + (\Delta x)^{n-1}$$

But

$$f'(x) = \lim_{\Delta x\to 0}\frac{f(x+\Delta x)-f(x)}{\Delta x}$$

$$= \lim_{\Delta x\to 0}\left[nx^{n-1} + \frac{n(n-1)}{2}x^{n-2}(\Delta x) + \cdots + (\Delta x)^{n-1}\right] = nx^{n-1}$$

since all terms containing Δx drop out. Hence, $f':x \mapsto nx^{n-1}$. ∎

Example 3 The derivative of the function $f:x \mapsto 7x+2$ is $f':x \mapsto 7$. The derivative of $f:x \mapsto x^{317}$ is $f':x \mapsto 317x^{316}$. The derivative of the identity $id:x \mapsto x$ is the constant function $x \mapsto 1$.

Since the derivative is the slope of the tangent to the curve at each point, there is a readily observable geometric significance to the sign of the derivative. Just as a straight line with positive slope is monotone increasing, so a function with positive derivative is a monotone increasing function (Figure 1-6). Similarly, a line with a negative slope is monotone decreasing,

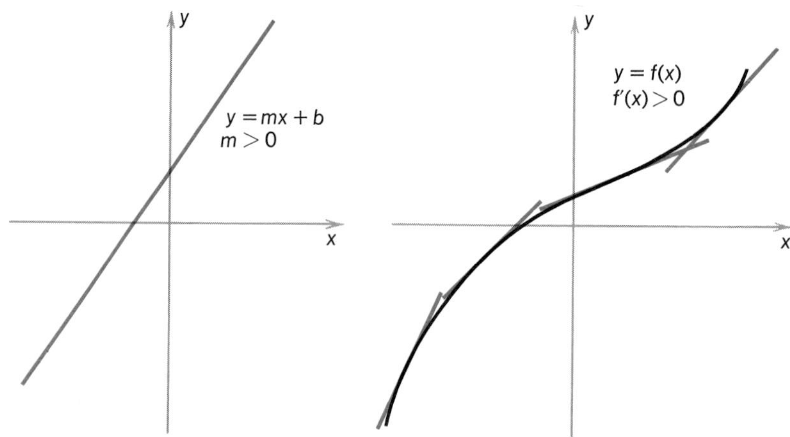

FIGURE 1-6

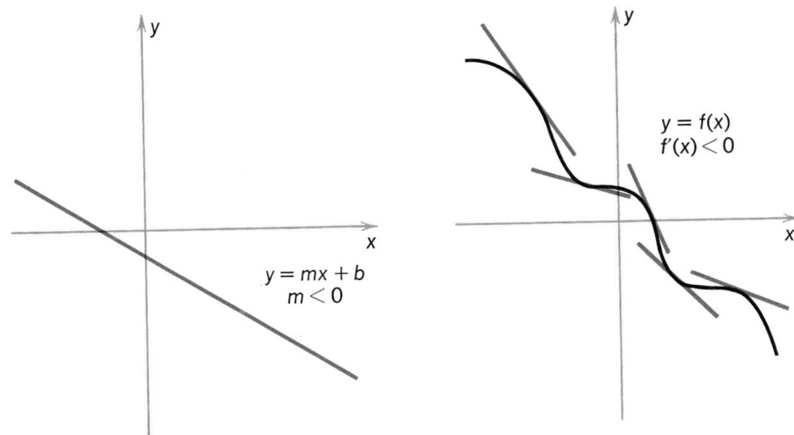

FIGURE 1-7

and a function with a negative derivative is a monotone decreasing function (Figure 1-7).

The sign of the derivative can also be used to find information about functions which are monotone increasing on some intervals, but monotone decreasing on others: on intervals where the derivative is positive, the function "rises"; in places where the derivative is negative, the function "falls." (See Figure 1-8.)

Example 4 The function $\sin:\theta \mapsto \sin\theta$ (Figure 1-8) has derivative $\sin':\theta \mapsto \cos\theta$, which is defined everywhere. This derivative is positive

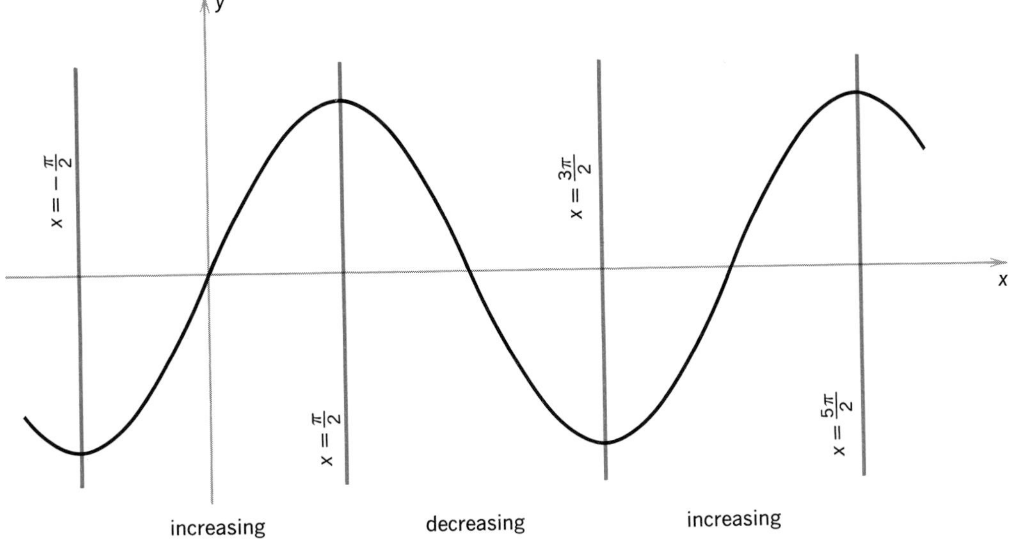

FIGURE 1-8

when $\cos\theta > 0$, and negative when $\cos\theta < 0$. The sine function is monotone increasing on the intervals $[\pi(2n-\frac{1}{2}), \pi(2n+\frac{1}{2})]$, and monotone decreasing on the intervals $[\pi(2n+\frac{1}{2}), \pi(2n+\frac{3}{2})]$.

We state these results formally in the next theorem.

Theorem 3 **First Derivative Test** Let f be differentiable on the interval I with $f'(x) \neq 0$ for all $x \in I$ except possibly at the endpoints of I. Then

(i) if $f'(x) > 0$ f is monotone increasing on I;
(ii) if $f'(x) < 0$ f is monotone decreasing on I.

The proof of this theorem is given in §4.7.

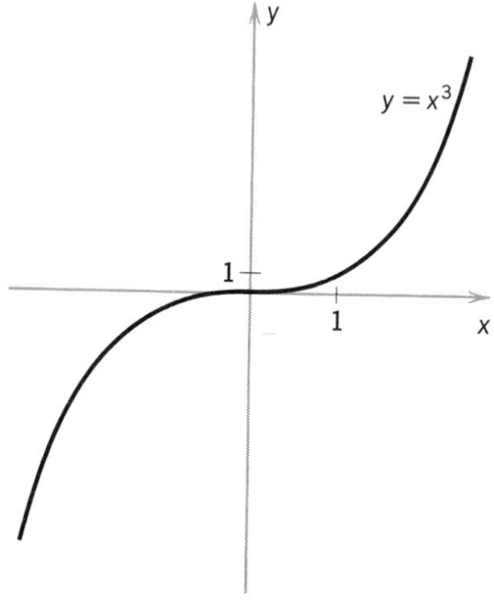

FIGURE 1-9

Example 5 The function $f\colon x \mapsto x^3$ (Figure 1-9) has derivative $f'\colon x \mapsto 3x^2$ which is positive except at zero. Therefore, f is monotone increasing on $]-\infty, 0]$ and on $[0, +\infty[$; and so f is monotone increasing on $]-\infty, +\infty[$.

Example 6 Let $g\colon x \mapsto x^3 - 12x$. We have $g'\colon x \mapsto 3x^2 - 12$. So $g'(x) < 0$ when $3x^2 < 12$; that is, when $x^2 < 4$ or $x \in]-2, 2[$. Hence, g is monotone decreasing on the closed interval $[-2, 2]$. For $x \in]2, +\infty[$ and $x \in]-\infty, -2[$, we have $g'(x) > 0$. Therefore, g is monotone increasing on the intervals $]-\infty, -2]$ and $[2, +\infty[$.

Points in the domain of f where $f'(x) = 0$ are called **critical points** of f. The function $f\colon x \mapsto x^3$ has a critical point at 0, while $\sin\colon \theta \mapsto \sin\theta$ has critical points at $(n+\frac{1}{2})\pi$. At a critical point, the tangent line to the graph of the function is horizontal (Figure 1-10).

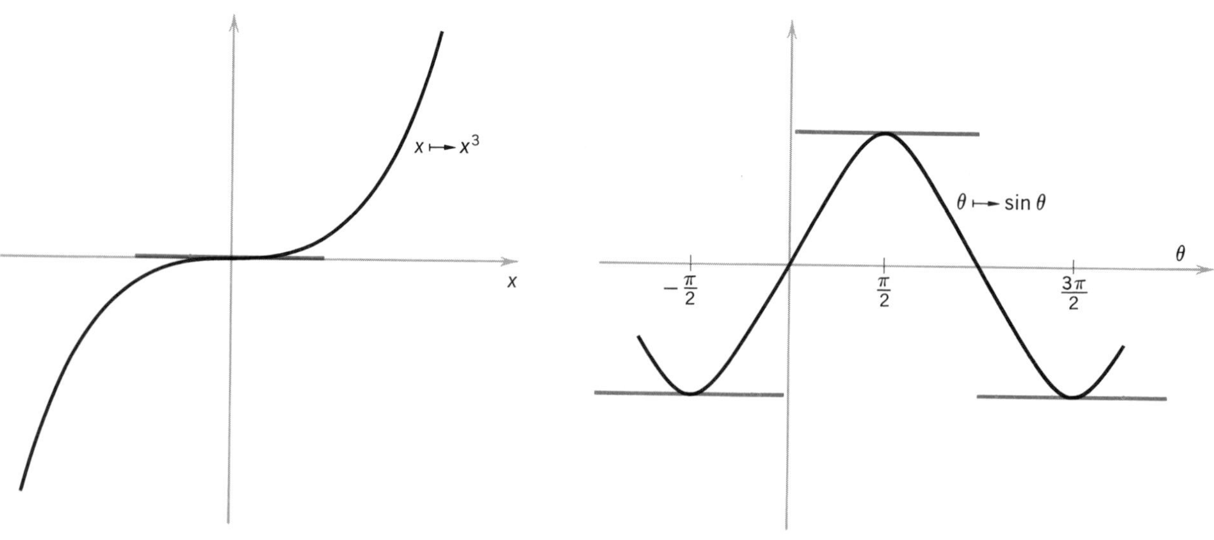

FIGURE 1-10

EXERCISES

1. Find $f'(x)$ for the following functions f.
   (a) $f:x \mapsto x^2$ (b) $f:x \mapsto x^5$
   (c) $f:x \mapsto x+6$ (d) $f:x \mapsto 4-3x$
   (e) $f:x \mapsto 2x+3$

2. Calculate dy/dx for each of the following functions, working directly from the definition of the derivative.
   (a) $y = x^2 - x + 1$ (b) $y = 1/x$
   (c) $y = 1/x^2$ (d) $y = \dfrac{1}{2x+1}$
   (e) $y = 2x^2 - x + 5$ (f) $y = x^3 - 12x + 11$
   (g) $y = ax^2 + bx + c$ (a, b, c constants) (h) $y = \sin^2 x$
   (i) $y = \cos^2 x + \sin x$ (j) $y = x + 7/x$

3. For each of the following functions, find the intervals where f is monotone increasing, and the intervals where f is monotone decreasing. Also, locate all critical points of f. Sketch the graphs [except for (d)].
   (a) $f:x \mapsto x^2 - x + 1$ (b) $f:x \mapsto 1/x$
   (c) $f:x \mapsto x^3 - 12x + 11$ (d) $f:x \mapsto ax^2 + bx + c$ (a, b, c constants)
   (e) $f:x \mapsto x^2 - 2x - 3$ (f) $f:x \mapsto 4 - x^2$

4. If f is differentiable at x, show that
$$\lim_{\Delta x \to 0} \frac{f(x_0+\Delta x) - f(x_0-\Delta x)}{\Delta x} = 2f'(x_0)$$
   [Hint: Add and subtract $f(x_0)$ from the numerator on the left-hand side.]

5. Establish that $\tan'(x) = \sec^2 x$.

6. The **optical property of parabolas** says that if a source of light is set at the focus of a parabola, all the light will be reflected off the parabola in lines parallel to its axis [Figure 1–11(a)]. Geometrically, this means that in the figure each α_i is equal to the corresponding β_i. Prove the **optical property of the parabola**; that is, show that the angle α, between the line connecting $(0, 1)$ and $(x, \frac{1}{4}x_0^2)$ and the tangent line to $y = \frac{1}{4}x^2$ at $(x_0, \frac{1}{4}x_0^2)$, is equal to the angle β between this tangent line and the vertical line $x = x_0$ [Figure 1–11(b)].

7. Let f be differentiable at mx_0+b; let $F(x) = f(mx+b)$. Show that $F'(x_0) = mf'(mx_0+b)$. Interpret geometrically.

8. A function $f: \boldsymbol{R} \to \boldsymbol{R}$ is **periodic** if there is a number a such that $f(x+a) = f(x)$ for all x. For example, all functions of the form $x \mapsto A\cos\alpha x + B\sin\alpha x$ are periodic, with $a = 2\pi/\alpha$. Prove that the derivative of a differentiable periodic function is also periodic.

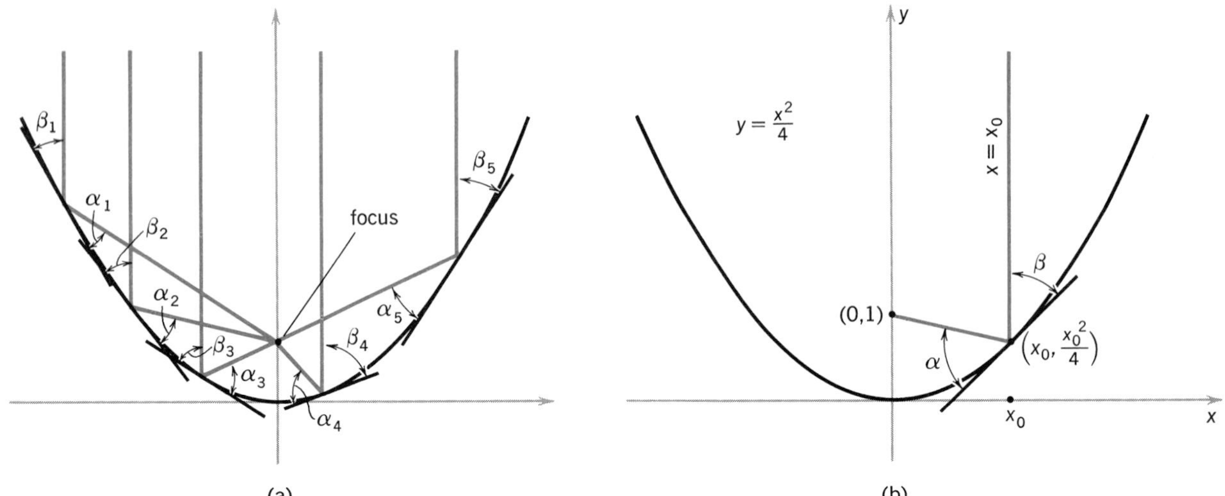

FIGURE 1-11

2 CONTINUITY OF DIFFERENTIABLE FUNCTIONS

Imagine a point Q on the graph of a function f moving toward a fixed point P on the graph. In Figure 2-1(a) the secant line through Q and P approaches the tangent line to the graph of f at P, as Q comes arbitrarily close to P. However, in Figure 2-1(b) the situation is quite different. There Q cannot approach arbitrarily near to P. Correspondingly, the secant line through Q and P cannot approach the tangent to the graph of f at P and, hence, the function cannot be differentiable at P. The observation that the function in Figure 2-1(b) is discontinuous suggests an intimate connection between differentiability and continuity. This is in fact true and we have the following theorem.

Theorem 1 **Continuity of a Differentiable Function** If a function is differentiable at a point, then it is continuous at the point.

Proof Let $f: x \mapsto f(x)$ be differentiable at the point x_0; a simple computation will be used to show that

$$\lim_{\Delta x \to 0} (f(x_0 + \Delta x) - f(x_0)) = 0$$

Now, the limit

$$f'(x_0) = \lim_{\Delta x \to 0} \frac{f(x_0 + \Delta x) - f(x_0)}{\Delta x}$$

exists by hypothesis. Therefore, using the product rule for limits, we can compute

$$\lim_{\Delta x \to 0} (f(x_0+\Delta x) - f(x_0)) = \lim_{\Delta x \to 0} \frac{f(x_0+\Delta x) - f(x_0)}{\Delta x} \cdot \lim_{\Delta x \to 0} \Delta x$$

$$= f'(x_0) \lim_{\Delta x \to 0} \Delta x = f'(x_0) \cdot 0 = 0$$

Theorem 1 is often stated as "Differentiability implies continuity." Notice that continuity is only a necessary, and not a sufficient, condition for differentiability. A continuous function need not be differentiable, as the next example shows.

Example 1 The absolute value function $x \mapsto |x|$ is continuous but not differentiable at 0 (Figure 2-2). It is easy to see that $x \mapsto |x|$ is continuous at 0. Now to verify that the absolute value function is not differentiable at 0,

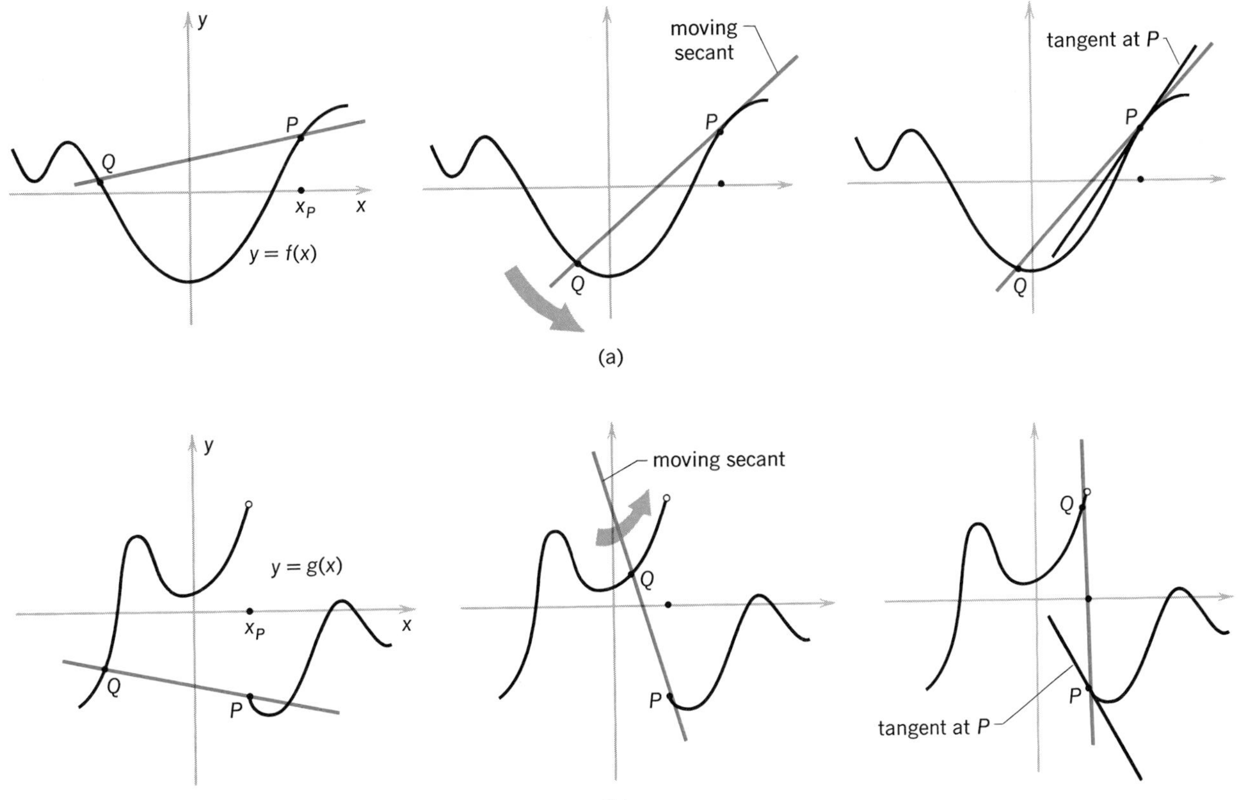

FIGURE 2-1

consider the ratio

$$\frac{|0+\Delta x| - |0|}{\Delta x} = \frac{|\Delta x|}{\Delta x} = \operatorname{sgn}(\Delta x)$$

In Chapter 2 we saw that $\lim_{\Delta x \to 0} \operatorname{sgn}(\Delta x)$ does not exist. Therefore the above ratio does **not** approach a limit as $\Delta x \to 0$ and it follows that the function $x \mapsto |x|$ is not differentiable at 0.

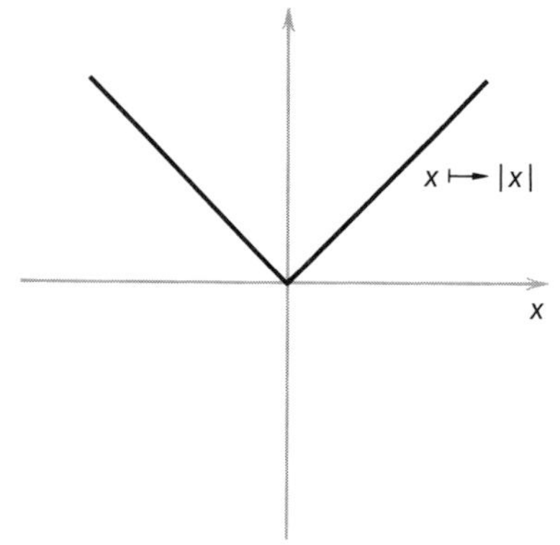

FIGURE 2-2

Geometrically, the graph of $x \mapsto |x|$ has no tangent at the origin since it has a cusp there. We can give intuitive explanation of why this curve has no tangent at the origin in terms of a simple mechanical example. Consider a line attached to a spool; if the line is pulled taut, the point Q where the line is held taut and the point P where the line leaves the spool determine the tangent to the spool at P. Therefore, as we change the location of the point Q the point P will also change since P and Q must lie on the tangent to the spool (Figure 2-3). On the other hand, if a line is attached to a triangular device as in Figure 2-4, the line can be pulled taut from any number of points Q and still pivot about the vertex P.

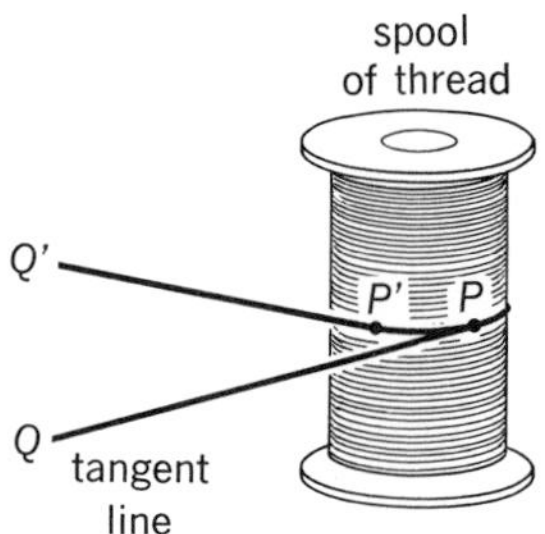

FIGURE 2-3

In §2.4, we saw that the following are all equivalent ways of saying that f is continuous at x_0:

(1) $\lim_{\Delta x \to 0} f(x_0+\Delta x) = f(x_0)$

(2) $\lim_{\Delta x \to 0} (f(x_0+\Delta x) - f(x_0)) = 0$

(3) $\lim_{x \to x_0} f(x) = f(x_0)$

(4) $\lim_{x \to x_0} (f(x) - f(x_0)) = 0$

Analagously, the following are equivalent ways of saying that the derivative of f at x_0 is equal to m:

(1) $\lim_{\Delta x \to 0} \dfrac{f(x_0+\Delta x) - f(x_0)}{\Delta x} = m$

(2) $\lim_{\Delta x \to 0} \dfrac{f(x_0+\Delta x) - f(x_0) - m\Delta x}{\Delta x} = 0$

(3) $\lim_{x \to x_0} \dfrac{f(x) - f(x_0)}{(x - x_0)} = m$

(4) $\lim_{x \to x_0} \dfrac{f(x) - f(x_0) - m(x-x_0)}{(x - x_0)} = 0$

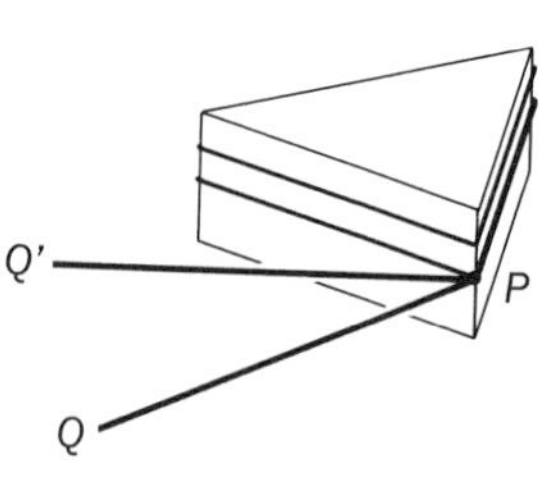

FIGURE 2-4

What distinguishes a differentiable function from a merely continuous one is that at every point on the graph of a differentiable function a tangent line is defined. If f is differentiable at x_0, then $f'(x_0)$ is the slope of the tangent line to the graph of f at $(x_0, f(x_0))$. Therefore, using the point-slope formula, this tangent line has equation

$$\frac{y - f(x_0)}{x - x_0} = f'(x_0)$$

or

$$y = f(x_0) + f'(x_0)(x - x_0)$$

If we consider the affine function $T: x \mapsto f(x_0) + f'(x_0)(x - x_0)$, the graph of T coincides with the tangent line to the graph of f at the point $(x_0, f(x_0))$. Since $f(x_0) = T(x_0)$, the continuity of f implies that

$$\lim_{x \to x_0} (f(x) - T(x)) = 0$$

However, when f is differentiable at x_0, we also have

$$\lim_{x \to x_0} \frac{(f(x) - T(x))}{(x - x_0)} = 0$$

This last limit can be verified by a straightforward computation. We leave the details to the student (exercise 10).

Example 2 Compute the tangent to the graph of $x \mapsto x^3$ at $x = 1$. By Theorem 2 of §1, the derivative is $f': x \mapsto 3x^2$ which equals 3 at $x = 1$. Thus the slope of the tangent line is 3. Its equation is thus

$$\frac{y - 1}{x - 1} = 3 \quad \text{or} \quad y = 3x - 2$$

EXERCISES

Give the domain of f' in each case, and find $f'(x)$.

1. $f(x) = x/(x+1)$
2. $f(x) = (x+1)/(x^2 - 4x + 3)$
3. $f(x) = \operatorname{sgn} x$

For exercises 4–8, indicate whether the given function is continuous and whether differentiable at the point indicated.

4. $x \mapsto x^2 \operatorname{sgn} x,$ at $x = 0$
5. $\theta \mapsto \begin{cases} \sin \theta, & \theta < \frac{1}{4}\pi \\ \cos \theta, & \theta \geqslant \frac{1}{4}\pi \end{cases}$ at $\theta = \frac{1}{4}\pi$
6. $\phi \mapsto \begin{cases} \sin \phi, & \phi < 0 \\ \tan \phi, & \phi \geqslant 0 \end{cases}$ at $\phi = 0$
7. $y \mapsto y^{1/3}$ at $y = 0$
8. $x \mapsto \begin{cases} \sin(1/x), & x > 1 \\ \frac{1}{2}, & x \leqslant 1 \end{cases}$ at $x = 1$

9. Suppose f is differentiable at x_0. Which of the following limits must exist, which might not. Justify your answers.

(a) $\lim_{\Delta x \to 0} f(x_0 + \Delta x^2) - f(x_0)$ (b) $\lim_{\Delta x \to 0} \frac{f(x_0+\Delta x) - f(x_0)}{\Delta x^2}$

(c) $\lim_{\Delta x \to 0} \frac{f(x_0-\Delta x) - f(x_0)}{\Delta x}$ (d) $\lim_{\Delta x \to 0} \frac{f(x_0+\sin\Delta x) - f(x_0)}{\sin \Delta x}$

10. Prove: If f is differentiable at x_0, then

$$\lim_{x \to x_0} \frac{(f(x) - T(x))}{x - x_0} = 0$$

where $T(x) = f(x_0) + f'(x_0)(x - x_0)$.

In exercises 11–15, compute the equation of the tangent line to the graph of the function at the indicated point.

11. $x \mapsto \sin x, \quad x = \frac{1}{4}\pi$
12. $x \mapsto x^5, \quad x = 2$
13. $x \mapsto mx + b, \quad x = x_0$
14. $x \mapsto 3x^2 + 5, \quad x = 3$
15. $x \mapsto (x+1)/(x^2 - 4x + 3), \quad x = 1$

3 ARITHMETIC OF DERIVATIVES

In this section we compute the derivatives of sums, products, and quotients of functions, and thereby determine the derivatives of all algebraic functions.

We first show that "the derivative of a sum equals the sum of the derivatives."

Theorem 1 Addition Rule of Differentiation Let f and g be two functions which are differentiable at x. Then the sum function $f+g$ is differentiable at x, and has derivative $(f+g)'(x) = f'(x) + g'(x)$. Thus we have

$$\frac{d(f(x) + g(x))}{dx} = \frac{d(f(x))}{dx} + \frac{d(g(x))}{dx}$$

Proof By the definition of the sum of functions,

$$\frac{(f+g)(x+\Delta x) - (f+g)(x)}{\Delta x} = \frac{f(x+\Delta x) + g(x+\Delta x) - f(x) - g(x)}{\Delta x}$$

$$= \frac{f(x+\Delta x) - f(x)}{\Delta x} + \frac{g(x+\Delta x) - g(x)}{\Delta x}$$

Since f and g are both differentiable at x, the limits as $\Delta x \to 0$ of the two terms in the last expression above are $f'(x)$ and $g'(x)$, and so we have

$$(f+g)'(x) = \lim_{\Delta x \to 0} \frac{(f+g)(x+\Delta x) - (f+g)(x)}{\Delta x}$$

$$= f'(x) + g'(x)$$

Corollary 1 If $f_1, \ldots, f_n$ are n functions all differentiable at x, then the sum function $(f_1+f_2+\cdots+f_n)$ is also differentiable at x, and has derivative $(f_1+\cdots+f_n)'(x) = f_1'(x)+\cdots+f_n'(x)$.

Example 1 Compute the derivative of the function $f: x \mapsto x^2 - 4x + 3$. Let $f_1: x \mapsto x^2$, $f_2: x \mapsto -4x+3$. Then $f'(x) = 2x$, $f_2'(x) = -4$, and by the addition rule

$$\begin{aligned} f'(x) &= (f_1+f_2)'(x_0) \\ &= f_1'(x_0) + f_2'(x_0) \\ &= 2x - 4 \end{aligned}$$

Example 2 The derivative of $f: x \mapsto x^{17} - 7x^3 + 3x$ is

$$f': x \mapsto 17x^{16} - 21x^2 + 3$$

We next treat the derivative of the product of two functions.

Theorem 2 Product Rule of Differentiation Let f and g be two functions which are differentiable at x. Then the product function fg is differentiable at x, and has derivative $(fg)'(x) = g(x)f'(x) + f(x)g'(x)$. Thus, we have

$$\frac{d(f(x)g(x))}{dx} = g(x)\frac{d(f(x))}{dx} + f(x)\frac{d(g(x))}{dx}$$

Proof We set $\Delta z = g(x+\Delta x) - g(x)$ and $\Delta y = f(x+\Delta x) - f(x)$. Now let $g(x+\Delta x)$ and $f(x+\Delta x)$ represent adjacent sides of a rectangle, as in Figure 3-1. From the figure or by simple algebra, we see that

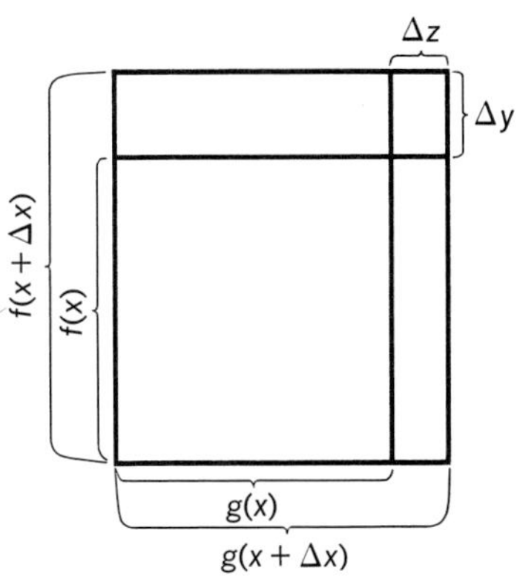

FIGURE 3-1

$$g(x+\Delta x)f(x+\Delta x) - f(x)g(x) = g(x)\Delta y + f(x)\Delta z + \Delta y \Delta z$$

Therefore

$$\frac{g(x+\Delta x)f(x+\Delta x) - f(x)g(x)}{\Delta x} = g(x)\frac{\Delta y}{\Delta x} + f(x)\frac{\Delta z}{\Delta x} + \frac{\Delta y}{\Delta x}\Delta z$$

(In Figure 3-1, all quantities are represented as positive, however, the above equation is easily seen to be valid in all cases.) Continuing the proof, we evaluate

$$\lim_{\Delta x \to 0} \frac{\Delta y}{\Delta x} = f'(x), \qquad \lim_{\Delta x \to 0} \frac{\Delta z}{\Delta x} = g'(x)$$

and

$$\lim_{\Delta x \to 0} \frac{\Delta y}{\Delta x} \Delta z = f'(x) \cdot \lim_{\Delta x \to 0} [g(x+\Delta x) - g(x)] = f'(x) \cdot 0 = 0$$

since g is continuous at x. Hence, $(fg)'(x) = g(x)f'(x) + f(x)g'(x)$. ∎

Corollary 2 Let f, g, and h be functions which are differentiable at x. Then $(fgh)'(x) = g(x)h(x)f'(x) + f(x)h(x)g'(x) + f(x)g(x)h'(x)$.

The reader should supply a proof. That is $(fghk)'(x)$?

Corollary 3 Let f be a function differentiable at x, and let c be a constant. Then $(cf)'(x) = cf'(x)$. Thus

$$\frac{d(cf(x))}{dx} = c\frac{d(f(x))}{dx}$$

Proof By the previous theorem we have

$$\frac{d(cf(x))}{dx} = c\frac{d(f(x))}{dx} + f(x)\frac{d(c)}{dx} = c\frac{d(f(x))}{dx} + 0$$ ∎

Corollary 4 Subtraction Rule of Differentiation If f and g are two functions differentiable at x, then the difference function $f-g$ is also differentiable at x, and has derivative $(f-g)'(x) = f'(x) - g'(x)$. Thus,

$$\frac{d((f-g)(x))}{dx} = \frac{d(f(x))}{dx} - \frac{d(g(x))}{dx}$$

Proof Apply Corollary 3 with $c = -1$, and then use the addition rule. ∎

The utility of all these theorems should be clear; we can compute derivatives of the sum, difference, and product of functions as soon as we know the derivatives of the individual functions themselves. For example, we can now compute the derivative of any polynomial function using the above rules and Theorem 2 of §1.

Theorem 3 Derivative of a Polynomial The polynomial function $f: x \mapsto a_n x^n + a_{n-1} x^{n-1} + \cdots + a_1 x + a_0$ is differentiable, and its derivative is $f': x \mapsto na_n x^{n-1} + (n-1) a_{n-1} x^{n-2} + \cdots + 2a_2 x + a_1$. Thus,

$$\frac{d(a_n x^n + a_{n-1} x^{n-1} + \cdots + a_0)}{dx} = na_n x^{n-1} + (n-1) a_{n-1} x^{n-2} + \cdots + a_1$$

Example 3 The derivative of $f: x \mapsto x^3 - 3x^2$ is $f': x \mapsto 3x^2 - 3(2x^1) = 3x^2 - 6x$.

Example 4 Compute the derivative of

$$f : x \mapsto 8x^4 - 22x^3 - 91x^2 + 276x + 63$$

at $x = 0$ and at $x = -1$. We can work directly from Theorem 3 to obtain $f' : x \mapsto 32x^3 - 66x^2 - 182x + 276$. Thus, $f'(0) = 276$, $f'(-1) = 360$.

Alternatively, if we were given the polynomial in its factored form $f : x \mapsto (x-3)(x-3)(2x+7)(4x-1)$, the first corollary to the product rule yields

$$\begin{aligned}\frac{d(f(x))}{dx} &= \frac{d(x-3)}{dx}[(x-3)(2x+7)(4x-1)] \\ &\quad + (x-3)\frac{d(x-3)}{dx}[(2x-7)(4x-1)] \\ &\quad + (x-3)^2\frac{d(2x+7)}{dx}[(4x-1)] \\ &\quad + (x-3)^2(2x-7)\frac{d(4x-1)}{dx} \\ &= 2(x-3)(2x+7)(4x-1) + (x-3)^2 \cdot 2 \cdot (4x-1) \\ &\quad + (x-3)^2(2x-7) \cdot 4\end{aligned}$$

Therefore $f'(0) = (-6)\cdot 7\cdot(-1) + 2\cdot 9\cdot(-1) + 4\cdot 9\cdot 7 = 276$, and $f'(-1) = (-8)\cdot 5\cdot(-5) + 2\cdot 16\cdot(-5) + 4\cdot 16\cdot 5 = 360$. Naturally, the values agree with those derived by direct use of the rule for the derivative of a polynomial.

The next derivative we wish to determine is that of the quotient of differentiable functions.

Theorem 4 Quotient Rule of Differentiation Let f and g be two functions differentiable at x, and suppose $g(x) \neq 0$. Then the quotient function f/g is differentiable at x, and has derivative

$$(f/g)'(x) = \frac{f'(x)g(x) - f(x)g'(x)}{(g(x))^2}$$

Thus,

$$\frac{d\left(\dfrac{f(x)}{g(x)}\right)}{dx} = \frac{g(x)\dfrac{d(f(x))}{dx} - f(x)\dfrac{d(g(x))}{dx}}{(g(x))^2}$$

Proof If we knew that f/g were differentiable, the result would be an easy application of the product rule (see exercise 24). However, since we do not know that the function f/g is differentiable, we must go back to the definition of the derivative and work from there. Thus we have

$$\frac{(f/g)(x+\Delta x)-(f/g)(x)}{\Delta x}$$

$$=\frac{\dfrac{f(x+\Delta x)}{g(x+\Delta x)}-\dfrac{f(x)}{g(x)}}{\Delta x}$$

$$=\frac{g(x)f(x+\Delta x)-f(x)g(x+\Delta x)}{\Delta x\, g(x)g(x+\Delta x)}$$

Adding and subtracting $f(x)g(x)$ we obtain

$$=\frac{[f(x+\Delta x)g(x)-f(x)g(x)]-[f(x)g(x+\Delta x)-f(x)g(x)]}{\Delta x\, g(x)g(x+\Delta x)}$$

$$=\frac{1}{g(x+\Delta x)}\frac{f(x+\Delta x)-f(x)}{\Delta x}-\frac{f(x)}{g(x)g(x+\Delta x)}\frac{g(x+\Delta x)-g(x)}{\Delta x}$$

But, since f and g are differentiable at x (and so g is continuous at x), all the necessary limits as $\Delta x \to 0$ exist, and

$$(f/g)'(x)=\lim_{\Delta x\to 0}\frac{(f/g)(x+\Delta x)-(f/g)(x)}{\Delta x}$$

$$=\lim_{\Delta x\to 0}\left[\frac{1}{g(x+\Delta x)}\frac{f(x+\Delta x)-f(x)}{\Delta x}-\frac{f(x)}{g(x)g(x+\Delta x)}\frac{g(x+\Delta x)-g(x)}{\Delta x}\right]$$

$$=\frac{1}{\lim\limits_{\Delta x\to 0} g(x+\Delta x)}\lim_{\Delta x\to 0}\frac{f(x+\Delta x)-f(x)}{\Delta x}-\frac{\lim\limits_{\Delta x\to 0} f(x)}{(\lim\limits_{\Delta x\to 0} g(x))(\lim\limits_{\Delta x\to 0} g(x+\Delta x))}\left(\lim_{\Delta x\to 0}\frac{g(x+\Delta x)-g(x)}{\Delta x}\right)$$

$$=\frac{1}{g(x)}f'(x)-\frac{f(x)}{(g(x))^2}g'(x)$$

$$=\frac{g(x)f'(x)-f(x)g'(x)}{(g(x))^2}$$

Example 5 Differentiate $y=(x^2+1)/\sin x$. First, $d(x^2+1)/dx=2x$ and $d(\sin x)/dx=\cos x$. Therefore,

$$\frac{dy}{dx}=\frac{(\sin x)(2x)-(x^2+1)\cos x}{\sin^2 x}$$

We can apply the quotient rule to compute the derivative of $\tan x$, because $\tan x = \sin x/\cos x$, and we already know that $\sin'(x) = \cos x$ and $\cos'(x) = -\sin x$. Thus

$$\tan' x = \frac{d\left(\dfrac{\sin x}{\cos x}\right)}{dx} = \frac{\cos x(\cos x) - (-\sin x)\sin x}{\cos^2 x}$$

$$= \frac{\cos^2 x + \sin^2 x}{\cos^2 x} = \frac{1}{\cos^2 x} = \sec^2 x$$

Hence,

$$\frac{d(\tan x)}{dx} = \frac{1}{\cos^2 x} = \sec^2 x$$

Thus, the slope of the tangent line to the tangent function at 0 is 1 (Figure 3-2). The computation of the derivatives of the other trigonometric functions is left as an exercise.

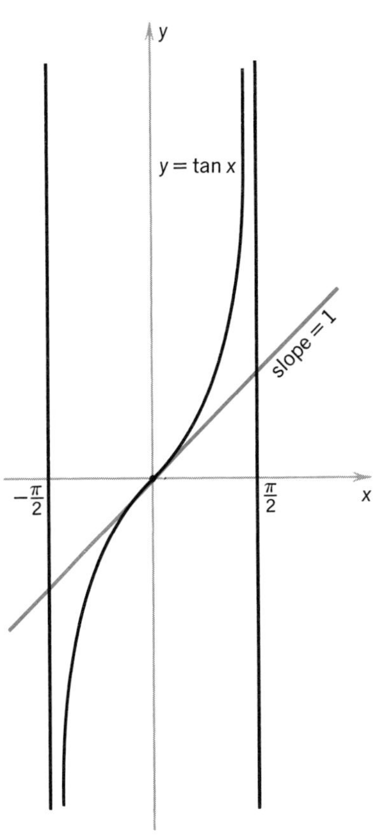

FIGURE 3-2

Corollary 5 If the function g is differentiable at x and $g(x) \neq 0$, then the function $1/g$ is differentiable at x and

$$\left(\frac{1}{g}\right)'(x) = \frac{-g'(x)}{(g(x))^2}$$

Proof Let f be the constant function $x \mapsto 1$, and apply Theorem 4. 

Example 6 Find the derivative of

$$h: x \mapsto \frac{x^3 + 3x^2 + 2}{x^4 + 1}$$

Since $x^4 + 1 > 0$ for all $x \in \boldsymbol{R}$, the domain of h is all of $\boldsymbol{R}$. If we define $f: x \mapsto x^3 + 3x^2 + 2$, $g: x \mapsto x^4 + 1$, then $h = f/g$, so

$$h'(x) = \frac{f'(x)g(x) - f(x)g'(x)}{(g(x))^2}$$

$$= \frac{(3x^2 + 6x)(x^4 + 1) - 4x^3(x^3 + 3x^2 + 2)}{(x^4 + 1)^2}$$

$$= \frac{-x(x^5 + 6x^4 + 8x^2 - 3x - 6)}{(x^4 + 1)^2}$$

We have already shown that $d(x^n)/dx = nx^{n-1}$, where n is a non-negative integer. We now extend this rule to include negative n.

Corollary 6 For any integer n,

$$\frac{d(x^n)}{dx} = nx^{n-1}$$

Proof We need only consider the case where $n < 0$. Then $-n > 0$, and so $d(x^{-n})/dx = -nx^{-n-1}$. The above corollary now yields

$$\frac{d(x^n)}{dx} = \frac{d\left(\dfrac{1}{x^{-n}}\right)}{dx} = \frac{-(-nx^{-n-1})}{x^{-2n}}$$

This last expression simplifies to $nx^{(-n-1-(-2n))} = nx^{n-1}$. ∎

EXERCISES

In exercises 1–18, calculate the derivative of the given function, using the theorems of this section.

1. $s = t^2 - 4t + 3$
2. $s = 2t^3 - 5t^2 + 4t - 3$
3. $s = \frac{1}{2}gt^2 + v_0 t + s_0$ (g, v_0, and s_0 are constants)
4. $s = 3 + 4t - t^2$
5. $s = (2t+3)^2$
6. $y = x^4 - 7x^3 + 2x^2 + 5$
7. $y = 5x^3 - 3x^5$
8. $y = 1/x$
9. $y = \frac{1}{4}x^4 - \frac{1}{3}x^3 + \frac{1}{2}x^2 - x + 3$
10. $y = 3x^7 - 7x^3 + 21x^2$
11. $y = x^2(x^3 - 1)$
12. $y = (x-2)(x+3)$
13. $y = (3x-1)(2x+5)$
14. $y = (x-1)^2(x+2)$
15. $y = \dfrac{2x + 5}{3x - 2}$
16. $y = \dfrac{2x + 1}{x^2 - 1}$
17. $r = q^2(q + 1)^{-1}$
18. $r = (q + q^{-1})^2$
19. Find $\dfrac{d(\cot x)}{dx}$, $\dfrac{d(\sec x)}{dx}$, $\dfrac{d(\csc x)}{dx}$, using the Quotient Rule.
20. Calculate
$$\frac{d(\sin\theta + \cos\theta)^2}{d\theta}, \frac{d(2\sin\theta\cos\theta)}{d\theta}.$$
Simplify $(\sin\theta + \cos\theta)^2 - 2\sin\theta\cos\theta$, and relate your answer to the derivatives just calculated.
21. In exercises 1–9 and 12–16 above, discuss the intervals where the function is monotone increasing and decreasing, and locate all critical points.

22. Where are the trigonometric functions monotone increasing? monotone decreasing?

23. A proof of the formula

$$\frac{d\left(\dfrac{f(x)}{g(x)}\right)}{dx} = \frac{g(x)\dfrac{d(f(x))}{dx} - f(x)\dfrac{d(g(x))}{dx}}{(g(x))^2}$$

can be motivated in a geometric fashion similar to the one given for the product rule (Theorem 2) as follows: Let $f(x)$ represent the area of a rectangle with adjacent sides $g(x)$ and $f(x)/g(x)$. Thus $f(x+\Delta x)$ is the area of a slightly larger rectangle such as the one in Figure 3–3. We have

$$\Delta A = f(x+\Delta x) - f(x)$$
$$\Delta z = g(x+\Delta x) - g(x)$$
$$\Delta y = \frac{f(x+\Delta x)}{g(x+\Delta x)} - \frac{f(x)}{g(x)}$$

Now, $(f/g)'(x) = \lim_{\Delta x \to 0} (\Delta y)/(\Delta x)$ by definition of Δy. Hence, to obtain the derivative of the quotient, solve for Δy in terms of ΔA, Δz, $g(x)$, $f(x)$, $g(x+\Delta x)$, $f(x+\Delta x)$ and then compute $\lim_{\Delta x \to 0} \Delta y/\Delta x$.

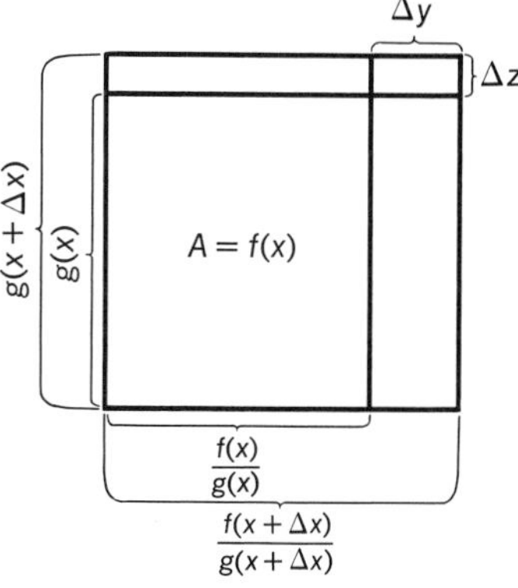

FIGURE 3-3

24. Suppose that the quotient function f/g is differentiable at x. Using the product rule for derivatives, show that

$$(f/g)'(x) = \frac{g(x)f'(x) - f(x)g'(x)}{[g(x)]^2}$$

[Hint: Let $h = f/g$ and write $f = hg$.]

4 HIGHER-ORDER DERIVATIVES

The derivative f' of a real function $f: x \mapsto f(x)$ is again a real function. As we pointed out in §1, the sign of $f'(x)$ can be used to help determine the shape of the graph of f. We shall see in §5.1 that further analysis of f' will yield additional information about the function f. An obvious step in studying the derivative f' is to determine whether f' itself is differentiable and, if it is, to find its derivative. If f' is differentiable at a point x, the derivative of f' at x is denoted by $f''(x)$.

The number $f''(x)$ is called the **second derivative** of f at x. The function $f'': x \mapsto f''(x)$ is called the second derivative of f.

Example 1 The function $f: x \mapsto x^3 + 7x + 1$ has derivative $f': x \mapsto 3x^2 + 7$ and second derivative $f'': x \mapsto 6x$. The second derivative of f at 0 is $f''(0) = 0$.

Example 2 The sine function $x \mapsto \sin x$ has first derivative $x \mapsto \cos x$, which is the cosine function. Since the derivative of $x \mapsto \cos x$ is $x \mapsto -\sin x$, the second derivative of the sine function is $x \mapsto -\sin x$. The second derivative of the sine function at $\frac{1}{2}\pi$ is $-\sin \frac{1}{2}\pi = -1$.

If f' is differentiable throughout the domain of f, we say f is **twice differentiable**. Thus, for example, the sine and cosine functions are twice differentiable. If f is differentiable and if f' is continuous, we say that f is **continuously differentiable**. By the theorem of §2, we see that a twice differentiable function is necessarily continuously differentiable. The converse does not necessarily hold. That is, there are continuously differentiable functions which are not twice differentiable; an example of such a function is given in exercise 21.

An alternative notation for the second derivative $f''(x)$ is $d^2(f(x))/dx^2$. Thus, if $f(x) = \sin x$, we write

$$f''(x) = \frac{d^2(f(x))}{dx^2} = \frac{d^2(\sin x)}{dx^2} = -\sin x$$

When we set $y = f(x)$, then we also denote the second derivative by d^2y/dx^2; the situation is then

$$y = f(x)$$

$$\frac{dy}{dx} = \frac{d(f(x))}{dx} = f'(x)$$

$$\frac{d^2y}{dx^2} = \frac{d^2(f(x))}{dx^2} = f''(x)$$

The value $f''(x_0)$ is in this case denoted by

$$\left.\frac{d^2y}{dx^2}\right|_{x=x_0} \quad \text{or} \quad \left.\frac{d^2(f(x))}{dx^2}\right|_{x=x_0}$$

Thus we have

$$y_0 = f(x_0)$$

$$\left.\frac{dy}{dx}\right|_{x=x_0} = \left.\frac{d(f(x))}{dx}\right|_{x=x_0} = f'(x_0)$$

$$\left.\frac{d^2y}{dx^2}\right|_{x=x_0} = \left.\frac{d^2(f(x))}{dx^2}\right|_{x=x_0} = f''(x_0)$$

Example 3 Let $y = \sin x$. Then we have $dy/dx = \cos x$. Since $d(\cos x)/dx = -\sin x$, we obtain $d^2y/dx^2 = -\sin x$. We also have

$$\left.\frac{d^2y}{dx^2}\right|_{x=0} = 0 \quad \text{and} \quad \left.\frac{d^2y}{dx^2}\right|_{x=\pi/2} = -1$$

Example 4 We list some further examples of this notation.

(a) $$\frac{d(x^3)}{dx} = 3x^2 \quad \text{and} \quad \frac{d^2(x^3)}{dx^2} = \frac{d(3x^2)}{dx} = 6x$$

(b) $\dfrac{d(mx+b)}{dx} = m$ and $\dfrac{d^2(mx+b)}{dx^2} = \dfrac{d(m)}{dx} = 0$

(c) $\dfrac{d(x\sin x)}{dx} = x\cos x + \sin x$ and $\dfrac{d^2(x\sin x)}{dx^2} = \dfrac{d(x\cos x + \sin x)}{dx}$

$$= 2\cos x - x\sin x$$

hence,

$$\left.\frac{d^2(x\sin x)}{dx^2}\right|_{x=\pi/2} = -\frac{\pi}{2}$$

If we continue the process of differentiating the function f, then the nth derivative of f, denoted by $f^{(n)}(x)$ or $d^n(f(x))/dx^n$, is defined in terms of the $(n-1)$st derivative by

$$\frac{d^n(f(x))}{dx^n} = \frac{d(f^{(n-1)}(x))}{dx}$$

By way of illustration, we have

$$\cos'(x) = \frac{d(\cos x)}{dx} = -\sin x$$

$$\cos''(x) = \frac{d^2(\cos x)}{dx^2} = \frac{d(\cos'(x))}{dx} = \frac{d(-\sin x)}{dx} = -\cos x$$

$$\cos'''(x) = \frac{d^3(\cos(x))}{dx^3} = \frac{d(\cos''(x))}{dx} = \frac{d(-\cos x)}{dx} = \sin x$$

$$\cos^{(iv)}(x) = \frac{d^4(\cos x)}{dx^4} = \frac{d(\cos'''(x))}{dx} = \frac{d(\sin x)}{dx} = \cos x$$

$$\cos^{(v)}(x) = \frac{d^5(\cos(x))}{dx^5} = \frac{d(\cos^{(iv)}(x))}{dx} = \frac{d(\cos x)}{dx} = -\sin x$$

In the next section we apply the first and second derivative to the analysis of motion along straight line paths. The derivatives serve to make precise the concepts of "instantaneous velocity" and "instantaneous acceleration."

We conclude this section with an instructive example where we compute the first three derivatives of a function.

Example 5 Compute the first three derivatives of the function

$$f\colon x \mapsto \frac{x^2 + 3x + 1}{x + 3}$$

and evaluate them at 0 and -2. We have,

$$f'(x) = \frac{(2x+3)(x+3) - (x^2+3x+1)}{(x+3)^2}$$

$$= \frac{x^2 + 6x + 8}{(x + 3)^2} = \frac{x^2 + 6x + 8}{x^2 + 6x + 9}$$

$$= 1 - \frac{1}{x^2 + 6x + 9}$$

$$= 1 - \frac{1}{(x + 3)^2}$$

Next,

$$f''(x) = 0 - \frac{d}{dx}\left(\frac{1}{x^2 + 6x + 9}\right)$$

$$= \frac{2x + 6}{(x + 3)^4} = \frac{2}{(x + 3)^3}$$

and finally,

$$f'''(x) = \frac{d}{dx}\left(\frac{d^2(f(x))}{dx^2}\right) = \frac{d\left(\dfrac{2}{x^3 + 9x^2 + 27x + 27}\right)}{dx}$$

$$= \frac{-2(3x^2 + 18x + 27)}{(x^3 + 9x^2 + 27x + 27)^2}$$

$$= \frac{-6(x + 3)^2}{(x + 3)^6}$$

$$= \frac{-6}{(x + 3)^4}$$

Thus

$$f'(x) = \frac{(x+2)(x+4)}{(x+3)^2}, \qquad f''(x) = \frac{2}{(x+3)^3}, \qquad f'''(x) = \frac{-6}{(x+3)^4}$$

and $f'(0) = 8/9$, $f''(0) = 2/27$, $f'''(0) = -2/27$, $f'(-2) = 0$, $f''(-2) = 2$, $f'''(-2) = -6$. Also notice that f is defined and three times differentiable for all values of x except $x = -3$.

EXERCISES

In exercises 1–17, calculate the second and third derivatives of each function.

1. $s = 5t^2 - 4t + 3$
2. $s = 2t^3 - 5t^2 + 4t - 3$
3. $s = \frac{1}{2}gt^2 + v_0t + s_0$
4. $s = (2t+3)^2$
5. $y = 5x^3 - 3x^5$
6. $y = 1/x$

7. $y = \frac{1}{4}x^4 - \frac{1}{3}x^3 + \frac{1}{2}x^2 - x + 3$
8. $y = (x-2)(x+3)$
9. $y = x^2(x^3-1)$
10. $y = \dfrac{2x+5}{3x-2}$
11. $r = (q+q^{-1})^2$
12. $r = \sin\theta$
13. $r = \cos\theta$
14. $r = \tan\theta$
15. $r = \sec\theta$
16. $f:x \mapsto x^n$
17. $f:x \mapsto 1/x^n$
18. Prove that the function

$$f:x \mapsto \begin{cases} x\sin^2(1/x), & x \neq 0 \\ 0, & x = 0 \end{cases}$$

is differentiable, but not continuously differentiable at $x = 0$. To do this, work directly from the definition of derivative to show that

$$f'(0) = \lim_{\Delta x \to 0} \frac{f(\Delta x)}{\Delta x} = 0$$

and then show that $f'(x)$ does not approach any limit as x approaches zero.

19. Prove that any twice differentiable function must be continuously differentiable.
20. Let $p(x) = a_n x^n + \cdots + a_1 x + a_0$ and $q(x) = b_n x^n + \cdots + b_1 x + b_0$ be two polynomials of degree n. Suppose $p(0) = q(0)$ and

$$p^{(k)}(x) = \frac{d^k(p(x))}{dx^k} = \frac{d^k(q(x))}{dx^k} = q^{(k)}(x)$$

for all $k \leqslant n$. Show $p(x) = q(x)$; that is, show $a_0 = b_0, a_1 = b_1, \ldots, a_n = b_n$.

21. Let $h: \boldsymbol{R} \to \boldsymbol{R}$ be defined by

$$h(x) = \begin{cases} \frac{1}{2}x^2, & x \geqslant 0 \\ -\frac{1}{2}x^2, & x \leqslant 0 \end{cases}$$

Work directly from the definition of derivative to show that $h'(x) = |x|$ for all x. Why is h not twice differentiable?

5 VELOCITY AND ACCELERATION; RATES OF CHANGE

In this section we treat one of the earliest and most important of the applications of differential calculus, the analysis of speed, velocity, and acceleration. What is **speed?** If a car travels 100 miles in two hours, we say that its average speed is 50 miles per hour. We find this speed by dividing the total distance traveled by the time spent traveling. It is possible that at different moments in its two hour journey, the car may be traveling faster or slower than the fixed speed of 50 miles per hour. If the trip is terminated by the automobile hitting a brick wall, the extent of the driver's injury will depend on the speed of the car at the moment of impact, and not on the average speed of the car for the trip up to that point. The concept of **instantaneous speed** will be clarified by defining it in terms of derivatives.

We also talk about **average** and **instantaneous velocity**. To see the distinction between velocity and speed, consider a stone thrown straight up into the air. Its speed decreases as it rises against the pull of gravity. Finally, it reaches its peak and begins to fall again, first slowly then faster. The speed of the stone at any height when rising is about the same as the speed at the same height when falling. The difference in direction is conveyed by agreeing that while the stone is rising it has **positive velocity** and while it is falling it has **negative velocity**. "Fifty miles an hour" is a speed, while "fifty miles an hour due north" is a velocity. Speed must be expressed as a positive number, while velocity is a "signed" number. For instance, we can say "—50 miles an hour due north" to indicate the same velocity as "50 miles per hour due south."

We would like to find a mathematical way to make the intuitive notion of instantaneous velocity precise. Suppose an object is moving along a straight line at varying speed. We fix the time $t = 0$ to be the moment the object leaves its starting point. We can graph the motion of the object by plotting the distance of the object from its starting point against time. It is customary to denote this distance by the letter s. This gives rise to a function which is called the **position function**. Denoting the position function by f we have $s = f(t)$. If $[t_1, t_2]$ is an interval of time, then the average velocity over this interval is

$$\text{average velocity} = \frac{f(t_2) - f(t_1)}{t_2 - t_1} = \frac{\Delta s}{\Delta t}$$

This is the average velocity, not average speed, because $f(t_2) < f(t_1)$ is possible, in which case $\Delta s/\Delta t$ would be negative.

To get an approximate idea of the velocity at the instant t_0 we can consider the average velocity during a time interval which is a small neighborhood of t_0. The smaller the neighborhood, the more reliable will be the approximation. We therefore define the **instantaneous velocity** at t_0 to be the limit of the average velocity during intervals containing t_0, as the length of the intervals approaches zero. Since the average velocity is given by $\Delta s/\Delta t$, we see that the instantaneous velocity at t_0 is equal to $(ds/dt)|_{t=t_0}$. Denoting the velocity at time t by $v(t)$, we have

$$v(t) = \frac{ds}{dt}$$

$$v(t_0) = \left.\frac{ds}{dt}\right|_{t=t_0}$$

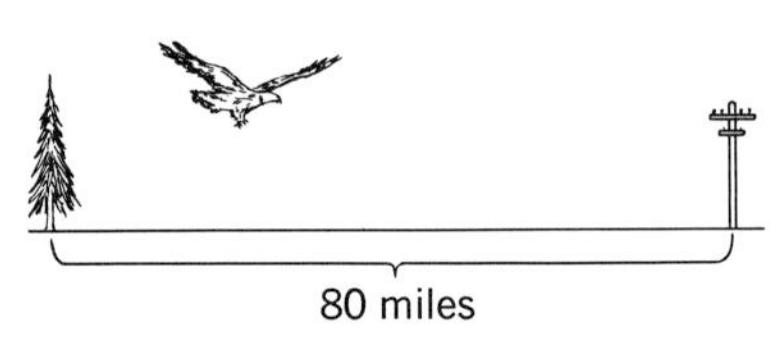

FIGURE 5-1

Example 1 An eagle takes off from a treetop and flies in a straight line for two hours landing on a telephone pole 80 miles away (Figure 5-1). Thus the eagle flies at an average rate of 40 miles per hour. However, the eagle will speed up at takeoff and slow down at landing, so he will not maintain a constant speed throughout his journey. Suppose we are given that the eagle's distance in miles from the treetop is

$$s = \frac{t^2}{60} - \frac{t^3}{10{,}800}$$

at t seconds after takeoff. The graph of the position function is shown in Figure 5-2. Let us compute the velocity. We obtain

$$\begin{aligned} v(t) &= \frac{ds}{dt} \\ &= \frac{2t}{60} - \frac{3t^2}{10{,}800} \\ &= \frac{t}{30} - \frac{t^2}{3600} \end{aligned}$$

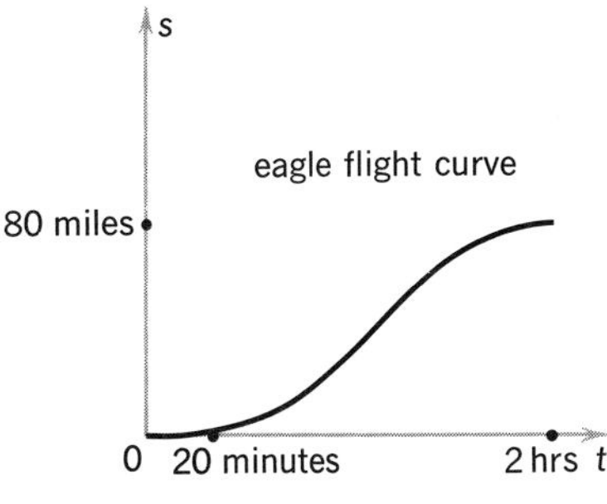

FIGURE 5-2

The velocity function is graphed in Figure 5-3. Notice the eagle has velocity 0 at the beginning and the end of his trip.

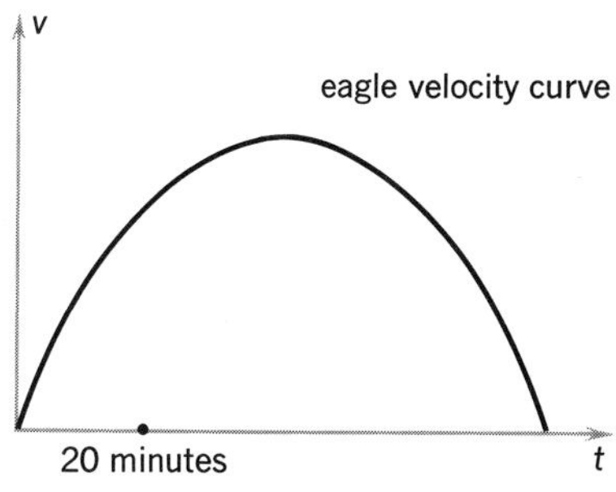

FIGURE 5-3

Example 2 A cannon is aimed straight upwards and is charged so as to fire a small ball. As we shall see in §5.6, Example 8, if we neglect air resistance, the motion of the ball is given by

$$s = v_0 t - 16t^2$$

where s is the distance straight up from the mouth of the cannon, measured in feet; t is the number of seconds after the cannon is fired; and v_0 is the "muzzle velocity" of the cannon, that is, the initial velocity of the ball.

We can ask several questions about the motion of the ball: How fast is it moving and in which direction, after one second? How long is it rising and how high does it go?

Suppose $v_0 = 50$ ft/sec. Let us try to answer the above questions. Since the position function is given by $s = 50t - 16t^2$, the velocity function is $v(t) = 50 - 32t$. Therefore the speed at $t = 1$ is the absolute value of $v(1)$. Since $v(1) = 50 - 32 = 18$, the speed at $t = 1$ is 18 ft/sec. Since the velocity $v(1)$ is positive, the ball is still rising. To find out how long the ball continues to rise, we must determine at what time it starts to fall. It will continue to rise as long as the velocity is positive, and it will begin to fall when the velocity changes sign. The velocity is given by

$$v(t) = 50 - 32t$$

Solving the equation $0 = 50 - 32t$, we see that $v(t)$ is negative for $t > 50/32$ sec and positive for $t < 50/32$ sec. At $t = 50/32$ sec the ball has stopped rising and is about to fall. We find out how high the ball goes by substituting $50/32$ into the position function to obtain

$$s = 50\left(\frac{50}{32}\right) - 16\left(\frac{50}{32}\right)^2 = \frac{625}{16} \text{ ft}$$

Derivatives also serve to make precise the concept of instantaneous acceleration. Acceleration is to velocity what velocity is to distance. By this we mean that

$$\text{average acceleration} = \frac{\text{change in velocity}}{\text{change in time}}$$

If $v(t)$ gives velocity as a function of time, then in the interval from t_0 to $t_0 + \Delta t$, the average acceleration is given by

$$\text{average acceleration} = \frac{v(t_0+\Delta t) - v(t_0)}{\Delta t}$$

As $\Delta t \to 0$, we obtain the instantaneous acceleration at t_0:

$$\begin{aligned} a(t_0) &= \lim_{\Delta t \to 0} \frac{v(t_0+\Delta t) - v(t_0)}{\Delta t} = \left.\frac{dv}{dt}\right|_{t=t_0} \\ &= \left.\frac{d^2 s}{dt^2}\right|_{t=t_0} \end{aligned}$$

Thus, the acceleration at any instant is the derivative of the velocity at that instant. We also see that acceleration is the second derivative of the distance. In the example of our flying eagle, we compute

$$\begin{aligned} a(t) &= \frac{dv(t)}{dt} \\ &= \frac{1}{30} - \frac{t}{180} \end{aligned}$$

Example 3 Suppose a ball is tossed in the air at a starting point five feet above the ground. If its initial velocity is 20 ft/sec, then its height at any time t is

$$h(t) = 5 + 20t - 16t^2$$

where h is in feet, and t in seconds. (This equation which is essentially the same as the one in Example 2 will be derived in §5.6.) Then the velocity as a function of time is

$$v(t) = h'(t) = 20 - 32t \text{ ft/sec}$$

and the acceleration is

$$a(t) = v'(t) = h''(t) = -32 \text{ ft/sec}^2$$

Notice that the acceleration is a negative quantity. This reflects the fact that the force of gravity is pulling the ball down.

If $f: t \mapsto f(t)$ is any function and t is interpreted as time, we naturally interpret f' as a rate of change. For example, while filling a bathtub, we can graph the volume of water in the tub versus the time. This defines a function whose derivative at t_0 represents the rate at which the tub is filling at time t_0.

We often think of $t \mapsto f'(t)$ as a rate of change even if t represents a quantity other than time: f' gives the change in the function value as t increases at a uniform rate. Hence, for any function f we speak of

$$\frac{f(b) - f(a)}{b - a}$$

as the **average rate of change** of f between a and b and $f'(a)$ as the **instantaneous rate of change** of f at a (with respect to t).

Example 4 The fish tank pictured in Figure 5-4 is filling with water at such a rate that the level rises $\frac{1}{4}$ ft/min. What is the instantaneous rate of change of the volume of water with respect to time?

The volume of water after t minutes is $t \times 4 \times \frac{1}{4}$ ft^3, so the volume function is $V: t \mapsto t$. Then $V': t \mapsto 1$, so that at any instant the volume is increasing at the rate of 1 ft^3/min.

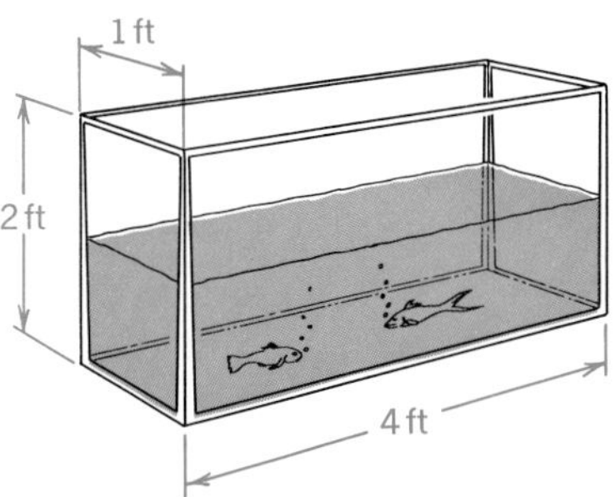

FIGURE 5-4

Example 5 The trough shown in Figure 5-5 is being filled with liquid in such a way that the depth of the liquid t minutes after the water begins to flow is $2t$ inches. What is the rate of change of the volume? How fast is the volume changing after 5 minutes?

When the depth of the liquid is $2t$ inches, the cross section is as shown in Figure 5-6 ($\frac{1}{6}t$ feet). Evidently $\frac{1}{36}t^2$ square feet of the end surface is under water, so the volume of the liquid in the trough is $\frac{1}{2}t^2$ cubic feet. The volume is thus given by $V: t \mapsto \frac{1}{2}t^2$ and the rate of change is $V': t \mapsto t$. At $t = 5$ min the volume is changing at the rate of 5 ft^3/min.

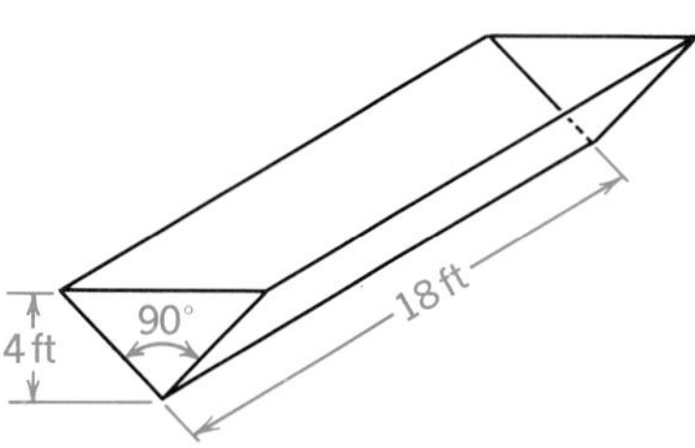

FIGURE 5-5

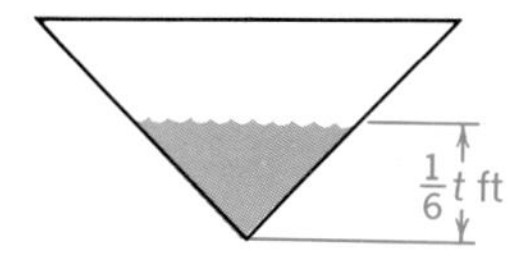

FIGURE 5-6

Example 6 Let A be the function which assigns to each positive number r the area of a circle of radius r. Find the rate of change of A with respect to r, and interpret the answer geometrically.

We have $A: r \mapsto \pi r^2$, and $A': r \mapsto 2\pi r$. Since $2\pi r$ is the circumference of the circle of radius r, we conclude that the rate of change of the area with respect to the radius is equal to the circumference. We interpret this to mean that if the circumference is s inches and the radius is increased by Δr inches, where Δr is very small, the change in area will be about $s \cdot \Delta r$ square inches, which should be clear geometrically.

EXERCISES

1. An aquarium is being filled with water. If Q represents the number of cubic inches of water in the tank, and t the number of minutes after filling begins, then $Q = 100(t + 20)^2$. At what rate is water entering the aquarium after 5 minutes? What is the average rate of filling during the first 5 minutes?
2. Find the rate of change of the volume of a sphere with respect to its radius. Interpret geometrically.

3. A ball is thrown directly upward with a speed of 108 ft/sec, and it rises according to the equation
$$h = 108t - 16t^2$$
   (a) How fast is it rising after 2 seconds?
   (b) After how many seconds does the ball begin to fall? How high is it then?
   (c) How fast is the ball traveling when it hits the ground?
   (d) What is the acceleration of the ball throughout its trip?

4. A mouse runs in a straight tunnel (the positive direction is to the right) according to a given law. For each of the laws below, determine the speed of the mouse at the given time, and when the mouse turns around.
   (a) $s = t^2 - 3t + 5, \quad t = 0$
   (b) $s = t^3 - 3t^2 - 7t - 2, \quad t = 2$
   (c) $s = t^3 - 2t^2 - t + 1, \quad t = 1$
   (d) $s = 6 + 2t - 16t^2, \quad t = 2$

In exercises 5–10, find the velocity and acceleration.

5. $s = t^2 - 4t + 2$
6. $s = t^3 + 2t^2 - t - 1$
7. $s = \dfrac{1+t}{4+t^2}$
8. $s = (\cos t)/t^2$
9. $s = 3(\cos t)(t^2+7)$
10. $s = 9t$
11. A pebble is dropped into a pool, causing an expanding circular ripple. If the radius of the ripple increases at 4 ft/sec, determine the area of the circle as a function of time, and find at what rate the area is increasing 10 seconds after the stone hits the water.

*12. A right circular cylinder of radius x is inscribed in a right circular cone of radius 8 and height 18. Find the volume of the cylinder as a function of the radius of the base ($0 \leqslant x \leqslant 8$). Over what interval is this function increasing? Decreasing? Can you determine the maximum possible volume for the cylinder?

*13. A car is moving along a level street at a rate of 12 ft/sec when the ignition is turned off. If the car experiences a constant deceleration (due to friction) and stops after 180 ft, find the deceleration.

6 THE CHAIN RULE

In this section we consider the problem of differentiating composite functions. The function $f: x \mapsto (x^2+7x+13)^{14} - 31$ can be differentiated by multiplying out and then differentiating the resulting polynomial. This procedure becomes quite tedious as the number of terms increases, and the risk of error is high. However, suppose we look at this function as the composite of $y \mapsto y^{14} - 31$ on $x \mapsto x^2 + 7x + 13$, each part of which is easy to differentiate. The computation of the derivative of a composite function can be accomplished using a simple formal device justified by the following important mathematical result.

Theorem 1 **The Chain Rule** Let f and g be real functions, and suppose f is differentiable at x_0, and g is differentiable at $f(x_0)$. Then the composite function $g \circ f$ is differentiable at x_0, and we have

$$(g \circ f)'(x) = g'(f(x_0))f'(x_0)$$

Proof Let $y = f(x)$, $z = g(y)$, and suppose $y_0 = f(x_0)$. Let $\Delta y = f(x_0 + \Delta x) - f(x_0)$, $\Delta z = g(y_0 + \Delta y) - g(y_0)$. The limit to be computed is $\lim_{\Delta x \to 0} \Delta z/\Delta x$, and we must show that

$$\lim_{\Delta x \to 0} \frac{\Delta z}{\Delta x} = g'(f(x_0))f(x_0)$$

The proof is suggested by the equation

$$\frac{\Delta z}{\Delta x} = \frac{\Delta z}{\Delta y}\frac{\Delta y}{\Delta x} \tag{1}$$

which is valid whenever $\Delta y \neq 0$. If $\Delta y \neq 0$ whenever $\Delta x \neq 0$, then for all $\Delta x \neq 0$, equation (1) holds and, hence,

$$\lim_{\Delta x \to 0} \frac{\Delta z}{\Delta x} = \lim_{\Delta x \to 0} \frac{\Delta z}{\Delta y} \cdot \lim_{\Delta x \to 0} \frac{\Delta y}{\Delta x}$$

We know that

$$\lim_{\Delta x \to 0} \frac{\Delta y}{\Delta x} = f'(x_0)$$

Moreover, $\lim_{\Delta x \to 0} \Delta y = 0$, and so

$$\lim_{\Delta x \to 0} \frac{\Delta z}{\Delta y} = \lim_{\Delta y \to 0} \frac{\Delta z}{\Delta y} = g'(f(x_0))$$

by the theorem on the composition of limits (Theorem 5, §2.4). Hence,

$$\lim_{\Delta x \to 0} \frac{\Delta z}{\Delta x} = g'(f(x_0))f'(x_0)$$

However, it is not in general the case that $\Delta y \neq 0$ whenever $\Delta x \neq 0$, so we must give a proof which avoids division by Δy when Δy might equal zero. Therefore, we define a function

$$\varepsilon : \Delta x \mapsto \begin{cases} 0, & \Delta y = 0 \\ \dfrac{\Delta z}{\Delta y} - g'(y_0), & \Delta y \neq 0 \end{cases}$$

Since $\Delta z = 0$ if $\Delta y = 0$, it follows that $\Delta z = \Delta y(g'(y_0) + \varepsilon(\Delta x))$. Dividing by Δx, we get

$$\frac{\Delta z}{\Delta x} = \frac{\Delta y}{\Delta x}(g'(y_0) + \varepsilon(\Delta x))$$

But f is differentiable at x_0, and $\lim_{\Delta x \to 0} \varepsilon(\Delta x) = 0$; hence

$$\lim_{\Delta x \to 0} \frac{\Delta z}{\Delta x} = \lim_{\Delta x \to 0} \frac{\Delta y}{\Delta x} \cdot \lim_{\Delta x \to 0} [g'(y_0) + \varepsilon(\Delta x)]$$

$$= f'(x_0)[g'(y_0) + 0] = g'(f(x_0)) \cdot f'(x_0)$$

Theorem 1 is sometimes called the Composite Function Theorem. If our functions are given in the form $y = f(x)$, $z = g(y)$, the Chain Rule takes a convenient form.

FORMAL DEVICE

If $y = f(x)$ and $z = g(y)$, then we have $z = (g \circ f)(x)$ and

$$\frac{dz}{dx} = \frac{dz}{dy}\frac{dy}{dx}, \qquad \left.\frac{dz}{dx}\right|_{x=x_0} = \left.\frac{dz}{dy}\right|_{y=f(x_0)} \left.\frac{dy}{dx}\right|_{x=x_0}$$

Example 1 Differentiate $z = (x^4 + \sin x)^3 + 3$.
Substituting $y = x^4 + \sin x$, we have $z = y^3 + 3$, and

$$\frac{d[(x^4 + \sin x)^3 + 3]}{dx} = \frac{d[y^3 + 3]}{dy} \frac{d[x^4 + \sin x]}{dx}$$

$$= (3y^2)(4x^3 + \cos x)$$

$$= 3(x^4 + \sin x)^2 (4x^3 + \cos x)$$

Example 2 Differentiate $z = (x^2 + 7x + 13)^{14} - 31$.
Substituting $y = x^2 + 7x + 13$, we have $z = y^{14} - 31$, and

$$\frac{d[(x^2 + 7x + 13)^{14} - 31]}{dx} = \frac{d[y^{14} - 31]}{dy} \frac{d[x^2 + 7x + 13]}{dx}$$

$$= (14y^{13})(2x + 7)$$

$$= 14(x^2 + 7x + 13)^{13}(2x + 7)$$

The thought of having to perform this differentiation by expanding the polynomial is appalling.

In terms of rates of change, the Chain Rule has a simple interpretation: If y is moving dy/dx times as fast as x and if z is moving dz/dy times as fast as y, then z is moving $(dz/dy)(dy/dx)$ times as fast as x; thus

$$\frac{dz}{dx} = \frac{dz}{dy}\frac{dy}{dx}$$

The Chain Rule also gives important information about the shapes of graphs. For example, if f has a critical point at x_0, then the composite $g \circ f$ must also have a critical point at x_0. Thus, we know at once that all differentiable curves of the form $z = g(x^2)$ have horizontal tangents at the origin, since $x \mapsto x^2$ has zero derivative at 0 and

$$\left.\frac{dz}{dx}\right|_{x=0} = g'(0) \cdot 0 = 0$$

We can also employ the Chain Rule to differentiate powers of trigonometric functions.

Example 3 Differentiate $f: x \mapsto \cos^3 x$ and $g: x \mapsto \sin^n x$. To differentiate f, let $h: x \mapsto \cos x$, $k: y \mapsto y^3$; then $f = k \circ h$, and we have

$$\begin{aligned}\frac{d(f(x))}{dx} &= \frac{d(k(y))}{dy}\frac{d(h(x))}{dx} \\ &= (3y^2)(-\sin x) \\ &= -3\cos^2 x \sin x\end{aligned}$$

To differentiate g, let $h: x \mapsto \sin x$, $k: y \mapsto y^n$, so $g = k \circ h$, and we have

$$\begin{aligned}\frac{d(g(x))}{dx} &= \frac{d(k(y))}{dy}\frac{d(h(x))}{dx} \\ &= (ny^{n-1})(\cos x) \\ &= n \sin^{n-1} x \cos x\end{aligned}$$

Obviously, in a similar way we find that

$$\begin{aligned}\frac{d(\tan^n x)}{dx} &= n \tan^{n-1} x \frac{d(\tan x)}{dx} \\ &= n \tan^{n-1} x \sec^2 x\end{aligned}$$

Example 4 Find the first and second derivatives of $f: x \mapsto \tan^2 x + \cos^5 x$, and evaluate them at $x = 0$. We have

$$\begin{aligned}f'(x) &= \frac{d(\tan^2 x + \cos^5 x)}{dx} \\ &= \frac{d(\tan^2 x)}{dx} + \frac{d(\cos^5 x)}{dx} \qquad \text{(Addition Rule)} \\ &= 2\tan x \cdot \frac{d(\tan x)}{dx} + 5\cos^4 x \frac{d(\cos x)}{dx} \qquad \text{(Chain Rule)} \\ &= 2\tan x \sec^2 x - 5 \sin x \cos^4 x\end{aligned}$$

Then $f'(0) = 2\tan(0)\sec^2(0) - 5\sin(0)\cos^4(0) = 2\cdot 0\cdot 1 - 5\cdot 0\cdot 1 = 0$. Moreover,

$$\begin{aligned}
\frac{d^2(f(x))}{dx^2} &= \frac{d(2\tan x\sec^2 x - 5\sin x\cos^4 x)}{dx} \\
&= \frac{d(2\tan x\sec^2 x)}{dx} - \frac{d(5\sin x\cos^4 x)}{dx} \\
&= 2\left[\frac{d(\tan x)}{dx}\sec^2 x + \frac{d(\sec^2 x)}{dx}\tan x\right] \\
&\quad -5\left[\frac{d(\sin x)}{dx}\cos^4 x + \frac{d(\cos^4 x)}{dx}\sin x\right] \qquad \text{(Product Rule)} \\
&= 2\left[\sec^2 x\sec x + 2\tan x\sec x\frac{d(\sec x)}{dx}\right] \\
&\quad -5\left[\cos x\cos^4 x + 4\sin x\cos^3 x\frac{d(\cos x)}{dx}\right] \qquad \text{(Chain Rule)} \\
&= 2[\sec^4 x + 2\tan^2 x\sec^2 x] - 5[\cos^5 x - 4\sin^2 x\cos^3 x] \\
&= 6\sec^4 x - 4\sec^2 x - 25\cos^5 x + 20\cos^3 x
\end{aligned}$$

Therefore,

$$\begin{aligned}
f''(0) &= 6\sec^4(0) - 4\sec^2(0) - 25\cos^5(0) + 20\cos^3(0) \\
&= 6 - 4 - 25 + 20 = -3
\end{aligned}$$

EXERCISES

In exercises 1–5, use the Chain Rule to compute dy/dx.

1. $y = u^2 + 3u - 7, \quad u = 2x + 1$
2. $y = \dfrac{u^2}{u^2+1}, \quad u = 2x + 1$
3. $y = w^3, \quad w = x^2 + 1$
4. $y = 2v^3 + 2/v^3, \quad v = (3x+2)^2$
5. $y = \sin 3\theta, \quad \theta = \cos 2x$

In exercises 6–8, compute $\dfrac{d((h\circ g)(x))}{dx}$.

6. $g{:}x \mapsto \sin x, \quad h{:}y \mapsto 1/y$
7. $g{:}x \mapsto \cos x, \quad h{:}\alpha \mapsto 1/\alpha$
8. $g{:}x \mapsto 2x + 4, \quad h{:}z \mapsto |z|$

In exercises 9–25, compute dy/dx. Indicate where you use the Chain Rule.

9. $y = (3x+5)^{10}$
10. $y = (6-3x)^7$

11. $y=(x+5)^{-3}$

12. $y=(3x+2x-6x)^{-4}$

13. $y=(x^2+1)^2(x^3-2x)^2$

14. $y=(x^3+2x-6)^3(x^2-4x+5)^7$

15. $y=(x^2-x^{-1}+1)(x^3+2x-6)^4$

16. $y=\dfrac{(3x-2)^2}{(2x-3)^2}$

17. $y=\dfrac{(2x-6)^4}{(x^2+5)^2}$

18. $y=2\sin^2 x+6\cos x$

19. $y=(\tan^3 x+3\cos x-7)^2$

20. $y=\dfrac{(\sin^2 x+2\cos x-1)^4}{\cos 3x}$

21. $y=\cos(ax^n)$ $\quad$ (a constant)

22. $y=\dfrac{(x^2+2x-1)^4(x^2+1)^2}{(3x-2x^{-1})^5(3x^2-6)^2}$

23. $y=\dfrac{\sin x}{x}$

24. $y=3\cos^2 2x-3\sin^2 2x$

25. $y=\dfrac{\sin x-x\cos x}{\cos x}$.

26. Show that the derivative of a differentiable periodic function is periodic using the Chain Rule. (See exercise 8 of §3.1.)

REVIEW EXERCISES FOR CHAPTER 3

1. Define the following terms.

continuously differentiable	First Derivative Test
critical point	instantaneous acceleration
derivative	instantaneous velocity
differentiable function	*n*th derivative

In exercises 2–13, compute the first and second derivatives of the given function and evaluate them at the point indicated. Find the equation of the line tangent to the graph at the indicated point.

2. $x \mapsto x^4(4-x^2)(16-x^4)$ $\quad$ at $x=2$

3. $x \mapsto \dfrac{2ax^3-x^4}{x-a}$ $\quad$ at $x=-a$

4. $x \mapsto 2x^{-3}-3x^{-2}-36x+20$ $\quad$ at $x=1$

5. $x \mapsto 3x^{-3}+2x^{-2}-5x^{-1}+3$ $\quad$ at $x=0$

6. $x \mapsto \cos^2(2x)\sin^3(x^2)$ $\quad$ at $x=0$

7. $x \mapsto \dfrac{x}{8}-\dfrac{\sin 4ax}{32a}$ $\quad$ at $x=0$

8. $x \mapsto 2\cos ax+\frac{1}{2}\sin^2 ax\sin ax$ $\quad$ at $x=\frac{1}{2}\pi$, a a positive integer

9. $x \mapsto \sin mx\sin^m x$ $\quad$ at $x=\frac{1}{4}\pi$, m a positive integer.

10. $x \mapsto \dfrac{1-\cos x}{2x^2}$ $\quad$ at $x=\frac{1}{3}\pi$

11. $x \mapsto \dfrac{4x^3}{1 - \cos^2(\frac{1}{2}x)}$ at $x = \frac{1}{2}\pi$

12. $x \mapsto 2x + \cot 2x$ at $x = -\frac{1}{3}\pi$

13. $x \mapsto (3x^2+1)(x^3+6x)$ at $x = -3$

In exercises 14–24, use the First Derivative Test to determine where the given function is increasing and where it is decreasing. Locate all the critical points of the function.

14. $x \mapsto (3x-2)(9x^2+4)(27x^3-8)$

15. $x \mapsto \dfrac{1 - x^2}{1 + x^2}$

16. $x \mapsto \dfrac{x^2(a+x)}{a - x}$

17. $x \mapsto \dfrac{x^2 - 4x + 2}{x - 4}$

18. $x \mapsto 2\sin(5x-7)$

19. $x \mapsto (x+2)(x-2)$

20. $x \mapsto \dfrac{x^3}{1 - x^2}$

21. $x \mapsto \dfrac{(x^2-x)^2}{(1-2x)^4}$

22. $x \mapsto \sin x \cos^2 x$

23. $x \mapsto \dfrac{\sin x}{1 - \sin x}$

24. $x \mapsto (1+\sin^2 x)$

In each exercise 25–35 enough data is given to find the position, velocity, and acceleration functions of a particle moving along the s axis. Find these functions.

25. $s = t^2 - t - 2$

26. $v = 2t + 3;\quad s = 3$ when $t = 1$

27. $s = t^3 - 5t + 7$

28. $s = 5 - 7t^2$

29. $a = -t^2 + 2t + 1;$ when $t = 0,\ v = 5,$ and $s = 0$

30. $v = 6 - 2t - 3t^2;\quad s = 0$ when $t = 0$

31. $s = -96t - 16t^2$

32. $v = \frac{3}{2}t^2 - 3t + 1;\quad s = -4$ when $t = 2$

33. $a = t;\quad v = s = 0$ when $t = 0$

34. $v = 6t - 8;\quad s = 4$ when $t = -1$

35. $a = 0;\quad v = 12;$ and $s = 7$ when $t = 2$

36. Let f be a differentiable even function with zero in its domain. Show that $f'(0) = 0$. [Recall that a function is even if it satisfies $f(x) = f(-x)$.]

37. For any real number m, show there is an odd function g satisfying $g'(0) = m$. [Recall that g is odd if it satisfies $g(x) = -g(-x)$.]

38. Let f be a differentiable function with domain $\boldsymbol{R}$. Define f_E, the even part of f, by $f_E(x) = \frac{1}{2}(f(x)+f(-x))$; similarly, define f_O, the odd part of f, by $f_O(x) = \frac{1}{2}(f(x)-f(-x))$. (a) Compute the derivatives of f_E and f_O and conclude that the derivative of the even part of f is equal to the odd part of the derivative of f and that the derivative of the odd part of f is equal to the even part of the derivative of f. (b) Conclude that the derivative of an odd function is even and that the derivative of an even function is odd. Interpret this result geometrically.

*39. Do exercise 21(b) of §2.1 for a curve $y = f(x)$ where $f'(0) = 0 = f(0)$ and $f''(0) > 0$. Show that the limiting position of $(0, C)$ is $(0, 1/f''(0))$. Interpret the condition $f''(0) > 0$ geometrically.

40. Consider the function f defined by the rule

$$f : x \mapsto \begin{cases} x^2, & x \geqslant 1 \\ ax + b & x \leqslant 1 \end{cases}$$

For what values of a and b is f continuous? For what values of a and b is f differentiable? Twice differentiable?

41. Consider the function g defined by the rule

$$g : x \mapsto \begin{cases} \sin x, & x \leqslant 0 \\ ax + b, & x \geqslant 0 \end{cases}$$

For what values of a and b is g differentiable? Twice differentiable? Twice continuously differentiable?

4

FURTHER RESULTS OF DIFFERENTIAL CALCULUS

In this chapter we introduce the **differential** and apply it to the tangent problem for relations and for parametric curves. In §§5, 6 we return to the theory of limits and continuous functions. Then we prove the important Mean Value Theorem and use this theorem to justify the First Derivative Test. This chapter includes three optional sections: §§3, 4 which are devoted to vector analysis, and §8 which treats inverse functions and implicit functions. The proof of Theorem 2, §6 is also optional.

1 DIFFERENTIALS AND IMPLICIT DIFFERENTIATION

In §3.1 we promised an explanation of the dy/dx and $d(f(x))/dx$ notations for derivatives. In this section we give that explanation and then exploit further the power of this system of notation.

Symbols such as dx, dy, dz, dA, dx, and so on denote variables in exactly the same way as x and Δx do. These variables are used exclusively in working with special linear functions which arise in the study of differential calculus.

A pair of variables dx and dy can be used to label a set of rectangular coordinate axes (just as the variables x and y label the x and y axes). Thus, in a plane whose axes are labeled dx and dy the straight line through the origin with slope m is the graph of the linear function $dx \mapsto m\,dx$. Just as we define linear functions using x and y by equations $y = mx$, so a linear function can be defined by an equation $dy = m\,dx$.

Suppose a function f is differentiable at x_0. Let us now superimpose the $dx\,dy$ plane on the xy plane in such a way that

(1) the $dx\,dy$ origin coincides with the point $(x_0, f(x_0))$;

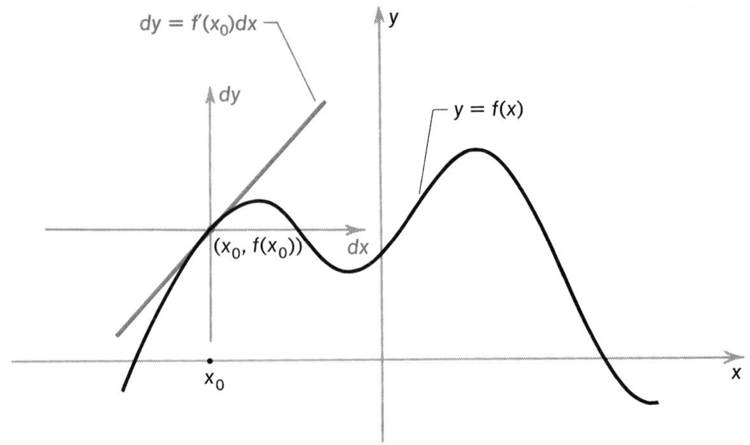

FIGURE 1-1

(2) the dx and dy axes are parallel to the x and y axes, respectively, and point in the same positive directions;
(3) the dx and dy axes have the same scale as the x and y axes, respectively (see Figure 1-1).

Now consider the linear function given by $dy = f'(x_0)\,dx$. The graph of this function in the $dx\,dy$ plane coincides with the tangent line to the graph of f at $(x_0, f(x_0))$ in the xy plane. This tangent line is the graph of the affine function $x \mapsto f'(x_0)(x - x_0) + f(x_0)$ in the xy plane; it has now become the graph of the linear function $dx \mapsto f'(x_0)\,dx$ in the $dx\,dy$ plane.

We always use the variables dx and dy in the context of the above discussion, and because of this usage, these variables are called **differentials**. The rule for using differentials is always the same: If x and y are related by the equation $y = f(x)$, then dy and dx are related, at the point (x, y), by the equation

$$dy = f'(x)\,dx$$

Thus for all $dx \neq 0$, we have

$$\frac{dy}{dx} = f'(x)$$

In particular, when $x = x_0$, we write

$$\left.\frac{dy}{dx}\right|_{x=x_0} = f'(x_0)$$

This is, of course, exactly the notation that we introduced earlier and have been using all along.

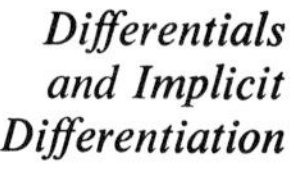

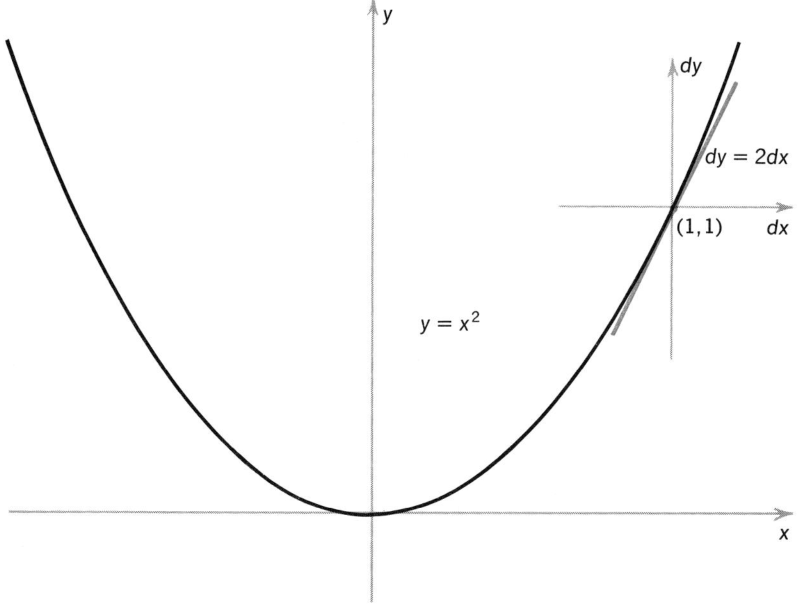

FIGURE 1-2

Example 1 If $y = x^2$ then $dy = 2x\,dx$. In particular the tangent to the curve at $(1, 1)$ has equation $dy = 2dx$ and slope $2 = (dy/dx)|_{x=1}$. (See Figure 1-2.)

Example 2 Suppose $u = \sin\alpha$. Then $du = \cos\alpha\,d\alpha$. In particular, the tangent to the graph of the sine function at the point $(-\frac{1}{3}\pi, \sin(-\frac{1}{3}\pi))$ has equation $du = \frac{1}{2}d\alpha$ (Figure 1-3).

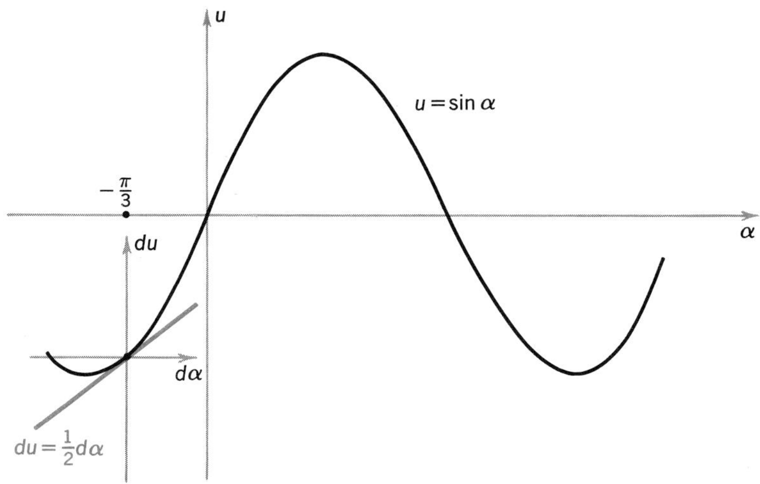

FIGURE 1-3

Example 3 Consider the constant function given by $y = c$. This function has constant derivative 0, so for every point on the graph the tangent line has equation $dy = 0$. (A picture for this example would show the dx axis superimposed on the line $y = c$.)

When given equations $y = f(x)$ and $dy = f'(x)\,dx$, it is often useful to set

$$d(f(x)) = dy$$

obtaining

$$d(f(x)) = f'(x)\,dx$$

This yields the familiar notation

$$\frac{d(f(x))}{dx} = f'(x)$$

The notation $d(f(x))$ is especially useful when a function is given in the form $x \mapsto f(x)$. The equation of the line tangent to the graph of f at $(x, f(x))$ is then

$$d(f(x)) = f'(x)\,dx$$

Note that like dy the symbol $d(f(x))$ denotes a variable; one can therefore substitute numerical values for $d(f(x))$ in an expression.

Example 4 Given the function $\theta \mapsto \tan\theta$, we have $d(\tan\theta) = \sec^2\theta\,d\theta$. Similarly, given the function $u \mapsto 6+u^2+u^3$, we have $d(6+u^2+u^3) = (2u+3u^2)\,du$. For the constant function $x \mapsto c$, we obtain $d(c) = 0 \cdot dx = 0$; for the identity function $x \mapsto x$, we find $d(x) = 1 \cdot dx = dx$. In similar fashion we obtain such formulas as

$$\begin{aligned} d(x^2) &= 2x\,dx \\ d(\sin u) &= \cos u\,du \\ d(x\sin x) &= (\sin x + x\cos x)\,dx \end{aligned}$$

The arithmetic formulas for taking derivatives all have differential counterparts. For example, if $v = g(x)$ and $u = f(x)$, then we have

$$u + v = (f+g)(x)$$

and

$$\begin{aligned} d(u+v) &= (f+g)'(x)\,dx \\ &= (f'(x) + g'(x))\,dx \\ &= f'(x)\,dx + g'(x)\,dx \\ &= du + dv \end{aligned}$$

In a similar manner, we derive the formulas

$$d(u+v) = du + dv$$

$$d(cu) = c\,du \qquad (c \text{ a constant})$$

$$d(c) = 0$$

$$d(uv) = u\,dv + v\,du$$

$$d\left(\frac{u}{v}\right) = \frac{v\,du - u\,dv}{v^2}$$

Example 5 Suppose $u = 5x^2$ and $v = 2x^3$. Then $du = d(5x^2) = 5d(x^2) = 10x\,dx$, $dv = d(2x^3) = 2d(x^3) = 6x^2\,dx$. Therefore

$$\begin{aligned} d(u+v) &= 10x\,dx + 6x^2\,dx \\ &= (10x+6x^2)\,dx \\ &= d(5x^2+2x^3) \\ &= d(u+v) \end{aligned}$$

Moreover, $uv = 10x^5$, so $d(uv) = 50x^4\,dx$. Using the product formula, we obtain

$$\begin{aligned} d(uv) &= u\,dv + v\,du \\ &= (5x^2)(6x^2\,dx) + (2x^3)(10x\,dx) \\ &= 30x^4\,dx + 20x^4\,dx = 50x^4\,dx \end{aligned}$$

in agreement with the calculation made directly from the product uv.

The Chain Rule states that the derivative of a composite function $x \mapsto g(f(x))$ is $x \mapsto g'(f(x))f'(x)$. In terms of differentials the Chain Rule can be stated as follows. If $v = g(u)$ and $u = f(x)$, then

$$dv = \left.\frac{dv}{du}\right|_{u=f(x)} \frac{du}{dx}\,dx$$

Example 6 Consider the function $x \mapsto 3\sin(x^2)$. Let us evaluate the differential $d(3\sin(x^2))$. We set $u = x^2$ and $v = 3\sin u$. We then have $dv/du = 3\cos u$ and $du/dx = 2x$. By the Chain Rule for differentials we obtain

$$\begin{aligned} d(3\sin(x^2)) = dv &= \left.\frac{dv}{du}\right|_{u=x^2} \frac{du}{dx}\,dx \\ &= 3\cos(x^2)\cdot 2x\,dx \\ &= 6x\cos(x^2)\,dx \end{aligned}$$

We now turn to the problem of computing the tangent line to a curve which is given as the locus of a relation of the form $F(x,y)=0$. Let us suppose that the relation is such that near each point it defines y as a function of x or vice versa (this point is discussed in §8 below). Then the differentials dx and dy will be related and this relationship will yield the tangent line to the locus of $F(x,y)=0$ at (x,y).

Given a relation $F(x,y)=0$, the first step in computing the tangent line to its locus is to use the formulas for differentials to compute the **differential relation**

$$dF(x,y,dx,dy) = 0$$

To do this computation, we write out the expression $F(x,y)$ and take differentials; this yields an expression in the four variables x, y, dx, and dy, which will take the form (expression in x and y) $dx+$(expression in x and y) dy. (A formal definition of the differential relation is given in §14.4.) We illustrate this step with a number of examples.

Example 7 Consider the relation $x^2+y^2-r^2=0$ (where r is a constant). Now we have $d(x^2)=2x\,dx$, $d(y^2)=2y\,dy$, and $d(r^2)=0$; so by the addition rule we have

$$d(x^2+y^2+r^2) = 2x\,dx + 2y\,dy - 0$$

and the differential relation is

$$2x\,dx + 2y\,dy = 0 \quad \text{or} \quad x\,dx + y\,dy = 0$$

Example 8 Differentiate the relation $Ax\sin\theta+Bx\cos\theta=0$ where A and B are constants.

Now $d(\sin\theta)=\cos\theta\,d\theta$, $d(\cos\theta)=-\sin\theta\,d\theta$, and $d(x)=dx$; so

$$\begin{aligned} d(Ax\sin\theta + Bx\cos\theta) &= d(Ax\sin\theta) + d(Bx\cos\theta) \\ &= A\,d(x\sin\theta) + B\,d(x\cos\theta) \\ &= A(x\,d(\sin\theta) + \sin\theta\,d(x)) \\ &\qquad + B(x\,d(\cos\theta) + \cos\theta\,d(x)) \\ &= Ax\cos\theta\,d\theta + A\sin\theta\,dx \\ &\qquad - Bx\sin\theta\,d\theta + B\cos\theta\,dx \end{aligned}$$

Collecting terms, we have the differential relation

$$(A\cos\theta - B\sin\theta)\,x d\theta + (A\sin\theta + B\cos\theta)\,dx = 0$$

Example 9 Suppose the equation $y=f(x)$ defines a differentiable function. Then $y-f(x)=0$ is a relation whose locus is the graph of f. Since $d(f(x))=f'(x)\,dx$, the differential relation is $dy-f'(x)\,dx=0$.

If a relation has the form

$$y - f(x) = 0$$

then we have just seen that the differential relation is

$$dy - f'(x)\,dx = 0$$

Setting $x = x_0$ we find

$$dy - f'(x_0)\,dx = 0$$

which is the equation for the tangent line to the curve $y - f(x) = 0$ at the point $(x_0, f(x_0))$ in the $dx\,dy$ coordinate system. If we have a relation $F(x, y) = 0$ and if we can compute the differential relation

$$dF(x, y, dx, dy) = 0$$

then the equation

$$dF(x_0, y_0, dx, dy) = 0$$

is the equation of the tangent to the locus of the relation at (x_0, y_0) in the $dx\,dy$ coordinate system (Figure 1-4). Moreover, solving this equation for dy/dx gives the slope of the tangent line at (x_0, y_0). This method of finding dy/dx is known as **implicit differentiation**.

The justification of this technique is discussed in §8. In brief, the method uses the fact that if $dF(x_0, y_0, dx, dy) = 0$ does not reduce identically to $dx = 0$ or $0 = 0$, then a portion of the locus of the relation containing the point (x_0, y_0) coincides with the graph of a differentiable function $f: x \mapsto f(x)$. One then differentiates $x \mapsto F(x, f(x))$ to obtain the result.

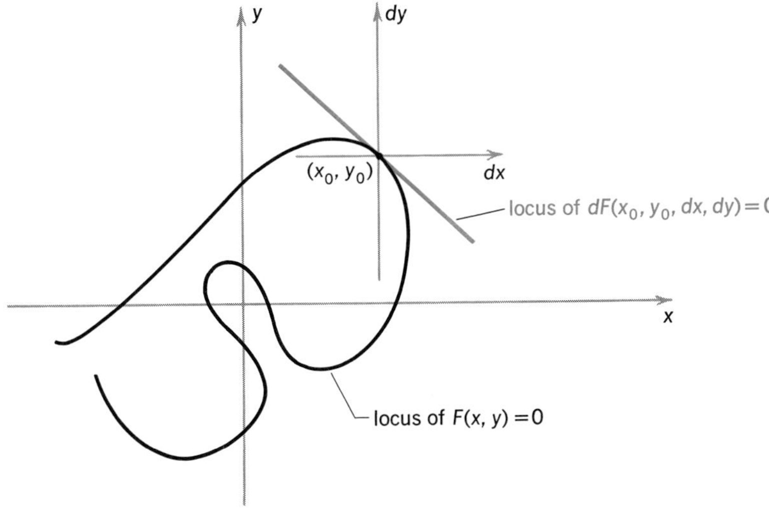

FIGURE 1-4

This method is called implicit differentiation because it enables one to compute the slope dy/dx without explicitly solving for y in terms of x; that is, without finding the function f.

Example 10 Let us consider again the relation $x^2+y^2-r^2=0$, where we now assume that the constant r is greater than zero. The locus of this relation is the circle of radius r with center at the origin. If (x_0, y_0) is a point on the circle, then from analytic geometry we can compute that the tangent to the circle at this point has slope $-x_0/y_0$ (Figure 1-5). In Example 7, we found the differential relation of this relation to be $x\,dx+y\,dy=0$. Solving for dy/dx yields

$$\frac{dy}{dx} = -\frac{x}{y}$$

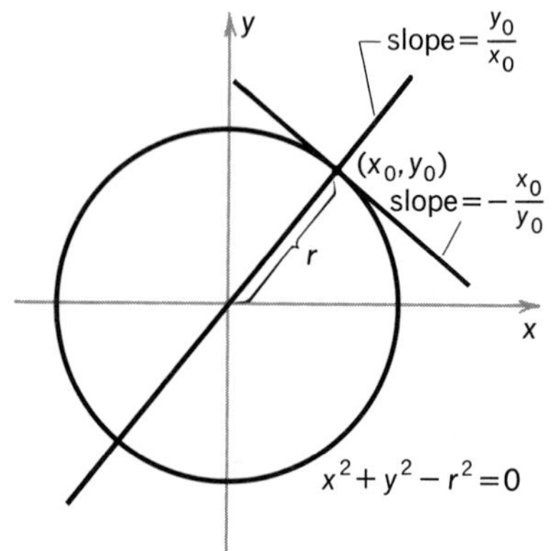

FIGURE 1-5

Evaluating dy/dx when $x=x_0$ and $y=y_0$ (where $x_0^2+y_0^2=r^2$), we again have

$$\left.\frac{dy}{dx}\right|_{(x_0, y_0)} = -\frac{x_0}{y_0}$$

which is the same as the slope computed from geometry. We must remark that the slope given above is only defined when $y_0 \neq 0$. When $y_0=0$, $x_0=\pm r$, and the tangent to the circle is a vertical line; in this case the equation $dF(x_0, y_0, dx, dy)=0$ reduces identically to $dx=0$, which is the equation of the (vertical) tangent line in the $dx\,dy$ coordinate system.

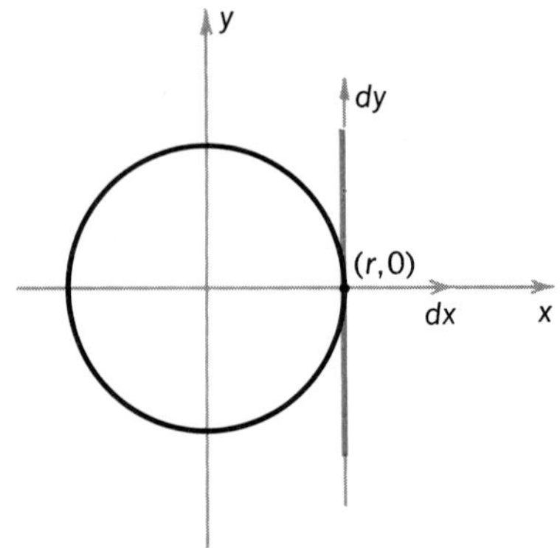

FIGURE 1-6

In general, the quotient $(dy/dx)|_{(x_0, y_0)}$ is defined exactly when the expression $dF(x_0, y_0, dx, dy)=0$ does not reduce identically to $dx=0$, or to $0=0$. When $dF(x_0, y_0, dx, dy)$ reduces to $dx=0$, the tangent to the locus of the relation is vertical and $dx=0$ is still the equation describing it in the $dx\,dy$ coordinate system (Figure 1-6).

Example 11 The relation $x^3-3xy+y^3=0$ describes the locus known as the **folium of Descartes** (Figure 1-7). We can find the slope of the tangent to the folium at an arbitrary point (x_0, y_0) on the curve by computing the differential relation

$$3x^2\,dx - 3y\,dx - 3x\,dy + 3y^2\,dy = 0$$

or

$$(x^2-y)\,dx + (y^2-x)\,dy = 0$$

Solving for dy/dx yields

$$\frac{dy}{dx} = \frac{x^2+y}{x-y^2}$$

and evaluating at (x_0, y_0), we get

$$\left.\frac{dy}{dx}\right|_{(x_0, y_0)} = \frac{x_0^2-y_0}{x_0-y_0^2} \tag{1}$$

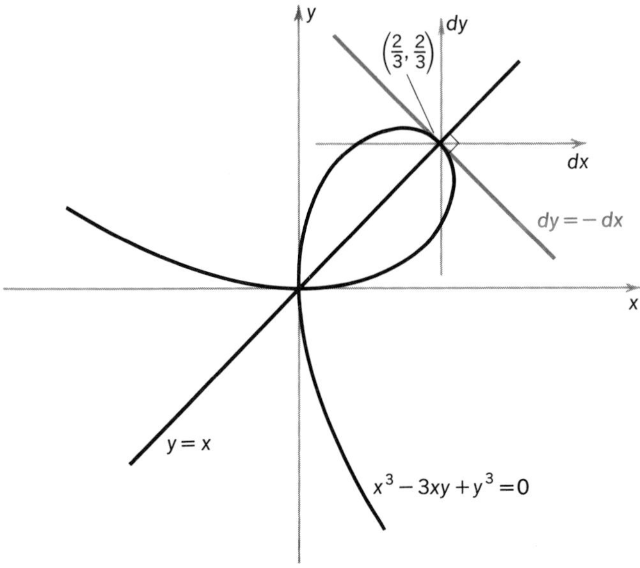

FIGURE 1-7

The folium intersects the line $y=x$ at the origin and at the point $(\frac{2}{3}, \frac{2}{3})$. From Figure 1-7 it is clear that the tangent to the folium at $(\frac{2}{3}, \frac{2}{3})$ is perpendicular to the line $y=x$ and hence has slope -1. We verify this immediately from equation (1) by evaluating

$$\left.\frac{dy}{dx}\right|_{(2/3,2/3)} = \frac{(\frac{2}{3})^2 - \frac{2}{3}}{\frac{2}{3} - (\frac{2}{3})^2} = -1$$

At the origin, the folium crosses itself, and there the differential relation reduces identically to $0=0$, demonstrating that a unique tangent line is not defined.

As a further application of implicit differentiation, we shall extend the formula

$$\frac{d(x^r)}{dx} = rx^{r-1}$$

to all rational numbers r. First, suppose q is a positive odd natural number. Then each real number x has a unique real qth root, denoted $x^{1/q}$. Therefore the rule $x \mapsto x^{1/q}$ defines a function whose domain is all of $\boldsymbol{R}$. This function can be described by the equation $y = x^{1/q}$ or, equivalently, by $y^q = x$, so the graph of the function $x \mapsto x^{1/q}$ is the locus of the relation $y^q - x = 0$ (Figure 1-8). The corresponding differential relation is given by $qy^{q-1}\,dy - dx = 0$. Therefore, if $(x, x^{1/q})$ is a point on the graph, the tangent line at this point has the equation

$$q(x^{1/q})^{q-1}\,dy - dx = 0$$

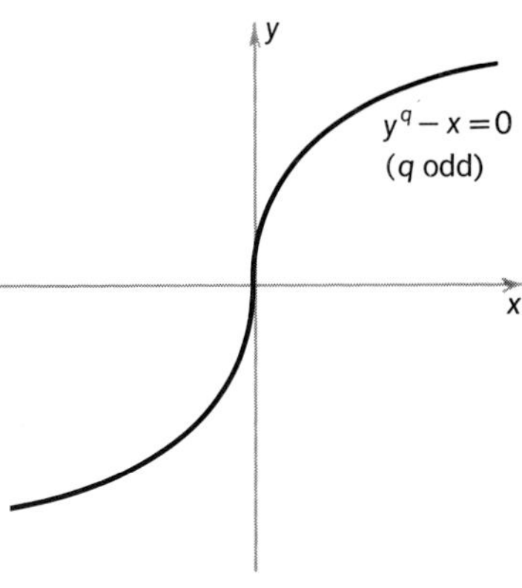

FIGURE 1-8

or

$$qx^{(q-1)/q}\,dy - dx = 0$$

When $x \neq 0$, this tangent line is nonvertical and the slope is

$$\frac{dy}{dx} = \frac{1}{q}x^{-(q-1)/q}$$

or [since $-(q-1)/q = (1/q) - 1$]

$$\frac{dy}{dx} = \frac{1}{q}x^{(1/q)-1}$$

Therefore, we have established the formula

$$\frac{d(x^{1/q})}{dx} = \frac{1}{q}x^{(1/q)-1}$$

for all positive odd natural numbers q

Now we must consider the case when q is a positive even natural number; then each positive real number x has two qth roots, one positive and one negative, denoted by $x^{1/q}$ and $-x^{1/q}$, respectively. (Negative numbers do not have real qth roots when q is even.) Thus, the symbol $x^{1/q}$ always denotes the nonnegative qth root of x (for $x \geqslant 0$), and so the rule $x \mapsto x^{1/q}$ defines a function whose domain is $[0, \infty[$. The graph of this function is the upper half of the locus of the relation $y^q - x = 0$ (Figure 1-9). As in the case of q odd, we can use implicit differentiation to compute that the derivative is

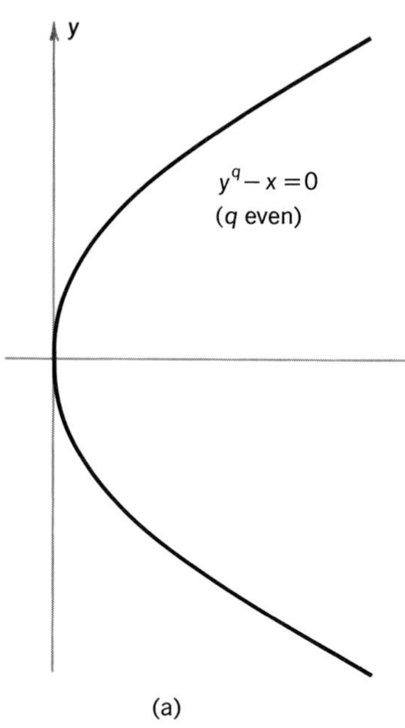

(a)

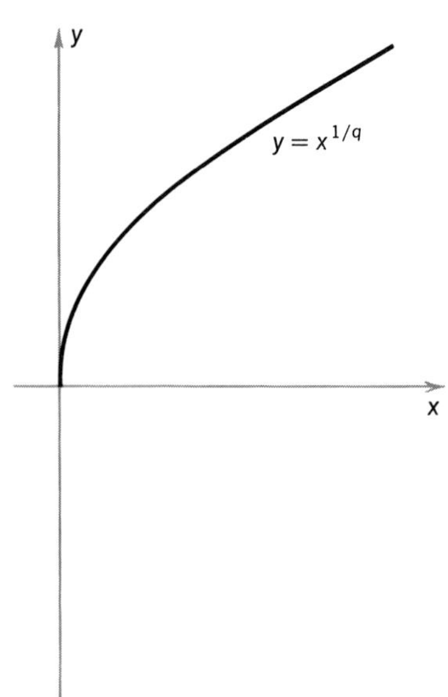

(b)

FIGURE 1-9

$$\frac{dy}{dx} = \frac{1}{q}x^{(1/q)-1}$$

Thus the formula

$$\frac{d(x^{1/q})}{dx} = \frac{1}{q}x^{(1/q)-1}$$

is extended to all natural numbers q.

It is now easy to establish the formula

$$\frac{d(x^r)}{dx} = rx^{r-1} \tag{2}$$

for all rational numbers r. First recall that a real number is rational if it is equal to the quotient of two integers. Thus if r is rational, it is a number of the form p/q, where p and q are integers. Note that for any rational number r, we can always find p and q such that $r = p/q$ and $q > 0$. (How?) To establish equation (2), we use the Chain Rule: First we write $x^{p/q} = (x^{1/q})^p$. Since p, is an integer we can differentiate and the derivative is, by our above work,

$$\frac{d((x^{1/q})^p)}{dx} = p(x^{1/q})^{p-1}\frac{d(x^{1/q})}{dx}$$

$$= p(x^{1/q})^{p-1}\frac{1}{q}x^{(1/q)-1}$$

$$= \frac{p}{q}x^{(p-1)/q+(1/q)-1}$$

$$= \frac{p}{q}x^{(p/q)-1}$$

We have proved the following theorem.

Theorem 1 If $r = p/q$ is any rational number, then

$$\frac{d(x^r)}{dx} = rx^{r-1}$$

Example 12 Compute the derivative of $f: x \mapsto 1/\sqrt{x^2+3}$. We notice that $x^2+3 > 0$ for all x, so the domain of f is all of $\boldsymbol{R}$. We write $f(x) = (x^2+3)^{-1/2}$, and use both the previous theorem and the Chain Rule:

$$f'(x) = -\tfrac{1}{2}(x^2+3)^{-3/2}\frac{d(x^2+3)}{dx}$$

$$= -\tfrac{1}{2}(x^2+3)^{-3/2}(2x)$$

$$= -\frac{x}{(x^2+3)^{3/2}}$$

Example 13 Differentiate the function $g: x \mapsto (3x^3 - \sin x)^{4/3}\tan^2 x$. We compute

$$\frac{d(g(x))}{dx} = \frac{d(3x^3-\sin x)^{4/3}}{dx}\tan^2 x + (3x^3-\sin x)^{4/3}\frac{d(\tan^2 x)}{dx} \text{ (Product Rule)}$$

$$= \tfrac{4}{3}(3x^3-\sin x)^{1/3}\frac{d(3x^3-\sin x)}{dx}\tan^2 x$$

$$+ (3x^3-\sin x)^{4/3}\, 2\tan x\frac{d(\tan x)}{dx} \text{ (Chain Rule)}$$

$$= \tfrac{4}{3}(3x^3-\sin x)^{1/3}(9x^2-\cos x)\tan^2 x$$

$$+ 2(3x^3-\sin x)^{4/3}\tan x\sec^2 x$$

$$= 2(3x^3-\sin x)^{1/3}\tan x(6x^2\tan x - \tfrac{2}{3}\sin x + 3x^3\sec^2 x - \tan x\sec x)$$

As a final application of implicit differentiation, we shall discuss the **reflected wave property of the ellipse**. Consider an ellipse E centered at the origin, with foci $(-c,0)$ and $(c,0)$, and equation

$$\frac{x^2}{a^2}+\frac{y^2}{b^2}-1=0$$

If a light wave is sent out from the focus at $(-c,0)$ and strikes the ellipse, then the reflective wave property says that the wave will bounce back and pass through the other focus at $(c,0)$. Geometrically, this means that the angles α and β in Figure 1-10 must be equal [T is the tangent line to E at (x_0, y_0)]. Let us now demonstrate this property using implicit differentiation. The differential relation of the ellipse is

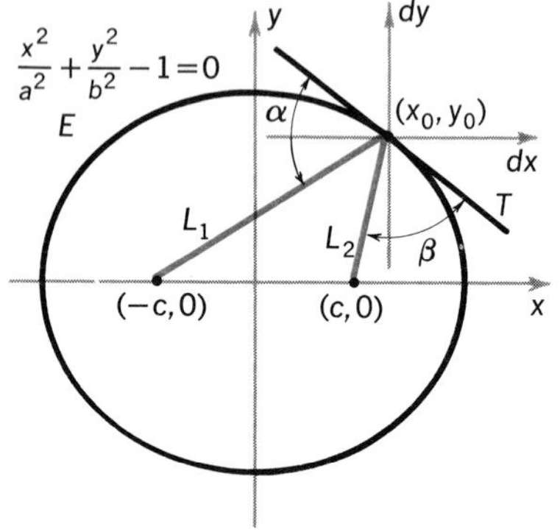

FIGURE 1-10

$$\frac{2x}{a^2}dx+\frac{2y}{b^2}dy=0$$

Therefore, the tangent line through the point (x_0, y_0) on E has equation

$$\frac{x_0}{a^2}dx+\frac{y_0}{b^2}dy=0$$

or

$$dy=-\frac{b^2x_0}{a^2y_0}dx$$

whenever $y_0 \neq 0$. (The case where $y_0=0$ is obvious.) In the $dx\,dy$ system, the line L_1 through $(-c,0)$ and (x_0, y_0) has equation $dy=y_0/(x_0+c)\,dx$. Likewise, the line L_2 through $(c,0)$ and (x_0,y_0) has equation $dy=y_0/(x_0-c)\,dx$. If we denote the angle of inclination of the tangent line T by $\pi-\gamma$, we have

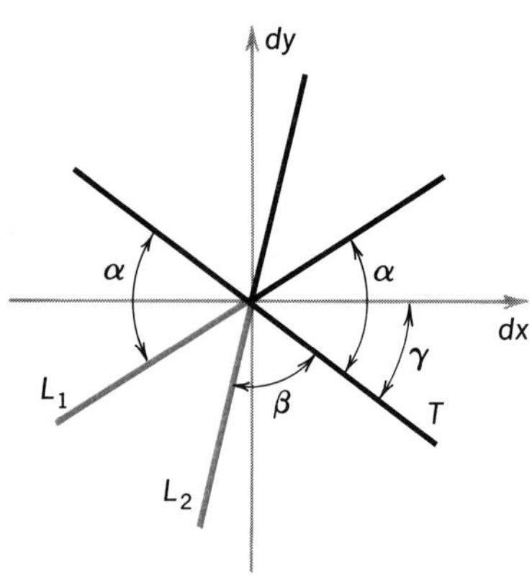

FIGURE 1-11

$$\alpha-\gamma = \text{angle of inclination of } L_1$$

$$\pi-(\beta+\gamma) = \text{angle of inclination of } L_2$$

(Figure 1-11).

We then have

$$\tan(\alpha-\gamma) = y_0/(x_0+c) \tag{3}$$

$$\tan(\beta+\gamma) = -\tan(\pi-(\beta+\gamma))$$

$$= y_0/(c-x_0) \tag{4}$$

$$\tan\gamma = -b^2x_0/a^2y_0 \tag{5}$$

We must now show that $(\alpha-\gamma)+\gamma=(\beta+\gamma)-\gamma$. But the formula

$$\tan(\theta+\phi)=\frac{\tan\theta+\tan\phi}{1-\tan\theta\tan\phi}$$

applied to equations (3)–(5) yields

$$\tan((\alpha-\gamma)+\gamma)=\tan((\beta+\gamma)-\gamma) \qquad \text{(Verify!)}$$

Thus $\beta=\alpha$, since α and β are both in the first quadrant.

EXERCISES

In exercises 1–7, compute the differential.

1. $d(x-x^3)$
2. $d(\sqrt{1-3x})$
3. $d(2\sin^2\theta)$
4. $d(\tan 2\phi)$
5. $d\left(\dfrac{ax}{\sqrt{a-x}}\right)$
6. $d(\sec^4 3x-\tan^4 3x)$
7. $d\left(\dfrac{x-a}{2}\sqrt{2ax-x^2}\right)$

In exercises 8–14, compute the differential relation and solve for dy/dx. Also, give the equation of the tangent line to the curve at the point indicated, both in the $dx\,dy$ and xy coordinate systems.

8. $y^2-2ax-4a^2=0$ at $(-3a/2, a)$
9. $\dfrac{x^2}{a^2}-\dfrac{y^2}{b^2}-1=0$ at $(5a/3, 4b/3)$
10. $3xy-2y^2-5=0$ at $(7/3, 5/2)$
11. $\sqrt[3]{xy}-4=0$ at $(8, 8)$
12. $x^4-3xy^3-x+2y=0$ at $(1, -\sqrt{\tfrac{2}{3}})$
13. $\sqrt{x/y^3}-2\sqrt{x^3/y}=0$ at $(\tfrac{1}{2}, 1)$
14. $x^2+\dfrac{x-y}{x+y}=0$ at $(2, -\tfrac{10}{3})$

Differentiate each of the functions of exercises 15–18.

15. $x \mapsto 4(1-x)^{4/3}+2(2x^2-3x+5)^{1/3}-(2x-x^3)^{-1/3}$
16. $t \mapsto \sqrt[5]{\left(\dfrac{1-t}{1+t}\right)^3}$
17. $y \mapsto \dfrac{y-3}{\sqrt{6y-y^2}}$
18. $x \mapsto \sqrt[3]{1-x^2}$
19. The total surface area of a right circular cone is given by
$$S=\pi(r^2+r\sqrt{r^2+h^2})$$
where r is the base radius and h is the height. If S is constant, find dr/dh when $r=7$ and $h=24$.
20. Show that the tangent line to the ellipse
$$\frac{x^2}{a^2}+\frac{y^2}{b^2}-1=0$$
at the point directly above the focus on the positive x axis has slope $-\sqrt{a^2-b^2}/a$. Where does this tangent line intersect the x and y axes?

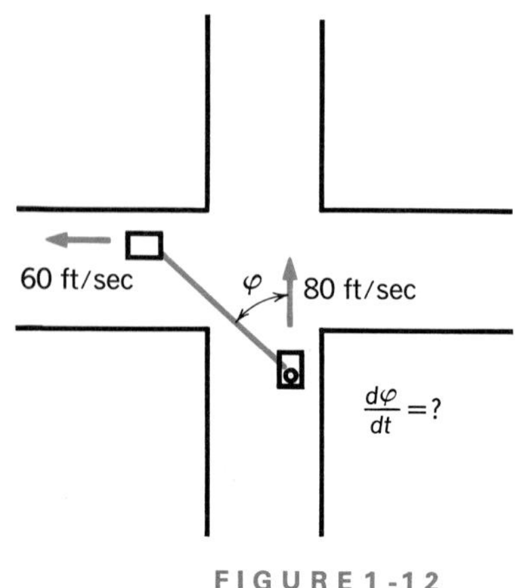

FIGURE 1-12

21. A police car approaching an intersection at 80 ft/sec spots a suspicious vehicle on the cross street. When the police car is 210 feet from the intersection, he turns his spotlight on the vehicle, which is at that time just crossing the intersection at a constant rate of 60 ft/sec. How fast must the light beam be turning 3 seconds later in order to follow the suspicious vehicle (Figure 1-12)?

2 PARAMETRIC EQUATIONS; RELATED RATES

Suppose we are given two equations $x=f(t)$ and $y=g(t)$, where f and g are continuous functions with the same domain. With every point t_0 in the common domain of f and g, we can associate the point $(f(t_0), g(t_0))$ in the xy plane. As t changes, the points $(f(t), g(t))$ will trace out a curve in the xy plane.

The equations $x=f(t)$ and $y=g(t)$ are called **parametric equations**, and t is called the **parameter**. The locus of points $(f(t), g(t))$ is the **parametrized curve**.

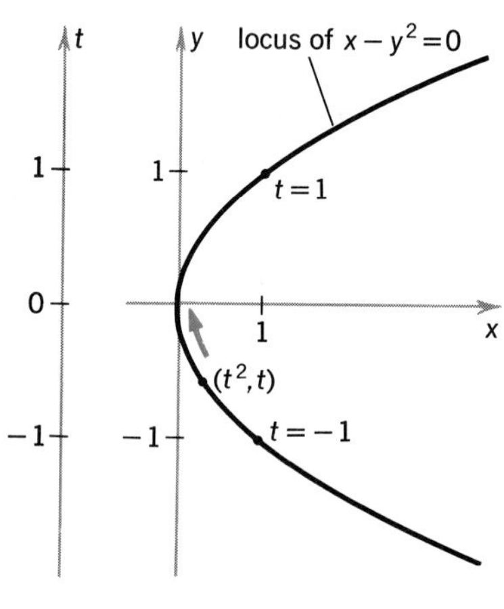

FIGURE 2-1

By way of example, consider the parametric equations

$$x = f(t) = t^2$$

$$y = g(t) = t$$

We note that $f(t)-g(t)^2=0$ for all t, and so all points $(f(t), g(t))=(t^2, t)$ lie on the locus of $x-y^2=0$ (Figure 2-1). Now as t goes from -1 to 0, the point (t^2, t) moves along the locus from $(1, -1)$ up to $(0,0)$; as t goes from 0 to 1, (t^2, t) moves along the upper half of the locus to the point $(1, 1)$.

Since parametric equations are almost always given in the form

$$x = \text{expression in } t$$

$$y = \text{expression in } t$$

(t is usually time) it is convenient to call the two functions defined by the equations X and Y, and to write

$$x = X(t), \qquad y = Y(t)$$

Thus for example, if the equations are $x=\cos t$ and $y=t^2$, we have $X(t)=\cos t$ and $Y(t)=t^2$, and $X'(t)=dx/dt=-\sin t$, $Y'(t)=dy/dt=2t$.

An important feature of parametric equations is that they distinguish among the ways in which a point can move along a curve. For example, the equations

$$x = t^2, \qquad y = -t \tag{1}$$

and the equations

$$x = t^2, \qquad y = t \tag{2}$$

both describe a point which moves along the locus $x - y^2 = 0$, but the points determined by the two sets of equations move in different directions. Equations (1) describe a point moving along the upper half of the locus when $t < 0$, reaching the origin when $t = 0$, and then moving along the lower half of the locus when $t > 0$. On the other hand, equations (2) describe the point as moving on the lower half when $t < 0$, reaching the origin at $t = 0$, and then moving along the upper half of the locus when $t > 0$.

Suppose $x = X(t)$ and $y = Y(t)$ are parametric equations such that $t \mapsto X(t)$ and $t \mapsto Y(t)$ are continuously differentiable functions. Then we have well-defined rates of change dx/dt of x with respect to t and dy/dt of y with respect to t. We may inquire what is the rate of change of y with respect to x? This is equivalent to asking for the slope of the line tangent to the parametrized curve (Figure 2-2). To compute the equation of the line tangent to the parametrized curve, we consider the differential relations

$$dx = X'(t_0)\,dt, \qquad dy = Y'(t_0)\,dt$$

Provided that $X'(t_0)$ and $Y'(t_0)$ are not both equal to zero, we have either

$$dt = \frac{dx}{X'(t_0)} \quad \text{or} \quad dt = \frac{dy}{Y'(t_0)}$$

Both cases yield

$$X'(t_0)\,dy = Y'(t_0)\,dx \tag{3}$$

which is the equation of the tangent line in $dx\,dy$ coordinates. That this is indeed the case is demonstrated in §8. In brief the argument is that the condition $X'(t_0) \neq 0$ means that a portion of the parametrized curve containing the point $(X(t_0), Y(t_0))$ is the graph of a differentiable function $f{:}\,x \mapsto f(x)$; one then computes that

$$\left.\frac{d(f(x))}{dx}\right|_{x=X(t_0)} = \frac{Y'(t_0)}{X'(t_0)}$$

Thus, when $t = t_0$ the line tangent to the parametrized curve at the point $(X(t_0), Y(t_0))$ has the equation

$$X'(t_0)\,dy = Y'(t_0)\,dx$$

in the $dx\,dy$ coordinate system, and the equation

$$(y - Y(t_0))\,X'(t_0) = (x - X(t_0))\,Y'(t_0)$$

in the xy system. When $X'(t_0) = 0$, equation (3) reduces to $dx = 0$, and the tangent line is vertical. When $X'(t_0) \neq 0$, the tangent line has a slope which is given by

$$\left.\frac{dy}{dx}\right|_{t=t_0} = \frac{Y'(t_0)}{X'(t_0)}$$

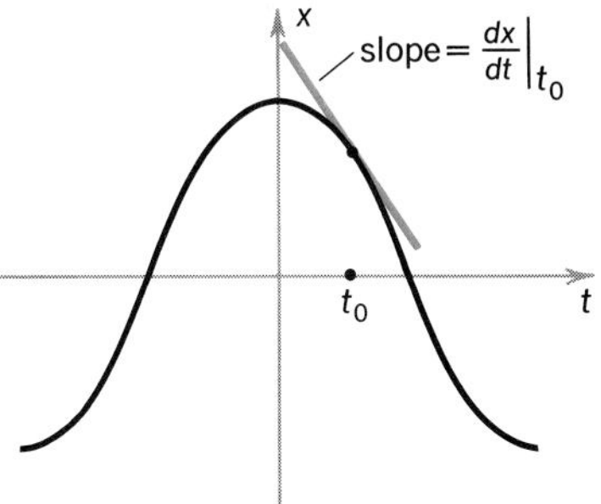

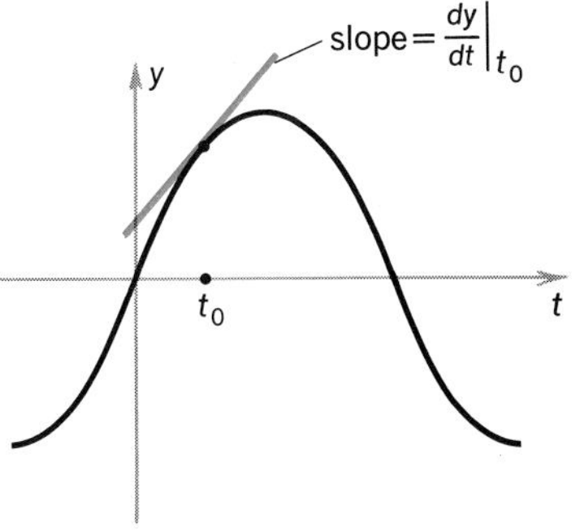

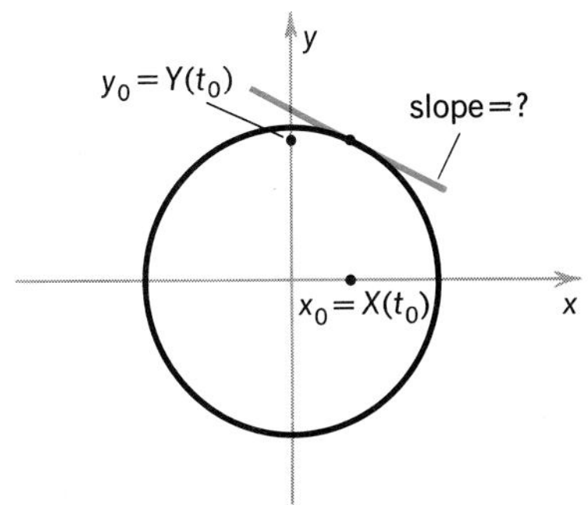

FIGURE 2-2

In general, when $X'(t) \neq 0$, we have the formula

$$\frac{dy}{dx} = \frac{Y'(t)}{X'(t)} = \frac{dy/dt}{dx/dt}$$

Example 1 Consider the parametric equations

$$x = t^2, \quad y = -t$$

(See Figure 2-3.) Using the formula just derived, we compute that the tangent to the curve at the point $(x_0, y_0) = (X(t_0), Y(t_0))$ is

$$2t_0\, dy = -dx$$

Since $y_0 = Y(t_0) = -t_0$, we obtain

$$-2y_0\, dy = -dx$$

or

$$2y_0\, dy = dx \tag{4}$$

as the equation for the tangent line in the $dx\,dy$ system.

We have already remarked that the parametric equations of this example describe a point moving along the locus of $y^2 - x = 0$. Using differentials, we can compute that the differential relation corresponding to this equation is

$$2y\, dy - dx = 0$$

Therefore, the tangent to the locus at (x_0, y_0) has equation $2y_0\, dy = dx$, which agrees with equation (4).

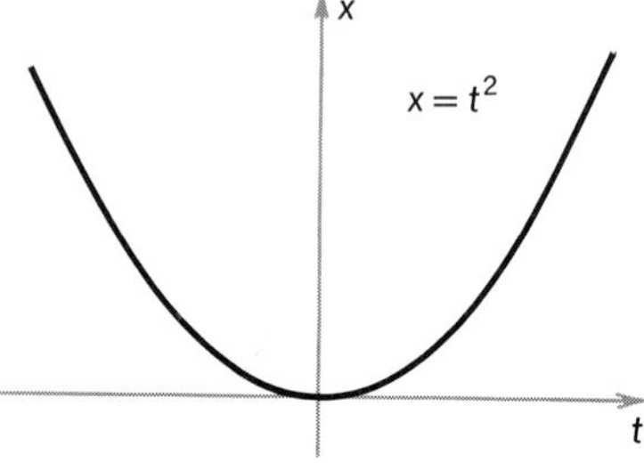

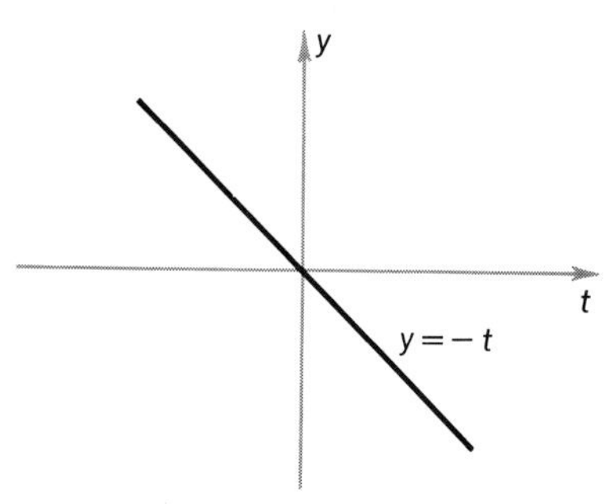

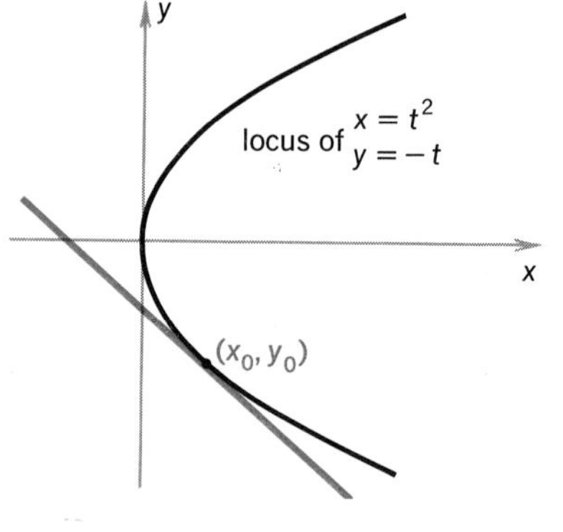

FIGURE 2-3

Example 2 Consider the equations

$$x = 5\cos t, \qquad y = 5\sin t$$

These equations parametrize the circle of radius 5 centered at the origin (Figure 2-4). To see this, note that for all t,

$$X(t)^2 + Y(t)^2 = 25(\cos^2 t + \sin^2 t) = 25$$

so all points $(X(t), Y(t))$ lie on the locus of $x^2 + y^2 - 25 = 0$, which is the circle in question. Conversely, if (x_0, y_0) lies on the circle, we can choose as t_0 the angle of inclination of the line through the origin and (x_0, y_0) (Figure 2-4). Then $x_0 = 5\cos t_0$, $y_0 = 5\sin t_0$, and $(x_0, y_0) = (X(t_0), Y(t_0))$.

Now the tangent line at (x_0, y_0) must have equation

$$Y'(t_0)\, dx = X'(t_0)\, dy$$

which in this case is

$$5\cos t_0\, dx = -5\sin t_0\, dy$$

However, $5\cos t_0 = x_0$ and $5\sin t_0 = -y_0$, so the equation of the tangent line is

$$x_0\,dx = -y_0\,dy$$

Recall that this formula was derived in §1, Example 10.

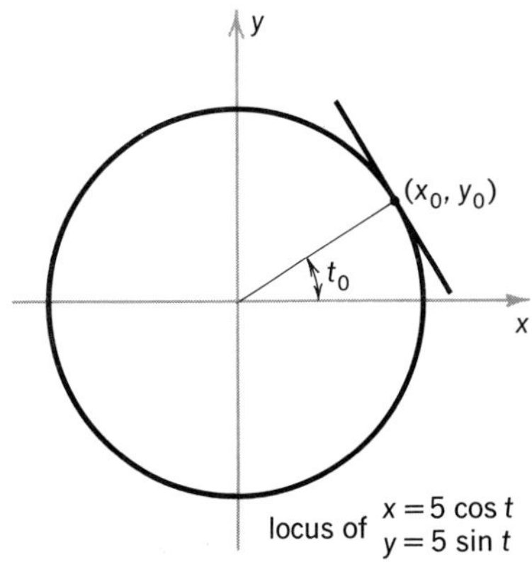

FIGURE 2-4

Example 3 Calculate the rate of change of y with respect to x, where $y = \sin t \tan t$ and $x = t^2+2$. We can write

$$\frac{dy}{dt} = \sin t \sec^2 t + \cos t \tan t$$

$$= \tan t \sec t + \sin t$$

and

$$\frac{dx}{dt} = 2t$$

Therefore, we have

$$\frac{dy}{dx} = \frac{dy/dt}{dx/dt} = \frac{\tan t \sec t + \sin t}{2t}$$

Example 4 If a ball falls according to the equation for height $h = 100-16t^2$, and the distance of a fielder from the point of impact is $s = 7(3-t)^{5/4}$, find the rate of change of the ball's height with respect to the fielder's approach, at time $t=2$. Hence, we calculate

$$\frac{dh}{dt} = -32t, \qquad \frac{ds}{dt} = -\frac{35}{4}(3-t)^{1/4}$$

so

$$\frac{dh}{ds} = \frac{128t}{35(3-t)^{1/4}}$$

Therefore, at $t=2$, the relative rate of change is $256/35 = 7\frac{11}{35}$.

In the previous section, we considered relations $F(x,y)=0$ and computed the differential relation $dF(x,y,dx,dy)=0$. The technique of calculating differentials also applies to the equation $F(x,y)=G(x,y)$. In fact, the method can be applied to equations involving more than just two variables. For example, consider the equation

$$s^2 = u^2 + v^2$$

Differentiating both sides gives the differential relation

$$2s\,ds = 2u\,du + 2v\,dv$$

or

$$s\,ds = u\,du + v\,dv \tag{5}$$

Now if s, u, and v are all defined in terms of a parameter t, we can solve for ds/dt in terms of s, du/dt, dv/dt, u, and v; hence, dividing both sides of (5) by dt gives

$$s\frac{ds}{dt} = u\frac{du}{dt} + v\frac{dv}{dt}$$

or

$$\frac{ds}{dt} = \frac{u}{s}\frac{du}{dt} + \frac{v}{s}\frac{dv}{dt} \tag{6}$$

whenever $s \neq 0$.

This technique of computing rates of change has many applications. We now employ the method to solve problems of **related rates**.

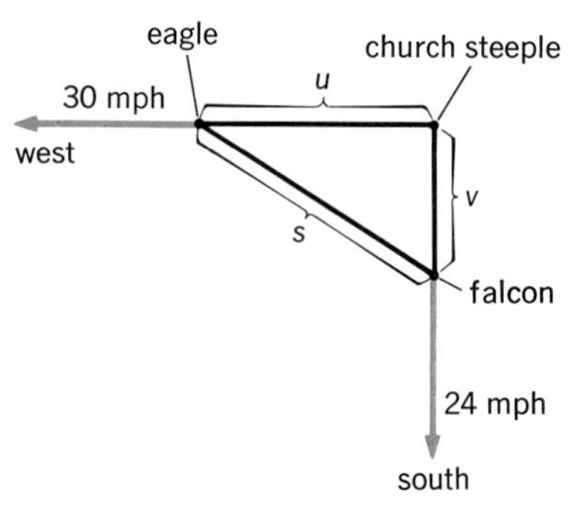

FIGURE 2-5

Example 5 An eagle flying due west at 30 mph passes over a church steeple at time $t = 0$. A falcon heading due south at 24 mph passes over the steeple 10 min later. Both birds fly at the same constant altitude. At what speed are the two birds separating from each other 20 min after the falcon passes over the steeple (Figure 2-5)?

Let us analyze the information contained in the problem. The eagle is flying west at 30 mph, so if we let u denote the distance from the eagle's position to the steeple, we have $du/dt = 30$. Similarly, if we let v denote the falcon's distance from the point over the steeple, we have $dv/dt = 24$. As shown in the figure, we denote the distance from the falcon to the eagle by s. We also know that 20 min after the falcon flew over the steeple is 30 min after the eagle flew over it; hence, $t = 30$ min $= \frac{1}{2}$ hr, and so we have to find $(ds/dt)|_{t=1/2}$.

The directions south and west are at right angles to one another, so u, v, and s are related by the equation

$$s^2 = u^2 + v^2$$

for all values of the parameter t. Differentiating, we obtain

$$2s\,ds = 2u\,du + 2v\,dv$$

or

$$s\,ds = u\,du + v\,dv$$

Observe that this is exactly the expression we obtained in equation (5). Therefore, to evaluate ds/dt, we again divide both sides by dt which yields

$$\frac{ds}{dt} = \frac{u}{s}\frac{du}{dt} + \frac{v}{s}\frac{dv}{dt} \tag{7}$$

We seek $(ds/dt)|_{t=1/2}$. We know that $(du/dt)|_{t=1/2} = 30$ and $(dv/dt)|_{t=1/2} = 24$. Moreover, it is easy to compute u, v, and s when $t = \frac{1}{2}$—the eagle has been flying west for $\frac{1}{2}$ hr, so $u = 30 \cdot \frac{1}{2} = 15$; the falcon has been flying south for 20 min, or $\frac{1}{3}$ hr from the time he passed over the steeple, so $v = 24 \cdot \frac{1}{3} = 8$.

Finally, $s = \sqrt{u^2 + v^2} = \sqrt{289} = 17$ when $t = \frac{1}{2}$ hr. Substituting these values into equation (7) yields

$$\left.\frac{ds}{dt}\right|_{t=1/2} = \frac{15 \cdot 30 + 8 \cdot 24}{17} = \frac{642}{17} \text{ mph} = 37\tfrac{13}{17} \text{ mph}$$

Example 6 A ball is hit along the third-base line with a speed of 100 ft/sec. At what rate is the ball's distance from the first base changing when it crosses third base?

Let u denote the position of the ball; u depends on time t, and we have $du/dt = 100$ ft/sec. If s denotes the distance from the ball to the first base, we have $s^2 = 90^2 + u^2$, since the infield is a square 90 ft on each side (Figure 2-6). Passing to the differential equation, we have $2s\,ds = 0 + 2u\,du$, which reduces to

$$s\,ds = u\,du \tag{8}$$

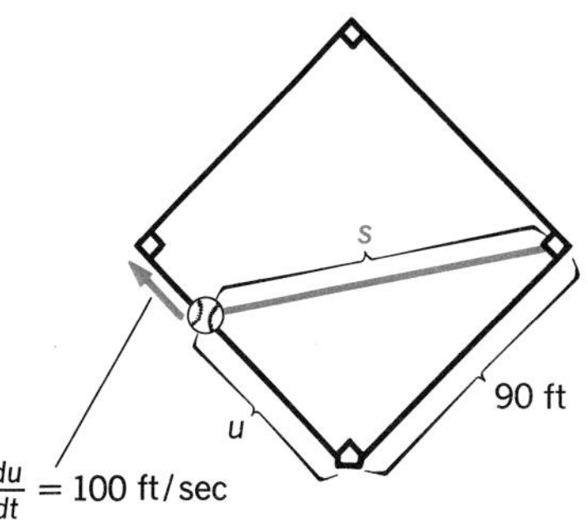

FIGURE 2-6

We want to find ds/dt when $u = 90$ ft, so we divide both sides of equation (8) by dt to arrive at

$$s\frac{ds}{dt} = u\frac{du}{dt} \tag{9}$$

The distance from the first base to the third base is (approximately) 127.2 ft (by the Pythagorean theorem), so when $u = 90$ ft, then $s = 127.2$ ft. We know that $du/dt = 100$ ft/sec, and so we can solve for ds/dt when $u = 90$ ft by substituting into equation (9), finding

$$\frac{ds}{dt} = \frac{90}{127.2} \cdot 100 = 70 \text{ ft/sec}$$

Example 7 Coffee is dripping from a conical filter at the rate of 0.3 in^3/sec. The height of the filter is 6 in and the radius at the top is 3 in (Figure 2-7). How fast is the coffee level falling when the depth of the coffee in the filter is 3 in?

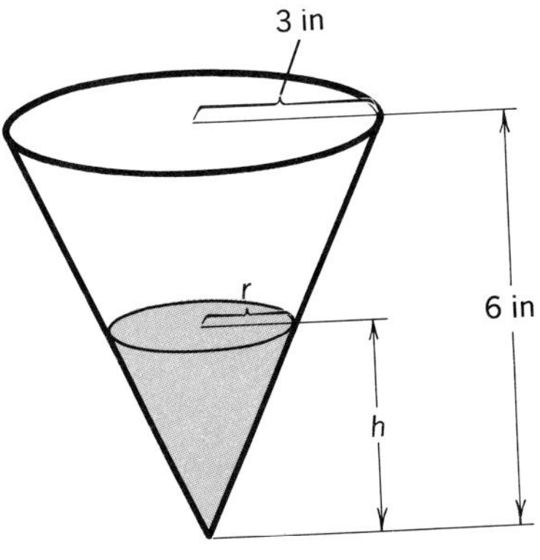

FIGURE 2-7

Let V denote the volume of coffee in the filter; then V depends on time and $dV/dt = -0.3$ in^3/sec (the minus sign indicates that V decreases with time). Let h be the depth of the coffee—h also depends on t. The problem is to find dh/dt when $h = 3$ in. Now, if r denotes the radius of the section of the filter at the coffee level h, by similar triangles we have $r/h = \frac{3}{6} = \frac{1}{2}$, so $r = \frac{1}{2}h$ for all time t. Next, the formula for the volume of a cone is $V = (\pi r^2 h/3)$ for all t, so we obtain

$$V = \frac{\pi}{3}h\left(\frac{h}{2}\right)^2 = \frac{\pi h^3}{12}$$

Taking differentials, we have

$$dV = \frac{\pi}{4}h^2\,dh$$

Since we seek dh/dt when $h = 3$ in, we divide by dt to obtain

$$\frac{dV}{dt} = \frac{\pi}{4}h^2\frac{dh}{dt}$$

Substituting -0.3 for dV/dt and 3 for h, we can evaluate

$$\frac{dh}{dt} = \frac{-0.3}{9}\frac{4}{\pi} = \frac{-0.4}{3\pi} = \frac{-2}{15\pi} \text{ in/sec}$$

when $h = 3$ in. (The negative sign appears because h is decreasing with time, and dh/dt measures the rate at which the coffee level is rising.)

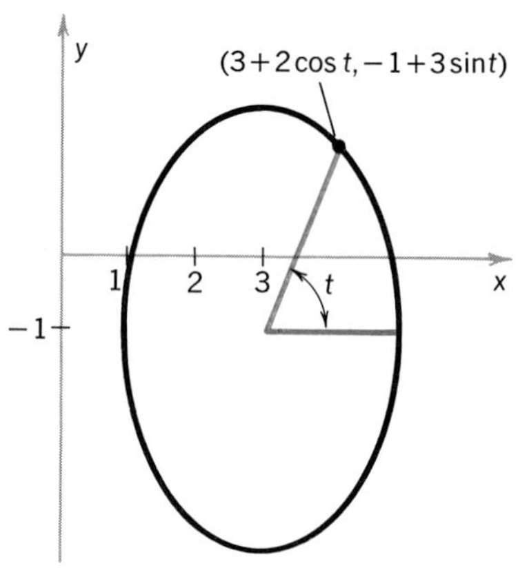

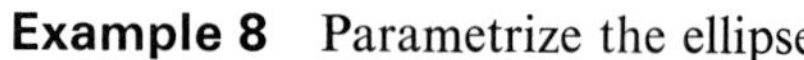

FIGURE 2-8

We conclude this section with a discussion of how to find parametric equations for a given curve. If we have a relation $F(x, y) = 0$, it is often convenient to parametrize both x and y in terms of some variable t. For example, motion along a circle is often studied by means of parametric equations. Other conics can also be parametrized, as in the following example.

Example 8 Parametrize the ellipse

$$\frac{(x-3)^2}{4} + \frac{(y+1)^2}{9} - 1 = 0$$

(See Figure 2-8.) We notice that we have the sum of two squares equaling 1. This suggests that the trigonometric formula $\cos^2 t + \sin^2 t = 1$ might be useful here. We therefore let

$$\frac{x-3}{2} = \cos t, \qquad \frac{y+1}{3} = \sin t$$

obtaining the parametrization $x = 3 + 2\cos t$, $y = -1 + 3\sin t$, $0 \leqslant t \leqslant 2\pi$.

Parametric equations in geometry can result from a word description of a problem.

Example 9 The **cycloid** can be described as the curve traced out by the point P on the circumference of a circle of radius 1 as the circle rolls along the x axis. By examining Figure 2-9, we see that the coordinates of the point P are given by

$$x = t - \sin t, \qquad y = 1 - \cos t$$

The cycloid is an example of a curve whose description as the graph of a function is rather complicated, but whose description by parametric equations is relatively simple. From the figure, we see that the cycloid has cusps at points corresponding to $t = 2n\pi$, where n is an integer. At all other points a nonvertical tangent line exists and its equation is

$$(1 - \cos t)\,dy = \sin t\,dx$$

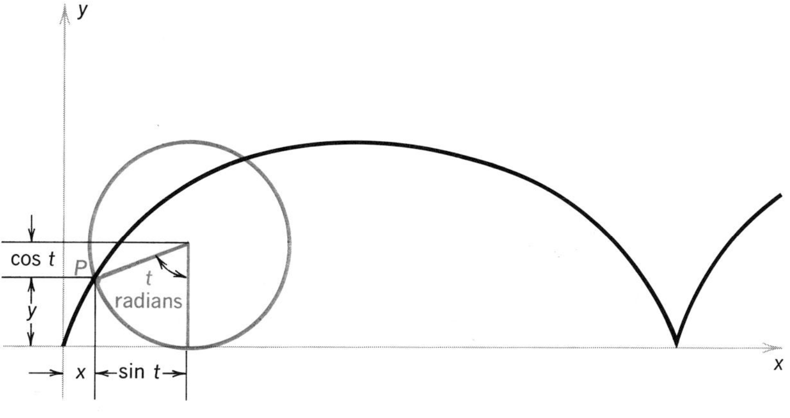

FIGURE 2-9

This equation reduces to $0 = 0$ when $t = 2n\pi$. The tangent is horizontal when t is an odd integral multiple of π.

EXERCISES

In exercises 1–5, compute dy/dx for each pair of parametric equations. Also, compute the equation of the tangent line to the parametrized curve at the point indicated. Sketch each parametrized curve, and the indicated tangent line.

1. $x = t^3, \quad y = t^4$ at $t = 2$
2. $x = u^2 + 1, \quad y = u^2 - u$ at $u = 3$
3. $x = r\sin^2\theta, \quad y = r\cos^2\theta$ at $\theta = -\frac{1}{3}\pi$
4. $x = \cos\theta, \quad y = \sin\theta$ at $\theta = -\frac{1}{3}\pi$
5. $x = \phi - \sin\phi, \quad y = 1 - \cos\phi$ at $\phi = \pi$
6. A falcon flies up from its trainer at an angle of 60° until it has flown 200 ft (Figure 2–10). It then levels off and continues to fly away. If the speed of the bird is 132 ft/sec, how fast is the falcon moving away from the falconer after 6 sec.

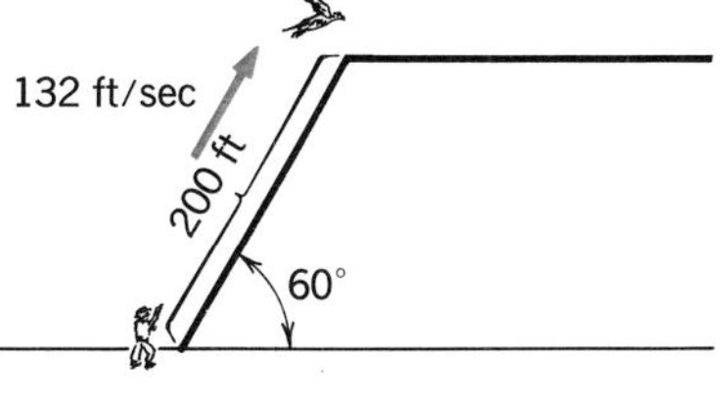

FIGURE 2-10

7. Two airplanes fly on parallel courses 20 mi apart, at speeds of 450 mph and 500 mph. How fast is the distance between the planes changing when the slower plane is 12 mi ahead of the faster one?
8. A lamp post is 10 ft high. A man six feet tall walks away from the lamp post at a rate of 4 mph. How fast does the end of his shadow move?
9. As sand is being poured at the rate of 2 ft/sec, it forms a conical pile whose altitude is equal to the radius of the base. How fast is the altitude increasing when the radius is 10 ft?

10. An object moves along the curve $y = x^3$. At what points on the curve are the x and y coordinates of the particle changing at the same rate?

11. A ferris wheel is 50 ft in diameter and has center located 30 ft above the ground. If the wheel revolves once every two minutes, how fast is a passenger rising when he is 42.5 ft above the ground? How fast is he moving horizontally?

12. The dome of an observatory is 60 ft in diameter. A boy is playing near the observatory at sunset. He throws a ball upward so that its shadow climbs to the highest point on the dome. How fast is the shadow moving along the dome $\frac{1}{2}$ sec after the ball begins to fall? How did you use the fact that it was sunset in solving the problem? Note: A ball falling from rest covers a distance $s = 16t^2$ ft after t sec.

13. An elevated train on a track 20 ft above the ground crosses a street at a rate of 25 ft/sec and at an angle of 30°. Five seconds later a car crosses under the tracks (Figure 2–11), going 40 ft/sec. How fast are the train and the car separating 3 sec later?

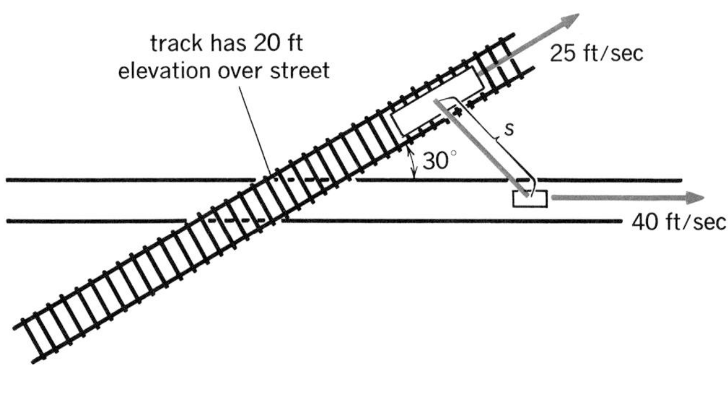

FIGURE 2-11

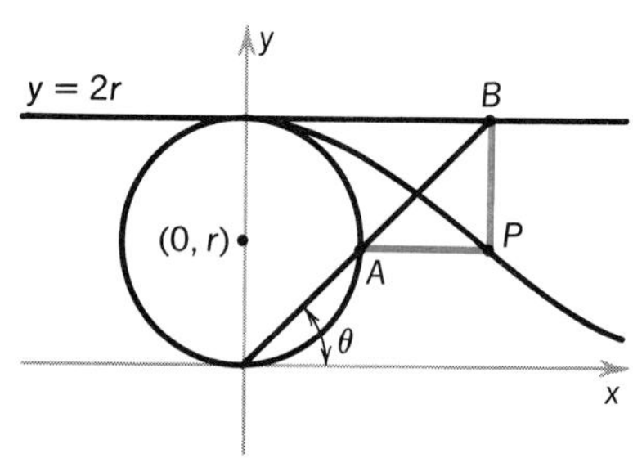

FIGURE 2-12

14. A ladder 15 ft long leans against an 8 ft high wall so the end of the ladder juts over the top of the wall. If the base of the ladder slides away from the wall at the rate of 3 ft/sec, find the rate at which the height of the top end of the ladder is decreasing when 2 feet of the ladder projects over the wall.

15. A clock has a minute hand 5 in long and an hour hand 3 in long. How fast are the hands separating from each other at 10:00?

16. Let the point P be obtained as in Figure 2–12. Find the locus of P as θ varies from 0 to π, and determine dy/dx in terms of θ. This curve is called the **witch of Agnesi.**

*17. A circular disk of radius R rolls without slipping along the positive x axis. (See Figure 2–13.) Describe the locus of a point located on the disk at a distance r from the center.

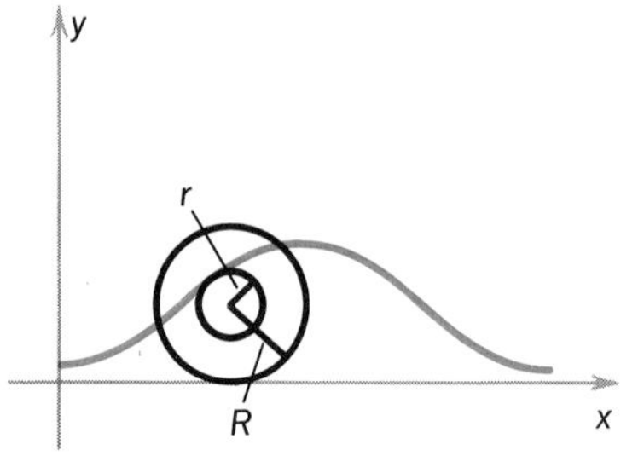

FIGURE 2-13

*18. A circular disk of radius $\frac{1}{4}a$ rolls without slipping inside and along the circumference of the circle $x^2+y^2=a^2$. With the aid of Figure 2–14, show that the locus described by the point on the disk which passes through $(a, 0)$ has equation $x^{2/3}+y^{2/3}=a^{2/3}$. This locus is called the **hypocycloid of four cusps.**

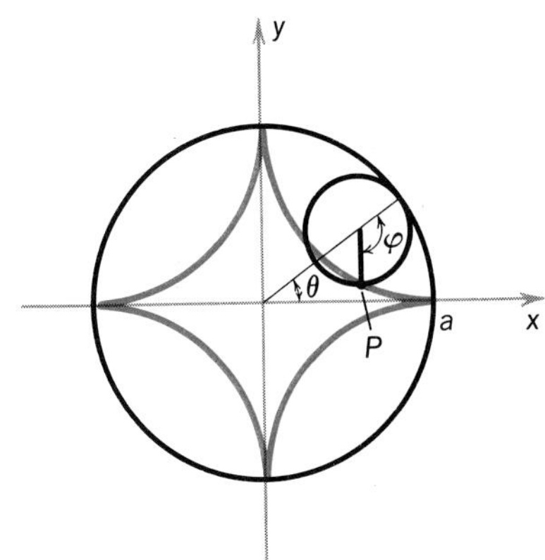

FIGURE 2-14

3 VECTORS IN THE PLANE

In §3.5 we applied the differential calculus to problems of velocity and acceleration. There we restricted ourselves to analyzing the motion of objects along **straight line paths**. In order to analyze motion in the plane or in space, the **calculus of vectors** is required. In this section we introduce the concept of vector and discuss elementary algebraic and geometric properties of vectors. In the section that follows we discuss **vector-valued functions**. There we establish some elementary results of vector calculus which can be applied to the study of the motion of a particle moving in the plane.

A basic distinction exists among the mathematical quantities used to represent mechanical or geometrical variables. A real number may represent a time or temperature or possibly the distance between two points; but a force or velocity cannot be represented by just a single real number because these quantities include a notion of direction together with an idea of magnitude. We speak not only of how strong a force is, but also of the direction in which it is applied; with velocity we need to know the direction of a motion in addition to how fast it is. Both the magnitude and direction of a quantity may be represented by an **arrow**, whose direction gives the direction of the quantity and whose length gives the magnitude. (At this time we shall only be interested in quantities whose representative arrows lie in the plane.) Such an arrow is also called a **directed line segment**.

Geometrically, a **vector** in the plane is a directed line segment originating at the origin of a rectangular coordinate system. Figure 3-1 shows several vectors. Vectors may thus be thought of as arrows beginning at the origin. Each such arrow determines a point in the plane, namely the point where it terminates, its tip. Conversely, given a point P in the plane, we have the vector which terminates at P; this vector is called the **position vector** of P.

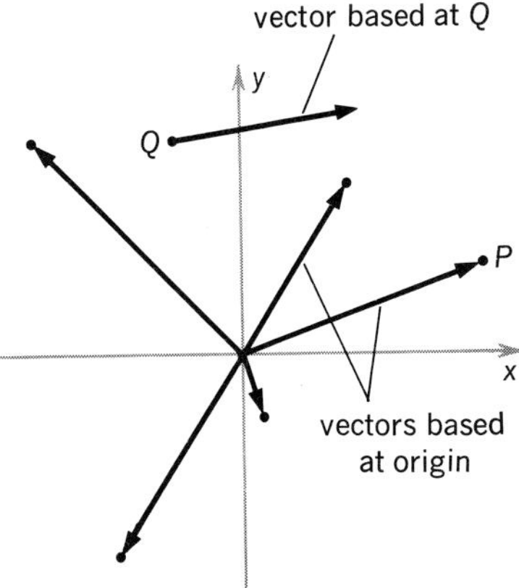

FIGURE 3-1

Occasionally it is useful to think of a vector as beginning at a point Q other than the origin (see Figure 3-1). We then say that the vector is **based** at Q. However, unless otherwise specified, we shall think of vectors as being based at the origin.

Therefore, with every vector we can associate the coordinates of its tip. Conversely, given the coordinates of a point we have the position vector of this point. We thus have two equivalent ways of thinking of vectors (in the plane). One way is to think of arrows of given magnitude, pointing in a specified direction. The other way is to think of ordered pairs of real numbers, which correspond to the coordinates of the tip of a position vector.

It may often be the case that the arrow description furnishes more geometric insight into a problem than the ordered pair description, but on the other hand, the ordered pair description is almost always the easier to work with analytically. We therefore make our formal definition of a vector correspond to this mode of description.

Definition A **vector (in the plane)** is an ordered pair (v_1, v_2) of real numbers, called the first and second **components** of the vector (or x and y components, where appropriate).

Thus geometrically, the components of a vector are just the rectangular coordinates of its tip. The vector $(0, 0)$ is called the **zero vector** and is denoted by $\mathbf{0}$.

The concept of vector is a generalization of the concept of real number in the sense that it is possible to define addition of two vectors and multiplication of a vector by a real number (also called a **scalar**). We now define these arithmetic operations and then proceed to discuss their geometric significance.

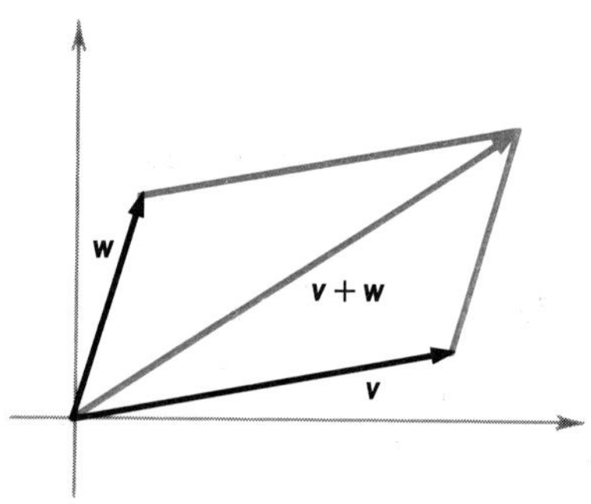

FIGURE 3-2

Definition Given two vectors $\boldsymbol{v} = (v_1, v_2)$ and $\boldsymbol{w} = (w_1, w_2)$, the **sum** of $\boldsymbol{v}$ and $\boldsymbol{w}$ is the vector $\boldsymbol{v} + \boldsymbol{w} = (v_1 + w_1, v_2 + w_2)$; the **difference** is the vector $\boldsymbol{v} - \boldsymbol{w} = (v_1 - w_1, v_2 - w_2)$. Given a vector $\boldsymbol{v}$ and a real number α then the **scalar multiple** of $\boldsymbol{v}$ by α is the vector $\alpha\boldsymbol{v} = (\alpha v_1, \alpha v_2)$.

From the definition of the sum of two vectors, we see that for any vector $\boldsymbol{v}$,

$$\boldsymbol{v} + \mathbf{0} = \boldsymbol{v}$$

The student should have no difficulty verifying that vector addition is **commutative**

$$\boldsymbol{v} + \boldsymbol{w} = \boldsymbol{w} + \boldsymbol{v}$$

and **associative**

$$\boldsymbol{u} + (\boldsymbol{v} + \boldsymbol{w}) = (\boldsymbol{u} + \boldsymbol{v}) + \boldsymbol{w}$$

and that scalar multiplication is **distributive** over vector addition

$$\alpha(\boldsymbol{v} + \boldsymbol{w}) = \alpha\boldsymbol{v} + \alpha\boldsymbol{w}$$

It is also evident that

$$\boldsymbol{v} - \boldsymbol{w} = \boldsymbol{v} + (-1)\boldsymbol{w}$$

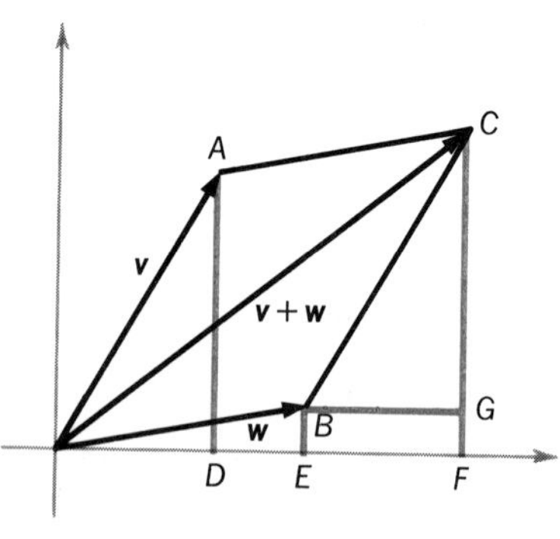

FIGURE 3-3

The vector $(-1)\boldsymbol{w}$ is usually written $-\boldsymbol{w}$.

Geometrically, to find the sum of two vectors $\boldsymbol{v}$ and $\boldsymbol{w}$, we form the parallelogram with adjacent sides $\boldsymbol{v}$ and $\boldsymbol{w}$; the sum $\boldsymbol{v} + \boldsymbol{w}$ is then the directed diagonal of this parallelogram (Figure 3-2). To prove this we set up notation as in Figure 3-3: let $\boldsymbol{v}$ be the vector ending at the point $A = (v_1, v_2)$ and let $\boldsymbol{w}$ be the vector ending at $B = (w_1, w_2)$. We want to show that the vertex C has coordinates $(v_1 + w_1, v_2 + w_2)$.

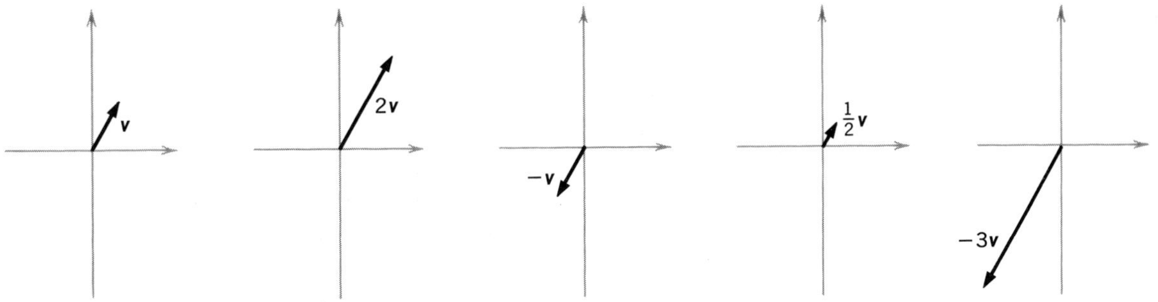

FIGURE 3-4

From the Figure 3-3 the reader can observe that the triangle OAD is congruent to the triangle CBG. Also observe that length $OD = v_1$, length $OE = w_1$. By the congruence relation length $OD =$ length BG, and since $BGFE$ is a rectangle, we have length $OD =$ length EF. But length $OF =$ length $OE +$ length EF, which means that length $OF = v_1 + w_1$. This shows that the x coordinate of C is $v_1 + w_1$. The proof for the y coordinate is analogous.

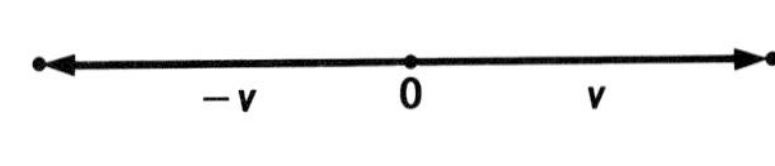

FIGURE 3-5

Scalar multiplication also has a natural geometric interpretation. If α is a scalar and $\boldsymbol{v}$ a vector, then $\alpha\boldsymbol{v}$ is the vector which is $|\alpha|$ times as long as v and having the same direction as $\boldsymbol{v}$ if $\alpha > 0$, but the opposite direction if $\alpha < 0$. Figure 3-4 shows several examples. This interpretation can be justified by an argument using similar triangles.

Suppose we add $\boldsymbol{v}$ and $-\boldsymbol{v}$ by constructing the parallelogram with adjacent sides $\boldsymbol{v}$ and $-\boldsymbol{v}$ (Figure 3-5). The construction yields a degenerate parallelogram whose diagonal has magnitude 0. In general, $\boldsymbol{v} - \boldsymbol{w}$ is the vector obtained by adding $-\boldsymbol{w}$ to $\boldsymbol{v}$; geometrically it corresponds to the arrow drawn from the tip of $\boldsymbol{w}$ to the tip of $\boldsymbol{v}$ (Figure 3-6).

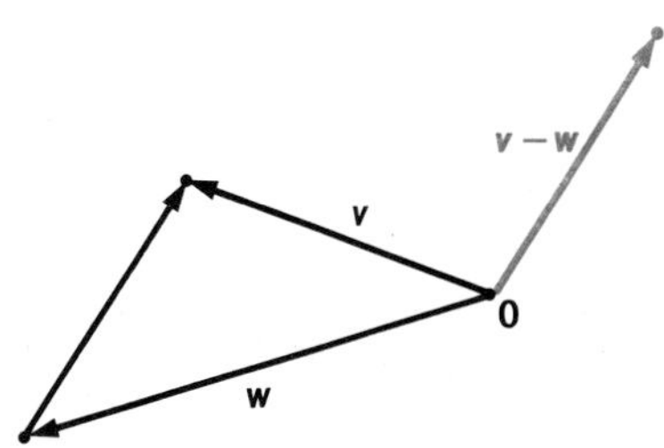

FIGURE 3-6

The magnitude of a vector, geometrically speaking, is its length, that is, the distance from the origin to the tip of the vector. Hence, if $\boldsymbol{v}$ is a vector with components v_1 and v_2, then we denote the **magnitude**, **norm**, or **absolute value** of $\boldsymbol{v}$ by $\|\boldsymbol{v}\|$, defined as

$$\|\boldsymbol{v}\| = \sqrt{v_1^2 + v_2^2}$$

Figure 3-7 illustrates this geometrically, and we see that the definition is based on the Pythagorean theorem.

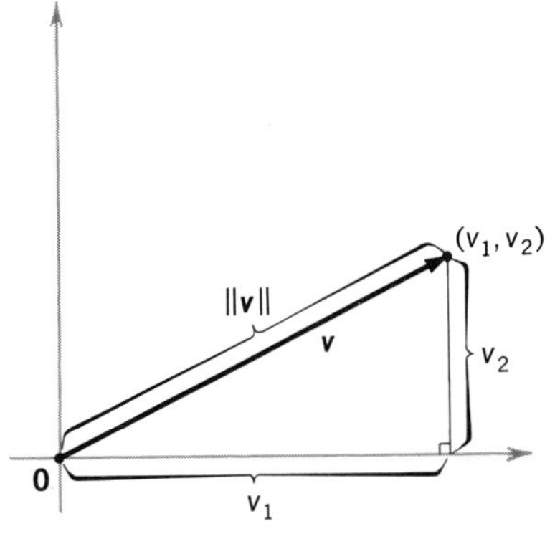

FIGURE 3-7

Theorem 1 The norm function $\boldsymbol{v} \mapsto \|\boldsymbol{v}\|$ sends vectors to real numbers and has the following properties: For all vectors $\boldsymbol{v}$ and $\boldsymbol{w}$, and for any real number λ,

(i) (Triangle inequality) $\|\boldsymbol{v} + \boldsymbol{w}\| \leqslant \|\boldsymbol{v}\| + \|\boldsymbol{w}\|$,

(ii) (Positive definiteness) $\|\boldsymbol{v}\| \geqslant 0$, and $\|\boldsymbol{v}\| = 0$ if and only if $\boldsymbol{v} = \boldsymbol{0}$,

(iii) (Homogeneity) $\|\lambda\boldsymbol{v}\| = |\lambda|\,\|\boldsymbol{v}\|$.

Properties (ii) and (iii) follow easily from the definitions (exercise 2). Property (i) is evident geometrically from Figure 3-2 and the fact that one side of a triangle is always shorter than the sum of the lengths of the other two sides.

The most important method of "multiplying" two vectors $\boldsymbol{v}$ and $\boldsymbol{w}$ yields a scalar as the result. The operation is referred to as the **inner product** (also called the **dot product** or **scalar product**). Suppose $\boldsymbol{v} = (v_1, v_2)$ and $\boldsymbol{w} = (w_1, w_2)$ are vectors (in the plane). Then the **inner product** of $\boldsymbol{v}$ and $\boldsymbol{w}$ is defined as

$$\boldsymbol{v} \cdot \boldsymbol{w} = v_1 w_1 + v_2 w_2$$

Theorem 2 Properties of the Inner Product Let $\boldsymbol{u} = (u_1, u_2)$, $\boldsymbol{v} = (v_1, v_2)$, and $\boldsymbol{w} = (w_1, w_2)$ be vectors, and let λ be a real number. Then

(i) (Commutativity) $\boldsymbol{u} \cdot \boldsymbol{v} = \boldsymbol{v} \cdot \boldsymbol{u}$

(ii) (Distributive Property)

$$\boldsymbol{u} \cdot (\boldsymbol{v} + \boldsymbol{w}) = \boldsymbol{u} \cdot \boldsymbol{v} + \boldsymbol{u} \cdot \boldsymbol{w}, \quad (\boldsymbol{u} + \boldsymbol{v}) \cdot \boldsymbol{w} = \boldsymbol{u} \cdot \boldsymbol{w} + \boldsymbol{v} \cdot \boldsymbol{w}$$

(iii) (Linearity) $\boldsymbol{u} \cdot (\lambda \boldsymbol{v}) = \lambda(\boldsymbol{u} \cdot \boldsymbol{v})$

(iv) (Normality) $\boldsymbol{v} \cdot \boldsymbol{v} = \|\boldsymbol{v}\|^2$

(v) (Law of Cosines) In the arrow representation, let θ $(0 \leqslant \pi)$ be the angle whose sides are $\boldsymbol{u}$ and $\boldsymbol{v}$. Then

$$\boldsymbol{u} \cdot \boldsymbol{v} = \|\boldsymbol{u}\| \, \|\boldsymbol{v}\| \cos \theta \quad \text{(Figure 3-8).}$$

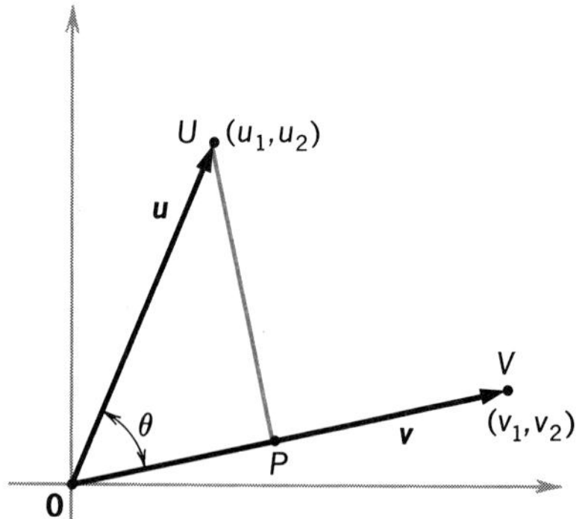

FIGURE 3-8

Proof The proofs of (i)–(iv) are immediate consequences of the definition and are left as exercises. To prove (v), we label points as in Figure 3-8 and proceed as follows. If we drop a perpendicular UP from the tip U of $\boldsymbol{u}$ to the line OV, then it is clear that

$$\cos \theta = \frac{\overline{OP}}{\overline{OU}}$$

Evidently, $\overline{OU} = \|\boldsymbol{u}\|$. To compute $\overline{OP}$ it is necessary to find the coordinates of P; this can be done by noting that P lies at the intersection of lines OV and UP. Now, line OV has equation

$$y = \frac{v_2}{v_1} x \tag{1}$$

Line UP is perpendicular to OV and passes through (u_1, u_2); hence, UP has equation

$$y - u_2 = -\frac{v_1}{v_2}(x - u_1) \tag{2}$$

Solving equations (1) and (2) simultaneously yields

$$x_P = \frac{v_1(u_1 v_1 + u_2 v_2)}{v_1^2 + v_2^2}, \qquad y_P = \frac{v_2(u_1 v_1 + u_2 v_2)}{v_1^2 + v_2^2}$$

as the coordinates of P. (Verify!) Then the length $\overline{OP}$ is

$$\overline{OP} = \sqrt{(x_P-0)^2 + (y_P-0)^2}$$

$$= \sqrt{v_1^2+v_2^2}\,\frac{u_1 v_1 + u_2 v_2}{v_1^2 + v_2^2} \qquad \text{(Verify!)}$$

$$= \frac{u_1 v_1 + u_2 v_2}{\sqrt{v_1^2+v_2^2}}$$

$$= \frac{\boldsymbol{u}\cdot\boldsymbol{v}}{\|\boldsymbol{v}\|}$$

Returning to the original expression for $\cos\theta$, we have

$$\cos\theta = \frac{(\boldsymbol{u}\cdot\boldsymbol{v})/\|\boldsymbol{v}\|}{\|\boldsymbol{u}\|}$$

or

$$\boldsymbol{u}\cdot\boldsymbol{v} = \|\boldsymbol{u}\|\,\|\boldsymbol{v}\|\cos\theta$$

When the vectors $\boldsymbol{v}$ and $\boldsymbol{w}$ are perpendicular, the angle between them is a right angle. Since $\cos\frac{1}{2}\pi = 0$, if $\boldsymbol{v}$ and $\boldsymbol{w}$ are perpendicular, then $\boldsymbol{v}\cdot\boldsymbol{w} = 0$. Conversely, if $\boldsymbol{v}\cdot\boldsymbol{w} = 0$ and if neither $\boldsymbol{v}$ nor $\boldsymbol{w}$ is the $\mathbf{0}$ vector, then $\boldsymbol{v}$ and $\boldsymbol{w}$ are perpendicular. Two vectors are said to be **orthogonal** if their inner product is zero. Hence we have the following corollary.

Corollary 1 Two nonzero vectors are orthogonal if and only if they are perpendicular. The zero vector is orthogonal to every vector.

Geometrically, we define the **projection** of $\boldsymbol{u}$ onto $\boldsymbol{v}$ to be the vector whose tip is at the foot of the perpendicular dropped from the tip of $\boldsymbol{u}$ onto the line in the direction of $\boldsymbol{v}$ (Figure 3-9).

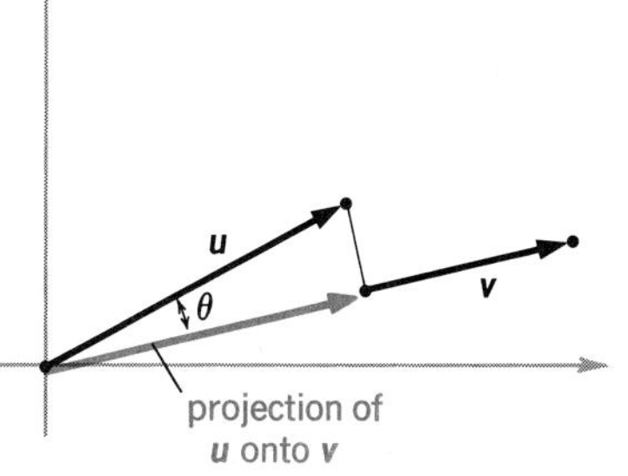

FIGURE 3-9

Algebraically, the projection of $\boldsymbol{u}$ onto $\boldsymbol{v}$ is the vector

$$\frac{\boldsymbol{u}\cdot\boldsymbol{v}}{\|\boldsymbol{v}\|^2}\boldsymbol{v} \tag{3}$$

The justification of equation (3) closely follows the pattern of proof of Theorem 2 (v), so it left as an exercise (5).

A **unit vector** $\boldsymbol{u}$ is a vector with norm 1, that is, $\|\boldsymbol{u}\| = 1$. Unit vectors are useful for specifying directions, since there is exactly one unit vector in any given direction. In particular, if $\boldsymbol{v}$ is any nonzero vector, then the norm of the vector $\boldsymbol{v}/\|\boldsymbol{v}\|$ is 1 (verify!), and this vector points in the same direction as $\boldsymbol{v}$. If a direction is specified by an angle ϕ above the x axis, the unit vector in the direction ϕ is $(\cos\phi, \sin\phi)$. For any vector $\boldsymbol{v}$ and unit vector $\boldsymbol{u}$ we have $\boldsymbol{v}\cdot\boldsymbol{u} = \|\boldsymbol{v}\|\cos\theta$, where θ is the angle between $\boldsymbol{u}$ and $\boldsymbol{v}$; this is exactly the length of the projection of $\boldsymbol{v}$ on the line in the direction of $\boldsymbol{u}$.

If $\boldsymbol{u}$ is a unit vector, then the vector $(\boldsymbol{v}\cdot\boldsymbol{u})\boldsymbol{u}$ is called the **component of $\boldsymbol{v}$ along $\boldsymbol{u}$** or **in the direction of $\boldsymbol{u}$**. By homogeneity of the norm,

$$\|(\boldsymbol{v}\cdot\boldsymbol{u})\boldsymbol{u}\| = |\boldsymbol{v}\cdot\boldsymbol{u}|\,\|\boldsymbol{u}\| = |\boldsymbol{v}\cdot\boldsymbol{u}|$$

that is, the norm of the component of $\boldsymbol{v}$ along $\boldsymbol{u}$ is the length of the projection of $\boldsymbol{v}$ on the line in the direction of $\boldsymbol{u}$.

With respect to a given rectangular coordinate system, any vector $\boldsymbol{v} = (v_1, v_2)$ can be decomposed into the sum of real multiples of the vectors $(1,0)$ and $(0,1)$. The vectors $(1,0)$ and $(0,1)$ are called the **standard basis vectors** of the plane, and they are denoted by the special symbols

$$\boldsymbol{i} = (1,0)$$

$$\boldsymbol{j} = (0,1)$$

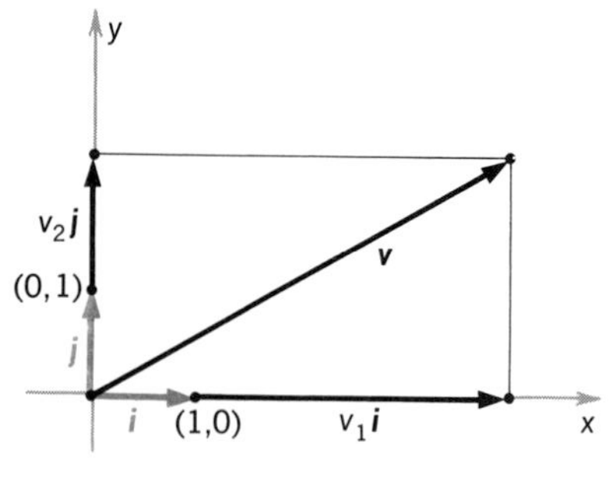

FIGURE 3-10

Clearly $\boldsymbol{i}$ and $\boldsymbol{j}$ are unit vectors (Figure 3-10). This figure also shows how we accomplish the desired decomposition. Hence, for any vector $\boldsymbol{v}$ with components v_1 and v_2 we can write

$$\boldsymbol{v} = v_1\boldsymbol{i} + v_2\boldsymbol{j}$$

Notice also that

$$\boldsymbol{i}\cdot\boldsymbol{j} = (1,0)\cdot(0,1) = (1\cdot 0) + (0\cdot 1) = 0$$

so $\boldsymbol{i}$ and $\boldsymbol{j}$ are orthogonal. Since neither is the zero vector, this means (as is obvious from the figure) that $\boldsymbol{i}$ and $\boldsymbol{j}$ are mutually perpendicular.

EXERCISES

1. For each pair of vectors, find $\boldsymbol{u}+\boldsymbol{v}$, $\boldsymbol{u}-\boldsymbol{v}$, and $\boldsymbol{u}\cdot\boldsymbol{v}$.
   (a) $\boldsymbol{u} = (3,2)$, $\boldsymbol{v} = (1,3)$
   (b) $\boldsymbol{u} = (2,1)$, $\boldsymbol{v} = (-7,4)$
   (c) $\boldsymbol{u} = (1,-1)$, $\boldsymbol{v} = (-7,7)$
   (d) $\boldsymbol{u} = (-4,3)$, $\boldsymbol{v} = (0,5)$
   (e) $\boldsymbol{u} = 5\boldsymbol{i} + 2\boldsymbol{j}$, $\boldsymbol{v} = 3\boldsymbol{i} + 2\boldsymbol{j}$
   (f) $\boldsymbol{u} = 3\boldsymbol{i} + 4\boldsymbol{j}$, $\boldsymbol{v} = 2\boldsymbol{i} - 25\boldsymbol{j}$
   (g) $\boldsymbol{u} = \dfrac{\sqrt{2}}{2}(\boldsymbol{i}+\boldsymbol{j})$, $\boldsymbol{v} = \dfrac{\sqrt{3}\,\boldsymbol{i}+\boldsymbol{j}}{2}$
2. Prove Theorem 1, (ii), (iii).
3. Prove Theorem 2(i)–(iv).
4. In each part, find $\|\boldsymbol{u}\|$ and $c\boldsymbol{u}$; also find the unit vector in the direction of $\boldsymbol{u}$, and a unit vector orthogonal to $\boldsymbol{u}$.
   (a) $\boldsymbol{u} = (-3,-2)$, $c = 3$
   (b) $\boldsymbol{u} = (1,1)$, $c = -7$
   (c) $\boldsymbol{u} = (-1,7)$, $c = \sqrt{5}$
   (d) $\boldsymbol{u} = (\sqrt{2}/2, -\sqrt{2}/2)$, $c = \pi$
   (e) $\boldsymbol{u} = 5\boldsymbol{i} + 4\boldsymbol{j}$, $c = 2$
   (f) $\boldsymbol{u} = 5\boldsymbol{i} + 12\boldsymbol{j}$, $c = 3$
   (g) $\boldsymbol{u} = 25\boldsymbol{i} - 4\boldsymbol{j}$, $c = -\frac{1}{2}$

5. Verify equation (3).
6. Prove Theorem 2(v) using the Law of Cosines.
7. Does the following inequality hold?
$$|\boldsymbol{v}\cdot\boldsymbol{w}| \leqslant \|\boldsymbol{v}\|\,\|\boldsymbol{w}\|$$
Justify your answer.
8. Thinking of a vector as a **force** applied to an object at a point, give an interpretation of vector addition and scalar multiplication.

4 VECTOR-VALUED FUNCTIONS

A function whose values are vectors is called a **vector-valued function**. We have already seen vector-valued functions in the form of **parametrized curves**. Recall that a parametrized curve is described by a pair of functions

$$x = X(t), \qquad y = Y(t) \tag{1}$$

where variable t is called the parameter. A parametrized curve assigns to t the point $P = (X(t), Y(t))$ in the plane. Now if $X(t)$ and $Y(t)$ are regarded as the components of the position vector of the point P (Figure 4-1), then $t \mapsto (X(t), Y(t))$ is a vector-valued function. If we let bold-faced letters $\boldsymbol{f}, \boldsymbol{g}, \boldsymbol{h}, \ldots$ denote vector-valued functions, then the equation

$$\boldsymbol{v} = \boldsymbol{f}(t) \tag{2}$$

describes a function $\boldsymbol{f}$ which sends real numbers to vectors. Equation (1) and equation (2) both represent exactly the same situation. Thus a vector-valued function $\boldsymbol{f}$ can always be decomposed into a pair of real valued functions. If $\boldsymbol{f}$ decomposes into f_1 and f_2, which give, respectively, the x

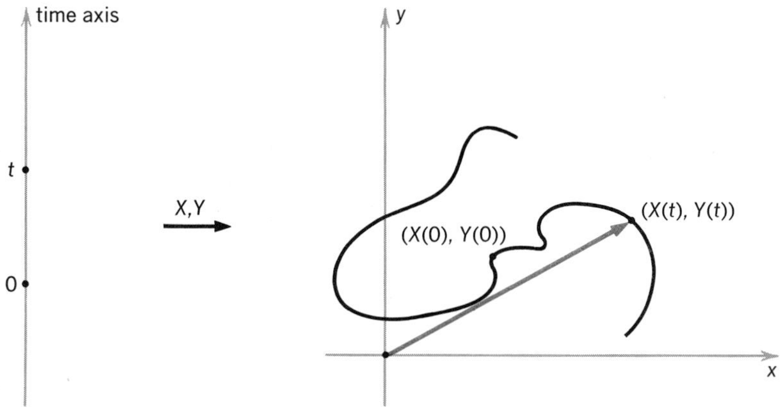

FIGURE 4-1

and y coordinates of $\boldsymbol{f}(t)$, it is convenient to write $\boldsymbol{f}=(f_1,f_2)$. We say that f_1 and f_2 are the first and second, or x and y, **components of the vector-valued function** $\boldsymbol{f}$.

Example 1 The vector-valued function

$$\boldsymbol{f}: t \mapsto (t-\sin t,\, 1-\cos t)$$

is the position function of a point on a circle rolling along the x axis. (See §4.2, Example 9.) The circle rolls along the x axis at a constant speed; that is, the center of the circle is moving to the right along the line $y=1$ at a constant speed of 1 distance unit per unit of time. The motion of the point $\boldsymbol{f}(t)$ is more complex and is analyzed in Example 3 of this section.

We have seen that vector functions also arise in purely mathematical situations. Hence, the function $t \mapsto (t,t^2)$ could simply describe the parametrization $x=t$, $y=t^2$ of the curve $y=x^2$, or it could describe the motion of a point moving along the parabola.

For vector-valued functions, **limits** are defined in terms of the component functions. We say that $\boldsymbol{f}(t)$ approaches $\boldsymbol{v}=(v_1,v_2)$ as t approaches t_0 if and only if $f_1(t)$ approaches v_1 and $f_2(t)$ approaches v_2 as t approaches t_0. Hence, we have

$$\lim_{t\to t_0} \boldsymbol{f}(t) = \boldsymbol{v} \Leftrightarrow \left(\lim_{t\to t_0} f_1(t) = v_1 \quad \text{and} \quad \lim_{t\to t_0} f_2(t) = v_2\right)$$

Example 2 If $\boldsymbol{f}$ is the function of Example 1 we have

$$\lim_{t\to 0} \boldsymbol{f}(t) = \left(\lim_{t\to 0} f_1(t), \lim_{t\to 0} f_2(t)\right) = (0,0)$$

and more generally

$$\lim_{t\to t_0} \boldsymbol{f}(t) = \left(\lim_{t\to t_0} f_1(t), \lim_{t\to t_0} f_2(t)\right) = (t_0-\sin t_0, 1-\cos t_0)$$

We say that $\boldsymbol{f}$ is **continuous** at t_0 if and only if both f_1 and f_2 are continuous at t_0, or, in vector notation, if and only if

$$\lim_{t\to t_0} \boldsymbol{f}(t) = \boldsymbol{f}(t_0)$$

A vector-valued function is said to be **differentiable** (at t_0) if its components are differentiable (at t_0). Then the **derivative** of a vector-valued function $\boldsymbol{f}=(f_1,f_2)$ is the vector-valued function $\boldsymbol{f}'=(f_1',f_2')$.

If we use the notations dx/dt and dy/dt for $f_1'(t)$ and $f_2'(t)$, then we write

$$\frac{d(\boldsymbol{f}(t))}{dt} = \left(\frac{dx}{dt}, \frac{dy}{dt}\right)$$

If we situate the $dx\,dy$ coordinate system at the point $\boldsymbol{f}(t_0)$ on the curve described by $t \mapsto \boldsymbol{f}(t)$, then the tangent line has equation (§2, equation (3))

$$f_1'(t_0)\,dy = f_2'(t_0)\,dx$$

Hence in $dx\,dy$ coordinates the vector $(f_1'(t_0), f_2'(t_0))$ is an arrow which is tangent to the curve at $\boldsymbol{f}(t_0)$ [see Figure 4-2]. Moreover, this arrow will point in the direction of motion along the curve.

We are now prepared to apply vector calculus to problems of velocity and acceleration. First we define the notion of **average velocity over an interval of time**. Then the instantaneous velocity vector will be defined by means of a limiting process, just as we did in §2.5.

Let the position of a particle be given by $t \mapsto \boldsymbol{f}(t)$. If $\Delta t > 0$, then the **average velocity** over the interval $[t, t+\Delta t]$ is

$$\boldsymbol{v}_{\text{av}} = \frac{\boldsymbol{f}(t+\Delta t) - \boldsymbol{f}(t)}{\Delta t}$$

$$= \frac{\text{change in position}}{\text{change in time}}$$

The physical significance of $\boldsymbol{v}_{\text{av}}$ is clear: A particle starting at $\boldsymbol{f}(t)$, moving in the direction of $\boldsymbol{v}_{\text{av}}$ with speed equal to the magnitude of $\boldsymbol{v}_{\text{av}}$ will reach the point $\boldsymbol{f}(t+\Delta t)$ after Δt units of time (see Figure 4-3).

Therefore we define the **instantaneous velocity** at t to be the vector

$$\boldsymbol{v} = \boldsymbol{v}(t) = \lim_{\Delta t \to 0} \frac{\boldsymbol{f}(t+\Delta t) - \boldsymbol{f}(t)}{\Delta t} = \boldsymbol{f}'(t)$$

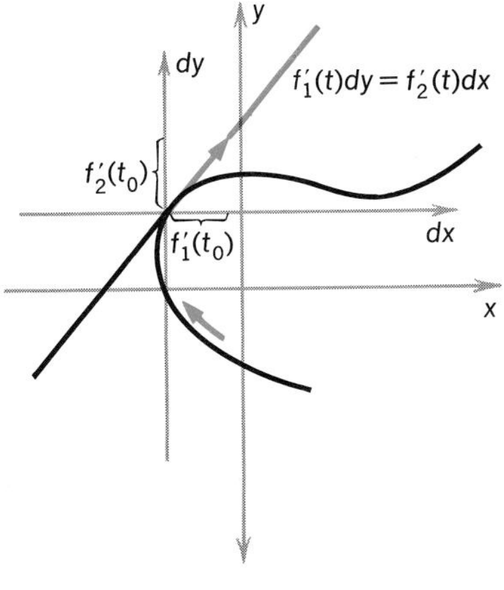

FIGURE 4-2

The **instantaneous speed** is by definition the magnitude of the velocity vector. Thus

$$\text{instantaneous speed} = \|\boldsymbol{v}(t)\| = \|\boldsymbol{f}'(t)\|$$

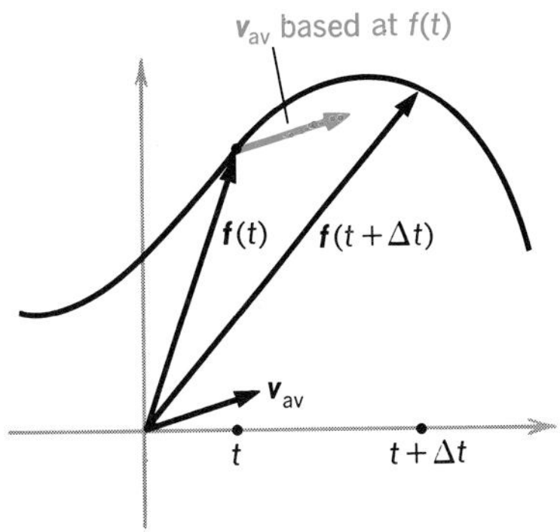

FIGURE 4-3

Example 3 If as in Example 1, $\boldsymbol{f}: t \mapsto (t - \sin t, 1 - \cos t)$ is the position function, the velocity vector is then

$$\boldsymbol{v}(t) = (1 - \cos t, \sin t)$$

The speed is

$$\|\boldsymbol{v}(t)\| = 2 - 2\cos t$$

Therefore, the point with position vector $\boldsymbol{f}(t)$ moves at variable speed. At time $t = 0$, the moving point is at rest; it attains its maximum speed when $\cos t = -1$; that is, at the times $t = (2n+1)\pi$, where n is an integer. In general the point slows down to speed zero as it approaches each of the points corresponding to $t = 2n\pi$.

We remark here that since $\boldsymbol{v}(t) = \boldsymbol{f}'(t)$, the velocity vector when translated so as to be based at the point $\boldsymbol{f}(t)$ is tangent to the curve (see Figure 4-2). Therefore the velocity vector is also called a **tangent vector**.

Example 4 A clock has a minute hand 5 in long and an hour hand 3 in long. At what speed are the tips of the hands moving apart at 10 o'clock?

When t is measured in minutes after 12 o'clock, the position of the tip of the minute hand is given by

$$\boldsymbol{m}:t \mapsto \left[5\cos\left(\frac{\pi}{2}-\frac{2\pi t}{60}\right), 5\sin\left(\frac{\pi}{2}-\frac{2\pi t}{60}\right)\right]$$

that is

$$\boldsymbol{m}:t \mapsto \left(5\sin\frac{\pi t}{30}, 5\cos\frac{\pi t}{30}\right)$$

Similarly, the position of the tip of the hour hand is given by

$$\boldsymbol{h}:t \mapsto \left(3\sin\frac{\pi t}{360}, 3\cos\frac{\pi t}{360}\right)$$

Then the velocities are

$$\boldsymbol{m}'(t) = \left(\frac{\pi}{6}\cos\frac{\pi t}{30}, -\frac{\pi}{6}\sin\frac{\pi t}{30}\right)$$

$$\boldsymbol{h}'(t) = \left(\frac{\pi}{120}\cos\frac{\pi t}{360}, -\frac{\pi}{120}\sin\frac{\pi t}{360}\right)$$

At $t = -120$ (10 o'clock) we therefore have

$$\boldsymbol{m}'(-120) = \left(\frac{\pi}{6}\cos(-4\pi), -\frac{\pi}{6}\sin(-4\pi)\right) = \left(\frac{\pi}{6}, 0\right)$$

$$\boldsymbol{h}'(-120) = \left[\frac{\pi}{120}\cos\left(-\frac{\pi}{3}\right), -\frac{\pi}{120}\sin\left(-\frac{\pi}{3}\right)\right] = \left(\frac{\pi}{240}, \frac{\pi\sqrt{3}}{240}\right)$$

The vector which gives the separation of the hands is $\boldsymbol{m}-\boldsymbol{h}$ and so the velocity of separation at $t = -120$ is therefore

$$\boldsymbol{v} = \boldsymbol{m}'(-120) - \boldsymbol{h}'(-120) = \left(\frac{\pi}{6}-\frac{\pi}{240}, -\frac{\pi\sqrt{3}}{240}\right)$$

$$= \left(\frac{39\pi}{240}, -\frac{\pi\sqrt{3}}{240}\right)$$

and the speed of separation is

$$\|\boldsymbol{v}\| = \sqrt{\left(\frac{39\pi}{240}\right)^2 + \left(\frac{-\pi\sqrt{3}}{240}\right)^2} = \frac{\pi}{240}\sqrt{39^2+3} = \frac{\pi}{240}\sqrt{1524}$$

$$= \frac{\pi}{120}\sqrt{381} \approx \frac{\pi}{6} \text{ in/min}$$

At the start of §3, we observed that it is not always possible to assign a single real number to represent certain variables. This is especially true of physical quantities, so let us now turn to some further examples. We have already noted that the position of a point P in the plane has associated with it the position vector of P, and this vector has components equal to the x and y coordinates of P.

The velocity of a point P moving in the plane is a vector whose components are the rates of change of the x and y coordinates of P with respect to time. The norm of the velocity vector is the speed of the point. The velocity vector of a point P is normally depicted as an appropriate arrow with its tail at P, directed along the line of motion (Figure 4-4).

By definition, the **momentum** vector $\boldsymbol{p}$ of a point P with mass m moving in the plane is $m\boldsymbol{v}_P$ where $\boldsymbol{v}_P$ is the velocity vector of P. The **acceleration** vector $\boldsymbol{a}$ of a point P moving in the plane has as components the rates of change with respect to time of the components of $\boldsymbol{v}_P$. The **force** vector $\boldsymbol{F}$ of a point P with mass m moving in the plane is the vector $m\boldsymbol{a}$, the scalar multiple of m with the acceleration vector. Thus the acceleration vector is

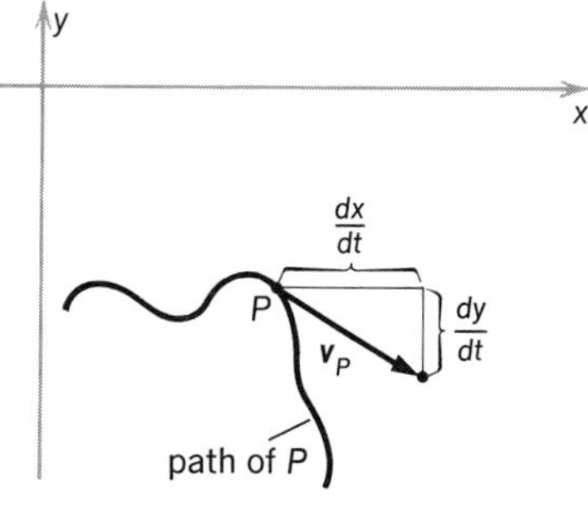

FIGURE 4-4

$$\boldsymbol{a} = \left(\frac{d^2x}{dt^2}, \frac{d^2y}{dt^2}\right)$$

and the force vector is

$$\boldsymbol{F} = \left(m\frac{d^2x}{dt^2}, m\frac{d^2y}{dt^2}\right)$$

In physics one usually specifies what $\boldsymbol{F}$ must be according to some physical law. For example, near the surface of the earth a particle of mass m experiences a force equal (approximately) in magnitude $32m$ (m measured in pounds) and directed downwards. Rewriting the above formula in terms of components, we get

$$F_1 = m\frac{d^2x}{dt^2}, \; F_2 = m\frac{d^2y}{dt^2}$$

which is **Newton's law of motion**, or **Newton's second law.**

Example 5 The vector-valued function

$$t \mapsto (10t, 5t - 16t^2)$$

is the position function of a ball thrown with initial velocity $(10, 5)$, measured in ft/sec. The velocity function is

$$t \mapsto (10, 5 - 32t)$$

and the acceleration function is $t \mapsto (0, -32)$. We also write $\boldsymbol{v}(t) = (10, 5 - 32t)$, $\boldsymbol{a}(t) = (0, -32)$. If the ball weighs one-half pound, the momentum function is $\boldsymbol{p}(t) = (5, \frac{5}{2} - 16t)$ ($\boldsymbol{p}$ measured in ft-lbs/sec) and the ft-lbs/sec force function is $\boldsymbol{F}(t) = (0, -16)$ (measured in ft-lbs/sec^2). The **speed** of the ball is a real valued (**not** vector-valued) function of time:

$$t \mapsto \|\boldsymbol{v}(t)\| = \sqrt{10^2 + (5 - 32t)^2}$$

that is,

$$t \mapsto \sqrt{1024t^2 - 320t + 125}$$

EXERCISES

1. For each part, find the components of the functions, and evaluate the function at the given points. Sketch the curves described.
   (a) $f:t \mapsto (8t, 96-16t^2)$; $t = 5, \sqrt{3}$
   (b) $g:\theta \mapsto (r\cos\theta, r\sin\theta)$; $\theta = \frac{1}{4}\pi, \frac{4}{3}\pi$
   (c) $h:p \mapsto (p-\sin p, 1-\cos p)$; $p = 0, -\frac{3}{4}\pi$
   (d) $f:t \mapsto (4\cos t+\cos 4t, 4\sin t-\sin 4t)$; $t = -\frac{1}{8}\pi, \frac{1}{8}\pi$
   (e) $g:t \mapsto (t, t^{3/5})$; $t = -1, -32$
2. Compute the derivatives of each of the functions of exercise 1, and evaluate these derivatives at the points given.
3. A particle moves counterclockwise around the ellipse $9x^2+y^2=81$, with a constant speed of 5 ft/sec. Find the velocity vector at the point $(2\sqrt{2}, 3)$.
4. A particle moves around the circle $x^2+y^2=25$ in a clockwise direction with a constant speed of 25 ft/sec. Find the velocity and acceleration vectors at the point $(4, 3)$.
5. If f and g are differentiable vector-valued functions, prove that
   (a) $(f+g)'(t) = f'(t)+g'(t)$
   (b) $(cf)'(t) = cf'(t)$
   (c) $(f\cdot g)'(t) = f(t)\cdot g'(t)+f'(t)\cdot g(t)$
6. Complete the following table ($m=$ mass, $\boldsymbol{v}=$ velocity, $s=$ speed, $\boldsymbol{a}=$ acceleration, $\boldsymbol{p}=$ momentum, $\boldsymbol{F}=$ force).

	m	$\boldsymbol{v}$	s	$\boldsymbol{a}$	$\boldsymbol{p}$	$\boldsymbol{F}$
(a)	3	(7, 1)	?	?	?	(18, 0)
(b)	2	?	?	?	$(4\cos\theta, 4\sin\theta)$	$(-4\sin\theta, 4\cos\theta)$
(c)	?	?	25	$\mathbf{0}$	$(-\frac{7}{8}, 3)$	?
(d)	?	?	?	(4, 5)	(−4, 6)	(16, 20)

5 ONE-SIDED AND INFINITE LIMITS

For further applications of the differential calculus we must broaden our use of the limit concept and study some of the deeper properties of continuous functions. In this section we make several straightforward adaptations of the techniques of Chapter 2 to define various types of limits. With these definitions we define the important concept of "function continuous on a closed interval." Then in the section that follows, we discuss the properties of such continuous functions.

The limits we have studied thus far have been "two sided" in the following sense: When we say $\lim_{x\to x_0} f(x) = l$, we mean that for every neighborhood N of l, there is a neighborhood M of x_0 such that

$$x \in M(x_0) \Rightarrow f(x) \in N$$

The deleted neighborhood $M(x_0)$ always contains points both to the right and to the left of x_0. Thus $\lim_{x\to x_0} f(x) = l$ implies that $f(x)$ is close to l for values of x near x_0 on either side of x_0.

Often it is necessary to consider situations where x approaches x_0, but remains on one side of x_0. By way of example, look at the function $g: x \mapsto \sqrt{x}$. The domain of g is the interval $[0, \infty[$. Now if x is positive and close to 0, $g(x) = \sqrt{x}$ will also be close to 0. We therefore can say that "as x remains positive and approaches 0, $g(x)$ also approaches 0." It is very simple to give a precise meaning to the expression in quotes and we make the following definition:

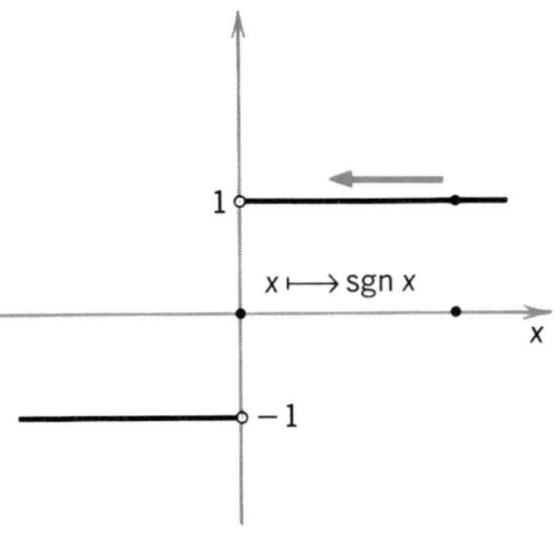

FIGURE 5-1

Definition Given a function f, we say that l is **the (one-sided) limit of f as x approaches x_0 from above** (respectively, **below**) if for every neighborhood N of l there is an interval $]x_0, b[$ (respectively, $]a, x_0[$) such that $f(x) \in N$ for all $x \in]x_0, b[$ (respectively, $]a, x_0[$); in this case we write $\lim_{x\to x_0^+} f(x) = l$ (respectively, $\lim_{x\to x_0^-} f(x) = l$). In this definition it is understood that f is defined on the appropriate interval.

Example 1 Consider the function $f: x \mapsto \operatorname{sgn}(x)$ as x approaches 0 from above. (See Figure 5-1.) If N is any neighborhood of 1, $\operatorname{sgn}(x) = 1 \in N$ for all $x > 0$. Thus, $\lim_{x\to x_0^+} \operatorname{sgn}(x) = +1$. Now let us look at the limit as x approaches 0 from below. Clearly, if N is any neighborhood of -1, then $\operatorname{sgn}(x) = 1 \in N$ for all $x < 0$. Thus, $\lim_{x\to x_0^-} \operatorname{sgn}(x) = -1$. Hence, we see that as we approach 0 through different points, two different limits are obtained. This result is expected, since we have already seen that this function is not continuous at 0.

Example 2 If $\lim_{x\to x_0} f(x) = l$, then $\lim_{x\to x_0^+} f(x) = l$. This result follows at once from the definitions.

One-sided limits can be used to define the important notion of a function which is continuous on a closed interval.

Definition Suppose the function f is defined for every x in the closed interval $[A, B]$. Then f is **continuous on $[A, B]$** if

(i) f is continuous at every point of the open interval $]A, B[$; and
(ii) $\lim\limits_{x\to A^+} f(x) = f(A)$ and $\lim\limits_{x\to B^-} f(x) = f(B)$.

Example 3 If f is continuous on all of $\boldsymbol{R}$, then f is continuous on every closed interval $[A, B]$. This should be clear from Example 2.

Example 4 If f is the function

$$f: x \mapsto \begin{cases} x^2 + 1, & x > 0 \\ -x^2, & x \leqslant 0 \end{cases}$$

(see Figure 5-2) then f is continuous on $[-1,0]$ but is not continuous on $[0,1]$. Moreover, f is continuous on $[\frac{1}{2},1]$ but not continuous on $[-1,1]$. In fact, f is continuous on all closed intervals $[A,B]$ with $B \leqslant 0$ and all closed intervals $[a,b]$ with $0 < a$.

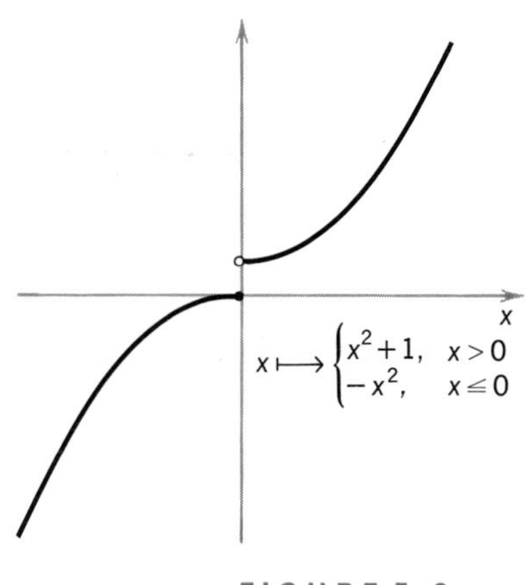

FIGURE 5-2

If $\lim_{x\to x_0} f(x)$ exists, then we know that both one sided limits $\lim_{x\to x_0^+} f(x)$ and $\lim_{x\to x_0^-} f(x)$ exist, and all three limits are equal. The converse of this is also true.

Theorem 1 If $\lim_{x\to x_0^+} f(x)$ and $\lim_{x\to x_0^-} f(x)$ exist and are equal, then $\lim_{x\to x_0} f(x)$ exists and equals the one-sided limits.

Proof Let $l = \lim_{x\to x_0^+} f(x) = \lim_{x\to x_0^-} f(x)$, and suppose N is a neighborhood of l. By our hypothesis, there are points $a < x_0$ such that $f(x) \in N$ for all $x \in]a, x_0[$ and $b > x_0$ such that $f(x) \in N$ for all $x \in]x_0, b[$. Now let $M =]a,b[$; hence, M is a neighborhood of x_0 and for all $x \neq x_0$ in M, we have $f(x) \in N$, by the choice of a and b. Therefore, $f(x)$ is eventually in N as x approaches x_0. ∎

Example 5 Consider the function

$$f: x \mapsto \begin{cases} \sin x, & x < 0 \\ \cos x - 1, & x \geqslant 0 \end{cases}$$

Then, f is continuous at 0, because $\lim_{x\to 0^+} f(x) = \lim_{x\to 0^-} f(x) = 0$, and hence, by Theorem 1, $\lim_{x\to 0} f(x) = 0 = f(0)$.

The ideas we have just discussed can be combined into a useful technique for dealing with limits.

FORMAL DEVICE

If $\lim_{x\to x_0^+} f(x) \neq \lim_{x\to x_0^-} f(x)$, $\lim_{x\to x_0} f(x)$ does not exist.

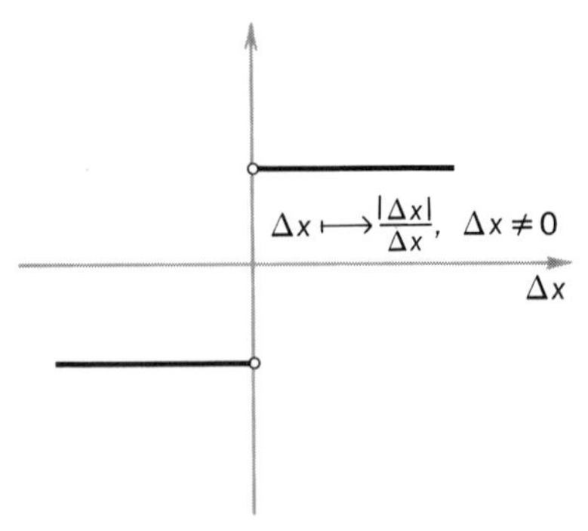

FIGURE 5-3

Example 6 Consider the function

$$\Delta x \mapsto \frac{|\Delta x|}{\Delta x}, \quad \Delta x \neq 0$$

(see Figure 5-3). Then

$$\lim_{\Delta x\to 0^+} \frac{|\Delta x|}{\Delta x} = 1 \quad \text{and} \quad \lim_{\Delta x\to 0^-} \frac{|\Delta x|}{\Delta x} = \lim_{\Delta x\to 0^-} -\frac{\Delta x}{\Delta x} = -1$$

Therefore, $\lim_{\Delta x\to 0} |\Delta x|/\Delta x$ does not exist.

Example 7 Let k be the function

$$x \mapsto \begin{cases} x, & x \leqslant 0 \\ 1-x, & 0 < x < 1 \\ x, & 1 \leqslant x \end{cases}$$

which is shown in Figure 5-4 (see §1.3, Example 10). We have $\lim_{x\to 0^-} k(x) = 0$, $\lim_{x\to 0^+} k(x) = 1$, $\lim_{x\to 1^-} k(x) = 0$, and $\lim_{x\to 1^+} k(x) = 1$. Therefore, $\lim_{x\to 0} k(x)$ and $\lim_{x\to 1} k(x)$ do not exist.

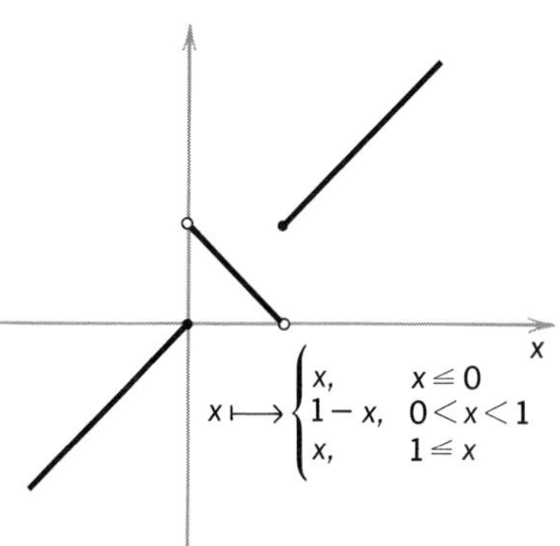

FIGURE 5-4

If both one-sided limits exist as x approaches x_0, but $\lim_{x\to x_0^-} f(x) \neq \lim_{x\to x_0^+} f(x)$, then f is said to have a **jump discontinuity** at x_0. (Figure 5-5.) Thus sgn (x) has a jump discontinuity at 0, and it is continuous everywhere else. The function k of Example 6 has jump discontinuities at 0 and 1, and is continuous everywhere else.

Two further types of limits are needed to complete our list:

(1) Limits where $|x|$ increases without bound.

(2) Limits where $|f(x)|$ increases without bound.

These limits are known as **improper** or **infinite limits**.

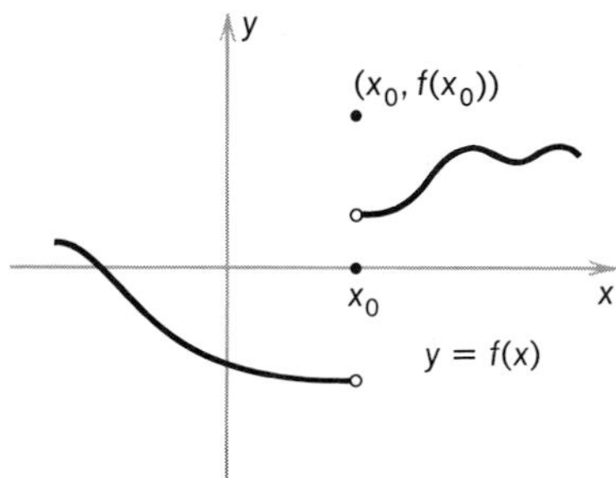

FIGURE 5-5

An interval of the form $]a, +\infty[$ is called a **neighborhood of** $+\infty$. We say

$$\lim_{x\to+\infty} f(x) = l$$

if for every neighborhood N of l, there is a neighborhood $]a, +\infty[$ such that $f(x) \in N$ for all $x \in]a, +\infty[$.

Example 8 Let $f: x \mapsto 1/x$, and let us show that $\lim_{x\to+\infty} f(x) = 0$ (see Figure 5-6). Suppose $]A, B[$ is a neighborhood of 0. Then $B > 0$ and so $1/x < B$ whenever $x > 1/B$. So for all x in the neighborhood $]1/B, +\infty[$, $f(x) \in]A, B[$.

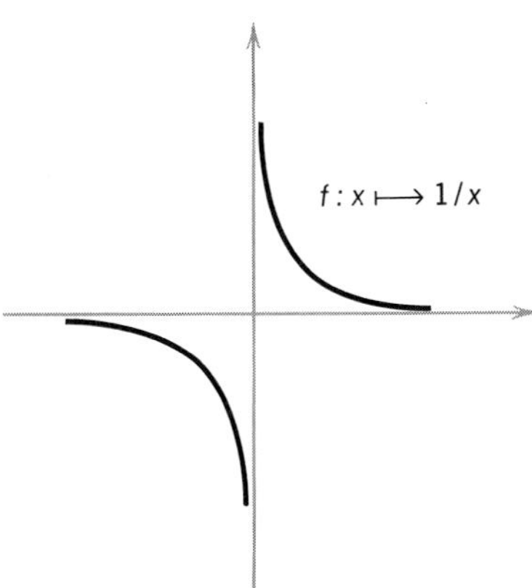

FIGURE 5-6

Example 9 Let

$$g: t \mapsto \frac{t^2+1}{t^2-1}$$

(See Figure 5-7.) From the graph it is clear that $\lim_{x\to+\infty} g(t) = 1$. Let us verify this algebraically. Suppose $]A, B[$ is a neighborhood of 1. We may divide t^2+1 by t^2-1 to obtain $g(t) = 1 + 2/(t^2-1)$. For $t > 1$, $g(t) > 1$, so we need to choose a sufficiently large so that $(2/(t^2-1)) < B-1$ whenever $t > a$. Now, $(2/(t^2-1)) < B-1$ is equivalent to $t^2 > 2/(B-1)+1$. Since $B-1 > 0$, we only need to choose $a = \sqrt{2/(B-1)+1}$; then for all $t \in]a, +\infty[$, we get $A < 1 < g(t) < B$, and so we have shown that $\lim_{t\to\infty} g(t) = 1$.

The graph of $g:t \mapsto (t^2+1)/(t^2-1)$ shows that "as t approaches $-\infty$, $g(t)$ also approaches 1." We can easily formalize the expression in quotes by defining a **neighborhood of** $-\infty$ to be an interval of the form $]-\infty, a[$. Then, given a function f and a real number l, we say that

$$\lim_{x \to -\infty} f(x) = l$$

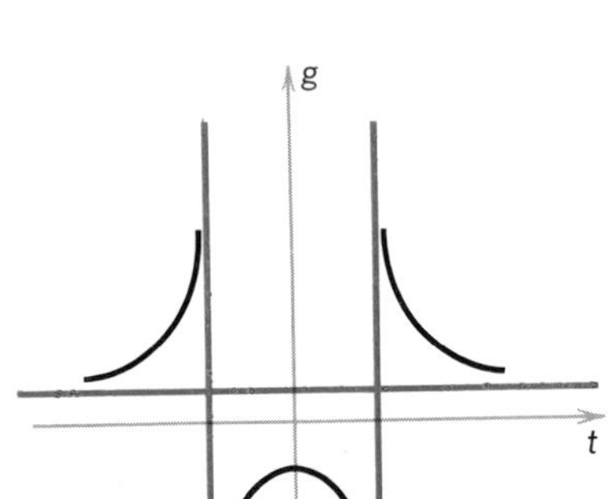

FIGURE 5-7

if for every neighborhood N of l, there is a neighborhood $]-\infty, a[$ such that $f(x) \in N$ whenever $x \in]-\infty, a[$. Let us now show that $\lim_{t \to -\infty} g(t) = 1$. Suppose $]A, B[$ is a neighborhood of 1. For all $t < -1$, $g(t) > 1 > A$. Therefore, we want to find $a < -1$ such that $g(t) < B$ whenever $t < a$. However, $1+2/(t^2-1) < B$ is the same as $t^2 > 2/(B-1)+1$. By hypothesis, $B-1 > 0$, so we can choose $a = -\sqrt{2/(B-1)+1}$.

If $\lim_{x \to +\infty} f(x)$ and $\lim_{x \to +\infty} g(x)$ exist (and, therefore, are finite real numbers), the arithmetic of limits applies and so

$$\lim_{x \to +\infty} (f+g)(x) = \lim_{x \to +\infty} f(x) + \lim_{x \to +\infty} g(x)$$

and

$$\lim_{x \to +\infty} (fg)(x) = \lim_{x \to +\infty} f(x) \cdot \lim_{x \to +\infty} g(x)$$

Similar formulas apply with ∞ replaced by $-\infty$.

We now consider a type of limit for which the theorems of the arithmetic of limits do not hold. The graph of $g:t \mapsto (t^2+1)/(t^2-1)$ shows that as t approaches 1 from above, $g(t)$ becomes positively infinite and as t approaches 1 from below, $g(t)$ becomes negatively infinite. These intuitive notions can be made precise as follows.

Definition Given a function f and a point x, we say

$$\lim_{x \to x^+} f(x) = +\infty$$

if for every neighborhood N of $+\infty$, there is an interval $]x_0, b[$ such that $f(x) \in N$ for all $x \in]x_0, b[$. We say $\lim_{x \to x_0^-} f(x) = +\infty$, if for every neighborhood N of $+\infty$, there is an interval $]a, x_0[$ such that $f(x) \in N$ for all $x \in]a, x_0[$.

If $\lim_{x \to x_0^+} f(x) = +\infty$ and if $\lim_{x \to x_0^-} f(x) = +\infty$, we say

$$\lim_{x \to x_0} f(x) = +\infty$$

We define $\lim_{x \to x_0^+} f(x) = -\infty$, $\lim_{x \to x_0^-} f(x) = -\infty$, and $\lim_{x \to x_0} f(x) = -\infty$ in a perfectly analogous manner. Also, the limits $\lim_{x \to +\infty} f(x) = +\infty\,(-\infty)$ and $\lim_{x \to -\infty} f(x) = +\infty\,(-\infty)$ can be defined in a straightforward fashion. (The reader should write out the details of these definitions as a useful exercise.)

Example 10

(a) $\lim_{t\to 1^+} g(t) = +\infty$ and $\lim_{t\to 1^-} g(t) = -\infty$, where $g(t) = (t^2+1)/(t^2-1)$

(b) $\lim_{x\to 0^+} 1/x = +\infty$ and $\lim_{x\to 0^-} 1/x = -\infty$

(c) $\lim_{x\to\infty} x^3 = +\infty$ and $\lim_{x\to -\infty} x^3 = -\infty$

(d) $\lim_{x\to\infty} \sin x$ is not defined, $\lim_{x\to\infty} x\sin x$ is also not defined

The arithmetic of limits cannot apply to this last type of limit because $+\infty$ and $-\infty$ are not real numbers. Hence, we cannot add or multiply them either with each other or with real numbers. The following examples list some of the possibilities for $\lim_{x\to x_0}[f(x)+g(x)]$ and $\lim_{x\to x_0} f(x)g(x)$ when $\lim_{x\to x_0} f(x) = +\infty\,(-\infty)$ and/or $\lim_{x\to x_0} g(x) = +\infty\,(-\infty)$. Further possibilities are explored in the exercises.

Example 11

(a) $\lim_{x\to 0} \dfrac{1}{x^2} = +\infty;\ \lim_{x\to 0} \dfrac{-1}{x^2} = -\infty;$

$$\text{however, } \lim_{x\to 0}\left(\frac{1}{x^2} + \frac{-1}{x^2}\right) = \lim_{x\to 0} 0 = 0.$$

(b) $\lim_{x\to 0} x^2 = 0;\ \lim_{x\to 0} \dfrac{1}{x^2} = +\infty;$

$$\text{however, } \lim_{x\to 0}\left(x^2 \cdot \frac{1}{x^2}\right) = \lim_{x\to 0} 1 = 1.$$

(c) $\lim_{x\to 0^+} \sin^2 x = 0;\ \lim_{x\to 0^+} \dfrac{1}{x} = +\infty;$

$$\text{however, } \lim_{x\to 0^+} \frac{\sin^2 x}{x} = \lim_{x\to 0^+}\left(\frac{\sin x}{x}\cdot \sin x\right) = 1\cdot 0 = 0.$$

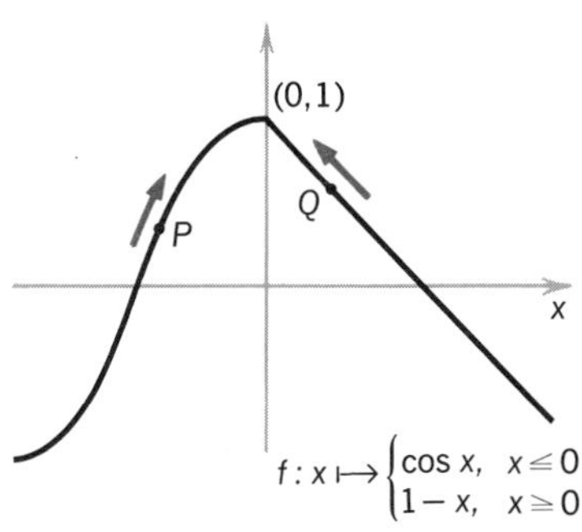

FIGURE 5-8

Now that we have discussed one-sided limits and continuity, as well as how the concept of infinity can be included in our formal apparatus, it is possible to apply these ideas to the derivative. In Figure 5-8 we have the graph of the function

$$f: x \mapsto \begin{cases} \cos x, & x \leqslant 0 \\ 1-x, & x \geqslant 0 \end{cases}$$

Consider a point P on the curve and to the left of the y axis. As P moves along the curve toward $(0, 1)$, the slope of the secant through P and $(0, 1)$ approaches $0 = \cos'(0)$. On the other hand, if Q approaches $(0, 1)$ along the right-hand part of the curve, the secant slope is constantly equal to -1.

We thus have

$$\lim_{\Delta x \to 0^-} \frac{f(0+\Delta x) - f(0)}{\Delta x} = 0$$

and

$$\lim_{\Delta x \to 0^+} \frac{f(0+\Delta x) - f(0)}{\Delta x} = -1$$

Therefore, f is **not** differentiable at 0. However, we say that f has a right-hand derivative, namely -1, at 0 and a left-hand derivative, namely 0, at 0. More generally, we have the following definition.

Definition If $\lim_{\Delta x \to 0^+} (f(x+\Delta x) - f(x))/\Delta x$ exists, the value of the limit is the **right-hand derivative** of f at x. If $\lim_{\Delta x \to 0^-} (f(x+\Delta x) - f(x))/\Delta x$ exists, this value is the **left-hand derivative** of f at x.

Clearly, a function f is differentiable at x if and only if both one-sided derivatives exist and are equal.

We also state the following definition.

Definition A function f is said to be **differentiable on the closed interval $[A, B]$** if f is differentiable on $]A, B[$ and if f has a right-hand derivative at A and a left-hand derivative at B.

Example 12

(a) The function $f: x \mapsto \sqrt{x}$ is *not* differentiable on $[0, 1]$, since it does not have a right-hand derivative at 0.

(b) The function

$$F(t) = \begin{cases} +1, & t \geqslant 0 \\ -1, & t < 0 \end{cases}$$

is differentiable on $[0, 1]$ but not on $[-1, 0]$, because the left-hand derivative does not exist at 0.

(c) The function $f: x \mapsto x^{3/2}$ is differentiable on $[0, 1]$.

(d) The absolute value function $t \mapsto |t|$ is differentiable on $[0, 1]$ and differentiable on $[-1, 0]$, but it is not differentiable on $[-1, 1]$. In fact, this function is not differentiable on any interval containing zero as an interior point; that is, not an endpoint.

If f is differentiable on $[A, B]$, then the following function has domain $[A, B]$:

$$x \mapsto \begin{cases} \lim_{\Delta x \to 0^+} \dfrac{f(A+\Delta x) - f(A)}{\Delta x}, & x = A \\ f'(x), & A < x < B \\ \lim_{\Delta x \to 0^-} \dfrac{f(B+\Delta x) - f(B)}{\Delta x}, & x = B \end{cases}$$

If this function is continuous on $[A, B]$, we say that f is **continuously differentiable on $[A, B]$**. Thus, for example, $x \mapsto |x|$ is continuously differentiable on $[0, 1]$.

The notion of **infinite derivative** corresponds to that of vertical tangent. For example, the function $g: x \mapsto x^{1/3}$ has a vertical tangent at 0 (Figure 5-9). As the figure indicates, we have

$$\lim_{\Delta x \to 0} \frac{g(0+\Delta x) - g(0)}{\Delta x} = \infty$$

The function $h: x \mapsto -(x^{1/3})$ also has a vertical tangent at 0 (Figure 5-10), but in this case we have

$$\lim_{\Delta x \to 0} \frac{h(0+\Delta x) - h(0)}{\Delta x} = -\infty$$

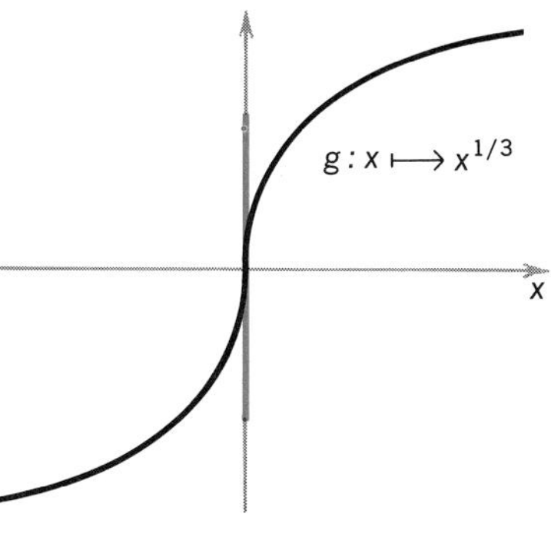

FIGURE 5-9

Definition If f is continuous at x_0 and if

$$\lim_{\Delta x \to 0} \frac{f(x_0+\Delta x) - f(x_0)}{\Delta x} = +\infty \text{ (or } -\infty)$$

f is said to have an **infinite derivative** at x_0.

The functions g and h above both have infinite derivative at 0. We require that f be continuous near x_0 in order not to assign an infinite derivative at 0 to a function such as

$$x \mapsto \begin{cases} 1/x^2, & x \neq 0 \\ 0, & x = 0 \end{cases}$$

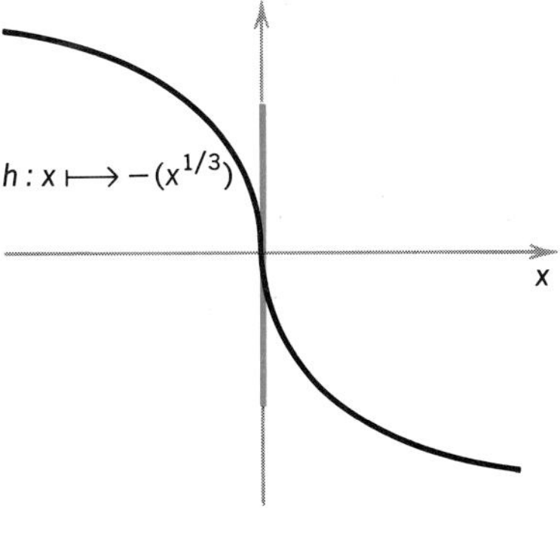

FIGURE 5-10

Moreover, this definition does not assign an infinite derivative at 0 to $x \mapsto |x^{1/3}|$. (Why?) (See Figure 5-11.)

We leave it to the reader to give a definition of infinite one-sided derivatives which will handle functions such as $x \mapsto |x^{1/3}|$.

EXERCISES

1. Evaluate the following limits.

(a) $\displaystyle\lim_{x \to \infty} \frac{\sqrt{x^2+1}}{x+1}$ (b) $\displaystyle\lim_{x \to \infty} \frac{3x^2 - 2x + 4}{x^2 - 2x - 3}$

(c) $\displaystyle\lim_{x \to -\infty} \frac{\sqrt{x^2+1}}{x+1}$ (d) $\displaystyle\lim_{x \to 1^-} \tan x$

(e) $\displaystyle\lim_{x \to 1^+} \tan x$ (f) $\displaystyle\lim_{x \to \infty} (x - \sqrt{x^2 + a^2})$

(g) $\displaystyle\lim_{x \to 0^+} (x\sqrt{1 + 1/x^2})$ (h) $\displaystyle\lim_{x \to 2^+} \frac{\sqrt{x^2 - 4}}{x - 2}$

(i) $\displaystyle\lim_{x \to 2^-} \frac{\sqrt{9 - x^2}}{\sqrt{10 - 7x + x^2}}$ (j) $\displaystyle\lim_{t \to \frac{1}{2}^+} \cot 2t$

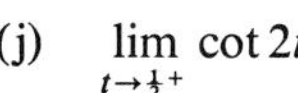

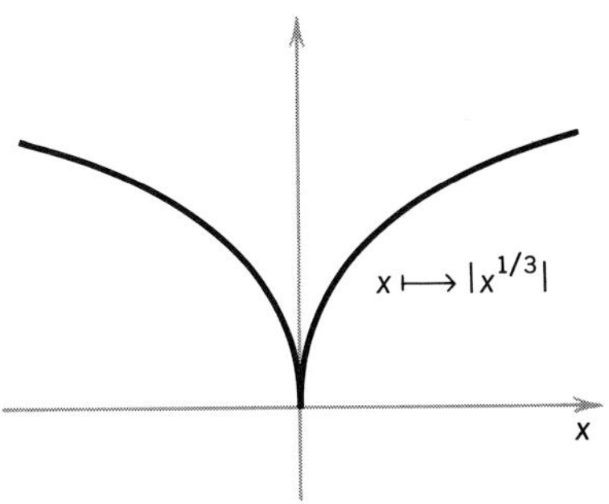

FIGURE 5-11

2. Suppose $\lim_{x\to x_0} f(x) = +\infty$ and $\lim_{x\to x_0} g(x) = l \neq 0$. Show that $\lim_{x\to x_0} [f(x)+g(x)] = +\infty$ and that
$$\lim_{x\to x_0} f(x)g(x) = \begin{cases} +\infty, & l > 0 \\ -\infty, & l < 0 \end{cases}$$

3. Suppose $\lim_{x\to x_0} f(x) = +\infty$ and $\lim_{x\to x_0} h(x) = -\infty$. Show that $\lim_{x\to x_0} [f(x)h(x)] = -\infty$.

4. Give an example of a function G satisfying $\lim_{x\to 0}(G(x))/x = 0$ and $\lim_{x\to 0}(G(x))/x^2 = +\infty$.

5. Compute the right-hand derivative of the following functions at the points indicated.
   (a) $x \mapsto |x|$ at 2
   (b) $x \mapsto |\tan x|$ at 0
   (c) $x \mapsto (\sin x)^{3/2}$ at $x = 2\pi$
   (d) $x \mapsto (\cos x)^{5/2}$ at $x = -\frac{1}{2}\pi$

6. Determine whether the following functions have infinite derivatives at the points indicated.
   (a) $x \mapsto \begin{cases} \sqrt{x}, & x \geq 0 \\ -\sqrt{|x|}, & x < 0 \end{cases}$ at 0
   (b) $x \mapsto \sqrt{|x|}$ at 0
   (c) $x \mapsto x^{1/5}$ at 0
   (d) $x \mapsto (\sin x)^{1/3}$ at 0
   (e) $x \mapsto x^{1/n}$, n an odd natural number, at 0
   (f) $x \mapsto |x|^{1/n}$, n an even natural number, at 0

7. Let f be an odd function which has both left- and right-handed derivatives at zero. Prove that these one-sided derivatives must be equal and conclude that f is differentiable at zero.

8. Suppose f has a (finite) right-hand derivative at x_0. Show
$$\lim_{x\to x_0^+} f(x) = f(x_0).$$

9. Do exercise 21 of §2.1 for the curve $y = x^3$. Show that as P approaches the origin from the right D and E approach $+\infty$; as P approaches the origin from the left D and E approach $-\infty$.

*10. Let f satisfy $f(0) = 0 = f'(0) = f''(0)$. Do exercise 21 of §2.1 for the curve $y = f(x)$. Show that as P approaches the origin from the right, both D and E become infinite (that is, approach either $+\infty$ or $-\infty$). Show the same thing happens as P approaches the origin from the left.

11. Suppose f has an infinite derivative at x_0. Let $C(\Delta x)$ denote the length of the chord from $(x_0, f(x_0))$ to $(x_0+\Delta x, f(x_0+\Delta x))$. Show that $\lim_{\Delta x\to 0^+}(C(\Delta x))/\Delta x = +\infty$. Interpret geometrically.

12. Let f and g be two continuous functions on $[A, B]$.
   (a) Suppose $f(x) = g(x)$ for all $x \in]A, B[$; show that $f(A) = g(A)$ and $f(B) = g(B)$.
   (b) Suppose f is monotone (increasing or decreasing) on $]A, B[$; show that f is monotone (increasing or decreasing) on $[A, B]$.

13. Denote the greatest integer less than or equal to x by $[x]$. (Thus, $[\frac{1}{2}] = 0$, $[3\frac{1}{4}] = [\pi] = [3] = 3$, $[-7\frac{1}{4}] = -8$.) Show that $x \mapsto [x]$ has a jump discontinuity at every integer. Determine the points of continuity and discontinuity of $x \mapsto x - [x]$.

6 THE EXTREME VALUE THEOREM

In this section we discuss two very important theorems about the values of a continuous function, the Intermediate Value Theorem and the Extreme Value Theorem. Together, these theorems imply that if a non-constant function with domain a closed interval is continuous on its domain, then its range is also a closed interval.

To begin this section we discuss an important property of the real numbers. By constructing a real number line we identify the set of real numbers with the set of points on the line. A straight line has many geometric properties; since the real numbers are representable by a straight line, we can expect the real numbers to have properties which correspond to certain geometric properties of the line.

A straight line has no "holes"; it is solid and continuous with no breaks. Is there a property of the real numbers which corresponds to this continuity? The answer is yes. However, in order to state this property of the real numbers we need to make some preliminary definitions.

Definition Suppose X is a set of real numbers. Then the real number a is an **upper bound** for X if $x \leqslant a$ for all numbers $x \in X$.

Example 1 The number 1 is an upper bound for the interval $[0, 1]$. In fact, all numbers in the infinite interval $[1, \infty[$ are upper bounds for $[0, 1]$. On the other hand, no element of the open interval $]-\infty, 1[$ is an upper bound for $[0, 1]$.

It follows that if a set has an upper bound, then it has infinitely many upper bounds.

Example 2 The set N of natural numbers has no upper bound.

Since upper bound is defined with the $\leqslant$ sign, it is possible for an upper bound of a set to be an element of the set. In Example 1 above, the number 1 is an element of $[0, 1]$ and it is an upper bound for $[0, 1]$. However, it can also happen that no upper bound for a given set is an element of that set. An open interval $]a, b[$ is an example of such a set.

Definition Suppose a is an upper bound for the set X. We say that a is the **least upper bound** for X if every upper bound for X is greater than or equal to a.

Example 3 The number 1 is the least upper bound for $[0, 1]$. It is also the least upper bound for $[0, 1[$.

Example 4 Let X be the set of rational numbers whose squares are less than 3. [That is, X consists of all rational numbers p/q such that $(p/q)^2 < 3$.] The real number $\sqrt{3}$ is the least upper bound for X.

Example 5 If a is an upper bound for X and if $a \in X$, then a is the least upper bound for X.

It is clear that a set has at most one least upper bound. If a is the least upper bound for X, we write

$$a = \text{l.u.b.}\, X$$

We now state the **Completeness Property for the Real Numbers**: Every set of real numbers which has an upper bound has a least upper bound.

The completeness property is an **assumption** we make about the real numbers. For example, one can show that it amounts to the same thing as requiring every infinite decimal fraction to be represented as a real number (see EMS 1AB, §8.1 for this approach). Intuitively we can see that the completeness property suggests that there are no "breaks" in the continuity of the real numbers.

Example 6 As in Example 4, let X be the set of rational numbers whose squares are less than 3. The least upper bound of this set is $\sqrt{3}$. As is well known, $\sqrt{3}$ is **not** a rational number. Therefore, the rational numbers do not have the Completeness Property.

We know that if a function f is continuous on a closed interval $[A, B]$, then the graph of f between A and B does not have any "gaps," "breaks," or "jumps." The graph of f connects the points $(A, f(A))$ and $(B, f(B))$ with an unbroken path. As we go from A to B, the values of $f(x)$ go from $f(A)$ to $f(B)$. The Intermediate Value Theorem says that all numbers y between $f(A)$ and $f(B)$ are of the form $f(x)$ for some x between A and B.

Theorem 1 **Intermediate Value Theorem** Suppose f is continuous on $[A, B]$. If either $f(A) \leqslant y_0 \leqslant f(B)$ or $f(B) \leqslant y_0 \leqslant f(A)$, then for some c in the interval $[A, B]$, we have $f(c) = y_0$.

Proof For definiteness, suppose that $f(A) < f(B)$, and let $y_0 \in]f(A), f(B)[$. Consider the set S of points in $[A, B]$ which are sent by f to values less than or equal to y_0. Clearly $A \in S$ and $B \notin S$. The set S is bounded from above by B; therefore there is a number which is the least upper bound of S. Let us denote this l.u.b. by c; we shall show that $f(c) = y_0$.

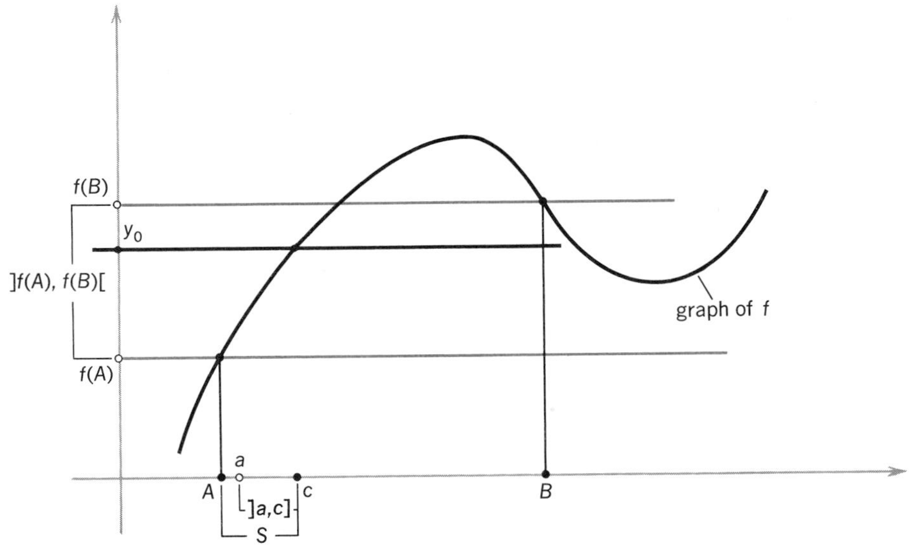

FIGURE 6-1

Now since c is the l.u.b. of S, every interval $]a, c[$ to the left of c contains elements x such that $f(x) \leqslant y_0$ (Figure 6-1). Hence,

$$\lim_{x \to c^-} f(x) \leqslant y_0$$

On the other hand, $f(x) \geqslant y_0$ for all x in $]c, B]$ by the selection of c; hence

$$\lim_{x \to c^+} f(x) \geqslant y_0$$

But since f is continuous on $[A, B]$, and in particular at c, we have

$$\lim_{x \to c^-} f(x) = \lim_{x \to c^+} f(x) = \lim_{x \to c} f(x) = f(c)$$

by Theorem 1, §5. Therefore, the two above inequalities yield

$$y_0 \leqslant \lim_{x \to c} f(x) \leqslant y_0$$

so

$$\lim_{x \to c} f(x) = y_0$$

The case where $f(A) > f(B)$ is left to the student. ∎

All polynomial functions are continuous, and so the Intermediate Value Theorem applies to them. For example, the function $f: x \mapsto x^3 - 3x^2 + 3x - 1$ takes the values $f(2) = 1$ and $f(0) = -1$. Thus, since $f(0) < 0 < f(2)$, the Intermediate Value Theorem indicates that there is a root of the polynomial between 0 and 2; that is, for some c such that $0 < c < 2$, we have $f(c) = 0$. In fact, $f(1) = 0$, since $f: x \mapsto (x-1)^3$.

Consider the function $g: x \mapsto x^2$ on the interval $[-1, 1]$; $g(-1) = 1 = g(1)$, and for all other $x \in [-1, 1]$, we have $0 \leqslant g(x) < 1$. Therefore, 1 is the largest value that g assigns to elements of $[-1, 1]$. We say that $1 = g(1)$ is the **maximum** of g on $[-1, 1]$. Similarly, $g(2) = 4$ is the largest value that g assigns to elements of $[-1, 2]$, and so $g(2)$ is the maximum value of g on $[-1, 2]$. On the other hand, $g(0) = 0$ is the smallest value that g assigns to elements of this interval, and we say that $g(0)$ is the **minimum** of g on $[-1, 2]$. Also, $g(2) = 4$ is the minimum of g on $[2, 3]$.

Definition Let f be a function defined on a closed interval $[A, B]$. Then $y_0 = f(x_0)$ is the **maximum** of f on $[A, B]$ if $x_0 \in [A, B]$, and $y_0 \geqslant f(x)$ for all $x \in [A, B]$. Similarly, $y_1 = f(x_1)$ is the **minimum** of f on $[A, B]$ if $x_1 \in [A, B]$, and $y_1 \leqslant f(x)$ for all $x \in [A, B]$. (See Figure 6-2.)

A maximum (or minimum) of a function is also called an **extreme value** of the function.

Example 7 The maximum of the sine function on $[-\frac{1}{2}\pi, \frac{1}{2}\pi]$ is $\sin(\frac{1}{2}\pi) = 1$; the minimum of the function on that interval is $\sin(-\frac{1}{2}\pi) = -1$. The maximum of the sine on $[-\frac{1}{2}\pi, 0]$ is $\sin 0 = 0$, and the minimum is again $\sin(-\frac{1}{2}\pi) = -1$.

The maximum and minimum values of a function do not always exist. The function

$$h: x \mapsto \begin{cases} 1/x, & x \neq 0 \\ 0, & x = 0 \end{cases}$$

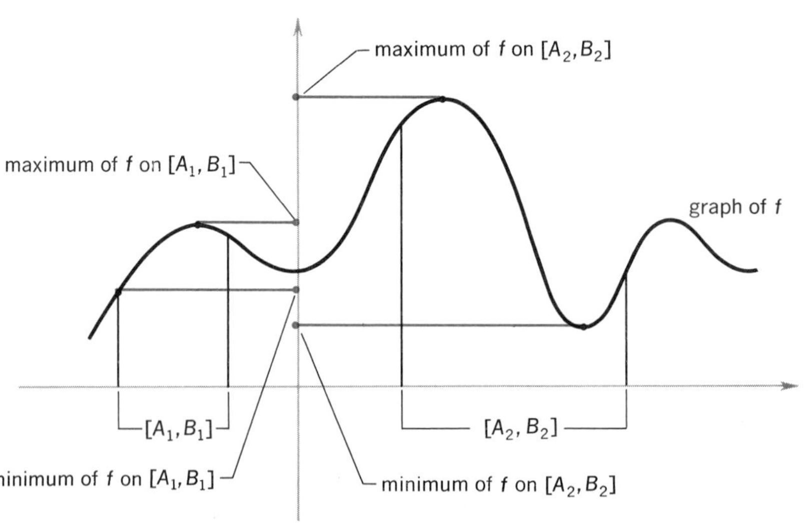

FIGURE 6-2

has no maximum or minimum on $[-1, 1]$ (Figure 6-3). However, observe that h is not continuous at 0. This is the crucial point, because when a function is defined and continuous on a closed interval, then at some point of the interval the function attains its maximum, and, likewise at some point of the interval it attains its minimum.

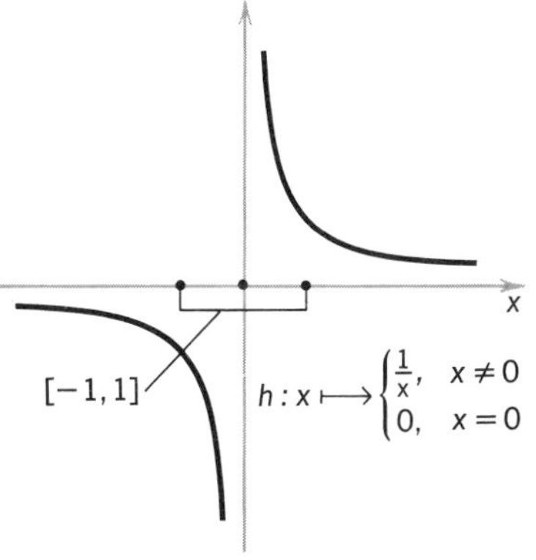

FIGURE 6-3

Theorem 2 Extreme Value Theorem If f is defined and continuous on $[A, B]$, then the maximum of f on $[A, B]$ and the minimum of f on $[A, B]$ are both attained. That is, there are points $x_0 \in [A, B]$ and $x_1 \in [A, B]$ such that for all $x \in [A, B]$

$$f(x_1) \leqslant f(x) \leqslant f(x_0)$$

***Proof** The proof of this theorem is long, and it makes extensive use of the Completeness Property of the real numbers. We say that m is a **bound** for f on the closed interval J if $f(x) \leqslant m$ holds for all points $x \in J$. To prove the theorem we first show that f has a bound on $[A, B]$; that done, we prove that the maximum value of f is the smallest of the possible bounds for f on $[A, B]$. The case of the minimum of f may be deduced by applying this result to the function $-f$, whose maximum is the minimum of f.

We claim that there are $x_0 \in]A, B]$ such that f is bounded on $[A, x_0]$. This is not hard to verify: $f(A)$ is a real number and we have $\lim_{x \to A^+} f(x) = f(A)$. Therefore there exists x_0 such that $f(x) < f(A) + 1$ for all $x \in [A, x_0]$.

Now let x_0 be any point in $[A, B]$. We shall prove that f is bounded on the interval $[A, x_0]$ (and hence on $[A, B]$). We use an argument by contradiction: We suppose that x exist such that f is not bounded on $[A, x]$ and we take a to be the greatest lower bound (see exercise 7) of all such x. Clearly $a > A$, since $a \geqslant x_0 > A$. From the definition of the point a it follows that for $A \leqslant x < a$, f is bounded on $[A, x]$; hence for $x < a$ the function f must not be bounded on the interval $]x, a]$. This contradicts the fact that $\lim_{x \to a^-} f(x) = f(a)$.

We define M to be the least upper bound of the set of values $f(x)$ where $A \leqslant x \leqslant B$. The number M exists, since f is bounded on $[A, B]$. Since M is the least upper bound of bounds for f on $[A, B]$ we must have $M \geqslant f(x)$ for all $x \in [A, B]$. We must now find a point c in $[A, B]$ satisfying $f(c) = M$. If $f(A) = M$, we are done. So we consider the case where $f(A) < M$. For $B \geqslant x > A$ we define $m(x)$ to be the least upper bound of the values of f on $[A, x]$. (Thus, $m(B) = M$.) Supposing $f(A) < M$, there exist intervals $[A, x_0]$ such that

$$x \in [A, x_0] \Rightarrow m(x) < M$$

We take c to be the least upper bound of the x such that $m(x) < M$. The point $c \in [A, B]$ exists since $m(b) = M$. By the above remark, we know that $A < c$; in fact, we have $m(x) < M$ for all x in $[A, c[$. Finally since f is continuous on $[A, B]$, it follows that $m(c) = M$. Next we shall show that $\lim_{x \to c^-} f(x)$ must equal M. Since $f(c) = \lim_{x \to c^-} f(x)$, this will complete the proof. Let I be a neighborhood of the number M. From the above properties of c, we see that there are points in $[A, c[$ satisfying $f(x) \in I$. But we can say more: For every $x_0 < c$, there is a point $x \in]x_0, c]$ satisfying $f(x) \in I$. This can be established as follows: Let I' be a neighborhood of M such that $m(x) \notin I'$ and take $I'' = I \cap I'$. I'' is a neighborhood of M and we have $f(x) \notin I''$ for all x in $[A, x_0]$. Since I'' is a neighborhood of M, there must be points x in $[A, c]$ such that $f(x) \in I''$. Such x can only be in $]x_0, c]$. Therefore, we conclude that $]x_0, c]$ contains points x such that $f(x) \in I$. Since I is any neighborhood of M, it follows that $f(c) = \lim_{x \to c^-} f(x) = M$. ∎

Example 8 (a) The function $x \mapsto 1/x$ has a maximum and a minimum value on every closed interval not containing zero. This function is continuous on $]0, 1]$ but has no maximum value on this interval; however, this interval is not a closed interval; it is only half-closed.

(b) The function $t \mapsto \sin(\cos t)$ has a maximum value and a minimum value on $[0, 2]$. The Extreme Value Theorem implies the **existence** of the values. However, it does not provide a method for finding where they occur. In the section that follows we will develop a method for locating the extreme values of a function.

The Extreme Value Theorem and the Intermediate Value Theorem give us the following theorem.

Theorem 3 Suppose f has domain $[A, B]$, is continuous on $[A, B]$, and nonconstant. Then the range of f is a closed interval.

Proof Let $M = f(c_1)$ be the maximum value of f and let $m = f(c_2)$ be the minimum value of f. Clearly the range of f is a subset of the interval $[m, M]$. From the Intermediate Value Theorem, it follows that $[m, M]$ is equal to the range of f. ∎

EXERCISES

1. Find the maximum and minimum of each function on the specified interval.
   (a) $x \mapsto -2x + 1, \quad [-2, 1]$ (b) $x \mapsto x^2 + 4x + 4, \quad [-4, 4]$
   (c) $x \mapsto 1/x, \quad [-6, -2]$ (d) $x \mapsto \sqrt{1 + x}, \quad [-1, 2]$
   (e) $t \mapsto \sin t, \quad [0, 2\pi]$ (f) $t \mapsto \tan t, \quad [-1, 1]$
2. In the text of this section the Intermediate Value Theorem is proved for the case $f(B) > f(A)$. Using this result, show the theorem holds for the case $f(B) < f(A)$, by considering $g = -f$.
3. Let h be monotone (increasing or decreasing) and continuous on $]-1, 1[$. Suppose $h(0) = 0$ and $\lim_{x \to 0} f(h(x)) = f(0)$. Show that f is continuous at 0. [Hint: Use Theorem 3.]

4. Let g be continuous on $[0,1]$ and suppose $0 \leqslant g(t) \leqslant 1$ for all $t \in [0,1]$. Prove that there is a point $t_0 \in [0,1]$ such that $g(t_0) = t_0$.
5. Prove
   (a) $\lim_{x \to +\infty} (x^3 + Bx^2 + Cx + D) = +\infty$
   (b) $\lim_{x \to +\infty} (x^3 + Bx^2 + Cx + D) = -\infty$
   (c) Conclude that the cubic polynomial $p(x) = x^3 + Bx^2 + Cx + D$ has a real root.
   (d) Conclude that the polynomial $g(x) = Ax^3 + Bx^2 + Cx + D$ $(A \neq 0)$ has a real root.
6. Prove that a continuous periodic function has a minimum and a maximum value.
7. (a) Give three upper bounds, including the least upper bound, for each of the following sets: the set whose elements are 1, 2, 5, and 8; $[0, \pi[$; $[a, b]$; $]a, b[$; $]-\infty, q[$; the set of all numbers of the form $n/(n+1)$ where n is a natural number; the range of $f: x \mapsto 0$.
   (b) A number x_1 is said to be a lower bound for a set X if x_1 is less than or equal to every element of X. The number y_1 is the greatest lower bound of X, abbreviated $y_1 = \text{g.l.b. } X$, if y_1 is the largest possible lower bound for X. Using the Completeness Property, prove that every set of real numbers which has a lower bound has a greatest lower bound.
   (c) For which sets of part (a) is there no g.l.b.? Why not? Give the greatest lower bound of all the other sets of part (a).
8. Show by example that the points x_0 and x_1 in Theorem 2 need not be unique.

7 THE MEAN VALUE THEOREM

The Extreme Value Theorem that we studied in the previous section applies to any function f continuous on a closed interval $[A, B]$. This theorem becomes much more powerful when we add the condition that f be differentiable on the open interval $]A, B[$. The differentiability allows a method for actually finding a point $x \in [A, B]$ such that $f(x)$ is an extreme value of f on $[A, B]$.

Theorem 1 **The Extreme Value Test** Let f be continuous on $[A, B]$ and differentiable on $]A, B[$. If $A < x_0 < B$ and if $f(x_0)$ is either the maximum or minimum of f on $[A, B]$, then $f'(x_0) = 0$.

Proof To fix ideas, let us suppose that $f(x_0)$ is the maximum of f on $[A, B]$. Then $f(x_0) \geqslant f(x_0 + \Delta x)$ for all $(x_0 + \Delta x) \in [A, B]$. Therefore, if $\Delta x > 0$ and $x_0 + \Delta x \in \,]A, B[$, we have

$$\frac{f(x_0 + \Delta x) - f(x_0)}{\Delta x} \leqslant 0$$

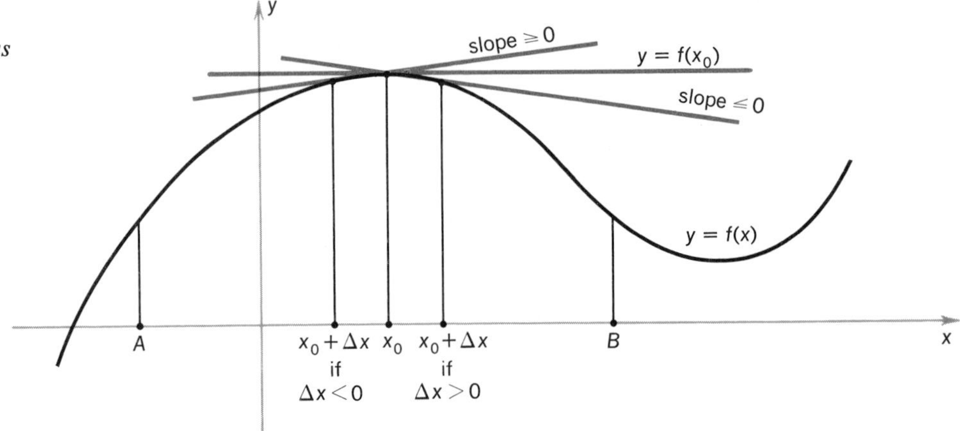

FIGURE 7-1

(Figure 7-1). Hence

$$\lim_{\Delta x \to 0^+} \frac{f(x_0 + \Delta x) - f(x_0)}{\Delta x} \leqslant 0$$

so $f'(x_0) \leqslant 0$. On the other hand, if $\Delta x < 0$ and $(x_0 + \Delta x) \in]A, B[$, we have

$$\frac{f(x_0 + \Delta x) - f(x_0)}{\Delta x} \geqslant 0$$

and so

$$\lim_{\Delta x \to 0^-} \frac{f(x_0 + \Delta x) - f(x_0)}{\Delta x} \geqslant 0$$

Thus $f'(x_0) \geqslant 0$. But then $0 \leqslant f'(x_0) \leqslant 0$ so we conclude that $f'(x_0) = 0$. The case when $f(x_0)$ is the minimum of f on $[A, B]$ is analogous and is left as an exercise. ■

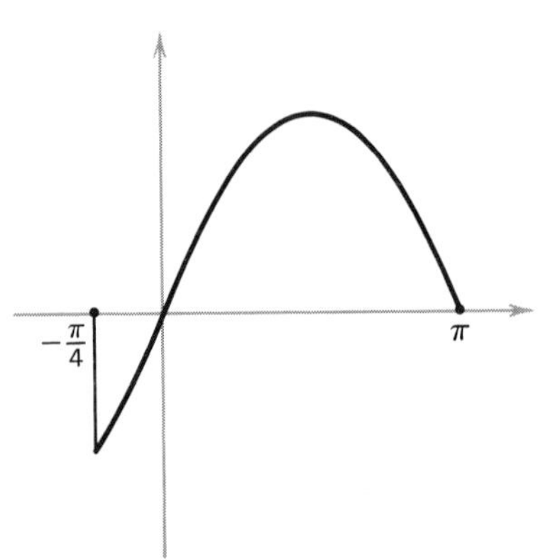

FIGURE 7-2

Theorem 1 states that if the maximum or minimum of f on $[A, B]$ occurs at a point in $]A, B[$, it occurs at a critical point. **That is, an extreme value occurs either at A, or at B, or at a critical point.**

Example 1 Find the extreme values of the sine function in the interval $[-\frac{1}{4}\pi, \pi]$ (Figure 7-2). The conditions of Theorem 1 apply, so the extreme values must occur at $-\frac{1}{4}\pi$, at π, or at a point in the interval where the derivative is zero. To find the critical points, we solve the equation

$$\frac{d(\sin x)}{dx} = 0$$

or

$$\cos x = 0$$

which occurs in $[-\frac{1}{4}\pi, \pi]$ only at $x = \frac{1}{2}\pi$. We compute $\sin\frac{1}{2}\pi = 1$, $\sin(-\frac{1}{4}\pi) = -1/\sqrt{2}$, and $\sin\pi = 0$. The maximum is the greatest of these values, that is, $\sin\frac{1}{2}\pi = 1$; the minimum is the smallest, that is, $\sin(-\frac{1}{4}\pi) = -1/\sqrt{2}$. Hence, the minimum occurs at the left endpoint of the interval.

Example 2 Find the extreme values of the function $f: x \mapsto x^3$ on $[-1, 2]$ (Figure 7-3). The extreme values must occur at $x = -1$, at $x = 2$, or at a critical point of the function. Since $f'(x) = 3x^2$, the only critical point in the given interval is at $x = 0$; we compute $(-1)^3 = -1$, $2^3 = 8$, and $0^3 = 0$. Therefore, the minimum on $[-1, 2]$ occurs at $x = -1$, and the maximum occurs at $x = 2$. Notice that here no extreme value occurs at the critical point. This shows that the converse of Theorem 1 need not hold. That is, a critical point does not have to be a maximum or minimum, even in a small neighborhood of this point.

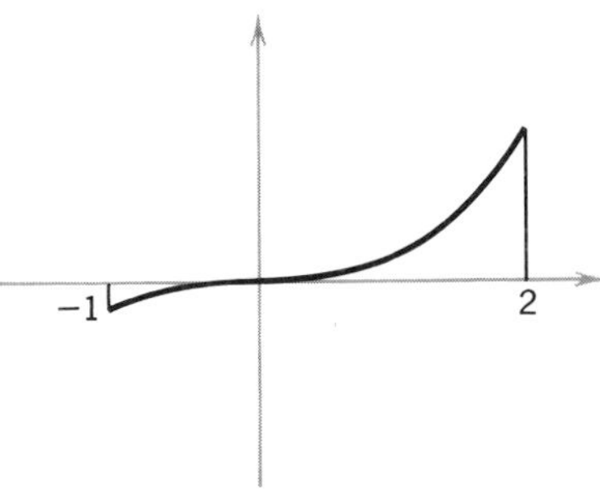

FIGURE 7-3

Example 3 Find the extreme values of $f: x \mapsto 3x^4 - 28x^3 + 66x^2 - 60x$ on the interval $[0, 6]$.

The extreme values must occur at $x = 0$, $x = 6$, or at a solution to

$$\begin{aligned} 0 &= f'(x) \\ &= 12x^3 - 84x^2 + 132x - 60 \\ &= 12(x^3 - 7x^2 + 11x - 5) \\ &= 12(x-1)^2(x-5) \end{aligned}$$

that is, at $x = 1$ or $x = 5$. We therefore evaluate

$$\begin{aligned} f(0) &= 0 \\ f(1) &= -19 \\ f(5) &= -275 \\ f(6) &= -144 \end{aligned}$$

The maximum value of f on $[0, 6]$ is $f(0) = 0$, and the minimum value is $f(5) = -275$.

The Extreme Value Test establishes an important relation between extreme values and critical points. Geometric intuition indicates that if a smooth curve crosses the x axis at $x = A$ and again at $x = B$, then there should be at least one critical point between A and B, or, in other words, a point where the tangent to the curve is parallel to the x axis (Figure 7-4(a)). This intuition fails if the curve does not have a tangent at every point, for then the curve can cross the axis at A, rise to a cusp where there is no tangent, and come back down to B (Figure 7-4(b)). However, when the curve is the graph of a differentiable function, then the tangent must always exist, and we have the following result.

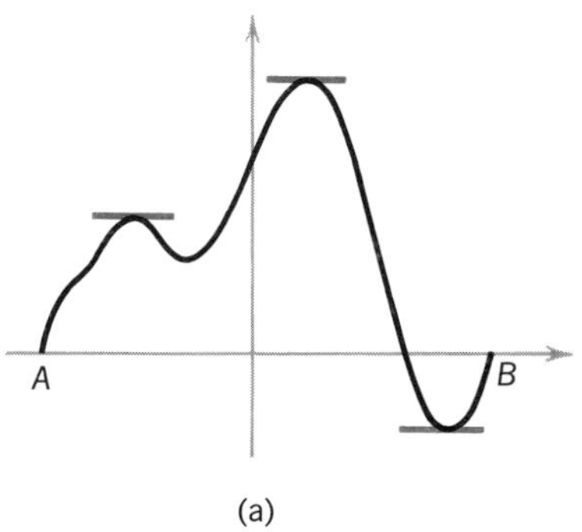

(a)

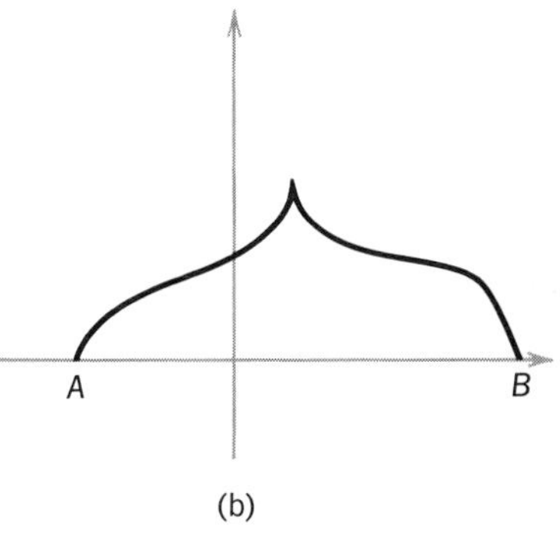

(b)

FIGURE 7-4

Theorem 2 Rolle's Theorem Suppose that f is continuous on $[A, B]$, differentiable on $]A, B[$, and that $f(A) = f(B) = 0$. Then there is a number c such that $A < c < B$ and $f'(c) = 0$.

Proof We distinguish three cases (see Figure 7-5): (i) If $f(x) = 0$ for all x in $[A, B]$, then $f'(c) = 0$ for any $c \in]A, B[$. (ii) If $f(x) > 0$ for some x in $[A, B]$, then f has a maximum positive value. Let $f(c)$ be the maximum of f on $[A, B]$, then $f(c) > 0$, so c is neither A nor B. Therefore $A < c < B$, and by the Extreme Value Test, $f'(c) = 0$. (iii) If $f(x) < 0$ for some x in $[A, B]$, then f has a negative minimum value. Let $f(c)$ be the minimum of f in $[A, B]$, then $f(c) < 0$, so again c is neither A nor B, and again we must have $f'(c) = 0$. ∎

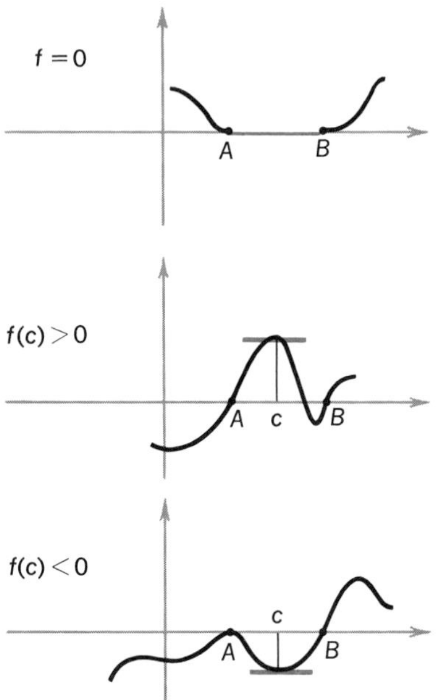

FIGURE 7-5

We remark that there may be more than one place between A and B where the derivative is zero (see Example 3 above); Rolle's Theorem guarantees us that there is at least one such place. Rolle's Theorem has several important corollaries.

Corollary 1 Suppose f and g are continuous on $[A, B]$ and differentiable on $]A, B[$. If $f(A) = g(A)$ and $f(B) = g(B)$, then for some c in the open interval $]A, B[$, we have $f'(c) = g'(c)$ (Figure 7-6).

Proof The function $f - g$ satisfies the hypotheses of Rolle's Theorem, so let $A < c < B$ be such that $(f - g)'(c) = 0$. But $(f - g)'(c) = f'(c) - g'(c)$, so then $f'(c) = g'(c)$. ∎

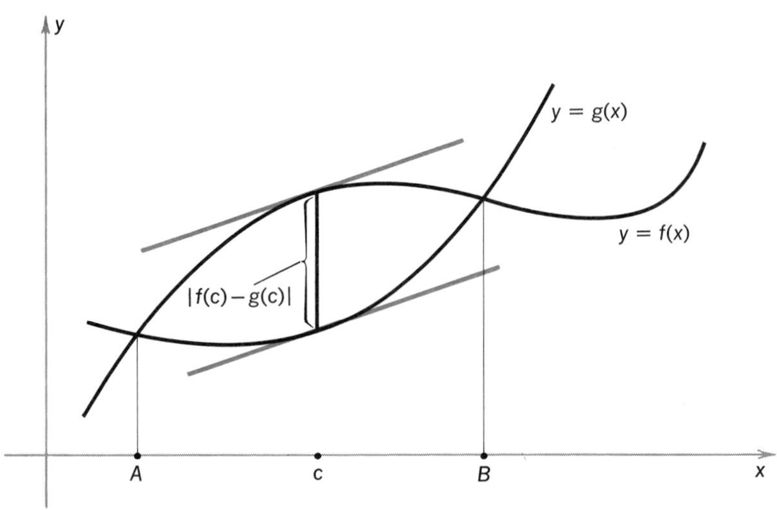

FIGURE 7-6

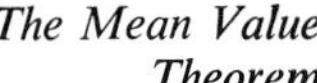

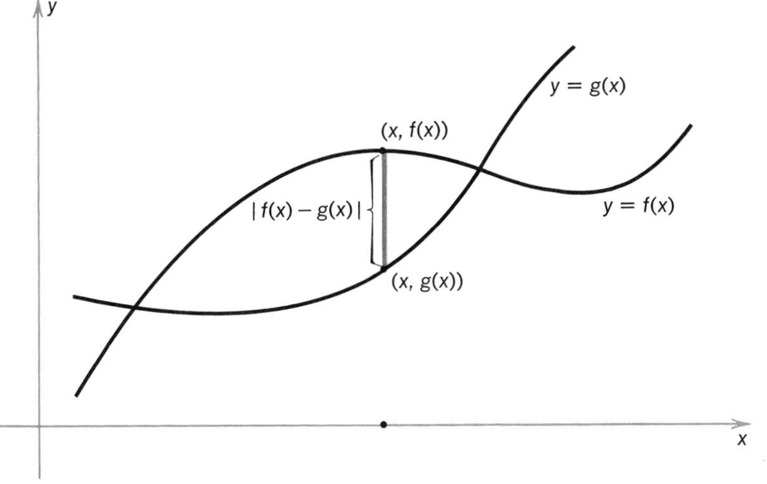

FIGURE 7-7

There is an interesting geometric significance to this corollary. The quantity $f(x)-g(x)$ measures the distance between the points $(x, f(x))$ and $(x, g(x))$; (Figure 7-7). Therefore, the curves are farthest apart at an extreme value of $f-g$ on $[A, B]$, which can occur only when $(f-g)'(x)=0$. Hence, this point must have $f'(x)=g'(x)$; thus the tangent lines are parallel at the point where the graphs of f and g are farthest apart.

A very important special case of Corollary 1 is known as the **Mean Value Theorem.**

Theorem 3 Mean Value Theorem Suppose f is continuous on the closed interval $[A, B]$ and differentiable on the open interval $]A, B[$. Then for at least one number c in $]A, B[$, we have

$$\boldsymbol{f'(c) = \frac{f(B)-f(A)}{B-A}}$$

Proof Consider the straight line passing through the points $(A, f(A))$ and $(B, f(B))$ (Figure 7-8). This line has the slope

$$m = \frac{f(B)-f(A)}{B-A}$$

and is therefore (by the point slope formula) the graph of the affine function $g: x \mapsto m(x-A)+f(A)$. The derivative of g is the constant function $g': x \mapsto m$. Now $g(A)=f(A)$ and $g(B)=f(B)$, so by Corollary 1 there is a number $c \in]A, B[$ (i.e., $A < c < B$) for which

$$f'(c) = m = \frac{f(B)-f(A)}{B-A}$$

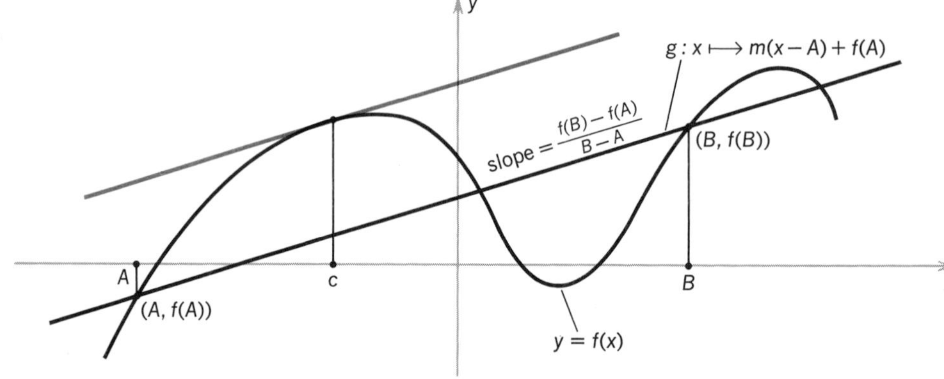

FIGURE 7-8

The result of this theorem is often written as

$$f(B) - f(A) = f'(c)(B - A)$$

The geometric significance of the Mean Value Theorem is that for at least one point c between A and B, the tangent to f at $(c, f(c))$ is parallel to the straight line connecting $(A, f(A))$ and $(B, f(B))$ (Figure 7-8).

If we are given a differentiable function f, the Mean Value Theorem can be applied to any pair of points x_1, x_2 in the domain of f, provided of course that $f(x)$ is defined for all x in the closed interval with endpoints x_1 and x_2. This makes the Mean Value Theorem flexible and easy to apply. In particular, if the domain of f is all of $\boldsymbol{R}$, we can apply the theorem to any pair of numbers x_1, x_2 and be sure that there is always a point c between x_1 and x_2 such that

$$f'(c) = \frac{f(x_2) - f(x_1)}{x_2 - x_1}$$

(See Figure 7-9.)

The Mean Value Theorem also applies to the analysis of motion, and yields the result that **a moving body must attain its average velocity over an interval of time at some instant during the interval.** To see this, notice that we have, for the interval $[t_1, t_2]$,

$$\text{average velocity} = \frac{s(t_2) - s(t_1)}{t_2 - t_1}$$

$$\text{instantaneous velocity } v(t) = \frac{ds}{dt} = s'(t)$$

By the Mean Value Theorem, there exists a time τ such that $t_1 < \tau < t_2$ and $s'(\tau) = [s(t_2) - s(t_1)]/(t_2 - t_1)$, therefore, $v(\tau)$ is the average velocity over the interval.

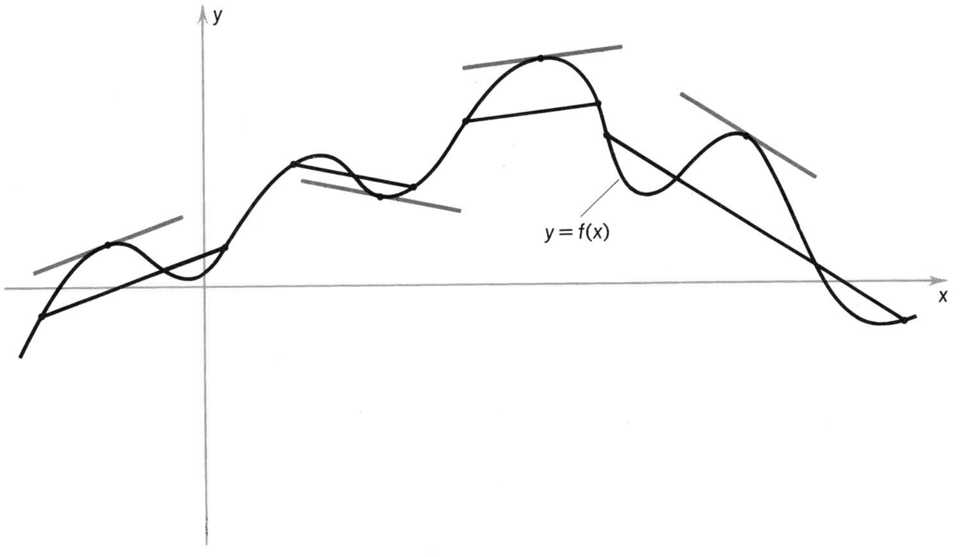

FIGURE 7-9

Example 4 Returning to §3.5, Example 1, we know that our flying eagle must attain his average velocity of 40 mph at some moment during his flight.

We have

$$s = \frac{t^2}{60} - \frac{t^3}{10800}, \quad s(0) = 0, \quad s(120) = 80$$

$$\frac{ds}{dt} = \frac{t}{30} - \frac{t^2}{3600}$$

so, by the Mean Value Theorem there is a moment τ when

$$\frac{\tau}{30} - \frac{\tau}{3600} = \frac{80}{120}$$

or

$$\tau^2 - 120\tau + 2400 = 0$$

Solving by the quadratic formula we obtain

$$\tau = 60 \pm 20\sqrt{3}$$

so the eagle attains his average velocity twice during his flight, at $\tau_1 = 60 - 20\sqrt{3}$ min and at $\tau_2 = 60 + 20\sqrt{3}$ min. (See Figure 7-10.)

We can use the Mean Value Theorem to prove the First Derivative Test from §3.1. Recall that this test says that the sign of the derivative indicates where the function is increasing or decreasing.

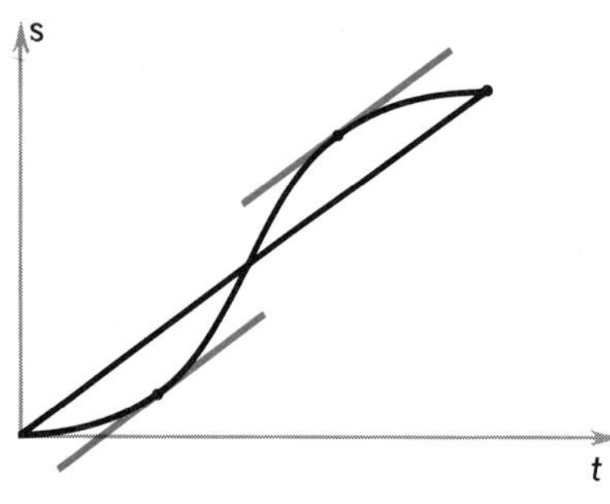

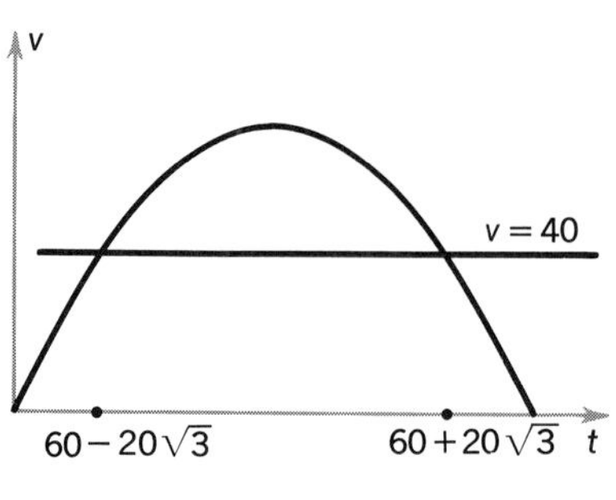

FIGURE 7-10

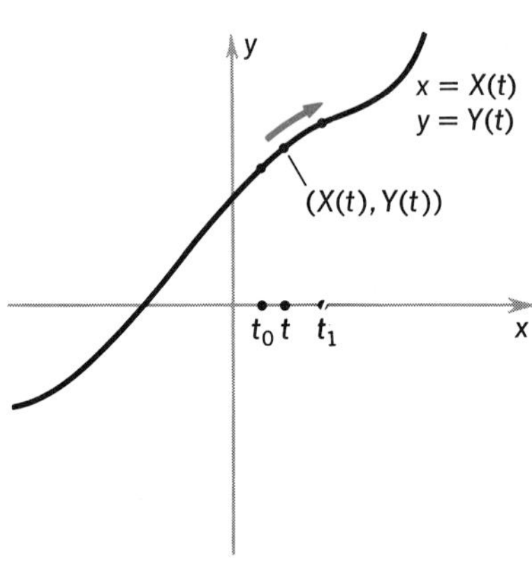

FIGURE 7-11

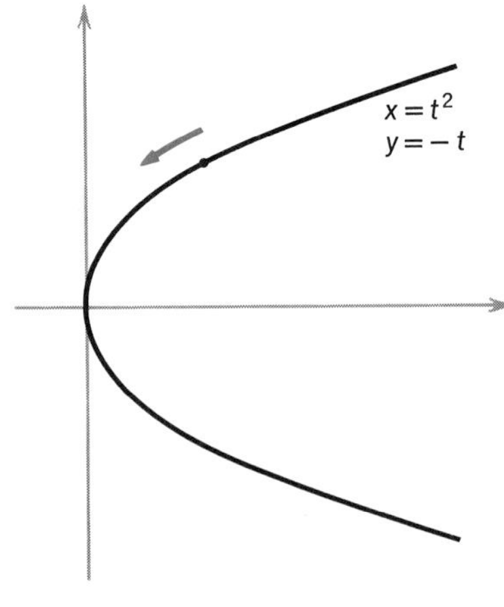

FIGURE 7-12

Proof First Derivative Test Suppose f is continuous on $[A, B]$ and differentiable on $]A, B[$, with $f'(x) > 0$ throughout $]A, B[$. (We leave as an exercise the proof for $f'(x) < 0$.) Let x_1 and x_2 be any two points in $[A, B]$ such that $A \leqslant x_1 < x_2 \leqslant B$; we shall show that $f(x_2) > f(x_1)$. Applying the Mean Value Theorem to f on the interval $[x_1, x_2]$, we see that there is a number c such that $x_1 < c < x_2$ and $f(x_2) - f(x_1) = f'(c)(x_2 - x_1)$. But $f'(c) > 0$ and $(x_2 - x_1) > 0$, so we must have $f(x_2) > f(x_1)$. ∎

We defer intensive exploitation of this result until we discuss curve sketching in the next chapter. For the moment we content ourselves with applying it to the motion described by parametric equations.

Suppose we are given parametric equations $x = X(t)$, $y = Y(t)$, and suppose that $X'(t) > 0$ for all t in the interval $]t_0, t_1[$. Then as t goes from t_0 to t_1, the point $(X(t), Y(t))$ must be moving to the right, since $X(t)$ is monotone increasing on $[t_0, t_1]$. If in addition $Y'(t) > 0$ for t in $]t_0, t_1[$, the point $(X(t), Y(t))$ must be moving upward, since Y is also monotone increasing on $[t_0, t_1]$ (Figure 7-11). Similar remarks apply when $X'(t) < 0$ and/or $Y'(t) < 0$. By way of example, consider the parametric equations $x = t^2$, $y = -t$. Then $X'(t) = 2t$, so $X'(t)$ is positive when t is positive and negative when t is negative. On the other hand, $Y'(t) = -1 < 0$ for all t. Thus, looking at Figure 7-12, we see that when t is approaching 0 through negative values, $(X(t), Y(t))$ is moving downward and to the left (top half of the curve). As t increases through positive values, $(X(t), Y(t))$ moves downward and to the right (bottom half of the curve).

We conclude this section with two important corollaries to the Mean Value Theorem. As we already know, the derivative of a constant function $x \mapsto k$ is $d(k)/dx = 0$. Conversely, we shall now prove that if $f: x \mapsto f(x)$ is a differentiable function such that $d(f(x))/dx = 0$, then f is a constant function.

Corollary 1 If f is continuous on $[A, B]$, differentiable on $]A, B[$, and if $f'(x) = 0$ for all $x \in]A, B[$, then $f(x) = f(A)$ for all $x \in [A, B]$.

Proof If $A < x \leqslant B$, we apply the Mean Value Theorem to f on the interval $[A, x]$ and obtain

$$f(x) - f(A) = f'(c)(x - A)$$

for some point c in the open interval $]A, x[$. But $f'(c) = 0$ by hypothesis, hence, $f(x) - f(A) = 0$, and so $f(x) = f(A)$. ∎

Suppose f and g are two functions such that

$$f(x) = g(x) + k$$

for some constant k. If g is differentiable, then so is f, and as we know,

$$\frac{d(f(x))}{dx} = \frac{d(g(x))}{dx}$$

The converse to this is also true, namely, two functions with the same derivative differ at most by a constant.

Corollary 2 If f and g are two functions both defined and differentiable on $[A, B]$, and if $f'(x) = g'(x)$ for all $x \in]A, B[$, then there is a constant k such that $f(x) = g(x) + k$ for all $x \in [A, B]$.

Proof The difference function $f - g$ satisfies the hypothesis of Corollary 1. Thus $f(x) - g(x) = (f - g)(A) = f(A) - g(A)$ for all $x \in [A, B]$, and so with $k = f(A) - g(A)$, we have $f(x) = g(x) + k$ for all $x \in [A, B]$, and k is the constant sought. ∎

The geometric significance of this corollary is found by extending the notion of parallel from lines to general curves. As usual, it is easier to give the definition in terms of functions rather than in terms of their graphs. The first theorem of this book said that two nonvertical straight lines are parallel if and only if they have the same slope. We now **define** two differentiable functions f and g to be **parallel** if they have the same domain and if $f' = g'$. Then Corollary 2 says that two differentiable functions are parallel if and only if they differ by a constant (Figure 7-13).

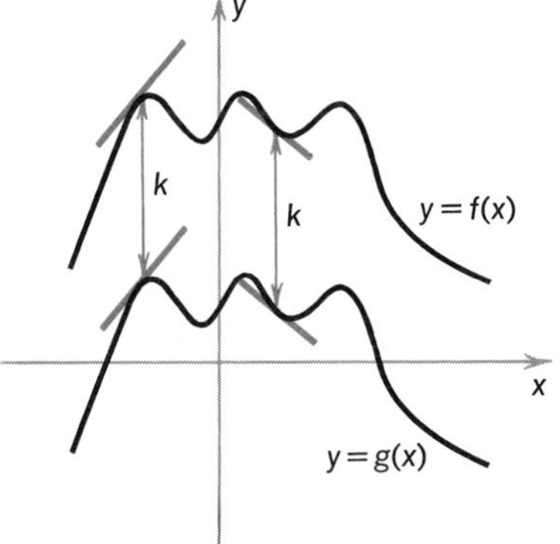

FIGURE 7-13

EXERCISES

1. Find the maximum and minimum value of f on the indicated interval.
   (a) $f: x \mapsto 1 - 4x - x^2$ on $[-2, 3]$
   (b) $f: x \mapsto x^4 + 2x^3 - 4x^2$ on $[0, 4]$
   (c) $f: 2x \mapsto \frac{1}{2}x^2 + \frac{2}{3}x^3 - \frac{1}{4}x^4$ on $[-1, 6]$
   (d) $f: x \mapsto \sin x + \cos x$ on $[0, 2\pi]$
   (e) $f: x \mapsto \sin(\cos x)$ on $[0, \pi]$
   (f) $f: x \mapsto \begin{cases} \dfrac{\sin x}{x}, & x \neq 0 \\ 1, & x = 0 \end{cases}$ on $[-\pi, \pi]$

2. The function $x \mapsto |x|$ attains its minimum at $x = 0$. Why is the derivative not zero there?

3. Find a number $c \in [A, B]$ such that $(f'(c) = (f(B) - f(A))/(B - A)$, where f, A, and B are as indicated:
   (a) $f: x \mapsto x^3 - 3x^2 + 3x + 1$, $A = 0$, $B = 1$
   (b) $f: x \mapsto x^{2/3}$, $A = 0$, $B = 1$
   (c) $f: x \mapsto x + 1/x$, $A = \frac{1}{3}$, $B = 3$
   (d) $f: x \mapsto (x+7)^{1/2}$, $A = -3$, $B = 9$

4. Rolle's Theorem can be used to help locate the zeros of a polynomial or other continuously differentiable function as follows: If (i) f is continuously differentiable on $[A, B]$, (ii) $(\operatorname{sgn} f(A)) \cdot (\operatorname{sgn} f(B)) = -1$, and (iii) $f'(x) \neq 0$ for all $x \in [A, B]$, then there is exactly one zero of f in $[A, B]$, that is, there is one and only one root of the equation $f(x) = 0$ in $[A, B]$.
   (a) That there is at least one such root follows from hypotheses (i) and (ii) and the Intermediate Value Theorem. Prove, using Rolle's Theorem, that there is at most one root of f in $[A, B]$.
   (b) Find two consecutive integers between which there is exactly one root of the polynomial $x \mapsto x^4 + 3x + 1$.
   (c) Repeat part (b) for $x \mapsto x^4 + 2x^3 - 2$.
   (d) Repeat part (b) for $x \mapsto 2x^3 - 3x^2 - 12x - 6$.

5. Let $x = X(t)$ and $y = Y(t)$ be parametric equations and suppose X and Y are continuously differentiable on $]t_0, t_1[$. Show that, if $X'(t) \neq 0$ for all t in $]t_0, t_1[$, then the parametrized curve is the graph of a function. That is, show that if P and Q lie on the curve, $x_P \neq x_Q$.

8 THE INVERSE FUNCTION THEOREM

Our discussion of *inverse functions* first began in Chapter 1. At this point, the student should read (or reread) the latter part of §1.3. In this section we show that the inverse of a differentiable function is itself differentiable, and that an important relation exists between the derivative of a function and the derivative of its inverse. The third and fourth theorems of this section are the Local Inverse Function Theorem and the Implicit Function Theorem. These are two fundamental results of mathematical analysis and they underlie the calculus of differentials of §4.1 and §4.2. Here we apply the Local Inverse Function Theorem to parametric equations, and we justify the use of differentials to compute the tangent to the parametrized curve. To do this, we reduce the tangent problem for parametrized curves to the tangent problem for graphs of real functions. The latter problem we have solved by means of **the derivative**. We then proceed to discuss the Implicit Function Theorem and the method of **implicit differentiation**, extending the ideas of §4.1.

In Chapter 1, we remarked that if f and f^{-1} are inverses of one another, then the graph of $y = f^{-1}(x)$ is obtained by "flipping" the graph of $y = f(x)$ about the line $y = x$. We now analyze the relation between the graphs of f and f^{-1} and the process of "flipping" in some detail. Given a point $P=(x_0, y_0)$, we denote by P^{-1} (read "P-inverse") the point $P^{-1} = (y_0, x_0)$. If $P = P^{-1}$, then P lies on the line $y = x$ and conversely. If $P \neq P^{-1}$, then the line $y = x$ is the perpendicular bisector of the line segment from P to P^{-1}. To see this, we first note that the midpoint of the segment PP^{-1} has coordinates

$$\left(\frac{x_0+y_0}{2}, \frac{y_0+x_0}{2}\right)$$

and so the midpoint lies on $y = x$; hence, $y = x$ bisects the segment. To check that PP^{-1} is perpendicular to $y = x$, we note that the straight line through P and P^{-1} has slope

$$m = \frac{x_0 - y_0}{y_0 - x_0} = -1$$

Since the slopes of the two lines are negative reciprocals, the lines are perpendicular.

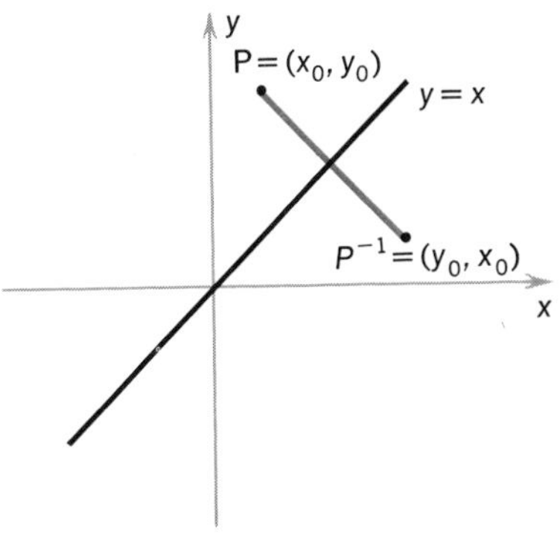

FIGURE 8-1

We therefore say that P and P^{-1} are **symmetric with respect to the line $y = x$** (Figure 8-1). Now suppose f is a one-one function and that f^{-1} is its inverse. Then $f(x_0) = y_0$ is equivalent to $f^{-1}(y_0) = x_0$. Thus a point $P = (x_0, y_0)$ is on the graph of f if and only if the point $P^{-1} = (y_0, x_0)$ is on the graph of f^{-1}. Therefore, we say that the graphs of f and f^{-1} are **symmetric to one another with respect to the line $y = x$** (Figure 8-2). The graph of f^{-1} is obtained by replacing all points P on $y = f(x)$ by their inverses P^{-1}. This is what we mean when we say that the curve $y = f^{-1}(x)$ is obtained by "flipping" the curve $y = f(x)$ about the line $y = x$.

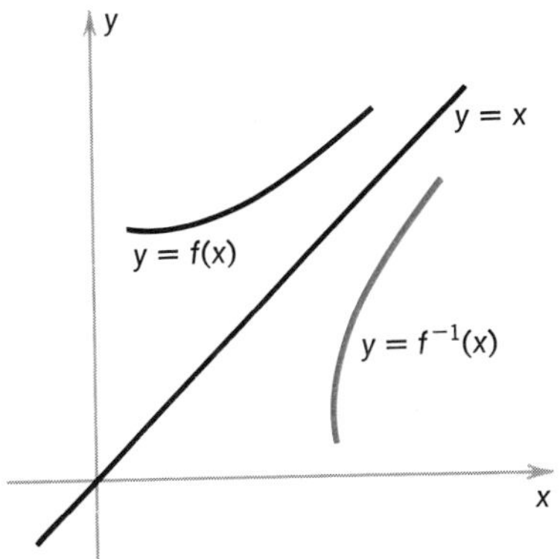

FIGURE 8-2

Example 1 Consider the function $f: x \mapsto mx$, where $m > 0$. Then the inverse function is $f^{-1}: x \mapsto x/m$, since $f^{-1}(f(x)) = (1/m)(mx) = x$. Let us denote by α the angle of inclination of $y = x/m$ and by β the angle of inclination of $y = mx$. By the symmetry of the lines with respect to $y = x$, we should have $\alpha + \beta = \frac{1}{2}\pi$ (Figure 8-3). We can verify this analytically as follows: $\cot\alpha = m$ and $\cot\beta = 1/m$. So

$$\cot(\alpha+\beta) = \frac{\cot\alpha\cot\beta - 1}{\cot\alpha + \cot\beta} = \frac{0}{\left(m + \dfrac{1}{m}\right)} = 0$$

Thus $\alpha + \beta = \frac{1}{2}\pi$, since both α and β are in the first quadrant.

(b) The function $x \mapsto -x$ is its own inverse. If $P = (x_0, -x_0)$, then $P^{-1} = (-x_0, x_0) = (-x_0, -(-x_0))$ is also on the line $y = -x$. The line $y = -x$ is perpendicular to $y = x$ and is symmetric with respect to $y = x$ (Figure 8-4).

(c) If $m > 0$, then the function $f: x \mapsto -mx$ has as its inverse $f^{-1}: x \mapsto -x/m$. The reader can show that the lines $y = -mx$ and $y = -x/m$ make equal acute angles with the line $y = x$. (That is, the angles α and β of Figure 8-5 are equal.)

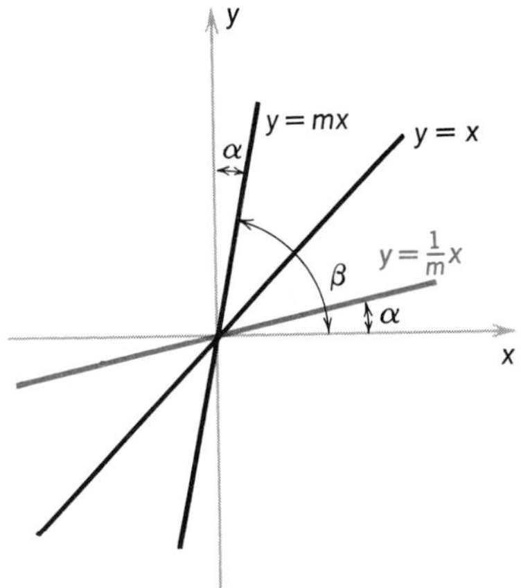

FIGURE 8-3

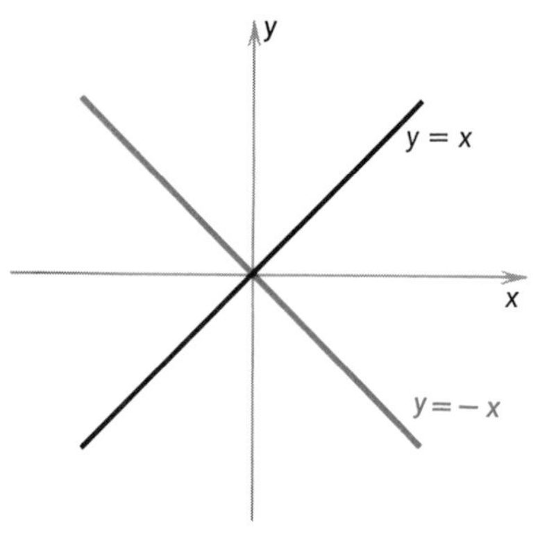

FIGURE 8-4

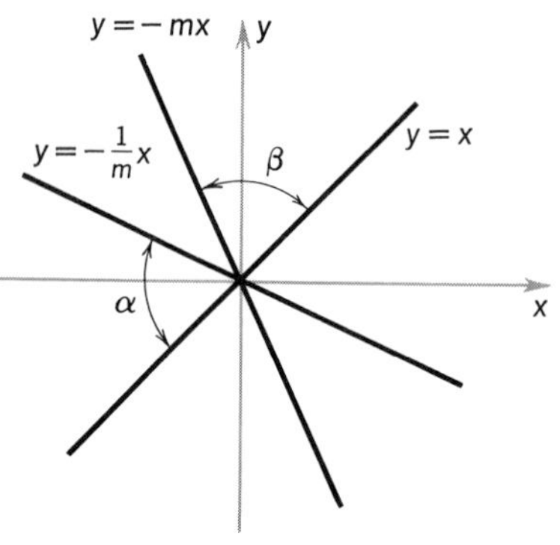

FIGURE 8-5

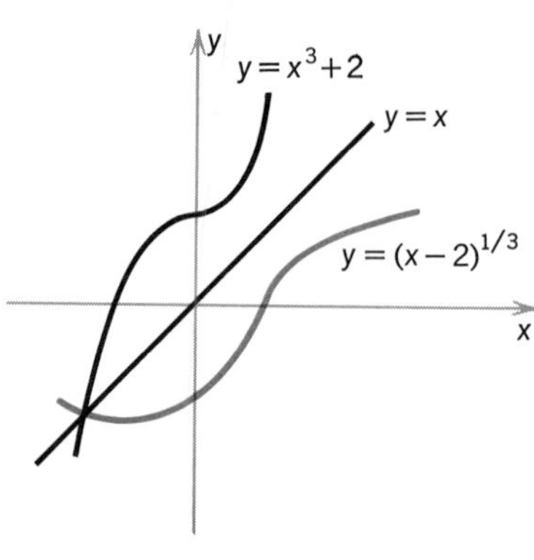

FIGURE 8-6

Example 2 (a) The inverse of $x \mapsto 1/x$ is $x \mapsto 1/x$. The graph of this function, the hyperbola $y = 1/x$, is symmetric with respect to $y = x$.

(b) If we have $f\colon x \mapsto x^3 + 2$, then we obtain $f^{-1}\colon x \mapsto (x-2)^{1/3}$ (Figure 8-6).

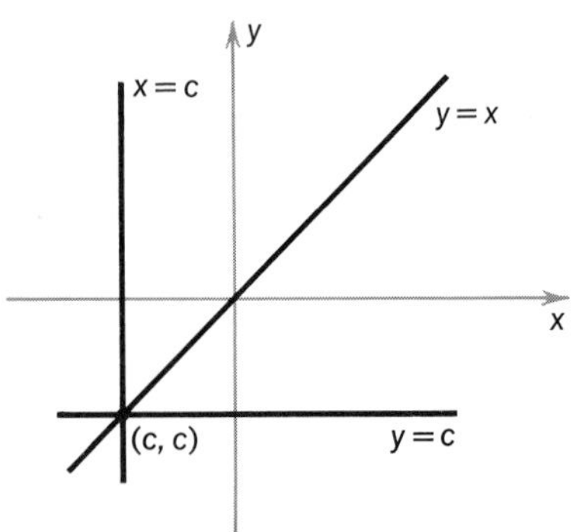

FIGURE 8-7

Now consider a horizontal line $y = c$, where c is a constant. If $P = (x_0, c)$ is on this line, then $P^{-1} = (c, x_0)$ lies on the vertical line $x = c$. So we say that the lines $y = c$ and $x = c$ are **symmetric to one another with respect to the line** $y = x$ (Figure 8-7). The three lines $x = c$, $y = c$, and $y = x$ intersect at the point (c, c). We see that $x = c$ and $y = c$ are perpendicular and that they both form angles of 45° with the line $y = x$.

From the figures, and our understanding of the process of flipping, it is clear that if f is continuous, then f^{-1} is also continuous. (We give an analytic proof, without pictures, of this fact in a moment.) It is also clear from the pictures that, if f is differentiable (and therefore has a smooth graph), then the graph of f^{-1} is also smooth and without sharp changes of direction. Hence, f^{-1} is also differentiable.

Moreover, when f is differentiable, the tangent to $y = f(x)$ at P is related to the tangent to $y = f^{-1}(x)$ at P^{-1} as follows. Flipping the graph of f about $y = x$ sends P to P^{-1}, and the tangent to $y = f(x)$ at P is sent to the tangent to $y = f^{-1}(x)$ at P^{-1} (Figure 8-8). So, if the tangent at P to the graph of f has equation $y = mx + b\,(m \neq 0)$, then the tangent to the graph of f^{-1} at P^{-1} has equation

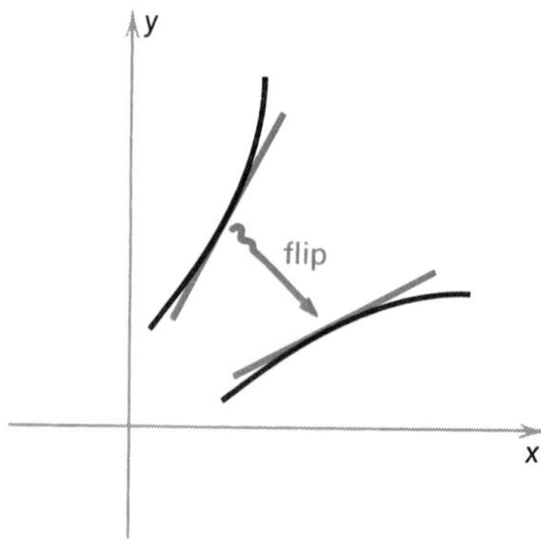

FIGURE 8-8

$$y = \frac{1}{m}x - \frac{b}{m}$$

When the tangent to f at P is horizontal (respectively, vertical), then the tangent to f^{-1} at P^{-1} is vertical (respectively, horizontal). See Figure 8-9. In terms of derivatives, this means that if $f(x_0) = y_0$ and $f'(x_0) = m \neq 0$, then $(f^{-1})'(y_0) = 1/m$. This last statement is, in effect, the **Inverse Function**

Theorem. We now proceed to give a complete analytic proof of this theorem and a discussion of its implications and corollaries.

The first theorem we prove seems restricted in that it is stated only for monotone functions. However, a more general result follows quickly from this special case (see exercise 4).

Theorem 1 Continuity of Inverses Suppose f is a continuous function with domain an open interval. Then if f is monotone (increasing or decreasing), f^{-1} is also monotone and continuous.

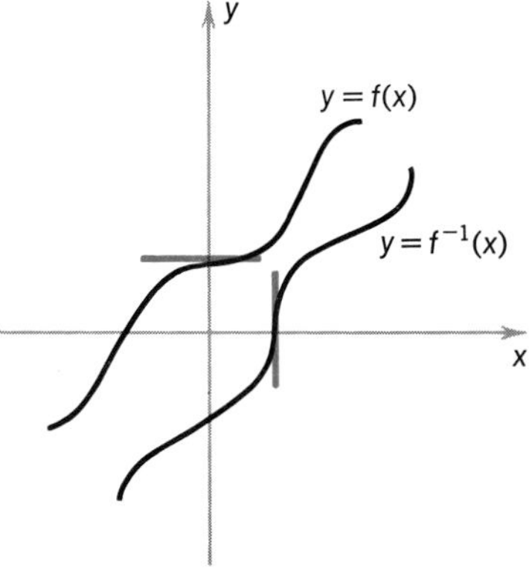

FIGURE 8-9

Proof The inverse function f^{-1} is easily seen to be monotone (the same way as f); that part of the proof is left to the student. Let us set $y_0 = f(x_0)$ and let $]A, B[$ be a neighborhood of x_0 such that all points in the closed interval $[A, B]$ are in the domain of f. The continuity of f^{-1} will be established if we can show that $f^{-1}(y)$ is eventually in every neighborhood of x_0 such as $]A, B[$ as y approaches y_0. (Why?) Now, if f is monotone increasing, $]f(A), f(B)[$ is a neighborhood of y_0. By applying the Intermediate Value Theorem to f on $[A, B]$, we see that $f^{-1}(y)$ is defined for all $y \in]f(A), f(B)[$. Moreover, for all y in this neighborhood, we must have $f^{-1}(y) \in]A, B[$. Similarly, if f is monotone decreasing, $]f(B), f(A)[$ is a neighborhood of y_0 and for all y in this neighborhood of y_0 we must have $f^{-1}(y) \in]A, B[$. In both cases, we have shown that $f^{-1}(y)$ is eventually in $]A, B[$ as y approaches y_0. ∎

Our study of differential calculus has already given some consideration to monotone functions. In fact, we have seen that a differentiable function whose derivative is positive (respectively, negative) throughout some interval is monotone increasing (respectively, decreasing) on that interval.

Now consider a continuously differentiable function f whose domain is an open interval, and such that $f'(x)$ is either always positive or always negative; thus, f is monotone. Hence, f^{-1} exists, and by the previous theorem is continuous. In the Inverse Function Theorem, we show that f^{-1} is also continuously differentiable and establish a relation between the two derivatives f' and $(f^{-1})'$.

Theorem 2 The Inverse Function Theorem Suppose f is a continuously differentiable function whose domain is an open interval, and that for all x in the domain of f, $f'(x) \neq 0$. Then f is a monotone function, and the inverse f^{-1} is also monotone and continuously differentiable. Moreover, when $f(x_0) = y_0$, we have

$$(f^{-1})'(y_0) = \frac{1}{f'(x_0)}$$

Thus setting $y = f(x)$ and $x = f^{-1}(y)$ yields

$$\left.\frac{dx}{dy}\right|_{y=f(x_0)} = \left.\frac{1}{dy/dx}\right|_{x=x_0}$$

Proof We first show that either $f'(x) > 0$ for all x or $f'(x) < 0$ for all x. By hypothesis, f' is continuous; hence, by the Intermediate Value Theorem, if $f'(x)$ changed sign, there would have to be a point c such that $f'(c) = 0$. By assumption such a point c does not exist, so $f'(x)$ cannot change sign; thus either $f'(x) > 0$ for all x or $f'(x) < 0$ for all x. By the remarks preceding this theorem, this fact establishes that f is monotone. Hence, f^{-1} exists and is monotone also.

By Theorem 1, we also know that f^{-1} is continuous. To show that f^{-1} is differentiable, we explicitly compute

$$\lim_{\Delta y \to 0} \frac{f^{-1}(y_0+\Delta y) - f^{-1}(y_0)}{\Delta y}$$

Establish notation as follows:

$$f(x_0) = y_0 \qquad f^{-1}(y_0) = x_0$$

$$f(x_0+\Delta x) = y_0 + \Delta y \qquad f^{-1}(y_0+\Delta y) = x_0 + \Delta x$$

The function f is continuous at y_0, and so as Δy approaches zero, $\Delta x = f^{-1}(y_0+\Delta y) - f^{-1}(y_0)$ must also approach zero, that is,

$$\lim_{\Delta y \to 0} \Delta x = 0$$

Since f is monotone and one-one, $\Delta y \neq 0$ implies $\Delta x \neq 0$. We have the equation

$$\frac{\Delta x}{\Delta y} = \frac{1}{\Delta y/\Delta x} = \frac{1}{\dfrac{f(x_0+\Delta x) - f(x_0)}{\Delta x}}$$

We wish to evaluate

$$\lim_{\Delta y \to 0} \frac{\Delta x}{\Delta y} = \lim_{\Delta y \to 0} \frac{1}{\dfrac{f(x_0+\Delta x) - f(x_0)}{\Delta x}}$$

We can apply the theorem on limits of composite functions (§2.4, Theorem 5) to the composition

$$\Delta y \mapsto f^{-1}(y_0+\Delta y) - f^{-1}(y_0) = \Delta x \mapsto \frac{1}{\dfrac{f(x_0+\Delta x) - f(x_0)}{\Delta x}}$$

which yields

$$\lim_{\Delta y \to 0} \frac{1}{\dfrac{f(x_0+\Delta x) - f(x_0)}{\Delta x}} = \lim_{\Delta x \to 0} \frac{1}{\dfrac{f(x_0+\Delta x) - f(x_0)}{\Delta x}} = \frac{1}{f'(x_0)}$$

The continuity of $(f^{-1})'$ is now easy to establish and we leave it for the student to verify (use Theorems 2 and 3 of §2.4). ∎

Example 3 (a) The inverse of the linear function $f: x \mapsto mx$ $(m \neq 0)$ is the linear function $f^{-1}: x \mapsto x/m$. Clearly, whenever $y_0 = f(x_0)$, we have $(f^{-1})'(y_0) = 1/m = 1/f'(x_0)$.

(b) The function $f: x \mapsto \sin^2 x + 5x$ with domain all real x has derivative

$$f': x \mapsto 2\sin x \cos x + 5 = \sin(2x) + 5$$

Thus $f'(x) > 0$ for all x, and so f satisfies the hypotheses of the Inverse Function Theorem. The inverse f^{-1} is difficult to compute, but $(f^{-1})'(y_0)$ can be easily evaluated if we know y_0 to equal $f(x_0)$. For example, if $y_0 = 0$, then $y_0 = f(0)$ and so $(f^{-1})'(0) = 1/f'(0) = \frac{1}{5}$; similarly, $5\pi = f(\pi)$ and $(f^{-1})'(5\pi) = \frac{1}{5}$.

Example 4 (a) The function with domain $]0, \infty[$ and rule $x \mapsto x^3$ satisfies the hypotheses of the Inverse Function Theorem. Its inverse has domain $]0, \infty[$ and rule $y \mapsto y^{1/3}$. Therefore, when $y_0 = x_0^3$, we have (by the Inverse Function Theorem)

$$\left.\frac{d(y^{1/3})}{dy}\right|_{y=y_0} = \frac{1}{\left.\dfrac{d(x^3)}{dx}\right|_{x=x_0}}$$

or

$$\left.\frac{d(y^{1/3})}{dy}\right|_{y=x_0^3} = \frac{1}{\left.\dfrac{d(x^3)}{dx}\right|_{x=x_0}}$$

As a check, let us evaluate both sides of this last equation. Clearly,

$$\frac{1}{\left.\dfrac{d(x^3)}{dx}\right|_{x=x_0}} = \frac{1}{3x_0^2}$$

On the other hand,

$$\left.\frac{d(y^{1/3})}{dy}\right|_{y=x_0^3} = \left.\left(\frac{1}{3}y^{-2/3}\right)\right|_{y=x_0^3} = \frac{1}{3}(x_0^3)^{-2/3} = \frac{1}{3x_0^2}$$

and all is well.

(b) The principal sine function with domain $]-\frac{1}{2}\pi, \frac{1}{2}\pi[$ and rule $\sin: \theta \mapsto \sin\theta$ satisfies the hypotheses of the Inverse Function Theorem. And the function arc sine is defined to be the inverse of sine. We thus have the formula

$$\frac{d(\arcsin y)}{dy} = \frac{1}{\dfrac{d(\sin\theta)}{d\theta}} = \frac{1}{\cos\theta}$$

From Figure 8-10, we see that $\cos\theta = \sqrt{1-y^2}$; hence,

$$\frac{d(\arcsin y)}{dy} = \frac{1}{\sqrt{1-y^2}}$$

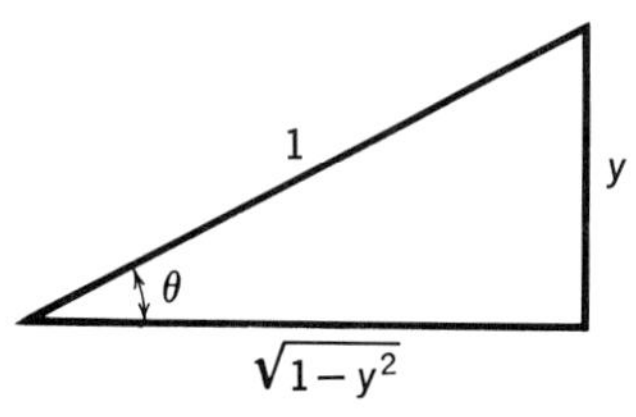

FIGURE 8-10

We have seen many functions which, though not monotone, are monotone on an interval; the sine function is one example. Suppose f is monotone (increasing or decreasing) on an interval $]A, B[$. Then we can define a function g whose domain is $]A, B[$ and whose rule is the rule of f, namely $g: x \mapsto f(x)$; g is called the **restriction** of f to $]A, B[$. Now since f is monotone on $]A, B[$, g is monotone and one-one, and so has an inverse g^{-1}. The function g^{-1} is called a **local inverse** of f. If the interval $]A, B[$ is a neighborhood of the point x_0, we say that g^{-1} is **local inverse of f at x_0**.

Example 5 (a) The arc sine function is a local inverse of the sine. In particular, arc sine is a local inverse of the sine function at $\frac{1}{4}\pi$.

(b) The function $f: x \mapsto x^2$ is monotone increasing on $]1, 2[$. So if we let g have domain $]1, 2[$ and rule $g: x \mapsto x^2$, g is a monotone increasing function. The function g^{-1}, the inverse of g, has domain $]1, 4[$ and rule $g^{-1}: y \mapsto \sqrt{y}$. Then g^{-1} is a local inverse of f, and in particular is a local inverse of f at $\frac{3}{2}$. Of course, g^{-1} is also a local inverse of f at $\frac{4}{3}$ and at every other point in $]1, 2[$. Similarly, $y \mapsto -\sqrt{y}$ is a local inverse of f at any point $x_0 < 0$. On the other hand, there is no neighborhood $]A, B[$ of 0 such that f is monotone on $]A, B[$ (why?), so f does not have a local inverse at 0. However, f does have a local inverse at every point except 0.

The most general form of the Inverse Function Theorem for real functions gives a condition for the existence of a local inverse.

Theorem 3 The Local Inverse Function Theorem Suppose f is continuously differentiable in a neighborhood of x_0, and that $f'(x_0) \neq 0$. Then there exists a continuously differentiable local inverse g^{-1} to f at x_0, and

$$(g^{-1})'(f(x_0)) = \frac{1}{f'(x_0)}$$

Proof From the hypotheses, there exists a neighborhood of x_0 throughout which $f'(x) \neq 0$ (Why?). Then we define a function g whose domain is this neighborhood and whose rule is $g: x \mapsto f(x)$. Clearly, g^{-1} is a local inverse of f at x_0. Moreover, g satisfies the hypotheses of the Inverse Function Theorem. Applying the Inverse Function Theorem to g yields

$$(g^{-1})'(f(x_0)) = \frac{1}{f'(x_0)}$$

Example 6 (a) The arc sine function $x \mapsto \arcsin x$ is a local inverse of $x \mapsto \sin x$ at 0. Since $\arcsin(0) = 0$ and $\sin'(0) = 1$, we have $\arcsin'(0) = 1/\sin'(0) = 1$.

(b) The function $h:x \mapsto -\sqrt{x}$ is a local inverse of $f:x \mapsto x^2$ at -2; and $h(f(-2)) = h(4) = -\sqrt{4} = -2$. By the theorem we have $h'(4) = 1/f'(-2)$. Since we know $h':x \mapsto -1/(2\sqrt{x})$ and $f':x \mapsto 2x$, we can compute $h'(4)$ and $f'(-2)$:

$$h'(4) = \frac{-1}{4} = \frac{1}{f'(-2)}$$

which is in agreement with the Local Inverse Function Theorem.

(c) Let $h:x \mapsto x + \sin x$. Then we have $h':x \mapsto 1 + \cos x$. The function h does not satisfy the hypotheses of the Inverse Function Theorem since $h'(\pi) = 0$; in fact $h'(\pi \pm 2n\pi) = 0$ for all natural numbers n. However, h' is continuous and $h'(0) = 2 > 0$. Therefore, by the Local Inverse Function Theorem, h has a local inverse k at 0 and $k'(0) = 1/h'(0) = 1/2$.

In §4.2, we studied parametric equations and computed a formula for the slope of the tangent to a parametrized curve,

$$\frac{dy}{dx} = \frac{Y'(t)}{X'(t)}$$

This formula was arrived at by manipulating differentials. However, at that time we did not show that dy/dx was obtained by a limiting process involving smaller and smaller secants. In order to justify that $Y'(t)/X'(t)$ is the limit of the slopes of secants, and therefore the slope of the tangent, we need the Local Inverse Function Theorem. With this theorem, we can reduce the tangent problem for parametrized curves to the tangent problem for the graph of a function—a problem already solved by limits and the derivative.

Let us carry out this program. Suppose that $X'(t_0) \neq 0$ and that X' is continuous. Then there is a neighborhood M of t_0 such that $X:t \mapsto X(t)$ is monotone on M. We restrict attention to the segment of the curve traced out while t is in M. Now this segment is the graph of a function $f:x \mapsto f(x)$; in fact, f is given by

$$f = Y \circ X^{-1}$$

where X^{-1} is the local inverse of X on M. By the Local Inverse Function Theorem, we know that X^{-1} is differentiable and that $(X^{-1})'(x) = 1/X'(t)$, where $x = X(t)$. Since Y is differentiable, so is the composite $f = Y \circ X^{-1}$. Applying the Chain Rule to $y = f(x) = Y(X^{-1}(x))$, we compute

$$\frac{dy}{dx} = f'(x) = Y'(X^{-1}(x))\frac{1}{X'(t)} = \frac{Y'(t)}{X'(t)}$$

Example 7 Consider the parametric equations

$$X:t \mapsto \cos t = x$$

$$Y:t \mapsto \sin t = y$$

whose locus is the unit circle (Figure 8-11). A local inverse of X is the function $x \mapsto \arccos x$, where $-1 < x < 1$. Then composing Y on X^{-1}, we have

$$f(x) = Y \circ X^{-1}(x) = \sin(\arccos x)$$
$$= \sqrt{1-x^2}$$

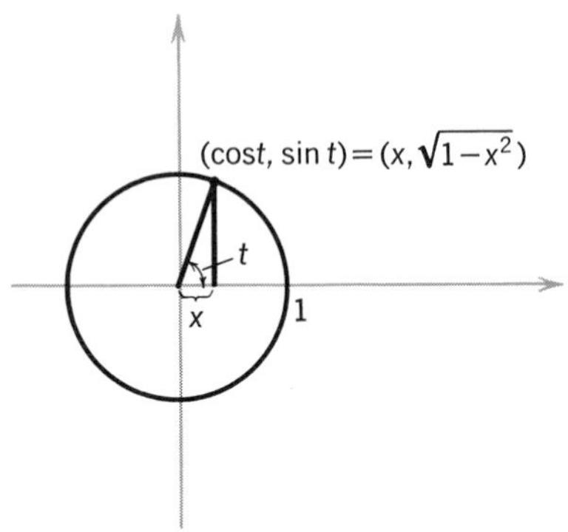

FIGURE 8-11

It is now easy to verify that $f'(x) = Y'(t)/X'(t)$ [where $x = X(t)$], for we have

$$\frac{Y'(t)}{X'(t)} = \frac{\cos t}{-\sin t} = \frac{x}{-y} = -\frac{x}{\sqrt{1-x^2}} = f'(x)$$

Having discussed inverse functions, we make some remarks on implicit functions and differentials. Recall that in §4.1 our main concern was developing techniques for differentiating implicit functions, whose existence and differentiability were taken for granted. At this time, we would like to examine these problems more closely.

Suppose we have a relation in two variables given by

$$F(x, y) = 0 \tag{1}$$

This equation expresses a condition which a point may or may not satisfy. The set of all points which satisfy this condition (the set may be empty) is the locus of equation (1). We know that this locus is often some sort of curve, For example, if

$$F(x, y) = x^2 + y^2 - 1$$

the locus is the unit circle. Now suppose $f(x)$ is a function defined for certain values of x. Then we say that $y = f(x)$ is a **solution** of equation (1) (or that we can solve this equation for y in terms of x) if all the points $(x, f(x))$ are part of the locus of (1), that is, if

$$F(x, f(x)) = 0$$

In some cases we find several functions, $y = f_1(x)$, $y = f_2(x), \ldots$ and the graphs of all these functions taken together form the locus of the relation. The relation $F(x, y) = 0$ is then said to define y **implicitly** in terms of x. [In a similar manner, $F(x, y) = 0$ might also define x implicitly in terms of y.]

The questions we would like to consider in this section are: "Under what conditions is a solution possible?" and "What can be said about the differentiability of the function f in terms of what may be known about the relation $F(x, y) = 0$."

Example 8 Solve the equation

$$\frac{(x-3)^2}{4} + \frac{(y+1)^2}{9} - 1 = 0$$

(Figure 8-12) for y in terms of x, and for x in terms of y. First solving for y, we obtain

$$\frac{(y+1)^2}{9} = 1 - \frac{(x-3)^2}{4}$$

$$(y+1)^2 = 9 - \tfrac{9}{4}(x-3)^2$$

$$y + 1 = \pm 3\sqrt{1 - \tfrac{1}{4}(x-3)^2}$$

$$y = -1 \pm \tfrac{3}{2}\sqrt{4 - (x-3)^2}$$

$$y = -1 \pm \tfrac{3}{2}\sqrt{-x^2 + 6x - 5} \tag{2}$$

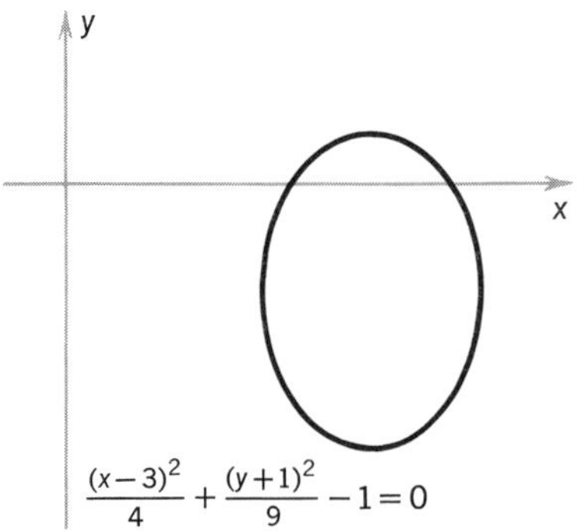

FIGURE 8-12

The $\pm$ sign shows that the solution for y defines **two** functions of x (the top and bottom halves of the ellipse). Similarly, we could solve for x in terms of two functions of y (the right and left halves of the ellipse):

$$x - 3 = \pm 2\sqrt{1 - \tfrac{1}{9}(y+1)^2}$$

$$= \pm \tfrac{2}{3}\sqrt{9 - (y+1)^2}$$

$$x = 3 \pm \tfrac{2}{3}\sqrt{-y^2 - 2y + 8}$$

Suppose that in solving $F(x,y) = 0$ for y in terms of x, we obtain functions $f_1, f_2, \ldots$, and that f_1 is defined **throughout an open neighborhood** of the point x_0. If $f_1(x_0) = y_0$, we have $F(x_0, f_1(x_0)) = F(x_0, y_0) = 0$, and f_1 is called a **local representation** of the relation $F(x,y) = 0$ at the point (x_0, y_0). More explicitly, a **local representation** of $F(x,y) = 0$ at (x_0, y_0) is a function f, whose domain is an open interval $]a,b[$ containing x_0, such that $f(x_0) = y_0$, and $F(x_0, f(x_0)) = 0$ for all x in $]a,b[$.

Example 9 Find a local representation for

$$\frac{(x-3)^2}{4} + \frac{(y+1)^2}{9} - 1 = 0$$

at $(2, \frac{3}{2}\sqrt{3} - 1)$ and at $(1, -1)$. All we have to do is determine which of equations (2) for y in terms of x gives a neighborhood of each point in question. For $(2, \frac{3}{2}\sqrt{3} - 1)$, this is the equation using the $+$ part of the $\pm$ sign; thus, the desired local representation is

$$y = f(x) = -1 + \tfrac{3}{2}\sqrt{-x^2 + 6x - 5}$$

For $(1, -1)$, however, there is no local representation at $x_0 = 1$ since the relation is not defined for any value of x less than 1; therefore, no **open** interval containing x_0 could be a domain for a local representation (Figure 8-12).

If $F(x,y) = 0$ defines y implicitly in terms of x, we may want to calculate the rate of change of y with respect to x at a point (x_0, y_0) in the locus of F. In §4.1, we accomplished this by deriving the relation

$dF(x, y, dx, dy) = 0$, solving for dy/dx, and evaluating the resulting expression at (x_0, y_0). If we have a local representation $y = f(x)$ for F at (x_0, y_0), it is not unreasonable to expect that $f'(x_0)$ would also give the desired rate of change. That the method we used in §4.1, and the method of local representations do in fact produce the same result is the content of the **Implicit Function Theorem**. This theorem provides a basis for the method of implicit differentiation, because the existence of the local representation $f(x)$ means that the tangent problem for $F(x, y) = 0$ is reduced to that for $y = f(x)$. This means that implicit differentiation is based on the method of computing tangents by means of limits. For technical reasons, we state the theorem without being completely rigorous about the hypotheses.

Theorem 4 **The Implicit Function Theorem.** Let $F(x, y) = 0$ be a relation and let the point (x_0, y_0) be in the locus of the relation. If the differential relation $dF(x, y, dx, dy) = 0$ exists and if $dF(x_0, y_0, dx, dy)$ does not reduce identically either to $dx = 0$ or to $0 = 0$, then there exists a differentiable local representation f for $F(x, y) = 0$ at (x_0, y_0), and the quotient of differentials dy/dx, found by solving the differential relation, is equal to $f'(x)$.

The proof of this theorem is beyond the scope of this course. The theorem assures that with sufficient hypotheses on $F(x, y)$, the graph of $F(x, y) = 0$ "breaks up" into pieces of graphs of differentiable functions.

Example 10 Consider the ellipse

$$\frac{(x-3)^2}{4} + \frac{(y+1)^2}{9} = 0$$

The differential relation is

$$\frac{2(x-3)}{4}\,dx + \frac{2(y+1)}{9}\,dy = 0 \tag{3}$$

whence,

$$\frac{dy}{dx} = -\frac{9(x-3)}{4(y+1)}$$

At $(2, \frac{3}{2}\sqrt{3} - 1)$, we evaluate

$$\left.\frac{dy}{dx}\right|_{(2, 3/2\sqrt{3}-1)} = -\frac{9(-1)}{4(\frac{3}{2}\sqrt{3})} = \frac{\sqrt{3}}{2}$$

Similarly, differentiating the local representation

$$f(x) = -1 + \tfrac{3}{2}\sqrt{-x^2 + 6x - 5}$$

see Example 9), yields

$$f'(x) = \frac{3}{2} \cdot \frac{-2x+6}{2\sqrt{-x^2+6x-5}}$$

$$= \frac{-3x+9}{2\sqrt{-x^2+6x-5}}$$

and evaluating at $x = 2$ gives

$$f'(2) = \frac{-6+9}{2\sqrt{3}} = \frac{\sqrt{3}}{2}$$

At $(1, -1)$, we have already remarked that there is no local representation, but there equation (2) becomes

$$\frac{2(1-3)}{4}\,dx + \frac{2(-1+1)}{9}\,dy = 0$$

which reduces to

$$-dx + 0 \cdot dy = 0$$

or just $dx = 0$. Therefore, one of the conditions of the theorem is violated.

So far in our discussion of local representations we have "preferred" the variable x over the variable y. If we have a relation $F(x, y) = 0$ and if (x_0, y_0) is in the locus of F, it may happen that there is a function g such that

(1) g is defined in a neighborhood of y_0; and
(2) $F(g(y), y) = 0$ for all y in the domain of g.

Such a function is also called a **local representation** of F at (x_0, y_0). We leave it to the reader to reword the Implicit Function Theorem in terms of local representations of the form $y = g(x)$.

We can consider local inverse functions in the light of local representations. If $y = f(x)$ specifies a function, then $F(x, y) = y - f(x) = 0$ is a relation. A local representation of this relation at (x_0, y_0) of the form $x = g(y)$ is just a local inverse of f at x_0.

Example 11 Find a local inverse of the function given by $y = x^2 + 3x + 2$, at the point $x = -1$. We solve the quadratic equation $x^2 + 3x + (2 - y) = 0$,

$$x = \frac{-3 \pm \sqrt{9 - 4(2-y)}}{2}$$

$$= \frac{-3 \pm \sqrt{1+4y}}{2}$$

Here x has actually been solved for two functions of y (corresponding to the $\pm$ sign). Substituting values reveals that the $+$ is the sign desired here, so the local inverse is given by

$$x = -\frac{3}{2} + \frac{\sqrt{1+4y}}{2}$$

EXERCISES

1. Prove that the inverse of a one-one odd function is also odd. [Hint Examine the graph of the inverse function.] What about even functions?
2. In Figure 8-5 (page 200), show that $\alpha = \beta$.
3. Prove that the inverse of a monotone function is monotone.
4. (a) Prove that a continuous one-one function with domain an interval is monotone.
   (b) Show that the inverse of a one-one continuous function is continuous. [Hint: Use part (a) together with Theorem 1.]
5. In parts (a)–(c), find the derivative of f^{-1} at the point given in two ways, first using the Inverse Function Theorem, then by explicit computation and differentiation of the inverse at the point in question.
   (a) $f: x \mapsto 1/(1-x)$, at $f(x) = 1$, $x = 0$
   (b) $f: x \mapsto \sqrt{1 + x^2}$ $(x > 0)$, at $f(x) = \sqrt{10}$, $x = 3$
   (c) $f: x \mapsto \sqrt[3]{x^2 + 3x}$ $(x > -\frac{3}{2})$, at $f(x) = 0$, $x = 0$
6. Compute the derivatives of the local inverses to the trigonometric functions (the sine was done in Example 4). [Hint: $(\sec^{-1})'(y) = 1/(|y|\sqrt{1+y^2})$.]
7. In parts (a)–(d), find local representations for the given relations at the points indicated. Then compute the rate of change of y with respect to x at these points, first by implicit differentiation of the relation and then by differentiating the local representation.
   (a) $x^{2/3} + y^{2/3} - 1 = 0$, at $(\frac{1}{8}, 3\sqrt{3/8})$
   (b) $x^4 + x^2 y^2 - 100 = 0$, at $(3, -\sqrt{19/3})$
   (c) $2x^2 - xy + 3y^2 - 18 = 0$, at $(3, 1)$
   (d) $x^2 - xy + y^2 - 4 = 0$, at $(-2, 2)$

REVIEW EXERCISES FOR CHAPTER 4

1. Define the following terms.

acceleration
components of a vector
continuity on a closed interval
differential
differential relation
force
implicit differentiation
inner product
left-hand derivative
local inverse
local representation
maximum of f on $[a, b]$
minimum of f on $[a, b]$
momentum
neighborhood of infinity
norm
one-sided limit
orthogonal
parallel functions
parameter
parametric equations
parametrized curve
right-hand derivative
scalar
scalar multiplication
standard basis vector
unit vector
vector
vector-valued function
velocity
zero vector

2. Restate the arithmetic theorems of differentiation, and the Chain Rule, in terms of differentials.
3. Give examples of neighborhoods of ∞ and $-\infty$.

In exercises 4–9, compute the (possibly infinite) limits.

4. $\lim_{x\to 3}\dfrac{1}{(x-3)^2}$
5. $\lim_{x\to\infty}\dfrac{1}{(x-3)^2}$
6. $\lim_{x\to\infty}\dfrac{x^2+7x+3}{3x^2+17x-2}$
7. $\lim_{x\to-\infty}\dfrac{2x-1}{x}$
8. $\lim_{x\to\infty}\dfrac{x^2+7x+3}{3x^3+17x-2}$
9. $\lim_{x\to-\infty}\dfrac{\sqrt{x^2-1}}{x}$

In exercises 10–14, indicate why finite limits do not exist.

10. $\lim_{x\to 0}\dfrac{1}{x^4}$
11. $\lim_{x\to 2}\dfrac{2}{x-2}$
12. $\lim_{x\to\infty} x^2+4$
13. $\lim_{x\to-\infty} x^2+4$
14. $\lim_{x\to 0}\dfrac{|x|}{x}$

In exercises 15–20, compute the (possibly infinite) one-sided limits.

15. $\lim_{x\to 0^+}\dfrac{1}{x^5}$
16. $\lim_{x\to 0^-}\dfrac{1}{x^3}$
17. $\lim_{x\to 0^+}\dfrac{1}{x^4}$
18. $\lim_{x\to 1^+}\dfrac{\sqrt{x^2-1}}{\sqrt{x-1}}$
19. $\lim_{x\to 1^+}\dfrac{|x-1|}{x-1}$
20. $\lim_{x\to 1^-}\dfrac{|x-1|}{x-1}$
21. Prove that if $\lim_{x\to a^+} f(x)$ and $\lim_{x\to a^-} f(x)$ both exist and are equal, then $\lim_{x\to a} f(x)$ exists and equals the one-sided limits.
22. Prove that if f is continuous on $[a,b]$ and on $[b,c]$, then f is continuous on $[a,c]$.
23. Suppose f is continuous on $[a,b]$ and g is continuous on $[b,c]$. Why might the function

$$h: x \mapsto \begin{cases} f(x), & a \leqslant x \leqslant b \\ g(x), & b < x \leqslant c \end{cases}$$

not be continuous on the interval $[a,c]$? What further conditions on f and g are needed?

24. Suppose that f is continuous on $[a,b]$ and differentiable on $]a,b[$. Prove that f is constant on $[a,b]$ if and only if f' is identically zero on $]a,b[$. Is the continuity of f necessary for the theorem to hold?

In exercises 25–29, compute the first and second derivatives of the function and evaluate at the point indicated.

25. $x \mapsto \left(\dfrac{1-x}{1+x}\right)^{2/3}$, at $x = 1$
26. $x \mapsto (x^2 - \frac{2}{3}a^2)^{1/5}(a^2-x^2)^{3/2}$, at $x = a$

27. $x \mapsto 3x^{1/3} - 4x^{-1/4} + \frac{3}{7}x^{7/3}$, at $x = 1$
28. $x \mapsto \frac{1}{5}(x^2+x-3)\sqrt{x^2+x-3}$, at $x = 4$
29. $x \mapsto \dfrac{\cot \frac{1}{2}x}{\sqrt{1-\cot^2(\frac{1}{2}x)}}$, at $x = \pi$

In exercises 30–33, compute dy/dx and indicate where the tangent to the locus is horizontal and where it is vertical.

30. $x^3 + y^3 - 3xy = 0$
31. $y^2 - 2x + 1 = 0$
32. $y^2 - \sin x = 0$
33. $x^2 y + 3xy - 4y^2 + 1 = 0$

In exercises 34–48, compute the equation of the tangent line at (x_0, y_0) to the locus described by the function, relation, or parametric equations. Express your results in both xy and $dx\,dy$ coordinates.

34. $x^3 + y^3 = r^3$
35. $|x|^{1/2} + |y|^{1/2} = r^{1/2}$
36. $x = 2t, \quad y = -5t$
37. $y = (9-x^{2/3})^{3/2}$
38. $x^5 - 2x^3y^2 + 3xy^4 - y^5 - 5 = 0$
39. $y = \cos(\tan x)$
40. $2x^2 - 3xy - 4y^2 - 5 = 0$
41. $y = \sin(\cos t), \quad x = \sin t$
42. $x^4 + y^4 = r^4$
43. $x = t^2 - 2, \quad y = t^3 - 3t + 1$
44. $x^2 + y^2 = xy$
45. $x = t^2 - 2t + 3, \quad y = t^2 + t + 1$
46. $x = \cos^3\theta, \quad y = \sin^3\theta$
47. $x = 2 + \frac{1}{2}s^2, \quad y = -1 + \frac{1}{8}s^3$
48. $x = 4\cos\theta, \quad y = 2\sin^2\theta$

In exercises 49–52, find the maximum value of the function on the interval indicated.

49. $x \mapsto x^3 - 12x + 2$, on $[-5, 4]$
50. $x \mapsto x^2(x-4)^3$, on $[-3, 1]$
51. $x \mapsto x^3(4-x)^3$, on $]-\infty, +\infty[$
52. $x \mapsto x(x-2)^2(x-4)^3$, on $[0, 3]$
53. Show that the function $x \mapsto 1/x$ attains neither a maximum nor a minimum on $]0, +\infty[$.
54. Does the Mean Value Theorem apply to $x \mapsto |x|$ on the interval $[-5, 2]$? Explain.
55. A particle moves along the locus of the relation $x^2 + y^2 - y = \sqrt{x^2 + y^2}$ in such a way that its position is described by a continuously differentiable parametrization. What are the values of the derivatives of the functions describing the particles position as it passes through the origin? Explain.

56. The equation for adiabatic expansion of air is $PV^{1.4} = C$, where P is pressure, V is volume, and C is a constant. At a certain time the pressure is 50 lbs/in² in a volume of 70 cubic inches and at this certain time the volume is decreasing at a rate of 10 cubic inches per second. How fast is the pressure increasing at this time?

57. A ten-foot lamp illuminates a six-foot high wall from a distance of twenty feet. An owl perched atop the wall spies a moth hovering at the lamp and swoops directly towards it at a speed of 20 feet per second. How fast is the height above the ground of the owl's shadow changing when the owl is three feet from the lamp?

58. (a) Prove the following, under suitable conditions:

$$\lim_{x \to x_0} \frac{f(x) - f(x_0)}{\sqrt{x} - \sqrt{x_0}} = 2f'(x_0)\sqrt{x_0}$$

(b) Evaluate $\lim_{x \to 1} \frac{x^2 - 1}{\sqrt{x} - 1}$

(c) Evaluate $\lim_{x \to 1^+} \frac{\sqrt{x^2 - 1}}{\sqrt{x - 1}}$

5

APPLICATIONS OF DIFFERENTIAL CALCULUS

In this chapter we present important applications of the techniques developed in Chapters 3 and 4. We shall discover that derivatives yield extensive information that can be used for the graphing of functions, and the location of their extreme values. The method of differential approximation is developed and L'Hôpital's Rule is used to solve difficult limit problems. Finally, we introduce the concept of the antiderivative, which is the basis for our later study of integral calculus and differential equations. The normal to a curve and the radius of curvature are discussed in an optional section.

1 CURVE SKETCHING

We have already seen how the first derivative gives information about the shape of the graph of a function. In this section, we carry this analysis further and show how the second derivative also provides data on the graph. We begin by consolidating the graphing information gleaned from the first derivative. The **First Derivative Test** (§3.1) indicates that on intervals where the derivative is positive, the function is monotone increasing; where the derivative is negative, the function is monotone decreasing. To state the First Derivative Test, we defined what it meant for a function to be "monotone (increasing or decreasing) on an interval."

The concept of maximum and minimum value of a function can also be relativized to intervals. (In fact, we have already relativized it to closed intervals; see §4.6.) We say that $f(x_0)$ is the **maximum of f on $]A, B[$** if $x_0 \in]A, B[$ and if $f(x_0) \geq f(x)$ for all $x \in]A, B[$. When $f(x_0)$ is the maximum of f on some neighborhood of x_0, we say that $f(x_0)$ is a **relative** or **local maximum** of f. When $f(x_0)$ is a relative maximum of f, we also say that **f has a relative maximum at x_0**. Similarly, we say that f has a **local** or

relative minimum at x_0, if $f(x_0)$ is the minimum of f on **some** neighborhood of x_0. (See Figure 1-1.) It is then an immediate consequence of the Extreme Value Test (§4.7) that any local maximum or minimum of a differentiable function occurs at a critical point of the function. Therefore, all the local maxima and minima of a differentiable function f occur at solutions of the equation $f'(x) = 0$. However, as we have already pointed out (§4.7), not all solutions of this equation necessarily give local maxima or minima (Figure 1-2).

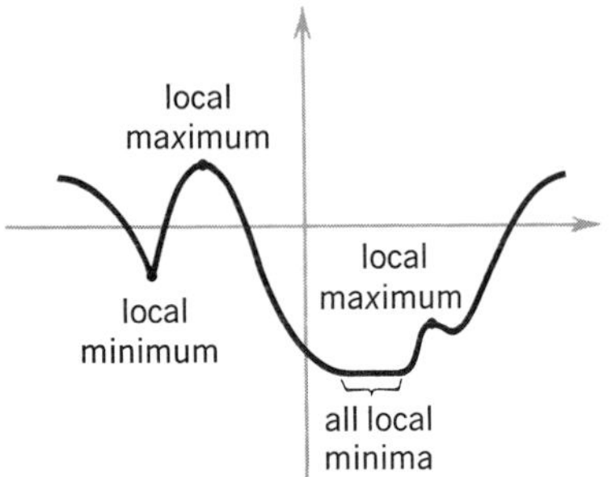

FIGURE 1-1

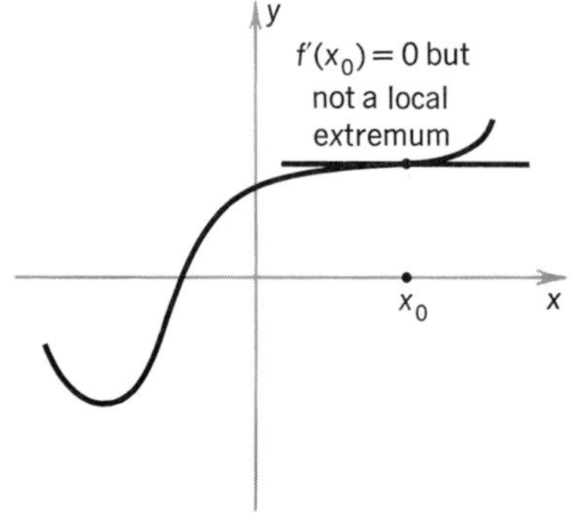

FIGURE 1-2

We now turn our attention to the second derivative. Basically, the sign of the second derivative tells where the derivative is increasing and where it is decreasing. This in turn gives a remarkable amount of information about the shape of the graph of the function itself.

Let us start by considering the function $x \mapsto x^2$. In §3.1 we demonstrated that the graph of this function, a parabola, had a very interesting property: every line tangent to the graph intersects it only at the point of tangency and otherwise lies entirely below the graph (Figure 1-3). The derivative of $x \mapsto x^2$ is $x \mapsto 2x$, which is a monotone increasing function. The first theorem of this section will reveal that all differentiable functions with monotone increasing derivatives have the property just described, and this property is important enough to be named.

Definition A differentiable function f is said to be **concave upward** if every tangent to the graph of f intersects the graph only at the point of tangency and otherwise lies entirely below the graph of f.

The function $x \mapsto x^2$ is concave upward as we have seen. In Figure 1-4, we have an example of a function which is concave upward but which does not have a critical point.

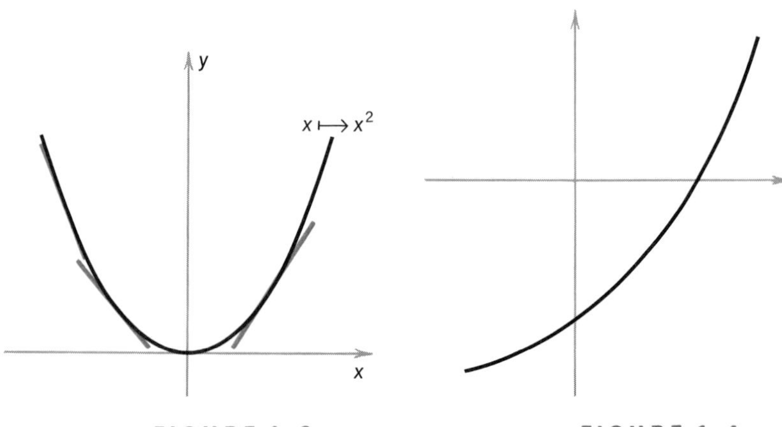

FIGURE 1-3

FIGURE 1-4

Changing the phrase "entirely below" to "entirely above," we have a related property.

Definition A differentiable function f is said to be **concave downward** if the tangent to the graph of f intersects the graph only at the point of tangency and otherwise lies entirely above the graph of f.

By way of example, $x \mapsto -x^4$ is concave downward [Figure 1-5(a)]; in this example, the derivative $x \mapsto -4x^3$ is monotone decreasing [Figure 1-5(b)]. We now state and prove the theorem which relates downward concavity to decreasing derivatives and upward concavity to increasing derivatives.

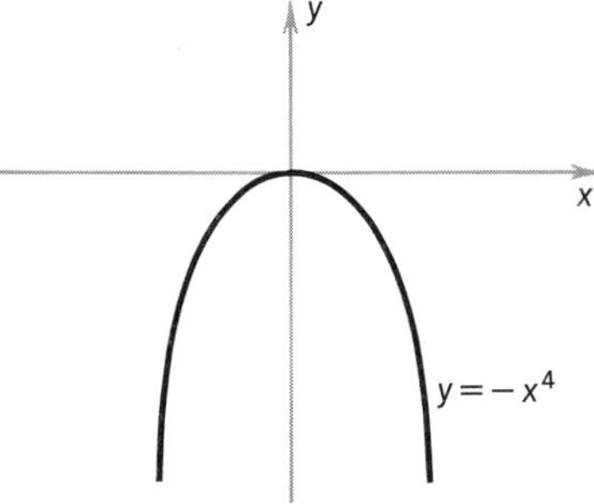

FIGURE 1-5(a)

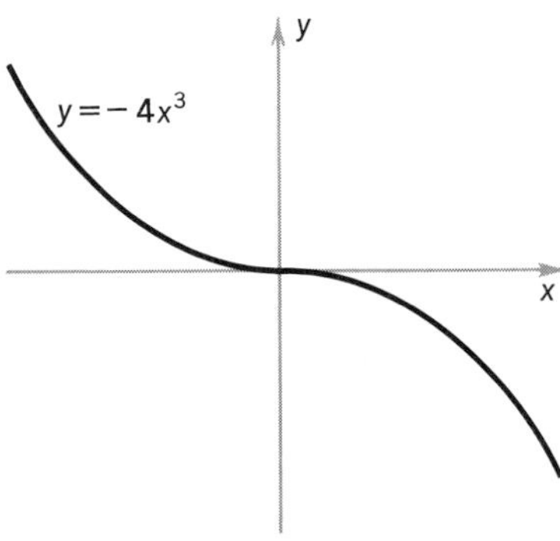

FIGURE 1-5(b)

Theorem 1 Let f be a differentiable function whose domain is an open interval. If f' is monotone increasing, then f is concave upward; if f' is monotone decreasing, then f is concave downward.

Proof We shall prove the theorem for the case f' monotone increasing and leave the decreasing case for the student. So suppose f' is increasing. Given a point x_0, the tangent line to the graph of f at $(x_0, f(x_0))$ is itself the graph of the affine function $T: x \mapsto f(x_0) + f'(x_0)(x - x_0)$. We have to show that $x \neq x_0$ implies $f(x) > T(x)$. First consider the case $x > x_0$. We know that $f'(c) > f'(x_0)$ for all points c in the interval $[x_0, x]$. Applying the Mean Value Theorem to f on the interval $[x_0, x]$ we conclude that for some c in $]x_0, x[$,

$$f(x) - f(x_0) = f'(c)(x - x_0)$$

or

$$f(x) = f(x_0) + f'(c)(x - x_0)$$

Since $f'(c) > f'(x_0)$ and $(x - x_0) > 0$, we conclude that $f(x) > T(x)$. Now let us consider the case $x < x_0$. Applying the Mean Value Theorem to f on $[x, x_0]$, we obtain a point c in $]x, x_0[$ such that

$$f(x) - f(x_0) = f'(c)(x - x_0)$$

or

$$f(x) = f(x_0) + f'(c)(x - x_0)$$

But now $(x - x_0) < 0$ and $f'(c) < f'(x_0)$, and so

$$f'(c)(x - x_0) > f'(x_0)(x - x_0)$$

Hence, once again $T(x) < f(x)$. ∎

We have the following important corollary to this theorem.

Corollary 1 Let f be a twice differentiable function whose domain is an open interval. If $f''(x) > 0$ for all x, then f is concave upward. If $f''(x) < 0$ for all x, then f is concave downward.

Proof We apply the First Derivative Test to the function f'. If $f''(x) > 0$ for all x, then f' is monotone increasing. If $f''(x) < 0$ for all x, then f' is monotone decreasing. The result now follows immediately from Theorem 1. ∎

Example 1 (a) The function $g: x \mapsto 5x^2 - \cos x$ has derivative

$$g': x \mapsto 10x + \sin x$$

and second derivative

$$g'': x \mapsto 10 + \cos x$$

Since $g''(x) > 0$ for all x, g' is increasing and g is concave upward.

(b) Consider $h: x \mapsto \sqrt{x^2+1}$. The derivative is $h': x \mapsto x/\sqrt{x^2+1}$. The reader might be able to see directly that h' is increasing. Alternatively, one can check that $h''(x) > 0$ for all x. We have

$$h''(x) = \frac{1}{(x^2+1)^{3/2}}$$

Since $(x^2+1)^{3/2} > 0$ for all x, one has $h''(x) > 0$ for all x. The graph of h is sketched in Figure 1-6.

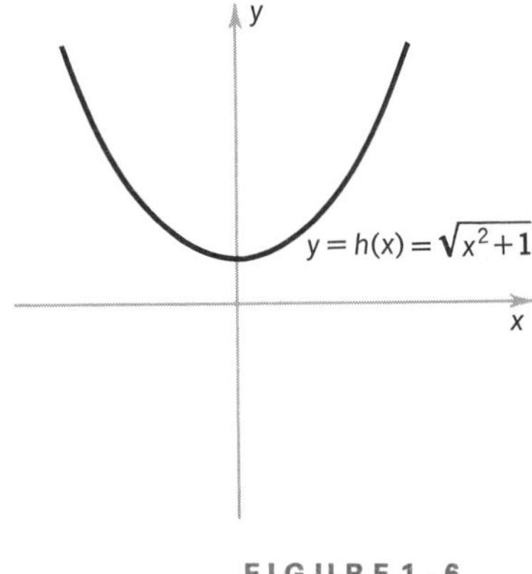

FIGURE 1-6

(c) If f is concave upward (respectively, downward), then the affine substitution $F: x \mapsto f(mx+b)$ $(m \neq 0)$ is also concave upward (respectively, downward). This is clear from the geometric interpretation of affine substitution; however, we can also verify this using Corollary 1. By the Chain Rule

$$F'(x) = mf'(mx+b)$$

and

$$F''(x) = m^2 f''(mx+b)$$

now $m^2 > 0$, so $F''(x) > 0$ if $f''(mx+b) > 0$. Thus if f'' is always positive, F'' is also always positive. Similarly, if f'' is always negative, F'' is always negative.

The concavity of a function indicates whether a critical point gives a maximum or a minimum value of the function. If x_0 is a critical point of f, the tangent to the graph of f at $(x_0, f(x_0))$ is horizontal and has equation $y = f(x_0)$. Suppose f is concave up and that $f'(x_0) = 0$. Then the graph of f intersects the line $y = f(x_0)$ only at $(x_0, f(x_0))$ and otherwise the graph lies entirely above this line. So $(x_0, f(x_0))$ is the lowest point on the graph and $f(x_0)$ must be the minimum value of f.

By way of example, consider the function $g: x \mapsto (x-1)^4 + 3$ which is concave up and has a critical point at 1. Therefore, $g(1) = 3$ is the minimum value of g.

By an analogous argument, if the function f is concave down and if $f'(x_0) = 0$, then $(x_0, f(x_0))$ is the highest point on the graph of f and $f(x_0)$ is the maximum value of f.

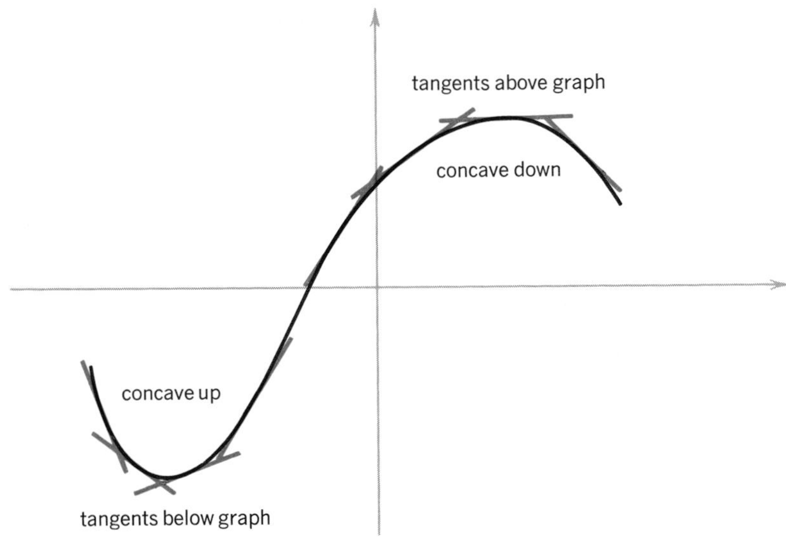

FIGURE 1-7

The concept of concavity can also be relativized to intervals. We say that a differentiable function f is **concave upward on $]A, B[$** if we have $f(x_1) > f(x_0) + f'(x_0)(x_1 - x_0)$ whenever $x_0 \neq x_1$ are points in $]A, B[$. The tangent to $y = f(x)$ at $(x_0, f(x_0))$ has equation $y = f(x_0) + f'(x_0)(x - x_0)$. Hence, f concave upwards on $]A, B[$ means that in the region between $x = A$ and $x = B$, the graph of f lies entirely above every tangent line except for the point of tangency itself (Figure 1-7). Similarly, we say that f is **concave downward on $]A, B[$** if we have $f(x_1) < f(x_0) + f'(x_0)(x_1 - x_0)$ whenever $x_0 \neq x_1$ are point in $]A, B[$. Clearly f concave downward on $]A, B[$ means that in the region bounded by $x = A$ and $x = B$, the graph of f lies entirely below every tangent line except for the point of tangency (Figure 1-7).

As is to be expected we have the following theorem and corollary.

Theorem 1′ Let f be differentiable on $]A, B[$. Then

(i) f is concave upward on $]A, B[$ if f' is monotone increasing on $]A, B[$;

(ii) f is concave downward on $]A, B[$ if f' is monotone decreasing on $]A, B[$.

Corollary 1′ Let f be twice differentiable on $]A, B[$. Then

(i) f is concave upward on $]A, B[$ if $f''(x) > 0$ for all $x \in]A, B[$;

(ii) f is concave downward on $]A, B[$ if $f''(x) < 0$ for all $x \in]A, B[$.

Proof This follows at once from Theorem 1 and its corollary applied to f restricted to $]A, B[$. ∎

We can paraphrase Corollary 1′ as follows: In regions where $f''(x)$ is positive, f is concave upward; where $f''(x)$ is negative, f is concave downward. Figure 1-8 illustrates this situation.

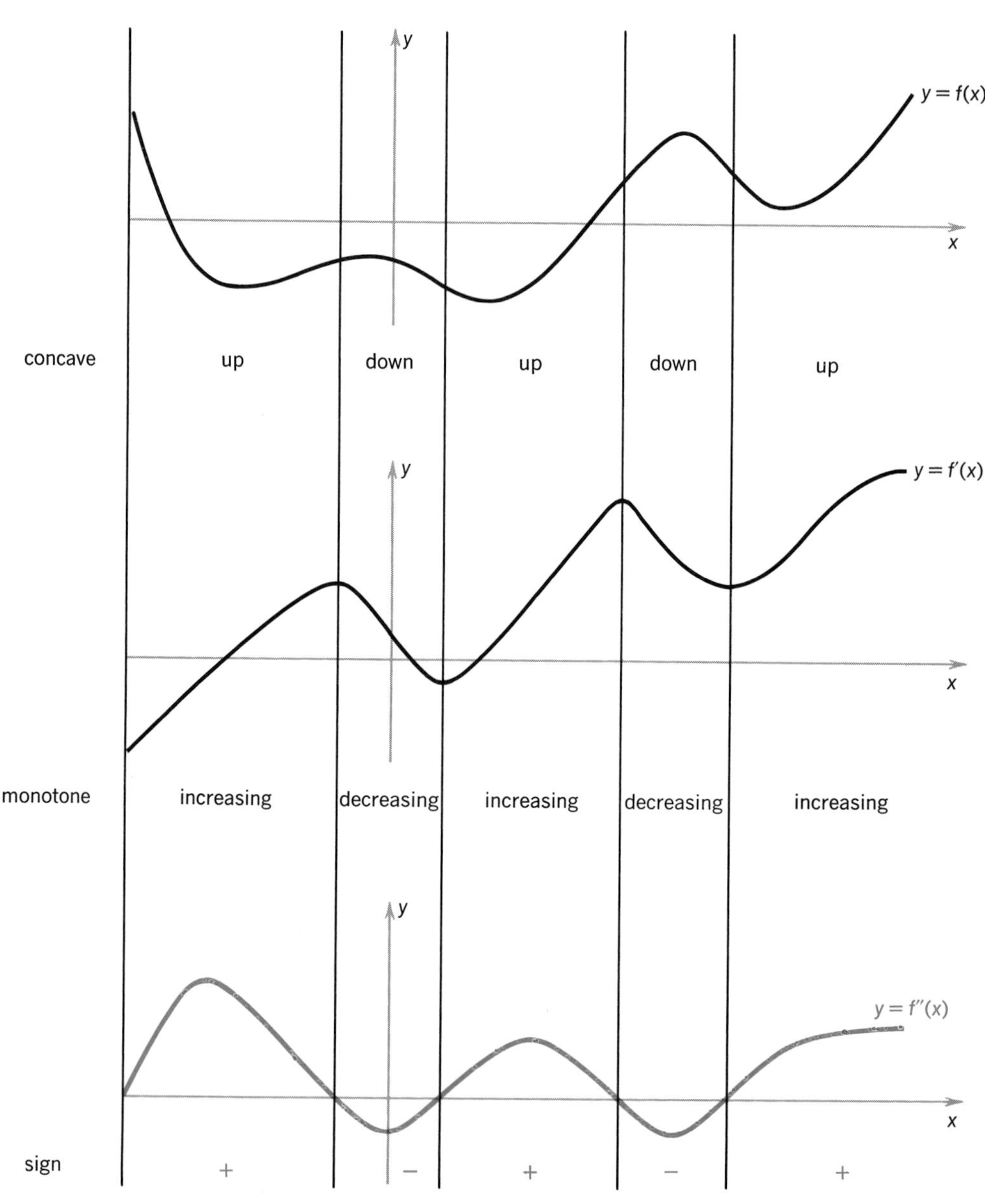

FIGURE 1-8

Example 2 To determine the concavity of the sine function, we compute $d^2(\sin x)/dx^2 = -\sin x$. We notice that $-\sin x > 0$ on the intervals $](2n-1)\pi, 2n\pi[$, so the sine function is concave upward there. Similarly, $-\sin x < 0$ on $]2n\pi, (2n+1)\pi[$, where the sine function is concave downward (Figure 1-9).

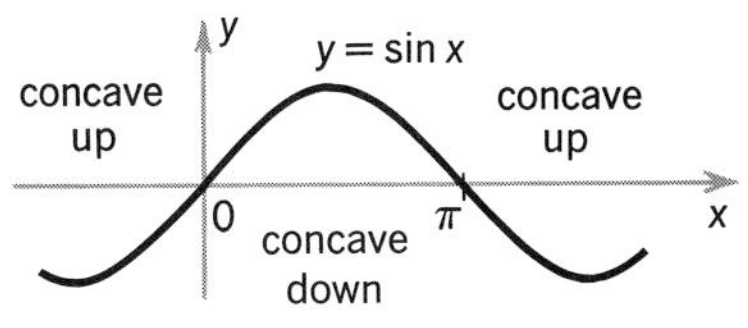

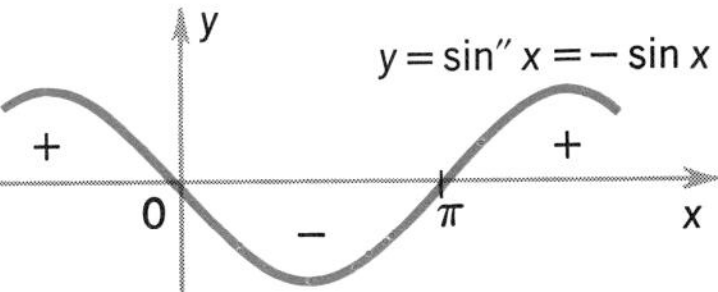

FIGURE 1-9

Example 3 (a) The second derivative of $y = 7x^3 + x^2 - 8x$ is $d^2y/dx^2 = 42x + 2$. Now $42x + 2$ is greater than zero when $x > -1/21$, and less than zero when $x < -1/21$; hence the function is concave upward on $]-1/21, \infty[$ and concave downward on $]-\infty, -1/21[$ (Figure 1-10).

(b) Consider $f: x \mapsto 5x - \sin x$. We have $f': x \mapsto 5 - \cos x$ and $f'': x \mapsto \sin x$. Since $f'(x) > 0$ for all x, the function f is monotone increasing. However, the concavity of f changes often—$f''(x) = \sin x$ and so f is concave upward when $\sin x > 0$ and concave downward when $\sin x < 0$. Hence, the graph of f is as in Figure 1-11.

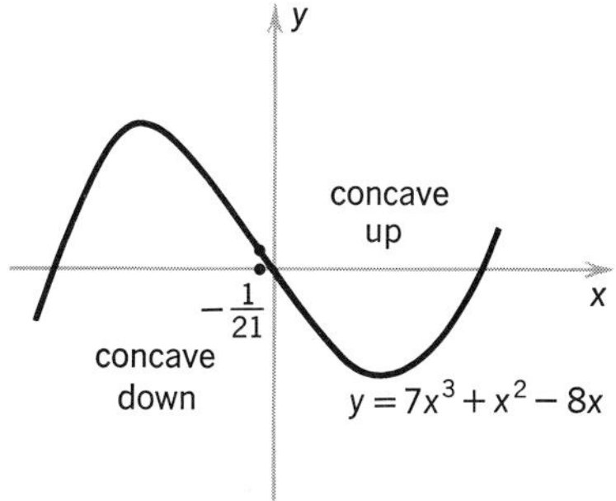

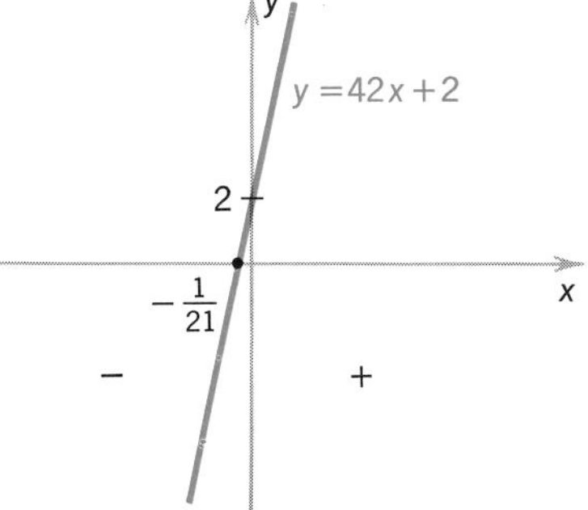

FIGURE 1-10

The direction of the concavity is the key to whether a critical point indicates a local maximum or minimum. Clearly, if $f'(x_0) = 0$ and f is concave downward on some neighborhood of x_0, then $f(x_0)$ is a local maximum (Figure 1-12). Likewise, if $f'(x_0) = 0$ and f is concave upward on some neighborhood of x, then $f(x_0)$ is a local minimum (Figure 1-13). These observations together with Corollary 1′ yield a result which is called the Second Derivative Test.

Theorem 2 **Second Derivative Test** Suppose x_0 is a critical point of f and f'' is continuous at x_0. Then

(i) if $f''(x_0) < 0$, f has a relative maximum at x_0;
(ii) if $f''(x_0) > 0$, f has a relative minimum at x_0.

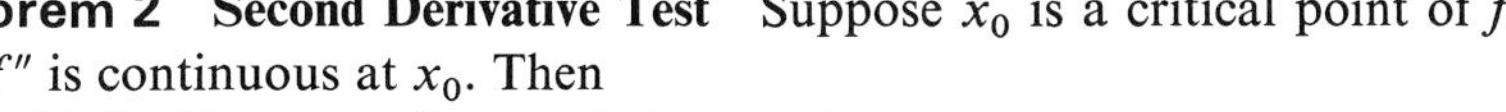

Proof We are given that $\lim_{x \to x_0} f''(x) = f''(x_0)$. Then if $f''(x_0) \neq 0$, there is a neighborhood M of x_0 such that $f''(x)$ does not change sign on M—

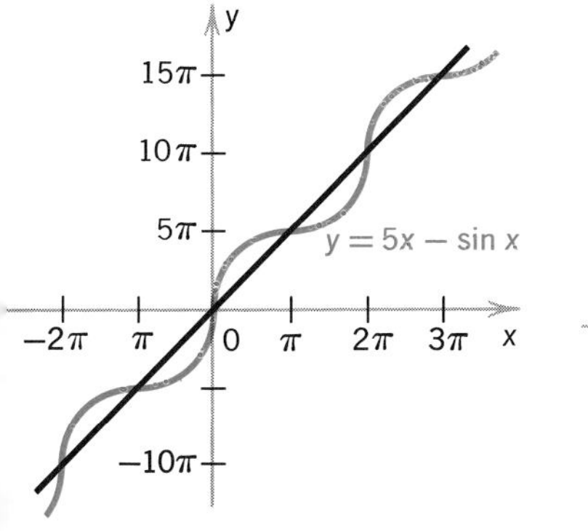

FIGURE 1-11

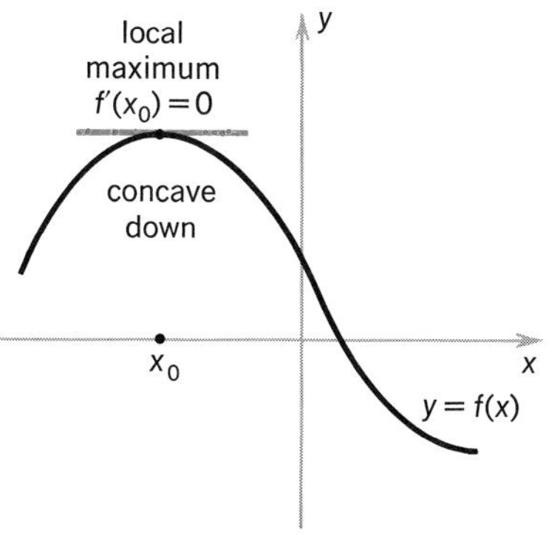

FIGURE 1-12

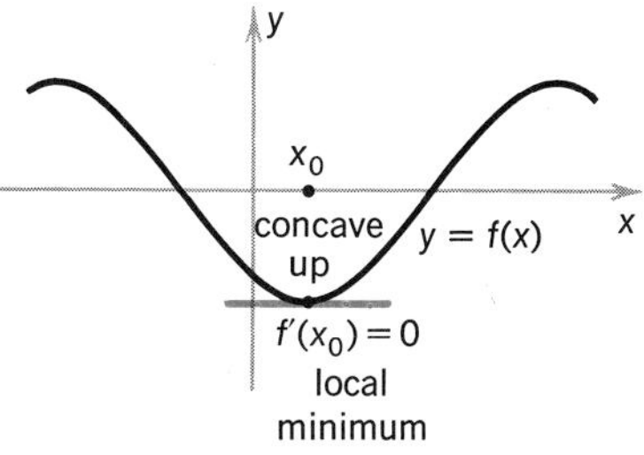

FIGURE 1-13

that is, $\operatorname{sgn}(f''(x)) = \operatorname{sgn}(f''(x_0))$ for all $x \in M$. If $f''(x_0) < 0$, f is concave downward on M and so $f(x_0)$ is a relative maximum. If $f''(x_0) > 0$, the same reasoning shows f is concave upward on M and $f(x_0)$ is a relative minimum. Here we are using the remarks on page 218. ∎

Example 4 The critical points of the sine function (Example 2) occur where $d(\sin x)/dx = \cos x = 0$, that is, at $x = (n+\frac{1}{2})\pi$. To discover at which of these critical points there are local maxima and minima, we compute

$$\left.\frac{d^2(\sin x)}{dx^2}\right|_{x=(n+\frac{1}{2})\pi} = -\sin(n+\tfrac{1}{2})\pi = \begin{cases} -1, & n \text{ even} \\ 1, & n \text{ odd} \end{cases}$$

Therefore, the second derivative test shows that for even n, $\sin(n+\frac{1}{2})\pi = 1$ is a relative maximum of the sine function, while for odd n, $\sin(n+\frac{1}{2})\pi = -1$ is a relative minimum, in agreement with what we know about $\sin x$.

Example 5 Consider the function of Example 3, $y = 7x^3 + x^2 - 8x$. The critical points are the solutions of

$$\begin{aligned} 0 &= dy/dx \\ &= 21x^2 + 2x - 8 \\ &= (3x+2)(7x-4) \end{aligned}$$

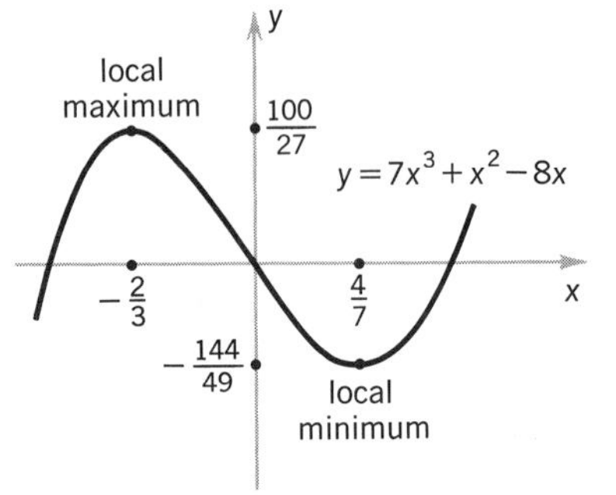

FIGURE 1-14

which are $x = -\frac{2}{3}$ and $x = \frac{4}{7}$. We can see whether these are extreme values using the Second Derivative Test. We compute

$$\frac{d^2y}{dx^2} = 42x + 2$$

so

$$\left.\frac{d^2y}{dx^2}\right|_{x=-2/3} < 0 \quad \text{and} \quad \left.\frac{d^2y}{dx^2}\right|_{x=4/7} > 0$$

Therefore, $7(-\frac{2}{3})^3 + (-\frac{2}{3})^2 - 8(-\frac{2}{3}) = \frac{100}{27}$ is a local maximum of the function, and $7(\frac{4}{7})^3 + (\frac{4}{7})^2 - 8(\frac{4}{7}) = -\frac{144}{49}$ is a local minimum (Figure 1-14).

If $f'(x_0) = f''(x_0) = 0$, we have to check directly whether f' is monotone increasing or decreasing on some neighborhood of x_0. For example, consider the function $k: x \mapsto x^4$. Then $k'(0) = k''(0) = 0$. But the derivative $k': x \mapsto 4x^3$ is an increasing function and so k is concave upward. Therefore, $k(0)$ is the minimum of k.

Suppose a function f is continuous on a closed interval $[A, B]$ and differentiable on $]A, B[$. If f' is monotone (increasing or decreasing) on

$]A, B[$, then f has at most one critical point on the interval. (Why?). Moreover, this critical point gives the maximum of f on $[A, B]$ if f' is decreasing and the minimum of f on the interval if f' is increasing.

We have shown that at points where the concavity of a curve is defined, the tangent does not cross the curve but instead remains on one side of it. Let us now consider the opposite case, where the tangent crosses the curve.

A point where the direction of concavity of a function changes is called a **point of inflection** of the function. Thus, x_0 is a point of inflection of the function f if f is concave upward to the left of x_0 and concave downward to the right of x_0, or vice versa. In other words, x_0 is a point of inflection of f if f is concave one way on some interval $]A, x_0[$ and concave the other way on some interval $]x_0, B[$. For example, the origin is a point of inflection of the sine function, because the sine is concave downward on $]-\pi, 0[$ and concave upward on $]0, \pi[$. The point $\frac{1}{2}\pi$ is a point of inflection of the cosine function, and 1 is an inflection point of $x \mapsto (x-1)^3$. Inflection points are an important aid in graphing functions (Figure 1-15).

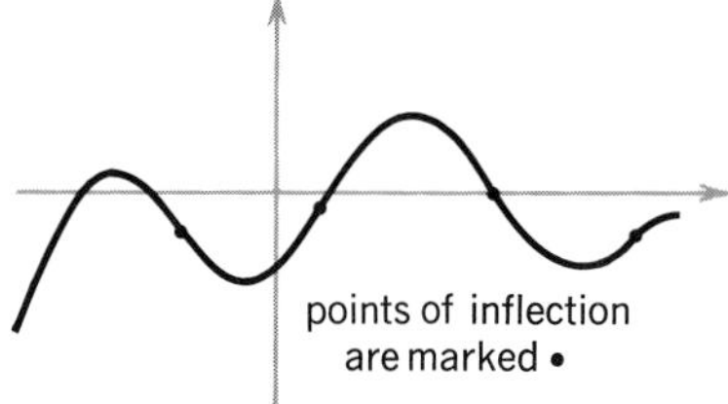

FIGURE 1-15

If x_0 is a point of inflection of f and if f'' is continuous at x_0, we must have $f''(x_0) = 0$. This is easy to see for if $f''(x_0) \neq 0$, then f must either be concave upward or concave downward in some neighborhood of x_0, making it impossible for f to change concavity at x_0. Hence, if f changes concavity at x_0, $f''(x_0) = 0$. For example, we notice that $\sin''(0) = -\sin(0) = 0$; $\cos''(\frac{1}{2}\pi) = -\cos(\frac{1}{2}\pi) = 0$; and

$$\left.\frac{d^2((x-1)^3)}{dx^2}\right|_{x=1} = \left.\frac{d(3(x-1)^2)}{dx}\right|_{x=1} = 0$$

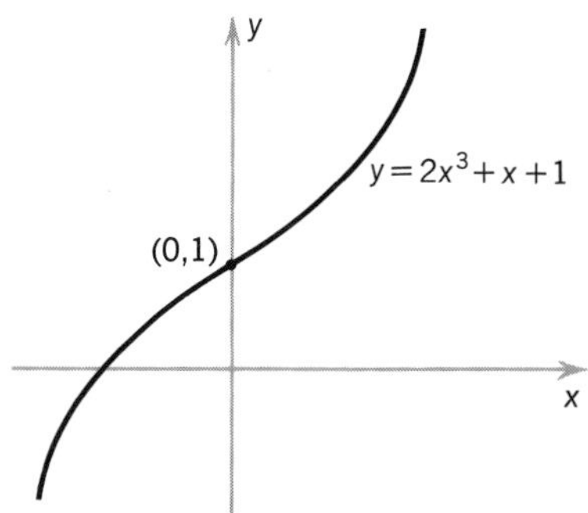

FIGURE 1-16

Example 6 (a) Consider the function $g: x \mapsto 2x^3 + x + 1$. The first and second derivatives are $g': x \mapsto 6x^2 + 1$ and $g'': x \mapsto 12x$. So $g''(x)$ changes sign at the origin and g has a point of inflection there. Since $g''(x) < 0$ when $x < 0$, g is concave downward on $]-\infty, 0[$; and since $g''(x) > 0$ when $x > 0$, g is concave upward on $]0, +\infty[$. Thus the graph of g is as in Figure 1-16.

(b) Since

$$\tan''(x) = \frac{2 \sin x \cos x}{\cos^4 x}$$

$\tan''(x)$ changes sign at 0 and the tangent function has a point of inflection there; in fact this function has an inflection point at all points $x = n\pi$, n an integer (Figure 1-17). Note that $\tan'(0) \neq 0$.

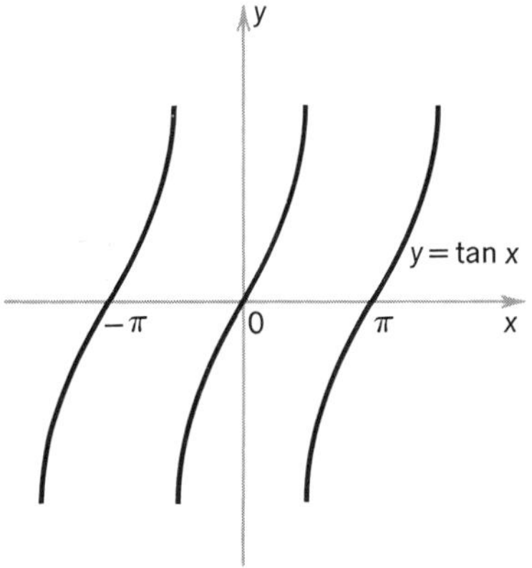

FIGURE 1-17

However, even if $f''(x_0) = 0$, we cannot be certain that f has a point of inflection at x_0. It is still necessary to verify that $f''(x)$ changes sign at x_0. For instance, $f: x \mapsto x^4 + x$ has $f''(0) = 0$, but $f': x \mapsto 4x^3 + 1$ is monotone increasing so f is always concave upward; hence the origin cannot be the point of inflection of f. However, we notice that $f'': x \mapsto 12x^2$ does not change sign at 0 (Figure 1-18).

At a point where both $f'(x_0) = 0$ and $f''(x_0) = 0$, there may be a local extreme value or a point of inflection or neither. However, if the sign of $f'(x)$ changes at most once on some neighborhood of x_0, then there must be a local extreme value or point of inflection at x_0 (see exercise 2). That is, let us suppose there is a neighborhood of x_0 such that x_0 is the only critical point of f on this neighborhood. By examining the sign of $f'(x)$ for x in this neighborhood we can determine which of the following three possible cases occurs:

(1) $f'(x)$ has the same sign for $x > x_0$ and $x < x_0$; in this case there is a point of inflection at x_0.
(2) $f'(x) < 0$ for $x < x_0$ and $f'(x) > 0$ for $x > x_0$; in this case $f(x_0)$ is a local minimum.
(3) $f'(x) > 0$ for $x < x_0$ and $f'(x) < 0$ for $x > x_0$; in this case $f(x_0)$ is a local maximum.

Figure 1-19 illustrates each of the above cases.

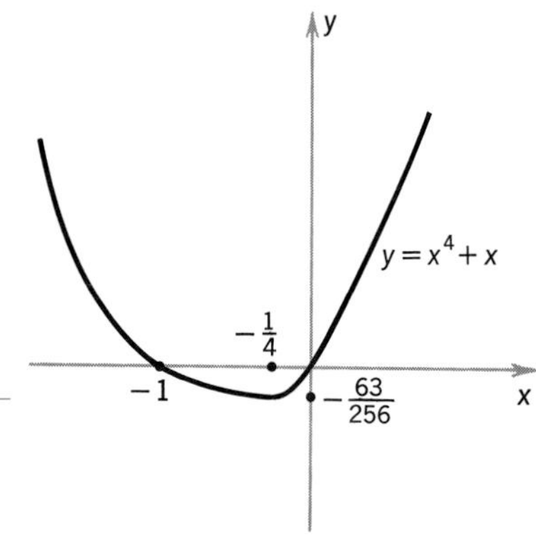

FIGURE 1-18

Example 7 The function $x \mapsto x^4$ has a local minimum at 0 and $x \mapsto x^3$ has a point of inflection at 0. These functions thus come under categories

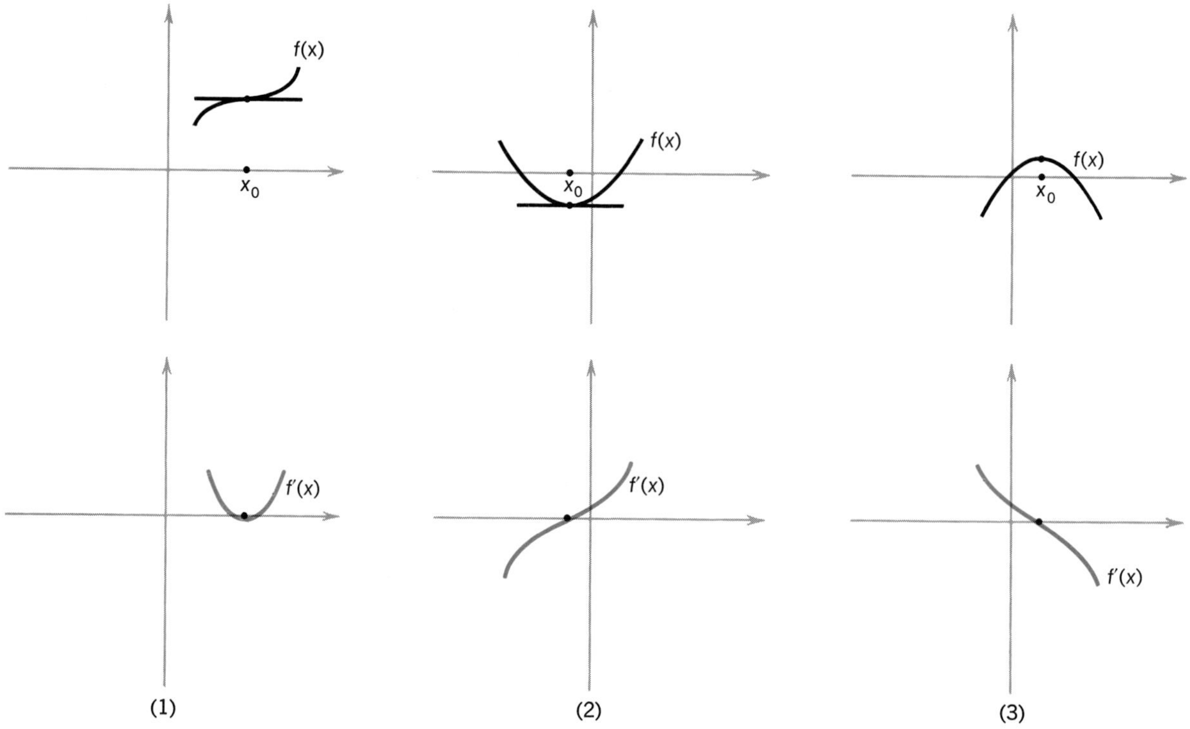

FIGURE 1-19

(2) and (1), respectively. Now consider the function given by $y = 5x^6 - 24x^5 + 30x^4$. We first compute

$$\frac{dy}{dx} = 30x^5 - 120x^4 + 120x^3$$

$$= 30x^3(x-2)^2$$

and

$$\frac{d^2y}{dx^2} = 150x^4 - 480x^3 + 360x^2$$

$$= 30x^2(x-2)(5x-6)$$

Examining these derivatives we see that

$$\left.\frac{d^2y}{dx^2}\right|_{x=6/5} = 0 \neq \left.\frac{dy}{dx}\right|_{x=6/5}$$

and we know that there is a point of inflection at $x = \frac{6}{5}$ (specifically, the point $(\frac{6}{5}, 17\frac{1307}{3125})$) because the second derivative changes sign there. At $x = 0$ and $x = 2$, we have both the first and second derivatives equal to zero. We therefore examine the sign of the first derivative around these values of x. Near $x = 0$, we see that dy/dx is positive for $x > 0$ and negative for $x < 0$; hence by the rule above, there is a relative minimum at $x = 0$ (namely, the value $y = 0$). On the other hand, we find that dy/dx is positive on both sides of $x = 2$; hence by the rule, there is a point of inflection at $x = 2$ [namely, the point $(2, 32)$]. Alternatively, we can verify directly that d^2y/dx^2 changes sign at $x = 2$. (See Figure 1-20.)

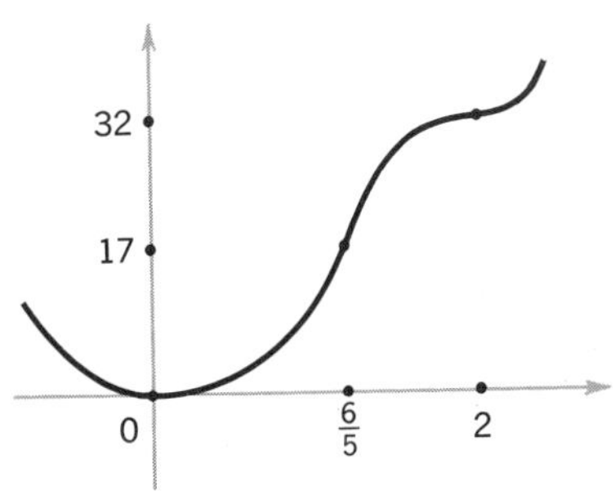

FIGURE 1-20

The above discussion provides a method for graphing functions with continuous first and second derivatives, provided we can determine the zeroes and changes of sign of these derivatives. The method is as follows.

FORMAL DEVICE

If f is a twice continuously differentiable function, then in order to sketch the graph of f we must do the following:

(1) Compute f' and f''. Set $f'(x) = 0$ and solve for critical points. Use the Second Derivative Test at critical points where the second derivative is not zero to find local maxima and local minima. Plot these values.

(2) Set $f''(x) = 0$ and solve for x. If $f''(x_0) = 0$, determine whether $f''(x)$ changes sign at x_0 to see if f has a point of inflection at x_0. If $f''(x_1) = 0 = f'(x_1)$, examine the sign of $f'(x)$ for x near x_1 to determine what happens at x_1. This procedure will yield the points of inflection and will complete the list of local maxima and local minima. Plot these values.

(3) Plot a few additional points if necessary. Draw a smooth curve through these points consistent with the information obtained from the signs of $f'(x)$ and $f''(x)$.

Example 8 Graph the cubic $f\colon x \mapsto x^3+\frac{1}{2}x^2-14x$.

(1) We compute $f'(x)=3x^2+x-14$, and $f''(x)=6x+1$. Critical points occur where $0=f'(x)=(x-2)(3x+7)$, that is, at $x=2$ and $x=-\frac{7}{3}$. Using the Second Derivative Test, we find $f''(2)=13>0$, so $f(2)=-18$ is a local minimum; $f''(-\frac{7}{3})=-13<0$, so $f(-\frac{7}{3})=22\frac{37}{54}$ is a local maximum. We plot these points in Figure 1-21(a).

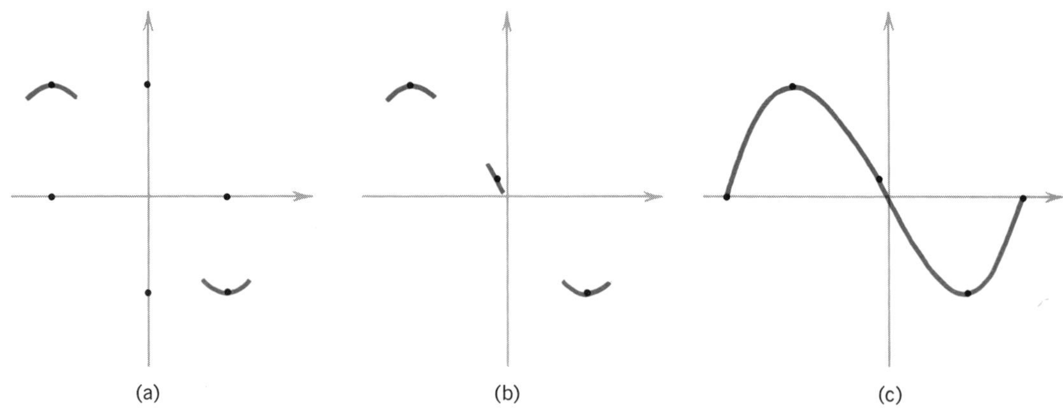

FIGURE 1-21

(2) Setting $f''(x)=0$, we find a single point of inflection of f to be at $x=-\frac{1}{6}$, and plot the point $(-\frac{1}{6}, 2\frac{37}{108})$ in Figure 1-21(b).

(3) We locate the zeroes of f at $x=0$, $x=-4$, and $x=\frac{7}{2}$. We notice that f is increasing where $f'(x)=(x-2)(3x+7)>0$, so that f increases where $x-2$ and $3x+7$ have the same sign; that is, on the intervals $]-\infty, -\frac{7}{3}[$ and $]2, \infty[$. On $]-\frac{7}{3}, 2[$, $f'(x)<0$ and f is decreasing. Also, f is concave downward where $f''(x)<0$, on $]-\infty, -\frac{1}{6}[$, and concave upward where $f''(x)>0$, on $]-\frac{1}{6}, \infty[$. The finished sketch is shown in Figure 1-21(c).

The above discussion gives a technique for graphing twice differentiable functions whose domain is all of $\boldsymbol{R}$. However, functions often arise in our study which are not defined, not continuous, or not differentiable at certain points. In this case, more information is needed for graphing.

The first step in graphing a function is to determine exactly what the domain of the function is, and where discontinuities or points of nondifferentiability occur.

Example 9 Consider the function

$$g\colon x \mapsto \begin{cases} \dfrac{(x-1)^2}{(x+1)(x-4)}, & x \neq -1, 4 \\ 0, & x = -1 \end{cases}$$

We can see from inspection that g is defined for every real number except 4, and is continuous at every real number except 4 and -1. By the quotient rule, we know that g is differentiable at points other than $x=4$ and $x=-1$. Let us check.

$$g'(x) = \frac{2(x-1)(x+1)(x-4)-(x-1)^2(2x-3)}{(x+1)^2(x-4)^2}$$

$$= \frac{(x+11)(1-x)}{(x+1)^2(x-4)^2}$$

Clearly, g' is defined everywhere except at $x=-1$ and $x=4$, so function g is continuous and differentiable everywhere except at -1 and 4.

Example 10 The tangent function $x \mapsto \tan x$ is defined, continuous, and differentiable everywhere except at points $(n+\frac{1}{2})\pi$, $n=0, \pm 1, \pm 2, \ldots$.

Example 11 The function $k: x \mapsto 2x^{1/3}+x^{2/3}$ is defined and continuous over all of $\boldsymbol{R}$, but the derivative $k'(x)=\frac{2}{3}(x^{-2/3}+x^{-1/3})$ is not defined at $x=0$.

After determining the domain of definition, continuity, and differentiability of a function, we use one-sided limits to investigate the nature of the function in the vicinity of its discontinuities. Suppose a function $f(x)$ is not continuous (possibly, not even defined) at the point $x=c$. Then as x approaches c from above or from below, f may approach a finite real number, or it may approach $\pm\infty$, or it might do none of these things.

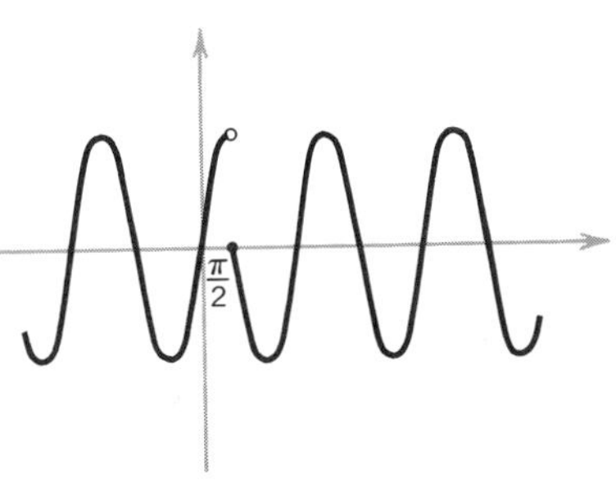

FIGURE 1-22

Example 12 The function

$$x \mapsto \begin{cases} \sin x, & x < \frac{1}{2}\pi \\ \cos x, & x \geqslant \frac{1}{2}\pi \end{cases}$$

is discontinuous at $\frac{1}{2}\pi$. As $x \to \frac{1}{2}\pi^-$, $f(x) \to 1$. As $x \to \frac{1}{2}\pi^+$, $f(x) \to 0$ (Figure 1-22).

Example 13 In Example 9, we found the function was discontinuous at $x=4$ and $x=-1$, where we have these one-sided limits:

$$\lim_{x\to 4^+} g(x) = +\infty \qquad \lim_{x\to 4^-} g(x) = -\infty$$

$$\lim_{x\to 1^+} g(x) = -\infty \qquad \lim_{x\to -1^-} g(x) = +\infty$$

Using this information, we sketch the graph in Figure 1-23.

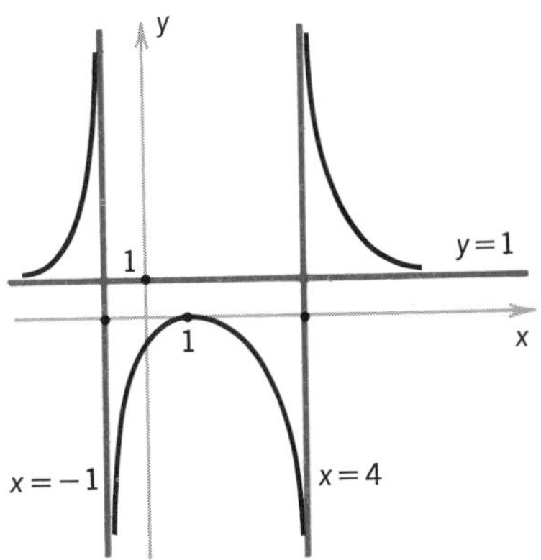

FIGURE 1-23

Example 14 At the discontinuities $x=(n+\frac{1}{2})\pi$ (see Example 10), the tangent function has the one-sided limits

$$\lim_{x\to(n+\frac{1}{2})\pi^+} \tan x = -\infty, \qquad \lim_{x\to(n+\frac{1}{2})\pi^-} \tan x = +\infty$$

If both of the one-sided limits of a function f are infinite (positive or negative) as x approaches a point of discontinuity c, the graph of f comes arbitrarily close to the vertical line $x = c$, and $f'(x)$ approaches (positive or negative) infinity. We then call the line $x = c$ an **asymptote** of f. In Example 9, the function g has asymptotes $x = 4$ and $x = -1$. The tangent function (Example 10) has asymptotes $x = (n + \frac{1}{2})\pi$.

The graph of a function f may also come arbitrarily close to a non-vertical line $y = mx + b$ as x goes to $+\infty$ or $-\infty$. In order for this to happen, the condition that must be satisfied is

$$\lim_{x\to\infty} [f(x) - (mx+b)] = 0$$

or

$$\lim_{x\to-\infty} [f(x) - (mx+b)] = 0$$

If in addition, the tangent to the graph of f at $(x, f(x))$ approaches the line $y = mx + b$ as $x \to \infty$ or as $x \to -\infty$, then the line $y = mx + b$ is also called an **asymptote** of f (Figure 1-24). The latter condition on the tangent is equivalent to saying that $f'(x) \to m$ as $x \to \infty$ or as $x \to -\infty$.

Example 15 Returning to Example 9, we see that by dividing out, we can write the rule for the function g (when $x \neq 4$, $x \neq -1$) as

$$g(x) = 1 + \frac{x-3}{x^2 - 3x - 4}$$

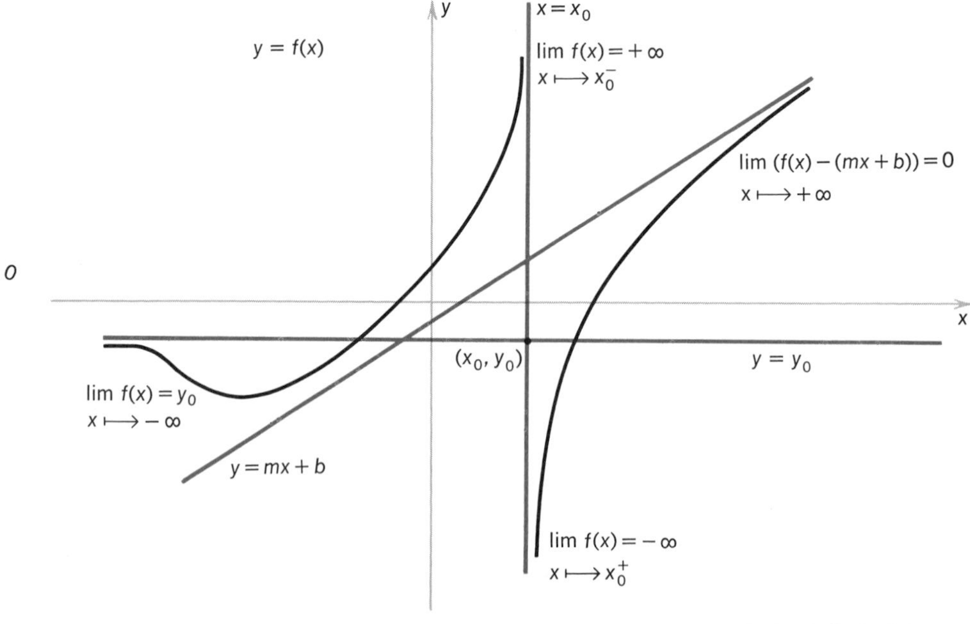

FIGURE 1-24

As $x \to \pm\infty$, the remainder term $(x-3)/(x^2-3x-4)$ approaches zero, and so $g(x) \to 1$. Therefore $\lim_{x\to\pm\infty}(g(x)-1)=0$. Moreover, $g'(x) \to 0$ as $x \to \pm\infty$, and so the line $y=1$ is an asymptote to g (Figure 1-23).

Example 16 Consider the hyperbola

$$\frac{x^2}{a^2} - \frac{y^2}{b^2} = 1$$

Solving for y in terms of x, we have

$$y = \pm b\sqrt{\frac{x^2}{a^2} - 1}$$

Now as $x \to \pm\infty$, the 1 under the radical sign becomes insignificant relative to the term x^2/a^2, and y becomes close to $\pm b\sqrt{x^2/a^2}$, that is, close to $\pm(b/a)x$. In fact, we have

$$\lim_{x\to\pm\infty}\left[\pm b\sqrt{\frac{x^2}{a^2} - 1} - \left(\pm\frac{b}{a}x\right)\right] = 0$$

Also

$$\frac{dy}{dx} = \pm\frac{b}{a}\cdot\frac{x/a}{\sqrt{(x/a)^2-1}} \to \pm\frac{b}{a} \text{ as } x \to \pm\infty$$

Hence the lines $y = \pm(b/a)x$ are asymptotes to the hyperbola (Figure 1-25).

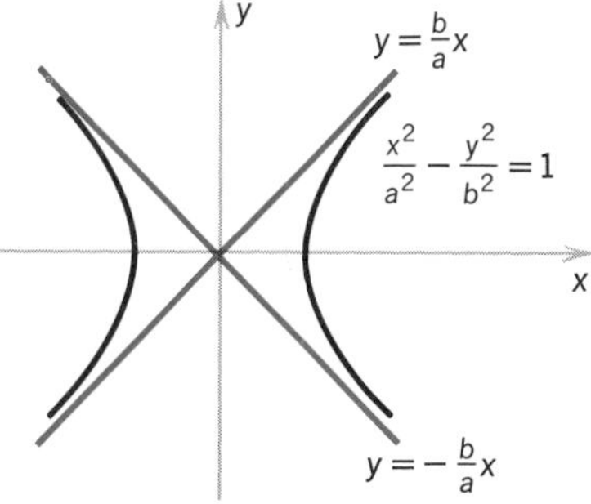

FIGURE 1-25

It is always easy to tell whether the graph of a function f eventually lies above or below its asymptote as x approaches $\pm\infty$. In particular, if $y = mx+b$ gives the asymptote and if $f(x)-(mx+b) > 0$ for very large positive (or negative) x, then f eventually lies above its asymptote as x approaches positive (or negative) infinity. On the other hand, if $f(x) < mx+b$ for very large positive (or negative) x, then f eventually lies below its asymptote.

Example 17 Returning to Example 9, we see that for large positive x, $g(x)-1 = (x-3)/(x^2-3x-4)$ is positive; hence as $x \to +\infty$, g approaches $y=1$ from above. For large negative x, $g(x)-1$ is negative; hence as $x \to -\infty$, g approaches $y=1$ from below (Figure 1-23).

After finding asymptotes and observing behavior around discontinuities, the graphing of general functions proceeds as with the everywhere twice differentiable functions. The results about local extrema, increasing and decreasing, and concavity, all hold for **any** interval on which the function and its first and second derivatives are defined. There is a further oberservation to make about points of inflection: If at some point on the graph of a function, the tangent line to the graph is vertical, then the behavior of f at that point must be determined by inspection, to see if it is a point of inflection.

Example 18 The function k from Example 11 has a vertical tangent at the origin (Figure 1-26). We can verify by inspection that k changes concavity at $x=0$. Notice that $k''(x) = -\frac{2}{9}(2x^{-5/3}+x^{-4/3}) = -\frac{2}{9}x^{-5/3}(2+x^{1/3})$, and so for points x near 0, $k''(x)$ is positive for $x<0$ and negative for $x>0$. Hence the concavity changes at $x=0$, and the origin is a point of inflection.

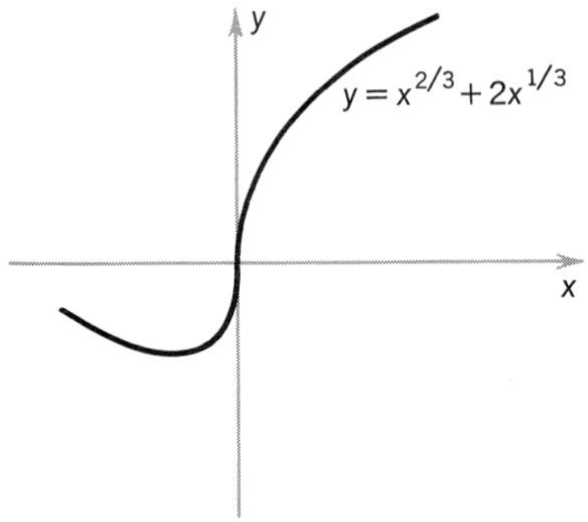

FIGURE 1-26

We conclude this section with detailed analyses of two graphing problems.

Example 19 Sketch the graph of the quartic

$$f\colon x \mapsto 3x^4 - 20x^3 + 42x^2 - 36x + 10$$

A polynomial is continuous and differentiable at all $x \in \boldsymbol{R}$, and we have

$$f'(x) = 12x^3 - 60x^2 + 84x - 36$$

and

$$f''(x) = 36x^2 - 120x + 84$$

As $x \to \pm\infty$, $f'(x) \to \pm\infty$; hence there are no nonvertical asymptotes. (There are no vertical asymptotes since f is continuous on all of $\boldsymbol{R}$.) Critical points of f occur where $f'(x) = 12(x-1)^2(x-3) = 0$; the points are $x=1$, $x=3$. At $x=3$, $f''(x)=48>0$, so $f(3)=-17$ is a local minimum, and we plot $(3, -17)$. At $x=1$, $f''(x)=0$, so we examine the sign of the first derivative on either side of $x=1$, noticing that $f'(x)<0$ on both sides; therefore, there is a point of inflection at $(1, -1)$. We locate other points of inflection by solving $f''(x) = 12(x-1)(3x-7) = 0$, discovering that the only other inflection point occurs at $x=\frac{7}{3}$, which yields the point $(\frac{7}{3}, -10\frac{13}{27})$. We now plot a few additional points:

x	$f(x)$
0	10
2	-6
4	26

Checking the signs of f' and f'', we observe that f is increasing on $[3, \infty[$ and decreasing on $]-\infty, 3]$. Moreover, f is concave downward on the interval $]1, \frac{7}{3}[$ and upward elsewhere. We draw the final sketch consistent with all this information (Figure 1-27).

Example 20 Sketch the graph of the function

$$f\colon x \mapsto \frac{x^3 - 4x^2}{(x-2)^2}$$

The function is undefined only at $x=2$, and elsewhere is continuous and differentiable, with

$$f'(x) = \frac{x^3 - 6x^2 + 16x}{(x-2)^3}, \quad f''(x) = \frac{-8x - 32}{(x-2)^4}$$

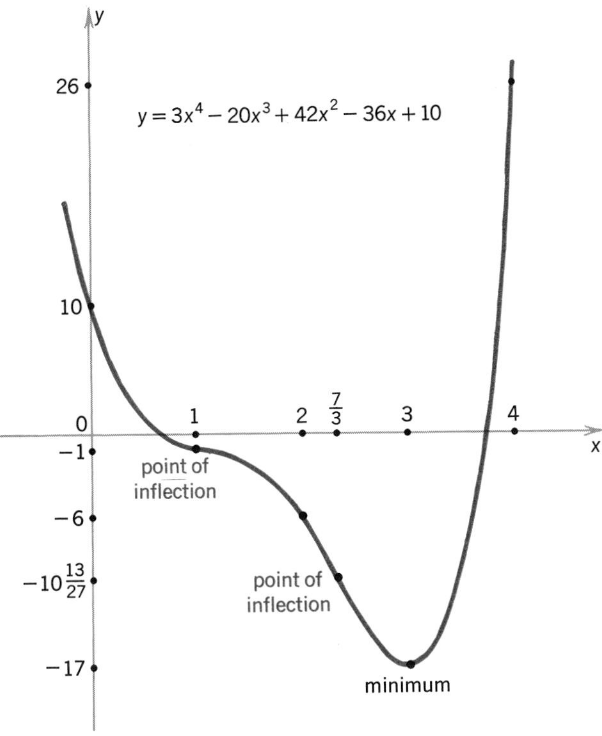

FIGURE 1-27

(Verify!) Observing the behavior of f near $x=2$, we find that

$$\lim_{x\to 2^+} f(x) = -\infty \quad \text{and} \quad \lim_{x\to 2^-} f(x) = -\infty$$

so the line $x=2$ is a vertical asymptote of f. To discover nonvertical asymptotes, we divide out $f(x) = x - 4x/(x^2-4x+4)$, and it follows that

$$\lim_{x\to\pm\infty} (f(x)-x) = \lim_{x\to\pm\infty}\left(-\frac{4x}{x^2-4x+4}\right) = 0$$

The reader can check that $\lim_{x\to\pm\infty} f'(x) = 1$, and so f has the nonvertical asymptote $y=x$. For large positive x, we have $-4x/(x-2)^2 < 0$, so the graph of f lies below the line $y=x$ as x approaches $+\infty$. For large negative x, we have $-4x/(x-2)^2 > 0$, so the graph of f lies above the asymptote as x approaches $-\infty$. Critical points of f are found by solving

$$f'(x) = \frac{x(x^2-6x+16)}{(x-2)^3} = 0$$

which yields $x = 0$ as the only real root. Now $f''(0) = -2 < 0$, so $f(0) = 0$ is a relative minimum, and we plot the point $(0, 0)$. Points of inflection are found by solving

$$f''(x) = \frac{-8x - 32}{(x-2)^2} = 0$$

yielding $x = -4$. We plot the single inflection point $(-4, -3\frac{5}{9})$.

We now evaluate and plot f at a few more points:

x	$f(x)$
-2	$-\frac{3}{2}$
1	-3
3	-3
4	0
6	$\frac{9}{2}$

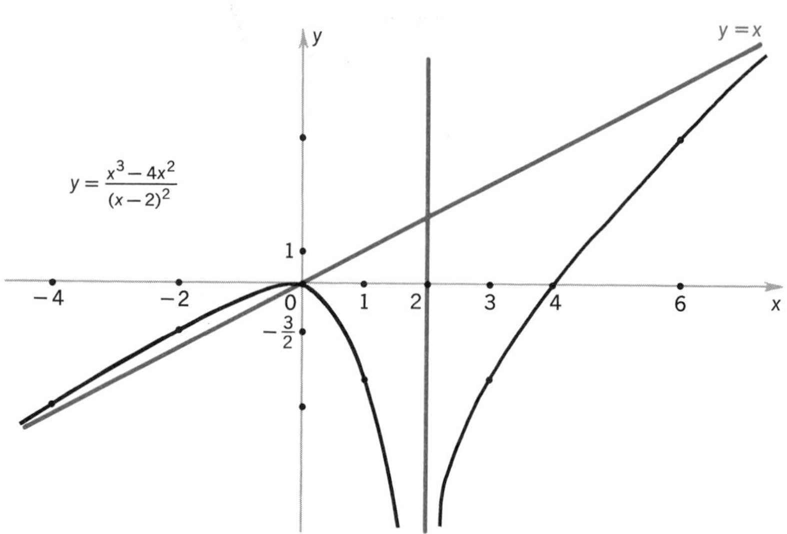

FIGURE 1-28

Examining the signs of the first and second derivatives, we find that $f'(x) < 0$ on the interval $]0, 2[$, so f is decreasing there. On $]-\infty, 0[$ and $]2, \infty[$, f is increasing. Since $f''(x) > 0$ on the interval $]-\infty, -4[$, f is concave upward there; $f''(x) < 0$ on $]-4, 2[$ and $]2, \infty[$, where f is concave downward. All this information having been gathered, we draw the final sketch (Figure 1-28).

EXERCISES

1. (a) Show that f is concave upward if and only if $-f$ is concave downward.
   (b) Use (a) to complete the proof of Theorem 1.

2. Explain why a twice differentiable function f with a critical point at x_0 must have either a local extreme value or a point of inflection at x_0 provided x_0 is the only critical point of f on some open interval. [Hint: The tangent to the graph of f at $(x_0, f(x_0))$ is horizontal; either the graph does or does not cross this tangent line at x_0. Consider the cases.]

3. Show that the asymptotes of an arbitrary hyperbola,
$$Ax^2 + Bxy + Cy^2 + Dx + Ey + F = 0$$
(where $B^2 - 4AC > 0$; see §1.7) are given by
$$2Cy = \pm\left(\mathscr{D}x + \frac{BE-2CD}{\mathscr{D}}\right) - \left(Bx+E\right)$$
where $\mathscr{D} = \sqrt{B^2 - 4AC}$. ($\mathscr{D}$ is called the **square root of the discriminant.**) Verify that if the hyperbola is given by $x^2/a^2 - y^2/b^2 = 1$, then this equation reduces to $y = \pm(b/a)x$.

4. Show that a polynomial function of degree 2 or higher cannot have any asymptotes.

5. When will the quotient
$$x \mapsto \frac{P(x)}{Q(x)}$$
of two polynomial functions have asymptotes? [Hint: Compare the degrees of P and Q.]

6. If a function is twice differentiable on all of R and has at least one local extreme value, then knowing the location of all relative maxima, relative minima, and points of inflection is sufficient to determine on precisely what intervals the function is increasing, decreasing, concave upward or downward. Explain why this is so and why there must be at least one extreme value for the method to hold. Then apply the method to determine the monotonicity and concavity of functions with extreme values and points of inflection as indicated. Sketch a curve with these characteristics.
   (a) Maxima at $(0, 9)$ and $(-9, 3)$; minimum at $(-4, -2)$; points of inflection at $(-6, 0)$ and $(-2, 7)$.
   (b) Maximum at $(-10/3, -1)$; minimum at $(-1/7, -7/2)$; points of inflection at $(-1, -2)$, $(0, 0)$, and $(3, 4)$.
   (c) Maximum at $(-9, 8)$; minima at $(-14, 4)$ and $(3, -5)$; points of inflection at $(-11, 6)$, $(-7, 5)$, $(-7/2, 2)$, and $(\sqrt{2}, -3)$.

7. Prove that the graph of a cubic is symmetric with respect to its point of inflection.

In exercises 8–30, graph the functions, remarking all discontinuities, asymptotes, local maxima and minima, and points of inflection.

8. $y = x^3 - 2x^2 + x + 1$
9. $y = x^5 - 30x^3 + 160x$
10. $y = \cos(1/x)$
11. $x \mapsto (x-1)^{2/3}$
12. $x \mapsto (\sin 2x)^{1/3}$
13. $x \mapsto \dfrac{x^4}{4x^3 + 1}$
14. $y = \sin x \cos x$
15. $y = \dfrac{3x^3 - 2x^2 - 8}{(x-1)^2}$
16. $y = \dfrac{3x^3 - 2x^2 - 8}{x^2 - 1}$
17. $y = (x^2+1)(x^2-1)$
18. $x \mapsto \sqrt{x^2+2x} - 3$
19. $x \mapsto \dfrac{3x^2 + x - 1}{x^2 - 1}$
20. $y = x + \sin x$
21. $y = x^4 - 32x + 25$
22. $x = 1 + \sqrt{x^3+x}$
23. $y = (\tan x + \sin x)$
24. $y = 3x^5 + 5x^3 + 1$
25. $y = \cot x$
26. $y = \sec x$
27. $y = x^{1/3}(x-4)$
28. $y = \dfrac{1}{\sqrt{x}} + \dfrac{\sqrt{x}}{9}$
29. $x \mapsto \dfrac{x + a}{\sqrt{x^2 + 1}}$
30. $y = x \sin x$

*31. Let $y = f(x)$ have $y = mx + b$ as an asymptote as $x \to \infty$ so that in particular $f'(x) \to m$ as $x \to \infty$.
   (a) Try to think of an example of a function g where limit $[g(x) - (mx+b)] = 0$ but $g'(x)$ does not go to m. Hence discuss the reason for making this extra assumption on f'.
   (b) By analogy with (a), try to argue that $f''(x)$ need not $\to 0$ as $x \to \infty$.

2 EXTREME VALUE PROBLEMS

The results on extreme values from the last section can be applied to a variety of problems that ask "What is the largest ...?" or "What is the least ...?" The first statement usually requires maximizing a function, while the second calls for minimizing the function.

For example, one might ask, "If an isosceles triangle has perimeter 24 inches, what is the greatest possible area it can have?" The problem can be solved by setting up an equation for the area A of the triangle in terms of the length l of the base of the form $A = f(l)$. We can then find the maximum area A by finding the maximum value of the function f.

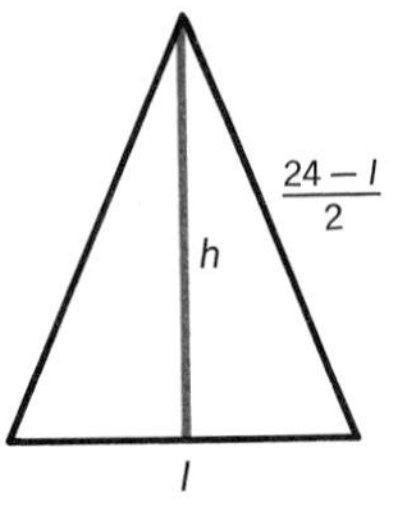

FIGURE 2-1

The altitude of the triangle is found by the Pythagorean Theorem (Figure 2-1):

$$h^2 = \left(\frac{24-l}{2}\right)^2 - \left(\frac{l}{2}\right)^2$$

$$h = \sqrt{144 - 12l}$$

and using $A = \frac{1}{2}lh$, we arrive at

$$A = \frac{l}{2}(144-12l)^{1/2}$$
$$= (36l^2 - 3l^3)^{1/2}$$

Then

$$\frac{dA}{dl} = \frac{72l - 9l^2}{2(36l^2 - 3l^3)^{1/2}}$$
$$= \frac{9l(8-l)}{2(36l^2 - 3l^3)^{1/2}}$$

and setting $dA/dl = 0$, we obtain $l = 8$ (dA/dl is undefined for $l = 12$). Now

$$\left.\frac{d^2A}{dl^2}\right|_{l=8} = \left(\frac{9}{2}\right)\left.\frac{(8-2l)(36l^2-3l^3) - \frac{1}{2}(8l-l^2)(72l-9l^2)}{(36l^2-3l^3)^{3/2}}\right|_{l=8} < 0$$

so the area is maximum when $l = 8$, at which time the triangle is equilateral, and $A = 16\sqrt{3}$.

Actually, all we knew about $l = 8$ was that it was the location of a **local** maximum. The **absolute maximum**, that is, the greatest value obtained by the function in its entire domain, might not have been at $l = 8$. However, if the function is defined over some interval (in this case [0, 12]), the absolute maximum must occur at a relative maximum or at an endpoint of the interval. In our problem, the area for $l = 0$ or $l = 12$ is zero, so the relative maximum is in fact absolute.

The Extreme Value Test (§4.7) justifies a procedure for solving extreme value problems.

FORMAL DEVICE

To solve an extreme value problem:

(1) Set up a function expressing the variable whose extreme value is to be determined in terms of a variable whose conditions are known. It is also helpful to draw a figure where possible.

(2) Equate the derivative of the function to zero and locate the critical points.

(3) Evaluate the function at its critical points and at the endpoints of the domain. If the derivative fails to exist at some point, examine this point as a possible maximum or minimum.

(4) The greatest value thus obtained is the absolute maximum; the least value is the absolute minimum.

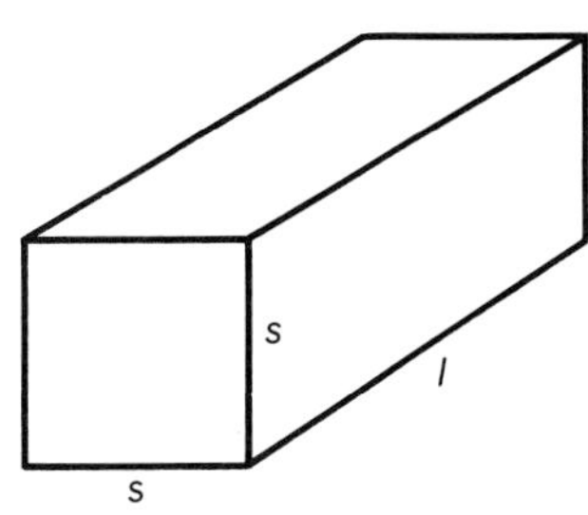

FIGURE 2-2

Example 1 What is the maximum volume that can be obtained by a rectangular solid with square ends, whose surface area is 96 square feet (Figure 2-2)? Let the length of the side of a square end be s. Then the rectangular faces of the solid each have area $A = (96 - 2s^2)/4$ for $0 \leqslant s \leqslant \sqrt{48}$, so the length of a rectangular face is $l = (24/s) - (s/2)$, and the volume is

$$\begin{aligned} V &= l \cdot s \cdot s \\ &= \left(\frac{24}{s} - \frac{s}{2}\right)s^2 \\ &= 24s - \frac{s^3}{2} \end{aligned}$$

Hence, the volume has been expressed in terms of the single variable s. We now seek the value of s which maximizes V. Differentiating

$$\frac{dV}{ds} = 24 - \frac{3}{2}s^2$$

and setting $dV/ds = 0$ yields $s = \pm 4$. However, the rectangular solid must have square ends with sides of length greater than zero and less than $\sqrt{96/2} = 4\sqrt{3}$, so $s = 4$ is the only critical point in the domain. At $s = 0$ and $s = 4\sqrt{3}$, the volume is 0; while at $s = 4$, $V = 64$ (cubic feet), which is therefore the maximum possible volume. Notice that the rectangular solid of maximum volume is a cube.

Example 2 A straight piece of wire of length L is to be cut into two pieces. One piece is to be formed into a circle and the other piece into a square (Figure 2-3). Where should the wire be cut so that the sum of the areas of the circle and the square will be greatest? Least? Let the wire be cut so that a piece of length x is made into the circle. Then x is the circumference of the circle, so

$$x = 2\pi r$$

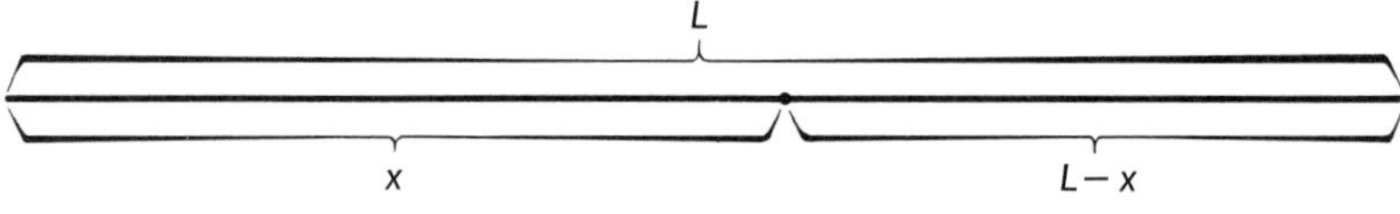

FIGURE 2-3

Thus

$$r = \frac{x}{2\pi}$$

and the area of the circle is

$$\begin{aligned} A_c &= \pi r^2 \\ &= \pi\left(\frac{x}{2\pi}\right)^2 \\ &= \frac{x^2}{4\pi}, \qquad 0 \leqslant x \leqslant L \end{aligned}$$

On the other hand, the square is made out of the remaining piece of wire, and so has perimeter $L-x$. One side of the square therefore has length

$$s = \frac{L-x}{4}$$

and the area of the square is

$$\begin{aligned} A_s = s^2 &= \left(\frac{L-x}{4}\right)^2 \\ &= \frac{x^2-2Lx+L^2}{16} \end{aligned}$$

The sum of the two areas is

$$\begin{aligned} A &= A_c + A_s \\ &= \frac{x^2}{4\pi} + \frac{x^2-2Lx+L^2}{16} \end{aligned}$$

We wish to find the maximum and minimum values of A for x such that $0 \leqslant x \leqslant L$. We therefore set

$$\begin{aligned} 0 &= \frac{dA}{dx} \\ &= \frac{x}{2\pi} + \frac{x-L}{8} \end{aligned}$$

which yields $x = L\pi/(4+\pi)$ as the only critical point of the function. We evaluate A at this critical point and at the endpoints of $[0, L]$. At $x=0$, $A = A_s = L^2/16$. At $x = L$, $A = A_c = L^2/4\pi$. At $x = \pi/(4+\pi)\,L$, we have

$$\begin{aligned} A &= \frac{\left(\frac{\pi}{4+\pi}\right)^2 L^2}{4\pi} + \frac{L^2\left(1 - \frac{\pi}{4+\pi}\right)^2}{16} \\ &= L^2\left[\frac{\pi}{4}\left(\frac{1}{4+\pi}\right)^2 + \left(\frac{1}{4+\pi}\right)^2\right] \\ &= L^2\frac{\pi+4}{4(\pi+4)^2} = L^2\frac{1}{4(\pi+4)} \end{aligned}$$

Clearly $L^2/4(\pi+4) < L^2/16 < L^2/4\pi$. Therefore the minimum area occurs where the wire is cut at $x = \pi L/(4+\pi)$; the maximum area is obtained when the wire is not cut at all, but rather when all the wire is used for the circle.

Example 3 A long lake has banks 200 feet apart. A man on one side of the lake would like to reach a tree on the other side, 500 feet from the point just opposite him (Figure 2-4). If he swims at a speed of 2 ft/sec and walks at 5 ft/sec, what is the shortest possible time in which he can reach the tree? Let the distance the man walks be x feet; his walking time is then $x/5$ sec. If he swims y feet, his swimming time is $y/2$ sec. Therefore the total time T is given by the equation

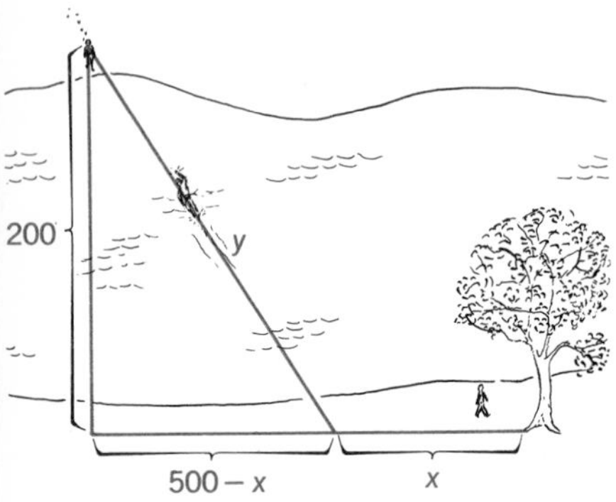

FIGURE 2-4

$$T = \frac{x}{5} + \frac{y}{2}$$

To apply the formal device, we must express T in terms of a single variable. We observe from the figure that $y^2 = (500-x)^2 + 200^2$, so

$$y = \sqrt{200^2 + (500-x)^2} = \sqrt{x^2 - 1000x + 290{,}000}$$

Substituting this expression for y into the equation for T yields

$$T = \frac{x}{5} + \left(\frac{x^2}{4} - 250x + 72{,}500\right)^{1/2}$$

where $0 \leqslant x \leqslant 500$. Differentiating, we have

$$\frac{dT}{dx} = \frac{1}{5} + \frac{x/2 - 250}{2(x^2/4 - 250x + 72{,}500)^{1/2}}$$

Setting dT/dx equal to zero and solving for x gives

$$-2(x^2/4 - 250x + 72{,}500)^{1/2}/5 = x/2 - 250$$

or

$$4(x^2/4 - 250x + 72{,}500)/25 = x^2/4 - 250x + 62{,}500$$

or

$$\frac{21}{100}x^2 - 210x + 50{,}900 = 0$$

Using the quadratic formula, we find

$$x = 500 \pm 400\frac{\sqrt{21}}{21}$$

The only value in the domain is $x = 500 - 400\sqrt{21}/21 \approx 412.7$. We evaluate T here and at the endpoints of the domain, $x = 0$ and $x = 500$. At $x = 0$, $T = \sqrt{72{,}500} = 50\sqrt{29} \approx 269$ (seconds). At $x = 500$, $T = 100 + 100 = 200$ (seconds). At $x = 500 - 400\sqrt{21}/21$, $T \approx 190$ (seconds), which is therefore the minimum possible time, by step (4) of the formal device.

Example 4 Find the coordinates of the points on the graph of $x \mapsto x^2 - 1$ which are nearest the origin (see Figure 2-5).

The distance from the point $(x, x^2 - 1)$ to the origin is given by

$$D(x) = \sqrt{x^2 + (x^2 - 1)^2}$$

The problem is simplified if we notice that the distance $D(x)$ is nonnegative and so takes on its least value precisely when the square of the distance takes on its least value. The square of the distance is given by

$$[D(x)]^2 = S(x) = x^2 + (x^2 - 1)^2 = x^4 - x^2 + 1$$

The domain of S is all of $\boldsymbol{R}$; however, since

$$\lim_{x \to -\infty} S(x) = +\infty = \lim_{x \to +\infty} S(x)$$

the minimum value of $S(x)$ must exist and must occur at a critical point.

We find that $S' : x \mapsto 4x^3 - 2x$ has roots 0 and $\pm\frac{1}{2}\sqrt{2}$. Since $S(0) = 1$ and $S(\frac{1}{2}\sqrt{2}) = \frac{3}{4}$, the minimum value of S occurs at both $\pm\frac{1}{2}\sqrt{2}$. We see, therefore, that the points on the graph of $x \mapsto x^2 - 1$ closest to the origin are $(\frac{1}{2}\sqrt{2}, -\frac{1}{2})$ and $(-\frac{1}{2}\sqrt{2}, -\frac{1}{2})$. The distance from either point to the origin is $\frac{1}{2}\sqrt{3}$.

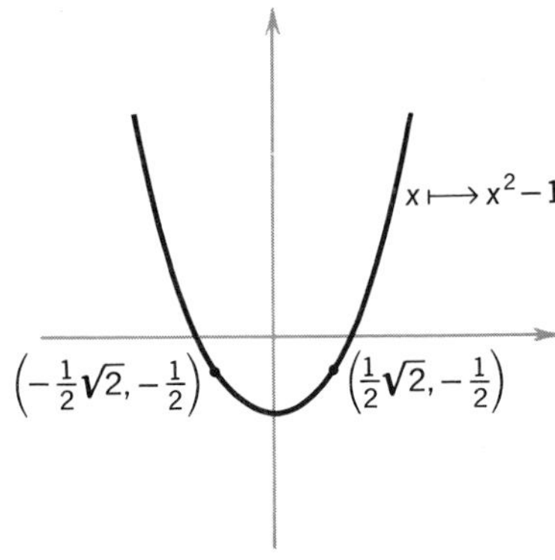

FIGURE 2-5

Example 5 Two long, straight railroad tracks cross each other at right angles. Train A, traveling at a constant velocity of 75 mph, crosses the intersection at 5 p.m. on track R. Train B, traveling at 60 mph, crosses the intersection 10 minutes later on track S. At what time are the trains closest together? (See Figure 2-6.)

Let 5 p.m. be time $t = 0$. If we measure time in minutes, A's distance from the intersection at time t is

$$r = \tfrac{5}{4}t$$

B's distance from the intersection is

$$s = t - 10$$

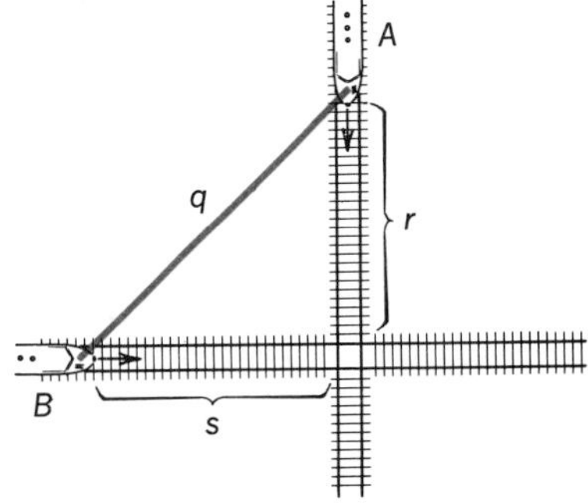

FIGURE 2-6

Therefore, the distance q between A and B satisfies

$$\begin{aligned} q^2 &= r^2 + s^2 \\ &= (\tfrac{5}{4}t)^2 + (t - 10)^2 \end{aligned}$$

Now the distance q is always positive and so q is minimum when q^2 is also minimum; hence, we solve for the value of t that minimizes q^2. We have

$$\frac{d(q^2)}{dt} = \frac{5}{2} \cdot \frac{5}{4}t + 2(t - 10)$$

Setting $d(q^2)/dt$ equal to zero yields

$$0 = 25t + 16t - 160 = 41t - 160$$

or

$$t = 160/41 = 3\tfrac{37}{41} \approx 4 \text{ minutes}$$

The critical value occurs at $5.03\frac{37}{41}$ p.m. From physical considerations we see that this must give the minimum value of q—at first the trains are far apart, then they reach the intersection at almost the same time, and soon after the distance between them begins to grow indefinitely. Alternatively, the Second Derivative Test can be used here; we have

$$\frac{d^2(q^2)}{dt^2} = \frac{41}{8} > 0$$

for all t, and so the critical point of q is the location of a minimum.

Example 6 An underdeveloped country whose only export product is coffee can sell x tons of coffee per month on the international market at the price of $(300-x/1000)$ dollars per ton. The cost of shipping x tons is 10 dollars per ton plus 1000 dollars overhead. Assuming the country has nationalized its coffee industry, what level of export will maximize the dollar income of the country? The total income or revenue on x tons of coffee is

$$R(x) = \left(300 - \frac{x}{1000}\right)x = 300x - \frac{x^2}{1000}$$

The cost of exporting x tons is

$$C(x) = 1000 + 10x$$

Therefore the net dollar income from x tons of coffee is

$$N(x) = R(x) - C(x) = \frac{-x^2}{1000} + 290x - 1000$$

We seek the maximum value of $x \mapsto N(x)$ on the interval $[0, \infty[$. This function is concave downward and achieves its maximum at a solution to

$$\begin{aligned} 0 &= N'(x) \\ &= \frac{-x}{500} + 290 \end{aligned}$$

that is, at the point $x = 145{,}000$. The country will therefore have maximum dollar income if it exports 145,000 tons of coffee per month. This will net the country somewhat over 21 million dollars a month.

In geometric terms, the country's dollar income is maximized when the distance between the revenue curve and the cost curve is maximized. From the discussion following Rolle's Theorem in §4.7 we see that this occurs when the two curves have parallel tangents. That is, profit is maximized when the rate of increase of cost "catches up" to the rate of increase of revenue. In economic theory, the rate of increase in revenue per unit of production is called the **marginal revenue**; the rate of increase in cost per unit of production is called the **marginal cost**. Whence the dictum of mathematical economics: Profit is maximized when marginal revenue equals marginal cost.

EXERCISES

1. Find two numbers whose sum is $S > 0$, if the product is to be maximum. What can we infer about the dimensions of a rectangle with given perimeter, and whose area we wish to maximize?

2. A box without a top is to be made by cutting small squares of equal size from the corners of an 8×5 inch piece of cardboard and then turning up the sides. Find the maximum possible volume for the box (Figure 2-7).

3. A cylindrical can is to be made with a tin lateral surface and copper top and bottom. If the can is to hold volume V, and if copper is five times as expensive as tin, find the dimensions of the most economical possible can.

4. A right circular cylindrical log of radius 2 feet is to be cut to make a rectangular beam. If the strength of the beam is proportional to the width times the square of the depth, find the dimensions for the strongest possible beam (Figure 2-8).

5. A trough is to be made from a piece of metal 15 inches wide, with cross section as shown in Figure 2-9. What should be the separation of the sides at the top of the trough in order to maximize holding capacity?

6. Find the volume of the largest right circular cone which can be inscribed in a sphere of radius r.

7. Figure 2-10 shows two posts, 8 and 12 feet high, which stand 15 feet apart. A wire is to join these posts and a stake on the ground between them. Where should the stake be placed to minimize the amount of wire needed?

8. An editor wants the page for his new math book to contain 80 square inches of print. If the page is to have two inch margins at the top and bottom and $\frac{3}{2}$ inch margins on the sides, what is the smallest area of paper that can be used?

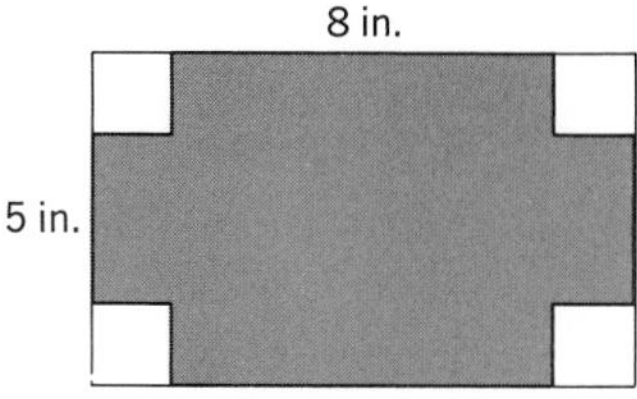

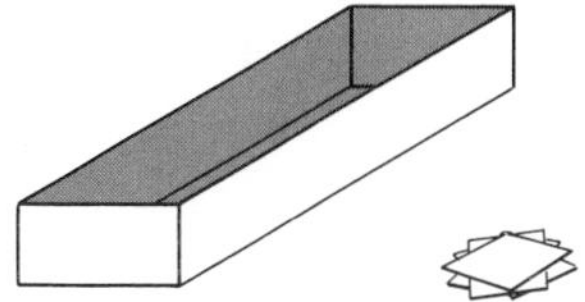

FIGURE 2-7

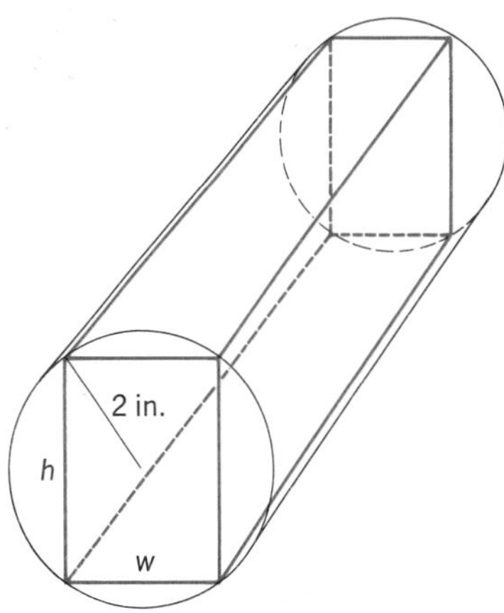

FIGURE 2-8

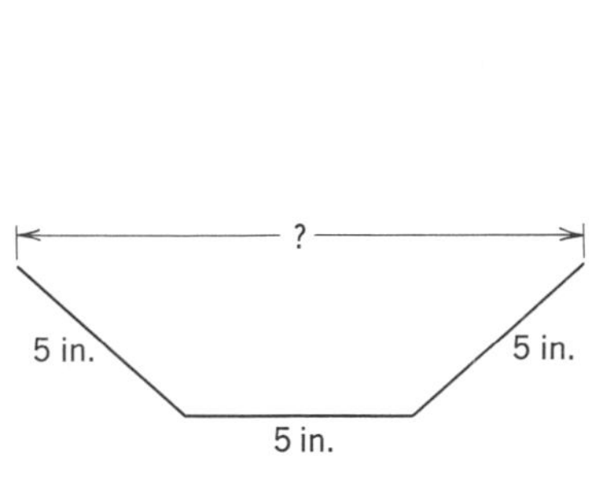

FIGURE 2-9

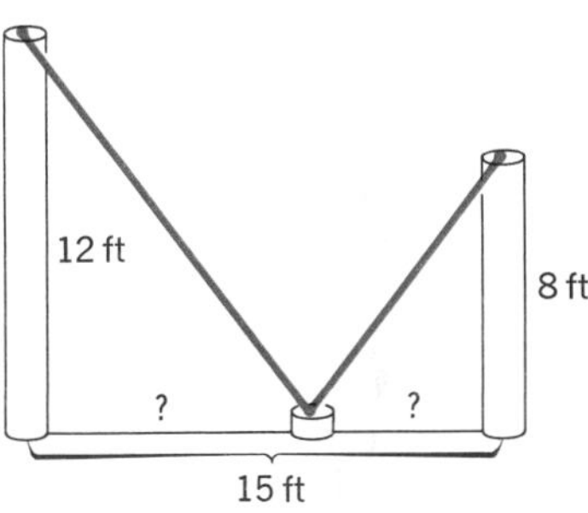

FIGURE 2-10

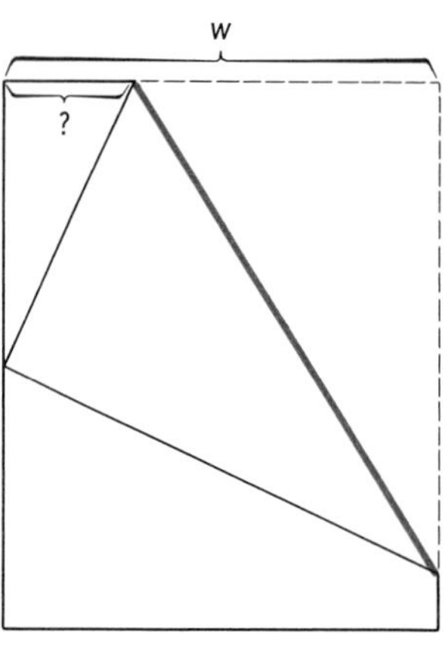

FIGURE 2-11

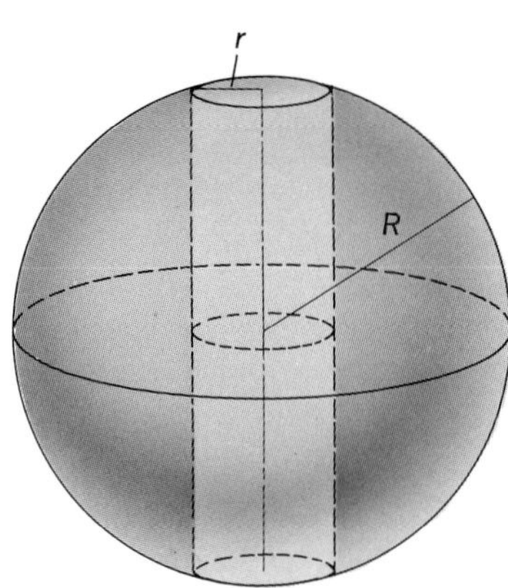

FIGURE 2-12

9. A rectangle is inscribed in an acute triangle, so that one side of the rectangle lies on one side of the triangle. Show that the area of the rectangle cannot exceed half the area of the triangle.
10. Find the positive number with the minimum sum of its square and 10 times the cube of its reciprocal.
11. The corner of a sheet of paper of width w is folded over so that it touches the opposite side (Figure 2-11). Where should the paper be folded along the width to minimize the length of the crease?
12. A cylindrical hole of radius r is bored in a sphere of radius R, the axis of the hole passing through the center of the sphere (Figure 2-12). What should r be so that the total surface area of the remaining solid will be maximum? Show that this area is $3\pi\sqrt{3}R$.
13. What is the greatest possible area of a rectangle inscribed in the ellipse $x^2/a^2+y^2/b^2=1$?
14. What is the shortest distance from the point $(0,0)$ to the hyperbola $xy-x^2=12$?
15. At what point on the ellipse $x^2/a^2+y^2/b^2=1$ does the tangent line segment cut off by the x and y axes have the least length? What is this length?
16. A triangle has vertices at $(0,0)$, $(a,0)$, and (b,c) (Figure 2-13). What are the coordinates of the point the sum of the squares of whose distances from the vertices of the triangle is smallest?
17. **Fermat's Principle** states that light travels from one point to another along the path requiring the shortest time of transit. Light is to travel from point A in Figure 2-14 to point B. At P, the light enters a new medium and is refracted as shown. Snell's Law says that if v_1 is the velocity in the first medium and v_2 is the velocity in the second medium, then

$$\frac{\sin\theta_1}{\sin\theta_2}=\frac{v_1}{v_2}$$

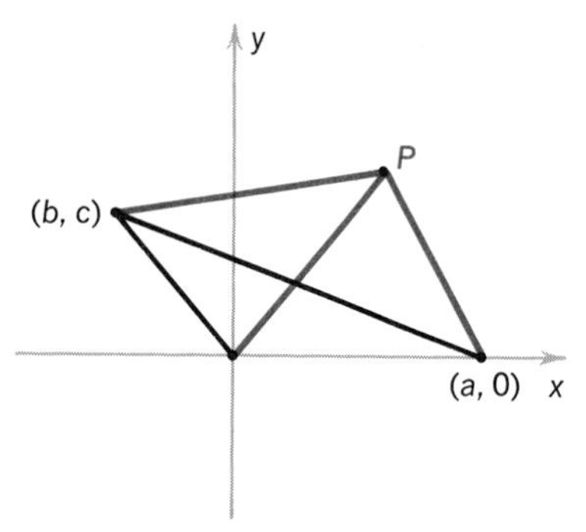

FIGURE 2-13

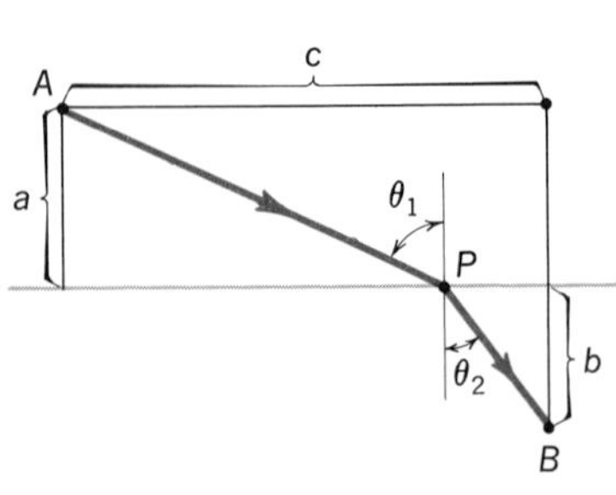

FIGURE 2-14

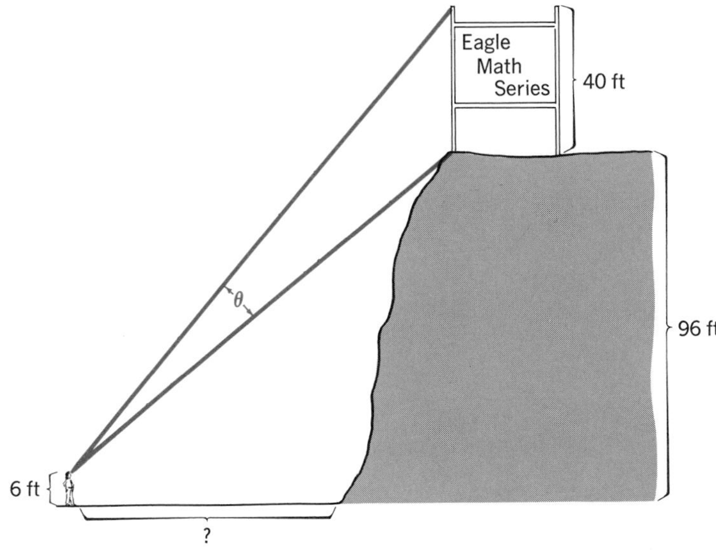

FIGURE 2-15

Also notice from the figure that $c = a \tan \theta_1 + b \tan \theta_2$. If a,b, c, and the physical properties of the media are fixed, establish Snell's Law from Fermat's Principle by locating a P which minimizes the time of transit.

18. A billboard 40 feet high stands atop a 96 foot cliff. How far should a man stand from the foot of the cliff for the billboard to subtend the greatest possible angle to his eyes, which are six feet above the ground? (See Figure 2-15.)

19. It costs \$25 per hour for fuel and \$100 per hour additional costs to operate a locomotive at 25 mph. The cost of fuel is proportional to the square of the speed, while additional costs are the same for any speed. At what speed will the cost per mile of operating the train be minimum?

20. Of all right circular cones of slant height 10 inches, what is the altitude of the one with maximum volume?

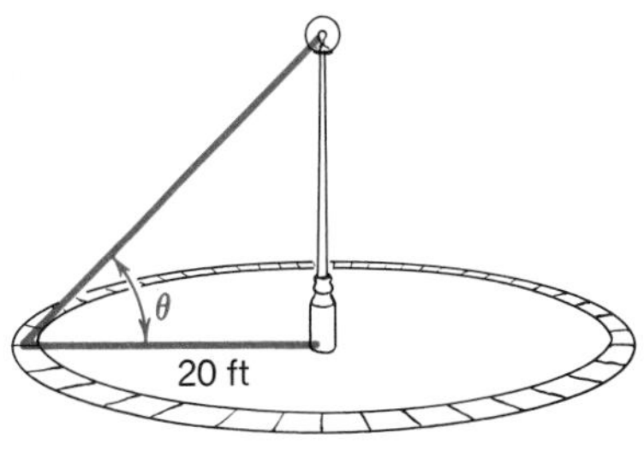

FIGURE 2-16

21. In Figure 2-16 a circular area of radius 20 feet is surrounded by a walk. A light is placed above the center of the area. What height most strongly illuminates the walk? (The intensity of illumination is given by $I = \sin \theta / s$, where s is the distance from the source and θ is the angle at which the light strikes the surface.)

22. A toy manufacturer is planning on mass-producing a new addition to their line of stuffed animals—an eagle. An efficiency expert estimates that if the price per eagle is x dollars, the profit is

$$P = K\left(\frac{x-9}{x^2} - 0.01\right)$$

where $K > 0$ for $x > 9$. Find the value of x that maximizes the profit.

3 DIFFERENTIAL APPROXIMATION

The tangent to a curve is obtained by approximating it with secant lines. In this section we turn this process around and use tangent lines to approximate secant lines. This leads to one of the earliest applications of calculus which is a method of approximating a function near a particular point. This method is called **differential** or **linear approximation.**

Suppose we know the value of a function given by $y=f(x)$ at the point x_0, and we want its value at $x_0+\Delta x$, where Δx is some small real number. Let $\Delta y=f(x_0+\Delta x)-f(x_0)$. Consider the equation $dy=f'(x)\,dx$. Setting $x=x_0$ and $dx=\Delta x$, this becomes $dy=f'(x_0)\,\Delta x$, and subtracting from the expression for Δy, we get

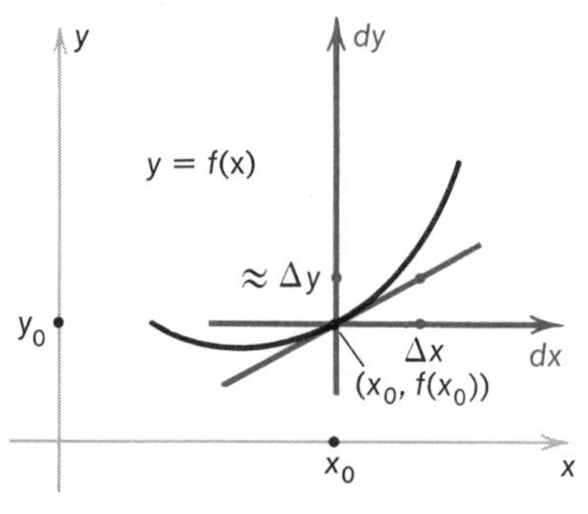

FIGURE 3-1

$$\Delta y - dy = f(x_0+\Delta x) - f(x_0) - f'(x_0)\,\Delta x$$

Therefore, $\lim_{\Delta x \to 0} \Delta y - dy = 0$, and so for very small Δx, Δy, and dy are approximately equal (we write $\Delta y \approx dy$). Therefore we have

$$f(x_0+\Delta x) - f(x_0) = \Delta y \approx dy = f'(x_0)\,\Delta x$$

and we approximate

$$\boldsymbol{f(x_0+\Delta x) \approx f(x_0) + f'(x_0)\,\Delta x}$$

(Figure 3-1).

We expect this approximation to be a good one because not only do we have

$$\lim_{\Delta x \to 0} \Delta y - dy = 0$$

but we also have

$$\lim_{\Delta x \to 0} \frac{\Delta y - dy}{\Delta x} = \lim_{\Delta x \to 0} \left(\frac{f(x_0+\Delta x) - f(x_0)}{\Delta x} - f'(x_0) \right) = 0$$

Therefore, the difference $\Delta y - dy$ goes to 0 "very quickly" as Δx goes to 0.

FORMAL DEVICE Linear Approximation

To approximate the value of a function at a point x_1 near a point x_0, where $f(x_0)$ is known, let $\Delta x = x_1 - x_0$ and approximate $f(x_0+\Delta x) \approx f(x_0)+f'(x_0)\,\Delta x$, that is,

$$f(x_1) \approx f(x_0) + f'(x_0)(x_1 - x_0)$$

Example 1 Approximate the value of the function $f(x) = x^5 - 3x^3 + x + 2$ at $x = 0.998$. We take $x_0 = 1$, $\Delta x = -0.002$, $f(x_0) = 1$, and $f'(x_0) = 5x_0^4 - 9x_0^2 + 1 = -3$, so

$$f(0.998) \approx f(1) + f'(1)(-0.002)$$
$$\approx 1 - 3(-0.002) = 1.006$$

Incidentally, actually calculating the value yields

$$f(0.998) = 1.006003896080032$$

Hence, our approximation is quite good here.

Example 2 Approximate $\sin 32°$. Letting $f(x) = \sin x$, $x_0 = 30° = \frac{1}{6}\pi$, $\Delta x = 2° = \frac{1}{90}\pi$ we have $f(x_0) = \sin\frac{1}{6}\pi = \frac{1}{2}$, $f'(x_0) = \cos\frac{1}{6}\pi = \sqrt{3}/2$, and so

$$\sin 32° \approx \frac{1}{2} + \frac{\sqrt{3}}{2}\left(\frac{\pi}{90}\right)$$
$$\approx 0.5 + 0.0302 = 0.5302$$

A four-place table gives $\sin 32° = 0.5299$, so the approximation is only off by 0.0003 (about $\frac{57}{1000}$ of one percent).

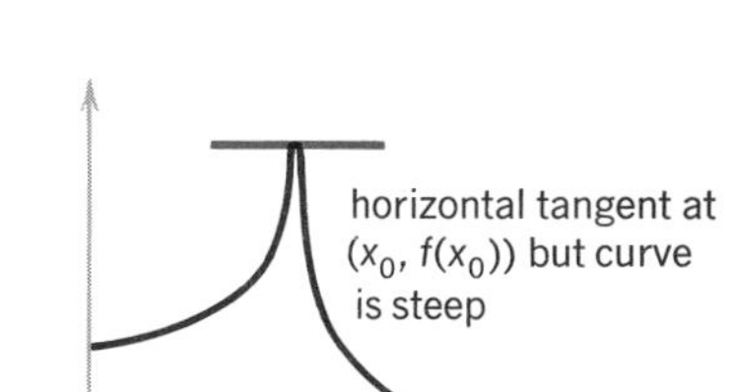

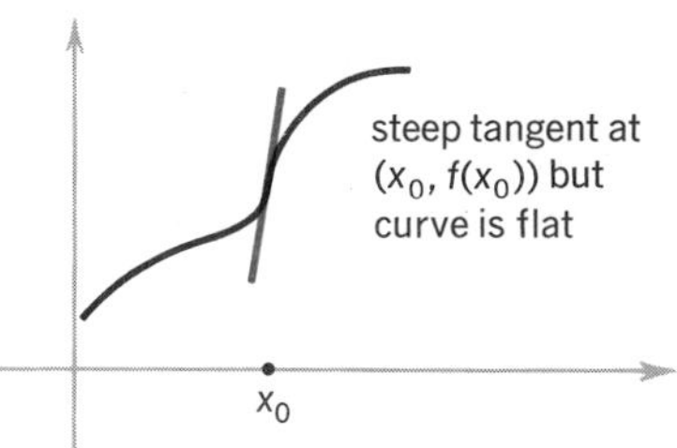

FIGURE 3-2

In the above examples, differential approximation produced excellent results. The tangent line to a curve at a point "fits" the curve very closely and that is the reason why differential (or **tangential**) approximation is effective. With this type of approximation, we are using the fact that when Δx is small, the secant line through $(x_0, f(x_0))$ and $(x_0 + \Delta x, f(x_0 + \Delta x))$ and the tangent line at $(x_0, f(x_0))$ are close together. However, if the curve is bending too sharply at $(x_0, f(x_0))$ then Δx will have to be very small indeed before these lines approach each other (see Figure 3-2). The bending of a curve is measured by the second derivative and it is in terms of the second derivative that we can give an explicit expression of the **error** involved in differential approximation. By the **error** we mean the difference between the true value and the approximate value:

$$\text{error} = \text{true value} - \text{approximate value}$$

As we shall see in §12.1, when f is twice differentiable we have

$$f(x_0 + \Delta x) = f(x_0) + f'(x_0)\Delta x + \tfrac{1}{2}f''(c)(\Delta x)^2 \tag{1}$$

where c is some point between x_0 and $x_0 + \Delta x$. (This proof is not difficult and the interested reader may consult it at this time.)

Equation (1) thus yields the error E in the form

$$E = \tfrac{1}{2}f''(c)(\Delta x)^2$$

When f'' is bounded on the interval with endpoints x_0 and $x_0 + \Delta x$, this means that we have

$$|E| \leqslant \tfrac{1}{2}M(\Delta x)^2$$

where M is a bound for $f''(x)$. In this case, as $\Delta x \to 0$, E tends to 0 *like* $(\Delta x)^2$.

Example 3 In the case of the sine function we have $|\sin'' x| \leqslant 1$ for all x. Therefore the error in differential approximation is bounded by $|E| \leqslant \frac{1}{2}(\Delta x)^2$. The approximation will be best near points at which the value of the sine is small.

Example 4 Let $f: x \mapsto \sin(9000x)$. Let us use differential approximation to estimate $f(\frac{1}{90}\pi)$. Setting $\Delta x = \frac{1}{90}\pi$ (the same value of Δx as in Example 2), we have

$$f(\tfrac{1}{90}\pi) \approx f(0) + f'(0)\tfrac{1}{90}\pi = 0 + 9000\tfrac{1}{90}\pi = 100\pi$$

The correct value is $f(\frac{1}{90}\pi) = \sin(9000 \cdot \frac{1}{90}\pi) = \sin(1000\pi) = 0$, so the error is gross. Here we have $|f''(x)| = 9000^2 |\sin x|$ which takes values as large as 2,835,000 on the interval $[0, \frac{1}{90}\pi]$.

EXERCISES

In exercises 1–7, use linear approximation to approximate the value of the function at the given point. Then find the actual value (by direct computation, looking in a table, or the like) and compute the error.

1. $x \mapsto \sqrt[3]{x}$ at $x = 100.1$
2. $x \mapsto \sqrt{x}$ at $x = 3.97$
3. $x \mapsto x^{5/6}$ at $x = 63.7$
4. $x \mapsto \dfrac{2}{x+5}$ at $x = 4.92$
5. $x \mapsto x\sqrt[3]{x^2+4}$ at $x = 1.97$
6. $x \mapsto x^n$ at $x = 1.001$
7. $x \mapsto \sin x$ at $x = 46°$
8. If two spheres with surface areas A_1 and A_2 are each slightly increased in radius, show that the ratio of the increase in their volume approximately equals A_1/A_2.
9. A box is $10 \times 9 \times 8$ ft^3. Find the approximate change in volume of the interior when the box is lined with paper $\frac{1}{16}$ in thick.
10. How can the graph of a function be used to determine whether an approximated value is too high or too low? Which order derivative of the function is relevant here? Compare with the formula given for approximating error.

4 THE NORMAL TO A CURVE; CURVATURE

Recall from analytic geometry that: (1) a line perpendicular to a horizontal line is vertical; (2) a line perpendicular to a vertical line is horizontal; and (3) if a line is neither horizontal nor vertical, and thus has slope $m \neq 0$, then a line perpendicular to it has slope $-1/m$. Recall also from elementary plane geometry that through any given point on a line, there is exactly one line perpendicular to the given line. We can extend these observations to the graphs of differentiable functions, by replacing the idea of "perpendicular" with the idea of "normal."

Definition Suppose a function given by $y = f(x)$ is differentiable at x_0, and therefore has a unique tangent line to its graph at $(x_0, f(x_0))$. Then the **normal** to the function at $(x_0, f(x_0))$ is the line passing through $(x_0, f(x_0))$ and perpendicular to the tangent there.

Thus, the normal line at $(x_0, f(x_0))$ has equation

$$f'(x_0)\,dy + dx = 0$$

in the $dx\,dy$ coordinate system. In the xy system, the equation of the normal at $(x_0, f(x_0))$ is

$$f'(x_0)(y - f(x_0)) + (x - x_0) = 0$$

Example 1 Consider the function $f: x \mapsto 2x^{1/3} + x^{2/3}$ (Figure 4-1). Then $f'(x) = \frac{2}{3}x^{-2/3} + \frac{2}{3}x^{-1/3}$. Thus at $x = -1$, $f(x) = -1$ and $f'(x) = 0$. Therefore the tangent line at $(-1, -1)$ is horizontal, with equation $y = -1$, and so the normal line is the vertical line $x = -1$. At $x = 0$, the tangent line is vertical (the y axis), and the normal is horizontal (the x axis). At $x = 1$, $f(x) = 3$ and $f'(x) = \frac{4}{3}$. Therefore the tangent line is $y - 3 = \frac{4}{3}(x - 1)$ or, $3y = 4x + 5$. The normal line is $y - 3 = -\frac{3}{4}(x - 1)$, which reduces to $4y = 15 - 3x$.

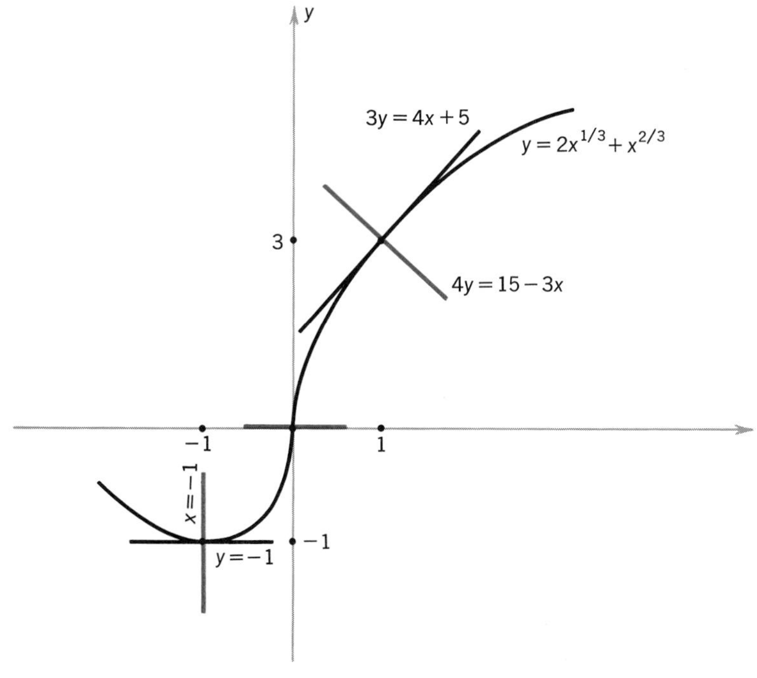

FIGURE 4-1

Example 2 Compute the equations for the tangent and normal lines to $y=\sin x+2\cos\frac{1}{2}x$ at $x=\pi$, $x=2\pi$, and $x=\frac{1}{3}\pi$ (Figure 4-2). We compute $dy/dx=\cos x-\sin\frac{1}{2}x$. Therefore we have:

x	$f(x)$	$f'(x)$	tangent at x	normal at x
π	0	-2	$y=-2(x-\pi)$	$y=\frac{1}{2}(x-\pi)$
2π	-2	1	$y-2=x+2\pi$	$y-2=-x+2\pi$
$\frac{1}{3}\pi$	$\frac{3}{2}\sqrt{3}$	0	$y=\frac{3}{2}\sqrt{3}$	$x=\frac{1}{3}\pi$

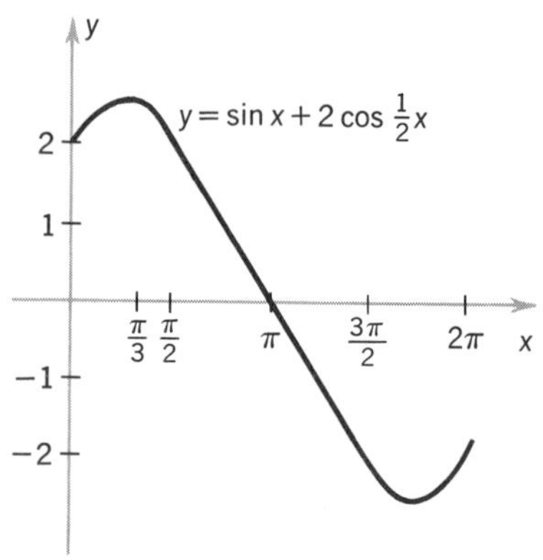

FIGURE 4-2

The technique of implicit differentiation can be used to compute the normal line to the locus of a relation. If $F(x,y)=0$ defines a relation, and if (x_0,y_0) is a point on the locus (so that $F(x_0,y_0)=0$), recall that the equation of the tangent at (x_0,y_0) is

$$dF(x_0,y_0,dx,dy)=0$$

which is an equation of the form $A\,dx+B\,dy=0$. Then the normal to $F(x,y)=0$ at (x_0,y_0) is perpendicular to the tangent, and so has equation $B\,dx-A\,dy=0$.

Example 3 Compute the equations of the tangent and normal lines to the locus of $x^3+y^2=0$ at the points $(-1,1)$ and $(-4,-8)$ (Figure 4-3). We compute the differential relation to be $3x^2dx+2y\,dy=0$, so the tangent at $(-1,1)$ has the equation

$$3\,dx+2\,dy=0$$

The normal line, therefore, is

$$2\,dx-3\,dy=0$$

At $(-4,-8)$, the tangent has the equation

$$48\,dx-16\,dy=0$$

or

$$3\,dx-dy=0$$

Therefore, the normal has the equation

$$dx+3\,dy=0$$

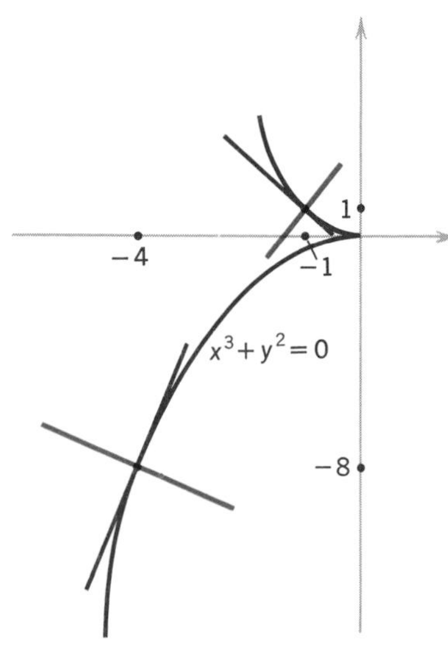

FIGURE 4-3

In the previous section we showed how the tangent line could be used to approximate a function near the point of tangency. If the curve is sloping rapidly, however, a better approximation might be offered by a circle. We can use the normal to a curve to assist us in finding the **closest circular approximation** to a curve at a given point.

Two curves (in this case a curve and a circle) are said to be **tangent** at some point if they interesect at that point, and if the tangents to the two curves at the point are identical. It follows from elementary geometry that the center of any circle tangent to a function at a given point on its graph is on the normal to the graph at that point.

Let $y = f(x)$ be the function, whose normal at the point (x_0, y_0) has the equation $y - y_0 = [-1/f'(x_0)](x - x_0)$. Consider the circle tangent to the graph of f at (x_0, y_0) with center at (x_1, y_1). Then (x_1, y_1) lies on this normal, and the equation of the circle is

$$(x - x_1)^2 + (y - y_1)^2 = R^2 = (x_1 - x_0)^2 + (y_1 - y_0)^2 \qquad (1)$$

Moreover, x_1 and y_1 are related by

$$y_1 - y_0 = -\frac{1}{f'(x_0)}(x_1 - x_0) \qquad (2)$$

Now since any circle tangent to the curve at (x_0, y_0) has the same line of tangency, dy/dx for any tangent circle at (x_0, y_0) will be just $f'(x_0)$. In general, however, there will be only one circle with

$$\left.\frac{d^2y}{dx^2}\right|_{(x_0,\, y_0)} = f''(x_0)$$

This circle is called the **closest circular approximation** to the function at (x_0, y_0), and its radius is known as the **radius of curvature** of f at (x_0, y_0).

We can calculate the value x_1 (and thus also y_1 and R) for this circle by reducing equations (1) and (2) above: Substituting equation (2) into equation (1) yields

$$(x - x_1)^2 + (y - y_1)^2 = (x_1 - x_0)^2 + \left(-\frac{1}{f'(x_0)}(x_1 - x_0)\right)^2$$

$$R^2 = (x_1 - x_0)^2\left(1 + \frac{1}{[f'(x_0)]^2}\right) \qquad (3)$$

Differentiating (1), we have

$$2(x - x_1)\,dx + 2(y - y_1)\,dy = 0$$

or

$$\frac{dy}{dx} = -\frac{x - x_1}{y - y_1}$$

Differentiating again yields

$$\frac{d^2y}{dx^2} = \frac{-(y - y_1) + (x - x_1)\,dy/dx}{(y - y_1)^2}$$

and at (x_0, y_0), we obtain

$$\left.\frac{d^2y}{dx^2}\right|_{(x_0,\,y_0)} = \frac{-(y_0-y_1)+(x_0-x_1)f'(x_0)}{(y_0-y_1)^2}$$

$$= \frac{(y_1-y_0)-(x_1-x_0)f'(x_0)}{(y_1-y_0)^2}$$

$$= \frac{(x_0-x_1)/f'(x_0)-(x_1-x_0)f'(x_0)}{((x_1-x_0)/f'(x_0))^2}$$

(using equation (2))

$$= \frac{f'(x_0)+(f'(x_0))^3}{x_0-x_1}$$

Then to find the closest circular approximation, we set

$$\left.\frac{d^2y}{dx^2}\right|_{(x_0,\,y_0)} = f''(x_0)$$

giving

$$f''(x_0) = \frac{f'(x_0)+(f'(x_0))^3}{x_0-x_1}$$

or

$$x_1 = x_0 - \frac{f'(x_0)+(f'(x_0))^3}{f''(x_0)} \tag{4}$$

Equations (4) and (2) combine to give

$$y_1 = y_0 + \frac{f'(x_0)+(f'(x_0))^3}{f'(x_0)f''(x_0)}$$

Equations (4) and (3) yield the radius of curvature:

$$R = \left|\frac{f'(x_0)+(f'(x_0))^3}{f''(x_0)}\right|\sqrt{1+\frac{1}{(f'(x_0))^2}}$$

$$= \frac{(1+(f'(x_0))^2)^{3/2}}{|f''(x_0)|}$$

Definition If f is a function twice differentiable at x_0 and if $f''(x_0) \neq 0$, then the **radius of curvature** of f at $(x_0, f(x_0))$ is

$$R = \frac{[1+(f'(x_0))^2]^{3/2}}{|f''(x_0)|}$$

The circle with center on the normal to the graph of f at $(x_0, f(x_0))$ which is tangent to the graph there, has radius R, and lies on the concave side of the graph, is called the **closest circular approximation** to f at x_0.

Example 4 Compute the radius of curvature and closest circular approximation to $y = 2x^{1/3} + x^{2/3}$ at the point $(1, 3)$. (See Example 1.) We know from Example 1 that $f'(x) = \frac{2}{3}(x^{-2/3} + x^{-1/3})$ and so $f'(1) = \frac{4}{3}$. We compute

$$\begin{aligned} f''(x) &= \tfrac{2}{3}\left(-\tfrac{2}{3}x^{-5/3} - \tfrac{1}{3}x^{-4/3}\right) \\ &= -\tfrac{2}{9}x^{-5/3}(2 - x^{1/3}) \end{aligned}$$

Therefore

$$f''(1) = -\tfrac{2}{9}$$

Then the radius of curvature is

$$R = \left(1 + \tfrac{16}{9}\right)^{3/2} \Big/ \left|-\tfrac{2}{9}\right| = \tfrac{125}{27}\Big/\tfrac{2}{9} = \tfrac{125}{6} = 20\tfrac{5}{6}$$

and we also derive

$$x_1 = 1 - \frac{\frac{4}{3} + \frac{64}{27}}{-\frac{2}{9}} = 17\tfrac{2}{3}$$

$$y_1 = 3 + \frac{\frac{4}{3} + \frac{64}{27}}{\left(\frac{4}{3}\right)\left(-\frac{2}{9}\right)} = -9\tfrac{1}{2}$$

Therefore the closest circular approximation is

$$\left(x - 17\tfrac{2}{3}\right)^2 + \left(y + 9\tfrac{1}{2}\right)^2 = \left(20\tfrac{5}{6}\right)^2 = 434\tfrac{1}{36}$$

Notice that the larger the radius of curvature, the larger the circle necessary to approximate the function, and thus the smaller the actual bending of the curve. We therefore define the **curvature** K of f at x_0 to be

$$K(x_0) = \begin{cases} 1/R, & f''(x_0) > 0 \\ -1/R, & f''(x_0) < 0 \\ 0, & f''(x_0) = 0 \end{cases}$$

Thus the curvature at x_0 is defined to be

$$K(x_0) = \frac{f''(x_0)}{[1 + f'(x_0)^2]^{3/2}}$$

A straight line has curvature 0 at all points; the curve of example 4 has curvature $-\frac{6}{125}$ at $(1, 3)$. The graph of $x \mapsto x^4$ is very flat around the origin and has curvature 0 at $x = 0$. Also notice that $\operatorname{sgn}(K) = \operatorname{sgn}(f''(x))$, and so the sign of the curvature is determined by the concavity of the graph of the function when $f''(x_0) \neq 0$.

Example 5 Analyze the curvature and radius of curvature for the sine function. Since $d(\sin\theta)/d\theta = \cos\theta$ and $d^2(\sin\theta)/d\theta^2 = -\sin\theta$, the radius of curvature is

$$R(\theta) = \frac{(1 + \cos^2\theta)^{3/2}}{|-\sin\theta|} = \frac{(2 - \sin^2\theta)^{3/2}}{|\sin\theta|}$$

and the curvature is

$$K(\theta) = \frac{-\sin\theta}{(2-\sin^2\theta)^{3/2}}$$

To find the critical points of curvature and radius of curvature, set

$$\begin{aligned} 0 &= K'(\theta) \\ &= \frac{-\cos\theta\,(2-\sin^3\theta)^{3/2} - 3\sin^2\theta\cos\theta\,(2-\sin^2\theta)^{1/2}}{(2-\sin^2\theta)^3} \\ &= \frac{\cos\theta\,(2-\sin^2\theta) + 3\sin^2\theta\cos\theta}{-(2-\sin^2\theta)^{5/2}} \\ &= \frac{2\cos\theta\,(1+\sin^2\theta)}{-(2-\sin^2\theta)^{5/2}} \end{aligned}$$

which has solutions $\theta = (n+\frac{1}{2})\pi$. The student should verify that the curvature has minima at $\theta = (\frac{1}{2}+2n)\pi$, where $K = -1$; maxima at $\theta = (\frac{3}{2}+2n)\pi$, where $K = 1$; and points of inflection at $\theta = n\pi$, where $K = 0$. The radius of curvature is therefore least at $\theta = (n+\frac{1}{2})\pi$, where $R = 1$. The radius of curvature becomes infinite at $n\pi$, the points of inflection of the sine function (Figure 4-4).

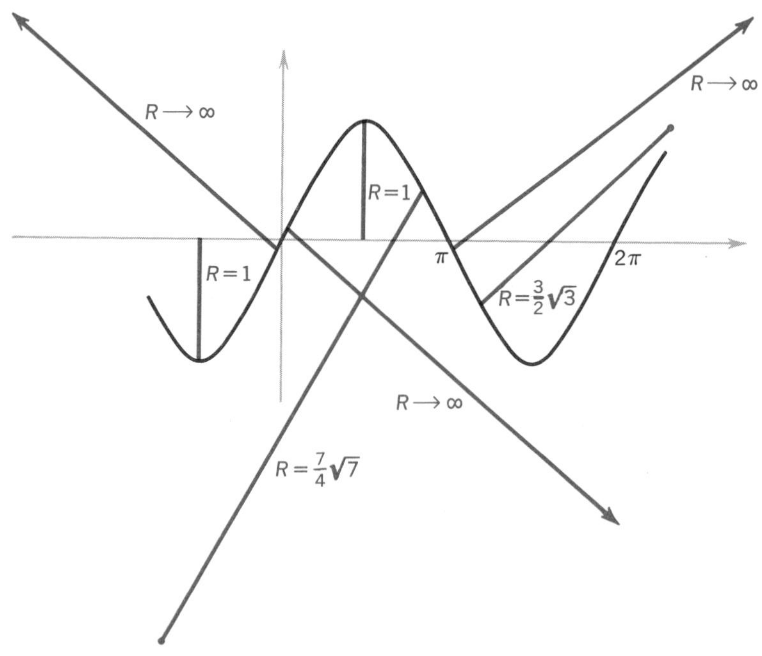

FIGURE 4-4

EXERCISES

1. Find the tangent and normal lines to the graph of each function at the point indicated.
   (a) $y = 3x^2 + 5x$ at $x = 2$
   (b) $y = x^2 + 3x$ at $x = 4$
   (c) $y = x^{3/2}/a$ at $y = 8a$
   (d) $(3x-y)^2 + 2x + y = 9$ at $(1,3)$
   (e) $y^2 - 2x - 4y - 1 = 0$ at $(-2,1)$

2. Find the normal to
   (a) $9x^2 - 4y^2 = 36$, perpendicular to the line $9x + 2y = 3$
   (b) $y = 2x^3$ with slope -2
   (c) $xy + 2x - y = 0$, parallel to $x + 2y = 0$
   (d) $y = 1 - x^2$, through the point $(3,1)$ [Hint: The normal is the minimum distance from a point to a curve.]
   (e) $y = \frac{1}{12}(x-1)^2(3x+1)$, through the point $(1,0)$

3. Find the curvature and radius of curvature to the following loci and compute these quantities at the point indicated. In each case, draw a figure analogous to the one in Example 5.
   (a) $y = \sqrt{4-x}$ at $(0,2)$
   (b) $y = \cot x$ at $(\frac{1}{6}\pi, \sqrt{3})$
   (c) $y = \cos t$ at $(0,1)$
   (d) $y = x^3$ at $(1,1)$
   (e) $y = f(mx+b)$ at $(0, f(b))$
   (f) $y = \sqrt{36-x^2}$ at $(\sqrt{11}, 5)$

4. For parts (a), (c), (d), and (f) of exercise 3, indicate where extrema of the curvature occur, thus finding where the radius of curvature has local minima. Sketch the graphs of the curvature and radius of curvature functions.

5. The **curvature** of a parametric curve is defined to be

$$K(t) = \frac{(dx/dt)(d^2y/dt^2) - (d^2x/dt^2)(dy/dt)}{[(dx/dt)^2 + (dy/dt)^2]^{3/2}}$$

   and the **radius of curvature** to be $R = 1/|K|$. Justify this definition partially by showing that if $x = t$, $y = f(t)$, then the two expressions for curvature coincide.

*6. Prove that the curvature of a curve C at a point P is equal to the rate of turning of the tangent line at the point. Thus, suppose a point moves along a parametrized curve

$$x = X(t), \qquad y = Y(t)$$

   at a constant speed, so that $\sqrt{X'(t)^2 + Y'(t)^2} = 1$. Let ϕ be the angle of inclination of the tangent to the curve (Figure 4-5). Prove that $K = d\phi/dt$.

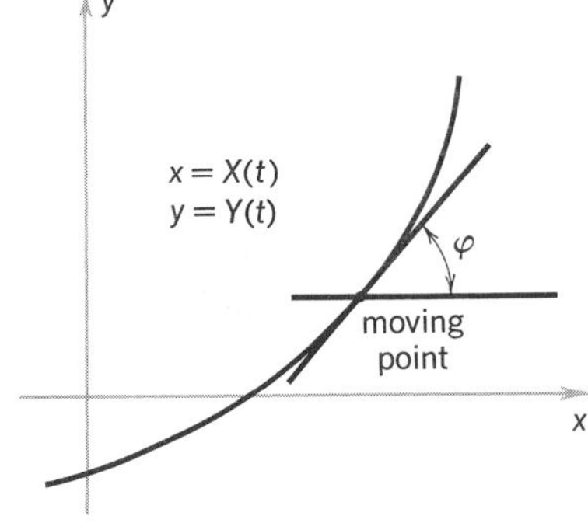

FIGURE 4-5

*7. Two curves are said to be **orthogonal** if they intersect, and if at the point of intersection the tangents to the two curves are perpendicular. Prove that an ellipse and a hyperbola having the same foci are orthogonal.

8. Let f be twice differentiable at 0 and let f satisfy $f(0) = 0, f'(0) = 0$, and $f''(0) > 0$. Do exercise 21 of §2.1 for the curve $y = f(x)$ and show that the limiting value of C is equal to the radius of curvature of f at 0. Interpret geometrically. Express the limit value of D in terms of $f'(0)$ and $f''(0)$.

5 L'HÔPITAL'S RULE

The elementary properties of limits tell us that if

$$\lim_{t\to a} f(t) = A, \qquad \lim_{t\to a} g(t) = B \neq 0$$

then

$$\lim_{t\to a} \frac{f(t)}{g(t)} = \frac{\lim_{t\to a} f(t)}{\lim_{t\to a} g(t)} = \frac{A}{B}$$

However, there still remains a variety of cases in which the above rule does not apply, either because $\lim_{t\to a} g(t) = 0$ or one of the limits is undefined. For example, consider the two cases:

$$\lim_{t\to 0} \frac{(1 - \cos t - \frac{1}{2}t^2)}{t^4}$$

$$\lim_{x\to\infty} \frac{x^3 + 3x^2}{2x^3 - x}$$

If we substitute $t = 0$ in the first expression we get the meaningless form $0/0$, while the second case we are led to ∞/∞. These expressions are called **indeterminate forms**.

Yet the limits in each of these cases are defined, and equal to $-\frac{1}{24}$ and $\frac{1}{2}$, respectively, as we shall soon demonstrate. The method for computing such limits is called L'Hôpital's Rule. Generally speaking, the rule is applicable if both the numerator and denominator of the function approach zero, or if they both approach $\pm\infty$.

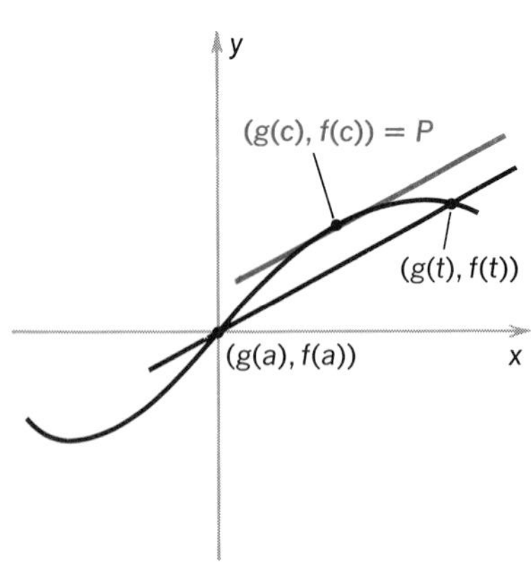

FIGURE 5-1

Theorem 1 **L'Hôpital's Rule** Suppose f and g are differentiable functions such that $f(a) = g(a) = 0$. If $\lim_{t\to a}(f'(t)/g'(t))$ exists, then $\lim_{t\to a}(f(t)/g(t))$ also exists, and

$$\lim_{t\to a} \frac{f(t)}{g(t)} = \lim_{t\to a} \frac{f'(t)}{g'(t)}$$

Proof Consider the parametric equations

$$x = g(t), \qquad y = f(t)$$

Since $g(a) = f(a) = 0$, the curve described by these equations passes through the origin of the xy plane when $t = a$ (Figure 5-1). If $t \neq a$, then the ratio $f(t)/g(t)$ is equal to the slope of the secant to the curve which passes through the origin and the point $(g(t), f(t))$. Moreover, we see that there is a point $P = (g(c), f(c))$ on the curve between the origin and $(g(t), f(t))$, such that the tangent line to the curve at P is parallel to the secant. (To show this analytically, apply Rolle's Theorem to the function $x \mapsto g(t)f(x) - f(t)g(x)$ on the interval $[a, t]$ if $t > a$ and on $[t, a]$ if $t < a$.) The slope of the tangent at P is $f'(c)/g'(c)$. Therefore

$$\frac{f(t)}{g(t)} = \frac{f'(c)}{g'(c)}$$

Now as t approaches a, c also approaches a, since c is always between t and a. Hence, passing to limits, we have

$$\lim_{t\to a} \frac{f(t)}{g(t)} = \lim_{c\to a} \frac{f'(c)}{g'(c)}$$ ∎

In practice, several applications of L'Hôpital's Rule may be required to reach a form which is not indeterminate. We simply keep applying the rule until at least one of the zeros disappears.

Example 1 Compute

$$\lim_{t\to 0} \frac{1 - \cos t - \frac{1}{2}t^2}{t^4}$$

We have

$$\lim_{t\to 0} \frac{1 - \cos t - \frac{1}{2}t^2}{t^4} \qquad [0/0]$$

$$= \lim_{t\to 0} \frac{\sin t - t}{4t^3} \qquad [0/0]$$

$$= \lim_{t\to 0} \frac{\cos t - 1}{12t^2} \qquad [0/0 \text{ again!}]$$

$$= \lim_{t\to 0} \frac{-\sin t}{24t} \qquad [\text{good grief, still } 0/0]$$

$$= \lim_{t\to 0} \frac{-\cos t}{24} \qquad [\text{whew!}]$$

$$= -\frac{1}{24}$$

Example 2 Compute

$$\lim_{x\to 0} \frac{(1+x)^{1/3} - (1 + \frac{1}{3}x)}{x^2}$$

We have

$$\lim_{x\to 0}\frac{(1+x)^{1/3}-(1+\frac{1}{3}x)}{x^2} \qquad [0/0]$$

$$= \lim_{x\to 0}\frac{\frac{1}{3}(1+x)^{-2/3}-\frac{1}{3}}{2x} \qquad [0/0]$$

$$= \lim_{x\to 0}\frac{-\frac{2}{9}(1+x)^{-5/3}}{2}$$

$$= -\frac{1}{9}$$

Although we shall not prove it, L'Hôpital's Rule also applies to finding the limit when $\lim_{t\to a} f(t) = \pm\infty$ and $\lim_{t\to a} g(t) = \pm\infty$. The rule is the same:

$$\lim_{t\to a}\frac{f(t)}{g(t)} = \lim_{t\to a}\frac{f'(t)}{g'(t)}$$

provided

$$\lim_{t\to a}\frac{f'(t)}{g'(t)}$$

exists. The limit may also be taken as t approaches a from one side or as t approaches $\pm\infty$. (For a proof of L'Hôpital's Rule for the case ∞/∞ see *Advanced Calculus* by R. C. Buck, McGraw-Hill, 2nd ed., 1964.)

Example 3

$$\lim_{n\to\infty}\frac{n^3+3n^2}{2n^3-n} \qquad [\infty/\infty]$$

$$= \lim_{n\to\infty}\frac{3n^2+6n}{6n^2-1} \qquad [\infty/\infty]$$

$$= \lim_{n\to\infty}\frac{6n+6}{12n} \qquad [\infty/\infty]$$

$$= \lim_{n\to\infty}\frac{6}{12} = \frac{1}{2}$$

In conclusion, we explicitly warn the student against two very common mistakes made when using L'Hôpital's Rule. First, **do not** differentiate the fraction $f(t)/g(t)$. The new fraction is obtained by dividing the derivative of the numerator by the derivative of the denominator. Second, **do not** apply L'Hôpital's Rule unless the limit is of the type 0/0 or $\pm\infty/\infty$. For example,

$$\lim_{t\to 0}\frac{t^3-7}{2t+7} \neq \lim_{t\to 0}\frac{3t^2}{2}$$

It is sometimes possible to find a limit of the type $\infty-\infty$ by transforming it into one of the types 0/0 or ∞/∞ and applying L'Hôpital's Rule.

Example 4 Find

$$\lim_{x\to 0^+}\left(\frac{1}{x}-\frac{1}{\sin x}\right)$$

Since both

$$\lim_{x\to 0^+}\frac{1}{x}=+\infty=\lim_{x\to 0^+}\frac{1}{\sin x}$$

we cannot apply the arithmetic of limits to determine this limit. However, we have

$$\begin{aligned}\lim_{x\to 0^+}\left(\frac{1}{x}-\frac{1}{\sin x}\right)&=\lim_{x\to 0^+}\frac{\sin x-x}{x\sin x} && [0/0]\\ &=\lim_{x\to 0^+}\frac{\cos x-1}{\sin x+x\cos x} && [0/0]\\ &=\lim_{x\to 0^+}\frac{-\sin x}{\cos x+\cos x-x\sin x}\\ &=\frac{0}{2}=0\end{aligned}$$

EXERCISES

Use L'Hôpital's Rule to compute the following limits.

1. $\lim_{\theta\to 0}\frac{\sin^2\theta}{\theta}$
2. $\lim_{\theta\to 0}\frac{\sin 3\theta}{\sin\theta}$
3. $\lim_{x\to\pi}\frac{1+\cos x}{(\pi-x)^2}$
4. $\lim_{y\to 0}\frac{\sqrt{1-y}-\sqrt{1+y}}{y}$
5. $\lim_{\alpha\to 0}\frac{\tan\alpha-\sin\alpha}{\alpha^2}$
6. $\lim_{x\to\infty}\frac{3x^3-2x^2+1}{2x^3+x^2}$
7. $\lim_{t\to\theta}\frac{\sqrt[3]{1+t}+(1+\frac{1}{3}t)}{t^2}$
8. $\lim_{\theta\to\pi/2}(\theta-\frac{1}{2}\pi)\tan 3\theta$
9. $\lim_{z\to 0}\frac{1-\sqrt{z+1}}{z}$
10. $\lim_{x\to\infty}\frac{1-x^2}{x+2x^2}$
11. $\lim_{x\to\infty}\frac{a_n x^n+a_{n-1}x^{n-1}+\cdots+a_0}{b_m x^m+b_{m-1}x^{m-1}+\cdots+b_0}$
12. $\lim_{\alpha\to 0}\frac{\sin(\alpha^2)-\sin^2\alpha}{\alpha^4}$
13. Two functions are **tangent** at a point if their function values and derivatives are equal at that point. Show that if f and g are tangent at x_0, then

$$\lim_{\Delta x\to 0}\frac{f(x_0+\Delta x)-g(x_0+\Delta x)}{\Delta x}=0$$

Show that if f and g are twice continuously differentiable functions which are tangent at x_0 with $f''(x_0)=g''(x_0)$, then

$$\lim_{\Delta x\to 0}\frac{f(x_0+\Delta x)-g(x_0+\Delta x)}{\Delta x^2}=0$$

6 ANTIDERIVATIVES

The preceding two chapters have been devoted mainly to the problem: Given a function f, find the derivative f'. We can also turn this question around and ask: Given a function f, find a function F such that $F' = f$. Such a function F is called an **antiderivative** of f, and the process of finding F is called **antidifferentiation**. Thus, the sine function is an antiderivative of the cosine function. The concept of antidifferentiation is closely linked to the integral calculus, which we shall begin to study in the next chapter.

Definition A function F is an **antiderivative** of a function f if F and f have the same domain and $F' = f$.

Example 1 Find an antiderivative of $f: x \mapsto 3x^4 + 8x^3 - 6$. Since we know that differentiating lowers the degree of a monomial by one, and multiplies its coefficient,

$$\frac{d(cx^n)}{dx} = cnx^{n-1}$$

antidifferentiation should therefore divide the coefficient and raise the degree. Let us guess at an antiderivative of f; consider

$$F: x \mapsto \tfrac{3}{5}x^5 + \tfrac{8}{4}x^4 - 6x = \tfrac{3}{5}x^5 + 2x^4 - 6x$$

To see if our guess is right, we differentiate

$$\begin{aligned} \frac{d(F(x))}{dx} &= 5(\tfrac{3}{5})x^4 + 4 \cdot 2x^3 - 6 \\ &= 3x^4 + 8x^3 - 6 \\ &= f(x) \end{aligned}$$

so F is in fact an antiderivative of f. Another antiderivative would be

$$x \mapsto \tfrac{3}{5}x^5 + 2x^4 - 6x + 11$$

(Verify!), or in fact,

$$x \mapsto \tfrac{3}{5}x^5 + 2x^4 - 6x + C$$

where C is any constant number. This follows because when we differentiate any constant, we obtain the same result—zero.

In Example 1, we guessed at and then verified that it was indeed a correct antiderivative. Antidifferentiation is essentially a problem of enlightened guesswork; there is no guaranteed process for finding antiderivatives of an arbitrary function. However, for polynomial functions there is a dependable process, which consists of applying the following formal devices.

FORMAL DEVICE

An antiderivative of the function $x \mapsto cx^n$ is

$$x \mapsto \frac{c}{n+1} x^{n+1}, \qquad \text{if } n \neq -1$$

FORMAL DEVICE

If F and G are antiderivatives of functions f and g, respectively, then $F+G$ is an antiderivative of $f+g$.

Later we shall develop some more methods which will extend our powers of antidifferentiation.

Example 2 The function $x \mapsto 7x^3$ has $x \mapsto \frac{7}{4}x^4$ as an antiderivative, and $x \mapsto 2x^2$ has $x \mapsto \frac{2}{3}x^3$ as an antiderivative. Thus, by the formal devices above, an antiderivative of $x \mapsto 2x^2 + 7x^3$ is $x \mapsto \frac{2}{3}x^3 + \frac{7}{4}x^4$.

It is not difficult to verify that an antiderivative of the nth-degree polynomial function

$$f: x \mapsto a_n x^n + a_{n-1} x^{n-1} + \cdots + a_0$$

is the $(n+1)$st-degree polynomial function

$$F: x \mapsto \frac{a_n x^{n+1}}{n+1} + \frac{a_{n-1} x^n}{n} + \cdots + \frac{a_1 x^2}{2} + a_0 x + C$$

where C is an arbitrary constant.

A guide to classifying antiderivatives has already been provided by the second corollary to the Mean Value Theorem (§4.7), which we restate here in the language of antidifferentiation.

Theorem 1 Two antiderivatives of the same function differ by at most a constant. That is, if F and G are both antiderivatives of a function f, with domain an interval, then there is a real number C such that

$$F(x) = G(x) + C$$

for all x in the domain of f. Moreover, if G is an antiderivative of f, and $F = G + C$, then F is also an antiderivative of f.

In effect, this theorem states that the antiderivative of a function is unique up to a constant term. Hence, every antiderivative of a function is the sum of any one antiderivative and some constant. (For example, x^2, x^2+3, $x^2-\sqrt{2}$, $x^2+2\pi/7$ are all antiderivatives of $2x$.) For this reason, we sometimes refer to **the** antiderivative of a function f as the **family** of functions written $F(x)+C$, where F is an antiderivative of f and C denotes an arbitrary constant. (Thus C is different for each different antiderivative of f.)

Example 3 Find the antiderivative of $f: x \mapsto 2x-1$. Using the formal devices, we set $F(x)=x^2-x+C$, and we verify that

$$F'(x) = \frac{d(x^2)}{dx} - \frac{d(x)}{dx} + \frac{d(C)}{dx} = 2x - 1 + 0 = 2x - 1 = f(x)$$

Example 4 The antiderivative of $x \mapsto 30x^5-30x^4-24x^3+24x^2+7$ is $x \mapsto \frac{30}{6}x^6 - \frac{30}{5}x^5 - \frac{24}{4}x^4 + \frac{24}{3}x^3 + 7x + C = 5x^6 - 6x^5 - 6x^4 + 8x^3 + 7x + C$. The antiderivative of $x \mapsto x^3-5x^2+7x-2$ is $x \mapsto \frac{1}{4}x^4 - \frac{5}{3}x^3 + \frac{7}{2}x^2 - 2x + C$.

Sometimes a problem involving antiderivatives is stated as a **differential equation**, an equation which expresses a relationship between a function and one or more of its derivatives. The antiderivative is the **solution** to the differential equation. Thus, if $dy/dx = f(x)$ is the equation and $F(x)+C$ is the antiderivative of f, then $y = F(x)+C$ is the solution to the equation, for we have

$$\begin{aligned} \frac{dy}{dx} &= \frac{d(F(x)+C)}{dx} \\ &= \frac{d(F(x))}{dx} + 0 \\ &= f(x) \end{aligned}$$

Differential equations are often difficult to solve, and the theory of differential equations is a large and important branch of mathematics. However, there are some elementary forms of differential equations which we can deal with here.

Example 5 Solve the differential equation

$$\frac{dy}{dx} = (x+1)^2$$

We have

$$\frac{dy}{dx} = x^2 + 2x + 1$$

and so the solution is

$$y = \tfrac{1}{3}x^3 + x^2 + x + C$$

as we could verify by differentiating y with respect to x.

If we are just given a differential equation $dy/dx = f(x)$, then there is a whole family of solutions. However, sometimes an additional condition is added, which allows us to determine the arbitrary constants involved in the antidifferentiation process. Such a condition is called an **initial condition**.

Example 6 Solve the equation of Example 5, $dy/dx=(x+1)^2$ for $y=F(x)$, where $F(1)=0$. All the solutions are of the form $y=\frac{1}{3}x^3+x^2+x+C$, so to find C we use the fact that $y=0$ when $x=1$:

$$0=\tfrac{1}{3}+1+1+C \Rightarrow C=-\tfrac{7}{3}$$

Therefore the desired solution is $y=\frac{1}{3}x^3+x^2+x-\frac{7}{3}$.

In Chapter 3, we discussed the fact that velocity is the derivative of the position function, and acceleration is the derivative of velocity (both derivatives are taken with respect to time). It follows that position is an antiderivative of velocity, and velocity is an antiderivative of acceleration. Therefore, given the velocity of an object, we can find the position of the object as a function of time by taking an appropriate antiderivative of the velocity function.

Example 7 A ball is thrown straight up into the air from an initial height of 6 feet. The velocity is $v=20-32t$ (ft/sec). After how many seconds will the ball hit the ground? The height of the ball is $h=20t-16t^2+C$ (the antiderivative of v, since $v=dh/dt$). But at $t=0$, we are told that $h=6$ (ft) which means $C=6$. Therefore, $h=20t-16t^2+6$. When the ball hits the ground $h=0$ and so we obtain a quadratic equation which we can solve for t:

$$t=\frac{-20\pm\sqrt{400+384}}{-32}=\frac{-20\pm\sqrt{(28)^2}}{-32}$$

$$=\frac{5\pm 7}{8}$$

We are only interested in time after the toss, so the + part of the sign gives the correct solution, $t=\frac{3}{2}$ (sec).

Example 8 A car brakes to a stop with constant acceleration. If the braking time is 20 seconds, during which the car travels 800 feet, find the acceleration. Let A be the constant acceleration. Then

$$a(t)=\frac{dv}{dt}=A$$

and so

$$v(t)=\frac{ds}{dt}=tA+C_1$$

and

$$s(t)=\tfrac{1}{2}At^2+C_1t+C_2$$

But we know that $v(20)=0$. Moreover, $s(20)-s(0)=800$. Substituting this information into the equations, we obtain

$$v(20)=0=20A+C_1 \Rightarrow C_1=-20A$$

and

$$s(20) - s(0) = 800$$
$$= \tfrac{1}{2}A(400) + 20(-20)A + C_2 - \tfrac{1}{2}A(0) - 0(-20A) - C_2$$
$$= -200A$$

so

$$A = -4(\text{ft/sec}^2)$$

Example 9 The position function of a body in free fall is

$$s = -16t^2 + v_0 t + s_0$$

where v_0 is the initial velocity and s_0 is the initial height above the ground. This equation can be derived from empirical data using antidifferentiation. A free falling body in the earth's gravitational field falls with a constant downward acceleration of g, the gravitational constant, approximately 32 ft/sec^2. If upward is taken as the positive direction, the acceleration is $-g$ and the velocity at time t is therefore

$$v = -gt + C_1$$

Since v must be v_0 at time $t = 0$, C_1 can only be v_0. Then the position is

$$s = -\tfrac{1}{2}gt^2 + v_0 t + C_2$$

Since this expression must equal s_0 at time $t = 0$, we see that C_2 is s_0.

We note that the result just obtained is not true on a planet where the gravitational field is different from that of the earth. One would need to change g appropriately.

EXERCISES

1. Find the antiderivative for each of the following functions.
   (a) $x \mapsto x^4$ (b) $\theta \mapsto \sin 2\theta$
   (c) $y \mapsto (2y+1)^4$ (d) $\phi \mapsto \csc\phi \cot\phi$
   (e) $v \mapsto v^{-3}$ (f) $x \mapsto 3x^4 - 5x^3 + 4x^2 + 7$
   (g) $z \mapsto 3z(z-1)^2$ (h) $x \mapsto x^6 - \sin x$
2. Solve each of the following differential equations, with initial conditions indicated.
   (a) $dy/dx = \cos x$ $y = \tfrac{1}{2}$ when $x = 0$
   (b) $dy/dx = 3x^2$ $y = 2$ when $x = 3$
   (c) $d^2y/dx^2 = 4$ $y = 1$ and $dy/dx = 0$ when $x = 1$
   (d) $d^2y/dx^2 = 6x - 6$ with critical point at $(1,0)$
   (e) $d^2y/dx^2 = 2$ $y = 1$ when $x = 2$, $y = 2$ when $x = 3$
   (f) $dy/dx = x(x-1)(x-2)$ $y = 4$ when $x = 2$
3. A ball rolls across a level field with initial velocity 35 ft/sec. If friction slows the ball at a constant rate of 10 ft/sec^2, how long will it take for the ball to come to a stop? How far will it have gone by then? Give the distance traveled by the ball as a function of time.

4. Suppose the x axis sticks out of the earth with origin at the center of the earth. An object of mass m on the x axis experiences a gravitational attraction mgR^2/x^2 to the earth, where R is the radius of the earth and g is a gravitional constant. If distance units are chosen as miles and the time units as seconds, then $R \approx 4000$ and $g \approx 1/165$. If an object's acceleration toward the earth is equal to the gravitional force of attraction, and if the effect of atmosphere and of other celestial bodies is ignored, with what initial velocity must a rocket be shot in order for it to travel out 4000 miles before falling to earth? To keep going forever?

5. Find the relations which correspond to the following differential relations.
   (a) $y\,dy - x\,dx = 0$ (b) $y^{-1/3}\,dy - x^{-1/3}\,dx = 0$
   (c) $dy = \sqrt[3]{y/x}\,dx$ (d) $dy = 2xy^2\,dx$
   (e) $dy/\sqrt{y} + \sqrt{x}\,dx = 0$

REVIEW EXERCISES FOR CHAPTER 5

In exercises 1–10
- (a) sketch the curves described;
- (b) indicate relative extrema;
- (c) indicate points of inflexion;
- (d) indicate regions of increase, decrease, and concavity;
- (e) note any asymptotes;
- *(f) determine the curvature and equation of the normal line at the; point indicated.

1. $y = x^3 - 8x^2 + 64x + 8$ $(0, 8)$
2. $y = 40x^6 - 16x^5 - x^4$ $(0, 0)$
3. $y = \dfrac{80}{3x^2 + 80}$ $(0, 1)$

*4. $y = x - \sin x$ $(0, 0)$

5. $y = \dfrac{x}{1 - x^2}$ $(0, 0)$
6. $y = \dfrac{x^2 - 4}{x^3}$ $(1, -3)$
7. $y = \dfrac{x^2 - x - 2}{x - 1}$ $(0, 2)$
8. $y = \dfrac{(x+2)^2(x+1)}{x(x-1)}$ $(-1, 0)$
9. $y = (3x-4)(1+3x)^{1/2}$ $(0, -4)$
10. $y = \sin 2x + \sqrt{3}\cos 2x$ $(0, \sqrt{3})$
11. For a right triangle with fixed hypotenuse, show that maximum area is attained when the triangle is isosceles.

12. Find the dimensions of a cylindrical can with volume V and minimal surface area.

13. A field is bounded by a long straight brick wall. A farmer wishes to fence a rectangular portion of the field using 200 meters of fencing. He need not fence the side against the wall. What is the largest area he can fence off?

14. What is the smallest positive number of the form $x+y$, where
$$\frac{2}{x}+\frac{4}{y}=3$$

15. An airline company will sell tickets on a charter flight for 100 dollars if fewer than 60 people fly, or for one-half dollar per ticket less for each person above 60. The plane holds 120 people. What is the maximum possible intake of the airline for the flight?

16. What is the area of the largest rectangle which can be inscribed in an isosceles right triangle with side s, if one angle of the rectangle is to be at the right angle of the triangle? If one side of the rectangle lies along the hypotenuse?

17. In a psychology experiment mice are to be placed in adjacent congruent rectangular cages with a common side. If 12 feet of caging material is available for the sides, what is the greatest total area which can be enclosed?

18. The owners of two houses, one 60 feet from the road and the other 40 feet from the road and 200 feet farther down, agree to share a single mailbox to be built on the road between them. Where on the road should the mailbox be built in order to minimize the sum of its distances from the houses? How could this problem be solved without calculus?

19. A cylindrical tank with hemispherical ends is to be built from a fixed amount of material in order to house the maximal volume. What should the shape of the tank be?

20. Legend has it that the Greek playwright Aeschylus died when an eagle, mistaking the learned gentleman's bald pate for a shiny stone, dropped a tortoise on his head in order to break it open (the tortoise, not the head). Suppose that the tortoise was dropped from an altitude of 221 ft and took 4 sec to fall. If Aeschylus was 5 ft tall, how rapidly was the eagle rising the moment it released the tortoise?

In exercises 21–26, compute antiderivatives for the given functions.

21. $x \mapsto x^3 - 8x^2 + 64x + 8$

22. $x \mapsto 40x^6 - 16x^5 - x^4$

23. $x \mapsto 80/3x^2$

24. $x \mapsto \dfrac{x-4}{x^3}$

25. $x \mapsto \sqrt{1+3x}$

26. $x \mapsto \sin 2x + \sqrt{3}\cos 2x$

In exercises 27–32, compute the limits.

27. $\lim_{x \to 0} \frac{x^2 \cot^2 x}{(x+1)^3}$

28. $\lim_{x \to \infty} \frac{\sqrt{6x^4 + 8x^3 + 7x + 3}}{x^2 - 4 + 2x}$

29. $\lim_{\theta \to \pi} \frac{\tan 5\theta}{\sin 3\theta}$

30. $\lim_{z \to 2} \frac{4 - \sqrt{z+2}}{z - 2}$

31. $\lim_{z \to n^2} \frac{n - \frac{1}{2}\sqrt{2z + 2n^2}}{z - n^2}$

32. $\lim_{\theta \to 0} \frac{\tan \theta - \sin \theta}{\theta^3}$

6

INTEGRAL CALCULUS

Here we analyze the final problem presented at the end of Chapter 1; that of finding the area under a curve.

We shall proceed as follows: In §1 we give a heuristic discussion based on intuitive notions of area. This is followed by a geometric derivation of the Fundamental Theorem of Calculus, which reduces the problem of finding an integral to one of finding an antiderivative. This result sets up the basic link between differential and integral calculus. In the remaining sections of the chapter we develop some applications of integration and further consequences of the Fundamental Theorem.

In Chapter 7 we reexamine the definition of the integral from a rigorous point of view. The idea is that in the present chapter we accept as intuitively clear the notion of the area of a region and the familiar geometric properties of area. This has the advantage of leading quickly to the Fundamental Theorem and an appreciation of the power of integration. In Chapter 7 we are faced with the problem of rigorously defining area and establishing properties of integrals corresponding to intuitively clear properties of area. This is done by the method of Riemann sums. Chapter 7 also contains further applications.

1 INTEGRALS AND AREA

As we mentioned in the introduction above, we begin our study with an intuitive discussion of integrals based on the assumption that the regions in question have an area and that the area satisfies the usual geometrical properties (for example, if a region R_1 is contained inside a region R_2, that is $R_1 \subset R_2$, then the area of R_2 is at least as large as that of R_1). If we accept these basic ideas, then we can develop a powerful method of

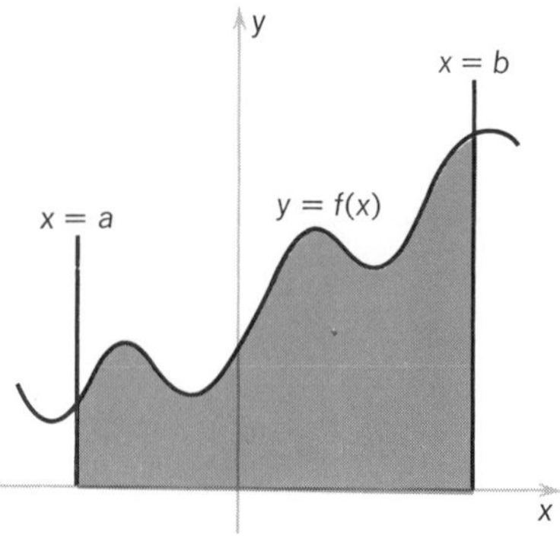

FIGURE 1-1

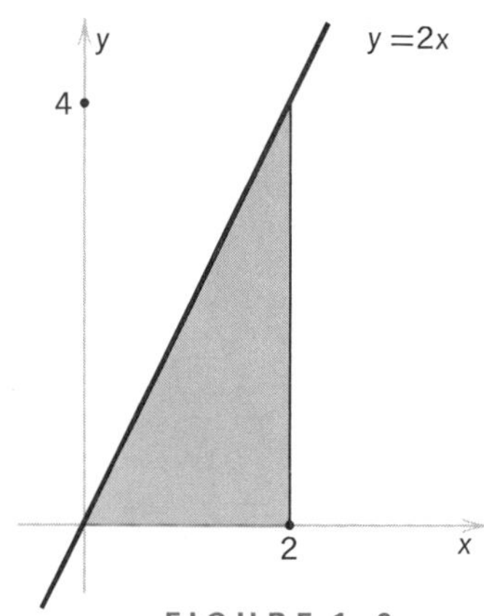

FIGURE 1-2

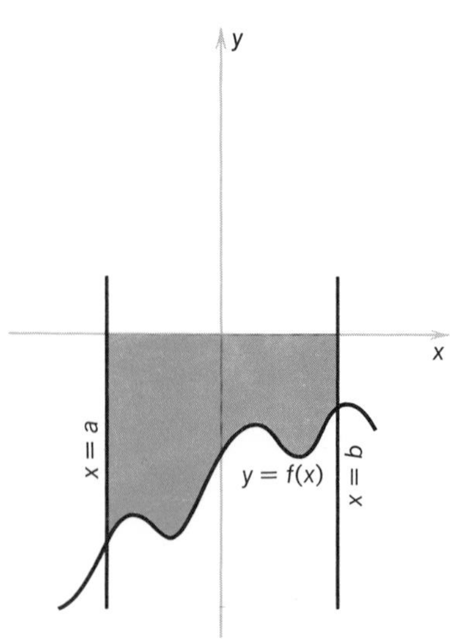

FIGURE 1-3

computing areas using the Fundamental Theorem of Calculus, which we state and prove. The justification for this initial approach and the longer and more technical rigorous development is deferred until the next chapter.

Again we emphasize that the proofs in this section are to be understood as informal, being based on (intuitively clear) notions of areas of regions.

Suppose f is a continuous function on an interval $[a, b]$ and suppose $f(x) \geqslant 0$ for $x \in [a, b]$. Then the graph of f, the vertical lines $x = a$ and $x = b$, and the x axis determine a region as illustrated in Figure 1-1. This region is called the **region under the graph of *f* from *a* to *b***. (One requires f to be continuous for technical reasons which are explained in §2 and in Chapter 7. This hypothesis is not important for the moment.)

We define $\int_a^b f$, called the **integral of *f* from *a* to *b***, to be the area of this region under the graph of f from a to b. The explanation of the notation $\int$ (standing for an elongated S for sum) and the alternative notation $\int_a^b f(x)\,dx$ is given later. We shall use the two notations interchangeably.

Example 1 Let $f: [0, 2] \to \boldsymbol{R}$, $x \mapsto 2x$. Then the area under the graph of f from 0 to 2 is the area of a triangle with base 2 and height 4 which is $\frac{1}{2} \cdot 2 \cdot 4 = 4$. Thus $\int_0^2 f = \int_0^2 2x\,dx = 4$. See Figure 1-2.

We must also deal with the case when negative values of f occur. Suppose first that $f(x) \leqslant 0$ for $x \in [a, b]$. Then as before we have the **region above *f* between *a* and *b*** (see Figure 1-3).

In this case we define $\int_a^b f(x)\,dx$ to be the **negative** of this area. The reason for choosing the negative can be made plausible as follows. Suppose f is a constant function $x \mapsto c$ on $[a, b]$. Then if $c \geqslant 0$,

$$\int_a^b f(x)\,dx = c(b-a)$$

since this is the area of the corresponding rectangle. The same formula is valid for $c \leqslant 0$ if the integral is defined as the negative of the area. Defining the integral in this way also leads to other simplifications as we shall see.

If f is general, that is, neither positive nor negative everywhere, we define $\int_a^b f(x)\,dx$ to be

$$\int_a^b f(x)\,dx = (\text{area below } f \text{ where } f(x) \text{ is} \geqslant 0) \\ -(\text{area above } f \text{ where } f(x) \leqslant 0)$$

(See Figure 1-4.) The numbers a and b are called, respectively, the **lower** and **upper limits of integration**. The function f is called the **integrand**.

Below we show that $\int_a^b f(x)\,dx$ can be evaluated by using antiderivatives. This is a very important result because it enables us to compute areas of regions with curved edges which otherwise would be difficult (or even impossible) to determine. Before doing this, however, we consider some additional examples and develop the basic properties of integrals.

Example 2 Suppose $g:x \mapsto mx$ where $m > 0$. Then

$$\int_0^b g = \int_0^b mx\,dx = \frac{mb \cdot b}{2} = \frac{m}{2}b^2$$

(Figure 1-5). More generally, $\int_a^b mx\,dx = m(b^2 - a^2)/2$, whenever $a < b$. Note that this formula is true whether $a < 0$, $a = 0$, or $a > 0$; and whether $b > 0$ or $b \leqslant 0$. The reader can verify that $\int_a^b mx\,dx = m(b^2 - a^2)/2$ also holds when $m \leqslant 0$.

Example 3 Since $\sin x = -\sin(-x)$, the two shaded regions in Figure 1-6 have the same area. In computing the integral $\int_{-\pi}^{\pi} \sin x\,dx$, the area on the left receives a negative sign and that on the right a positive sign; since the areas are equal and receive opposite signs, we have $\int_{-\pi}^{\pi} \sin x\,dx = 0$.

The following **additivity property** of the definite integral is clear from the definition: If $a < b < c$, and if f is continuous on $[a, c]$, we have

$$\int_a^b f + \int_b^c f = \int_a^c f$$

This just expresses the fact that the area of the union of disjoint regions is the sum of the areas (see Figure 1-7).

Suppose M is a constant such that $f(x) \leqslant M$ for all x in $[a, b]$. Then, when f is continuous on $[a, b]$, we must have

$$\int_a^b f \leqslant M(b-a)$$

since the area under f is less than that of the enclosed rectangle (Figure 1-8). Note that this inequality holds even when $f(x) \leqslant 0$ on parts of $[a, b]$. On the other hand, if m is a constant such that $m \leqslant f(x)$ for all $x \in [a, b]$, we have

$$m(b-a) \leqslant \int_a^b f$$

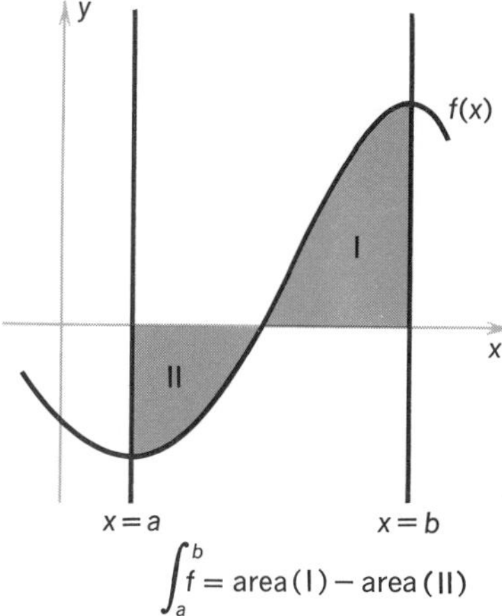

FIGURE 1-4

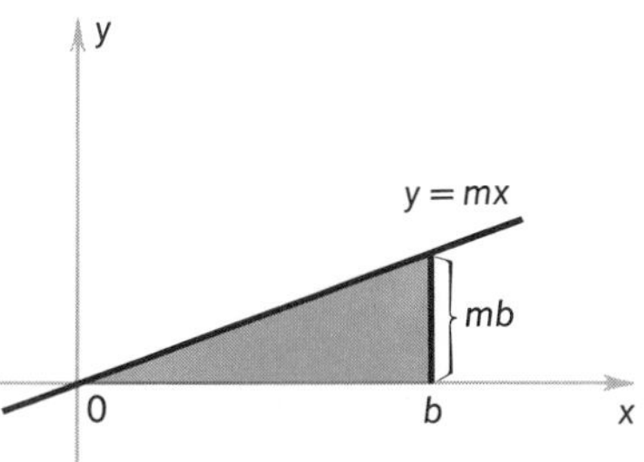

FIGURE 1-5

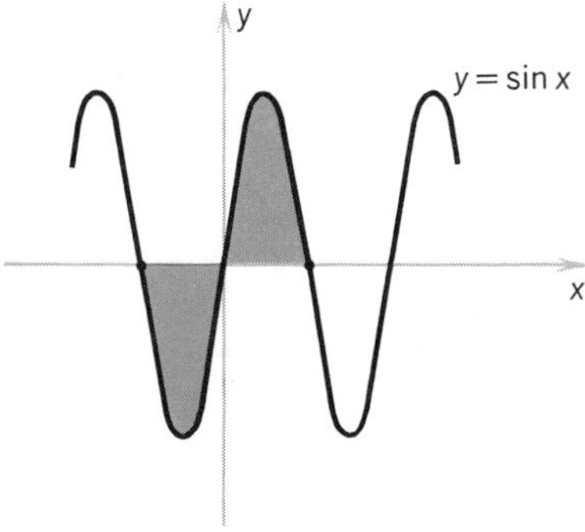

FIGURE 1-6

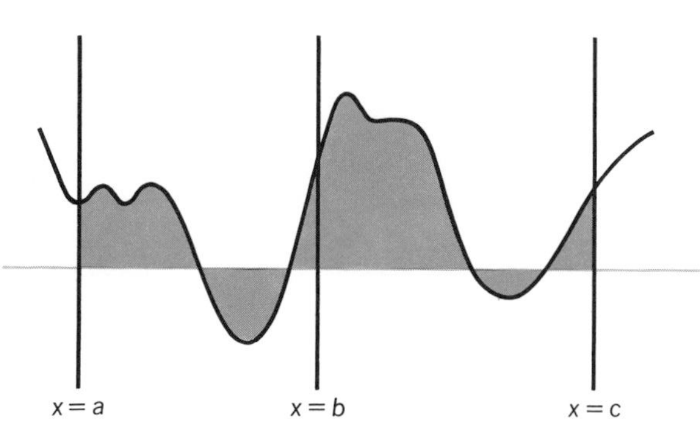

FIGURE 1-7

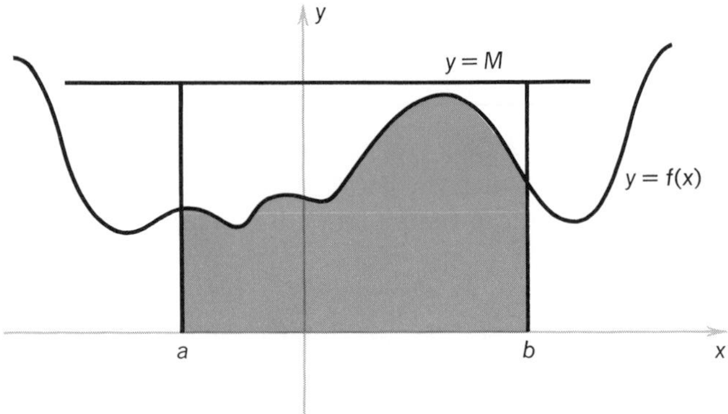

FIGURE 1-8

(See Figure 1-9.) In particular, if M_0 is the maximum of f on $[a, b]$ and m_0 the minimum, we have

$$m_0(b-a) \leqslant \int_a^b f \leqslant M_0(b-a)$$

or

$$m_0 \leqslant \frac{\int_a^b f}{b-a} \leqslant M_0$$

Therefore, the number $(\int_a^b f)/(b-a)$ lies between m_0 and M_0. By the Intermediate Value Theorem (§4.6) there must be a point $c \in [a, b]$, such that

$$f(c) = \frac{\int_a^b f}{(b-a)}$$

or

$$f(c)(b-a) = \int_a^b f$$

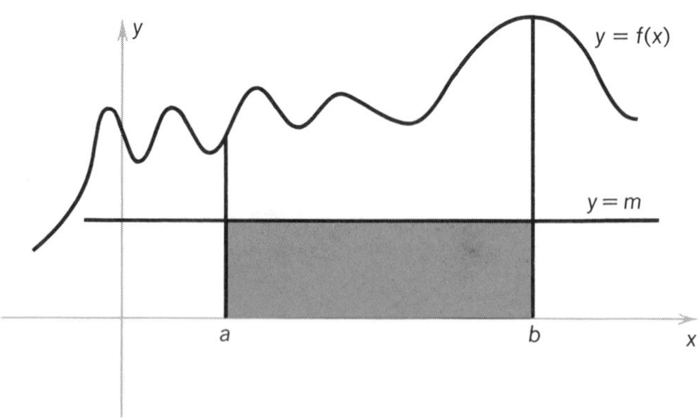

FIGURE 1-9

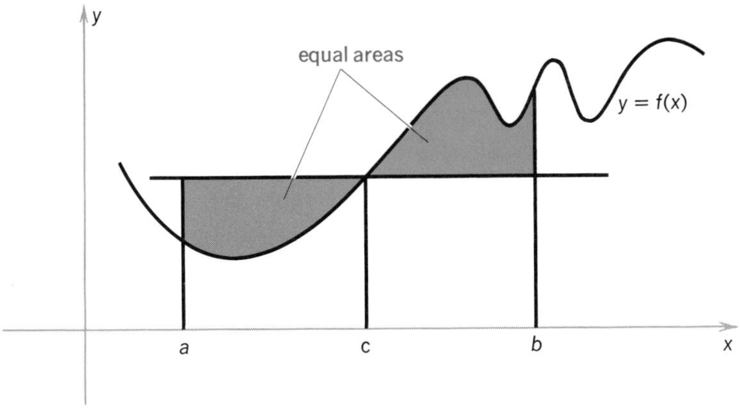

FIGURE 1-10

This last result is known as the **Mean Value Theorem of Integral Calculus.** The value $\int_a^b f/(b-a)$ is called the **average value** or **mean value** of f on $[a, b]$. The average value is such that the constant function $F: x \mapsto f(c)$ has the same integral from a to b as does f; that is,

$$\int_a^b F = \int_a^b f$$

Example 4 (a) If $f(x) \geqslant 0$ for all $x \in [a, b]$, then the average value $f(c)$ has a natural geometric meaning: It is the height of the rectangle of base $(b-a)$ whose area is equal to the area under the graph of f from a to b (Figure 1-10).

(b) In the case of an affine function $h: x \mapsto mx+k$, we see that the average value of h on $[a, b]$ occurs at the point $c = \frac{1}{2}(a+b)$, which is the average of a and b. Moreover, $h(c) = \frac{1}{2}[m(a+b)]+k = \frac{1}{2}[h(a)+h(b)]$, which is the average of $h(a)$ and $h(b)$ (Figure 1-11).

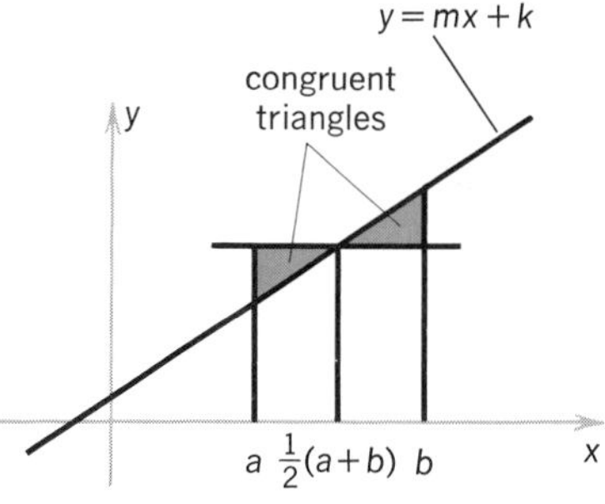

FIGURE 1-11

So far we have only defined $\int_a^b f$ when $b > a$. When $b < a$, and f is continuous on $[b, a]$, we define

$$\int_a^b f = -\int_b^a f$$

When $b = a$, we define $\int_b^a f = 0$. These conventions assure that the following formula is valid for all a and b, whether $a < b$, $a = b$, or $b < a$:

$$\int_a^b f = -\int_b^a f$$

Furthermore, the **additivity property**

$$\int_a^c f + \int_c^b f = \int_a^b f$$

continues to hold whenever $\int_a^c f$ and $\int_c^b f$ are defined; and it is not necessary that $a < c < b$ or even that $a < b$.

Example 5 Let us show explicitly that

$$\int_0^{-\pi} \sin x\, dx + \int_{-\pi}^{\pi} \sin x\, dx = \int_0^{\pi} \sin x\, dx$$

We have

$$\int_0^{-\pi} \sin x\, dx = -\int_{-\pi}^{0} \sin x\, dx$$

and

$$\int_{-\pi}^{\pi} \sin x\, dx = \int_{-\pi}^{0} \sin x\, dx + \int_0^{\pi} \sin x\, dx$$

Summing the right-hand sides of the two expressions above yields the result we want:

$$-\int_{-\pi}^{0} \sin x\, dx + \int_{-\pi}^{0} \sin x\, dx + \int_0^{\pi} \sin x\, dx = \int_0^{\pi} \sin x\, dx$$

With the above definitions, the Mean Value Theorem of Integral Calculus is valid when $b < a$; that is, as long as $\int_a^b f$ is defined we have (whether $b < a$ or $a < b$)

$$\int_a^b f = f(c)(b-a)$$

for some point c between a and b. Let us verify this for the case $b < a$. By definition, we have $\int_a^b f = -\int_b^a f$. Since $b < a$, there is a point c in $[b, a]$ such that $f(c)$ is the average value of f on $[b, a]$. Thus,

$$f(c) = \frac{\int_b^a f}{a-b}$$

However,

$$\frac{\int_b^a f}{a-b} = \frac{\int_a^b f}{b-a}$$

and so

$$f(c)(b-a) = \int_a^b f$$

which is what we wanted to show.

To summarize, whenever the integrals are defined, we have

(Additivity) $$\int_a^b f = \int_a^c f + \int_c^b f$$

(Mean Value Theorem) $$\int_a^b f = f(c)\,(b-a)$$

for some c between a and b. These are the two main properties of integrals which we will need in the following sections, and much of the development can be based solely on them. This remark will prove valuable in Chapter 7.

EXERCISES

1. Compute the following integrals.
   (a) $\int_0^1 3\,dx$
   (b) $\int_2^3 5x\,dx$
   (c) $\int_1^2 (2x+3)\,dx$
2. For constants c and d prove that
$$\int_a^b (c+d)\,dx = \int_a^b c\,dx + \int_a^b d\,dx$$
3. Compute $\int_0^{2\pi} \cos x\,dx$.
4. Prove that $0 \leqslant \int_0^\pi \sin x\,dx \leqslant \pi$.
5. Plot the graph of $f: x \mapsto |x|$ for $-3 \leqslant x \leqslant 3$.
   (a) Compute $\int_{-3}^3 f$.
   (b) What is the average value of f over $[-3, 3]$? On $[-3, 0]$? On $[-1, 2]$?
6. Plot the graph of the function $F: t \mapsto \int_0^t x\,dx$ for $0 \leqslant t \leqslant 1$. Compute the derivative of F.
7. Plot the graph of $G: t \mapsto \int_0^t |x|\,dx$ for $-3 \leqslant t \leqslant 3$. Compute $G'(t)$.
8. (a) If $0 \leqslant g(x) \leqslant h(x)$ for $x \in [a, b]$ argue that
$$\int_a^b g(x)\,dx \leqslant \int_a^b h(x)\,dx$$
   (b) Prove that
$$\int_0^1 x^n\,dx \leqslant \int_0^1 x^m\,dx \qquad \text{if } n \geqslant m$$
9. Prove that $\int_a^b cx^2\,dx \leqslant \frac{1}{2}c(a^2+b^2)(b-a)$ by comparing $x \mapsto cx^2$ with a linear function ($c \geqslant 0$).
10. Plot the graph of $f: x \mapsto 3x^2$ on $[0, 1]$ and estimate the average value of f between 0 and 1.

2 THE FUNDAMENTAL THEOREM OF INTEGRAL CALCULUS

In this section we justify the fundamental theorem which relates differential and integral calculus through the following important formula: If F is an antiderivative of the continuous function f, then

$$\int_a^b f = F(b) - F(a)$$

To demonstrate this result we shall construct a special antiderivative of f. If f is continuous on $[a, b]$, then for every $x \in [a, b]$, f is continuous on $[a, x]$, and so the integral $\int_a^x f$ is defined. This gives rise to a new function $x \mapsto \int_a^x f$, whose domain is $[a, b]$. In particular this function sends a to $\int_a^a f = 0$ and b to $\int_a^b f$. When $f(x) \geqslant 0$ for all $x \in [a, b]$ the function $x \mapsto \int_a^x f$ simply sends x to the area under the graph of f between a and x (Figure 2-1).

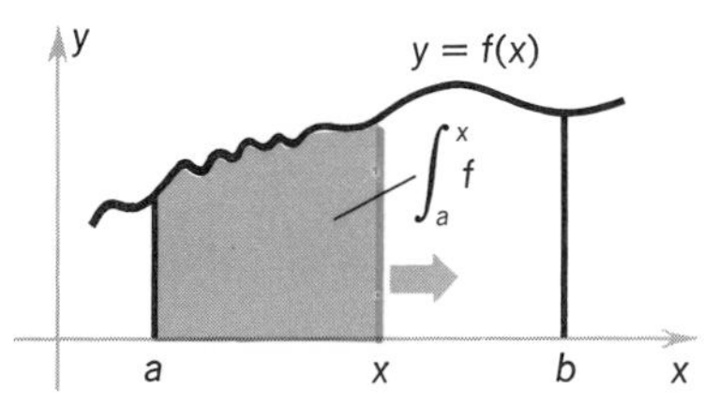

FIGURE 2-1

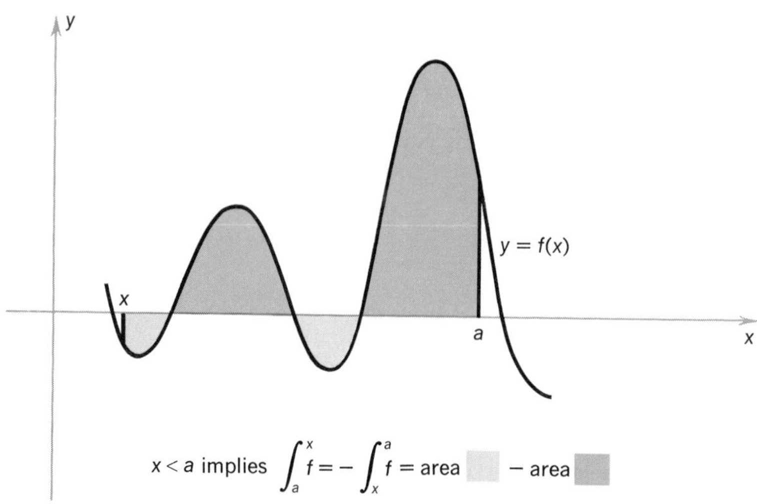

FIGURE 2-2

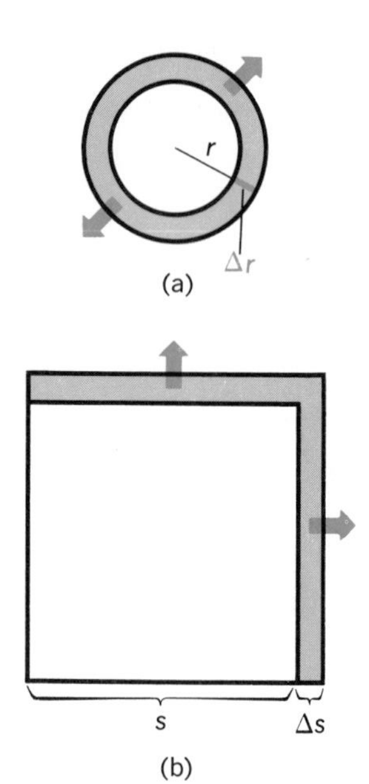

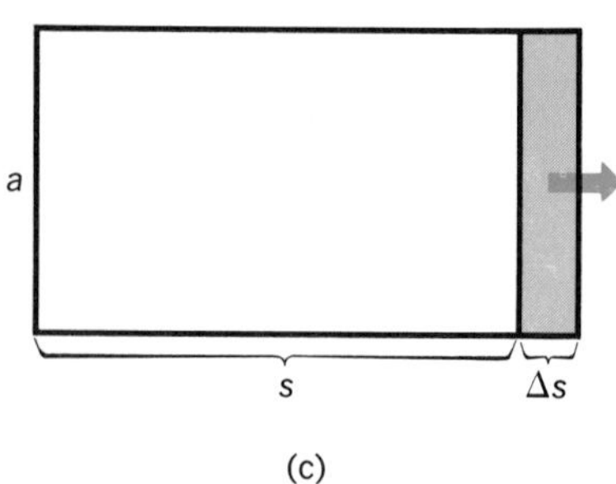

FIGURE 2-3

More generally, if f is continuous on an interval I and if $a \in I$, then $x \mapsto \int_a^x f$ is defined for all $x \in I$ (Figure 2-2).

We can now ask if the function $x \mapsto \int_a^x f$ is continuous and if it is differentiable. Before answering these questions in general, let us look at some familiar examples of area functions (compare Example 6, §3.5).

Consider a circle with increasing radius r. The area of the circle is

$$A = \pi r^2$$

Since

$$\frac{dA}{dr} = 2\pi r$$

we see that the rate of change of the area is equal to the circumference [Figure 2-3(a)].

Next consider a square of increasing side s. The area of the square is $A = s^2$ and so $dA/ds = 2s$. Here, the rate of change of the area is the semi-perimeter $2s$. Notice that while the circle grows by expanding simultaneously on all sides at once, the square, pictured in Figure 2-3(b), grows by expanding along two sides and this difference is reflected in the rates of change.

Let us now examine the case of a rectangle with fixed side a and increasing side s [Figure 2-3(c)]. This time the area is $A = as$ and we have $dA/ds = a$. Here, the rectangle grows by pushing out along a side of constant length a, and the length of this side is the rate of change of the area.

Finally, we turn to the area under the graph of a nonnegative function f. Here the area $\int_a^x f$ is increasing as x moves to the right (Figure 2-4) and the area expands by pushing out along a side of length $f(x)$.

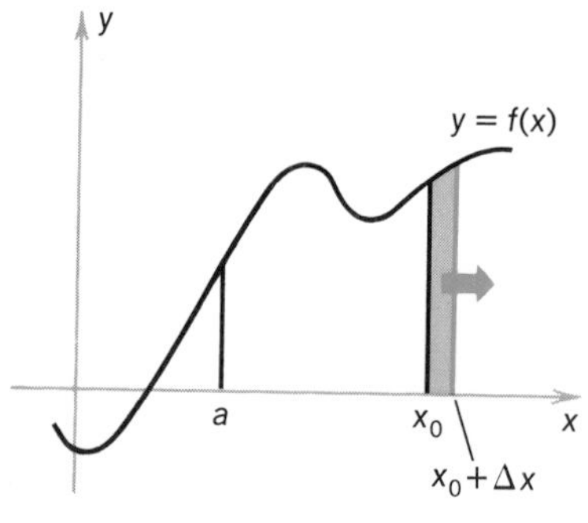

FIGURE 2-4

By now, we should suspect that the rate of change of $\int_a^x f$ with respect to x would be just $f(x)$. This is, in fact, the case and this result is known as the Fundamental Theorem of Calculus.

Theorem 1 The Fundamental Theorem of Calculus Let f be continuous on the interval I and let $a \in I$. Let G be the function with domain I and rule $G: x \to \int_a^x f$. Then:

(i) G is differentiable on I, and $G' = f$.

(ii) If F is any function with $F'(x) = f(x)$ (that is, any antiderivative of f), then

$$F(x) - F(a) = \int_a^x f = G(x)$$

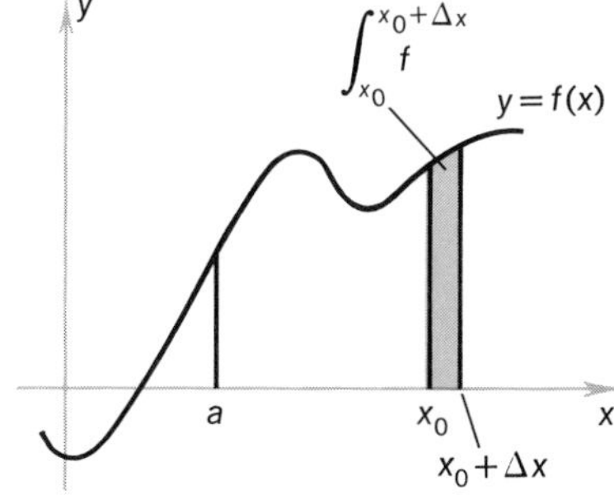

FIGURE 2-5

Proof The proof of (i) is computational. We explicitly calculate

$$\lim_{\Delta x \to 0} \frac{G(x+\Delta x) - G(x)}{\Delta x}$$

showing that this limit exists and equals $f(x)$. Now, by additivity, we have (see Figure 2-5)

$$G(x+\Delta x) - G(x) = \int_a^{x+\Delta x} f - \int_a^x f = \int_x^{x+\Delta x} f$$

By the Mean Value Theorem of Integral Calculus, there is a point c between x and $x+\Delta x$ such that

$$\int_x^{x+\Delta x} f = f(c) \cdot \Delta x$$

Thus

$$\frac{G(x+\Delta x) - G(x)}{\Delta x} = f(c) \cdot \frac{\Delta x}{\Delta x} = f(c)$$

We therefore have

$$\lim_{\Delta x \to 0} \frac{G(x+\Delta x) - G(x)}{\Delta x} = \lim_{\Delta x \to 0} f(c)$$

But c is between x and $x+\Delta x$, and as Δx approaches 0, the point c is squeezed toward x. The function f is continuous, so $f(c)$ approaches $f(x)$ as c approaches x. Thus $\lim_{\Delta x \to 0} f(c) = f(x)$, and we have

$$G'(x) = \lim_{\Delta x \to 0} \frac{G(x+\Delta x) - G(x)}{\Delta x} = f(x)$$

The first part of the theorem is demonstrated. To establish part (ii), we let F be any antiderivative of f. Then $G' = f = F'$, by (i), and so G and F differ by at most a constant. That is, there is a number C such that

$$G(x) = F(x) + C$$

for all $x \in I$. We now compute C: We know that $G(a) = 0$, since by definition $G(a) = \int_a^a f$. But $G(a) - F(a) = C$, hence $-F(a) = C$. Therefore, $G(x) = F(x) - F(a)$. In particular, setting $x = b$, $\int_a^b f = G(b) = F(b) - F(a)$. ∎

As part of the theorem observe that one always has an antiderivative for a continuous function. If f is not continuous this need not be true (see Exercise 1).

In summary we have the following basic computational method.

FORMAL DEVICE

If F is any antiderivative of f then the integral of f may be computed as follows:

$$\int_a^b f(x)\,dx = F(b) - F(a)$$

Example 1 Consider a linear function $f: x \mapsto mx$. As we have seen $\int_0^x f = mx^2/2$. According to the Fundamental Theorem $F: x \mapsto mx^2/2$ is an antiderivative of f, and of course this is the case since $d(mx^2/2)^2/dx = mx$.

We have also seen that, more generally, $\int_a^x f = (m/2)(x^2 - a^2)$. The Fundamental Theorem implies that $G: x \mapsto (m/2)(x^2 - a^2)$ is also an antiderivative of f. Since a is a constant, we have $G'(x) = mx$ which is as it should be.

Example 2 Evaluate $\int_0^{\frac{1}{2}\pi} \cos x\,dx$. We know that $\sin'(x) = \cos x$; hence $\sin x$ is an antiderivative of $\cos x$ and so $\int_0^{\frac{1}{2}\pi} \cos x\,dx = \sin(\frac{1}{2}\pi) - \sin(0) = 1 - 0 = 1$. Another antiderivative of $\cos x$ is $F(x) = \sin x + 10$. By the Fundamental Theorem

$$\int_0^{\frac{1}{2}\pi} \cos x\,dx = F(1) - F(0) = (1+10) - (0+10)$$
$$= 1 - 0 + (10-10) = 1$$

We see that in computing the integral with $F(x)$, the constants 10 and -10 both appear and cancel one another.

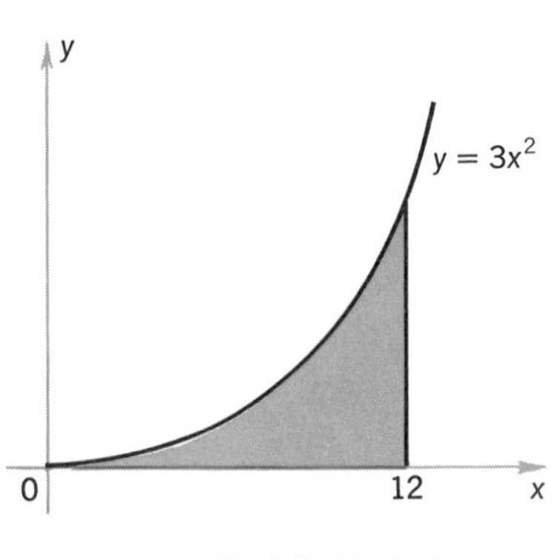

FIGURE 2-6

Notice that here we have computed the area of a region with a curved side. Using elementary geometry alone such a computation would be very difficult.

Example 3 Find the area of the region bounded by the x axis, the line $x = 12$, and the parabola $y = 3x^2$. Since the curve $y = 3x^2$ never dips below the x axis (Figure 2-6), the area sought is given by

$$A = \int_0^{12} 3x^2\,dx$$

We see that $F(x)=x^3$ is an antiderivative of $f(x)=3x^2$, and so $A=F(12)-F(0)=12^3-0=1728$.

It is convenient to introduce two forms of notation for the difference $F(b)-F(a)$:

$$F(x)\big|_a^b = F(b) - F(a) = [F(x)]_a^b$$

By way of example, if $F(x)=x^2$, we have $F(x)|_a^b = x^2|_a^b = b^2-a^2 = [x^2]_b^a$. If $F(x)=\sin x$, we have $F(x)|_0^\pi = \sin x|_0^\pi = 0-0=0=[\sin x]_0^\pi$. If $F(x)=1/x^2$, we have $F(x)|_1^2 = 1/x^2|_1^2 = \frac{1}{4}-1 = -\frac{3}{4} = [1/x^2]_1^2$.

Thus the Fundamental Theorem provides the following formula for evaluating integrals:

$$\int_a^b f(x)\,dx = F(x)\Big|_a^b, \quad \text{whenever} \quad \frac{dF(x)}{dx} = f(x)$$

Example 4 (a) Since $F(x)=\frac{1}{3}x^3$ is an antiderivative of x^2, we have

$$\int_0^2 x^2\,dx = \tfrac{1}{3}x^3\Big|_0^2 = \tfrac{8}{3}$$

(b) Since $F(x)=x^3+x^2+x$ is an antiderivative of $3x^2+2x+1$, we have

$$\int_0^{-1} (3x^2+2x+1)\,dx = [(x^3+x^2+x)]_0^{-1} = -1 - 0 = -1$$

(c) Similarly,

$$\int_0^2 \sin x\,dx = -\cos x\Big|_0^2 = -\cos 2 + 1$$

and

$$\int_0^{\frac{1}{4}\pi} \frac{1}{\cos^2 u}\,du = [\tan u]_0^{\frac{1}{4}\pi} = 1 - 0 = 1$$

(d) Also,

$$\int_1^2 \frac{1}{u^2}\,du = \frac{-1}{u}\Big|_1^2 = \frac{-1}{2} + 1 = \frac{1}{2}$$

and

$$\int_1^2 \frac{-1}{t^3}\,dt = \frac{1}{2t^2}\Big|_1^2 = \frac{1}{8} - \frac{1}{2} = \frac{-3}{8}$$

We have the following corollary to the Fundamental Theorem.

Corollary 1 If $F'(x)=f(x)$, the average value of f on $[a,b]$ is given by

$$f(c) = \frac{F(b)-F(a)}{b-a}$$

for some point c in $[a,b]$.

Proof Since $\int_a^b f = F(b) - F(a)$, the formula for the average value, $f(c) = \int_a^b f/(b-a)$ immediately transforms into

$$f(c) = \frac{F(b) - F(a)}{b - a}$$

∎

Example 5 Find the average value of $f: x \mapsto 3x^2$ on the interval $[2, 3]$. The function $F: x \mapsto x^3$ is an antiderivative of f and, by Corollary 1, the average value is

$$\frac{F(3) - F(2)}{3 - 2} = \frac{27 - 8}{1} = 19$$

Example 6 Find the average value of the cosine on $[-\frac{1}{2}\pi, \frac{1}{2}\pi]$. We know that $d(\sin x)/dx = \cos x$, so we compute

$$\frac{\sin(\frac{1}{2}\pi) - \sin(-\frac{1}{2}\pi)}{\frac{1}{2}\pi - (-\frac{1}{2}\pi)} = \frac{1 - (-1)}{\frac{1}{2}\pi + \frac{1}{2}\pi} = \frac{2}{\pi}$$

The above Corollary suggests that there is a relationship between the Mean Value Theorem of Differential Calculus and the Mean Value Theorem of Integral Calculus. If $F'(x) = f(x)$, we can apply the Mean Value Theorem of Differential Calculus to F on the interval $[a, b]$. This gives a point c in $]a, b[$ such that

$$F'(c) = \frac{F(b) - F(a)}{b - a}$$

But $F'(c) = f(c)$ and so that applying this theorem of differential calculus to F yields the same conclusion as the Mean Value Theorem of Integral Calculus applied to $f = F'$.

We recall that the Mean Value Theorem of Differential Calculus had the following implication for the analysis of distance and velocity: If a body is moving (along a straight line) during an interval of time $[t_0, t_1]$, there is a point τ in the interval such that the instantaneous velocity at τ is equal to the average velocity on the interval $[t_0, t_1]$. Recall the proof. If we denote the distance traveled by $s(t)$ and the velocity by $v(t)$, the average velocity on $[t_0, t_1]$ is

$$\text{average velocity} = \frac{s(t_1) - s(t_0)}{t_1 - t_0}$$

Since $v = ds/dt$, the theorem, applied to $s(t)$ on $[t_0, t_1]$, states that at some point τ between t_0 and t_1,

$$v(\tau) = \left.\frac{ds}{dt}\right|_{t=\tau} = \frac{s(t_1) - s(t_0)}{t_1 - t_0}$$

On the other hand, $s(t)$ is an antiderivative of $v(t)$ and so by the Fundamental Theorem, we have

$$s(t_1) - s(t_0) = \int_{t_0}^{t_1} v(t)\, dt \qquad (1)$$

The Mean Value Theorem of Integral Calculus, applied to $v(t)$ on $[t_0, t_1]$, provides a point τ such that

$$v(\tau)(t_1 - t_0) = \int_{t_0}^{t_1} v(t)\,dt$$

or

$$v(\tau) = \frac{s(t_1) - s(t_0)}{t_1 - t_0}$$

The value $v(\tau)$ is (by definition) the average value of $v(t)$ on $[t_0, t_1]$. We conclude that the average value of the velocity function on $[t_0, t_1]$ is equal to the average velocity on $[t_0, t_1]$.

Equation (1) shows that the Fundamental Theorem provides a formula for computing the distance traveled once the velocity function is known. The formula, based on equation (1) and the Fundamental Theorem, is

$$s(t_1) - s(t_0) = \int_{t_0}^{t_1} v(t)\,dt = V(t_1) - V(t_0)$$

Example 7 A body moves along a straight line with velocity $v(t) = 6\sin t$. What is the distance traveled during the interval $[0, \frac{1}{2}\pi]$? During the interval $[\frac{1}{4}\pi, \frac{1}{2}\pi]$? The distance traveled on $[0, \frac{1}{2}\pi]$ is

$$\begin{aligned} s(\tfrac{1}{2}\pi) - s(0) &= \int_0^{\frac{1}{2}\pi} v(t)\,dt = \int_0^{\frac{1}{2}\pi} 6\sin t\,dt \\ &= -6\cos t \Big|_0^{\frac{1}{2}\pi} = 0 - (-6) = 6 \end{aligned}$$

The distance traveled during the interval $[\frac{1}{4}\pi, \frac{1}{2}\pi]$ is

$$\begin{aligned} s(\tfrac{1}{2}\pi) - s(\tfrac{1}{4}\pi) &= \int_{\frac{1}{4}\pi}^{\frac{1}{2}\pi} v(t)\,dt = -6\cos t \Big|_{\frac{1}{4}\pi}^{\frac{1}{2}\pi} \\ &= 0 - \left(-6\frac{\sqrt{2}}{2}\right) = 3\sqrt{2} \end{aligned}$$

Consult §5.6 for some related problems.

We conclude this section with some further corollaries to the Fundamental Theorem. Observe that the proofs depend only on properties of the integral and the Fundamental Theorem and not on the geometric properties of area.

Corollary 2 **Linearity** If f is continuous on $[a, b]$, then for all $x \in [a, b]$

$$\int_a^x cf = c\int_a^x f$$

where c is a constant. In particular,

$$\int_a^b cf = c\int_a^b f$$

Proof Set $F: x \mapsto \int_a^x cf$ and $G: x \mapsto c\int_a^x f$. By the Fundamental Theorem $F'(x) = cf(x) = G'(x)$ for all x and so G and F differ by at most a constant. Since $G(a) = F(a) = 0$, the constant is 0 and $G = F$. ∎

As an exercise the reader should explore the geometric meaning of this result in terms of area.

Corollary 3 **Additivity** If f and g are continuous on $[a, b]$, then for all $x \in [a, b]$, $\int_a^x (f+g) = \int_a^x f + \int_a^x g$. In particular

$$\int_a^b f + g = \int_a^b f + \int_a^b g$$

Proof Let $F(x) = \int_a^x f$, $G(x) = \int_a^x g$, and $H(x) = \int_a^x (f+g)$. Then $F'(x) = f(x)$, $G'(x) = g(x)$ and $H'(x) = (f+g)(x)$. Thus $H' = F' + G' = (F+G)'$. Hence H and $F+G$ differ by a constant. The constant must be zero as we see by taking $x = a$. ∎

Corollary 4 Suppose that g and h are continuous on $[a, b]$ and that $g(x) \leqslant h(x)$ for all $x \in [a, b]$. Then $\int_a^b g \leqslant \int_a^b h$.

Proof Consider the function $h - g: x \mapsto h(x) - g(x)$. Since $h(x) - g(x) \geqslant 0$ for all x we have

$$\int_a^b (h - g) \geqslant 0$$

The theorem now follows easily from Corollaries 2 and 3. ∎

Theorem 2 **Second Mean Value Theorem of Integral Calculus** Suppose f and g are continuous on $[a, b]$, and that $g(x) \geqslant 0$ for all $x \in [a, b]$. Then for some point $c \in [a, b]$, we have

$$\int_a^b f(x)g(x)\,dx = f(c)\int_a^b g(x)\,dx$$

Proof Let M be the maximum value of f on $[a, b]$ and let m be the minimum value of f on $[a, b]$. Then for all $x \in [a, b]$, we have

$$mg(x) \leqslant f(x)g(x) \leqslant Mg(x)$$

Integrating, we obtain

$$m\int_a^b g(x)\,dx \leqslant \int_a^b f(x)g(x)\,dx \leqslant M\int_a^b g(x)\,dx$$

The rest of the proof is analogous to the proof of the Mean Value Theorem of Integral Calculus, and is left as an exercise. ∎

The reader should also verify that the above theorem is true if we assume $g(x) \leqslant 0$ on $[a, b]$. Furthermore, the reader should construct examples to show the theorem is not valid if $g(x)$ is allowed to change sign on $[a, b]$.

The final theorem of this section gives a criterion for determining when an integral is nonzero, and it is therefore useful in conjunction with the Second Mean Value Theorem of Integral Calculus.

Theorem 3 Suppose that f is continuous on $[a,b]$, that $f(x) \geqslant 0$ for all $x \in [a,b]$, and that for some $x_0 \in [a,b]$, we have $f(x_0) > 0$. Then

$$\int_a^b f(x)\,dx > 0$$

Proof First, we shall assume that x_0 is in $]a,b[$. Since f is continuous at x_0, if we take $\varepsilon = \frac{1}{2}f(x_0)$, there must be a neighborhood $]A,B[$ of x_0 such that $f(x) > \varepsilon$ for all $x \in [A,B]$. The interval $[c,d] = [\max(a,A), \min(B,b)]$ is a subinterval of $[a,b]$; and $f(x) > \varepsilon$ for $x \in [c,d]$ We have

$$\int_a^b f(x)\,dx \geqslant \int_c^d f(x)\,dx \geqslant \int_c^d \varepsilon\,dx \geqslant \varepsilon(d-c) > 0$$

The case when x_0 is an endpoint can be handled similarly. ∎

From this it follows that for a continuous function f which is non-negative on $[a,b]$, that is, $f(x) \geqslant 0$ for $x \in [a,b]$, then $\int_a^b f = 0$ implies that $f = 0$.

EXERCISES

1. Let f be continuous on $[a,b]$ and let g be continuous on $[b,c]$ with $g(b) \neq f(b)$. Define a function H by the rule

$$H: x \mapsto \begin{cases} \int_a^x f, & x \leqslant b \\ \int_a^b f + \int_b^x g, & x \geqslant b \end{cases}$$

   (a) Interpret H as the area under the graph of

$$h: x \mapsto \begin{cases} f(x), & x \leqslant b \\ g(x), & x > b \end{cases}$$

   from a to x.

   (b) Show that H is continuous on $[a,c]$.

   (c) Show that H is differentiable on $[a,c]$ except at $x = b$.

   (d) Write out H, h explicitly if $a = 0$, $b = 1$, $c = 2$, $f = 0$, $g = 1$.

2. What is the area under the graph of $f: x \mapsto 12(x^3 + x^2 + x + \frac{1}{2})$ between 1 and 10?
3. Show that the average value of the sine function between 0 and π is $2/\pi$.
4. The velocity of a particle at time t is given by $v(t) = t^2 + 3$. Find the distance traveled between times $t = 1$ and $t = 3$.

In exercises 5–12 find

(a) the integral of the given function on the given interval;
(b) the value of the function at the midpoint of the interval; and
(c) the average value of the function on the interval.

5. $x \mapsto 12x^3 + 3x^2 + 6x + 3$; $[0, 1]$
6. $x \mapsto 12x^3 - 3x^2 + 6x - 1$; $[2, 4]$
7. $x \mapsto 3\sqrt{2}x^2 + 4x + 2\sqrt{2}$, $[\sqrt{2}, 2\sqrt{2}]$
8. $x \mapsto 3x^2 + 6\sqrt[3]{3}x - 3$, $[\sqrt[3]{3}, 3\sqrt[3]{3}]$
9. $x \mapsto x^3 - 6x^2 + 11x - 4$; $[1, 3]$
10. $x \mapsto x^2$; $[-1, 1]$
11. $x \mapsto \cos x$; $[-\frac{1}{2}\pi, \pi]$
12. $x \mapsto 3\sin 2x$; $[0, \frac{1}{2}\pi]$
13. Find the area bounded by the parabola passing through $(-1, 11)$, $(1, 1)$, and $(2, 2)$ and the lines $y = 0$, $x = -2$, $x = 1$.
14. Find the area bounded by the parabola passing through $(-2, 0)$, $(0, 3)$, and $(1, 1)$ and the lines $y = 0$, $x = -1$, and $x = 0.5$.
15. Show by example that the Second Mean Value Theorem is false if we allow $g(x)$ to change sign on $[a, b]$.
16. For any continuous f on $[0, 1]$ prove there is a point c for which
$$\int_0^1 xf(x)\,dx = \tfrac{1}{2}f(c)$$
Find such a point if $f(x) = x^2$.
17. Give an example to show that the point c such that $f(c)$ is the average of f on $[a, b]$ need not be unique.
18. Deduce the Mean Value Theorem of Integral Calculus as a special case of the Second Mean Value Theorem.

3 CALCULATION OF AREAS

We now make use of the Fundamental Theorem to compute areas of regions bounded by continuous curves.

Recall that if $f(x) \geqslant 0$, then $\int_a^b f(x)\,dx$ was defined as the area under f from a to b (see §1). The significance of the Fundamental Theorem is that it gives us a method for computing the integral, namely by the formula $\int_a^b f(x)\,dx = F(b) - F(a)$, where F is an antiderivative of f. Hence, for non-negative f, this formula gives the area under the graph of f from a to b. Again, we emphasize that the present discussion is somewhat intuitive, since it depends on certain properties of area which we take for granted. A more rigorous approach is given in Chapter 7.

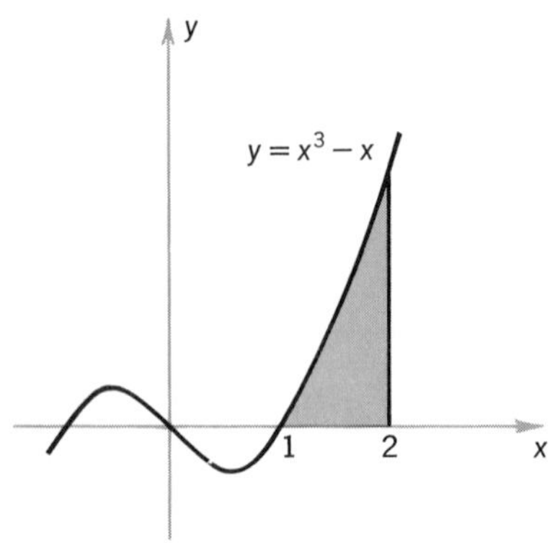

FIGURE 3-1

Example 1 (a) Find the area A under the graph of $f: x \mapsto x^6 + 5x^4$ from 0 to 2. Clearly, $f(x) \geqslant 0$ for all x in the interval, so the preceding remarks apply. An antiderivative of f is $F: x \mapsto (x^7/7) + x^5$. Therefore, we can compute the area to be

$$A = F(x)\Big|_0^2 = \left(\frac{x^7}{7} + x^5\right)\Big|_0^2 = \frac{128}{7} + 32 = \frac{352}{7}$$

(b) Find the area of the shaded portion in Figure 3-1, where $f: x \mapsto x^3 - x$.

Since $F: x \mapsto \frac{1}{4}x^4 - \frac{1}{2}x^2$ is an antiderivative of f, we have $A = F(x)|_1^2 = 2-(-\frac{1}{4}) = 2\frac{1}{4}$.

Similarly, given a function f continuous on $[a, b]$ such that $f(x) \leqslant 0$ for $x \in [a, b]$, then by definition of the definite integral, the area above the graph of f from a to b is

$$A = -\int_a^b f(x)\,dx$$

Therefore, in this case, if F is an antiderivative of f, we have

$$A = -[F(x)]_a^b = [-F(x)]_a^b$$

Example 2 Find the area above the graph of $f: x \mapsto x^2 - 4$ from -2 to 2. Since $F: x \mapsto \frac{1}{3}x^3 - 4x$ is an antiderivative of f, the area is

$$\begin{aligned}[-F(x)]_{-2}^2 &= (8-\tfrac{8}{3}) - (\tfrac{8}{3}-8)\\ &= 16 - \tfrac{16}{3} = \tfrac{32}{3}\end{aligned}$$

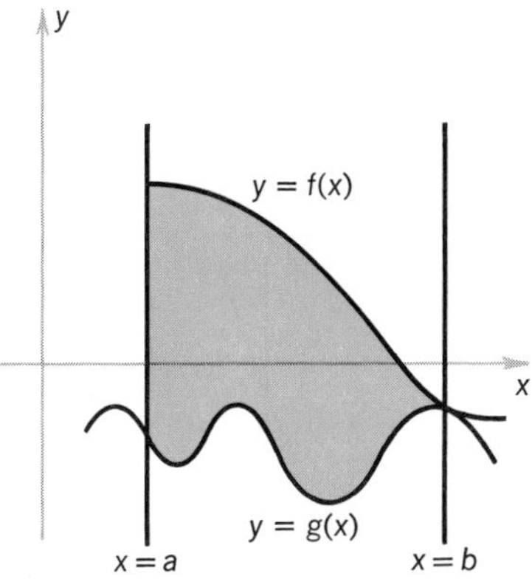

FIGURE 3-2

Consider now two functions g and f defined on $[a, b]$ with $f(x) \geqslant g(x)$. We claim that the area of the region between the graphs is given by

$$\int_a^b (f-g)$$

The region under consideration is sketched in Figure 3-2.

To justify this conclusion let us first suppose that $f(x) \geqslant g(x) \geqslant 0$. Then the area under the graph of f is $\int_a^b f$ and that under g is $\int_a^b g$. The difference is

$$\int_a^b f - \int_a^b g = \int_a^b (f-g) \qquad (1)$$

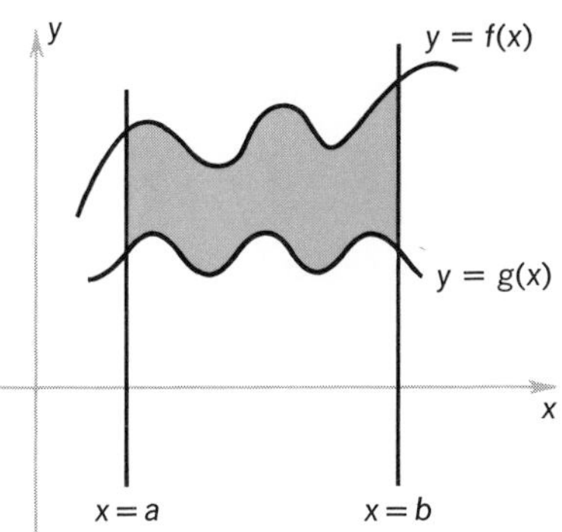

FIGURE 3-3

which represents the area of the difference between the regions (the shaded portion in Figure 3-3).

Formula (1) is also valid for the cases where one or both of f and g are negative. The reader should examine the previous figures to convince himself of this (break up $[a, b]$ into regions where f and g are $\leqslant 0$ or $\geqslant 0$). Keep in mind that $f(x) \geqslant g(x)$.

Example 3 Find the area of the region bounded by the curves $y = x^2 + 1$ and $y = x^3$, and the lines $x = 0$, $x = 1$ (Figure 3-4). From the figure, it is clear that the area of the shaded region can be obtained by subtracting the area under $y = x^3$ from the area under $y = x^2 + 1$. Thus,

$$A = \int_0^1 (x^2+1)\,dx - \int_0^1 x^3\,dx$$

Applying the Fundamental Theorem, we obtain

$$A = (\tfrac{1}{3}x^3 + x)\,\Big|_0^1 - \tfrac{1}{4}x^4\,\Big|_0^1 = \tfrac{4}{3} - \tfrac{1}{4} = \tfrac{13}{12}$$

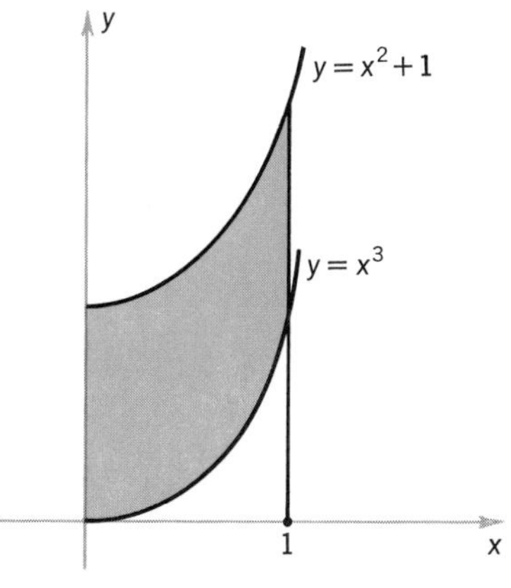

FIGURE 3-4

Example 4 Find the area of the region bounded by the y axis, the line $x=1$, and the graphs of $g:x \mapsto x^2+1$ and $f:x \mapsto x^4-x^3-1$ (Figure 3-5). By the formula above the area is

$$\int_0^1 (x^2+1)-(x^4-x^3-1)\,dx = \tfrac{1}{3}+1-\tfrac{1}{5}+\tfrac{1}{4}+1 = \tfrac{143}{60}$$

Let us verify this another direct way.

The area sought is the sum of the area A_1 of region I and the area A_2 of region II. The function $G:x \mapsto \frac{1}{3}x^3+x$ is an antiderivative of g and so $A_2 = (\frac{1}{3}x^3+x)|_0^1 = \frac{4}{3}$. An antiderivative of f is $F:x \mapsto \frac{1}{5}x^5-\frac{1}{4}x^4-x$. Since $f(x) \leqslant 0$ for all x in $[0, 1]$, we have

$$A = -F(x)\,|_0^1 = -F(1) = \tfrac{21}{20}$$

Hence,

$$A = \frac{4}{3}+\frac{21}{20} = \frac{80+63}{60} = \frac{143}{60}$$

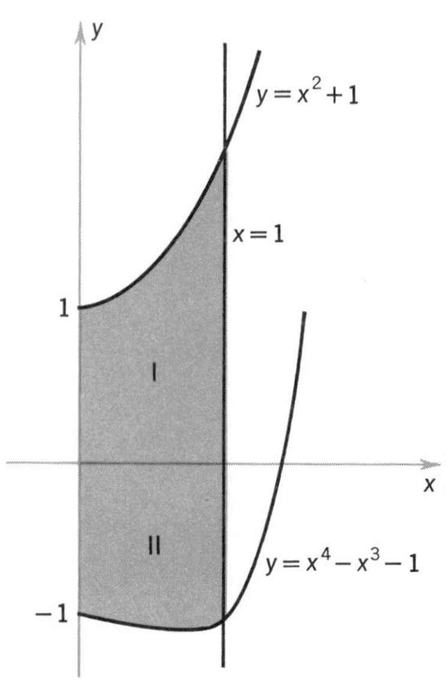

FIGURE 3-5

To find the area between two graphs which intersect, as in Figure 3-6, we must solve for the points of intersection of the two graphs before we can use the above methods. We illustrate with some examples.

Example 5 Find the area between the graphs of $f:x \mapsto \frac{1}{3}x^2+1$ and $g:x \mapsto -\frac{1}{3}x^2+3$ between 0 and 3.

The first step in solving such a problem is to sketch the graphs (Figure 3-7). Before we can apply the techniques of this section, we must find the point P by solving the equation

$$\tfrac{1}{3}x^2+1 = -\tfrac{1}{3}x^2+3$$

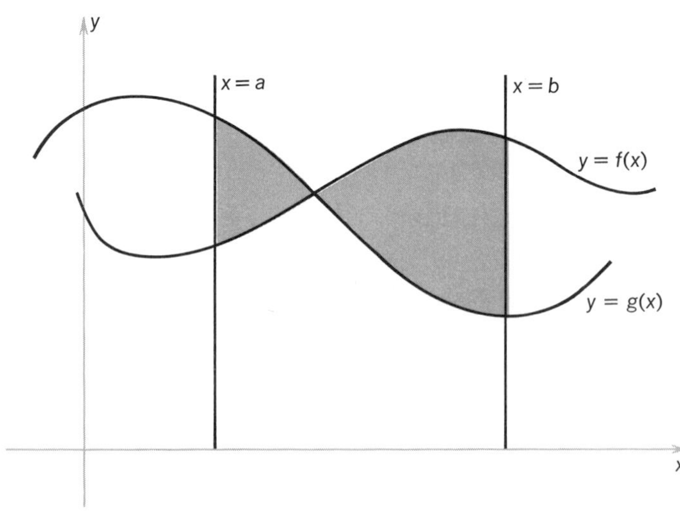

FIGURE 3-6

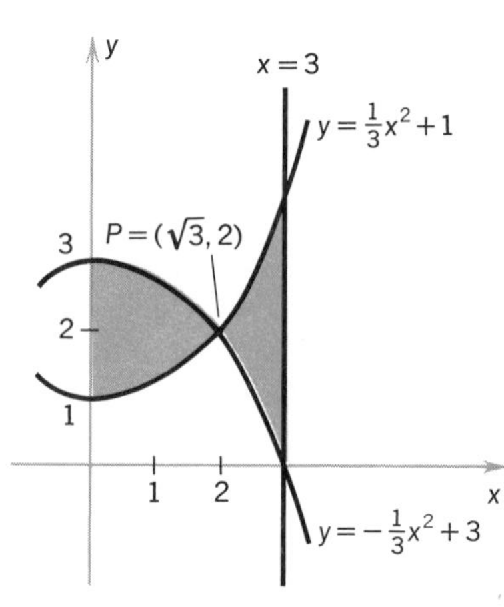

FIGURE 3-7

The two solutions are $\pm\sqrt{3}$, corresponding to the two points $(-\sqrt{3},2)$ and $(\sqrt{3},2)$ where the graphs of f and g intersect. By inspection $P=(\sqrt{3},2)$.

An antiderivative of f is $F:x\mapsto \frac{1}{9}x^3+x$ and an antiderivative of g is $G:x\mapsto -\frac{1}{9}x^3+3x$, so the area of the left-hand region is

$$G(x)\Big|_0^{\sqrt{3}} - F(x)\Big|_0^{\sqrt{3}} = \tfrac{4}{3}\sqrt{3}$$

The area of the right-hand region is

$$F(x)\Big|_{\sqrt{3}}^{3} - G(x)\Big|_{\sqrt{3}}^{3} = \tfrac{4}{3}\sqrt{3}$$

Hence, the total area equals the sum of these two areas and we have

$$\tfrac{4}{3}\sqrt{3}+\tfrac{4}{3}\sqrt{3} = \tfrac{8}{3}\sqrt{3}$$

Example 6 Find the area between the graph of $f:x\mapsto x^3-3x^2+2x$ and the x axis between 0 and 3 (Figure 3-8).

This can be thought of as the area between the graphs of f and the zero function. Finding the intersections of the two graphs amounts to finding the roots of f. These are 0, 1, 2. Since $F:x\mapsto \frac{1}{4}x^4-x^3+x^2$ is an antiderivative of f, the area sought is:

$$\begin{aligned}
&F(x)\Big|_0^1 - F(x)\Big|_1^2 + F(x)\Big|_2^3\\
&\quad = F(1)-F(0)-F(2)+F(1)+F(3)-F(2)\\
&\quad = -F(0)+2F(1)-2F(2)+F(3)\\
&\quad = 0+2(\tfrac{1}{4})-2(0)+\tfrac{9}{4}\\
&\quad = \tfrac{11}{4}
\end{aligned}$$

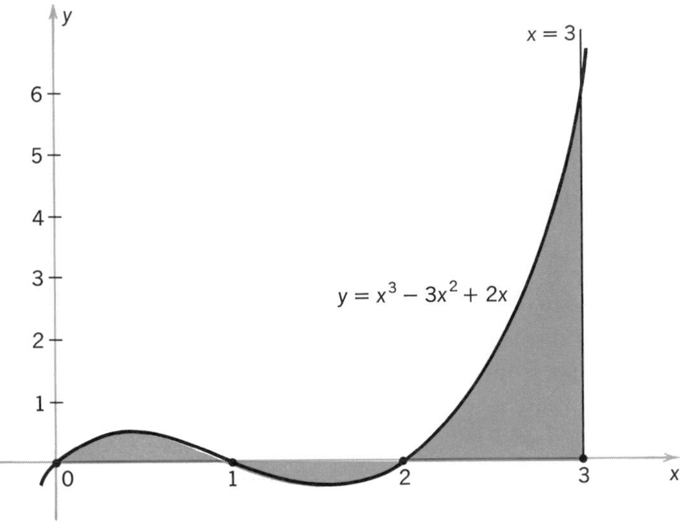

FIGURE 3-8

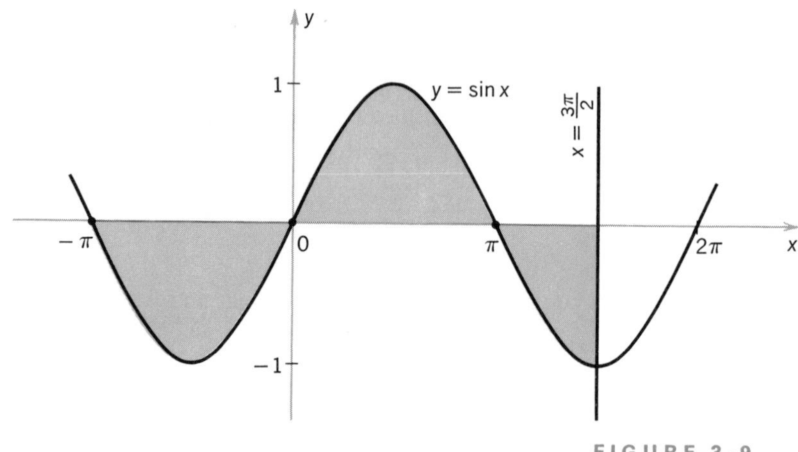

FIGURE 3-9

Example 7 The shaded area shown in Figure 3-9 between the x axis and the graph of the sine from $-\pi$ to $\frac{3}{2}\pi$ is

$$-(-\cos x)\Big|_{-\pi}^{0} + (-\cos x)\Big|_{0}^{\pi} - (-\cos x)\Big|_{\pi}^{3/2\pi} = 5$$

In §1, we established the formula

$$s(t_1) - s(t_0) = \int_{t_0}^{t_1} v(t)\,dt$$

for the distance traveled by a body moving along a straight line with velocity $v(t)$. If the velocity $v(t)$ changes sign during the interval $[t_0, t_1]$, then the above formula gives the net distance traveled. For example, using this formula the result would be 7 miles as the distance traveled if the body were to move forward 100 miles and then backward 93 miles during the interval. If we want to compute the actual total displacement of the body, we want not the integral $\int_{t_0}^{t_1} v(t)\,dt$ but rather the area of the region bounded by the t axis and the curve $y = v(t)$ from t_0 to t_1 This would be the integral of the speed rather than the velocity (explain).

Example 8 A body travels with velocity $v(t) = \sin t$. What is the total displacement of the body during $[0, 2\pi]$? (See Figure 3-10.)

The total displacement is equal to the area bounded by the x axis and the sine curve from 0 to 2π. Thus the total displacement is

$$\int_0^{\pi} \sin t\,dt - \int_{\pi}^{2\pi} \sin t\,dt = -\cos t\Big|_0^{\pi} - \left(-\cos t\Big|_0^{\pi}\right) = 2 - (-2) = 4$$

On the other hand, the net distance is

$$s(2\pi) - s(0) = \int_0^{2\pi} \sin t\,dt = -\cos t \int_0^{2\pi} = 2 - 2 = 0$$

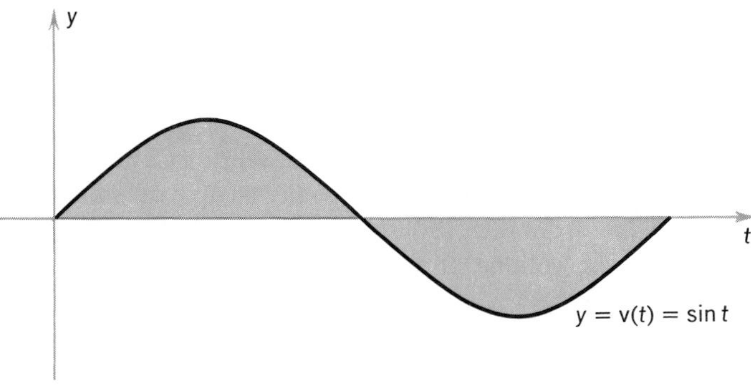

FIGURE 3-10

We see that the body moves forward 2 units of distance during $[0, \pi]$ and then retreats back to the starting point during $[\pi, 2\pi]$. Thus the total displacement is 4 and the net distance is 0.

EXERCISES

1. Find the area bounded by the x axis, the graph of f, and the two given vertical lines.
   (a) $f: x \mapsto x - 1,\ x = 0,\ x = 1$
   (b) $f: x \mapsto x^2,\ x = 2,\ x = 5$
   (c) $f: x \mapsto x^3,\ x = 1,\ x = 3$
   (d) $f: x \mapsto 4x - x^2,\ x = 1,\ x = 3$
   (e) $f: x \mapsto 2\sqrt{x},\ x = 4,\ x = 16$
   (f) $f: x \mapsto -x^2 + 6x - 8,\ x = 2,\ x = 4$
   (g) $f: x \mapsto x^2 + 4x,\ x = -1,\ x = 1$
   (h) $f: x \mapsto x^3 - 9x,\ x = 0,\ x = 3$
2. Find the area lying above the x axis and under the graph of $f: x \mapsto 4x - x^2$.
3. Find the area bounded by the graph of $f: x \mapsto x^2 - 7x + 6$, the x axis, and the lines $x = 2$, $x = 6$.
4. Find the area bounded by the graphs of each of the following pairs of functions.
   (a) $f: x \mapsto \frac{1}{2}x^2,\ g: x \mapsto \frac{1}{8}x^3$
   (b) $f: x \mapsto x^3,\ g: x \mapsto 4x$
   (c) $f: x \mapsto \frac{1}{4}x^2,\ g: x \mapsto 16 - x^2$
   (d) $f: x \mapsto 6x - x^2,\ g: x \mapsto x^2 - 2x$
   (e) $f: x \mapsto x^4 - 4x^2,\ g: x \mapsto 4x^2$
   (f) $f: x \mapsto x^2 - 6x + 8,\ g: x \mapsto -x^2 + 6x - 8$
5. Find the area bounded by the parabola $y^2 = 4x$ and the chord through $(1, -2)$ and $(4, 4)$.
6. Find the area bounded by the curves $y^2 = a^2 x$ and $y^2 = x^3$.

4 VOLUMES AND SOLIDS OF REVOLUTION

In the seventeenth century the Jesuit Bonaventura Cavalieri announced the principle which bears his name: Two bodies have equal volumes if plane sections equidistant from their lowest points have equal areas (Figure 4-1). In other words, the volume of a solid is completely determined by its cross-sectional areas.

We shall now derive a formula for the volume of a solid in terms of the cross-sectional areas. Cavalieri's principle will be an immediate consequence of the formula. Our proof is based on intuitively clear ideas about volumes. After developing these methods, we apply them to the particular case of solids of revolution. The methods will be put on a more rigorous foundation later in Chapter 15.

Consider then a solid body and let $A(x)$ be the cross-sectional area measured a distance x from some fixed reference plane (Figure 4-2). We claim that the volume is given by

$$\textbf{Volume} = \int_a^b A(x)\,dx$$

where a and b are the minimum and maximum distances from the reference plane. We refer to this as **Cavalieri's formula**. It is assumed that $x \mapsto A(x)$ is a continuous function.

To demonstrate this result, we proceed in a way similar to the Fundamental Theorem. Let $V(x)$ be the volume of the solid up to a distance x. We shall argue that $V'(x) = A(x)$. Then the desired result follows from the Fundamental Theorem since

$$\int_a^b A(x)\,dx = V(b) - V(a) = V(b)$$

because $V(a) = 0$ and $V(b)$ is the volume we seek.

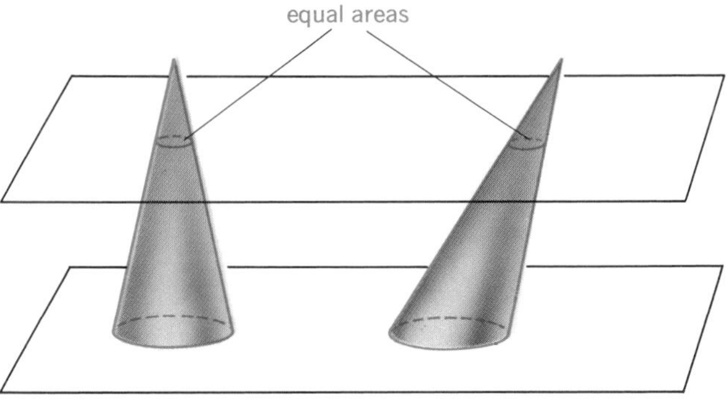

FIGURE 4-1

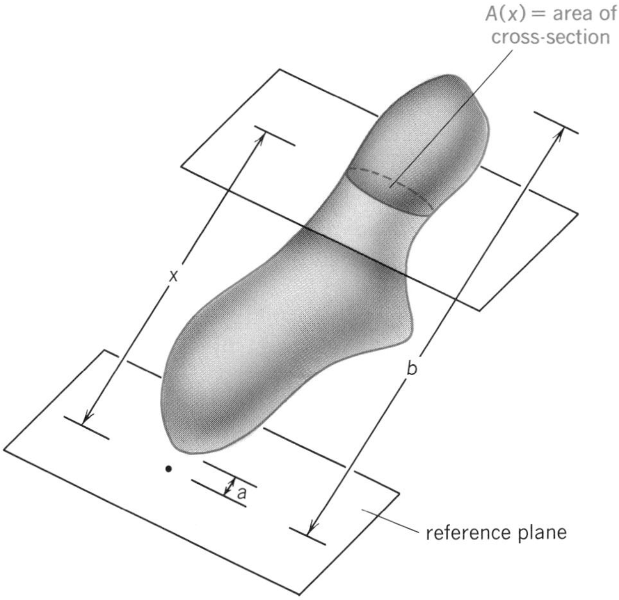

FIGURE 4-2

Consider a small increment Δx (which we assume is > 0; the other case is similar). Then

$$V(x+\Delta x) - V(x)$$

is the volume between two cross sections. A good approximation to this is $A(x)\Delta x$ which is the volume of the slab shown in Figure 4-3. Here we are taking for granted the fact that the volume of a solid with base area $A(x)$ and vertical walls of height h is $A(x)h$.

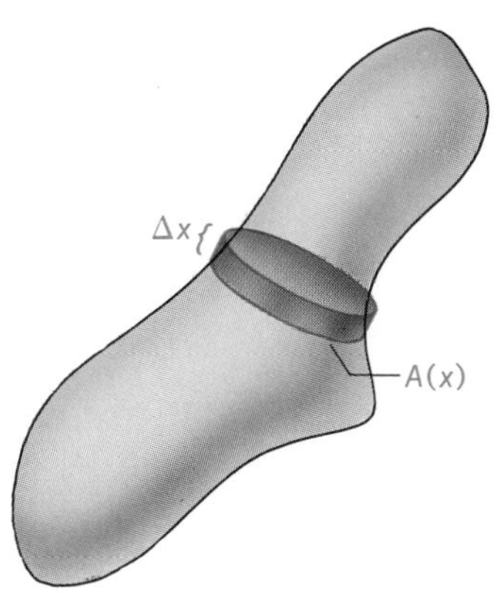

FIGURE 4-3

To compute the derivative $V'(x)$, assume $\Delta x \neq 0$ is given and as above, suppose $\Delta x > 0$. We let $A(c)$ and $A(C)$ be the minimum and maximum, respectively, of the function A on $[x, x+\Delta x]$. Hence (Figure 4-4),

$$A(c)\Delta x \leqslant V(x+\Delta x) - V(x)$$

and

$$V(x+\Delta x) - V(x) \leqslant A(C)\Delta x \qquad \text{(Why?)}$$

Division by Δx yields

$$A(c) \leqslant \frac{V(x+\Delta x) - V(x)}{\Delta x} \leqslant A(C)$$

This inequality is also valid when $\Delta x < 0$ and $A(c)$ and $A(C)$ are the minimum and maximum of A on the interval $[x+\Delta x, x]$.

Now A is a continuous function, and as Δx approaches 0, c and C are squeezed toward x and so

$$\lim_{\Delta x \to 0} A(c) = \lim_{\Delta x \to 0} A(C) = A(x)$$

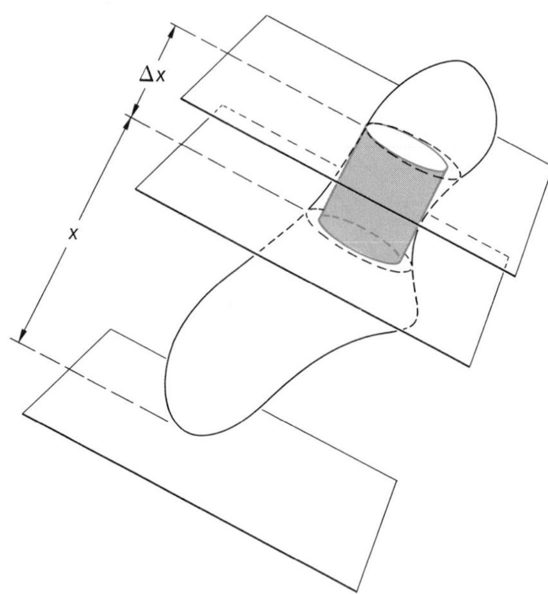

FIGURE 4-4

Therefore, we also have

$$\lim_{\Delta x \to 0} \frac{V(x+\Delta x) - V(x)}{\Delta x} = A(x)$$

or

$$V'(x) = A(x)$$

By the Fundamental Theorem of Calculus we obtain our result

$$V = V(b) - V(a) = \int_a^b A(x)\,dx$$

Let us verify these results in a simple case.

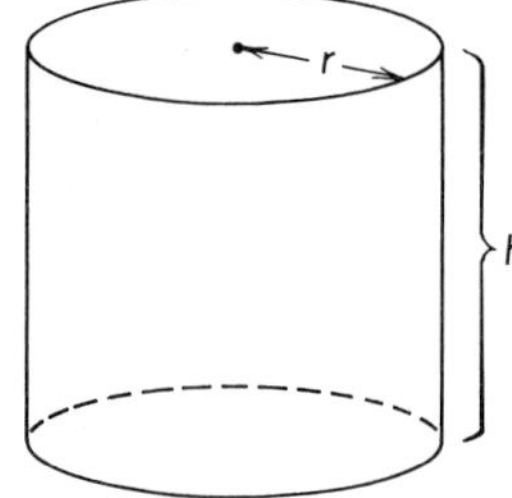

FIGURE 4-5

Example 1 Consider a cylinder with base radius r and height h (Figure 4-5). The volume of the cylinder is $\pi r^2 h$. The rate of change of volume with respect to the height is πr^2 which is exactly the cross-sectional area.

By Cavalieri's principle we would conclude that the two bodies in Figure 4-6 (a right cylinder and a skew cylinder) have the same volume.

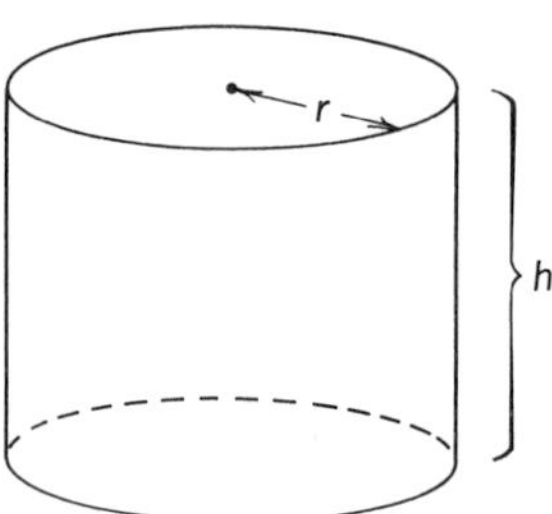

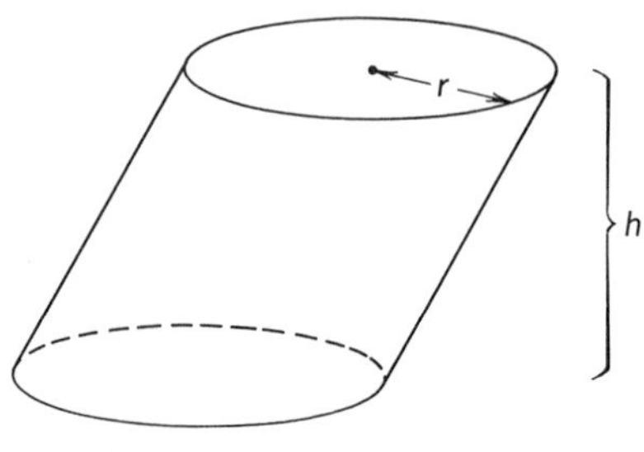

FIGURE 4-6

This fact should also be intuitively obvious. Indeed, think of this body as being a pile of records. One shape can be pushed into the other without changing the total volume.

The reader should go through similar arguments for a cube or a cone (see exercises 1 and 2). The similarity with the arguments in §6.2 on the Fundamental Theorem should also be noted.

Next we turn to a special type of solid, namely a **solid of revolution**. They are defined as follows.

Let f be a function continuous on $[a, b]$ and suppose that $f(x) \geqslant 0$ for all $x \in [a, b]$. We can construct a solid by rotating the graph of f on $[a, b]$ about the x axis (Figure 4-7). This type of solid is called a **solid of revolution**.

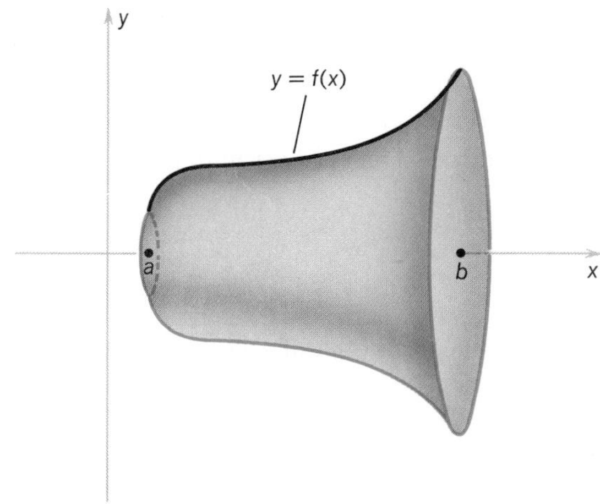

FIGURE 4-7

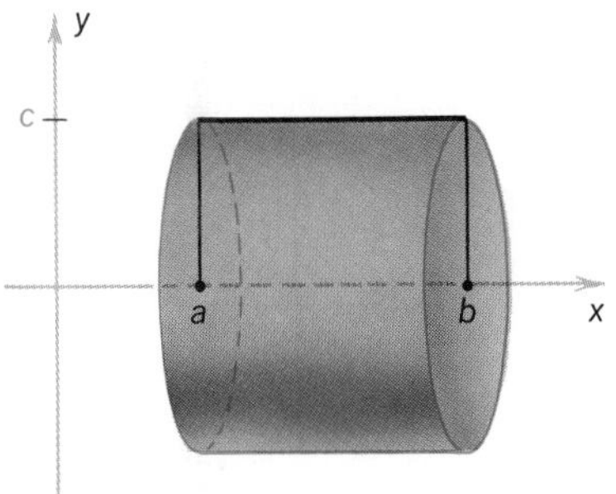

FIGURE 4-8

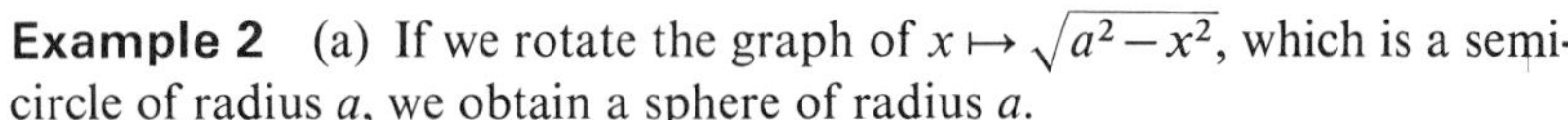

Example 2 (a) If we rotate the graph of $x \mapsto \sqrt{a^2 - x^2}$, which is a semicircle of radius a, we obtain a sphere of radius a.

(b) Consider the graph of the constant function $x \mapsto c$, $c > 0$, on $[a, b]$. Rotating about the x axis generates a cylinder of height $b - a$; and the base of the cylinder has radius c (Figure 4-8).

(c) Rotating the graph of the linear function $x \mapsto cx$, $c > 0$, on $[0, b]$ generates a cone with height b and with base of radius cb (Figure 4-9).

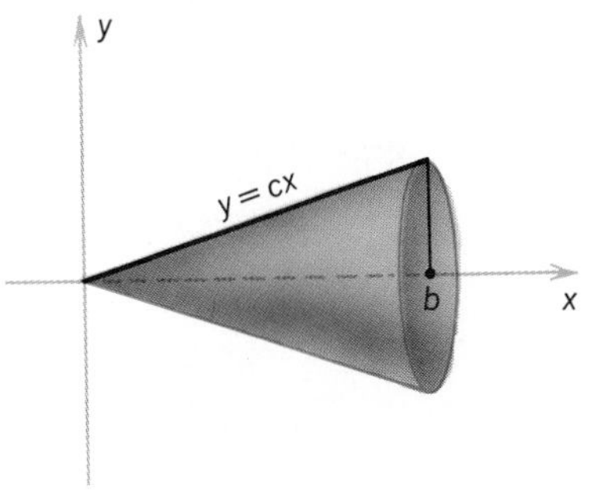

FIGURE 4-9

Suppose we now consider the general case of a solid of revolution generated by rotating the function f on $[a, b]$ about the x axis. If we cut this solid with a plane perpendicular to the x axis and passing through the point x_0 (Figure 4-10), the cross section cut out is a circle of radius $f(x_0)$. Thus, the cross section has area $\pi f(x_0)^2$. Therefore, the function $A: x \mapsto \pi f(x)^2$ gives the area of the cross section of the solid at x.

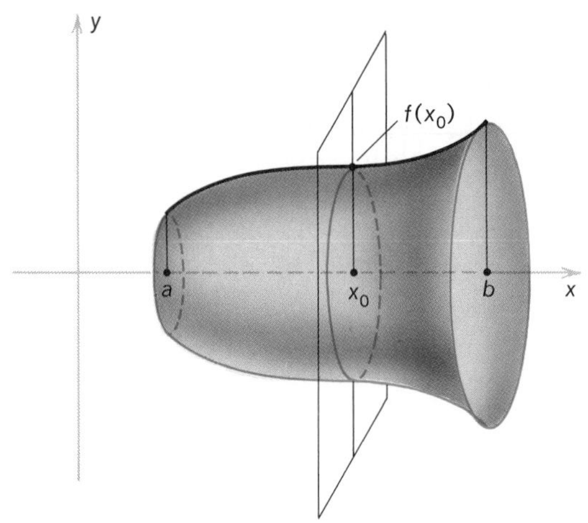

FIGURE 4-10

Therefore by Cavalieri's principle we obtain this result: The volume of a solid of revolution obtained by rotating the nonnegative function f on $[a, b]$ about the x axis is

$$\textbf{Volume} = \pi \int_a^b [f(x)]^2\, dx$$

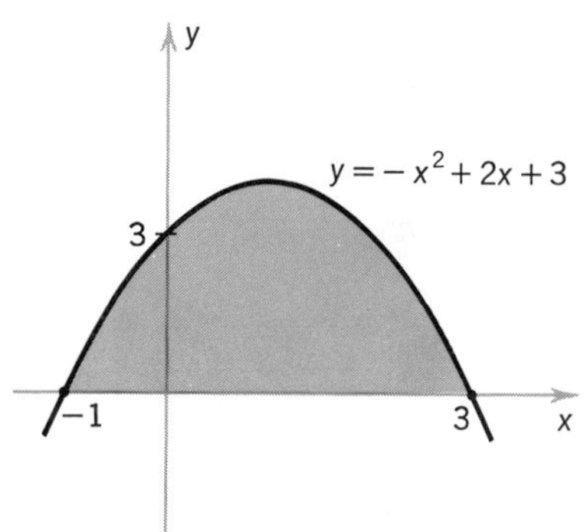

FIGURE 4-11

Example 3 Find the volume of the solid generated by revolving $y = \sqrt{\sin x}$ from $x = 0$ to $x = \pi$ about the x axis.

The area of a cross section at a point x is $A(x) = \pi(\sqrt{\sin x})^2 = \pi \sin x$. So the volume is

$$\begin{aligned} V = \pi \int_0^\pi \sin x\, dx &= \pi[-\cos x]_0^\pi = \pi[1 - (-1)] \\ &= 2\pi \end{aligned}$$

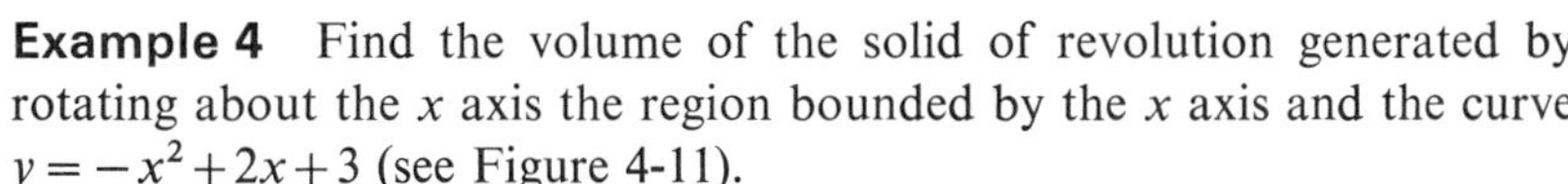

Example 4 Find the volume of the solid of revolution generated by rotating about the x axis the region bounded by the x axis and the curve $y = -x^2 + 2x + 3$ (see Figure 4-11).

The region is bounded at $x = -1$ and $x = 3$, so

$$\begin{aligned} V &= \pi \int_{-1}^{3} (-x^2 + 2x + 3)^2\, dx \\ &= \pi \int_{-1}^{3} (x^4 - 4x^3 - 2x^2 + 12x + 9)\, dx \\ &= \pi\left(\tfrac{1}{5}x^5 - x^4 - \tfrac{2}{3}x^3 + 6x^2 + 9x\right)\Big|_{-1}^{3} \\ &= \pi\left(\tfrac{512}{15}\right) = 34\tfrac{2}{15}\pi \end{aligned}$$

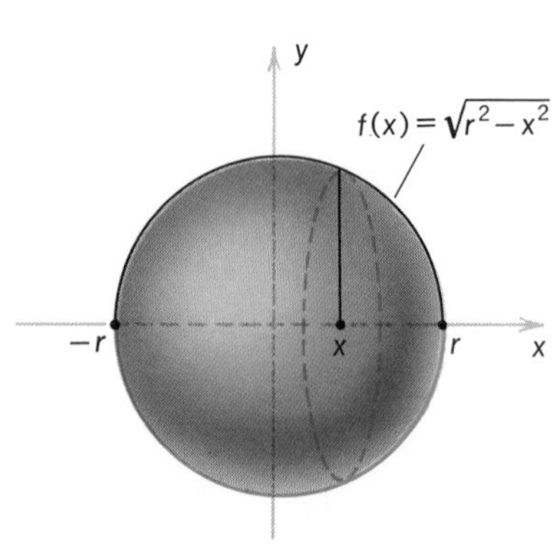

FIGURE 4-12

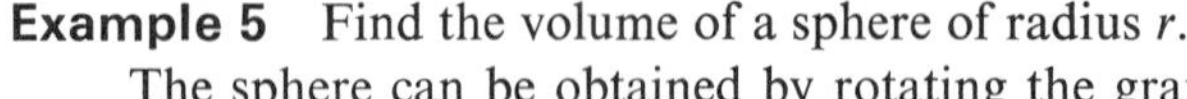

Example 5 Find the volume of a sphere of radius r.

The sphere can be obtained by rotating the graph of the semicircle $f: x \mapsto \sqrt{r^2 - x^2}$ around the x axis (Figure 4-12). At a point x in the interval $[-r, r]$, the cross section is a circle of radius $f(x) = \sqrt{r^2 - x^2}$. Hence the cross section has area $\pi f(x) = \pi(r^2 - x^2)$. So the volume is

$$\begin{aligned} V &= \int_{-r}^{r} \pi(r^2 - x^2)\,dx \\ &= \pi(r^2 x - \tfrac{1}{3}x^3)\Big|_{-r}^{r} \\ &= \pi((r^3 - \tfrac{1}{3}r^3) - (-r^3 - -\tfrac{1}{3}r^3)) \\ &= \pi(\tfrac{2}{3}r^3 - (-\tfrac{2}{3}r^3)) = \tfrac{4}{3}\pi r^3 \end{aligned}$$

A solid of revolution can also be obtained by revolving the graph of a function $y = f(x)$ about the y axis. For example, if we rotate the graph of $x \mapsto \sqrt{r^2 - x^2}, 0 \leqslant x \leqslant r$, about the y axis we obtain a hemisphere of radius r. The top half of a doughnut can be generated in the following way: Consider the graph of $x \mapsto \sqrt{r^2 - (x-a)^2}$ where $a > r > 0$. The graph of this function is a semicircle lying in the first quadrant with center at a and with radius r. When this graph is rotated about the y axis the top half of a **torus** or doughnut is obtained (Figure 4-13).

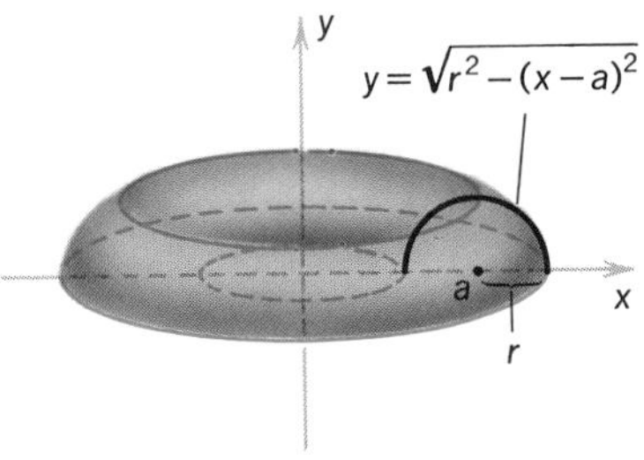

FIGURE 4-13

The problem is to compute the volume of solids obtained in this manner. That is, given a function f continuous on a closed interval $[a, b]$ with $a \geqslant 0$ and with $f(x) \geqslant 0$ on $[a, b]$, what is the volume generated by rotating f about the y axis? The answer is provided by the formula

$$\boldsymbol{V = 2\pi \int_a^b x f(x)\,dx}$$

To establish this formula, we argue much as we did for Cavalieri's principle; namely we consider the volume function $V(x)$ which gives the volume generated by rotating the graph of f from a to x about the y axis. The geometric significance of the difference $V(x+\Delta x) - V(x)$ leads one to look at the **cylindrical shell** obtained by rotating a rectangle with base the interval $[x, x+\Delta x]$ and height equal to $f(c)$, where c is a point in $[x, x+\Delta x]$ (Figure 4-14). Such a cylindrical shell has volume

$$\pi f(c)(x+\Delta x)^2 - \pi f(c)x^2 = \pi f(c)[2x\Delta x + \Delta x^2]$$

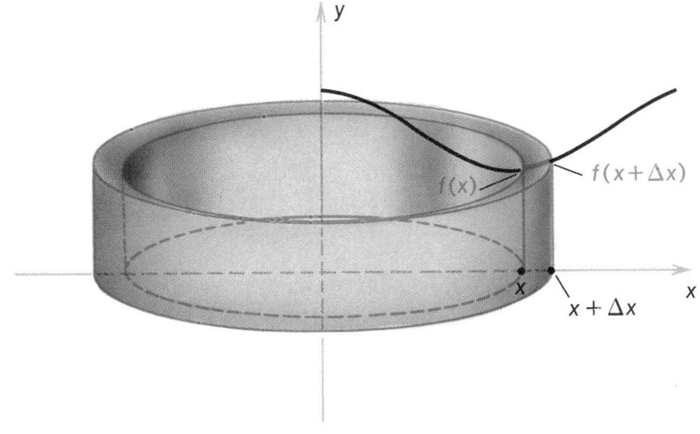

FIGURE 4-14

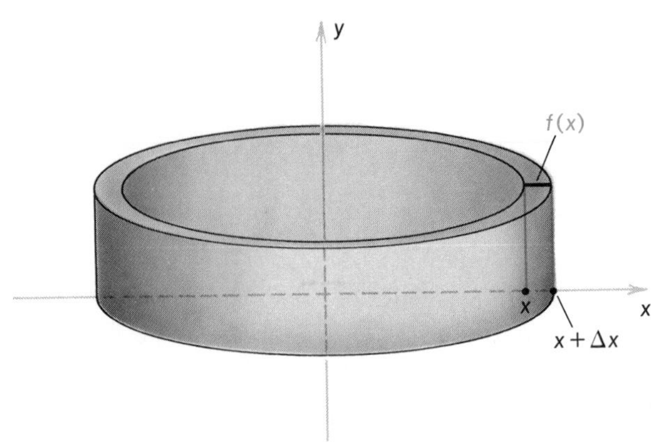

FIGURE 4-15

The difference $V(x+\Delta x)-V(x)$ represents the volume obtained by rotating only the portion of f from x to $x+\Delta x$ about the y axis. This volume is approximated by the volume of the shell shown in Figure 4-15.

The volume of this shell for Δx small is approximately $2\pi\Delta x f(x)$. Hence we expect, intuitively, the rate of change of $V(x)$ to be $2\pi x f(x)$. Let us verify this more carefully.

The actual volume of a shell with radii x and $x+\Delta x$ and height h is

$$V = \pi(x+\Delta x)^2 h - \pi x^2 h = 2\pi h x\Delta x + \pi h(\Delta x)^2$$

Now, letting m, M be the minimum and maximum, respectively, of f on $[x, x+\Delta x]$, we have

$$2\pi m x\Delta x + \pi m(\Delta x)^2 \leqslant V(x+\Delta x) - V(x) \leqslant 2\pi M x\Delta x + \pi M(\Delta x)^2$$

since the actual volume $V(x+\Delta x)-V(x)$ lies between the two shells with heights m and M. Dividing by Δx and letting $\Delta x \to 0$ yields

$$\lim_{\Delta x\to 0} \frac{V(x+\Delta x) - V(x)}{\Delta x} = 2\pi x f(x)$$

since m and M approach $f(x)$. The error terms $\pi m(\Delta x)^2$ and $\pi M(\Delta x)^2$ do not contribute in the limit since they go to 0. Hence, we have proven that $V'(x) = 2\pi x f(x)$; thus by the Fundamental Theorem it follows that

$$V = 2\pi \int_a^b x f(x)\, dx$$

Example 6 Find the volume generated by rotating $y = x^2$ from 0 to 1 about the y axis.

By the above formula, we have

$$V = 2\pi \int_0^1 x \cdot x^2\, dx = 2\pi \left(\tfrac{1}{4}x^4 \Big|_0^1\right) = \tfrac{1}{2}\pi$$

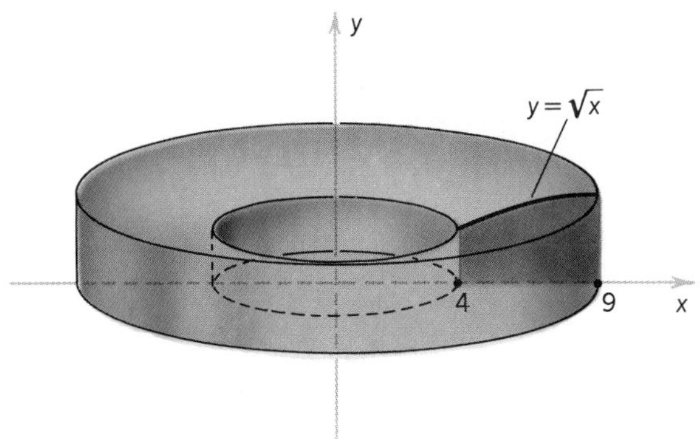

FIGURE 4-16

Example 7 The graph of $x \mapsto \sqrt{x}$ from 4 to 9 is rotated about the y axis. Find the volume generated.

We sketch the figure (Figure 4-16) and we see that the volume generated is the volume of a shell whose height varies from $2 = \sqrt{4}$ to $3 = \sqrt{9}$. By the formula, we have

$$V = 2\pi \int_4^9 x\sqrt{x}\,dx = 2\pi \int_4^9 x^{3/2} = 2\pi \left(\tfrac{2}{5}x^{5/2}\Big|_4^9\right)$$
$$= \tfrac{4}{5}\pi(243-32) = \tfrac{844}{5}\pi$$

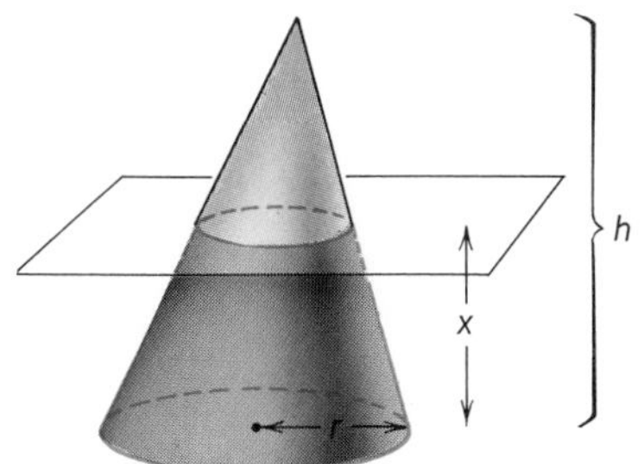

FIGURE 4-17

EXERCISES

1. Compute the rate of change of volume under a plane with height x for the cone in Figure 4-17.
2. Apply Cavalieri's principle to conclude that a cone and a skew cone with the same height and base radius have equal volumes.
3. What is the rate of change of the volume of the cone in exercise 1 with respect to r?
4. Show that the volume of a pyramid of height h whose base is a square of side s is $\frac{1}{3}s^2 h$.
5. The volume of a sphere with radius r is $\frac{4}{3}\pi r^3$. Compute the rate of change of volume with respect to r. Give a geometric reason for the answer in terms of surface area. If the surface area of a sphere were known, could you use this to obtain a formula for the volume or vice versa?

In exercises 6–12, find the volume generated by revolving the region bounded by the indicated graphs about the x axis.

6. $x^2 - y^2 = 16$, $y = 0$, and $x = 8$
7. $y^2 = x^3$, $y = 0$, and $x = 2$
8. $y = 4x^2$, $x = 0$, and $y = 16$

9. $y = x^2$, $y = 4x$
10. $y = x$, $y = x^2$
11. $y = (x-1)^{-3}$, $x = -1$, $x = 0$, and $y = 0$
12. $y = \sqrt{4+x}$, $x = 0$, and $y = 0$

In exercises 13–17, find the volume generated by revolving the region bounded by the indicated graphs about the y axis.

13. $y = x^3 + 10$, $x = 1$, and $x = 2$
14. $y = x^3 + Bx^2 + Cx + D$, $x = 0$, and $x = 1$
15. $y = \sqrt{x^2+10}$, $x = 3$, and $x = 4$
16. $y = \sin(x^2) + 5$, $x = 0$, and $x = \sqrt{\frac{1}{2}\pi}$
17. $y = 1/\cos^2(x^2)$, $x = 0$, and $x = \sqrt{\frac{1}{4}\pi}$
18. Revolving the region enclosed by the ellipse $x^2/a^2 + y^2/b^2 - 1 = 0$ about the x axis generates an egg-shaped solid. (a) Find the volume of the solid. (b) Without computing any further integrals, find the volume generated by revolving this region about the y axis.
19. An irregularly shaped well has a cross sectional area given by
$$A = \sqrt{100-x}, \qquad 0 \leqslant x \leqslant 100$$
where x is the depth of the cross section (all in feet). A pump is removing water from the well at a steady rate of one cubic foot per minute. How fast is the level dropping when the water level is 40 feet below the surface? When the level is 0 feet from the surface? Derive a general formula for which this is a special case.

5 FUNCTIONS DEFINED BY INTEGRALS

In §2 we began our study of functions of the form

$$x \mapsto \int_a^x f$$

In this section, we study a greater variety of functions defined by means of integrals. Here we bring the Chain Rule and the Fundamental Theorem together to establish the important **Change of Variables Theorem**.

Suppose the function $f: t \mapsto f(t)$ is continuous on $[a, b]$. Then we can form the function

$$F: x \mapsto \int_a^x f$$

Thus, if $f: t \mapsto t^2$, we have

$$F: x \mapsto \int_a^x t^2\, dt$$

We can also define F by the equation

$$F(x) = \int_a^x t^2\, dt$$

When defining functions using integrals, it is desirable to avoid expressions such as

$$G(x) = \int_a^x x^2\,dx$$

because the variable x is then being used to define two different functions. Instead, we use two variables; one to define the function being integrated (for example, $t \mapsto t^2$) and the other to define the integral function (thus, $F: x \mapsto \int_a^x t^2\,dt$). Or, in other words, we use t as the variable of integration, and x as a limit of integration.

Example 1 (a) The function $G: x \mapsto \int_0^x \cos$ can also be defined by the rule $G: x \mapsto \int_0^x \cos t\,dt$ and by the equation $G(x) = \int_0^x \cos t\,dt$. Moreover, $t \mapsto \sin t$ is an antiderivative of $t \mapsto \cos t$ and so by the Fundamental Theorem,

$$G(x) = \int_0^x \cos t\,dt = \sin t\,\Big|_0^x = \sin x$$

(b) If $f: t \mapsto t^2$, then the rule $F: x \mapsto \int_a^x t^2\,dt$ defines the same function as do $F: x \mapsto \int_a^x f$ and $F(x) = \int_a^x t^2\,dt$. Furthermore, $t \mapsto \frac{1}{3}t^3$ is an antiderivative of f, so

$$F(x) = \int_a^x t^2\,dt = \tfrac{1}{3}t^3\,\Big|_a^x = \tfrac{1}{3}x^3 - \tfrac{1}{3}a^3$$

(c) Suppose F is defined by $F: u \mapsto \int_0^u \sqrt{t}\,dt$. Since $t \mapsto \frac{2}{3}t^{3/2}$ is an antiderivative of $t \mapsto \sqrt{t}$, we obtain

$$F(u) = \tfrac{2}{3}t^{3/2}\,\Big|_0^u = \tfrac{2}{3}u^{3/2}$$

(d) Suppose $F(u) = \int_0^u |t|\,dt$. Then we have

$$F(u) = \begin{cases} -\tfrac{1}{2}u^2, & u \leqslant 0 \\ \tfrac{1}{2}u^2, & u \geqslant 0 \end{cases}$$

Let us return to the general case of a continuous f with F defined by $F(x) = \int_a^x f(t)\,dt$; then we can ask what is the derivative $F'(x)$? Since $\int_a^x f(t)\,dt = \int_a^x f$, we know by the Fundamental Theorem that $F' = f$, and so $F'(x) = f(x)$ for $x \in [a, b]$. This gives us the following important formula:

$$\frac{d\left(\int_a^x f(t)\,dt\right)}{dx} = f(x)$$

The formula can be state verbally as: The derivative of the definite integral of a continuous function with respect to the upper limit of integration is equal to the value of the integrand function at this upper limit.

Example 2 (a) If $F(x)=\int_0^x |t|\,dt$, then $F'(x)=d(\int_0^x |t|\,dt)/dx=|x|$.

(b) The following results are obtained by using the above formula:

$$\frac{d\left(\int_a^x \sin t\,dt\right)}{dx}=\sin x$$

$$\frac{d\left(\int_a^x t^2+t^3\,dt\right)}{dx}=x^2+x^3$$

$$\frac{d\left(\int_a^u \sin t\,dt\right)}{du}=\sin u$$

$$\frac{d\left(\int_0^u \sqrt{t}\,dt\right)}{du}=\sqrt{u}$$

$$\frac{d\left(\int_0^x c\,dt\right)}{dx}=c, \qquad c \text{ a constant.}$$

If we have a function $F:x\mapsto \int_a^x f(t)\,dt$, the domain of F is the set of all x such that $\int_a^x f(t)\,dt$ is defined. In the case that $x>a$, the integral $\int_a^x f(t)\,dt$ is defined if f is continuous on $[a,x]$. When $x<a$, the integral $\int_a^x f(t)\,dt$ is defined when f is continuous on $[x,a]$. And, of course, when $x=a$, $\int_a^x f(t)\,dt$ is defined and equals 0. We give some examples.

Example 3 (a) The domain of $x\mapsto \int_a^x mt\,dt$ is all of $\boldsymbol{R}$, since $t\mapsto mt$ is continuous on all intervals $[a,x]$ and on all intervals $[x,a]$.

(b) The domain of $x\mapsto \int_1^x 1/t\,dt$ is $]0,+\infty[$, the set of positive real numbers. This follows from the fact that $t\mapsto 1/t$ is continuous on $[x,1]$ only if $x>0$.

(c) The domain of $u\mapsto \int_{-1}^u 1/t\,dt$ is $]-\infty,0[$, the set of all negative real numbers.

(d) The domain of $u\mapsto \int_0^u \sqrt{t}\,dt$ is $[0,+\infty[$, the set of all non-negative real numbers.

(e) If the function f is continuous and has domain $\boldsymbol{R}$, then $F:x\mapsto \int_a^x f(t)\,dt$ is defined, continuous, and even differentiable on all of $\boldsymbol{R}$. If f has domain $]A,B[$ and is continuous, then for any a in $]A,B[$, the function $x\mapsto \int_a^x f$, is continuous and differentiable on $]A,B[$.

Given a function $F:u\mapsto \int_a^u f(t)\,dt$ and another function h we can, if a certain condition is satisfied, form $F\circ h$, the composite of F on h. The condition to be satisfied is that for all x in the domain of h, $h(x)$ must be in the domain of F. That is, for all x in the domain of h, the integral $\int_a^{h(x)} f(t)\,dt$ must be defined. When this condition is met, the composite is defined by the equation $(F\circ h)(x)=\int_a^{h(x)} f(t)\,dt$. Let us look at some examples.

Example 4 (a) Suppose $F(u)=\int_0^u |t|\,dt$ and $h(x)=x^2$. Then, since $\int_0^{x^2} |t|\,dt$ is defined for all x in $\boldsymbol{R}$, the composite $F\circ h$ is defined and has domain $\boldsymbol{R}$. And, we see that $(F\circ h)(x)=F(h(x))=\int_0^{x^2}|t|\,dt$.

(b) If $F(u)=\int_0^u mt\,dt$ and if $h(x)=\sin x$, the composite $F\circ h$ is defined for all $x\in\boldsymbol{R}$. Moreover, we have $F(\sin x)=\int_0^{\sin x} mt\,dt$. We recall that $\int_0^b mt\,dt=\frac{1}{2}mb^2$. Therefore, $F(u)=\frac{1}{2}mu^2$ and so we obtain

$$F(\sin x) = \tfrac{1}{2}m(\sin^2 x)$$

(c) If $F(u)=\int_{-1}^u 1/t\,dt$ and if $h(x)=x^2$, the composite $F\circ h$ cannot be formed since $t\mapsto 1/t$ is never continuous on $[-1,x^2]$.

(d) If $F(x)=\int_0^{\tan x} 3t^2\,dt$, the domain of F is equal to the domain of the tangent function. We know that the tangent is a periodic function with period π. Since F is the composite of $u\mapsto\int_0^u 3t^2dt$ on $x\mapsto\tan x$, F is also periodic. Explicitly, F is given by $F{:}\,x\mapsto\tan^3 x$ and so F also has period π.

Theorem 1 **Generalized Fundamental Theorem** Suppose that f and h are continuous functions such that the composite of $F{:}\,u\mapsto\int_a^u f(t)\,dt$ on h is defined. If in addition h is differentiable, then $F\circ h$ is also differentiable and we have

$$(F\circ h)'(x) = f(h(x))\,h'(x)$$

Since the composite $F\circ h$ is given by $(F\circ h)(x)=\int_a^{h(x)} f(t)\,dt$, we obtain the formula

$$\frac{d\left(\int_a^{h(x)} f(t)\,dt\right)}{dx} = f(h(x))\,h'(x)$$

Proof First we note that by the Fundamental Theorem, $F'(u)=f(u)$ and so $F'(h(x))=f(h(x))$. The composite $F\circ h$ is differentiable since both F and h are. We apply the Chain Rule to obtain

$$(F\circ h)'(x) = F'(h(x))\,h'(x)$$

Since $F'(h(x))=f((h(x))$ as we have just seen, we have

$$(F\circ h)'(x) = f\,(h(x))\,h'(x)$$ ∎

Example 5 (a) Find the derivative of $x\mapsto\int_0^{x^2} 1/(1+\sqrt{t})\,dt$.

In the notation of the above formula, we have $h(x)=x^2$ and $f(t)=1/(1+\sqrt{t})$. Thus $f(h(x))=f(x^2)=1/(1+\sqrt{x^2})=1/(1+|x|)$. Hence, we have

$$\frac{d\left(\int_0^{x^2}\frac{1}{1+\sqrt{t}}\,dt\right)}{dx} = \frac{1}{1+|x|}\cdot 2x = \frac{2x}{1+|x|}$$

(b) If $F{:}\,u\mapsto\int_0^u mt\,dt$ and if $h(x)=\sin x$, we can apply the above formula [with $f(t)=mt$] to obtain

$$\frac{d\left(\int_0^{\sin x} mt\,dt\right)}{dx} = m(\sin x)(\cos x)$$

On the other hand, in Example 4(b), we saw that

$$\int_0^{\sin x} mt\,dt = F(\sin x) = \tfrac{1}{2}m\sin^2 x$$

Differentiating $x \mapsto \frac{1}{2}m\sin^2 x$, we obtain

$$\frac{d\left(\frac{1}{2}m\sin^2 x\right)}{dx} = \tfrac{1}{2}2m(\sin x)(\cos x) = m(\sin x)(\cos x)$$

which agrees with the expression arrived at using the formula.

(c) Find $T'(x)$ if

$$T(x) = \int_{-x^2}^{x^2+1} \frac{dt}{\sqrt{t^4+t^2+1}}$$

We first note that the domain of T is $]-\infty, +\infty[$. We then fix a point c in the domain of T (so c can be any real number) and we write

$$\begin{aligned} T(x) &= \int_{-x^2}^{c} \frac{dt}{\sqrt{t^4+t^2+1}} + \int_{c}^{x^2+1} \frac{dt}{\sqrt{t^4+t^2+1}} \\ &= -\int_{c}^{-x^2} \frac{dt}{\sqrt{t^4+t^2+1}} + \int_{c}^{x^2+1} \frac{dt}{\sqrt{t^4+t^2+1}} \end{aligned}$$

We set

$$T_1(x) = \int_{c}^{-x^2} \frac{dt}{\sqrt{t^4+t^2+1}}, \quad \text{and} \quad T_2(x) = \int_{c}^{x^2+1} \frac{dt}{\sqrt{t^4+t^2+1}}$$

Thus, $T(x) = -T_1(x) + T_2(x)$ and so $T'(x) = -T_1'(x) + T_2'(x)$. Applying the Generalized Fundamental Theorem to $T_1(x)$ and $T_2(x)$ we obtain

$$T_1'(x) = \frac{1}{\sqrt{(-x^2)^4 + (-x^2)^2 + 1}}(-2x)$$

and

$$T_2'(x) = \frac{1}{\sqrt{(x^2+1)^4 - (x^2+1)^2 + 1}}(2x)$$

Hence,

$$\begin{aligned} T'(x) &= \frac{2x}{\sqrt{x^8+x^4+1}} + \frac{2x}{\sqrt{(x^2+1)^4 + (x^2+1)^2 + 1}} \\ &= \frac{2x}{\sqrt{x^8+x^4+1}} + \frac{2x}{\sqrt{x^8+4x^6+7x^4+6x^2+3}} \end{aligned}$$

More generally if we are given a continuous function f and differentiable functions g and h we can consider the function

$$F : x \mapsto \int_{g(x)}^{h(x)} f(t)\,dt$$

[it is not necessary that $h(x)$ be greater than $g(x)$]. The derivative of F is, as we shall show,

$$F'(x) = f(h(x))h'(x) - f(g(x))g'(x)$$

Indeed, pick an arbitrary point c (in the domain under consideration) and write, by additivity of the integral,

$$\begin{aligned} F(x) &= \int_{c}^{h(x)} f(t)\,dt + \int_{g(x)}^{c} f(t)\,dt \\ &= \int_{c}^{h(x)} f(t)\,dt - \int_{c}^{g(x)} f(t)\,dt \end{aligned}$$

Now the formula of the Generalized Fundamental Theorem applies and yields the desired result.

We next apply the Generalized Fundamental Theorem to prove a very important result of integral calculus.

Theorem 2 Change of Variables; The Chain Rule for Integral Calculus Let f be a continuous function and let h be a continuously differentiable function with domain an open interval. Suppose also that the composite function $f \circ h$ is defined. Then for all points a and x in the domain of h, we have

$$\int_{a}^{x} (f \circ h)\,h' = \int_{h(a)}^{h(x)} f$$

Proof We set $G : x \mapsto \int_{h(a)}^{h(x)} f$ and $F : x \mapsto \int_{a}^{x} (f \circ h)\,h'$. We must show that $G = F$. To do this, we first show that $G' = F'$. By the Fundamental Theorem,

$$F'(x) = (f \circ h)(x)\,h'(x) = f\big(h(x)\big)\,h'(x)$$

On the other hand, by the Generalized Fundamental Theorem,

$$G'(x) = f\big(h(x)\big)\,h'(x)$$

Hence, G and F differ by a constant. To evaluate this constant, we note that $G(a) = 0 = F(a)$; hence the constant is 0 and $G = F$. ∎

In the above proof, the hypothesis that the domain of h is an open interval is necessary to insure that the integral $\int_{h(a)}^{h(x)} f$ is defined for all x in the domain of h. To see this, apply the Intermediate Value Theorem to h on $[a, x]$ or $[x, a]$ as the case may be, to determine that f is defined on the closed interval with endpoints $h(a)$ and $h(x)$.

The following **Change of Variables Formula** is helpful for applying the Change of Variable Theorem to the evaluation of integrals:

$$\int_a^b \left(f(u)\frac{du}{dx} \right) dx = \int_{u(a)}^{u(b)} f(u)\, du$$

This formula is derived from Theorem 2 as follows: We specialize x to b to obtain

$$\int_a^b (f \circ h)\, h' = \int_{h(a)}^{h(b)} f$$

We then set $h(x) = u$ and $h'(x) = du/dx$. Thus,

$$\int_a^b (f \circ h)\, h' = \int_a^b f(h(x))\, h'(x)\, dx = \int_a^b f(u)\frac{du}{dx}\, dx \tag{1}$$

Finally, we define $u(a) = h(a)$ and $u(b) = h(b)$, which yields

$$\int_{h(a)}^{h(b)} f = \int_{h(a)}^{h(b)} f(u)\, du = \int_{u(a)}^{u(b)} f(u)\, du \tag{2}$$

Putting equations (1) and (2) together gives the desired formula. The use of this formula is often called **integration by substitution**.

Example 6 (a) Evaluate $\int_0^{\frac{1}{4}\pi} \sin^2 x \cos x\, dx$. We set $u = \sin x$ and we have $du/dx = \cos x$. Therefore,

$$\int_0^{\frac{1}{4}\pi} \sin^2 x \cos x\, dx = \int_0^{\frac{1}{4}\pi} u^2 \frac{du}{dx}\, dx$$

Now, we see that $u(0) = 0$ and $u(\frac{1}{4}\pi) = \sqrt{2}/2$, so

$$\int_0^{\frac{1}{4}\pi} u^2 \frac{du}{dx}\, dx = \int_0^{\sqrt{2}/2} u^2\, du = \frac{u^3}{3}\bigg|_0^{\sqrt{2}/2} = \frac{\sqrt{2}}{12}$$

(b) Evaluate

$$\int_0^1 \frac{2x+1}{(x^2+x+1)^2}\, dx$$

If we set $u = x^2+x+1$, we obtain $du/dx = 2x+1$. Thus,

$$\int_0^1 \frac{2x+1}{(x^2+x+1)^2}\, dx = \int_1^3 \frac{1}{u^2}\, du = \frac{-1}{u}\bigg|_1^3 = -\tfrac{1}{3} + 1 = \tfrac{2}{3}$$

(c) Evaluate

$$\int_0^1 \frac{x}{\sqrt{x^2+1}}\, dx$$

If we set $u = x^2+1$, then we obtain $du/dx = 2x$. To manipulate the integral into the proper form, we note that

$$\int_0^1 \frac{x}{\sqrt{x^2+1}}\, dx = \frac{1}{2}\int_0^1 \frac{2x}{\sqrt{x^2+1}}\, dx$$

Now with $u=x^2+1$, we have

$$\int_0^1 \frac{x}{\sqrt{x^2+1}}\,dx = \frac{1}{2}\int_0^1 \frac{1}{\sqrt{u}}\frac{du}{dx}\,dx$$

$$= \frac{1}{2}\int_1^2 \frac{1}{\sqrt{u}}\,du = \tfrac{1}{2}\left(2\sqrt{u}\Big|_1^2\right)$$

$$= \tfrac{1}{2}(2\sqrt{2}-2)$$

$$= \sqrt{2}-1$$

The Change of Variables Theorem enables us to see that certain regions have the same area; this is useful in cases where we are not able to compute the areas numerically.

Example 7 Show that the area under the curve $y=\tan x$ from 0 to $\frac{1}{4}\pi$ is equal to the area under the curve $y=1/x$ from $\sqrt{2}/2$ to 1. The first area is equal to $\int_0^{\frac{1}{4}\pi}\tan x\,dx$. Now $\tan x=(\sin x)/(\cos x)$; so if we set $u=\cos x$, we have

$$\int_0^{\frac{1}{4}\pi}\tan x\,dx = \int_0^{\frac{1}{4}\pi}\frac{-1}{u}\frac{du}{dx}\,dx$$

$$= \int_1^{\sqrt{2}/2}\frac{-1}{u}\,du$$

$$= \int_{\sqrt{2}/2}^1 \frac{1}{u}\,du$$

But this last integral is equal to the area under the curve $y=1/x$ from $\sqrt{2}/2$ to 1.

EXERCISES

Differentiate each of the following functions.

1. $x \mapsto \int_0^x t^3\,dt$
2. $x \mapsto \int_0^x \sin t\,dt$
3. $x \mapsto \int_x^0 t\,dt$
4. $x \mapsto \int_0^{x^2} t\,dt$
5. $x \mapsto \int_0^{\cos x}\sin t\,dt$
6. $x \mapsto \int_0^x xt\,dt$

*7. $x \mapsto \int_0^{x^2}\cos xt\,dt$

State the domain for each of the following functions and differentiate.

8. $x \mapsto \int_1^x \frac{1}{\sqrt{t}}\,dt$
9. $x \mapsto \int_0^x \frac{1}{1-\sqrt{t}}\,dt$
10. $x \mapsto \int_0^{x^2}\frac{1}{\cos t}\,dt$
11. $x \mapsto \int_x^{x^2+1} t\,dt$
12. $x \mapsto \int_{\sqrt{x}}^x \frac{1}{\sqrt{t}}\,dt$

Evaluate the following integrals.

13. $\int_0^\pi x^2 \cos x^3\, dx$

14. $\int_7^8 \frac{3x^2+6}{(x^3+6x)^7}\, dx$

15. $\int_6^2 (3x+1)\, dx$

16. $\int_{\pi/4}^{\pi/3} \sin x \cos x\, dx$

17. $\int_0^{\pi/4} \cos^5 x \sin x\, dx$

18. $\int_2^4 [(x+3)^5 - 7(x+3)^4 + 3(x+3)^2 + 2]\, dx$

19. $\int_0^1 x\sqrt{1-x^2}\, dx$

6 THE INDEFINITE INTEGRAL

In Chapter 5 we studied antiderivatives. Recall that for a function f an antiderivative is a function F such that $F' = f$. Furthermore, given one antiderivative F, any other antiderivative is of the form $F + C$, where C is a constant. In the previous chapter, we also discovered that the Fundamental Theorem gave the connection between integration and antiderivatives:

$$\int_a^b f(t)\, dt = F(b) - F(a)$$

Because of this connection one uses a special notation for the antiderivative. Namely, we write

$$\int f(x)\, dx$$

for the antiderivative of f. To include all antiderivatives we add on an arbitrary constant. For historical reasons one generally uses x as the variable in this notation, hence, one typical example is

$$\int x\, dx = \frac{x^2}{2} + C$$

where C is a constant.

We call $\int f(x)\, dx$ the **indefinite integral** of f; it is a function of x (with an arbitrary constant). We call $\int_a^b f(x)\, dx$ the **definite integral** of f; it is a real number. They are related by

$$\int_a^b f(x)\, dx = \int f(x)\, dx \Big|_a^b$$

In this section we shall gain further practice with integrals and methods of evaluation.

Example 1

$$\int x^2\, dx = \frac{x^3}{3} + C$$

$$\int \cos x\, dx = \sin x + C$$

$$\int \sin u\, du = -\cos u + C$$

The symbol C can be replaced by any other symbol for an arbitrary constant.

$$\int \sqrt{x}\,dx = \tfrac{2}{3}x^{3/2} + C_1$$

$$\int (3u^2+2\sin u)\,du = u^3 - 2\cos u + C_2$$

$$\int 1\,dx = \int dx = x + C_1$$

Because of the equation $\int du = u + C$, the indefinite integral operates as the inverse of differentiation. This fact is of great importance in computing antiderivatives. In fact, every differentiation formula immediately produces a corresponding formula for indefinite integrals. Below is a list of some of the important differentiation formulas we have established so far, along with their corresponding integral formulas:

$d(u^r) = ru^{r-1}\,du$	$\int u^{r-1}\,du = u^r/r, \quad r \neq 0$
$d(\sin u) = \cos u\,du$	$\int \cos u\,du = \sin u + C$
$d(\cos u) = -\sin u\,du$	$\int \sin u\,du = -\cos u + C$
$d(\tan u) = \sec^2 u\,du$	$\int \sec^2 u\,du = \tan u + C$

Theorem 1 The indefinite integral has the following properties:

(i) **Additivity** $\int f(x)\,dx + \int g(x)\,dx = \int (f(x) + g(x))\,dx$

(ii) **Linearity** $\int cf(x)\,dx = c\int f(x)\,dx$

Proof To establish (i), let $\int f(x)\,dx = F(x) + C_1$ and $\int g(x)\,dx = G(x) + C_2$. Then $d(F(x)+C_1+G(x)+C_2)/dx = f(x)+g(x)$, so

$$\int (f(x)+g(x))\,dx = F(x) + C_1 + G(x) + C_2 = F(x) + G(x) + C$$

This follows because the sum of two arbitrary constants is simply another arbitrary constant. To establish (ii), one uses the fact that a multiple of an arbitrary constant is also an arbitrary constant. The proof of (ii) is left to the reader. ∎

Example 2

(a) $$\int (3x^2+2x)\,dx = \int 3x^2\,dx + \int 2x\,dx = (x^3+C_1) + (x^2+C_2) = x^3 + x^2 + C$$

(b) $$\int (\cos u + \sin u)\,du = \int \cos u\,du + \int \sin u\,du = (\sin u + C_1) + (-\cos u + C_2) = \sin u - \cos u + C$$

(c) $\int 2\sec^2 u\,du = 2\int \sec^2 u\,du = 2(\tan u + C_1) = 2\tan u + C$

(d) $\int (3x - 3\sin x)\,dx = \int 3(x - \sin x)\,dx = 3\int (x - \sin x)\,dx$

$= 3(\tfrac{1}{2}x^2 + \cos x + C_1) = \tfrac{3}{2}x^2 + 3\cos x + C$

(e) $\int \frac{2}{\sqrt{x}}\,dx = \int 2x^{-1/2}\,dx = 2\int x^{-1/2}\,dx = 2\left(\frac{x^{1/2}}{\frac{1}{2}} + C_1\right) = 4x^{1/2} + C$

Theorem 2 If $\int f(u)\,du = G(u) + C$, then

$$\int f(mx+b)\,dx = \frac{1}{m}G(mx+b) + C$$

Proof Let $H: x \mapsto (1/m)G(mx+b)$. H is the composition of $u \mapsto (1/m)G(u)$ on the affine function $x \mapsto mx+b$. So we have $H'(x) = (1/m)G'(mx+b)\cdot m = G'(mx+b) = f(mx+b)$. Thus

$$\int f(mx+b)\,dx = \frac{1}{m}G(mx+b) + C$$ ∎

Example 3 (a) $\int \cos 2x\,dx = \tfrac{1}{2}\sin 2x + C$

(b) $\int \sqrt{x+1}\,dx = \tfrac{2}{3}(x+1)^{3/2} + C$

(c) $\int (2x+1)^2\,dx = \tfrac{1}{6}(2x+1)^3 + C$

(d) $\int \frac{1}{\sqrt{2x}}\,dx = \tfrac{1}{2}\cdot 2\sqrt{2x} + C = \sqrt{2x} + C$

Many times we are given a function whose antiderivative is not obvious. In such a case, a useful technique is to transform the function into an equivalent form and then try to find the antiderivative of the new form. Trigonometric identities are therefore often helpful in evaluating integrals. We shall illustrate their use while establishing the important formulas:

$$\int \sin^2 x\,dx = \frac{1}{2}\left(x - \frac{\sin 2x}{2}\right) + C$$

$$\int \cos^2 x\,dx = \frac{1}{2}\left(x + \frac{\sin 2x}{2}\right) + C$$

The identities we need are

$$\cos 2x = \cos^2 x - \sin^2 x$$

and

$$\sin^2 x + \cos^2 x = 1$$

These combine to yield

$$\cos^2 x = \frac{1+\cos 2x}{2}$$

and

$$\sin^2 x = \frac{1-\cos 2x}{2}$$

Using this last identity we have

$$\int \sin^2 x\,dx = \frac{1}{2}\int(1-\cos 2x\,dx) = \frac{1}{2}\left(x-\int\cos 2x\,dx\right) = \frac{1}{2}\left(x-\frac{\sin 2x}{2}\right)+C$$

Similarly,

$$\int \cos^2 x\,dx = \frac{1}{2}\left(x+\int\cos 2x\,dx\right) = \frac{1}{2}\left(x+\frac{\sin 2x}{2}\right)+C$$

The student is warned that although there was a simple formula for differentiating a product or a quotient, there is *no* such simple formula for integrating products or quotients except in certain circumstances. In this sense, integration is not as straightforward as differentiation and special tricks are often involved.

The Chain Rule provides us with another important technique for indefinite integration. Recall that the Chain Rule asserts that the derivative of a composition of functions is a product of derivatives. This means that the antiderivative of certain products will be a composition of functions.

Example 4 (a) Let us find $\int 3\sin^2 x\cos x\,dx$. We know that $d(\sin x)/dx = \cos x$ and that $d(\sin^3 x)/dx = 3\sin^2 x\cos x$. Therefore, $\int 3\sin^2 x\cos x\,dx = \sin^3 x + C$. (Note that $x \mapsto \sin^3 x$ is the composition of $u \mapsto u^3$ on $x \mapsto \sin x$.)

(b) Compute

$$\int \frac{7\tan^6\theta}{\cos^2\theta}\,d\theta$$

We have

$$\frac{d(\tan\theta)}{d\theta} = \frac{1}{\cos^2\theta}; \qquad \text{so} \qquad \frac{d(\tan^7\theta)}{d\theta} = \frac{7\tan^6\theta}{\cos^2\theta}$$

Therefore,

$$\int \frac{7\tan^6\theta}{\cos^2\theta}\,d\theta = \tan^7\theta + C$$

Suppose we are given that $\int G(u)\,du = H(u)+C$; let us evaluate $\int G(h(x))\,h'(x)\,dx$. By the Chain Rule, we obtain

$$(H\circ h)'(x) = \frac{dH(h(x))}{dx} = H'(h(x))\,h'(x) = G(h(x))\,h'(x)$$

Hence

$$\int G(h(x))\,h'(x)\,dx = H(h(x)) + C$$

If we substitute

$$u = h(x)$$

the last equation becomes

$$\int G(u)\frac{du}{dx}\,dx = H(h(x)) + C$$

FORMAL DEVICE Method of Substitution

To evaluate $\int f(x)\,dx$, attempt to put the integral into the form $\int G(h(x))\,h'(x)\,dx$. Then substitute $u = h(x)$ and find $\int G(u)\,du = H(u) + C$. Then $\int f(x)\,dx = H(h(x)) + C$.

Example 5 (a) Let us compute $\int \sin^{10} x \cos x\,dx$. This integral does not appear in our list of formulas and cannot be obtained either by linearity or additivity. However, we note that $d(\sin x)/dx = \cos x$. Letting $u = \sin x$, the integral becomes $\int u^{10}(du/dx)dx$. Since $\int u^{10}\,du = \frac{1}{11}u^{11} + C$, we conclude that

$$\int \sin^{10} x \cos x\,dx = \frac{\sin^{11} x}{11} + C$$

Let us verify our claim that $\int \sin^{10} x \cos x\,dx = (\sin^{11} x)/11 + C$. We differentiate $x \mapsto (\sin^{11} x)/11 + C$ and we find, by the Chain Rule, that

$$\frac{d\left(\dfrac{\sin^{11} x}{11} + C\right)}{dx} = \sin^{10} x \cos x$$

(b) Compute $\int 4x\sqrt{1+2x^2}\,dx$. Since $d(1+2x^2) = 4x\,dx$, we set $u = 1 + 2x^2$ and we have

$$\int 4x\sqrt{1+2x^2} = \int \sqrt{u}\,du = \tfrac{2}{3}u^{3/2} + C = \tfrac{2}{3}(1+2x^2)^{3/2} + C$$

Again, we check

$$\frac{d\left(\frac{2}{3}(1+2x^2)^{3/2} + C\right)}{dx} = (1+2x^2)^{1/2} \cdot 4x\,,$$

(c) Compute $\int x\sqrt{1+2x^2}\,dx$. Since $d(1+2x^2) = 4x\,dx$. we first write

$$\int x\sqrt{1+2x^2}\,dx = \tfrac{1}{4}\int 4x\sqrt{1+2x^2}\,dx$$

Combining the result in (b) with linearity, we obtain

$$\int x\sqrt{1+2x^2} = \tfrac{1}{4}\left(\tfrac{2}{3}(1+2x^2)^{3/2}\right) + C = \tfrac{1}{6}(1+2x^2)^{3/2} + C$$

We conclude this section with an area problem which we can solve using the formula for $\int \cos^2 x\,dx$.

Example 6 Compute the area under the cycloid parametrized by

$$x = \phi - \sin\phi, \qquad y = 1 - \cos\phi, \qquad 0 \leqslant \phi \leqslant 2\pi$$

(see Figure 6-1).

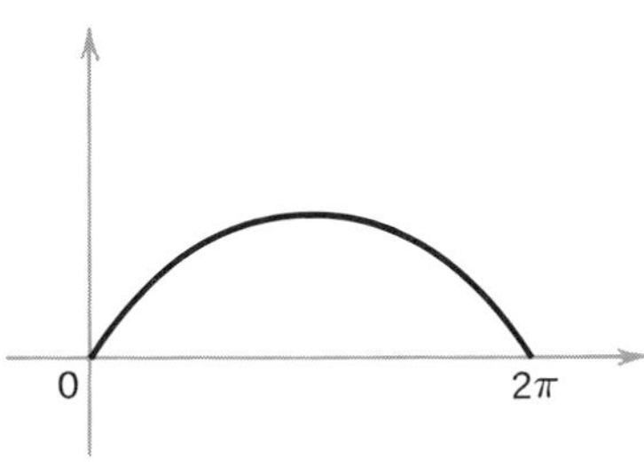

FIGURE 6-1

The cycloid is the graph of a continuous function $f: x \mapsto f(x)$ whose domain is $[0, 2\pi]$. So the area sought is equal to

$$\int_0^{2\pi} f(x)\,dx$$

Using the change of variable $x = \phi - \sin\phi$, we have $f(x) = 1 - \cos\phi$. So the area is

$$\int_0^{2\pi} f(x)\,dx = \int_0^{2\pi} f(x)\frac{dx}{d\phi}\,d\phi = \int_0^{2\pi} (1-\cos\phi)^2\,d\phi$$

We expand the integrand on the right and we find

$$\begin{aligned}\int_0^{2\pi} (1-\cos\phi)^2\,d\phi &= \int_0^{2\pi} (1 - 2\cos\phi + \cos^2\phi)\,d\phi \\ &= \left[\phi - 2\sin\phi + \frac{1}{2}\left(\phi - \frac{\sin 2\phi}{2}\right)\right]_0^{2\pi} \\ &= 2\pi + \pi = 3\pi\end{aligned}$$

EXERCISES

Evaluate the following indefinite integrals.

1. $\int 3x\,dx$
2. $\int 12(x^3 - x^2 + x - 1)\,dx$
3. $\int (-10x^2 + 1)\,dx$
4. $\int (3000x^{14} + 0.5x^3)\,dx$
5. $\int (x+10)^2\,dx$
6. $\int (2x^4 - x^3 + 10x^2 + x - 1)\,dx$
7. $\int (5x^2 + x + 1)^3\,dx$
8. $\int 45x^4\,dx$
9. $\int \frac{-1}{25}x^3\,dx$
10. $\int \cos x \sin x\,dx$
11. $\int (x+10)^9\,dx$
12. $\int \sin^4 x \cos x\,dx$
13. $\int \sin(3x+4)\,dx$
14. $\int \frac{x}{\sqrt{x^2+1}}\,dx$

In exercises 15–18 find the volume of the solid of revolution generated by revolving the given curve about the given axis.

15. $y = \sin x$ from 0 to π about the x axis
16. $y = \cos x$ from π to 2π about the x axis
17. $y = \sqrt{x}$ from 0 to 1 about the x axis
18. $y = \sqrt{1+x^2}$ from 1 to 2 about the y axis

19. Find the area between $y = \cos^2 x$ and $y = -\sin^2 x$ from 0 to 2π.
20. The graph of $x \mapsto \sqrt{r^2 - x^2}$ is a semicircle of radius r. Compute $\int_{-r}^{r} \sqrt{r^2 - x^2}\, dx$ by substituting $x = \cos\theta$.
21. The equations

$$x = a\cos t, \qquad y = b\sin t$$
$$0 \leqslant t \leqslant 2\pi, \qquad a > 0, \quad b > 0$$

describe an ellipse with semi-axes of length a and b. Find the area enclosed by the ellipse.

REVIEW EXERCISES FOR CHAPTER 6

1. Define the following terms.

antiderivative	indefinite integral
average value of a function	integrand
definite integral	limit of integration

In exercises 2–21, evaluate the given expression.

2. $\int (x^2 - 2x + 4x - 1)\, dx$
3. $\int_0^{\pi} \sin(\cos x)\sin x\, dx$
4. $\int_3^2 x^3\, dx$
5. $\dfrac{d(\int_a^{x^2} t^2\, dt)}{dx}$
6. $\int (x+1)(2x-1)\, dx$
7. $\int_a^b (\sin t - \cos t)\, dt$
8. $\int_a^b (\sin t - \cos t)\, dx$
9. $\int (3x^2 - 1)\sqrt{x^3 - x + 2}\, dx$
10. $\int x^2(x^3-1)^{1/2}\, dx$
11. $\int \sin^4 3x \cos 3x\, dx$
12. $\int \dfrac{x^3}{(x^4+1)^3}\, dx$
13. $\int_0^0 \dfrac{x-3}{x^2 - 6x + 7}\, dx$
14. $\int (\sin x)^{1/2} \cos x\, dx$
15. $\int_0^{\sqrt[3]{\pi}} x^2 \cos x^3\, dx$
16. $\int_8^7 \dfrac{3x^2 + 6}{(x^3 + 6x)^7}\, dx$
17. $\int_{1/4\pi}^{1/3\pi} \sin x \cos x\, dx$
18. $\int_4^2 [(x+3)^5 - 7(x+3)^4 + 3(x+3)^2 + 2]\, dx$
19. $\int_0^{\sqrt{t^2+1}} (1-u^2)^{-1/2}\, du$
20. $\dfrac{d(\int_3^{u^2} \sin\sqrt{t}\, dt)}{du}$
21. $\int_a^b \left(\dfrac{d(\int_0^x t^2\, dt)}{dx}\right) dx$

What is the domain of the function defined by each of the following rules? Where is the function differentiable?

22. $x \mapsto \int_0^{\sin x} \cos t\, dt$

23. $x \mapsto \int_0^{x^2} \tan u\, du$

24. $x \mapsto \int_1^{x^2+1} \frac{1}{u^2}\, du$

25. $x \mapsto \frac{d(\int_0^x |t|\, dt)}{dx}$

26. The function $F: x \mapsto \int_1^x 1/t\, dt$ has domain $]0, +\infty[$ and its derivative has rule $x \mapsto 1/x$. Use these facts to show that $F(at) = F(a) + F(t)$ for all positive real numbers a and t. Show that the only differentiable functions satisfying $f(ax) = f(x) + f(a)$ for all positive reals a and x are the constant multiples of F.

27. Find the area enclosed by the figure eight whose equation is $y^2 = x^2 - x^4$.

28. Let f be continuous on $[a, b]$. Show that for any $c \in\,]a, b[$, the average value of f on $[a, b]$ is the "weighted average" of the average values of f on $[a, c]$ and $[c, b]$, that is

$$av_{[a,b]} = \frac{c-a}{b-a} av_{[a,c]} + \frac{b-c}{b-a} av_{[c,b]}$$

In exercises 29–35, find the volume generated by revolving the region bounded by the indicated graphs about the x axis.

29. $y = 4/(x+1),\ x = -5,\ x = -2,\ y = 0$

30. $y = \tan x,\ x = \frac{1}{3}\pi,\ y = 0$

31. $y = x,\ y = x^3$

32. $y = x + 2,\ y^2 - 3y = 2x$

33. $x = 0,\ y = 0,\ 2x + 3y = 1$

34. $y = 2,\ x = 0,\ x = y^2$

35. $y = 0,\ y = 4 - x^2$

In exercises 36–40, find the volume generated by revolving the region bounded by the indicated graphs about the y axis.

36. $y = 2,\ y = x^2 + 1,\ x = 0$

37. $y = x^3 + x,\ x = 1,\ x = 0$

38. $y = x^3 + x,\ x = 1,\ x = a\ (a > 1)$

39. $y = x^5,\ x = -1,\ x = -2,\ y = -1$

40. $y = \sin(x^2),\ y = 0,\ x = 0,\ x = \sqrt{\pi}$

In exercises 41–44, calculate the area between the graphs of the given functions on the given intervals.

41. $x \mapsto 60(x^5 - x^4 + x^3),\ x \mapsto -2x;\ [-1, 1]$

42. $x \mapsto \sin x,\ x \mapsto \cos x;\ [0, 2\pi]$

43. $x \mapsto -\cos(\pi x),\ x \mapsto x^2 - 1;\ [-1, 1]$

44. $x \mapsto x^2 - x,\ x \mapsto x - x^2;\ [-1, 2]$

45. Use the definite integral to show that for every natural number n there is a function which is n times continuously differentiable but is not $n+1$ times differentiable at zero.

7

APPLICATIONS OF THE DEFINITE INTEGRAL

We begin this chapter with a discussion of sequences; this is done to develop ideas which are necessary for studying the next important topic of integration, Riemann sums. Using Riemann sums, we are able to formulate rigorous definitions of area and the definite integral, which can then be applied to yield formulas for arc length, surface area, and work. The final section of this chapter is devoted to improper integrals.

1 SEQUENCES OF REAL NUMBERS

We think of a **sequence**, informally, as a list of objects given in a definite order. A sequence is in fact a particular kind of function.

Definition A **sequence** s is a real-valued function whose domain is the positive integers. The number $s(n)$ is denoted by s_n and is called the **nth term** of the sequence.

We often describe a sequence s by displaying its first few terms, for example, $s_1, s_2, \ldots, s_n, \ldots$, which we abbreviate $\{s_n\}$. This can be done if it is clear what the general rule is. Graphically, a sequence may be thought of as the set of points of the graph of the form $(n, s(n))$ (Figure 1-1).

No doubt the reader is already familiar with certain kinds of sequences. One very simple example is an **arithmetic progression**, a sequence such as $1, 3, 5, 7, 9, \ldots$. This sequence $\{s_n\}$ is defined by $s_n = 2n - 1$. Thus $s_{20} = 2 \cdot 20 - 1 = 39$.

Another common type of sequence is a geometric progression, such as $1, 2, 4, 8, 16, \ldots$. This sequence is defined by $r_n = 2^{n-1}$. Thus $r_{20} = 2^{20-1} = 2^{19}$. Other examples of sequences follow.

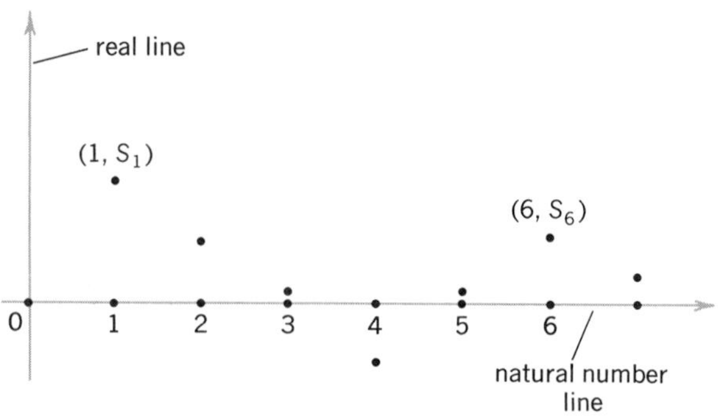

FIGURE 1-1

Example 1 Let us define $\{s_n\}$ by $s_n = (n+1)/n$, $\{r_n\}$ by $r_n = n^{1/n}$, and $\{t_n\}$ by $t_n = (1+1/n)^n$. Then we have

$$\{s_n\} = 2, 1\tfrac{1}{2}, 1\tfrac{1}{3}, 1\tfrac{1}{4}, \ldots$$

$$\{r_n\} = 1, \sqrt{2}, \sqrt[3]{3}, \sqrt[4]{4}, \ldots$$

$$\{t_n\} = 2, \tfrac{9}{4}, \tfrac{64}{27}, \tfrac{625}{256}, \ldots$$

Since sequences are real-valued functions, we can define addition, scalar multiplication, quotients, and products of sequences in the same manner as these operations are normally defined for functions. For example, the nth term of the sum of two sequences is the sum of the two nth terms.

Example 2 Define $\{s_n\}$ by $s_n = 2n$, $\{r_n\}$ by $r_n = n^2$. Then,

$$\{s_n\} = 2, 4, 6, 8, \ldots$$

$$\{r_n\} = 1, 4, 9, 16, \ldots$$

We now have

$$(r+s)_n = r_n + s_n = n^2 + 2n$$

$$(rs)_n = r_n s_n = n^2 \cdot 2n = 2n^3$$

$$(cr)_n = c(r_n) = cn^2, \qquad c \text{ is any real number}$$

$$\left(\frac{r}{s}\right)_n = \frac{r_n}{s_n} = \frac{n}{2}$$

and hence

$$\{(r+s)_n\} = 3, 8, 15, 24, \ldots$$

$$\{(rs)_n\} = 2, 16, 54, 128, \ldots$$

$$\{(cr)_n\} = c, 4c, 9c, 16c, \ldots$$

$$\left\{\left(\frac{r}{s}\right)_n\right\} = \tfrac{1}{2}, 1, \tfrac{3}{2}, 2, \ldots$$

One of the goals in studying sequences is to determine how $\{s_n\}$ behaves as n tends toward infinity.

Example 3 The sequence $\{s_n\}$ given by $s_n = 2n^2$ tends towards infinity as n increases. The sequence $\{r_n\}$ defined by $r_n = (1/3n)$ approaches zero as n becomes large, while the sequence $\{t_n\}$ given by $t_n = (-1)^n$ merely oscillates between $+1$ and -1 as n increases.

A sequence $\{s_n\}$ is said to **converge** to a number α if the values s_n become arbitrarily close to α as n gets large, and we write $\lim_{n\to\infty} s_n = \alpha$. Let us define the concept of convergence formally.

Definition A sequence $\{s_n\}$ is **eventually in a neighborhood** M if there is a number N such that $s_n \in M$ for all $n > N$. A sequence $\{s_n\}$ **converges to a real number** α if $\{s_n\}$ is eventually in every neighborhood of α. We then write $\lim_{n\to\infty} s_n = \alpha$ (or $s_n \to \alpha$ as $n \to \infty$). A sequence is **convergent** if it converges to some α. If a sequence does not converge, it is said to be **divergent**.

The following theorem follows easily from the definition of limit of a sequence.

Theorem 1 A sequence $\{s_n\}$ has a finite limit α if and only if given any $\varepsilon > 0$, there is a $N \in N$ such that

$$|s_n - \alpha| < \varepsilon, \text{ for all } n > N$$

Geometrically we can illustrate the concept of limit in the following way. Looking at Figure 1-2, we observe that as n becomes large the points representing the sequence seem to cluster about the line $y = \alpha$.

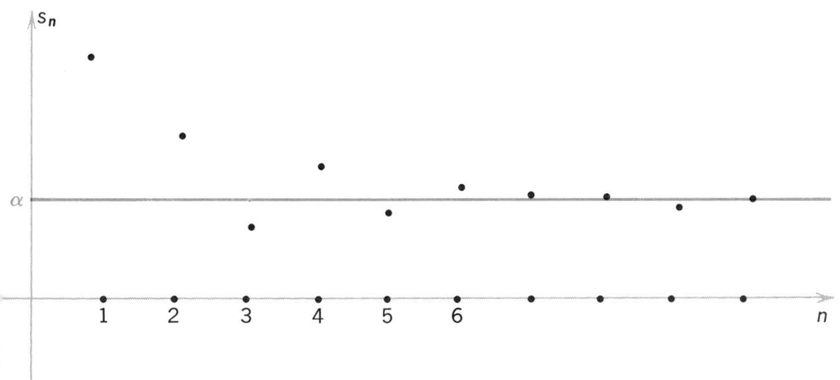

FIGURE 1-2

Example 4 (a) Let us verify that $\lim_{n\to\infty} 1/n = 0$. If $]A, B[$ is a neighborhood of 0, we choose N such that $N > 1/B$. Then for all $n > N$, we have $0 < 1/n < 1/N < B$ and so $\{1/n\}$ is eventually in $]A, B[$ as $n \to \infty$.

(b) Let f be a function which is continuous on $[0, 1]$. Define a sequence $\{s_n\}$ by $s_n = f(1/n)$. Let us show that $\lim_{n\to\infty} s_n = f(0)$. By hypothesis, we have $\lim_{x\to 0^+} f(x) = f(0)$. Therefore, for any neighborhood $]A, B[$ of $f(0)$ we know that there is an interval $[0, b]$ such that $x \in [0, b]$ implies $f(x) \in]A, B[$. So we choose N such that $1/N < b$. For $n > N$, we have $0 < 1/n < 1/N < b$ and therefore $s_n = f(1/n)$ is in $]A, B[$. Hence s_n is eventually in $]A, B[$ as $n \to \infty$. By way of example, $f: x \mapsto \sqrt{x}$ is continuous on $[0, 1]$ and so $\lim_{n\to\infty} \sqrt{1/n} = f(0) = 0$.

(c) If $\lim_{x\to\infty} f(x) = \alpha$, then $\lim_{n\to\infty} f(n) = \alpha$. (The student should have no difficulty verifying this.) For example, $\lim_{x\to\infty} 1/x^2 = 0$ and so $\lim_{n\to\infty} 1/n^2 = 0$. Also, by L'Hôpital's Rule, $\lim_{x\to\infty} 3x^3/(x^3 + 10x) = 3$, so $\lim_{n\to\infty} 3n^3/n^3 + 10n = 3$.

It is clear from the concept of eventuality that convergence is a property of the "last terms" of a sequence. In other words, the alteration of a finite number of terms of a sequence does not affect its convergence or divergence. Thus, we can explore the convergence of $a_1, a_2, a_3, \ldots, a_n, \ldots$ by considering $a_4, a_5, \ldots, a_n, \ldots$, or $a_7, a_8, \ldots, a_n, \ldots$.

Example 5 (a) Consider the sequence $\{s_n\}$ with rule $s_n = 1/2^{n-1}$. Then

$$s_n = 1, \tfrac{1}{2}, \tfrac{1}{4}, \tfrac{1}{8}, \ldots$$

Observe that whether one considers $\frac{1}{2}, \frac{1}{4}, \frac{1}{8}, \ldots$, or $\frac{1}{16}, \frac{1}{32}, \ldots$ the behavior is the same: $\lim_{n\to\infty} s_n = 0$.

(b) Consider the sequence $\{s_n\}$ given by $s_n = n^2$:

$$\{s_n\} = 1, 4, 9, 16, \ldots$$

It does not matter whether one considers $4, 9, 16, \ldots$ or $9, 16, 25, \ldots$ since the behavior is the same: The sequence diverges.

When for every real number B, the values s_n are eventually all greater than B [as in Example 5(b)], we say

$$\lim_{n\to\infty} s_n = +\infty$$

Formally, $\lim_{n\to\infty} s_n = +\infty$ if, for every real number B, there exists an $N \in N$ such that $s_n > B$ for every $n > N$. We define $\lim_{n\to\infty} s_n = -\infty$ in the same way, except the inequality $s_n > B$ is reversed.

It should be pointed out, however, that a sequence may not be convergent and yet not approach $\pm\infty$.

Example 6 Consider the sequence $\{s_n\}$ defined by $s_n = (-1)^n$. Then $\{s_n\} = -1, 1, -1, \ldots$. This sequence does not converge to a limit nor do we have $\lim_{n\to\infty} s_n = +\infty$ (or $-\infty$).

Let us next consider certain properties of limits of sequences. Since sequences are functions, these properties may be obtained from the corresponding properties of limits of functions.

Theorem 2 If $\lim_{n\to\infty} s_n = \alpha$, $\lim_{n\to\infty} r_n = \beta$, then

(i) $\lim_{n\to\infty} (s+r)_n = \alpha + \beta$

(ii) $\lim_{n\to\infty} (sr)_n = \alpha\beta$

(iii) for c a real number,

$$\lim_{n\to\infty} (cs_n) = c\alpha$$

(iv) $\lim_{n\to\infty} \left(\frac{s}{r}\right)_n = \frac{\alpha}{\beta}, \qquad r_n \neq 0 \text{ for all } n, \qquad \beta \neq 0$

Proof Each of these proofs is perfectly analogous to the corresponding proof for the limits of functions. In fact, the only real changes that need to be made in the wording of the corresponding proofs is that "there is a neighborhood M of x_0" is systematically replaced by "there is a natural number N" and "for all x in $M(x_0)$" by "for all $n > N$." ▨

Example 7 We list some limits which can be obtained by combining the above theorem with the fact that $\lim_{n\to\infty} 1/n = 0$:

(a) $\lim_{n\to\infty} \frac{2}{n} = 0$

(b) $\lim_{n\to\infty} \left(1 - \frac{7}{n}\right) = 1$

(c) $\lim_{n\to\infty} \left(1 + \frac{1}{n} + \frac{1}{n^2}\right) = 1$

(d) $\lim_{n\to\infty} \frac{1+n+n^2}{n^2} = 1 \qquad \left(\text{Note that } \frac{1+n+n^2}{n^2} = \frac{1}{n^2} + \frac{1}{n} + 1.\right)$

(e) $\lim_{n\to\infty} \frac{2A}{n^3} = 0, \qquad A$ a constant

The reader can also check the following limits:

(f) $\lim_{n\to\infty} \frac{1}{2^n} = 0$

(g) $\lim_{n\to\infty} \frac{\sqrt{n}+1}{n+10} = 0,$

(h) $\lim_{n\to\infty} \frac{(-1)^n}{n} = 0 \qquad$ [Note that $\lim_{n\to\infty} (-1)^n$ does not exist.]

(i) $\lim_{n\to\infty} \left(1 + \frac{1}{2^n}\right)\left(2 + \frac{1}{2^n}\right) = 2$

An important class of sequences of real numbers are those which are monotone.

Definition A sequence $\{s_n\}$ is **monotone increasing** if $n > m$ implies $s_n \geqslant s_m$; $\{s_n\}$ is **monotone decreasing** if $n > m$ implies $s_n \leqslant s_m$. A sequence $\{s_n\}$ is **bounded** if there is a number $K > 0$ such that $|s_n| \leqslant K$ for all n.

Note that we have used $s_n \geqslant s_m$, $s_n \leqslant s_m$. By our definition, a sequence does not have to be strictly increasing to be considered monotone increasing. Thus, $1, 1, 2, 2, 3, 3, \ldots$ is a monotone increasing sequence. Similarly, $-1, -1, -2, -2, -3, -3, \ldots$ is a monotone decreasing sequence.

If we compare this definition with that for monotone functions (§1.3) we see that there we meant strictly increasing (or decreasing). Because of this standard ambiguity of terminology, mathematicians sometimes add additional phrases to make their meanings more precise. For example, if $s_n > s_m$ when $n > m$ we might say $\{s_n\}$ is strictly monotone increasing, while if $s_n \geqslant s_m$ we might say $\{s_n\}$ is monotone nondecreasing. However, in this chapter let us use the stated definition, as it is most convenient for present purposes.

Example 8 The sequence $\frac{1}{2}, \frac{2}{3}, \frac{3}{4}, \ldots, n/(n+1), \ldots$ is monotone increasing and bounded (let $K = 1$). The sequence $1, \frac{1}{2}, \frac{1}{3}, \frac{1}{4}, \ldots, 1/n, \ldots$ is monotone decreasing and bounded. The sequence $1, 4, 9, 16, \ldots, n^2, \ldots$ is monotone increasing and unbounded.

It follows at once that every convergent sequence is bounded. Although the converse of this is false (there are bounded divergent sequences—see Example 6). What we do expect to be true is that bounded monotone sequences converge, since the terms cannot grow arbitrarily and they steadily move in one direction. This is proved in the following theorem.

Theorem 3 Monotone Convergence Theorem If $\{s_n\}$ is a bounded monotone sequence, then $\{s_n\}$ converges.

Proof We prove the theorem for the case where $\{s_n\}$ is increasing, and leave the formulation of the decreasing case as an exercise. The set of numbers $\{s_1, s_2, \ldots\}$ has an upper bound since it is bounded, and therefore it has a least upper bound. Let $\alpha = \text{l.u.b.}\,\{s_n\}$. We will show that $\lim_{n\to\infty} s_n = \alpha$.

Let $]A, B[$ be a neighborhood of α. We always have $s_n \leqslant \alpha < B$, so to show that $\{s_n\}$ is eventually in $]A, B[$ as $n \to \infty$, we have only to show that there is an N such that $n > N$ implies $A < s_n$. But $A < \alpha$ and since α is the l.u.b. of $\{s_n\}$, there must be some n_0 such that $A < s_{n_0} \leqslant \alpha$ (otherwise α would not be the **least** upper bound of the sequence). We now set $N = n_0$. Since $\{s_n\}$ is monotone increasing, for all $n > N$ we have $A < s_N \leqslant s_n < \alpha$. ∎

The following theorem is important in the next section. It states that if a sequence $\{R_n\}$ is "hemmed in" between two sequences converging to the same limit, then all three sequences converge to the same limit.

Theorem 4 Squeezing Lemma Suppose sequences $\{s_n\}$, $\{R_n\}$, and $\{S_n\}$ have the following properties: (i) $s_n \leqslant R_n \leqslant S_n$ for all n; (ii) $\{s_n\}$ is bounded and monotone increasing, and $\{S_n\}$ is bounded and monotone decreasing; (iii) the sequence $\{(S-s)_n\}$ converges to 0. Then $\{s_n\}$, $\{R_n\}$, and $\{S_n\}$ all converge to the same limit.

Proof The proof is in two parts; first we show $\lim_{n\to\infty} s_n = \lim_{n\to\infty} S_n$, and then we prove $\lim_{n\to\infty} s_n = \lim_{n\to\infty} R_n$. Then to begin, let $\alpha = \text{l.u.b. } \{s_n\}$, $\beta = \text{g.l.b. } \{S_n\}$ (see exercise 7, §4.6 for definition of g.l.b. = greatest lower bound). Then by the Monotone Convergence Theorem, $\lim_{n\to\infty} s_n = \alpha$ and $\lim_{n\to\infty} S_n = \beta$. We claim that $\alpha = \beta$. For if $\alpha \neq \beta$, either $\beta > \alpha$ or $\alpha > \beta$. Assume $\beta > \alpha$. Then $S_n \geqslant \beta$ for all n, and $s_n \leqslant \alpha$ for all n, so $S_n - s_n \geqslant \beta - \alpha > 0$ for all n, contradicting assumption (iii) of the hypothesis. On the other hand, if $\beta < \alpha$, then there must be a $k \in N$ with $S_k < \alpha$ (otherwise β could not be the g.l.b. of $\{S_n\}$), and a $j \in N$ with $S_k < s_j < \alpha$ (since α is the l.u.b. of $\{s_n\}$). Then by the monotonicity assumption, $S_n \leqslant S_k$ for all $n \geqslant k$, and $s_n \geqslant s_j$ for all $n \geqslant j$, so

$$s_n - S_n > s_j - S_k > 0 \qquad n \geqslant \max(k, j)$$

which contradicts assumption (ii) of the hypothesis. Therefore $\alpha = \beta$, and $\lim_{n\to\infty} s_n = \alpha = \lim_{n\to\infty} S_n$.

For the second part let $\varepsilon > 0$ be given. Then there is an N such that for $n > N$, we have $0 \leqslant \alpha - s_n < \varepsilon$, $0 \leqslant S_n - \alpha < \varepsilon$. But from assumption (i), which is that $s_n \leqslant R_n \leqslant S_n$ for all n, we conclude that $-\varepsilon < s_n - \alpha \leqslant R_n - \alpha \leqslant S_n - \alpha < \varepsilon$. Therefore $|R_n - \alpha| < \varepsilon$ for all $n > N$, which implies that $\lim_{n\to\infty} R_n = \alpha$. ∎

Example 9 Let

$$s_n = 1 - \frac{1}{n}, \qquad R_n = \frac{\sqrt{n^2 - n + 1}}{n}, \qquad S_n = 1 + \frac{1}{n}$$

Clearly, S_n is decreasing, s_n is increasing, and $S_n - s_n = 2/n$ converges to 0. Moreover,

$$R_n = \sqrt{1 - \frac{1}{n} + \frac{1}{n^2}} < \sqrt{1 + \frac{2}{n} + \frac{1}{n^2}} = \sqrt{\left(1 + \frac{1}{n}\right)^2} = 1 + \frac{1}{n} = S_n$$

Also,

$$R_n > \sqrt{1 - \frac{2}{n} + \frac{1}{n^2}} = \sqrt{\left(1 - \frac{1}{n}\right)^2} = 1 - \frac{1}{n} = s_n$$

Thus, $S_n > R_n > s_n$ for all n and by the Squeezing Lemma,

$$\lim_{n\to\infty} R_n = \lim_{n\to\infty} s_n = 1$$

Suppose we take the terms of a sequence of real numbers and insert + signs between them, thus forming a *sum*. For example, the sum of the first one hundred integers is written

$$1 + 2 + 3 + \cdots + 99 + 100$$

If N is a natural number and if $s_1, s_2, \ldots, s_N$ are real numbers, then the sum $s_1 + s_2 + \cdots + s_N$ can be abbreviated by writing

$$s_1 + s_2 + \cdots + s_N = \sum_{k=1}^{N} s_k$$

using the Greek capital sigma for **sum**. This symbol is read "the sum of s_k for $k = 1, 2, \ldots, N$." The k appears because 1 and N are fixed and k is thought of as being in turn each natural number from 1 to N (here, k is called the **index**, and the letter used as the index is always arbitrary).

Example 10 (a) If $s_1 = 1, s_2 = 2, s_3 = 3, s_1 + s_2 + s_3 = \sum_{k=1}^{3} s_k = 1+2+3$. More generally, if $s_k = k$, we have $\sum_{k=1}^{N} s_k = 1+2+3+\cdots+N$. Similarly, if $s_k = k^2$, we have $\sum_{k=1}^{N} k^2 = 1+4+9+\cdots+N^2$. If $s_k = 1$ for all k, then

$$\sum_{k=1}^{N} s_k = \sum_{k=1}^{N} 1 = \underbrace{1 + 1 + \cdots + 1}_{N \text{ times}} = N$$

(b) As further examples of this notation, we have formulas such as the following:

$$\sum_{n=1}^{6} n^2 = 1 + 4 + 9 + \cdots + 36$$

$$\sum_{n=1}^{N} t_n = t_1 + t_2 + \cdots + t_N$$

$$\sum_{i=1}^{n} t_i = t_1 + t_2 + \cdots + t_n$$

$$\sum_{i=1}^{3} i^2 = 1 + 4 + 9$$

$$\sum_{j=1}^{4} j^3 = 1 + 8 + 27 + 64$$

$$\sum_{n=1}^{3} \sin(n\pi) = 0 + 0 + 0 = 0$$

In the next section, we shall need the formulas

$$(1)\quad \sum_{k=1}^{N} k = 1 + 2 + \cdots + N = \frac{N(N+1)}{2}$$

$$(2)\quad \sum_{k=1}^{N} k^2 = 1 + 2^2 + \cdots + N^2 = \frac{N(N+1)(2N+1)}{6}$$

ʌs an example, we use (1) to find

$$\sum_{k=1}^{5} k = \frac{5 \cdot 6}{2} = 15 = 1 + 2 + 3 + 4 + 5$$

ʒy (2), we have

$$\sum_{k=1}^{3} k^2 = \frac{3 \cdot 4 \cdot 7}{6} = 14 = 1 + 4 + 9$$

These formulas can be established using the principle of mathematical ıduction. However, we shall obtain (1) and (2) by ad hoc methods.

'roof of (1) We have

$$\sum_{k=1}^{N} k = 1 + 2 + \cdots + (N-1) + N$$

ɴd, backwards,

$$\sum_{k=1}^{N} k = N + (N-1) + \cdots + 1$$

.dding the two rows term by term, we obtain

$$2\sum_{k=1}^{N} k = \underbrace{(N+1) + (N+1) + \cdots + (N+1)}_{N \text{ times}}$$

$$= N(N+1)$$

Ience,

$$\sum_{k=1}^{N} k = \frac{N(N+1)}{2}$$

roof of (2) If k is a real number, we have

$$k^3 - (k-1)^3 = k^3 - (k^3 - 3k^2 + 3k - 1) = 3k^2 - 3k + 1$$

eplacing k in the above expression successively by $1, 2, \ldots, N$ gives N ղuations:

$$\begin{aligned}
1^3 - 0^3 &= 3 \cdot 1^2 - 3 \cdot 1 + 1 \\
2^3 - 1^3 &= 3 \cdot 2^2 - 3 \cdot 2 + 1 \\
3^3 - 2^3 &= 3 \cdot 3^2 - 3 \cdot 3 + 1 \\
&\vdots \\
(N-1)^3 - (N-2)^3 &= 3(N-1)^2 - 3(N-1) + 1 \\
N^3 - (N-1)^3 &= 3N^2 - 3N + 1
\end{aligned}$$

we add and notice how the terms on the left cancel, we obtain

$$\begin{aligned}
N^3 &= 3[1^2 + 2^2 + \cdots + N^2] - 3[1 + 2 + \cdots + N] + N \\
&= 3\sum_{k=1}^{N} k^2 - 3\sum_{k=1}^{N} k + N
\end{aligned}$$

So

$$3\sum_{k=1}^{N} k^2 = N^3 + 3\sum_{k=1}^{N} k - N$$

$$= N^3 + 3\frac{N(N+1)}{2} - N$$

$$= \frac{2N^3+3N^2+3N-2N}{2}$$

$$= \frac{N}{2}(N^2+3N+1)$$

$$= \frac{N(N+1)(2N+1)}{2}$$

We can use the sigma notation to define sequences. For exampl the equation

$$s_n = \sum_{k=1}^{n} k$$

defines the sequence

$$s_n = \frac{n(n+1)}{2}$$

Similarly, the equation

$$r_n = 6\sum_{k=1}^{n} k^2$$

defines the sequence $r_n = n(n+1)(2n+1)$.

Example 11 (a) Evaluate $\lim_{n\to\infty}(\sum_{k=1}^{n} k)/n^2$. Using formula (1), w have

$$\frac{\sum_{k=1}^{n} k}{n^2} = \frac{n(n+1)}{2n^2} = \frac{n+1}{2n} = \frac{1}{2} + \frac{1}{2n}$$

Thus,

$$\lim_{n\to\infty} \frac{\sum_{k=1}^{n} k}{n^2} = \lim_{n\to\infty}\left(\frac{1}{2}+\frac{1}{2n}\right) = \frac{1}{2}$$

(b) Let $s_n = (1/n^3)\sum_{k=1}^{n} k^2$ and evaluate $\lim_{n\to\infty} s_n$. We have

$$s_n = \frac{1}{n^3}\left(\frac{n}{6}\right)(n+1)(2n+1) = \frac{1}{6}\left(\frac{2n^2+3n+1}{n^2}\right) = \frac{1}{6}\left(2+\frac{3}{n}+\frac{1}{n^2}\right)$$

We see that $\lim_{n\to\infty} s_n = \frac{1}{3}$.

EXERCISES

1. Prove that the limit of a convergent sequence is unique, that is, if $\lim_{n\to\infty} s_n = s$, $\lim_{n\to\infty} s_n = t$, then $s = t$.

2. Prove Theorem 1. Prove Theorem 2(i).

3. In each of the following, the nth element s_n of a sequence is given. Write down the first five numbers in the sequence, and then find $\lim_{n\to\infty} s_n$ if it exists.

   (a) $s_n = \dfrac{7}{n}$ (b) $s_n = \dfrac{3n}{n+5}$

   (c) $s_n = \dfrac{n^2-1}{n^2+1}$ (d) $s_n = \dfrac{(-1)^n}{n^2}$

   (e) $s_n = \dfrac{n^2}{(n+1)^2}$ (f) $s_n = (-1)^n\dfrac{n+1}{n^2}$

   (g) $s_n = r^n$, $|r| < 1$ (h) $s_n = r^n$, $r > 1$

4. Find the limits of the sequences defined by the following:

   (a) $s_n = \dfrac{n^2+5n-2}{2n^2}$ (b) $s_n = \dfrac{3n}{n+2} - \dfrac{n+5}{3n}$

   (c) $s_n = n\sin\dfrac{1}{n}$ (d) $s_n = \dfrac{1+(-1)^n}{n}$

   (e) $s_n = e^{1/n}$ (f) $s_n = \sqrt{n} - \sqrt{n+1}$

   (g) $s_n = \dfrac{\sqrt{n+1}}{\sqrt{2n+1}}$ (h) $s_n = \dfrac{6n^3-2n}{n^4+3n}$

5. Determine if each sequence defined below converges, and if it does find the limit.

   (a) $s_n = 1 + (-1)^n$ (b) $s_n = \cos n$

   (c) $s_n = \begin{cases} 1 + \dfrac{1}{n}, & n \text{ a positive even integer} \\ 1, & n \text{ a positive odd integer} \end{cases}$

   (d) $s_n = 2^n$ (e) $s_n = 1 + \dfrac{(-1)^n}{n}$

6. Show that the sequences defined by

   (a) $s_n = 1 - \dfrac{1}{n}$

   (b) $s_n = \dfrac{1\cdot3\cdot5\cdot7\cdots(2n-1)}{2\cdot4\cdot6\cdot8\cdots(2n)}$

   are convergent without computing their limits.

7. Using the definition of a limit of a sequence, prove that if $\{s_n\}$ is a sequence of nonzero terms and $\lim_{n\to\infty} s_n = \infty$, then

$$\lim_{n\to\infty} \frac{1}{s_n} = 0$$

8. Suppose $\lim_{n\to\infty} r_n = \alpha = \lim_{n\to\infty} s_n$. Define a sequence $\{u_n\}$ by

$$u_n = \begin{cases} r_n, & n \text{ odd} \\ s_n, & n \text{ even} \end{cases}$$

Show that $\lim_{n\to\infty} u_n = \alpha$. Show also that the sequence $\{v_n\}$ defined by

$$v_n = \begin{cases} r_n, & n \text{ a multiple of } 7 \\ s_n, & n \text{ not a multiple of } 7 \end{cases}$$

also converges to α.

9. Suppose that $f(x)$ is defined for all x in $[0, 1]$, $f(0) = 0$, and that f is differentiable at 0. Define a sequence $\{s_n\}$ by $s_n = nf(1/n)$. Prove that

$$\lim_{n\to\infty} s_n = f'(0)$$

10. Using the result of exercise 9, evaluate the following limits: (a) $\lim_{n\to\infty} n \sin(1/n)$; (b) $\lim_{n\to\infty} n \tan(1/n)$; (c) $\lim_{n\to\infty} (n \cos(1/n) - n)$.

11. Suppose f is continuous and monotone increasing on $]A, B[$ and that $f(x) \leqslant M$ for all x in $]A, B[$. Define the sequence $\{r_n\}$ by $r_n = f(B - 1/n)$. (a) Show that $\{r_n\}$ converges. (b) Let $\alpha = \lim_{n\to\infty} r_n$; show that $\lim_{x\to B^-} f(x) = \alpha$.

12. Suppose f is continuous and monotone increasing on $]B, C[$ and that $M \leqslant f(x)$ for all x in $]B, C[$. Consider $s_n = f(B + 1/n)$. (a) Show that $\{s_n\}$ converges. (b) Show that $\lim_{x\to B^+} f(x) = \lim_{n\to\infty} s_n$.

13. Prove that if f is monotone increasing on $]A, C[$ and continuous on $]A, B[$ and $]B, C[$, then either f is continuous at B or f has a jump discontinuity at B.

14. Prove that $\lim_{n\to\infty} s_n = 0$ if and only if $\lim_{n\to\infty} |s_n| = 0$.

15. Prove that $\lim_{n\to\infty} s_n = 0$ if and only if for every $\varepsilon > 0$ there is an N such that $n > N$ implies $|s_n| < \varepsilon$.

16. Find the sum of the first one hundred numbers. Find the sum of the squares of the first one hundred numbers.

17. Prove the following:

(a) $\sum_{i=1}^{n} (u_i + v_i) = \sum_{i=1}^{n} u_i + \sum_{i=1}^{n} v_i$ (b) $\sum_{i=1}^{n} cu_i = c \sum_{i=1}^{n} u_i$

(c) $\left|\sum_{i=1}^{n} u_i\right| \leqslant \sum_{i=1}^{n} |u_i|$

18. Show that for $f: \boldsymbol{R} \to \boldsymbol{R}$, if $f(n) \to 0$ then $\lim_{x\to\infty} f(x)$ need not exist. [Hint: Consider $f: x \mapsto \sin \pi x$.]

2 RIEMANN SUMS

In this section we turn to the foundations of integral calculus and define the definite integral rigorously as a limit of sequences of approximating sums known as **Riemann sums**. We also treat the problem of assigning area to regions bounded by the graphs of continuous functions. This time we base our definition of area on the definite integral and use the Principle of Exhaustion as a justification for doing this.

In general the Principle of Exhaustion is the following. Given a region R in the plane, consider a sequence of polygons inscribed in R, say $\{P_n\}$ and another sequence of circumscribed polygons $\{Q_n\}$. If the areas of $\{P_n\}$ and $\{Q_n\}$ both converge to the same limit, the limit will be defined to be the area of R. (We usually take elements of $\{P_n\}$, $\{Q_n\}$ to consist of unions of rectangles, as in the example below, so computing their areas is no problem.) The method is called the **Principle of Exhaustion** because the regions $\{P_n\}$ eventually exhaust all of R. The method was used first by Greek mathematicians, long before calculus itself was invented by Newton and Leibnitz.

This technique has theoretical as well as practical value. It is important that we have available a method by which the area of a region can be rigorously defined (previously we accepted that regions had areas). We find that our geometrical intuition breaks down for some sets: For example, what is the area of the set of points P whose coordinates satisfy $0 \leqslant x_P \leqslant 1$, $0 \leqslant y_P \leqslant 1$ and where x_P and y_P are rational numbers? Therefore, to put areas on a more satisfactory theoretical basis, the Principle of Exhaustion is needed. It also has many practical uses, as we shall see.

Our study begins with a detailed examination of the region under $y = x^2$ from 0 to b ($b > 0$). If we call the region R, then we shall show that there exist sequences $\{P_n\}$ and $\{Q_n\}$ of polygons such that

(1) Each Q_n is circumscribed about R
(2) Each P_n is inscribed in R

and

$$(3)\quad \lim_{n\to\infty} A(P_n) = \lim_{n\to\infty} A(Q_n) = \frac{b^3}{3}$$

where $A(P_n)$ denotes the area of P_n and $A(Q_n)$ that of Q_n. This result then shows that $b^3/3$ is the only possible number that can be assigned to the region R as its area, for the area of R must be less than or equal to $A(Q_n)$ for every n and greater than or equal to $A(P_n)$ for every n. After this example we extend the procedure to general regions bounded by the graphs of continuous functions. In doing this, we shall develop the machinery which will enable us to define the integral as a limit of sequences of Riemann sums.

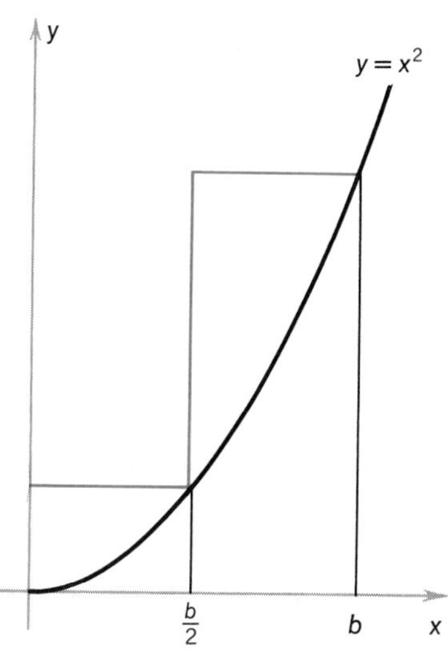

FIGURE 2-1

The region R under $y = x^2$ from 0 to b is sketched in Figure 2-1. We can construct a circumscribed polygon by dividing the interval $[0, b]$ into the two parts $[0, b/2]$ and $[b/2, b]$ and the region R into two parts—the first part being the region under $y = x^2$ from 0 to $b/2$, and the second part being the region under $y = x^2$ from $b/2$ to b. As shown in Figure 2-1, we can cover the first part of R with a rectangle of area $(b/2)(b/2)^2 = (b/2)^3$ and the second part with a rectangle of area $(b/2)b^2 = b^3/2$. Thus R is now circumscribed by a polygon of area

$$A_1 = \left(\frac{b}{2}\right)^3 + \frac{b^3}{2} = \frac{b^3}{8} + \frac{b^3}{2} = \frac{5b^3}{8}$$

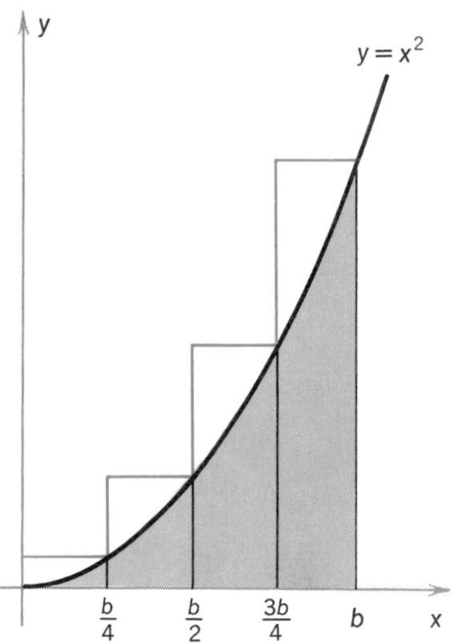

FIGURE 2-2

We next apply the division process to both $[0, b/2]$ and $[b/2, b]$ and we partition $[0, b]$ into four equal parts $[0, b/4]$, $[b/4, b/2]$, $[b/2, 3b/4]$, and $[3b/4, b]$. Then we circumscribe each of the four corresponding parts of R with a rectangle as in Figure 2-2. When constructing circumscribed rectangles, we always use the right endpoint of each interval to determine the height of the rectangle lying over that interval. The areas of these rectangles are (from left to right)

$$\left(\frac{b}{4}\right)^2 \frac{b}{4}, \quad \left(\frac{b}{2}\right)^2 \frac{b}{4}, \quad \left(\frac{3b}{4}\right)^2 \frac{b}{4}, \quad b^2 \frac{b}{4}$$

The sum of these areas $A_2 = \frac{15}{32} b^3$ is the area of a second circumscribed polygon. Notice that $A_1 > A_2$ and that the second polygon is contained in the first. The idea, of course, is to continue this process indefinitely obtaining a decreasing sequence $A_1 > A_2 > A_3 > \cdots$. It should then turn out that $\lim_{n\to\infty} = A_n = \frac{1}{3}b^3$. Let us therefore compute A_n and try to evaluate $\lim_{n\to\infty} A_n$.

At step 1, we divided $[0, b]$ into 2 equal parts. At step 2, each of these intervals was in turn divided into equal parts giving $4 = 2^2$ intervals of equal length. The reader can verify that at stage n, we will have 2^n intervals each of length $b/2^n$ and at stage $n+1$ we will have $2 \cdot 2^n = 2^{n+1}$ intervals each of length $\frac{1}{2}(b/2^n) = b/2^{n+1}$. Moreover, it is easy to list the endpoints of the intervals obtained at each stage. At step 1, they are $0, b/2, b$. At step 2, they are 0, $b/2^2$, $2b/2^2$, $3b/2^2$, and $4b/2^2$; at step n, they are 0, $b/2^n$, $2b/2^n, \ldots$ $kb/2^n, \ldots, (2^n - 1)b/2^n$, and $2^n b/2n$. In other words, at stage n, the endpoints are $0 \cdot b/2^n$, $1 \cdot b/2^n$, $2 \cdot b/2n$, $3 \cdot b/2^n$, and so on up to $2^n b/2^n$. Furthermore, for each integer k from 1 to 2^n the rectangle that we construct over the interval

$$\left[\frac{(k-1)}{2^n} b, \frac{kb}{2^n}\right]$$

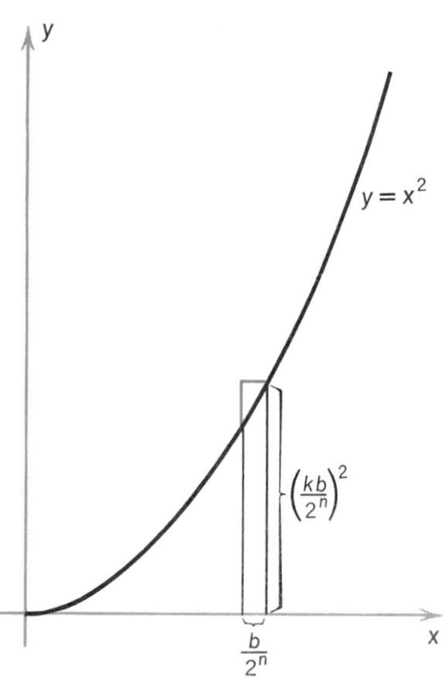

FIGURE 2-3

has base equal to $b/2^n$ and height $(kb/2^n)^2$ (Figure 2-3). The total area of the rectangles constructed at stage n is

$$A_n = \frac{b}{2^n}\left(\frac{0}{2^n}\right)^2 + \frac{b}{2^n}\left(\frac{2b}{2^n}\right)^2 + \frac{b}{2^n}\left(\frac{3b}{2^n}\right)^2 + \cdots + \frac{b}{2^n}\left(\frac{2^n b}{2^n}\right)^2$$

Each summand contains the factor $(b/2^n)(b/2^n)^2$, so we have

$$A_n = \frac{b}{2^n}\left(\frac{b}{2^n}\right)^2 [1+2^2+3^2+\cdots+(2^n)^2]$$

or

$$A_n = b^3\left(\frac{1}{2^n}\right)^3 \sum_{k=1}^{2^n} k^2$$

We set $N = 2^n$ and we have

$$\begin{aligned} A_n &= b^3\left(\frac{1}{N}\right)^3 \sum_{k=1}^{N} k^2 \\ &= \frac{b^3}{6}\frac{1}{N^3}N(N+1)(2N+1) \\ &= \frac{b^3}{6}\frac{N}{N}\frac{N+1}{N}\frac{2N+1}{N} \\ &= \frac{b^3}{6}\left(1+\frac{1}{N}\right)\left(2+\frac{1}{N}\right) \end{aligned}$$

We have therefore obtained

$$A_n = \frac{b^3}{6}\left(1+\frac{1}{2^n}\right)\left(2+\frac{1}{2^n}\right)$$

Hence

$$\lim_{n\to\infty} A_n = \frac{b^3}{6}\cdot 2 = \frac{b^3}{3}$$

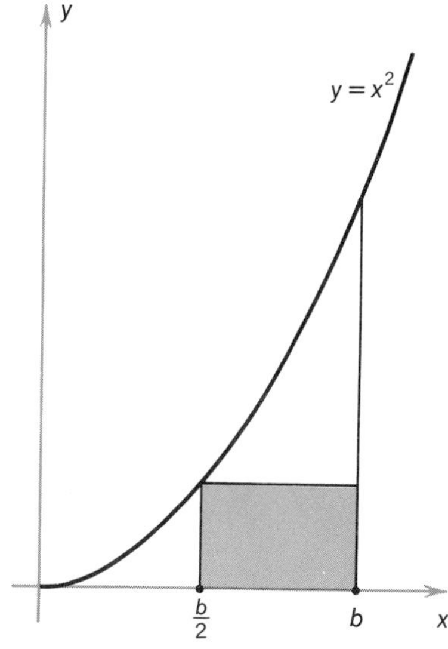

FIGURE 2-4

Let us now exhaust R by **inscribing** rectangles and polygons. This time at step 1, after partitioning $[0, b]$ into $[0, b/2]$ and $[b/2, b]$, we inscribe the rectangle with base $[b/2, b]$ and height $(b/2)^2$. This rectangle has area $B_1 = (b/2)(b/2)^2 = b^3/8$ (Figure 2-4). This time when constructing inscribed rectangles we will always use the left endpoint of the interval to determine the height of the rectangle lying over that interval.

At step 2, we again divide $[0, b]$ into four equal parts: $[0, b/4]$, $[b/4, b/2]$, $[b/2, 3b/4]$, $[3b/4, b]$ and we construct inscribed rectangles as shown in Figure 2-5.

The inscribed polygon has area

$$\begin{aligned} B_2 &= \frac{b}{4}\left[\left(\frac{b}{4}\right)^2 + \left(\frac{2b}{4}\right)^2 + \left(\frac{3b}{4}\right)^2\right] \\ &= \frac{b}{4}\left(\frac{b}{4}\right)^2 (1^2+2^2+3^2) \\ &= \frac{b^3}{4^3}14 = \frac{7b^3}{32} \end{aligned}$$

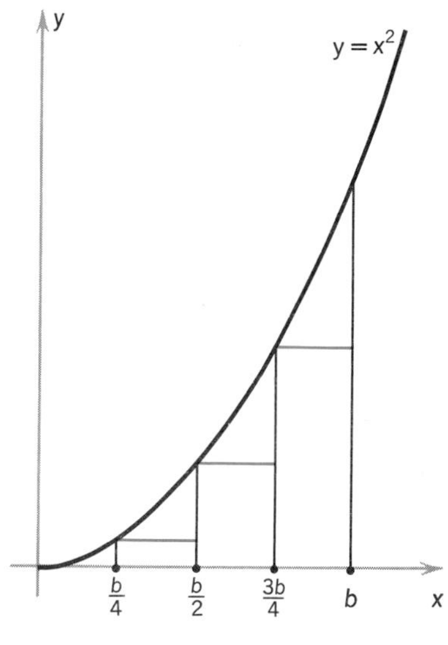

FIGURE 2-5

We see that $B_1 < B_2$, and that the second inscribed polygon exhausts more of R than did the first. Continuing in this fashion, at stage n, we construct $2^n - 1$ rectangles each of base $b/2^n$ and successively (from left to right) of height

$$0, \left(\frac{b}{2^n}\right)^2, \left(\frac{2b}{2^n}\right)^2, \ldots, \left(\frac{kb}{2^n}\right)^2, \ldots, \left[\frac{(2n-1)b}{2^n}\right]^2$$

So we have

$$\begin{aligned} B_n &= \frac{b}{2^n}\left[\left(\frac{b^2}{2^n}\right)^2 + \left(\frac{2b}{2^n}\right)^2 + \cdots + \left(\frac{(2^n-1)b}{2^n}\right)^2\right] \\ &= \frac{b}{2^n}\left(\frac{b}{2^n}\right)^2 (1^2 + 2^2 + \cdots + (2^{n-1})^2) \end{aligned}$$

So, if $N = 2^n - 1$, we have

$$\begin{aligned} B_n &= \frac{b^3}{(N+1)^3}\sum_{k=1}^{N} k^2 = \frac{b^3}{(N+1)^3}\,\frac{N(N+1)(2N+1)}{6} \\ &= \frac{b^3}{6}\left(\frac{N}{N+1}\right)\left(\frac{2N+1}{N+1}\right)\left(\frac{N+1}{N+1}\right) \\ &= \frac{b^3}{6}\left(1 - \frac{1}{N+1}\right)\left(2 - \frac{1}{N+1}\right) \end{aligned}$$

Clearly, as $n \to \infty$,

$$B_n \to \frac{b^3}{3}$$

We have established the Principle of Exhaustion for the region under $y = x^2$. Our program is to extend this result to all regions bounded by graphs of continuous functions. To accomplish this we now set up machinery for relating area and limit and the definite integral and limit. We define a **regular partition of order N** of a closed interval $[a, b]$ to be a division of $[a, b]$ into N closed intervals each of length $(b-a)/N$ (Figure 2-6).

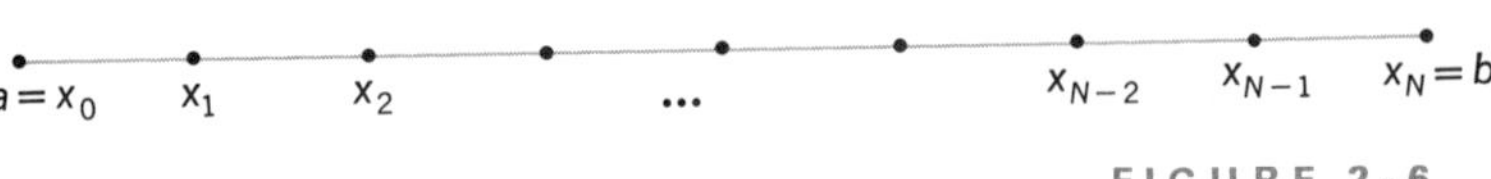

FIGURE 2-6

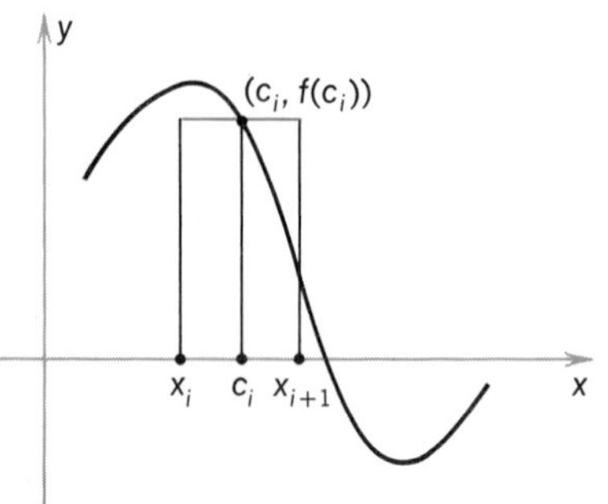

FIGURE 2-7

Suppose f is defined on $[a, b]$ and that the numbers $x_0 = a, x_1, x_2, \ldots, x_N =$ subdivide $[a, b]$ as a regular partition of order N. For each i, $0 \leqslant i \leqslant N-1$ we have an interval $[x_i, x_{i+1}]$. Let c_i be a point in $[x_i, x_{i+1}]$. We can construct the rectangle with base the interval $[x_i, x_{i+1}]$ and side the line segment from $(x_i, 0)$ to $(x_i, f(c_i))$ (see Figure 2-7). We note that the product

$f(c_i)(x_{i+1}-x_i)$ is equal to $\pm$ the area of this rectangle, depending on the sign of $f(c_i)$. If $f(c_i)=0$, the construction yields a straight line segment which we can consider a "degenerate rectangle" with area equal to zero.

If we select a point c_i for each i, we can form the sum

$$\begin{aligned} R_N &= \sum_{i=0}^{N-1} f(c_i)(x_{i+1}-x_i) \\ &= \sum_{i=0}^{N-1} f(c_i)\left(\frac{b-a}{N}\right) \\ &= \frac{b-a}{N}\sum_{i=0}^{N-1} f(c_i) \end{aligned}$$

A sum of this form is called an **Nth Riemann sum** for f on $[a,b]$. Evidently by choosing different c_i's we can form different Nth Riemann sums for the same f and N and same interval $[a,b]$.

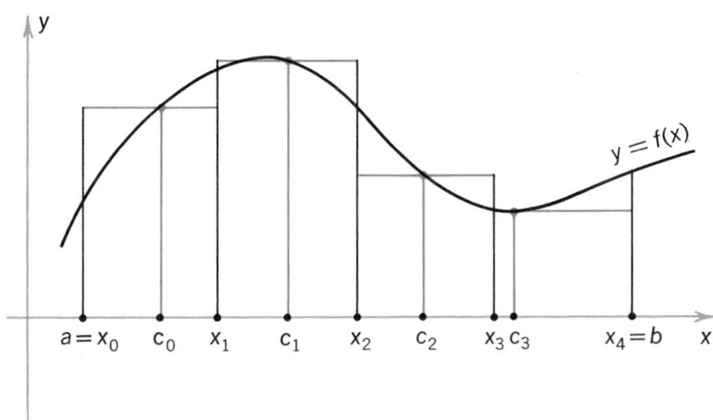

FIGURE 2-8

To get a firm idea of what the sum $R_N=\sum_{i=0}^{N-1} f(c_i)(x_{i+1}-x_i)$ represents, let us interpret it in the three following cases. For the first case we suppose that $f(x)\geqslant 0$ on $[a,b]$. Then the sum $R_N=\sum_{i=0}^{N-1} f(c_i)(x_{i+1}-x_i)$ is equal to the area of the polygon made up of the rectangles constructed over the intervals $[x_i,x_{i+1}]$ (Figure 2-8). Each such rectangle has area $f(c_i)(x_{i+1}-x_i)$ and so the polygon has area equal to the Riemann sum R_N. For the second case we suppose that $f(x)\leqslant 0$ on $[a,b]$. Here the rectangles again form a polygon although this time the polygon lies below the x axis. We see that the sum $R_N=\sum_{i=0}^{N-1} f(c_i)(x_{i+1}-x_i)$ is equal to **minus** the area of the polygon (see Figure 2-9).

In the third case we suppose that $f(x)$ changes sign on $[a,b]$. If $f(c_i)\geqslant 0$, the rectangle over $[x_i,x_{i+1}]$ lies above the x axis; if $f(c_i)\leqslant 0$, the rectangle over $[x_i,x_{i+1}]$ lies below the x axis. Therefore the sum $R_N=\sum_{i=0}^{N-1} f(c_i)(x_{i+1}-x_i)$ will represent the sum of the areas of the rectangles

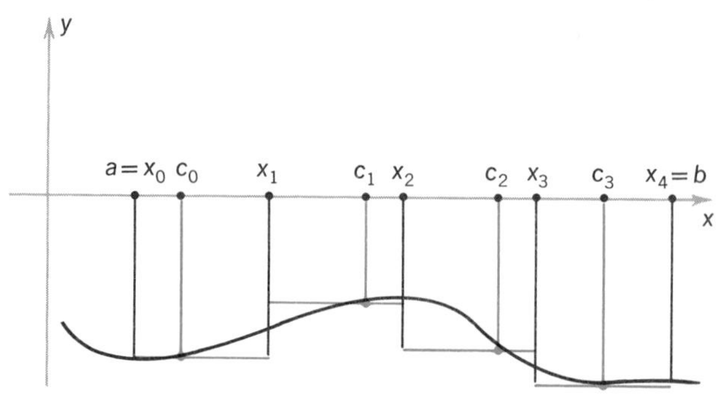

FIGURE 2-9

lying above the x axis **minus** the sum of the areas of the rectangles lying below the axis (Figure 2-10).

The above discussion should suggest that there is a relationship between Riemann sums and the area under a curve. In the theorems that follow we see that this is indeed the case.

For $N = 2^n$, a Riemann sum $R_N = R_{2^n}$ will be called a 2^nth Riemann sum for f on $[a, b]$. We shall now study sequences of 2^nth Riemann sums—more general sequences are considered in Theorem 5. That is, we now study sequences $\{r_n\}$ of the form $r_n = R_{2^n}$. For example, if f is $f: x \mapsto x^2$,

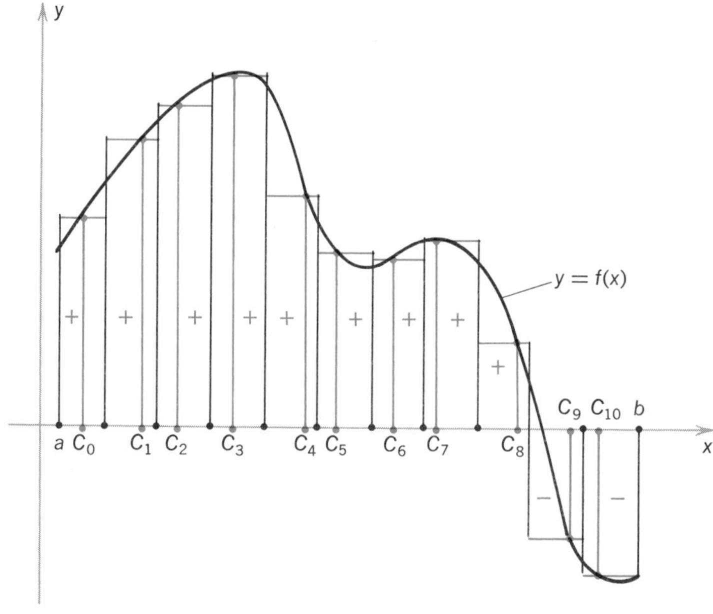

FIGURE 2-10

the sequences $\{A_n\}$ and $\{B_n\}$ are both sequences of 2^nth Riemann sums for f on $[0, b]$. Moreover, the sequences $A_1, B_1, A_2, B_2, \ldots$ and $B_1, A_2, B_3, A_4, \ldots$ are also sequences of 2^nth Riemann sums for $x :\mapsto x^2$ on $[0, b]$. Note that all of these sequences converge to $b^3/3$. We have singled out 2^nth Riemann sums because they are a little easier to handle, since the partitions are obtained by successive bisection.

We now prove three important interrelated theorems.

Theorem 1 Suppose f is continuous on $[a, b]$. Then every sequence of 2^nth Riemann sums for f on $[a, b]$ converges, and all such sequences converge to the same limit. This common limit is defined to be

$$\int_a^b f$$

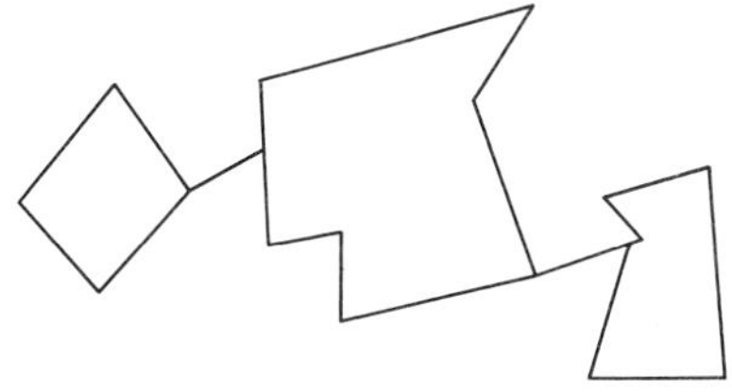

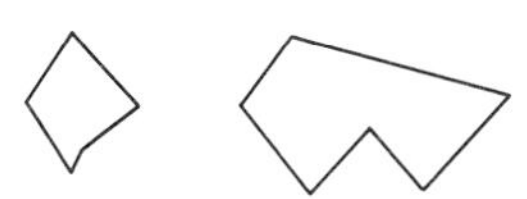

FIGURE 2-11

For nonnegative f, this limit represents the area under the graph of f. We see this further on in the proof of Theorem 2. For general f we interpret this limit as the area under the graph of f above the x axis minus the area below the axis.

The second theorem establishes the Principle of Exhaustion. We shall call a **polygonal figure** any plane figure which is made up of a finite number of polygons (possibly connected by straight line segments). Some examples of polygonal figures are shown in Figure 2-11. In our examples, the polygons will be made up of rectangles.

Theorem 2 **The Principle of Exhaustion** Suppose f and g are continuous on $[a, b]$ and that $f(x) \geqslant g(x)$ for all $x \in [a, b]$. Let R be the region bounded by the graphs of f and g from a to b. Then there exist sequences $\{P_n\}$ and $\{Q_n\}$ of polygonal figures such that

(i) Each P_n is inscribed in R
(ii) Each Q_n is circumscribed about R
(iii) $\lim_{n\to\infty} A(P_n) = \lim_{n\to\infty} A(Q_n)$

This common limit is given by the difference $\int_a^b f - \int_a^b g$.

We define the **area** of R to be this real number:

$$A(R) = \int_a^b f - \int_a^b g$$

Our third result is that the "area" of a graph of a continuous function is equal to zero. This will turn out to be a corollary of the proof of Theorem 1.

Theorem 3 Let f be continuous with domain $[a, b]$. Then there exists a sequence of polygonal figures $\{P_n\}$ such that

(i) The graph of f is a subset of each P_n
(ii) $\lim_{n\to\infty} A(P_n) = 0$

Next we turn to the proofs of Theorems 1–3. The algebraic details of the discussion now become somewhat technical. The reader should make the effort, however, to understand the geometric significance of the sequences and limits involved. The proof of Lemma 4 is rather difficult but it can be omitted without affecting the continuity of the geometric discussion.

The Principle of Exhaustion motivates us to look at two special 2^nth Riemann sums. (In the case of $x \mapsto x^2$ on $[0, b]$, one will be A_n and the other B_n. More generally, if $f(x) \geqslant 0$ on $[a, b]$, one of these special sums will represent the area of an inscribed polygon and the other the area of a circumscribed polygon.)

To define these special sums, we introduce some further notation. Consider the regular partition $a = x_0, x_1, \ldots, x_{2^n} = b$ of $[a, b]$. Since the function f is continuous on $[x_i, x_{i+1}]$, it has a maximum and a minimum on this interval. Let $f(m_i^n)$ be the minimum of f on $[x_i, x_{i+1}]$ and let $f(M_i^n)$ be the maximum of f on the same interval $[x_i, x_{i+1}]$. We use the superscript n in m_i^n and M_i^n to emphasize that m_i^n and M_i^n both lie in the interval

$$[x_i, x_{i+1}] = \left[a + \frac{i(b-a)}{2^n}, a + \frac{(i+1)(b-a)}{2^n}\right]$$

If we choose $c_i = m_i^n$, we have the 2^nth Riemann sum

$$\begin{aligned} s_n &= \sum_{i=0}^{2^n-1} f(m_i^n)(x_{i+1} - x_i) \\ &= \frac{b-a}{2^n} \sum_{i=0}^{2^n-1} f(m_i^n) \end{aligned}$$

The Riemann sum s_n is called the nth **lower sum** for f on $[a, b]$.

If we choose $c_i = M_i^n$, we have the sum

$$\begin{aligned} S_n &= \sum_{i=0}^{2^n-1} f(M_i^n)(x_{i+1} - x_i) \\ &= \frac{b-a}{2^n} \sum_{i=0}^{2^n-1} f(M_i^n) \end{aligned}$$

The sum S_n is called the nth **upper sum** for f on $[a, b]$.

Since $f(M_i^n) \geqslant f(m_i^n)$ for all i, $i = 0, 1, \ldots, 2^n - 1$, we have

$$s_n \leqslant S_n$$

In fact, for any choice of c_i we have

$$f(m_i^n) \leqslant f(c_i) \leqslant f(M_i^n)$$

and so

$$s_n \leqslant R_{2^n} \leqslant S_n$$

for all 2^nth Riemann sums R_{2^n} (Figure 2-12).

The plan now is to show (1) that $\{S_n\}$ is a bounded decreasing sequence and therefore a convergent sequence, (2) that $\{s_n\}$ is a bounded increasing

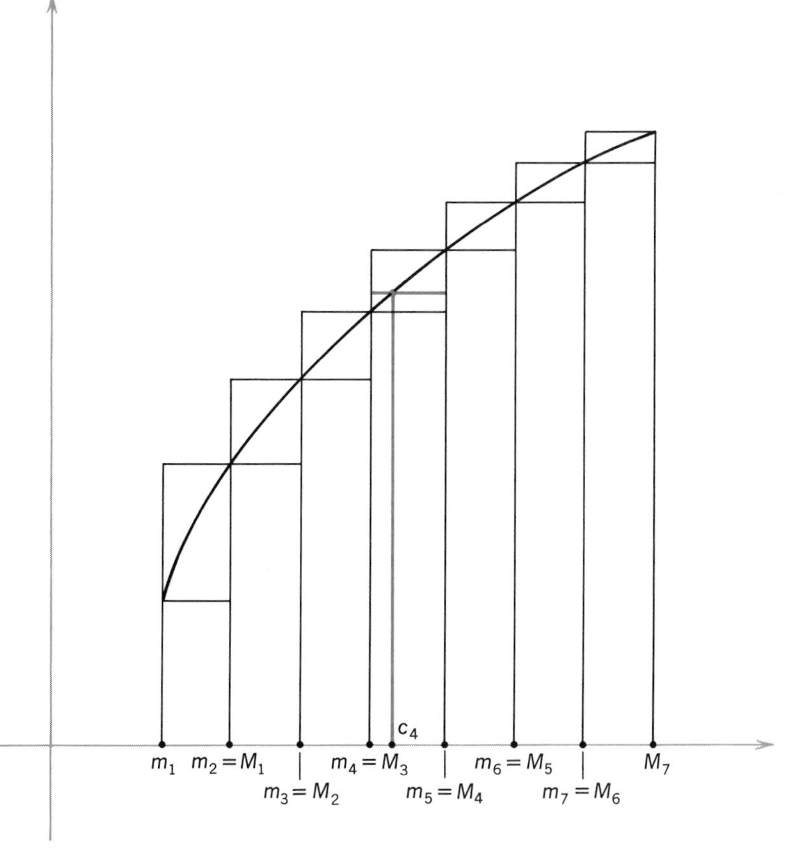

FIGURE 2-12

sequence and therefore a convergent sequence, and (3) both sequences converge to the same limit. It will then follow from the Squeezing Lemma that all sequences of 2^nth Riemann sums $\{R_{2^n}\}$ converge to the same limit. This will prove Theorem 1.

Lemma 1 The sequences $\{s_n\}$ and $\{S_n\}$ are bounded.

Proof Let M be the maximum value of f on $[a, b]$ and let m be the minimum value of f on $[a, b]$. We have $m \leqslant m_i^n$ and $M_i^n \leqslant M$ for all i. Thus,

$$\begin{aligned} m(b-a) &= \sum_{i=0}^{2^n-1} m\left(\frac{b-a}{2^n}\right) \leqslant \sum_{i=0}^{2^n-1} m_i^n\left(\frac{b-a}{2^n}\right) \\ &\leqslant \sum_{i=0}^{2^n-1} M_i^n\left(\frac{b-a}{2^n}\right) \leqslant \sum_{i=0}^{2^n-1} M\left(\frac{b-a}{2^n}\right) = M(b-a) \end{aligned}$$

So

$$m(b-a) \leqslant s_n \leqslant S_n \leqslant M(b-a)$$ ∎

Lemma 2 The sequence $\{s_n\}$ is monotone increasing and the sequence $\{S_n\}$ is monotone decreasing.

Proof We prove only that $\{s_n\}$ is monotone increasing, leaving the statement about $\{S_n\}$ as an exercise. Geometrically, it should be clear that $s_{n+1} \geqslant s_n$ for all n (see Figure 2-13). To show this analytically, consider a typical interval $A_i^n = [x_i, x_{i+1}]$ of the nth partition. Then to find s_{n+1} we subdivide this interval into two equal intervals,

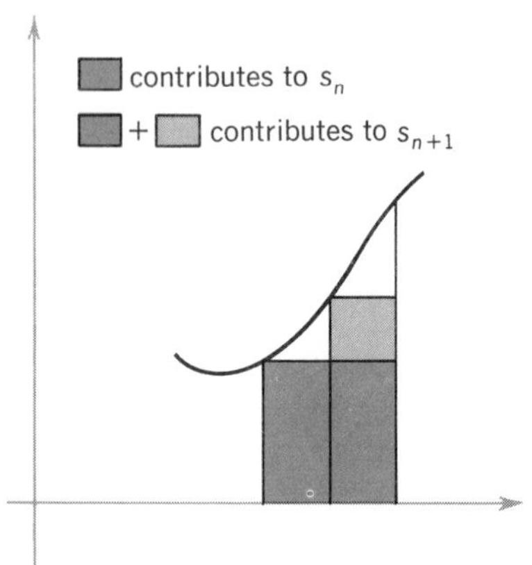

FIGURE 2-13

$$A_{2i}^{n+1} = \left[x_i, \frac{x_i + x_{i+1}}{2}\right] \quad \text{and} \quad A_{2i+1}^{n+1} = \left[\frac{x_i + x_{i+1}}{2}, x_{i+1}\right]$$

Since A_{2i}^{n+1} is included in A_i^n, we must have $f(m_{2i}^{n+1}) \geqslant f(m_i^n)$ (Why?). Similarly, we have $f(m_{2i+1}^{n+1}) \geqslant f(m_i^n)$. Therefore,

$$f(m_i^n) \leqslant \tfrac{1}{2}(f(m_{2i}^{n+1}) + f(m_{2i+1}^{n+1}))$$

and

$$\sum_{i=0}^{2^n-1} f(m_i^n) \leqslant \tfrac{1}{2} \sum_{i=0}^{2^n-1} (f(m_{2i}^{n+1}) + f(m_{2i+1}^{n+1}))$$

Multiplying both sides by $(b-a)/2^n$ we obtain

$$s_n = \frac{b-a}{2^n} \sum_{i=0}^{2^n-1} f(m_i^n) \leqslant \frac{b-a}{2^{n+1}} \sum_{i=0}^{2^n-1} (f(m_{2i}^{n+1}) + f(m_{2i+1}^{n+1}))$$

But the right-hand sum is just s_{n+1}, proving

$$s_n \leqslant s_{n+1}$$

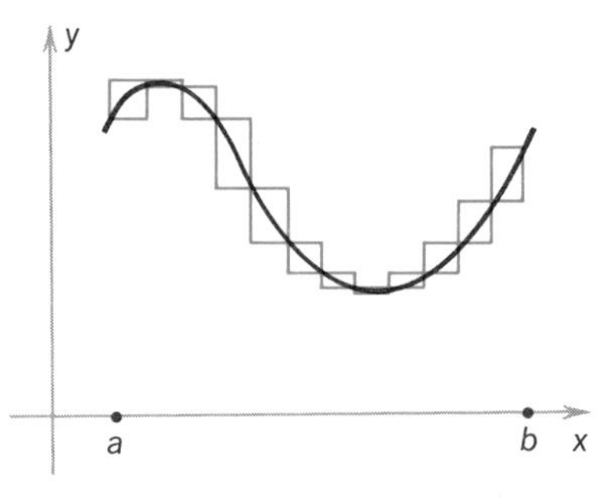

FIGURE 2-14

Next, look at the difference $S_n - s_n$. This number is nonnegative and has an important geometric significance: $S_n - s_n$ represents the area of a polygon which completely contains the graph of f from a to b (Figure 2-14). This polygon is composed of rectangles of base $x_{i+1} - x_i$ and height $f(M_i^n) - f(m_i^n)$. The area of the polygon is

$$\sum_{i=1}^{2^n-1} (f(M_i^n) - f(m_i^n))(x_{i+1} - x_i) = S_n - s_n$$

This is the case whether or not $f(x)$ changes sign on $[a, b]$. Therefore, in the following lemma, when we prove that $\lim_{n\to\infty} (S_n - s_n) = 0$, we will also be showing that there is a sequence of polygonal figures which cover the graph of f on $[a, b]$ and whose areas tend to 0. Thus, we will have proved Theorem 3.

Lemma 3 Condition (iii) of the Squeezing Lemma is satisfied; that is, $\lim_{n\to\infty} (S_n - s_n) = 0$.

Proof We have

$$S_n - s_n = \sum_{i=0}^{2^n-1} (f(M_i^n) - f(m_i^n))(x_{i+1} - x_i)$$
$$= \frac{b-a}{2^n} \sum_{i=0}^{2^n-1} (f(M_i^n) - f(m_i^n))$$

which yields

$$S_n - s_n = |S_n - s_n| \leqslant \frac{b-a}{2^n} \sum_{i=0}^{2^n-1} |f(M_i^n) - f(m_i^n)|$$

At this point we need a technical lemma before continuing with the proof of Lemma 3.

Lemma 4 Uniformity Lemma Let f be continuous with domain $[a, b]$. Then for every $\varepsilon > 0$, there exists a $\delta > 0$ such that $|u - v| < \delta \Rightarrow |f(u) - f(v)| < \varepsilon$.

The conclusion of this lemma is stronger than simple continuity because the **same** δ must work uniformly for all u, v in $[a, b]$.

*** Proof of the Uniformity Lemma** Let $\varepsilon > 0$ be given and let us consider the set A of points $x \in [a, b]$ for which there exists a $\delta > 0$ such that

$$(x_1, x_2 \in [a, x],\ |x_1 - x_2| < \delta) \text{ implies } (|f(x_1) - f(x_2)| < \varepsilon)$$

Now $a \in A$ so A is not empty. Also, A is bounded above by b, so we know A has a least upper bound; call it c. Since f is continuous at c, there is an $\eta > 0$ such that $|y - c| < \eta$, $y \in [a, b]$ implies $|f(y) - f(c)| < \varepsilon/2$. By definition of least upper bound, there exists $x \in A$ such that $c - \eta/2 < x \leqslant c$. By defition of A there is a δ such that $|f(x_1) - f(x_2)| < \varepsilon$ whenever $x_1, x_2 \in [a, x]$ and $|x_1 - x_2| < \delta$. Let ζ be the minimum of $\eta/2$ and δ. Next let $x_1, x_2 \in [a, c + \eta/2] \cap [a, b]$ and $|x_1 - x_2| < \zeta$. If $x_1, x_2 \leqslant x$ then we know $|f(x_1) - f(x_2)| < \varepsilon$. If one of x_1, x_2 is greater than x then both lie in the interval $]c - \eta, c + \eta[$ since $|x_1 - x_2| < \zeta \leqslant \eta/2$. Hence by continuity of f at c, whenever $x_1, x_2 \in [a, c + \eta/2] \cap [a, b]$ and $|x_1 - x_2| < \zeta$, we have

$$|f(x_1) - f(x_2)| \leqslant |f(x_1) - f(c)| + |f(c) - f(x_2)|$$
$$< \varepsilon/2 + \varepsilon/2 = \varepsilon$$

This argument shows that $c \in A$. Moreover, we conclude that $c = b$, for otherwise if $c < b$ there would be elements of A greater than c. Hence $b \in A$ and so by definition of A, f is uniformly continuous on $[a, b]$. ∎

The reader wishing to gain further insight into the idea of uniform continuity may consider this example: The function $x \mapsto 1/x$ is continuous on $]0, 1]$ but is not uniformly continuous. This does not contradict the lemma because $]0, 1]$ is not a closed interval.

Proof of Lemma 3 (continued) We shall apply the Uniformity Lemma to show that for every $e>0$ there exists n such that $|S_n-s_n|<e$. Since $|S_n-s_n|$ is decreasing, this will imply that $\lim_{n\to\infty}(S_n-s_n)=0$.

So let $e>0$ be given. Applying the Uniformity Lemma with $\varepsilon=e/(b-a)$, let $1/2^n<\delta$ be such that

$$|u-v| < \frac{b-a}{2^n} \Rightarrow |f(u)-f(v)| < \frac{e}{b-a}$$

Since

$$|M_i^n - m_i^n| < \frac{b-a}{2^n}$$

we find that

$$|f(M_i^n) - f(m_i^n)| < \frac{e}{b-a}$$

This yields

$$|S_n - s_n| \leqslant \frac{b-a}{2^n}\sum_{i=0}^{2^n-1} |f(M_i^n) - f(m_i^n)| < \frac{b-a}{2^n}\frac{2^n e}{b-a} = e$$

This completes the proof of Theorem 1. In the course of the argument, we have proven Theorems 1 and 3. Now we prove Theorem 2.

Proof of Theorem 2 We let U_n be the nth upper sum for f on $[a,b]$ and u_n the nth lower sum for f on $[a,b]$. Similarly, we let V_n be the nth upper sum for g on $[a,b]$ and v_n the nth lower sum for g on $[a,b]$. As the reader can easily verify, the number u_n-V_n is equal to the area of an inscribed polygonal figure P_n (Figure 2-15). [This is true if $u_n-V_n>0$, which we shall

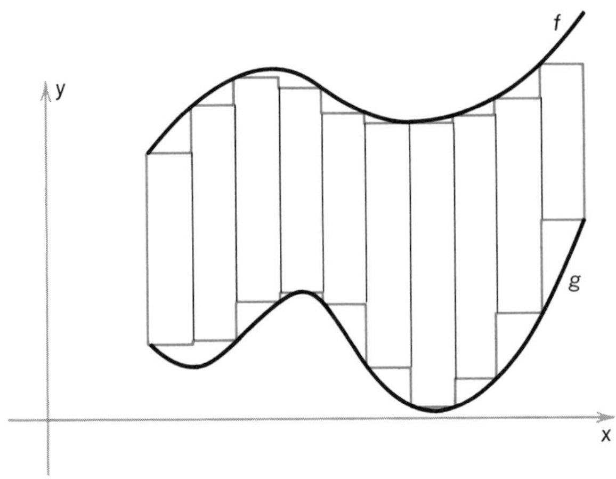

FIGURE 2-15

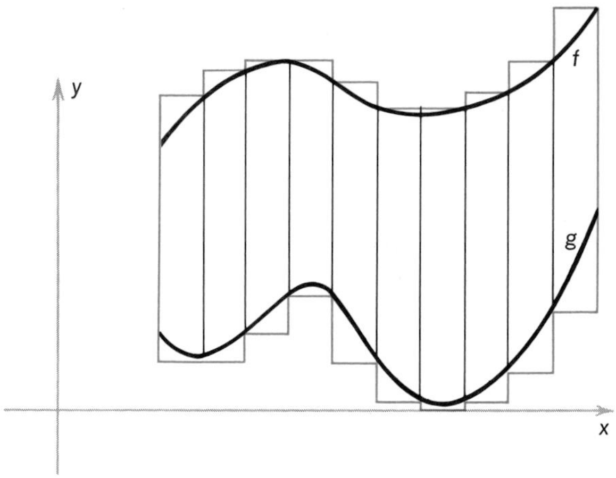

FIGURE 2-16

ıssume (see exercise 12).] On the other hand, $U_n - v_n$ represents the area ›f a circumscribed polygonal figure Q_n (Figure 2-16). From Theorem 1, ve have $\lim_{n\to\infty}(u_n - V_n) = \lim_{n\to\infty}(U_n - v_n)$ and so we conclude that

$$\begin{aligned}\lim_{n\to\infty} A(P_n) &= \lim_{n\to\infty}(u_n - V_n)\\ &= \lim_{n\to\infty}(U_n - v_n)\\ &= \lim_{n\to\infty} A(Q_n)\end{aligned}$$

Moreover, each expression equals

$$\lim_{n\to\infty}(u_n - V_n) = \lim_{n\to\infty} u_n - \lim_{n\to\infty} V_n = \int_a^b f(x)\,dx - \int_a^b g(x)\,dx \qquad ◩$$

Let us now recap what we have done so far. We have shown that for . continuous function $f(x)$ on $[a, b]$, the Riemann sums for $f(x)$ all converge o the same limit and this limit is denoted $\int_a^b f(x)\,dx$. The Principle of :xhaustion and the definition of the Riemann sum justify our interpretation ›f the integral in terms of area, a process with which we are already familiar.

Now we wish to extend our development to include the Fundamental 'heorem. Recall that our proof in §6.2 depended only on several key ›roperties of the integral. Hence, if we establish these properties in the ›resent context, the same proof is then valid here. Let us therefore prove hese results.

'heorem 4 Let f be continuous on $[a, b]$. Then

(i) For $c \in [a, b]$ we have

$$\int_a^b f(x)\,dx = \int_a^c f(x)\,dx + \int_c^b f(x)\,dx$$

(ii) If $m \leqslant f(x) \leqslant M$ for all $x \in [a, b]$ then

$$m(b-a) \leqslant \int_a^b f(x)\,dx \leqslant M(b-a)$$

Proof Part (ii) follows at once from the definition of the integral as a limit of sequences of Riemann sums. We now prove (i), leaving it to the reader to fill in the details of the proof of (ii). If f is defined on an interval $[\alpha, \beta]$, then we will denote by $s_n(\alpha, \beta)$ and $S_n(\alpha, \beta)$ the nth lower sum for f on $[\alpha, \beta]$ and the nth upper sum for f on $[\alpha, \beta]$ respectively. For the interval $[a, b]$, $[a, c]$, and $[c, b]$ we have

$$s_n(a, b) \leqslant S_n(a, c) + S_n(c, b)$$

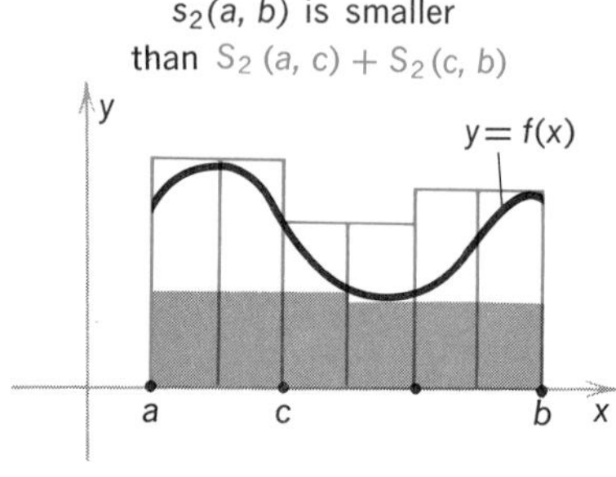

FIGURE 2-17

This inequality can be seen geometrically (Figure 2-17) and can also be verified by a direct examination of the sums involved (see the proof of Theorem 5 below for the algebraic demonstration of a similar fact; see also exercise 13). Taking the limit as $n \to \infty$ we find by Theorem 1 that

$$\int_a^b f = \lim_{n \to \infty} s_n(a, b) \leqslant \lim_{n \to \infty} S_n(a, c) + \lim_{n \to \infty} S_n(c, b) = \int_a^c f + \int_c^b f$$

Thus,

$$\int_a^b f \leqslant \int_a^c f + \int_c^b f$$

On the other hand we have

$$S_n(a, b) \geqslant s_n(a, c) + s_n(c, b)$$

Taking the limit as $n \to \infty$, we now find that

$$\int_a^b f \geqslant \int_a^c f + \int_c^b f$$

This completes the proof of (i). ∎

If $b < a$ we define $\int_a^b f(x)\,dx = -\int_b^a f(x)\,dx$ as we did in Chapter 6. The additivity property (i) then holds for all a, b, c.

Having established the above basic properties of the integral, our earlier proofs of the Mean Value Theorem and the Fundamental Theorem of Calculus carry over. Thus our discussion has come full circle. We began by developing properties of the integral using geometric intuition based on area. Next, we gave a precise definition of the integral using Riemann sums and verified (after considerable effort) that our intuition was consistent. All subsequent developments, such as the Change of Variables Theorem for integrals, of course remain valid. We shall therefore use them freely in the future.

Next we shall see how Riemann sums can be used to give an alternative approach to volumes of solids of revolution. This method is more rigorous in that volume is defined in terms of a definite procedure—the Principle of Exhaustion—rather than relying on our intuition about volumes.

In §6.4 we found that the volume of the solid obtained by revolving the graph of f from a to b about the x axis is

$$V = \int_a^b \pi f(x)^2\,dx$$

where f is assumed continuous on $[a,b]$. By Theorem 1, the integral $\int_a^b \pi f(x)^2\,dx$ is a limit of upper and lower sums and we have

$$\begin{aligned} V &= \lim_{n\to\infty} \sum_{i=0}^{2^n-1} \pi f(m_i^n)^2 (x_{i+1}-x_i) \\ &= \lim_{n\to\infty} \sum_{i=0}^{2^n-1} \pi f(M_i^n)^2 (x_{i+1}-x_i) \end{aligned}$$

Geometrically, the lower sum $\sum_{i=0}^{2^n-1} \pi f(m_i^n)^2 (x_{i+1}-x_i)$ represents the volume of an inscribed solid which is made up of cylindrical disks (Figure 2-18). The upper sum $\sum_{i=0}^{2^n-1} \pi f(M_i^n)^2 (x_{i+1}-x_i)$ represents the volume of a circumscribed solid composed of cylindrical disks (Figure 2-19). As $n\to\infty$, the volumes of the inscribed and circumscribed solids approach a common limit and this limit is the volume of the solid of revolution. We thus have a Principle of Exhaustion for solids of revolution which demonstrates that volume is indeed defined for such solids. This justifies our somewhat cavalier approach to the problem of volume in Chapter 6.

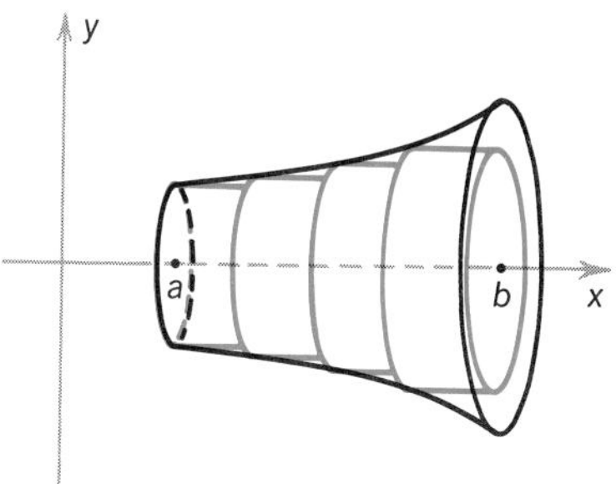

FIGURE 2-18

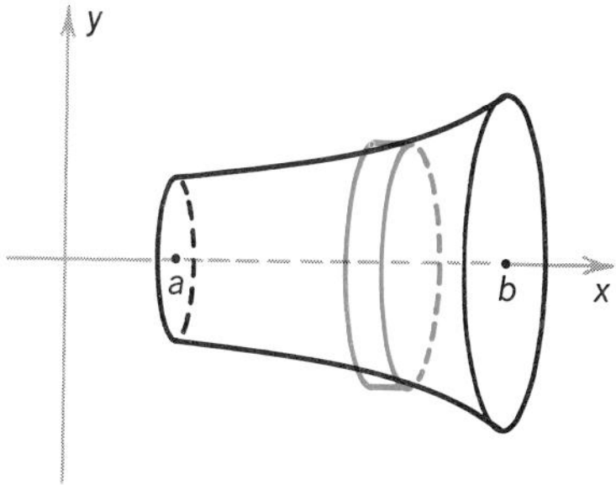

FIGURE 2-19

The Riemann sum also affords further insight into the notion of the average value of a function. In §6.1 we defined the average value of f on $[a,b]$ to be the number $(\int_a^b f)/(b-a)$, where f is assumed continuous on $[a,b]$. By Theorem 1, we have

$$\int_a^b f = \lim_{n\to\infty} \frac{b-a}{2^n} \sum_{i=1}^{2^n} f(x_i)$$

and

$$\frac{\int_a^b f}{b-a} = \lim_{n\to\infty} \frac{1}{2^n} \sum_{i=1}^{2^n} f(x_i)$$

where the x_i are the points of a regular partition of $[a,b]$ of order 2^n. The number $(1/2^n)\sum_{i=1}^{2^n} f(x_i)$ is the average of $f(x_1), f(x_2), \ldots, f(x_{2^n})$ and the average value of f is obtained by taking the limit of these averages as $n\to\infty$.

The two examples we have just seen illustrate how the Riemann sum can be applied in situations far removed from the problem of finding the area under a curve.

Thus far in our discussion of the integral as a limit of sums, we have used only 2^nth-Riemann sums. However, much more general sequences of Riemann sums converge to the integral. For completeness, therefore, we state and prove a result known as **Darboux's Theorem.**

Theorem 5 Let $a = x_0 < x_1 < \cdots < x_n = b$ be a (not necessarily regular) partition of $[a,b]$ and let $c_i \in [x_i, x_{i+1}]$. Form the Riemann sum

$$R = \sum_{i=1}^{n} f(c_i)(x_{i+1}-x_i)$$

where f is a continuous function on $[a, b]$.

Call the maximum of $|x_{i+1}-x_i|$ the **size** of the partition. If $\{R_n\}$ is any sequence of Riemann sums corresponding to partitions whose size $\to 0$ then

$$\lim_{n\to\infty} R_n = \int_a^b f(x)\,dx$$

*** Proof** We first define the notion of a **refinement** of a partition $P: x_0 < x_1 < \cdots < x_n$. This is simply another partition obtained by inserting additional points in the given partition. It should be clear that the size of the refined partition must be less than or equal to the size of the original partition.

If for a given partition P, c_i is chosen so that $f(c_i)$ is the minimum of f on $[x_i, x_{i+1}]$ the corresponding Riemann sum is called a lower sum and is denoted L_P. Similarly we have an upper sum U_P. Therefore, for any Riemann sum R for P, we have

$$L_P \leqslant R \leqslant U_P$$

Now let Q be a refinement of the partition P. Then we claim that

$$L_P \leqslant L_Q \leqslant U_Q \leqslant U_P$$

Indeed, let us examine the assertion $L_P \leqslant L_Q$. Now for simplicity, we insert one point c in $[x_i, x_{i+1}]$. If m_i is the minimum of f on $[x_i, x_{i+1}]$, l_i the minimum of f on $[x_i, c]$, and k_i the minimum on $[c, x_{i+1}]$, then $m_i \leqslant l_i$ and $m_i \leqslant k_i$ so

$$m_i(x_{i+1}-x_i) = m_i\{(x_{i+1}-c)+(c-x_i)\} \leqslant k_i(x_{i+1}-c)+l_i(c-x_i)$$

Thus we conclude that the sum for L_Q is greater than that for L_P. The general case is analogous.

Now let $\{R_n\}$ be a sequence of Riemann sums as hypothesized and let the corresponding partitions be P^n. Since $L_{P^n} \leqslant R_n \leqslant U_{P^n}$ it suffices to show L_{P^n} and U_{P^n} both converge to the integral. Consider the partition Q^n which is the refinement obtained by combining all the points of P^n and a regular partition of order 2^n. Then, by the above,

$$s_n \leqslant L_{Q^n} \leqslant U_{Q^n} \leqslant S_n$$

Hence, since s_n and S_n both converge to $\int_a^b f(x)\,dx$ (Theorem 1), we find

$$\lim_{n\to\infty} L_{Q^n} = \lim_{n\to\infty} U_{Q^n} = \int_a^b f(x)\,dx$$

It remains to prove that L_{Q^n} and L_{P^n} converge to the same limit, as do U_{Q^n} and U_{P^n}. We do this using the Uniformity Lemma. Given $\varepsilon > 0$ choose $\delta > 0$ so that for $|x'-x''| < \delta$, $|f(x')-f(x'')| < \varepsilon$. Also, choose N so $n \geqslant N$ implies that the size of P^n is $< \delta$. Now we consider the refinement Q^n of P^n as above, and let $n \geqslant N$. For simplicity suppose Q^n is obtained by inserting a point c in $[x_i, x_{i+1}]$. The corresponding terms in the Riemann sums for L_{P^n} and L_{Q^n} are

$$m_i(x_{i+1}-x_i) \quad \text{and} \quad k_i(x_{i+1}-c)+l_i(c-x_i)$$

where m_i, l_i, and k_i have the same meaning as before. The difference is

$$\begin{aligned} &k_i(x_{i+1}-c)+l_i(c-x_i)-m_i(x_{i+1}-x_i) \\ &= k_i(x_{i+1}-c)+l_i(c-x_i)-m_i(x_{i+1}-c)-m_i(c-x_i) \\ &= (k_i-m_i)(x_{i+1}-c)+(l_i-m_i)(c-x_i) \\ &\leqslant \varepsilon(x_{i+1}-x_i) \end{aligned}$$

since $0 \leqslant k_i - m_i \leqslant \varepsilon$ by hypothesis. The case where more points are added to $[x_i, x_{i+1}]$ is entirely similar.

We find that, on adding,

$$0 \leqslant L_{Q^n} - L_{P^n} \leqslant \varepsilon(b-a)$$

In a similar way

$$0 \leqslant U_{P^n} - U_{Q^n} \leqslant \varepsilon(b-a)$$

Thus our assertion follows. ∎

We conclude this section with an example of a sequence of nth Riemann sums which does not converge. We define a function G by the rule

$$G(x) = \begin{cases} 1/x^2, & x \neq 0 \\ 0, & x = 0 \end{cases}$$

Hence G is defined on $[0, 1]$ but is not continuous on $[0, 1]$. The points $0, 1/2^n, 2/2^n, 3/2^n, \ldots, (2^n-1)/2^n, 2^n/2^n = 1$ give a regular partition of $[0, 1]$ of order 2^n. Taking $c_i = (i+1)/2^n, 0 \leqslant i \leqslant 2^{n-1}$ we form the following 2^nth Riemann sum for G on $[0, 1]$:

$$\begin{aligned} R_n &= \sum_{i=0}^{2^n-1} G(c_i)(x_{i+1}-x_i) \\ &= \frac{1}{2^n}\sum_{i=0}^{2^n-1} \frac{(2^n)^2}{(i+1)^2} \\ &= 2^n \sum_{i=0}^{2^n-1} \frac{1}{(i+1)^2} \geqslant 2^n \end{aligned}$$

The sequence $\{2^n\}$ diverges to $+\infty$; hence $\{R_n\}$ also diverges to $+\infty$.

EXERCISES

*1. Let $f:[a,b] \to \boldsymbol{R}$ be continuous except perhaps at a single point $c \in [a,b]$. Suppose f is bounded. Prove that the Riemann sums for f converge (as in Theorem 3).

2. Evaluate $\int_0^2 x\,dx$ using Riemann sums as we did in the text for $x \mapsto x^2$.

Evaluate the following integrals.

3. $\int_0^1 x(x^2+1)^{10}\,dx$
4. $\int_0^2 x\sin(x^2)\,dx$
5. $\int_0^1 \frac{x}{(1+x^2)^{1/2}}\,dx$
6. Review the Mean Value Theorem and the Fundamental Theorem and verify that the proofs given depend only on the properties developed in this section.
7. For each N let L_N be the lower sum for the Nth regular partition of $[a,b]$. Give an example of a continuous function f such that the sequence $\{L_N\}$ is not monotone increasing. However, conclude from Theorem 5 that the sequence $\{L_N\}$ converges for all continuous f.
8. Prove Theorem 2 for monotone functions, without recourse to the Uniformity Lemma.
9. Prove the Uniformity Lemma for continuously differentiable functions. [Hint: Use the Extreme Value Theorem and the Mean Value Theorem.]
10. Let f be continuous on $[a,b]$. Show that for every N, there is a Riemann sum R_N for the Nth regular partition of $[a,b]$ which satisfies $R_N = \int_a^b f$. [Hint: Use the Mean Value Theorem of Integral Calculus.] Show that this is true for any partition.
11. Use Theorem 1 to prove the following result without recourse to the Fundamental Theorem: Let f and g be continuous on $[a,b]$. Then $\int_a^b f+g = \int_a^b f + \int_a^b g$ and $\int_a^b cf = c\int_a^b f$, where c is a constant.

*12. (a) Give examples to show that $u_n - V_n$ in the proof of Theorem 3 can be < 0.
(b) By omitting such rectangles, show that the remaining ones define a sequence of polygons P_n with the required properties. (Note that such omissions might be necessary at each stage n if f and g touch.)

*13. (a) Let P and Q be any partitions on $[a,b]$. Using the notation of Theorem 5, prove that $L_P \leqslant U_Q$. [Hint: Consider a refinement R of P and Q.]
(b) Give an analytical proof of the inequalities required in the proof of Theorem 4.

3 LENGTH OF A CURVE

Recall that the length of the straight line segment joining two points $P = (x_P, y_P)$ and $Q = (x_Q, y_Q)$ in the plane is

$$d(P,Q) = \sqrt{(x_Q - x_P)^2 + (y_P - y_Q)^2} \tag{1}$$

(Figure 3-1). In this section we derive a formula for the length of the graph of a continuously differentiable real function $f:[a,b] \to \boldsymbol{R}$. We shall also treat parametrized curves and apply our results to the analysis of the motion of a moving particle. This discussion should give some idea of the power and wide applicability of the method of Riemann sums.

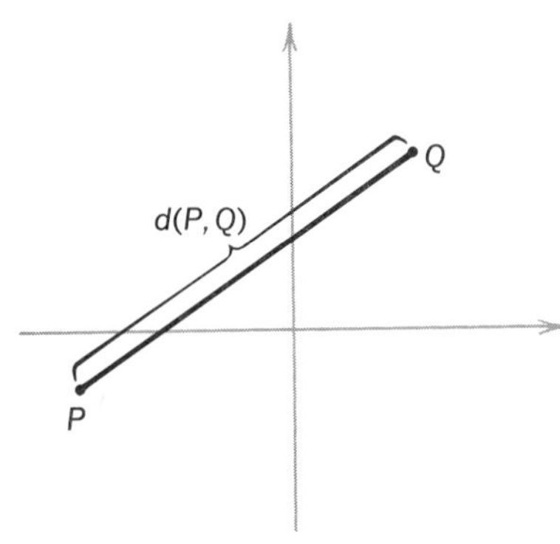

FIGURE 3-1

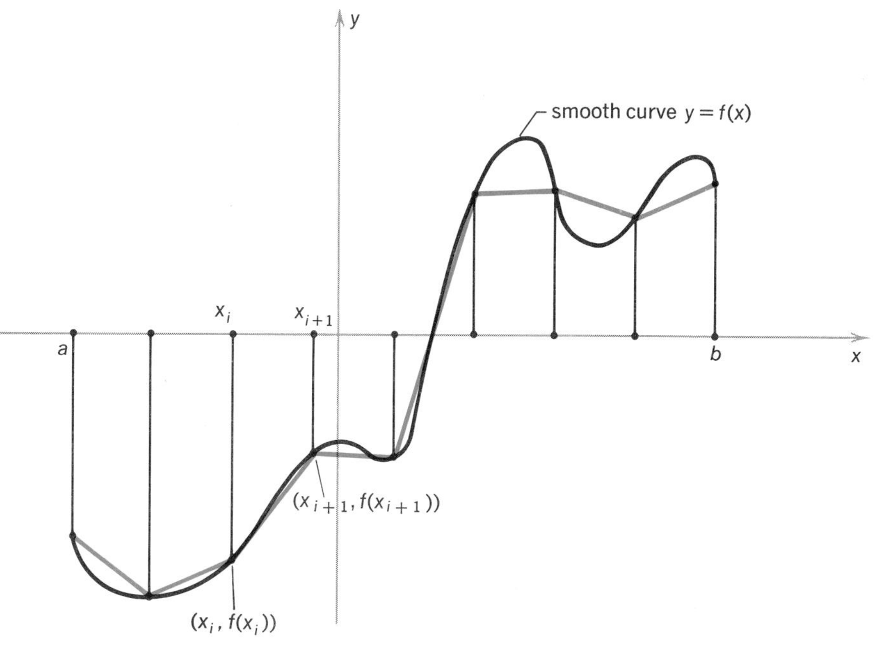

FIGURE 3-2

Assume now that the real function f is not only continuous but also continuously differentiable on $[a, b]$. Let $[a, b]$ be subdivided by the points $a = x_0, \ldots, x_{2^n} = b$ of a regular partition of order $N = 2^n$. We associate with this partition an **approximate curve length** of f on $[a, b]$ by (1) connecting each point $(x_i, f(x_i))$ to $(x_{i+1}, f(x_{i+1}))$ with a line segment for $i = 0, \ldots, 2^n - 1$, and then (2) adding together the lengths of these 2^n line segments (Figure 3-2). This sum is

$$\sum_{i=0}^{N-1} d((x_i, f(x_i)), (x_{i+1}, f(x_{i+1}))) \tag{2}$$

Using (1), we may rewrite (2) as

$$\sum_{i=0}^{N-1} \sqrt{(x_{i+1} - x_i)^2 + (f(x_{i+1}) - f(x_i))^2} \tag{3}$$

Since f is continuous on $[x_i, x_{i+1}]$ and differentiable on $]x_i, x_{i+1}[$ the hypotheses of the Mean Value Theorem are satisfied. Hence, there is a point $c_i \in]x_i, x_{i+1}[$ such that

$$f(x_{i+1}) - f(x_i) = f'(c_i)(x_{i+1} - x_i) \tag{4}$$

Squaring both sides of (4), we substitute $f'(c_i)^2 (x_{i+1} - x_i)^2$ for $(f(x_{i+1}) - f(x_i))^2$ in (3) to obtain the following expression for the approximate curve length:

$$\sum_{i=0}^{N-1} \sqrt{(x_{i+1}-x_i)^2 + f'(c_i)^2(x_{i+1}-x_i)^2}$$

or

$$\sum_{i=0}^{N-1} \sqrt{1+f'(c_i)^2}(x_{i+1}-x_i) \tag{5}$$

after taking $(x_{i+1}-x_i)^2$ outside the radical sign.

Now (5) is an Nth Riemann sum for the function $x \mapsto \sqrt{1+f'(x)^2}$ which is continuous (since f' is continuous). Thus a sequence of Riemann sums of the form (5) has a limit as $n \to \infty$, namely the definite integral

$$\int_a^b \sqrt{1+f'(x)^2}\,dx$$

Since it is clear that the polygonal path whose length is the sum (2) approximates the curve more closely as the number n increases, we take the limit of these sums as $n \to \infty$ as the definition of the length of the curve. That is, the **curve length** of f on $[a, b]$ is

$$\int_a^b \sqrt{1+f'(x)^2}\,dx$$

The curve length is also called the **arc length**.

Example 1 Find the length of the line segment given by $x \mapsto 2x+3$ over $[-2, 2]$ by using the formula developed in this section.

If we did not have the new formula, we could easily obtain the answer using the expression for the distance

$$\sqrt{(-2-2)^2+(-1-7)^2} = \sqrt{16+64} = \sqrt{80} = 4\sqrt{5}$$

Using the arc length formula, we evaluate

$$\int_{-2}^{2} \sqrt{1+(2)^2}\,dx = \int_{-2}^{2} \sqrt{5}\,dx = \sqrt{5}x\Big|_{-2}^{2} = 4\sqrt{5}$$

Example 2 Find the arc length of $f: x \mapsto x^{3/2}$ on $[\frac{5}{9}, 4]$. Here we have $f'(x) = \frac{3}{2}x^{1/2}$ and so the arc length is

$$L = \int_{5/9}^{4} \sqrt{1+\tfrac{9}{4}x}\,dx$$

To find this integral, we substitute $u = 1+\frac{9}{4}x$. We have $du = \frac{9}{4}\,dx$ and $\int u^{1/2}\,du = \frac{2}{3}u^{3/2}+c$. By the method of substitution, we deduce that

$$\int \sqrt{1+\tfrac{9}{4}x}\,dx = (\tfrac{4}{9})(\tfrac{2}{3})(1+\tfrac{9}{4}x)^{3/2} + c$$

Therefore, by the Fundamental Theorem, we have

$$L = \int_{5/9}^{4} \sqrt{1+\tfrac{9}{4}x}\,dx = (\tfrac{4}{9})(\tfrac{2}{3})[(1+\tfrac{9}{4}x)^{3/2}]_{5/9}^{4}$$

$$= \tfrac{8}{27}(10^{3/2} - \tfrac{27}{8})$$

We remark here that the formula for arc length can be derived without recourse to Riemann sums. Recall that to prove the Fundamental Theorem of Calculus we used certain geometrically evident facts about area to show that the derivative of $x \mapsto \int_a^x f$ is f itself. Similarly, with Cavaliere's Principle for volume, we showed that the derivative of the volume function $V(x)$ was the cross-sectional area function $A(x)$; in this case we used certain simple intuitive facts about volume and the assumption that $A(x)$ was continuous.

Analogous simple, intuitive facts about arc length are:

(1) "A straight line is the shortest distance between two points." The arc length of f on $[a,b]$ is greater than or equal to the length of the (straight) line segment from $(a,f(a))$ to $(b,f(b))$ (Figure 3-3).

(2) "The steeper the curve, the longer the arc length." More precisely, if g and f are both continuously differentiable on $[a,b]$, and if $|g'(x)| \geqslant |f'(x)|$ for all x in $[a,b]$, then the arc length of g on $[a,b]$ is greater than or equal to the arc length of f on $[a,b]$ (Figure 3-4).

(3) "Additivity." The arc length of f on $[a,b]$ is equal to the arc length of f on $[a,c]$ plus the arc length of f on $[c,b]$ whenever $a \leqslant c \leqslant b$.

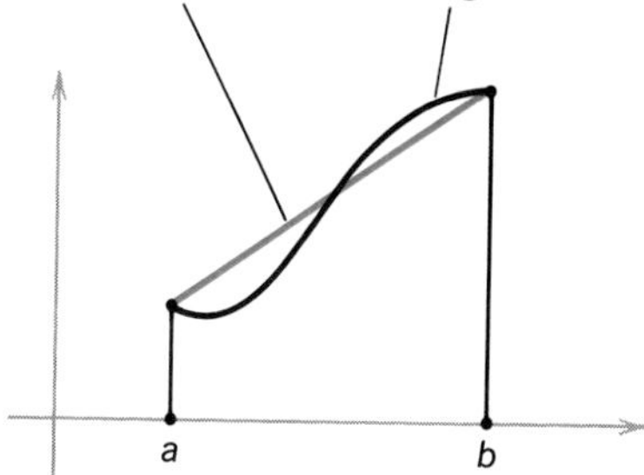

FIGURE 3-3

Now suppose f is continuously differentiable on $[a,b]$; let us denote by $S(x)$ the arc length of f on the interval $[a,x]$. Using (1), (2), and (3), we shall show that $S'(x)=\sqrt{1+f'(x)^2}$. Let $x_0 \in [a,b]$; we want to show that

$$\lim_{\Delta x\to 0} \frac{S(x_0+\Delta x)-S(x_0)}{\Delta x} = \sqrt{1+f'(x_0)^2}$$

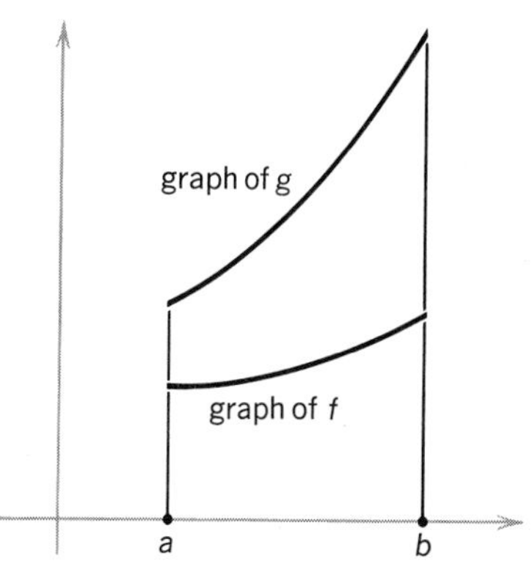

FIGURE 3-4

Let $\Delta x \neq 0$ be given and to fix ideas let us assume $\Delta x > 0$. By (3) $S(x_0+\Delta x)-S(x_0)$ is the arc length of f on $[x_0, x_0+\Delta x]$. Let us set $\Delta y = f(x_0+\Delta x)-f(x_0)$. The distance from $(x_0, f(x_0))$ to $(x_0+\Delta x, f(x_0)+\Delta y)$ is $\sqrt{\Delta x^2+\Delta y^2}$. By (1),

$$\sqrt{\Delta x^2+\Delta y^2} \leqslant S(x_0+\Delta x)-S(x_0)$$

Let $M=|f'(c)|$ be the maximum value of $|f'|$ on $[x_0, x_0+\Delta x]$. The straight line passing through the point $(x_0, f(x_0))$ with slope M is the graph of the affine function $A: x \mapsto f(x_0)+M(x-x_0)$ (Figure 3-5). Clearly, $A'(x) = M \geqslant |f'(x)|$ on $[x_0, x_0+\Delta x]$ so we have by (2),

$$S(x_0+\Delta x)-S(x_0) \leqslant \sqrt{M^2\Delta x^2+\Delta x^2} = \Delta x\sqrt{1+f'(c)^2}$$

Combining this result with the inequality above, and dividing by Δx, we obtain

$$\frac{\sqrt{\Delta x^2+\Delta y^2}}{\Delta x} \leqslant \frac{S(x_0+\Delta x)-S(x_0)}{\Delta x} \leqslant \sqrt{1+f'(c)^2}$$

or

$$\sqrt{1+\left(\frac{\Delta y}{\Delta x}\right)^2} \leqslant \frac{S(x_0+\Delta x)-S(x_0)}{\Delta x} \leqslant \sqrt{1+f'(c)^2} \qquad (6)$$

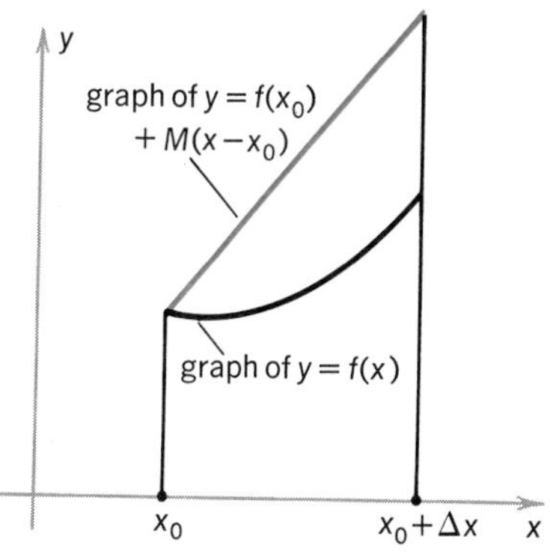

FIGURE 3-5

As $\Delta x \to 0$, c is squeezed to x_0 and, by continuity, $\sqrt{1+f'(c)^2}$ approaches $\sqrt{1+f'(x_0)^2}$. On the other hand, we also have

$$\lim_{\Delta x \to 0} \sqrt{1+\left(\frac{\Delta y}{\Delta x}\right)^2} = \sqrt{1+f'(x_0)^2}$$

Therefore, it follows that

$$\lim_{\Delta x \to 0} \frac{S(x_0+\Delta x) - S(x_0)}{\Delta x} = \sqrt{1+f'(x_0)^2}$$

or $S'(x_0) = \sqrt{1+f'(x_0)^2}$. Hence using the Fundamental Theorem we conclude that the arc length is $\int_a^b \sqrt{1+f'(x)^2}\,dx$ as before.

Not all curves are graphs of functions. A **parametrized curve** is a set of points in the plane with coordinates x and y which depend on a **parameter** t

$$x = f(t), \qquad y = g(t) \tag{7}$$

See Figure 3-6. You may wish to review §4.2 at this time.

To develop a formula for the arc length of a parametrized curve, assume we are given two continuously differentiable functions $f: t \mapsto x = f(t)$ and $g: t \mapsto y = g(t)$ which are defined on the closed interval $[\alpha, \beta]$ (we may think of $[\alpha, \beta]$ as an interval of time, and of $(f(t), g(t))$ as the position at time $t \in [\alpha, \beta]$ of a particle moving in the plane).

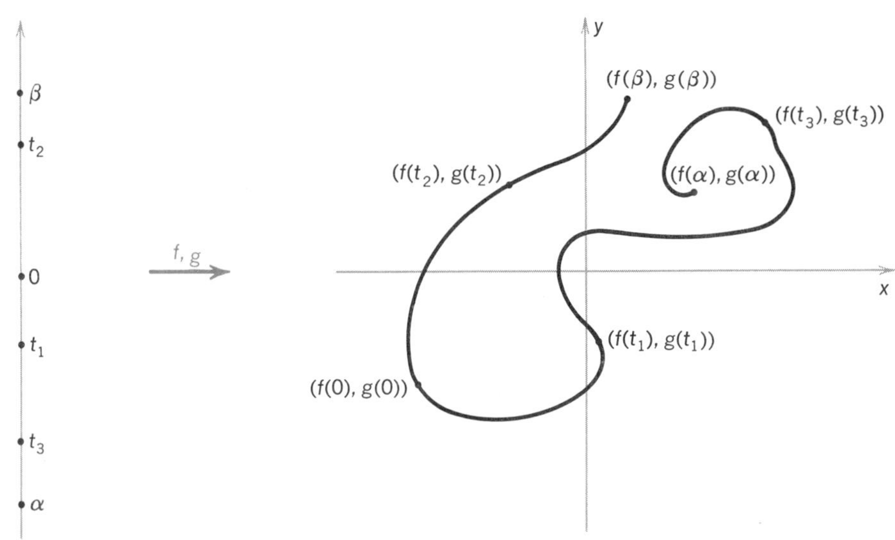

FIGURE 3-6

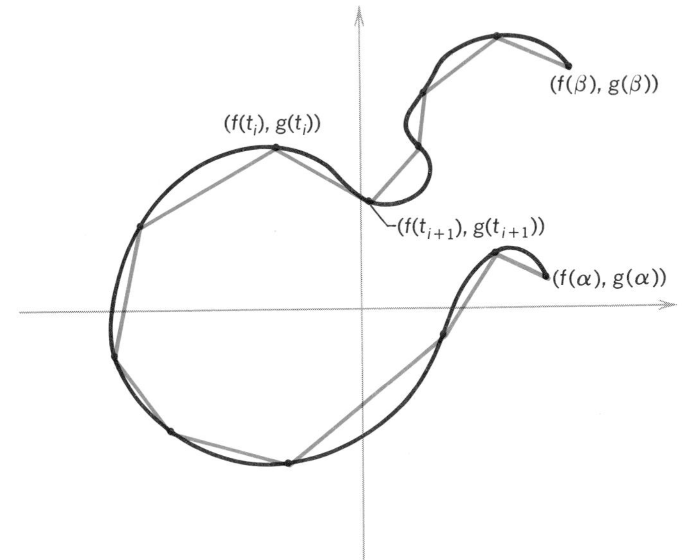

FIGURE 3-7

Let $\alpha = t_0, \ldots, t_{2^n} = \beta$ be the subdivision points of a regular partition of $[\alpha, \beta]$ of order $N = 2^n$. Then we may connect the points $(f(t_i), g(t_i))$ and $(f(t_{i+1}), g(t_{i+1}))$ by line segments, for $i = 0, \ldots, N$, so that the resulting polygonal curve fits along the given "smooth" curve (Figure 3-7).

We regard the length of this polygonal curve as an approximation to the length of the given curve which improves as $n \to \infty$. Applying the distance formula to each of these line segments, the sum of their length is

$$\sum_{i=0}^{N-1} \sqrt{(f(t_{i+1}) - f(t_i))^2 + (g(t_{i+1}) - g(t_i))^2} \tag{8}$$

We wish to bring (8) into the form of the general term of a sequence which clearly converges to a definite integral. As before, we apply the Mean Value Theorem to the differences $f(t_{i+1}) - f(t_i)$ and $g(t_{i+1}) - g(t_i)$.

Since f and g do satisfy the hypotheses of the Mean Value Theorem (Why?), there are numbers c_i and $\bar{c}_i$ in $[t_i, t_{i+1}]$ such that

$$f(t_{i+1}) - f(t_i) = f'(c_i)(t_{i+1} - t_i)$$

and

$$g(t_{i+1}) - g(t_i) = g'(\bar{c}_i)(t_{i+1} - t_i)$$

Substituting in (8) and bringing $(t_{i+1} - t_i)$ outside the radical sign, we can rewrite (8) as

$$\sum_{i=0}^{N-1} \sqrt{f'(c_i)^2 + g'(\bar{c}_i)^2}\,(t_{i+1} - t_i) \tag{9}$$

Now (9) is **not** an Nth Riemann sum, because c_i and $\bar{c}_i$ are not necessarily equal. However, it can be proved (the interested reader should consult the remarks at the end of §4), on the basis of the continuity of f' and g', that (9) has the same limit as

$$\sum_{i=0}^{N-1} \sqrt{f'(c_i)^2+g'(c_i)^2}\,(t_{i+1}-t_i) \tag{10}$$

as $n \to \infty$. The expression (10) **is** an Nth Riemann sum for the continuous function $t \mapsto \sqrt{f'(t)^2+g'(t)^2}$, so its limit as $n \to \infty$ is the definite integral

$$\int_\alpha^\beta \sqrt{f'(t)^2+g'(t)^2}\,dt$$

This number is by definition the **arc length** of the parametrized curve (7).

Example 3 Find the arc length of the parametrized curve described by the parametric equations $x = t^3+1$ and $y = 2t^{9/2}-4$ over $[1,3]$. Writing $f: t \mapsto t^3+1$ and $g: t \mapsto 2t^{9/2}-4$, we have the derivatives $f'(t) = 3t^2$ and $g'(t) = 9t^{7/2}$, both of which are continuous functions on $[1,3]$. Hence, the arc length is

$$\begin{aligned} L &= \int_1^3 \sqrt{(3t^2)^2+(9t^{7/2})^2}\,dt \\ &= \int_1^3 \sqrt{9t^4+81t^7}\,dt \\ &= 3\int_1^3 t^2\sqrt{1+9t^3}\,dt \end{aligned}$$

Let us substitute $u = 1+9t^3$, so $du = 27t^2\,dt$; and we evaluate

$$3\int \tfrac{1}{27}\sqrt{u}\,du = \tfrac{2}{27}u^{3/2} + C$$

By the method of substitution, we have

$$3\int t^2\sqrt{1+9t^3}\,dt = \tfrac{2}{27}(1+9t^3)^{3/2} + C$$

Hence,

$$L = \left[\tfrac{2}{27}(1+9t^3)^{3/2}\right]_1^3 = \tfrac{2}{27}(488\sqrt{61}-10\sqrt{10})$$

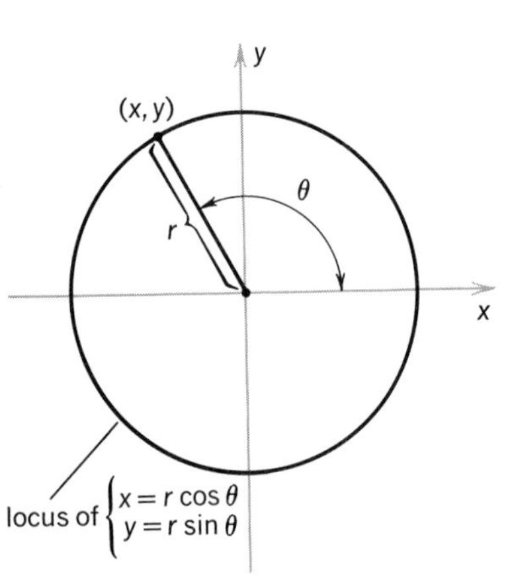

FIGURE 3-8

Example 4 Find the length of the parametrized curve given by $x = r\cos\theta$, $y = r\sin\theta$, over the interval $[0, 2\pi]$.

This curve is just the circle of radius r parametrized by an angle θ as shown in Figure 3-8. If we compute the length of this curve by a definite integral, we should obtain the well-known circumference $2\pi r$.

To proceed, therefore, we find the derivatives of the parametric functions to be $\theta \mapsto -r\sin\theta$ and $\theta \mapsto r\cos\theta$, respectively, which are both continuous on $[0, 2\pi]$. Hence, the length is given by

$$L = \int_0^{2\pi} \sqrt{(-r\sin\theta)^2 + (r\cos\theta)^2}\, d\theta$$

$$= r\int_0^{2\pi} \sqrt{\sin^2\theta + \cos^2\theta}\, d\theta$$

$$= r\int_0^{2\pi} 1\, d\theta = r(2\pi - 0) = 2\pi r$$

It is inexact to say that the integral $\int_a^b \sqrt{X'(t)^2 + Y'(t)^2}\, dt$ is the length of the arc described by the equations

$$x = X(t), \qquad y = Y(t), \qquad a \leqslant t \leqslant b$$

Rather, the above integral gives the total displacement or total distance traveled by a particle whose motion is described by these equations. Let us consider an example. Suppose our equations are

$$x = t^2 = X(t), \qquad y = t^2 = Y(t), \qquad -1 \leqslant t \leqslant 1$$

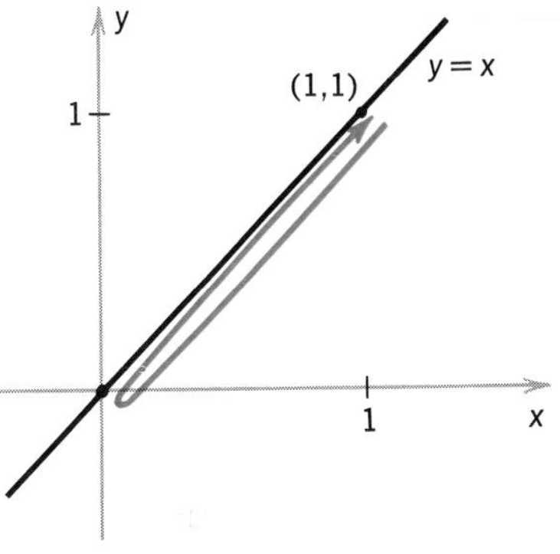

FIGURE 3-9

As t goes from -1 to 1, the point $(X(t), Y(t))$ travels along the line $y = x$ from the point $(1, 1)$ down to the origin and then back up again to $(1, 1)$ (Figure 3-9). So the distance traveled is $d = 2\sqrt{2}$, while the length of the line segment from $(0,0)$ to $(1,1)$ is $\sqrt{2}$. The formula for arc length (or better, distance traveled) yields

$$d = \int_{-1}^{1} \sqrt{4t^2 + 4t^2}\, dt$$

$$= \int_{-1}^{1} 2\sqrt{2}(\sqrt{t^2})\, dt = \int_{-1}^{1} 2\sqrt{2}|t|\, dt$$

$$= 2\sqrt{2}\left(\int_{-1}^{0} -t\, dt + \int_0^1 t\, dt\right) = 2\sqrt{2}\left(\tfrac{1}{2} + \tfrac{1}{2}\right) = 2\sqrt{2}$$

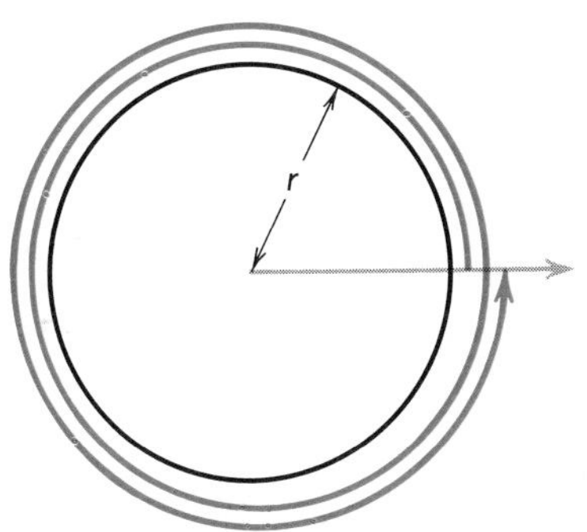

FIGURE 3-10

As another example, consider

$$x = r\cos\theta, \qquad y = r\sin\theta, \qquad 0 \leqslant \theta \leqslant 4\pi$$

These equations describe a point moving **twice** around a circle of radius r. So the integral $\int_0^{4\pi} r\, d\theta = 4\pi r$ is equal to **twice** the circumference of the circle (Figure 3-10).

We may associate with a given function $f:[a, b] \to \boldsymbol{R}$ the function $S:[a, b] \to \boldsymbol{R}$ which assigns to $x \in [a, b]$ the arc length of the graph of f from a to x. (Two different functions f_0 and f_1 may give rise to the same associated arc length function S. Can you give an example?) If we use the notations $y = f(x)$ and $s = S(x)$ we may write

$$S(x) = \int_a^x \sqrt{1 + f'(u)^2}\, du$$

and

$$\frac{ds}{dx} = \sqrt{1 + f'(x)^2}$$

so that $ds = \sqrt{1 + f'(x)^2}\, dx$ (Figure 3-11).

Therefore, the tangent to the graph of $s=S(x)$ at $(x,S(x))$ has slope $\sqrt{1+f'(x)^2}$. In the $dx\,ds$ coordinate system, the tangent line has equation

$$ds = \sqrt{1+f'(x)^2}\,dx$$

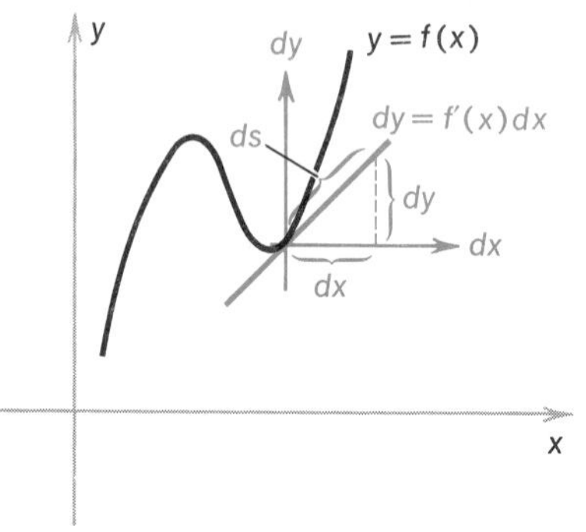

Furthermore, $f'(x)=dy/dx$ implies $f'(x)^2=(dy/dx)^2$, so that

$$ds = \sqrt{1+\left(\frac{dy}{dx}\right)^2}\,dx, \quad \text{and} \quad ds^2 = \left[1+\left(\frac{dy}{dx}\right)^2\right]dx^2$$

that is,

$$ds^2 = dx^2 + dy^2$$

This last equation can be interpreted geometrically as follows: Consider the tangent to the graph of f at $(x,f(x))$. In $dx\,dy$ coordinates this line has equation $dy=f'(x)\,dx$ (Figure 3-12). The variable ds represents the length of the hypotenuse of the right triangle with sides of length dx and dy. The triangle so formed—with sides dx and dy and hypotenuse ds—is called the **differential triangle.**

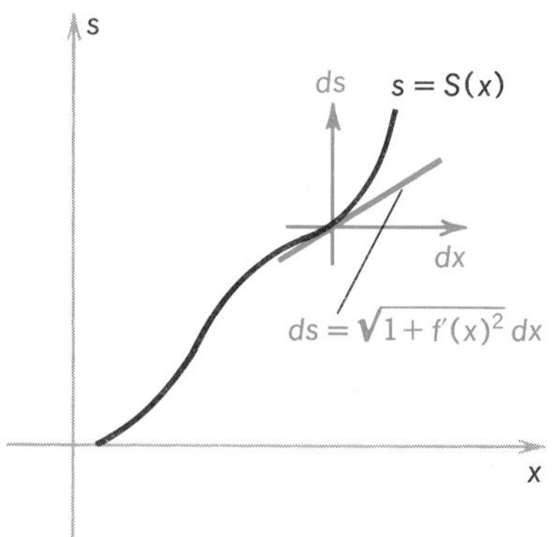

FIGURE 3-11

If a curve is described by parametric equations

$$x = f(t), \qquad y = g(t), \qquad a \leqslant t \leqslant b$$

then the function $S: t \mapsto \int_a^t \sqrt{[f'(u)]^2+[g'(u)]^2}\,du$ is called the **arc length function.** The number $S(t_0)$ is the length of the path traced out by the point $(f(t),g(t))$ as t goes from a to t_0. Setting $s=S(t)$, we have

$$\frac{ds}{dt} = \sqrt{[f'(t)]^2+[g'(t)]^2} = \sqrt{\left(\frac{dx}{dt}\right)^2+\left(\frac{dy}{dt}\right)^2}$$

and we again find

$$ds^2 = dx^2 + dy^2$$

This equation can again be interpreted geometrically by the differential triangle.

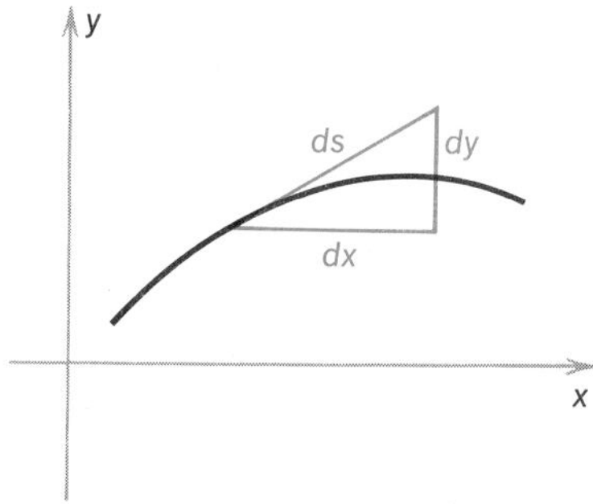

FIGURE 3-12

The derivative ds/dt measures the rate of change of distance traveled with respect to time t:

$$\frac{ds}{dt} = \lim_{\Delta t\to 0}\frac{S(t+\Delta t)-S(t)}{\Delta t} = \lim_{\Delta t\to 0}\frac{\text{change in distance traveled}}{\text{change in time}}$$

Therefore the number

$$\left.\frac{ds}{dt}\right|_{t=t_0} = S'(t_0)$$

is defined to be the **speed** of the particle at time t_0. We say speed and not velocity because ds/dt is always a nonnegative number and is the rate of change of total distance traveled with respect to time. Even if a particle "turns around" and "retraces its steps," the total distance traveled $S(t)$ is still increasing; hence $S'(t) = ds/dt \geqslant 0$.

Example 5 A particle moves in an elliptical path according to the equations

$$x = a\cos 2t, \qquad y = b\sin 2t, \qquad a, b > 0, \qquad 0 \leqslant t \leqslant \pi$$

What is the speed of the particle at the points $(a, 0)$, $(0, b)$, $(-a, 0)$, and $(0, -b)$? We have

$$dx = -2a\sin 2t\,dt, \qquad dy = 2b\cos 2t\,dt$$

$$ds^2 = dx^2 + dy^2$$

$$= (4a^2\sin^2 2t + 4b^2\cos^2 2t)\,dt^2$$

or

$$\left(\frac{ds}{dt}\right)^2 = 4a^2\sin^2 2t + 4b^2\cos^2 2t$$

The particle passes through the designated points at $t = 0$, $t = \frac{1}{4}\pi$, $t = \frac{1}{2}\pi$, and $t = \frac{3}{4}\pi$, so we evaluate

$$\left(\left.\frac{ds}{dt}\right|_{t=0}\right)^2 = 4b^2, \qquad \left.\frac{ds}{dt}\right|_{t=0} = 2b$$

$$\left(\left.\frac{ds}{dt}\right|_{t=\frac{1}{4}\pi}\right)^2 = 4a^2, \qquad \left.\frac{ds}{dt}\right|_{t=\frac{1}{4}\pi} = 2a$$

$$\left(\left.\frac{ds}{dt}\right|_{t=\frac{1}{2}\pi}\right)^2 = 4b^2, \qquad \left.\frac{ds}{dt}\right|_{t=\frac{1}{2}\pi} = 2b$$

$$\left(\left.\frac{ds}{dt}\right|_{t=\frac{3}{4}\pi}\right)^2 = 4a^2, \qquad \left.\frac{ds}{dt}\right|_{t=\frac{3}{4}\pi} = 2a$$

Example 6 Establish the following formula for the perpendicular distance from the origin to the tangent to a parametrized curve:

$$\mathbf{d} = \left|x\frac{dy}{ds} - y\frac{dx}{ds}\right|$$

Figure 3-13 shows the differential triangle and two other triangles similar to it. In the notation of the figure we have

$$x_1 + x_2 = x, \qquad \frac{\mathbf{d}}{x_1} = \frac{dy}{ds}, \qquad \frac{y}{x_2} = \frac{dy}{dx}$$

Also,

$$x_1 = x - y\frac{dx}{dy}$$

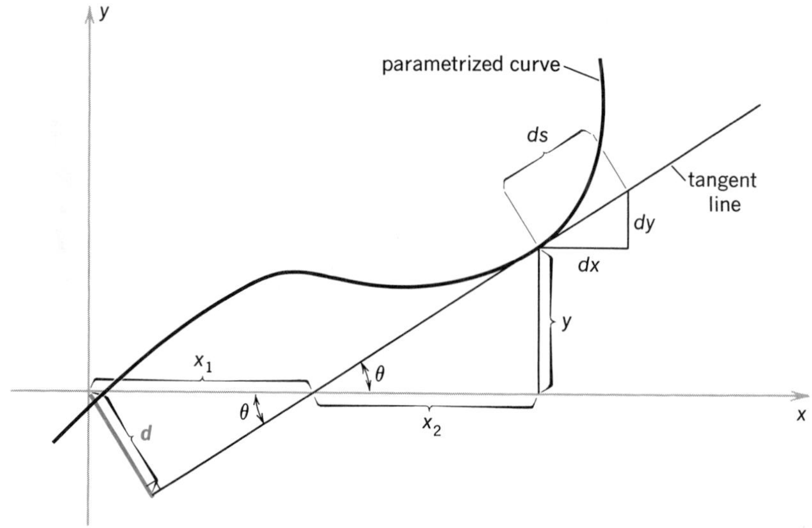

FIGURE 3-13

and hence,

$$\boldsymbol{d} = \left(x - y\frac{dx}{dy}\right)\frac{dy}{ds} = x\frac{dy}{ds} - y\frac{dx}{dy}\frac{dy}{ds}$$

Therefore,

$$\boldsymbol{d} = x\frac{dy}{ds} - y\frac{dx}{ds}$$

In cases other than that in Figure 3-13 we might get a negative sign, so the general situation is covered by taking the absolute value of the expression. Note that if the tangent is horizontal, division by dy is not possible. However, in this case we have $dx/ds = 1$ and the formula $\boldsymbol{d} = |y(ds/dx)|$ is trivial. The tangent line is not defined if both $dx/dt = 0$ and $dy/dt = 0$.

EXERCISES

1. Find the length of the arc described by the following equations:

   (a) $y = \dfrac{x^3}{6} + \dfrac{1}{2x}$ from $x = 2$ to $x = 5$

   (b) $y = \dfrac{x^5}{5} + \dfrac{1}{12x^3}$ from $x = \frac{1}{2}$ to $x = 1$

   (c) $y = \int_0^x \sqrt{u}\,du$ from $x = 0$ to $x = \frac{1}{4}\pi$

   (d) $x = a\cos^3\theta$, $y = a\sin^3\theta$, $0 \leqslant \theta \leqslant \frac{1}{2}\pi$

   (e) $x = \cos t + t\sin t$, $y = \sin t - t\cos t$, $0 \leqslant t \leqslant \frac{1}{2}\pi$

   (f) $x = a(t - \sin t)$, $y = a(1 - \cos t)$, $a > 0$, $0 \leqslant t \leqslant 2\pi$

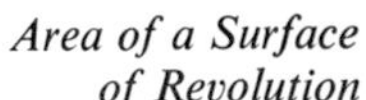

2. Using the definition of the arc length of h on $[a, b]$ as the definite integral $\int_a^b \sqrt{1+h'(x)^2}\,dx$ (where h is continuously differentiable on $[a, b]$), check properties (1)–(3) for arc length.

In exercises 3–6, f is a function continuously differentiable on $[a, b]$ and $S(x)$ denotes the arc length of f on $[a, x]$.

3. Show that S is continuously differentiable and has no critical points on $[a, b]$. Conclude that S is monotone increasing. Also show that x_0 is a point of inflection for S if and only if $f'(x)\cdot f''(x)$ changes sign at x_0. Show that $S'(x) \geq 1$ and interpret geometrically.
4. Let S^{-1} denote the inverse of S. What is the domain of S^{-1}? Show that S^{-1} is differentiable. Suppose $S_0 = S(x_0)$; express $(S^{-1})'(S_0)$ in terms of $f(x_0)$.
5. Consider the parametric equations
$$x = S^{-1}(s) = X(s), \qquad y = f(S^{-1}(s)) = Y(s)$$
Express dx/ds and dy/ds in terms of $f'(x)$. [Hint: $dx/ds = 1/(ds/dx)$] Show that $(dx/ds)^2+(dy/ds)^2 = 1$.
6. Suppose also that f is monotone increasing on $[a, b]$. Show that S is concave upward on $[a, b]$ if f is concave upward on $[a, b]$; show that S is concave downward on $[a, b]$ if f is concave downward on $[a, b]$.
7. Suppose a curve is defined by the parametric equations
$$x = X(t), \qquad y = Y(t)$$
where X and Y are continuously differentiable. Suppose $P = (X(t_0), Y(t_0))$ is a point on the curve and let $Q = (X(t_0+\Delta t), Y(t_0+\Delta t))$ be another point on the curve. Denote the length of the arc of the curve from P to Q by $\widehat{PQ}$ and the length of the chord by $\overline{PQ}$. Show that
$$\lim_{\Delta t \to 0} \frac{\widehat{PQ}-\overline{PQ}}{\Delta t} = 0$$
[Hint: Use the Mean Value Theorem.]

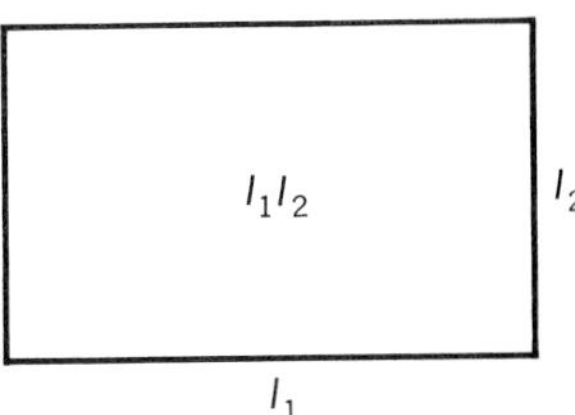

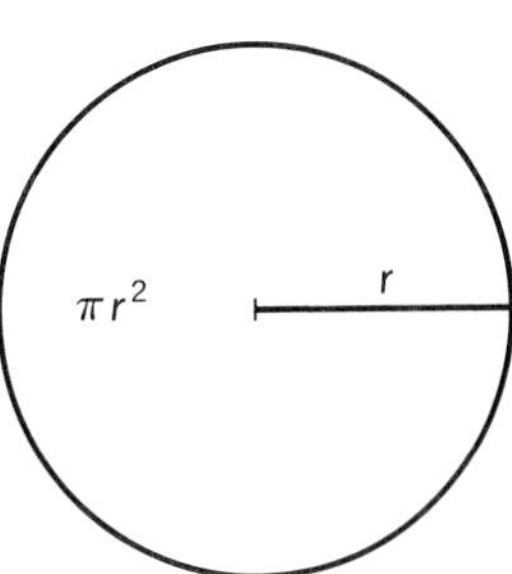

FIGURE 4-1

4 AREA OF A SURFACE OF REVOLUTION

The area of a rectangle is the product of the length of its sides, and the circumference of a circle of radius r is $2\pi r$ (Figure 4-1). From these facts it is not difficult to prove that the lateral area of a right circular cylinder of base radius r and altitude l is $2\pi rl$. Similarly, it is easy to show that the area of the lateral surface of a right circular cone of base radius r and altitude h is equal to half the product of the slant height $l = \sqrt{r^2+h^2}$ and the circumference $c = 2\pi r$ of the base, that is, $A = \pi rl$ (Figure 4-2).

The lateral surfaces of the cylinder and cone are simple examples of **surfaces of revolution**. This means they can be obtained by revolving the graph of a function around an axis (Figure 4-3). It is natural to ask whether we can define the area of the surface generated by revolving the graph of a function f about a coordinate axis (Figure 4-4).

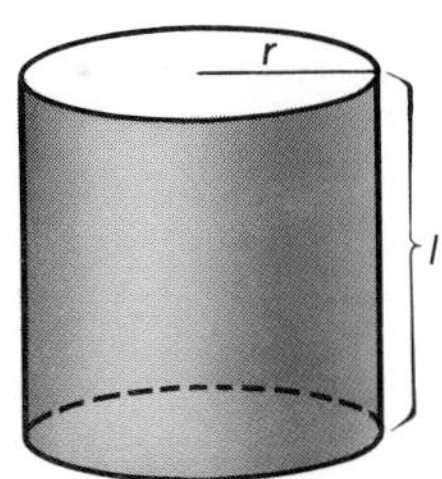

$A = 2\pi rl$

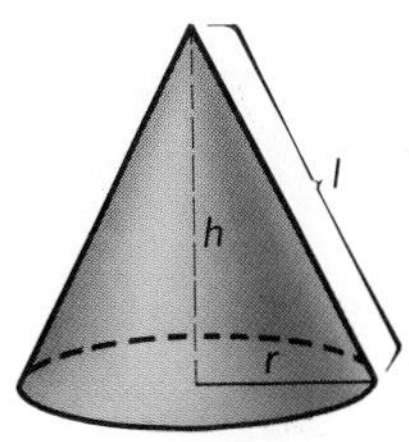

$A = \pi rl$

FIGURE 4-2

Employing the same method of approximation we used to define arc length, let us assume that f is a nonnegative continuous function on $[a, b]$ which is also continuously differentiable on $[a, b]$. Let $[a, b]$ be subdivided by the points $a = x_0, \ldots, x_{2^n} = b$ of a regular partition of order $N = 2^n$. Connect $(x_i, f(x_i))$ to $(x_{i+1}, f(x_{i+1}))$ by a line segment for $i = 0, \ldots, 2^n - 1$, so that the graph of f is approximated by a polygonal curve (Figure 4-5).

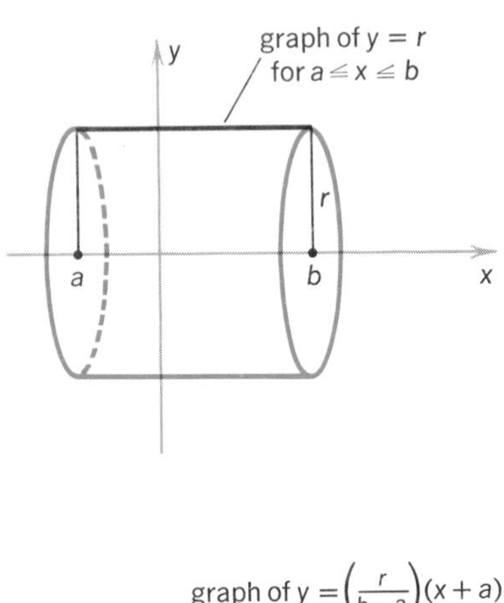

FIGURE 4-3

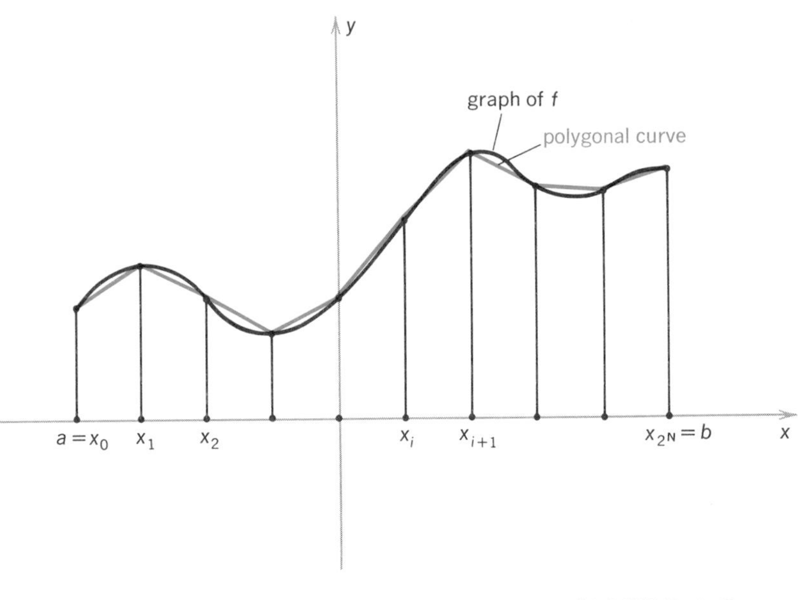

FIGURE 4-5

When this entire figure is revolved around the x axis, the graph of f generates a surface, and the polygonal approximation to f also generates a surface. The area of the latter surface is an approximation to the surface area of the revolved graph of f.

The area of the approximation is the sum of 2^n areas, any one of which is the lateral area of a frustum of a cone (Figure 4-6). In exercise 1 the reader is asked to prove the result from geometry that the area of a frustum of a cone with slant height l and radii r_1, and r_2 of its faces is

$$\pi l(r_1 + r_2) \tag{1}$$

Using (1) on each of the frustra of the polygonal surface, the following sum represents an approximation to the area of the surface of revolution of the graph of f:

$$\sum_{i=0}^{N-1} \pi(f(x_i) + f(x_{i+1}))\sqrt{(x_{i+1} - x_i)^2 + (f(x_{i+1}) - f(x_i))^2} \tag{2}$$

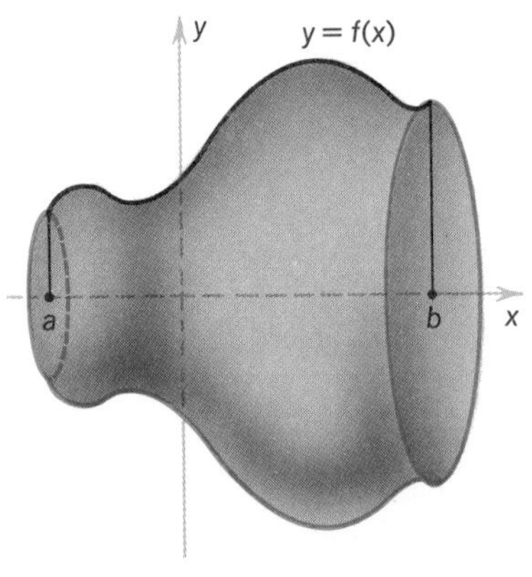

FIGURE 4-4

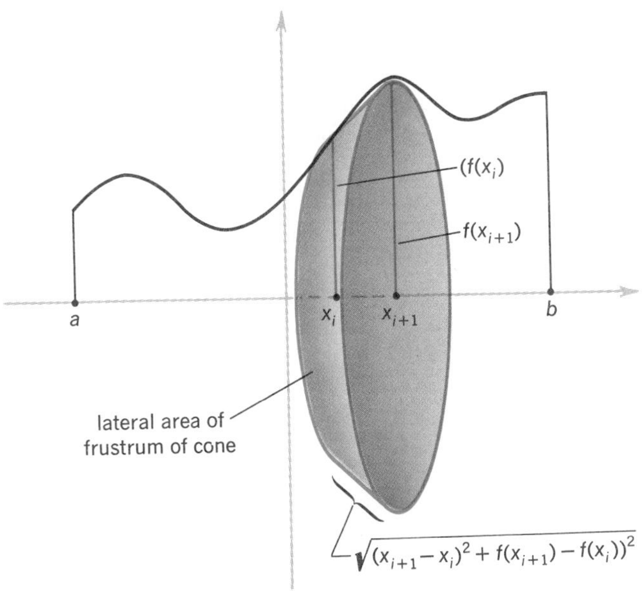

FIGURE 4-6

In each term in the summation (2) we apply the Mean Value Theorem, yielding

$$f(x_{i+1}) - f(x_i) = f'(c_i)(x_{i+1} - x_i), \qquad x_i \leqslant c_i \leqslant x_{i+1}$$

and factor the term $(x_{i+1} - x_i)^2$ from the radical, yielding

$$\sum_{i=0}^{N-1} \pi(f(x_i) + f(x_{i+1}))\sqrt{1 + [f'(c_i)]^2}\,(x_{i+1} - x_i)$$

$$= \sum_{i=0}^{N-1} \pi f(x_i)\sqrt{1 + [f'(c_i)]^2}\,(x_{i+1} - x_i)$$

$$+ \sum_{i=0}^{N-1} \pi f(x_{i+1})\sqrt{1 + [f'(c_i)]^2}\,(x_{i+1} - x_i) \qquad (3)$$

Because of the continuity hypothesis, each of the summands above has the same limit as $n \to \infty$ as the Riemann sum

$$\sum_{i=0}^{N-1} \pi f(c_i)\sqrt{1 + [f'(c_i)]^2}\,(x_{i+1} - x_i)$$

(The interested reader should consult the remarks at the end of this section.)

So the limit of (3) as $n \to \infty$ is

$$\int_a^b \pi f(x)\sqrt{1 + [f'(x)]^2}\,dx + \int_a^b \pi f(x)\sqrt{1 + [f'(x)]^2}\,dx$$

$$= 2\pi \int_a^b f(x)\sqrt{1 + [f'(x)]^2}\,dx$$

This number is the **(lateral) area** of the surface of revolution generated by revolving the graph of f around the x axis.

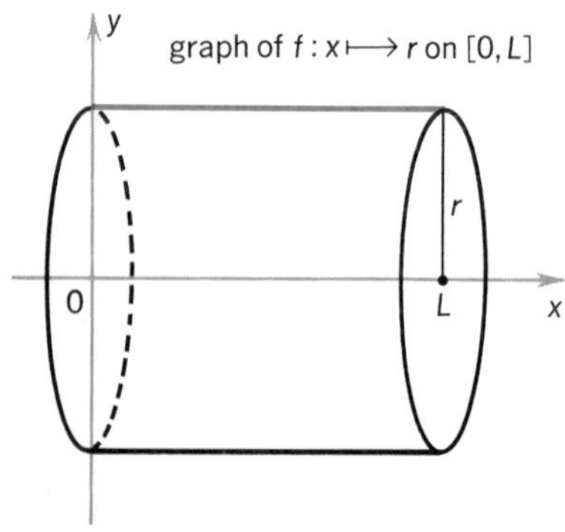

FIGURE 4-7

Example 1 Let us verify our formula for surface area in a simple known case, the lateral area of a cylinder of length L and radius r. We regard this surface as a surface of revolution whose "radius function" is the constant function $f: x \mapsto r$ on the interval $[0, L]$ (Figure 4-7). Then the surface area of revolution is

$$\begin{aligned}
&\int_0^L 2\pi f(x)\sqrt{1+[f'(x)]^2}\,dx \\
&= \int_0^L 2\pi r\sqrt{1+0}\,dx \\
&= \int_0^L 2\pi r\,dx \\
&= 2\pi r \int_0^L dx \\
&= 2\pi r(L-0) = 2\pi rL
\end{aligned}$$

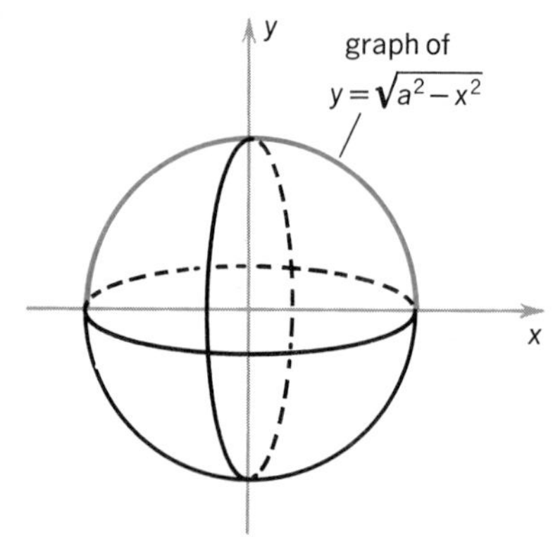

FIGURE 4-8

Example 2 Use the formula of this section to find the surface area of a sphere of radius a.

We consider the surface of the sphere to be the surface of revolution generated by the semicircle $y = \sqrt{a^2 - x^2}$ (Figure 4-8). Then the area is given by

$$\begin{aligned}
&\int_{-a}^{a} 2\pi\sqrt{a^2-x^2}\sqrt{1+\left(\frac{x^2}{a^2-x^2}\right)}\,dx \\
&= \int_{-a}^{a} 2\pi\sqrt{a^2-x^2+x^2}\,dx \\
&= \int_{-a}^{a} 2\pi a\,dx = 2\pi a x\Big|_{-a}^{a} = 4\pi a^2
\end{aligned}$$

Revolving the graph of a function $f: x \mapsto f(x)$ about the y axis also gives rise to a surface of revolution (see Figure 4-9). If we apply the analysis of this surface in terms of polygonal approximations to the graph of f, we find the following formula for surface area (the student should work this case out for himself).

$$A = \int_a^b 2\pi x\sqrt{1+[f'(x)]^2}\,dx$$

For this formula to apply, we do not need to assume that $f(x) \geqslant 0$ on $[a, b]$; but we do assume $a \geqslant 0$.

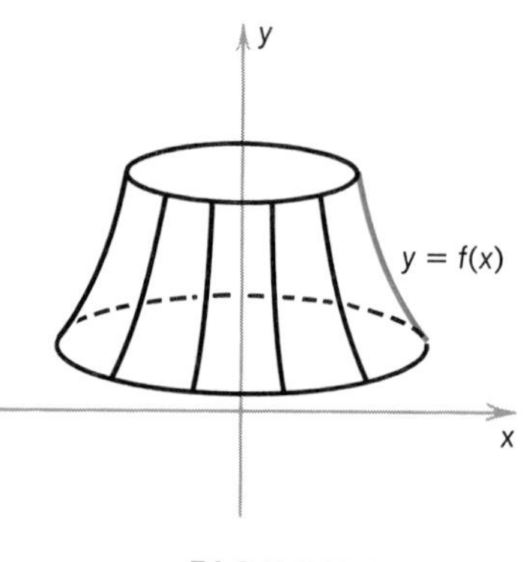

FIGURE 4-9

Example 3 Find the surface area generated by revolving the graph of $y = \frac{1}{2}x^2$ from 0 to 1 about the y axis. Applying the above formula we obtain

$$A = \int_0^1 2\pi x\sqrt{1+x^2}\,dx$$

We make the substitution $u = 1 + x^2$ which yields $du = 2x\,dx$. Then we have $\int \pi u^{1/2}\,du = \frac{2}{3}\pi u^{3/2} + C$ and substituting back we obtain $\int \pi 2x\sqrt{1+x^2}\,dx = \frac{2}{3}\pi(1+x^2)^{3/2} + C$. We now evaluate

$$A = \int_0^1 2\pi x\sqrt{1+x^2}\,dx = \tfrac{2}{3}\pi[(1+x^2)^{3/2}]_0^1$$
$$= \tfrac{2}{3}\pi(\sqrt{8}-1) = \tfrac{2}{3}\pi(2\sqrt{2}-1)$$

In §3 we saw that the formula for arc length could be derived from reasonable geometric assumptions about arc length as an alternative to Riemann sums. One can do a similar thing for surface area (exercise 8).

As a final remark, we note that in §3 and §4 we have encountered approximating sums such as

$$\sum_{i=0}^{N-1} \pi\sqrt{[f'(c_i)]^2 + [g'(\bar{c}_i)]^2}\,(x_{i+1}-x_i) \tag{4}$$

and

$$\sum_{i=0}^{N-1} \pi f(x_i)\sqrt{1+[f'(c_i)]^2}\,(x_{i+1}-x_i) \tag{5}$$

which correspond to a partition $a = x_0, x_1, \ldots, b = x_N$ of the interval $[a, b]$, but which are not Riemann sums, because the "function part" of the sum is not a single function evaluated at some point c_i with $x_i \leqslant c_i \leqslant x_{i+1}$. When all the functions involved are continuous on $[a, b]$, as they are in the above applications, it can be proved that the sums (4) and (5) have the same limits respectively as $N \to \infty$ as the corresponding Riemann sums:

$$\sum_{i=0}^{N} \pi\sqrt{f'(c_i) + g'(c_i)}\,(x_{i+1}-x_i) \tag{4'}$$

and

$$\sum_{i=0}^{N} \pi f(c_i)\sqrt{1+[f'(c_i)]^2}\,(x_{i+1}-x_i) \tag{5'}$$

namely the integrals

$$\int_a^b \pi\sqrt{[f'(x)]^2 + [g'(x)]^2}\,dx \tag{6}$$

and

$$\int_a^b \pi f(x)\sqrt{1+[f'(x)]^2}\,dx \tag{7}$$

We refer, for example, to B. Mitchell, *Calculus Without Analytic Geometry*, D. C. Heath and Co., Lexington, Mass., 1969, Chapter IV, §9.

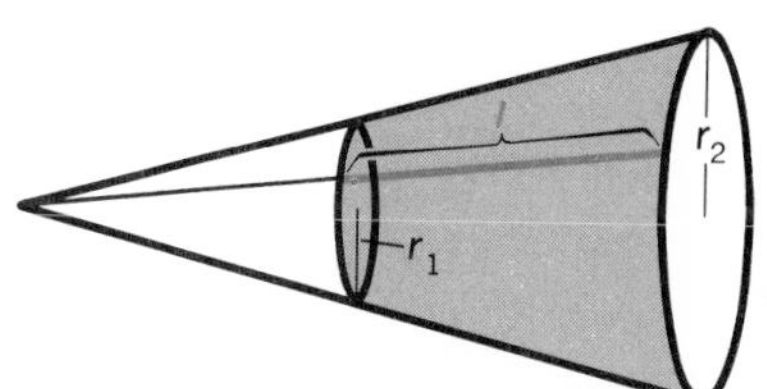

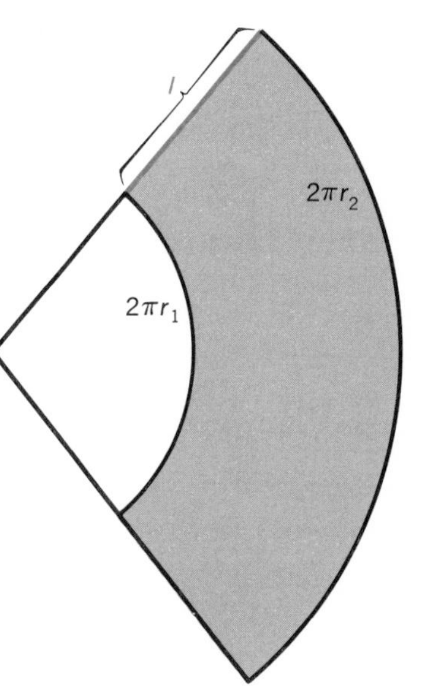

FIGURE 4-10

EXERCISES

1. Argue that the surface area of a frustrum of a cone is $\pi l(r_1+r_2)$ (l = slant height, r_1, r_2 are the radii of the faces). Do this as follows: Take the cone and slit it along a line from the base to the vertex and spread the cone flat. It will be a sector of a circle with radius the slant height of the whole cone and arc length $2\pi r_2$ (here $r_2 > r_1$). (See Figure 4-10.)

2. Find the lateral surface area obtained by revolving each of the following curves about the y axis:
   (a) $y = x^2$ from $x = 0$ to $x = 2$
   (b) $y = \frac{1}{3}(x^2+2)^{3/2}$ from $x = 0$ to $x = 3$
3. Find the lateral surface area of the solid obtained by revolving each of the following graphs about the x axis.
   (a) $y = x^3$ from $x = 0$ to $x = 1$
   (b) $y = mx$, $m > 0$, from $x = 0$ to $x = 2$
   (c) $y = \sqrt{a^2-x^2}$ from $x = \dfrac{-a}{2}$ to $x = \dfrac{a}{2}$
4. Let $f:x \mapsto mx$, $m > 0$, and let $g:x \mapsto 2m$. Revolving f on $[0,2]$ about the x axis generates a cone, and revolving g on $[0,2]$ generates a cylinder which circumscribes the cone. What are the values of m such that the cone has lateral area less than or equal to the lateral area of the circumscribed cylinder?

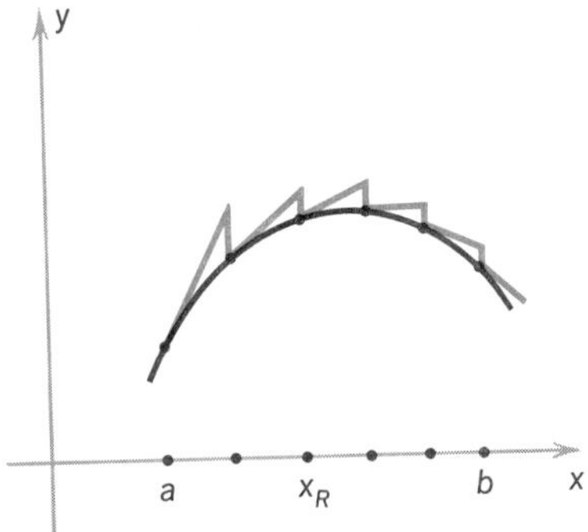

FIGURE 4-11

5. (a) Let f be continuously differentiable on $[a,b]$ and let $a = x_0, \ldots, x_k, \ldots, x_{2^n} = b$ be a regular partition of order $N = 2^n$. Set $\Delta x = (b-a)/2^n$. The tangent to f at $(x_k, f(x_k))$ has slope $f'(x_k)$, and the point $(x_{k+1}, f(x_k)+f'(x_k)\,\Delta x)$ lies on this tangent line. The tangents to f at the points $(x_k, f(x_k))$ lead to a "toothed" polygonal approximation to $y = f(x)$ as in Figure 4-11. The vertical segment over x_{k+1} of this polygon has length $\Delta s_k = |f'(x_k)\,\Delta x - (f(x_{k+1}) - f(x_k))|$. Let $r_n = \sum_{k=0}^{2^n-1} \Delta s_k$. Show $\lim_{n\to\infty} r_n = 0$.
   (b) Compute the area of the surface of revolution generated by the "toothed" polygonal approximation to $y = f(x)$ of (a). Show that these areas converge to the surface area generated by revolving f on $[a,b]$.
6. The curve parametrized by $x = \cos^3 t$, $y = \sin^3 t$, $0 \leqslant t \leqslant \frac{1}{2}\pi$ is rotated about the x axis. Find the lateral surface area generated. [Hint: This curve is the graph of a function $f:[0,1] \to \boldsymbol{R}$.]
7. Suppose the equations $x = x(t)$, $y = y(t)$, $a \leqslant t \leqslant b$, parametrize a curve which is also the graph of a continuously differentiable function $f:[x(a), x(b)] \to \boldsymbol{R}$. (a) Show that the surface area generated by revolving this curve about the x axis is $S = \int_a^b 2\pi\, y(t)\sqrt{x'(t)^2 + y'(t)^2}\, dt$ [Hint: Apply a change of variables]. (b) Suppose further that $x(a) \geqslant 0$. Find a formula for the lateral surface area obtained by revolving the curve about the y axis.

*8. As with arc length, derive the formula for the surface area of the graph of f revolved about the x axis using the following assumptions. Let $A_{x_1}^{x_2}(f)$ denote the area for f restricted to $[x_1, x_2]$. Assume
   (1) Additivity: $A_{x_1}^{x_2}(f) = A_{x_1}^{x_3}(f) + A_{x_3}^{x_2}(f)$.
   (2) Monotonicity: If $f(x_1) \geqslant g(x_2) \geqslant 0$ for all x_1, x_2 in $[c,d]$ and if
   $$\int_c^d \sqrt{1+[f'(x)]^2}\, dx \geqslant \int_c^d \sqrt{1+[g'(x)]^2}\, dx$$
   then $A_c^d(f) \geqslant A_c^d(g)$, whenever f and g are continuously differentiable on $[c,d]$.
   (3) The usual formula for the area of a frustrum of a cone holds.

5 WORK

Suppose a force F is applied to an object to move it along a straight line in one direction. If this force, usually measured in pounds, is a constant force, and it moves the object a distance of s feet, then the **work** done on the object is the product

$$W = Fs \tag{1}$$

Observe that if the unit of force is pounds, and the unit of distance is feet, then the unit of work will be foot-pounds. Other units of work may be inch-tons, or inch-pounds, and so on.

Example 1 Suppose we move an object 3 feet along a straight line with a constant force of 5 pounds applied in the direction of motion. Then the work done on the object is

$$W = 5 \cdot 3 = 15 \text{ ft-lb}$$

Suppose, however, that the force used to move the object is a variable force, that is, suppose the force is exerted in a fixed direction and at a given instant of time depends on the distance of the object from an initial point. (For example, stretching or compressing a spring.) In this case, equation (1) cannot be used directly to give the work done.

Let us consider the line of motion as an x coordinate axis with starting points at $x = a$. To simplify reality we idealize the object as a point moving along the line. Then the applied force is a function of x (x in turn may be a function of time t), and when the object is at the point x, the force applied is $F(x)$ units.

Let us suppose that the object moves from $x = a$ to $x = b$ and that in this interval $[a, b]$ the force function $x \mapsto F(x)$ is continuous. We would like to define the work done in moving the object from a to b.

To do this we first partition the interval, with $a = x_0, x_1, \ldots, x_{2^n} = b$ the subdivision points of a regular partition of $[a, b]$ of order $N = 2^n$. We assume for the moment that **additivity** is part of our definition of work; the work done in moving an object from a to b is the sum of the works done in moving the object from one intermediate position to the next. Thus if $W(u, v)$ denotes the work done in moving the object from u to v,

$$W(a, b) = W(a, x_1) + W(x_1, x_2) + \cdots + W(x_{2^n - 1}, b) \tag{2}$$

We consider the subinterval $[x_i, x_{i+1}]$, and choose $c_i \in [x_i, x_{i+1}]$. For large n it is reasonable to use

$$F(c_i)(x_{i+1} - x_i)$$

as an approximation of the work $W(x_i, x_{i+1})$, even though F is not constant on $[x_i, x_{i+1}]$. Thus we get an approximation to the total work by adding these individual approximations

$$\sum_{i=0}^{2^n-1} F(c_i)(x_{i+1}-x_i)$$

and since this is a Riemann sum for the continuous function $F(x)$ on $[a,b]$, it is natural to take its limit as $n\to\infty$, $\int_a^b F(x)\,dx$ as our definition of $W(a,b)$.

Definition The **work** $W(a,b)$ done by a continuous force function F exerting a variable force on an object along a coordinate line equal to $F(x)$ units at the point x, moving the object from point a to point b is

$$W(a,b) = \int_a^b F(x)\,dx$$

Observe that by defining work in this manner, the additivity in the sense of (2) follows from the properties of the definite integral.

We see, too, that by the Mean Value Theorem for definite integrals, there is a point $c\in[a,b]$ such that

$$F(c)(b-a) = \int_a^b F(x)\,dx = W(a,b)$$

Thus, as far as the above definition of work goes, the work is the same as if an appropriate constant value of the force function $F(c)$ were applied to move the object from a to b.

This definition can be used to show that for forces acting in a fixed direction, the work done in moving a particle (our idealized object) is simply the change in the kinetic energy of the particle, whether the force is constant or variable. (See exercise 11.)

Example 2 A container full of gas weighing G pounds is lifted and the valve is set so that gas escapes at a uniform rate. At height H feet the container weighs $\frac{1}{2}G$ pounds. Find the work done in lifting the container to height H.

We take the x axis vertical from the starting point. Then the force function $F(x)$ is the weight of the container

$$F(x) = G - g(x)$$

where $g(x)$ is the weight of the escaped gas when the container is at height x. Since the rate of escape is uniform, a simple proportion yields $g(x)$:

$$\frac{g(x)}{x} = \frac{\frac{1}{2}G}{H}$$

or

$$g(x) = \frac{G}{2H}x$$

Thus

$$W = \int_0^H\left(G-\frac{G}{2H}x\right)dx = GH - \frac{GH}{4} = \frac{3}{4}GH \text{ ft-lb}$$

The amount of work is, therefore, the same as if a constant force of $\frac{3}{4}G$ pounds were used to move an object a distance H feet.

Example 3 Find the work done in stretching a spring from its natural length of 8 inches to a length of 11 inches, when a force of 20 pounds stretches the spring $\frac{1}{2}$ inch.

The force $F(x)$ required to hold a spring extended x units beyond its natural length is given by Hooke's law,

$$F(x) = kx, \qquad k \text{ constant}$$

where k depends on the material, thickness of the wire, and other physical properties.

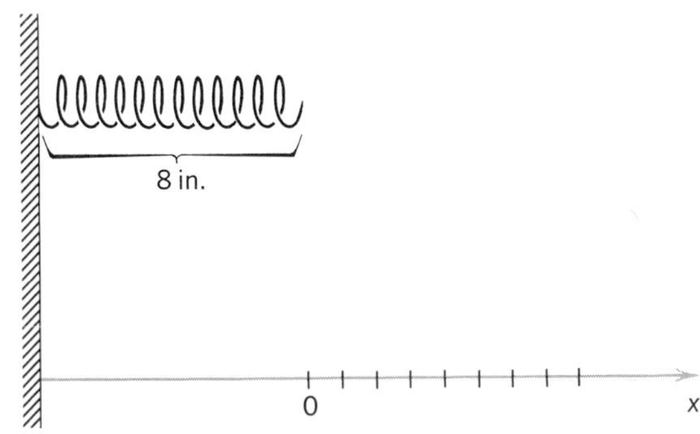

FIGURE 5-1

We first consider the spring as placed along the x axis, with the origin at the position where we start stretching (Figure 5-1). We then solve for k using the fact that the force is 20 pounds when $x = \frac{1}{2}$:

$$20 = k \cdot \tfrac{1}{2}$$

$$40 = k$$

and we have $F(x) = 40x$.

Finally, from the definition of work we have

$$W = \int_0^3 F(x)\,dx = \int_0^3 40x\,dx = 20x^2 \Big|_0^3 = 180 \text{ in-lb}$$

We can also use this same general technique to solve problems of the following type.

Example 4 A tank with the shape of a right circular cylinder of altitude 10 feet and radius of the base 4 feet is full of water. Find the amount of work done in pumping all the water in the tank to a level 10 feet above the top of the tank.

To do this we consider the x axis as vertically downward (Figure 5-2). We divide the tank into 2^n discs, forming a regular partition of $[0, 10]$ with $0 = x_0, x_1, x_2, \ldots, x_{2^n} = 10$ the subdivision points. The plan is to find the total work by finding the sum of the amounts of work needed to raise each little disc to that height.

Let us consider the ith disc with thickness $x_{i+1} - x_i$. The force which we must exert on this disc to raise it must be a force equal to that of gravity. This force is equal to the weight of the disc. We find the weight by noting that water has a density of 62.4 lb/ft^3, and that the volume of the disc is

$$\pi(4^2)(x_{i+1} - x_i) = 16\pi(x_{i+1} - x_i)$$

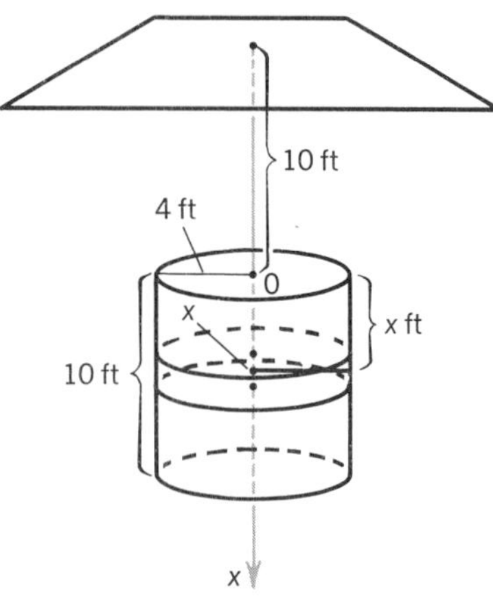

FIGURE 5-2

Thus the force required must be

$$(62.4)\,16\pi(x_{i+1} - x_i)$$

Observe that in this problem the force is constant, and the work done on each disc is merely the product of the force and the distance that it moves. But, each disc moves a different distance.

If, for the ith disc, we choose $x_i < c_i < x_{i+1}$, where the intermediate point c_i represents the distance of that disc to the origin, then the total distance the disc moved will be $10 + c_i$ feet. The work done in moving this disc is

$$(62.4)\,16\pi(x_{i+1} - x_i)(10 + c_i), \qquad x_i < c_i < x_{i+1}$$

Then, the total work is

$$W = \sum_{i=0}^{2^n - 1} (62.4)\,16\pi(x_{i+1} - x_i)(10 + c_i)$$

When we take the limit as $n \to \infty$ this sum approaches the integral

$$W = \int_0^{10} (62.4)\,16\pi(10 + x)\,dx = (62.4)\,16\pi\left(10x + \frac{x}{2}\right)^2 \Big|_0^{10}$$

$$= (62.4)\,16\pi(150) = 149{,}760\pi \text{ ft-lb}$$

EXERCISES

1. A particle is moving along the x axis from $x = 1$ to $x = 2$ according to the force law $F(x) = x^3 + 2x^2 + 6x - 1$. Find the work done.
2. If a spring of normal length 10 inches requires a force of 25 pounds to stretch it $\frac{1}{4}$ inch, find the work done in stretching it from 11 to 12 inches.
3. If a spring of normal length 10 inches requires 30 pounds to stretch it to $11\frac{1}{2}$ inches, find the work done in stretching it

   (a) from 10 to 12 inches (b) from 12 to 14 inches
4. A 100 foot cable weighing 5 lb/ft supports a safe weighing 500 pounds. Find the work done in winding 80 feet of the cable on a drum.

5. Suppose a hollow cylindrical pipe connects the surface of the earth with the center of the earth and that a particle of mass m in the pipe would be attracted toward the center of the earth with force $F(r) = mg(r/R)$ when it is at a distance r from the center of the earth, where R is the radius of the earth and g is the acceleration due to gravity at the surface of the earth. What would be the total work done on the particle in lifting it all the way from the center of the earth to the surface of the earth?
6. A vertical cylindrical tank 6 feet in diameter and 10 feet high is half full of water. Find the amount of work done in pumping all the water out at the top of the tank.
7. A swimming pool full of water is in the form of a rectangular parallelepiped 5 feet deep, 15 feet wide, and 25 feet long. Find the work required to pump the water up to a level 1 foot above the surface of the pool.
8. A hemispherical tank of radius 3 feet is full of water.
   (a) Find the work done in pumping the water over the edge of the tank.
   (b) Find the work done in emptying the tank by an outlet pipe 2 feet above the top of the tank.
9. Two particles repel each other with a force inversely proportional to the square of the distance between them. If one particle remains fixed at $(2, 0)$, find the work done in moving the second along the x axis from $(-3, 0)$ to $(0, 0)$.
10. Find the work done in pumping all the water out of a conical reservoir of radius 10 feet at the top, altitude 8 feet, to a height of 6 feet above the top of the reservoir.
11. Using the definition of $W(a, b)$ and that the velocity v is equal to dx/dt, use Newton's Second Law $F = m(dv/dt)$ to show that

$$W(a, b) = \tfrac{1}{2}mv_b^2 - \tfrac{1}{2}mv_a^2$$

where $v_u =$ value of v when $x = u$. The quantity $\frac{1}{2}mv^2$ is called the **kinetic energy** of the particle of mass m, and the content of the equation is that the work is the difference of the kinetic energies of the particle.

6 IMPROPER INTEGRALS

So far the symbol $\int_a^b f(x)\,dx$ has only been meaningful when (1) a and b are finite real numbers and (2) f is continuous on the closed interval with endpoints a and b. In this section we shall use the concept of limit to assign meaning to $\int_a^b f(x)\,dx$ in cases where (1) and/or (2) might not hold.

Let us consider the branch of the graph of $y = 1/x^2$ in the first quadrant (Figure 6-1). We observe that the curve approaches the x axis as x approaches $+\infty$. In fact, the curve hugs the x axis very closely for large values of x and is asymptotic to the axis. When a and b satisfy $0 < a < b$, then the area under $y = 1/x^2$ from a to b is equal to

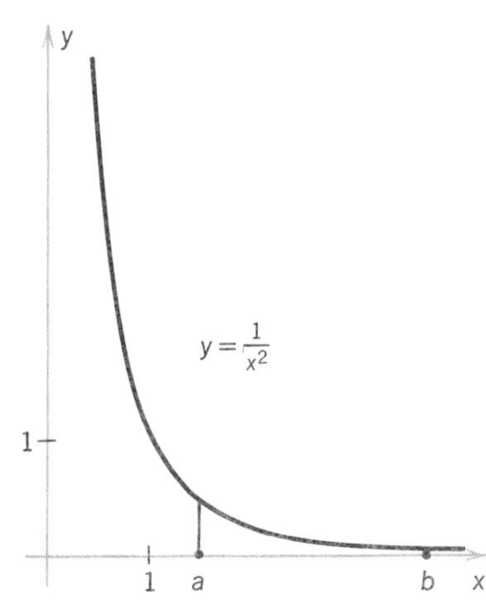

FIGURE 6-1

$$\int_a^b \frac{1}{x^2}\,dx = \left[-\frac{1}{x}\right]_a^b = \frac{1}{a} - \frac{1}{b}$$

Let us now treat this definite integral as a function of b; that is, let us consider the function $b \mapsto \int_a^b 1/x^2\,dx$ and study the behavior of this function as $b \to +\infty$. We observe that

$$\lim_{b\to+\infty}\left(\frac{1}{a} - \frac{1}{b}\right)$$

exists and is equal to $1/a$. Thus, as b approaches $+\infty$, the area under the curve approaches the limit $1/a$. We can say therefore that "the area under $y = 1/x^2$ from a to $+\infty$ is equal to $1/a$." And, we will assign to the symbol $\int_a^{+\infty} 1/x^2\,dx$ the value $1/a$. With this in mind, we make the following definition.

Definition If f is continuous on $[a, \infty[$, and if

$$\lim_{b\to\infty}\int_a^b f(x)\,dx$$

exists, then we define

$$\int_a^\infty f(x)\,dx = \lim_{b\to\infty}\int_a^b f(x)\,dx$$

Similarly, if f is continuous on $]-\infty, b]$ and if $\lim_{a\to-\infty}\int_a^b f(x)\,dx$ exists, we define

$$\int_{-\infty}^b f(x)\,dx = \lim_{a\to-\infty}\int_a^b f(x)\,dx$$

We say that $\int_a^\infty f(x)\,dx$ is **convergent** when the limit exists. If the limit does not exist, then we say that $\int_a^\infty f(x)\,dx$ is **divergent**.

Example 1 Show that the integral $\int_1^\infty dx/x^3$ converges and find its value. We have

$$\int_1^\infty \frac{dx}{x^3} = \lim_{b\to\infty}\int_1^b \frac{dx}{x^3}$$

if this limit exists.

By the Fundamental Theorem, we obtain

$$\int_1^b \frac{dx}{x^3} = -\frac{1}{2x^2}\bigg|_1^b$$

$$= \frac{-1}{2b^2} + \frac{1}{2}$$

so

$$\lim_{b\to\infty}\left(-\frac{1}{2b^2} + \frac{1}{2}\right) = \frac{1}{2}$$

Thus $\int_1^\infty dx/x^3$ converges to the value $\frac{1}{2}$.

Example 2 Determine whether

$$\int_2^\infty \frac{1}{x^2+2x+1}\,dx$$

converges or diverges.

Since $x^2+2x+1=(x+1)^2$, we obtain for $2 \leqslant b < \infty$

$$\int_2^b \frac{1}{x^2+2x+1}\,dx = \int_2^b \frac{1}{(x+1)^2}\,dx = \left.\frac{-1}{(x+1)}\right|_2^b = \frac{1}{3} - \frac{1}{(b+1)}$$

Clearly

$$\lim_{b\to\infty}\left(\frac{1}{3} - \frac{1}{(b+1)}\right) = \frac{1}{3}$$

and so the integral converges.

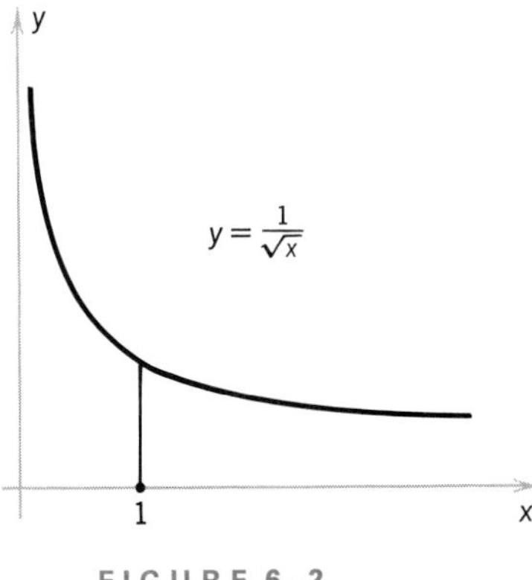

FIGURE 6-2

We now turn to the problem of defining $\int_a^b f(x)\,dx$ when f is not necessarily continuous on $[a,b]$. The first example we shall study is one where $\lim_{x\to a^+} f(x) = +\infty$.

Consider $\int_0^1 dx/\sqrt{x}$. Let us first look at the graph of this function (Figure 6-2). Note that as $x\to 0$ from the right side, the curve rises steadily, and it is unclear whether the area under the curve between 0 and 1 is well defined. However, let us consider the area between $x=a$ and $x=1$ for $1>a>0$. This area is defined and is equal to $\int_a^1 dx/\sqrt{x}$. If the limit of $\int_a^1 dx/\sqrt{x}$ exists as a approaches 0 from the right, we define this limit to be the value of the integral $\int_0^1 dx/\sqrt{x}$.

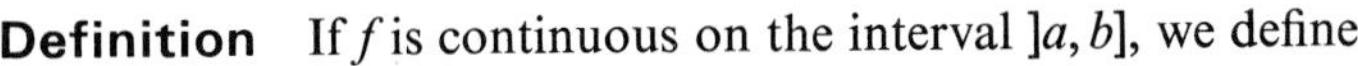

Definition If f is continuous on the interval $]a,b]$, we define

$$\int_a^b f(x)\,dx = \lim_{c\to a^+}\int_c^b f(x)\,dx$$

provided this limit exists. Similarly, for f continuous on $[a,b[$ we define

$$\int_a^b f(x)\,dx = \lim_{c\to b^-}\int_a^c f(x)\,dx$$

when the limit exists. Again, we say that the integral converges when the limit exists and diverges when the limit does not exist.

Example 3 Determine whether $\int_0^1 dx/\sqrt{x}$ converges and if it does, find its value. We have

$$\int_0^1 \frac{dx}{\sqrt{x}} = \lim_{c\to 0^+}\int_c^1 \frac{dx}{\sqrt{x}}$$

if the limit exists. We have

$$\int_c^1 \frac{dx}{\sqrt{x}} = 2\sqrt{x}\,\bigg|_c^1 = 2 - 2\sqrt{c}$$

and so,

$$\lim_{c \to 0^+} (2 - 2\sqrt{c}) = 2$$

Thus, the integral converges and its value is 2.

Example 4 Determine whether

$$\int_1^2 \frac{1}{x^2 - 2x + 1}\, dx$$

converges or diverges.

The function $g: x \mapsto 1/x^2 - 2x + 1$ is not continuous at 1 and in fact

$$\lim_{x \to 1^+} g(x) = \lim_{x \to 1^+} \frac{1}{(x-1)^2} = +\infty$$

So we consider

$$\int_b^2 \frac{1}{x^2 - 2x + 1}\, dx, \qquad 1 < b \leqslant 2$$

From the Fundamental Theorem we conclude that

$$\int_b^2 \frac{1}{x^2 - 2x + 1}\, dx = \int_b^2 \frac{1}{(x-1)^2}\, dx = \left.\frac{-1}{(x-1)}\right|_b^2 = \frac{1}{b-1} - 1$$

However,

$$\lim_{b \to 1^+} \left(\frac{1}{b-1} - 1 \right) = +\infty$$

and so the integral diverges.

We next consider the integration of functions with discontinuity at an interior point of the interval.

Definition Suppose $f(x)$ is continuous in $[a, b]$ except at $x = c$ where $a < c < b$. Then $\int_a^b f(x)\, dx$ is **convergent** and we set

$$\int_a^b f(x)\, dx = \int_a^c f(x)\, dx + \int_c^b f(x)\, dx$$

if and only if both the integrals $\int_a^c f(x)\, dx$ and $\int_c^b f(x)\, dx$ are convergent.

All we have done here is to split our integral into a sum of smaller intervals and then consider our original integral as a sum of convergent integrals.

Example 5 Determine whether $\int_{-1}^1 dx/x^2$ converges and if it does find its value.

We first sketch the graph (Figure 6-3). Observe that at $x=0$, both branches of the graph become infinite. We consider the interval of integration $x=-1$ to $x=1$ as the two smaller intervals, $x=-1$ to $x=0$, and $x=0$ to $x=1$. We then have,

$$\int_{-1}^{1} \frac{dx}{x^2} = \lim_{c_1 \to 0-} \int_{-1}^{c_1} \frac{dx}{x^2} + \lim_{c_2 \to 0+} \int_{c_2}^{1} \frac{dx}{x^2}$$

if these limits exist. However

$$\lim_{c_1 \to 0-} \int_{-1}^{c_1} \frac{dx}{x^2} = \lim_{c_1 \to 0-} \left(-\frac{1}{x} \right) \Bigg|_{-1}^{c_1} = \lim_{c_1 \to 0-} \left(-\frac{1}{c_1} + 1 \right)$$

and

$$\lim_{c_2 \to 0+} \int_{c_2}^{1} \frac{dx}{x^2} = \lim_{c_2 \to 0+} \left(-\frac{1}{x} \right) \Bigg|_{c_2}^{1} = \lim_{c_2 \to 0+} \left(\frac{1}{c_2} - 1 \right)$$

Since neither of these limits exists, the integral $\int_{-1}^{1} dx/x^2$ is divergent.

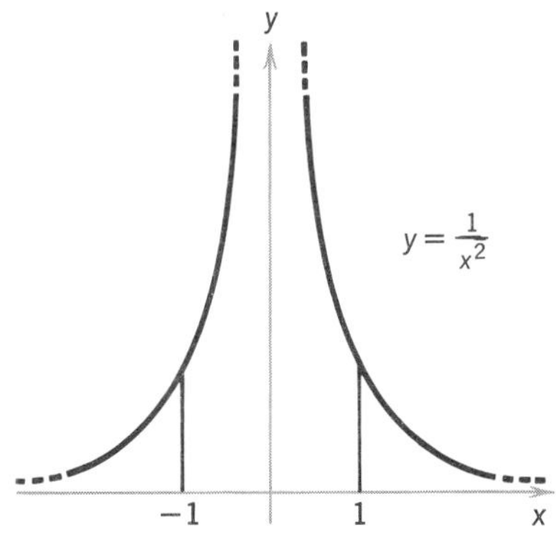

FIGURE 6-3

It is often possible to determine whether an improper integral converges without computing it, by comparing it with another simpler integral which is known to converge.

Suppose we are given a function f whose graph is as in Figure 6-4. Suppose we also can find a function g as in Figure 6-4 such that $\int_a^\infty g(x)\,dx$ converges.

This means that the area under the graph of g approaches a finite value as $x \to \infty$. We would then expect that $\int_a^\infty f(x)\,dx$ also converges, that is, the area under f also approaches a finite value since $f(x) \leqslant g(x)$ everywhere. In fact, we have the following theorem called the **comparison test.**

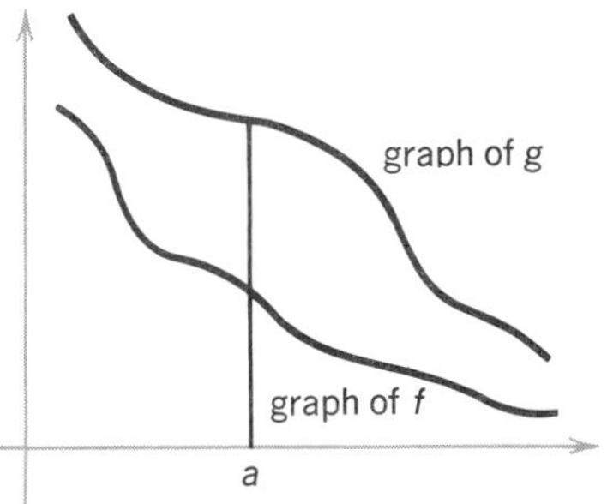

FIGURE 6-4

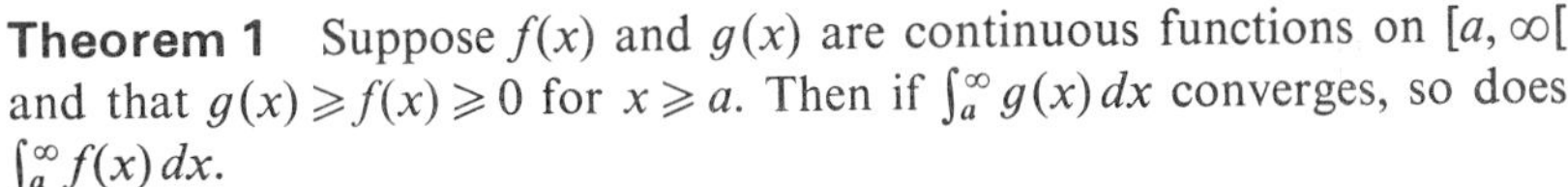

Theorem 1 Suppose $f(x)$ and $g(x)$ are continuous functions on $[a, \infty[$ and that $g(x) \geqslant f(x) \geqslant 0$ for $x \geqslant a$. Then if $\int_a^\infty g(x)\,dx$ converges, so does $\int_a^\infty f(x)\,dx$.

Proof From the hypothesis $0 \leqslant f(x) \leqslant g(x)$, we have, for $x \geqslant a$,

$$0 \leqslant \int_a^b f(x)\,dx \leqslant \int_a^b g(x)\,dx \leqslant \int_a^\infty g(x)\,dx \tag{1}$$

Thus by the inequality (1) the function

$$\psi(b) = \int_a^b f(x)\,dx$$

is monotone and bounded, and hence, as with monotone sequences (§1), has a limit as $b \to \infty$. ▨

Example 6 Determine whether $\int_1^\infty x/(x^3+1)\,dx$ converges. Observe that $x^3 \leqslant x^3+1$ for $x \geqslant 1$ and, hence, that

$$\frac{x}{x^3+1} \leqslant \frac{x}{x^3}$$

So we have

$$\frac{x}{x^3+1} \leqslant \frac{1}{x^2}$$

Both of these functions are greater than 0 for $x \geqslant 1$, and since we already know that $\int_1^\infty (dx/x^2)$ converges we conclude that $\int_1^\infty x/(x^3+1)\,dx$ converges.

EXERCISES

Show that each of the following integrals converge and evaluate them.

1. $\int_1^\infty \frac{dx}{x^4}$

2. $\int_0^\infty \frac{dx}{(x+1)^{3/2}}$

3. $\int_0^\infty \frac{x\,dx}{(x^2+9)^2}$

4. $\int_0^1 \frac{dx}{x^{1/3}}$

5. $\int_0^3 \frac{dx}{\sqrt{9-x^2}}$ [Hint: Substitute $u = 3\sin x$.]

6. $\int_0^4 \frac{dx}{\sqrt{4-x}}$

7. $\int_0^4 \frac{dx}{\sqrt[3]{x-1}}$

In exercises 8-11, determine whether the integral converges or diverges.

8. $\int_0^\infty \frac{dx}{x^3}$

9. $\int_{-2}^0 \frac{dx}{\sqrt[3]{x+1}}$

10. $\int_0^4 \frac{dx}{(x-2)^{2/3}}$

11. $\int_0^1 \frac{dx}{x^n}$, where n is an integer $\geqslant 2$

Show that the following integrals exist by comparing them to a known integral.

12. $\int_1^\infty \frac{dx}{\sqrt{x^4+2x+6}}$

13. $\int_1^\infty \frac{dx}{x^r}$, where r is a rational number $\geqslant 2$

REVIEW EXERCISES FOR CHAPTER 7

In exercises 1–10, sketch the regions bounded by the given curves and find their areas.

1. $x = 0, \quad x = 2, \quad y = x^2, \quad y = 9x$
2. $x = 1, \quad x = 2, \quad x^2 y = 2, \quad x + y = 4$
3. $y = \sqrt{x}, \quad y = -\sqrt{x}, \quad x = 4$
4. $x = 1 + y^2, \quad x = 10, \quad y = 0$
5. $y = -1, \quad y = 1, \quad x - y + 1 = 0, \quad x = 1 - y^2$
6. $x = 3y^2 - 9, \quad x = 0, \quad y = 0, \quad y = 1$
7. $x^2 y = 4, \quad 3x + y - 7 = 0$
8. $y = x^4 - 4x^2, \quad y = 4x^2$
9. $y = x^2 - 4x + 1, \quad x + y - 5 = 0$
10. $y^2 = 2 - x, \quad y = 2x + 2$

In exercises 11–13, find the mean value of the function over the indicated intervals.

11. $y = x^2 - 2x + 1, \quad [0, 1]$
12. $y = 4 - x^2, \quad [-2, 2]$
13. $y = \sqrt{2x+1}, \quad [4, 12]$

Find the lengths of the arcs described in exercises 14–17.

14. $y = \frac{2}{3}x^{3/2} - \frac{1}{2}x^{1/2}$, from $x = 0$ to $x = 4$
15. $x = \frac{3}{5}y^{5/3} - \frac{3}{4}y^{1/3}$, from $y = 0$ to $y = 1$
16. $x = 4\cos t, \quad y = 4\sin t$, from $t = 0$ to $t = \frac{1}{2}\pi$
17. $y = (9 - x^{2/3})^{3/2}$ from $x = 1$ to $x = 2$

For exercises 18–22, find the area of each surface obtained by revolving the given arc about the x axis.

18. $y^2 = 2 - x$, from $x = 0$ to $x = 2$
19. $y = \dfrac{x^4}{8} + \dfrac{1}{4x^2}$, from $x = 1$ to $x = 2$
20. $y = mx$, from $x = 0$ to $x = 2$
21. $y = \frac{1}{3}x^3$, from $x = 0$ to $x = 3$
22. $y = Ax^2 + Bx + C$ from $x = -a$ to $x = a$

In exercises 23–28, find the volumes generated when the areas bounded by the given curves and lines are rotated about the x axis.

23. $y = \sin x, \quad y = 0, \quad 0 \leqslant x \leqslant \pi$

24. $y = 3x - x^2, \quad y = x$
25. $y = 2x^2, \quad y = 0, \quad x = 0, \quad x = 5$
26. $x^2 - y^2 = 16, \quad y = 0, \quad x = 8$
27. $x = 2y - y^2, \quad x = 0$
28. $y = 3 + x^2, \quad y = 4$
29. Find the volume generated when the area bounded by the curve $y = x^2$ and the line $y = x$ is revolved around the y axis.
30. Find the volume generated when the area bounded by the curve $y = x^2$ and the line $y = 4$ is rotated about the line $x = 2$.
31. The base of a solid is the circle $x^2 + y^2 = a^2$. Each plane section of the solid cut out by a plane perpendicular to the x axis is a square with one edge of the square in the base of the solid. Find the volume of the solid.
32. A solid is generated by rotating $f(x)$, $0 \leqslant x \leqslant a$, about the x axis. Its volume for all a is $a^2 + a$; find $f(x)$.
33. Find the volume of the solid generated by rotating the larger area bounded by $y^2 = x - 1$, $x = 3$, and $y = 1$ about the y axis.
34. A round hole of radius $\sqrt{3}$ feet is bored through the center of a solid sphere of radius 2 feet. Find the volume cut out.
35. The area bounded by the curve $y^2 = 4ax$ and the line $x = a$ is rotated about the line $x = 2a$. Find the volume generated.
36. The expansion of a gas in a cylinder causes a piston to move so that the volume of the enclosed gas increases from 15 to 25 cubic inches. Assuming the relation between the pressure p and the volume v to be $pv = 60$, find the work done.
37. The force with which the earth attracts a weight w pounds at a distance s miles from its center is

$$F = (4000)^2 \frac{w}{s^2}$$

where the radius of the earth has been taken as 4,000 miles. Find the work done against the force of gravity in moving a 1 pound mass from the surface of the earth to a point 1000 miles above the surface.
38. A vertically hanging cable weighing 3 lb/ft is unwinding from a cylindrical drum. If 50 feet are already unwound, find the work done by the force of gravity as an additional 250 feet are unwound.
39. A cistern is 10 feet square and 8 feet deep. Find the work done in emptying it over the top if it is $\frac{3}{4}$ full of water.
40. A conical vessel is 12 feet across the top and 15 feet deep. If it contains a liquid weighing w lb/ft^3 to a depth of 10 feet, find the work done in pumping the liquid to a height 3 feet above the top of the vessel.

8

TRANSCENDENTAL FUNCTIONS

A real function $f: x \mapsto f(x)$ is **algebraic** if, for some positive integer n, there exist $n+1$ polynomials $A_0(x), A_1(x), \ldots, A_n(x)$, where neither $A_0(x)$ nor $A_n(x)$ is the zero polynomial, such that

$$A_0(x)f^n(x) + A_1(x)f^{n-1}(x) + \cdots + A_{n-1}(x)f(x) + A_n(x) = 0$$

holds for all x. Thus polynomial functions are algebraic functions; sums, products, quotients, powers, and roots of algebraic functions are also algebraic. A function that is not algebraic is called **transcendental**. The elementary transcendental functions include the trigonometric, logarithmic, exponential, and inverse trigonometric functions. In this chapter, we study these functions, their derivatives, graphs, and certain applications. We begin with the trigonometric functions.

In the final two sections, we introduce the system of **polar coordinates** for the plane. We then see how to compute tangent lines, arc lengths, and areas in terms of these coordinates.

1 TRIGONOMETRIC FUNCTIONS AND THEIR INVERSES

We briefly recall that the trigonometric functions express relations between the sides of a right triangle in terms of the base angle, or the properties of an arc of the unit circle (Figure 1-1). The student is already familiar with the basic characteristics of the trigonometric functions; we shall confine our remarks here to restating several key identities, and to a discussion of periodicity and graphing.

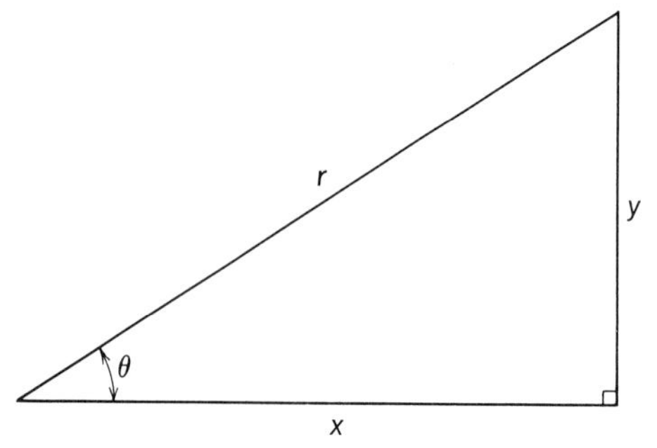

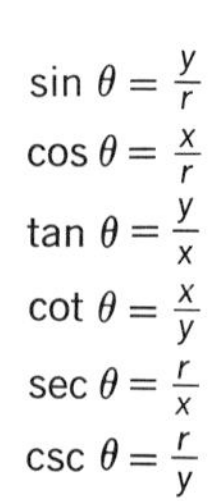

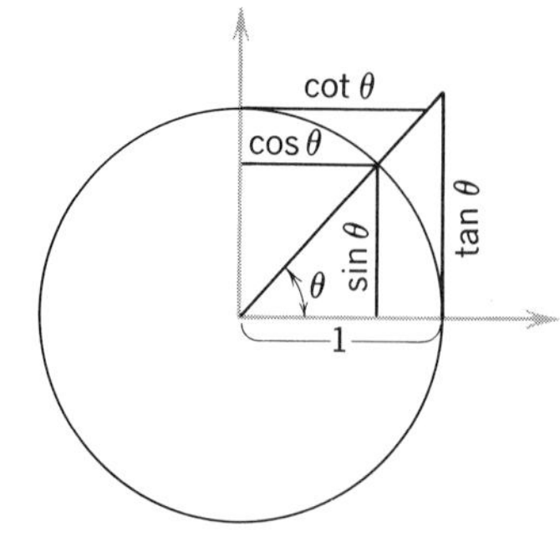

FIGURE 1-1

The following identities are easily established:

$$\sin(-\theta) = -\sin\theta$$

$$\cos(-\theta) = \cos\theta$$

$$\sin(\tfrac{1}{2}\pi - \theta) = \cos\theta$$

$$\sin^2\theta + \cos^2\theta = 1$$

$$\tan^2\theta + 1 = \sec^2\theta$$

$$\sin 2\theta = 2\sin\theta\cos\theta$$

$$\cos 2\theta = \cos^2\theta - \sin^2\theta$$

$$\sin(\theta + \phi) = \sin\theta\cos\phi + \cos\theta\sin\phi$$

$$\cos(\theta + \phi) = \cos\theta\cos\phi - \sin\theta\sin\phi$$

$$\tan(\theta + \phi) = \frac{\tan\theta + \tan\phi}{1 - \tan\theta\tan\phi}$$

$$\sin\tfrac{1}{2}\theta = \sqrt{\frac{1-\cos\theta}{2}}$$

$$\cos\tfrac{1}{2}\theta = \sqrt{\frac{1+\cos\theta}{2}}$$

It is obvious from the graphs of the trigonometric functions that these functions "repeat themselves." Consider for example the sine function; clearly

$$\sin\theta = \sin(\theta + 2\pi n)$$

for any angle θ and integer n. The smallest positive number after which a function starts to repeat itself is called the **period** of the function; thus the period of the sine function is 2π. A function with a period is called **periodic**.

Clearly, all the trigonometric functions are periodic. The periods of the cosine, secant, and cosecant functions are also 2π, while those of the tangent and cotangent functions are π.

The following chart summarizes the important information about trigonometric functions (see Figure 1-2):

Function	Domain	Range	Symmetry	Period	Asymptotes
sine	$\boldsymbol{R}$	$[-1, 1]$	odd	2π	—
cosine	$\boldsymbol{R}$	$[-1, 1]$	even	2π	—
tangent	$x \neq \pi(n+\frac{1}{2})$	$\boldsymbol{R}$	odd	π	$x = \pi(n+\frac{1}{2})$
cotangent	$x \neq n\pi$	$\boldsymbol{R}$	odd	π	$x = n\pi$
secant	$x \neq \pi(n+\frac{1}{2})$	$\|y\| \geqslant 1$	even	2π	$x = \pi(n+\frac{1}{2})$
cosecant	$x \neq n\pi$	$\|y\| \geqslant 1$	odd	2π	$x = n\pi$

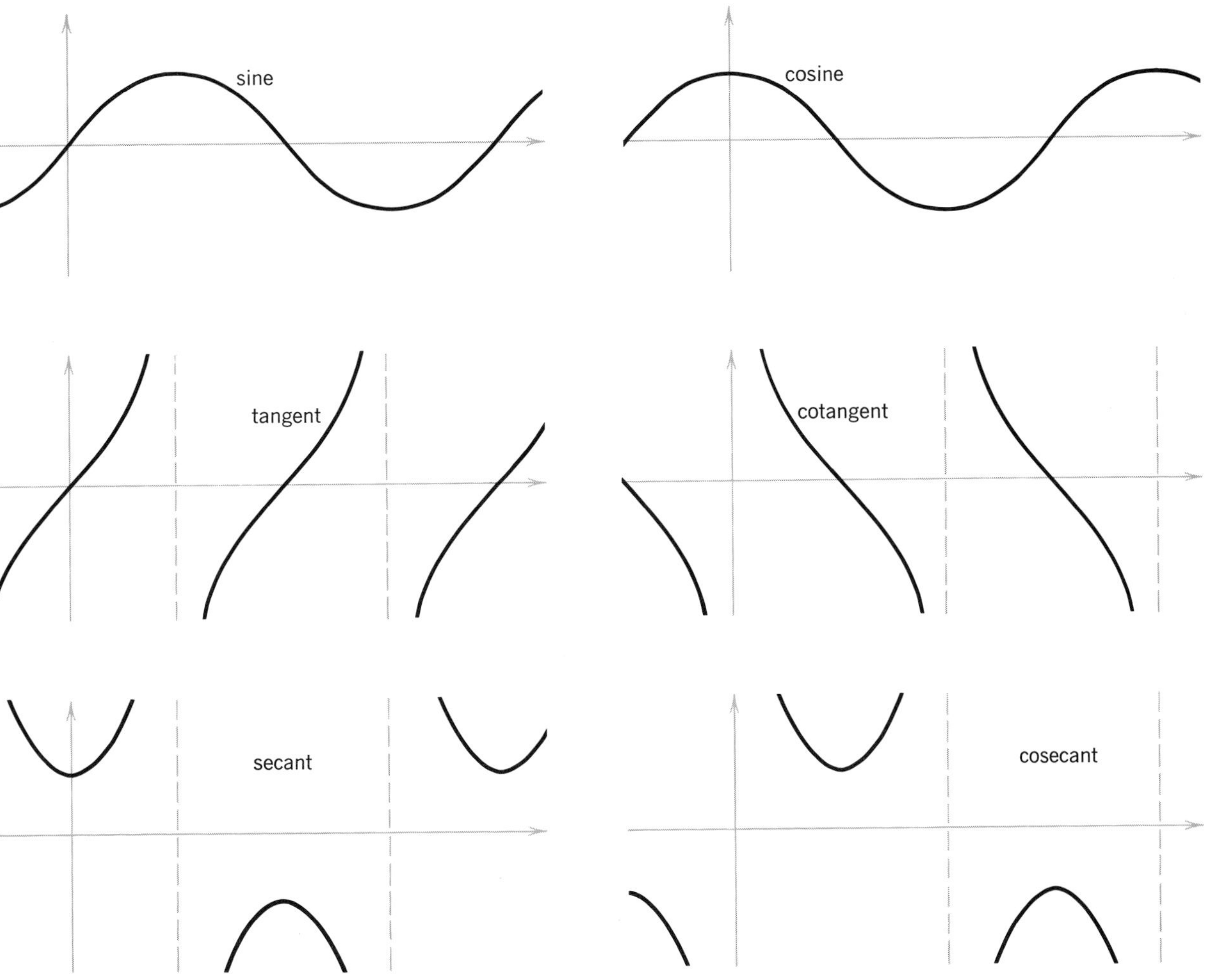

FIGURE 1-2

If the function $x \mapsto \sin x$ is multiplied by some continuous function f, which is not periodic or which has a large period, the graph of the resulting function $F: x \mapsto f(x) \sin x$ has an interesting characteristic.

FORMAL DEVICE

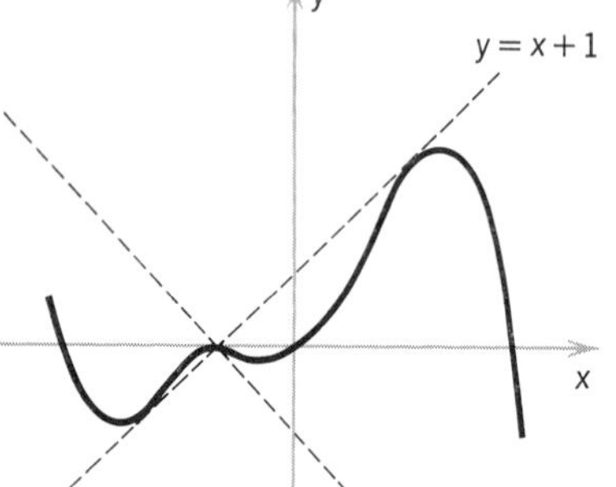

FIGURE 1-3

Under the conditions just stated, the graph of F oscillates between the graphs of f and $-f$, touching the former curve at $x = (2n + \frac{1}{2})\pi$ and the latter at $x = (2n - \frac{1}{2})\pi$, and intersecting the x axis at $x = n\pi$.

Example 1 Sketch the graph of $g: x \mapsto (x+1)\sin x$. We note that the function is defined and differentiable throughout $\boldsymbol{R}$. The information provided by the formal device is really sufficient to sketch the graph (Figure 1-3). The student might also wish to check more precisely the location of critical and inflection points, but in this case it is rather awkward.

Example 2 Sketch the graph of $x \mapsto \sin x \cos(\frac{1}{4}x)$. Again the graph oscillates back and forth, this time between $\cos(\frac{1}{4}x)$ and $-\cos(\frac{1}{4})$ (Figure 1-4).

Clearly, the trigonometric functions are not one-one (in fact, no periodic function is). Therefore, we cannot expect to find inverses to the trigonometric functions. Nevertheless, it is still meaningful to ask for an angle whose cosine is $\frac{1}{2}$ or for an angle whose tangent is $\sqrt{3}$. The reason that we cannot define inverses of the trigonometric functions is that these questions each have many answers, whereas a function must be single-valued. In particular, the answers to the first question are $\theta = (2n \pm \frac{1}{3})\pi$, n an integer; the answers to the second question are $\phi = (n + \frac{1}{3})\pi$, n an integer.

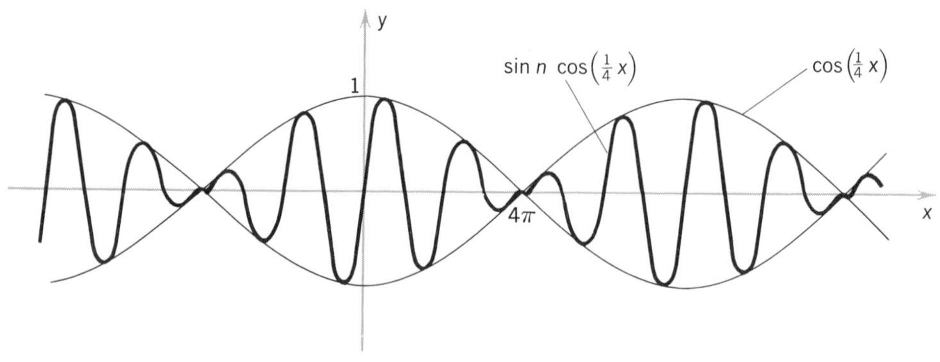

FIGURE 1-4

Defining functions which "look like" inverses to the trigonometric functions is equivalent to somehow providing single answers to the above type of questions. This can be done if we make a convention to restrict our attention for each trigonometric function to some interval on which that function is one-one. The convention is chosen to include the origin and part of the positive x axis in this interval, and if possible to make the function continuous on the interval. With this convention the **principal intervals** of the trigonometric functions are defined as follows (Figure 1-5):

Function	Principal Domain
sine	$[-\frac{1}{2}\pi, \frac{1}{2}\pi]$
cosine	$[0, \pi]$
tangent	$]-\frac{1}{2}\pi, \frac{1}{2}\pi[$
cotangent	$]0, \pi[$
secant	$[0, \pi]$
cosecant	$]-\frac{1}{2}\pi, \frac{1}{2}\pi[$

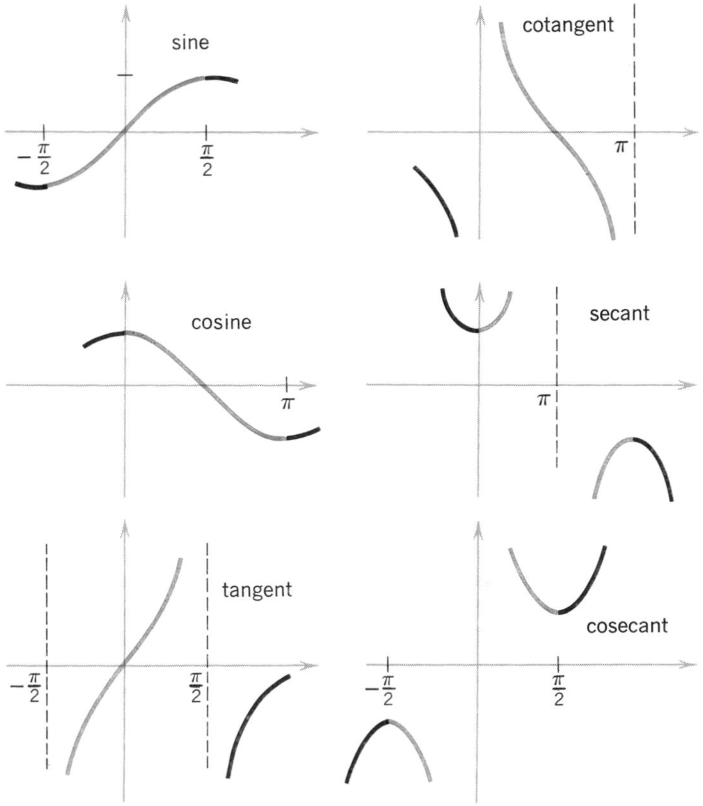

FIGURE 1-5

On the principal intervals, the trigonometric functions are defined (except at $\frac{1}{2}\pi$ for the secant and 0 for the cosecant) and one-one. Therefore, a function whose rule is that of a trigonometric function and whose domain is the principal interval of that trigonometric function has an inverse. This inverse is called the **principal inverse trigonometric function**, or sometimes (by abuse of language) just the **inverse trigonometric function**. The domains of the principal inverse trigonometric functions are precisely the ranges of the original functions; the ranges of the inverses are the principal intervals of the originals. The inverse functions are denoted by $\sin^{-1}$ or arcsin, $\cos^{-1}$ or arccos, $\tan^{-1}$ or arctan, and so on. The following chart summarizes information on the principal inverse trigonometric functions whose graphs are shown in Figure 1-6.

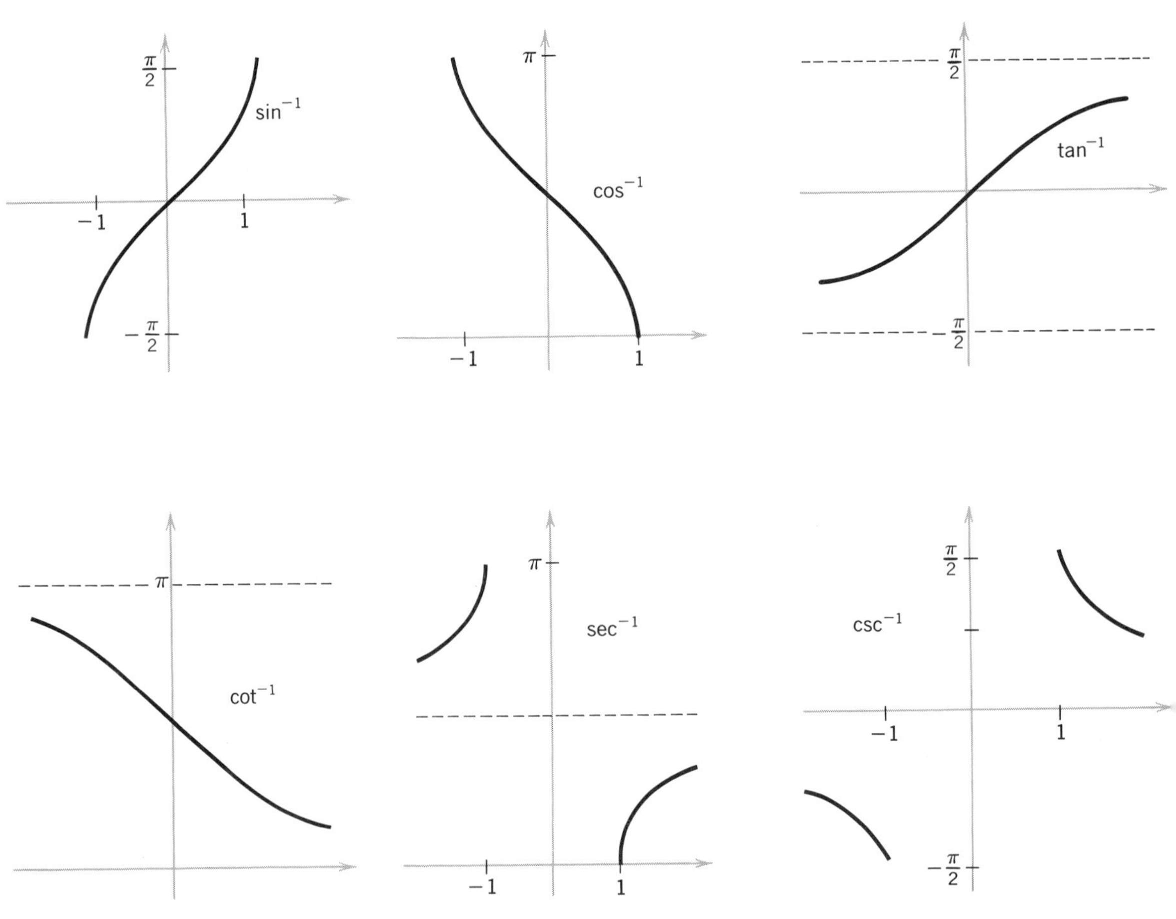

FIGURE 1-6

Function	Domain	Range	Symmetry	Asymptotes
$\sin^{-1}$	$[-1, 1]$	$[-\frac{1}{2}\pi, \frac{1}{2}\pi]$	odd	—
$\cos^{-1}$	$[-1, 1]$	$[0, \pi]$	—	—
$\tan^{-1}$	$\boldsymbol{R}$	$]-\frac{1}{2}\pi, \frac{1}{2}\pi[$	odd	$y = \pm\frac{1}{2}\pi$
$\cot^{-1}$	$\boldsymbol{R}$	$]0, \pi[$	—	x axis, $y = \pi$
$\sec^{-1}$	$\lvert x\rvert \geqslant 1$	$[0, \pi](y \neq \frac{1}{2}\pi)$	—	$y = \frac{1}{2}\pi$
$\csc^{-1}$	$\lvert x\rvert \geqslant 1$	$[-\frac{1}{2}\pi, \frac{1}{2}\pi](y \neq 0)$	odd	x axis

To differentiate the inverse trigonometric functions, we notice that the composition of a trigonometric function on its inverse is the identity function on the domain of the inverse; the Chain Rule can therefore be applied.

Theorem 1 The derivative of the inverse sine is

$$\frac{d(\sin^{-1} x)}{dx} = \frac{1}{\sqrt{1-x^2}}, \qquad (x \neq \pm 1)$$

Proof We have

$$id(x) = \sin(\sin^{-1} x)$$

where $id(x) = x$. Hence

$$\frac{d(id(x))}{dx} = \sin'(\sin^{-1} x)\cdot\frac{d(\sin^{-1} x)}{dx}$$

$$1 = \cos(\sin^{-1} x)\cdot\frac{d(\sin^{-1} x)}{dx}$$

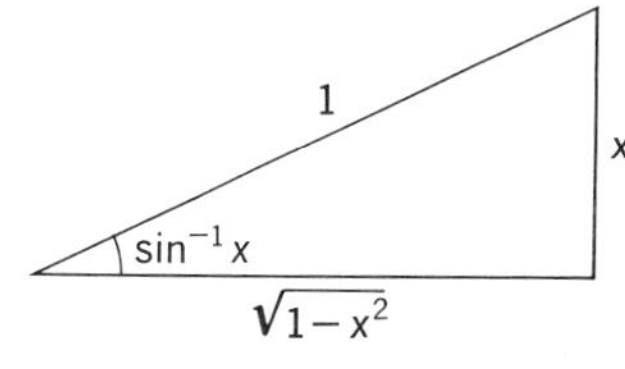

FIGURE 1-7

But $\cos(\sin^{-1} x) = \sqrt{1-x^2}$ (Figure 1-7). Therefore

$$1 = \sqrt{1-x^2}\,\frac{d(\sin^{-1} x)}{dx}$$

or

$$\frac{d(\sin^{-1} x)}{dx} = \frac{1}{\sqrt{1-x^2}}$$

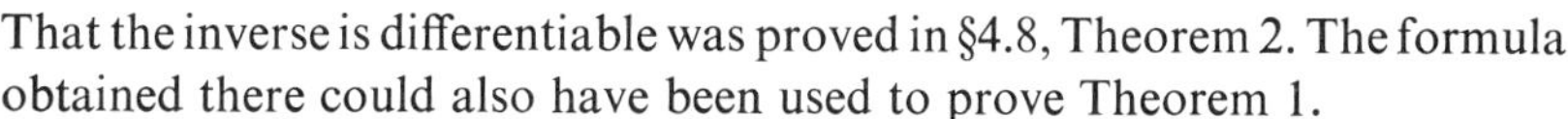
That the inverse is differentiable was proved in §4.8, Theorem 2. The formula obtained there could also have been used to prove Theorem 1.

Theorem 2 The derivatives of the remaining inverse trigonometric functions are

$$\frac{d(\cos^{-1} x)}{dx} = \frac{-1}{\sqrt{1-x^2}}, \qquad (x \neq \pm 1)$$

$$\frac{d(\tan^{-1} x)}{dx} = \frac{1}{1+x^2}$$

$$\frac{d(\cot^{-1} x)}{dx} = \frac{-1}{1+x^2}$$

$$\frac{d(\sec^{-1} x)}{dx} = \frac{1}{|x|\sqrt{x^2-1}}, \qquad (x \neq \pm 1)$$

$$\frac{d(\csc^{-1} x)}{dx} = \frac{-1}{|x|\sqrt{x^2-1}}, \qquad (x \neq \pm 1)$$

Proof The proofs of the first three formulas are analogous to that of the inverse sine. A word of explanation is needed for the absolute values in the inverse secant and cosecant. We shall therefore prove the formula for the secant. Thus

$$\begin{aligned} id(x) &= \sec(\sec^{-1} x) \\ 1 &= \sec'(\sec^{-1} x) \cdot \frac{d(\sec^{-1} x)}{dx} \\ &= \sec(\sec^{-1} x)\tan(\sec^{-1} x) \cdot \frac{d(\sec^{-1} x)}{dx} \end{aligned}$$

Clearly $\sec(\sec^{-1} x) = x$. To find $\tan(\sec^{-1} x)$, let $\theta = \sec^{-1} x$, so $x = 1/\cos\theta$. Now

$$\begin{aligned} \tan(\sec^{-1} x) &= \tan\theta \\ &= \frac{\sqrt{1-\cos^2\theta}}{\cos\theta} \\ &= \sec\theta\sqrt{1-\frac{1}{\sec^2\theta}} \\ &= x\sqrt{1-1/x^2} \\ &= \frac{x}{|x|}\sqrt{x^2-1} \\ &= (\operatorname{sgn} x)\sqrt{x^2-1} \end{aligned}$$

Therefore

$$\begin{aligned} 1 &= x[(\operatorname{sgn} x)\sqrt{x^2-1}] \cdot \frac{d(\sec^{-1} x)}{dx} \\ &= |x|\sqrt{x^2-1} \cdot \frac{d(\sec^{-1} x)}{dx} \end{aligned}$$

and

$$\frac{d(\sec^{-1} x)}{dx} = \frac{1}{|x|\sqrt{x^2-1}}, \qquad x \neq \pm 1$$

Example 3 Compute the derivative of $\sin^{-1} 2x \cos^{-1} x$. We have

$$\frac{d(\sin^{-1} 2x \cos^{-1} x)}{dx} = \frac{d(\sin^{-1} 2x)}{dx} \cos^{-1} x + \sin^{-1} 2x \frac{d(\cos^{-1} x)}{dx}$$

$$= \frac{2\cos^{-1} x}{\sqrt{1-x^2}} - \frac{\sin^{-1} 2x}{\sqrt{1-x^2}}$$

We already know that the integrals of the sine and cosine functions are the negative cosine and the sine functions, respectively. Integrating the other trigonometric functions and all of the inverse trigonometric functions involves techniques which we have not yet acquired. Therefore, consideration of these integrals is deferred until we have the apparatus necessary for computation. However, we note here the following integration formulas which are furnished by Theorems 1 and 2:

$$\int \frac{du}{\sqrt{1-u^2}} = \sin^{-1} u + C$$

$$\int \frac{du}{1+u^2} = \tan^{-1} u + C$$

$$\int \frac{du}{|u| \sqrt{u^2-1}} = \sec^{-1} u + C$$

Example 4 Evaluate

$$\int_0^{\frac{1}{2}\pi} \frac{\cos x}{1+\sin^2 x}\, dx$$

If we make the substitution $u = \sin x$, we have $du = \cos x\, dx$.

Since $\int 1/(1+u^2)\, du = \tan^{-1} u + C$, we obtain by the method of substitution

$$\int \frac{\cos x}{1+\sin^2 x}\, dx = \tan^{-1}(\sin x) + C$$

Therefore, we can compute

$$\int_0^{\frac{1}{2}\pi} \frac{\cos x}{1+\sin^2 x}\, dx = [\tan^{-1}(\sin x)]_0^{\frac{1}{2}\pi} = \tan^{-1}(1) - \tan^{-1}(0)$$

$$= \tfrac{1}{4}\pi - 0 = \tfrac{1}{4}\pi$$

EXERCISES

1. Explain why the formal device for graphing functions of the form $x \mapsto f(x) \sin x$ is valid. Produce a similar formal device for functions $x \mapsto f(x) \cos x$.
2. Graph the following functions: (a) $x \mapsto (\sin x)/x$; (b) $x \mapsto \sin x \cos(x/n)$; (c) $x \mapsto (\frac{1}{2}x-2) \cos x$

3. Why can a periodic function not be one-one?
4. Evaluate $\sin^{-1}(\frac{1}{2})$, $\cos^{-1}(-\frac{1}{2})$, $\sin^{-1}(1)-\sin^{-1}(-1)$, $\tan^{-1}(1)-\tan^{-1}(-1)$, $\cos(\sin^{-1}0.8)$, $\sin(2\sin^{-1}0.8)$, $\cos^{-1}(0)-\sin^{-1}(\frac{1}{2})$.
5. Does $\tan^{-1}x=\sin^{-1}x/\cos^{-1}x$? Does $\operatorname{cosec}^{-1}x=1/\cos^{-1}x$?
6. Prove the differentiation formulas for the inverse trigonometric functions which are not proved in the text.
7. Compute the derivative of each function.
   (a) $x\mapsto\sin^{-1}(x/2)$
   (b) $x\mapsto\sin^{-1}\left(\dfrac{x-1}{x+1}\right)$
   (c) $x\mapsto x\sec^{-1}(2/x)-2\sqrt{1+4x^2}$
   (d) $x\mapsto\cot^{-1}(2/x)+\tan^{-1}(x/2)$
   (e) $x\mapsto\tan^{-1}x+\cot^{-1}x$
   (f) $x\mapsto\sin^{-1}(1/x)+\sec^{-1}x$
8. Integrate each of the following.
   (a) $\displaystyle\int_0^{1/2}\frac{dx}{\sqrt{1-x^2}}$ (b) $\displaystyle\int_{\sqrt{2}}^{2}\frac{dx}{x\sqrt{x^2-1}}$ (c) $\displaystyle\int_{-2}^{-\sqrt{2}}\frac{dx}{x\sqrt{x^2-1}}$
9. Establish Theorem 1 using Theorem 2 of §4.8.
10. Prove directly, and from Theorems 1 and 2, that $\cos^{-1}x+\sin^{-1}x=$ constant.

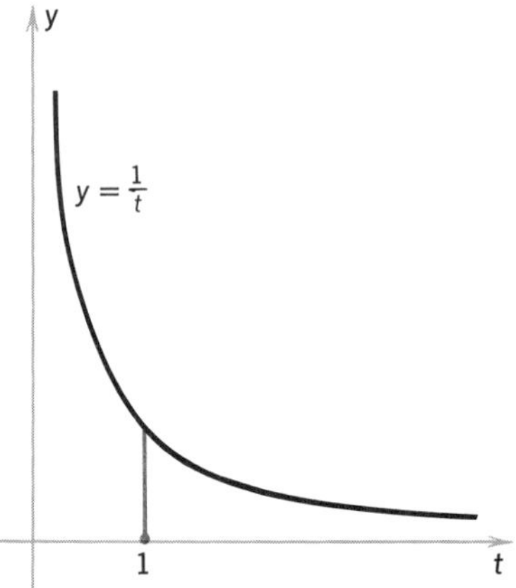

FIGURE 2-1

2 LOGARITHMIC AND EXPONENTIAL FUNCTIONS

We have already observed that for $n\neq-1$,

$$\int t^n dt=\frac{t^{n+1}}{n+1}+C$$

However, this formula cannot be used to compute $\int dt/t$ since it leads to the meaningless expression $1/0$. The equation $y=1/t$, for positive values of t, describes one branch of an hyperbola (Figure 2-1). The definite integral

$$\int_1^x\frac{dt}{t}\qquad(x>0)$$

gives the area under the hyperbola between the lines $t=1$ and $t=x$ (Figure 2-2). This area is used to define a new function in the following way: For $x>0$, define the **logarithmic function** ln by

$$\ln : x\mapsto\int_1^x\frac{1}{t}\,dt$$

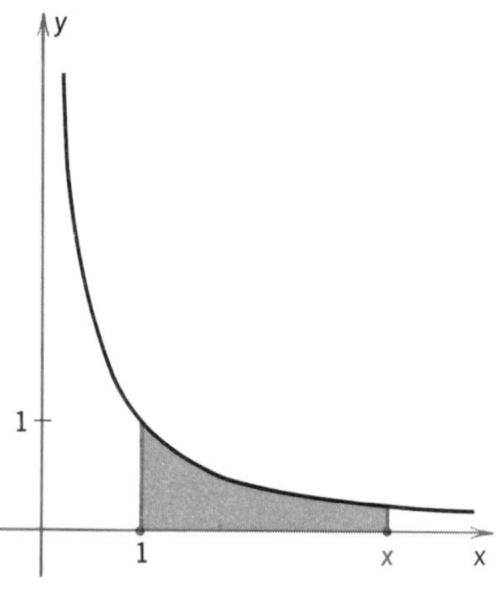

FIGURE 2-2

Observe that for $x>1$, $\ln x$ represents the area bounded by the curve $y=1/t$, the t axis, and the lines $t=1$ and $t=x$. For $x=1$, the upper and lower limits of integration are identical and the area is zero; thus

$\ln 1 = \int_1^1 dt/t = 0$. For $0 < x < 1$, the left boundary is the line $t = x$ and the right boundary is $t = 1$ (Figure 2-3). We have

$$\ln x = \int_1^x dt/t = -\int_x^1 dt/t$$

Thus for $0 < x < 1$, $\ln x$ is the negative of the area under the curve between x and 1. In summary,

$$\ln x < 0 \quad \text{when } 0 < x < 1$$

$$\ln x = 0 \quad \text{when } x = 1$$

$$\ln x > 0 \quad \text{when } x > 1$$

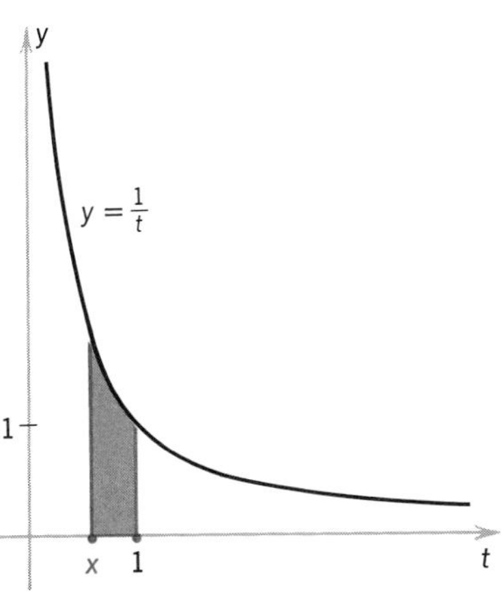

FIGURE 2-3

Computing the derivative of the logarithmic function is an easy application of the Fundamental Theorem of Calculus. By that theorem, we know that

$$\frac{d\left(\int_a^x f(t)\, dt\right)}{dx} = f(x)$$

Hence

$$\frac{d(\ln x)}{dx} = \frac{d\int_1^x dt/t}{dx} = \frac{1}{x} \tag{1}$$

Equation (1) can be used in conjunction with the Chain Rule to yield a number of interesting results.

Example 1 Compute $d(\ln(\cos x))/dx$, where $-\frac{1}{2}\pi < x < \frac{1}{2}\pi$. Using equation (1) and the Chain Rule, we have

$$\frac{d(\ln(\cos x))}{dx} = \frac{1}{\cos x}\frac{d(\cos x)}{dx}$$

$$= \frac{-\sin x}{\cos x} = -\tan x$$

Now $\ln x$ is only defined for $x > 0$. However, for $x < 0$, we may still speak of $\ln|x|$. Differentiating this function yields

$$\frac{d(\ln|x|)}{dx} = \frac{1}{|x|}\frac{d(|x|)}{dx}$$

$$= \begin{cases} -1/|x|, & x < 0 \\ 1/|x|, & x > 0 \end{cases}$$

$$= \frac{1}{x} \qquad (x \neq 0)$$

Thus $\ln|x|$ is an antiderivative of $1/x$, and we have the indefinite integral

$$\int \frac{du}{u} = \ln|u| + C \tag{2}$$

We next turn to analyzing the graph of the logarithmic function (Figure 2-4). The function has domain $]0, \infty[$ and, as we shall see in Example 2, range all of $\boldsymbol{R}$, and $\ln x \to -\infty$ as $x \to 0^+$ and $\ln x \to \infty$ as $x \to \infty$. As $x \to 0^+$, $\ln x \to -\infty$, so the y axis is a vertical asymptote of the logarithmic function. As $x \to \infty$, although $\ln x \to \infty$, there is no nonvertical asymptote (see exercise 1). We have already noted that $\ln 1 = 0$, $\ln x < 0$ for $x < 1$, and $\ln x > 0$ for $x > 1$. Furthermore, $d(\ln x)/dx = 1/x > 0$ for all $x > 0$; therefore $x \mapsto \ln x$ is **monotone increasing.** We also compute the second derivative

$$\frac{d^2(\ln x)}{dx^2} = -\frac{1}{x^2} < 0$$

for all x, so the function ln is **concave downward.**

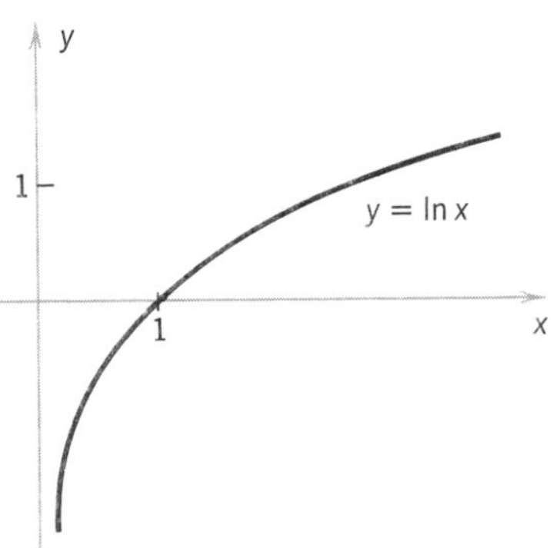

FIGURE 2-4

The next theorem establishes a fundamental property of the ln function.

Theorem 1 If $a, b > 0$, then $\ln(ab) = \ln a + \ln b$.

Proof Consider the function given by $y = \ln ax$, defined for $x > 0$. Differentiating using the Chain Rule, we have

$$\frac{dy}{dx} = \frac{1}{ax} \cdot a = \frac{1}{x}$$

Thus, the functions $x \mapsto \ln ax$ and $x \to \ln x$ have the same derivative; consequently they differ by at most a constant:

$$\ln ax = \ln x + C$$

Since this is true for all $x > 0$, it is true in particular for $x = 1$; hence

$$\ln(a \cdot 1) = \ln 1 + C = 0 + C$$

which implies

$$\ln a = C$$

so constant C is equal to $\ln a$. But the above relation is also true if $x = b$, therefore

$$\ln(ab) = \ln b + \ln a$$

This theorem indicates that the ln function has the basic property of all logarithmic functions. Shortly we shall relate the concepts of log and ln.

Corollary 1 If $a, b > 0$, then $\ln(b/a) = \ln b - \ln a$.

Proof We have $b = (b/a)\cdot a$, so $\ln b = \ln((b/a)\cdot a) = \ln(b/a) + \ln a$ by the above theorem. Hence $\ln(b/a) = \ln b - \ln a$. ▨

Corollary 2 For any rational number q, and $x > 0$, $\ln x^q = q \ln x$.

This proof is left as an exercise. [Hint: First consider the case of q an integer.]

Example 2 Establish the limits

$$\lim_{x\to\infty} \ln x = +\infty, \qquad \lim_{x\to 0^+} \ln x = -\infty$$

Since ln is a continuous function, these limits imply that the range of ln is all of $\boldsymbol{R}$ (Why?). To establish the first limit, we note that

$$\ln 2 = \int_1^2 \frac{1}{t}\,dt \geqslant \tfrac{1}{2}(2-1) = \tfrac{1}{2}$$

since $1/t \geqslant \frac{1}{2}$ on the interval $[1, 2]$ (see Figure 2-5).

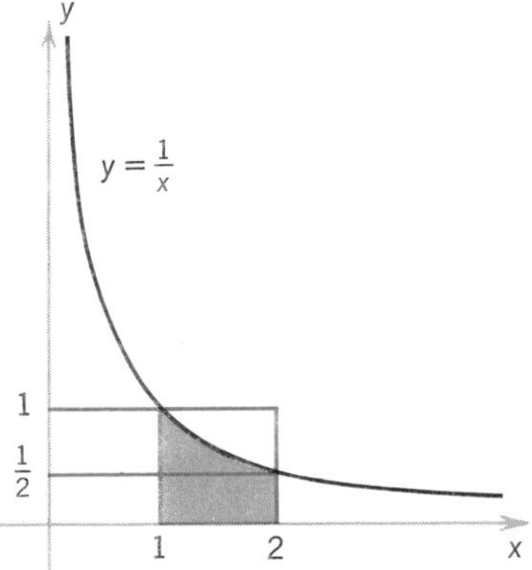

FIGURE 2-5

By Corollary 2 above, we have

$$\ln 2^n = n \ln 2 \geqslant \frac{n}{2}$$

Therefore the sequence of values $\{\ln 2^n\}$ diverges to $+\infty$ as $n \to \infty$. Since ln is monotone increasing, this shows that $\lim_{x\to\infty} \ln x = +\infty$. Moreover, by Corollary 1, the sequence $\{\ln \frac{1}{2^n}\}$ diverges to $-\infty$ as $n \to \infty$, which proves that $\lim_{x\to 0^+} \ln x = -\infty$.

The antiderivative of $u \mapsto \ln u$ is easily found by inspection: Differentiating $u \mapsto u \ln u - u$ gives us the logarithmic function ln, and so,

$$\int \ln u\,du = u \ln u - u + C \tag{3}$$

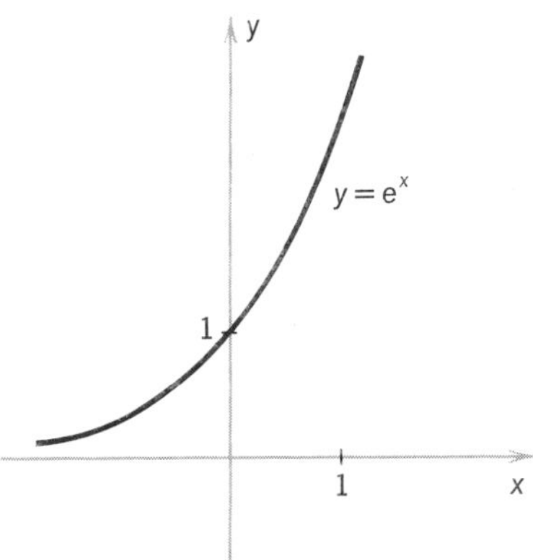

FIGURE 2-6

We have already remarked that the logarithmic function is monotone increasing; hence, it is one-one and so has an inverse (see §1.3). The domain of this inverse is the range of the ln function, that is, all of $\boldsymbol{R}$. The inverse of the logarithmic function is called the **exponential function**, and is denoted **exp**. For every real number x, the exponential function assigns to x the unique (positive) real number y such that $\ln y = x$ (Figure 2-6). Thus if $\ln y = x$, then $y = \exp x$. We denote by e the real number $\exp 1$ (e is an irrational number, whose decimal expansion begins 2.71828 ...). Then

$$\ln e = 1$$

and for any rational x,

$$\ln e^x = x \ln e = x$$

hence,

$$\exp x = e^x$$

so that $e^x = \exp x$ for all rational x. Since the logarithmic function is continuous, for *any* real number x there is a positive number y such that $x = \ln y$, and $y = \exp x$; we therefore **define** e^x for any real number x to be the unique real number $y = \exp x$. We shall henceforth use the notations "$\exp x$" and "e^x" interchangeably.

For a typical x, it is impossible to compute either $\ln x$ or e^x by elementary methods. A table of logarithms and exponentials is given in the appendix of this book, to be consulted when approximate numerical answers to problems are needed. Special numerical techniques are used to construct these tables (see Chapter 12).

We can differentiate the exponential function using the Chain Rule:

$$x = \ln e^x$$

so

$$\frac{d(x)}{dx} = \frac{d(\ln e^x)}{dx}$$

or

$$\begin{aligned} 1 &= \left.\frac{d(\ln y)}{dy}\right|_{y=e^x} \frac{d(e^x)}{dx} \\ &= \frac{1}{e^x}\frac{d(e^x)}{dx} \end{aligned}$$

Hence

$$\boldsymbol{\frac{d(e^x)}{dx} = e^x} \qquad (4)$$

This also leads to the integration formula

$$\boldsymbol{\int e^u\,du = e^u + C} \qquad (5)$$

Example 3 Find $d(e^{1/x})/dx$. By the Chain Rule

$$\begin{aligned} \frac{d(e^{1/x})}{dx} &= e^{1/x}\frac{d(1/x)}{dx} \\ &= e^{1/x}\left(-\frac{1}{x^2}\right) \\ &= \frac{-e^{1/x}}{x^2} \end{aligned}$$

Example 4 Integrate $\int xe^{x^2}\,dx$. Let $u = x^2$; $du = 2x\,dx$. Then

$$\int x\,e^{x^2}\,dx = \tfrac{1}{2}\int e^{x^2}\,2x\,dx$$
$$= \tfrac{1}{2}\int e^u du = \tfrac{1}{2}e^u + C$$
$$= \tfrac{1}{2}e^{x^2} + C$$

The following properties of the exponential function are easily derived using properties of the logarithmic function:

$$e^{x_1} \cdot e^{x_2} = e^{x_1 + x_2} \tag{6}$$

$$e^{-x} = 1/e^x \tag{7}$$

To verify equation (6), let $y_1 = e^{x_1}$, $y_2 = e^{x_2}$. Then $x_1 = \ln y_1$, $x_2 = \ln y_2$, and $x_1 + x_2 = \ln y_1 + \ln y_2 = \ln y_1 y_2$, so by definition,

$$e^{x_1} \cdot e^{x_2} = y_1 y_2 = \exp(x_1 + x_2) = e^{x_1 + x_2}$$

Establishing equation (7) is left as an exercise.

Finally, we examine the graph of $x \mapsto e^x$ (Figure 2-6). Since $e^x > 0$ for all x, the graph lies entirely above the x axis. The y intercept of the function is $(0, 1)$. As $x \to 0$, $\ln x \to -\infty$; therefore as $x \to -\infty$, $e^x \to 0$ and $d(e^x)/dx = e^x \to 0$, so the exponential function has the x axis as an asymptote. As $x \to \infty$, $d(e^x)/dx = e^x \to \infty$, so there is no asymptote as $x \to \infty$. It is clear that both the first and second derivatives of the exponential function are always positive, so that $x \mapsto e^x$ is monotone increasing and concave upward throughout all of $\boldsymbol{R}$.

Since $x \mapsto e^x$ and $x \mapsto \ln x$ are inverse functions, their graphs are symmetric to one another with respect to the line $y = x$. Thus the graph of the exponential function is obtained by "flipping" the graph of the logarithm function about the line $y = x$ (see Figure 2-7).

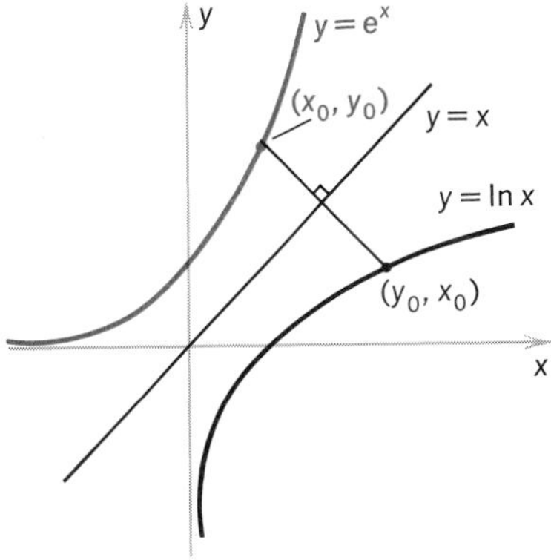

FIGURE 2-7

EXERCISES

1. Prove that $\ln x$ does not approach an asymptote as $x \to \infty$. This is done in several steps: (a) Show that $\lim_{x\to\infty} d(\ln x)/dx = 0$, and explain why this means that any asymptote as $x \to \infty$ would have to be horizontal. (b) Since $\ln x$ is unbounded as $x \to \infty$, explain why this result shows that $x \mapsto \ln x$ cannot have a horizontal asymptote as $x \to \infty$. (c) Use parts (a) and (b) to reach the desired conclusion.
2. Prove that for any rational number q, $\ln x^q = q \ln x$.
3. Establish equation (7) of this section.
4. Simplify each of the following expressions.

   (a) $\ln e^{1/x}$ (b) $e^{\ln 2 + \ln x}$

   (c) $e^{-\ln(1/x)}$ (d) $\ln(x^2 e^{-3x})$

5. Express each of the following in terms of $a = \ln 3$ and $b = \ln 5$. For example, $\ln(0.6) = \ln(\frac{3}{5}) = \ln 3 - \ln 5 = a - b$.
   (a) $\ln 625$ (b) $\ln 0.04$ (c) $\ln \frac{25}{27}$
   (d) $\ln \sqrt[3]{9}$ (e) $\ln 5\sqrt{5}$ (f) $\ln 45$
   (g) $\ln 1.8$ (h) $\ln \sqrt{5.4}$
6. Compute dy/dx and d^2y/dx^2.
   (a) $y = \ln(x^2+2x)$ (b) $y = 5\ln(x^2-9)$
   (c) $y = \ln(\sin x)$ (d) $y = \ln(\tan x + \sec x)$
   (e) $y = \frac{1}{3}\ln \dfrac{x^3}{x^3+1}$ (f) $y = \ln(\ln x)$
   (g) $y = \cos(\ln x)$ (h) $y = \ln\sqrt{(1+x)/(1-x)}$
   (i) $y = x\ln(a^2+x^2) - 2x + 2a\tan^{-1}(x/a)$
   (j) $y = x^2 e^x$ (k) $y = e^{2x}(2\cos 3x + 3\sin 3x)$
   (l) $y = (ax-1)e^{ax}/a^3$ (m) $y = e^x \cos(e^x)$
   (n) $y = \ln \dfrac{e^x}{1+e^x}$ (o) $y = \cos^{-1}(e^{-\frac{1}{2}x})$
7. Integrate each of the following.
   (a) $\int e^{2x}\,dx$ (b) $\int x^3 e^{x^4}\,dx$
   (c) $\int \dfrac{4\,dx}{e^{3x}}$ (d) $\int \dfrac{(\ln x)^2\,dx}{x}$
   (e) $\int e^{\sin x}\cos x\,dx$ (f) $\int \dfrac{dx}{2x+5}$
   (g) $\int \dfrac{\sin x\,dx}{\cos x} = \int \tan x\,dx$ (h) $\int \dfrac{\cos x\,dx}{\sin x} = \int \cot x\,dx$
   (i) $\int \dfrac{dx}{\sqrt{x}+x}$
8. Prove equation (4) using Theorem 2 of §4.8.

3 MORE LOGARITHMIC AND EXPONENTIAL FUNCTIONS

In earlier courses one has already learned the meaning of a^x for any positive real number a and rational number x. If we express x in the form p/q, where p and q are integers and q is positive, then $a^x = a^{p/q}$ means "the (positive) qth root of a, raised to the pth power"; that is, $a^{p/q} = (\sqrt[q]{a})^p$.

We would like to be able to talk about the expression a^x for *any* real number x and positive real number a. Now, if a is any positive real number and x is any rational number, we have

$$a^x = \exp(\ln a^x)$$

But

$$\ln a^x = x\ln a$$

so

$$\exp(x \ln a) = a^x \tag{1}$$

The left-hand side of equation (1) is defined for x any real number, and so may be taken to be the definition of a^x for any real number x.

Definition If $x \in \boldsymbol{R}$ and $a > 0$, we define

$$\boldsymbol{a^x = \exp(x \ln a) = e^{x\ln a}} \tag{2}$$

Using the properties of the exponential and logarithmic functions together with the above definition, the following properties, already known for rational x and y, can be established:

$$a^x \cdot a^y = a^{x+y} \tag{3}$$

$$a^x/a^y = a^{x-y} \tag{4}$$

$$(a^x)^y = a^{xy} \tag{5}$$

$$(ab)^x = a^x \cdot b^x \tag{6}$$

We prove equation (3) and leave the others as exercises. Now $a^x \cdot a^y = e^{x\ln a}e^{y\ln a}$. But

$$e^{x\ln a}\, e^{y\ln a} = e^{(x+y)\ln a}$$

and by definition

$$a^{x+y} = e^{(x+y)\ln a}$$

Therefore

$$a^x \cdot a^y = e^{(x+y)\ln a} = a^{x+y}$$

Let us examine the graph of $y = a^x$.

Example 1 Sketch the graph of $y = 2^x$. By equation (2), $y = e^{x(\ln 2)}$, and so the graph has the x axis as an asymptote as $x \to -\infty$. Since $dy/dx = (\ln 2)e^{x(\ln 2)} \to \infty$ as $x \to \infty$, there is no asymptote as $x \to \infty$. The function is monotone increasing and concave upward. We compute several values and then sketch the curve (Figure 3-1).

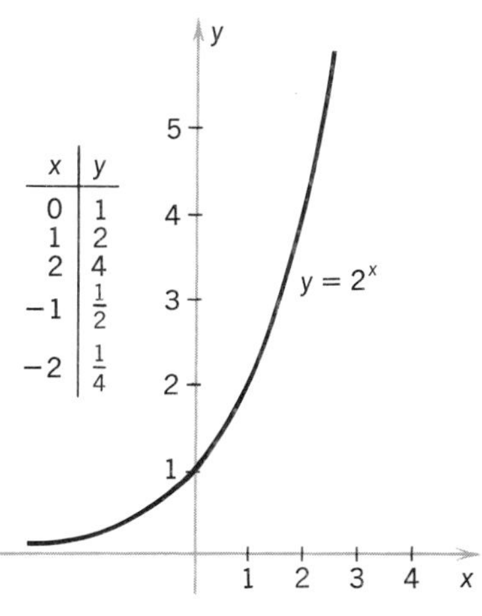

x	y
0	1
1	2
2	4
−1	$\frac{1}{2}$
−2	$\frac{1}{4}$

FIGURE 3-1

Any exponential function $x \mapsto a^x\,(a > 0, a \neq 1)$ is one-one, and therefore has an inverse, which is a corresponding logarithmic function.

Definition If a and x are positive real numbers $(a \neq 1)$ such that $a^y = x$, then we say that y is the **logarithm of x to the base a**, and write

$$y = \log_a x$$

For example, $\log_3 9 = 2$, since $3^2 = 9$. Suppose that $y = \log_a x$ and $a \neq 1$. Then $a^y = x$, so

$$y \ln a = \ln x$$

and

$$y = \ln x/\ln a$$

Thus we have the general formula

$$\log_a x = \frac{\ln x}{\ln a} \tag{7}$$

for all $a \neq 1$. In particular, if $a = e$, equation (7) reduces to

$$\log_e x = \ln x / \ln e$$

or

$$\log_e x = \ln x \tag{8}$$

Hence the ln function is just the logarithm to the base e. It is sometimes called the **natural logarithm**.

The following properties of logarithms to any base are easily established from the properties of the ln function and the above formula:

$$\log_a (x \cdot y) = \log_a x + \log_a y, \qquad a \neq 1 \tag{9}$$

$$\log_a (x/y) = \log_a x - \log_a y, \qquad a \neq 1 \tag{10}$$

$$\log_a x = y \log_a x \quad x > 0, \qquad a \neq 1 \tag{11}$$

We prove equation (9) and leave the other two as exercises. Thus

$$\begin{aligned} \log_a (x \cdot y) &= \ln (x \cdot y)/\ln a \\ &= (\ln x + \ln y)/\ln a \\ &= \ln x/\ln a + \ln y/\ln a \\ &= \log_a x + \log_a y \end{aligned}$$

As the reader undoubtedly knows, it is this basic property (9) which led to the introduction of logarithms, generally credited to John Napier. They have been used extensively as an aid in numerical multiplication.

Let us examine the graph of $y = \log_a x$.

Example 2 Sketch the graph of $y = \log_3 x$. We leave to the student to show that there is a vertical asymptote at $x = 0$ and no nonvertical asymptote. By equation (7),

$$y = \frac{1}{\ln 3} \ln x$$

so

$$\frac{dy}{dx} = \frac{1}{x \ln 3} > 0$$

for all $x > 0$, and

$$\frac{d^2y}{dx^2} = \frac{1}{-x^2 \ln 3} < 0$$

hence, the function is monotone increasing and concave downward. We compute several values and then sketch the graph (Figure 3-2).

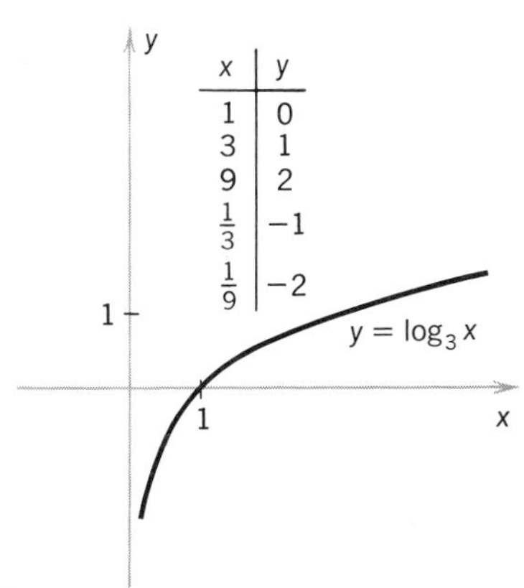

FIGURE 3-2

Having defined a^x and $\log_a x$, we are now ready to compute their derivatives. We first compute the derivative of $x \mapsto a^x$ where $a > 0$. Now $a^x = e^{x \ln a}$ by definition, so

$$\frac{d(a^x)}{dx} = \frac{d(e^{x \ln a})}{dx}$$
$$= e^{x \ln a} \ln a$$
$$= a^x \ln a$$

Thus,

$$\frac{d(a^x)}{dx} = a^x \ln a \tag{12}$$

Equation (12) also yields the integration formula

$$\int a^u\, dx = a^u/\ln a + C \tag{13}$$

Next, to compute the derivative of $\log_a x (a \neq 1)$, recall that

$$\log_a x = \frac{\ln x}{\ln a}$$

so

$$\frac{d(\log_a x)}{dx} = \frac{d}{dx}\left(\frac{\ln x}{\ln a}\right)$$
$$= \frac{d(\ln x/\ln a)}{dx}$$

Thus,

$$\frac{d(\log_a x)}{dx} = \frac{1}{x \ln a} \tag{14}$$

We have thus far differentiated functions of the form $x \mapsto a^x$ and $x \mapsto x^a$ (see also exercise 12). It is sometimes necessary to differentiate functions of the form

$$x \mapsto [f(x)]^{g(x)}$$

This can be accomplished by the following technique.

FORMAL DEVICE Logarithmic Differentiation

To compute the derivative of a function

$$F: x \mapsto [f(x)]^{g(x)}$$

(1) Take the logarithm of both sides of the equation $F(x) = [f(x)]^{g(x)}$ yielding

$$\ln(F(x)) - g(x)\ln(f(x)) = 0$$

(2) Differentiate both sides of this equation to find

$$\frac{F'(x)}{F(x)} - \left[g'(x) \cdot \ln(f(x)) + g(x)\frac{f'(x)}{f(x)}\right] = 0$$

(3) Solve for $F'(x)$

$$F'(x) = F(x)\left[g'(x) \cdot \ln(f(x)) + g(x)\frac{f'(x)}{f(x)}\right]$$

In practice it is far easier to remember the technique and apply it to individual cases, than it is to memorize the final formula.

Example 3 Find $d(x^x)/dx$. Let $y = x^x$. Then $\ln y = x \ln x$, or

$$\ln y - x \ln x = 0$$

The corresponding differential relation is

$$dy/y - (\ln x\, dx + dx) = 0$$

or

$$dy/dx = y(\ln x + 1)$$

Hence

$$\frac{d(x^x)}{dx} = x^x(\ln x + 1)$$

Example 4 Compute $d((\cos^2 x)^{\sin x})/dx$. Let $y = (\cos^2 x)^{\sin x}$. Then

$$\ln y - 2\sin x \ln(\cos x) = 0$$

Computing the differential relation yields

$$dy/y - 2(\cos x \ln(\cos x) - \sin^2 x/\cos x)\, dx = 0$$

or

$$dy/dx = 2y(\cos x \ln(\cos x) - \sin x \tan x)$$

Therefore

$$d((\cos^2 x)^{\sin x})/dx = 2(\cos^2 x)^{\sin x}(\cos x \ln(\cos x) - \sin x \tan x)$$

We have already utilized graphing techniques from Chapter 5 to study several exponential and logarithmic functions. These techniques can also be applied to more complex transcendental functions.

Example 5 Sketch the graph of $f{:}x \mapsto xe^x$. The function f is defined, continuous, and differentiable on all of $\boldsymbol{R}$, and

$$f'(x) = e^x(x+1)$$

$$f''(x) = e^x(x+2) \qquad \text{(Verify!)}$$

We see that f' is unbounded as $x \to \infty$, so there is no asymptote in that direction. However, as $x \to -\infty$, we find that $f(x)$ and $f'(x)$ both approach 0, so the x axis is an asymptote.

Because $e^x > 0$ for all x, $f'(x) = 0$ only when $x+1=0$, that is, at $x=-1$. Using the Second Derivative Test, we calculate $f''(-1) = e^{-1} > 0$, so the point $(-1, -e^{-1})$ locates a relative minimum. Similarly $f''(x) = 0$ only when $x+2=0$, and $(-2, -e^{-2})$ is the only point of inflection.

We now plot a few additional points:

x	y
0	0
1	$e \approx 2.7$
2	$2e^2 \approx 14.8$

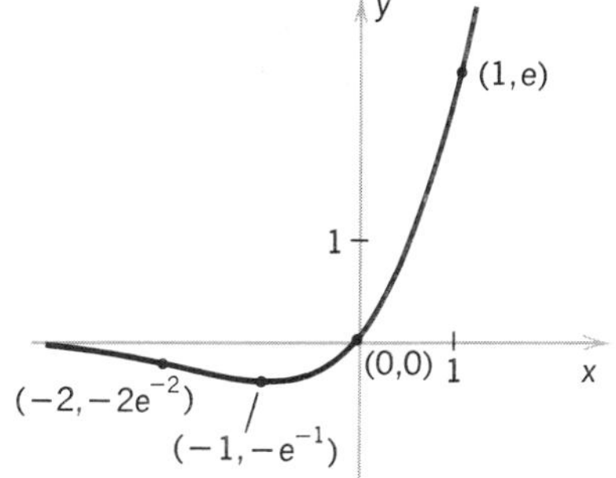

FIGURE 3-3

Using the derivative tests we see that f decreases on $]-\infty, -1]$, increases on $[-1, \infty]$, is concave downward on $]-\infty, -2[$, and concave upward on $]-2, \infty[$. We now sketch the final curve (Figure 3-3).

Example 6 Sketch the graph of $f{:}x \mapsto e^{-x}\sin x$. The domain of f is all of $\boldsymbol{R}$. By inspection we see that f is continuous and differentiable, and using the Chain Rule we compute

$$f'(x) = e^{-x}(\cos x - \sin x)$$

$$f''(x) = -2e^{-x}\cos x \qquad \text{(Verify!)}$$

Since $-1 \leqslant \sin x \leqslant 1$ for all x, we have $-e^{-x} \leqslant f(x) \leqslant e^{-x}$ for all x. But both $-e^{-x}$ and e^{-x} approach zero as $x \to \infty$; hence, f also approaches zero. Moreover, $|f'(x)| \leqslant \sqrt{2}\,e^{-x}$ (Why?), so $f'(x) \to 0$ as $x \to \infty$. Therefore, the x axis is an asymptote of f as $x \to \infty$. On the other hand, $f''(x)$ is not bounded as $x \to -\infty$; hence, there is no asymptote in that direction.

Since $e^{-x} > 0$ for all x, critical points of f occur where $\cos x - \sin x = 0$, that is, at $x = (n+\frac{1}{4})\pi$. The Second Derivative Test then shows that for even n, these critical points are maxima; for odd n they are minima. Points of inflection occur when $\cos x = 0$ (Why?), which is at $x = (n+\frac{1}{2})\pi$. Monotonicity and concavity are evident from the locations of critical points and points of inflection (see §5.1, exercise 6).

The zeros of f occur when $\sin x = 0$, at $x = \pi n$. Recall from §1 that in this case the graph of f oscillates back and forth between the graphs of $y = -e^{-x}$ and $y = e^{-x}$, each of which it touches at alternate points $x = (n+\frac{1}{2})\pi$; here it happens to be that these are the points of inflection of f. This is sufficient information to sketch the function (Figure 3-4).

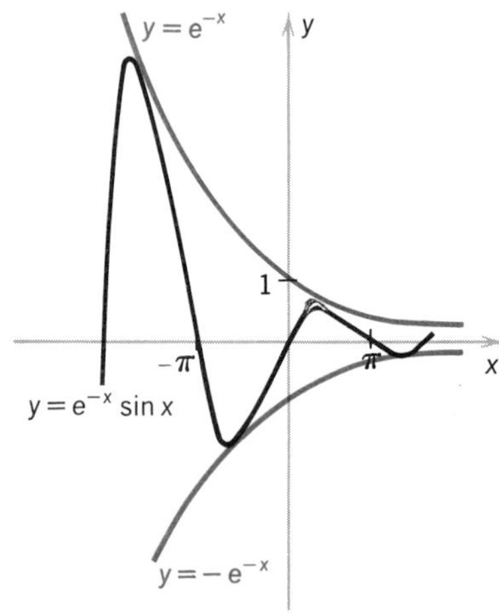

FIGURE 3-4

If f is a twice differentiable function, the graph of $F: x \mapsto e^{f(x)}$ considerably reflects the graph of f. We have

$$F'(x) = f'(x)\, e^{f(x)}$$

$$F''(x) = \left(f'(x)^2 + f''(x)\right) e^{f(x)} \qquad \text{(Verify!)}$$

Since $e^{f(x)} > 0$ for all x, the graph of F lies entirely above the x axis. Moreover, $\operatorname{sgn}(F'(x)) = \operatorname{sgn}(f'(x))$; hence, critical points and regions of monotonicity of F and f are identical. (As a result, F is clearly one-one if and only if f is one-one; the same holds true for periodicity.) With regard to concavity, we see that F is concave upward wherever f is. Downward concavity must be determined for each individual case.

Now consider nonvertical asymptotes. If f has an asymptote $y = mx + b$ as $x \to \infty$, then we know that $[f(x) - (mx+b)] \to 0$, and $f'(x) \to m$. If $m > 0$, then $F'(x)$ becomes close to $m\, e^{mx+b}$, which is unbounded; hence F has no asymptote. If $m = 0$, then $F(x) = e^{f(x)} \to e^b$, as $x \to \infty$, and $F'(x) = f'(x) e^{f(x)} \to 0 \cdot e^b = 0$ as $x \to \infty$; hence, F has the horizontal asymptote $y = e^b$. If $m < 0$, then $F(x) \to 0$ and $F'(x) \to 0$ (Why?); hence, F has the x axis as an asymptote. We leave consideration of asymptotes as $x \to -\infty$ to the exercises.

Example 7 Sketch the graph of $g: x \mapsto e^{\sin x}$. The function is clearly defined and differentiable everywhere, with

$$g'(x) = \cos x\, e^{\sin x}$$

$$g''(x) = (\cos^2 x - \sin x)\, e^{\sin x}$$

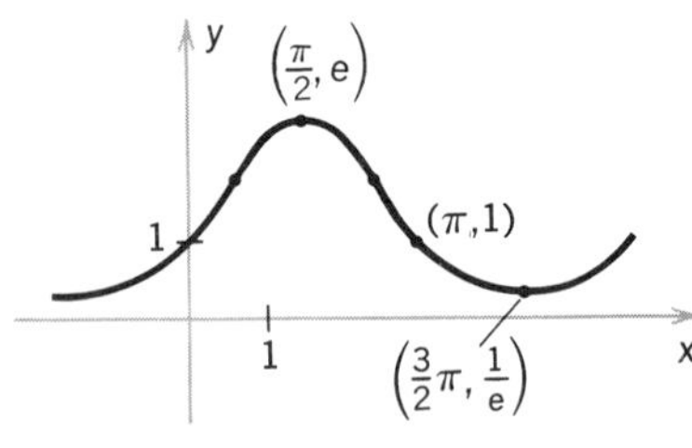

FIGURE 3-5

Since $g'(x)$ does not approach a limit as $x \to \pm\infty$, there are no asymptotes. Relative maxima and minima occur at the critical points of $x \mapsto \sin x$. Maxima are of the form $((2n+\frac{1}{2})\pi, e)$; minima are of the form $((2n-\frac{1}{2})\pi, 1/e)$. Points of inflection occur when $\cos^2 x - \sin x = 0$, that is, where $\cos x = \tan x$, which occurs approximately at $x = (2n+\frac{1}{2})\pi \pm 0.90$. [One obtains this point approximately by examining where the graphs of $\cos x$ and $\tan x$ intersect.] The points of inflection therefore have the form $((2n+\frac{1}{2})\pi \pm 0.90, 1.85)$. Finally, we note that the graph of g crosses the line $y = 1$ at $x = n\pi$. This information is sufficient to sketch the curve (Figure 3-5). Notice that g is periodic with period 2π.

We remark finally that l'Hôpital's Rule is frequently useful in determining limits of functions involving exponentials.

Example 8 Compute $\lim_{x\to 0}(\sec^3 2x)^{\cot^2 3x}$. Let $y=(\sec^3 2x)^{\cot^2 3x}$. Then $\ln y = \cot^2 3x \ln(\sec^3 2x) = \ln(\sec^3 2x)/\tan^2 3x$. Therefore,

$$\lim_{x\to 0} \ln y = \lim_{x\to 0} \frac{\ln(\sec^3 2x)}{\tan^2 3x} \qquad [=0/0]$$

$$= \lim_{x\to 0} \frac{6\sec^3 2x \tan 2x/\sec^3 2x}{6\tan 3x \sec^2 3x}$$

$$= \lim_{x\to 0} \frac{\tan 2x}{\tan 3x \sec^2 3x} \qquad [=0/0]$$

$$= \lim_{x\to 0} \frac{2\sec^2 2x}{3\sec^4 3x + 6\tan^2 3x \sec^2 3x} = \frac{2}{3}$$

Therefore, $\lim_{x\to 0} y = e^{2/3}$.

Example 9 Compute $\lim_{x\to 0}(e^x+e^{-x}-x^2-2)/(\sin^2 x - x^2)$.

$$\lim_{x\to 0} \frac{e^x+e^{-x}-x^2-2}{\sin^2 x - x^2} \qquad [=0/0]$$

$$= \lim_{x\to 0} \frac{e^x-e^{-x}-2x}{2\sin x \cos x - 2x} \qquad [=0/0]$$

$$= \lim_{x\to 0} \frac{e^x+e^{-x}-2}{2\cos 2x - 2} \qquad [=0/0]$$

$$= \lim_{x\to 0} \frac{e^x-e^{-x}}{-4\sin 2x} \qquad [=0/0]$$

$$= \lim_{x\to 0} \frac{e^x+e^{-x}}{-8\cos 2x} = -\frac{1}{4}$$

EXERCISES

1. Solve for x. Check all answers.
   (a) $\log_5(2x+5) + \log_5(2x-5) = 2$
   (b) $\log_2(x^2+1) + \log_2(x^2-1) = 0$
   (c) $12^x - 2^{x+1} = 0$
   (d) $\sqrt[3]{5^{2x+1}} = 2$
   (e) $2^x + 2^{-x} = 1$
2. Sketch the graphs of the following functions.
   (a) $x \mapsto 4^x$
   (b) $x \mapsto 4^{2-x}$
   (c) $x \mapsto \log_9 x$
   (d) $x \mapsto \log_9(x-5)$
3. Prove equations (4)–(6) of this section.
4. Prove equation (8) of this section.
5. Prove equations (10) and (11) of this section.

6. Differentiate the following functions.
   (a) $x \mapsto y^x$ (b) $x \mapsto (\sqrt{2})^{2x}$ (c) $x \mapsto 2^x e^x$
   (d) $x \mapsto 12^{3x} 5^{-x}$ (e) $x \mapsto z^{kx} w^{mx}$ (f) $x \mapsto e^x \ln x$
   (g) $x \mapsto e^{-e^x}$ (h) $x \mapsto x^{\sin x}$ (i) $x \mapsto x^{\ln x}$
   (j) $x \mapsto x^{x^{\exp x}}$

7. Integrate.
   (a) $\int_0^{1.2} 3^x \, dx$ (b) $\int_0^{\frac{1}{6}\pi} \sin\theta \, 3^{-\cos\theta} \, d\theta$ (c) $\int_1^{\sqrt{3}} x \, 3^{-x^2} \, dx$

8. Show that if $0 < a < 1$, then $x \mapsto a^x$ is monotone decreasing. What happens when $a = 1$?

9. Sketch the graphs of the following functions.
   (a) $x \mapsto x^2 e^x$ (b) $x \mapsto e^{\cos x}$
   (c) $x \mapsto 1/\ln x$ (d) $x \mapsto e^x \sin(e^x)$

10. Suppose a function f has an asymptote $y = mx + b$ as $x \to -\infty$. Analyze the asymptotes of $x \mapsto e^{f(x)}$ as $x \to -\infty$.

11. Compute the following limits.
    (a) $\lim\limits_{x \to \frac{1}{2}} \dfrac{\ln 2x}{2x - 1}$ (b) $\lim\limits_{x \to 0} \dfrac{e^{2x} - \cos x}{\tan 3x}$
    (c) $\lim\limits_{x \to \infty} \dfrac{\ln(1 + e^{2x})}{x}$ (d) $\lim\limits_{x \to \infty} \dfrac{x^2 \ln x}{e^x}$
    (e) $\lim\limits_{x \to \infty} a^x/b^x \quad (a, b > 1)$ (f) $\lim\limits_{x \to \infty} \dfrac{(1 + x^5)^{1/3}}{\ln x}$
    (g) $\lim\limits_{x \to 1} x^{1/(1-x)}$ (h) $\lim\limits_{x \to 0+} x^x$
    (i) $\lim\limits_{x \to 0} (e^x + x)^{1/x}$ (j) $\lim\limits_{x \to 0} (1 + 1/x)^x$

12. In Theorem 1, §4.1, we proved that $d(x^a)/dx = ax^{a-1}$ for a rational. Extend this now to hold for all real a.

4 HYPERBOLIC FUNCTIONS AND THEIR INVERSES

In this section we consider certain combinations of the functions $x \mapsto e^x$ and $x \mapsto e^{-x}$, which are called **hyperbolic functions**. These combinations appear so often in physics and engineering that they have been given special names. The hyperbolic functions are also useful in solving differential equations; our application is to the problem of finding the tension at any point in a cable suspended by its ends.

The two basic hyperbolic functions are defined in terms of exponentials by the formulas:

$$\mathbf{sinh}\, x = \tfrac{1}{2}(e^x - e^{-x})$$

$$\mathbf{cosh}\, x = \tfrac{1}{2}(e^x + e^{-x})$$

The equations define the **hyperbolic sine** and **hyperbolic cosine**, so-called because they have many properties analogous to the trigonometric sine and cosine functions.

Consider a curve parametrized by the equations

$$x = \cosh t, \qquad y = \sinh t$$

Then

$$\begin{aligned} x^2 - y^2 &= \cosh^2 t - \sinh^2 t \\ &= \tfrac{1}{4}[(e^x + 2 + e^{-x}) - (e^x - 2 + e^{-x})] \\ &= 1 \end{aligned}$$

Notice also that $\cosh t > 0$ for all t; hence $x > 0$. We thus see that the hyperbolic sine and cosine give rise to a parametric representation of the right-hand branch of the hyperbola $x^2 - y^2 = 1$ (Figure 4-1). In an analogous manner, the equations $x = \cos t$ and $y = \sin t$ are a parametric representation of the circle $x^2 + y^2 = 1$.

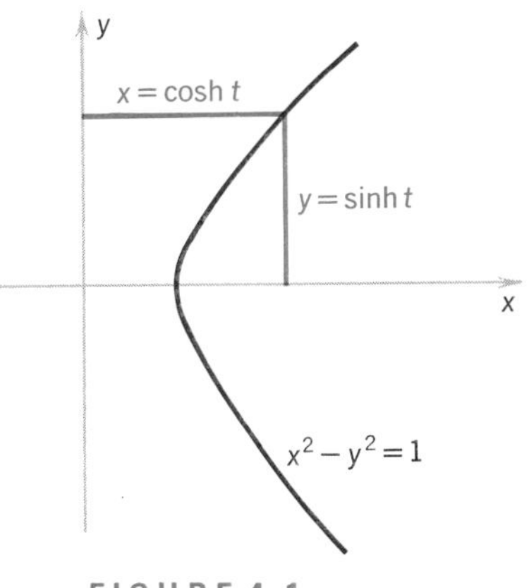

FIGURE 4-1

The remaining hyperbolic functions are defined in terms of $\sinh x$ and $\cosh x$.

$$\tanh x = \frac{\sinh x}{\cosh x} = \frac{e^x - e^{-x}}{e^x + e^{-x}}$$

$$\coth x = \frac{\cosh x}{\sinh x} = \frac{e^x + e^{-x}}{e^x - e^{-x}}$$

$$\operatorname{sech} x = \frac{1}{\cosh x} = \frac{2}{e^x + e^{-x}}$$

$$\operatorname{csch} x = \frac{1}{\sinh x} = \frac{2}{e^x - e^{-x}}$$

Let us now consider the graphs of the hyperbolic functions. The following chart summarizes information about these graphs, which appear in Figure 4-2(a)–(f).

Function	Domain	Range	Symmetry	Asymptotes
$\sinh x$	R	R	odd	—
$\cosh x$	R	$[1, \infty[$	even	—
$\tanh x$	R	$]-1, 1[$	odd	$y = \pm 1$
$\coth x$	$x \neq 0$	$]-\infty, -1]$ and $[1, \infty[$	odd	$y = \pm 1$
$\operatorname{sech} x$	R	$]0, 1]$	even	$y = 0, 1$
$\operatorname{csch} x$	$x \neq 0$	$y \neq 0$	odd	$y = 0$ $x = 0$

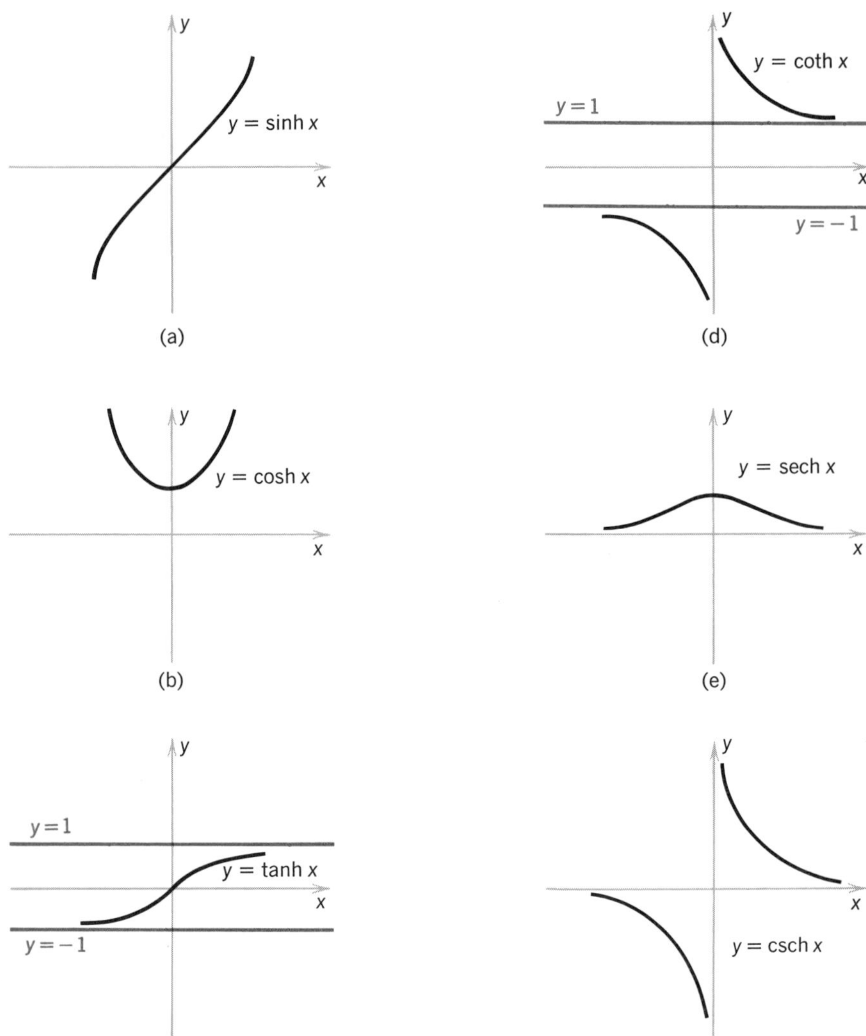

FIGURE 4-2

To differentiate the hyperbolic functions, recall that $d(e^x)/dx = e^x$ and $d(e^{-x})/dx = -e^{-x}$. Therefore we have

$$\frac{d(\sinh x)}{dx} = \frac{e^x + e^{-x}}{2} = \cosh x$$

and

$$\frac{d(\cosh x)}{dx} = \frac{e^x - e^{-x}}{2} = \sinh x$$

To find $d(\tanh x)/dx$, we use the quotient rule:

$$\frac{d(\tanh x)}{dx} = \frac{d}{dx}\left(\frac{\sinh x}{\cosh x}\right)$$

$$= \frac{\cosh x(\cosh x) - \sinh x(\sinh x)}{\cosh^2 x}$$

$$= \frac{\cosh^2 x - \sinh^2 x}{\cosh^2 x} = \frac{1}{\cosh^2 x}$$

$$= \operatorname{sech}^2 x$$

In a similar manner we can derive the following formulas:

$$\frac{d(\coth x)}{dx} = -\operatorname{csch}^2 x$$

$$\frac{d(\operatorname{sech} x)}{dx} = -\operatorname{sech} x \tanh x$$

$$\frac{d(\operatorname{csch} x)}{dx} = -\operatorname{csch} x \coth x$$

Observe that except for differences in sign, these formulas are exactly analogous to those of the trigonometric functions. Also note that each of these formulas can be used to produce a matching integration formula:

$$d(\sinh u) = \cosh u\, du \qquad \int \cosh u\, du = \sinh u + C$$

$$d(\cosh u) = \sinh u\, du \qquad \int \sinh u\, du = \cosh u + C$$

$$d(\tanh u) = \operatorname{sech}^2 u\, du \qquad \int \operatorname{sech}^2 u\, du = \tanh u + C$$

$$d(\coth u) = -\operatorname{csch}^2 u\, du \qquad \int \operatorname{csch}^2 u\, du = -\coth u + C$$

$$d(\operatorname{sech} u) = -\operatorname{sech} u \tanh u\, du \qquad \int \operatorname{sech} u \tanh u\, du = -\operatorname{sech} u + C$$

$$d(\operatorname{csch} u) = -\operatorname{csch} u \coth u\, du \qquad \int \operatorname{csch} u \coth u\, du = -\operatorname{csch} u + C$$

We have seen that the relation of the hyperbolic functions to their derivatives is analogous to the relationship of the trigonometric functions to their derivatives. There are also various identities which can be established with hyperbolic functions that are analogous to trigonometric identities. Using the fact that

$$\cosh^2 x - \sinh^2 x = 1$$

the student can easily derive the identities

$$1 - \tanh^2 x = \operatorname{sech}^2 x$$

and

$$\coth^2 x - 1 = \operatorname{csch}^2 x$$

which are left as exercises.

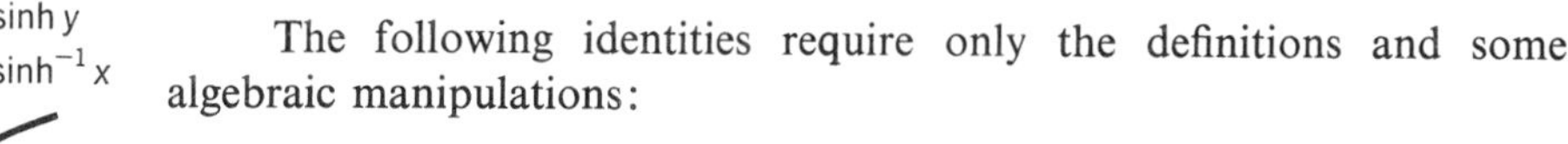

The following identities require only the definitions and some algebraic manipulations:

$$\sinh(x + y) = \sinh x \cosh y + \cosh x \sinh y$$

$$\cosh(x + y) = \cosh x \cosh y + \sinh x \sinh y$$

These in turn can be used to derive the formulas

$$\sinh 2x = 2 \sinh x \cosh x$$

$$\cosh 2x = \cosh^2 x + \sinh^2 x$$

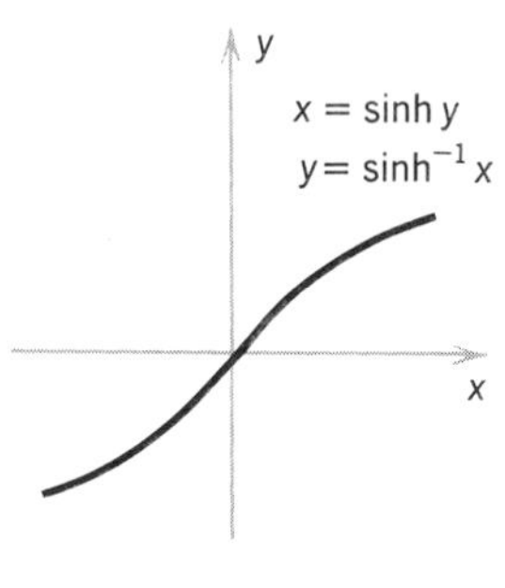

FIGURE 4-3

Thus, almost all the trigonometric identities have hyperbolic counterparts.

It is also reasonable to ask if the hyperbolic functions have inverses. Hence, suppose we consider the locus of $x = \sinh y$ (Figure 4-3). Since $d(\sinh y)/dy = \cosh y > 0$ for all y, we verify immediately that the hyperbolic sine is monotone increasing and, therefore, one-one which implies that the inverse function exists. We define $\sinh^{-1} x$, the **inverse hyperbolic sine** of x, as the unique real number y such that $x = \sinh y$. There is thus an inverse hyperbolic sine function $x \mapsto \sinh^{-1} x$.

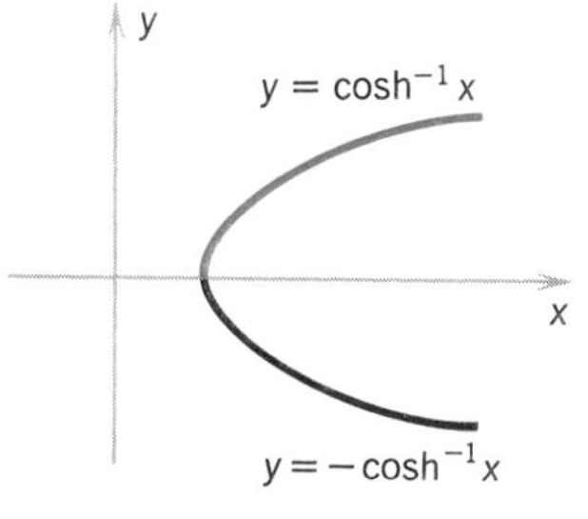

FIGURE 4-4

Now let us examine the locus of $x = \cosh y$ (Figure 4-4). Observe that for any given x, there are two values of y such that $x = \cosh y$. We therefore cannot define $y = \cosh^{-1} x$ without any restrictions, since for a given value of x there would be two values of y. Hence, we take a "principle branch" of the locus; that is, we consider only positive y in the relation $x = \cosh y$. We thus define **inverse hyperbolic cosine** $\cosh^{-1} x$ to be the unique **nonnegative** real number y such that $x = \cosh y$. Hence, $x \mapsto \cosh^{-1} x$ now describes the upper half of the locus $x = \cosh y$, while $x \mapsto -\cosh^{-1} x$ describes the lower half.

The only other hyperbolic function with which we need worry about a principle branch is the hyperbolic secant, and we again select this principal branch to be the one where $y > 0$. Thus for $0 < x \leqslant 1$, we define $\operatorname{sech}^{-1} x$ to be the unique positive real number y such that $x = \operatorname{sech} y$ (Figure 4-5).

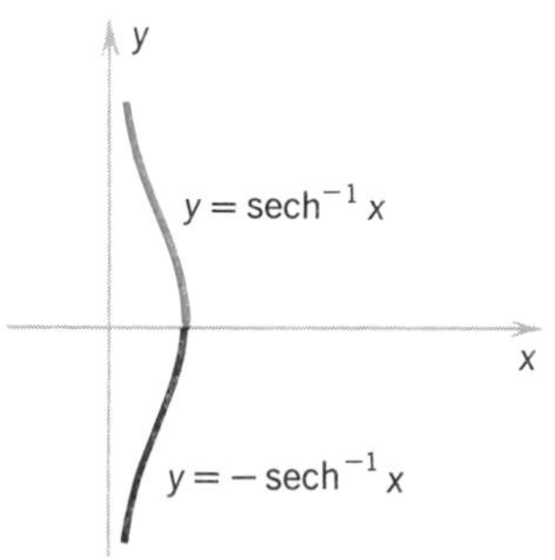

FIGURE 4-5

For the remaining inverse functions $\tanh^{-1} x$, $\coth^{-1} x$, and $\operatorname{csch}^{-1} x$, there is no need to consider principal branches, and the definitions are straightforward [Figure 4-6(a)–(c)].

The following chart summarizes the information about the graphs of the inverse hyperbolic functions.

Function	Domain	Range	Symmetry	Asymptotes
$\sinh^{-1} x$	$\boldsymbol{R}$	$\boldsymbol{R}$	odd	—
$\cosh^{-1} x$	$[1, \infty[$	$[0, \infty[$	—	—
$\tanh^{-1} x$	$]-1, 1[$	$\boldsymbol{R}$	odd	$x = \pm 1$
$\coth^{-1} x$	$]-\infty, -1]$ and $[1, \infty[$	$y \neq 0$	odd	$x = \pm 1$
$\operatorname{sech}^{-1} x$	$]0, 1]$	$[0, \infty[$	—	$x = 0$
$\operatorname{csch}^{-1} x$	$x \neq 0$	$y \neq 0$	odd	$x = 0$

Since the hyperbolic functions are defined in terms of exponentials, and the exponential and logarithmic functions are inverses of one another, it is useful to determine the relationship that exists between the inverse hyperbolic functions and the logarithmic function.

Suppose $x = \sinh y = \frac{1}{2}(e^y - e^{-y})$. Then $e^y - (1/e^y) - 2x = 0$, or

$$(e^y)^2 - 2x(e^y) - 1 = 0$$

Using the quadratic formula, we obtain

$$e^y = x \pm \sqrt{x^2 + 1}$$

Since $e^y > 0$ we must choose the plus sign; then, taking the natural logarithm of both sides yields

$$y = \ln(x + \sqrt{x^2 + 1})$$

Hence

$$\sinh^{-1} x = \ln(x + \sqrt{x^2 + 1}), \qquad -\infty < x < \infty$$

Using similar methods, the following formulas are established:

$$\cosh^{-1} x = \ln(x + \sqrt{x^2 - 1}), \qquad x \geqslant 1$$

$$\tanh^{-1} x = \tfrac{1}{2}\ln\left(\frac{1+x}{1-x}\right), \qquad |x| < 1$$

$$\coth^{-1} x = \tfrac{1}{2}\ln\left(\frac{x+1}{x-1}\right) = \tanh^{-1}\left(\frac{1}{x}\right), \qquad |x| > 1$$

$$\operatorname{sech}^{-1} x = \ln\left(\frac{1+\sqrt{1-x^2}}{x}\right) = \cosh^{-1}\left(\frac{1}{x}\right), \qquad 0 < x \leqslant 1$$

$$\operatorname{csch}^{-1} x = \ln\left(\frac{1}{x} + \frac{\sqrt{1+x^2}}{|x|}\right) = \sinh^{-1}\left(\frac{1}{x}\right) \qquad x \neq 0$$

We can use these representations of the inverse hyperbolic functions to compute their derivatives. For example,

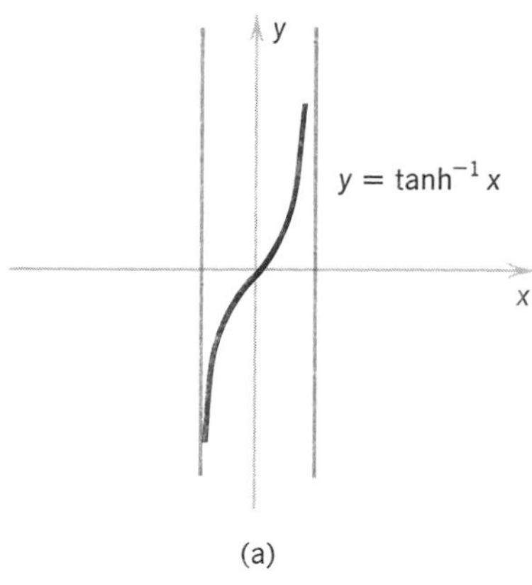

(a)

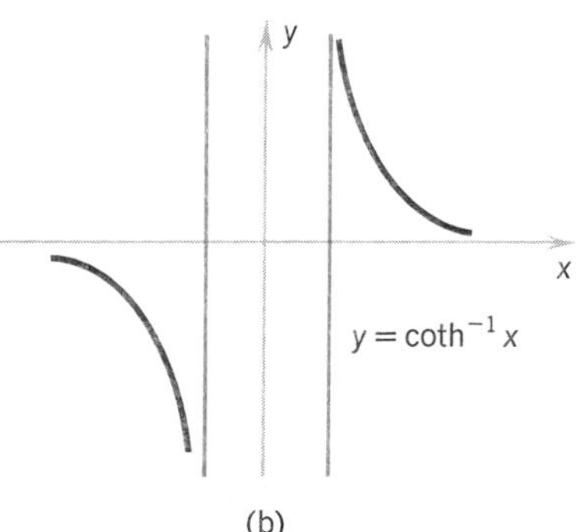

(b)

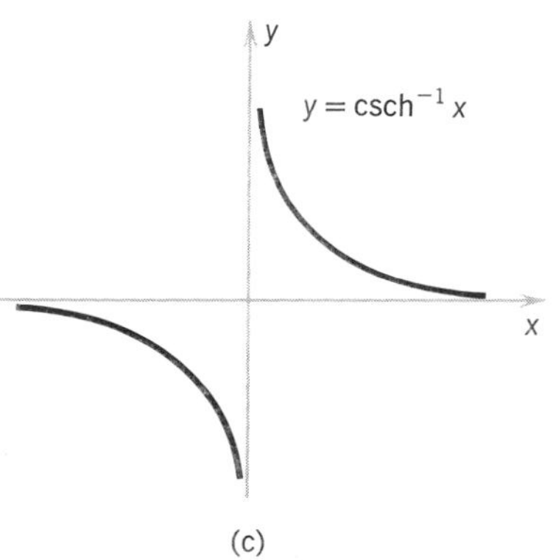

(c)

FIGURE 4-6

$$\frac{d(\sinh^{-1} x)}{dx} = \frac{d(\ln[x+\sqrt{x^2+1}])}{dx}$$

$$= \frac{1}{x+\sqrt{x^2+1}} \frac{d(x+\sqrt{x^2+1})}{dx}$$

$$= \frac{1}{x+\sqrt{x^2+1}} \left(1 + \frac{x}{\sqrt{x^2+1}}\right)$$

$$= \frac{1}{\sqrt{x^2+1}}$$

The remaining formulas are established in a similar fashion:

$$\frac{d(\cosh^{-1} x)}{dx} = \frac{1}{\sqrt{x^2-1}}, \qquad x > 1$$

$$\frac{d(\tanh^{-1} x)}{dx} = \frac{1}{1-x^2}, \qquad |x| < 1$$

$$\frac{d(\coth^{-1} x)}{dx} = \frac{1}{1-x^2}, \qquad |x| > 1$$

$$\frac{d(\operatorname{sech}^{-1} x)}{dx} = -\frac{1}{x\sqrt{1-x^2}}, \qquad 0 < x < 1$$

$$\frac{d(\operatorname{csch}^{-1} x)}{dx} = -\frac{1}{|x|\sqrt{1+x^2}}, \qquad x \neq 0$$

Again note that every differentiation formula gives rise to an integration formula, and moreover, these are new formulas, since the integrands have not occurred before.

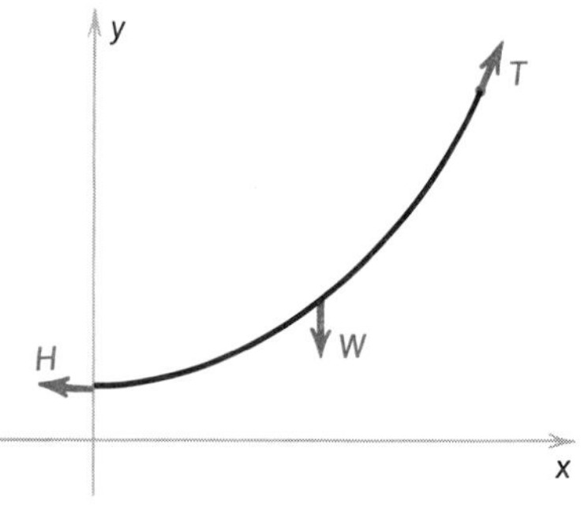

FIGURE 4-7

We conclude this section with an application of hyperbolic functions to the solution of a differential equation. Let us consider the problem of a hanging cable or *catenary*. We suppose that the cable is freely suspended by its ends and we wish to describe the curve along which it will lie.

The forces acting on the cable are displayed in Figure 4-7. The tension in the cable (a force along the cable) is denoted T and its horizontal component is denoted H. The horizontal tension H may be assumed constant. The weight W of the cable is given by $W = ws$, where $w =$ weight/foot and $s =$ arc length measured in feet. Consider the point (x, y) on the cable. Since the cable is in equilibrium, the opposite forces acting on the point counterbalance each other (Figure 4-8). Thus

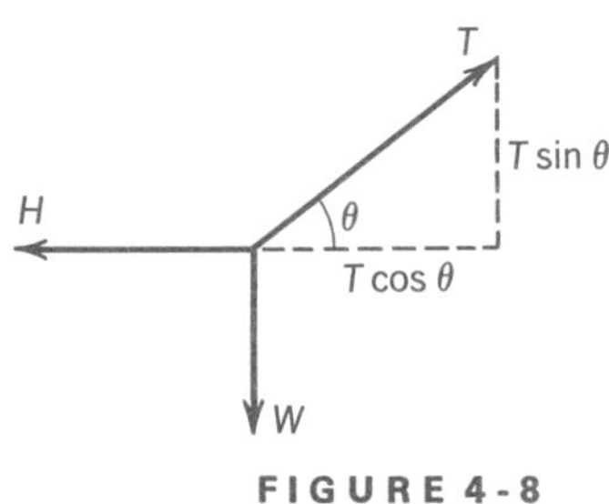

FIGURE 4-8

$$H = T\cos\theta$$

$$W = ws = T\sin\theta$$

and

$$\frac{W}{H} = \frac{ws}{H} = \frac{T\sin\theta}{T\cos\theta} = \tan\theta$$

But $\tan\theta = dy/dx$; hence

$$\frac{ws}{H} = \frac{dy}{dx} \tag{1}$$

Differentiating both sides of equation (1) with respect to x yields

$$\frac{w}{H}\frac{ds}{dx} = \frac{d^2y}{dx^2} \tag{2}$$

We recall from §7.3 that

$$\frac{ds}{dx} = \sqrt{1 + \left(\frac{dy}{dx}\right)^2}$$

Substituting this expression for ds/dx into equation (2) gives

$$\frac{w}{H}\sqrt{1 + \left(\frac{dy}{dx}\right)^2} = \frac{d^2y}{dx^2} \tag{3}$$

It is necessary to solve the differential equation (3). Since this equation involves the second derivative, we will need two initial conditions to determine constants of integration. These conditions are provided by describing the situation at one end of the cable. In Figure 4-7 the lowest point of the cable is on the y axis, so

$$\frac{dy}{dx} = 0 \quad \text{when } x = 0$$

To obtain the other initial condition we set $y = y_0$ for $x = 0$; y_0 will be selected so that the final equation for the cable is in simplest form.

For notational convenience, let

$$p = \frac{dy}{dx}$$

Then

$$\frac{d^2y}{dx^2} = \frac{dp}{dx}$$

and so equation (3) becomes

$$\frac{w}{H}\sqrt{1 + p^2} = \frac{dp}{dx}$$

or

$$\frac{w}{H}dx = \frac{dp}{\sqrt{1+p^2}} \tag{4}$$

Integrating both sides of equation (4) yields

$$\frac{w}{H}x + C_1 = \sinh^{-1} p$$

When $x = 0$, $p = dy/dx = 0$, and the constant C_1 is determined by

$$\frac{w}{H} \cdot 0 + C_1 = \sinh^{-1} 0$$

which implies

$$C_1 = 0$$

Therefore, equation (4) yields

$$\frac{w}{H} x = \sinh^{-1} p$$

Since $p = dy/dx$, this means

$$\frac{dy}{dx} = \sinh\left(\frac{w}{H} x\right) \tag{5}$$

or, integrating both sides,

$$\int \sinh\left(\frac{w}{H} x\right) dx = \int dy$$

$$\frac{H}{w} \int \sinh\left(\frac{w}{H} x\right)\left(\frac{w}{H} dx\right) = y + C_2$$

$$\frac{H}{w} \cosh\left(\frac{w}{H} x\right) = y + C_2$$

The constant C_2 is found by using the second initial condition, that $y = y_0$ when $x = 0$. Hence,

$$y_0 + C_2 = \frac{H}{w} \cosh 0 = \frac{H}{w}$$

so

$$C_2 = \frac{H}{w} - y_0$$

and

$$\frac{H}{w} \cosh\left(\frac{w}{H} x\right) = y + \frac{H}{w} - y_0$$

To simplify this result, we choose $y_0 = H/w$. The curve described by the hanging cable is then

$$y = \frac{H}{w} \cosh\left(\frac{w}{H} x\right)$$

or

$$y = a \cosh \frac{x}{a} \tag{6}$$

where $a = H/w$.

EXERCISES

1. Verify that $x \mapsto \coth x$ and $x \mapsto \operatorname{csch} x$ are odd functions, and that $x \mapsto \operatorname{sech} x$ is an even function.
2. Prove the differentiation formulas for the coth, sech, and csch functions.
3. Prove that
$$\sinh(x+y) = \sinh x \cosh y + \cosh x \sinh y$$
$$\cosh(x+y) = \cosh x \cosh y + \sinh x \sinh y$$
and use these formulas to find a formula for $\tanh(x+y)$.
4. Prove that $(\cosh x + \sinh x)^n = \cosh nx + \sinh nx$.
5. Given one hyperbolic function value, determine the other five.
(a) $\cosh x = \frac{13}{12}$ (b) $\coth x = \frac{17}{15}$
(c) $\sinh x = -\frac{3}{4}$ (d) $\tanh x = -\frac{5}{12}$
(e) $\operatorname{csch} x = \frac{7}{25}$ (f) $\operatorname{sech} x = \frac{3}{5}$
6. Prove the formulas for the inverse hyperbolic functions in terms of the logarithmic function as stated in the text, modeling your proof on the one given for the inverse hyperbolic sine.
7. Prove the differentiation formulas for the inverse hyperbolic functions. Use the method suggested and/or Theorem 2 of §4.8.
8. Differentiate each of the following.
(a) $x \mapsto \coth(\tan x)$ (b) $x \mapsto \operatorname{sech}^3 x$
(c) $x \mapsto \sqrt{\cosh x - 1}$ (d) $x \mapsto \ln(\cosh 2x)$
(e) $x \mapsto \tanh^{-1}(1-2x)$ (f) $x \mapsto \tan^{-1}(\sinh x)$
(g) $x \mapsto \tanh^{-1}(\operatorname{sech} x)$ (h) $x \mapsto \sinh^{-1}(\tan x)$
9. Evaluate the integrals.
(a) $\int \frac{\sinh x}{\cosh^4 x} dx$ (b) $\int \tanh x \, dx$
(c) $\int \frac{\sinh \sqrt{x} \, dx}{\sqrt{x}}$
10. Determine the second derivatives of the hyperbolic functions. Which of them have points of inflection, and where?
11. A cable weighing $\frac{1}{2}$ pound per foot has a horizontal tension of 2 pounds when suspended from two points on the same level a distance 100 feet apart. How far below this point will the cable hang?

5 GROWTH, DECAY, AND OSCILLATION

There are many applications of transcendental functions to physical situations. In any problem where the rate of change (usually with respect to time) of a physical quantity is proportional to the magnitude of the quantity itself, an exponential function is called for. To see why this is the case, suppose that y denotes the function which expresses the physical quantity in terms of time. Then we have the differential equation

$$\frac{dy}{dt} = ky \tag{1}$$

where k is the constant of proportionality. Equation (1) is equivalent to the differential relation

$$\frac{dy}{y} = k\,dt \tag{2}$$

both sides of which we integrate, yielding

$$\ln|y| = kt + C \tag{3}$$

or

$$\begin{aligned} |y| &= e^{kt+C} \\ &= K\,e^{kt} \end{aligned}$$

where $K = e^C$ is a fixed nonnegative constant. The constant K can be determined by the initial condition of the problem (that is, the value of y when $t = 0$). If we let $y = y_0$ at $t = 0$, we obtain

$$y_0 = K\,e^{k \cdot 0} = K\,e^0 = K$$

so the final equation for the problem is

$$y = y_0\,e^{kt} \tag{4}$$

FORMAL DEVICE

If a problem gives rise to a differential equation of the form

$$\frac{dy}{dt} = ky$$

with $y = y_0$ at $t = 0$, we can solve for y in terms of t:

$$y = y_0\,e^{kt} \tag{4}$$

or for t in terms of y:

$$t = \frac{1}{k}(\ln|y| - \ln|y_0|) = \frac{1}{k}\ln\left|\frac{y}{y_0}\right| \tag{5}$$

In the following examples and in the exercises, we display the significance of equations (4) and (5) in a variety of contexts. First let us consider some examples of **growth.**

Example 1 A certain bacterial culture doubles in size after 20 seconds. If the culture contains 100 specimens at time $t = 0$ (seconds), when will the number have increased to 5000 specimens?

The rate of change in size of the culture is proportional to the size of the culture, so we have by equation (5)

$$t = c \ln \frac{y}{y_0}$$

where $y_0 = 100$. At $t = 20$, $y = 200$; hence,

$$20 = c \ln \frac{200}{100}$$

or, using a table of natural logarithms

$$c = \frac{20}{\ln 2} \approx \frac{20}{0.6931} \approx 28.9$$

Therefore, we have

$$t \approx 28.9 \ln \frac{y}{100}$$

so $y = 5000$ at time

$$\begin{aligned} t &\approx 28.9 \ln 50 \\ &\approx 28.9\,(3.912) \\ &\approx 113.4 \text{ sec} \end{aligned}$$

Example 2 The prime minister of a small Asian country is appalled that his country's population is increasing at a rate of 3.4% per year. After the institution of an intensive birth control program, the rate of increase drops to 2.1% per year. The prime minister, in a relieved frame of mind, issues a statement saying: "At our current rate of growth the population of our country in 50 years will be only X% of what it would have been at the old rate." If the statement is true, what is X?

Suppose the population of the country at the present ($t = 0$) is p_0. The old growth rate is derived from the equation

$$t = \frac{1}{k_1} \ln \frac{y}{p_0}$$

where $y = p_0$ at $t = 0$, and $y = 1.034p_0$ at $t = 1$ (year). Hence,

$$1 = \frac{1}{k_1} \ln 1.034$$

so

$$k_1 \approx 0.0334$$

and we have

$$y = p_0 e^{0.0334t}$$

as the population projected from the old growth rate. For the new rate, we have

$$t = \frac{1}{k_2} \ln \frac{y}{p_0}$$

where $y = p_0$ at $t = 0$, and $y = 1.021 p_0$ at $t = 1$ (year). Hence,

$$1 = \frac{1}{k_2} \ln 1.021$$

so

$$k_2 \approx 0.0208$$

and

$$y = p_0 e^{0.0208t}$$

is the population projected from the new growth rate. The ratio of these after 50 years is

$$\frac{p_0 e^{0.0208(50)}}{p_0 e^{0.0334(50)}} = e^{-0.63} \approx 0.532$$

so the population is only 53.2% what it might have been, and $X = 53.2$.

Example 3 If bank A compounds interest biannually and bank B compounds interest continually, what is Jones' profit after one year in placing his \$1000 in bank B instead of A, if both banks advertise 5% interest?

Bank A pays 2.5% twice in the course of the year, so the actual amount of Jones' investment after 1 year is

$$T_A = (1.025)^2 (1000) \approx (1.0506) \cdot (1000) = \$1050.6$$

At bank B, the growth of the money is represented by

$$\frac{dT_B}{dt} = 0.05 T_B$$

or

$$T_B = T_{B_0} e^{0.05t}$$
$$= 1000\, e^{0.05} \approx \$1051.3$$

Therefore Jones realizes $T_B - T_A \approx \$0.70$ more after one year by investing at bank B.

Now let us look at an example of **decay**.

Example 4 In a certain condenser the instantaneous rate of discharge is two percent of the number of volts in the condenser per second. Find the voltage in the condenser as a function of time.

We have

$$\frac{dV}{dt} = -0.02V$$

or

$$V = V_0 e^{-0.02t}$$

Note that in the examples of growth, the constant k in equation (4) was positive; in the example of decay, the value of k was negative. This is always the case:

$$k > 0 \qquad \text{means growth}$$

$$k < 0 \qquad \text{means decay}$$

Also notice that in all our examples, the quantity measured was strictly positive, so we dispensed with the absolute value sign in equation (5), writing $\ln y$ instead of $\ln|y|$. If the measure of the quantity varies continuously and changes sign, at some point we have $y = 0$, and the term $\ln|y| = \ln 0$ from equation (3) is not defined. This situation gives rise to an "improper integral," the phenomenon which we considered in §7.6.

Another situation which arises frequently in physics is that of a body accelerating in a straight line toward a fixed point, the magnitude of the acceleration being proportional to the distance from the fixed point. The differential equation for this motion is

$$\frac{d^2x}{dt^2} = -k^2x \tag{6}$$

where k is constant, and the minus sign is chosen because the acceleration is always directed in the direction opposite to the displacement of the moving object with respect to the fixed point (Figure 5-1). Because an object obeying equation of motion (6) oscillates back and forth about the fixed point, equation (6) is said to describe **simple harmonic motion**.

It can be shown that the general solution to equation (6) has the form

$$x = A\cos kt + B\sin kt \tag{7}$$

where A and B are constants determined by the initial conditions for x and $v = dx/dt$. We can easily verify that equation (7) is a solution for (6):

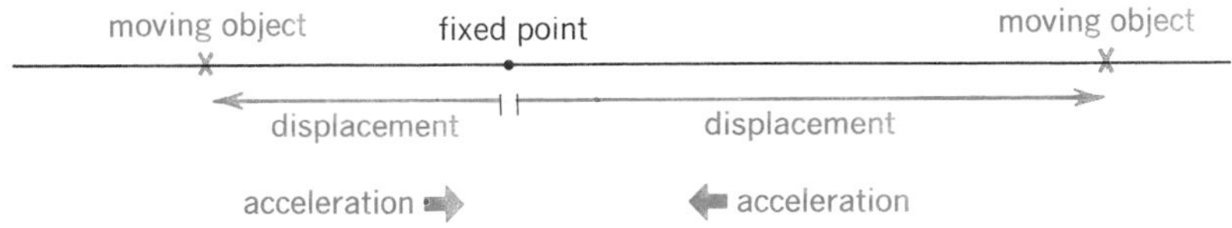

FIGURE 5-1

$$v = \frac{dx}{dt} = \frac{d}{dt}(A\cos kt + B\sin kt)$$

$$= -Ak\sin kt + Bk\cos kt$$

$$a = \frac{d^2x}{dt^2} = -Ak^2\cos kt - Bk^2\sin kt$$

$$= -k^2(A\cos kt + B\sin kt)$$

$$= -k^2 x$$

The student should check that if a problem describes initial displacement x_0 and velocity v_0 at $t = 0$, then in equation (7), $A = x_0$ and $B = v_0/k$.

FORMAL DEVICE

Suppose a problem gives rise to an equation of motion of the form

$$\frac{d^2x}{dt^2} = -k^2x, \qquad \text{with} \quad x = x_0 \quad \text{and} \quad \frac{dx}{dt} = v_0 \text{ at } t = 0$$

Then we can solve for x in terms of t:

$$x = x_0\cos kt + \frac{v_0}{k}\sin kt \tag{8}$$

Also, we have

$$v = \frac{dx}{dt} = -kx_0\sin kt + v_0\cos kt \tag{9}$$

Example 5 **Hooke's Law** states "When a spring of (unstretched) length l is stretched to a length $l+x$, the force tending to restore the spring to its natural length is proportional to the extension x." Suppose that a spring of negligible natural length and mass is fixed at one end and attached to a steel ball at the other end. The spring is stretched to a length of 10 cm and released. If the proportionality constant from Hooke's law is -16, find the position and velocity of the ball as functions of time.

At $t = 0$, we have $x_0 = 10$ (cm), $v_0 = 0$. We also have the equation

$$\frac{d^2x}{dt^2} = -16x$$

Then from equations (8) and (9) we get

$$kx = 10\cos 4t, \qquad v = -40\sin 4t$$

The farthest distance the object of oscillation moves from the fixed point is called the **amplitude** of the oscillation. When the oscillating object is at this distance, its velocity is 0 (Why?). We can therefore determine the amplitude from equations (8) and (9): Setting $dx/dt = 0$ gives

$$x_0 \sin kt = v_0 \cos kt$$

If $x_0 \neq 0$, we have

$$\sin kt = \frac{v_0}{kx_0} \cos kt \tag{10}$$

or substituting into (8),

$$x = \left(x_0 + \frac{v_0^2}{k^2 x_0}\right) \cos kt \tag{11}$$

But from (10) we have

$$kt = \tan^{-1} \frac{v_0}{kx_0}$$

so (11) becomes

$$\begin{aligned} x &= \left(x_0 + \frac{v_0^2}{k^2 x_0}\right) \cos\left(\tan^{-1} \frac{v_0}{kx_0}\right) \\ &= \left(x_0 + \frac{v_0^2}{k^2 x_0}\right)\left(\frac{x_0}{\sqrt{k^2 x_0^2 + v_0^2}}\right) \\ &= \frac{x_0^2 + (v_0/k)^2}{\sqrt{x_0^2 + (v_0/k)^2}} \\ &= \sqrt{x_0^2 + (v_0/k)^2} \end{aligned}$$

The student should work out the details to show that the same formula applies when $x_0 = 0$, in which case we assume $v_0 \neq 0$. (Clearly, if $x_0 = v_0 = 0$, the amplitude is 0.)

FORMAL DEVICE

The amplitude of an oscillation given by

$$x = x_0 \cos kt + \frac{v_0}{k} \sin kt$$

is

$$A = \sqrt{x_0^2 + (v_0/k)^2}$$

In particular, if $v_0 = 0$, $A = |x_0|$; if $x_0 = 0$, $A = |v_0/k|$.

In Example 5 above, $v_0 = 0$, and $A = |x_0| = 10$ cm.

The time for the oscillation to go through a complete cycle is called the **period** of the oscillation and is given by the formula

$$p = \frac{2\pi}{k} \tag{12}$$

where k is the proportionality constant of equations (8) and (9). Observe that the period of an oscillation is independent of the initial conditions. In Example 5, the period is $\frac{1}{2}\pi$.

EXERCISES

1. A sample of isotope X weighs 320 milligrams. After 12 minutes, only 20 milligrams remain. What is the half-life of isotope X? How much of the sample will be left after 30 minutes? (Hint: The rate of decay of an isotope is proportional to the weight of the sample. The **half-life** of the isotope is the time necessary for $\frac{1}{2}$ of any given sample of the isotope to decay.)
2. The half-life of uranium is about 0.45 billion years. What percent of a given sample of uranium will be left after 1 billion years?
3. The population of a city was 10,000 in 1900, 20,000 in 1930, and 30,000 in 1960. Could the rate of increase of the population have been proportional to the population? Explain.
4. In 1940 the population of Eagleville was 80,000. In 1967 it was 100,000, If the rate of population increase is proportional to the population, what will the population be in 1984?
5. In a certain chemical reaction, the concentration of substance S increases at a rate proportional to the concentration. If the concentration is 1 part per 100 at the beginning of the reaction and 10 parts per 100 after 10 seconds, when will the concentration be 50 parts per hundred? 95? 99?
6. **Newton's law of cooling** says that the difference between the temperature of a body and that of the surrounding medium decreases at a rate proportional to this difference. Suppose a frying pan with temperature 150°F is placed in a 40°F constant temperature room to cool, and that after two minutes the temperature of the pan is 120°F. Find the temperature of the pan after 5 minutes; after 10 minutes.
7. A thermometer which reads 80°F is plunged into a swimming pool full of water. After 10 minutes the thermometer reads 65°F. After 30 minutes it reads 50°F. What is the temperature of the water in the pool.
8. A large tank contains 5000 gallons of sea water. In order to reduce the salinity (proportion of mineral salts in the water to volume of water) of the water, fresh water is poured into the tank at the rate of 50 gallons per minute. Show that the salinity of the mixture decreases exponentially, and find by what factor the salinity has been reduced after 1 hour.
9. A point moves on the x axis in simple harmonic motion of period 4π. If $x = 3$ and $dx/dt = -2$ when $t = 0$, find x as a function of t. What is the amplitude of the motion? What is the first time after $t = 0$ when $x = 0$?
10. Consider a point in simple harmonic motion. How long does it take for the speed of the point to decrease from its maximum value to half its value? What fraction of the amplitude of the motion does the point travel in the interval?

6 POLAR COORDINATES

A coordinate system in the plane allows us to associate a pair of real numbers with each point in the plane. So far we have considered rectangular coordinate systems exclusively. We shall now describe the system of **polar coordinates.**

We begin by selecting a point in the plane which we call the origin 0, and an initial ray from 0 (or half line with one end at 0), usually drawn horizontally and to the right of 0.

If we are given a point P in the plane, its position is determined by its distance r from the origin, and by the counterclockwise angle θ made by the line segment $0P$ and the initial ray (Figure 6-1). We say in this case that the point P has the **polar coordinates** (r, θ). As in trigonometry, the angle θ is positive when measured counterclockwise from the initial ray and negative when measured clockwise. The distance r from the origin to the point P is taken to be positive. Observe that while in a rectangular coordinate system each point may be represented in just one way by a pair of rectangular coordinates, the same point may be represented in many ways by polar coordinates, since the angle associated with a given point is not unique.

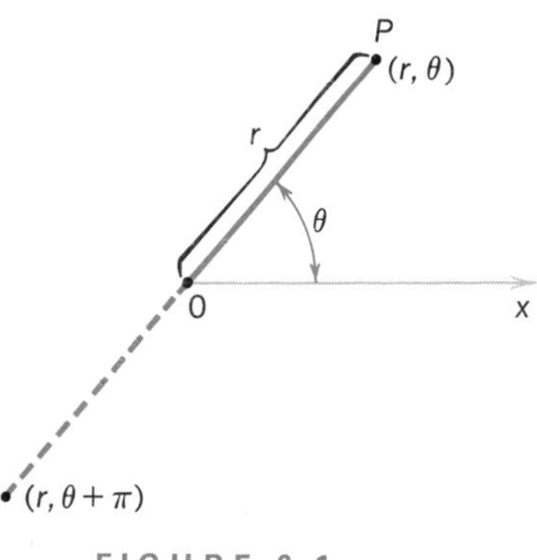

FIGURE 6-1

For instance, the point $(3, \pi/3)$ also has the coordinates $(3, \pi/3+2\pi)$, $(3, \pi/3+4\pi)$, $(3, \pi/3+6\pi)$, $(3, \pi/3-2\pi)$, and in general $(3, \pi/3\pm 2n\pi)$. The origin has the coordinates $(0, \theta)$ for all θ.

It is convenient to allow r to be negative with the convention that $(-r, \theta)$ is another representation of the point with coordinates $(r, \theta+\pi)$. Thus the points (r, θ) and $(-r, \theta)$ are symmetric with respect to the origin.

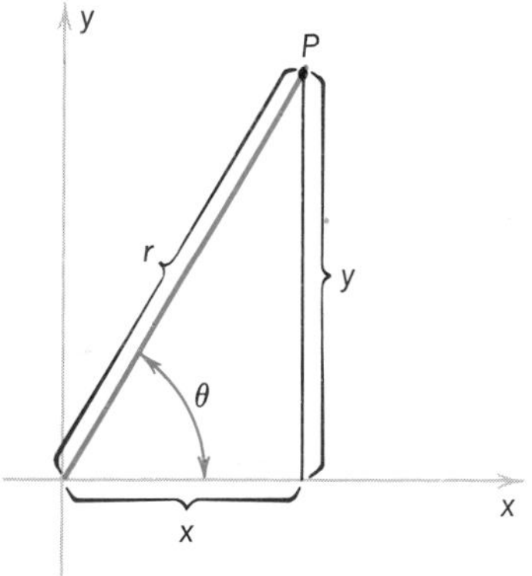

FIGURE 6-2

With both rectangular and polar coordinate systems available, it is important to be able to change from one coordinate system to the other. To obtain equations for doing this, consider a common origin and take the initial ray as the positive x axis (see Figure 6-2). From the definitions of sine and cosine, we have the following relationships for converting the polar coordinates (r, θ) to the rectangular coordinates (x, y):

$$x = r\cos\theta, \qquad y = r\sin\theta$$

Similarly, we convert the rectangular coordinates into polar coordinates using the relationships

$$r = \sqrt{x^2+y^2}, \qquad r > 0$$

$$\theta = \tan^{-1}\left(\frac{y}{x}\right)$$

Many mathematical curves are most conveniently described by equations involving polar coordinates. The locus of the relation $F(r, \theta) = 0$

is the set of points with coordinates (r_0, θ_0) such that

$$F(r_0, \theta_0) = 0 \tag{1}$$

When r depends on θ, that is, $r = f(\theta)$, we have the relation $r - f(\theta) = 0$, which is a special case of (1).

When we consider the loci of relations that are expressed in terms of polar coordinates, we can adopt two different points of view. First, we can start with a given relation involving r and θ and consider the locus of all the points (r_0, θ_0) which satisfy the relation. Alternatively, we can begin with a locus defined by geometric properties (for example, the circle) and try to find a relation involving r and θ such that every point of the locus has at least one set of polar coordinates which satisfy the relation. (Recall that representation in polar coordinates is not unique.)

Example 1 Suppose we are given $r = f(\theta)$, that is, the relation $r - f(\theta) = 0$. To plot the locus of such a relation we must find all ordered pairs (r_0, θ_0) which satisfy the given relation, and then plot the points (r_0, θ_0). Usually, all we want is to sketch the locus of $r = f(\theta)$, and to do this we should first locate those points which are most helpful in visualizing the curve—for example, the points where $f(\theta)$ is a maximum or minimum, or where $f(\theta) = 0$. One can also note where $f(\theta)$ is increasing and where it is decreasing.

An especially helpful observation is that of symmetry, which is usually easily detected. The following cases should be noted: The curve is

(1) symmetric about the origin if the relation is unchanged when r is replaced by $-r$;

(2) symmetric about the x axis if the relation is unchanged when θ is replaced by $-\theta$;

(3) symmetric about the y axis if the relation is unchanged when θ is replaced by $\pi - \theta$.

[Note that we have used x and y axes in this description. The positive x axis corresponds to the initial ray and thus the x axis is the line $\theta = 0$. The y axis is the line $\theta = \frac{1}{2}\pi$.]

Example 2 Sketch the locus of $r = a(1 - \cos\theta)$ where a is a positive constant.

Observe that since

$$\cos\theta = \cos(-\theta)$$

the relation is unchanged when θ is replaced by $-\theta$; hence the locus is symmetric about the x axis. Since

$$-1 \leqslant \cos\theta \leqslant 1$$

the values of r vary from 0 to $2a$. The minimum value $r=0$ occurs at $\theta=0$, and the maximum value $r=2a$ occurs at $\theta=\pi$. As θ varies from 0 to π, $\cos\theta$ decreases from 1 to -1, and $1-\cos\theta$ increases from 0 to 2, that is, r increases from 0 to $2a$ as the radius ranges from $\theta=0$ to $\theta=\pi$.

It helps to compute several points:

θ	r
0	0
$\frac{1}{3}\pi$	$\frac{1}{2}a$
$\frac{1}{2}\pi$	a
$\frac{2}{3}\pi$	$\frac{3}{2}a$
π	$2a$

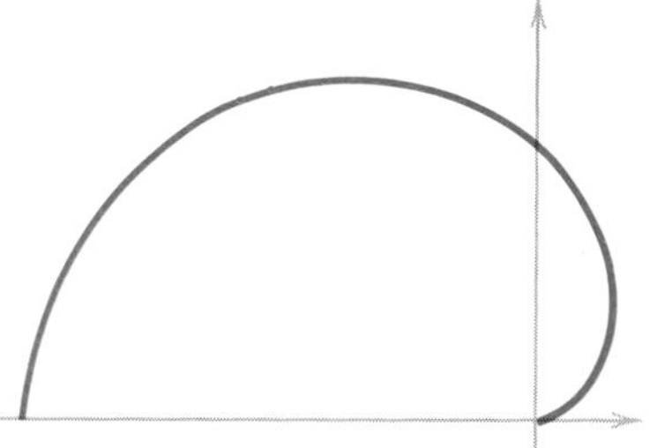

FIGURE 6-3

We also note that $\theta \mapsto a(1-\cos\theta)$ is increasing on $]0,\pi[$ since the derivative $dr/d\theta = a\sin\theta$ is positive on $]0,\pi[$.

Our curve appears as in Figure 6-3.

We can then use the symmetry of the relation and reflect this portion across the x axis. The result is the curve shown in Figure 6-4. This locus is called a **cardioid** because of its heart-shaped appearance.

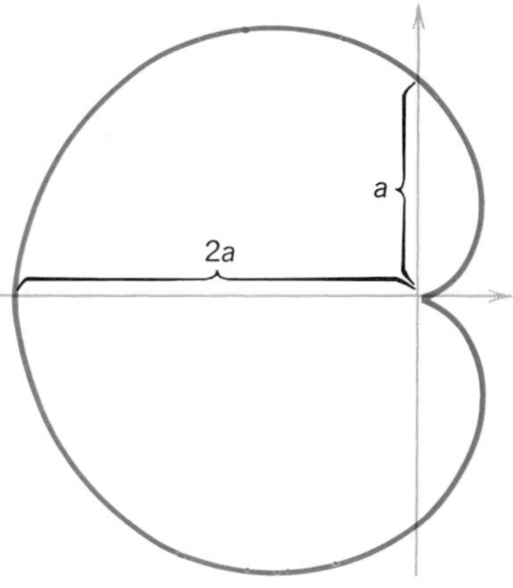

FIGURE 6-4

Example 3 Sketch the locus of $r^2=\sin\theta$. When we replace r by $-r$ the relation is unchanged, and hence the locus is symmetric with respect to the origin. If we replace θ by $\pi-\theta$, $\sin(\pi-\theta)=\sin\theta$ and so the locus is symmetric with respect to the y axis. If $\sin\theta$ is negative, r^2 cannot equal $\sin\theta$, so we restrict θ to the interval $[0,\pi]$. The function $\theta\mapsto\sqrt{\sin\theta}$ is increasing on $[0,\frac{1}{2}\pi]$ and decreasing on $[\frac{1}{2}\pi,\pi]$. It has a maximum at $\theta=\frac{1}{2}\pi$. Computing some values, we obtain the following table:

θ	r
0	0
$\frac{1}{6}\pi$	$\pm\sqrt{2}/2 \approx \pm 0.71$
$\frac{1}{3}\pi$	$\pm\sqrt{\frac{1}{2}\sqrt{3}} \approx \pm 0.93$
$\frac{1}{2}\pi$	± 1
$\frac{2}{3}\pi$	$\pm\sqrt{\frac{1}{2}\sqrt{3}} \approx \pm 0.93$
$\frac{5}{6}\pi$	$\pm\sqrt{2}/2 \approx 0.71$
π	0

Because the locus is symmetric with respect to the origin and the y axis, it is enough to plot points in the first quadrant. This locus is called a **lemniscate** and is shown in Figure 6-5.

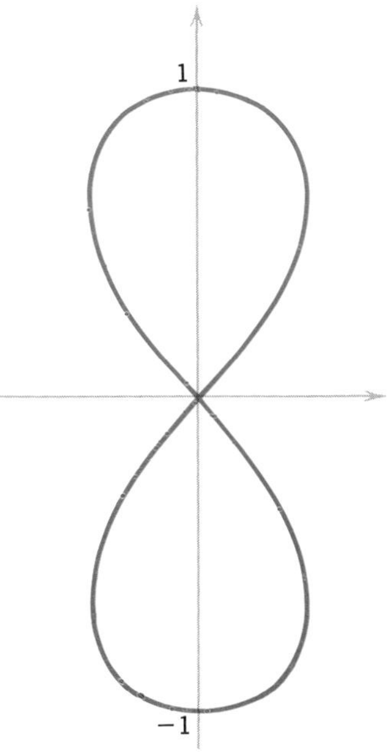

FIGURE 6-5

Let us now turn to our second approach to the study of loci of polar relations. Consider the locus of points each of which is at distance a from the origin, which is the circle shown in Figure 6-6. Can we find a relation in polar form that is represented by this locus? One way to solve such a problem is to first find the relation in rectangular coordinates. We know that the relation in Cartesian coordinates is

$$x^2 + y^2 = a^2 \tag{2}$$

We then use the equations

$$x = r\cos\theta, \qquad y = r\sin\theta$$

to convert the relation into polar form. Thus (2) becomes

$$r^2\cos^2\theta + r^2\sin^2\theta = a^2$$

$$r^2 = a^2$$

$$r = \pm a$$

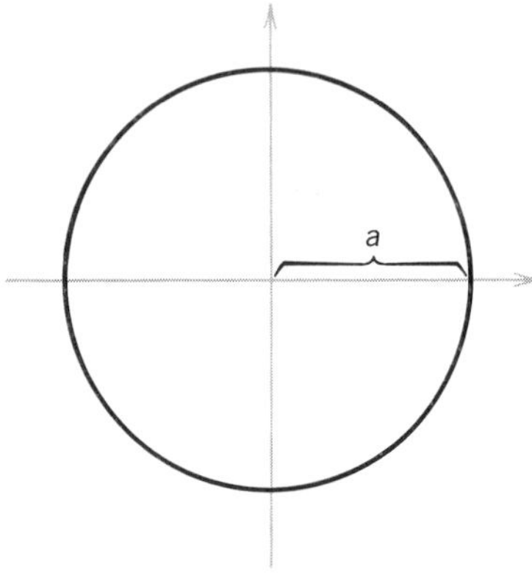

FIGURE 6-6

In this case, however, and in many other cases, it is easier to derive the polar relation directly from the geometric properties of the locus. Here we know that every point of the locus is at distance a from the origin, so it follows that for every point of the locus we have $r = a$. Thus, the given locus can be represented by the relation $r - a = 0$.

Observe that a locus in the plane may be represented by more than one relation; for example, $r + a = 0$ is also a relation for the given circle.

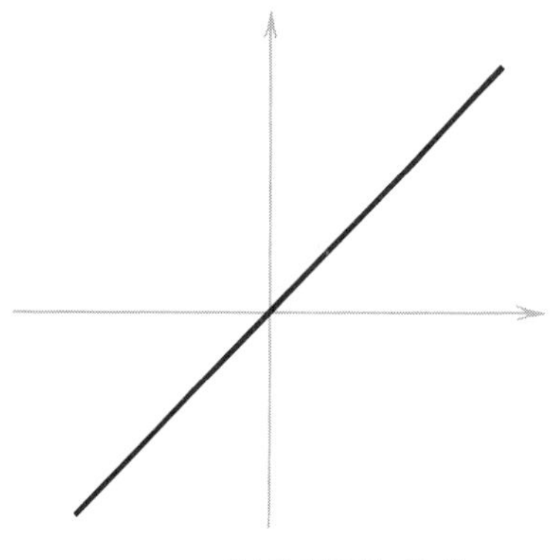

FIGURE 6-7

Example 4 Find a polar relation which represents a line through the origin with slope 1 (Figure 6-7).

Clearly, in this case, r varies from $-\infty$ to $+\infty$, but θ is constant. For all points (r, θ) of the locus we have $\theta = \frac{1}{4}\pi$. One relation whose locus is the given line is $\theta - \frac{1}{4}\pi = 0$. However, each point $(r, \frac{1}{4}\pi)$ on the line also has coordinates $(r, \frac{1}{4}\pi \pm 2n\pi)$; and so $\theta - \frac{9}{4}\pi = 0$ and $\theta - \frac{17}{4}\pi = 0$ are other relations whose locus is the given line.

Example 5 The geometric definition of a parabola is the set of all points in the plane which are equidistant from a fixed line and a fixed point not on the line. We call the line the directrix and the point the focus.

Consider the parabola with focus at the origin and directrix the line $x = -2p$, $p > 0$ (see Figure 6-8). Find a relation in polar form that is represented by this locus.

By the definition, $\overline{OP}$ must equal $\overline{MP}$. But

$$\overline{MP} = \overline{MQ} + \overline{QP}$$

and

$$\overline{MQ} = 2p, \qquad \overline{QP} = r\cos\theta$$

so

$$\overline{MP} = 2p + r\cos\theta$$

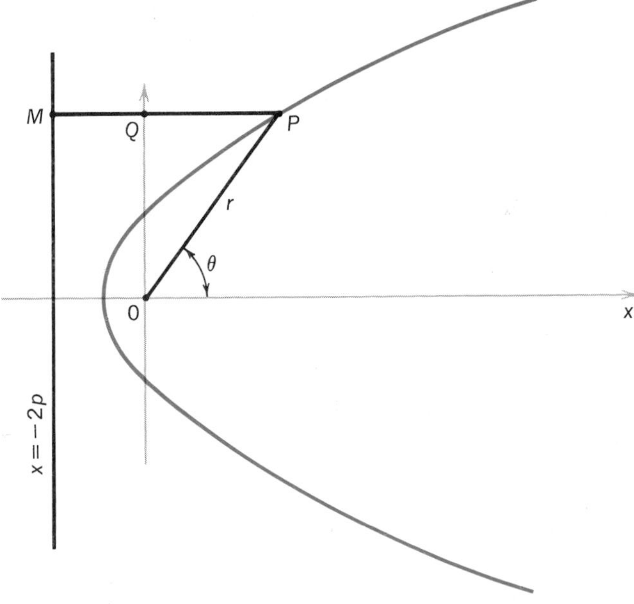

FIGURE 6-8

Since $\overline{OP} = r$, we see that

$$r = 2p + r\cos\theta$$

$$r - r\cos\theta = 2p$$

$$r = 2p/(1 - \cos\theta)$$

Thus $r = 2p/(1-\cos\theta)$ is one relation which is represented by the given locus. In Cartesian coordinates, this equation is $4p(x+p/2)-y^2 = 0$. The reader can check that this agrees with what we derived in §1.5 for the parabola.

Occasionally it is necessary to find intersections of loci which are defined by relations in polar form. This problem cannot be completely solved by solving the two relations simultaneously, since a point can be on each of two curves and yet not have a pair of polar coordinates which satisfies both relations simultaneously. For example, the point $(2, \pi)$ is on the locus of

$$r^2 = 4\cos\theta$$

even though its coordinates as given do not satisfy the relation, since the same point can be represented as $(-2, 0)$ and these coordinates do satisfy the relation. The same point $(2, \pi)$ is on the locus of

$$r = 1 - \cos\theta$$

and thus this point should be included among the points of intersection of the two loci. But solving the equations simultaneously does not give the point $(2, \pi)$ as a point of intersection. To find all of these points one must carefully plot and examine the loci of the relations.

EXERCISES

1. Plot the following points which are given in polar form.
   (a) $(2, \frac{1}{2}\pi)$ (b) $(3, -\frac{1}{3}\pi)$ (c) $(-1, \frac{1}{3}\pi)$
   (d) $(-2, \frac{1}{2}\pi)$ (e) $(2, \frac{13}{4}\pi)$
2. Find the Cartesian coordinates of the points in exercise 1.
3. Plot the following points. Find all polar coordinates of each point.
   (a) $(-3, \frac{1}{4}\pi)$ (b) $(2, \frac{1}{6}\pi)$ (c) $(-1, -\frac{1}{4}\pi)$
4. Find a pair of polar coordinates for each of the following points given in Cartesian coordinates and plot.
   (a) $(1, -1)$ (b) $(-1, -1)$ (c) $(2, 0)$
   (d) $(\sqrt{3}, 1)$ (e) $(-\sqrt{2}, -\sqrt{2})$
5. Graph the locus of points whose polar coordinates satisfy the given relation or inequality.
   (a) $r = 3$ (b) $r \geqslant 2$ (c) $2 < r \leqslant 3$
   (d) $r < -1,\ \theta = \frac{1}{2}\pi$ (e) $r > -1,\ 0 < \theta < \frac{1}{2}\pi$
6. Graph the loci of each of the following.
   (a) $r\cos\theta = 1$ (b) $r\sin\theta = -2$
   (c) $r = -3\sec\theta$ (d) $r = 4\csc\theta$
7. Plot the loci.
   (a) $r = 4\sin\theta$ (b) $r = -4\cos\theta$
   (c) $r = 1$ (d) $r = -2\sin\theta$
8. Graph each of the following cardioids.
   (a) $r = a(1+\cos\theta)$ (b) $r = a(1+\sin\theta)$
   (c) $r = a(1-\cos\theta)$ (d) $r = a(1-\sin\theta)$
9. Graph the loci of the following, called **limaçons**.
   (a) $r = 4-2\sin\theta$ (b) $r = 2+\sin\theta$
   (c) $r = 1+2\cos\theta$ (*d*) $r = 3+\sin\theta$
10. Plot the following lemniscates.
    (a) $r^2 = 4\sin 2\theta$ (b) $r^2 = \cos 2\theta$
11. Plot each of the following curves, called **roses**. (Note that these are of the form $r = a\cos n\theta$ or $r = a\sin n\theta$. When n is an odd integer, the rose has n petals; when n is an even integer, the rose has $2n$ petals.)
    (a) $r = 3\sin 3\theta$ (b) $r = 2\cos 3\theta$
    (c) $r = 2\sin 2\theta$ (d) $r = 3\cos 2\theta$
12. Discuss and sketch.
    (a) $r = 2(1+\sin\theta)$ (b) $r = 2+4\cos\theta$
    (c) $r = 2-4\sin\theta$ (d) $r^2 = 2\cos\theta$
    (e) $r = 5\cos 2\theta$ (f) $r = 4\sin^2 \frac{1}{2}\theta$
    (g) $r(2-\cos\theta) = 2$ (h) $r(1+3\sin\theta) = 2$

13. Find the polar equation.
    (a) $x = 3$ (b) $2x + y = 3$
    (c) $xy = 4$ (d) $x^2 + y^2 + 2x - 4y = 0$
    (e) $x^2 + y^2 - 2ay = 0$ (f) $(x^2+y^2)^2 = x^2 - y^2$
14. Determine the Cartesian equation and sketch the locus.
    (a) $r = 4\cos\theta$ (b) $r = \sin 2\theta$
    (c) $r = 2\sec\theta\tan\theta$ (d) $r = \dfrac{3}{2-\cos\theta}$
    (e) $r = \dfrac{2}{1+\cos\theta}$ (f) $r = \dfrac{9}{4-5\sin\theta}$
15. Sketch the locus of $r\cos(\theta-\frac{1}{3}\pi) = 3$. [Hint: Rotate the axes so that $\theta' = \theta - \frac{1}{3}\pi$.]
16. Sketch the curves:
    (a) $r = 2\cos(\theta+\frac{1}{4}\pi)$ (b) $r = 2\sin(\theta-\frac{1}{3}\pi)$
17. Show that the distance between the points (r_1, θ_1) and (r_2, θ_2), where $r_1 \geqslant 0$ and $r_2 \geqslant 0$, is $\sqrt{r_1^2+r_2^2-2r_1 r_2\cos(\theta_1-\theta_2)}$.
18. Find the polar equation of the parabola whose focus is at the origin and whose vertex is at $(C, 0)$. Consider the cases $C > 0$ and $C < 0$.
19. Show that $r = \cos\theta + 1$, and $r = \cos\theta - 1$ represent the same curve.
20. Find all the points of intersection of each of the following pairs of loci.
    (a) $r = 2\cos 2\theta$, $r = 2(1+\cos\theta)$
    (b) $r = a(1+\cos\theta)$ $r = a(1-\sin\theta)$

7 TANGENTS AND ARC LENGTH

When working in polar coordinates, the calculation of tangent lines and rates of change is best done in terms of the angle between the tangent to a curve at a point P and the line $\overline{OP}$ through P (Figure 7-1).

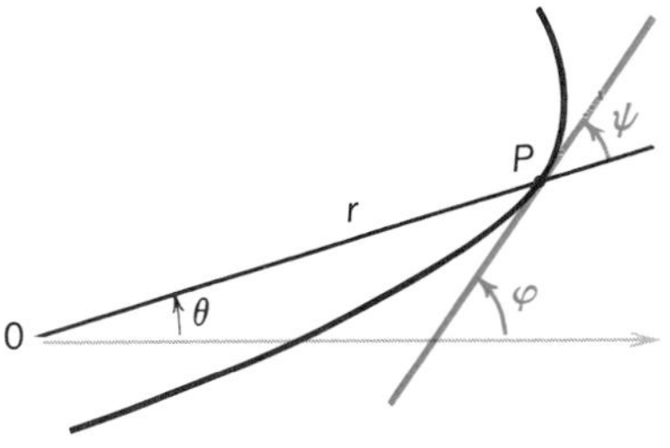

FIGURE 7-1

We know that $\varphi = \theta + \psi$. Suppose a relation which represents the given locus is in the form $r - f(\theta) = 0$, where f is a differentiable function of θ. From the previous section we know that

$$x = r\cos\theta, \qquad y = r\sin\theta$$

so

$$\frac{dx}{d\theta} = -r\sin\theta + \cos\theta\frac{dr}{d\theta}$$

$$\frac{dy}{d\theta} = r\cos\theta + \sin\theta\frac{dr}{d\theta}$$

Since $\psi = \varphi - \theta$, it follows that

$$\tan\psi = \tan(\varphi - \theta) = \frac{\tan\varphi - \tan\theta}{1+\tan\varphi\tan\theta}$$

But $\tan\varphi$ is the slope of the tangent line and so

$$\tan\varphi = \frac{dy}{dx} = \frac{dy/d\theta}{dx/d\theta}$$

Also, $\tan\theta$ is the slope of OP and so $\tan\theta = y/x$. Thus,

$$\tan\psi = \frac{\dfrac{dy/d\theta}{dx/d\theta} - \dfrac{y}{x}}{1 + \dfrac{y}{x}\dfrac{dy/d\theta}{dx/d\theta}} = \frac{x\dfrac{dy}{d\theta} - y\dfrac{dx}{d\theta}}{x\dfrac{dx}{d\theta} + y\dfrac{dy}{d\theta}} \tag{1}$$

The numerator on the right of (1) is found to be

$$x\frac{dy}{d\theta} - y\frac{dx}{d\theta} = (r\cos\theta)\left(r\cos\theta + \sin\theta\frac{dr}{d\theta}\right) - r\sin\theta\left(-r\sin\theta + \cos\theta\frac{dr}{d\theta}\right)$$
$$= r^2$$

while the denominator is

$$x\frac{dx}{d\theta} + y\frac{dy}{d\theta} = r\cos\theta\left(-r\sin\theta + \cos\theta\frac{dr}{d\theta}\right) + r\sin\theta\left(r\cos\theta + \sin\theta\frac{dr}{d\theta}\right)$$
$$= r\frac{dr}{d\theta}$$

Substituting these results into equation (1), we obtain the formula

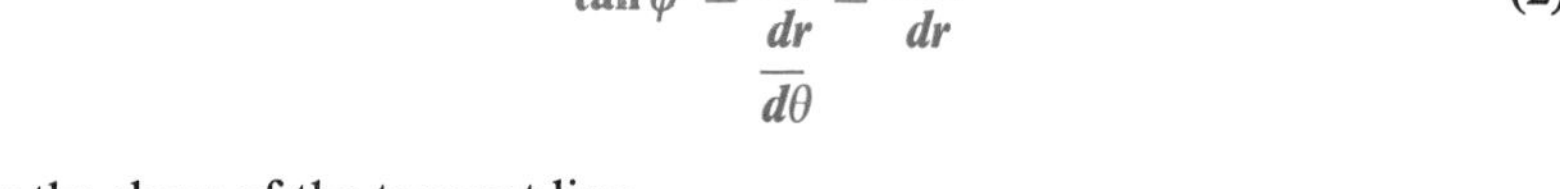

$$\tan\psi = \frac{r}{\dfrac{dr}{d\theta}} = \frac{r\,d\theta}{dr} \tag{2}$$

for the slope of the tangent line.

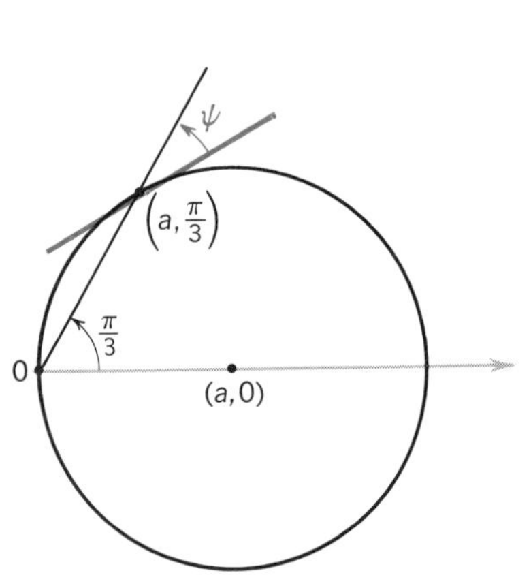

FIGURE 7-2

Example 1 Find the angle ψ at the point $(a, \frac{1}{3}\pi)$ for $r = 2a\cos\theta$ (see Figure 7-2). We have

$$r = 2a\cos\theta$$

$$\frac{dr}{d\theta} = -2a\sin\theta$$

and from the formula

$$\tan\psi = \frac{r}{dr/d\theta} = \frac{2a\cos\theta}{-2a\sin\theta} = -\cot\theta$$

At $(a, \frac{1}{3}\pi)$,

$$\cot\theta = \frac{1}{\sqrt{3}}$$

so

$$\tan\psi = -\frac{1}{\sqrt{3}}$$

$$\psi = \tan^{-1}\left(-\frac{1}{\sqrt{3}}\right)$$

$$\psi = \tfrac{5}{6}\pi$$

Suppose we now consider the locus of the relation $r - f(\theta) = 0$, where f is continuously differentiable. Let s denote the arc length measured along the curve.

We know that

$$ds^2 = dx^2 + dy^2$$

Recall that we have

$$dx = -r\sin\theta\, d\theta + \cos\theta\, dr$$

$$dy = r\cos\theta\, d\theta + \sin\theta\, dr$$

so if we square and add the differentials, we obtain

$$\mathbf{ds^2 = r^2\, d\theta^2 + dr^2} \tag{3}$$

Hence, to find the length of a polar curve, we use this last formula and integrate ds between appropriate limits.

Example 2 Find the length of the arc of $r = a\theta^2$ between $\theta = 0$ and $\theta = \pi$.

We have

$$r = a\theta^2, \qquad dr = 2a\theta\, d\theta$$

so

$$ds^2 = r^2\, d\theta^2 + dr^2$$

$$= a^2\theta^4\, d\theta^2 + 4a^2\theta^2\, d\theta^2$$

Taking the square root yields

$$ds = \sqrt{a^2\theta^4\, d\theta^2 + 4a^2\theta^2\, d\theta^2}$$

$$= a\theta\sqrt{\theta^2 + 4}\, d\theta$$

Then

$$s = \int_{\theta=0}^{\theta=\pi} ds = \int_0^\pi a\theta\sqrt{\theta^2+4}\, d\theta$$

$$= \frac{1}{2}\frac{a(\theta^2+4)^{3/2}}{\frac{3}{2}}\Bigg|_0^\pi = \tfrac{1}{3}a(\theta^2+4)^{3/2}\Big|_0^\pi$$

$$= \frac{a}{3}(\pi^2+4)^{3/2} - \tfrac{8}{3}a$$

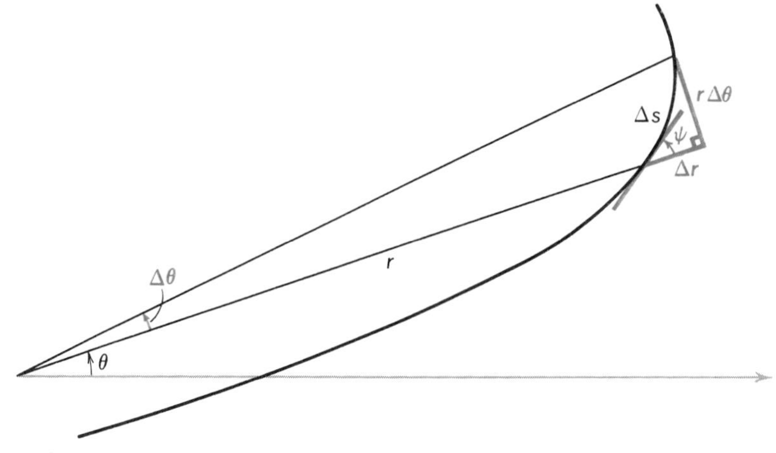

FIGURE 7-3

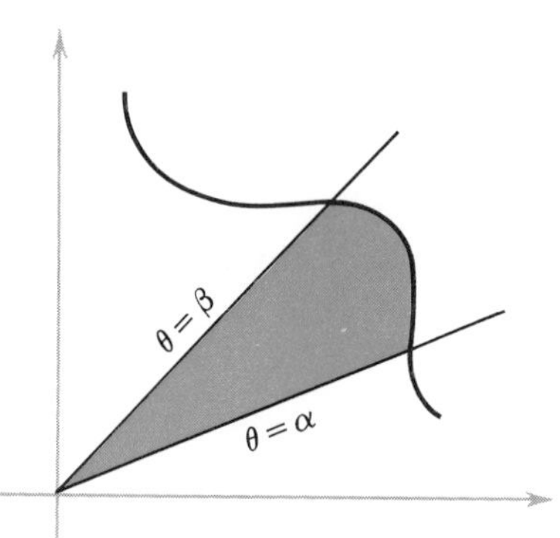

FIGURE 7-4

We can see a geometric interpretation of formulas (2) and (3) derived above if we look at the triangle shown in Figure 7-3. We can consider Δr and $r\Delta\theta$ as two "legs" and Δs as the "hypotenuse" of a "right triangle" with angle ψ opposite side $r\Delta\theta$, since we have

$$\tan\psi \approx \frac{r\Delta\theta}{\Delta r}$$

and

$$\Delta s^2 \approx (r\Delta\theta)^2 + (\Delta r)^2$$

where these approximations become better and better as $\Delta\theta \to 0$.

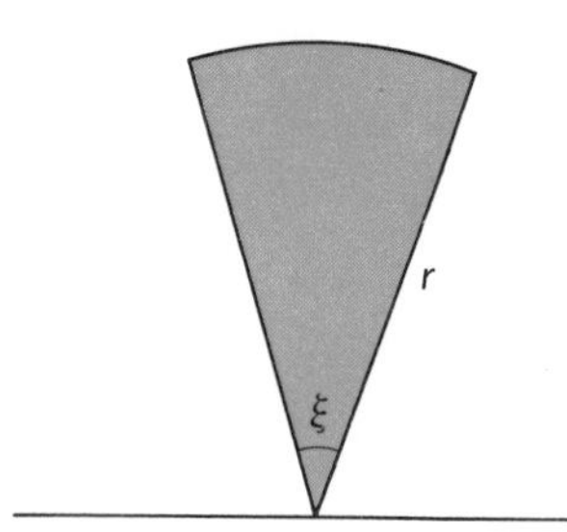

FIGURE 7-5

We next turn to the problem of computing the area of a region whose boundary is given by an equation in polar coordinates. Now suppose $\theta \mapsto f(\theta)$ is continuous on $[\alpha, \beta]$ where $0 \leqslant \alpha < \beta \leqslant 2\pi$; and suppose further that $f(\theta) \geqslant 0$ for all $\theta \in [\alpha, \beta]$. Let us now find a formula for the area of the region R bounded by the rays $\theta = \alpha$, $\theta = \beta$ and the locus of $r - f(\theta) = 0$ (Figure 7-4). We shall need the following fact: The area of a circular sector of central angle ζ and radius r is $\frac{1}{2}r^2\zeta$ (Figure 7-5).

We consider a regular partition $\alpha = \theta_0 < \theta_1 = < \cdots < \theta_{2^n} = \beta$ of $[\alpha, \beta]$. This partition corresponds to subdividing the angle AOB into 2^n subangles each of magnitude $\Delta\theta = (\beta - \alpha)/2^n$. Let $f(\Gamma_i^n)$ be the maximum value of f on $[\theta_i, \theta_{i+1}]$, $0 \leqslant i \leqslant 2^n - 1$, and let $f(\gamma_i^n)$ be the minimum value of f on $[\theta_i, \theta_{i+1}]$, $0 \leqslant i \leqslant 2^n - 1$. The angle ξ_i formed by the rays $\theta = \theta_i$ and $\theta = \theta_{i+1}$ and the radius $r_i = f(\Gamma_i^n)$ determine a circular sector of area $\frac{1}{2}r_i^2\xi_i = \frac{1}{2}f(\Gamma_i^n)^2(\beta - \alpha)/2^n$ (Figure 7-6). By the monotonicity and additivity of area, we have

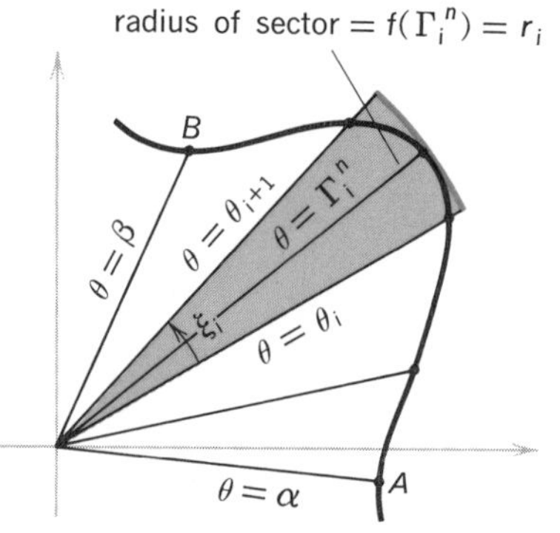

FIGURE 7-6

$$A(R) \leqslant \sum_{i=0}^{2^n-1} \tfrac{1}{2}f(\Gamma_i^n)^2 \frac{(\beta-\alpha)}{2^n} = S_n$$

Similarly we have

$$s_n = \sum_{i=0}^{2^n-1} \tfrac{1}{2}f(\gamma_i^n)^2 \frac{(\beta-\alpha)}{2^n} \leqslant A(R)$$

Since the above sums S_n and s_n are upper and lower nth Riemann sums for the function $\theta \mapsto \frac{1}{2}f(\theta)^2$ on $[\alpha, \beta]$, we conclude that

$$A(R) = \lim_{n\to\infty} S_n = \lim_{n\to\infty} s_n = \int_\alpha^\beta \tfrac{1}{2}f(\theta)^2\, d\theta$$

Thus we have the desired formula

$$A = \tfrac{1}{2}\int_\alpha^\beta f(\theta)^2\, d\theta$$

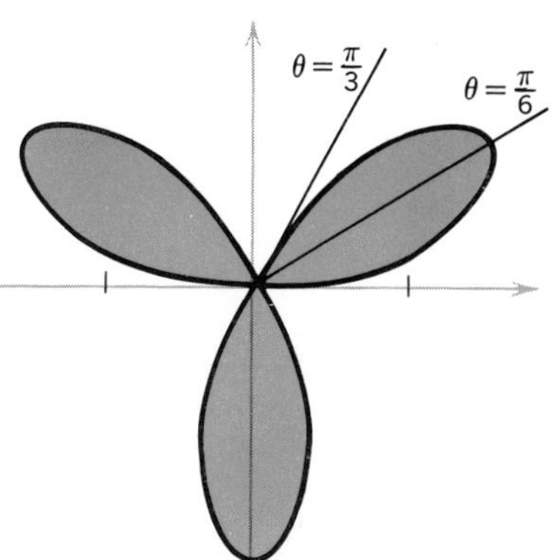

FIGURE 7-7

Example 3 Find the area enclosed by $r = 2\sin 3\theta$. We first draw the locus, which is a rose with three petals (Figure 7-7). We observe that we need only find the area of the loop in the first quadrant. Since $r = 0$ when $\theta = 0$, r increases as θ increases from 0 to $\frac{1}{6}\pi$, and r decreases as θ increases from $\frac{1}{6}\pi$ to $\frac{1}{3}\pi$, the area of the loop in the first quadrant can be found using the limits $\theta = 0$ and $\theta = \frac{1}{3}\pi$.

This area is one-third the entire area and so we have

$$\begin{aligned}
\tfrac{1}{3}A &= \tfrac{1}{2}\int_0^{1/3\pi} (2\sin 3\theta)^2\, d\theta \\
&= 2\int_0^{1/3\pi} \sin^2 3\theta\, d\theta \\
&= \int_0^{1/3\pi} (1 - \cos 6\theta)\, d\theta \\
&= [\theta - \tfrac{1}{6}\sin 6\theta]_0^{1/3\pi} \\
&= \tfrac{1}{3}\pi
\end{aligned}$$

Hence

$$A = 3 \cdot \tfrac{1}{3}\pi = \pi$$

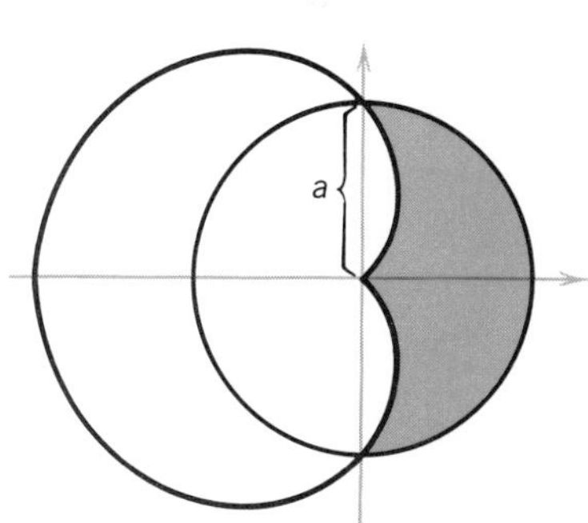

FIGURE 7-8

Example 4 Find the area that is inside the circle $r = a$ and outside the cardioid $r = a(1 - \cos\theta)$ (see Figure 7-8).

After sketching the locus, we note that we need only consider the area in the first quadrant, which represents one-half the desired area.

In the first quadrant the area between the loci equals

$$\int_0^{1/2\pi} \tfrac{1}{2}a^2\, d\theta - \int_0^{1/2\pi} \tfrac{1}{2}a^2(1 - \cos\theta)^2\, d\theta$$

Thus we have

$$\tfrac{1}{2}A = \tfrac{1}{2}a^2 \int_0^{\frac{1}{2}\pi} d\theta - \tfrac{1}{2}a^2 \int_0^{\frac{1}{2}\pi} d\theta + a^2 \int_0^{\frac{1}{2}\pi} \cos\theta\, d\theta - \tfrac{1}{2}a^2 \int_0^{\frac{1}{2}\pi} \cos^2\theta\, d\theta$$

$$= a^2 \int_0^{\frac{1}{2}\pi} \cos\theta\, d\theta - \tfrac{1}{4}a^2 \int_0^{\frac{1}{2}\pi} (1 + \cos 2\theta) d\theta$$

$$= a^2 \sin\theta \Big|_0^{\frac{1}{2}\pi} - \left[\tfrac{1}{4}a^2 (\theta + \tfrac{1}{2}\sin 2\theta)\right]_0^{\frac{1}{2}\pi}$$

$$= a^2 - a^2 \tfrac{1}{8}\pi = a^2(1 - \tfrac{1}{8}\pi)$$

Therefore

$$A = a^2(2 - \tfrac{1}{4}\pi)$$

EXERCISES

1. Find the angle ψ between the radius and the tangent to the locus at the points indicated.

   (a) $r = \dfrac{2}{\cos\theta}$ at $(2\sqrt{2}, \tfrac{1}{4}\pi)$

   (b) $r = 2 + \sin\theta$ at $(\tfrac{5}{2}, \tfrac{1}{6}\pi)$

   (c) $r = 4\cos\theta$ at $(2, \tfrac{1}{3}\pi)$

2. Show that the angle β between the tangents to two loci at a point of intersection may be found using the formula

$$\tan\beta = \frac{\tan\psi_2 - \tan\psi_1}{1 + \tan\psi_1 \tan\psi_2}$$

3. (a) If the loci of $r = f_1(\theta)$, $r = f_2(\theta)$ intersect (not at the origin) at a common value of θ, show that they intersect orthogonally when $\tan\psi_1 \tan\psi_2 = -1$.

   (b) Sketch the loci of $r(1 - \cos\theta) = a$ and $r = a(1 - \cos\theta)$ and show that they intersect at right angles.

4. Find the angles at which the following pairs of loci intersect.

   (a) $r = 4\cos\theta, \quad r = \dfrac{1}{\sin\theta}$ (b) $r = \dfrac{4}{1 - \cos\theta}, \quad r = 8$

5. Show that the loci of

$$r = \frac{1}{1 + \cos\theta}, \qquad r = \frac{1}{1 - \cos\theta}$$

   intersect at right angles.

In exercises 6–10, find the length of the arc.

6. $r = 3\theta^2$ $\quad 1 \leqslant \theta \leqslant 2$

7. $r = 3\cos\theta$ $\quad 0 \leqslant \theta \leqslant \tfrac{1}{4}\pi$

8. $r = 4e^{\theta}$ $\quad 0 \leqslant \theta \leqslant \pi$

9. $r = \dfrac{2}{1 + \cos\theta}$ $\quad 0 \leqslant \theta \leqslant \tfrac{1}{2}\pi$

10. $r = a\sin^2 \tfrac{1}{2}\theta$ $\quad 0 \leqslant \theta \leqslant \pi$

In exercises 11–15, find the area of the region bounded by the given curve and the given radii.

11. $r = \theta, \quad \theta = 0, \quad \theta = \frac{1}{2}\pi$
12. $r = e^{\theta/2}, \quad \theta = 0, \quad \theta = \frac{1}{2}\pi$
13. $r = \tan\theta, \quad \theta = 0, \quad \theta = \frac{1}{4}\pi$
14. $r = \dfrac{1}{\cos\theta}, \quad \theta = -\frac{1}{4}\pi, \quad \theta = \frac{1}{4}\pi$
15. $r = \sin\frac{1}{2}\theta + \cos\frac{1}{2}\theta, \quad \theta = 0, \quad \theta = \frac{1}{2}\pi$

Find the area enclosed by each of the curves in exercises 16–20.

16. $r = 2\cos 3\theta$
17. $r = \sin^2\frac{1}{2}\theta$
18. $r = a(1+\cos\theta)$
19. $r^2 = 2\sin 2\theta$
20. $r^2 = 2a^2\cos 2\theta$

21. Find the area inside the circle $r = 3$ and outside the cardioid $r = 1+\cos\theta$.
22. Find the area inside the loci of $r = 3\cos\theta$ and $r = 2-\cos\theta$.
23. Find the area inside the loci of $r = 4(1+\cos\theta)$ and $r = -4\sin\theta$.
24. Derive the formula $A = \int\frac{1}{2}[f(\theta)]^2\,d\theta$ for the area bounded by $r = f(\theta)$ by considering the appropriate area function and computing its derivative.
25. If $f(\theta)$ changes sign on $[\alpha,\beta]$, what does the integral $\int\frac{1}{2}[f(\theta)]^2 d\theta$ represent? Consider the example $r = \sin\theta$, $0 \leqslant \theta \leqslant 2\pi$.

REVIEW EXERCISES FOR CHAPTER 8

1. Define the following terms.

   amplitude
   e
   exponential function
   hyperbolic functions
   inverse hyperbolic functions
   inverse trigonometric functions
   logarithmic function
   logarithm to the base a
   oscillation
   period
   periodic function
   principal interval
   simple harmonic motion

2. Describe a process for graphing the product of two twice differentiable functions defined on all of $\boldsymbol{R}$, one of which is bounded and periodic, and the other of which is not periodic.
3. All the inverse trigonometric functions are monotone. Explain why.
4. Describe the process of logarithmic differentiation.
5. Show that $y = y_0 e^{kt}$ is the solution to the differential equation $dy/dt = ky$.

6. Differentiate the following functions
   (a) $x \mapsto 3^x$
   (b) $x \mapsto x^3$
   (c) $x \mapsto 3^3$
7. Which is larger: e^π or π^e?

In exercises 8–17, compute dy/dx and d^2y/dx^2, and sketch the curve.

8. $y = 8\sin(\frac{1}{4}\pi + x)$
9. $x = \tan\frac{1}{2}\pi y$
10. $y = xe^{-x}$
11. $y = x\ln x$
12. $y = e^{-x}\sin 2x$
13. $y = \ln(x+\sqrt{x^2+1})$
14. $y = x\tan^{-1}\frac{1}{2}x$
15. $y = \frac{1}{2}\ln\left(\frac{1+\tanh x}{1-\tanh x}\right)$
16. $\cosh y = 1 + \frac{1}{2}x^2$
17. $y = \tan(\frac{1}{2}\pi\tanh x)$

In exercises 18–31, compute the derivative.

18. $x \mapsto x^2e^{2x}\sin 3x$
19. $x \mapsto \sin^{-1}x/\sqrt{1+x^2}$
20. $x \mapsto \ln\left(\frac{x^5}{1+x^2}\right) + 3^{(x^{2/3})}$
21. $x \mapsto \tan^{-1}\left(\frac{3\sin x}{4+5\cos x}\right)$
22. $x \mapsto \ln\left(\frac{\sec x+\tan x}{\sec x-\tan x}\right)$
23. $x \mapsto \ln\sqrt{(1+\cos x)/(1-\cos x)}$
24. $x \mapsto \sin^{-1}(x^2) + xe^{-x^2}$.
25. $x \mapsto e^{\tan x}$
26. $x \mapsto \sqrt[3]{\frac{x(x-2)}{x-1}}$
27. $x \mapsto (\sin x)^x$
28. $x \mapsto e^{\frac{1}{2}x}\ln x$
29. $x \mapsto (\ln x)^x$
30. $x \mapsto x^{-1}e^{\tanh^{-1}(1-x)}$
31. $x \mapsto xe^{-x}\cosh^{-1}(1-x)$
32. Prove that for **any real number** n, $d(x^n)/dx = nx^{n-1}$.
33. The **gudermannian** function is defined by $\mathrm{gd}: x \mapsto \tan^{-1}(\sinh x)$. Sketch the graph of the gudermannian and compute its derivative. Show that the function is one-one and find its inverse.
34. If from any point on the graph of the function $f(x)$, perpendiculars are dropped to the coordinate axis, the upper of the two regions into which the graph divides the rectangle thus determined (by the perpendiculars and the coordinate axes) has twice the area of the lower region. Find $f(x)$.
35. A particle starts at the origin and moves along the x axis in such a way that its velocity at $(x, 0)$ is $\cos^2\pi x$. After how long does the point reach $x = \frac{1}{4}$? Does it ever reach $x = \frac{1}{2}$? Explain.
36. Solve the differential equation $dy/dx = y^3 \cdot e^{-2x}$ if $y = 4$ when $x = 0$.
37. Integrate:
   (a) $\int \frac{dx}{3-5x}$ (b) $\int \frac{x\,dx}{2-x^2}$ (c) $\int \frac{x^2\,dx}{x^3-2}$
   (d) $\int \sinh^3 x\,dx$ (e) $\int \frac{x\,dx}{(2-x^2)^2}$ (f) $\int \frac{x+1}{x}\,dx$
   (g) $\int \frac{d\theta}{\sinh\theta+\cosh\theta}$ (h) $\int \frac{e^t\,dt}{\sqrt{1+e^{2t}}}$

38. Crystalization of a certain chemical takes place at a rate proportional to the amount of crystal present. If there are initially 2 grams of crystal and there are 5 grams one hour later, about how much crystal will there be 12 hours after the initial 2 grams is introduced, assuming an unlimited supply of the chemical is present.

39. The half-life of radium is 1690 years. Find the approximate time necessary for 90% of a given sample of radium to decay.

40. If 200 pounds of saline solution are dissolved in 400 gallons of water, and pure fresh water is added at a rate of 10 gallons per minute, what is the percentage decrease of salinity after one hour?

41. A snake is moved from a 120° desert into a 70° laboratory. One hour later the snake's temperature is 100°. After how long will his temperature be 75°? (Recall that snakes are "cold-blooded," and therefore tend to assume the temperature of the atmosphere in which they are placed.)

42. Sugar dissolves in water at a rate proportional to the undissolved amount. If thirty pounds of sugar are mixed with fifty gallons of water, and twenty pounds are left after one day, how long will it take for only one pound of sugar to remain?

43. A car is traveling along a road at a velocity of 70 mph when it runs out of gas. Friction slows the car at a rate proportional to the velocity of the car. When the car is going 60 mph, it is decelerating at a rate of 2 miles per hour per minute. Finally, when the car is going only two mph, the driver pulls it over to the side of the road and begins to walk to the nearest gas station. How far does the car travel after it runs out of gas?

44. Discuss and sketch each of the following curves.
(a) $r = a\cos\theta - a\sin\theta$ (b) $r\cos\frac{1}{2}\theta = a$
(c) $r = a(1-2\sin 3\theta)$ (d) $r = 2/(1-\cos\theta)$
(e) $r^2 = 4a^2\sin 4\theta$ (f) $r = 1/(1+\cos\theta)$

45. Sketch each of the following pairs of curves and find all points of intersection.
(a) $r = a,\quad r = 2a\sin\theta$
(b) $r = a\cos\theta,\quad r = a(1+\cos\theta)$
(c) $r^2 = 4\cos 2\theta,\quad r^2 = \sec 2\theta$

46. Find $\tan\psi$ for the given curve at the given point.
(a) $r = 2+\cos\theta,\quad \theta = \frac{1}{3}\pi$
(b) $r = 2\sin 3\theta,\quad \theta = \frac{1}{4}\pi$
(c) $r^2 = 4\sin 2\theta,\quad \theta = \frac{2}{3}\pi$

47. Show that each of the following pairs of curves intersect at right angles at all points of intersection.
(a) $r = 4\cos\theta,\quad r = 4\sin\theta$
(b) $r^2\cos 2\theta = 4,\quad r^2\sin 2\theta = 9$
(c) $r = 2(1+\sin\theta),\quad r = 2(1-\sin\theta)$

48. Find the length of the cardioid $r = a(1-\cos\theta)$.

49. Find the length of the curve $r = a\cos^4\frac{1}{4}\theta$.

50. Find the length of $r = e^{\theta/2}$ from $\theta = 0$ to $\theta = 8$.

51. Find the area
    (a) inside $r = \cos\theta$ and outside $r = 1 - \cos\theta$;
    (b) between the inner and outer ovals of $r^2 = a^2(1 + \sin\theta)$;
    (c) bounded by $r^2 = a^2 \sin\theta(1 - \cos\theta)$.

52. Find the polar equation of a parabola with focus at the origin and vertex at $(a, \frac{1}{4}\pi)$.

53. Let P be a point on the hyperbola $r^2 \sin 2\theta = 2a^2$. Show that the triangle formed by OP, the tangent at P, and the initial line is isosceles.

9

TECHNIQUES OF INTEGRATION

In this chapter we introduce various techniques for finding an indefinite integral, or antiderivative, of a given function. Using these methods we can integrate many more types of functions than we have been able to handle up to now, and we shall also expand our list of standard integration formulas.

Basic trigonometric identities and substitutions are used to reduce the integrand to familiar derivative form; the method of integration by parts treats the problem of finding an antiderivative for a product of functions. The method of partial fractional decomposition of a proper rational function is used to solve the problem of integrating rational functions.

1 BASIC FORMULAS AND SUBSTITUTION

The problem of computing an indefinite integral $\int f(x)\,dx$ is that of finding a function F such that $d(F(x)) = f(x)\,dx$. If we can manipulate the integrand into a form which represents the differential of some function $F(x)$, we have $\int f(x)\,dx = \int d(F(x)) = F(x) + C$. Since integration is the inverse of differentiation, when we integrate the differential of a function, we get back the original function (plus an arbitrary constant.) Therefore, the fundamental technique for computing integrals is to reduce the integrand to the desired differential form $F'(x)\,dx$. In this section we shall review the most basic techniques which we learned in §§6.5 and 6.6.

Example 1 Integrate $\int 4x^3\,dx$. We know from differential calculus that $4x^3\,dx$ represents the differential $d(x^4)$, that is,

$$d(x^4) = 4x^3\,dx$$

Thus,

$$\int 4x^3\,dx = \int d(x^4) = x^4 + C$$

This method of finding integrals requires the ability to recognize the answer, that is, to guess the function whose differential is represented by the given integrand. To do this, the student must be able to convert the integrands into familiar forms, forms for which he can use tables or basic formulas that he has learned. Our list of standard forms is already quite long as a result of our study of the differential calculus.

For example, we know that $d(\sin u) = \cos u\,du$. Hence we know that

$$\int \cos u\,du = \int d(\sin u) = \sin u + C \tag{1}$$

The following formulas have been derived in this way:

$$\int \sin u\,du = -\cos u + C \tag{2}$$

$$\int \sec^2 u\,du = \tan u + C \tag{3}$$

$$\int \csc^2 u\,du = -\cot u + C \tag{4}$$

$$\int \sec u \tan u\,du = \sec u + C \tag{5}$$

$$\int \csc u \cot u\,du = -\csc u + C \tag{6}$$

$$\int u^n\,du = \frac{u^{n+1}}{n+1} + C, \qquad n \neq -1 \tag{7}$$

$$\int \frac{du}{u} = \ln|u| + C \tag{8}$$

$$\int e^u\,du = e^u + C \tag{9}$$

$$\int a^u\,du = \frac{a^u}{\ln a} + C \tag{10}$$

$$\int \frac{du}{1+u^2} = \tan^{-1} u + C \tag{11}$$

We may apply these formulas if the given integrand is in a correct form to begin with, or if we can rearrange the integrand so that it becomes apparent which is the formula to use. The most elementary technique for accomplishing this rearrangement is the method of **substitution,** a method whose differential basis is the Chain Rule.

Suppose S is the composition of G on h; that is, $S(x) = G[h(x)]$. By the Chain Rule, the derivative of S is given by

$$S'(x) = G'[h(x)]\,h'(x)$$

Writing this expression in differential form, we have

$$d(S(x)) = G'[h(x)]\,h'(x)\,dx$$

Let us denote $h(x)$ by u. Hence,

$$\int d(S(x)) = \int G'(u)\,du = G(u) + C$$

and we also have

$$\int G[h(x)]\,h'(x)\,dx = G[h(x)] + C$$

FORMAL DEVICE Substitution

To compute $\int F(x)\,dx$, where $F(x)$ can be put into the form $G(h(x))\,h'(x)$, and we know how to integrate G, substitute $u = h(x)$ and $du = h'(x)\,dx$, so $\int F(x)\,dx$ becomes $\int G(u)\,du$. Compute this latter integral, and substitute $h(x) = u$ back into the result, yielding $\int F(x)\,dx$.

Example 2 Integrate $\int (x+3)^{1/3}\,dx$. The function $x \mapsto (x+3)^{1/3}$ is the composition of $u \mapsto u^{1/3}$ on $x \mapsto x+3$, so we let $G(u) = u^{1/3}$ and $h(x) = x+3$. Then

$$(x+3)^{1/3} = G[h(x)]$$

Therefore,

$$d(h(x)) = h'(x)\,dx = 1 \cdot dx = dx$$

and

$$(x+3)^{1/3}\,dx = G[h(x)]\,h'(x)\,dx$$

If we set $u = x+3$, this equation becomes $(x+3)^{1/3}\,dx = u^{1/3}\,du$, and so

$$\int (x+3)^{1/3}\,dx = \int u^{1/3}\,du$$

We can now use formula (7) to obtain

$$\int u^{1/3}\,du = \tfrac{3}{4}u^{4/3} + C$$

and substituting back $u = x+3$, we get

$$\int (x+3)^{1/3}\,dx = \tfrac{3}{4}(x+3)^{4/3} + C$$

Example 3 Integrate $\int (2x+3)^{1/3}\,dx$. If we let

$$u = h(x) = 2x + 3$$

then

$$du = h'(x)\,dx = 2\,dx$$

We would like to put our integrand into the form $u^{1/3}\,du$ but $u^{1/3}\,du = (2x+3)^{1/3}2\,dx$ and the given integrand is just $(2x+3)^{1/3}\,dx$. However, we can obtain the desired form of the integrand if we multiply inside the integral by 2 and outside the integral by $\frac{1}{2}$ (this does not alter the final value, since the integral is linear and what we have done is therefore equivalent to multiplying by 1). Thus,

$$\int (2x+3)^{1/3}\,dx = \tfrac{1}{2}\int (2x+3)^{1/3}2\,dx$$

$$= \tfrac{1}{2}\int u^{1/3}\,du = \tfrac{3}{8}u^{4/3} + C$$

Substituting back we have

$$\int (2x+3)^{1/3}\,dx = \tfrac{3}{8}(2x+3)^{4/3} + C$$

To check this answer, we differentiate $\frac{3}{8}(2x+3)^{4/3}+C$ using the Chain Rule:

$$d(\tfrac{3}{8}(2x+3)^{4/3} + C) = \tfrac{1}{2}(2x+3)^{1/3}2\,dx$$

$$= (2x+3)^{1/3}\,dx$$

Example 4 Integrate

$$\int \frac{(x+1)\,dx}{x^2+2x+3}$$

Let $u = x^2+2x+3$, so $du = (2x+2)\,dx$. Then

$$\int \frac{(x+1)\,dx}{x^2+2x+3} = \frac{1}{2}\int \frac{2(x+1)\,dx}{x^2+2x+3} = \frac{1}{2}\int \frac{du}{u}$$

Since this is one of our standard forms, we have

$$\frac{1}{2}\int \frac{du}{u} = \tfrac{1}{2}\ln|u| + C$$

Substituting $u = x^2+2x+3$ yields

$$\int \frac{(x+1)\,dx}{x^2+2x+3} = \tfrac{1}{2}\ln|x^2+2x+3| + C$$

Example 5 Integrate

$$\int \frac{\sin\sqrt{x}\,dx}{\sqrt{x}}$$

Let $u = \sqrt{x} = x^{1/2}$, so

$$du = \tfrac{1}{2}x^{-1/2}\,dx = \frac{1}{2}\frac{dx}{\sqrt{x}}$$

Hence,

$$\int \frac{\sin\sqrt{x}}{\sqrt{x}}\,dx = 2\int \frac{1}{2}\frac{\sin\sqrt{x}}{\sqrt{x}}\,dx = 2\int \sin u\,du$$

$$= -2\cos u + C$$

Thus,

$$\int \frac{\sin\sqrt{x}}{\sqrt{x}}\,dx = -2\cos\sqrt{x} + C$$

Example 6 Integrate $\int \tan^3 2x \sec^2 2x\,dx$. Let $u = \tan 2x$, so $du = 2\sec^2 2x\,dx$. Then we have

$$\int \tan^3 2x \sec^2 2x\,dx = \frac{1}{2}\int (\tan^3 2x)(\sec^2 2x)\,2\,dx$$

$$= \frac{1}{2}\int u^3\,du = \frac{u^4}{8} + C$$

Hence

$$\int \tan^3 2x \sec^2 2x\,dx = \tfrac{1}{8}(\tan^4 2x) + C$$

Observe that the method of substitution for integration can also be used to evaluate definite integrals by the Change of Variable Theorem (§6.5). Suppose we are given

$$\int_a^b f[h(x)]\,h'(x)\,dx$$

and we let $u = h(x)$ then

$$\int_a^b f[h(x)]\,h'(x)\,dx = \int_{h(a)}^{h(b)} f(u)\,du$$

For example, consider

$$\int_0^1 (2x+3)\,dx$$

Let $u = 2x+3$, so that $du = 2\,dx$. We may then proceed in two different ways. First, we can solve the integral in terms of u, and substitute back $u = 2x+3$, keeping the same limits of integration. Thus,

$$\int_0^1 (2x+3)\,dx = \frac{1}{2}\int_0^1 (2x+3)\,2dx = \frac{1}{2}\int_0^1 u\,du$$

$$= \tfrac{1}{4}u^2\Big|_{x=0}^{1} = \tfrac{1}{4}(2x+3)^2\Big|_0^1 = 4$$

Alternatively, we can leave the solution of the integral in terms of u and set new limits of integration; when $x = 0$, $u = 3$ and when $x = 1$, $u = 5$. Therefore, our new limits of integration are 3 and 5, and our integral then becomes

$$\int_0^1 (2x+3)\,dx = \frac{1}{2}\int_3^5 u\,du = \frac{1}{2}\frac{u^2}{2}\Big|_3^5 = 4$$

EXERCISES

In exercises 1–20, perform the indicated integration using substitution.

1. $\int \sqrt{3x+2}\,dx$
2. $\int (2x-1)^{3/2}\,dx$
3. $\int \frac{dx}{(x-1)^3}$
4. $\int \frac{dx}{(3x+7)^3}$
5. $\int x\sqrt{16-x^2}\,dx$
6. $\int (x^3+2)^2\,3x^2\,dx$
7. $\int \frac{x^2\,dx}{\sqrt[4]{x^3+2}}$
8. $\int \frac{(2x+1)\,dx}{x^2+x+1}$
9. $\int \frac{\cos x}{\sin x}\,dx$
10. $\int \sin^7 x \cos x\,dx$
11. $\int \sec 3x \tan 3x\,dx$
12. $\int x \sec^2 (x^2)\,dx$
13. $\int \frac{\ln x}{x}\,dx$
14. $\int x e^{x^2}\,dx$
15. $\int \frac{dx}{\sqrt{x}(1+\sqrt{x})}$
16. $\int e^x \cos (e^x)\,dx$
17. $\int e^{(3\cos 2x)} \sin 2x\,dx$
18. $\int e^x \sqrt{1-4e^x}\,dx$
19. $\int \frac{x\ln(1+x^2)}{1+x^2}\,dx$
20. $\int \frac{\sec^2 3x\,dx}{(4+\cot 3x)}$

Evaluate the integrals in exercises 21–23, first by keeping same limits of integration, and then by changing limits.

21. $\int_0^2 \frac{dx}{(x+1)^2}$
22. $\int_{-1}^1 (x^3+1)^3\,x^2\,dx$
23. $\int_0^2 x\sqrt{4-x^2}\,dx$
24. Find the area bounded by the curve
$$y = \frac{2x+1}{\sqrt{x^2+x+1}}$$
the x axis, and the lines $x=0$, $x=2$.
25. Find the area bounded by the curve $y = \sec^2 x \tan x$, the x axis, and the lines $x=0$, $x=\frac{1}{3}\pi$.

2 TRIGONOMETRIC INTEGRALS

In this section we develop techniques for integrating various combinations of products and powers of trigonometric functions.

The simplest forms which we encounter are

$$\int \sin ax \cos bx\, dx$$

$$\int \sin ax \sin bx\, dx$$

or

$$\int \cos ax \cos bx\, dx$$

When $b = a$, these integrals can be evaluated by previous techniques; when $b \neq a$, the easiest way to evaluate these forms is to use the identities

$$\sin ax \cos bx = \tfrac{1}{2}(\sin(a-b)x + \sin(a+b)x) \tag{1}$$

$$\sin ax \sin bx = \tfrac{1}{2}(\cos(a-b)x - \cos(a+b)x) \tag{2}$$

$$\cos ax \cos bx = \tfrac{1}{2}(\cos(a-b)x + \cos(a+b)x) \tag{3}$$

These identities are derived from the well-known trigonometric formulas:

$$\cos(A+B) = \cos A \cos B - \sin A \sin B$$

$$\cos(A-B) = \cos A \cos B + \sin A \sin B$$

$$\sin(A+B) = \sin A \cos B + \cos A \sin B$$

$$\sin(A-B) = \sin A \cos B - \cos A \sin B$$

Example 1 Integrate $\int \sin x \sin 6x\, dx$. Using identity (2) we have

$$\sin x \sin 6x = \tfrac{1}{2}(\cos(-5x) - \cos 7x)$$

so

$$\int \sin x \sin 6x\, dx = \tfrac{1}{2}\int \cos(-5x)\, dx - \tfrac{1}{2}\int \cos 7x\, dx$$

$$= \tfrac{1}{2}\int \cos 5x\, dx - \tfrac{1}{2}\int \cos 7x\, dx$$

Let $u = 5x$, $du = 5dx$, $v = 7x$, and $dv = 7dx$. Therefore,

$$\int \sin x \sin 6x\, dx = \tfrac{1}{2}\cdot\tfrac{1}{5}\int \cos u\, du - \tfrac{1}{2}\cdot\tfrac{1}{7}\int \cos v\, dv$$

$$= \tfrac{1}{10}\sin u - \tfrac{1}{14}\sin v + C$$

$$= \tfrac{1}{10}\sin 5x - \tfrac{1}{14}\sin 7x + C$$

Example 2 Integrate $\int \cos 3x \cos x\, dx$. Using the appropriate identity yields

$$\cos 3x \cos x = \tfrac{1}{2}(\cos 2x + \cos 4x)$$

so

$$\int \cos 3x \cos x\, dx = \tfrac{1}{2}\int \cos 2x\, dx + \tfrac{1}{2}\int \cos 4x\, dx$$
$$= \tfrac{1}{4}\sin 2x + \tfrac{1}{8}\sin 4x + C$$

We next consider integrals involving powers of trigonometric functions, which can be put into the form $\int u^n\, du$. We first discuss the integration of expressions involving powers of sine and cosine.

Example 3 Find $\int \sin^2 2x \cos 2x\, dx$. Letting $u = \sin 2x$, $du = 2\cos 2x\, dx$. Therefore,

$$\int \sin^2 2x \cos 2x\, dx = \tfrac{1}{2}\int \sin^2 2x \cos 2x\, 2\, dx$$
$$= \tfrac{1}{2}\int u^2\, du$$
$$= \frac{1}{2}\frac{u^3}{3} + C = \frac{u^3}{6} + C$$

Substituting back we have,

$$\int \sin^2 2x \cos 2x\, dx = \frac{\sin^3 2x}{6} + C$$

In general,

$$\int \sin^n ax \cos ax\, dx = \frac{\sin^{n+1} ax}{(n+1)a} + C, \qquad n \neq -1 \tag{4}$$

and

$$\int \cos^n ax \sin ax\, dx = -\frac{\cos^{n+1} ax}{(n+1)a} + C, \qquad n \neq -1 \tag{5}$$

Example 4 Integrate $\int \cos^3 2x\, dx$.

In the previous example our method worked because we had a power of $\sin 2x$ and a $\cos 2x\, dx$ in the integrand to form du. In this example, if we let $u = \cos 2x$ then $du = -2\sin 2x\, dx$ and there is no $\sin 2x$ in the integrand. However, if we write the integrand $\cos^3 2x$ as

$$\cos^2 2x \cos 2x$$

we can then use the substitution $\cos^2 2x = 1 - \sin^2 2x$ to introduce the $\sin 2x$ into the integrand.

Then,

$$\int \cos^3 2x\, dx = \int \cos^2 2x \cos 2x\, dx$$
$$= \int (1 - \sin^2 2x)\cos 2x\, dx$$

Let $u = \sin 2x$, so $du = 2\cos 2x\, dx$. Then

$$\tfrac{1}{2}\int (1-\sin^2 2x)\,2\cos 2x\, dx = \tfrac{1}{2}\int (1-u^2)\, du$$

$$= \frac{1}{2}\left(u - \frac{u^3}{3}\right) = \frac{u}{2} - \frac{u^3}{6}$$

$$= \frac{\sin 2x}{2} - \frac{\sin^3 2x}{6} + C$$

Observe that this method will work for integrals of expressions involving sine and cosine, with at least one of them to a positive odd power.

Example 5 Integrate $\int \cos^3 x \sin^2 x\, dx$.

$$\int \cos^3 x \sin^2 x\, dx = \int \cos x \cos^2 x \sin^2 x\, dx$$

$$= \int \cos x (1-\sin^2 x) \sin^2 x\, dx$$

Let $u = \sin x$, so $du = \cos x\, dx$, and we have

$$\int (1-\sin^2 x)\sin^2 x \cos x\, dx = \int (1-u^2)u^2\, du$$

$$= \int (u^2 - u^4)\, du$$

$$= \frac{u^3}{3} - \frac{u^5}{5} + C$$

$$= \frac{\sin^3 x}{3} - \frac{\sin^5 x}{5} + C$$

Although we can use the identity $\sin^2 x + \cos^2 x = 1$ to integrate expressions which contain sines and cosines when at least one of the exponents is a positive odd integer, we would also like to be able to integrate expressions containing both sine and cosine to even powers. To do this we use the following trigonometric identities:

$$\sin^2 x = \frac{1-\cos 2x}{2} \qquad (6)$$

$$\cos^2 x = \frac{1+\cos 2x}{2} \qquad (7)$$

These identities are easily derived by adding or subtracting the equations

$$\cos^2 x + \sin^2 x = 1$$

$$\cos^2 x - \sin^2 x = \cos 2x$$

and then dividing by 2.

We take as our first example of this technique the simplest case, $\int \sin^2 x\, dx$.

Example 6

$$\int \sin^2 x\, dx = \int \frac{1-\cos 2x}{2}\, dx$$
$$= \tfrac{1}{2}\int dx - \tfrac{1}{2}\int \cos 2x\, dx$$
$$= \tfrac{1}{2}x - \tfrac{1}{4}\int \cos 2x\, 2dx$$
$$= \tfrac{1}{2}x - \tfrac{1}{4}\sin 2x + C$$

Example 7

$$\int \cos^4 x\, dx = \int (\cos^2 x)^2\, dx$$
$$= \int \tfrac{1}{4}(1 + \cos 2x)^2\, dx$$
$$= \tfrac{1}{4}\int (1 + 2\cos 2x + \cos^2 2x)\, dx$$
$$= \tfrac{1}{4}\int (1 + 2\cos 2x + \tfrac{1}{2}(1+\cos 4x))\, dx$$
$$= \tfrac{1}{4}x + \tfrac{1}{4}\sin 2x + \tfrac{1}{8}x + \tfrac{1}{32}\sin 4x + C$$
$$= \tfrac{3}{8}x + \tfrac{1}{4}\sin 2x + \tfrac{1}{32}\sin 4x + C$$

In Example 7, we see how to find integrals of sine or cosine to positive, even powers greater than two by using the identities (6) and (7) and the binomial expansion. As our last example of the use of these identities, let us consider the integration of an expression which contains both sine and cosine to a positive even power.

Example 8 Integrate $\int \sin^2 x \cos^2 x\, dx$.

$$\int \sin^2 x \cos^2 x\, dx = \int (1-\cos^2 x)\cos^2 x\, dx$$
$$= \int \cos^2 x\, dx - \int \cos^4 x\, dx$$

These last two integrals can be found using the method of Examples 6 and 7. Hence, integrals of even powers of both sine and cosine can be broken into a sum of integrals, each of which can be evaluated using identities (6) and (7). One gets $\frac{1}{8}x - \frac{1}{32}\sin 4x + C$.

Let us now consider the other trigonometric functions. Suppose we are asked to find

$$\int \tan x\, dx$$

If we were to let

$$u = \tan x, \qquad du = \sec^2 x\, dx$$

we see that we do not have *du* in the integrand. Often, in integrating more complex combinations of trigonometric functions it is useful to express the integrand in terms of sines and cosines.

Example 9 Since

$$\tan x = \frac{\sin x}{\cos x}$$

we have

$$\int \tan x \, dx = \int \frac{\sin x}{\cos x} \, dx$$

Letting $u = \cos x$, $du = -\sin x \, dx$ gives

$$-\int \frac{du}{u}$$

$$= -\ln |u| + C$$

$$= -\ln |\cos x| + C$$

Hence,

$$\int \tan x \, dx = -\ln |\cos x| + C \tag{8}$$

As a second example of this technique consider $\int \csc x \cot x \, dx$. Although we already can guess the answer since $d(-\csc x) = \csc x \cot x \, dx$, let us express the integrand in terms of sines and cosines.

Example 10

$$\int \csc x \cot x \, dx = \int \frac{1}{\sin x} \frac{\cos x}{\sin x} \, dx = \int \frac{\cos x \, dx}{\sin^2 x}$$

Let $u = \sin x$, $du = \cos x \, dx$, so that

$$\int \frac{\cos x \, dx}{\sin^2 x} = \int \frac{du}{u^2} = -u^{-1} + C$$

$$= -\frac{1}{\sin x} = -\csc x + C$$

We see that this technique is helpful in dealing with certain combinations of $\tan x \sec x$ and so on. However, a different method is used to integrate powers of $\tan x$ or $\sec x$.

Example 11 Integrate $\int \tan^3 x \, dx$.

If we were to let $u = \tan x$, then $du = \sec^2 x \, dx$ and it would be necessary to have a $\sec^2 x$ in the integrand. Hence, what we do is to use the identity

$$\tan^2 x + 1 = \sec^2 x \tag{9}$$

in a manner analogous to that used for sines and cosines. Thus,

$$\begin{aligned}\int \tan^3 x\,dx &= \int \tan x \tan^2 x\,dx\\ &= \int \tan x(\sec^2 x - 1)\,dx\\ &= \int \tan x \sec^2 x\,dx - \int \tan x\,dx\end{aligned}$$

We can solve the first integral by letting $u = \tan x$, and we have already found the second integral. Hence

$$\int \tan^3 x\,dx = \frac{\tan^2 x}{2} + \ln|\cos x| + C$$

Example 12 Integrate $\int \tan^n x\,dx$.

$$\begin{aligned}\int \tan^n x\,dx &= \int \tan^{n-2} x \tan^2 x\,dx\\ &= \int \tan^{n-2} x(\sec^2 x - 1)\,dx\\ &= \int \tan^{n-2} x \sec^2 x\,dx - \int \tan^{n-2} x\,dx\end{aligned}$$

Therefore,

$$\int \tan^n x\,dx = \frac{\tan^{n-1} x}{n-1} - \int \tan^{n-2} x\,dx$$

This is what is known as a **reduction formula**, in which the problem of finding $\int \tan^n x\,dx$ is reduced to finding that of $\int \tan^{n-2} x\,dx$. For higher powers of $\tan x$ we can use this formula repeatedly, each time reducing the exponent of $\tan x$ by 2. Similar reduction formulas for the sine and cosine functions are derived in the next section on integration by parts. The student is advised that it is more important to master the general techniques rather than to learn these formulas by rote.

Identity (9) can also be used to integrate even powers of $\sec x$.

Example 13 Integrate $\int \sec^4 3x\,dx$. We have

$$\begin{aligned}\int \sec^4 3x\,dx &= \int \sec^2 3x \sec^2 3x\,dx\\ &= \int (1 + \tan^2 3x)\sec^2 3x\,dx\\ &= \int \sec^2 3x\,dx + \int \tan^2 3x \sec^2 3x\,dx\\ &= \tfrac{1}{3}\tan 3x + \tfrac{1}{9}\tan^3 3x + C\end{aligned}$$

Note, however, that we use a simple trick to find the next integral.

Example 14

$$\int \sec x\,dx = \int \frac{\sec x(\tan x + \sec x)}{\sec x + \tan x}\,dx$$

Letting $u = \sec x + \tan x$, $du = (\sec x \tan x + \sec^2 x)\,dx$, so in this form the numerator is the derivative of the denominator. Therefore

$$\int \frac{\sec x \tan x + \sec^2 x}{\sec x + \tan x}\,dx = \int \frac{du}{u}$$

$$= \ln|u| + C$$

Thus,

$$\int \sec x\,dx = \ln|\sec x + \tan x| + C$$

In summary, we have learned how to integrate odd powers of sine and cosine using the identity $\cos^2 x + \sin^2 x = 1$; we have learned how to integrate even powers of sine and cosine using the identities $\sin^2 x = \frac{1}{2}(1 - \cos 2x)$, $\cos^2 x = \frac{1}{2}(1 + \cos 2x)$; and we have learned how to integrate certain powers of $\tan x$ and $\sec x$ using the identity $1 + \tan^2 x = \sec^2 x$.

EXERCISES

Evaluate the following integrals.

1. $\int \sin 3x \sin 2x\,dx$
2. $\int \sin 3x \cos 5x\,dx$
3. $\int \cos 4x \cos 2x\,dx$
4. $\int \cos^{53} x\,dx$
5. $\int \cos^4 2x \sin^3 2x\,dx$
6. $\int \sqrt{\sin x} \cos^5 x\,dx$
7. $\int \sin^3 x (\cos x)^{5/2}\,dx$
8. $\int \frac{\cos^3 x\,dx}{\sin^2 x}$
9. $\int \sin^2 2x\,dx$
10. $\int \cos^2 3x\,dx$
11. $\int \sin^4 x\,dx$
12. $\int \cos^2 x \sin^4 x\,dx$
13. $\int \cos^6 3x\,dx$
14. $\int \tan^4 x\,dx$
15. $\int \tan^3 3x \sec^4 3x\,dx$
16. $\int \cot^3 2x\,dx$
17. $\int \cot 3x \csc^4 3x\,dx$
18. $\int \sec^3 x \tan^2 x\,dx$
19. $\int \sec^6 x \sqrt{\tan x}\,dx$
20. $\int \frac{dx}{\sec x \tan x}$
21. By a method analogous to the one used to find $\int \sec x\,dx$, show that

$$\int \csc x\,dx = -\ln|\csc x + \cot x| + C$$

22. Show that for any integers m, n

(a) $\int_{-\pi}^{\pi} \sin mx \cos nx\,dx = 0$

(b) $\int_{-\pi}^{\pi} \sin mx \sin nx\,dx = 0, \qquad m \neq n$

23. Find the area bounded by the curve
$$y = \frac{\cos^3 x}{\sqrt{\sin x}}$$
the x axis, and the lines $x = \frac{1}{6}\pi$, $x = \frac{1}{2}\pi$.

24. Find the area bounded by the curve $y = \cot^2 2x \csc^2 2x$, the x axis, and the lines $x = \frac{1}{6}\pi$, $x = \frac{1}{3}\pi$.

3 INTEGRATION BY PARTS

Consider the formula for the differential of a product

$$d(uv) = u\,dv + v\,du$$

When we integrate this we obtain

$$\int d(uv) = \int u\,dv + \int v\,du + C$$

$$uv = \int u\,dv + \int v\,du + C$$

or

$$\int u\,dv = uv - \int v\,du + C \tag{1}$$

This last formula is the one used for **integration by parts.** If we are given an expression to integrate and can choose part of that expression to be u and part of it to be dv where we know $\int dv$ and where the integral $\int v\,du$ is simpler to obtain than $\int u\,dv$ then integration by parts is useful.

Example 1 Integrate $\int x \ln x\,dx$.

We would like to express $x \ln x\,dx$ as $u\,dv$ so that $v\,du$ will be easier to integrate. Suppose we choose

$$u = \ln x, \qquad dv = x\,dx$$

Then we know that

$$du = \frac{1}{x}\,dx, \qquad v = \frac{x^2}{2}$$

(Note that when we integrate dv to get v we would normally add a constant C', that is, $v = x^2/2 + C'$. In exercise 1 of this section, the reader is asked to show that setting this constant equal to 0 does not affect the final answer.)

We now form the integral

$$\int v\,du = \int \frac{x^2}{2}\frac{1}{x}\,dx = \int \frac{x}{2}\,dx$$

which we can easily solve, and so we have

$$\int v\,du = \frac{x^2}{4} + C$$

Substituting our original quantities into formula (1) yields the result

$$\int x \ln x\,dx = \frac{x^2}{2}\ln x - \frac{x^2}{4} + C$$

Example 2 Integrate $\int e^x \sin x\,dx$. Let

$$u = e^x, \qquad dv = \sin x\,dx$$

Then

$$du = e^x\,dx, \qquad v = -\cos x$$

and

$$\int e^x \sin x\,dx = e^x(-\cos x) - \int (-\cos x)e^x\,dx + C_1$$

$$\int e^x \sin x\,dx = -e^x \cos x + \int e^x \cos x\,dx + C_1 \tag{2}$$

It does not appear that we have simplified matters since $\int e^x \cos x\,dx$ seems just as difficult to integrate as $\int e^x \sin x\,dx$; however, it is often necessary to use the formula for integration by parts twice or more before one has complete success. Hence, let us use integration by parts on

$$\int e^x \cos x\,dx$$

Let

$$u = e^x, \qquad dv = \cos x\,dx$$

so then

$$du = e^x\,dx, \qquad v = \sin x$$

and

$$\int e^x \cos x\,dx = e^x \sin x - \int e^x \sin x\,dx + C_2$$

We can now substitute this expression into formula (2):

$$\int e^x \sin x\,dx = -e^x \cos x + e^x \sin x - \int e^x \sin x\,dx + C_1 + C_2$$

Note that we now have $\int e^x \sin x\,dx$ on both sides and we can combine them so that

$$2\int e^x \sin x\,dx = e^x \sin x - e^x \cos x + C_1 + C_2$$

which yields the final result

$$\int e^x \sin x\,dx = \frac{e^x \sin x - e^x \cos x}{2} + C$$

The method of integration by parts is especially useful in finding the integrals of the inverse trigonometric functions.

Example 3 Integrate $\int \sin^{-1} x\,dx$. Let

$$u = \sin^{-1} x, \qquad dv = dx$$

Then

$$du = \frac{dx}{\sqrt{1-x^2}}, \qquad v = x$$

and

$$\int \sin^{-1} x\,dx = x\sin^{-1} x - \int \frac{x\,dx}{\sqrt{1-x^2}}$$

$$= x\sin^{-1} x + \frac{1}{2}\int \frac{-2x\,dx}{\sqrt{1-x^2}}$$

Therefore, we have the formula

$$\boldsymbol{\int \sin^{-1} x\,dx = x\sin^{-1} x + \sqrt{1-x^2} + C}$$

As was mentioned in the preceding section, we can use integration by parts to derive reduction formulas for various trigonometric functions.

Example 4 Find $\int \sin^n x\,dx$. We rewrite the integral as

$$\int \sin^n x\,dx = \int \sin^{n-1} x \sin x\,dx$$

Let

$$u = \sin^{n-1} x, \qquad dv = \sin x\,dx$$

Then

$$du = (n-1)\sin^{n-2} x \cos x\,dx, \qquad v = -\cos x$$

so we have

$$\int \sin^n x\,dx = -\sin^{n-1} x\cos x + (n-1)\int \sin^{n-2} x\cos^2 x\,dx$$

Replacing $\cos^2 x$ by $1-\sin^2 x$ gives

$$\int \sin^n x\,dx = -\sin^{n-1} x\cos x + (\mathrm{n}-1)\int \sin^{n-2} x(1-\sin^2 x)\,dx$$

$$\int \sin^n x\,dx = -\sin^{n-1} x\cos x + (n-1)\int \sin^{n-2} x\,dx - (n-1)\int \sin^n x\,dx$$

Since $\int \sin^n x\,dx$ appears on both sides we can combine and get

$$n\int \sin^n x\,dx = -\sin^{n-1} x\cos x + (n-1)\int \sin^{n-2} x\,dx$$

$$\boldsymbol{\int \sin^n x\,dx = -\frac{\sin^{n-1} x\cos x}{n} + \frac{\mathrm{n}-1}{n}\int \sin^{n-2} x\,dx}$$

The analogous reduction formula for cosine is left to the exercises.

We see that the success of the integration by parts method depends on strategic choices for the functions u and v. Since the idea is to replace an integral containing u as a factor in the integrand by one containing du as a factor, it is reasonable to look for a function whose derivative is simpler (in some sense) than the original function. Experience has shown that the most suitable choices for u are logarithmic functions and inverse trigonometric functions, followed by functions which are positive integral powers of x. Least suitable for u are the exponential and trigonometric functions. The reader should be able to see the reasoning behind this classification, which is a useful working rule in many practical problems.

EXERCISES

1. By substituting $v+C'$ for v in the integration by parts formula, verify that the constant C' does not appear in the final answer.
2. Using the method followed in Example 2, derive the more general formula

$$\int e^{ax}\sin bx\,dx = \frac{e^{ax}}{a^2+b^2}(a\sin bx - b\cos bx) + C$$

In exercises 3–20, perform the indicated integrations.

3. $\int \ln x\,dx$
4. $\int \tan^{-1} x\,dx$
5. $\int (\ln x)^{-12}\,x\,dx$
6. $\int xe^x\,dx$
7. $\int x^2 e^{-x}\,dx$
8. $\int x\sin x\,dx$
9. $\int \sin(\ln x)\,dx$
10. $\int \cos(\ln x)\,dx$
11. $\int x^2\ln x\,dx$
12. $\int x^2\cos x\,dx$
13. $\int x\sec^2 x\,dx$
14. $\int x\cos(2x+1)\,dx$
15. $\int x\sin^{-1} x\,dx$
16. $\int x\sqrt{1+x}\,dx$
17. $\int \cos^{-1} 2x\,dx$
18. $\int x\tan^{-1} x\,dx$
19. $\int \ln(x^2+2)\,dx$
20. $\int (x^3-7x+2)\sin x\,dx$

Derive the following reduction formulas.

21. $\displaystyle\int \cos^n x\,dx = \frac{\cos^{n-1} x\sin x}{n} + \frac{n-1}{n}\int \cos^{n-2} x\,dx$

22. $\displaystyle\int \sec^n x\,dx = \frac{\sec^{n-2} x\tan x}{n-1} + \frac{n-2}{n-1}\int \sec^{n-2} x\,dx, \quad n \neq 1$

23. (a) Integrate $\int \sin^3 x\,dx$ using the fact that the exponent of $\sin x$ is an odd, positive integer.
    (b) Integrate $\int \sin^3 x\,dx$ using the reduction formula derived in this section.
    (c) Show that the answers obtained in (a), (b) differ only by a constant.

24. (a) Integrate $\int \cos^2 x\,dx$ using the identity $\cos^2 x = \frac{1}{2}(1+\cos 2x)$.
    (b) Integrate $\int \cos^2 x\,dx$ by the method of parts.
    (c) Show that the answers in (a), (b) differ only by a constant.

Evaluate each of the following integrals.

25. $\int_0^{\frac{1}{3}\pi} x \tan^{-1} 2x\,dx$
26. $\int_0^{\frac{1}{4}\pi} e^{3x} \sin 4x\,dx$
27. $\int_{-\pi}^{\pi} x^2 \cos x\,dx$
28. $\int_0^1 x^3 e^{-x^2}\,dx$

4 TRIGONOMETRIC SUBSTITUTIONS

Various integrals involving the terms a^2-u^2, a^2+u^2, and u^2-a^2 (where a is a constant) can be computed most simply by using certain trigonometric substitutions.

Suppose we are asked to find an integral which has in its integrand an expression of the form a^2-u^2. Let us use the substitution $u = a \sin\theta$, $a > 0$, and see if it might simplify anything:

$$a^2 - u^2 = a^2 - a^2 \sin^2\theta = a^2(1-\sin^2\theta)$$

Since

$$1 - \sin^2\theta = \cos^2\theta$$

We have

$$a^2 - u^2 = a^2\cos^2\theta$$

Thus we have replaced the binomial a^2-u^2 with a single term $a^2\cos^2\theta$. The advantages of this replacement will become apparent as we use the technique to solve various problems.

Let us now use $u = a\tan\theta$ to simplify u^2+a^2:

$$u^2 + a^2 = a^2\tan^2\theta + a^2 = a^2(\tan^2\theta+1)$$

Recall that

$$\tan^2\theta + 1 = \sec^2\theta$$

Hence,

$$u^2 + a^2 = a^2\sec^2\theta$$

Similarly, for u^2-a^2, we use $u = a\sec\theta$ to obtain

$$u^2 - a^2 = a^2\sec^2\theta - a^2 = a^2(\sec^2\theta-1)$$

and since

$$\sec^2\theta - 1 = \tan^2\theta$$

it follows that

$$u^2 - a^2 = a^2\tan^2\theta$$

Thus, the plan of attack in using these substitutions is as follows:

Expression	Substitution	Result
$a^2 - u^2$	$u = a\sin\theta, a > 0$	$a^2\cos^2\theta$
$a^2 + u^2$	$u = a\tan\theta, a > 0$	$a^2\sec^2\theta$
$u^2 - a^2$	$u = a\sec\theta, a > 0$	$a^2\tan^2\theta$

These substitutions can be easily remembered by recalling the Pythagorean Theorem, and then taking a and u as sides, and θ as an angle in a right triangle. Thus the first case, a^2-u^2, suggests we let a equal the hypotenuse, and let θ be the angle opposite the leg u [Figure 4-1(a)]. Similarly, for a^2+u^2, we let a and u be legs, with θ opposite u [Figure 4-1(b)] and for u^2-a^2, consider u as the hypotenuse and a as a leg, with θ the included angle [Figure 4-1(c)].

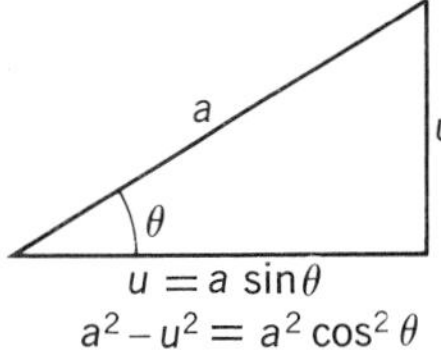

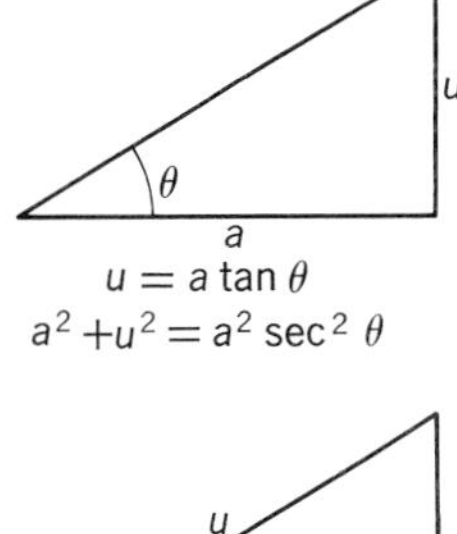

u

θ

a

$u = a\sec\theta$

$u^2 - a^2 = a^2\tan\theta$

FIGURE 4-1

Example 1 Integrate $\int dx/\sqrt{4+x^2}$. Let

$$x = 2\tan\theta, \qquad dx = 2\sec^2\theta\, d\theta$$

Then

$$\int \frac{dx}{\sqrt{4+x^2}} = \int \frac{2\sec^2\theta\, d\theta}{\sqrt{4+4\tan^2\theta}} = \int \frac{2\sec^2\theta\, d\theta}{2\sqrt{1+\tan^2\theta}} = \int \frac{\sec^2\theta\, d\theta}{\sqrt{\sec^2\theta}}$$

$$= \pm \int \sec\theta\, d\theta$$

the sign depending on the sign of $\sec\theta$.

Since $x = 2\tan\theta$,

$$\theta = \tan^{-1}\tfrac{1}{2}x, \qquad -\tfrac{1}{2}\pi < \theta < \tfrac{1}{2}\pi$$

and for these values of θ, $\sec\theta$ is positive. Hence,

$$\int \frac{dx}{\sqrt{4+x^2}} = \int \sec\theta\, d\theta$$

$$= \ln|\sec\theta + \tan\theta| + C$$

Since $\sec\theta = \frac{1}{2}\sqrt{4+x^2}$, and $\tan\theta = \frac{1}{2}x$, we substitute back to obtain

$$\int \frac{dx}{\sqrt{4+x^2}} = \ln\left|\frac{\sqrt{4+x^2}}{2} + \frac{x}{2}\right| + C$$

$$= \ln|\sqrt{4+x^2} + x| - \ln 2 + C$$

$$= \ln(\sqrt{4+x^2} + x) + C'$$

where $C' = C - \ln 2$. We can remove the absolute value sign because $\sqrt{x^2+4}+x$ is always positive.

The reader should verify that by using the substitution $u = a\tan\theta$, one can generalize Example 1 to get the following result:

$$\int \frac{du}{\sqrt{a^2 + u^2}} = \ln(\sqrt{a^2 + u^2} + u) + C, \qquad a > 0$$

The reader will also recall that this result was established in §8.4 where we found the formulas

$$\sinh^{-1} x = \ln(x + \sqrt{1+x^2})$$

and

$$\int \frac{du}{\sqrt{1+u^2}} = \sinh^{-1} u + C$$

Example 2 Integrate $\int du/\sqrt{a^2-u^2}$. We use the substitution

$$u = a\sin\theta, \qquad du = a\cos\theta\, d\theta$$

Therefore,

$$\int \frac{du}{\sqrt{a^2-u^2}} = \int \frac{a\cos\theta\, d\theta}{\sqrt{a^2-a^2\sin^2\theta}}$$

$$= \int \frac{a\cos\theta\, d\theta}{a\sqrt{\cos^2\theta}} = \pm \int d\theta$$

where the sign of the integral depends on the sign of $\cos\theta$. Since we use only the principal value of $\sin^{-1} u/a$, we have

$$\theta = \sin^{-1}\frac{u}{a}, \qquad -\frac{\pi}{2} \leqslant \theta \leqslant \frac{\pi}{2}$$

hence, $\cos\theta \geqslant 0$ and we choose the $+$ sign. Continuing,

$$\int \frac{du}{\sqrt{a^2-u^2}} = \int d\theta = \theta + C$$

so

$$\int \frac{du}{\sqrt{a^2 - u^2}} = \sin^{-1}\frac{u}{a} + C$$

which again agrees with what we found in §8.1.

Since this is a general formula, if we are asked to integrate

$$\int \frac{dx}{\sqrt{1-4x^2}}$$

we let

$$a = 1, \qquad u = 2x, \qquad du = 2\,dx$$

and then

$$\int \frac{dx}{\sqrt{1-4x^2}} = \tfrac{1}{2}\sin^{-1} 2x + C$$

Example 3 Integrate

$$\int \frac{x^3\,dx}{\sqrt{4-x^2}}$$

Let

$$x = 2\sin\theta, \qquad dx = 2\cos\theta\,d\theta$$

$$\int \frac{x^3\,dx}{\sqrt{4-x^2}} = \int \frac{16\sin^3\theta\cos\theta\,d\theta}{\sqrt{4-4\sin^2\theta}}$$

$$= \pm 8 \int \frac{\sin^3\theta\cos\theta\,d\theta}{\cos\theta}$$

but for $-\frac{1}{2}\pi < \theta < \frac{1}{2}\pi$, $\cos\theta > 0$, so the last integral is

$$8 \int \sin^3\theta\,d\theta$$

From §3 we know that

$$\int \sin^3\theta\,d\theta = -\frac{\sin^2\theta\cos\theta}{3} - \frac{2}{3}\cos\theta + C$$

Hence,

$$\int \frac{x^3\,dx}{\sqrt{4-x^2}} = -\frac{8\sin^2\theta\cos\theta}{3} - \frac{16}{3}\cos\theta + C$$

Since $\sin\theta = \frac{1}{2}x$, and $\cos\theta = \frac{1}{2}\sqrt{4-x^2}$, we have

$$\int \frac{x^3\,dx}{\sqrt{4-x^2}} = -\frac{x^2\sqrt{4-x^2}}{3} - \frac{8\sqrt{4-x^2}}{3} + C$$

$$= \frac{\sqrt{4-x^2}}{3}(-x^2 - 8) + C$$

We can also use trigonometric substitutions in problems involving the definite integral, where a change of limits often simplifies matters.

Example 4 Evaluate

$$\int_2^{2\sqrt{3}} \frac{x^2\,dx}{\sqrt{16-x^2}}$$

Let

$$x = 4\sin\theta, \qquad dx = 4\cos\theta\,d\theta$$

In this case we also want to change the limits. Now, when $x = 2$, $\sin\theta = \frac{1}{2}$, and $\theta = \sin^{-1}\frac{1}{2}$. Using the principal value, we find $\theta = \frac{1}{6}\pi$. When $x = 2\sqrt{3}$, $\sin\theta = \frac{1}{2}\sqrt{3}$ and $\theta = \frac{1}{3}\pi$. So we have

$$\int_2^{2\sqrt{3}} \frac{x^2\,dx}{\sqrt{16-x^2}} = \int_{\frac{1}{6}\pi}^{\frac{1}{3}\pi} \frac{64\sin^2\theta\cos\theta\,d\theta}{4\cos\theta}$$

$$= 16\int_{\frac{1}{6}\pi}^{\frac{1}{3}\pi} \sin^2\theta\,d\theta$$

We know from §2 that $\int \sin^2\theta\,d\theta = \frac{1}{2}\theta - \frac{1}{4}\sin 2\theta + C$ so

$$16\int_{\frac{1}{6}\pi}^{\frac{1}{3}\pi} \sin^2\theta\,d\theta = 16\left[\tfrac{1}{2}\theta - \tfrac{1}{4}\sin 2\theta\right]_{\frac{1}{6}\pi}^{\frac{1}{3}\pi}$$

$$= 16(\tfrac{1}{6}\pi - \tfrac{1}{8}\sqrt{3}) - 16(\tfrac{1}{12}\pi - \tfrac{1}{8}\sqrt{3})$$

$$= \tfrac{8}{3}\pi - \tfrac{4}{3}\pi = \tfrac{4}{3}\pi$$

At this time we want to consider an algebraic technique for transforming certain complex expressions into expressions which can be integrated using trigonometric substitutions. This method involves completing the square of a quadratic expression.

Recall that given the general quadratic $f(x) = ax^2 + bx + c$ $(a \neq 0)$, we can make the binomial part $(ax^2 + bx)$ a perfect square by adding to it the square of one-half b, the coefficient of x:

$$ax^2 + bx + c = a\left(x^2 + \frac{b}{a}x\right) + c$$

$$= a\left(x^2 + \frac{b}{a}x + \frac{b^2}{4a^2}\right) + \frac{c}{2} - \frac{b^2}{4a}$$

$$= a\left(x + \frac{b}{2a}\right)^2 + \frac{4ac - b^2}{4a}$$

Example 5 Find

$$\int \frac{dx}{\sqrt{24-2x-x^2}}$$

First, consider the expression $24-2x-x^2$. We can write

$$24 - 2x - x^2 = -(x^2 + 2x - 24)$$

$$= -(x^2 + 2x + 1 - 1 - 24)$$

$$= -[(x+1)^2 - 5^2]$$

$$= 5^2 - (x+1)^2$$

and

$$\int \frac{dx}{\sqrt{24-2x-x^2}} = \int \frac{dx}{\sqrt{5^2-(x+1)^2}}$$

If we substitute

$$u = x + 1, \qquad du = dx$$

this integral is in the form

$$\int \frac{du}{\sqrt{a^2-u^2}}$$

We can then use a trigonometric substitution to find the integral. In this case we already know that

$$\int \frac{du}{\sqrt{a^2-u^2}} = \sin^{-1}\frac{u}{a} + C$$

Hence

$$\int \frac{du}{\sqrt{5^2-u^2}} = \sin^{-1}\frac{u}{5} + C$$

$$= \sin^{-1}\frac{x+1}{5} + C$$

Example 6 Find

$$\int \frac{dx}{\sqrt{9x^2+6x+10}}$$

First we complete the square

$$9x^2 + 6x + 10 = 9(x^2+\tfrac{2}{3}x) + 10$$

$$= 9(x^2+\tfrac{2}{3}x+\tfrac{1}{9}) + 10 - \tfrac{9}{9}$$

$$= 9(x+\tfrac{1}{3})^2 + 9$$

so

$$\int \frac{dx}{\sqrt{9x^2+6x+10}} = \int \frac{dx}{\sqrt{9(x+\frac{1}{3})^2+9}}$$

Substituting

$$u = x + \tfrac{1}{3}, \qquad du = dx$$

we have

$$\int \frac{dx}{\sqrt{9(x+\frac{1}{3})^2+9}} = \int \frac{du}{\sqrt{9}\sqrt{u^2+1}}$$

$$= \tfrac{1}{3}\int \frac{du}{\sqrt{u^2+1}}$$

Again, we could use a trigonometric substitution. But we have already found that $\int du/\sqrt{u^2+a^2} = \ln(\sqrt{a^2+u^2}+u) + C$. Hence,

$$\frac{1}{3}\int \frac{du}{\sqrt{u^2+1}} = \tfrac{1}{3}\ln(\sqrt{1+u^2}+u) + C$$

and

$$\int \frac{dx}{\sqrt{9x^2+6x+10}} = \tfrac{1}{3}\ln(\sqrt{1+(x+\tfrac{1}{3})^2}+x+\tfrac{1}{3}) + C$$

EXERCISES

In exercises 1–2, derive the given formulas.

1. $\int \frac{du}{a^2+u^2} = \frac{1}{a}\tan^{-1}\frac{u}{a} + C, \qquad a > 0$

2. $\int \frac{du}{\sqrt{u^2-a^2}} = \ln|u + \sqrt{u^2-a^2}| + C, \qquad a > 0$

In exercises 3–17, perform the indicated integration.

3. $\int \frac{dx}{\sqrt{2-5x^2}}$

4. $\int \frac{dx}{x^2\sqrt{4+x^2}}$

5. $\int \frac{dx}{\sqrt{4+x^2}}$

6. $\int \frac{x^2\,dx}{\sqrt{x^2-4}}$

7. $\int \frac{\sqrt{x^2-9}}{x}\,dx$

8. $\int x\sqrt{16-x^2}\,dx$

9. $\int \frac{dx}{x^2-9}$

10. $\int \frac{x^3\,dx}{\sqrt{4-x^2}}$

11. $\int \frac{\sqrt{9-4x^2}}{x}\,dx$

12. $\int \frac{\sqrt{x^2+a^2}}{x}\,dx$

13. $\int \frac{dx}{x^2+2x+5}$

14. $\int \frac{dx}{2x^2+12x+20}$

15. $\int \frac{x\,dx}{\sqrt{x^2-2x+5}}$

16. $\int \frac{(x-1)\,dx}{\sqrt{x^2-4x+3}}$

17. $\int \frac{dx}{(x^2+2x+2)^3}$

Evaluate the definite integral.

18. $\int_0^4 \frac{dx}{\sqrt{9+x^2}}$

19. $\int_0^2 \frac{dx}{\sqrt{4+x^2}}$

20. $\int_0^1 \frac{dx}{4-x^2}$

21. $\int_1^3 \frac{dx}{\sqrt{x^2-2x+5}}$

22. $\int_1^2 \frac{x^2\,dx}{\sqrt{2x-x^2}}$

23. Find the area of the region enclosed by the ellipse

$$\frac{x^2}{4} + \frac{y^2}{16} = 1$$

24. Using a trigonometric substitution, find the area of a circle $x^2+y^2=r^2$.

25. Find the area of the region bounded by the curve

$$y = \frac{x^2}{(4-x^2)^{3/2}}$$

the x axis, and the lines $x=0$, $x=1$.

5 PARTIAL FRACTIONS

In this section we investigate integrals of functions which can be expressed as $x \mapsto f(x)/g(x)$, where $f(x)$ and $g(x)$ are polynomials, $g(x) \neq 0$. Such functions are called **rational functions.** A result from algebra states that any given rational function can be expressed as the sum of a polynomial and another rational function whose numerator has smaller degree than the denominator. That is,

$$\frac{f(x)}{g(x)} = Q(x) + \frac{R(x)}{g(x)}$$

where the degree of $R(x)$ is less than that of $g(x)$.

For example, given $f(x) = x^3 - x + 1$ and $g(x) = x^2 + 1$, then using division we have

$$\frac{f(x)}{g(x)} = x^2 + 1 + \frac{-2x+1}{x^2+1}$$

Notice that in the last term the degree of the numerator (1) is less than the degree of the denominator (2).

Since every rational function can be broken down into a sum this way and since the integral is linear, we shall restrict our attention to the integration of functions which are expressed as $h(x)/g(x)$ and in which the numerator $h(x)$ is of lesser degree than the denominator.

Suppose we are asked to find $\int (x\,dx)/(x^2 - 3x + 2)$. Recall that in algebra one learns how to combine expressions such as $A/(x-c) + B/(x-d)$ over a common denominator. If we can use a reverse process on $x/(x^2 - 3x + 2)$, that is, if we can split it into the sum of rational functions having simpler denominators, then we will often simplify the problem of integrating.

Since the denominator $x^2 - 3x + 2 = (x-1)(x-2)$ let us try to find real numbers A, B such that

$$\frac{x}{x^2-3x+2} = \frac{A}{x-1} + \frac{B}{x-2}$$

Thus, we wish to solve for A and B. We have

$$\frac{x}{x^2-3x+2} = \frac{A(x-2) + B(x-1)}{(x-1)(x-2)}$$

or

$$x = A(x-2) + B(x-1)$$

Now, we want to know when the last expression above is an identity, and that occurs when coefficients of like powers of x on the two sides of the equation are equal. Using this condition then, we have

$$x = Ax - 2A + Bx - B$$

$$x = (A+B)x - (2A+B)$$

Matching coefficients, we find

$$A + B = 1, \qquad 2A + B = 0$$

Hence,

$$A = -1, \qquad B = 2$$

So we have

$$\frac{x}{x^2-3x+2} = \frac{-1}{x-1} + \frac{2}{x-2}$$

It is clear that the integrals of the two terms on the right are easier to evaluate than the single integral on the left. This technique is known as the method of **partial fractions.**

Another algebraic fact about polynomials is that any polynomial (of degree at least one) with real coefficients can be written as a product of linear and quadratic factors each with real coefficients. Thus, the first step in using the method of partial fractions to integrate an expression of the form $f(x)/g(x), g(x) \neq 0$ [degree of $f(x) <$ degree of $g(x)$], is to factor the denominator $g(x)$ into its linear and quadratic factors. We consider three cases.

In the first case, suppose $g(x)$ can be factored into distinct linear factors; that is, $g(x) = (x-p_1)(x-p_2)\cdots(x-p_n)$, where $p_i \neq p_j$ for $i \neq j$. Then $f(x)/g(x)$ is the sum of fractions whose numerators are constants and whose denominators are the factors $(x-p_i)$ of $g(x)$. For example, if

$$\frac{f(x)}{g(x)} = \frac{x+1}{6x^3-5x^2-12x-4} = \frac{x+1}{(2x+1)(3x+2)(x-2)}$$

then

$$\frac{f(x)}{g(x)} = \frac{A}{2x+1} + \frac{B}{3x+2} + \frac{C}{x-2}$$

and A, B, and C may be found using the method already described.

For the second case, suppose $g(x)$ contains a repeated factor $(x-a)^m$. For this factor we use the expression which is the sum of m partial fractions:

$$\frac{A}{x-a} + \frac{B}{(x-a)^2} + \cdots + \frac{C}{(x-a)^m}$$

Example 1 Find $\int dx/x(x+1)^2$. Let

$$\frac{1}{x(x+1)^2} = \frac{A}{x} + \frac{B}{x+1} + \frac{C}{(x+1)^2}$$

then

$$1 = (A+B)x^2 + (2A+B+C)x + A$$

and we find

$$A = 1, \qquad B = -1, \qquad C = -1$$

Hence, we can write

$$\frac{1}{x(x+1)^2} = \frac{1}{x} - \frac{1}{x+1} - \frac{1}{(x+1)^2}$$

and so

$$\int \frac{dx}{x(x+1)^2} = \int \frac{dx}{x} - \int \frac{dx}{x+1} - \int \frac{dx}{(x+1)^2}$$

$$= \ln|x| - \ln|x+1| + \frac{1}{x+1} + C$$

$$= \ln\left|\frac{x}{x+1}\right| + \frac{1}{x+1} + C$$

For the third and final case, suppose $g(x)$ has a factor of the form $(ax^2+bx+c)^m$. Then for this factor we use the expression

$$\frac{Ax+B}{ax^2+bx+c} + \frac{Cx+D}{(ax^2+bx+c)^2} + \cdots + \frac{Ex+F}{(ax^2+bx+c)^m}$$

Example 2 Evaluate $\int dx/x(x^2+1)^2$. First we find the partial fraction decomposition

$$\frac{1}{x(x^2+1)^2} = \frac{A}{x} + \frac{Bx+C}{x^2+1} + \frac{Dx+E}{(x^2+1)^2}$$

$$1 = (A+B)x^4 + Cx^3 + (2A+B+D)x^2 + (C+E)x + A$$

which implies

$$A + B = 0, \qquad C = 0, \qquad 2A + B + D = 0,$$

$$C + E = 0, \qquad A = 1$$

Hence,

$$A = 1, \qquad B = -1, \qquad C = 0, \qquad D = -1, \qquad E = 0$$

and we have

$$\frac{dx}{x(x^2+1)^2} = \frac{1}{x} - \frac{x}{x^2+1} - \frac{x}{(x^2+1)^2}$$

Therefore

$$\int \frac{dx}{x(x^2+1)^2} = \int \frac{dx}{x} - \int \frac{x\,dx}{x^2+1} - \int \frac{x\,dx}{(x^2+1)^2}$$

$$= \ln|x| - \tfrac{1}{2}\ln|x^2+1| + \frac{1}{2(x^2+1)} + C$$

$$= \ln\left|\frac{x}{\sqrt{x^2+1}}\right| + \frac{1}{2(x^2+1)} + C$$

Observe that we can simplify any rational function with the method of partial fractions by reducing the function to some combination of these standard cases.

Example 3 Find

$$\int \frac{dx}{(x+1)(x^2+1)}$$

Using partial fractions, we write

$$\frac{1}{(x+1)(x^2+1)} = \frac{A}{x+1} + \frac{Bx+C}{x^2+1}$$

$$1 = (A+B)x^2 + (B+C)x + (A+C)$$

$$A + C = 1, \qquad A + B = 0, \qquad B + C = 0$$

Hence

$$A = \tfrac{1}{2}, \qquad B = -\tfrac{1}{2}, \qquad C = \tfrac{1}{2}$$

and we have

$$\frac{1}{(x+1)(x^2+1)} = \frac{1}{2}\frac{1}{x+1} - \frac{1}{2}\frac{x-1}{x^2+1}$$

$$\int \frac{dx}{(x+1)(x^2+1)} = \frac{1}{2}\int \frac{dx}{x+1} - \frac{1}{2}\int \frac{(x-1)\,dx}{x^2+1}$$

$$= \tfrac{1}{2}\ln|x+1| - \tfrac{1}{4}\ln|x^2+1| + \tfrac{1}{2}\tan^{-1}x + C$$

Example 4 Find

$$\int \frac{dx}{(x+1)(x+2)^2}$$

First we decompose the integrand

$$\frac{x}{(x+1)(x+2)^2} = \frac{A}{x+1} + \frac{B}{x+2} + \frac{C}{(x+2)^2}$$

$$= \frac{A(x+2)^2 + B(x+1)(x+2) + C(x+1)}{(x+1)(x+2)^2}$$

so

$$x = Ax^2 + 4Ax + 4A + Bx^2 + 3Bx + 2B + Cx + C$$

and it follows that

$$A + B = 0$$

$$4A + 3B + C = 1$$

$$4A + 2B + C = 0$$

which yields

$$A = -1, \qquad B = 1, \qquad C = 2$$

Integrating, we obtain

$$\int \frac{x\,dx}{(x+1)(x+2)^2} = -\int \frac{dx}{x+1} + \int \frac{dx}{x+2} + \int \frac{2\,dx}{(x+2)^2}$$

$$= -\ln|x+1| + \ln|x+2| - \frac{2}{x+2} + C$$

EXERCISES

Evaluate the following integrals.

1. $\int \frac{dx}{x^2-4}$
2. $\int \frac{dx}{x^2+7x+6}$
3. $\int \frac{2x^3+3x^2-2}{2x^2+3x+1}\,dx$
4. $\int \frac{(x+2)\,dx}{(2x+1)(x+1)}$
5. $\int \frac{x\,dx}{x^2+4x-5}$
6. $\int \frac{x\,dx}{(x+1)^2}$
7. $\int \frac{x\,dx}{(x+1)(x+2)^2}$
8. $\int \frac{(3x+5)\,dx}{x^3-x^2-x+1}$
9. $\int \frac{(x^3+x^2+x+2)}{x^4+3x^2+2}\,dx$
10. $\int \frac{dx}{x(x^2+x+1)}$
11. $\int \frac{(6x^2-x+13)\,dx}{(x+1)(x^2+4)}$
12. $\int \frac{dx}{(x^2+1)^2}$
13. $\int \frac{(2x^2+3)\,dx}{(x^2+1)^2}$
14. $\int \frac{(2x^3+x^2+4)\,dx}{(x^2+4)^2}$

6 OTHER SUBSTITUTIONS

We have already seen that it is possible to integrate any rational function $x \mapsto f(x)/g(x)$ by the method of partial fractions. In this section we show that any rational function of sine and cosine can be integrated using the substitution $y = \tan(x/2)$.

A rational function of sine and cosine is defined by replacing each x in the polynomials $f(x), g(x)$ by sines and cosines. As an example consider

$$F(x) = \frac{\sin^2 x + 2\sin^3 x + \sin x \cos x}{\cos^2 x + \sin x \cos^2 x}$$

This would correspond to the rational function

$$\frac{x^2 + 2x^3 + x^2}{x^2 + x^3}$$

To integrate these functions, first we define y, a new variable of integration, by the equation

$$y = \tan\frac{x}{2}$$

Next, we express $\cos x$ in terms of y:

$$\begin{aligned}\cos x &= \cos 2\tfrac{1}{2}x = \cos^2 \tfrac{1}{2}x - \sin^2 \tfrac{1}{2}x \\ &= \frac{\cos^2 \frac{1}{2}x - \sin^2 \frac{1}{2}x}{1} = \frac{\cos^2 \frac{1}{2}x - \sin^2 \frac{1}{2}x}{\cos^2 \frac{1}{2}x + \sin^2 \frac{1}{2}x} \\ &= \frac{\cos^2 \frac{1}{2}x - \sin^2 \frac{1}{2}x}{\cos^2 \frac{1}{2}x + \sin^2 \frac{1}{2}x} \cdot \frac{1/(\cos^2 \frac{1}{2}x)}{1/(\cos^2 \frac{1}{2}x)} = \frac{1 - \tan^2 \frac{1}{2}x}{1 + \tan^2 \frac{1}{2}x}\end{aligned}$$

from which it follows that

$$\cos x = \frac{1-y^2}{1+y^2}$$

Similarly,

$$\begin{aligned}\sin x &= 2\sin \tfrac{1}{2}x \cos \tfrac{1}{2}x \\ &= \frac{2\sin \frac{1}{2}x \cos^2 \frac{1}{2}x}{\cos \frac{1}{2}x} = \frac{2\tan \frac{1}{2}x}{\sec^2 \frac{1}{2}x} \\ &= \frac{2\tan \frac{1}{2}x}{1 + \tan^2 \frac{1}{2}x}\end{aligned}$$

so

$$\sin x = \frac{2y}{1+y^2}$$

Finally, since $y = \tan \frac{1}{2}x$

$$x = 2\tan^{-1} y$$

$$dx = \frac{2\,dy}{1+y^2}$$

Using these substitutions we are able to integrate any rational function of sine and cosine.

Example 1

$$\begin{aligned}\int \frac{dx}{1+\cos x} &= \int \frac{2\,dy}{1+y^2}\cdot\frac{1}{1+(1-y^2)/(1+y^2)}\\ &= \int \frac{2\,dy}{1+y^2}\cdot\frac{1+y^2}{2}\\ &= \int dy = y + C\\ &= \tan\tfrac{1}{2}x + C\end{aligned}$$

Example 2

$$\begin{aligned}\int \frac{dx}{1+\sin x} &= \int \frac{1}{1+2y/(1+y^2)}\cdot\frac{2\,dy}{1+y^2}\\ &= \int \frac{1+y^2}{1+y^2+2y}\cdot\frac{2\,dy}{1+y^2}\\ &= \int \frac{2\,dy}{y^2+2y+1}\\ &= 2\int \frac{dy}{(y+1)^2}\\ &= -2(y+1)^{-1} + C\\ &= -2(\tan\tfrac{1}{2}x+1)^{-1} + C\end{aligned}$$

Certain integrals involving fractional powers of x may be simplified by using the substitution $y = x^n$.

Example 3 Find

$$\int \frac{1-\sqrt{x}}{1+\sqrt{x}}\,dx$$

Let

$$y = \sqrt{x}, \qquad dy = \frac{dx}{2\sqrt{x}} = \frac{dx}{2y}$$

Therefore

$$\begin{aligned}\int \frac{1-\sqrt{x}}{1+\sqrt{x}}\,dx &= \int \frac{1-y}{1+y}\,2y\,dy\\ &= \int \frac{2y-2y^2}{1+y}\,dy\end{aligned}$$

Dividing through the integrand we have,

$$\int \frac{2y-2y^2}{1+y}\,dy = \int \left(-2y+4-\frac{4}{1+y}\right)dy$$

$$= -\int 2y\,dy + \int 4\,dy - \int \frac{4}{1+y}\,dy$$

$$= -y^2 + 4y - 4\ln|1+y| + C$$

$$= -x + 4\sqrt{x} - 4\ln|1+\sqrt{x}| + C$$

EXERCISES

Perform the indicated integrations.

1. $\int \frac{\sin x\,dx}{1+\sin x}$
2. $\int \frac{3\,dx}{\sin x+\cos x}$
3. $\int \frac{dx}{1+\sin x-\cos x}$
4. $\int \frac{dx}{3-2\cos x}$
5. $\int \frac{dx}{5+4\sin x}$
6. $\int \frac{\sin x\cos x}{1-\cos x}\,dx$
7. $\int \frac{dx}{\sqrt{x}-\sqrt[4]{x}}$
8. $\int \frac{dx}{x\sqrt{1-x}}$
9. $\int x^5\sqrt{1-x^3}\,dx$
10. $\int \frac{\sqrt{x}\,dx}{1+x}$
11. $\int \frac{dx}{3+\sqrt{x+2}}$
12. $\int_{\frac{1}{2}\pi}^{\pi} \frac{dx}{1-\cos x}$
13. $\int_0^{\frac{1}{2}\pi} \frac{dx}{2+\cos x}$
14. $\int_0^3 \frac{x\,dx}{\sqrt{1+x}}$
15. $\int_1^{64} \frac{dx}{\sqrt[3]{x}+2\sqrt{x}}$

REVIEW EXERCISES FOR CHAPTER 9

In exercises 1–54, perform the indicated integration.

1. $\int \frac{(x+1)\,dx}{\sqrt{x^2+2x-4}}$
2. $\int \frac{1}{x^3}\left(\frac{1-x^2}{x^2}\right)dx$
3. $\int \frac{\cos(\ln x)}{x}\,dx$
4. $\int x\cos x\,dx$

5. $\int \frac{\sqrt{25-x^2}}{x}\,dx$

6. $\int \frac{3x\,dx}{\sqrt[3]{x^2+3}}$

7. $\int \frac{(x^2+3x-4)}{x^2-2x-8}\,dx$

8. $\int \frac{dx}{\sqrt{x}(1+\sqrt{x})}$

9. $\int x^2\sqrt{1-x}\,dx$

10. $\int \frac{dx}{x^2\sqrt{a^2-x^2}}$

11. $\int \frac{x\,dx}{(x-2)^2}$

12. $\int \frac{dx}{(x-2)\sqrt{x+2}}$

13. $\int \frac{e^{2x}-1}{e^{2x}+3}\,dx$

14. $\int b\sec ax\tan ax\,dx$

15. $\int \cos^6\frac{x}{2}\,dx$

16. $\int x^2e^{-3x}\,dx$

17. $\int \frac{dx}{2+\sin x}$

18. $\int \frac{\sqrt{x^2+a^2}\,dx}{x}$

19. $\int \frac{x^3+x^2+x+3}{(x^2+1)(x^2+3)}\,dx$

20. $\int e^{\tan 2x}\sec^2 2x\,dx$

21. $\int \sin^3 x\cos^3 x\,dx$

22. $\int \cos^3 x\cos 2x\,dx$

23. $\int e^{2x}\cos 3x\,dx$

24. $\int \frac{x^2\,dx}{\sqrt{x^2-16}}$

25. $\int \frac{dx}{e^{2x}-3e^x}$

26. $\int \frac{dx}{\sin x-\cos x-1}$

27. $\int x\sin^{-1}x^2\,dx$

28. $\int \frac{x\,dx}{\sqrt{2x+1}}$

29. $\int x(\cos^3 x^2-\sin^3 x^2)\,dx$

30. $\int \frac{x^3+x-1}{(x^2+1)^2}\,dx$

31. $\int \frac{dx}{\sqrt{x^2-4x+13}}$

32. $\int \frac{(2x-3)\,dx}{x^2+6x+13}$

33. $\int \tan^4 x\sec^4 x\,dx$

34. $\int \cot^3 x\csc^3 x\,dx$

35. $\int \frac{dx}{x\sqrt{x^2+9}}$

36. $\int \frac{(3x-1)\,dx}{4x^2-4x+1}$

37. $\int \ln(x^2+1)\,dx$

38. $\int \frac{\sqrt{\tan^{-1}x}\,dx}{1+x^2}$

39. $\int \tan x\sqrt{\sec x}\,dx$

40. $\int \frac{x^2\,dx}{\sqrt{x^2-a^2}}$

41. $\int \frac{x^2\,dx}{x^4-16}$

42. $\int \frac{(3x+1)}{(x^2-4)^2}\,dx$

43. $\int \frac{\sin x\,dx}{1+\sin^2 x}$

44. $\int \sin\sqrt{2x}\,dx$

45. $\int \frac{\sin^2 x\,dx}{\cos^4 x}$

46. $\int \frac{(x^2-x+1)\,dx}{x^4-5x^3+5x^2+5x-6}$

47. $\int 6x^2 \sin^{-1} 2x\, dx$

48. $\int x^3 \sqrt{a^2x^2+b^2}\, dx$

49. $\int \tan^{3/2} x \sec^4 x\, dx$

50. $\int \frac{x^3\, dx}{\sqrt{9-x^2}}$

51. $\int \frac{x^2-2x+3}{(x-1)^2(x^2+4)}\, dx$

52. $\int 9x \tan^2 3x\, dx$

53. $\int \frac{\cot^3 x}{\csc x}\, dx$

54. $\int \frac{dx}{(x^2-4x+5)^2}$

Determine whether or not each of the following integrals converge, and if it does, evaluate it.

55. $\int_0^1 \frac{dx}{x}$

56. $\int_{-1}^{8} \frac{dx}{\sqrt[3]{x}}$

57. $\int_{-1}^{\infty} \frac{x}{e^{x^2}}\, dx$

58. $\int_{-\infty}^{\infty} \frac{1}{e^x+e^{-x}}\, dx$

59. $\int_0^4 \frac{dx}{\sqrt{4x-x^2}}$

60. $\int_{-\infty}^{\infty} x^2 e^{-x^3}\, dx$

61. $\int_2^{\infty} \frac{dx}{x \ln^2 x}$

62. $\int_{-\infty}^{6} \frac{dx}{(4-x)^2}$

63. Find the area bounded by the curve $y^2 = x^4/(4-x^2)$ and its asymptotes.

64. Find the area bounded by the curve $y^2 = 1/x(1-x)$ and its asymptotes.

65. Show that the area under $y = e^{-2x}$, $x \geqslant 0$, is $\frac{1}{2}$ and find the volume generated by revolving the area about the x axis.

Derive the following reduction formulas.

66. $\int x^n \sin bx\, dx = \frac{-x^n}{b} \cos bx + \frac{n}{b} \int x^{n-1} \cos bx\, dx$

67. $\int x^n \cos bx\, dx = \frac{x^n}{b} \sin bx - \frac{n}{b} \int x^{n-1} \sin bx\, dx$

10

INFINITE SERIES

In this chapter we present a comprehensive treatment of **infinite series** of real numbers, as well as a discussion of the main aspects of power series and the Taylor expansion of a function. The final (starred) section is an introduction to convergence of sequences and series of functions on an interval; this section contains the definition of uniform convergence for sequences and series, as well as the main theorem on continuity of the limit function and term-wise differentiation and integration.

After the elements of the subject are introduced in §1, we proceed to derive the comparison, ratio, and integral tests in §2. Section 3 treats absolute convergence and alternating series. The discussion of power series in §4 is self-contained, except for reference to optional §6 for term-by-term differentiation and integration. In §5 on the Taylor expansion, both the integral and derivative forms of the remainder are considered.

1 SERIES OF REAL NUMBERS

In §7.1 we defined a sequence S to be a real-valued function whose domain is the natural numbers. We also know that given a sequence of real numbers $a_1, a_2, a_3, \ldots, a_n$ the sum $a_1+a_2+\cdots+a_n$ is always defined if n is a positive integer and the a_i are finite numbers. Now, suppose that we have an infinite sequence of real numbers $a_1, a_2, a_3, \ldots$. How can we define the sum

$$a_1 + a_2 + \cdots + a_n + \cdots \tag{1}$$

consisting of an infinite number of terms? An expression such as (1) is called an **infinite series**, which is also denoted by

$$\sum_{n=1}^{\infty} a_n$$

where a_n is the nth term of the sequence.

Let us construct another sequence in the following way. We let

$$s_1 = a_1$$
$$s_2 = a_1 + a_2$$
$$s_3 = a_1 + a_2 + a_3$$
$$\vdots$$
$$s_n = a_1 + a_2 + a_3 + \cdots + a_n = \sum_{i=1}^{n} a_i$$

This sequence is called the **sequence of partial sums** of the series in (1) and s_n is then the ***n*th partial sum** of the infinite series.

Example 1 We can create an infinite series from the arithmetic sequence $1, 3, 5, 7, 9, \ldots$ by taking

$$s_1 = 1$$
$$s_2 = 1 + 3$$
$$s_3 = 1 + 3 + 5$$
$$\vdots$$
$$s_n = 1 + 3 + 5 + 7 + 9 + \cdots + (2n - 1)$$

so that our infinite series is

$$1 + 3 + 5 + 7 + 9 + \cdots + (2n - 1) + \cdots$$

In order to generalize the concept of a sum to the case where we are adding infinitely many terms, we consider the sequence of partial sums s_n for the infinite series $\sum_{n=1}^{\infty} a_n$. We then define the infinite sum as the limit of s_n as $n \to \infty$, if the limit exists.

Definition A series

$$\sum_{n=1}^{\infty} a_n = a_1 + a_2 + a_3 + \cdots + a_n + \cdots$$

is said to **converge to the sum S** whenever the sequence of partial sums $\{s_n\}$ converges to S. A series that does not converge is said to **diverge**.

Hence, if the sequence $\{s_n\}$ of partial sums of a series converges, we take its limit to be the value of the infinite series, that is, if

$$\lim_{n \to \infty} s_n = S$$

we write

$$\sum_{n=1}^{\infty} a_n = S$$

If the series represented by $\sum_{n=1}^{\infty} a_n$ converges to the sum S, we call

$$r_n = S - s_n$$

the **nth remainder**.

The reasoning behind the next theorem should be clear to the reader.

Theorem 1 If an infinite series $\sum_{n=1}^{\infty} a_n$ converges, then

(i) $$\lim_{n\to\infty} a_n = 0$$

(ii) $$\lim_{n\to\infty} r_n = 0$$

Proof We prove (i) and leave (ii) as an exercise. Let $s_n = a_1 + a_2 + \cdots + a_n$ denote the nth partial sum of the series, and suppose the series converges to S, that is,

$$\lim_{n\to\infty} s_n = S$$

Then, corresponding to any number $\varepsilon > 0$, there is an $N \in \mathbf{N}$ such that all terms of the sequence $\{s_n\}$ lie in the interval $]S-\frac{1}{2}\varepsilon, S+\frac{1}{2}\varepsilon[$ for $n > N$. Hence, the difference between any two terms does not exceed ε; that is, if m and n are both greater than N, then

$$|s_n - s_m| < \varepsilon$$

In particular, if $m = n-1$, and $n > N+1$, we obtain

$$s_n - s_{n-1} = (a_1 + a_2 + \cdots + a_n) - (a_1 + a_2 + \cdots + a_{n-1}) = a_n$$

Therefore,

$$|a_n| < \varepsilon, \qquad n > N+1$$

Since ε was any positive number, we have

$$\lim_{n\to\infty} a_n = 0$$

Observe that part (i) of the theorem states a **necessary condition** for a series to converge. Since a necessary condition is one that must be satisfied if the series converges, it follows that any convergent series has $\lim_{n\to\infty} a_n = 0$. However, this is **not** a **sufficient condition** for a series to converge—a series may very well have $\lim_{n\to\infty} a_n = 0$ and yet not converge (see Example 3). Because we can conclude that a series whose nth term does not approach 0 is divergent, this provides a useful test for establishing divergence. For example, the series

$$\frac{1}{2}+\frac{2}{3}+\frac{3}{4}+\cdots+\frac{n}{n+1}+\cdots$$

diverges since

$$\lim_{n\to\infty} a_n = \lim_{n\to\infty}\frac{n}{n+1} = \lim_{n\to\infty}\frac{1}{1+1/n} = 1$$

As a straightforward consequence of the fact that the nth term of a convergent series tends to zero, we have the following theorem.

Theorem 2 If the series $\sum_{n=1}^{\infty} a_n$ converges, then there is a positive number C such that for all n, $|a_n| < C$.

Proof Since $a_n \to 0$ as $n \to \infty$, there is an N in $\mathbf{N}$ such that for $n \geqslant N$, $|a_n| < 1$. Let

$$C_1 = \max|a_k|, \qquad k = 1, \ldots, N$$

Then if $C = C_1 + 1$, we have $|a_n| < C$, for all n. ∎

Hence, all the terms of a convergent series are bounded.

Example 2 A **geometric series** is of the form

$$a + ar + ar^2 + ar^3 + \cdots + ar^{n-1} + \cdots, \qquad a \neq 0$$

where the ratio of any term to the one before it is r. The sum of the first n terms is

$$s_n = a + ar + ar^2 + \cdots + ar^{n-1} \tag{2}$$

Multiplying both sides of (2) by r, we have

$$rs_n = ar + ar^2 + ar^3 + \cdots + ar^n \tag{3}$$

Subtracting (3) from (2) gives

$$s_n(1-r) = a(1-r^n)$$

If $r \neq 1$, we may divide by $1-r$ to obtain

$$s_n = \frac{a(1-r^n)}{1-r}, \qquad r \neq 1 \tag{4}$$

In the case that $r = 1$, $s_n = na$. If $a \neq 0$ this expression has no finite limit as $n \to \infty$. If $a = 0$, then the series is $0+0+0+\cdots$ and clearly it converges to zero.

If $r \neq 1$, we use equation (4). When $|r| < 1$, then $r^n \to 0$ as $n \to \infty$ (see exercises in §7.1) and

$$\lim_{n\to\infty}\frac{a(1-r^n)}{1-r} = \frac{a}{1-r}, \qquad |r| < 1$$

Hence, as we may recall from algebra, the geometric series with $|r|<1$ converges to the sum $a/(1-r)$. If $|r|>1$, $|r^n|\to\infty$ as $n\to\infty$, and therefore the series diverges. In the remaining case of $r=-1$, the series is

$$a-a+a-a+a-a+\cdots$$

This series has no limit as $n\to\infty$, since the partial sums form the oscillating sequence $a, 0, a, 0, \ldots$ which does not converge when $a\neq 0$, and hence the series is divergent.

Example 3 Consider the **harmonic series**

$$\sum_{n=1}^{\infty}\frac{1}{n}=1+\frac{1}{2}+\frac{1}{3}+\frac{1}{4}+\cdots$$

Note that $\lim_{n\to\infty}a_n=\lim_{n\to\infty}1/n=0$, but we shall see that this series is divergent. To prove this, observe that s_n, the nth partial sum of the series, is given by

$$s_n=1+\frac{1}{2}+\cdots+\frac{1}{n}$$

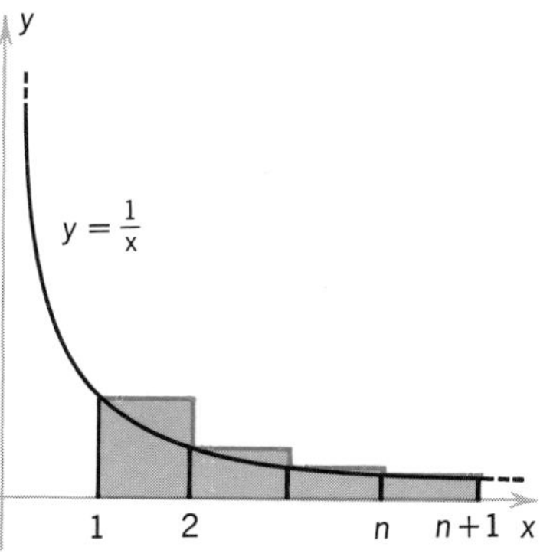

FIGURE 1-1

Next, consider Figure 1-1, which is the graph of $y=1/x$ between $x=1$ and $x=n+1$. Using the partition points $1, 2, \ldots, n+1$, we form the sum of the areas of the shaded rectangles; this sum is given by

$$A=1+\frac{1}{2}+\cdots+\frac{1}{n}$$

Hence, $A=s_n$. Since the sum A is greater than or equal to the corresponding definite integral, we have

$$s_n=A\geqslant\int_1^{n+1}\frac{1}{x}\,dx=\ln(n+1)$$

We know that $\ln(n+1)$ increases without bound as n increases, hence the same must be true of s_n. Thus,

$$\lim_{n\to\infty}s_n=\infty$$

and the harmonic series diverges.

Perhaps a more intuitive way to see that the harmonic series diverges is by grouping terms into blocks containing 2^n terms, $n=1,2,\ldots$. For example,

$$\begin{aligned}s_{16}&=1+\tfrac{1}{2}+\tfrac{1}{3}+\tfrac{1}{4}+\tfrac{1}{5}+\tfrac{1}{6}+\tfrac{1}{7}+\tfrac{1}{8}+\tfrac{1}{9}+\cdots+\tfrac{1}{16}\\&=1+\tfrac{1}{2}+\underbrace{\tfrac{1}{3}+\tfrac{1}{4}}+\underbrace{\tfrac{1}{5}+\tfrac{1}{6}+\tfrac{1}{7}+\tfrac{1}{8}}+\underbrace{\tfrac{1}{9}+\tfrac{1}{10}+\cdots+\tfrac{1}{16}}\end{aligned}$$

Each bracketed group above has a sum greater than $\frac{1}{2}$, so it follows that

$$s_{16}>1+\tfrac{1}{2}+\tfrac{1}{2}+\tfrac{1}{2}+\tfrac{1}{2}=3$$

Thus if we choose both k and n large enough and set $N = k2^n$, we can make the Nth partial sum

$$s_N > 1 + \frac{k}{2}$$

as large as we please.

We now define certain algebraic operations for series.

Definition (i) The **sum** of the series $\sum_{n=1}^{\infty} a_n$ and $\sum_{n=1}^{\infty} b_n$ is the series $\sum_{n=1}^{\infty} (a_n + b_n)$.

(ii) The **product** of the series $\sum_{n=1}^{\infty} a_n$ by the number c is the series $\sum_{n=1}^{\infty} ca_n$.

(iii) The **Cauchy product** of the series $\sum_{n=1}^{\infty} a_n$ and $\sum_{n=1}^{\infty} b_n$ is the series $\sum_{n=1}^{\infty} c_n$ where

$$c_n = a_1 b_n + a_2 b_{n-1} + \cdots + a_n b_1$$

The method of multiplying series is motivated by the way one multiplies finite sequences. [Consider, for example, the product $(a_1 + a_2 + \cdots + a_m) \cdot (b_1 + b_2 + \cdots + b_m) = (a_1 b_1) + (a_1 b_2 + a_2 b_1) + (a_1 b_3 + a_2 b_2 + a_3 b_1) + \cdots$.]

Example 4 The sum of the series

$$\sum_{n=1}^{\infty} \frac{1}{n} = 1 + \frac{1}{2} + \frac{1}{3} + \frac{1}{4} + \cdots$$

and the series

$$\sum_{n=1}^{\infty} \frac{1}{2^n} = 1 + \frac{1}{2} + \frac{1}{4} + \frac{1}{8} + \cdots$$

is the new series

$$(1 + 1) + (\tfrac{1}{2} + \tfrac{1}{2}) + (\tfrac{1}{3} + \tfrac{1}{4}) + (\tfrac{1}{4} + \tfrac{1}{8}) + \cdots = 2 + 1 + \tfrac{7}{12} + \tfrac{3}{8} + \cdots$$

If the series

$$\sum_{n=1}^{\infty} (2n - 1) = 1 + 3 + 5 + 7 + 9 + \cdots$$

is multiplied by 2 we get the new series

$$\sum_{n=1}^{\infty} 2n = 2 + 6 + 10 + 14 + 18 + \cdots$$

The Cauchy product of the series $1 + \frac{1}{2} + \frac{1}{3} + \frac{1}{4} + \cdots$ and the series $-\frac{1}{2} - \frac{1}{3} - \frac{1}{4} - \frac{1}{5} \cdots$ is the new series

$$-\tfrac{1}{2} + (-\tfrac{1}{3} - \tfrac{1}{2}\tfrac{1}{2}) + (-\tfrac{1}{4} - \tfrac{1}{2}\tfrac{1}{3} - \tfrac{1}{3}\tfrac{1}{2}) + \cdots = -\tfrac{1}{2} - \tfrac{7}{12} - \tfrac{7}{12} - \tfrac{101}{180} - \cdots$$

EXERCISES

1. Determine whether or not each of the following infinite series converges, and find its sum if it does.

   (a) $5 + \frac{5}{7} + \cdots + \frac{5}{7^{n-1}} + \cdots$

   (b) $\frac{a}{5} + \frac{a}{25} + \frac{a}{125} + \cdots + \frac{a}{5^n} + \cdots$

   (c) $\sin\left(\frac{\pi}{2}\right) + \sin\left(\frac{2\pi}{2}\right) + \cdots + \sin\left(\frac{n\pi}{2}\right) + \cdots$

   (d) $(1-\frac{1}{2}) + (\frac{1}{2}-\frac{1}{3}) + (\frac{1}{3}-\frac{1}{4}) + \cdots + \left(\frac{1}{n} - \frac{1}{n+1}\right) + \cdots$

   (e) $\ln\frac{1}{2} + \ln\frac{2}{3} + \ln\frac{3}{4} + \cdots + \ln\frac{n}{n+1} + \cdots$

   (f) $\frac{1}{\sqrt{1+100}} + \frac{2}{\sqrt{1+400}} + \cdots + \frac{n}{\sqrt{1+100n^2}} + \cdots$

2. Find the sum of each of the following convergent series.

   (a) $0.612612612\ldots$

   [Hint: Consider $1000s_n - s_n$.]

   (b) $\sum_{n=1}^{\infty} \left(\frac{1}{2^n} - \frac{1}{3^n}\right)$

   (c) $5 - \frac{5}{2} + \frac{5}{4} - \frac{5}{8} + \cdots + \frac{(-1)^{n-1}5}{2^{n-1}} + \cdots$

   [Hint: Consider $s_n + \frac{1}{2}s_n$.]

   (d) $\sum_{n=1}^{\infty} \frac{1}{(4n-3)(4n+1)}$

3. Show that each of the following series diverges.

   (a) $\sum_{n=1}^{\infty} \frac{2n}{3n+5}$

   (b) $\sum_{n=1}^{\infty} \frac{3n^2+72n-3}{2n^2+n-1}$

   (c) $\frac{1}{3} + \frac{2}{5} + \frac{3}{7} + \frac{4}{9} + \cdots$

   (d) $\frac{3}{2} + \frac{9}{4} + \frac{27}{8} + \frac{81}{16} + \cdots$

   (e) $\sum_{n=1}^{\infty} \frac{1}{\sqrt{n}+\sqrt{n-1}}$

4. Prove that if $\sum_{n=1}^{\infty} a_n$ and $\sum_{n=1}^{\infty} b_n$ are convergent series with $\sum_{n=1}^{\infty} a_n = \alpha$, $\sum_{n=1}^{\infty} b_n = \beta$ and if c is a real number, then the series $\sum_{n=1}^{\infty} (a_n + b_n)$ and $\sum_{n=1}^{\infty} c\,a_n$ are also convergent and

   (a) $\sum_{n=1}^{\infty} (a_n + b_n) = \alpha + \beta$

   (b) $\sum_{n=1}^{\infty} c\,a_n = c\alpha$

5. Using the results of exercise 4, show that if $\sum_{n=1}^{\infty} a_n$ converges and $\sum_{n=1}^{\infty} b_n$ diverges, then $\sum_{n=1}^{\infty} (a_n + b_n)$ diverges.
6. Show that if $\sum_{n=1}^{\infty} a_n$ converges, then any series obtained from this one by deleting a finite number of terms also converges. [Hint: Observe that s_n, the nth partial sum of $\sum_{n=1}^{\infty} a_n$, converges to a limit. Find the value to which s_n', the nth partial sum of the deleted series, must tend.]

2 TESTS FOR CONVERGENCE

We would like to be able to determine whether or not a given series converges, but it is often difficult to do this directly from the definition of convergence, especially, when we do not have a simple expression for the sum of the first n terms of a series as a function of n. In this section, we shall establish several tests for deciding whether or not a given series converges.

So far we have considered two basic types of series, those in which all terms of the series are positive numbers, and those in which the terms are alternately positive and negative. The tests which we shall study here are applicable to the first type, **positive series**.

Recall that a series converges if the sequence of partial sums $\{s_n\}$ converges, that is, if

$$\lim_{n \to \infty} s_n = S$$

exists and is finite. We have seen that a series diverges if $\lim_{n \to \infty} s_n = \pm\infty$. However, we have also encountered cases where a series diverges, but which has a bounded sequence of partial sums. For example, the series $1-1+1-1+1-\cdots$ has as its sequence of partial sums the oscillating sequence $1, 0, 1, 0, 1, \ldots$ which is divergent yet bounded since it does not approach $\pm\infty$.

We assert that for series with nonnegative terms, there are only two alternatives.

Theorem 1 Every series $\sum_{n=1}^{\infty} a_n$, with $a_n \geqslant 0$ for all n, either converges or $\lim_{n \to \infty} s_n = \infty$, where $\{s_n\}$ is the sequence of partial sums.

Proof Consider the partial sums

$$\begin{aligned} s_1 &= a_1 \\ s_2 &= a_1 + a_2 \\ s_3 &= a_1 + a_2 + a_3 \\ &\vdots \\ s_n &= a_1 + a_2 + a_3 + \cdots + a_n \\ s_{n+1} &= s_n + a_{n+1} \geqslant s_n \end{aligned}$$

since $a_n \geqslant 0$. Therefore

$$s_1 \leqslant s_2 \leqslant s_3 \leqslant \cdots \leqslant s_n \leqslant s_{n+1}$$

and it follows that $\{s_n\}$ is a monotone increasing sequence. Either this sequence is bounded or it is not bounded.

If $\{s_n\}$ is bounded, then by the monotone convergence theorem of §7.1, it has a limit S,

$$S = \lim_{n\to\infty} s_n = \sum_{n=1}^{\infty} a_n$$

and so the series is convergent.

If the sequence is not bounded, then for every real number B, there exists an integer N such that $s_N > B$. Since the sequence is increasing, $s_n \geqslant s_N > B$ for every $n > N$. Hence

$$\lim_{n\to\infty} s_n = \infty$$

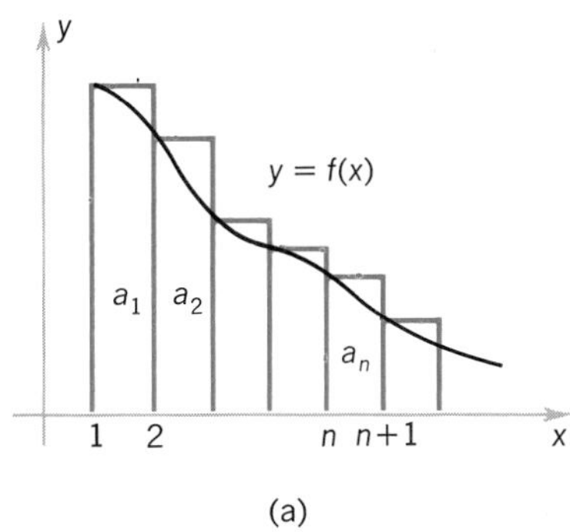

(a)

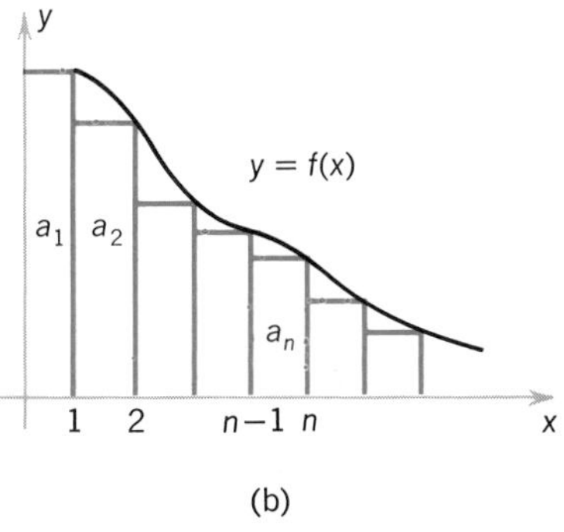

(b)

FIGURE 2-1

Therefore, we see that in the case of a series with nonnegative terms, divergence by oscillation is ruled out. A positive series either converges or becomes infinite.

The following test, called the **integral test**, is a generalization of the method that was used to prove the divergence of the harmonic series.

Theorem 2 The Integral Test If f is continuous, nonnegative, and monotone decreasing on $[1, +\infty[$, then $\sum_{n=1}^{\infty} f(n)$ and $\int_1^{\infty} f(x)\,dx$ converge or diverge together.

Proof In Figure 2-1(a) the rectangles of areas $a_1, a_2, \ldots, a_n$ enclose more area than that under the curve from $x = 1$ to $x = n+1$. Therefore, we have

$$a_1 + a_2 + \cdots + a_n \geqslant \int_1^{n+1} f(x)\,dx$$

If we now consider Figure 2-1(b) and take the area from $x = 1$ to $x = n$ we have

$$a_2 + a_3 + \cdots + a_n \leqslant \int_1^n f(x)\,dx$$

Adding a_1 to both sides

$$a_1 + a_2 + a_3 + \cdots + a_n \leqslant a_1 + \int_1^n f(x)\,dx$$

Combining our two results, yields

$$\int_1^{n+1} f(x)\,dx \leqslant a_1 + a_2 + \cdots + a_n \leqslant a_1 + \int_1^n f(x)\,dx$$

If the integral $\int_1^{\infty} f(x)\,dx$ is finite, then the right-hand inequality implies that the series $\sum_{n=1}^{\infty} a_n$ is also finite. But if $\int_1^{\infty} f(x)\,dx$ is infinite, the left-hand inequality shows that the series is also infinite. Hence, the series and the integral converge or diverge together.

Example 1 Let us determine whether or not $\sum_{n=1}^{\infty} n/(n^2+1)$ converges. Observe that for $x \geqslant 1$, $f(x) = x/(x^2+1)$ is positive and continuous. Since $f'(x) = (-x^2+1)/(x^2+1)^2 \leqslant 0$, f is monotone decreasing.

$$\int_1^{\infty} \frac{x\,dx}{x^2+1} = \lim_{b\to\infty} \int_1^b \frac{x\,dx}{x^2+1}$$

$$= \lim_{b\to\infty} [\tfrac{1}{2}\ln(x^2+1)]_1^b$$

$$= \lim_{b\to\infty} \tfrac{1}{2}\ln((b^2+1)/2)$$

But, as $b \to \infty$, $\frac{1}{2}\ln((b^2+1)/2) \to \infty$, and so the series diverges by the integral test.

We can use the integral test to determine the convergence or divergence of series of the form

$$\sum_{n=1}^{\infty} \frac{1}{n^p}$$

where p is a positive real number. This type of series is called a ***p*-series.**

Example 2 For which values of p does the p-series $\sum_{n=1}^{\infty} 1/n^p$ converge, and for which values of p does it diverge?

Observe that $1/x^p > 0$ on $[1, \infty[$ and the function is monotone decreasing on that interval since $p > 0$. For $p \neq 1$, we have

$$\int \frac{dx}{x^p} = \frac{x^{-p+1}}{-p+1}$$

If $p > 1$ then

$$\int_1^{\infty} \frac{dx}{x^p} = \lim_{b\to\infty} \left.\frac{x^{-p+1}}{-p+1}\right|_1^b = \frac{1}{1-p} \lim_{b\to\infty} \left(\frac{1}{b^{p-1}} - 1\right) = \frac{1}{p-1}$$

which is finite. Thus, **for $p > 1$, the p-series converges.** If $p < 1$, then

$$\int_1^{\infty} \frac{dx}{x^p} = \frac{1}{1-p} \lim_{b\to\infty} \left. x^{1-p} \right|_1^b = \frac{1}{1-p} \lim_{b\to\infty} (b^{1-p} - 1) = \infty$$

Hence, **for $p < 1$ the p-series diverges.** Finally, consider $p = 1$:

$$\int_1^{\infty} \frac{dx}{x} = \lim_{b\to\infty} \ln|x| \Big|_1^b = \lim_{b\to\infty} \ln b = \infty$$

Hence, **for $p = 1$ the p-series also diverges.** This last case is the harmonic series (Example 1.3).

The theorems above only tell us whether or not a given series converges; it does not show us how to find the sum S when it exists. There are a number of tests for convergence which do not deal with the question of finding the sum but are concerned only with determining its existence. The **comparison test** consists of comparing the given series term by term with a series that is known to converge.

Theorem 3 The Comparison Test If there is an $N \in \boldsymbol{N}$ such that $n \geqslant N$ implies $0 \leqslant a_n \leqslant b_n$ and $\sum_{n=1}^{\infty} b_n$ converges, then $\sum_{n=1}^{\infty} a_n$ converges.

Proof Recall that the convergence of a series does not depend on a finite number of terms at the beginning. Hence, we may suppose $\sum_{n=1}^{\infty} a_n$ and $\sum_{n=1}^{\infty} b_n$ are such that $a_n \leqslant b_n$ for $n \geqslant 1$ and suppose that the latter series is convergent with sum $T = \sum_{n=1}^{\infty} b_n$.

Now the sequence of partial sums $\{s_n\}$ of the series $\sum_{n=1}^{\infty} a_n$ is a monotone increasing sequence. Since $a_n \leqslant b_n$, we have that $s_n \leqslant t_n$, where t_n is the nth partial sum of the series $\sum_{n=1}^{\infty} b_n$. Also, since the terms of $\sum_{n=1}^{\infty} b_n$ are positive, $\{t_n\}$ is monotone increasing, and it follows that $t_n \leqslant T$. Thus $s_n \leqslant t_n \leqslant T$, and the monotone increasing sequence $\{s_n\}$ is bounded, and hence has a limit. ∎

Observe that in applying Theorem 2 to test for convergence of a given series $\sum_{n=1}^{\infty} a_n$ we need only find another series $\sum_{n=1}^{\infty} b_n$ in which $b_n \geqslant a_n$ after a finite number of terms.

Example 3 Let us determine whether the series $\sum_{n=1}^{\infty} (n-2)/n^3$ is convergent. Observe that the first term of this series is negative; in order to use the comparison test we must have a positive series, so let us consider the series for $n > 2$. We know that for $n > 2$, $(n-2)/n^3 < n/n^3$, hence $(n-2)/n^3 < 1/n^2$ for $n > 2$. But by Example 2, $\sum_{n=1}^{\infty} 1/n^2$ converges, hence $\sum_{n=1}^{\infty} (n-2)/n^3$ converges.

We should note that there is a helpful corollary to the above theorem.

Corollary 1 If for $n \geqslant N$, $0 \leqslant b_n \leqslant a_n$ and the series $\sum_{n=1}^{\infty} b_n$ diverges, then $\sum_{n=1}^{\infty} a_n$ diverges.

The proof of the corollary is left as an exercise.

Observe that in the case where $\sum_{n=1}^{\infty} b_n$ converges and $a_n \geqslant b_n \geqslant 0$, the comparison test does not yield any information about $\sum_{n=1}^{\infty} a_n$; it may or may not converge.

Example 4 Consider the series $\sum_{n=1}^{\infty} 2/(n+1)$. We would like to kno whether or not it converges. Let us compare this series to the harmoni series $\sum_{n=1}^{\infty} 1/n$. Observe that $2/(n+1) \geqslant 1/n$ since $2n/(n+1) \geqslant 1$ fo $n \geqslant 1$. Since we know that the harmonic series diverges, we conclud that the series

$$\sum_{n=1}^{\infty} \frac{2}{n+1} = 1 + \frac{2}{3} + \frac{1}{2} + \frac{2}{5} + \cdots$$

diverges, also.

The **ratio test** is another useful criterion for convergence whic concerns only the terms of the given series.

Theorem 4 **The Ratio Test** If $0 < a_n$ and

$$r = \lim_{n \to \infty} \frac{a_{n+1}}{a_n}$$

exists, then

(i) $\sum_{n=1}^{\infty} a_n$ converges if $r < 1$.

(ii) $\sum_{n=1}^{\infty} a_n$ diverges if $r > 1$.

(iii) $\sum_{n=1}^{\infty} a_n$ may converge or diverge if $r = 1$.

Proof (i) Suppose $r < 1$. Let t be a number between r and 1, for example $t = \frac{1}{2}(1+r)$. Then the number ε defined by the equation $\varepsilon = t - r$ i positive. Since

$$r = \lim_{n \to \infty} \frac{a_{n+1}}{a_n}$$

we know that a_{n+1}/a_n is eventually in $]r-\varepsilon, r+\varepsilon[$ as $n \to \infty$. That is there is an N such that

$$r - \varepsilon < \frac{a_{n+1}}{a_n} < r + \varepsilon = t$$

for all $n \geqslant N$. Therefore, we have

$$a_{N+1} < ta_N$$

$$a_{N+2} < ta_{N+1} < t^2 a_N$$

$$\vdots$$

$$a_{N+m} < ta_{N+m-1} < t^m a_N$$

Thus, for $m \geqslant 1$, $a_{N+m} < a_N t^m$. Compare the given series with the series $\sum_{n=1}^{\infty} b_n$ where

$$a_1 = b_1, a_2 = b_2, \ldots, a_N = b_N, a_{N+m} = a_N t^m, \qquad m = 1, 2, \ldots$$

Since $|t| < 1$, the geometric series $1 + t + t^2 + \cdots$ converges, therefore $\sum_{n=1}^{\infty} b_n = \sum_{n=1}^{N} a_n + a_N \sum_{m=1}^{\infty} t^m$ and by the comparison test, the given series converges.

(ii) Suppose $r > 1$. Then, for $n \geqslant m$ we have

$$\frac{a_{n+1}}{a_n} \geqslant 1$$

or

$$a_{m+1} \geqslant a_m, \qquad a_{m+2} \geqslant a_{m+1} \geqslant a_m \cdots$$

Since $a_m > 0$, the sum of the terms

$$a_m + a_{m+1} + a_{m+2} + \cdots + a_{m+q} \geqslant a_m(1 + 1 + \cdots + 1) = (q+1)\, a_m$$

which tends to ∞ as $q \to \infty$. Hence, the series

$$a_{m+1} + a_{m+2} + \cdots = \sum_{q=1}^{\infty} a_{m+q}$$

has unbounded partial sums and must diverge. Therefore the series $\sum_{n=1}^{\infty} a_n$ must diverge also.

(iii) We prove this case by displaying two examples of series for which $r = 1$, one which diverges and one which converges. First, consider the harmonic series $\sum_{n=1}^{\infty} 1/n$. We have already seen that it is divergent and it is not difficult to see that $r = 1$. Next, consider the series $\sum_{n=1}^{\infty} 1/n^2$. The reader should verify here again $r = 1$, but this time the series converges. ∎

Example 5 Does the series $\sum_{n=1}^{\infty} 1/(2^n 3^n)$ converge?

Let us look at $\lim_{n\to\infty} a_{n+1}/a_n$. We have

$$\frac{a_{n+1}}{a_n} = \frac{1/2^{n+1}3^{n+1}}{1/2^n 3^n} = \frac{2^n 3^n}{2^{n+1}3^{n+1}} = \frac{1}{2}\frac{1}{3} = \frac{1}{6}$$

and so

$$r = \lim_{n\to\infty} \frac{a_{n+1}}{a_n} = \frac{1}{6}$$

Since $r < 1$ the ratio test guarantees that the series converges. We could, alternatively, observe that our series is a geometric series with ratio $1/6$ and so converges.

Example 6 Consider the series $\sum_{n=1}^{\infty} 2^n/(n^3+1)$. First we form the ratio a_{n+1}/a_n:

$$\frac{a_{n+1}}{a_n} = \frac{\dfrac{2^{n+1}}{(n+1)^3+1}}{\dfrac{2^n}{n^3+1}}$$

$$= \frac{2^{n+1}}{n^3+3n^2+3n+2} \cdot \frac{n^3+1}{2^n}$$

$$= 2\left(\frac{n^3+1}{n^3+3n^2+3n+2}\right)$$

$$= 2\left[\frac{1+\dfrac{1}{n^3}}{1+\dfrac{3}{n}+\dfrac{3}{n^2}+\dfrac{2}{n^3}}\right]$$

and so

$$\lim_{n\to\infty} \frac{a_{n+1}}{a_n} = 2$$

Thus this series diverges, by the ratio test.

EXERCISES

In exercises 1–18, determine if the series converge or diverge.

1. $\sum_{n=1}^{\infty} \frac{1}{7n-2}$

2. $\sum_{n=1}^{\infty} \frac{1}{\sqrt{2n+1}}$

3. $\sum_{n=1}^{\infty} \frac{1}{3^{n-1}+2}$

4. $\sum_{n=1}^{\infty} \frac{n!}{3^n}$

5. $\sum_{n=1}^{\infty} \frac{\sin^2 n}{2^n}$

6. $\sum_{n=1}^{\infty} \frac{3n+1}{n} \cdot \frac{3^n}{4^n-1}$

7. $\sum_{n=1}^{\infty} \frac{1}{\sqrt{k^2+1}}$

8. $\sum_{n=1}^{\infty} \frac{1}{(n+1)(n+3)}$

9. $\sum_{n=1}^{\infty} \frac{n!}{1\cdot 3\cdot 5\cdots(2n-1)}$

10. $\sum_{n=3}^{\infty} \frac{1}{n\ln n}$

11. $\sum_{n=1}^{\infty} \frac{2^n+n}{3^n-n}$

12. $\sum_{n=1}^{\infty} \frac{n+1}{\ln(n+2)}$

13. $\sum_{n=1}^{\infty} \frac{n}{e^n}$

14. $\sum_{n=1}^{\infty} \frac{(n+1)2^n}{n!}$

15. $\sum_{n=1}^{\infty} \frac{n+1}{n\sqrt{3n-2}}$

16. $\sum_{n=1}^{\infty} \frac{n^3}{(\ln 2^2)^n}$

17. $\sum_{n=1}^{\infty} \frac{\ln n}{\sqrt{n}}$

18. $\sum_{n=1}^{\infty} \frac{n^n}{n!}$

19. Prove that if $\sum_{n=1}^{\infty} b_n$ is a positive series, and if $\lim_{n\to\infty} a_n/b_n = 1$, then the series $\sum_{n=1}^{\infty} a_n$ converges if and only if $\sum_{n=1}^{\infty} b_n$ does.

20. Use the same method as in the proof of the integral test to show that if $p > 1$ then

$$\frac{1}{(p-1)(m+1)^{p-1}} < \sum_{n=1}^{\infty} \frac{1}{n^p} - \sum_{n=1}^{m} \frac{1}{n^p} < \frac{1}{(p-1)m^{p-1}}$$

21. Show that the ratio test yields known results in case it is applied to a geometric sequence.

3 ABSOLUTE CONVERGENCE

All the convergence tests we have discussed so far apply to positive series; to use them for negative series we can simply factor out a minus sign. However, we would also like to be able to determine the convergence of **alternating series** such as

$$\sum_{n=1}^{\infty} (-1)^{n-1}\frac{1}{n} = 1 - \frac{1}{2} + \frac{1}{3} - \frac{1}{4} + \frac{1}{5} - \cdots$$

To do this we introduce an important concept.

Definition An infinite series $\sum_{n=1}^{\infty} a_n$ is said to be **absolutely convergent** if the corresponding series of absolute values, $\sum_{n=1}^{\infty} |a_n|$, is convergent. If a series $\sum_{n=1}^{\infty} a_n$ converges, but $\sum_{n=1}^{\infty} |a_n|$ does not, then we say that $\sum_{n=1}^{\infty} a_n$ is **conditionally convergent**.

We have the following theorem.

Theorem 1 If the infinite series $\sum_{n=1}^{\infty} a_n$ is absolutely convergent, then it is convergent.

Proof Let the nth partial sum of $\sum_{n=1}^{\infty} a_n$ be denoted as usual by s_n, and let the nth partial sum of $\sum_{n=1}^{\infty} |a_n|$ be denoted by p_n. By hypothesis,

$$\lim_{n\to\infty} p_n = p$$

exists. We want to prove that $\lim_{n\to\infty} s_n$ also exists. Consider the sum

$$q_n = (a_1 + |a_1|) + (a_2 + |a_2|) + \cdots + (a_n + |a_n|) = s_n + p_n$$

Each term of the series $\sum_{k=1}^{n} (a_k + |a_k|)$ is positive or zero, so the sequence of numbers $\{q_n\}$ is a monotone increasing sequence. But q_n is less than or equal to

$$(|a_1| + |a_1|) + (|a_2| + |a_2|) + \cdots + (|a_n| + |a_n|)$$

that is,

$$q_n \leqslant 2p_n$$

Since $\{p_n\}$ is a monotone increasing sequence whose limit is p, it follows that

$$p_n \leqslant p \qquad \text{for all } n$$

Therefore, $q_n \leqslant 2p$, and since $\{q_n\}$ is thus a bounded monotonic sequence it converges to a limit. We set $q = \lim_{n\to\infty} q_n$. Thus, $s_n = q_n - p_n$ also has a limit as $n \to \infty$:

$$\lim_{n\to\infty} s_n = \lim_{n\to\infty} q_n - \lim_{n\to\infty} p_n = q - p$$

We next observe that to determine the convergence or divergence of a series we can apply the ratio test to the series of absolute values. Hence, for

$$r = \lim_{n\to\infty} \left|\frac{a_{n+1}}{a_n}\right|, \qquad a_n \neq 0$$

the series converges absolutely for $r < 1$, diverges for $r > 1$, and may do either for $r = 1$.

Example 1 Determine all values of x for which $\sum_{n=1}^{\infty} x^n/(n!)$ converges absolutely.

We have

$$a_{n+1} = \frac{x^{n+1}}{(n+1)!}, \qquad a_n = \frac{x^n}{n!}$$

Hence,

$$\left|\frac{a_{n+1}}{a_n}\right| = \left|\frac{\dfrac{x^{n+1}}{(n+1)!}}{\dfrac{x^n}{n!}}\right| = \left|\frac{x^{n+1}}{(n+1)!} \cdot \frac{n!}{x^n}\right|$$

$$= \left|\frac{x}{n+1}\right| = \frac{|x|}{n+1}$$

Therefore,

$$\lim_{n\to\infty} \left|\frac{a_{n+1}}{a_n}\right| = \lim_{n\to\infty} \frac{|x|}{n+1} = |x| \lim_{n\to\infty} \frac{1}{n+1} = 0$$

Hence, this series converges absolutely for all $-\infty < x < +\infty$ and so converges for all $-\infty < x < +\infty$.

Since a series that converges absolutely also converges, we can apply tests for positive series to the series of absolute values of the terms of an alternating series. If the alternating series is absolutely convergent, we can conclude that it is convergent. But if the series of absolute values diverges, we can conclude nothing about the convergence of the original series. Fortunately, we have the following criterion for testing the convergence of an alternating series.

Theorem 2 **Alternating Series Test** If the following conditions are all satisfied, then the series converges.

(i) The series $\sum_{n=1}^{\infty} a_n$ is **strictly alternating**, that is, it is of the form

$$a_1 - a_2 + a_3 - a_4 + \cdots + (-1)^{n-1}a_n + \cdots, \text{ where } a_i > 0$$

(ii) $\lim_{n\to\infty} a_n = 0$

(iii) $a_n \geqslant a_{n+1} \geqslant 0$, for all n

Proof If all the above conditions are satisfied, then the partial sums of $\sum_{n=1}^{\infty}(-1)^{n-1}a_n$ which have an even number of terms may be written as follows:

$$s_2 = a_1 - a_2, \qquad s_4 = (a_1 - a_2) + (a_3 - a_4)$$

and, in general,

$$s_{2n} = (a_1 - a_2) + (a_3 - a_4) + \cdots + (a_{2n-1} - a_{2n})$$

Since $a_n \geqslant a_{n+1}$, it follows that

$$a_1 - a_2 \geqslant 0, \qquad a_3 - a_4 \geqslant 0$$

and, in general,

$$a_{2n-1} - a_{2n} \geqslant 0$$

Therefore s_{2n} is a sequence of nonnegative terms and, clearly,

$$0 \leqslant s_2 \leqslant s_4 \leqslant \cdots \leqslant s_{2n} \leqslant \cdots$$

Alternatively, s_{2n} may be written in the form

$$s_{2n} = a_1 - (a_2 - a_3) - (a_4 - a_5) - \cdots - (a_{2n-2} - a_{2n-1}) - a_{2n}$$

Each expression in parentheses is positive or zero, from which we conclude that for all n

$$s_{2n} \leqslant a_1$$

Since $\{s_{2n}\}$ is monotonically increasing, and we have just shown that it is bounded, by the Monotone Convergence Theorem $\lim_{n\to\infty} s_{2n}$ exists. Since $s_{2n+1} = s_{2n} + a_{2n+1}$,

$$\lim_{n\to\infty} s_{2n+1} = \lim_{n\to\infty} s_{2n} + \lim_{n\to\infty} a_{2n+1}$$

By hypothesis (ii) $\lim_{n\to\infty} a_n = 0$, so

$$\lim_{n\to\infty} s_{2n+1} = \lim_{n\to\infty} s_{2n}$$

Thus,

$$\lim_{n\to\infty} s_n = \lim_{n\to\infty} s_{2n} = \lim_{n\to\infty} s_{2n+1} = S \quad ■$$

Example 2 Show that the alternating series

$$1-\frac{1}{1}+\frac{1}{2}-\frac{1}{3}+\cdots+\frac{(-1)^n}{n}+\cdots$$

is convergent.

Since

$$a_n=\frac{1}{n}, \qquad a_{n+1}=\frac{1}{(n+1)}$$

we see that $a_n > a_{n+1}$. Furthermore, $\lim_{n\to\infty} 1/n = 0$, so Theorem 2 guarantees that the series converges. Notice that this series is not absolutely convergent.

We must be certain that all the conditions in the theorem are satisfied before we can conclude that a given series converges.

Example 3 Determine whether or not

$$1-\frac{2}{3}+\frac{3}{5}-\cdots+\frac{(-1)^{n-1}n}{2n-1}+\cdots$$

converges.

The series diverges since $\lim_{n\to\infty} a_n \neq 0$, even though the other hypotheses of the alternating series test are satisfied:

$$a_n=\frac{n}{2n-1}>\frac{n+1}{2n+1}=a_{n+1}$$

and the series is strictly alternating.

EXERCISES

1. Determine whether each of the following alternating series converges or diverges.

(a) $\sum_{n=1}^{\infty}(-1)^n\frac{1}{\sqrt{n}}$

(b) $\sum_{n=1}^{\infty}\frac{(n+1)(-1)^{n-1}}{3n}$

(c) $\sum_{n=2}^{\infty}\frac{(-1)^{n-1}}{\ln n}$

(d) $\sum_{n=1}^{\infty}\frac{\sqrt{n}(-1)^{n-1}}{2n+1}$

(e) $\sum_{n=1}^{\infty}(-1)^{n-1}\frac{n+1}{n}$

(f) $\sum_{n=2}^{\infty}(-1)^n e^{-n}$

(g) $\sum_{n=1}^{\infty}(-1)^{n-1}\frac{1}{\sqrt[n]{3}}$

(h) $\sum_{k=1}^{\infty}(-1)^{n-1}\frac{n!}{10^n}$

(i) $\sum_{n=1}^{\infty}(-1)^{n-1}\frac{\ln^n n}{3n+2}$

2. Classify each of the following infinite series as absolutely convergent, conditionally convergent, or divergent.

(a) $1-\frac{1}{3!}+\frac{1}{5!}-\frac{1}{7!}+\cdots$ (b) $\sum_{n=1}^{\infty}\frac{n!}{10^n}$

(c) $\sum_{n=1}^{\infty}(-1)^n\frac{1}{(n+1)^{2/3}}$ (d) $1-\frac{1}{\sqrt{2}}+\frac{1}{\sqrt{3}}-\frac{1}{\sqrt{4}}+\cdots$

(e) $\sum_{n=1}^{\infty}\frac{2^{2n-1}}{(2n-1)!}$ (f) $\sum_{n=1}^{\infty}\frac{(-1)^{n-1}(\frac{4}{3})^n}{n^2}$

(g) $\sum_{n=1}^{\infty}(-1)^n e^{-n^2}$ (h) $\sum_{k=1}^{\infty}\frac{(1+4k)(-1)^k}{7k^2-1}$

(i) $\sum_{n=1}^{\infty}\frac{(-5)^{n-1}}{n\cdot n!}$ (j) $\sum_{n=1}^{\infty}(-1)^{n+1}\frac{\ln(n+1)}{n+1}$

(k) $\sum_{n=1}^{\infty}\frac{(-1)^{n-1}}{\sqrt{n(n+1)}}$ (l) $\sum_{n=1}^{\infty}\frac{(-1)^{n-1}}{(n!)^3}$

3. Prove that the series $\sum_{n=1}^{\infty}a_n$ diverges if there exists a number N such that $|a_{n+1}/a_n|>1$ for every $n>N$.

4. Prove the following extension of the comparison test: If $\sum_{n=1}^{\infty}a_n$ is absolutely convergent, and if $|b_n|\leqslant|a_n|$ for every $n>N$, then $\sum_{n=1}^{\infty}b_n$ is absolutely convergent.

5. Prove the extended ratio test, that is, if $r=\lim_{n\to\infty}|a_{n+1}/a_n|$, then the series $\sum_{n=1}^{\infty}a_n$
   (i) converges absolutely for $r<1$,
   (ii) diverges for $r>1$,
   (iii) may do either for $r=1$.

*6. Suppose $\sum_{n=1}^{\infty}a_n=A$ and $\sum_{n=1}^{\infty}b_n=B$ converge absolutely. Show that the Cauchy product converges to AB.

4 POWER SERIES

Although polynomials constitute a wide class of functions, we have already seen that not all functions are polynomials (of finite degree). Using series we can represent certain functions as "polynomials of infinite degree," called **power series**. We may then ask the question, are all functions polynomials of either finite or infinite degree? The answer is still no, but a very wide class of functions can be defined in terms of power series, making this one of the most useful mathematical techniques. In this section we investigate the properties of power series and functions defined by power series; in the following section we establish a criterion for deciding when a function can be represented by a converging power series on some interval of real numbers.

Definition The general form of a **power series** is

$$\sum_{n=0}^{\infty}a_n x^n=a_0+a_1x+a_2x^2+a_3x^3+\cdots$$

where a_n is a real constant and x is a real variable. The numbers $a_0, a_1, a_2, \ldots$ are called the **coefficients** of the power series.

Notice that if we assign a particular value to x, we obtain an infinite series of numbers of the type we have already considered. The convergence of a power series is determined by the convergence of these infinite series of numbers. Ordinarily, the convergence or divergence of this series of constants depends on the value of x; hence, if a power series converges for $x = c$, we say that the **power series converges at** c.

Naturally, we shall be most interested in finding the values of x for which a given power series converges. Observe that every power series converges at 0, since

$$a_0 + a_1 \cdot 0 + a_2 \cdot 0^2 + a_3 \cdot 0^3 + \cdots = a_0$$

Theorem 1 If the power series $\sum_{n=0}^{\infty} a_n x^n$ converges for some real number c, then it converges absolutely for every real number x such that $|x| < |c|$.

Proof We can assume $c \neq 0$. Since the series $\sum_{n=0}^{\infty} a_n c^n$ converges, we know

$$\lim_{n \to \infty} a_n c^n = 0$$

Thus, there exists an $N \in N$ such that

$$|a_n c^n| \leqslant 1, \qquad n \geqslant N$$

Since

$$|a_n c^n| = |a_n|\,|c|^n$$

this means that

$$|a_n| \leqslant \frac{1}{|c|^n}, \qquad n \geqslant N$$

Thus, for $n \geqslant N$

$$|a_n|\,|x|^n \leqslant \frac{|x|^n}{|c|^n}$$

Now suppose that $|x| < |c|$. Let $r = |x|/|c|$; then $r < 1$ and

$$|a_n x^n| \leqslant r^n, \qquad n \geqslant N$$

Since the geometric series $\sum_{n=0}^{\infty} r^n$ converges for $r < 1$, it follows by the comparison test that $\sum_{n=0}^{\infty} |a_n x^n|$ also converges. ∎

As a consequence of this theorem, it is easy to show that there are only three possibilities for a power series:

(1) The series converges only at $x = 0$.

(2) The series converges absolutely in an interval symmetrically centered about the origin; an interval containing both, or only one, or neither of its end-points.

(3) The series converges absolutely for all values of x.

We call the set of numbers at which a power series converges the **interval of convergence**. If this interval is bounded and has the symmetrical endpoints $-R$ and R, R is called the **radius of convergence** of the series. In case (3) the interval of convergence is the set of all real numbers, so the radius of convergence is said to be infinite, and we write $R = \infty$.

Observe that, in general, we cannot say whether the series converges or diverges at the endpoints of the interval of convergence. This must be determined independently for each specific case. Figure 4-1 summarizes our results.

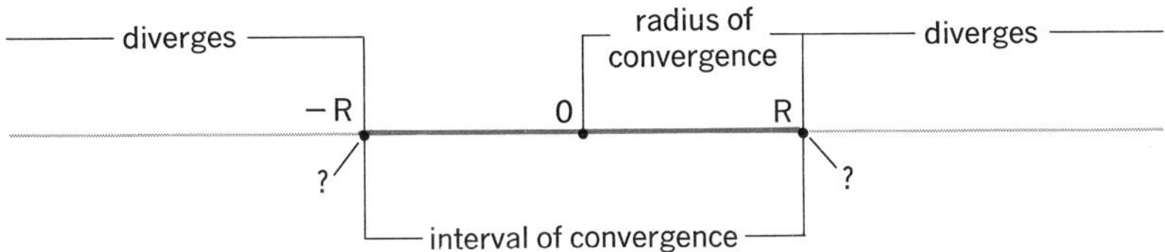

FIGURE 4-1

Example 1 Find the interval of convergence of the series $\sum_{n=1}^{\infty} (1/n)x^n$.

Here, as in many other examples of power series, the interval of convergence can be found easily using the ratio test. If u_n is the nth term of the series, we have

$$u_n = \frac{1}{n}x^n, \qquad u_{n+1} = \frac{1}{n+1}x^{n+1}$$

Then,

$$\left|\frac{u_{n+1}}{u_n}\right| = \frac{|x|^{n+1}}{n+1} \cdot \frac{n}{|x|^n} = |x|\frac{n}{n+1}$$

and

$$\lim_{n\to\infty}\left|\frac{u_{n+1}}{u_n}\right| = \lim_{n\to\infty}|x|\frac{n}{n+1} = |x|\lim_{n\to\infty}\frac{1}{1+(1/n)} = |x|$$

Therefore, the ratio test tells us that the series converges when $|x| < 1$ and diverges for $|x| > 1$. However, for $|x| = 1$, the endpoints of the interval, the ratio test gives no information.

If $x = 1$, the series becomes $\sum_{n=1}^{\infty} 1/n$, the harmonic series, which is divergent. If $x = -1$, the series becomes $\sum_{n=1}^{\infty}(-1)^n/n$, an alternating series which converges by the alternating series test (Theorem 3.1).

Combining results, we see that the interval of convergence is $[-1, 1[$, a half-open interval.

Example 2 Find the interval of convergence of the series

$$\sum_{n=0}^{\infty} \frac{(-1)^n x^n}{n!}$$

Again we apply the ratio test

$$u_n = \frac{(-1)^n x^n}{n!}, \qquad u_{n+1} = \frac{(-1)^{n+1} x^{n+1}}{(n+1)!}$$

and

$$\frac{|u_{n+1}|}{|u_n|} = \frac{|x|^{n+1}}{(n+1)!} \cdot \frac{n!}{|x|^n} = |x| \frac{1}{n+1}$$

Hence,

$$\lim_{n\to\infty} \frac{|u_{n+1}|}{|u_n|} = |x| \lim_{n\to\infty} \frac{1}{n+1} = 0$$

Thus, the series converges for all values of x and the interval of convergence is $]-\infty, +\infty[$.

Next, we generalize the definition of power series.

Definition A **power series in $x-a$** is a series of the form

$$\sum_{n=0}^{\infty} a_n (x-a)^n = a_0 + a_1 (x-a) + a_2 (x-a)^2 + \cdots$$

Observe that the power series we have been studying up to now have been a special case of the more general form, the case where $a = 0$. The proofs which have been given for the series $\sum_{n=0}^{\infty} a_n x^n$ can easily be modified to apply to the power series in $x-a$. We observe that the interval of convergence of $\sum_{n=0}^{\infty} a_n (x-a)^n$ is a translation of the interval of convergence of $\sum_{n=0}^{\infty} a_n y^n$. Thus, the interval of convergence of

$$\sum_{n=0}^{\infty} a_n (x-a)^n$$

where $a > 0$ is centered at a [Figure 4-2(a)], and the interval of convergence of $\sum_{n=0}^{\infty} a_n (x+a)^n$, $a > 0$, is centered at $-a$ [Figure 4-2(b)]. The power series $\sum_{n=0}^{\infty} a_n (x-a)^n$ is frequently called a **power series about a**.

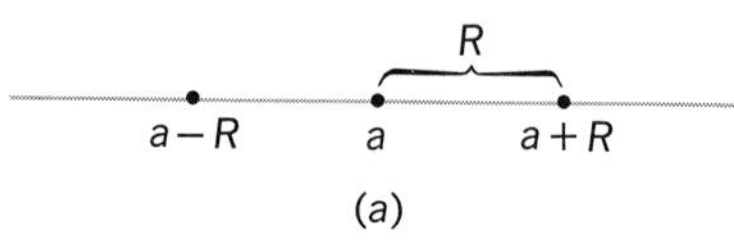

(a)

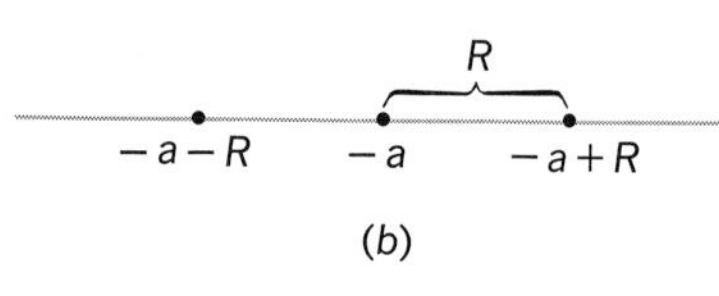

(b)

FIGURE 4-2

Example 3 Find the interval and radius of convergence of the power series $\sum_{n=1}^{\infty} n^2 (x-1)^n$.

We have

$$u_n = n^2 (x-1)^n, \qquad u_{n+1} = (n+1)^2 (x-1)^{n+1}$$

and

$$\left|\frac{u_{n+1}}{u_n}\right| = \frac{(n+1)^2\,|x-1|^{n+1}}{n^2\,|x-1|^n} = \frac{1+(2/n)+(1/n^2)}{1}\,|x-1|$$

Thus,

$$\lim_{n\to\infty}\left|\frac{u_{n+1}}{u_n}\right| = |x-1|\lim_{n\to\infty}\left(1+\frac{2}{n}+\frac{1}{n^2}\right) = |x-1|$$

Hence, the series converges for $|x-1|<1$. Since $|x-1|$ is the distance from any point x to the point 1 we see that the interval of convergence is centered at 1 with $R=1$ (Figure 4-3).

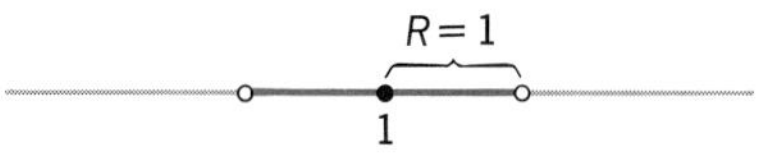

FIGURE 4-3

Next, let us examine the convergence at the endpoints. For $x=0$, the series becomes $\sum_{n=1}^{\infty} n^2(-1)^n$ which is divergent. For $x=2$, the series is $\sum_{n=1}^{\infty} n^2$ which is also divergent. Hence, our interval of convergence is the open interval $]0,2[$.

We now wish to begin our investigation of functions which are defined by power series.

Definition The **function defined by the power series** $\sum_{n=0}^{\infty} a_n(x-a)^n$ is the function $f\colon x\mapsto \sum_{n=0}^{\infty} a_n(x-a)^n$; the domain of the function is the interval of convergence of the power series.

We sometimes write $f(x)=\sum_{n=0}^{\infty} a_n(x-a)^n$. One of the most important properties of functions defined by power series is that every such function which has a positive or infinite radius of convergence has derivatives of all orders, and these derivatives are the functions represented by the power series **obtained by differentiating the original series term-by-term**. For example, if

$$f(x) = \sum_{n=0}^{\infty} a_n(x-a)^n = a_0 + a_1(x-a) + a_2(x-a)^2 + \cdots$$

is convergent when $|x-a|<R$, then

$$f'(x) = \sum_{n=1}^{\infty} na_n(x-a)^{n-1} = a_1 + 2a_2(x-a) + 3a_3(x-a)^2 + \cdots$$

$$f''(x) = \sum_{n=1}^{\infty} n(n-1)\,a_n(x-a)^{n-2} = 2a_2 + 6a_3(x-a) + 12(x-a)^2 + \cdots$$

and so on.

A similar result is true for **term-by-term integration** of power series:

$$\int_a^b f(x)\,dx = \int_a^b \sum_{n=0}^{\infty} a_n(x-a)^n\,dx = \sum_{n=0}^{\infty}\int_a^b a_n(x-a)^n\,dx$$

$$= \sum_{n=0}^{\infty}\frac{a_n}{n+1}(x-a)^{n+1}$$

The results on termwise differentiation and integration of power series are valid for x in the interval of convergence, and are consequences of more general theorems on differentiation and integration of sequences and series of functions (see Theorem 3, §6).

To obtain the result about differentiation from Theorem 3 of §6, we shall need the following theorem, which shows that the derived series $\sum_{n=1}^{\infty} n a_n x^{n-1}$ converges at points inside the interval of convergence of the original series $\sum_{n=0}^{\infty} a_n x^n$.

Theorem 2 Suppose $\sum_{n=0}^{\infty} a_n x^n$ converges for $|x| < R$. Then the derived series $\sum_{n=1}^{\infty} n a_n x^{n-1}$ converges for $|x| < R$.

Proof Let $0 < b < c < R$. It suffices to show that $\sum_{n=1}^{\infty} n a_n b^n$ converges absolutely. Since $\sum_{n=0}^{\infty} a_n c^n$ converges, there is a number $C > 0$ such that $|a_n c^n| < C$, for all n. Therefore,

$$|n a_n b^n| = \left| n a_n c^n \frac{b^n}{c^n} \right| = |a_n c^n| \left| n \frac{b^n}{c^n} \right| < C \left| n \frac{b^n}{c^n} \right|$$

By the ratio test, $\sum_{n=1}^{\infty} n(b^n/c^n)$ converges $(b/c < 1)$, so the power series converges absolutely, by the comparison test. ∎

Note that this theorem does not assert anything about the behavior of the derived series at endpoints of the interval of convergence. It is possible that the derived series will diverge at an endpoint, even though the original series may be convergent at that point.

Assuming the results stated above on termwise differentiation and integration, let us show how they can be used to express familiar functions as power series.

Example 4 First note that for $|x| < 1$, the geometric series $\sum_{n=0}^{\infty} x^n$ converges and

$$\frac{1}{1-x} = \sum_{n=0}^{\infty} x^n$$

Hence,

$$\frac{1}{1+x} = \frac{1}{1-(-x)} = \sum_{n=0}^{\infty} (-1)^n x^n$$

The function $f(x) = -1/(1+x)^2$ is the derivative of $g(x) = 1/(1+x)$. Let us express $f(x)$ as a power series. Term-by-term differentiation of the above power series expansion for $g(x)$, valid for $|x| < 1$, yields

$$\frac{-1}{(1+x)^2} = f(x) = g'(x) = \sum_{n=0}^{\infty} \frac{d((-1)^n x^n)}{dx} = \sum_{n=1}^{\infty} n(-1)^n x^{n-1}$$

Example 5 If we integrate the above power series expansion for $1/(1+x)$ term-by-term, then for $|x|<1$,

$$\sum_{n=0}^{\infty}\left(\int_0^x (-1)^n t^n\,dt\right) = \sum_{n=0}^{\infty}\frac{(-1)^n}{n+1}x^{n+1}$$

But $\int_0^x 1/(1+t)\,dt = \ln(1+x)$, hence we have the power series expansion, valid for $|x|<1$:

$$\ln(1+x) = \sum_{n=0}^{\infty}\frac{(-1)^n}{n+1}x^{n+1} = \sum_{n=1}^{\infty}\frac{(-1)^{n-1}}{n}x^n$$

EXERCISES

1. Find the radius of convergence of each of the following power series.

   (a) $\sum_{n=1}^{\infty}\frac{x^n}{n^2}$ (b) $1+x+\frac{x^2}{\sqrt{2}}+\frac{x^3}{\sqrt{3}}+\cdots$

   (c) $\sum_{n=0}^{\infty}(-1)^n(n+1)x^n$ (d) $\sum_{n=0}^{\infty}2^n x^n$

   (e) $\sum_{n=0}^{\infty}\frac{x^n}{(2n)!}$ (f) $\sum_{n=0}^{\infty}\frac{(3x)^n}{2^{n+1}}$

2. Find the interval of convergence of each of the following power series.

   (a) $\sum_{n=1}^{\infty}(-1)^{n-1}\frac{x^n}{n}$ (b) $\sum_{n=1}^{\infty}(-1)^{n-1}\frac{x^{2n-1}}{2n-1}$

   (c) $\sum_{n=1}^{\infty}(-1)^n n^2 x^n$ (d) $\sum_{n=1}^{\infty}\frac{x^{n+1}}{[\ln(n+1)]^{n+1}}$

3. Find the interval of convergence of each of the following power series.

   (a) $\sum_{n=1}^{\infty}\frac{(x-3)^{n-1}}{(n-1)^2}$ (b) $\sum_{n=0}^{\infty}\frac{n!}{2^n}(x-1)^n$

   (c) $\sum_{n=1}^{\infty}\frac{(x+1)^n}{\sqrt{n}}$ (d) $\sum_{n=1}^{\infty}\frac{(-1)^{n-1}(x+4)^n}{3^n n^2}$

   (e) $\sum_{n=0}^{\infty}\frac{(x-2)^n}{2^n\sqrt{n+1}}$

   (f) $\frac{x-3}{1\cdot 3}+\frac{(x-3)^2}{2\cdot 3^2}+\frac{(x-3)^3}{3\cdot 3^3}+\frac{(x-3)^4}{4\cdot 3^4}+\cdots$

   (g) $\sum_{n=1}^{\infty}(-1)^{n+1}\frac{(n!)^2(x-2)^n}{2^n(2n)!}$ (h) $\sum_{n=0}^{\infty}\frac{(-1)^{n+1}(x+1)^{2n}}{(n+1)^2 5^n}$

4. Let f be the function defined by

$$f(x) = \sum_{n=1}^{\infty}\frac{1}{n2^n}(x-2)^n$$

   (a) Find the domain of f.
   (b) Write down the derived series and find its domain.

5. Find the domains of the functions f defined by

   (a) $$f(x) = \frac{1}{1\cdot 2\cdot 3} + \frac{x^2}{2\cdot 3\cdot 4} + \frac{x^4}{3\cdot 4\cdot 5} + \frac{x^6}{4\cdot 5\cdot 6} + \cdots$$

   (b) $$f(x) = \frac{x}{5} - \frac{x^2}{2\cdot 5^2} + \frac{x^3}{3\cdot 5^3} - \frac{x^4}{4\cdot 5^4} + \cdots$$

6. Show that

$$\ln(1-x) = -x - \frac{x^2}{2} - \frac{x^3}{3} - \frac{x^4}{4} - \cdots, \qquad |x| < 1$$

7. Show that

$$\ln\left(\frac{1+x}{1-x}\right) = 2\left(x + \frac{x^3}{3} + \frac{x^5}{5} + \cdots\right), \qquad |x| < 1$$

8. (a) Find a power series for $1/(1+x^2)$ using Example 4.
   (b) Use integration to find a power series for $\tan^{-1} x$.

5 TAYLOR SERIES

Now that we have investigated functions defined by power series, we would like to know which functions can be represented by convergent power series, and if there is a systematic method for constructing these representations. In this section we find the answer by considering the *Taylor series.*

Recall that if a power series $\sum_{n=0}^{\infty} a_n(x-a)^n$ converges in some interval, the sum of the series has a value for each x in this interval and thus defines a function

$$f(x) = a_0 + a_1(x-a) + a_2(x-a)^2 + \cdots$$

defined for all x in the interval of convergence.

It would be helpful if we could discover a relationship between the coefficients a_n and the function f, for then we could take a function which is known to have a series representation and proceed to construct that power series. To find this relationship, we begin as follows.

Consider $f(x) = a_0 + a_1(x-a) + a_2(x-a)^2 + \cdots$. Evaluating at $x = a$ we get

$$f(a) = a_0 + a_1(0) + a_2(0) + a_3(0) + \cdots$$

so

$$f(a) = a_0$$

Hence, we have found the first coefficient in terms of f. To find a_1, let us use the result on termwise differentiation stated in the preceding section to find $f'(x)$, which we will then evaluate at a (this process is valid when the radius of convergence of the series is greater than 0):

$$f'(x) = a_1 + 2a_2(x-a) + 3a_3(x-a)^2 + \cdots$$

and

$$f'(a) = a_1 + 2a_2(0) + 3a_3(0) + \cdots$$

so

$$f'(a) = a_1$$

Hence, we have found a_1 in terms of f. Using this termwise differentiation repeatedly we have,

$$f''(x) = 2a_2 + 3 \cdot 2a_3(x-a) + 4 \cdot 3a_4(x-a)^2 + 5 \cdot 4a_5(x-a)^3 + \cdots$$

so

$$f''(a) = 2a_2$$

or

$$a_2 = \frac{f''(a)}{2!}$$

Similarly

$$f'''(x) = 3 \cdot 2a_3 + 4 \cdot 3 \cdot 2a_4(x-a) + 5 \cdot 4 \cdot 3a_5(x-a)^2 + 6 \cdot 5 \cdot 4a_6(x-a)^3 + \cdots$$

$$f'''(a) = 3 \cdot 2a_3$$

or

$$a_3 = \frac{f'''(a)}{3 \cdot 2} = \frac{f'''(a)}{3!}$$

Continuing in this manner we find the general formula for the coefficient a_n to be

$$a_n = \frac{f^{(n)}(a)}{n!}$$

Note that this result implies that the power series representing a function is **uniquely determined** by the function (when the radius of convergence of the series is positive, that is, when the domain of f is not just the number a).

Substituting the expression for a_n into our original power series yields

$$f(x) = \sum_{n=0}^{\infty} \frac{f^{(n)}(a)}{n!}(x-a)^n = f(a) + f'(a)(x-a) + \frac{f''(a)(x-a)^2}{2!} + \cdots$$

Definition If f is a function with derivatives of all orders at a, then its **Taylor series** about a is the power series

$$\sum_{n=0}^{\infty} \frac{f^n(a)}{n!}(x-a)^n = f(a) + f'(a)(x-a) + \frac{f''(a)(x-a)^2}{2!} + \cdots$$

If f is any function defined near a which has derivatives of all orders, we can compute the coefficients a_n and form the Taylor series for the function. However, the mere existence of the Taylor series for a given function does not imply that the series converges to the values of the function. Recall that when we found the expressions relating a_n to $f(x)$, we were assuming that the function f was, in fact, defined by a power series. The remainder of this discussion is devoted to determining when a function can be validly represented by its Taylor series. Such a function is said to be **analytic.**

We can approximate the value of a convergent infinite series by its partial sums. For a Taylor series the nth partial sum is a polynomial in $(x-a)$.

Definition If f is a function such that $f^{(n)}(a)$ exists, then the **nth Taylor approximation** to the function f about a is the polynomial T_n given by

$$T_n(x) = f(a) + f'(a)\,(x-a) + \cdots + \frac{1}{n!}f^{(n)}(a)\,(x-a)^n$$

If we compute the ith derivative of the polynomial $T_n(x)$, we find

$$T_n^{(i)}(a) = f^{(i)}(a), \qquad i = 1, \ldots, n$$

That is, the first n derivatives of our original function coincide with the first n derivatives of its nth Taylor approximation. We recall that the fact that an approximating function has certain order derivatives equal to those of the original function is related graphically to the degree of contact of the two loci, that is, as the two functions agree in higher and higher order derivatives, the degree of contact is greater, and the approximating curve approaches the original curve more closely. For example, in Figure 5-1 we sketch the graphs of $f: x \mapsto 1/x$ and the first four Taylor approximations to f at $x = 1$.

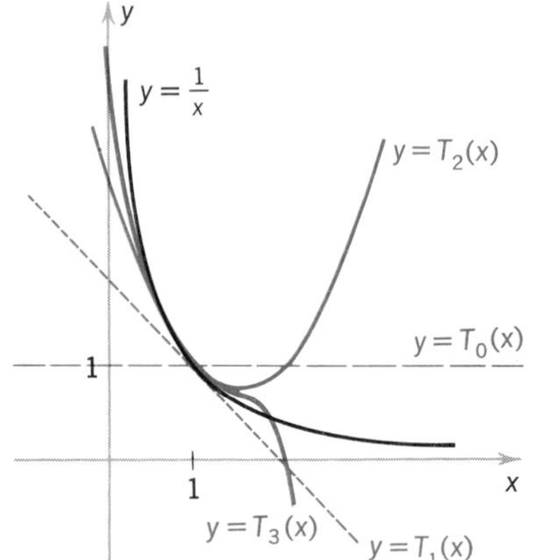

FIGURE 5-1

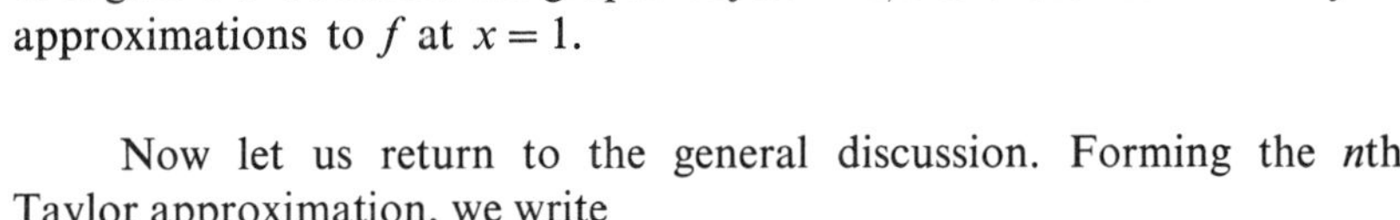

Now let us return to the general discussion. Forming the nth Taylor approximation, we write

$$f(x) = a_0 + a_1(x-a) + a_2(x-a)^2 + \cdots + a_n(x-a)^n + R_n(x,a)$$

or

$$f(x) = T_n(x) + R_n(x,a)$$

where $R_n(x,a)$ is the **nth remainder**. The nature of this remainder is revealed in Taylor's theorem.

Theorem 1 **Taylor's Theorem** Let f be differentiable at least $n+1$ times on an interval containing a and x, where $f^{(n+1)}$ is continuous. Then

$$f(x) = f(a) + f'(a)(x-a) + \frac{f''(a)}{2!}(x-a)^2 + \cdots$$

$$+ \frac{f^n(a)}{n!}(x-a)^n + R_n(x,a)$$

where

$$R_n(x,a) = \int_a^x \frac{(x-t)^n}{n!} f^{(n+1)}(t)\,dt$$

Proof We start the proof with the formula

$$\int_a^b f'(t)\,dt = [f(t)]_a^b = f(b) - f(a) \tag{1}$$

and integrate the left side of (1) by parts with

$$u = f'(t), \qquad dv = dt$$

$$du = f''(t)\,dt, \qquad v = t - b$$

(where the constant of integration $-b$ is introduced for reasons that will make themselves clear). Hence,

$$\int_a^b f'(t)\,dt = [(t-b)f'(t)]_a^b - \int_a^b (t-b)f''(t)\,dt$$

$$= (b-a)f'(a) + \int_a^b (b-t)f''(t)\,dt$$

Using integration by parts again with

$$u = f''(t), \qquad dv = (b-t)\,dt$$

$$du = f'''(t)\,dt, \qquad v = \frac{-(b-t)^2}{2}$$

we have

$$\int_a^b (b-t)f''(t)\,dt = \left[-\frac{(b-t)^2}{2}f''(t)\right]_a^b + \int_a^b \frac{(b-t)^2}{2}f'''(t)\,dt$$

$$= \frac{(b-a)^2}{2!}f''(a) + \int_a^b \frac{(b-t)^2}{2!}f'''(t)\,dt$$

Continuing in this manner we obtain

$$\int_a^b f'(t)\,dt = (b-a)f'(a) + \frac{(b-a)^2}{2!}f''(a) + \frac{(b-a)^3}{3!}f'''(a) + \cdots$$

$$+ \frac{(b-a)^n}{n!}f^n(a) + \int_a^b \frac{(b-t)^n}{n!}f^{(n+1)}(t)\,dt \tag{2}$$

Comparing equations (1) and (2), we see that

$$f(b) = f(a) + (b-a)f'(a) + \frac{(b-a)^2}{2!}f''(a) + \cdots$$

$$+ \frac{(b-a)^n}{n!}f^n(a) + \int_a^b \frac{(b-t)^n}{n!}f^{(n+1)}(t)\,dt$$

Replacing b by x we have

$$f(x) = f(a) + (x-a)f'(a) + \frac{(x-a)^2}{2!}f''(a) + \cdots + \frac{(x-a)^n}{n!}f^n(a) + R_n(x,a)$$

where

$$R_n(x,a) = \int_a^x \frac{(x-t)^n}{n!}f^{(n+1)}(t)\,dt$$ ▧

This theorem is of great value in approximations and numerical computations, since $R_n(x,a)$ is a measure of how much f differs from a certain approximating polynomial of degree n. For a treatment of this subject the reader is advised to see Chapter 12 on numerical methods.

Now, we know that given a function f with derivatives continuous through order $n+1$, we can express it as $f(x) = T_n(x) + R_n(x,a)$. We would still like to know when we can represent a function f by its Taylor series. This is equivalent to asking for what values of x does the Taylor series converge to the value $f(x)$?

We observe that this will be true when

$$f(x) = \lim_{n\to\infty} T_n(x) \tag{3}$$

since as $n \to \infty$, $T_n(x)$ approaches $\sum_{n=0}^{\infty}(f^n(a)/n!)(x-a)^n$. But we already know that $f(x) = T_n(x) + R_n(x,a)$; hence, statement (3) is true when and only when

$$\lim_{n\to\infty} R_n(x,a) = 0$$

This is the condition that we have been seeking, and we formally state it in a theorem.

Theorem 2 If a function f has derivatives of all orders on an interval containing a and x, then f can be represented by its Taylor series provided

$$\lim_{n\to\infty} R_n(x,a) = 0$$

Before we investigate some examples of functions with series expansions, we will state and prove an alternative form of Taylor's Theorem, a form which expresses the remainder $R_n(x,a)$ in terms of a derivative.

Theorem 3 Suppose we are given a function f with continuous derivatives through order $n+1$ in an interval containing a and x. Then there is a number c between a and x such that

$$f(x) = f(a) + f'(a)(x-a) + \frac{f''(a)}{2!}(x-a)^2 + \cdots + \frac{f^n(a)}{n!}(x-a)^n$$

$$+ \frac{f^{n+1}(c)(x-a)^{n+1}}{(n+1)!}$$

Thus,

$$R_n(x, a) = \frac{f^{n+1}(c)(x-a)^{n+1}}{(n+1)!}$$

This is known as **Lagrange's** form of the remainder.

Proof By Theorem 1 we have

$$R_n(x, a) = \int_a^x \frac{(x-t)^n}{n!} f^{n+1}(x)\, dx$$

We apply the Second Mean Value Theorem of Integral Calculus to this integral which yields

$$R_n(x, a) = \frac{f^{n+1}(c)}{n!} \int_a^x \frac{(x-t)^n}{n!}\, dt$$

where c is between a and x. By the Fundamental Theorem, we have

$$\int_a^x (x-t)^n\, dt = \left[-\frac{(x-t)^{n+1}}{n+1}\right]_a^x$$

$$= \frac{(x-a)^{n+1}}{n+1}$$

Hence

$$R_n(x, a) = f^{n+1}(c) \cdot \frac{1}{n!} \cdot \frac{(x-a)^{n+1}}{n+1} = \frac{f^{n+1}(c)(x-a)^{n+1}}{(n+1)!}$$ ∎

When a Taylor series is computed about the point zero, it is called a **Maclaurin series**. Thus, the Maclaurin series for a given function is a power series in x, the special case of the Taylor series in which $a = 0$.

Example 1 Express e^x as a power series.

Let $f(x) = e^x$. This function and all its derivatives are everywhere continuous, so we can apply Taylor's Theorem. We will use the case $a = 0$, that is, we will be writing the Maclaurin series.

Since $f'(x) = e^x$ and, in general, for this function $f^{(i)}(x) = e^x$, we have $f^{(i)}(0) = 1$. Therefore,

$$e^x = f(0) + f'(0)\,x + \frac{f''(0)\,(x)^2}{2!} + \frac{f'''(0)\,x^3}{3!} + \cdots + \frac{f^n(0)\,x^n}{n!} + R_n(x,0)$$

$$e^x = 1 + x + \frac{x^2}{2!} + \frac{x^3}{3!} + \cdots + \frac{x^n}{n!} + R_n(x,0)$$

where

$$R_n(x,0) = \int_0^x \frac{(x-t)^n}{n!}\,e^t\,dt$$

We wish to know when e^x can be expressed this way, therefore by Theorem 2 we ask when

$$\lim_{n\to\infty} R_n(x,0) = 0$$

For simplicity, let us first assume that $x > 0$. We estimate the remainder by observing that the integrand is positive for $0 < t < x$ and $e^t < e^x < 3^x$. Therefore

$$|R_n(x,0)| \leqslant \int_0^x \frac{(x-t)^n}{n!}\,3^x\,dt = 3^x \frac{x^{n+1}}{(n+1)!}, \qquad x > 0$$

For x negative we replace x by $|x|$ on the right side of the inequality. Then for all real values of x we have

$$|R_n(x,0)| \leqslant 3^{|x|} \frac{|x|^{n+1}}{(n+1)!}$$

In order to determine whether $\lim_{n\to\infty} R_n(x,0) = 0$, we consider two separate cases. First, suppose $|x| \leqslant 1$; then $3^{|x|} \leqslant 3$ and

$$|R_n(x,0)| \leqslant \frac{3}{(n+1)!}$$

As $n \to \infty$, it is obvious that $R_n(x,0) \to 0$ and the series

$$e^x = 1 + x + \frac{x^2}{2!} + \frac{x^3}{3!} + \cdots + \frac{x^n}{n!} + \cdots$$

converges to e^x for $|x| \leqslant 1$.

Next, assume $|x| > 1$; then both $|x|^{n+1}$ and $(n+1)!$ increase indefinitely as n does. However, the factorial increases so much faster for large n that the quotient approaches zero. This can easily be shown using the ratio test on the series $\sum_{n=0}^{\infty} |x|^n/n!$; since the series converges, it follows that

$$\lim_{n\to\infty} \frac{|x|^{n+1}}{(n+1)!} = 0$$

because the nth term must approach 0.

The factor $3^{|x|}$ remains constant as $n \to \infty$, so we have

$$\lim_{n\to\infty} 3^{|x|} \frac{|x|^{n+1}}{(n+1)!} = 0$$

Thus we have shown that $\lim_{n\to\infty} R_n(x,0) = 0$ for all real x, and we can assert

$$e^x = \sum_{n=0}^{\infty} \frac{x^n}{n!}, \qquad -\infty < x < +\infty$$

We note, too, that the series for e^{-x} is found by substituting $-x$ for x in the above expression; hence,

$$e^{-x} = 1 - x + \frac{x^2}{2!} - \frac{x^3}{3!} + \cdots = \sum_{n=1}^{\infty} (-1)^n \frac{x^n}{n!}$$

which is valid for all real x.

Example 2 Find the Maclaurin series expansion for $\sin x$.

We have $f(x) = \sin x$, $a = 0$. Then

$$\begin{aligned}
f(x) &= \sin x, & f(0) &= \sin 0 = 0\\
f'(x) &= \cos x, & f'(0) &= \cos 0 = 1\\
f''(x) &= -\sin x, & f''(0) &= -\sin 0 = 0\\
f'''(x) &= -\cos x, & f'''(0) &= -\cos 0 = -1\\
f''''(x) &= \sin x, & f''''(0) &= \sin 0 = 0\\
&\vdots & &\vdots\\
f^{(k)}(x) &= \sin(x + \tfrac{1}{2}k\pi), & f^{(k)}(0) &= \sin \tfrac{1}{2}k\pi
\end{aligned}$$

Thus,

$$f^{(k)}(0) = \sin \tfrac{1}{2}k\pi, \qquad k = 0, 1, 2, \ldots$$

For k an even integer, $\sin\frac{1}{2}k\pi = 0$. When k is of the form $4m+1$ (the integers $1, 5, 9, 13, \ldots$) then $\sin\frac{1}{2}k\pi = 1$, but if k is of the form $4m+3$ (the integers $3, 7, 11, \ldots$) then $\sin\frac{1}{2}k\pi = -1$. Substituting these values into the formula yields

$$\sin x = x - \frac{x^3}{3!} + \frac{x^5}{5!} - \frac{x^7}{7!} + \cdots + \frac{(-1)^{n-1}x^{2n-1}}{(2n-1)!} + 0 \cdot x^{2n} + R_{2n}(x,0)$$

where

$$R_{2n}(x,0) = \frac{x^{2n+1}}{(2n+1)!} \sin\left(c + \frac{(2n+1)\pi}{2}\right)$$

Since $|\sin\theta| \leqslant 1$, we have

$$|R_{2n}(x,0)| \leqslant \frac{|x|^{2n+1}}{(2n+1)!}$$

As before, it can be shown, using the ratio test on the appropriate series, that

$$\lim_{n\to\infty}\frac{|x|^{2n+1}}{(2n+1)!} = 0, \qquad -\infty < x < +\infty$$

Hence,

$$\sin x = x - \frac{x^3}{3!} + \frac{x^5}{5!} - \frac{x^7}{7!} + \cdots$$

and we write

$$\sin x = \sum_{n=1}^{\infty} \frac{(-1)^{n-1}x^{2n-1}}{(2n-1)!}, \qquad -\infty < x < +\infty$$

It can be shown in an analogous fashion that

$$\cos x = 1 - \frac{x^2}{2!} + \frac{x^4}{4!} - \frac{x^6}{6!} + \cdots$$

or

$$\cos x = \sum_{n=0}^{\infty} \frac{(-1)^n x^{2n}}{(2n)!}, \qquad -\infty < x < +\infty$$

EXERCISES

1. Differentiate term by term the power series representing $\sin x$ to derive the expansion of $\cos x$ stated at the end of the section.
2. For each of the following, write the Taylor series for the function about a.
   (a) $f(x) = (1-x)^{-2}, \quad a = 0$ (b) $f(x) = e^{-2x}, \quad a = 0$
   (c) $f(x) = \ln x, \quad a = 1$ (d) $f(x) = \sinh x, \quad a = 0$
   (e) $f(x) = \tan^{-1} x, \quad a = 0$ (f) $f(x) = \cos x, \quad a = \frac{1}{3}\pi$
3. Find the Taylor polynomial T_n for the following functions and the given values of n and a.
   (a) $f(x) = \dfrac{1}{1+x^2}; \quad a = 0, \quad n = 4$
   (b) $f(x) = e^{-x^2}; \quad a = 0, \quad n = 4$
   (c) $f(x) = \sec x; \quad a = \frac{1}{3}\pi, \quad n = 3$
   (d) $f(x) = \sqrt{x+1}; \quad a = 3, \quad n = 3$
4. Given the polynomial
   $$f(x) = 1 - 3x + 2x^2 + 5x^3 - x^4$$
   show that f may be written in the form
   $$f(x) = a_0 + a_1(x-1) + a_2(x-1)^2 + a_3(x-1)^3 + a_4(x-1)^4$$
5. If f is a polynomial in x of degree $\leqslant m$, that is,
   $$f(x) = a_0 + a_1 x + \cdots + a_m x^m,$$
   and T_n is the Taylor polynomial which approximates f near an arbitrary real number a, prove that $f(x) = T_n(x)$ for all x when $n \geqslant m$.

6. Find the power series $y = \sum_{n=1}^{\infty} a_n x^n$ satisfying the following conditions:
   (i) $y = 2$ when $x = 0$
   (ii) $\dfrac{dy}{dx} = 0$ when $x = 0$
   (iii) $\dfrac{d^2y}{dx^2} - y = 0$
7. Show that
$$\int_0^x e^{-y^2}\,dy = x - \frac{x^3}{3\cdot 1!} + \frac{x^5}{5\cdot 2!} - \frac{x^7}{7\cdot 3!} + \cdots$$
   for all values of x.
8. Find the Taylor's series for each of the following functions about the given value of a, and find the interval on which the series validly represents the function.
   (a) $f(x) = e^{bx}$, $a = 0$ (b) $f(x) = \cos x$, $a = k$
   (c) $f(x) = \ln(b+x)$, $a = 0$ (d) $f(x) = \dfrac{1}{1-x}$, $a = 0$
   (e) $f(x) = \sin 3x$, $a = 0$ (f) $f(x) = \sin^{-1} x$, $a = 0$.
9. Prove that the closest integer to $n!/e$ is divisible by $n-1$. [Hint: Consider the series for $1/e$.]
10. Suppose f has derivatives of all orders on an interval I containing a, and $|f^{(n)}(x)| \leqslant M^n$ on I.
    (a) Show that the Taylor series of f converges to f, i.e. show f is analytic.
    (b) Apply this criterion in Examples 1 and 2.

*6 SEQUENCES AND SERIES OF FUNCTIONS

In more advanced parts of analysis, one often encounters functions which are defined by means of infinite series. Hence, it is important to discuss the notion of convergence for sequences and series of functions. In essence, the earlier material of this chapter is a special case of these more general considerations. However, we shall find that our new definitions are based on the previous ones.

After the basic definitions are given and examples are discussed, the important notion of **uniform convergence** is introduced. The section concludes with the main theorem about functions defined by uniformly convergent sequences which concerns questions of term-by-term differentiation and integration of sequences and series.

Let $\mathscr{F}(I)$ denote the set of real-valued functions on the interval I. A sequence of functions on I is a function $F: N \to \mathscr{F}(I)$ and we denote $F(n)$ by F_n. Thus for every number n in N there is a function F_n with domain I. For each x_0 in I, the sequence $\{F_n(x_0)\}$ is a sequence of real numbers.

Example 1 Consider the following sequences of functions:

(a) $F_n(x) = x^n$; this is the sequence of powers of x,

$$1 = x^0,\ x,\ x^2,\ \ldots,\ x^n,\ \ldots,$$

(b) $F_n(x) = x^{-n}$; this is the sequence of negative powers of x,

$$1 = x^0,\ x^{-1},\ x^{-2},\ \ldots,\ x^{-n},\ \ldots$$

(c) $$F_n(x) = \begin{cases} \cos nx, & n \text{ even} \\ \sin nx, & n \text{ odd} \end{cases}$$

this is the sequence of trigonometric functions,

$$1,\ \sin x,\ \cos 2x,\ \sin 3x,\ \ldots$$

In the above example, each sequence could be considered on any interval I. However, the sequence

$$F_n(x) = \ln\left[\left(\frac{1}{1-x}\right)^{1/n}\right]$$

is restricted to the interval $0 < x < 1$, since the functions are not defined outside that interval.

Evaluating every function of the sequence $\{F_n\}$ at a number x_0 in the interval I, yields a sequence $\{F_n(x_0)\}$ of real numbers. Thus, for $x = \frac{1}{2}$, we obtain the sequence of real numbers

$$0,\ \ln 2,\ \ln(2^{1/2}),\ \ldots,\ \ln(2^{1/n}),\ \ldots$$

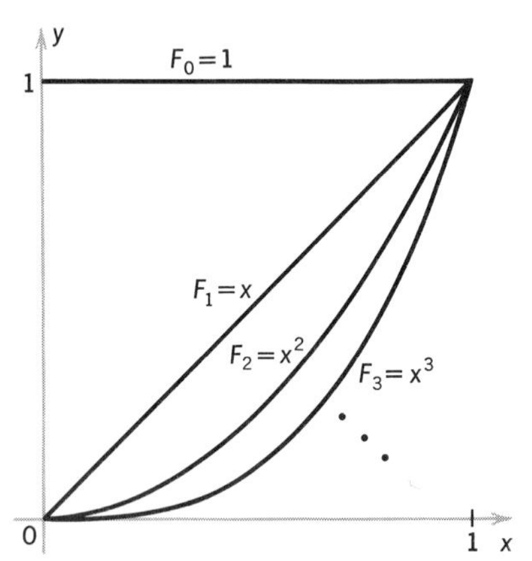

FIGURE 6-1

Definition If $\{F_n(x_0)\}$ converges as $n \to \infty$, then we say the **sequence of functions $\{F_n\}$ converges at x_0**. If $\{F_n(x)\}$ converges at each x_0 in the interval I, we say the **sequence $\{F_n\}$ converges on I**.

Thus in Example 1(a) above, the sequence $\{x^n\}$ converges on $[0, 1]$; for $x \neq 1$, the limit is 0, and the limit is 1, for $x = 1$ (Figure 6-1). The sequence of Example 1(b) converges on any interval of the form $[1, b]$, $b > 1$ (Figure 6-2), as the reader can easily verify.

If $\{F_n\}$ is a sequence of functions on the interval I, the infinite series of functions $\sum_{n=0}^{\infty} F_n$ associates to every x_0 in I an infinite series of real numbers $\sum_{n=0}^{\infty} F_n(x_0)$.

Definition The **infinite series of functions $\sum_{n=0}^{\infty} F_n$ converges at x_0** if the associated infinite series of real numbers $\sum_{n=0}^{\infty} F_n(x_0)$ converges. The **infinite series $\sum_{n=0}^{\infty} F_n$ converges on the interval I** if it converges at x_0, for every $x_0 \in I$.

In the same manner as for series of real numbers, we define the sequence s_n of partial sums of $\sum_{n=0}^{\infty} F_n$ by

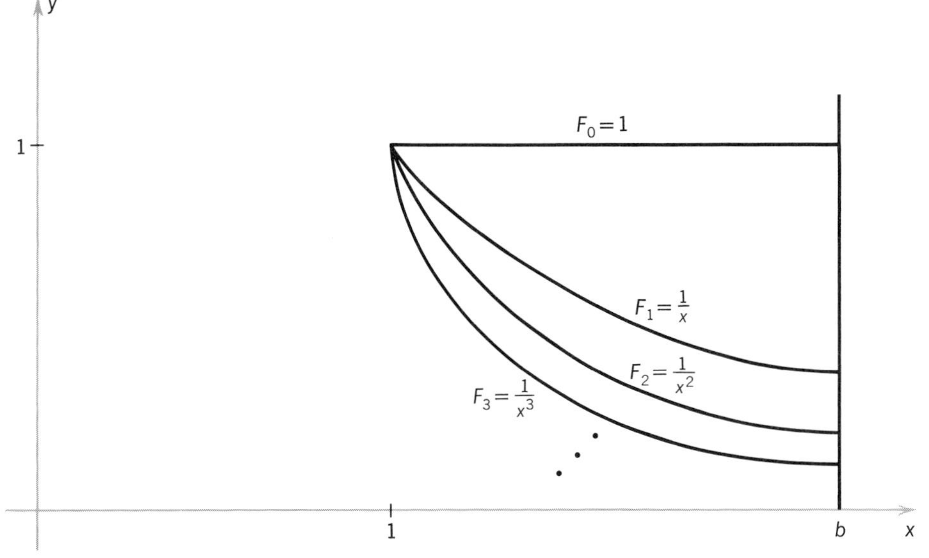

FIGURE 6-2

$$\begin{aligned} s_0 &= F_0 \\ s_1 &= F_0 + F_1 \\ &\vdots \\ s_{n+1} &= F_0 + F_1 + \cdots + F_n + F_{n+1} = s_n + F_{n+1} \end{aligned}$$

where, of course, each s_n is a function on the interval I. Thus, the sequence of partial sums is a sequence of functions on I. It is not difficult to show that an infinite series of functions converges on I if and only if the sequence of partial sums converges on I (see exercise 9).

Thus, the infinite series $\sum_{n=0}^{\infty} x^n$ has the sequence of partial sums $s_n = \sum_{k=0}^{n} x^k$, and since the geometric series $\sum_{n=0}^{\infty} r^n$ converges for $0 \leqslant r < 1$, it follows that the infinite series $\sum_{n=0}^{\infty} x^n$ converges on any interval $[0, r_0]$ with $0 \leqslant r_0 < 1$, but not on $[0, 1]$.

The series $\sum_{n=0}^{\infty} x^n$ is, of course, just one instance of a power series. Power series constitute a large and important class of infinite series of functions, and the material of §5 on power series should be referred to for examples and intuition concerning infinite series of functions. In a sense, the theory is nicest for power series, since many of the properties of power series are not shared by the more general series of functions.

Absolute convergence of a series of functions is defined in the natural way.

Definition If the infinite series of functions $\sum_{n=0}^{\infty} |F_n|$ converges on I, the **infinite series $\sum_{n=0}^{\infty} F_n$ converges absolutely on I.** (Here, of course, the function $|F_n|$ is defined by $|F_n| : x \mapsto |F_n(x)|$, for $x \in I$.)

Example 2 Show that $\sum_{n=1}^{\infty} \cos(nx)/n^2$ converges absolutely for all x.

Since $|\cos nx| \leqslant 1$ for all n, then for all x, the series converges absolutely by the comparison test, comparing with the convergent series $\sum_{n=1}^{\infty} 1/n^2$.

Now let us assume that the sequence of functions $\{F_n\}$ converges on an interval I. We define the **limit function** F of the sequence of functions $\{F_n\}$ by $F(x) = \lim_{n\to\infty} F_n(x)$, for $x \in I$. If an infinite series of functions $\sum_{n=0}^{\infty} F_n$ converges on the interval I, the limit function of the sequence of partial sums is called the **sum function**, $G = \sum_{n=0}^{\infty} F_n$, and has domain I.

For example, the limit function of $F_n(x) = x/(1+nx)$ is $F(x) = 0$. The sum function of $\sum_{n=0}^{\infty} x^n/n!$ (note $0! = 1!$) is $G(x) = e^x$, which is the same as saying that

$$e^x = \lim_{n\to\infty} \sum_{k=0}^{n} \frac{x^k}{k!}$$

which is true for all real numbers x.

It is important to know whether or not the limit function of a sequence of functions inherits the properties of the functions of the sequence. If the functions are continuous, is the limit function continuous? Similar questions for differentiation and integration can also be asked. The key to answering these questions lies in the notion of **uniform convergence**, which includes convergence, as defined above, with something more.

Recall the meaning of convergence at x_0, for the sequence $\{F_n\}$ of functions. If the limit is denoted by y_0, convergence at x_0 means that given any $\varepsilon > 0$, there is an $N_0 \in \boldsymbol{N}$, such that

$$|F_n(x_0) - y_0| < \varepsilon$$

whenever $n \geqslant N_0$.

For another number x_1 in the interval I $(x_1 \neq x_0)$, even if $\lim_{n\to\infty} F_n(x_1) = y_1$ exists, the number N_0 above may or may not guarantee the inequality $|F_n(x_1) - y_1| < \varepsilon$ for $n \geqslant N_0$. In a sense, the number N_0 is chosen out of consideration for ε and $F_n(x_0)$. The existence of $\lim F_n(x_1)$ guarantees that there is some number N_1 in $\boldsymbol{N}$ such that whenever $n \geqslant N_1$, $|F_n(x_1) - y_1| < \varepsilon$. Of course, a number N greater than either N_0 or N_1 would guarantee the inequality

$$|F_n(x) - y| < \varepsilon, \qquad \text{for} \quad n \geqslant N \tag{1}$$

where x could be x_0 or x_1, and y could be y_0, y_1, respectively.

If it is possible to choose one number N such that inequality (1) above holds simultaneously for all values of x in the interval I, then we have the "something more" of uniform convergence.

We shall now make our definition precise.

Definition The sequence of functions $\{F_n\}$ **converges to F uniformly on an interval I,** if for every $\varepsilon > 0$, there exists an $N \in \mathbf{N}$, such that

$$|F_n(x) - F(x)| < \varepsilon$$

whenever $n \geqslant N$ and x is in the interval I. Thus, the convergence is uniform if for each $\varepsilon > 0$, a single N can be chosen which will guarantee inequality (1) for all x in the interval I. The crucial point here is that the **same** N will work for all values of x in the given interval.

Example 3 Consider the sequence of partial sums $s_n(x)$ of the series

$$\sum_{n=0}^{\infty} x^n$$

on the interval $[s, r]$ where $0 < s < r < 1$. Since

$$s_n(x) = \sum_{k=0}^{n} x^k = \frac{1 - x^{n+1}}{1 - x}$$

by the formula for the sum of a geometric progression, and since the limit function is $F(x) = 1/(1-x)$, we have, for $y = F(x)$,

$$|s_n(x) - y| = \left|\frac{1 - x^{n+1}}{1 - x} - \frac{1}{1 - x}\right| = \left|\frac{x^{n+1}}{1 - x}\right| \leqslant \frac{r^{n+1}}{1 - r} \tag{2}$$

Therefore, given $\varepsilon > 0$, a number N_0 can be chosen such that $r^{n+1}/(1-r) < \varepsilon$ for $n \geqslant N_0$, for we know that $\lim_{n\to\infty} r^{n+1}/(1-r) = 0$ when $0 < r < 1$. Hence, for $n \geqslant N_0$ and for all x in $[s, r]$, we have

$$|s_n(x) - y| \leqslant \frac{r^{n+1}}{1 - r} < \varepsilon$$

This proves that the sequence of partial sums $s_n(x)$ of the series $\sum_{n=0}^{\infty} x^n$ converges uniformly on the interval $[s, r]$.

Uniform convergence may be interpreted graphically by considering the inequality of the definition

$$|F_n(x) - F(x)| < \varepsilon, \quad \text{for all } x \in I$$

This inequality is equivalent to requiring

$$F(x) - \varepsilon < F_n(x) < F(x) + \varepsilon, \quad \text{for all } x \in I$$

So we can imagine the graphs of the two functions $x \mapsto F(x) + \varepsilon$ and $x \mapsto F(x) - \varepsilon$. Then over the interval $I = [a, b]$, the portion of the xy plane between these two curves is a tubelike region of width 2ε. To have uniform convergence, the graph of $x \mapsto F_n(x)$ must lie entirely in the tube when n is sufficiently large ($n \geqslant N$). Moreover, for each $\varepsilon > 0$, such N must exist. (See Figure 6-3.)

So that we can discuss uniform convergence for series, we make the following definition.

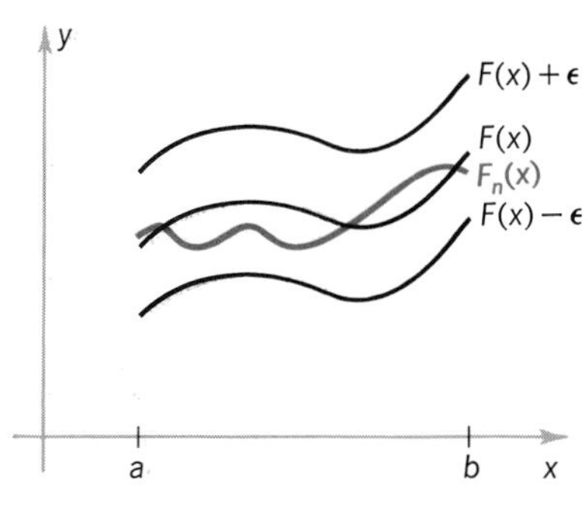

FIGURE 6-3

Definition The infinite series of functions $\sum_{n=0}^{\infty} F_n$ **converges uniformly on an interval** I if the sequence of partial sums of the series converges uniformly on I.

Example 3 above shows that on any interval $[s, r]$ where $0 \leqslant s < r < 1$, the infinite series $\sum_{n=0}^{\infty} x^n$ converges uniformly and the limit function is $y = F(x) = 1/(1-x)$. It is not difficult to demonstrate that the convergence is not uniform on the interval $0 \leqslant s \leqslant x < 1$. To show this suppose an $N \in \mathbf{N}$ is chosen, which works for a given $\varepsilon > 0$. By the equality in (2) we have

$$|s_N(x) - y| = \frac{x^{N+1}}{1-x} \geqslant \frac{s^{N+1}}{1-x} \tag{3}$$

But $0 < s^{N+1}/\varepsilon$, and therefore there is an $\bar{x}$ such that

$$0 < 1 - \bar{x} < \frac{s^{N+1}}{\varepsilon}$$

Thus, for such $\bar{x}$

$$\varepsilon < \frac{s^{N+1}}{1-\bar{x}}$$

Letting $\bar{y} = F(\bar{x})$, we use this result in (3) to obtain

$$|s_N(\bar{x}) - \bar{y}| \geqslant \frac{s^{N+1}}{1-\bar{x}} > \varepsilon$$

Therefore, the number N does not work, because we can produce an exceptional $\bar{x}$ (of course, all x such that $0 < 1 - x < 1 - \bar{x}$ are exceptions for this N) for which the necessary inequality does not hold.

The next theorem is a useful tool for proving that a series is uniformly convergent. Although a series may be uniformly convergent without satisfying the hypothesis of this theorem, in many important cases it is the simplest method for establishing this property.

Theorem 1 **Weierstrass Comparison Test** Suppose that $|F_n(x)| \leqslant c_n$ for all x in the interval I for $n = 0, 1, 2, \ldots$. Suppose further that $\sum_{m=0}^{\infty} c_m$ converges. Then the series $\sum_{n=0}^{\infty} F_n$ converges absolutely and uniformly on the interval I.

Proof Let $C = \sum_{n=0}^{\infty} c_n$, $C_n = \sum_{k=0}^{n} c_k$, and let the partial sums of the series be denoted by $s_n(x)$,

$$s_n(x) = \sum_{k=0}^{n} F_k(x)$$

Then if $F(x) = \sum_{n=0}^{\infty} F_n(x)$ converges, for x in the interval I,

$$|F(x) - s_n(x)| = \left|\sum_{k=n+1}^{\infty} F_k(x)\right| \leqslant \sum_{k=n+1}^{\infty} |F_k(x)| \leqslant \sum_{k=n+1}^{\infty} c_k$$

But $\sum_{k=n+1}^{\infty} c_k = C - \sum_{k=0}^{n} c_k = C - C_n$, and given $\varepsilon > 0$, there is an $N \in N$ such that for $n \geqslant N$, $C - C_n < \varepsilon$. Hence, for $n \geqslant N$,

$$|F(x) - s_n(x)| \leqslant C - C_n < \varepsilon$$

Clearly, N does not depend on x, since C_n does not. So $s_n(x)$ converges uniformly to $F(x)$ on the interval I, provided we can show that the series converges. But by the comparison test, the series $\sum_{n=0}^{\infty} |F_n(x)|$ converges for x in I, and therefore, since absolute convergence implies convergence, the series $\sum_{n=0}^{\infty} F_n(x)$ converges, for every x in the interval I. ∎

Note that the Weierstrass Comparison Test has the advantage that it can be applied without knowing the limit function $F(x)$ of the series.

Example 4 Reconsider the series $\sum_{n=0}^{\infty} x^n$ on the interval $I = [0, r]$ where $0 < r < 1$. Since for all $x \in I$, $F_n(x) = x^n$ satisfies the inequality

$$|F_n(x)| = x^n \leqslant r^n$$

and since the series $\sum_{n=0}^{\infty} r^n$ converges for $0 \leqslant r < 1$, the series $\sum_{n=0}^{\infty} x^n$ converges uniformly on the interval I, by the Weierstrass Comparison Test.

This example is a special case of a more general application of the Weierstrass Comparison Test to power series.

Theorem 2 If the power series $\sum_{n=0}^{\infty} a_n x^n$ converges at $x_1 \neq 0$, then the series converges absolutely for all x with $|x| < |x_1|$ and uniformly on any interval $[-h, h]$ with $h < |x_1|$.

Proof By the convergence of $\sum_{n=0}^{\infty} a_n x_1^n$, we know that $a_n x_1^n \to 0$ as $n \to \infty$. Therefore, there exists a constant $c > 0$ such that $|a_n x_1^n| < c$, for all n. Then, since

$$|a_n x^n| = \left| a_n x_1^n \frac{x^n}{x_1^n} \right| = |a_n x_1^n| \left| \frac{x}{x_1} \right|^n < c \left| \frac{x}{x_1} \right|^n$$

the geometric series $\sum_{n=0}^{\infty} c|x/x_1|^n$ converges for $|x/x_1| < 1$, so the given series converges absolutely. Furthermore, if $|x| \leqslant h < |x_1|$, the Weierstrass Comparison Test applies, with $C_n = c(h/x_1)^n$, therefore the series converges uniformly on any such interval $[-h, h]$ where $0 < h < |x_1|$. ∎

We now return to the question raised at the start of this section, namely, under what conditions does the limit function inherit properties of continuity and differentiability? It is also important for us to know when we are justified in differentiating and integrating a series or sequence of functions term-by-term.

The property of uniform convergence is what allows us to infer these desired results. Assuming uniform convergence and other suitable hypotheses, it can be shown that all goes well; on the other hand, we can produce examples of nonuniformly convergent sequences or series where all does not go well. The results for series follow from the corresponding proofs for sequences of functions, since convergence is always defined via sequences:

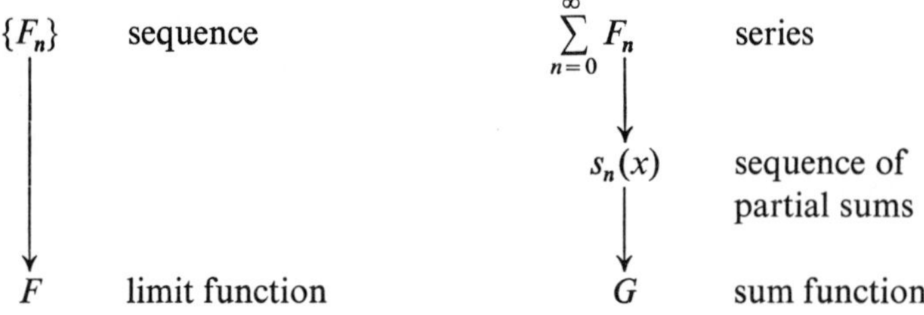

Theorem 3

Sequences	**Series**
(i) If $\{F_n\}$ converges uniformly to F on $[a, b]$ and each F_n is continuous on $[a, b]$, then $F = \lim_{n \to \infty} F_n$ is continuous on $[a, b]$.	If $\sum_{n=0}^{\infty} F_n$ converges uniformly to G on $[a, b]$ and each F_n is continuous on $[a, b]$, then $G = \sum_{n=0}^{\infty} F_n$ is continuous on $[a, b]$.
(ii) If $\{F_n\}$ converges uniformly to F on $[a, b]$ and each F_n is continuous on $[a, b]$, then $\int_a^b F_n(x)\,dx$ converges to $\int_a^b F(x)\,dx$.	If $\sum_{n=0}^{\infty} F_n$ converges uniformly to G on $[a, b]$ and each F_n is continuous on $[a, b]$, then $\sum_{n=0}^{\infty} (\int_a^b F_n(x)\,dx)$ converges to $\int_a^b G(x)\,dx$.
(iii) If $\{F_n\}$ converges uniformly to F on $[a, b]$, each F_n' is continuous on $[a, b]$, and $\{F_n'\}$ converges uniformly to the limit function G on $[a, b]$, then $F' = G$ on $[a, b]$	If $\sum_{n=0}^{\infty} F_n$ converges uniformly on $[a, b]$, each F' is continuous on $[a, b]$, and $\sum_{n=0}^{\infty} F_n'$ converges uniformly to the sum function G on $[a, b]$, then $F' = G$ on $[a, b]$.

To give some idea of how the result for series is obtained from the result for sequences, we will prove (i) for both sequences and series but omit the proofs for series of the other propositions.

Proof (i) For x_0 in $]a, b[$, we shall show that $F(x)$ is eventually in every neighborhood of the form $]F(x_0) - \varepsilon, F(x_0) + \varepsilon[$, where $\varepsilon > 0$, as x approaches x_0. For $x_0 = a$ or $x_0 = b$, the appropriate one-sided limits will be established by our argument.

So, let $\varepsilon > 0$ be given. By uniform convergence, there is an N such that for $n \geqslant N$,

$$|F_n(x) - F(x)| < \tfrac{1}{3}\varepsilon \tag{4}$$

for all x in $[a,b]$.

Now, since F_N is continuous on $[a,b]$, there is a neighborhood M of x_0 such that

$$|F_N(x) - F_N(x_0)| < \tfrac{1}{3}\varepsilon \tag{5}$$

whenever $x \in M$ and $x \in [a,b]$. Rewriting $F(x) - F(x_0)$ as

$$F(x) - F(x_0) = F(x) - F_N(x) + F_N(x) - F_N(x_0) + F_N(x_0) - F(x_0)$$

and using facts about absolute values, we have

$$|F(x) - F(x_0)| \leqslant |F(x) - F_N(x)| + |F_N(x) - F_N(x_0)| + |F_N(x_0) - F(x_0)|$$

By (4), the first and third terms on the right are less than $\frac{1}{3}\varepsilon$, and by (5), the second term is less than $\frac{1}{3}\varepsilon$, whenever $x \in M$ and $x \in [a,b]$. Thus

$$|F(x) - F(x_0)| < \tfrac{1}{3}\varepsilon + \tfrac{1}{3}\varepsilon + \tfrac{1}{3}\varepsilon = \varepsilon$$

for $x \in [a,b]$ and $x \in M$. This establishes that F is continuous on $[a,b]$.

The proof of (i) for series is immediate; since each F_n is continuous on $[a,b]$, the sum $s_n = \sum_{k=0}^{n} F_k$ is continuous, and the uniform convergence of $\sum_{n=0}^{\infty} F_n$ is, by definition, the uniform convergence of the sequence of functions $\{s_n\}$. Thus the result follows from (i) for sequences.

(ii) By (i) the limit function F is continuous on the interval $[a,b]$, so $\int_a^b F(x)\,dx$ is defined. By the uniform convergence on $[a,b]$, for any $\varepsilon > 0$, there is an N such that

$$|F(x) - F_n(x)| < \frac{\varepsilon}{b-a} \tag{6}$$

for $n \geqslant N$ and all x in $[a,b]$. To prove our theorem we use this and the basic inequality for estimating integrals:

$$\left| \int_a^b g(x)\,dx \right| \leqslant \int_a^b |g(x)|\,dx$$

We have, by the additivity of definite integrals

$$\left| \int_a^b F(x)\,dx - \int_a^b F_n(x)\,dx \right| = \left| \int_a^b (F(x) - F_n(x))\,dx \right|$$

and therefore,

$$\left| \int_a^b (F(x) - F_n(x))\,dx \right| \leqslant \int_a^b |F(x) - F_n(x)|\,dx$$

For $n \geqslant N$, this integrand is bounded by $\varepsilon/(b-a)$ by (6), so we combine these inequalities to obtain

$$\left| \int_a^b F(x)\,dx - \int_a^b F_n(x)\,dx \right| \leqslant \int_a^b |F(x) - F_n(x)|\,dx < \int_a^b \frac{\varepsilon}{b-a}\,dx = \varepsilon$$

which proves that

$$\lim_{n\to\infty} \int_a^b F_n(x)\,dx = \int_a^b F(x)\,dx$$

(iii) By (i) the limit function G is continuous on $[a, b]$ since $\{F_n'\}$ converges uniformly, and we can apply the above result on termwise integration to conclude that, for $a \leqslant x \leqslant b$,

$$\int_a^x G(t)\,dt = \lim_{n\to\infty} \int_a^x F_n'(t)\,dt$$

Applying the Fundamental Theorem of Calculus to the right-hand side, we have

$$\int_a^x G(t)\,dt = \lim_{n\to\infty} [F_n(x) - F_n(a)] = F(x) - F(a)$$

and using the Fundamental Theorem, we differentiate this equation to get

$$G(x) = F'(x)$$

The results of Theorem 3 are often rephrased as justifying the interchange of integration and sums, or differentiation and sums. Thus (ii) and (iii) may be written this way for series:

(ii) $\int_a^b \sum_{n=0}^{\infty} F_n(x)dx = \sum_{n=0}^{\infty} \int_a^b F_n(x)dx$

(iii) $d\left[\sum_{n=0}^{\infty} F_n(x)\right]/dx = \sum_{n=0}^{\infty} d\left[F_n(x)\right]/dx$

provided of course the hypotheses there are fulfilled.

To demonstrate the necessity of uniform convergence for the conclusion of (i) for sequences, consider the familiar sequence $F_n(x) = x^n$ on the interval $[0, 1]$. Clearly, each function is continuous on $[0, 1]$, and we know the limit function F is not continuous, $F(x) = 0$ for $x \neq 1$, $F(1) = 1$. Of course, the convergence is not uniform, as was shown earlier.

In §5 on power series, we remarked that termwise integration and differentiation can be used to find series expansions for functions from known series expansions.

Example 5 Find the Maclaurin expansion for $\arctan x^3$.

Since

$$\frac{1}{1+u} = \sum_{n=0}^{\infty} (-1)^n u^n, \qquad |u| < 1$$

we let $u = w^2$, and obtain

$$\frac{1}{1+w^2} = \sum_{n=0}^{\infty} (-1)^n w^{2n}, \qquad |w| < 1$$

Integrating term-by-term we have

$$\arctan w = \sum_{n=0}^{\infty} \frac{(-1)^n w^{2n+1}}{2n+1}, \qquad |w| < 1$$

and substituting $w = x^3$ yields

$$\arctan x^3 = \sum_{n=0}^{\infty} \frac{(-1)^n (x^3)^{2n+1}}{2n+1} = \sum_{n=0}^{\infty} \frac{(-1)^n x^{6n+3}}{2n+1}$$

EXERCISES

In exercises 1–6, show that the sequences of functions on the given interval I all converge to the zero function.

1. $F_n(x) = \dfrac{\cos(nx)}{n}, \quad I = [0, 1]$

2. $F_n(x) = \dfrac{x}{n}, \quad I = [-b, b], \quad b > 0$

3. $F_n(x) = \sin\left(\dfrac{x}{n}\right)\cos\left(\dfrac{x}{n}\right), \quad I = [0, 1]$

4. $F_n(x) = \dfrac{x^2}{1+nx^2}, \quad I = [0, 1]$

5. $F_n(x) = \dfrac{nx^2}{1+nx} - x, \quad I = [0, 1]$

6. $F_n(x) = \dfrac{\cos nx}{nx}, \quad I = [1, \infty[$

7. What is the largest interval of real numbers for which the series
$$\sum_{n=0}^{\infty} \frac{(-1)^n x^n}{n^2}$$
converges?

8. Show that if the series $\sum_{n=0}^{\infty} F_n$ converges on the interval I, then for all $x \in I$, $\lim_{n\to\infty} F_n(x) = 0$.

9. Show that the series $\sum_{n=0}^{\infty} F_n$ converges on I if and only if the sequence $G_n = \sum_{k=0}^{n} F_k$ converges on I.

10. Show that if $\sum_{n=0}^{\infty} F_n$ and $\sum_{n=0}^{\infty} G_n$ converge on I, then $\sum_{n=0}^{\infty} (F_n + G_n)$ converges on I to $\sum_{n=0}^{\infty} F_n + \sum_{n=0}^{\infty} G_n$.

11. Prove a ratio test for series of nonvanishing functions: If $F_n(x) \neq 0$ for all x in I and all n, and $|F_{n+1}(x)/F_n(x)| \to r(x)$ as $n \to \infty$, $0 \leqslant r(x) < 1$, then $\sum_{n=0}^{\infty} F_n$ converges absolutely on I.

12. Investigate the sequences of functions in exercises 1–6 and determine whether the convergence is uniform or not.

13. Prove that if $\{F_n\}$ converges uniformly to F on I and $\{G_n\}$ converges uniformly to G on I, then (a) $\{F_n + G_n\}$ converges uniformly to $F + G$ on I, and (b) $\{F_n \cdot G_n\}$ converges uniformly to $F \cdot G$ on I.

In exercises 14–18, determine h such that the given series of functions converges uniformly on the specified interval.

14. $\displaystyle\sum_{n=0}^{\infty} \frac{(2x)^n}{n!}, \quad 0 \leqslant |x| \leqslant h$

15. $\displaystyle\sum_{n=0}^{\infty} n^2 x^n, \quad 0 \leqslant |x| \leqslant h$

16. $\displaystyle\sum_{n=1}^{\infty} x(1-x)^n, \quad 0 < |x| \leqslant h$

17. $\sum_{n=1}^{\infty} \frac{1}{x}$, $\quad h \leqslant x < \infty$

18. $\sum_{n=0}^{\infty} e^{-nx}$, $\quad h \leqslant x < \infty$

19. Assuming that $\sum_{n=0}^{\infty} a_n$ converges, prove that $\sum_{n=0}^{\infty} a_n \cos nx$ converges uniformly for all x.

20. If $\sum_{n=0}^{\infty} a_n$ converges, and there is a $c > 0$ such that for all x and all n, $|F_n(x)| < c$, prove that $\sum_{n=0}^{\infty} a_n F_n$ converges uniformly for all x.

21. Find h such that $\sum_{n=1}^{\infty} (x \ln x)^n$ converges uniformly for $0 < x \leqslant h$.

22. Find h such that $\sum_{n=1}^{\infty} (\cos x)^n$ converges uniformly for $|x| \leqslant h$.

Find series expansions for the functions of exercises 23–28 using termwise differentiation or integration, and substitution where necessary.

23. $f(x) = e^{1+x^2}$

24. $f(x) = \ln(1+x^4)$

25. $f(x) = \cos\left(\frac{3x}{\pi}\right)$

26. $f(x) = \arctan(3x+2)$

27. $f(x) = \frac{1}{(1-x^3)^2}$

28. $f(x) = \cosh(x^2)$

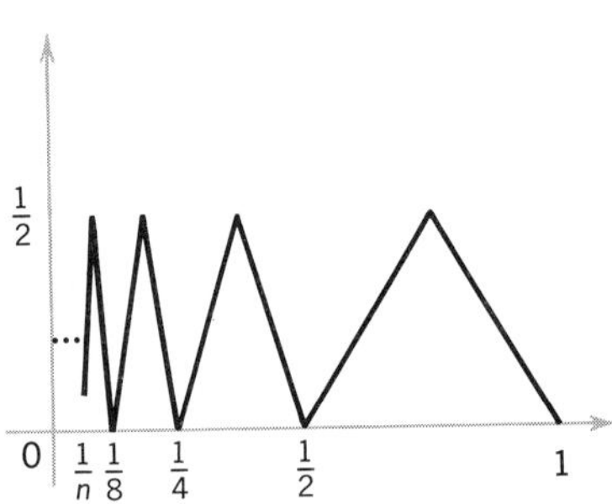

FIGURE 6-4

Many continuous, bounded functions of importance are not smooth enough to be treated by power series (Figure 6-4). An important tool for dealing with such functions is the **Fourier series,**

$$\frac{a_0}{2} + \sum_{n=1}^{\infty} a_n(\cos nx + b_n \sin nx)$$

The real numbers a_n, b_n are defined by the formulas

$$a_n = \frac{1}{\pi} \int_{-\pi}^{\pi} f(x) \cos nx \, dx$$

$$b_n = \frac{1}{\pi} \int_{-\pi}^{\pi} f(x) \sin nx \, dx$$

where the interval $[-\pi, \pi]$ is used because of interest in periodic functions f,

$$f(x+2\pi) = f(x), \qquad -\infty < x < \infty$$

Under appropriate assumptions it can be shown that when the Fourier series of f converges, it converges to f. Compute the Fourier series of the functions given in exercises 29–31.

29. $f(x) = x$

30. $f(x) = |x|$

31. $f(x) = \cos x$

32. Show that the sequence of functions defined for $n = 1, 2, \ldots$ on $[-1, 1]$ by

$$F_n(x) = \begin{cases} 0, & -1 \leqslant x \leqslant -\frac{1}{n} \\ nx + 1, & -\frac{1}{n} \leqslant x \leqslant 0 \\ -nx + 1, & 0 \leqslant x \leqslant \frac{1}{n} \\ 0, & \frac{1}{n} \leqslant x \leqslant 1 \end{cases}$$

(Figure 6-5) converges to the limit function

$$F(x) = \begin{cases} 0, & x \neq 0 \\ 1, & x = 0 \end{cases}$$

for x in $[-1, 1]$. Using Theorem 3(i), prove that the convergence is not uniform on $[-1, 1]$.

FIGURE 6-5

33. Show that $\int_0^1 F_n(x)\,dx$ does not converge to $\int_0^1 F(x)\,dx$ and conclude that F_n does not converge uniformly on $[0, 1]$, although $F_n \to F = 0$ on $[0, 1]$, where

$$F_n(x) = \begin{cases} n^2 x, & 0 \leqslant x \leqslant \dfrac{1}{n} \\ -n^2 x + 2n, & \dfrac{1}{n} \leqslant x \leqslant \dfrac{2}{n} \\ 0, & \dfrac{2}{n} \leqslant x \leqslant 1 \end{cases}$$

(see Figure 6-6).

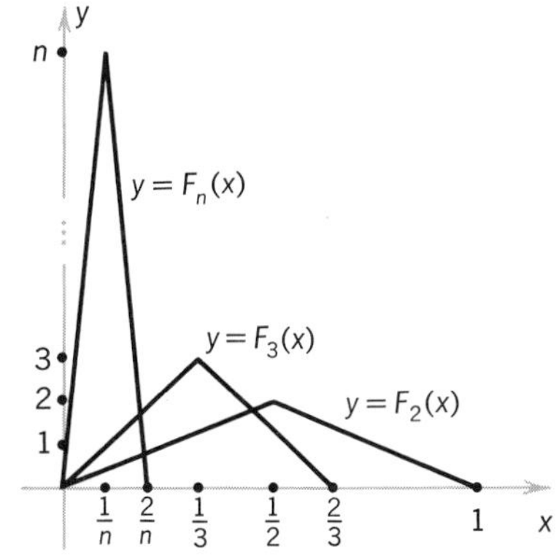

FIGURE 6-6

34. Suppose $f(x)$ is a differentiable real-valued function bounded for all x($|f(x)| < B$ for some $B > 0$), and suppose $f'(x) \geqslant 1$ for all x. Show that the series $\sum_{n=1}^{\infty} F_n$ with $F_n(x) = f(nx)/n^2$ (a) converges uniformly for all x; and (b) the series $\sum_{n=1}^{\infty} F_n'(x)$ does not converge for any x. Give an example of such a function f.

35. Prove Theorem 3(ii) and (iii) for series, assuming the results for sequences.

REVIEW EXERCISES FOR CHAPTER 10

1. Define the following.
   an infinite series of real numbers
   convergence of an infinite series
   absolute convergence
   conditional convergence
   a power series in x
   the Taylor series of a function f about a point a
   a Maclaurin series

2. Prove, without referring to the text, that if the infinite series $\sum_{n=1}^{\infty} a_n$ is convergent, then there is a $C > 0$ such that for all n, $|a_n| < C$.

3. Let $a_1, a_2, \ldots, a_n, \ldots$ be a sequence of nonzero numbers such that $\lim_{n\to\infty} a_n = +\infty$. Prove that the infinite series $\sum_{n=1}^{\infty}(a_{n+1} - a_n)$ diverges, and the series $\sum_{n=1}^{\infty}(1/a_n - 1/a_{n+1})$ converges.

4. Prove that every series of positive numbers either converges or the sequence of partial sums tends to infinity.

5. State and prove the comparison test for the convergence of a series of nonnegative terms, without referring to the text.

6. Prove the **root test** for convergence: A positive series $\sum_{n=1}^{\infty} a_n$ converges if $\lim_{n\to\infty} \sqrt[n]{a_n} < 1$, and diverges if $\lim_{n\to\infty} \sqrt[n]{a_n} > 1$. This test is inconclusive if $\lim_{n\to\infty} \sqrt[n]{a_n} = 1$. [Hint: If $\lim_{n\to\infty} \sqrt[n]{a_n} < 1$, then for $n > m$, $\sqrt[n]{a_n} < r < 1$ and hence $a_n < r^n$.]

7. Let r be the radius of convergence of the power series $\sum_{n=0}^{\infty} a_n x^n$. Show that if there exists a constant m such that $|a_n| \leqslant m$ for all n, then $r \geqslant 1$.

8. Prove that the representation of a function $f(x)$ in powers of x is unique, that is, if

$$f(x) = \sum_{n=0}^{\infty} s_n x^n, \quad \text{and} \quad f(x) = \sum_{n=0}^{\infty} t_n x^n$$

then $s_i = t_i$, $i = 0, 1, 2, \ldots$ (we assume that the series converge in an interval, and not merely for $x = 0$).

9. (a) Using the series for $1/(1+x^2)$ show that

$$\arctan x = \sum_{n=0}^{\infty} \frac{(-1)^n x^{2n+1}}{2n+1}$$

for $|x| < 1$.

(b) Use (a) to find a series expansion for $\pi/4$.

In exercises 10–13, find the sum of the given series.

10. $\displaystyle\sum_{n=1}^{\infty} \frac{1}{4n^2-1}$

11. $\displaystyle\sum_{n=1}^{\infty} \frac{2n-1}{(n^2+1)(n^2-2n+2)}$

12. $\displaystyle\sum_{n=1}^{\infty} \left(\frac{1}{n^p} - \frac{1}{(n+1)^p}\right), \quad p > 0$

13. $\displaystyle\sum_{n=1}^{\infty} \frac{1}{n(n+1)(n+2)}$

In exercises 14–17, show that each series diverges.

14. $3 + \frac{5}{2} + \frac{7}{3} + \frac{9}{4} + \cdots$

15. $\frac{1}{\ln 2} + \frac{1}{\ln 3} + \frac{1}{\ln 4} + \cdots$

16. $1 + \frac{1}{\sqrt{2}} + \cdots + \frac{1}{\sqrt{n}} + \cdots$

17. $\displaystyle\sum_{n=1}^{\infty} (-1)^{n+1} \frac{e^n}{n^3}$

In exercises 18–29, determine if the given series converges.

18. $\displaystyle\sum_{n=1}^{\infty} \frac{n^4+5}{n^5}$

19. $\displaystyle\sum_{n=3}^{\infty} \frac{1}{\sqrt[n]{2}}$

20. $\displaystyle\sum_{n=1}^{\infty} \frac{\ln(n+1)}{(n+1)^3}$

21. $\displaystyle\sum_{n=1}^{\infty} \frac{n+1}{n \cdot 2^n}$

22. $\displaystyle\sum_{n=1}^{\infty} \frac{n^3}{3^n}$

23. $\frac{2}{5} + \frac{2\cdot4}{5\cdot8} + \frac{2\cdot4\cdot6}{5\cdot8\cdot11} + \frac{2\cdot4\cdot6\cdot8}{5\cdot8\cdot11\cdot14} + \cdots$

24. $\displaystyle\sum_{n=1}^{\infty} \frac{2n}{(n+1)(n+2)(n+3)}$

25. $2 + \frac{3}{5} + \frac{4}{10} + \frac{5}{17} + \cdots$

26. $\displaystyle\sum_{n=1}^{\infty} (-1)^{n-1} \frac{\ln n}{3n+2}$

27. $\displaystyle\sum_{n=1}^{\infty} (-1)^{n-1} \frac{n^2}{n^4+2}$

28. $\sum_{n=1}^{\infty} \frac{(-1)^{n-1}(\frac{4}{3})^n}{n^2}$

29. $\sum_{n=1}^{\infty} (-1)^n \frac{2 \cdot 4 \cdot 6 \cdots 2n}{1 \cdot 4 \cdot 7 \cdots (3n-2)}$

Examine each of the series below for absolute or conditional convergence.

30. $\sum_{n=1}^{\infty} \frac{(-1)^n(6n^2-9n+4)}{n^3}$

31. $\sum_{n=1}^{\infty} \frac{(-1)^{n-1}}{(n+1)^2}$

32. $\sum_{n=1}^{\infty} (-1)^{n-1} \frac{n}{n^2+1}$

Find the intervals of convergence of the power series in exercises 33–39.

33. $\frac{x}{5} - \frac{x^2}{2 \cdot 5^2} + \frac{x^3}{3 \cdot 5^3} - \frac{x^4}{4 \cdot 5^4} + \cdots$

34. $\frac{x^2}{(\ln 2)^2} + \frac{x^3}{(\ln 3)^3} + \frac{x^4}{(\ln 4)^4} + \frac{x^5}{(\ln 5)^5} + \cdots$

35. $\sum_{n=1}^{\infty} \frac{(n-2)x^n}{n^2}$

36. $\sum_{n=1}^{\infty} \frac{(-1)^{n-1} n! (\frac{3}{2})^n x^n}{1 \cdot 3 \cdot 5 \cdots (2n-1)}$

37. $\sum_{n=2}^{\infty} \frac{(-1)^n x^n}{n(\ln n)^2}$

38. $\sum_{n=1}^{\infty} \frac{n^n}{n!} x^n$

39. $\sum_{n=1}^{\infty} \frac{(-1)^{n-1} x^n}{(n+1)\ln(n+1)}$

Find an infinite series expansion for each of the functions in exercises 40–43.

40. $f(x) = \frac{1-\cos x}{x}$, in powers of x

41. $f(x) = \ln x$, in powers of $x - 3$

42. $f(x) = \sqrt{x}$, in powers of $x - 4$

43. $f(x) = e^x \cos x$, in powers of x

In exercises 44–47, calculate the Taylor series for the given function and specify the interval on which the series is valid.

44. $f(x) = \ln x$, in powers of $x - 2$

45. $f(x) = e^{x/2}$, in powers of $x - 2$

46. $f(x) = \tan x$, in powers of x

47. $f(x) = \sin^{-1} x$, in powers of x

48. Give an example of a series which converges but which does not converge absolutely.

11

ELEMENTARY DIFFERENTIAL EQUATIONS

Problems involving differential equations often arise in physics, engineering, chemistry, and less frequently in such subjects as biology and economics. In this chapter, we explain some of the basic concepts of elementary differential equations and provide several solution techniques.

1 INTRODUCTION

In studying ordinary differential equations, it is important to be familiar with the kinds of physical problems in which they arise. Our first application of the calculus to physics in this course was in §3.4 where we studied the motion of a particle along a straight line path. We denote the position of a particle at time t by $s = s(t)$; the function $t \mapsto s(t)$ is therefore called the **position function**. We saw that the velocity of the particle was given by the derivative $s'(t)$ and the acceleration by the second derivative $s''(t)$.

The notation we shall use is as follows. Just as we write $s = s(t)$ for the position function, so we also write $s' = s'(t)$ and $s'' = s''(t)$ for the derivatives. To facilitate working with differential equations, throughout this chapter we use the convention that when a function is given by an equation, say $y = f(x)$, then we also write

$$y = y(x) = f(x)$$

and for the derivatives

$$y' = y'(x) = f'(x)$$

$$y'' = y''(x) = f''(x)$$

and so on.

Let us recall some ideas from physics. Newton's second law of motion asserts that a particle of mass m moving on a path described by a position function $s(t)$ and subject to a force function F satisfies the equation

$$\text{mass} \times \text{acceleration} = \text{force}$$

or

$$ms''(t) = F(s(t)) \tag{1}$$

Equation (1) is an example of a differential equation, because it involves $s(t)$ and the second derivative $s''(t)$. If friction is present the force may also depend on $s'(t)$ as well as on the position $s(t)$.

Our first illustration of Equation (1) is what is referred to as **simple harmonic motion.** We consider a particle on the line which is attracted to the origin by a force proportional to the distance from the origin, that is, $F(x) = -kx$, $k > 0$. An example of this type of motion occurs when there is an object on the end of a horizontal spring. The position of the object when the spring is neither stretched nor compressed (in equilibrium) is taken as the origin. The force acting on the object is proportional to the amount the spring is stretched or compressed. This is a physical property of a spring known as Hooke's law, and it holds to a good approximation whenever the spring is not stretched too much and when frictional forces are negligible. The force is a directed quantity, the sign indicating whether an object accelerates to the right or to the left (Figure 1-1). The minus sign indicates that the acceleration of the object is directed to the left if the object is located on the positive axis and to the right if the object is located on the negative axis. The acceleration is due to the force attempting to restore the spring to its equilibrium state. The object therefore oscillates back and forth (indefinitely, since we are neglecting various frictional forces). Observe that if the force had been given by $x \mapsto +kx$ with a positive coefficient the object would accelerate off in one direction and never come back.

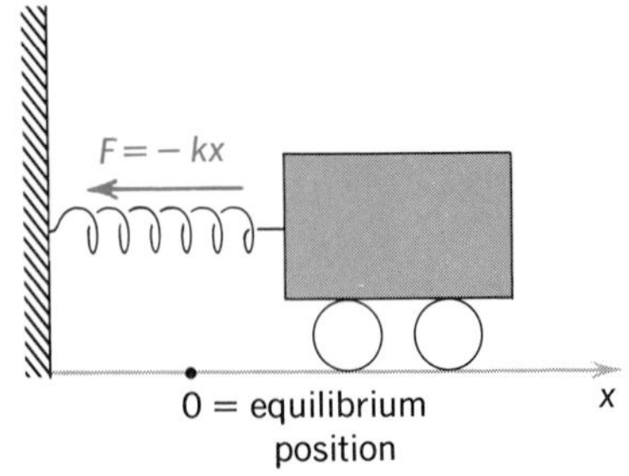

FIGURE 1-1

Other examples of simple harmonic motion are the swing of a pendulum over a small enough arc so that the curvature is negligible, vibrations of a stretched string (neglecting internal forces which "dampen" the motion), and various phenomena involving sound, electric currents, and the atom.

Now assume that a particle of mass m moves on the x axis in the presence of the force $x \mapsto -kx$. The characteristic property of the force says that the particle has the acceleration given by the function $x \mapsto -(k/m)x$. This function gives the acceleration at any position x, but does not give the acceleration of the particle as a function of time. Our problem is to find the position function $t \mapsto s(t)$ of the particle. To do this we write down an equation and "solve" for this function.

The acceleration as a function of time can be written in terms of the unknown function s as $a : t \mapsto -(k/m)s(t)$. Now we know that a is given by the second derivative s'' of the position function. Hence we have a differential equation to determine s:

$$ms''(t) = -ks(t), \qquad \text{or} \qquad ms'' + ks = 0 \tag{2}$$

Notice that even before we solve equation (2) we can find out a good deal about its solutions by obtaining a spring which will produce this motion and constructing an apparatus to record the path of the object. This will provide an accurate graph of the function s.

At present, we have no systematic way to solve a differential equation like the one above, but we happen to know of functions whose second derivatives are a negative number times the function itself: the functions sine and cosine. In order to get the factor k/m, we have to take the function $t \mapsto \sin\sqrt{k/m}\,t$ or $t \mapsto \cos\sqrt{k/m}\,t$ (Why?). Either of these functions is a solution of the differential equation, and is the position function for a particle moving under the force $x \mapsto -kx$. We have shown that simple harmonic motion can be described by trigonometric functions.

The problem has not been completely solved until we have found all solutions of the differential equation. Let us review (from §8.5) how this is done. The solutions above satisfy the **initial conditions**

$$s(0) = 0, \qquad s'(0) = \sqrt{\frac{k}{m}}$$

and

$$s(0) = 1, \qquad s'(0) = 0$$

respectively. These data can be regarded as the position and velocity of the particle when we begin to record its motion. It can be shown that there is precisely one solution of the equation for each choice of initial conditions.

Observe that if $s_1(t)$ and $s_2(t)$ are solutions of the equation $s'' + (k/m)s = 0$ then so is $c_1 s_1 + c_2 s_2$ for constants c_1 and c_2. This is simply because our equation is linear. Let us verify this; we want to prove

$$(c_1 s_1 + c_2 s_2)'' + \frac{k}{m}(c_1 s_1 + c_2 s_2) = 0$$

But

$$\begin{aligned}(c_1 s_1 + c_2 s_2)'' &= c_1 s_1'' + c_2 s_2'' \\ &= \left(c_1 \frac{k}{m} s_1 + c_2 \frac{k}{m} s_2\right) \\ &= -\frac{k}{m}(c_1 s_1 + c_2 s_2)\end{aligned}$$

which demonstrates our assertion.

Thus if we are seeking the solution with initial position x_0 and initial velocity v_0 we attempt to find c_1 and c_2 such that this will be the case for $s(t) = c_1 \cos\sqrt{k/m}\,t + c_2 \sin\sqrt{k/m}\,t$. This is quite easy; at $t = 0$, $s(0) = c_1$ so choose $c_1 = x_0$. Also, $s'(0) = c_2\sqrt{k/m}$ so choose $c_2 = \sqrt{m/k}\,v_0$. Thus our solution is

$$\boldsymbol{s(t) = x_0 \cos\sqrt{k/m}\,t + v_0\sqrt{m/k}\sin\sqrt{k/m}\,t} \tag{3}$$

In this procedure we say that we have **superimposed** the two solutions.

For graphing purposes, as well as for understanding the above solution, it is convenient to write equation (3) as $s(t) = A\cos(\omega t - \alpha)$. To see what A, ω, α should be, we write

$$A\cos(\omega t-\alpha) = A\cos\omega t\cos\alpha + A\sin\omega t\sin\alpha$$

Thus we should choose

$$\omega = \sqrt{k/m}, \qquad A = \sqrt{x_0^2+v_0^2 m/k}$$

and α such that $\cos\alpha = x_0/A$. Thus our solution may be rewritten

$$s(t) = \sqrt{x_0^2+v_0^2 m/k}\cos(\sqrt{m/k}\,t-\alpha) \tag{3'}$$

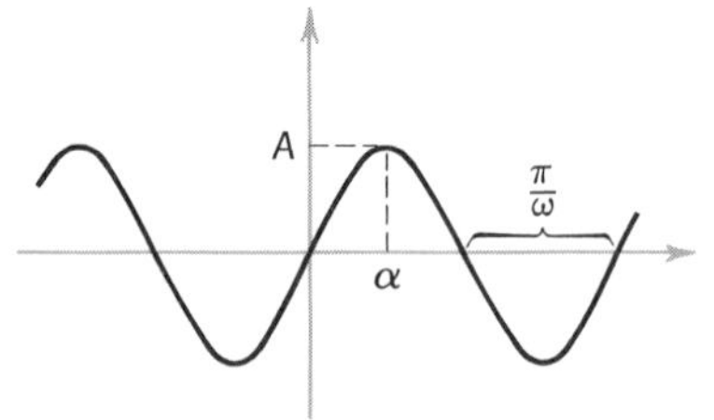

FIGURE 1-2

The general features of $t \mapsto A\cos(\omega t-\alpha)$ are sketched in Figure 1-2. We call A the **amplitude**, ω the **frequency**, and α the **phase shift**.

Thus the solution is harmonic (i.e., back and forth) motion and different choices of x_0, v_0 affect the amplitude and phase but not the frequency.

Notice that in order to fix a specific solution, initial conditions are necessary. This is natural because we expect that particles which start out with different initial positions and velocities will take different paths. The procedure is quite analogous to specifying the constant in an indefinite integral.

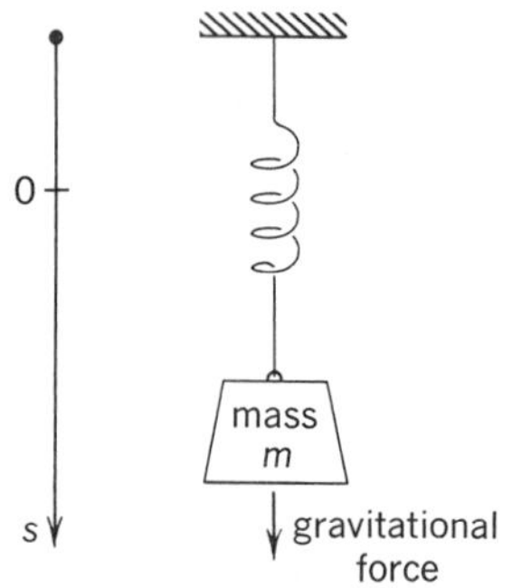

FIGURE 1-3

Let us investigate other equations arising from Newton's law. Suppose we have a weight on a spring which is suspended vertically (Figure 1-3). Now a constant force of gravity mg operates on this system. If the distance $s(t)$ is measured downwards from the equilibrium position of the spring, then the equation is, by (1),

$$ms'' = -ks + mg$$

This equation is easily put into the form of equation (2) above by letting $h = s - mg/k$, for then

$$mh'' = ms'' = -ks + mg = -kh$$

Thus the solution s is obtained simply by shifting the solution to equation (2). We see that gravity does not affect the behavior of the solution—rather it just pulls the weight down slightly. This is what we might expect intuitively.

Next, suppose we take our particle on a spring satisfying $ms'' = -ks$ [so the solution is given by (3)] and place it in some viscous fluid such as air or water, or allow frictional forces of sliding (see Figure 1-1). This introduces additional forces which we assume are proportional to the velocity. Then equation (1) becomes

$$ms'' = -ks - \mu s' \tag{4}$$

where $\mu > 0$ is a constant (depending on the materials). We should expect that the solutions to this equation behave rather differently than those of the equation for simple harmonic motion. Indeed, the particle should be slowed down by the friction and eventually come to rest. Thus, the term $-\mu s'$

should behave like a **damping** term or a **decay** term. We can see this directly by transforming the equation back into simple harmonic form as follows: Define the function h by $h(t) = e^{\gamma t}s(t)$ where γ is a constant to be determined. Then

$$h'(t) = \gamma e^{\gamma t}s(t) + e^{\gamma t}s'(t)$$

and so

$$h''(t) = \gamma^2 e^{\gamma t}s(t) + 2\gamma e^{\gamma t}s'(t) + e^{\gamma t}s''(t)$$

Now, substituting for $s''(t)$, the term $2\gamma s'(t)+s''(t)$ becomes

$$\frac{1}{m}(-ks-\mu s'(t)) + 2\gamma s'(t)$$

The term with $s'(t)$ will cancel if we chose

$$2\gamma = \frac{\mu}{m}$$

and we obtain

$$h'' = \gamma^2 e^{\gamma t}s(t) - \frac{k}{m}e^{\gamma t}s(t)$$

or

$$h'' = \left(\frac{\mu^2}{4m^2} - \frac{k}{m}\right)h(t) = -\frac{\tilde{k}}{m}h(t)$$

where $\tilde{k} = k - \mu^2/4m$. Now if $k > \mu^2/4m$ then h satisfies an equation which is a simple harmonic equation, and so it has solution (3) or (3′) with this new $\tilde{k}$. Thus

$$s(t) = e^{-\mu t/2m}h(t) = e^{-\mu t/2m}\left(c_1 \cos\sqrt{\frac{\tilde{k}}{m}}\,t + c_2 \sin\sqrt{\frac{\tilde{k}}{m}}\,t\right)$$

is the required solution. The term $e^{-\mu t/2m} \to 0$ as $t \to \infty$ and this damps out the solution. The solution may then be characterized as **damped oscillations.** The term $e^{\gamma t}$ used in finding the solution is called an **integrating factor**, and we deal with this technique again later in the chapter. The condition $k > \mu^2/4m$ has a simple meaning, which is that the frictional forces should not be too large; if the spring were put in thick molasses the situation would not be one of damped oscillations but rather a slow drifting to equilibrium. In this case the sign of $\tilde{k}$ changes and sines and cosines must be replaced by the hyperbolic functions sinh and cosh. We work out this case in §5; the various possibilities are sketched in Figure 1-4.

The equations studied so far are of **second order**, that is, the second derivative s'' is the highest derivative involved. An example of a first-order equation is the equation of growth or decay considered earlier (§8.5). This equation had the form $f' = kf$, to be solved for $f(t)$. The solution is

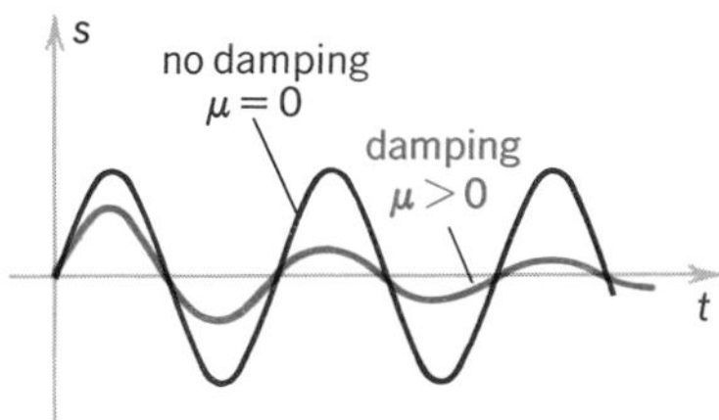

(a) damping, $\mu < \sqrt{2km}$

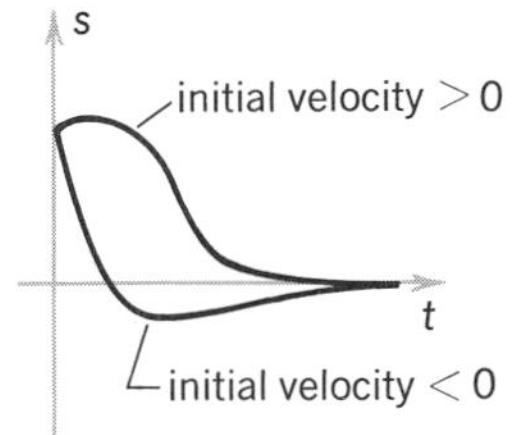

(b) overcritical damping, $\mu > \sqrt{2km}$

FIGURE 1-4

$f(t) = Ae^{kt}$ where the constant A is to be determined from the initial value $f(0) = A$. See exercise 5 for another example.

The above collection of examples is an indication of why differential equations are important and how they arise. More complicated physical situations quickly lead to more complicated equations. It is therefore desirable to have some general methods for dealing with such equations, and this is our primary goal in the ensuing sections.

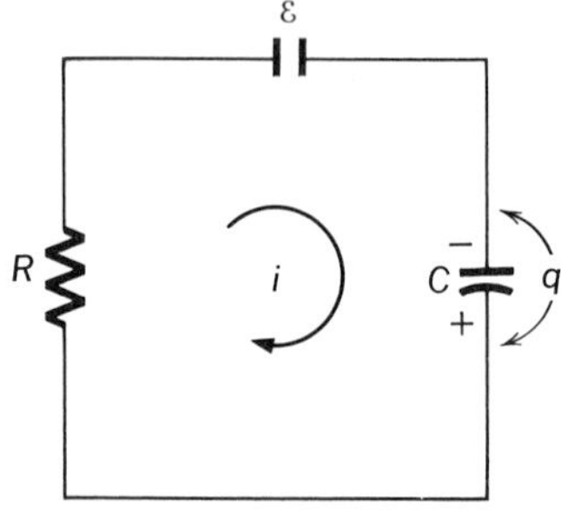

FIGURE 1-5

EXERCISES

Solve each of the following equations.

1. $s''(t) + 3s(t) = 0, \quad s(0) = 0, \quad s'(0) = 1$
2. $s''(t) + 4s(t) = 0, \quad s(0) = 1, \quad s'(0) = 0$
3. $s''(t) - s(t) = 0, \quad s(0) = 0, \quad s'(0) = 1$
4. $s'(t) + s(t) = 0, \quad s(0) = 1$
5. Figure 1-5 shows a typical RC-circuit which has the equation

$$\mathscr{E} = R\frac{dq}{dt} + \frac{q}{C}$$

(q represents charge and $i = dq/dt$, the current; $\mathscr{E}$, R, C are constants.) This equation is a modification of the equation of decay by the inclusion of the term $\mathscr{E}$. Consider the transformation $h(t) = e^{\gamma t}q(t)/C$, and solve the equation if $q(0) = 0$. Sketch the graph.

Solve the following equations and interpret physically.

6. $s''(t) + 3s(t) + s'(t) = 0, \quad s(0) = 0, \quad s'(0) = 1$
7. $s''(t) + 5s(t) + 25s'(t) = 0, \quad s(0) = 1, \quad s'(0) = 0$
8. Radium has a half-life of about 1600 years. If we start with 100 milligrams, show that about 65 milligrams are left after 1000 years.

2 CLASSIFICATION OF ORDINARY DIFFERENTIAL EQUATIONS

Now that we are familiar with some specific examples, let us begin to study more general aspects of differential equations. Recall that in the previous section we used the notation $s(t)$; however, at this point it is convenient to change and consider a differential equation for a function $y(x)$.

An **ordinary differential equation** is simply an equation involving $x, y(x)$, and derivatives of y. Such an equation has the form

$$F\left(x, y, \frac{dy}{dx}, \frac{d^2y}{dx^2}, \ldots, \frac{d^ny}{dx^n}\right) = 0 \tag{1}$$

Some examples of ordinary differential equations were given in §1; other examples are

(a) $\dfrac{dy}{dx} + \dfrac{d^3 y}{dx^3} = 1$

(b) $3x^2\,dx = 2y^3\,dy + 16$

(c) $3\dfrac{d^2 y}{dx^2} - 2\dfrac{dy}{dx} + 4x^2 y = 0$

(d) $\dfrac{x-y}{x+y}\,dx + \dfrac{y-x}{x-y}\,dy = 0$

(e) $\dfrac{dy}{dx} + y + 3 = 0$

(f) $\dfrac{dy}{dx} = 4\left(\dfrac{d^2 y}{dx^2}\right)^2$

An equation of the form (1), in which the derivative of highest order is the nth (that is, $(d^n y/dx^n)$ is the highest derivative occurring) is called an **nth-order differential equation.** Hence, for the equations above, (b), (d), and (e) are first-order, (c) and (f) are second-order, and (a) is third-order.

When the left-hand side of (1) has the form of a polynomial expression in the derivative involved, then the exponent of the highest-order derivative is called the **degree** of the equation. For example, (a)–(e) above have degree one, while (f) has degree two. Recognizing the order and degree of a differential equation is an important step in finding its solution.

A **linear differential equation** of order n is a differential equation of the form

$$p_0(x)\frac{d^n y}{dx^n} + p_1(x)\frac{d^{n-1} y}{dx^{n-1}} + \cdots + p_{n-1}(x)\frac{dy}{dx} + p_n(x)y + p_{n+1}(x) = 0 \quad (2)$$

where $p_0, p_1, \ldots, p_{n+1}$ are functions of x, and $p_0(x_0) \neq 0$ for some x_0. All other differential equations are called **nonlinear**. The first-order linear equation can always be written in the form

$$\frac{dy}{dx} + p(x)y = q(x), \qquad \text{if } p_0(x) \neq 0 \quad (3)$$

and similarly the second-order linear differential equation can always be put in the form

$$\frac{d^2 y}{dx^2} + p(x)\frac{dy}{dx} + q(x)y = r(x), \qquad \text{if } p_0(x) \neq 0 \quad (4)$$

If the coefficients $p(x)$, $q(x)$ are constants then, naturally enough, (4) is called a second-order linear differential equation with constant coefficients. Some illustrations of the terminology follow.

Equation	Order	Degree	Comments
$x^2 \dfrac{dy}{dx} - 3xy - 2x^2 + 4x^4 = 0$	one	one	linear
$(x+y)^2 \dfrac{dy}{dx} - 1 = 0$	one	one	nonlinear because of y^2
$\dfrac{d^2 y}{dx^2} + y - x^3 = 0$	two	one	linear, constant coefficients
$\dfrac{dy}{dx} - e^{x+y} = 0$	one	one	nonlinear
$\left(\dfrac{d^2 y}{dx^2}\right)^2 + xy \dfrac{dy}{dx} = 0$	two	two	nonlinear
$5\left(\dfrac{dy}{dx}\right)^3 + 6y = 0$	one	three	nonlinear
$\dfrac{dy}{dx} + 3y - 2\sin x = 0$	one	one	linear, constant coefficients

A function $y=f(x)$ is called a **solution** of a given differential equation on an interval I if the equation is satisfied for all $x \in I$ when y and its derivatives are substituted into the equation. For example, if we are given the second-order differential equation

$$F\left(x, y, \frac{dy}{dx}, \frac{d^2 y}{dx^2}\right) = 0 \tag{5}$$

then a solution of (5) is a function $y=f(x)$ such that f is at least twice differentiable and

$$F(x, f(x), f'(x) f''(x)) = 0 \tag{6}$$

for all values of x in some interval I. As a more specific example, a solution of the first-order equation

$$\frac{dy}{dx} - x = 0$$

is the function $y=f(x)$ where $y=\frac{1}{2}x^2+5$. This follows from the elementary fact that the derivative of $f: x \mapsto \frac{1}{2}x^2+5$ is $f': x \mapsto x$. Clearly, there are numerous other solutions of the given equation; indeed, any expression of the form $y=\frac{1}{2}x^2+c$, where c is a constant, is also a solution. We therefore refer to c as an **arbitrary constant**. Usually, the solution to an nth-order differential equation will have n arbitrary constants; this solution is called the **general solution**. Hence, $y=\frac{1}{2}x^2+c$ is the general solution of $dy/dx - x = 0$. The solution obtained by giving the c's some particular set of values is called a **particular solution** of the differential equation. Therefore,

$\frac{1}{2}x^2+5$ is a particular solution of $dy/dx-x=0$. As another example of the use of this terminology, consider the second-order differential equation

$$\frac{d^2y}{dx^2}+y=0 \tag{7}$$

We recall from §1 that $\sin x$ and $\cos x$ are particular solutions of (7). The general solution is $c_1 \sin x+c_2 \cos x$. As we expected from the remark above, equation (7) has a general solution involving two arbitrary constants; a particular solution of (7) is obtained by choosing values for these arbitrary constants, so, for example, $c_1=9$ and $c_2=-7$ yield the solution $f(x)=9\sin x-7\cos x$.

The main problem in working with differential equations is to find the general solution; once this is known it is relatively simple to determine the arbitrary constants (that is, find a particular solution) if **initial conditions** are specified. The physical significance of initial conditions was explained in §1. Sometimes the terminology **boundary conditions** is used.

Let us recall the example used above, that of finding a solution of $dy/dx-x=0$. However, suppose we require that the value of the solution is 3 when x has the value 7. Hence, we must find a differentiable function $f(x)$ such that $3=f(7)$, and $f'(x)-x=0$ for all x in the domain of f. We already know that $\frac{1}{2}x^2+c$ is the general solution for this equation; let us see if we can determine a particular solution $f(x)$ by giving c a special value such that f satisfies the given conditions. Substituting 3 and 7 into the general solution, we obtain

$$3=\tfrac{1}{2}7^2+c$$

so

$$c=-\tfrac{43}{2}$$

In other words, by specifying a value for the arbitrary constant in the general solution of this first-order differential equation, we can find a particular solution which satisfies a required **initial-value** condition; in this case the solution $y=\frac{1}{2}x^2-\frac{43}{2}$ satisfies $3=f(7)$. Other initial-value problems were considered in §1.

EXERCISES

Find the solution of each of the following differential equations with the stated initial conditions.

1. $y'+y=0, \quad y(0)=1$
2. $y''+y=0, \quad y(0)=1, \quad y'(0)=0$
3. $y'''+x=0, \quad y(0)=0, \quad y'(0)=1, \quad y''(0)=0$
4. $y''+y'+y=0, \quad y(0)=3, \quad y'(0)=0$

5. Show that each of the following solutions to $y''+y=0$ gives all solutions of the form $A\sin x+B\cos x$.
   (a) $C_1\sin(x+C_2)$
   (b) $C_1\cos x+3eC_2\sin x$
   (c) $C_1\cos(x+3C_2)$
6. Show that each of the following solutions to $y''+y=0$ is equivalent to a solution given in terms of only one arbitrary constant.
   (a) $C_1\sin x + 3eC_2\sin x$
   (b) $C_1\sin x + C_2\cos(x-\frac{9}{2}\pi)$
   (c) $C_1\sin x + C_2\sin(-x)$

3 SEPARABLE FIRST-ORDER EQUATIONS

Consider a first-order equation

$$F\left(x, y, \frac{dy}{dx}\right) = 0$$

As a simplification, let us suppose we can solve for dy/dx so the equation takes the form

$$\frac{dy}{dx} = f(x, y)$$

For example, the equation of growth or decay $dy/dx = ky$ has this form.

We shall find that most methods for solving differential equations rely on hypotheses about the expression $f(x, y)$; in this section we assume that the variables can be **separated** in the sense that we can write

$$f(x, y) = \frac{g(x)}{h(y)}$$

that is, the expression $f(x, y)$ can be manipulated into the form of a ratio of an expression involving x alone to an expression involving y alone. Hence the **separable equation** can always be put in the form

$$\frac{dy}{dx} = \frac{g(x)}{h(y)} \tag{1}$$

This is the same as

$$h(y)\,dy = g(x)\,dx \tag{2}$$

in which all y terms are collected with dy and all x terms with dx. If h and g are continuous functions then the differentials $h(y)\,dy$ and $g(x)\,dx$ are, by the Fundamental Theorem of Calculus, differentials of functions H and G, where $H'(y) = h(y)$ and $G'(x) = g(x)$. That is, the functions H and G are antiderivatives of the functions h and g. Since the derivative of a constant is zero, a general solution of (2) is found by using integration to express y in terms of x in the equation

$$\int h(y)\,dy = \int g(x)\,dx + c \qquad (3)$$

where c is an arbitrary constant.

Example 1 The equation of growth or decay $dy/dx = ky$ is separable, therefore integrating, we have

$$\ln|y| = kx + C_1$$

or

$$y = Ce^{kx}$$

Example 2 Consider the first-order equation

$$\frac{dy}{dx} = -\frac{(x^3+x^2)y}{x^2(y^3+2y)}$$

First we simplify the right-hand side and write

$$\frac{dy}{dx} = -\frac{(x^3+x^2)}{x^2}\frac{y}{(y^3+2y)} = \frac{x+1}{y^2+2}$$

Clearing fractions, this becomes

$$(y^2+2)\,dy = (x+1)\,dx$$

Antiderivatives of the functions y^2+2 and $x+1$ are $\frac{1}{3}y^3+2y+c_1$ and $\frac{1}{2}x^2+x+c_2$, so that $\frac{1}{3}y^3+2y = \frac{1}{2}x^2+x+c$, where $c = c_2 - c_1$ is an arbitrary constant. (This equation involving x and y is called an **implicit solution** to the problem, since we have not solved for y explicitly in terms of x.)

Example 3 Solve the first-order equation

$$(x^2+1)(y^2-1)\,dx + xy\,dy = 0$$

Dividing through by y^2-1 and x (with the agreement that we do not consider points where division by zero occurs) this is the same as

$$\left(\frac{x^2+1}{x}\right)dx + \left(\frac{y}{y^2-1}\right)dy = 0$$

In performing the above division by $x(y^2-1)$, we did this under the tacit assumptions that $x \neq 0$ and $y \neq \pm 1$. Since our purpose here is to gain facility in solving separable equations, we shall always make the assumption that any factors by which we divide are not zero.

After writing

$$y^2 - 1 = -(1-y^2)$$

and

$$\frac{x^2+1}{x} = x + \frac{1}{x}$$

we obtain

$$\left(x+\frac{1}{x}\right)dx = \left(\frac{y}{1-y^2}\right)dy$$

For the left and right-hand sides we have the antiderivatives

$$\int \left(x+\frac{1}{x}\right)dx = \tfrac{1}{2}x^2 + \ln|x| + c_1$$

and

$$\int \left(\frac{y}{1-y^2}\right)dy = -\tfrac{1}{2}\ln|1-y^2| + c_2$$

so that the implicit solution of this problem can be written in the form

$$\tfrac{1}{2}x^2 + \ln|x| + \tfrac{1}{2}\ln|1-y^2| = c$$

where $c = c_2 - c_1$ is an arbitrary constant.

Our next example shows how a differential equation can serve to solve a geometric locus problem.

Example 4 Let k be a positive constant. Find a curve with the property that for every point P on the curve the distance from P to the point where the tangent line at P intersects the y axis is equal to k. Let the curve we seek be $y = y(x)$. If $P = (x_0, y_0)$ is a point on the curve, the tangent line at P has slope y'_0, and the tangent line has equation $(x-x_0)y'_0 = y - y_0$. Setting $x = 0$ we compute the y intercept of the tangent line to be $(0, -x_0 y'_0 + y_0)$. Using the distance formula, we find that we must have

$$\begin{aligned} k^2 &= x_0^2 + (y_0 - (y_0 - x_0 y'_0))^2 \\ &= x_0^2 + x_0^2 y_0'^2 \end{aligned}$$

Since $P = (x_0, y_0)$ is an arbitrary point on the curve, we now drop the subscripts to obtain the differential equation that the curve must satisfy:

$$k^2 = x^2 + x^2(y')^2$$

or

$$k^2 = x^2\left[1+\left(\frac{dy}{dx}\right)^2\right]$$

We solve for dy/dx:

$$\frac{dy}{dx} = \pm\sqrt{\frac{k^2-x^2}{x^2}} = \pm\frac{\sqrt{k^2-x^2}}{x}$$

This equation has variables separable. The solution is, therefore,

$$\pm \int dy = \int \frac{\sqrt{k^2-x^2}}{x}\, dx$$

We can find the integral on the right-hand side by means of trigonometric substitutions (see exercise 14) or, alternatively, by consulting a table of integrals such as the one in the Appendix. Both methods yield

$$\int \frac{\sqrt{k^2-x^2}}{x}\, dx = \sqrt{k^2-x^2} - k\ln\frac{k+\sqrt{k^2-x^2}}{|x|} + C_1$$

Therefore we have

$$\pm y + C = \sqrt{k^2-x^2} - k\ln\frac{k^2+\sqrt{k^2-x^2}}{|x|}$$

The above solution gives pairs of curves of the form $y+C = \pm f(x)$. We see that f is symmetric in x and that $f(x)$ is undefined when $x = 0$ or when $|x| > k$. The solution curve, called a **tractrix**, is sketched in Figure 3-1.

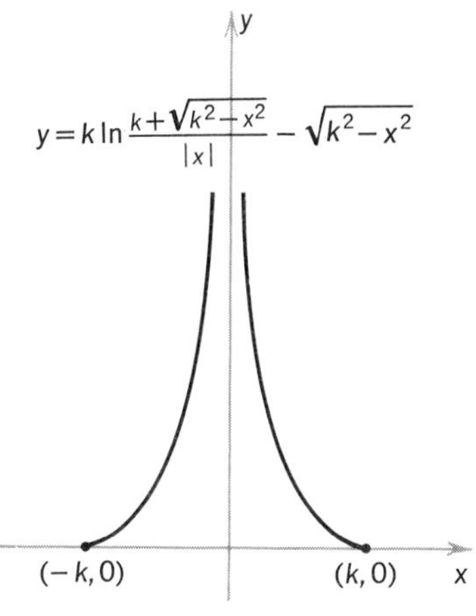

FIGURE 3-1

We shall conclude this section with a discussion of certain substitution techniques for reducing a second-order differential equation to one of first order. The solution of the first-order equation then leads to the solution of the original second-order equation.

Suppose a second-order differential equation takes the special form

$$F(x, y', y'') = 0$$

in which the variable y itself does not appear. In this case we say that the dependent variable is missing. If we make the substitution $u = y'$ we have $u' = y''$ and the equation becomes

$$F(x, u, u') = 0$$

which is an equation of the first order. Now if we can solve this first-order equation for $u = u(x)$, then we can find the solution $y = y(x)$ to the original equation performing the integration $y = \int u(x)\, dx$.

Example 5 In §8.4 we derived the differential equation for the hanging cable:

$$\frac{d^2y}{dx^2} = \frac{w}{H}\sqrt{1+\left(\frac{dy}{dx}\right)^2}$$

The locus of the cable is given by the solution to the equation which satisfies the initial conditions $y'(0) = 0$, $y(0) = H/w$. This is a second-order differential equation with the dependent variable missing. When we set $u = dy/dx$ the equation becomes

$$\frac{du}{dx} = \frac{w}{H}\sqrt{1+u^2}$$

We separate the variables to find

$$\frac{du}{\sqrt{1+u^2}} = \frac{w}{H}dx$$

which has solution

$$\sinh^{-1} u = \frac{w}{H}x + C_1$$

The initial condition $y'(0) = u(0) = 0$ determines that $C_1 = 0$. So we have

$$\sinh^{-1} u = \frac{w}{H}x$$

or

$$u = \sinh\frac{w}{H}x$$

We integrate once more to find $y(x)$:

$$y(x) = \int u(x)\,dx = \int \sinh\frac{w}{H}x\,dx = \frac{H}{w}\cosh\frac{w}{H}x + C$$

The initial condition $y(0) = H/w$ now determines that $C = 0$, so our solution is $y = (H/w)\cosh(w/H)x$.

If a differential equation takes the form

$$F(y, y', y'') = 0$$

we say that the independent variable is missing. Let us again set $u = y' = dy/dx$. Since

$$\left.\frac{d(dy/dx)}{dy}\right|_{y=y(x)}\frac{dy}{dx} = \frac{d(dy/dx)}{dx} = \frac{d^2y}{dx^2}$$

we have

$$y'' = \frac{d^2y}{dx^2} = u\frac{du}{dy}$$

The substitution $u = y' = dy/dx$ therefore transforms the equation into

$$F\left(y, u, u\frac{du}{dy}\right) = 0$$

which is a first-order equation. If we solve this new equation for $u = u(y)$, then we can solve $dy/dx = u(y)$ to find the solution of the original equation.

Example 6 Consider the equation for harmonic motion

$$\frac{d^2x}{dt^2} = -kx, \qquad k > 0$$

We observe that this equation has the independent variable missing; hence, by setting $u = dx/dt$ we transform our equation into

$$u\frac{du}{dx} = -kx$$

Separating the variables and integrating yields

$$u^2 = -kx^2 + C$$

or

$$u = \pm\sqrt{C - kx^2}$$

Since $u = dx/dt$ we must now solve the equation

$$\frac{dx}{dt} = \pm\sqrt{C - kx^2}$$

or

$$\frac{dx}{\sqrt{C - kx^2}} = \pm dt$$

Integrating we obtain

$$\frac{1}{\sqrt{k}}\sin^{-1}\left(\frac{x}{C_2}\right) = \pm(t + C_1), \qquad C_2 = \sqrt{C}$$

Taking sines of both sides, we find

$$x = C_2 \sin(\pm\sqrt{k}(t + C_1))$$

Since $\sin(-\alpha) = -\sin\alpha$, we can write this as

$$x = \pm C_2 \sin(\sqrt{k}\,t + C_3), \qquad C_3 = \sqrt{k}\,C_1$$

Finally since C_2 is an arbitrary positive constant, the solution is simply

$$x = C_4 \sin(\sqrt{k}\,t + C_3)$$

where C_4 and C_3 are arbitrary constants.

EXERCISES

In exercises 1–12 find the solution (implicit or explicit) to the differential equation.

1. $x\,dy + (2y - c)\,dx = 0$, $\quad c$ a constant
2. $dy/dx = e^{x-y}$
3. $(3x + 1)\,dx + e^{x+y}\,dy = 0$
4. $\ln x \dfrac{dx}{dy} = \dfrac{x}{y}$
5. $(\cos^2\theta - \sin^2\theta)\,dr + 2r\sin\theta\cos\theta\,d\theta = 0$

6. $y\,dx - x\,dy = y^2\,dx + dy$
7. $y^3\,dx + \sqrt{1-x^2}\,dy = 0$
8. $x^2(y^2+1)\,dx + y\sqrt{x^3+1}\,dy = 0$
9. $x\dfrac{d^2y}{dx^2} + \dfrac{dy}{dx} = 0$
10. $2ay'y'' = 1$
11. $(1+x^2)\dfrac{d^2y}{dx^2} + x\dfrac{dy}{dx} = 0$
12. $y\dfrac{d^2y}{dx^2} = 1 - \left(\dfrac{dy}{dx}\right)^2$
13. Find all curves such that the tangent line to every point on the curve passes through the origin.
14. Compute the integral
$$\int \frac{\sqrt{k^2-x^2}}{x}\,dx$$
by first making the substitution $x = k\sin\theta$ then $z = \tan\frac{1}{2}\theta$. This second substitution yields
$$\cos\theta = \frac{1-z^2}{1+z^2}, \quad \sin\theta = \frac{2z}{1+z^2} \quad \text{and} \quad d\theta = \frac{2\,dz}{1+z^2}$$
The integral in z is then solved by the method of partial fractions.
15. A speedy hare starts at the origin and runs south along the y axis. A playful hound starting at the same moment from the point $(1,0)$ pursues the hare. The hare and the hound run at exactly the same speed at every moment. Describe the path followed by the hound, if at every instant the hound is aiming towards the hare.
16. A differential equation of the form $y' = F(y/x)$ is called **homogeneous**. Show that the substitution $v = y/x$ reduces such an equation to one with variables separable. Then use this technique to solve the following differential equations:
    (a) $(x+y)\,dy + (x-y)\,dx = 0$
    (b) $(x^2+y^2)\,dx + 2xy\,dy = 0$
    (c) $(xe^{y/x}+y)\,dx - x\,dy = 0$

4 LINEAR FIRST-ORDER DIFFERENTIAL EQUATIONS

The linear first-order differential equation has the standard form

$$\frac{dy}{dx} + p(x)y = q(x) \tag{1}$$

Here the assumption we make about $dy/dx = f(x,y)$ is that $f(x,y) = q(x) - p(x)y$ for some continuous functions q and p. Equation (1) is called **homogeneous** if $q(x) = 0$ for all x in its domain; otherwise (1) is **nonhomogeneous**. As we shall see in this section, there is an important relationship between the general solution of equation (1) and the general solution of the **associated homogeneous equation**

$$\frac{dy}{dx} + p(x)y = 0 \tag{2}$$

First we shall find the solution to the homogeneous equation and then apply this result to find the solution in the nonhomogeneous case.

Note that equation (2) is separable:

$$\frac{dy}{y} + p(x)\,dx = 0$$

or

$$\frac{dy}{y} = -p(x)\,dx \tag{3}$$

We integrate to find

$$\ln|y| = -p(x)\,dx + C_1$$

Taking the exponential of both sides, we have

$$\begin{aligned} |y| &= \exp\left(-\int p(x)\,dx + C_1\right) \\ &= e^{C_1}\exp\left(-\int p(x)\,dx\right) \end{aligned}$$

or

$$y = \pm e^{C_1}\exp\left(-\int p(x)\,dx\right)$$

Since $y = 0$ is also a solution of (2), the general solution of (2) is therefore

$$y = C\exp\left(-\int p(x)\,dx\right) \tag{4}$$

We note an important relationship between the general solution of (2) and the solution of (1). If $y_1(x)$ and $y_2(x)$ are any two solutions of (1), then $y_1(x) - y_2(x)$ is a solution of the associated homogeneous equation (1) as is quickly verified by substitution:

$$\begin{aligned} \frac{d(y_1(x) - y_2(x))}{dx} + p(x)(y_1(x) - y_2(x)) &= \frac{d(y_1(x))}{dx} + p(x)y_1(x) \\ &\quad - \left(\frac{d(y_2(x))}{dx} + p(x)y_2(x)\right) \\ &= q(x) - q(x) = 0 \end{aligned}$$

Conversely, if $h_1(x)$ is a solution of the homogeneous equation (2), and if $y_1(x)$ is a solution of (1), then the sum $h_1(x) + y_1(x)$ is again a solution of the nonhomogeneous equation (1). Therefore, the general solution of (1) is

$$h(x) + y_1(x)$$

where $h(x) = C\exp(-\int p(x)\,dx)$ is the general solution of the homogeneous equation (2) and where $y_1(x)$ is any particular solution to the nonhomogeneous equation (1).

Next we employ a method called **variation of parameters** to derive a particular solution of the nonhomogeneous equation from the known solution (4) to the homogeneous equation.

The method is to look for a solution of the form

$$y_1(x) = g(x)\exp\left(-\int p(x)\,dx\right) \tag{5}$$

What we have done is to replace the constant which appears in the solution of the homogeneous equation (4) by a function $g(x)$. We shall see that the assumption that a solution of the form (5) exists serves to explicitly determine $g(x)$, and that this gives the solution. Let us write equation (1) in the form

$$dy + p(x)\,y\,dx = q(x)\,dx \tag{6}$$

and substitute (5) in (6):

$$d\left(g(x)\exp\left(-\int p(x)\,dx\right)\right) + p(x)g(x)\exp\left(-\int p(x)\,dx\right)dx = q(x)\,dx$$

Applying the rule that $d(uv) = udv + vdu$ to the first term on the left-hand side and using the fact that $d(\exp(H(x))) = H'(x)\exp(H(x))\,dx$, we obtain

$$\begin{aligned} &g(x)(-p(x))\exp\left(-\int p(x)\,dx\right)dx \\ &+g'(x)\exp\left(-\int p(x)\,dx\right)dx \\ &+p(x)g(x)\exp\left(-\int p(x)\,dx\right)dx = q(x)\,dx \end{aligned}$$

Since the first and third terms cancel, we now have the simpler differential equation

$$g'(x)\exp\left(-\int p(x)\,dx\right) = q(x) \tag{7}$$

after dividing through by dx. Observing that $\exp(-a) = 1/\exp a$, we rewrite (7) in the form

$$g'(x) = q(x)\exp\left(\int p(x)\,dx\right) \tag{8}$$

and integrate to obtain

$$g(x) = \int \left[q(x)\exp\left(\int p(x)\,dx\right)\right]dx$$

Hence we have solved for the function $g(x)$ in (5) in terms of the known functions $p(x)$ and $q(x)$; therefore a particular solution to (1) is

$$y_1(x) = \int \left[q(x)\exp\left(\int p(x)\,dx\right)\right]dx\,\exp\left(-\int p(x)\,dx\right)$$

and by the above remarks the general solution to (1) is

$$y = C\exp\left(-\int p(x)\,dx\right) + \exp\left(-\int p(x)\,dx\right)g(x) \tag{9}$$

where

$$g(x) = \int \left[q(x)\exp\left(\int p(x)\,dx\right)\right]dx$$

Therefore to solve the first-order linear differential equation (1), we must take the given functions $p(x)$ and $q(x)$ and consecutively find two antiderivatives; first $P(x)$ of $p(x)$ and second one of $q(x)\,e^{P(x)}$. The general solution is then

$$\boldsymbol{y = Ce^{-P(x)} + e^{-P(x)}\int q(x)\,e^{P(x)}\,dx} \tag{10}$$

where C denotes an arbitrary constant.

Example 1 Let us find the general solution of the first-order equation

$$x\frac{dy}{dx} + (x+1)\,y = x^3$$

Initially, this is not of the form (1), but we divide through by x to obtain

$$\frac{dy}{dx} + \left(\frac{x+1}{x}\right)y = x^2,$$

Here, $p(x) = (x+1)/x$ and $q(x) = x^2$. Since $(x+1)/x = 1 + (1/x)$ the antiderivative of $p(x)$ is $P(x) = x + \ln|x|$. We have $q(x) = x^2$, and the other antiderivative to be found is that of

$$\begin{aligned} q(x)\exp P(x) &= x^2\,e^{(x+\ln|x|)} \\ &= x^2\,e^x\,e^{\ln|x|} \\ &= \pm x^3\,e^x \end{aligned}$$

where we have used well-known properties of the exponential function. Performing three successive integrations by parts, we find the antiderivative $e^x(x^3 - 3x^2 + 6x - 6)$ for $x^3\,e^x$; hence, the general solution of (10), after simplifications, is

$$y = \frac{C}{x}e^{-x} + x^2 - 3x + 6 - \frac{6}{x}$$

Example 2 Let us solve the initial-value problem

$$(x^2+1)\frac{dy}{dx} + 4xy = x$$

where the initial condition is that the value of y is 1 when the value of x is 2.

Dividing through by x^2+1 we obtain the equation in the form (1):

$$\frac{dy}{dx} + \frac{4x}{x^2+1}y = \frac{x}{x^2+1} \tag{11}$$

Here $p(x) = 4x/(x^2+1)$ and $q(x) = x/(x^2+1)$. The reader can quickly verify that $P'(x) = p(x)$ if $P(x) = \ln(x^2+1)^2$. The other antiderivative we need is that of

$$q(x)\,e^{P(x)} = \frac{x}{x^2+1}\,e^{\ln(x^2+1)^2} = x(x^2+1)$$

but clearly $\int x(x^2+1)\,dx = x^4/4 + x^2/2$. So after simplification we can write the general solution of (11) as

$$y = \frac{C}{(x^2+1)^2} + \frac{1}{(x^2+1)^2}\left(\frac{x^4}{4} + \frac{x^2}{2}\right)$$

To determine the value of C so that $y = 1$ when $x = 2$ we substitute these values in the general solution and solve for C, obtaining $C = 19$. Therefore, the particular solution of (10) satisfying the given initial condition is

$$y = \frac{19}{(x^2+1)^2} + \frac{1}{(x^2+1)^2}\left(\frac{x^4}{4} + \frac{x^2}{2}\right)$$

Example 3 A small lake containing one million cubic feet of water is fed by streams and rainfall with fresh water at the rate of 3,500 cubic feet per day. Waste water from a newly constructed industrial plant makes its way into the lake at a rate of 1,500 cubic feet per day. This waste water is 2% noxious matter. When the level of noxious matter in the lake reaches 0.5% a serious ecological imbalance will result. Is the lake threatened in the near future?

Analyzing the given information, we see that every day 5,000 cubic feet of water enter the lake and 5,000 cubic feet of water are drained from the lake (since the total amount of water is constant); moreover, 30 cubic feet of noxious matter enter the lake each day. If we let y denote the quantity of noxious material in the lake at time t, where t is time measured in days, then every day $(5{,}000/1{,}000{,}000)y$ cubic feet of noxious matter is drained from the lake (assuming naturally that the waters entering the lake and the lake water mix). Hence y satisfies the differential equation

$$\frac{dy}{dt} = 30 - \frac{1}{200}y$$

We establish the initial condition $y(0) = 0$ and measure time from when the waste water from the plant first begins entering the lake. This first-order linear equation is easily solved. The antiderivative of $p(t) = 1/200$ is $t/200$ and the antiderivative of $q(t)\exp(\int p(t)\,dt) = 30\,e^{t/200}$ is $6{,}000\,e^{t/200}$. Hence, the general solution to our differential equation is

$$y = Ce^{-t/200} + e^{-t/200}(6{,}000\,e^{t/200}) = Ce^{-t/200} + 6{,}000$$

We determine the value of C from the initial coniition $y(0) = 0$:

$$0 = Ce^0 + 6{,}000 \quad \text{or} \quad C = -6{,}000$$

So the solution to the initial-value problem is

$$y = 6{,}000(1-e^{-t/200})$$

Pollution will occur when $y \geqslant 5{,}000$ or when $e^{-t/200} \leqslant 1/6$. Consulting a table for the exponential function (see the Appendix) we see that this will happen when $t/200 \geqslant 1.8$ or when $t \geqslant 360$. Therefore, if action is not taken, the lake will be polluted in about one year's time.

EXERCISES

In exercises 1–5 solve the differential equation; find the solution satisfying the initial conditions where given.

1. $x\dfrac{dy}{dx} - 2y = x^2 + x;\quad y = 1$ when $x = 1$

2. $\dfrac{dy}{dx} + 2y = e^{-x}$

3. $\dfrac{dy}{dx} + \dfrac{xy}{x^2+1} = x^3;\quad y(0) = 1$

4. $\dfrac{dy}{dx} - (\tan x)y = \sin x;\quad y = 1$ when $x = 4$

5. $\cosh x\,dy + (y\sinh x + e^x)\,dx = 0$

6. The equation $y'+p(x)y = q(x)y^n$ is known as **Bernoulli's equation**. Solve this equation using variation of parameters; that is, look for a solution of the form $y = g(x)h(x)$ where $Ch(x)$ is the solution of the homogeneous equation $y'+p(x)y = 0$.

7. Consider the equation $y' + (\cos x)y = e^{-\sin x}$. (a) Find a solution $y = f(x)$ such that $f(\pi) = \pi$. (b) Show that *any* solution $y = g(x)$ has the property that $g(\pi k) - g(0) = \pi k$, for all integers k.

8. (a) The plant involved in Example 3 installs purification equipment which reduces the quantity of noxious matter in its waste water to 1.5%. The company then claims that the lake is now not threatened with 0.5% pollution. Show that the company's claims are justified. (b) However, ecologists with strong public support judge that to maintain a margin of safety the pollution level in the lake should not be allowed to pass 0.25%. To what level should the company reduce the noxious material in the plant's waste water in order to maintain good relations with the community?

9. The water from the lake in Example 3 is carried downstream several miles until it enters Lake Cayahoga, a larger lake of two million cubic feet capacity. In addition to the water coming from the lake upstream, Lake Cayahoga is fed 5,000 cubic feet of fresh water daily from other sources. Working with the figures of Example 3, determine the level of pollution in Lake Cayahoga in some three years' time. (Use three years = 1,000 days.)

5 LINEAR SECOND-ORDER DIFFERENTIAL EQUATIONS

In §1 we studied the second-order equation describing harmonic motion

$$ms'' + ks = 0 \tag{1}$$

We found that the general solution to this equation is $s(t) = C_1 \sin\sqrt{k/m}\,t + C_2 \cos\sqrt{k/m}\,t$, where C_1 and C_2 are two arbitrary constants corresponding to the order two of (1). Also in §1 we considered the equation

$$ms'' = ks - s' \tag{2}$$

which describes a vibration subject to restraining forces such as friction. Both of these equations are examples of the homogeneous second-order linear equation with constant coefficients whose general form is

$$y'' + ay' + by = 0 \tag{3}$$

where a and b are real constants. In this section we develop the general solution to this equation. We then consider the nonhomogeneous second-order linear equation with constant coefficients whose general form is

$$y'' + ay' + by = R(x) \tag{4}$$

and find a method for its solution.

First, let us look at the special case of the homogeneous equation where $a = 0$; that is, we consider

$$y'' + by = 0 \tag{5}$$

This equation is easily solved by examining the different possibilities for the sign of b.

The first possibility is $b = 0$. In this case the general solution of (5) is easily seen to be

$$y = C_1 + C_2 x \tag{6}$$

where C_1 and C_2 denote arbitrary constants.

The second case is $b > 0$. If we set $b = K^2$, the equation becomes $y'' = -K^2 y$, which we recognize as the equation for harmonic motion (1). This equation therefore has general solution

$$y = C_1 \sin Kx + C_2 \cos Kx \tag{7}$$

where $K = \sqrt{b}$.

The remaining case is $b < 0$. Here we set $-b = K^2$, and equation (5) becomes $y'' = K^2 y$. This equation has the solution

$$y = C_1 e^{Kx} + C_2 e^{-Kx} \tag{8}$$

where $K = -b$.

To prove that we have indeed found all solutions to (5), we refer the reader to exercise 18.

Having solved equation (5), we shall now reduce equation (3) to equation (5) by means of a substitution. Let us write, as we did in §1,

$$u = y e^{\gamma x}$$

where γ is a real constant which we will determine presently. Since $y = u e^{-\gamma x}$, we have

$$y' = e^{-\gamma x} u' - \gamma e^{-\gamma x} u$$

$$y'' = e^{-\gamma x} u'' - 2\gamma e^{-\gamma x} u' + \gamma^2 e^{-\gamma x} u$$

So upon substitution equation (3) becomes

$$0 = \gamma^2 e^{-\gamma x} u - 2\gamma e^{-\gamma x} u' + u'' e^{-\gamma x} + a(u' e^{-\gamma x} - \gamma e^{-\gamma x} u) + b e^{-\gamma x} u \qquad (9)$$

or

$$0 = (\gamma^2 e^{-\gamma x} - a\gamma e^{-\gamma x} + b e^{-\gamma x}) u + (-2\gamma e^{-\gamma x} + a e^{-\gamma x}) u' + e^{-\gamma x} u'' \qquad (10)$$

We can force the coefficient of u' in (10) to be zero by requiring that $a e^{-\gamma x} - 2\gamma e^{-\gamma x} = 0$, which is equivalent to $\gamma = a/2$. With this choice of γ equation (10) becomes

$$0 = (\tfrac{1}{4}a^2 e^{-\frac{1}{2}ax} - \tfrac{1}{2}a^2 e^{-\frac{1}{2}ax} + b e^{-\frac{1}{2}ax}) u + e^{-\frac{1}{2}ax} u'' \qquad (11)$$

Dividing both sides of (11) by $e^{-\frac{1}{2}ax}$ and simplifying yields

$$0 = u'' + \left(\frac{4b - a^2}{4}\right) u \qquad (12)$$

We have thus shown that y satisfies (3) if and only if $u = e^{\frac{1}{2}ax} y$ satisfies (12). But equation (12) is an example of equation (5) and can be solved for a solution $u(x)$ according to the sign of $(4b - a^2)/4$; therefore, the solution of (3) is $y = e^{-\frac{1}{2}ax} u(x)$. We can thus classify the solutions of equation (3) according to the sign of $4b - a^2$. We summarize the results with a theorem.

Theorem 1 The general solution of $y'' + ay' + by = 0$ is

$$y(x) = e^{-\frac{1}{2}ax}(C_1\, u_1(x) + C_2\, u_2(x))$$

where C_1 and C_2 are arbitrary constants and where $u_1(x)$ and $u_2(x)$ are determined by the sign of $4b - a^2$:

(i) If $4b - a^2 = 0$, then $u_1(x) = 1$ and $u_2(x) = x$;

(ii) if $4b - a^2 < 0$, then $u_1(x) = e^{Kx}$ and $u_2(x) = e^{-Kx}$ where $K = \frac{1}{2}\sqrt{a^2 - 4b}$;

(iii) if $4b - a^2 > 0$, then $u_1(x) = \cos Kx$ and $u_2(x) = \sin Kx$ where $K = \frac{1}{2}\sqrt{4b - a^2}$.

Example 1 Let us again consider the equation

$$ms'' = -ks - \mu s', \qquad m > 0,\ k > 0,\ \mu \geqslant 0$$

In the notation of equation (3) we have $a = \mu/m$ and $b = k/m$. Hence, the solution will depend on the sign of $4b - a^2 = (4k/m) - (\mu^2/m^2)$. The case where $(4k/m) - (\mu^2/m^2) > 0$ or $4k > \mu^2/m$ has the solution

$$s(t) = e^{-\mu/2m}(C_1 \cos Kt + C_2 \sin Kt), \qquad K = \frac{1}{2}\sqrt{\frac{4k}{m} - \frac{\mu^2}{m^2}}$$

which represents damped harmonic oscillation as we saw in §1. Notice that the damping term $-\mu s'$ affects the frequency of the solution: If $\mu = 0$ then the solution has frequency $\sqrt{k/m}$, while if $\mu > 0$, the frequency is $\sqrt{(k/m) - (\mu^2/4m^2)}$. The case $(4k/m) - (\mu^2/m^2) < 0$ has the solution

$$s(t) = e^{-\mu t/2m}(C_1 e^{Kt} + C_2 e^{-Kt}), \qquad K = \frac{1}{2}\sqrt{\frac{\mu^2}{m^2} - \frac{4k}{m}}$$

Notice that we have $\mu/2m > K$ so that in this case

$$s(t) = C_1 e^{-rt} + C_2 e^{-r't}$$

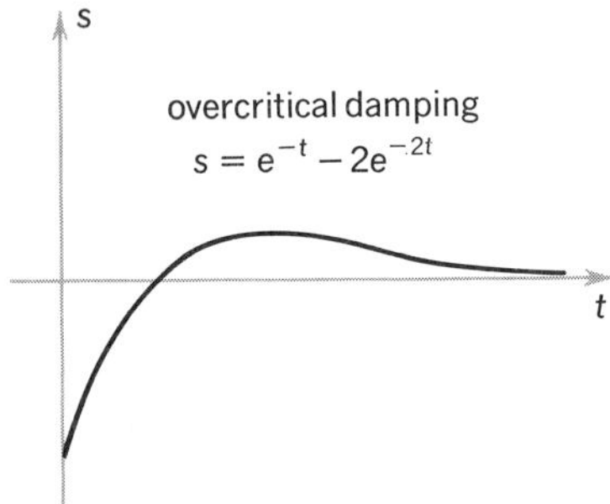

FIGURE 5-1

where r and r' are positive constants. Hence, $\lim_{t\to\infty} s(t) = 0$ and $s(t)$ can change sign at most once for $t \in [0, \infty[$. The motion described is therefore **not** oscillatory. This case is referred to as **overcritical damping** (see Figure 5-1). The final case $(4k/m) = (\mu^2/m^2)$ has the solution

$$s(t) = (C_1 + C_2 t)e^{-\mu t/2m}$$

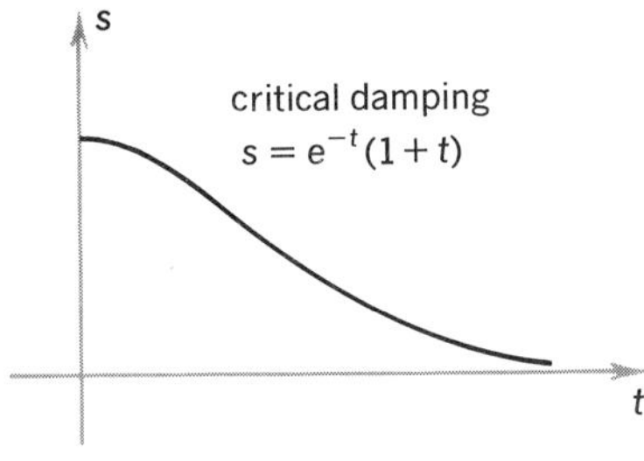

FIGURE 5-2

Here $\mu > 0$ (since $k > 0$) and so $s(t)$ again tends to zero as t tends to $+\infty$. This case is referred to as **critical damping** (see Figure 5-2).

Next we study the nonhomogeneous linear second-order equation with constant coefficients

$$y'' + ay' + by = R(x) \tag{13}$$

We note that there is the same important relationship between the solution of the homogeneous equation (3) and the nonhomogeneous equation as in the first-order case. If $y_1(x)$ and $y_2(x)$ are two solutions of the nonhomogeneous equation, the difference $y_1(x) - y_2(x)$ satisfies the homogeneous equation as is quickly verified by substitution. Conversely, if $h_1(x)$ is a solution of the homogeneous equation and if $y_1(x)$ is a solution of the nonhomogeneous equation, then $h_1(x) + y_1(x)$ is again a solution of the nonhomogeneous equation. Therefore, if we have a particular solution $y_1(x)$ of the nonhomogeneous equation and the general solution $h(x)$ of the homogeneous equation, then the general solution of the nonhomogeneous equation is $h(x) + y_1(x)$.

To complete the study of equation (13) we show how to find a particular solution to the nonhomogeneous equation (13) given the general solution to the homogeneous equation (3). The general solution to (13) is then taken to be the sum of this particular solution and the general solution of the homogeneous equation.

We employ the method of variation of parameters. The general solution of the homogeneous equation (3) is

$$e^{-\frac{1}{2}ax}(C_1 u_1(x)+C_2 u_2(x)) = C_1 v_1(x) + C_2 v_2(x)$$

where $v_1(x) = e^{-\frac{1}{2}ax} u_1(x)$ and $v_2(x) = e^{-\frac{1}{2}ax} u_2(x)$ and where $u_1(x)$ and $u_2(x)$ are determined by Theorem 1. We shall find a particular solution to (13) of the form

$$y_1(x) = g_1(x) v_1(x) + g_2(x) v_2(x) \tag{14}$$

The assumption that a solution of the form (14) exists will lead us to the correct choice of $g_1(x)$ and $g_2(x)$. Differentiating both sides of (14) yields

$$\begin{aligned} y_1'(x) &= g_1'(x) v_1(x) + g_1(x) v_1'(x) + g_2'(x) v_2(x) + g_2(x) v_2'(x) \\ &= g_1(x) v_1'(x) + g_2(x) v_2'(x) + (g_1'(x) v_1(x) + g_2'(x) v_2(x)) \end{aligned} \tag{15}$$

and

$$\begin{aligned} y_1''(x) &= g_1(x) v_1''(x) + g_2(x) v_2''(x) + (g_1'(x) v_1'(x) + g_2'(x) v_2'(x)) \\ &\quad + (g_1'(x) v_1(x) + g_2'(x) v_2(x))' \end{aligned} \tag{16}$$

Since $v_1(x)$ and $v_2(x)$ both satisfy the homogeneous equation (3) (Why?) we have

$$g_1(x)(v_1''(x) + av_1'(x) + bv_1(x)) = 0$$

and

$$g_2(x)(v_2''(x) + av_2'(x) + bv_2(x)) = 0$$

Therefore, substituting the expressions (15) and (16) for $y_1''(x)$ and $y_1''(x)$ into (3) yields

$$\begin{aligned} y_1''(x) + ay_1'(x) + by_1(x) &= (g_1'(x) v_1'(x) + g_2'(x) v_2'(x)) \\ &\quad + (g_1'(x) v_1(x) + g_2'(x) v_2(x))' \\ &\quad + a(g_1'(x) v_1(x) + g_2'(x) v_2(x)) \end{aligned} \tag{17}$$

We want to choose $g_1(x)$ and $g_2(x)$ so that the right-hand side of (17) is equal to $R(x)$. This will be achieved if we select $g_1(x)$ and $g_2(x)$ such that

$$g_1'(x) v_1(x) + g_2'(x) v_2(x) = 0$$

and

$$g_1'(x) v_1'(x) + g_2'(x) v_2(x) = R(x)$$

Solving these simultaneous linear equations for $g_1'(x)$ and $g_2'(x)$ we obtain

$$g_1'(x) = \frac{-v_2(x) R(x)}{W(x)}, \qquad g_2'(x) = \frac{v_1(x) R(x)}{W(x)}$$

where $W(x) = v_1(x)v_2'(x) - v_2(x)v_1'(x)$. Since $W(x) \neq 0$ for all x [as can be easily verified for each of the three sets $v_1(x), v_2(x)$ which arise as solutions to equation (3)] the functions $g_1'(x)$ and $g_2'(x)$ are therefore determined with the same domain as $R(x)$. The required functions $g_1(x)$ and $g_2(x)$ can now be obtained by integration:

$$g_1(x) = -\int \frac{v_2(x)R(x)}{W(x)}\,dx, \qquad g_2(x) = \int \frac{v_1(x)R(x)}{W(x)}\,dx$$

We summarize our findings by stating a theorem.

Theorem 2 The linear second-order differential equation with constant coefficients $y'' + ay' + by = R(x)$ has general solution

$$y(x) = C_1 v_1(x) + C_2 v_2(x) + y_1(x)$$

where $C_1 v_1(x) + C_2 v_2(x) = e^{-\frac{1}{2}ax}(C_1 u_1(x) + C_2 u_2(x))$ is the general solution to the homogeneous equation $y'' + ay' + by = 0$ as determined by Theorem 1, and where $y_1(x)$ is any particular solution of the given nonhomogeneous equation. Furthermore, there is a particular solution of the nonhomogeneous equation which is given by

$$y_1(x) = g_1(x)v_1(x) + g_2(x)v_2(x)$$

where

$$g_1(x) = -\int \frac{v_2(x)R(x)}{W(x)}\,dx, \qquad g_2(x) = \int \frac{v_1(x)R(x)}{W(x)}\,dx$$

and

$$W(x) = v_1(x)v_2'(x) - v_2(x)v_1'(x)$$

(One calls W the **Wronskian.**)

Example 2 Let us find the general solution of $y'' - y = 3e^{2x}$. First we solve the homogeneous equation $y'' - y = 0$: this is an example of equation (5) with $b = -1$ so the general solution to it is $h(x) = C_1 e^x + C_2 e^{-x}$. A particular solution to the nonhomogeneous equation is e^{2x} as can be found by inspection. Alternatively, we can compute a particular solution by the method given in Theorem 2: We have $v_1(x) = e^x$ and $v_2(x) = e^{-x}$, and we calculate $W(x) = e^x(-e^{-x}) - e^x(e^{-x}) = -1 - 1 = -2$. We next compute

$$g_1(x) = -\int \frac{e^{-x}(3e^{2x})}{-2}\,dx = \tfrac{3}{2}e^x \quad \text{and} \quad g_2(x) = \int \frac{e^x(3e^{2x})}{-2}\,dx = -\tfrac{1}{2}e^{3x}$$

So we find the particular solution

$$\begin{aligned} y_1(x) &= \tfrac{3}{2}e^x(e^x) + (-\tfrac{1}{2}e^{3x}(e^{-x})) \\ &= e^{2x} \end{aligned}$$

The general solution to our equation is, therefore

$$y = C_1 e^x + C_2 e^{-x} + e^{2x}$$

Theorem 2 has interesting applications to the theory of vibrations and oscillations. Physically, a solution of the homogeneous equation

$$ms'' = -\mu s' - ks \tag{18}$$

represents what is known as a **free oscillation**. A solution of the nonhomogeneous equation

$$ms'' = -\mu s' - ks + R(t) \tag{19}$$

represents a **forced oscillation** due to the external force $R(t)$ which is operating in addition to the vibrating force $-ks$ and the restraining force $-\mu s'$. Such an external force is also called an **exciting force**.

By Theorem 2, we see that if we have a forced oscillation $s_1(t)$ and superimpose a free oscillation $h_1(t)$ we obtain $s_1(t)+h_1(t)$ which satisfies the same nonhomogeneous equation as the original forced oscillation $s_1(t)$. If friction or some other restraining force is present ($\mu > 0$), the free oscillation $h_1(t)$ will eventually fade out because of the damping factor $e^{-\mu t/2m}$. Hence, for a given forced oscillation, with damping present, the motion will always tend to the same final state independently of the superimposition of any free oscillation.

A very important type of exciting force $R(t)$ is one which is periodic and of the form $R(t) = A\cos\omega t + B\sin\omega t$. In this case the differential equation

$$ms'' + \mu s' + ks = A\cos\omega t + B\sin\omega t, \qquad 4km - \mu^2 > 0,\ \mu > 0 \tag{20}$$

has the particular solution

$$s_1(t) = \beta^2(A\cos(\omega t - \varphi) + B\sin(\omega t - \varphi)) \tag{21}$$

where $\beta = 1/\sqrt{(k-m\omega^2)^2 + \mu^2\omega^2}$ and $\tan\varphi = \omega\mu/(k - m\mu^2)$. The positive number $\beta = \beta(\omega)$ is called the **distortion factor** and φ is called the **phase displacement**. By Theorem 2, the general solution to (20) is

$$\begin{aligned} s(t) = {} & e^{-\mu t/2m}(C_1\cos Kt + C_2\sin Kt) \\ & + \beta^2(A\cos(\omega t - \varphi) + B\sin(\omega t - \varphi)) \end{aligned} \tag{22}$$

where $K = \frac{1}{2}\sqrt{(k/m) - (\mu^2/m^2)}$. From (22) we see that if μ is small and if the exciting frequency ω is close to $\sqrt{k/m}$, which would be the frequency of the undamped free oscillation, then β^2 is large; hence, although $s(t)$ is bounded as $t \to +\infty$, still large amplitudes $|s(t)|$ occur. The appearance of these large amplitudes is known as the phenemenon of **resonance** (see exercise 19).

When $\mu = 0$, the particular solution to (20) does not take the form (21). [In fact (21) is meaningless when $\mu = 0$ and $\omega = \sqrt{k/m}$ since the distortion factor β is then not defined.] Instead we have the particular solution

$$s_1(t) = \frac{t}{2\sqrt{km}}(A \sin \omega t - B \cos \omega t) \tag{23}$$

and the general solution

$$s(t) = C_1 \cos \sqrt{\frac{k}{m}}\, t + C_2 \sin \sqrt{\frac{k}{m}}\, t + \frac{t}{2\sqrt{km}}(A \sin \omega t - B \cos \omega t) \tag{24}$$

Here the free oscillation does not fade away as t approaches $+\infty$ since it is undamped. Furthermore, the presence of the exciting force causes $|s(t)|$ to take on arbitrarily large values as t approaches $+\infty$.

EXERCISES

Solve each of the following equations.

1. $\dfrac{d^2 y}{dx^2} + 2\dfrac{dy}{dx} = 0$
2. $\dfrac{d^2 y}{dx^2} - 10\dfrac{dy}{dx} + 16y = 0$
3. $\dfrac{d^2 y}{dx^2} + w^2 y = 0 \quad (w = \text{constant} \neq 0)$
4. $\dfrac{d^2 y}{dx^2} - k^2 y = 0 \quad (k = \text{constant} \neq 0)$
5. $\dfrac{d^2 y}{dx^2} - 6\dfrac{dy}{dx} + 9y = 0$
6. $\dfrac{d^2 y}{dx^2} + 4y = 0; \quad y = 4, \quad \dfrac{dy}{dx} = 6$ when $x = 0$
7. $\dfrac{d^2 y}{dx^2} - 2\dfrac{dy}{dx} + 2y = 0; \quad y = 0, \quad \dfrac{dy}{dx} = \sqrt{2}$ when $x = \frac{1}{4}\pi$
8. $\dfrac{d^2 y}{dx^2} + 2\dfrac{dy}{dx} - 3y = 6$
9. $\dfrac{d^2 y}{dx^2} + y = \tan x$
10. $\dfrac{d^2 y}{dx^2} + 2\dfrac{dy}{dx} + y = e^x$
11. $\dfrac{d^2 y}{dx^2} + 4\dfrac{dy}{dx} + 5y = x + 2$
12. $\dfrac{d^2 x}{dy^2} - \dfrac{dx}{dy} = y + 2$
13. $\dfrac{d^2 y}{dx^2} + 4\dfrac{dy}{dx} + 4y = 2e^{-2x} + \sin 2x$
14. $\dfrac{d^2 x}{dt^2} + 2\lambda\dfrac{dx}{dt} + a^2 x = E \sin pt \quad (\lambda,\ a,\ E,\ p$ are constants)

15. $\dfrac{d^2x}{dt^2} + a^2x = E\cos pt \quad (p \neq a)$
16. Give a physical interpretation of the equations in exercises 8–12. Sketch the solutions for particular initial conditions.
17. Give a physical interpretation of the equation in exercise 14.
18. Let $f(x)$ be a solution to the equation $y'' + by = 0$. Let $u_1(x), u_2(x)$ be the solutions $\sin\sqrt{b}x, \cos\sqrt{b}x$ or $e^{\sqrt{-b}x}, e^{-\sqrt{-b}x}$ according to the sign of $b \neq 0$. (a) Define $W_1(x) = f(x)u_1'(x) - f'(x)u_1(x)$ and $W_2(x) = f(x)u_2'(x) - f'(x)u_2(x)$. Show that $W_1(x)$ and $W_2(x)$ are constant. (b) Show that there exist real numbers r_1 and r_2 such that

$$r_1u_1(x) + r_2u_2(x) = f(x)$$
$$r_1u_1'(x) + r_2u_2'(x) = f'(x)$$

Conclude that $C_1u_1(x) + C_2u_2(x)$ is indeed the general solution to the equation. (c) Prove that Theorem 2 gives the general solution to (5).

19. The graph of $\beta(\omega) = 1/\sqrt{(k - m\omega^2)^2 + \mu^2\omega^2}$ is known as the **resonance curve** of the exciting force $R(t) = A\cos\omega t + B\sin\omega t$. (a) Show that

$$\lim_{\omega\to\infty} \frac{\omega^2}{\beta(\omega)} = \frac{1}{m}.$$

(b) Assume $2km - \mu^2 > 0$ and $\mu > 0$. Show that the maximum value of $\beta(\omega)$ on $[0, +\infty[$ occurs at the point $\omega_1 = \sqrt{k/m - \mu^2/2m^2}$. Then show that $\lim_{\mu\to 0}\beta(\omega_1) = +\infty$. (c) Assume $2mk - \mu^2 \leqslant 0$ and $\mu > 0$. Show that the maximum value of $\beta(\omega)$ on $[0, \infty[$ occurs at $\omega = 0$. (d) Show that when $\mu = 0$ then $\beta(\omega)$ has an infinite discontinuity at $\sqrt{k/m}$.

6 SERIES SOLUTIONS OF LINEAR DIFFERENTIAL EQUATIONS

Consider the nonhomogeneous second-order linear equation with variable coefficients,

$$\frac{d^2y}{dx^2} + P(x)\frac{dy}{dx} + Q(x)y = R(x) \tag{1}$$

The preceding methods are not adequate for finding a general solution of (1). In this section we illustrate the application of a theorem which gives a sufficient condition that (1) has a solution defined by a power series. We begin by an important definition from Chapter 10.

Definition A real-valued function f is **analytic** at the point x_0 in its domain if and only if its Taylor series $\sum_{n=0}^{\infty}(f^{(n)}(x_0)/n!)(x-x_0)^n$ at x_0 has a nonzero radius of convergence and equals f on some neighborhood of x_0. A function is analytic on a set S if it is analytic at each point of S.

All polynomial functions are analytic everywhere, as are the sine, cosine, and exponential functions. A rational function $x \mapsto Q_1(x)/Q_2(x)$, where $Q_1(x)$ and $Q_2(x)$ are polynomials, is analytic everywhere except at those values of x where the denominator is zero.

We state the following theorem without proof.

Theorem 1 If the functions P and Q are analytic at x_0, then there are two linearly independent solutions of (1) which are analytic at x_0. (Two functions are **linearly independent** if neither is a real multiple of the other.)

It can be shown that a general solution of (1) is $C_1 g(x)+C_2 h(x)$ where $g(x)$ and $h(x)$ are the two solutions whose existence are guaranteed by the theorem. Hence, if we may now assume that (1) has an analytic solution $\sum_{n=0}^{\infty} c_n(x-x_0)^n$ at x_0 under the hypothesis of theorem, the question becomes how can we find such a solution? We must somehow produce a sequence of numbers, $c_0, c_1, c_2, \ldots$ with the required properties. The following examples illustrate the procedure for determining these constants.

Example 1 Consider the equation

$$\frac{d^2 y}{dx^2} - xy = 0 \tag{2}$$

Since the functions $x \mapsto 0$ and $x \mapsto x$ are analytic everywhere, we may seek analytic solutions at 0. Denote a power series solution of (2) at 0 by $f(x)=\sum_{n=0}^{\infty} c_n x^n$. Then

$$f''(x) - xf(x) = 0$$

or, in different notation,

$$\frac{d^2\left(\sum\limits_{n=0}^{\infty} c_n x^n\right)}{dx^2} - x\sum_{n=0}^{\infty} c_n x^n = 0 \tag{3}$$

Since Theorem 1 guarantees that the Taylor series of $f(x)$ has a nonzero radius of convergence, $f(x)=\sum_{n=0}^{\infty} c_n x^n$ is differentiable at 0 and its derivative $f'(x)$ is the derived series $f'(x)=\sum_{n=1}^{\infty} nc_n x^{n-1}$; similarly $f'(x)$ is differentiable at 0 and its derivative is $f''(x)=\sum_{n=2}^{\infty} n(n-1)c_n x^{n-2}$. Substituting the latter in (3), we have

$$\sum_{n=2}^{\infty} n(n-1)c_n x^{n-2} - x\sum_{n=0}^{\infty} c_n x^n = 0 \tag{4}$$

Since x is independent of the index of summation n, we may rewrite (4) as

$$\sum_{n=2}^{\infty} n(n-1)c_n x^{n-2} - \sum_{n=0}^{\infty} c_n x^{n+1} = 0 \tag{5}$$

In order to combine these terms we rewrite the summations so that x has the exponent n. Let us consider the first summation in (5)

$$\sum_{n=2}^{\infty} n(n-1)c_n x^{n-2} \tag{6}$$

Set $m=n-2$ in (6); then $n=m+2$, and since $m=0$ for $n=2$, the summation (6) takes the form

$$\sum_{m=0}^{\infty} (m+2)(m+1)\, c_{m+2}\, x^m \tag{7}$$

Now since the index of summation is a "dummy" variable, we may replace m by n in (7) to write the first summation in (5) as

$$\sum_{n=0}^{\infty} (n+2)(n+1)\, c_{n+2}\, x^n \tag{8}$$

In like manner, letting $m=n+1$, the other summation

$$\sum_{n=0}^{\infty} c_n\, x^{n+1} \tag{9}$$

in (5) takes the form

$$\sum_{m=1}^{\infty} c_{m-1}\, x^m \tag{10}$$

Then replacing m by n in (10), the second summation in (5) may be written as

$$\sum_{n=1}^{\infty} c_{n-1}\, x^n \tag{11}$$

Thus replacing (6) by its equivalent (8) and (9) by its equivalent (11), equation (5) may be written

$$\sum_{n=0}^{\infty} (n+2)(n+1)\, c_{n+2}\, x^n - \sum_{n=1}^{\infty} c_{n-1}\, x^n = 0 \tag{12}$$

Although x now has the same exponent in each summation in (12), the ranges of the summations are not the same. In the first summation the range is from 0 to ∞, and in the second it is from 1 to ∞. Observe that the common range is from 1 to ∞. Hence, we write out individually the terms in the first summation which do not belong to this common range, and we continue to employ the sigma notation to denote the remainder of the first summation. Therefore, we rewrite

$$\sum_{n=0}^{\infty} (n+2)(n+1)\, c_{n+2}\, x^n$$

in (12) as

$$2c_2 + \sum_{n=1}^{\infty} (n+2)(n+1)\, c_{n+2}\, x^n$$

and equation (12) becomes

$$2c_2 + \sum_{n=1}^{\infty} (n+2)(n+1)\, c_{n+2}\, x^n - \sum_{n=1}^{\infty} c_{n-1}\, x^n = 0$$

We can now combine like powers of x and write this equation as

$$2c_2 + \sum_{n=1}^{\infty} [(n+2)(n+1)\,c_{n+2} - c_{n-1}]\,x^n = 0 \tag{13}$$

For (13) to be valid for all x in the interval of convergence of $f(x)$, the coefficient of each power of x on the left-hand side of (13) must be equated to 0. Therefore

$$2c_2 = 0 \tag{14}$$

and

$$(n+2)(n+1)\,c_{n+2} - c_{n-1} = 0, \qquad n = 1, 2, 3, \ldots \tag{15}$$

Equation (15) is a recurrence relation which allows us to compute c_{n+2} in terms of c_{n-1} for $n = 1, 2, 3, \ldots$. That is, solving (15) for c_{n+2} we have

$$c_{n+2} = \frac{c_{n-1}}{(n+2)(n+1)} \tag{16}$$

For $n = 1$, equation (16) yields

$$c_3 = \frac{c_0}{3 \cdot 2}$$

and continuing in the same manner for $n = 2, 3, 4, 5, 6$, we obtain

$$c_4 = \frac{c_1}{4 \cdot 3}$$

$$c_5 = \frac{c_2}{5 \cdot 4} = 0 \quad \text{since } c_2 = 0$$

$$c_6 = \frac{c_3}{6 \cdot 5} = \frac{c_0}{6 \cdot 5 \cdot 3 \cdot 2}$$

$$c_7 = \frac{c_4}{7 \cdot 6} = \frac{c_1}{7 \cdot 6 \cdot 4 \cdot 3}$$

$$c_8 = \frac{c_5}{8 \cdot 7} = 0 \quad \text{since } c_5 = 0$$

Hence, using this process we can determine all the coefficients in the assumed solution $f(x) = \sum_{n=0}^{\infty} c_n x^n$ in terms of two arbitrary constants c_0 and c_1. Referring to the terms calculated above we may write

$$\begin{aligned} f(x) &= c_0 + c_1 x + \frac{c_0}{3 \cdot 2} x^3 + \frac{c_1}{4 \cdot 3} x^4 + \frac{c_0}{6 \cdot 5 \cdot 3 \cdot 2} x^6 + \frac{c_1}{7 \cdot 6 \cdot 4 \cdot 3} x^7 + \ldots \\ &= c_0 \left[\frac{x^3}{3 \cdot 2} + \frac{x^6}{6 \cdot 5 \cdot 3 \cdot 2} + \ldots \right] + c_1 \left[x + \frac{x^4}{4 \cdot 3} + \frac{x^7}{7 \cdot 6 \cdot 4 \cdot 3} + \ldots \right] \end{aligned}$$

Hence, $f(x)$ is the general solution to (2) expressed as a linear combination with arbitrary coefficients c_0, c_1 of two linearly independent convergent power series.

Example 2 Consider the second-order equation

$$\frac{d^2 y}{dx^2} + x^2 \frac{dy}{dx} + xy = 0 \tag{17}$$

Since x^2 and x are analytic expressions at 0, we seek a power series solution of (17) in the form

$$f(x) = \sum_{n=0}^{\infty} c_n x^n \tag{18}$$

Differentiating (18) term by term, we have

$$f'(x) = \sum_{n=1}^{\infty} nc_n x^{n-1} \tag{19}$$

and

$$f''(x) = \sum_{n=2}^{\infty} n(n-1)\, c_n x^{n-2} \tag{20}$$

Thus

$$\sum_{n=2}^{\infty} n(n-1)\, c_n x^{n-2} + x^2 \sum_{n=1}^{\infty} nc_n x^{n-1} + x \sum_{n=0}^{\infty} c_n x^n = 0 \tag{21}$$

after substituting (18), (19), and (20) into (17). Since x is independent of the index n, we may rewrite this as

$$\sum_{n=2}^{\infty} n(n-1)\, c_n x^{n-2} + \sum_{n=1}^{\infty} nc_n x^{n+1} + \sum_{n=0}^{\infty} c_n x^{n+1} = 0 \tag{22}$$

Next, we wish to rewrite the first summation in (22) so that x has exponent n. We let $m = n-2$; then $n = m+2$, and since $m = 0$ for $n = 2$, the first summation takes the form

$$\sum_{m=0}^{\infty} (m+2)(m+1)\, c_{m+2}\, x^m \tag{23}$$

The index of summation is a dummy variable, so we may replace m by n in (23) to obtain

$$\sum_{n=1}^{\infty} (n+2)(n+1)\, c_{n+2}\, x^n \tag{24}$$

Exactly the same procedure may be repeated for the second and third summations in (22) to yield the equivalent expressions

$$\sum_{n=2}^{\infty} (n-1)\, c_{n-1}\, x^n \tag{25}$$

and

$$\sum_{n=2}^{\infty} c_{n-1}\, x^n \tag{26}$$

Substituting expressions (24), (25), and (26) into (22), we obtain

$$\sum_{n=0}^{\infty}(n+2)(n+1)c_{n+2}x^n + \sum_{n=2}^{\infty}(n-1)c_{n-1}x^n + \sum_{n=1}^{\infty}c_{n-1}x^n = 0 \quad (27)$$

The common range of summation in (27) is from 2 to ∞. As in Example 1, we write out the terms from the first and third summation which do not belong to this common range, and we continue to employ the sigma notation to denote the remainder of the summations. After combining like powers of x, (27) becomes

$$2c_2 + (6c_3 + c_0)x + \sum_{n=2}^{\infty}[(n+2)(n+1)c_{n+2} + (n-1)c_{n-1} + c_{n-1}]x^n \quad (28)$$

For (27) to be valid for all values of x in the interval of convergence of $\sum_{n=0}^{\infty} c_n x^n$, the coefficient of each power of x in the left-hand side of (28) must be equated to zero. This leads immediately to the conditions

$$2c_2 = 0 \quad (29)$$

$$6c_3 + c_0 = 0 \quad (30)$$

and also

$$(n+2)(n+1)c_{n+2} + (n-1)c_{n-1} + c_{n-1} = 0, \qquad n = 2, 3, \ldots \quad (31)$$

After we simplify and solve for c_{n+2} in terms of c_{n-1}, equation (31) becomes

$$c_{n+2} = -\frac{nc_{n-1}}{(n+2)(n+1)} \quad (32)$$

From (29) we have $c_2 = 0$; from (30) we obtain $c_3 = -c_0/3!$. The coefficients $c_4, c_5, c_6, \ldots$ can be obtained by a methodic calculation which is illustrated below:

$$c_4 = -\frac{2c_1}{4\cdot 3} = -\frac{2^2 c_1}{4!}$$

$$c_5 = -\frac{3c_2}{5\cdot 4} = 0$$

$$c_6 = -\frac{4c_3}{6\cdot 5} = \left(-\frac{4}{6\cdot 5}\right)\left(-\frac{c_0}{3!}\right) = \frac{4^2 c_0}{6!}$$

$$c_7 = -\frac{5c_4}{7\cdot 6} = \left(-\frac{5}{7\cdot 6}\right)\left(-\frac{2^2 c_1}{4!}\right) = \frac{5^2 2^2 c_1}{7!}$$

$$c_8 = -\frac{6c_5}{8\cdot 7} = \left(-\frac{6}{8\cdot 7}\right)\cdot 0 = 0$$

$$c_9 = -\frac{7c_6}{9\cdot 8} = \left(-\frac{7}{9\cdot 8}\right)\left(\frac{4^2 c_0}{6!}\right) = -\frac{7^2 4^2 c_0}{9!}$$

$$c_{10} = -\frac{8c_7}{10 \cdot 9} = \left(-\frac{8}{10 \cdot 9}\right)\left(\frac{5^2\, 2^2\, c_1}{7!}\right) = -\frac{8^2\, 5^2\, 2^2\, c_1}{10!}$$

$$\vdots$$

Thus a power series solution for (17) is given by

$$f(x) = c_0 + c_1 x - \frac{c_0}{3!}x^3 - \frac{2^2 c_1}{4!}x^4 + \frac{4^2 c_0}{6!}x^6 + \frac{5^2\, 2^2\, c_1}{7!}x^7$$
$$-\frac{7^2\, 4^2\, c_0}{9!}x^9 - \frac{8^2\, 5^2\, 2^2\, c_1}{10!}x^{10} + \cdots \qquad (33)$$

Since the coefficients of the powers of x contain two arbitrary constants c_0 and c_1, the right-hand side of (33) can be separated into the sum of two power series, one a multiple of c_0 and the other a multiple of c_1:

$$f(x) = c_0\left[1 - \frac{x^3}{3!} + \frac{4^2}{6!}x^6 - \frac{7^2\, 4^2}{9!}x^9 + \cdots\right]$$
$$+c_1\left[x - \frac{2^2}{4!}x^4 + \frac{5^2\, 2^2}{7!}x^7 - \frac{8^2\, 5^2\, 2^2}{10!}x^{10} + \cdots\right] \qquad (34)$$

If these two series have nonzero radii of convergence, then they are the linearly independent solutions guaranteed by the existence theorem of this section (Theorem 1). We leave it as an exercise for the reader to use the ratio test on these series together with (32) to show that they converge in an interval $(-a, a)$ about 0 with $a > 0$.

EXERCISES

Find the general solution of each of the following differential equations using the power series method.

1. $\dfrac{d^2 y}{dx^2} - x^2 \dfrac{dy}{dx} + 2xy = 0$

2. $4\dfrac{d^2 y}{dx^2} + x^2 \dfrac{dy}{dx} - xy = 0$

3. $(x^2 - 4)\dfrac{d^2 y}{dx^2} - 3x\dfrac{dy}{dx} + 4y = 0$

4. $\dfrac{d^2 y}{dx^2} - y = 0$

5. $(1 - x^2)\dfrac{d^2 y}{dx^2} - 2x\dfrac{dy}{dx} + p(p+1)y = 0 \qquad (p \neq 0)$

   (This is known as **Legendre's equation.**)

6. $\dfrac{d^2 y}{dx^2} + (x+1)\dfrac{dy}{dx} - x^2 y = 0$

7. Determine the interval of convergence of each of the following power series. Then show that each series satisfies the given differential equation; use this fact to find the sum of each series.

   (a) $\sum_{n=0}^{\infty} \frac{x^{2n}}{n!}$; $y' = 2xy$

   (b) $\sum_{n=0}^{\infty} \frac{x^n}{n!}$; $y' = x + y$

REVIEW EXERCISES FOR CHAPTER 11

In exercises 1–10 solve the differential equation.

1. $yy' + xy^2 = x$
2. $(2\sqrt{xy} - x)\,dy + y\,dx = 0$
3. $y'' + y' - 2y = e^{2x}$
4. $xy' + y = y^2 \ln x$
5. $yy' + \frac{1}{2}y^2 = \sin x$
6. $d^2y/dx^2 - y = x$
7. $3y^2y' + y^3 = x - 1$
8. $(1 + e^{x/y}) = -e^{x/y}(1 - x/y)\,dy/dx$
9. $y'' - 4y - e^{2x} = 0$
10. $y'' + y = \tan x$
11. Solve the equation $y' - xy = x^3y^2$ by variation of parameters.
12. Obtain the solution to the first-order linear equation $y' + p(x)y = q(x)$ by setting $u = y\exp(\int p(x)\,dx)$ and reducing to an equation in u and x. Also, prove that the solution obtained is the general solution.
13. The equation

    $$x^2\frac{d^2y}{dx^2} + ax\frac{dy}{dx} + by = 0$$

    is known as **Euler's equation**. (a) Solve this equation by means of the substitution $x = e^t$ which reduces it to a linear equation with constant coefficients in y and t.

    $$\left[\text{Hint: } \left.\frac{d^2y}{dx^2}\right|_{x=x(t)}\left(\frac{dx}{dt}\right)^2 = \frac{d^2y}{dt^2} - \left.\frac{dy}{dx}\right|_{x=x(t)}\frac{d^2x}{dt^2}\right]$$

    (b) Find a condition on a and b such that the solution of Euler's equation is a rational function of x.
14. There is a sequence of N cups each of capacity Q cubic inches filled with water and arranged one below the other. Suppose that Q cubic inches of wine are poured into the topmost cup, the pouring being done in T minutes and at a constant rate. The overflow from each cup pours into the cup immediately below it. Assuming that a complete mixture of water and wine takes place instantaneously, determine the quantity of wine in each cup when the pouring is finished.

15. An electric circuit satisfies the differential equation

$$\mu \frac{dq}{dt} + \rho q = E_0 \sin(\omega t)$$

where $i = dq/dt$ denotes the electric current and where μ, ρ, ω, and E_0 are positive constants. Suppose that the initial condition $q(0) = 0$ is given. Solve this initial value problem and show that the solution takes the form

$$q(t) = \frac{E_0}{\sqrt{\rho^2 + \omega^2 \mu^2}} \sin(\omega t - \phi) + \frac{E_0 \omega \mu e^{-\rho t/\mu}}{\sqrt{\rho^2 + \omega^2 \mu^2}}$$

where $\tan \phi = \omega\mu/\rho$.

16. An equation of the form $y' + p(x)y + q(x)y^2 = R(x)$ is called a **Riccati equation**. (There is no known method for solving the general Riccati equation.) (a) Prove that if $g(x)$ is a given solution of this equation, then there are solutions of the form $g(x) + 1/v(x)$ where $v(x)$ is a solution of a first-order linear equation. (b) Solve the equation $y' - x^2 y^2 + x^4 - 1 = 0$ which has particular solution $y = x$.

17. The power series

$$y = \sum_{n=0}^{\infty} (-1)^n \frac{2^{2n} x^{2n}}{(2n)!}$$

satisfies the differential equation $y'' + 4y = 0$. Use this fact to compute the sum of the series.

18. Find the recurrence relation on the coefficients of series solutions to the equation $xy'' + y' + xy = 0$. Show that the solution satisfying the initial conditions $y(0) = 1$, $y'(0) = 0$ is

$$\sum_{n=0}^{\infty} (-1)^n \frac{x^{2n}}{2^{2n} (n!)^2}$$

12

NUMERICAL METHODS

Numerical methods have been an important part of the calculus since its inception. In this chapter we study interpolation by means of polynomials, techniques of approximate integration, interpolation by means of finite differences, and numerical methods for solving differential equations.

1 POLYNOMIAL INTERPOLATION

In §5.3, we discussed the method of differential approximation, which was based on the formula

$$f(x) \approx f(x_0) + f'(x_0)(x - x_0)$$

In this chapter we study other methods of approximation and numerical techniques.

It is most important in working with approximations to have control over the error associated with the approximation. By the **error** we mean the difference between the true value and the approximate value:

$$(\text{error}) = (\text{true value}) - (\text{approximate value})$$

For differential approximation let us denote the error by

$$E(x) = f(x) - [f(x_0) + f'(x_0)(x - x_0)]$$

Clearly, we have

$$\lim_{x \to x_0} E(x) = 0 \tag{1}$$

and

$$\lim_{x \to x_0} \frac{E(x)}{x - x_0} = 0 \tag{2}$$

Equations (1) and (2) lead us to expect that differential approximation will give accurate results when $(x-x_0)$ is small. If we make the further assumption that the second derivative f'' exists and is continuous on a neighborhood of x_0, then we can sharpen equation (2) to

$$\lim_{x\to x_0} \frac{E(x)}{(x-x_0)^2} \tag{3}$$

exists and is finite. Intuitively, (3) says that $E(x)$ tends to 0 **like** $(x-x_0)^2$ as x approaches x_0.

To establish equation (3) we first prove the Extended Mean Value Theorem for twice-differentiable functions. The proof is not difficult and is an example of a type of argument that is used several times later in this section to compute bounds for error terms. The proof is therefore included even though the result is a special case of Taylor's theorem (Theorem 3, §10.5).

Theorem 1 The Extended Mean Value Theorem Let f be continuously differentiable on $[a,b]$, and twice differentiable on $]a,b[$. Then there is a point c in $]a,b[$ satisfying

$$f(b)-f(a) = f'(a)(b-a)+\tfrac{1}{2}f''(c)(b-a)^2$$

Proof Consider the function

$$F(x) = f(x)-f(a)-f'(a)(x-a) - K(x-a)^2 \tag{4}$$

where K is a constant. Clearly $F(a)=0$. Now choose K so that $F(b)=0$ also:

$$K = \frac{f(b)-f(a)-f'(a)(b-a)}{(b-a)^2} \tag{5}$$

Then F satisfies the hypotheses of Rolle's Theorem, and so there is a point c_1 in $]a,b[$ such that

$$0 = F'(c_1) = f'(c_1)-f'(a)-2K(c_1-a)$$

By hypothesis, F' is continuous on $[a,c_1]$ and differentiable on $]a,c_1[$. Therefore, we may apply Rolle's Theorem to F' on $[a,c_1]$, yielding a point c in $]a,c_1[$ which satisfies

$$0 = F''(c) = f''(c)-2K$$

Thus

$$K = \tfrac{1}{2}f''(c)$$

Substituting this back into equation (4) yields the desired result. ∎

We next show how to obtain equation (3) under the hypothesis that f be twice continuously differentiable on some neighborhood of x_0. When f'' is continuous on $[x_0,x]$ (or $[x,x_0]$ when $x<x_0$), we can apply the above

theorem to obtain

$$f(x) = f(x_0) + f'(x_0)(x-x_0) + \tfrac{1}{2}f''(c)(x-x_0)^2$$

where c is a point between x_0 and x. Hence we have

$$\frac{E(x)}{(x-x_0)^2} = \frac{f(x)-f(x_0)-f'(x_0)(x-x_0)}{(x-x_0)^2}$$

$$= \tfrac{1}{2}f''(c)$$

Now let x approach x_0. The point c is squeezed toward x_0, and so

$$\lim_{x\to x_0} \frac{E(x)}{(x-x_0)^2} = \lim_{x\to x_0} \tfrac{1}{2}f''(c) = \tfrac{1}{2}f''(x_0)$$

Differential approximation is one example of an **extrapolation** method. That is, knowing $f(x_0)$ and $f'(x_0)$, we **predict** the approximate value of $f(x)$ for points x near x_0. In differential approximation we are interested in estimating the value of a function at a given point, that is, finding a single value of a function. We now turn to the problem of approximating a function over many points at the same time. The principal method we shall employ is **polynomial approximation**, in which a function f is approximated by means of polynomials which agree with f at a large number of points in its domain. We first show how to construct a polynomial of degree at most n which takes the same values as an arbitrary function f at the $n+1$ distinct points $x_0, \ldots, x_n$. In geometric terms, this means that for any $n+1$ points in the plane with distinct abscissas there is a polynomial of degree less than or equal to n whose graph passes through the $n+1$ given points. Let the points be (x_0, y_0), (x_1, y_1), ..., $(x_n y_n)$.

First let us remark that there can be at most one such polynomial: for if $p(x)$ and $q(x)$ are two polynomials of degree less than or equal to n satisfying

$$q(x_0) = y_0 = p(x_0)$$

$$\vdots$$

$$q(x_n) = y_n = p(x_n)$$

then $q(x)-p(x)$ is a polynomial of degree at most n having $n+1$ roots. This is possible only if $q(x)-p(x)$ is the zero polynomial, which means that $q(x) = p(x)$.

Now to find the desired polynomial, we first consider the polynomials

$$l_i(x) = \frac{(x-x_0)\cdots(x-x_{i-1})(x-x_{i+1})\cdots(x-x_n)}{(x_i-x_0)\cdots(x_i-x_{i-1})(x_i-x_{i+1})\cdots(x_i-x_n)}$$

where $0 \leqslant i \leqslant n$. These polynomials satisfy

$$l_i(x_j) = 1, \qquad i = j$$

$$l_i(x_j) = 0, \qquad i \neq j$$

and each $l_i(x)$ is of degree n. Then we simply set

$$p(x) = y_0 l_0(x) + y_1 l_1(x) + \cdots + y_n l_n(x) = \sum_{i=0}^{n} y_i l_i(x) \tag{6}$$

and $p(x)$ is the polynomial we seek. Equation (6) is known as **Lagrange's formula of interpolation**, the polynomials $l_i(x)$ are called the **Lagrange interpolation coefficients**, and the polynomial $p(x)$ is called the **Lagrangian interpolating polynomial**. If we set

$$\pi(x) = (x-x_0)(x-x_1)\cdots(x-x_n)$$

we have

$$\pi'(x_i) = (x_i-x_0)\cdots(x_i-x_{i-1})(x_i-x_{i+1})\cdots(x_i-x_n)$$

and we may rewrite

$$l_i(x) = \frac{\pi(x)}{\pi'(x_i)(x-x_i)}$$

Hence, the Lagrange formula becomes

$$p(x) = \sum_{i=0}^{n} \frac{\pi(x)f(x_i)}{\pi'(x_i)(x-x_i)}$$

Example 1 Construct a polynomial $p(x)$ of degree $\leqslant 3$ whose graph passes through the points $(-2,-5)$, $(-1,1)$, $(1,1)$, and $(2,7)$, and evaluate $p(\frac{3}{2})$.

Substituting the given values into Lagrange's interpolation formula, we have

$$\begin{aligned} p(x) &= \frac{(x+1)(x-1)(x-2)}{(-2+1)(-2-1)(-2-2)}(-5) + \frac{(x+2)(x-1)(x-2)}{(-1+2)(-1-1)(-1-2)}(1) \\ &\quad + \frac{(x+2)(x+1)(x-2)}{(1+2)(1+1)(1-2)}(1) + \frac{(x+2)(x+1)(x-1)}{(2+2)(2+1)(2-1)}(7) \\ &= \tfrac{5}{12}(x+1)(x-1)(x-2) + \tfrac{1}{6}(x+2)(x-1)(x-2) \\ &\quad - \tfrac{1}{6}(x+2)(x+1)(x-2) + \tfrac{7}{12}(x+2)(x+1)(x-1) \\ &= x^3 - x + 1 \end{aligned}$$

Clearly, $p(\frac{3}{2}) = 2\frac{7}{8}$. However, in practice, it is not necessary to obtain the algebraic form of $p(x)$ in order to find $p(x_0)$ for some value x_0. Direct substitution in the formula usually will give the result more quickly. For the polynomial under consideration:

$$\begin{aligned} p(\tfrac{3}{2}) &= \frac{(\frac{3}{2}+1)(\frac{3}{2}-1)(\frac{3}{2}-2)}{(-2+1)(-2-1)(-2-2)}(-5) + \frac{(\frac{3}{2}+2)(\frac{3}{2}-1)(\frac{3}{2}-2)}{(-1+2)(-1-1)(-1-2)} \\ &\quad + \frac{(\frac{3}{2}+2)(\frac{3}{2}+1)(\frac{3}{2}-2)}{(1+2)(1+1)(1-2)} + \frac{(\frac{3}{2}+2)(\frac{3}{2}+1)(\frac{3}{2}-1)}{(2+1)(2+2)(2+1)}(7) \\ &= -\frac{25}{96} - \frac{7}{48} + \frac{35}{48} + \frac{245}{96} = 2\tfrac{7}{8} \end{aligned}$$

Example 2 Find the minimum value of the polynomial $q(x)$ of degree 2 which satisfies $q(0) = 11, q(1) = 5$, and $q(5) = 21$. From Lagrange's formula we obtain

$$q(x) = \frac{(x-1)(x-5)}{(-1)(-5)}(11) + \frac{x(x-5)}{(1)(-4)}(5) + \frac{x(x-1)}{(5)(4)}(21)$$

Then differentiating we find,

$$q'(x) = \tfrac{11}{5}(2x-6) - \tfrac{5}{4}(2x-5) + \tfrac{21}{20}(2x-1)$$

Setting $q'(x) = 0$ gives

$$0 = 80x - 160$$

Thus, $q'(2) = 0$. Since $q''(2) > 0$, 2 is the minimum point of q, and the minimum value of $q(x)$ is $q(2) = 3$.

Since we wish to use the interpolating polynomial to approximate f at points other than the interpolating points, it is important that we obtain some estimate of the error. However, without additional hypotheses, nothing whatever can be said about this quantity, for we can change the function arbitrarily at points which are not interpolating points without changing $p(x)$ at all (see Figure 1-1). A definite statement can be made if we assume a qualitative knowledge of the derivative.

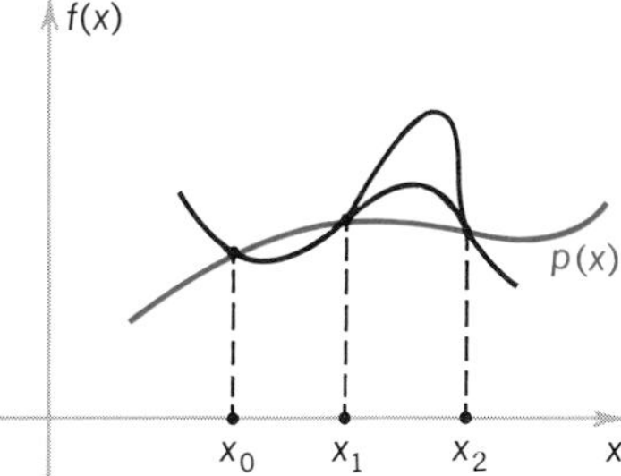

FIGURE 1-1

Theorem 2 Let the function f be approximated by the polynomial $p(x)$ which agrees with $f(x)$ at the $n+1$ distinct points $x_0, \ldots, x_n$. Denote by

$$E(x) = f(x) - p(x)$$

the error in the approximation. Then if f is $n+1$ times continuously differentiable, and x lies between the greatest and smallest of the x_i, we can express $E(x)$ in the form

$$E(x) = \frac{1}{(n+1)!} f^{(n+1)}(\xi)\,\pi(x) \tag{7}$$

where the point ξ also lies somewhere between the greatest and smallest of the x_i.

Proof We argue as follows. Consider the function

$$F(x) = f(x) - p(x) - K\pi(x)$$

where K is constant. Clearly, $F(x)$ vanishes at each of the points $x_0, \ldots, x_n$. Let I denote the closed interval whose endpoints respectively are the smallest and largest of the numbers $x_0, \ldots, x_n$. Let $\bar{x}$ be a point in I distinct from $x_0, \ldots, x_n$. Since $\pi(x)$ vanishes only at the points $x_0, \ldots, x_n$, K can be chosen so that $F(x)$ vanishes at $x_0, \ldots, x_n$, and $\bar{x}$. Let us choose this K, namely

$$K = \frac{f(\bar{x}) - p(\bar{x})}{\pi(\bar{x})}$$

Since $F(x)$ vanishes at least $n+2$ times on the interval I, then by Rolle's Theorem the derivative $F'(x)$ must vanish at least $n+1$ times on this interval. Similarly $F''(x)$ must vanish at least n times, and finally $F^{(n+1)}(x)$ must vanish at least once. Let ξ be a point in I satisfying $F^{(n+1)}(\xi)=0$. We have

$$0 = F^{(n+1)}(\xi) = f^{(n+1)}(\xi) - p^{(n+1)}(\xi) - K\pi^{(n+1)}(\xi)$$

Since $p(x)$ is of degree at most n, we see that $p^{(n+1)}(\xi)=0$. Also we have $\pi^{(n+1)}(\xi)=(n+1)!$ Therefore we obtain

$$0 = f^{(n+1)}(\xi) - K(n+1)!$$

or

$$K = \frac{f^{(n+1)}(\xi)}{(n+1)!}$$

The equation $F(\bar{x})=0$ becomes

$$f(\bar{x}) - p(\bar{x}) = \frac{1}{(n+1)!} f^{(n+1)}(\xi)\,\pi(\bar{x})$$

This equation reduces to $0=0$ when one of the x_i is substituted for $\bar{x}$, so as $\bar{x}$ was otherwise arbitrary, this equation reduces to (7). ∎

Note that the point ξ depends on x and changes as x varies over I. Hence, equation (7) cannot be used to calculate an exact value of the error, since ξ as a function of x is, in general, unknown. However, the formula can be used in many cases to find a **bound** for the error of the interpolating polynomial.

Example 3 Let us compute the error associated with linear or affine interpolation. If we are given the values $f(a)$ and $f(b)$, Lagrange interpolation (with $x_0=a, x_1=b$) yields the affine function

$$p(x) = f(a) + \frac{f(b)-f(a)}{b-a}\,x$$

with error

$$E(x) = f(x) - f(a) - \frac{f(b)-f(a)}{b-a}\,x$$

By the above analysis, assuming f twice differentiable on $]a,b[$, we have

$$E(x) = \frac{(x-a)(x-b)}{2} f''(\xi)$$

where ξ is a point in $]a,b[$. If f'' is bounded on $]a,b[$, with $|f''(\xi)| \leqslant M$ for all ξ in $]a,b[$, then

$$|E(x)| \leqslant \tfrac{1}{2}M\,|(x-a)(x-b)|$$

The function $x \mapsto |(x-a)(x-b)|$ has its maximum on $[a,b]$ at the midpoint $\frac{1}{2}(b-a)$, where it assumes the value $\frac{1}{4}(b-a)^2$. We thus find that on $[a,b]$,

$$|E(x)| = |f(x)-p(x)| \leqslant \tfrac{1}{8}M(b-a)^2$$

The Lagrange interpolating polynomial

$$p(x) = \sum_{i=0}^{n} l_i(x) f(x_i)$$

gives f exactly if f is a polynomial of degree at most n. If the values $f'(x_i)$, $0 \leqslant i \leqslant n$ are also known, then an interpolating formula can be found which will be exact if $f(x)$ is a polynomial of degree up to $2n+1$. The polynomial $h(x)$ to be defined will satisfy the $2n+2$ conditions

$$h'(x_i) = f'(x_i)$$

$$h(x_i) = f(x_i)$$

for all $i, 0 \leqslant i \leqslant n$. To construct $h(x)$, suppose that a solution can be found of the form

$$h(x) = \sum_{i=0}^{n} h_i(x) f(x_i) + \sum_{i=0}^{n} \bar{h}_i(x) f'(x_i)$$

where the $h_i(x)$ and $\bar{h}_i(x)$ are polynomials of degree at most $2n+1$, to be determined. We will clearly have $h(x_i) = f(x_i)$ if

$$h_i(x_j) = \delta_{ij}, \qquad \bar{h}_i(x_j) = 0$$

where δ_{ij}, the **Kronecker delta**, is defined by

$$\delta_{ij} = \begin{cases} 1, & i = j \\ 0, & i \neq j \end{cases}$$

Moreover, the requirement $h'(x_i) = f'(x_i)$ will be met if

$$h'(x_j) = 0, \qquad \bar{h}_i'(x_j) = \delta_{ij}$$

Note that the polynomial $[l_i(x)]^2$ which is of degree $2n$ satisfies

$$[l_i(x_j)]^2 = \delta_{ij}$$

and the derivative of $[l_i(x)]^2$ is zero for $x = x_j, j \neq i$.

The conditions on the zeros of $h_i(x)$ and $\bar{h}_i(x)$ then imply that

$$h_i(x) = u_i(x)\,[l_i(x)]^2$$

$$\bar{h}_i(x) = v_i(x)\,[l_i(x)]^2$$

where $u_i(x)$ and $v_i(x)$ are polynomials of degree at most 1. On the other hand, the conditions $h_i(x_j) = \delta_{ij} = \bar{h}_i'(x_j)$ imply that

$$u_i(x_i) = 1, \qquad u_i'(x_i) + 2l_i'(x_i) = 0$$

and

$$v_i(x_i) = 0, \qquad v_i'(x_i) = 1$$

We deduce that

$$u_i(x) = 1 - 2l_i'(x_i)(x-x_i)$$

$$v_i(x) = x - x_i$$

Collecting this information, we set

$$h_i(x) = (1-2l_i'(x_i)(x-x_i))[l_i(x)]^2$$

$$\bar{h}_i(x) = (x-x_i)[l_i(x)]^2$$

and the desired polynomial is

$$h(x) = \sum_{i=0}^{n} h_i(x)f(x_i) + \sum_{i=0}^{n} \bar{h}_i(x)f'(x_i) \tag{8}$$

Equation (8) is known as **Hermite's interpolation formula**.

Example 4 Find a cubic polynomial $p(x)$ satisfying $p(0) = A$, $p(1) = B$, $p'(0) = C$, $p'(1) = D$.

In this example, $n = 1$, $x_0 = 0$, and $x_1 = 1$. We have

$$l_0(x) = \frac{x-1}{-1}, \qquad l_0'(x) = -1$$

$$l_1(x) = \frac{x-0}{1}, \qquad l_1'(x) = 1$$

Therefore, we obtain

$$h_0(x) = (1+2x)(1-x)^2 = 1 - 3x^2 + 2x^3$$

$$h_1(x) = (1-2(x-1))x^2 = 3x^2 - 2x^3$$

$$\bar{h}_0(x) = x(1-x)^2 = x - 2x^2 + x^3$$

$$\bar{h}_1(x) = (x-1)x^2 = x^3 - x^2$$

The Hermitian interpolation formula yields

$$p(x) = A(2x^3-3x^2+1) + B(-2x^3+3x^2) + C(x^3-2x^2+x) + D(x^3-x^2)$$

$$= (2A-2B+C+D)x^3 + (-3A+3B-2C-D)x^2 + Cx + A$$

Inspection shows that indeed $p(0) = A$, $p(1) = B$, $p'(0) = C$, $p'(1) = D$.

For the error associated with Hermite interpolation, we have the result

$$E(x) = f(x) - h(x) = \frac{f^{(2n+2)}(\xi)}{(2n+2)!}\pi(x)^2 \tag{9}$$

where ξ is a point in the interval I whose endpoints are the smallest and largest, respectively, of the $x_0, \ldots, x_n$.

The result of equation (9) is established in a manner analogous to that employed to estimate the error in Lagrange interpolation. We form

$$F(x) = f(x) - h(x) - K[\pi(x)]^2$$

and choose K so that $F(x)$ vanishes at the point $\bar{x}$ in I, where $\bar{x}$ is different from $x_0, \ldots, x_n$. Then $F(x)$ vanishes at the $n+2$ points $\bar{x}, x_0, \ldots, x_n$ in I and so $F'(x)$ vanishes at $n+1$ points in I. However, $F'(x)$ also vanishes at the $n+1$ points $x_0, \ldots, x_n$. Hence $F'(x)$ vanishes at least $2n+2$ times in the interval I. Thus $F^{(2n+2)}(x)$ vanishes at least once in I. The rest of the derivation is strictly analogous to the Lagrangian case.

From the error formula (9) it also follows that the Hermitian interpolation polynomial is unique, that is, there is exactly one polynomial of degree at most $2n+1$ satisfying the $2n+2$ conditions

$$h(x_i) = y_i, \qquad h'(x_i) = y_i'$$

where $0 \leqslant i \leqslant n$.

EXERCISES

1. Find the Lagrange interpolating polynomial determined by the data $f(0) = 0, f(\frac{1}{5}) = \frac{126}{625}, f(\frac{1}{3}) = \frac{28}{81}, f(\frac{1}{2}) = \frac{65}{256}, f(1) = 2$.
2. Find the Lagrange interpolating polynomial determined by the data $f(0) = 1, f(1) = 5, f(2) = 19, f(3) = 49$.
3. Let $p(x)$ be a polynomial of degree less than or equal to n which satisfies $p(x_0) = y_0, p(x_1) = y_1, \ldots, p(x_n) = y_n$. Let $\pi(x) = (x-x_0)(x-x_1)\cdots(x-x_n)$. Obtain the Lagrangian form
$$p(x) = \sum_{i=0}^{i=n} \frac{\pi(x)\, y_i}{\pi'(x_i)(x-x_i)}$$
by determining the coefficients a_i in the partial fraction expansion of the ratio $p(x)/\pi(x) = \sum_{i=0}^{i=n} a_i/(x-x_i)$. [Hint: Multiply both sides of this equation by $(x-x_i)$ and let $x \to x_i$.]
4. Let $p(x)$ be a polynomial of degree less than or equal to 3. Show that
$$\int_h^{2h} p(x)\,dx = \frac{h}{24}(-p(0) + 13p(h) + 13p(2h) - p(3h))$$
5. Find the Hermite interpolation polynomial determined by the data $f(0) = 1, f'(0) = 1; f(1) = 6, f'(1) = 10$.
6. Find the Hermite interpolation polynomial determined by the data $f(0) = 0, f'(0) = 1; f(1) = 6, f'(1) = 6; f(2) = 34, f'(2) = 81$.
7. A table gives the values of $\sin x$ with spacing 0.02 radians between tabulated points. Affine interpolation is used to determine $\sin x$ at intermediate points. To how many decimal places will the values obtained for intermediate points be exact? (Assume for the purposes of this problem that the table lists the exact value of $\sin x$ at tabulated points.)

2 NEWTON–COTES QUADRATURE

The term **numerical quadrature** refers to the process of approximating definite integrals. The following general method is called **Newton–Cotes quadrature**. Suppose the continuous function f is well approximated on the interval $[a, b]$ by a Lagrangian polynomial

$$f(x) \approx p(x) = \sum_{i=1}^{n} f(x_i)\, l_i(x)$$

Then we approximate $\int_a^b f(x)\,dx$ by $\int_a^b p(x)$, yielding

$$\int_a^b f(x)\,dx \approx \sum_{i=1}^{n} f(x_i) \int_a^b l_i(x)$$

The simplest example of this method is to approximate f on $[a, b]$ by the affine function whose graph passes through $(a, f(a))$ and $(b, f(b))$. This affine function is given by

$$A: x \mapsto f(a) + m(x-a)$$

where

$$m = \frac{f(b)-f(a)}{b-a}$$

Here we obtain the approximation

$$\int_a^b f(x)\,dx \approx \int_a^b A(x)\,dx = \frac{b-a}{2}(f(a)+f(b)) \tag{1}$$

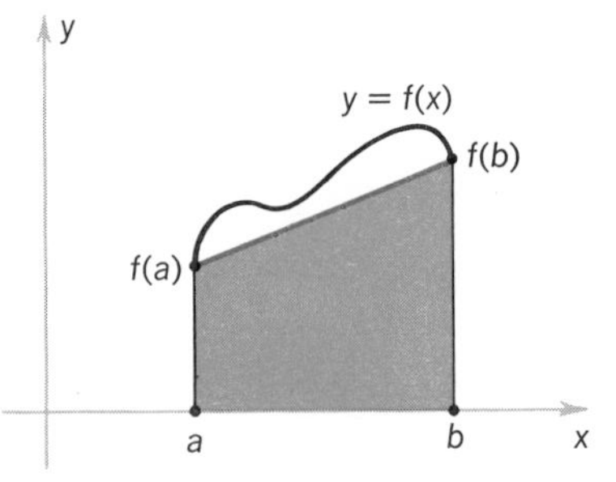

FIGURE 2-1

When $f(x) \geqslant 0$ on $[a, b]$, there is a natural geometric interpretation to this approximation formula. Referring to Figure 2-1, we see that the area under f from a to b is approximated by the shaded area. The shaded figure is a trapezoid, whose area is $\frac{1}{2}(b-a)(f(b)+f(a))$. Because of this geometric intuition, formula (1) is known as the **trapezoidal rule**.

The trapezoidal rule (1) is likely to give a very crude approximation to the value of the definite integral. We can hope to do much better by dividing $[a, b]$ into small subintervals, applying the trapezoidal rule to f in each subinterval and then summing. To carry out this program suppose we are given

$$y_0 = f(x_0)$$
$$\vdots$$
$$y_n = f(x_n)$$

where

$$a = x_0 < x_1 < \cdots < x_n = b$$

If we approximate f on $[x_i, x_{i+1}]$ by the affine function whose graph passes through (x_i, y_i) and (x_{i+1}, y_{i+1}), we have

$$\int_{x_i}^{x_{i+1}} f(x)\,dx \approx \frac{x_{i+1}-x_i}{2}(y_i+y_{i+1})$$

Summing,

$$\int_a^b f(x)\,dx \approx \tfrac{1}{2}(y_0+y_1)(x_1-x_0) + \cdots + \tfrac{1}{2}(y_{n-1}+y_n)(x_n-x_{n-1}) \qquad (2)$$

For equidistant spacing

$$x_1 - x_0 = x_2 - x_1 = \cdots = x_n - x_{n-1} = \frac{b-a}{n}$$

the formula (2) reduces to

$$\int_a^b f(x)\,dx \approx \frac{b-a}{n}(\tfrac{1}{2}y_0+y_1+\cdots+y_{n-1}+\tfrac{1}{2}y_n) \qquad (3)$$

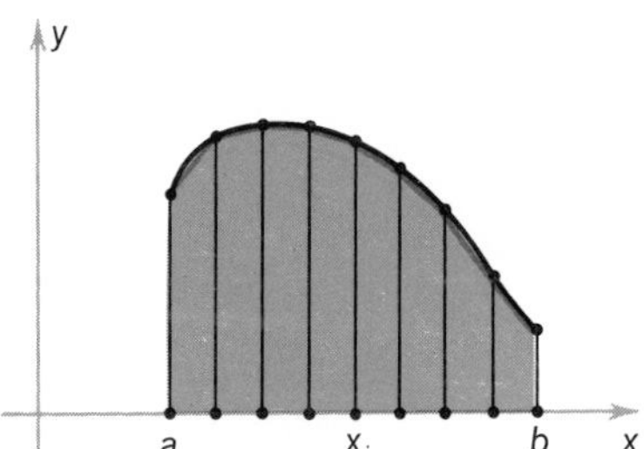

FIGURE 2-2

Thus (3) is obtained by applying the trapezoidal rule [equation (1)] n times. For simplicity, (3) is also often referred to as the trapezoidal rule.

Supposing $f(x) \geqslant 0$ on $[a,b]$, formula (3) can be interpreted geometrically (see Figure 2-2): The sum $\frac{1}{2}(b-a)(\frac{1}{2}y_0+y_1+\cdots+y_{n-1}+\frac{1}{2}y_n)$ is equal to the area of the polygon formed by the x axis, the lines $x=a$, $x=b$, and the straight line segments from (x_i, y_i) to (x_{i+1}, y_{i+1}). This polygon is composed of the trapezoids that are used to approximate f on $[x_i, x_{i+1}]$.

Using polygons to approximate integrals is, of course, closely related to the construction of the integral as the limit of approximating sums. In fact, by Darboux's Theorem (§7.2), we have

$$\lim_{n\to\infty} \frac{b-a}{n}(\tfrac{1}{2}y_0+y_1+\cdots+y_{n-1}+\tfrac{1}{2}y_n) = \int_a^b f(x)\,dx$$

when f is continuous. Therefore, in theory, by taking sufficiently many abscissas $x_0, \ldots, x_n$ we can approximate the integral $\int_a^b f(x)\,dx$ to any desired degree of accuracy. In practice, the trapezoidal rule gives satisfactory results, if we can determine $f(x_i)$ for sufficiently many x_i. The calculation is very simple and straightforward—the difficulty is to determine how many ordinates are needed to insure a given degree of accuracy. The error associated with this rule is discussed in some detail at the end of the section.

Example 1 Given the following data

x	$y = f(x)$
0	1.00000
0.5	1.64872
1.0	2.71828
1.5	4.48169
2.0	7.38906
2.5	12.18249
3.0	20.08554
3.5	33.11545
4.0	54.59815

let us approximate $\int_0^4 f(x)\,dx$ by means of the trapezoidal rule. Here we take $a=0$, $b=4$, $n=8$, and $(b-a)/n=0.5$. We compute

$$\begin{aligned}\int_0^4 f(x)\,dx &\approx 0.5(\tfrac{1}{2}\cdot 1+1.64872+2.71828+4.48169+\cdots\\&\quad +33.11545+\tfrac{1}{2}\cdot 54.59815)\\&= 54.71015\end{aligned}$$

The data for the above function are taken from a table of values of the exponential function $x \mapsto e^x$. We can evaluate

$$\int_0^4 e^x\,dx = e^4 - 1 = 53.59815$$

So, assuming $f(x)=e^x$, our approximation is off by 1.11200.

The trapezoidal rule is obtained by approximating a curve using straight line segments. To get a good approximation of a curve in this way, we need a great number of very small line segments. We can expect to be even more successful if we use parabolic arcs to approximate the curve. To do this, instead of connecting two consecutive given points on the curve by a line segment, we shall connect three consecutive given points by a parabolic arc, that is, by a segment of the graph of a polynomial of the second order. The result, known as **Simpson's rule**, is a formula of great accuracy and great practical importance.

The simplest case to consider is the following. Let us approximate f on $[a,b]$ by the Lagrange polynomial $p(x)$ whose graph passes through the points

$$(a,f(a)), \qquad (\tfrac{1}{2}(a+b), f(\tfrac{1}{2}(a+b))), \qquad (b,f(b))$$

We have the approximation

$$\int_a^b f \approx \int_a^b p(x)\,dx$$

To evaluate the latter integral, we note that by change of variable,

$$\int_0^{2h} p(x+a)\,dx = \int_a^b p(x)\,dx$$

where $h=\frac{1}{2}(b-a)$. However, $p(x+a)$ is a polynomial of degree $\leqslant 2$ which satisfies

$$p(0+a) = f(a), \qquad p(h+a) = f(\tfrac{1}{2}(a+b)), \qquad p(2h) = f(b)$$

Therefore, we must have, by Lagrange's interpolation formula [(6), §12.1],

$$p(x+a) = f(a)\,l_0(x) + f(\tfrac{1}{2}(a+b))\,l_1(x) + f(b)\,l_2(x)$$

where

$$l_0(x) = \frac{(x-h)(x-2h)}{(-h)(-2h)} = \frac{(x-h)(x-2h)}{2h^2}$$

$$l_1(x) = \frac{(x-0)(x-2h)}{h(-h)} = \frac{x(x-2h)}{-h^2}$$

$$l_2(x) = \frac{(x-0)(x-h)}{(2h)(h)} = \frac{x(x-h)}{2h^2}$$

Integrating, we obtain

$$\begin{aligned}
\int_a^b p(x)\,dx &= \int_0^{2h} p(x+a)\,dx \\
&= f(a)\int_0^{2h} \frac{(x-h)(x-2h)}{2h^2}\,dx + f(\tfrac{1}{2}(a+b))\int_0^{2h} \frac{x(x-2h)}{-h^2}\,dx \\
&\quad + f(b)\int_0^{2h} \frac{x(x-h)}{2h^2}\,dx \\
&= \tfrac{1}{3}h\,f(a) + \tfrac{4}{3}h\,f(\tfrac{1}{2}(a+b)) + \tfrac{1}{3}h\,f(b) \\
&= \tfrac{1}{3}h\big(f(a)+4f(\tfrac{1}{2}(a+b))+f(b)\big)
\end{aligned}$$

We thus arrive at **Simpson's rule**:

$$\int_a^b f(x)\,dx \approx \frac{b-a}{6}\left[f(a)+4f\left(\frac{a+b}{2}\right)+f(b)\right] \tag{4}$$

As with the trapezoidal rule, we can increase the effectiveness of Simpson's rule by dividing $[a, b]$ into subintervals, applying (4) on each of the subintervals, and then summing. This process leads to the rule

$$\begin{aligned}
\int_a^b f(x)\,dx \approx \frac{1}{3}\left(\frac{b-a}{n}\right)&(f(x_0)+4f(x_1)+2f(x_2)+4f(x_3)+2f(x_4)+\cdots \\
&+2f(x_{n-2})+4f(x_{n-1})+f(x_n))
\end{aligned} \tag{5}$$

where

$$\begin{aligned}
x_0 &= a \\
x_1 &= a + \frac{b-a}{n} \\
x_2 &= a + \frac{2(b-a)}{n} \\
&\vdots \\
x_n &= a + \frac{n(b-a)}{n} = b
\end{aligned}$$

Formula (5) is also commonly called Simpson's rule. The partition $a = x_0 < x_1 < \cdots < x_n = b$ divides $[a, b]$ into n subintervals each of length $h = (b-a)/n$. As we stated above, the rule (5) is obtained by applying (4) to each of the intervals $[x_0, x_2], [x_2, x_4], \ldots, [x_{2i}, x_{2i+2}], \ldots, [x_{n-2}, x_n]$. For this to work, n must, therefore, be even. Hence, an **odd number of equally spaced values** of x are required for Simpson's rule.

Example 2 Using the data of Example 1, let us approximate $\int_0^4 f(x)\,dx$. Applying Simpson's rule with $n=8$, we compute

$$\begin{aligned}\int_0^4 f(x)\,dx &\approx \tfrac{1}{6}(1+4(1.64812)+2(2.71828)+4(4.48169)+2(7.38906)\\&\quad+4(12.18249)+2(20.08554)+4(33.11545)+54.59815)\\&= 53.61622\end{aligned}$$

Assuming $f(x)=e^x$, we see that the error arising from Simpson's rule is

$$53.59815-53.61622 = -0.01807$$

which is much smaller than the error in Example 1, where the trapezoidal rule was used.

Example 3 Use Simpson's rule to estimate $\ln 7 = \int_1^7 dx/x$.

Taking $n=6$ and $(b-a)/n=(7-1)/6=1$, we have

$$\begin{aligned}\ln 7 &\approx \tfrac{1}{3}(1+4\cdot\tfrac{1}{2}+2\cdot\tfrac{1}{3}+4\cdot\tfrac{1}{4}+2\cdot\tfrac{1}{5}+4\cdot\tfrac{1}{6}+\tfrac{1}{7})\\&= 1.95873\end{aligned}$$

to five decimal places. Since the exact value of ln 7 to five places is 1.94591, the error occurs in the second decimal place, being 0.01282.

The trapezoidal rule was based on interpolation by affine functions, while Simpson's rule was based on interpolation by polynomial functions of degree two. The general form of Newton–Cotes quadrature is based on Lagrangian interpolation of arbitrary order n. We shall now derive the general formula for Newton–Cotes quadrature. Suppose the function f can be sufficiently well approximated by a polynomial of degree n on the interval $[a,b]$. Given the $n+1$ function values $y_0=f(x_0), \ldots, y_n=f(x_n)$, we can form the Lagrange approximating polynomial

$$p(x) = y_0 l_0(x) + \cdots + y_n l_n(x)$$

with the associated error

$$E_n(x) = \pi(x)\frac{f^{(n+1)}(\xi)}{(n+1)!}$$

Since $f(x)\approx p(x)$, we integrate $p(x)$ from a to b to find

$$\int_a^b f \approx \int_a^b p(x)\,dx = \sum_{i=0}^{n} y_i \int_a^b l_i(x) \tag{6}$$

with associated error

$$E_n = \int_a^b \pi(x) f^{(n+1)}(\xi)\,dx/(n+1)!$$

Now assuming the x_i are equally spaced points, we can readily evaluate $\int_a^b l_i(x)\,dx$. Making the change of variables

$$x = a + \frac{b-a}{n}u, \qquad x_i = a + \frac{i(b-a)}{n}$$

we have

$$\int_a^b l_i(x)\,dx = \frac{b-a}{n}\int_0^n l_i\left(a + \frac{b-a}{n}u\right)du$$

However, since

$$l_i(x) = \frac{(x-x_0)(x-x_1)\cdots(x-x_{i-1})(x-x_{i+1})\cdots(x-x_n)}{(x_i-x_0)\cdots(x_i-x_{i-1})(x_i-x_{i+1})\cdots(x_i-x_n)}$$

it is seen that the form of $l_i(x)$ is unchanged under an affine change of variables. Thus

$$l_i(x) = l_i\left(a + \frac{b-a}{n}u\right) = \frac{u(u-1)(u-2)\cdots(u-i+1)(u-i-1)\cdots(u-n)}{i(i-1)\cdots(i-i+1)(i-i-1)\cdots(i-n)}$$

We obtain

$$\int_a^b l_i(x) = \frac{b-a}{n}\int_0^n \frac{u(u-1)\cdots(u-i+1)(u-i-1)\cdots(u-n)}{i(i-1)\cdots(i-i+1)(i-i-1)\cdots(i-n)}\,du$$

Using the notation

$$C_i^{(n)} = \int_0^n \frac{u(u-1)\cdots(u-i+1)(u-i-1)\cdots(u-n)}{i(i-1)\cdots(i-i+1)(i-i-1)\cdots(i-n)}\,du$$

we can rewrite the formula for approximate quadrature,

$$\int_a^b \boldsymbol{f} \approx \frac{\boldsymbol{b-a}}{\boldsymbol{n}}\sum_{i=1}^{n} C_i^{(n)}\boldsymbol{f}(\boldsymbol{x}_i)$$

which is the general formula for Newton–Cotes quadrature.

For $n = 2$, we expect to find Simpson's rule. In fact,

$$C_0^{(2)} = \int_0^2 \frac{(u-1)(u-2)}{(-1)(-2)}\,du = \frac{1}{3}$$

$$C_1^{(2)} = \int_0^2 \frac{u(u-2)}{(1)(-1)}\,du = \frac{4}{3}$$

$$C_2^{(2)} = \int_0^2 \frac{u(u-1)}{(2)(1)}\,du = \frac{1}{3}$$

which indeed gives Simpson's rule (4).

For $n=1$, we of course find the simple form of the trapezoidal rule (1). For $n=3$, we find

$$C_0^{(3)} = \int_0^3 \frac{(u-1)(u-2)(u-3)}{(-1)(-2)(-3)}\,du = -\frac{3}{8}$$

$$C_1^{(3)} = \int_0^3 \frac{u(u-2)(u-3)}{(1)(-1)(-2)}\,du = \frac{9}{8}$$

$$C_2^{(3)} = \int_0^3 \frac{u(u-1)(u-3)}{(2)(1)(-1)}\,du = \frac{9}{8}$$

$$C_3^{(3)} = \int_0^3 \frac{u(u-1)(u-2)}{(3)(2)(1)}\,du = \frac{3}{8}$$

We thus obtain the **three-eighths rule**:

$$\int_a^b f(x)\,dx \approx \tfrac{3}{8}(b-a)\left(\frac{f(x_0)}{3} + f(x_1) + f(x_2) + \frac{f(x_3)}{3}\right)$$

Continuing this process one step further we can obtain **Villarceau's formula**

$$\int_a^b f \approx \tfrac{2}{45}(b-a)(7f(x_0) + 32f(x_1) + 12f(x_2) + 32f(x_3) + 7f(x_4))$$

(exercise 17).

Example 4 Applying the three-eighths rule twice, approximate ln 7.

We shall approximate $\int_1^4 dx/x$ and $\int_4^7 dx/x$, using the three-eighths rule for each integral; we obtain

$$\int_1^7 \frac{dx}{x} \approx 3\cdot\tfrac{3}{8}(\tfrac{1}{3}+\tfrac{1}{2}+\tfrac{1}{3}+\tfrac{1}{12}+\tfrac{1}{12}+\tfrac{1}{5}+\tfrac{1}{6}+\tfrac{1}{21}) = 1.96607$$

The error is thus 0.02016. This method of approximation required $f(x)$ for seven values of x, the same as were needed for three applications of Simpson's rule in Example 3. Moreover, the error is slightly larger than that obtained when Simpson's rule was applied. We shall remark further on these two rules in our discussion of the error associated with them.

We now wish to discuss the error involved in Newton–Cotes quadrature. We will see that for n even, the formula

$$\int_a^b f(x)\,dx = \sum_{i=0}^{n} f(x_i)\,C_i^{(n)}$$

is exact for polynomial functions of degree $n+1$. This result is striking in that the degree of the Lagrange interpolating polynomial is equal to n. Thus, for example, in Simpson's rule, while we use parabolic arcs to approximate the function, the formula gives an exact result for cubic polynomials!

It can be shown that when n is odd, E_n takes the form

$$E_n = \frac{((b-a)/n)^{n+2}}{(n+1)!} f^{(n+1)}(\xi) \int_0^n u(u-1)\cdots(u-n)\,du$$

On the other hand, when n is even, the error E_n takes the form

$$E_n = \frac{((b-a)/n)^{n+3}}{(n+1)!} f^{(n+2)}(\xi) \int_0^n (u-\tfrac{1}{2}n)\,u(u-1)\cdots(u-n)\,du$$

Therefore, when n is odd and an even number of ordinates are used, the $(n+1)$st derivative $f^{(n+1)}(\xi)$ appears in the error. When n is even and an odd number of ordinates are used, the $(n+2)$nd derivative $f^{(n+2)}(\xi)$ appears in the error (see exercise 20).

Applied to the simple form of the trapezoidal rule, the analysis of the error yields

$$\int_a^b f(x)\,dx = \frac{b-a}{2}(f(a)+f(b)) - \frac{(b-a)^3}{12} f''(\xi)$$

for some ξ in $]a,b[$. Since the general form of this rule

$$\int_a^b f(x)\,dx \approx \frac{b-a}{n}(\tfrac{1}{2}f(x_0)+f(x_1)+\cdots+f(x_{n-1})+\tfrac{1}{2}f(x_n))$$

is obtained by applying the simple rule to each of the intervals

$$[x_0, x_1], [x_1, x_2], \ldots, [x_{n-1}, x_n]$$

we find that the error in the general trapezoidal rule is given by

$$\int_a^b f(x)\,dx = \frac{b-a}{n}(\tfrac{1}{2}f(x_0)+\cdots+\tfrac{1}{2}f(x_n)) - \frac{1}{12}\left(\frac{b-a}{n}\right)^3 (f''(\xi_1)+\cdots+f''(\xi_n)) \tag{7}$$

where $x_i < \xi_i < x_{i+1}$. However, it follows from the Intermediate Value Theorem, assuming the continuity of f'', that there exists a point ξ in $[a,b]$ such that

$$f''(\xi) = \frac{f''(\xi_1)+\cdots+f''(\xi_n)}{n}$$

Hence, equation (7) becomes

$$\int_a^b f(x)\,dx = \frac{b-a}{n}(\tfrac{1}{2}f(x_0)+f(x_1)+\cdots+\tfrac{1}{2}f(x_n)) - \frac{(b-a)^3}{12n^2} f''(\xi)$$

Therefore, the magnitude of the error is

$$|E| = \frac{(b-a)^3}{12n^2}\,|f''(\xi)|$$

Hence, If f'' is continuous (from which it follows that f'' is bounded) on $[a,b]$, the error will tend to zero like $1/n^2$ as $n \to \infty$.

Applied to Simpson's rule, the analysis of the error yields

$$E_2 = \frac{((b-a)/2)^5 f^{(4)}(\xi)}{24} \int_0^2 u(u-1)^2(u-2)\,du$$

$$= -\frac{((b-a)/2)^5 f^{(4)}(\xi)}{90}$$

Therefore, we have an exact form of Simpson's rule

$$\int_a^b f(x)\,dx = \left(\frac{b-a}{2}\right)(f(x_0)+4f(x_1)+f(x_2)) - \frac{1}{90}\left(\frac{b-a}{2}\right)^5 f^{(4)}(\xi)$$

where ξ is some undetermined point in $]a, b[$.

When $f(x)$ is a polynomial of degree $\leqslant 3$, we have $f^{(4)}(x) = 0$ for all x and so in this case the error vanishes:

$$E_2 = -\frac{1}{90}\left(\frac{b-a}{2}\right)^5 f^{(4)}(\xi) = 0$$

The more general form of Simpson's rule is obtained by dividing the interval $[a, b]$ into $n/2$ subintervals each of length $2(b-a)/n$, and then applying the simple form of the rule to each of these subintervals and summing. Using the same reasoning as we did for the trapezoidal rule, we find

$$\int_a^b f(x)\,dx = \frac{b-a}{3}(f(x_0)+4f(x_1)+2f(x_2)+\cdots$$

$$+2f(x_{n-2})+4f(x_{n-1})+f(x_n)) - \frac{(b-a)^5}{180n^4} f^{(4)}(\xi)$$

Therefore, if $f^{(4)}$ is continuous on $[a, b]$, the error tends to zero like $1/n^4$ as n goes to infinity.

For the three-eighths rule, the error is

$$E_3 = -\frac{(b-a)}{6480} f^{(4)}(\xi)$$

We see that this rule, like Simpson's rule, is exact for cubic polynomials. In fact, it is roughly of the same accuracy as Simpson's rule. The coefficient of $-(b-a)f^{(4)}(\xi)$ in the error for this latter rule is $1/2880$. Hence, we expect the three-eighths rule to be but slightly more accurate than Simpson's rule. Since the three-eighths rule requires an additional tabulation of the value of the function and offers only slightly increased expectation of accuracy, in practice Simpson's rule is usually preferred. However, Simpson's rule suffers from the fact that to apply it an odd number of ordinates must be used. If an even number $N = 2k$ of values $f(x_i)$ are given, the three-eighths rule can be applied to f on $[x_0, x_3]$ and Simpson's rule can be applied to f on $[x_3, x_N]$, which will overcome this difficulty.

The Hermite interpolation formula also gives rise to a rule of quadrature. We recall the equation

$$f(x) = \sum_{i=0}^{n-1} h_i(x) f(x_i) + \sum_{i=0}^{n-1} \bar{h}_i(x) f'(x_i) + \frac{f^{(2n)}(\xi)\,\pi(x)^2}{(2n)!}$$

where ξ depends on x. Integrating both sides yields

$$\int_a^b f(x)\,dx = \sum_{i=0}^{n-1} H_i f(x_i) + \sum_{i=0}^{n-1} \bar{H}_i f'(x_i) + E$$

where

$$H_i = \int_a^b h_i(x) = \int_a^b (1-2l_i'(x_i)(x-x_i))\,[l_i(x)]^2\,dx$$

$$\bar{H}_i = \int_a^b (x-x_i)\,l_i(x)^2\,dx$$

$$E = \frac{1}{(2n)!}\int_a^b f^{(2n)}(\xi)\,\pi(x)^2\,dx$$

Since $\pi(x)^2$ does not change sign on $[a, b]$, the Second Mean Value Theorem of Integral Calculus can be applied to yield

$$E = \frac{f^{(2n)}(\eta)}{(2n)!}\int_a^b \pi(x)^2\,dx$$

where $a \leqslant \eta \leqslant b$. We see at once, therefore, that Hermitian quadrature yields exact results for polynomials of degree as high as $2n-1$.

Example 5 When we estimated $\ln 7 = \int_1^7 dx/x$ by Simpson's rule (Example 3), in effect we divided $[1, 7]$ into the subintervals

$$[1,3], \quad [3,5], \quad [5,7]$$

and approximated $x \mapsto 1/x$ on each of these by a parabolic arc. On $[1, 3]$, for instance, we used the parabola passing through $(1, 1), (2, \frac{1}{2}), (3, \frac{1}{3})$. The Hermitian interpolation formula gives a cubic polynomial $q_1(x)$ satisfying

$$q_1(1) = 1$$

$$q_1(3) = \tfrac{1}{3}$$

$$q_1'(1) = -1$$

$$q_1'(3) = -\tfrac{1}{9}$$

The method of Hermitian quadrature is to approximate

$$\int_1^3 \frac{dx}{x} \approx \int_1^3 q_1(x)\,dx$$

Since the Hermitian polynomial is a cubic satisfying four requirements, we can expect it to fit the function $x \mapsto 1/x$ somewhat better on $[1, 3]$ than does the parabolic arc used in Simpson's rule. If we form the Hermitian polynomials $q_2(x)$ and $q_3(x)$ satisfying

$$\begin{aligned} q_2(3) &= \tfrac{1}{3} & q_3(5) &= \tfrac{1}{5} \\ q_2(5) &= \tfrac{1}{5} & q_3(7) &= \tfrac{1}{7} \\ q_2'(3) &= -\tfrac{1}{9} & q_3'(5) &= -\tfrac{1}{25} \\ q_2'(5) &= -\tfrac{1}{25} & q_3'(7) &= -\tfrac{1}{49} \end{aligned}$$

we can estimate the integral by

$$\int_1^7 \frac{dx}{x} \approx \int_1^3 q_1(x)\,dx + \int_3^5 q_2(x)\,dx + \int_5^7 q_3(x)\,dx$$

The approximate result obtained using this method is 1.9526 to four decimal places, an error of 0.0067. The error is roughly half the error associated with Simpson's rule.

EXERCISES

In exercises 1–4, approximate the integral using the general form of the trapezoidal rule with the given value of n.

1. $\int_1^2 \frac{1}{x^2}\,dx, \quad n = 2$
2. $\int_0^3 (x^2+1)\,dx, \quad n = 6$
3. $\int_1^3 \frac{1}{x}\,dx, \quad n = 8$
4. $\int_{-1}^1 \frac{2}{1+x^2}\,dx, \quad n = 8$
5. $4\int_0^1 (1-x^2)\,dx, \quad n = 4$

In exercises 6–11, use the general form of Simpson's rule with the given value of n to approximate the integral.

6. $\int_1^3 \frac{1}{x}\,dx, \quad n = 4$
7. $\int_0^\pi \sqrt{t}\sin t\,dt, \quad n = 4$
8. $\int_0^1 \frac{1}{1+x^2}\,dx, \quad n = 4$
9. $\int_0^\pi \sqrt{\sin x}\,dx, \quad n = 2$
10. $\int_0^1 \frac{1}{1+x+x^3}\,dx, \quad n = 4$
11. $\int_0^\pi \frac{1}{2+\cos x}\,dx, \quad n = 2$

In exercises 12–16, approximate each number by expressing it as the definite integral of some function and then using Simpson's rule.

12. $\arctan \frac{1}{2}$
13. $\arcsin \frac{1}{4}$
14. $\ln 2$
15. e
16. π

17. Derive the formula for Newton–Cotes' quadrature with $n = 4$:

    $\int_a^b f = \frac{2}{45}(b-a)(7f(x_0) + 32f(x_1) + 12f(x_2) + 32f(x_3) + 7f(x_4))$

    (This rule is also known as Villarceau's formula.)

18. Estimate $\ln \frac{5}{3} = \int_3^5 dx/x$ by Newton–Cotes quadrature with $n = 1$, $n = 2$, $n = 3$, and $n = 4$; that is, use the simple form of the trapezoidal rule, the simple form of Simpson's rule, the three-eighths rule, and then Villarceau's formula. Compare each of your results with the correct value to six decimal places which is $\ln \frac{5}{3} = 0.510825$.
19. Estimate $\ln \frac{10}{3} = \int_3^{10} dx/x$ by Newton–Cotes quadrature with $n = 1$, $n = 2$, $n = 3$, and $n = 4$. Compare your results with the correct value to eight decimal places which is $\ln \frac{10}{3} = 1.203973$.
20. Verify that Simpson's rule is exact for cubic polynomials.

3 GAUSSIAN QUADRATURE

In our discussion of Newton–Cotes quadrature we made the simplifying assumption that the points $x_0, \ldots, x_{n-1}$ were equally spaced. However, equation (6) of the last section is valid without the assumption that the points x_i are equally spaced. Similarly, for Hermitian quadrature no assumption on the spacing of the points x_i was made and the formula is valid without any such hypothesis. In this section, we develop a very powerful method of numerical quadrature which chooses the points x_i by a more judicious method than that dictated by equal spacing.

We restrict attention for the time being to the interval $[-1, 1]$ and to integrals of the form $\int_{-1}^{1} f$. This will not affect the generality of results to be obtained, since, by a change of variables

$$\int_a^b f(x)\, dx = \frac{b-a}{2} \int_{-1}^{1} f\left(\frac{b-a}{2} x + \frac{b+a}{2}\right) dx$$

The **Legendre polynomials** $P_r(x)$ are defined by the equation

$$P_r(x) = \frac{1}{2^r r!} \frac{d^r [(x^2-1)^r]}{dx^r}$$

Thus

$$P_0(x) = 1$$

$$P_1(x) = x$$

$$P_2(x) = \tfrac{1}{2}(3x^2 - 1)$$

$$P_3(x) = \tfrac{1}{2}(5x^3 - 3x)$$

$$P_4(x) = \tfrac{1}{8}(35x^4 - 30x^2 + 3)$$

$$P_5(x) = \tfrac{1}{8}(63x^5 - 70x^3 + 15x)$$

These polynomials are easily seen to satisfy the recurrence formula

$$P_{r+1}(x) = \frac{2r+1}{r+1} x P_r(x) - \frac{r}{r+1} P_{r-1}(x)$$

Thus

$$P_6(x) = \tfrac{13}{7} x P_5(x) - \tfrac{6}{7} P_4(x)$$

and so on.

Theorem 1 The Orthogonality Property If $p(x)$ is a polynomial of degree less than r, then

$$\int_{-1}^{1} P_r(x)\, p(x)\, dx = 0$$

For the first few values of r the orthogonality property is easily checked (see exercise 7). The basic idea is to use the definition of P_n and integrate by parts. If the reader does exercise 7 he should also see the general proof.

Using the orthogonality property of the Legendre polynomials, we shall obtain a quadrature rule which involves n function values of the function and no further data, and yet is accurate up to polynomials of degree $2n-1$. Recall the formula for Hermite quadrature

$$\int_{-1}^{1} f(x)\, dx = \sum_{i=0}^{n-1} H_i f(x_i) + \sum_{i=0}^{n-1} \overline{H}_i f'(x_i) + E \tag{1}$$

where

$$E = \frac{f^{(2n)}(\xi)}{(2n)!} \int_{-1}^{1} \pi(x)^2\, dx$$

The weight $\overline{H}_i$ is given by

$$\overline{H}_i = \int_{-1}^{1} (x - x_i)\, [l_i(x)]^2\, dx$$

Since

$$(x - x_i)\, [l_i(x)]^2 = \frac{\pi(x)}{\pi'(x_i)}\, l_i(x)$$

where $\pi(x) = (x - x_0) \cdots (x - x_{n-1})$, we find that

$$\overline{H}_i = \frac{1}{\pi'(x_i)} \int_{-1}^{1} \pi(x)\, l_i(x)\, dx$$

Equation (1) is valid for all choices of n distinct points x_i in the interval $[a, b]$. We can now exploit this freedom in the choice of the points x_i to force the condition

$$\overline{H}_i = 0$$

Note that the degree of $l_i(x)$ is $n-1$, and the degree of $\pi(x)$ is n. Hence, simply take $\pi(x) = (x - x_0) \cdots (x - x_{n-1})$, where $x_0, \ldots, x_{n-1}$ are the n roots of $P_n(x)$. With this choice of points x_i, π is a multiple of P_n and by Theorem 1

$$\int_{-1}^{1} l_i(x)\, \pi(x)\, dx = 0$$

and the Hermite formula reduces to

$$\int_{-1}^{1} f(x)\,dx = \sum_{i=0}^{n-1} H_i f(x_i) + \frac{f^{(2n)}(\xi)}{(2n)!}\int_{-1}^{1} P_n(x)^2\,dx \tag{2}$$

Equation (2) is known as **Gauss' rule of quadrature.** It produces exact results for polynomials of degree less than or equal to $2n-1$ while making use of only n ordinates. However, the x_i's must be chosen as roots of $P_n(x)$.

For $2 \leqslant n \leqslant 5$, we list below the zeroes $x_0, \ldots, x_{n-1}$ of the nth Legendre polynomial and the numerical value of the weights $H_0, \ldots, H_{n-1}$.

When $n=2$,

$$x_0 = -1/\sqrt{3} = -0.577350 \qquad H_0 = 1$$
$$x_1 = 1/\sqrt{3} = 0.577350 \qquad H_1 = 1$$

When $n=3$,

$$x_0 = -\sqrt{\tfrac{3}{5}} = -0.774597 \qquad H_0 = \tfrac{5}{9}$$
$$x_1 = 0 \qquad H_1 = \tfrac{8}{9}$$
$$x_2 = \sqrt{\tfrac{3}{5}} = 0.774597 \qquad H_2 = \tfrac{5}{9}$$

When $n=4$,

$$x_0 = -\sqrt{\tfrac{3}{7} + \tfrac{2}{35}\sqrt{30}} = -0.861136 \qquad H_0 = \tfrac{1}{2} - \tfrac{1}{36}\sqrt{30} = 0.347855$$
$$x_1 = -\sqrt{\tfrac{3}{7} - \tfrac{2}{35}\sqrt{30}} = -0.339981 \qquad H_1 = \tfrac{1}{2} + \tfrac{1}{36}\sqrt{30} = 0.652145$$
$$x_2 = \sqrt{\tfrac{3}{7} - \tfrac{2}{35}\sqrt{30}} = 0.339981 \qquad H_2 = \tfrac{1}{2} + \tfrac{1}{36}\sqrt{30} = 0.652145$$
$$x_3 = \sqrt{\tfrac{3}{7} + \tfrac{2}{35}\sqrt{30}} = 0.861136 \qquad H_3 = \tfrac{1}{2} - \tfrac{1}{36}\sqrt{30} = 0.347855$$

When $n=5$, we obtain these approximate values:

$$x_0 = -0.906180 \qquad H_0 = 0.236927$$
$$x_1 = -0.538469 \qquad H_1 = 0.478629$$
$$x_2 = 0 \qquad H_2 = 0.568889$$
$$x_3 = 0.538469 \qquad H_3 = 0.478629$$
$$x_4 = 0.906180 \qquad H_4 = 0.236927$$

Example 1 Approximate the value of $\int_{-1}^{1} dt/(3+t)$.

Using three ordinates,

$$\int_{-1}^{1}\frac{dt}{3+t} = \frac{5}{9}\left(\frac{1}{3-\sqrt{\frac{3}{5}}}\right) + \frac{8}{9}\left(\frac{1}{3}\right) + \frac{5}{9}\left(\frac{1}{3+\sqrt{\frac{3}{5}}}\right)$$
$$= 0.69312165$$

Using five ordinates, the approximation becomes 0.693147157. The correct value is 0.69314718, so that using three ordinates, the error is in the fifth decimal place; using five ordinates, it is in the eighth decimal place!

Example 2 Using Gaussian quadrature, estimate $\int_0^4 e^u du$, and compare the result with approximations in §2.

We shall apply Gaussian quadrature with five ordinates. We first apply the change of variables

$$x = \frac{u-2}{2}, \qquad u = 2x+2$$

to bring the integral into the required form:

$$\int_0^4 e^u \, du = 2\int_{-1}^1 e^{2x+2} \, dx$$

We consult a table of values of the exponential function, and perform the necessary interpolations to find the data:

$x =$	$2x+2 =$	$e^{2x+2} =$
-0.906180	0.18764031	1.20639950
-0.538469	0.92306138	2.51698405
0	2	7.38905610
0.538469	3.07693862	21.69189349
0.906180	3.81235969	45.25710562

Multiplying each of the e^{2x_i+2} by H_i $(0 \leqslant i \leqslant 4)$, summing, and multiplying by two, we find

$$\int_0^4 e^u du \approx 53.59814$$

Therefore, the error is $E = 0.00001$, since the exact value to five places is 53.59815. When we used Simpson's rule with nine ordinates the error was 0.01807; the trapezoidal rule with nine ordinates yielded an error equal to 1.11200.

In the discussion that follows, we consider reasons for the remarkable accuracy of the Gaussian method. Since Gaussian quadrature with n ordinates is exact for polynomials of degree $2n-1$, we have the following theorem.

Theorem 2 If $p(x)$ and $q(x)$ are polynomials of degree at most $2n-1$ which satisfy

$$p(x_i) = q(x_i) \qquad 0 \leqslant i \leqslant n-1$$

for each root x_i of the nth Legendre polynomial, then

$$\int_{-1}^{1} p(x)\,dx = \int_{-1}^{1} q(x)\,dx$$

Using this result, we can gain insight into why Gaussian quadrature is a more powerful method than Newton–Cotes quadrature. In Gaussian quadrature we estimate $\int_{-1}^{1} f$ by

$$\sum_{i=0}^{n-1} y_i \int_{-1}^{1} h_i(x)\,dx$$

Since $h_i(x_j) = \delta_{ij}$, we see that for each of the points x_j we have

$$\sum_{i=0}^{n-1} y_i h(x_j) = y_j = f(x_j)$$

Therefore $\sum_{i=0}^{n-1} y_i h_i(x)$ is a polynomial of degree at most $2n-1$ which agrees with f at each of the points $x_0, x_1, \ldots, x_{n-1}$. Associated with these points and the ordinates $y_0, \ldots, y_{n-1}$, we also have the Lagrange interpolating polynomial of order $n-1$,

$$p(x) = \sum_{i=0}^{n-1} y_i l_i(x)$$

In view of the above theorem,

$$\sum_{i=0}^{n-1} y_i \int_{-1}^{1} h_i(x)\,dx = \int_{-1}^{1} p(x)\,dx$$

$$= \sum_{i=0}^{n-1} y_i \int_{-1}^{1} l_i(x)\,dx$$

Therefore, in the Gaussian method, we in fact approximate $\int_{-1}^{1} f(x)\,dx$ by $\sum_{i=0}^{n-1} y_i \int_{-1}^{1} l_i(x)\,dx$, the integral of the Lagrange interpolating polynomial which agrees with f at the n given points—very much the same as the procedure in Newton–Cotes quadrature.

However, in the Newton–Cotes case, we chose n equally spaced points. In the Gaussian case, we chose the points to be the zeroes of the nth Legendre polynomial. Moreover, as the above theorem shows, in the Gaussian case we could freely add any n points to our originally chosen ones in order to obtain a better polynomial approximation to f, but yet the value of the integral would not change. This is the great advantage of Gaussian quadrature. In effect we operate with $2n$ points while in fact we use only n ordinates. The power of Gaussian quadrature comes from the fact that we choose as the points x_i the zeroes of the members of an orthogonal family of polynomials. As $n \to \infty$, the interpolating polynomials $\sum_{i=0}^{n-1} y_i l_i(x)$ used in Gaussian quadrature "fit" the function f much more closely than do the polynomials associated with Newton–Cotes quadrature and equidistant spacing. The study of orthogonal families of functions is fundamental in modern numerical analysis. Unfortunately, this topic is beyond the scope of the present course.

EXERCISES

1. Estimate $\ln \frac{5}{3} = \int_3^5 dx/x$ by the method of Gaussian quadrature with three points and then with four points (the cases $n = 3$ and $n = 4$). Compare your results with the true value to nine decimal places which is $\ln \frac{5}{3} = 0.510825545$. Also compare your results with those obtained using Newton–Cotes quadrature in exercise 18 of §12.2.
2. Estimate $\ln \frac{10}{3} = \int_3^{10} dx/x$ by the method of Gaussian quadrature with three points and then with four points (the cases $n = 3$ and $n = 4$). Compare your results with the value $\ln \frac{10}{3} = 1.203973$ which is correct to seven decimal places. Also compare your results with those obtained using Newton–Cotes quadrature in exercise 19 of §12.2.
3. Estimate $\int_\pi^{8\pi} \sin x/x\, dx$ by the following method: Divide $[\pi, 8\pi]$ into seven subintervals of length π and to each subinterval apply Gaussian quadrature with two points. The correct value to six decimal places is 0.320805.
4. Estimate $\int_1^3 \sin x \sqrt{(x-1)(3-x)}\, dx$ by the method of Gaussian quadrature with two points and then with five points. The correct value to three places is 1.258.
5. Let $p(x) = A + Bx + Cx^2 + Dx^3$. Show that
$$\int_{-h}^{h} p(x)\, dx = h\left[p\left(\frac{h}{\sqrt{3}}\right) + p\left(\frac{-h}{\sqrt{3}}\right)\right]$$
6. Let $g(x) = A + Bx + Cx^2 + Dx^3 + Ex^4 + Fx^5$. Show that
$$\int_{-h}^{h} g(x)\, dx = \frac{h}{9}[5g(\sqrt{\tfrac{3}{5}}h) + 8g(0) + 5g(-\sqrt{\tfrac{3}{5}}h)]$$
7. Verify Theorem 1 for $r = 1, 2, 3, 4, 5$, and 6.

4 DIFFERENCE EQUATIONS

We now return to the subject of polynomial interpolation. With the aid of finite difference operators we shall develop alternative forms of the Lagrange interpolating polynomial, forms which are of great practical and theoretical importance. At the end of this section, we apply these new techniques to the problem of numerical differentiation.

Suppose we are given a fixed value for Δx and a function $f: x \mapsto f(x)$. We can define a new function $\mathbf{\Delta} f$ by the rule

$$\mathbf{\Delta} f: x \mapsto f(x + \Delta x) - f(x)$$

The function $\mathbf{\Delta} f$ is called the **first difference** of the function f corresponding to the increment Δx. The symbol $\mathbf{\Delta}$ is called the **difference operator.** It transforms the function f into a new function $\mathbf{\Delta} f$. If we apply the difference operator to the function $\mathbf{\Delta} f$ we obtain yet another function $\mathbf{\Delta}(\mathbf{\Delta} f)$, usually denoted $\mathbf{\Delta}^2 f$, whose rule is

$$\Delta^2 f : x \mapsto \Delta f(x + \Delta x) - \Delta f(x)$$

Since $\Delta f(x) = f(x + \Delta x) - f(x)$ and since $\Delta f(x + \Delta x) = f(x + \Delta x + \Delta x) - f(x + \Delta x)$, we find that

$$\Delta^2 f(x) = f(x + 2\Delta x) - 2f(x + \Delta x) + f(x)$$

The function $\Delta^2 f$ is called the **second difference** of f corresponding to the increment Δx. By repeating this process, we obtain in turn the functions $\Delta^3 f, \Delta^4 f, \ldots, \Delta^n f, \ldots$ which are called the third, fourth, ..., nth, ... differences of f corresponding to the increment Δx.

Example 1 In Figure 4-1 we have a difference table for the function $f : x \mapsto x^3$ with increment $\Delta x = 1$. The column $\Delta f(x)$ gives the first differences which are obtained by subtracting each entry of the column $f(x)$ from the entry following it. Thus, for $x = 114$, $x + \Delta x = 115$, we have $\Delta f(x) = (x + \Delta x)^3 - x^3 = 1520875 - 1481544 = 39331$. This result is set on a line midway between the lines determined by $x = 114$ and $x = 115$. In the same way the second differences $\Delta^2 f(x)$ are obtained from the first differences and so on. The entries in color are, therefore, from left to right, 110, $f(110)$, $\Delta f(110)$, $\Delta^2 f(110)$, $\Delta^3 f(110)$, $\Delta^4 f(110)$, $\Delta^5 f(110)$, $\Delta^6 f(110)$, $\Delta^7 f(110)$, and $\Delta^8 f(110)$.

x	$f(x) = x^3$	$\Delta f(x)$	$\Delta^2 f(x)$	$\Delta^3 f(x)$	$\Delta^4 f(x)$	$\Delta^5 f(x)$	$\Delta^6 f(x)$	$\Delta^7 f(x)$	$\Delta^8 f(x)$
110	**1331000**								
		36631							
111	1367631		**666**						
		37297		**6**					
112	1404928		672		**0**				
		37969		6		**0**			
113	1442897		678		0		**0**		
		38647		6		0		**0**	
114	1481544		684		0		0		**0**
		39331		6		0		0	
115	1520875		690		0		0		
		40021		6		0			
116	1560896		696		0				
		40717		6					
117	1601613		702						
		41419							
118	1643032								

FIGURE 4-1

We see that the third differences are constant and that the fourth differences are all zero. It will be shown shortly that the nth differences of a polynomial function of degree n are constant and that the differences of higher order are all zero.

The difference operator Δ has some fundamental algebraic properties:

(1) **Additivity** $\Delta(f+g) = \Delta f + \Delta g$

(2) **Linearity** $\Delta(cf) = c\Delta f$, c a real number

(3) **Law of Powers** $\Delta^r(\Delta^s f) = \Delta^{r+s} f$, r and s natural numbers

These properties of the difference operator are easily established and we leave this task to the reader. We note that properties (1) and (2) also hold for the powers Δ^n, n a natural number.

Let us now prove that the nth differences of a polynomial function of degree n are constant and that the $(n+1)$st differences are zero. Since the operator Δ^n is additive and linear, it suffices to establish the result for monomials $f(x)=x^n$. We have

$$\begin{aligned}\Delta f(x) &= (x+\Delta x)^n - x^n \\ &= x^n + nx^{n-1}(\Delta x) + \frac{n(n-1)}{2!}x^{n-2}(\Delta x)^2 + \cdots \\ &\quad + \frac{n(n-1)(n-2)\cdots(4)(3)(2)}{(n-1)!}x(\Delta x)^{n-1} + (\Delta x)^n - x^n \\ &= n(\Delta x)x^{n-1} + \cdots + (\Delta x)^n\end{aligned}$$

Thus $\Delta f(x)$ is a polynomial of degree $n-1$. Continuing in this manner we find that $\Delta^2 f(x)$ is a polynomial of degree $n-2$ and so on. Therefore $\Delta^n f(x)$ is a polynomial of degree zero; that is, a constant. In fact, we have $\Delta^n f(x) = (n!)(\Delta x)^n$; and we also have $\Delta^{n+1} f(x) = 0$.

The above result is very useful. It enables us, for example, to complete a mathematical table in which there are gaps. This is illustrated below.

Example 2 If

$$f(x_0) = 0.5563, \quad f(x_0+1) = 0.5682$$

$$f(x_0+3) = 0.5911, \quad f(x_0+5) = 0.6128$$

find $f(x_0+2)$ and $f(x_0+4)$ on the assumption that $f(x)$ is a polynomial of the third degree. We have seen that

$$\Delta^2 f(x) = f(x+2\Delta x) - 2f(x+\Delta x) + f(x)$$

Using $\Delta^4 f = \Delta^2(\Delta^2 f)$, we obtain

$$\Delta^4 f(x) = f(x+4\Delta x) - 4f(x+3\Delta x) + 6f(x+2\Delta x) - 4f(x+\Delta x) + f(x)$$

Since $f(x)$ is a polynomial of degree three, we must have $\Delta^4 f(x) = 0$. With $\Delta x = 1$, we obtain

$$0 = f(x_0+4) - 4f(x_0+3) + 6f(x_0+2) - 4f(x_0+1) + f(x_0)$$

and

$$0 = f(x_0+5) - 4f(x_0+4) + 6f(x_0+3) - 4f(x_0+2) + f(x_0+1)$$

Substituting the given values in these equation we get

$$f(x_0+4) + 6f(x_0+2) = 4.0809$$

$$f(x_0+4) + f(x_0+2) = 1.1819$$

whence

$$f(x_0+2) = 0.5798 \quad \text{and} \quad f(x_0+4) = 0.6021$$

We next consider some further operators and relate them to the Δ operator. The **identity operator**, denoted **1**, is defined by the rule

$$\mathbf{1}f : x \mapsto f(x)$$

Thus the operator **1** leaves a function unchanged. In particular, if we apply the identity operator to a function Δf, we obtain Δf. We define Δ to the 0*th* power, Δ^0, to be the identity operator **1**; thus $\Delta^0 = \mathbf{1}$.

Also associated with a fixed increment Δx is the **shifting operator** E which is defined by

$$Ef : x \mapsto f(x+\Delta x)$$

Thus, for example, if $\Delta x = 1$, we have $Ef(x) = f(x+1)$.

When we are given a fixed initial point x_0 and fixed increment Δx we shall employ the notations

$$x_1 = x_0 + \Delta x, \quad x_2 = x_1 + \Delta x, \quad \ldots, \quad x_{n+1} = x_n + \Delta x$$

We have therefore

$$Ef(x_0) = f(x_1), \quad Ef(x_1) = f(x_2), \quad \ldots, \quad Ef(x_n) = f(x_{n+1})$$

The powers E^2, E^3, ... are defined by iteration

$$E^2 f(x) = Ef(x+\Delta x), \ldots, E^{n+1}f(x) = E^n f(x+\Delta x)$$

We see that $E^2 f(x) = Ef(x+\Delta x) = f(x+2\Delta x)$, and in general we have $E^n f(x) = f(x+n\Delta x)$. We also define $E^0 = \mathbf{1}$.

In terms of the notation introduced above we have

$$E^n f(x_m) = f(x_{m+n}) \qquad n = 0, 1, 2, \ldots \quad m = 0, 1, 2, \ldots$$

We see that the operator E and its powers satisfy

(1) **Additivity** $E(f+g) = Ef + Eg$

(2) **Linearity** $E(cf) = cEf$

(3) **Law of Powers** $E^r(E^s f) = E^{r+s} f$

We have the relation

$$E = \mathbf{1} + \Delta$$

where the sum $\mathbf{1}+\Delta$ is the operator defined by

$$(\mathbf{1}+\Delta)f = \mathbf{1}f + \Delta f$$

The operators E, Δ, and $\mathbf{1}$ and their powers can be manipulated algebraically: We can add and subtract them, multiply them by real constants, and raise to powers. For natural numbers k we have

$$E^k = (\mathbf{1}+\Delta)^k$$

Applying the Binomial Theorem we find that

$$\begin{aligned} f(x_k) &= (\mathbf{1}+\Delta)^k f(x_0) \\ &= f(x_0) + k\Delta f(x_0) + \frac{k(k-1)}{2!}\Delta^2 f(x_0) + \cdots \\ &\quad + \frac{k(k-1)(k-2)\cdots(3)(2)(1)}{(k-1)!}\Delta^{k-1}f(x_0) + \Delta^k f(x_0) \quad (1) \\ &= \sum_{r=0}^{r=k}\binom{k}{r}\Delta^r f(x_0) \end{aligned}$$

where

$$\binom{k}{r} = \frac{k(k-1)(k-2)\cdots(k-(r-1))}{r!} \quad \text{when } r \neq 0 \text{ and} \binom{k}{r} = 1 \text{ when } r = 0$$

Equation (1) expresses $f(x_k)$ in terms of $f(x_0)$ and the successive differences $\Delta f(x_0), \Delta^2 f(x_0), \ldots, \Delta^k f(x_0)$. We see, therefore, that from $f(x_0)$ and these successive differences we can determine the $k+1$ function values $f(x_0), f(x_1), f(x_2), \ldots, f(x_k)$. Thus, for example, from the color entries in the table in Figure 4-1 we can reconstruct the entire table.

Since the data $f(x_0), \Delta f(x_0), \Delta^2 f(x_0), \ldots, \Delta^k f(x_0)$ determine $k+1$ function values, they also determine a Lagrange interpolating polynomial. It is natural to inquire whether the resulting polynomial can be expressed efficiently in terms of $f(x_0)$ and the successive differences.

Suppose we are given $f(x_0), \Delta f(x_0), \Delta^2 f(x_0), \ldots, \Delta^n f(x_0)$. We consider the following polynomial:

$$\begin{aligned} \psi(x) &= f(x_0) + \frac{(x-x_0)}{1!}\frac{\Delta f(x_0)}{\Delta x} + \frac{(x-x_0)(x-x_1)}{2!}\frac{\Delta^2 f(x_0)}{(\Delta x)^2} + \cdots \\ &\quad + \frac{(x-x_0)(x-x_1)\cdots(x-x_{n-1})}{n!}\frac{\Delta^n f(x_0)}{(\Delta x)^n} \end{aligned} \quad (2)$$

The degree of this polynomial is at most equal to n. Let us verify that we have $f(x_k) = \psi(x_k)$ for $k = 0, 1, \ldots, n$. At once from equation (2) we obtain

$$\psi(x_k) = f(x_0) + \frac{(x_k-x_0)}{1!}\frac{\Delta f(x_0)}{\Delta x} + \cdots + \frac{(x_k-x_0)\cdots(x_k-x_{k-1})}{k!}\frac{\Delta^k f(x_0)}{(\Delta x)^k}$$

Using the fact that

$$x_k - x_0 = k\Delta x$$

$$x_k - x_1 = (k-1)\,\Delta x$$

$$\vdots$$

$$x_k - x_{k-1} = \Delta x$$

we find that

$$\psi(x_k) = f(x_0) + \frac{k}{1!}\,\Delta f(x_0) + \frac{k(k-1)}{2!}\,\Delta^2 f(x_0) + \cdots$$

$$+ \frac{k(k-1)\cdots(3)(2)(1)}{k!}\,\Delta^k f(x_0)$$

However, by equation (1) the right-hand side of the above equation is equal to $f(x_k)$ and this establishes the result. The polynomial $\psi(x)$ given in equation (2) is known as the **Newton forward difference polynomial.** This polynomial is in fact the same as the Lagrange interpolating polynomial determined by the $n+1$ values $f(x_0), f(x_1), \ldots, f(x_n)$ since this polynomial is uniquely determined. The two polynomials differ in form, however, and in the way in which they are obtained from given data. The Newton polynomial is only defined when the points $x_0, x_1, \ldots, x_n$ are evenly spaced, while the Lagrange polynomial is always defined (provided the points are distinct). On the other hand, the Newton forward difference polynomial has an important recurrence property: To pass from the polynomial of degree n to the one of degree $n+1$, one simply adds the additional term

$$\frac{(x-x_0)(x-x_1)\cdots(x-x_n)}{(n+1)!}\,\frac{\Delta^{n+1}f(x_0)}{(\Delta x)^{n+1}}$$

The Lagrange polynomial, which is of the form

$$p(x) = \sum_{i=0}^{i=n} f(x_i)\,l_i(x)$$

does not enjoy such a property.

To facilitate working with the Newton polynomial, we make the affine substitution

$$s = \frac{x-x_0}{\Delta x}, \qquad x = x_0 + s\Delta x$$

We see that the integral values $0, 1, 2, \ldots$ of s correspond to the points $x_0, x_1, x_2, \ldots$. Moreover, we have $x - x_k = (s-k)\,\Delta x$. Therefore, with $x = x_0 + s\Delta x$, equation (2) becomes

$$\psi(x_0+s\Delta x) = f(x_0) + s\Delta f(x_0) + \frac{s(s-1)}{2!}\Delta^2 f(x_0) + \cdots + \frac{s(s-1)(s-2)\cdots(s-(n-1))}{n!}\Delta^n f(x_0) \tag{3}$$

$$= \sum_{k=0}^{k=n} \binom{s}{k} \Delta^k f(x_0)$$

The approximation

$$f(x_0+s\Delta x) \approx \sum_{k=0}^{k=n} \binom{s}{k} \Delta^k f(x_0) \tag{4}$$

is known as **Newton's forward difference formula.**

Example 3 From the data given in the table of Figure 4-2 estimate $f(1.5)$.

x	$f(x)$	$\Delta f(x)$	$\Delta^2 f(x)$	$\Delta^3 f(x)$
1	208460			
		29242		
2	237702		−213	
		29029		−27
3	266731		−240	
		28798		−26
4	295520		−266	
		28523		
5	324043			

FIGURE 4-2

We shall apply Newton's forward difference formula with $x_0 = 1$, $\Delta x = 1$. Since $s = (x - x_0)/\Delta x$, when $x = 1.5$ we have $s = \frac{1}{2}$. We compute

$$f(1.5) \approx \psi(1.5) = f(1) + \tfrac{1}{2}\Delta f(1) + \frac{\frac{1}{2}(-\frac{1}{2})}{2}\Delta^2 f(1) + \frac{\frac{1}{2}(-\frac{1}{2})(-\frac{3}{2})}{3\cdot 2}\Delta^3 f(1)$$

$$= 208460 + 14621 + 26.625 - 1.625$$

$$= 223106$$

From the data of Example 3, the Newton forward difference formula would serve badly to estimate $f(4.5)$: for with $x_0 = 4$ and $\Delta x = 1$, the approximation is $f(4.5) \approx f(4) + \frac{1}{2}\Delta f(4)$. The formula therefore yields an approximation by means of a polynomial of order one and uses very little of the available information about the function f. For working with difference tables and with values of the argument x which lie near the end of the entries in a table, we shall use **backward differences.** We define the **backward difference operator** ∇ by

$$\nabla f(x) = f(x) - f(x - \Delta x)$$

where Δx is a fixed increment. By iteration we define

$$\nabla^{r+1} f(x) = \nabla^r f(x) - \nabla^r f(x - \Delta x)$$

and by convention we define

$$\nabla^0 f(x) = f(x)$$

By way of example, we have

$$\begin{aligned}\nabla^2 f(x) &= \nabla f(x) - \nabla f(x - \Delta x)\\ &= f(x) - f(x - \Delta x) - f(x - \Delta x) - f(x - 2\Delta x)\\ &= f(x) - 2f(x - \Delta x) + f(x - 2\Delta x)\end{aligned}$$

It is easily checked that the operator ∇ satisfies the following properties:

(1) **Additivity** $\nabla(f+g) = \nabla f + \nabla g$

(2) **Linearity** $\nabla(cf) = c\nabla f$

(3) **Law of Powers** $\nabla^r(\nabla^n f) = \nabla^{r+n} f$

For working with backward differences, we shall use negative integers as indices, given a fixed spacing Δx and a point x_0, we set

$$\begin{aligned}x_{-1} &= x_0 - \Delta x\\ x_{-2} &= x_{-1} - \Delta x = x_0 - 2\Delta x\\ &\vdots\\ x_{-n-1} &= x_{-n} - \Delta x = x_0 + (-n-1)\,\Delta x\end{aligned}$$

It is easy to relate backward differences with forward differences. Note for example that we have

$$\begin{aligned}\nabla f(x_0) &= f(x_0) - f(x_0 - \Delta x)\\ &= f(x_0) - f(x_{-1})\\ &= \Delta f(x_{-1})\end{aligned}$$

And more generally we have

$$\nabla f(x_n) = \Delta f(x_{n-1}), \qquad n = 0, \pm 1, \pm 2, \pm 3, \ldots$$

We also see that

$$\begin{aligned}\nabla^2 f(x_n) &= \Delta^2 f(x_{n-2})\\ \nabla^3 f(x_n) &= \Delta^3 f(x_{n-3})\\ &\vdots\\ \nabla^r f(x_n) &= \Delta^r f(x_{n-r}), \qquad n = 0, \pm 1, \pm 2, \ldots, \quad r = 0, 1, 2, 3, \ldots\end{aligned}$$

These relationships enable us to read backward differences from a difference table of the type in Figure 4-1 and Figure 4-2. This is illustrated in the next example.

Example 4 Consider the table of Figure 4-1. We see that

$$\nabla f(118) = 41419 = \Delta f(117)$$

$$\nabla^2 f(118) = 702 = \Delta^2 f(116)$$

$$\nabla^3 f(118) = 6 = \Delta^3 f(115)$$

$$\nabla^4 f(118) = 0 = \Delta^4 f(114)$$

If we recast the Newton forward difference polynomial in terms of backward differences we obtain the following polynomial:

$$\Phi(x) = f(x_0) + \frac{x-x_0}{1!}\frac{\nabla f(x_0)}{\Delta x} + \frac{(x-x_0)(x-x_{-1})}{2!}\frac{\nabla^2 f(x_0)}{(\Delta x)^2} + \cdots$$

$$+ \frac{(x-x_0)(x-x_1)\cdots(x-x_{-(n-1)})}{n!}\frac{\nabla^n f(x_0)}{(\Delta x)^n} \tag{5}$$

This polynomial is of degree at most n and agrees with $f(x)$ at the $(n+1)$ points $x_0, x_{-1}, x_{-2}, \ldots, x_{-n}$. If we again set

$$s = \frac{x-x_0}{\Delta x}, \qquad x = x_0 + s\Delta x$$

we see that the values $0, -1, -2, -3, \ldots$ of s correspond to the points $x_0, x_{-1}, x_{-2}, x_{-3}, \ldots$. We obtain the formula of approximation

$$f(x_0 + s\Delta x) \approx \Phi(x_0 + s\Delta x)$$

$$= f(x_0) + s\nabla f(x_0) + \frac{s(s+1)}{2!}\nabla^2 f(x_0)$$

$$+ \frac{s(s+1)(s+2)}{3!}\nabla^3 f(x_0) + \cdots$$

$$+ \frac{s(s+1)(s+2)\cdots(s+n-1)}{n!}\nabla^n f(x_0) \tag{6}$$

This approximation formula is known as **Newton's backward difference formula**.

Example 5 From the data of the table in Figure 4-3, estimate $f(104.25)$.

x	$f(x)$	$\Delta f(x)$	$\Delta^2 f(x)$	$\Delta^3 f(x)$
101	1030198			
		30906		
102	1061104		612	
		31518		6
103	1092622		618	
		32136		6
104	1124758		624	
		32760		
105	1157518			

FIGURE 4-3

Here we take $x_0 = 105$, $\Delta x = 1$, and $s = -\frac{3}{4}$. We compute

$$f(104.25) \approx 1157518 + (-\tfrac{3}{4})(32760) + \frac{(-\frac{3}{4})(\frac{1}{4})}{2}(624) + \frac{(-\frac{3}{4})(\frac{1}{4})(1\frac{1}{4})}{6}(6)$$

$$= 1157518 - 24628\tfrac{47}{64}$$

$$= 1132889\tfrac{17}{64}$$

We shall consider one further type of difference operator and corresponding interpolation formula before going on to the problem of numerical differentiation.

The **central difference operator** δ is defined by the equation

$$\delta f(x) = f(x + \tfrac{1}{2}\Delta x) - f(x - \tfrac{1}{2}\Delta x)$$

We define the powers by iteration

$$\delta^{r+1} f(x) = \delta^r f(x + \tfrac{1}{2}\Delta x) - \delta^r f(x - \tfrac{1}{2}\Delta x), \qquad r = 1, 2, \ldots$$

Central differences are especially useful for estimating $f(x)$ when x lies close to the middle of a sequence of entries in a difference table. For working with central differences we shall use fractional indices: Given a fixed point x_0 and a fixed spacing Δx, we set

$$x_{1/2} = \frac{x_0 + x_1}{2}$$

$$x_{-1/2} = \frac{x_0 + x_{-1}}{2}$$

$$x_{1\frac{1}{2}} = \frac{x_1 + x_2}{2}$$

$$x_{-1\frac{1}{2}} = \frac{x_{-1} + x_{-2}}{2}$$

$$x_{n+1/2} = \frac{x_n + x_{n+1}}{2}, \qquad n = 0, \pm 1, \pm 2, \pm 3, \ldots$$

By way of illustration we note the following:

$$\delta f(x_0) = f(x_{1/2}) - f(x_{-1/2})$$

$$\delta f(x_{1/2}) = f(x_1) - f(x_0) = \Delta f(x_0)$$

$$\delta f(x_{-1/2}) = f(x_0) - f(x_{-1}) = \nabla f(x_0)$$

$$\delta^2 f(x_0) = \delta f(x_{1/2}) - \delta f(x_{-1/2}) = f(x_1) - f(x_0) - (f(x_0) - f(x_{-1}))$$

$$= f(x_1) - 2f(x_0) + f(x_{-1})$$

A problem associated with the method of central differences is that when k is an integer the difference $\delta f(x_k)$ does not involve tabulated values of f but is defined in terms of values at halfway points:

$$\delta f(x_k) = f(x_k + \tfrac{1}{2}\Delta x) - f(x_k - \tfrac{1}{2}\Delta x)$$

Thus, for example, the table of Figure 4-1 does not provide enough information to calculate $\delta f(110)$ or $\delta f(111)$ or $\cdots$ or $\delta f(118)$. However, the second central difference at a tabulated point is given in terms of tabular entries:

$$\delta^2 f(x_k) = f(x_{k+1}) - 2f(x_k) + f(x_{k-1})$$

This is, in fact, true for all central differences $\delta^{2m} f(x_k)$ of even order. Moreover, we may note that

$$\delta f(x_{k+1/2}) = f(x_{k+1}) - f(x_k)$$

and that in general the odd central differences $\delta^{2m+1} f(x_{k+1/2})$ at halfway points involve only tabulated values.

The following, known as **Stirling's formula**, is an interpolation formula using central differences. With $s = (x - x_0)/\Delta x$, we have

$$\begin{aligned} f(x) &= f(x_0 + s\Delta x) \\ &\approx f(x_0) + \tfrac{1}{2}s(\delta f(x_{1/2}) + \delta f(x_{-1/2})) + \tfrac{1}{2}s\delta^2 f(x_0) \\ &\quad + \frac{s(s^2-1)}{2\cdot 3!}(\delta^3 f(x_{1/2}) + \delta^3 f(x_{-1/2})) + \frac{s^2(s^2-1^2)}{4!}\delta^4 f(x_0) + \cdots \\ &\quad + \frac{s^2(s^2-1^2)\cdots(s^2-(m-1)^2)}{(2m)!}\delta^{2m} f(x_0) \\ &\quad + \frac{s(s^2-1^2)\cdots(s^2-m^2)}{2\cdot(2m+1)!}(\delta^{2m+1} f(x_{1/2}) + \delta^{2m+1} f(x_{-1/2})) \end{aligned} \quad (7)$$

Using the identities

$$\delta f(x_{1/2}) = \Delta f(x_0)$$

$$\delta f(x_{-1/2}) = \Delta f(x_{-1})$$

$$\delta^2 f(x_0) = \Delta^2 f(x_{-1})$$

$$\delta^3 f(x_{1/2}) = \Delta^3 f(x_{-1})$$

$$\delta^3 f(x_{-1/2}) = \Delta^3 f(x_{-2})$$

$$\delta^4 f(x_0) = \Delta^2 f(x_0) - 2\Delta^2 f(x_{-1}) + \Delta^2 f(x_{-2})$$

we can write Stirling's formula up to fourth differences as

$$\begin{aligned} f(x_0 + s\Delta x) \approx f(x_0) &+ s\left(\frac{\Delta f(x_0)+\Delta f(x_{-1})}{2}\right) + \frac{s^2}{2}(\Delta^2 f(x_{-1})) \\ &+ \frac{s(s^2-1)}{3!}\left(\frac{\Delta^3 f(x_{-1})+\Delta^3 f(x_{-2})}{2}\right) \\ &+ \frac{s^2(s^2-1)}{4!}(\Delta^2 f(x_0)-2\Delta^2 f(x_{-1})+\Delta^2 f(x_{-2})) \end{aligned} \tag{8}$$

Example 6 Using the data of Figure 4-4, approximate $f(3\frac{1}{2})$ by means of Stirling's formula.

x	$f(x)$	$\Delta f(x)$	$\Delta^2 f(x)$	$\Delta^3 f(x)$
1	0.208460			
		0.029242		
2	0.237702		−0.000213	
		0.029029		−0.000027
3	0.266731		−0.000240	
		0.028789		−0.000026
4	0.295520		−0.000266	
		0.028523		−0.000026
5	0.324043		−0.000292	
		0.028231		
6	0.352274			

FIGURE 4-4

In this example we take $x_0 = 3$, $\Delta x = 1$, = and $s = \frac{1}{2}$. Applying equation (8) above through third differences, we compute

$$\begin{aligned} f(3\tfrac{1}{2}) &\approx f(3) + \frac{1}{2}\left(\frac{\Delta f(3)+\Delta f(2)}{2}\right) + \frac{1}{8}\Delta^2 f(2) + \frac{\frac{1}{2}(-\frac{3}{4})}{3!}\left(\frac{\Delta^3 f(2)+\Delta^3 f(1)}{2}\right) \\ &= 0.266731 + 0.014455 - 0.000030 + 0.000001 \\ &= 0.281157 \qquad \text{(with roundoff in the sixth place)} \end{aligned}$$

We can obtain an important formula for numerical differentiation by differentiating Stirling's formula. If we differentiate both sides of (8) with respect to s, we find

$$f'(x_0 + s\Delta x)\,\Delta x \approx \frac{\Delta f(x_0)+\Delta f(x_{-1})}{2} + s\Delta^2 f(x_{-1})$$
$$+ \frac{3s^2-1}{3!}\left(\frac{\Delta^3 f(x_{-1})+\Delta^3 f(x_{-2})}{2}\right)$$
$$+ \frac{s(4s^2-2)}{4!}(\Delta^2 f(x_0)-2\Delta^2 f(x_{-1})+\Delta^2 f(x_{-2}))$$

Letting $s=0$ we obtain

$$f'(x_0) \approx \frac{1}{\Delta x}\left[\left(\frac{\Delta f(x_0)+\Delta f(x_{-1})}{2}\right) - \frac{1}{6}\left(\frac{\Delta^3 f(x_{-1})+\Delta^3 f(x_{-2})}{2}\right)\right] \tag{9}$$

If we differentiate Stirling's formula (8) twice with respect to s, we obtain

$$f''(x_0 + s\Delta x)\,(\Delta x)^2 \approx \Delta^2 f(x_{-1}) + s\left(\frac{\Delta^3 f(x_{-1})+\Delta^3 f(x_{-2})}{2}\right)$$
$$+ \frac{6s^2-1}{12}(\Delta^2 f(x_0)-2\Delta^2 f(x_{-1})+\Delta^2 f(x_{-2}))$$

Letting $s=0$, we now have

$$f''(x_0) \approx \frac{1}{(\Delta x)^2}\left[\Delta^2 f(x_{-1}) - \tfrac{1}{12}(\Delta^2 f(x_0) - 2\Delta^2 f(x_{-1}) + \Delta^2 f(x_{-2}))\right] \tag{10}$$

Formulas (9) and (10) provide us with a method of approximating the first and second derivatives of a function at a tabulated point. Let us consider an example of this technique.

Example 7 Oceanographers have made the following consecutive daily measurements of certain tidal phenomena:

0.099833	0.208460	0.314566	0.416871
0.514136	0.605186	0.688921	0.764329

Are the phenomena periodic and, if so, what is the period?

x	$f(x)$	$\Delta f(x)$	$\Delta^2 f(x)$	$\Delta^3 f(x)$	$\Delta^4 f(x)$
1	0.099833				
		0.108627			
2	0.208460		−0.002521		
		0.106106		−0.001280	
3	0.314566		−0.003801		0.000041
		0.102305		−0.001239	
4	0.416871		−0.005040		0.000064
		0.097265		−0.001175	
5	0.514136		−0.006215		0.000075
		0.091050		−0.001100	
6	0.605186		−0.007315		0.000088
		0.083735		−0.001012	
7	0.688921		−0.008327		
		0.075408			
8	0.764329				

FIGURE 4-5

From the given data we can construct the difference table of Figure 4-5. The function f determined by such tidal phenomena, if periodic, is often of the form

$$f(x) \approx A \cos \alpha x + B \sin \alpha x$$

and therefore satisfies the differential equation

$$f''(x) = -\alpha^2 f(x)$$

The difference table of Figure 4-5 gives sufficient information to apply (10) to estimate the second derivatives $f''(3)$, $f''(4)$, $f''(5)$, $f''(6)$. First we take $x_0 = 3$, $\Delta x = 1$, and we compute

$$\begin{aligned} f''(3) &\approx \Delta^2 f(2) - \tfrac{1}{12}(\Delta^2 f(3) - 2\Delta^2 f(2) + \Delta^2 f(1)) \\ &= -0.003801 - \tfrac{1}{12}(0.000041) \\ &= -0.003804 \qquad \text{(with roundoff in the sixth place)} \end{aligned}$$

Next we take $x_0 = 4$ and $x = 1$. Applying (10), we obtain

$$\begin{aligned} f''(4) &\approx \Delta^2 f(3) - \tfrac{1}{12}(\Delta^2 f(4) - 2\Delta^2 f(3) + \Delta^2 f(2)) \\ &= -0.005040 - \tfrac{1}{12}(0.000064) \\ &= -0.005045 \text{ (with roundoff in the sixth place)} \end{aligned}$$

In an analogous fashion we find

$$f''(5) \approx -0.006221 \quad \text{and} \quad f''(6) \approx -0.007322$$

again with roundoff in the sixth place. Computing the ratios $f''(x)/f(x)$ to five decimal places, we find that

$$\frac{f''(3)}{f(3)} = -0.01209$$

$$\frac{f''(4)}{f(4)} = -0.01210$$

$$\frac{f''(5)}{f(5)} = -0.01209$$

$$\frac{f''(6)}{f(6)} = -0.01209$$

Rounding off to four decimal places we have $f''(x)/f(x) = -0.0121$. Hence we can conclude that $\alpha = 0.11$ and that

$$f(x) \approx A \cos 0.11x + B \sin 0.11x$$

This result shows that f is periodic with period $2\pi/0.11$ or 57.12 days.

Formulas analogous to the ones obtained by differentiating Stirling's formula can be obtained by differentiating Newton's forward and backward difference formulas (see exercise 8). However, these formulas for numerical differentiation give much less satisfactory results than those obtained from Stirling's formula. Since Stirling's formula uses central differences we see that we can obtain a good approximation to the slope of the curve at a tabulated point which lies halfway between a sequence of observations. In advanced treatments, the reader can find a more careful discussion of the errors involved in these various procedures.

EXERCISES

1. Let f be twice continuously differentiable at x_0. Show that

$$\lim_{\Delta x \to 0} \frac{\Delta^2 f(x_0)}{(\Delta x)^2} = f''(x_0)$$

Generalize. [Hint: Use L'Hôpital's rule.]

2. Establish the formula

$$\sum_{k=0}^{n-1} f(x_k) = \sum_{k=0}^{n-1} \binom{n}{k+1} \Delta^k f(x_0)$$

3. The data in Figure 4-6 are taken from a five-place table of $f(x) = \sin x$; the differences have been computed with roundoff in the fifth decimal place. (a) Estimate $f(1.02)$ using Newton's forward difference formula. (b) Estimate $f(1.75)$ using Newton's backward difference formula. (c) Estimate $f(1.22)$ using Stirling's formula. (Note: The answers have been computed using first through fourth differences rounding off in the sixth decimal place.)

x	$f(x) = \sin x$	Δ	Δ^2	Δ^3	Δ^4	Δ^5
1.0	0.84147					
		0.04974				
1.1	0.89121		−0.00891			
		0.04083		−0.00040		
1.2	0.93204		−0.00931		0.00008	
		0.03152		−0.00032		0.00002
1.3	0.96356		−0.00963		0.00010	
		0.02189		−0.00022		0.00001
1.4	0.98545		−0.00985		0.00011	
		0.01204		−0.00011		−0.00003
1.5	0.99749		−0.00996		0.00008	
		0.00208		−0.00003		0.00004
1.6	0.99957		−0.00999		0.00012	
		−0.00791		0.00009		
1.7	0.99166		−0.00990			
		−0.01781				
1.8	0.97385					

FIGURE 4-6

4. Let $g = \Delta f$, $h = \nabla f$, and let $k = \boldsymbol{E}f$. Show that $\nabla g(x) = \Delta h(x)$, $\Delta k(x) = \boldsymbol{E}g(x)$, and $\nabla k(x) = \boldsymbol{E}h(x)$.
5. Let $h = fg$ and $k = f/g$. Show that $\Delta h(x) = f(x)\,\Delta g(x) + \boldsymbol{E}g(x)\,\Delta f(x)$ and that

$$\Delta k(x) = \frac{g(x)\,\Delta f(x) - f(x)\,\Delta g(x)}{g(x)\,\boldsymbol{E}g(x)}$$

6. Show that the following is true of a difference table: The sum of the entries in any column of differences is equal to the difference between the first and last entries in the preceding column.
7. From the following table compute $\delta^2 \arcsin x$ at $x = 0.95$, $x = 0.96$, $x = 0.97$, $x = 0.98$, $x = 0.99$.

x	$\arcsin x$
0.94	1.2226
0.95	1.2532
0.96	1.2870
0.97	1.3252
0.98	1.3705
0.99	1.4293
1.00	1.5708

8. Differentiate Newton's forward difference formula to obtain the approximations

$$f'(x_0) \approx \frac{1}{\Delta x}\left(\Delta f(x_0) - \tfrac{1}{2}\Delta^2 f(x_0) + \tfrac{1}{3}\Delta^3 f(x_0) - \tfrac{1}{4}\Delta^4 f(x_0)\right)$$

$$f''(x_0) \approx \frac{1}{(\Delta x)^2}\left(\Delta^2 f(x_0) - \Delta^3 f(x_0) + \tfrac{11}{12}\Delta^4 f(x_0)\right)$$

9. Let $p(x)$ be a polynomial of order n. Show that the nth backward differences $\nabla^n p(x)$ are constant.

10. Show that the following observations are those of a periodic function and find the period.

x (years)	$f(x)$
1	0.198669
2	0.295520
3	0.389418
4	0.479425
5	0.564642
6	0.644217
7	0.717356
8	0.783327

11. Determine the constants A, B in Example 7.

5 DIFFERENTIAL EQUATIONS

The numerical solution of differential equations is one of the most important branches of numerical analysis. In this section we briefly discuss two methods for obtaining approximate solutions of initial value problems. We begin with a simple technique known as **Euler's method.**

Suppose we are given a differential equation

$$\frac{dy}{dx} = F(x, y)$$

with the initial condition that $y = y_0$ when $x = x_0$ [which we write $y(x_0) = y_0$]. The solution curve $y = f(x)$ to this problem passes through the point (x_0, y_0) where it has slope $f'(x_0) = (dy/dx)|_{x=x_0} = F(x_0, y_0)$.

If we set $x_1 = x_0 + \Delta x$, we can approximate $f(x_1) = y_1$ by the method of differential approximation:

$$\begin{aligned} f(x_1) &= y_1 \approx f(x_0) + f'(x_0)\,\Delta x \\ &= y_0 + F(x_0, y_0)\,\Delta x = y_1^* \end{aligned}$$

We now have the approximation $f(x_1) \approx y_1^*$. The approximate value y_1^* can in turn be used to estimate $f'(x_1)$:

$$f'(x_1) = F(x_1, y_1) \approx F(x_1, y_1^*)$$

Let us again use differential approximation, this time with the estimates $y_1^* \approx f(x_1)$ and $F(x_1, y_1^*) \approx f'(x_1)$. We obtain

$$\begin{aligned} f(x_2) &= f(x_1 + \Delta x) \approx f(x_1) + f'(x_1)\,\Delta x \\ &\approx y_1^* + F(x_1, y_1^*)\,\Delta x = y_2^* \end{aligned}$$

If we perform this operation again we obtain

$$f(x_3) = f(x_2 + \Delta x) \approx f(x_2) + F(x_2, y_2)\,\Delta x$$
$$\approx y_2^* + F(x_2, y_2^*)\,\Delta x = y_3^*$$

Continuing in this manner we can obtain approximations y_4^*, y_5^* and so on.

Example 1 Consider the initial value problem

$$(x^2+1)\frac{dy}{dx} = x - 4xy, \qquad y(2) = 1$$

The solution $y = f(x)$ to this problem satisfies $f(2) = 1$. Let us apply Euler's method to estimate $f(x)$ at $x = 2.1$, $x = 2.2$, $x = 2.3$. Here we take $\Delta x = 0.1$ as our spacing, $x_0 = 2$ and $y_0 = f(x_0) = 1$. First we put the differential equation into the form

$$\frac{dy}{dx} = \frac{1}{x^2+1}(x - 4xy) = F(x, y)$$

We calculate

$$\left.\frac{dy}{dx}\right|_{x=2} = F(2, 1) = \tfrac{1}{5}(2 - 8) = -\tfrac{6}{5}$$

and

$$F(x_0, y_0)\,\Delta x = -\tfrac{6}{5}\left(\tfrac{1}{10}\right) = -\tfrac{3}{25}$$

This yields the estimate

$$y_1^* = 1 - \tfrac{3}{25} = \tfrac{22}{25} = 0.88$$

To obtain an approximate value for $f(x)$ at $x_2 = 2.2$ we compute

$$\left.\frac{dy}{dx}\right|_{x=2.1} \approx F(x_1, y_1^*) = \frac{2.1}{5.41}(1 - 3.52) = -0.978$$

and we evaluate

$$y_2^* = y_1^* + F(x_1, y_1^*)\,\Delta x = 0.88 + (-0.0978) = 0.7822$$

Continuing we find

$$\left.\frac{dy}{dx}\right|_{x=2.2} \approx F(x_2, y_2^*) = \frac{2.2}{5.84}(1 - 4(0.7822)) = -0.80190$$

This gives the estimate

$$y_3^* = y_2^* + F(x_2, y_2^*)\,\Delta x = 0.78220 - 0.08019 = 0.70201$$

The exact solution of the initial value problem of this example is

$$f(x) = \frac{1}{4}\left(1 + \frac{75}{(x^2+1)^2}\right)$$

(see Example 2 of §11.4). The estimated values y_1^*, y_2^*, and y_3^* that we have obtained can be compared to the true values:

$$y_1 = f(x_1) = 0.89, \qquad y_2 = f(x_2) = 0.799, \qquad y_3 = f(x_3) = 0.72225$$

We shall not go into an analysis of the error associated with Euler's method. We do note, however, that Euler's method is subject to several types of error: At each step there is the error arising from differential approximation and an additional error which comes in because $F(x_1, y_1^*)$, $F(x_2, y_2^*), \ldots$ are only estimates of $(dy/dx)|_{x=x_1}$, $(dy/dx)|_{x=x_2}, \ldots$ and not the true values. Reducing the spacing Δx will serve to reduce the above types of error; however, the smaller we take Δx the longer is the calculation necessary to determine the solution $y = f(x)$ over a given interval. In practice, if Δx is taken too small, a large roundoff error results.

A more sophisticated method of solving differential equations numerically makes use of the technique of polynomial interpolation. We again consider an initial value problem

$$\frac{dy}{dx} = F(x, y), \qquad y(x_0) = y_0$$

Let us suppose that we are given the $n+1$ values

$$y_0 = f(x_1), \quad y_1 = f(x_1), \quad \ldots, \quad y_n = f(x_n)$$

where $x_1 = x_0 + \Delta x$, $x_2 = x_1 + \Delta x$, $\ldots$, $x_n = x_{n-1} + \Delta x$ and where $y = f(x)$ is the solution to the initial value problem. We wish to estimate $y_{n+1} = f(x_{n+1})$. For this purpose we shall use the relation

$$f(x_{n+1}) = f(x_n) + \int_{x_n}^{x_{n+1}} f'(x)\, dx \tag{1}$$

The program is as follows: Knowing the function values $y_0 = f(x_0)$, $y_1 = f(x_1), \ldots, y_n = f(x_n)$, we can calculate the corresponding values $f'(x_0), f'(x_1), \ldots, f'(x_n)$ of the derivative from the equations

$$f'(x_k) = F(x_k, y_k), \qquad k = 0, 1, \ldots, n$$

If $f'(x)$ is known at $n+1$ points, then an interpolating polynomial of degree n can be constructed to approximate f'. We can then use this polynomial to approximate $f'(x)$ over the interval $[x_n, x_{n+1}]$ and we can estimate the integral $\int_{x_n}^{x_{n+1}} f'(x)\, dx$ by integrating this approximating polynomial from x_n to x_{n+1}.

The Newton backward difference formula yields the polynomial approximation

$$f'(x) = f'(x_n + s\Delta x) \approx f'(x_n) + s\nabla f'(x_n) + \frac{s(s+1)}{2!}\nabla^2 f'(x_n) + \cdots$$
$$+ \frac{s(s+1)\cdots(s+n-1)}{n!}\nabla^n f'(x_n) \tag{2}$$

where $s = (x - x_n)/\Delta x$. We then have by change of variable

$$\int_{x_n}^{x_{n+1}} f'(x)\,dx = \Delta x \int_0^1 f'(x_n + s\Delta x)\,ds$$

and using (2),

$$\int_0^1 f'(x_n + s\Delta x)\,ds \approx \sum_{k=0}^{k=n} \nabla^k f'(x_n) \int_0^1 \frac{s(s+1)\cdots(s+k-1)}{k!}\,ds$$

$$= \sum_{k=0}^{k=n} a_k \nabla^k f'(x_n) \tag{3}$$

where

$$a_k = \int_0^1 \frac{s(s+1)\cdots(s+k-1)}{k!}\,ds$$

Combining (1), (2), and (3) we have

$$f(x_{n+1}) = f(x_n) + \int_{x_n}^{x_{n+1}} f'(x)\,dx \approx f(x_n) + \Delta x \sum_{k=0}^{k=n} a_k \nabla^k f'(x_n) \tag{4}$$

Calculation of a_0, a_1, a_2, a_3, a_4, and a_5 yields

$$a_0 = 1, \quad a_1 = \tfrac{1}{2}, \quad a_2 = \tfrac{5}{12}, \quad a_3 = \tfrac{3}{8}, \quad a_4 = \tfrac{251}{720}, \quad a_5 = \tfrac{95}{288}$$

We can now rewrite (4) as

$$f(x_{n+1}) \approx f(x_n) + \Delta x[f'(x_n) + \tfrac{1}{2}\nabla f'(x_n) + \tfrac{5}{12}\nabla^2 f'(x_n) + \tfrac{3}{8}\nabla^3 f'(x_n) + \tfrac{251}{720}\nabla^4 f'(x_n) + \tfrac{95}{288}\nabla^5 f'(x_n) + \cdots + a_n \nabla^n f'(x_n)] \tag{5}$$

The technique for obtaining numerical solutions of initial value problems which is based on formula (5) is known as **Adam's method**. We see that for $n = 0$ this method reduces to Euler's method.

Example 2 Consider the initial value problem

$$\frac{dy}{dx} = x^2 - y, \qquad y(0) = 1$$

Suppose we wish to estimate the values $f(0.1), f(0.2), f(0.3), \ldots$ of the solution to this differential equation. Let us apply Adam's method with $x_0 = 0$ and $\Delta x = 0.1$. Applying (5) with $n = 0$ we have

$$y_1 = f(x_1) \approx f(x_0) + f'(x_0)\Delta x = 1 + (-1)(0.1) = 0.9 = y_1^*$$

where we have used $f'(x_0) = f'(0) = 0^2 - 1$. For the next step we first evaluate

$$f'(x_1) \approx F(x_1, y_1^*) = (0.1)^2 - 0.9 = -0.89$$

and

$$\nabla f'(x_1) = f'(x_1) - f'(x_0) \approx -0.89 - (-1) = 0.11$$

Applying (5) with $n = 1$ we find

$$\begin{aligned} y_2 = f(x_2) &\approx f(x_1) + \Delta x[f'(x_1) + \tfrac{1}{2}\nabla f'(x_1)] \\ &\approx 0.9 + 0.1(-0.89 + 0.055) \\ &= 0.8165 = y_2^* \end{aligned}$$

Continuing we find

$$f'(x_2) = F(x_2, y_2) \approx F(x_2, y_2^*) = (0.2)^2 - 0.8165 = -0.7765$$

and

$$\nabla f'(x_2) = f'(x_2) - f'(x_1) \approx -0.7765 - (-0.89) = 0.1135$$

$$\nabla^2 f'(x_2) = \nabla f'(x_2) - \nabla f'(x_1) \approx 0.1135 - 0.11 = 0.0035$$

Hence, applying (5) with $n = 2$ yields

$$\begin{aligned} y_3 = f(x_3) &\approx f(x_2) + \Delta x[f'(x_2) + \tfrac{1}{2}\nabla f'(x_2) + \tfrac{5}{12}\nabla^2 f'(x_2)] \\ &\approx 0.8165 + 0.1(-0.7765 + 0.05675 + 0.00146) \\ &= 0.744671 \end{aligned}$$

The initial value problem of this example is a linear differential equation with constant coefficients and has the exact solution

$$f(x) = 2 - 2x + x^2 - e^{-x}$$

We can therefore compare our estimates with the true values (with roundoff in the fourth decimal place):

$$y_1 = 0.9052, \qquad y_2 = 0.8213, \qquad y_3 = 0.7402$$

EXERCISES

1. Continue Example 1 to estimate $f(1.4)$. Compare your result with the true value.
2. Continue Example 2 to estimate $f(0.4)$. Compare your result with the true value.
3. Consider the initial value problem

$$\frac{dy}{dx} = G(x), \qquad y(x_0) = y_0$$

Suppose that the function G is twice differentiable with domain $\boldsymbol{R}$ and that $G''(x) < 0$ for all x. Show that Euler's method, applied to the initial value problem, would yield

$$0 < y_1^* - y_1 < y_2^* - y_2 < y_3^* - y_3 < \cdots < y_n^* - y_n$$

where y_k denotes the value of the exact solution at $x_0 + k\Delta x$ and y_k^* denotes the approximation obtained using of Euler's method.

4. Consider the initial value problem

$$\frac{dy}{dx} = q(x), \qquad y(x_0) = y_0$$

where $q(x)$ is a polynomial of degree k. Show that if Adam's method is applied to this initial value problem, one obtains

$$y_k^* - y_k = y_{k+1}^* - y_{k+1} = \cdots = y_{k+n}^* - y_{k+n}$$

that is, show that the error remains constant after k steps.

Find the exact solution to each of the initial value problems in exercises 5–8. Then use Euler's method and then Adam's method, with spacing $\Delta x = 0.1$, to find approximate values of the solution at four points. Compare these results with the exact values.

5. $(x^3 + y^3)\dfrac{dy}{dx} = 3x^2 y, \quad y(1) = 1$

6. $\dfrac{dy}{dx} + 2y = e^{-x}, \quad y(0) = 1$

7. $x\,dy + y\,dx = y\,dy, \quad y(1) = 1$

8. $(y^2 + 1)\,dx + (2xy + 1)\,dy = 0, \quad y(0) = 0$

REVIEW EXERCISES FOR CHAPTER 12

In exercises 1–3, find the Lagrange interpolating polynomial determined by the given set of data.

1. $f(0) = 10, f(-1) = 10, f(-2) = 6, f(-3) = -9$
2. $f(0) = 0, f(1) = 2, f(2) = 66, f(3) = 732, f(4) = 4100,$ $f(5) = 15630$
3. $f(100) = 10101, f(200) = 40201, f(300) = 90301$

In exercises 4–6, find the Hermite polynomial determined by the given set of data.

4. $f(0) = 0, f'(0) = 0; \quad f(-3) = -9, f'(-3) = 33$
5. $f(0) = 0, f'(0) = 0; \quad f(1) = 1, f'(1) = 5; \quad f(2) = 32, f'(2) = 80$
6. $f(0) = 1, f'(0) = 1; \quad f(1) = 4, f'(1) = 6$

7. Given the following data, estimate $f(3.927)$ using Newton's foward difference formula: $f(3.92) = 0.40044$, $f(3.94) = 1.41860$, $f(3.96) = 2.45733$, $f(3.98) = 3.51703$, $f(4.00) = 4.59815$, $f(4.02) = 5.70111$.

8. By differentiating Newton's backward difference formula, derive the following rules of approximate differentiation:

(a) $f'(x_0) \approx \dfrac{1}{\Delta x}(\nabla f(x_0) + \tfrac{1}{2}\nabla^2 f(x_0) + \tfrac{1}{3}\nabla^3 f(x_0))$

(b) $f''(x_0) \approx \dfrac{1}{(\Delta x)^2}(\nabla^2 f(x_0) + \nabla^3 f(x_0) + \tfrac{11}{12}\nabla^4 f(x_0))$

9. Define the differential operator **D** by $\mathbf{D}f(x) = f'(x)$ and the integration operator **J** by $\mathbf{J}f(x) = \int_x^{x+\Delta x} f$. (a) Show that the operators **D** and **J** are additive, linear, and satisfy the law of powers. (b) Show that $\mathbf{DJ}f = \mathbf{JD}f = \Delta f$.

10. (a) Write out Stirling's formula through sixth differences and through seventh differences. (b) Observe that the formula with sixth differences requires seven tabulated values and yields a polynomial of order six which agrees with $f(x)$ at each of these seven tabulated points. (c) Observe that the formula with seventh differences requires nine tabulated points $x_0, x_{\pm 1}, x_{\pm 2}, x_{\pm 3}, x_{\pm 4}$ and yields a polynomial of order seven agreeing with $f(x)$ at the seven points $x_0, x_{\pm 1}, x_{\pm 2}, x_{\pm 3}$. Argue that the approximating polynomial in this case need not agree with $f(x)$ at x_{+4} or x_{-4}.

11. (a) Show that for $n = 6$, the Newton–Cotes formula of approximate quadrature is

$$\int_a^b f \approx \sum_{i=0}^{i=6} f(x_i)\, C_i^{(6)}$$
$$= \tfrac{1}{140}[41f(x_0) + 216f(x_1) + 27f(x_2) + 272f(x_3) + 27f(x_4) + 216f(x_5) + 41f(x_6)]$$

(b) The following is known as **Weddle's rule of quadrature**:

$$\int_a^b f \approx \tfrac{3}{10}(f(x_0) + 5f(x_1) + f(x_2) + 6f(x_3) + f(x_4) + 5f(x_5) + f(x_6))$$

Show that the approximate value of $\int_a^b f$ given by Weddle's rule exceeds that given by the above Newton–Cotes formula by $\frac{1}{140}\delta^6 f(x_3)$.

(c) Explain why Weddle's rule gives exact results for polynomials of degree five but not for all polynomials of degree six. Discuss the error associated with Weddle's rule.

(d) Apply Weddle's rule to estimate $\int_1^7 dx/x = \ln 7$.

12. The **Bessel functions** are an important class of functions in pure and applied mathematics. The nth Bessel function is denoted J_n. The nth Bessel function $J_n : x \mapsto J_n(x)$ satisfies the differential equation

$$\frac{d^2 y}{dx^2} + \frac{1}{x}\frac{dy}{dx} + \left(1 - \frac{n^2}{x^2}\right) y = 0$$

The following data is taken from a table of values of one of the Bessel functions J_n:

x	$y = J_n(x)$
1.0	0.765198
1.1	0.719622
1.2	0.671133
1.3	0.620086
1.4	0.566855
1.5	0.511828
1.6	0.455402

Find approximate values of dy/dx and d^2y/dx^2 for $x = 1.3$ and, by substituting in the above differential equation, determine the value of n.

In exercises 13–16, apply the general form of Simpson's rule with $n = 4$ to approximate the given integral.

13. $\int_0^1 \frac{1}{1+x^3}\,dx$

14. $\int_0^1 \frac{1}{\sqrt{1+x^2}}\,dx$

15. $\int_0^1 \frac{x}{\sqrt{1+x^3}}\,dx$

16. $\int_0^1 (1+x^2)^{1/3}\,dx$

17. Estimate $\int_\pi^{8\pi} \sin x/x\,dx$ by the following method: Divide $[\pi, 8\pi]$ into seven subintervals of length π and to each subinterval apply Gaussian quadrature with three points. Apply the same technique using Gaussian quadrature with four points on each subinterval. Compare your results with the correct value to six decimal places which is 0.320805. Also compare with the result you obtained using Gaussian quadrature with two points (exercise 3 of §3).

In exercises 18–20, find the exact solution to the given initial value problem. Use Euler's method and then Adam's method with spacing $\Delta x = 0.1$ to find approximate values of the solution at four points. Compare these results with the true values.

18. $x\frac{dy}{dx} - 2y = x^2 + x, \quad y(1) = 1$

19. $\frac{dy}{dx} + 2y = e^{-x}, \quad y(0) = 0$

20. $(\cos^2 u - \sin^2 u)\,dr + 2r\sin u\cos u\,du = 0, \quad r(0) = 1$

13

VECTORS IN SPACE

In this chapter we translate the algebraic and geometric properties of two- and three-dimensional space into the language of vectors. The first section sets forth the basic definitions and properties of vectors and the remaining sections discuss vector operations that are essential for the remainder of this course.

1 VECTORS

Suppose you are in a room with one window and you want to specify the location of a fly sitting on a lamp hanging from the ceiling (see Figure 1-1). One way of giving such a precise location would be to fix a reference point O, say a corner of the room at the base of the wall containing the window, and then give measurements as if you were telling a spider how to locate the fly. You might instruct the spider to start from O and walk 10 feet along the wall with the window, then walk perpendicular to this wall for 15 feet and finally, climb up 8 feet (unfortunately the spider might not be able to perform this trick).

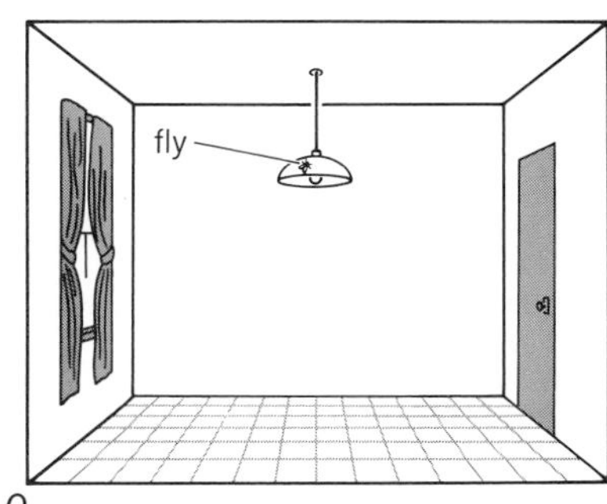

FIGURE 1-1

Considering this example carefully, one can see that the length, height, and width from a fixed reference point uniquely determines points in space. We can abstract this principle mathematically as follows.

Points in space may be represented as ordered triples of real numbers, just as points in the plane may be represented as ordered pairs of real numbers. To construct such a representation, we choose three mutually perpendicular lines which meet at a point in space. These lines are called the x axis, y axis, and z axis and the point at which they meet is called the **origin** (this is our reference point). The set of axes is often referred to as a **coordinate system**, and it is drawn as shown in Figure 1-2.

If we identify a real number with each point on these axis lines (as we did with the real line), then we may associate to each point P in space a unique triple of real numbers (x,y,z) and, vice versa, to each triple we may assign a unique point in space.

Let the triple $(0,0,0)$ correspond to the origin of the coordinate system and let the arrows on the axes indicate the positive directions. Then, for example, the triple $(4,-3,4)$ represents a point 4 units from 0 along the x axis in the positive direction, 3 units along the y axis in the negative direction, and 4 units along the z axis in the positive direction (Figure 1-3).

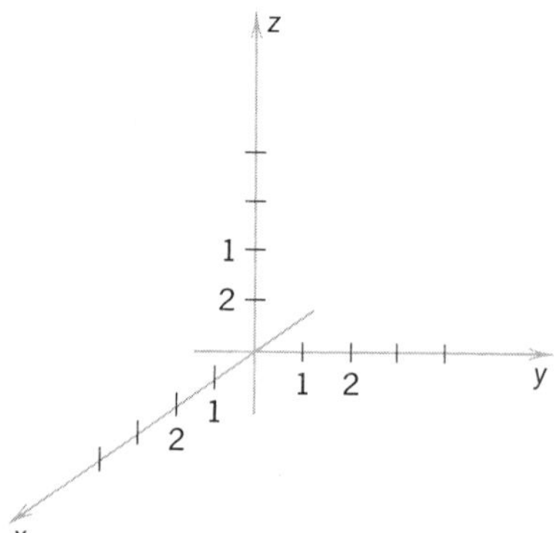

FIGURE 1-2

Because we can associate points in space and ordered triples in this way, we often use the expression "the point (x,y,z)" instead of the longer phrase "the point P which corresponds to the triple (x,y,z)." If the triple (x_0,y_0,z_0) corresponds to P, we say that x_0 is the x coordinate, y_0 the y coordinate, and z_0 the z coordinate of P.

With this method of representing points in mind, we see that the x axis consists of the points of the form $(\alpha,0,0)$ where α is any real number, the y axis consists of points $(0,\alpha,0)$, and the z axis consists of points $(0,0,\alpha)$.

What we have just constructed here is a **model** of three-dimensional space. In the next few paragraphs we examine the mathematical properties of this model and then we consider its geometric interpretation.

We employ the following notation for the line, the plane, and three-dimensional space with coordinate systems:

(1) The real line is denoted $\boldsymbol{R}^1$ (thus, $\boldsymbol{R}$ and $\boldsymbol{R}^1$ are identical).
(2) The set of all ordered pairs (x,y) of real numbers is denoted $\boldsymbol{R}^2$.
(3) The set of all ordered triples (x,y,z) of real numbers is denoted $\boldsymbol{R}^3$.

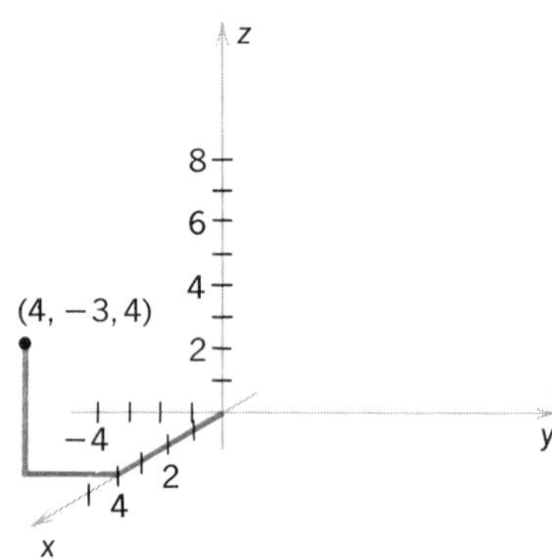

FIGURE 1-3

When speaking of $\boldsymbol{R}^1$, $\boldsymbol{R}^2$, and $\boldsymbol{R}^3$ collectively, we write $\boldsymbol{R}^n$, $n=1$, 2, or 3, or $\boldsymbol{R}^m$, $m=1$, 2, 3.

Since elements of $\boldsymbol{R}^3$ are ordered triples of real numbers, the operation of addition can be extended to $\boldsymbol{R}^3$. Given the two triples (x,y,z) and (x',y',z'), we define their **sum** by

$$(x,y,z)+(x',y',z') = (x+x',\, y+y',\, z+z')$$

Example 1

$$(1,1,1)+(2,-3,4) = (3,-2,5)$$

$$(x,y,z)+(0,0,0) = (x,y,z)$$

$$(1,7,3)+(2,0,6) = (3,7,9)$$

The element $(0,0,0)$ is called the **zero element** of $\boldsymbol{R}^3$. The element $(-x,-y,-z)$ is called the **additive inverse** of (x,y,z), and we write $(x,y,z)-(x',y',z')$ for $(x,y,z)+(-x',-y',-z')$.

There are two very important product operations in $\boldsymbol{R}^3$. One of these, called the **inner product**, sends pairs of elements of $\boldsymbol{R}^3$ to real numbers.

We discuss the inner product in detail in §13.2. The other important product operation for $\boldsymbol{R}^3$ is called **scalar multiplication** (the word "scalar" is a synonym for "real number"). This product combines scalars (real numbers) and elements of $\boldsymbol{R}^3$ (ordered triples) to yield elements of $\boldsymbol{R}^3$. Let us consider this operation first.

Given a scalar α and a triple (x, y, z), we define the **scalar multiple** or **scalar product** by

$$\alpha(x, y, z) = (\alpha x, \alpha y, \alpha z)$$

Example 2

$$\begin{aligned}
2(4, e, 1) &= (2\cdot 4, 2\cdot e, 2\cdot 1) = (8, 2e, 2) \\
6(1,1,1) &= (6,6,6) \\
1(x,y,z) &= (x,y,z) \\
0(x,y,z) &= (0,0,0) \\
(a+b)(x,y,z) &= ((a+b)x, (a+b)y, (a+b)z) \\
&= (ax+bx, ay+by, az+bz) \\
&= a(x,y,z) + b(x,y,z)
\end{aligned}$$

The real number 1 is called the **multiplicative identity** element. It is an immediate result of the definitions that scalar multiplication and addition for $\boldsymbol{R}^3$ satisfy the following identities:

$(ab)(x,y,z) = a(b(x,y,z))$ (associativity for scalar multiplication)

$(a+b)(x,y,z) = a(x,y,z) + b(x,y,z)$
$a[(x,y,z) + (x',y',z')] = a(x,y,z) + a(x',y',z')$ (distributivity for scalar multiplication)

$a(0,0,0) = (0,0,0)$
$0(x,y,z) = (0,0,0)$ (properties of zero elements)

$1(x,y,z) = (x,y,z)$ (property of identity element)

For $\boldsymbol{R}^2$, addition is defined by

$$(x,y) + (x',y') = (x+x', y+y')$$

and scalar multiplication by

$$\alpha(x,y) = (\alpha x, \alpha y)$$

We often identify $\boldsymbol{R}^2$ with the set of triples $(a,b,0)$ and speak of $\boldsymbol{R}^2$ as the plane (or more precisely, the xy plane).

Now let us turn to the geometry of our model. One of the more fruitful tools in mathematics has been the notion of a vector. We define a (geometric) **vector** to be a directed line segment beginning at the origin, that is, a line segment with magnitude and a specified direction with initial point at the origin. Figure 1-4 shows several vectors. Vectors may be thought of as arrows beginning at the origin.

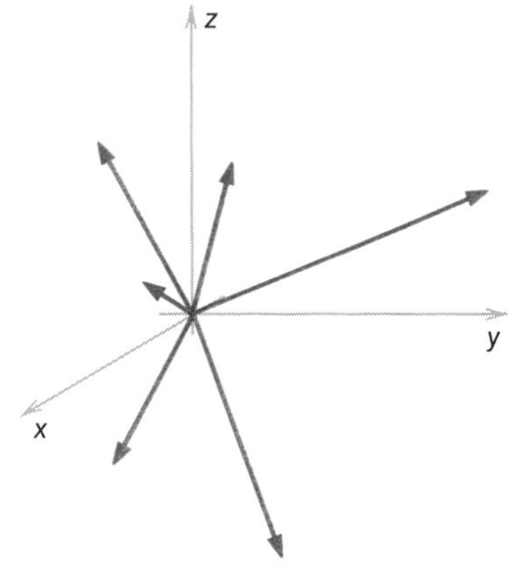

FIGURE 1-4

Using this definition of a vector, we may associate with each vector $\boldsymbol{v}$ the point in space (x, y, z) where $\boldsymbol{v}$ terminates, and write $\boldsymbol{v}(x, y, z)$ to indicate this correspondence. For this reason, the elements of $\boldsymbol{R}^3$ are also called **vectors**.

We say that two vectors are **equal** if and only if they have the same direction and the same magnitude. This condition may be expressed algebraically by saying that

$$\boldsymbol{v}_1(x, y, z) = \boldsymbol{v}_2(x', y', z') \text{ if and only if } x = x',\ y = y',\ z = z'$$

Geometrically, we define **vector addition** as follows. In the plane formed by the vectors $\boldsymbol{v}_1$ and $\boldsymbol{v}_2$ (see Figure 1-5), form the parallelogram having $\boldsymbol{v}_1$ as one side and $\boldsymbol{v}_2$ as its adjacent side. Then the sum $\boldsymbol{v}_1 + \boldsymbol{v}_2$ is the directed line segment along the diagonal of the parallelogram. To show that our geometric definition of addition does not contradict our algebraic definition, we must demonstrate that $\boldsymbol{v}(x, y, z) + \boldsymbol{v}(x', y', z') = \boldsymbol{v}(x+x', y+y', z+z')$.

We prove this result in the plane, and leave it to the reader to formulate the proposition in three-dimensional space. Thus we wish to show $\boldsymbol{v}(x, y) + \boldsymbol{v}(x', y') = \boldsymbol{v}(x+x', y+y')$.

In Figure 1-6 let $\boldsymbol{v}(x, y)$ be the vector ending at the point A, and let $\boldsymbol{v}(x', y')$ be the vector ending at point B. The vector $\boldsymbol{v}(x, y) + \boldsymbol{v}(x', y')$ ends at the vertex C of parallelogram $OBCA$. Hence, to verify that $\boldsymbol{v}(x, y) + \boldsymbol{v}(x', y') = \boldsymbol{v}(x+x', y+y')$, it is sufficient to show that the coordinates of C are $(x+x', y+y')$.

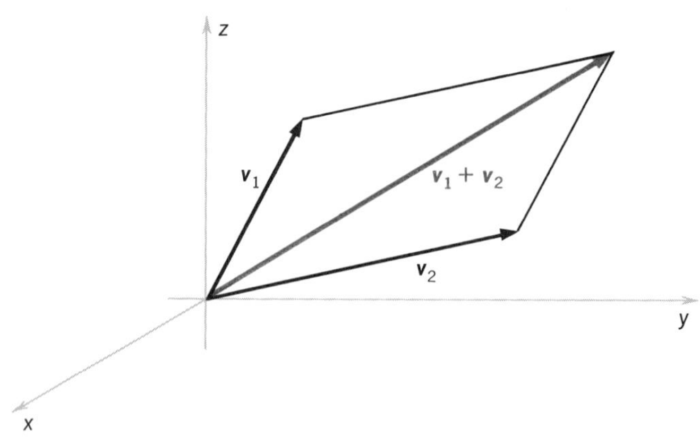

FIGURE 1-5

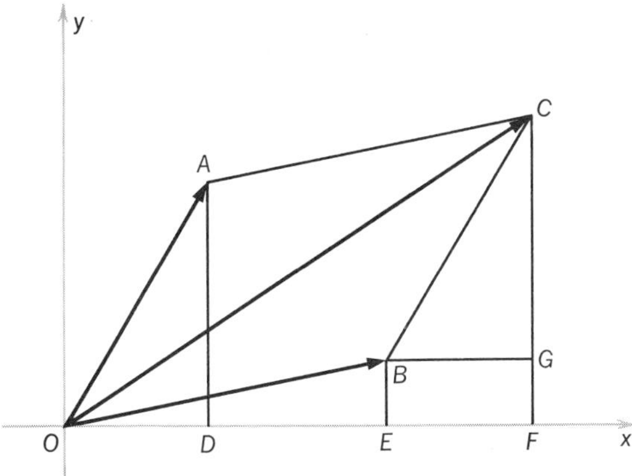

FIGURE 1-6

From the figure it is clear that triangle OAD is congruent to triangle BCG. Furthermore, $OD = x$ and $OE = x'$. By the congruence relation $OD = BG$, and since $BGFE$ is a rectangle, we have $OD = EF$.

Since $OF = OE + EF$, it follows that $OF = x + x'$. This shows that the x coordinate of C is $x + x'$. The proof for the y coordinate is analogous. Hence, we see that the geometric definition of vector addition is equivalent to the algebraic definition in which we add coordinates.

Figure 1-7 illustrates another way of looking at vector addition; that is, we translate (without rotation) the directed line segment representing the vector $\boldsymbol{v}_2$ so that it begins at the end of the vector $\boldsymbol{v}_1$. The endpoint of the resulting directed segment is the endpoint of the vector $\boldsymbol{v}_1 + \boldsymbol{v}_2$. We note that when $\boldsymbol{v}_1$ and $\boldsymbol{v}_2$ are collinear, the parallelogram collapses. This situation is illustrated in Figure 1-7(b).

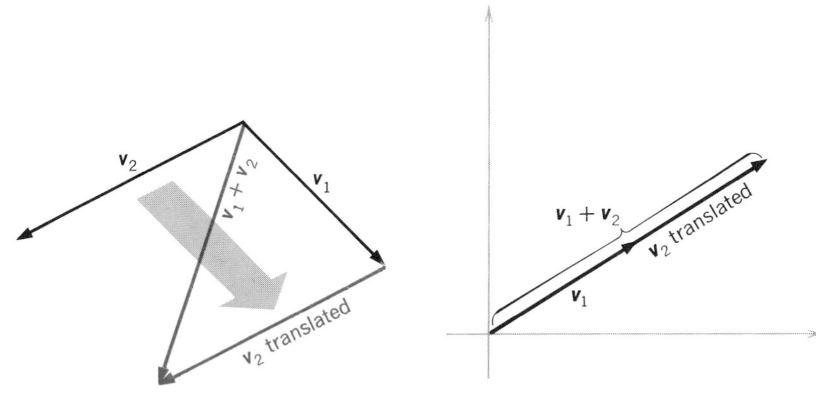

FIGURE 1-7

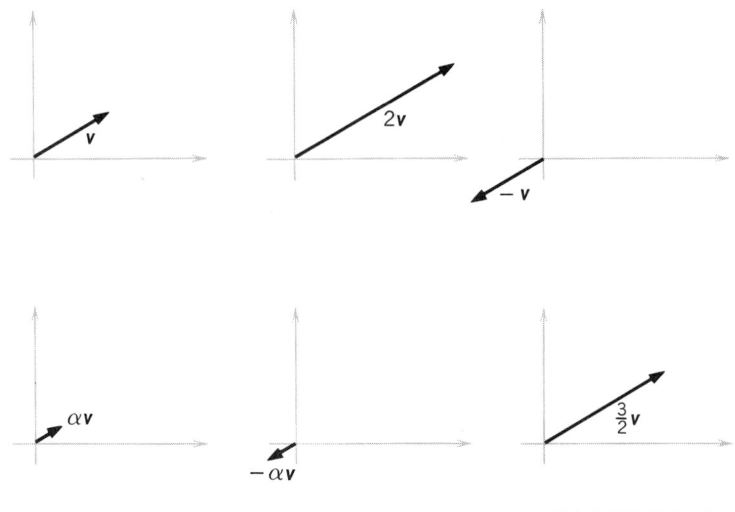

FIGURE 1-8

Scalar multiples of vectors have similar geometric interpretations. If α is a scalar and $\boldsymbol{v}$ a vector, we define $\alpha\boldsymbol{v}$ to be the vector which is α times as long as $\boldsymbol{v}$ with the same direction as $\boldsymbol{v}$ if $\alpha > 0$ but with the opposite direction if $\alpha < 0$. Figure 1-8 illustrates several examples.

Using an argument which depends on similar triangles we can prove that

$$\alpha\boldsymbol{v}(x, y, z) = \boldsymbol{v}(\alpha x, \alpha y, \alpha z) \tag{1}$$

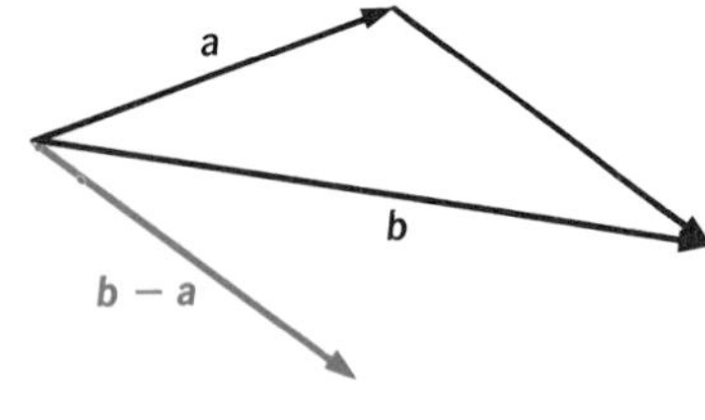

FIGURE 1-9

Again the geometric definition coincides with the algebraic one.

How do we represent the vector $\boldsymbol{b}-\boldsymbol{a}$ geometrically? Since $\boldsymbol{a}+(\boldsymbol{b}-\boldsymbol{a}) = \boldsymbol{b}$, $\boldsymbol{b}-\boldsymbol{a}$ is that vector which when added to $\boldsymbol{a}$ gives $\boldsymbol{b}$. In view of this, we may conclude that $\boldsymbol{b}-\boldsymbol{a}$ is the vector parallel to and with the same magnitude as the directed line segment beginning at the endpoint of $\boldsymbol{a}$ and terminating at the endpoint of $\boldsymbol{b}$ (see Figure 1-9).

Let us denote by $\boldsymbol{i}$ the vector which ends at (1,0,0), $\boldsymbol{j}$ the vector which ends at (0, 1, 0), and $\boldsymbol{k}$ the vector which ends at (0, 0, 1). Then using (1) above we find that

$$\begin{aligned} \boldsymbol{v}(x, y, z) &= \boldsymbol{v}(x, 0, 0) + \boldsymbol{v}(0, y, 0) + \boldsymbol{v}(0, 0, z) \\ &= x\boldsymbol{v}(1, 0, 0) + y\boldsymbol{v}(0, 1, 0) + z\boldsymbol{v}(0, 0, 1) \\ &= x\boldsymbol{i} + y\boldsymbol{j} + z\boldsymbol{k} \end{aligned}$$

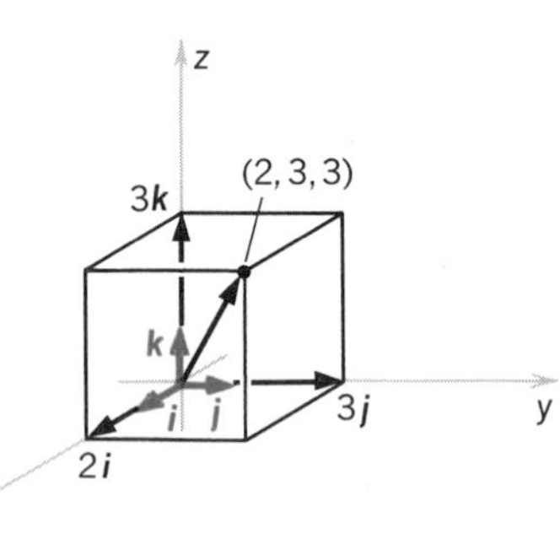

FIGURE 1-10

Hence, we can represent any vector in three-dimensional space in terms of the vectors, $\boldsymbol{i}$, $\boldsymbol{j}$, and $\boldsymbol{k}$.

For example, the vector ending at (2,3,3) is $2\boldsymbol{i}+3\boldsymbol{j}+3\boldsymbol{k}$, that ending at $(0, -1, 4)$ is $-\boldsymbol{j}+4\boldsymbol{k}$ (see Figure 1-10). The vectors $\boldsymbol{i}$, $\boldsymbol{j}$, and $\boldsymbol{k}$ are called the **standard basis** vectors for $\boldsymbol{R}^3$.

We also have the relationships

$$(x\boldsymbol{i}+y\boldsymbol{j}+z\boldsymbol{k}) + (x'\boldsymbol{i}+y'\boldsymbol{j}+z'\boldsymbol{k}) = (x+x')\boldsymbol{i} + (y+y')\boldsymbol{j} + (z+z')\boldsymbol{k}$$

and

$$\alpha(x\boldsymbol{i}+y\boldsymbol{j}+z\boldsymbol{k}) = (\alpha x)\boldsymbol{i} + (\alpha y)\boldsymbol{j} + (\alpha z)\boldsymbol{k}$$

Because of the correspondence between vectors and points, we may sometimes refer to a point $\boldsymbol{a}$ under circumstances in which $\boldsymbol{a}$ has been defined to be a vector. The reader should realize that by this statement we mean the endpoint of the vector $\boldsymbol{a}$.

As an example of the use of vectors, let us describe the points which lie in the parallelogram whose adjacent sides are the vectors $\boldsymbol{a}$ and $\boldsymbol{b}$.

Consider Figure 1-11. If P is any point in the given parallelogram and we construct lines l_1 and l_2 through P parallel to the vectors $\boldsymbol{a}$ and $\boldsymbol{b}$, respectively, we see that l_1 intersects the side of the parallelogram determined by the vector $\boldsymbol{b}$ at some point $t\boldsymbol{b}$, where $0 \leqslant t \leqslant 1$. Likewise, l_2 intersects the side determined by the vector $\boldsymbol{a}$ at some point $s\boldsymbol{a}$, where $0 \leqslant s \leqslant 1$.

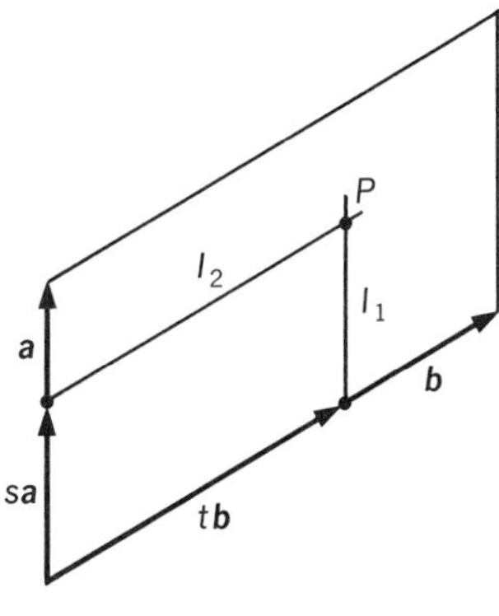

FIGURE 1-11

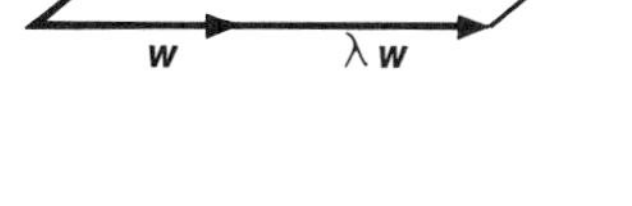

FIGURE 1-12

Since P is now the endpoint of the diagonal of a parallelogram having adjacent sides $s\boldsymbol{a}$ and $t\boldsymbol{b}$, if $\boldsymbol{v}$ denotes the vector ending at P, we see that $\boldsymbol{v} = s\boldsymbol{a} + t\boldsymbol{b}$. Thus, all the points in the given parallelogram are endpoints of vectors of the form $s\boldsymbol{a}+t\boldsymbol{b}$ for $0 \leqslant s \leqslant 1$ and $0 \leqslant t \leqslant 1$. By reversing our steps it is easy to see that all vectors of this form end within the parallelogram.

Since two line segments determine a plane, two vectors also determine a plane. If we apply the same reasoning as above, it is not hard to see that the plane formed by two vectors $\boldsymbol{v}$ and $\boldsymbol{w}$ consists of all points of the form $\alpha\boldsymbol{v}+\beta\boldsymbol{w}$ where α and β vary over the real numbers. This follows from the fact that any point P in the plane determined by $\boldsymbol{v}$ and $\boldsymbol{w}$ will lie in some parallelogram determined by $\lambda\boldsymbol{v}$ and $\lambda\boldsymbol{w}$ where λ is some scalar (see Figure 1-12). Thus $P = \alpha(\lambda\boldsymbol{v}) + \beta(\lambda\boldsymbol{w}) = (\alpha\lambda)\boldsymbol{v} + (\beta\lambda)\boldsymbol{w}$. The plane determined by $\boldsymbol{v}$ and $\boldsymbol{w}$ is called the plane **spanned** by $\boldsymbol{v}$ and $\boldsymbol{w}$. When $\boldsymbol{v} = \gamma\boldsymbol{w}$ $(\boldsymbol{w} \neq 0)$ the plane degenerates to a straight line. When $\boldsymbol{v} = \boldsymbol{w} = 0$ we obtain a single point.

There are three particular planes which arise naturally in the coordinate system and which will be of use to us later. We call the plane spanned by vectors $\boldsymbol{i}$ and $\boldsymbol{j}$ the xy plane, the plane spanned by $\boldsymbol{j}$ and $\boldsymbol{k}$ the yz plane, and the plane spanned by $\boldsymbol{i}$ and $\boldsymbol{k}$ the xz plane. These planes are illustrated in Figure 1-13.

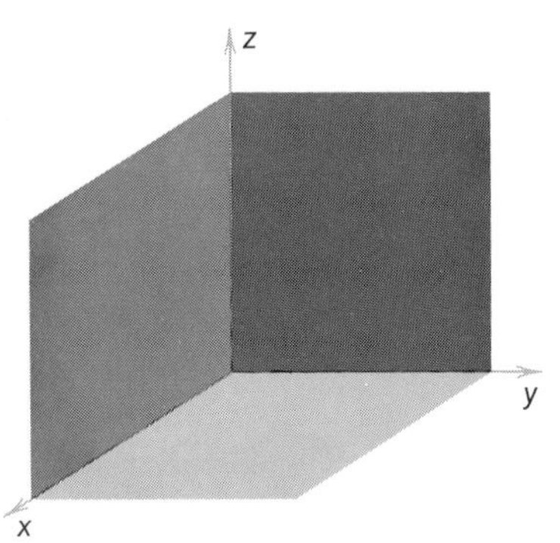

FIGURE 1-13

Using the geometric interpretation of vector addition and scalar multiplication, we may find the parametric equation of a line l passing through the endpoint of the vector $\boldsymbol{a}$ in the direction of a vector $\boldsymbol{v}$ (see Figure 1-14). As t varies through all real values, the points of the form $t\boldsymbol{v}$ are all scalar multiples of the vector $\boldsymbol{v}$, and therefore exhaust the points of the line passing through the origin in the direction of $\boldsymbol{v}$. Since every point on l is the endpoint of the diagonal of a parallelogram with sides $\boldsymbol{a}$ and $t\boldsymbol{v}$, for some suitable value of t, we see that all the points on l are of the form $\boldsymbol{a}+t\boldsymbol{v}$. Thus, the line l may be expressed parametrically in the form $l(t)=\boldsymbol{a}+t\boldsymbol{v}$. At $t=0$, $l(0)=\boldsymbol{a}$. As t increases the point $l(t)$ moves away from $\boldsymbol{a}$ in the direction of $\boldsymbol{v}$. As t decreases from $t=0$ through negative values, $l(t)$ moves away from $\boldsymbol{a}$ in the direction of $-\boldsymbol{v}$.

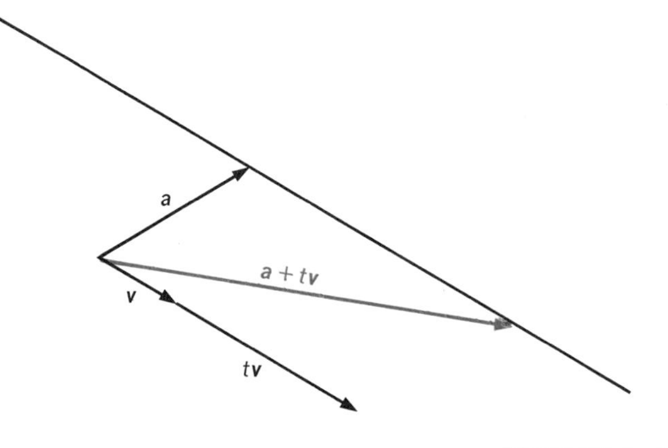

FIGURE 1-14

Of course, there are other parametrizations of the same line. These may be obtained by choosing a different point on the given line and forming the parametric equation of the line beginning at that point and in the direction of $\boldsymbol{v}$. For example, the endpoint of $\boldsymbol{a}+\boldsymbol{v}$ is on the line $\boldsymbol{a}+t\boldsymbol{v}$, and thus $l'(t)=\boldsymbol{a}+\boldsymbol{v}+t\boldsymbol{v}$ represents the same line.

Still other parametrizations may be obtained by observing that if $\alpha\neq 0$, the vector $\alpha\boldsymbol{v}$ has the same direction as $\boldsymbol{v}$. Thus $l''(t)=\boldsymbol{a}+\alpha t\boldsymbol{v}$ is another parametrization of $l(t)=\boldsymbol{a}+t\boldsymbol{v}$.

Example 3 Determine the equation of the line passing through (1,0,0) in the direction of $\boldsymbol{j}$. The desired line can be expressed parametrically as

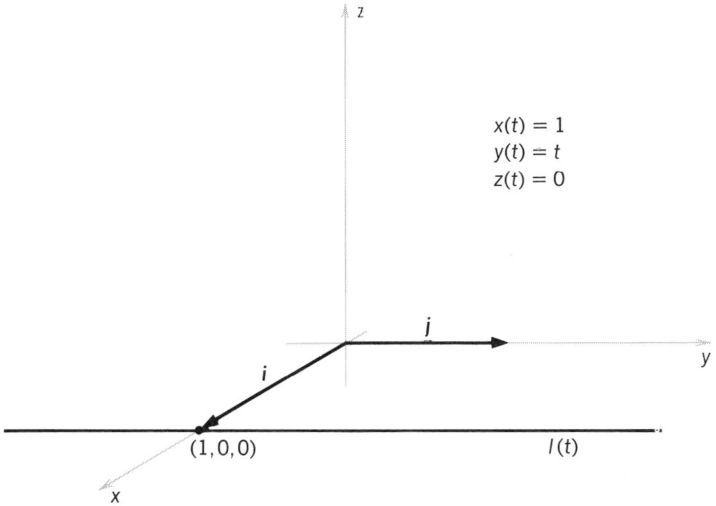

FIGURE 1-15

$l(t) = \boldsymbol{i} + t\boldsymbol{j}$ (Figure 1-15). In terms of coordinates we have

$$l(t) = (1,0,0) + t(0,1,0) = (1,t,0)$$

In this case the line is the intersection of the planes $z = 0$ and $x = 1$.

We may also derive the equation of a line passing through the endpoints of two given vectors $\boldsymbol{a}$ and $\boldsymbol{b}$.

Since the vector $\boldsymbol{b} - \boldsymbol{a}$ is parallel to the directed line segment from $\boldsymbol{a}$ to $\boldsymbol{b}$, what we really wish to do here is calculate the parametric equations of the line passing through $\boldsymbol{a}$ in the direction of $\boldsymbol{b} - \boldsymbol{a}$ (Figure 1-16). Thus $l(t) = \boldsymbol{a} + t(\boldsymbol{b} - \boldsymbol{a})$, or $l(t) = (1-t)\boldsymbol{a} + t\boldsymbol{b}$.

As t increases from 0 to 1, $t(\boldsymbol{b} - \boldsymbol{a})$ starts as the 0 vector and continues in the direction of $\boldsymbol{b} - \boldsymbol{a}$, increasing in length until at $t = 1$ it is the vector $\boldsymbol{b} - \boldsymbol{a}$. Thus, for $l(t) = \boldsymbol{a} + t(\boldsymbol{b} - \boldsymbol{a})$, as t increases from 0 to 1, $l(t)$ moves from the endpoint of $\boldsymbol{a}$ along the directed line segment from $\boldsymbol{a}$ to $\boldsymbol{b}$ to the endpoint of $\boldsymbol{b}$.

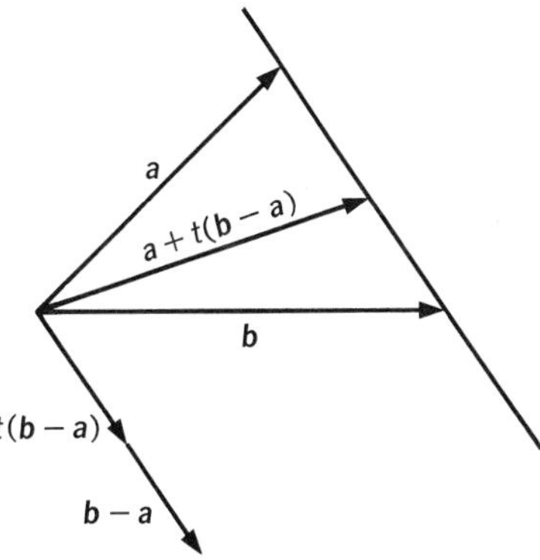

FIGURE 1-16

Example 4 Find the equation of the line passing through $(-1,1,0)$ and $(0,0,1)$ (see Figure 1-17). Letting $\boldsymbol{a} = -\boldsymbol{i} + \boldsymbol{j}$, $\boldsymbol{b} = \boldsymbol{k}$, we have

$$\begin{aligned} l(t) &= (1-t)(-\boldsymbol{i} + \boldsymbol{j}) + t\boldsymbol{k} \\ &= -(1-t)\boldsymbol{i} + (1-t)\boldsymbol{j} + t\boldsymbol{k} \end{aligned}$$

Before continuing this discussion, we note that any vector of the form $\boldsymbol{c} = \lambda\boldsymbol{a} + \mu\boldsymbol{b}$, where $\lambda + \mu = 1$, is on the line passing through the endpoints of $\boldsymbol{a}$ and $\boldsymbol{b}$. To see this, observe that $\boldsymbol{c} = (1-\mu)\boldsymbol{a} + \mu\boldsymbol{b} = \boldsymbol{a} + \mu(\boldsymbol{b} - \boldsymbol{a})$, and so $\boldsymbol{c}$ is on the line passing through the endpoints of $\boldsymbol{a}$ and $\boldsymbol{b}$.

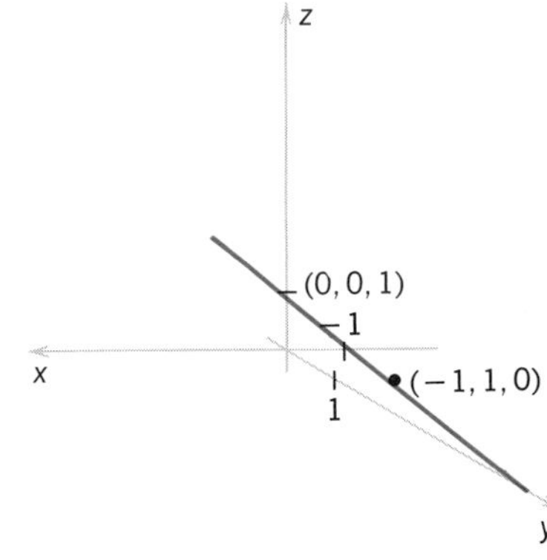

FIGURE 1-17

As another example of the power of vector methods, let us give a simple proof that the diagonals of a parallelogram bisect each other.

Let the adjacent sides of the parallelogram be represented by the vectors $\boldsymbol{a}$ and $\boldsymbol{b}$, as shown in Figure 1-18. We first calculate the vector to the midpoint of PQ. Since $\boldsymbol{b}-\boldsymbol{a}$ is parallel and equal in length to the directed segment from P to Q, $\frac{1}{2}(\boldsymbol{b}-\boldsymbol{a})$ is parallel and equal in length to the directed line segment from P to the midpoint of PQ. Thus, the vector $\boldsymbol{a}+\frac{1}{2}(\boldsymbol{b}-\boldsymbol{a}) = \frac{1}{2}(\boldsymbol{a}+\boldsymbol{b})$ ends at the midpoint of PQ.

Next, we calculate the vector to the midpoint of OR. We know $\boldsymbol{a}+\boldsymbol{b}$ ends at R, thus $\frac{1}{2}(\boldsymbol{a}+\boldsymbol{b})$ ends at the midpoint of OR.

Since we have shown that the vector $\frac{1}{2}(\boldsymbol{a}+\boldsymbol{b})$ ends at both the midpoint of OR and the midpoint of PQ, it follows that OR and PQ bisect each other.

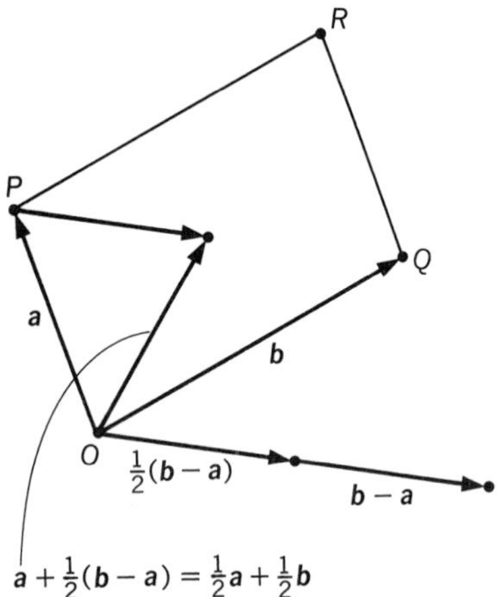

FIGURE 1-18

EXERCISES

1. What restrictions must be made on x, y, and z, so that the triple (x, y, z) will represent a point on the y axis? On the z axis? In the xz plane? In the yz plane?
2. Show by geometrical construction that $\boldsymbol{v}(x, y, z)+\boldsymbol{v}(x', y', z') = \boldsymbol{v}(x+x', y+y', z+z')$.
3. Supply the argument depending on similar triangles which proves that $\alpha\boldsymbol{v}(x, y, z) = \boldsymbol{v}(\alpha x, \alpha y, \alpha z)$.

Complete the following computations:

4. $(3, 4, 5) + (6, 2, -6) = ?$
5. $(-21, 23) - (?, 6) = (-25, ?)$
6. $3\boldsymbol{v}(133, -0.33, 0) + \boldsymbol{v}(-399, 0.99, 0) = ?$
7. $(8a, -2b, 13c) = (52, 12, 11) + \frac{1}{2}?$
8. $\boldsymbol{v}(2, 3, 5) - 4\boldsymbol{i} + 3\boldsymbol{j} = \boldsymbol{v}(?, ?, ?)$
9. $800\boldsymbol{v}(0.03, 0, 0) = ?\boldsymbol{i} + ?\boldsymbol{j} + ?\boldsymbol{k}$

In exercises 10–16 use vector notation to describe the points which lie in the given configurations.

10. The parallelogram whose adjacent sides are the vectors $\boldsymbol{i}+3\boldsymbol{k}$ and $-2\boldsymbol{j}$.
11. The plane determined by $\boldsymbol{v}(2, 7, 0)$ and $\boldsymbol{v}(0, 2, 7)$.
12. The line passing through $(0, 2, 1)$ in the direction of $2\boldsymbol{i}-\boldsymbol{k}$.
13. The line passing through $(-1, -1, -1)$ and $(1, -1, 2)$.
14. The parallelepiped with sides the vectors $\boldsymbol{a}$, $\boldsymbol{b}$, and $\boldsymbol{c}$
15. The parallelogram with one corner at (x_0, y_0, z_0) and sides meeting at that corner parallel to the vectors $\boldsymbol{a}$ and $\boldsymbol{b}$.
16. The plane determined by the three points (x_0, y_0, z_0), (x_1, y_1, z_1), and (x_2, y_2, z_2).

*17. Show that the medians of a triangle intersect at a point, and that this point divides each median in a ratio of 2:1.

2 THE INNER PRODUCT

In this section and the next we discuss two products of vectors which are often very useful in physical applications and which have interesting geometric interpretations. The first product we consider is called the **inner product**. The name **dot product** is often used instead.

Suppose we have two vectors $\boldsymbol{a}$ and $\boldsymbol{b}$ (Figure 2-1) and we wish to determine the angle between them; that is, the smaller angle subtended by $\boldsymbol{a}$ and $\boldsymbol{b}$ in the plane which they span. The inner product allows us to do this. Let us first develop the concept formally and then prove that this product does what we claim.

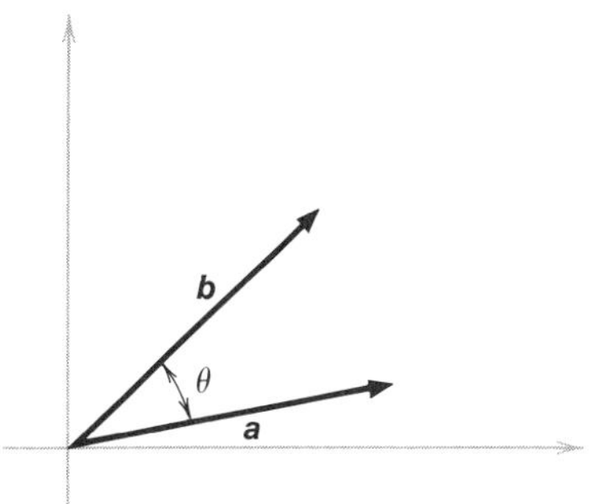

FIGURE 2-1

Let $\boldsymbol{a} = a_1\boldsymbol{i}+a_2\boldsymbol{j}+a_3\boldsymbol{k}$ and $\boldsymbol{b} = b_1\boldsymbol{i}+b_2\boldsymbol{j}+b_3\boldsymbol{k}$ be two vectors in $\boldsymbol{R}^3$. We define the **inner product** of $\boldsymbol{a}$ and $\boldsymbol{b}$, written $\boldsymbol{a}\cdot\boldsymbol{b}$, to be the real number

$$\boldsymbol{a}\cdot\boldsymbol{b} = a_1 b_1 + a_2 b_2 + a_3 b_3$$

Note that the inner product of two vectors is a scaler quantity.

Certain properties of the inner product follow immediately from the definition. If $\boldsymbol{a}$, $\boldsymbol{b}$, and $\boldsymbol{c}$ are vectors in $\boldsymbol{R}^3$, and α and β are real numbers, then

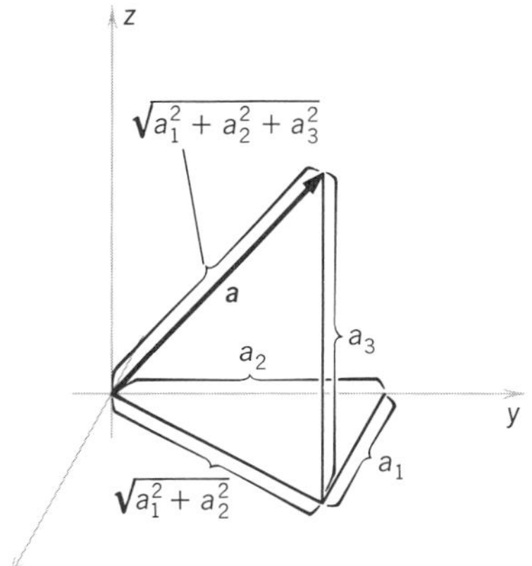

FIGURE 2-2

(1) $\boldsymbol{a}\cdot\boldsymbol{a} \geqslant 0$

$\boldsymbol{a}\cdot\boldsymbol{a} = 0$ if and only if $\boldsymbol{a} = 0$

(2) $\alpha\boldsymbol{a}\cdot\boldsymbol{b} = \alpha(\boldsymbol{a}\cdot\boldsymbol{b})$

$\boldsymbol{a}\cdot\beta\boldsymbol{b} = \beta(\boldsymbol{a}\cdot\boldsymbol{b})$

(3) $\boldsymbol{a}\cdot(\boldsymbol{b}+\boldsymbol{c}) = \boldsymbol{a}\cdot\boldsymbol{b} + \boldsymbol{a}\cdot\boldsymbol{c}$

$(\boldsymbol{a}+\boldsymbol{b})\cdot\boldsymbol{c} = \boldsymbol{a}\cdot\boldsymbol{c} + \boldsymbol{b}\cdot\boldsymbol{c}$

To prove (1), observe that if $\boldsymbol{a}=a_1\boldsymbol{i}+a_2\boldsymbol{j}+a_3\boldsymbol{k}$, $\boldsymbol{a}\cdot\boldsymbol{a}=a_1^2+a_2^2+a_3^2$. Since a_1, a_2, and a_3 are real numbers we know $a_1^2 \geqslant 0$, $a_2^2 \geqslant 0$, $a_3^2 \geqslant 0$. Thus, $\boldsymbol{a}\cdot\boldsymbol{a} \geqslant 0$. Moreover, if $a_1^2+a_2^2+a_3^2=0$, then $a_1 = a_2 = a_3 = 0$, therefore, $\boldsymbol{a}=0$. The proofs of the other properties of the inner product are easily obtained.

It follows from the Pythagorean theorem that the length of the vector $\boldsymbol{a}=a_1\boldsymbol{i}+a_2\boldsymbol{j}+a_3\boldsymbol{k}$ is $\sqrt{a_1^2+a_2^2+a_3^2}$ (see Figure 2-2). The length of the vector $\boldsymbol{a}$ is denoted by $|\boldsymbol{a}|$. This quantity is often called the **norm** of $\boldsymbol{a}$. Since $\boldsymbol{a}\cdot\boldsymbol{a}=a_1^2+a_2^2+a_3^2$, it follows that

$$|\boldsymbol{a}| = (\boldsymbol{a}\cdot\boldsymbol{a})^{1/2}$$

Vectors with norm 1 are called **unit vectors**. For example, the vectors $\boldsymbol{i}, \boldsymbol{j}, \boldsymbol{k}$ are unit vectors. Observe that for any nonzero vector $\boldsymbol{a}$, $\boldsymbol{a}/|\boldsymbol{a}|$ is a unit vector and we say that we have **normalized** $\boldsymbol{a}$.

In the plane, the vector $\boldsymbol{i}_\theta = \cos\theta\boldsymbol{i}+\sin\theta\boldsymbol{j}$ is the unit vector making an angle of θ degrees with the x axis (see Figure 2-3). Clearly, $|\boldsymbol{i}_\theta| = (\sin^2\theta+\cos^2\theta)^{1/2} = 1$.

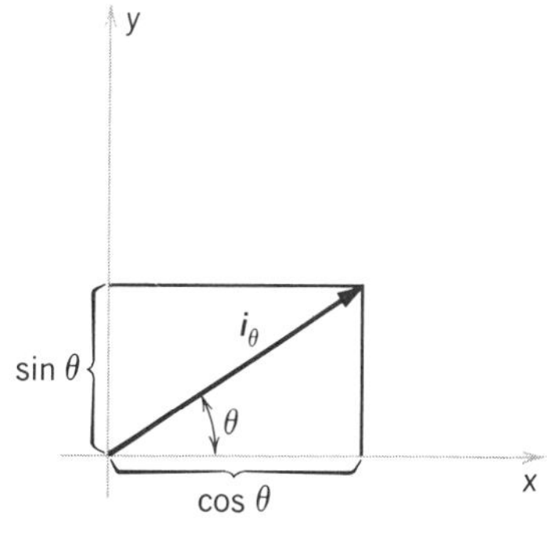

FIGURE 2-3

If $\boldsymbol{a}$ and $\boldsymbol{b}$ are vectors, we have seen that the vector $\boldsymbol{b}-\boldsymbol{a}$ is parallel to and has the same magnitude as the directed line segment from the endpoint of $\boldsymbol{a}$ to the endpoint $\boldsymbol{b}$. It follows that the distance from the endpoint of $\boldsymbol{a}$ to the endpoint $\boldsymbol{b}$ is $|\boldsymbol{b}-\boldsymbol{a}|$ (see Figure 2-4). For example, the distance from the endpoint of the vector $\boldsymbol{i}$, that is, the point $(1,0,0)$, to the endpoint of the vector $\boldsymbol{j}$, $(0,1,0)$, is $\sqrt{(1-0)^2+(0-1)^2+(0-0)^2}=\sqrt{2}$.

Let us now show that the inner product does indeed measure the angle between two vectors.

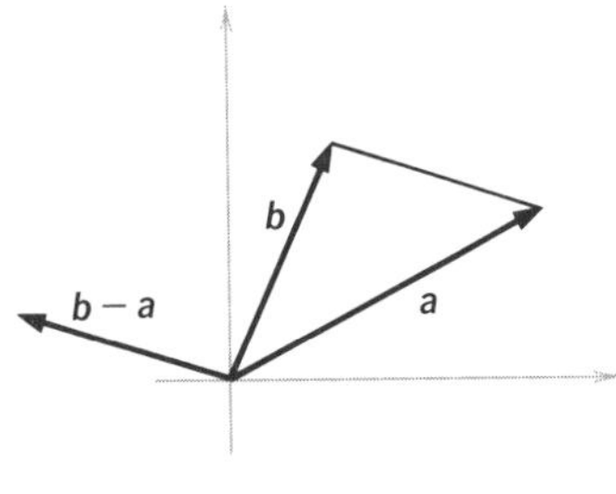

FIGURE 2-4

Theorem 1 Let $\boldsymbol{a}$ and $\boldsymbol{b}$ be two vectors in $\boldsymbol{R}^3$ and let θ, $0 \leqslant \theta \leqslant \pi$, be the angle between them (Figure 2-5). Then

$$\boldsymbol{a}\cdot\boldsymbol{b} = |\boldsymbol{a}|\,|\boldsymbol{b}| \cos \theta$$

Thus we may express the angle between $\boldsymbol{a}$ and $\boldsymbol{b}$ as

$$\theta = \cos^{-1}\frac{\boldsymbol{a}\cdot\boldsymbol{b}}{|\boldsymbol{a}|\,|\boldsymbol{b}|}$$

If $\boldsymbol{a}$ and $\boldsymbol{b}$ are unit vectors this simplifies to $\theta = \cos^{-1}(\boldsymbol{a}\cdot\boldsymbol{b})$.

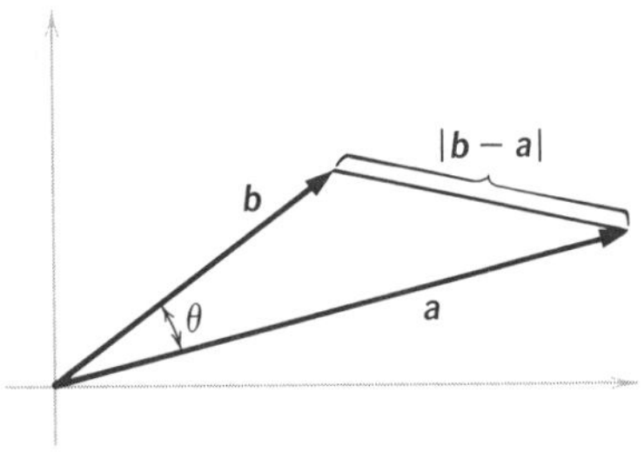

FIGURE 2-5

Proof If we apply the law of cosines from trigonometry to the triangle with one vertex at the origin and adjacent sides determined by the vectors $\boldsymbol{a}$ and $\boldsymbol{b}$, it follows that

$$|\boldsymbol{b}-\boldsymbol{a}|^2 = |\boldsymbol{a}|^2 + |\boldsymbol{b}|^2 - 2|\boldsymbol{a}|\,|\boldsymbol{b}| \cos\theta$$

Since $|\boldsymbol{b}-\boldsymbol{a}|^2 = (\boldsymbol{b}-\boldsymbol{a})(\boldsymbol{b}-\boldsymbol{a})$, $|\boldsymbol{a}|^2 = \boldsymbol{a}\cdot\boldsymbol{a}$, and $|\boldsymbol{b}|^2 = \boldsymbol{b}\cdot\boldsymbol{b}$, we can rewrite the above equation as

$$(\boldsymbol{b}-\boldsymbol{a})\cdot(\boldsymbol{b}-\boldsymbol{a}) = \boldsymbol{a}\cdot\boldsymbol{a} + \boldsymbol{b}\cdot\boldsymbol{b} - 2|\boldsymbol{a}|\,|\boldsymbol{b}| \cos\theta$$

Now

$$\begin{aligned}(\boldsymbol{b}-\boldsymbol{a})\cdot(\boldsymbol{b}-\boldsymbol{a}) &= \boldsymbol{b}\cdot(\boldsymbol{b}-\boldsymbol{a}) - \boldsymbol{a}\cdot(\boldsymbol{b}-\boldsymbol{a}) \\ &= \boldsymbol{b}\cdot\boldsymbol{b} - \boldsymbol{b}\cdot\boldsymbol{a} - \boldsymbol{a}\cdot\boldsymbol{b} + \boldsymbol{a}\cdot\boldsymbol{a} \\ &= \boldsymbol{a}\cdot\boldsymbol{a} + \boldsymbol{b}\cdot\boldsymbol{b} - 2\boldsymbol{a}\cdot\boldsymbol{b}\end{aligned}$$

Thus,

$$\boldsymbol{a}\cdot\boldsymbol{a} + \boldsymbol{b}\cdot\boldsymbol{b} - 2\boldsymbol{a}\cdot\boldsymbol{b} = \boldsymbol{a}\cdot\boldsymbol{a} + \boldsymbol{b}\cdot\boldsymbol{b} - 2|\boldsymbol{a}|\,|\boldsymbol{b}| \cos\theta$$

or

$$\boldsymbol{a}\cdot\boldsymbol{b} = |\boldsymbol{a}|\,|\boldsymbol{b}| \cos\theta$$

This result shows that the inner product of two vectors is the product of their lengths times the cosine of the angle between them. This relationship is often of value in problems of a geometric nature.

Corollary 1 **Cauchy–Schwarz Inequality** For any two vectors $\boldsymbol{a}$ and $\boldsymbol{b}$, we have

$$|\boldsymbol{a}\cdot\boldsymbol{b}| \leqslant |\boldsymbol{a}|\,|\boldsymbol{b}|$$

with equality if and only if $\boldsymbol{a}$ is a scalar multiple of $\boldsymbol{b}$.

Proof If $\boldsymbol{a}$ is not a scalar multiple of $\boldsymbol{b}$, then $|\cos\theta| < 1$ and the inequality holds. When $\boldsymbol{a}$ is a scalar multiple of $\boldsymbol{b}$ then $\theta = 0$ and $\cos\theta = 1$. ∎

Example 1 Find the angle between the vectors $\boldsymbol{i}+\boldsymbol{j}+\boldsymbol{k}$ and $\boldsymbol{i}+\boldsymbol{j}-\boldsymbol{k}$ (Figure 2-6). Using Theorem 1 we have

$$(\boldsymbol{i}+\boldsymbol{j}+\boldsymbol{k})\cdot(\boldsymbol{i}+\boldsymbol{j}-\boldsymbol{k}) = |\boldsymbol{i}+\boldsymbol{j}+\boldsymbol{k}|\,|\boldsymbol{i}+\boldsymbol{j}-\boldsymbol{k}|\cos\theta$$

So

$$1+1-1 = (\sqrt{3})(\sqrt{3})\cos\theta$$

Hence,

$$\cos\theta = \tfrac{1}{3}$$

or

$$\theta = \cos^{-1}(\tfrac{1}{3}) \approx 71°$$

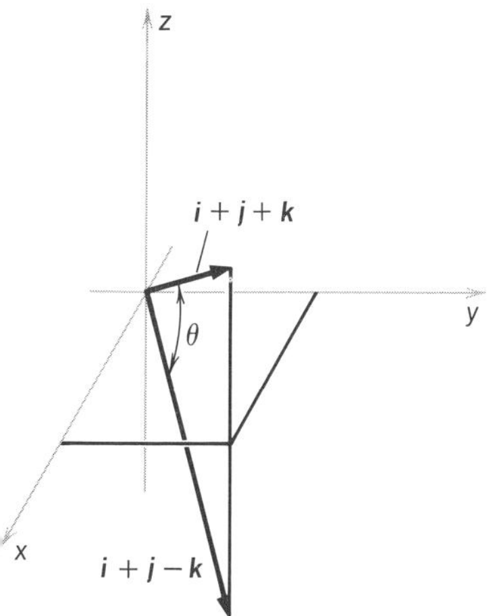

FIGURE 2-6

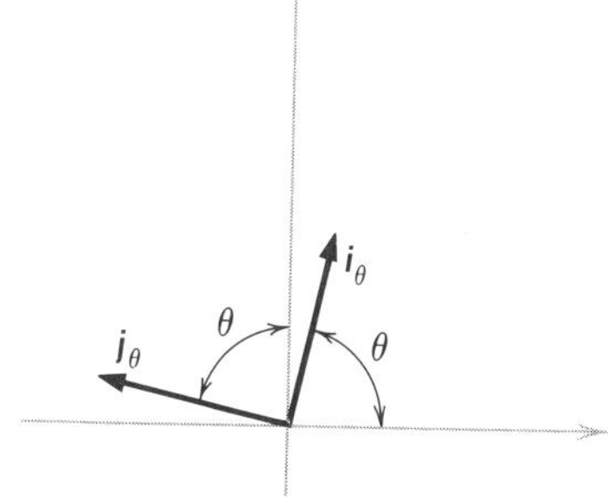

FIGURE 2-7

If $\boldsymbol{a}$ and $\boldsymbol{b}$ are nonzero vectors in $\boldsymbol{R}^3$ and θ is the angle between them, we see that $\boldsymbol{a}\cdot\boldsymbol{b} = 0$ if and only if $\cos\theta = 0$. From this it follows that the inner product of two nonzero vectors is zero if and only if the vectors are perpendicular. Often we say that perpendicular vectors are **orthogonal**. We adopt the convention that the zero vector is orthogonal to all vectors. Hence, the inner product provides us with a convenient method for determining if two vectors are orthogonal.

For example, the vectors $\boldsymbol{i}_\theta = \cos\theta\,\boldsymbol{i} + \sin\theta\,\boldsymbol{j}$ and $\boldsymbol{j}_\theta = -\sin\theta\,\boldsymbol{i} + \cos\theta\,\boldsymbol{j}$ are orthogonal, since

$$\boldsymbol{i}_\theta\cdot\boldsymbol{j}_\theta = -\cos\theta\sin\theta + \sin\theta\cos\theta = 0$$

(see Figure 2-7).

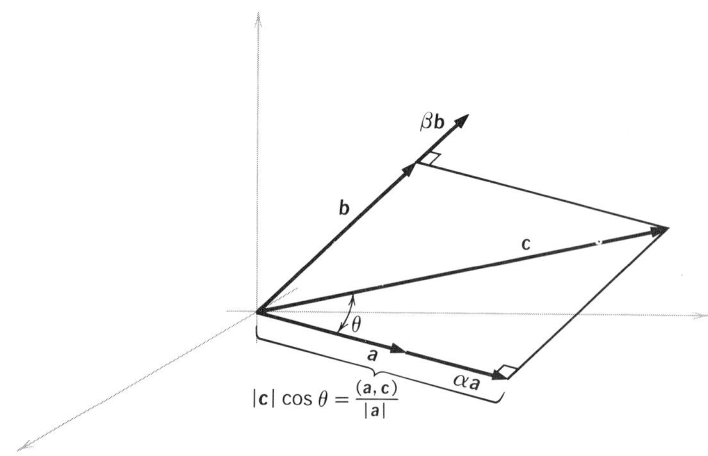

FIGURE 2-8

Example 2 Let $\boldsymbol{a}$ and $\boldsymbol{b}$ be two nonzero orthogonal vectors. Let $\boldsymbol{c}$ be another vector in the plane spanned by $\boldsymbol{a}$ and $\boldsymbol{b}$. As we have seen, there are scalars α and β, such that $\boldsymbol{c} = \alpha\boldsymbol{a} + \beta\boldsymbol{b}$. We use the inner product to determine α and β (see Figure 2-8). Taking the inner product of $\boldsymbol{a}$ and $\boldsymbol{c}$, we have

$$\boldsymbol{a}\cdot\boldsymbol{c} = \boldsymbol{a}\cdot(\alpha\boldsymbol{a}+\beta\boldsymbol{b}) = \alpha\boldsymbol{a}\cdot\boldsymbol{a} + \beta\boldsymbol{a}\cdot\boldsymbol{b}$$

Since $\boldsymbol{a}$ and $\boldsymbol{b}$ are orthogonal, $\boldsymbol{a}\cdot\boldsymbol{b} = 0$, and so,

$$\alpha = \frac{\boldsymbol{a}\cdot\boldsymbol{c}}{\boldsymbol{a}\cdot\boldsymbol{a}} = \frac{\boldsymbol{a}\cdot\boldsymbol{c}}{|\boldsymbol{a}|^2}$$

Similarly,

$$\beta = \frac{\boldsymbol{b}\cdot\boldsymbol{c}}{\boldsymbol{b}\cdot\boldsymbol{b}} = \frac{\boldsymbol{b}\cdot\boldsymbol{c}}{|\boldsymbol{b}|^2}$$

The same result may be obtained using the geometric interpretation of the scalar product. Let l be the distance measured from the origin to the point where the perpendicular from the endpoint of $\boldsymbol{c}$ intersects the line determined by extending $\boldsymbol{a}$. It follows that

$$l = |\boldsymbol{c}| \cos\theta$$

where θ is the angle between $\boldsymbol{a}$ and $\boldsymbol{c}$. Moreover, $l = \alpha|\boldsymbol{a}|$. Taken together these results yield

$$\alpha|\boldsymbol{a}| = |\boldsymbol{c}| \cos\theta, \text{ or } \alpha = \frac{|\boldsymbol{c}|\cos\theta}{|\boldsymbol{a}|} = \frac{|\boldsymbol{c}|}{|\boldsymbol{a}|}\left(\frac{\boldsymbol{a}\cdot\boldsymbol{c}}{|\boldsymbol{c}|\ |\boldsymbol{a}|}\right) = \frac{\boldsymbol{a}\cdot\boldsymbol{c}}{\boldsymbol{a}\cdot\boldsymbol{a}}$$

In Example 2, the vector $\alpha\boldsymbol{a}$ is called the **projection** of $\boldsymbol{c}$ onto $\boldsymbol{a}$. Similarly the vector $\beta\boldsymbol{b}$ is the projection of $\boldsymbol{c}$ onto $\boldsymbol{b}$ (Figure 2-9). In general, the length of the projection of a vector $\boldsymbol{b}$ onto a vector $\boldsymbol{a}$, where θ is the angle between $\boldsymbol{a}$ and $\boldsymbol{b}$, is given by

$$|\boldsymbol{b}| \cos\theta = \frac{\boldsymbol{a}\cdot\boldsymbol{b}}{|\boldsymbol{a}|}$$

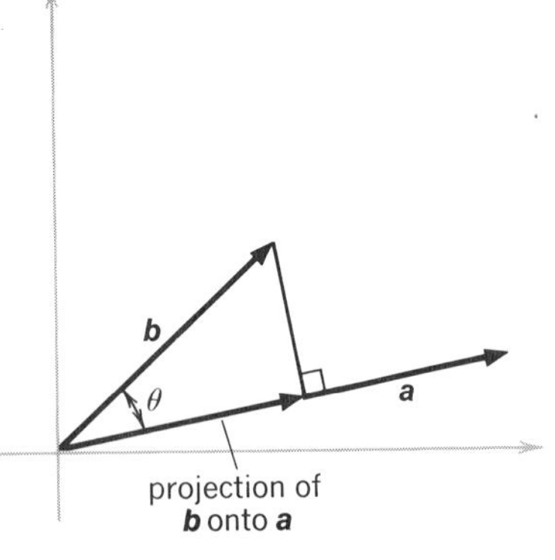

FIGURE 2-9

EXERCISES

1. (a) Prove properties (2) and (3) of the inner product.
   (b) Prove that $\boldsymbol{a} \cdot \boldsymbol{b} = \boldsymbol{b} \cdot \boldsymbol{a}$.
2. Calculate $\boldsymbol{a} \cdot \boldsymbol{b}$ where $\boldsymbol{a} = 2\boldsymbol{i} + 10\boldsymbol{j} - 12\boldsymbol{k}$ and $\boldsymbol{b} = -3\boldsymbol{i} + 4\boldsymbol{k}$.
3. Find the angle between $7\boldsymbol{j} + 19\boldsymbol{k}$ and $-2\boldsymbol{i} - \boldsymbol{j}$ (to the nearest degree).
4. Compute $\boldsymbol{u} \cdot \boldsymbol{v}$, where $\boldsymbol{u} = \sqrt{3}\boldsymbol{i} - 315\boldsymbol{j} + 22\boldsymbol{k}$ and $\boldsymbol{v} = \boldsymbol{u}/|\boldsymbol{u}|$.
5. What is $|8\boldsymbol{i} - 12\boldsymbol{k}| \cdot |6\boldsymbol{j} + \boldsymbol{k}| - |(8\boldsymbol{i} - 12\boldsymbol{k}) \cdot (6\boldsymbol{j} + \boldsymbol{k})|$ (to the nearest tenth)?

In exercises 6–11, compute $|\boldsymbol{u}|$, $|\boldsymbol{v}|$, and $\boldsymbol{u} \cdot \boldsymbol{v}$ for the given vectors.

6. $\boldsymbol{u} = 15\boldsymbol{i} - 2\boldsymbol{j} + 4\boldsymbol{k}$, $\boldsymbol{v} = \pi\boldsymbol{i} + 3\boldsymbol{j} - \boldsymbol{k}$
7. $\boldsymbol{u} = 2\boldsymbol{j} - \boldsymbol{i}$, $\boldsymbol{v} = -\boldsymbol{j} + \boldsymbol{i}$
8. $\boldsymbol{u} = 5\boldsymbol{i} - \boldsymbol{j} + 2\boldsymbol{k}$, $\boldsymbol{v} = \boldsymbol{i} + \boldsymbol{j} - \boldsymbol{k}$
9. $\boldsymbol{u} = -\boldsymbol{i} + 3\boldsymbol{j} + \boldsymbol{k}$, $\boldsymbol{v} = -2\boldsymbol{i} - 3\boldsymbol{j} - 7\boldsymbol{k}$
10. $\boldsymbol{u} = -\boldsymbol{i} + 3\boldsymbol{k}$, $\boldsymbol{v} = 4\boldsymbol{j}$
11. $\boldsymbol{u} = -\boldsymbol{i} + 2\boldsymbol{j} - 3\boldsymbol{k}$, $\boldsymbol{v} = -\boldsymbol{i} - 3\boldsymbol{j} + 4\boldsymbol{k}$
12. Normalize the vectors in exercises 6–8.

3 THE CROSS PRODUCT

In §13.2 we defined a product of vectors which was a scalar. In this section we develop a notion of **vector product**; that is, we show how given two vectors $\boldsymbol{a}$ and $\boldsymbol{b}$ we can produce a third vector $\boldsymbol{a} \times \boldsymbol{b}$ called the **cross product** of $\boldsymbol{a}$ and $\boldsymbol{b}$. This new vector will have the pleasing geometric property that is perpendicular to the plane spanned by $\boldsymbol{a}$ and $\boldsymbol{b}$.

As before, we shall first develop the somewhat lengthy mathematical formalism necessary to state a useful definition of this product. Once this has been accomplished, we can study the geometric implications of the mathematical structure we have built.

We begin by defining a 2×2 **matrix** to be an array

$$\begin{pmatrix} a_{11} & a_{12} \\ a_{21} & a_{22} \end{pmatrix}$$

where $a_{11}, a_{12}, a_{21}, a_{22}$ are four scalars. For example,

$$\begin{pmatrix} 2 & 1 \\ 0 & 4 \end{pmatrix}, \begin{pmatrix} -1 & 0 \\ 1 & 1 \end{pmatrix}, \text{ and } \begin{pmatrix} 13 & 7 \\ 6 & 11 \end{pmatrix}$$

are 2×2 matrices. We define the **determinant**

$$\begin{vmatrix} a_{11} & a_{12} \\ a_{21} & a_{22} \end{vmatrix}$$

of such a matrix by the equation

$$\begin{vmatrix} a_{11} & a_{12} \\ a_{21} & a_{22} \end{vmatrix} = a_{11}a_{22} - a_{12}a_{21} \tag{1}$$

Example 1

(a) $\begin{vmatrix} 1 & 1 \\ 1 & 1 \end{vmatrix} = 1-1 = 0$

(b) $\begin{vmatrix} 1 & 2 \\ 3 & 4 \end{vmatrix} = 4-6 = -2$

(c) $\begin{vmatrix} 5 & 6 \\ 7 & 8 \end{vmatrix} = 40-42 = -2$

A 3×3 matrix is an array

$$\begin{pmatrix} a_{11} & a_{12} & a_{13} \\ a_{21} & a_{22} & a_{23} \\ a_{31} & a_{32} & a_{33} \end{pmatrix}$$

again where each a_{ij} is a scalar. The sharp-eyed reader may have already observed that the subscripts i, j have some special meaning; a_{ij} denotes the entry in the array which is the ith row and the jth column. We define the determinant of a 3×3 matrix by the rule

$$\begin{vmatrix} a_{11} & a_{12} & a_{13} \\ a_{21} & a_{22} & a_{23} \\ a_{31} & a_{32} & a_{33} \end{vmatrix} = a_{11}\begin{vmatrix} a_{22} & a_{23} \\ a_{32} & a_{33} \end{vmatrix} - a_{12}\begin{vmatrix} a_{21} & a_{23} \\ a_{31} & a_{33} \end{vmatrix} + a_{13}\begin{vmatrix} a_{21} & a_{22} \\ a_{31} & a_{32} \end{vmatrix} \tag{2}$$

Without some mnemonic device, formula (2) would be difficult to memorize. The rule to learn here is that you move along the first row multiplying a_{1j} by the determinant of the 2×2 matrix obtained by cancelling out the first row and the jth column, and then you add these up remembering to put a minus in front of the a_{12} term. For example,

$$\begin{pmatrix} a_{21} & a_{23} \\ a_{31} & a_{33} \end{pmatrix}$$

is obtained by crossing out the first row and the second column of

$$\begin{pmatrix} \cancel{a_{11}} & \cancel{a_{12}} & \cancel{a_{13}} \\ a_{21} & \cancel{a_{22}} & a_{23} \\ a_{31} & \cancel{a_{32}} & a_{33} \end{pmatrix}$$

Example 2

(a) $\begin{vmatrix} 1 & 0 & 0 \\ 0 & 1 & 0 \\ 0 & 0 & 1 \end{vmatrix}$

$$= 1\begin{vmatrix} 1 & 0 \\ 0 & 1 \end{vmatrix} - 0\begin{vmatrix} 0 & 0 \\ 0 & 1 \end{vmatrix} + 0\begin{vmatrix} 0 & 1 \\ 0 & 0 \end{vmatrix} = 1$$

(b) $\begin{vmatrix} 1 & 2 & 3 \\ 4 & 5 & 6 \\ 7 & 8 & 9 \end{vmatrix}$

$$= 1\begin{vmatrix} 5 & 6 \\ 8 & 9 \end{vmatrix} - 2\begin{vmatrix} 4 & 6 \\ 7 & 9 \end{vmatrix} + 3\begin{vmatrix} 4 & 5 \\ 7 & 8 \end{vmatrix} = -3+12-9 = 0$$

An important property of determinants is that interchanging two rows or two columns results in a change of sign. This is an immediate consequence of the definition for the 2×2 case. For rows we have

$$\begin{vmatrix} a_{11} & a_{12} \\ a_{21} & a_{22} \end{vmatrix} = a_{11}a_{22} - a_{21}a_{12} = -(a_{21}a_{12} - a_{11}a_{22}) = -\begin{vmatrix} a_{21} & a_{22} \\ a_{11} & a_{12} \end{vmatrix}$$

and for columns

$$\begin{vmatrix} a_{11} & a_{12} \\ a_{21} & a_{22} \end{vmatrix} = -(a_{12}a_{21} - a_{11}a_{22}) = -\begin{vmatrix} a_{12} & a_{11} \\ a_{22} & a_{21} \end{vmatrix}$$

We leave it to the reader to verify this property for the 3×3 case.

A second fundamental property of determinants is that we can factor scalars out of any row or column. For 2×2 determinants, this means

$$\begin{vmatrix} \alpha a_{11} & a_{12} \\ \alpha a_{21} & a_{22} \end{vmatrix} = \begin{vmatrix} a_{11} & \alpha a_{12} \\ a_{21} & \alpha a_{22} \end{vmatrix}$$

$$= \alpha\begin{vmatrix} a_{11} & a_{12} \\ a_{21} & a_{22} \end{vmatrix} = \begin{vmatrix} \alpha a_{11} & \alpha a_{12} \\ a_{21} & a_{22} \end{vmatrix} = \begin{vmatrix} a_{11} & a_{12} \\ \alpha a_{21} & \alpha a_{22} \end{vmatrix}$$

Similarly for 3×3 determinants we have

$$\begin{vmatrix} \alpha a_{11} & \alpha a_{12} & \alpha a_{13} \\ a_{21} & a_{22} & a_{23} \\ a_{31} & a_{32} & a_{33} \end{vmatrix} = \alpha\begin{vmatrix} a_{11} & a_{12} & a_{13} \\ a_{21} & a_{22} & a_{23} \\ a_{31} & a_{32} & a_{33} \end{vmatrix} = \begin{vmatrix} a_{11} & \alpha a_{12} & a_{13} \\ a_{21} & \alpha a_{22} & a_{23} \\ a_{32} & \alpha a_{32} & a_{33} \end{vmatrix}$$

and so on. These results follow easily from the definitions.

A third fundamental fact about determinants is the following: If we change a row (respectively, column) by adding another row (respectively, column) to it, the value of the determinant remains the same. For the 2×2 case this means that

$$\begin{vmatrix} a_1 & a_2 \\ b_1 & b_2 \end{vmatrix} = \begin{vmatrix} a_1+b_1 & a_2+b_2 \\ b_1 & b_2 \end{vmatrix} = \begin{vmatrix} a_1 & a_2 \\ b_1+a_1 & b_2+b_2 \end{vmatrix} = \begin{vmatrix} a_1+a_2 & a_2 \\ b_1+b_2 & b_2 \end{vmatrix} = \begin{vmatrix} a_1 & a_1+a_2 \\ b_1 & b_1+b_2 \end{vmatrix}$$

For the 3×3 case, this means

$$\begin{vmatrix} a_1 & a_2 & a_3 \\ b_1 & b_2 & b_3 \\ c_1 & c_2 & c_3 \end{vmatrix} = \begin{vmatrix} a_1+b_1 & a_2+b_2 & a_3+b_3 \\ b_1 & b_2 & b_3 \\ c_1 & c_2 & c_3 \end{vmatrix} = \begin{vmatrix} a_1+a_2 & a_2 & a_3 \\ b_1+b_2 & b_2 & b_3 \\ c_1+c_2 & c_2 & c_3 \end{vmatrix}$$

and so on. Again, this property can be proved using the definition of determinant.

Example 3 Suppose

$$\boldsymbol{a} = \alpha\boldsymbol{b} + \beta\boldsymbol{c}, \ \boldsymbol{a} = (a_1, a_2, a_3) = \alpha(b_1, b_2, b_3) + \beta(c_1, c_2, c_3)$$

Let us show that

$$\begin{vmatrix} a_1 & a_2 & a_3 \\ b_1 & b_2 & b_3 \\ c_1 & c_2 & c_3 \end{vmatrix} = 0$$

We shall do the case $\alpha \neq 0$, $\beta \neq 0$. The case $\alpha = 0 = \beta$ is trivial and the case where one of α, β is zero is a simple modification of the one we do. Using the fundamental properties of determinants, we have

$$\begin{vmatrix} \alpha b_1+\beta c_1 & \alpha b_2+\beta c_2 & \alpha b_3+\beta c_3 \\ b_1 & b_2 & b_3 \\ c_1 & c_2 & c_3 \end{vmatrix}$$

$$= -\frac{1}{\alpha}\begin{vmatrix} \alpha b_1+\beta c_1 & \alpha b_2+c_2 & \alpha b_3+\beta c_3 \\ -\alpha b_1 & -\alpha b_2 & -\alpha b_3 \\ c_1 & c_2 & c_3 \end{vmatrix} \quad \text{(multiplying the second row by } -\alpha\text{)}$$

$$= \left(-\frac{1}{\alpha}\right)\left(-\frac{1}{\beta}\right)\begin{vmatrix} \alpha b_1+\beta c_1 & \alpha b_2+\beta c_2 & \alpha b_3+\beta c_3 \\ -\alpha b_1 & -\alpha b_2 & -\alpha b_3 \\ -\beta c_1 & -\beta c_2 & -\beta c_3 \end{vmatrix} \quad \text{(multiplying the third row by } -\beta\text{)}$$

$$= \frac{1}{\alpha\beta}\begin{vmatrix} \beta c_1 & \beta c_2 & \beta c_3 \\ -\alpha b_1 & -\alpha b_2 & -\alpha b_3 \\ -\beta c_1 & -\beta c_2 & -\beta c_3 \end{vmatrix} \quad \text{(adding the second row to the first row)}$$

$$= \frac{1}{\alpha\beta}\begin{vmatrix} 0 & 0 & 0 \\ -\alpha b_1 & -\alpha b_2 & -\alpha b_3 \\ -\beta c_1 & -\beta c_2 & -\beta c_3 \end{vmatrix} \quad \text{(adding the third row to the first row)}$$

$$= 0$$

Now that we have established the necessary results about determinants, we return to products of vectors. Let $\boldsymbol{a} = a_1\boldsymbol{i}+a_2\boldsymbol{j}+a_3\boldsymbol{k}$ and $\boldsymbol{b} = b_1\boldsymbol{i}+b_2\boldsymbol{j}+b_3\boldsymbol{k}$ be vectors in $\boldsymbol{R}^3$. The **cross product** of $\boldsymbol{a}$ and $\boldsymbol{b}$, denoted $\boldsymbol{a}\times\boldsymbol{b}$, is defined to be the vector

$$\boldsymbol{a}\times\boldsymbol{b} = \begin{vmatrix} a_2 & a_3 \\ b_2 & b_3 \end{vmatrix}\boldsymbol{i} - \begin{vmatrix} a_1 & a_3 \\ b_1 & b_3 \end{vmatrix}\boldsymbol{j} + \begin{vmatrix} a_1 & a_2 \\ b_1 & b_2 \end{vmatrix}\boldsymbol{k}$$

or, symbolically,

$$\boldsymbol{a}\times\boldsymbol{b} = \begin{vmatrix} \boldsymbol{i} & \boldsymbol{j} & \boldsymbol{k} \\ a_1 & a_2 & a_3 \\ b_1 & b_2 & b_3 \end{vmatrix}$$

Note that the cross product of two vectors is another vector. Just as the inner product is sometimes called the scalar product, the cross product is sometimes called the **vector product.**

Again, certain algebraic properties of the cross product follow immediately from the definition. If $\boldsymbol{a}$, $\boldsymbol{b}$, and $\boldsymbol{c}$ are vectors and α, β, and γ are scalars, then

(1) $\boldsymbol{a}\times\boldsymbol{b} = -(\boldsymbol{b}\times\boldsymbol{a})$

(2) $\boldsymbol{a}\times(\beta\boldsymbol{b}+\gamma\boldsymbol{c}) = \beta(\boldsymbol{a}\times\boldsymbol{b}) + \gamma(\boldsymbol{a}\times\boldsymbol{c})$

$\quad(\alpha\boldsymbol{a}+\beta\boldsymbol{b})\times\boldsymbol{c} = \alpha(\boldsymbol{a}\times\boldsymbol{c}) + \beta(\boldsymbol{b}\times\boldsymbol{c})$

Note that $\boldsymbol{a}\times\boldsymbol{a} = -(\boldsymbol{a}\times\boldsymbol{a})$ by (1). Thus, $\boldsymbol{a}\times\boldsymbol{a} = \boldsymbol{0}$. Also,

$$\boldsymbol{i}\times\boldsymbol{j} = \boldsymbol{k}, \qquad \boldsymbol{j}\times\boldsymbol{k} = \boldsymbol{i}, \qquad \boldsymbol{k}\times\boldsymbol{i} = \boldsymbol{j}$$

which can be remembered by cyclicly permuting $\boldsymbol{i},\boldsymbol{j},\boldsymbol{k}$ like this:

$$\boldsymbol{i} \to \boldsymbol{j} \to \boldsymbol{k} \to \boldsymbol{i}$$

For example,

$$(3\boldsymbol{i}-\boldsymbol{j}+\boldsymbol{k})\times(\boldsymbol{i}+2\boldsymbol{j}-\boldsymbol{k}) = \begin{vmatrix} \boldsymbol{i} & \boldsymbol{j} & \boldsymbol{k} \\ 3 & -1 & 1 \\ 1 & 2 & -1 \end{vmatrix} = -\boldsymbol{i}+4\boldsymbol{j}+7\boldsymbol{k}$$

Our next goal is to provide a geometric interpretation of the cross product. To do this, we first introduce the triple product. Given three vectors $\boldsymbol{a}$, $\boldsymbol{b}$, and $\boldsymbol{c}$, the real number

$$\boldsymbol{a}\cdot(\boldsymbol{b}\times\boldsymbol{c})$$

is called the **triple product** of $\boldsymbol{a}$, $\boldsymbol{b}$, and $\boldsymbol{c}$ (in that order). Let us obtain a formula for the triple product $\boldsymbol{a}\cdot(\boldsymbol{b}\times\boldsymbol{c})$. If $\boldsymbol{a}=a_1\boldsymbol{i}+a_2\boldsymbol{j}+a_3\boldsymbol{k}$, $\boldsymbol{b}=b_1\boldsymbol{i}+b_2\boldsymbol{j}+b_3\boldsymbol{k}$, and $\boldsymbol{c}=c_1\boldsymbol{i}+c_2\boldsymbol{j}+c_3\boldsymbol{k}$, then

$$\begin{aligned}\boldsymbol{a}\cdot(\boldsymbol{b}\times\boldsymbol{c}) &= (a_1\boldsymbol{i}+a_2\boldsymbol{j}+a_3\boldsymbol{k})\cdot\left(\begin{vmatrix} b_2 & b_3 \\ c_2 & c_3\end{vmatrix}\boldsymbol{i}-\begin{vmatrix} b_1 & b_3 \\ c_1 & c_3\end{vmatrix}\boldsymbol{j}+\begin{vmatrix} b_1 & b_2 \\ c_1 & c_2\end{vmatrix}\boldsymbol{k}\right)\\ &= a_1\begin{vmatrix} b_2 & b_3 \\ c_2 & c_3\end{vmatrix}-a_2\begin{vmatrix} b_1 & b_3 \\ c_1 & c_3\end{vmatrix}+a_3\begin{vmatrix} b_1 & b_2 \\ c_1 & c_2\end{vmatrix}\end{aligned}$$

This may be written more concisely as

$$\boldsymbol{a}\cdot(\boldsymbol{b}\times\boldsymbol{c}) = \begin{vmatrix} a_1 & a_2 & a_3 \\ b_1 & b_2 & b_3 \\ c_1 & c_2 & c_3\end{vmatrix}$$

Now suppose that $\boldsymbol{a}$ is a vector in the plane spanned by the vectors $\boldsymbol{b}$ and $\boldsymbol{c}$. This means that the first row in the determinant expression for $\boldsymbol{a}\cdot(\boldsymbol{b}\times\boldsymbol{c})$ is of the form $\boldsymbol{a}=\alpha b+\beta c$, and therefore $\boldsymbol{a}\cdot(\boldsymbol{b}\times\boldsymbol{c})=0$ by Example 3. In other words, the vector $\boldsymbol{b}\times\boldsymbol{c}$ is orthogonal to any vector in the plane spanned by $\boldsymbol{b}$ and $\boldsymbol{c}$, in particular to both $\boldsymbol{b}$ and $\boldsymbol{c}$.

Next, we calculate the magnitude of $\boldsymbol{b}\times\boldsymbol{c}$. Note that

$$\begin{aligned}|\boldsymbol{b}\times\boldsymbol{c}|^2 &= \begin{vmatrix} b_2 & b_3 \\ c_2 & c_3\end{vmatrix}^2+\begin{vmatrix} b_1 & b_3 \\ c_1 & c_3\end{vmatrix}^2+\begin{vmatrix} b_1 & b_2 \\ c_1 & c_2\end{vmatrix}^2\\ &= (b_2c_3-b_3c_2)^2+(b_1c_3-c_1b_3)^2+(b_1c_2-c_1b_2)^2\end{aligned}$$

Writing out this last expression, we see that it is equal to

$$\begin{aligned}&(b_1^2+b_2^2+b_3^2)(c_1^2+c_2^2+c_3^2)-(b_1c_1+b_2c_2+b_3c_3)^2\\ &= |\boldsymbol{b}|^2|\boldsymbol{c}|^2-(\boldsymbol{b}\cdot\boldsymbol{c})^2 = |\boldsymbol{b}|^2|\boldsymbol{c}|^2-|\boldsymbol{b}|^2|\boldsymbol{c}|^2\cos^2\theta\\ &= |\boldsymbol{b}|^2|\boldsymbol{c}|^2\sin^2\theta\end{aligned}$$

where θ is the angle between $\boldsymbol{b}$ and $\boldsymbol{c}$, $0\leqslant\theta\leqslant\pi$.

Combining our results, we conclude that $\boldsymbol{b}\times\boldsymbol{c}$ is a vector perpendicular to the plane spanned by $\boldsymbol{b}$ and $\boldsymbol{c}$ with length $|\boldsymbol{b}|\,|\boldsymbol{c}|\,|\sin\theta|$. However, there are two possible vectors which satisfy these conditions because there are two choices of directions perpendicular (or normal) to the plane P spanned by $\boldsymbol{b}$ and $\boldsymbol{c}$. This is clear from Figure 3-1 which shows the two choices $\boldsymbol{n}_1$ and $-\boldsymbol{n}_1$ perpendicular to P with $|\boldsymbol{n}_1|=|-\boldsymbol{n}_1|=|\boldsymbol{b}|\,|\boldsymbol{c}|\,|\sin\theta|$.

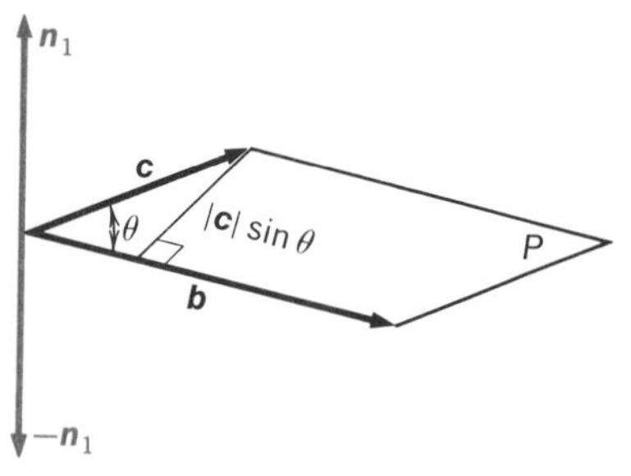

FIGURE 3-1

Which vector represents $\boldsymbol{b}\times\boldsymbol{c}$, $\boldsymbol{n}_1$ or $-\boldsymbol{n}_1$? The answer is $\boldsymbol{n}_1=\boldsymbol{b}\times\boldsymbol{c}$. This is not difficult to prove (see review exercise 17). The following "right-hand rule" determines the directions of $\boldsymbol{b}\times\boldsymbol{c}$.

Take the palm of your right hand and place it in such a way that your fingers curl in the direction of $\boldsymbol{c}$ through the angle θ. Then your thumb points in the direction of $\boldsymbol{b}\times\boldsymbol{c}$ (Figure 3-2).

If $\boldsymbol{b}$ and $\boldsymbol{c}$ are collinear, $\theta=0$, and so $\boldsymbol{b}\times\boldsymbol{c}=\boldsymbol{0}$; if $\boldsymbol{b}$ and $\boldsymbol{c}$ are not collinear, they span a plane and $\boldsymbol{b}\times\boldsymbol{c}$ is a vector perpendicular to this plane. The length of $\boldsymbol{b}\times\boldsymbol{c}$, $|\boldsymbol{b}|\ |\boldsymbol{c}|\ |\sin\theta|$, is just the area of the parallelogram with adjacent sides the vectors $\boldsymbol{b}$ and $\boldsymbol{c}$ (Figure 3-3).

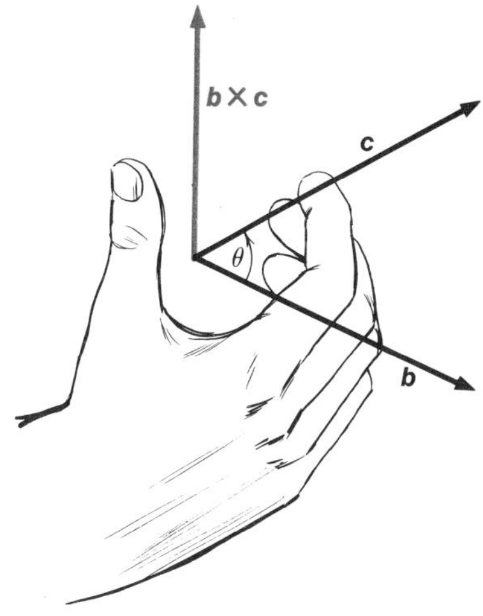

FIGURE 3-2

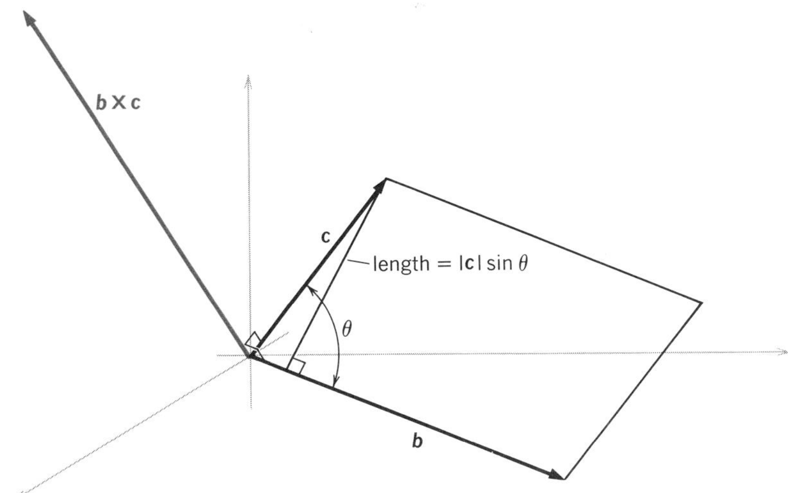

FIGURE 3-3

Example 4 Find a unit vector orthogonal to the vectors $\boldsymbol{i}+\boldsymbol{j}$ and $\boldsymbol{j}+\boldsymbol{k}$. A vector perpendicular to both $\boldsymbol{i}+\boldsymbol{j}$ and $\boldsymbol{j}+\boldsymbol{k}$ is the vector

$$(\boldsymbol{i}+\boldsymbol{j})\times(\boldsymbol{j}+\boldsymbol{k})=\begin{vmatrix}\boldsymbol{i} & \boldsymbol{j} & \boldsymbol{k}\\ 1 & 1 & 0\\ 0 & 1 & 1\end{vmatrix}=\boldsymbol{i}-\boldsymbol{j}+\boldsymbol{k}$$

Since $|\boldsymbol{i}-\boldsymbol{j}+\boldsymbol{k}|=\sqrt{3}$, the vector

$$\frac{1}{\sqrt{3}}(\boldsymbol{i}-\boldsymbol{j}+\boldsymbol{k})$$

is a unit vector perpendicular to $\boldsymbol{i}+\boldsymbol{j}$ and $\boldsymbol{j}+\boldsymbol{k}$.

Using the cross product, we may obtain the basic geometric interpretation of determinants. Let $\boldsymbol{b}=b_1\boldsymbol{i}+b_2\boldsymbol{j}$ and $\boldsymbol{c}=c_1\boldsymbol{i}+c_2\boldsymbol{j}$ be two vectors in the plane. If θ denotes the angle between $\boldsymbol{b}$ and $\boldsymbol{c}$, we have seen that $|\boldsymbol{b}\times\boldsymbol{c}|=|\boldsymbol{b}|\ |\boldsymbol{c}|\ |\sin\theta|$. As noted above, $|\boldsymbol{b}|\ |\boldsymbol{c}|\ |\sin\theta|$ is the area of the parallelogram with adjacent sides $\boldsymbol{a}$ and $\boldsymbol{b}$ (see Figure 3-1). Using the definition of the cross product,

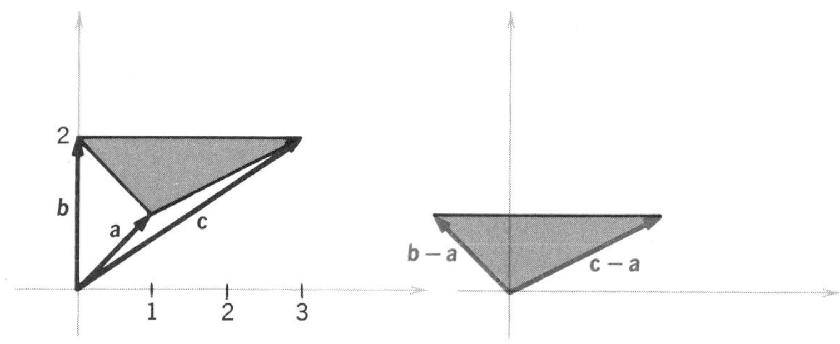

FIGURE 3-4

$$b \times c = \begin{vmatrix} i & j & k \\ b_1 & b_2 & 0 \\ c_1 & c_2 & 0 \end{vmatrix} = \begin{vmatrix} b_1 & b_2 \\ c_1 & c_2 \end{vmatrix} k$$

Thus $|b \times c|$ is the absolute value of the determinant

$$\begin{vmatrix} b_1 & b_2 \\ c_1 & c_2 \end{vmatrix} = b_1 c_2 - b_2 c_1$$

From this it follows that the absolute value of the above determinant is the area of the parallelogram with adjacent sides the vectors $b = b_1 i + b_2 j$ and $c = c_1 i + c_2 j$.

Example 5 Find the area of the triangle with vertices at the points (1,1), (0,2), and (3,2) (Figure 3-4).

Let $a = i + j$, $b = 2j$, and $c = 3i + 2j$. It is clear that the triangle whose vertices are the endpoints of the vectors a, b, and c has the same area as the triangle with vertices at 0, $b - a$, and $c - a$ (Figure 3-4). Indeed, the latter is merely a translation of the former triangle. Since the area of this translated triangle is one-half the area of the parallelogram with adjacent sides $b - a$ and $c - a$, we find that the area of the triangle with vertices (1,1), (0,2), and (3,2) is the absolute value of

$$\frac{1}{2}\begin{vmatrix} -1 & 1 \\ 2 & 1 \end{vmatrix} = -\frac{3}{2}$$

that is, 3/2.

There is an interpretation of determinants of 3×3 matrices as volumes which is analogous to the interpretation of determinants of 2×2 matrices as areas. Let $a = a_1 i + a_2 j + a_3 k$, $b = b_1 i + b_2 j + b_3 k$, and $c = c_1 i + c_2 j + c_3 k$ be vectors in R^3. We will show that the volume of the parallelepiped with adjacent sides a, b, and c (Figure 3-5) is the absolute value of the determinant

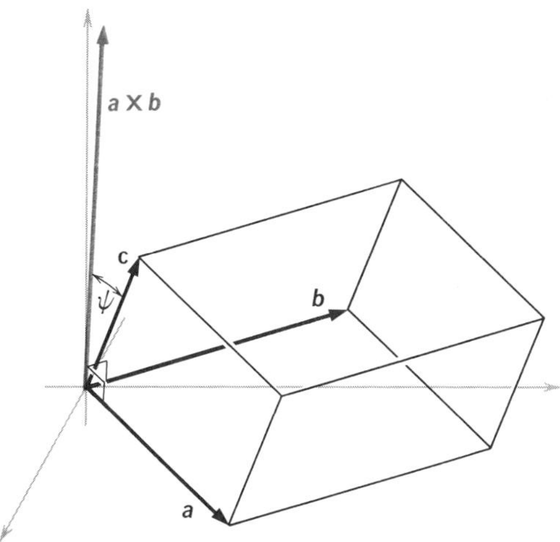

FIGURE 3-5

$$D = \begin{vmatrix} a_1 & a_2 & a_3 \\ b_1 & b_2 & b_3 \\ c_1 & c_2 & c_3 \end{vmatrix}$$

We know that $|\boldsymbol{a} \times \boldsymbol{b}|$ is the area of the parallelogram with adjacent sides $\boldsymbol{a}$ and $\boldsymbol{b}$. Moreover, $(\boldsymbol{a} \times \boldsymbol{b}) \cdot \boldsymbol{c} = |\boldsymbol{c}|\,|\boldsymbol{a} \times \boldsymbol{b}| \cos \psi$, where ψ is the angle which $\boldsymbol{c}$ makes with the normal to the plane spanned by $\boldsymbol{a}$ and $\boldsymbol{b}$. Since the volume of the parallelepiped with adjacent sides $\boldsymbol{a}$, $\boldsymbol{b}$, and $\boldsymbol{c}$ is the product of the area of the base $|\boldsymbol{a} \times \boldsymbol{b}|$ times the altitude $|\boldsymbol{c}| \cos \psi$, it follows that the volume is merely $|(\boldsymbol{a} \times \boldsymbol{b}) \cdot \boldsymbol{c}|$.

We saw earlier that $|(\boldsymbol{a} \times \boldsymbol{b}) \cdot \boldsymbol{c}|$ is the absolute value of the determinant D above; therefore, the absolute value of D is the volume of the parallelepiped with adjacent sides $\boldsymbol{a}$, $\boldsymbol{b}$, and $\boldsymbol{c}$.

To conclude this section, we shall use vector methods to determine the equation of a plane in space. Let P be a plane in space, $\boldsymbol{a}$ a vector ending on the plane, and $\boldsymbol{n}$ a vector normal to the plane (see Figure 3-6).

Now if $\boldsymbol{r}$ is a vector in $\boldsymbol{R}^3$, $\boldsymbol{r}$ ends on the plane P if and only if $\boldsymbol{r} - \boldsymbol{a}$ is parallel to P and, hence, if and only if $(\boldsymbol{r} - \boldsymbol{a}) \cdot \boldsymbol{n} = 0$ ($\boldsymbol{n}$ is perpendicular to any vector parallel to P—see Figure 3-6). Since the inner product is distributive, this last condition is equivalent to $\boldsymbol{r} \cdot \boldsymbol{n} = \boldsymbol{a} \cdot \boldsymbol{n}$. Therefore, if we let $\boldsymbol{a} = a_1 \boldsymbol{i} + a_2 \boldsymbol{j} + a_3 \boldsymbol{k}$, $\boldsymbol{n} = A\boldsymbol{i} + B\boldsymbol{j} + C\boldsymbol{k}$, and $\boldsymbol{r} = x\boldsymbol{i} + y\boldsymbol{j} + z\boldsymbol{k}$, it follows that $\boldsymbol{r}$ lies on P if and only if

$$Ax + By + Cz = \boldsymbol{r} \cdot \boldsymbol{n} = \boldsymbol{a} \cdot \boldsymbol{n} = Aa_1 + Ba_2 + Ca_3 \tag{3}$$

Since $\boldsymbol{n}$ is some fixed normal and $\boldsymbol{a}$ is also fixed, the right-hand side of equation (3) is a constant, say $-D$. Thus the equation which determines the plane P is

$$\boldsymbol{Ax + By + Cz + D = 0} \tag{4}$$

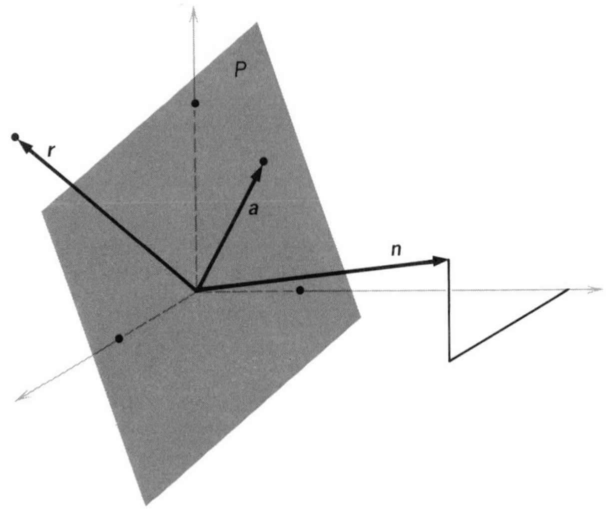

FIGURE 3-6

where $A\boldsymbol{i}+B\boldsymbol{j}+C\boldsymbol{k}$ is normal to P; conversely, if A, B, and C are not all zero the set of points (x, y, z) satisfying equation (4) is a plane with normal $A\boldsymbol{i}+B\boldsymbol{j}+C\boldsymbol{k}$.

The four numbers A, B, C, D are not determined uniquely by P. To see this, note that (x, y, z) satisfies equation (4) if and only if it also satisfies the relation

$$(\lambda A)x + (\lambda B)y + (\lambda C)z + (\lambda D) = 0$$

for $\lambda \neq 0$. Thus the quantities A, B, C, D are determined by a plane P up to a scalar multiple and, conversely, given A, B, C, D and A', B', C', D', they determine the same plane if $A=\lambda A'$, $B=\lambda B'$, $C=\lambda C'$, $D=\lambda D'$ for some scalar λ. This fact will become more apparent in Example 7 below.

Example 6 Determine the equation of the plane perpendicular to the vector $\boldsymbol{i}+\boldsymbol{j}+\boldsymbol{k}$ and containing the point (1,0,0).

From the above discussion it follows that the equation of the plane is of the form $x+y+z+D=0$. Since (1,0,0) is on the plane, $1+0+0+D=0$, or $D=-1$. Thus, the equation of the plane is $x+y+z=1$.

Example 7 Find the equation of the plane containing the points (1,1,1), (2,0,0), and (1,1,0).

Method 1. The equation of the plane is of the form $Ax+By+Cz+D=0$. Since the points (1,1,1), (2,0,0), and (1,1,0) lie on the plane, we have that

$$\begin{aligned} A+B+C+D &= 0 \\ 2A \qquad\quad +D &= 0 \\ A+B \qquad +D &= 0 \end{aligned}$$

Proceeding by elimination, we reduce the above system to the form

$$\begin{aligned} 2A + \quad D &= 0 \\ 2B + D &= 0 \\ C \quad &= 0 \end{aligned}$$

Since the numbers A, B, C, and D are determined only up to a scalar multiple, we can fix the value of one of them and then the others will be determined uniquely. If we let $D = -2$, then $A = +1$, $B = +1$, $C = 0$. Thus $x+y-2=0$ is the equation of the plane which contains the given points.

Method 2. Let $\boldsymbol{a} = \boldsymbol{i}+\boldsymbol{j}+\boldsymbol{k}$, $\boldsymbol{b} = 2\boldsymbol{i}$, $\boldsymbol{c} = \boldsymbol{i}+\boldsymbol{j}$. Any vector normal to the plane must be orthogonal to the vectors $\boldsymbol{a}-\boldsymbol{b}$ and $\boldsymbol{c}-\boldsymbol{b}$, which are parallel to the plane. Thus, $\boldsymbol{n} = (\boldsymbol{a}-\boldsymbol{b}) \times (\boldsymbol{c}-\boldsymbol{b})$ is normal to the plane and we have

$$\boldsymbol{n} = \begin{vmatrix} \boldsymbol{i} & \boldsymbol{j} & \boldsymbol{k} \\ -1 & 1 & 1 \\ -1 & 1 & 0 \end{vmatrix} = -\boldsymbol{i} - \boldsymbol{j}$$

Thus, the equation of the plane is of the form $-x-y+D=0$. Since (2,0,0) lies on the plane, $D = +2$. After substituting, we obtain $x+y-2=0$.

Example 8 Let $Ax+By+Cz+D=0$ be the equation of a plane in $\boldsymbol{R}^3$. Determine the distance from the point (x_1, y_1, z_1) to the plane (see Figure 3-7).

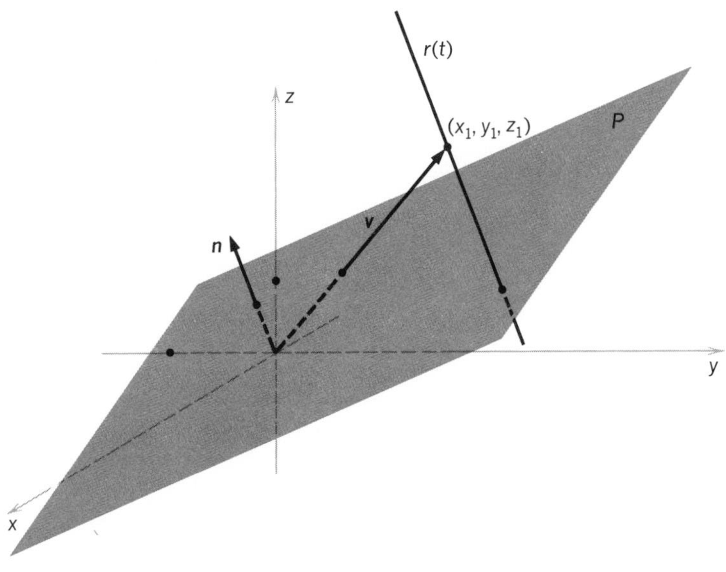

FIGURE 3-7

The vector

$$n = \frac{Ai + Bj + Ck}{\sqrt{A^2 + B^2 + C^2}}$$

is a unit vector normal to the plane. If

$$a = \frac{-Dn}{\sqrt{A^2 + B^2 + C^2}} = \frac{-DAi - DBj - DCk}{A^2 + B^2 + C^2}$$

then we assert that a ends on the given plane. To see this note that in coordinates a is represented by

$$\left(\frac{-DA}{A^2 + B^2 + C^2}, \frac{-DB}{A^2 + B^2 + C^2}, \frac{-DC}{A^2 + B^2 + C^2}\right) = (x, y, z)$$

Now (x, y, z) lies on P if and only if $Ax + By + Cz + D = 0$. But

$$A\left(\frac{-DA}{A^2 + B^2 + C^2}\right) + B\left(\frac{-DB}{A^2 + B^2 + C^2}\right) + C\left(\frac{-DB}{A^2 + B^2 + C^2}\right) + D$$

$$= -D\frac{(A^2 + B^2 + C^2)}{A^2 + B^2 + C^2} + D = 0$$

and so a does indeed end on P.

Now the distance from (x_1, y_1, z_1) to P will be the length of the line segment EF (Figure 3-7) through (x_1, y_1, z_1) perpendicular to P and ending on P. So if we can find where the line perpendicular to P passing through (x_1, y_1, z_1) intersects P, we can then compute the length of the line segment EF and this will complete the problem.

If the endpoint of the vector $r = xi + yj + zk$ lies on the plane, we know from before that $(r - a) \cdot n = 0$. Let $v = x_1 i + y_1 j + z_1 k$ be the vector ending at (x_1, y_1, z_1). Then, the line through (x_1, y_1, z_1) perpendicular to the plane is, in parametric form, $r(t) = v + tn$. This line intersects the plane for some $t = t_0$ when

$$(r(t_0) - a) \cdot n = 0$$

or

$$(v + t_0 n - a) \cdot n = 0 = v \cdot n + t_0(n \cdot n) - a \cdot n = t_0 + v \cdot n - a \cdot n$$

$$= t_0 - (a - v) \cdot n$$

so

$$t_0 = (a - v) \cdot n$$

The distance from the point (x_1, y_1, z_1) to the point at which $r(t)$ intersects the plane, i.e., the distance from (x_1, y_1, z_1) to the plane is

$$|v - r(t_0)| = |v - (v + t_0 n)| = |(a - v) \cdot n|$$

so

$$\textbf{distance} = \frac{|Ax_1 + By_1 + Cz_1 + D|}{\sqrt{A^2 + B^2 + C^2}}$$

EXERCISES

1. Verify that interchanging two rows or two columns of a 3×3 determinant changes the sign of the determinant.
2. Evaluate

(a) $\begin{vmatrix} 2 & -1 & 0 \\ 4 & 3 & 2 \\ 3 & 0 & 1 \end{vmatrix}$ (b) $\begin{vmatrix} 36 & 18 & 17 \\ 45 & 24 & 20 \\ 3 & 5 & -2 \end{vmatrix}$

(c) $\begin{vmatrix} 1 & 4 & 9 \\ 4 & 9 & 16 \\ 9 & 16 & 25 \end{vmatrix}$ (d) $\begin{vmatrix} 2 & 3 & 5 \\ 7 & 11 & 13 \\ 17 & 19 & 23 \end{vmatrix}$

3. Compute $\boldsymbol{a} \times \boldsymbol{b}$, where $\boldsymbol{a} = \boldsymbol{i} - 2\boldsymbol{j} + \boldsymbol{k}$, $\boldsymbol{b} = 2\boldsymbol{i} + \boldsymbol{j} + \boldsymbol{k}$.
4. Compute $\boldsymbol{a} \cdot (\boldsymbol{b} \times \boldsymbol{c})$, where $\boldsymbol{a}$ and $\boldsymbol{b}$ are as in exercise 3 and $\boldsymbol{c} = 3\boldsymbol{i} - \boldsymbol{j} + 2\boldsymbol{k}$.
5. Find the area of the parallelogram with sides the vectors $\boldsymbol{a}$ and $\boldsymbol{b}$, given in exercise 3.
6. What is the volume of the parallelepiped with sides $2\boldsymbol{i} + \boldsymbol{j} - \boldsymbol{k}$, $5\boldsymbol{i} - 3\boldsymbol{k}$, and $\boldsymbol{i} - 2\boldsymbol{j} - \boldsymbol{k}$?

In exercises 7–10, find unit vectors orthogonal to the given vectors. In each case, specify how many different answers are possible.

7. $\boldsymbol{i} + \boldsymbol{j}$
8. $-5\boldsymbol{i} + 9\boldsymbol{j} - 4\boldsymbol{k}$, $7\boldsymbol{i} + 8\boldsymbol{j} + 9\boldsymbol{k}$
9. $-5\boldsymbol{i} + 9\boldsymbol{j} - 4\boldsymbol{k}$, $7\boldsymbol{i} + 8\boldsymbol{j} + 9\boldsymbol{k}$, 0
10. $2\boldsymbol{i} - 4\boldsymbol{j} + 3\boldsymbol{k}$, $-4\boldsymbol{i} + 8\boldsymbol{j} - 6\boldsymbol{k}$

11. Determine the distance from the plane $12x + 13y + 5z + 2 = 0$ to the point $(1, 1, -5)$.
12. Find the distance from the line through the origin and $(1, 1, 1)$, to the line through $(1, 2, -2)$ parallel to $2\boldsymbol{i} - \boldsymbol{j} + 2\boldsymbol{k}$.
13. Compute $\boldsymbol{u} + \boldsymbol{v}$, $\boldsymbol{u} \cdot \boldsymbol{v}$, $|\boldsymbol{u}|$, $|\boldsymbol{v}|$, and $\boldsymbol{u} \times \boldsymbol{v}$ where $\boldsymbol{u} = \boldsymbol{i} - 2\boldsymbol{j} + \boldsymbol{k}$, $\boldsymbol{v} = 2\boldsymbol{i} - \boldsymbol{j} + 2\boldsymbol{k}$.
14. Repeat exercise 13 for $\boldsymbol{u} = 3\boldsymbol{i} + \boldsymbol{j} - \boldsymbol{k}$, $\boldsymbol{v} = -6\boldsymbol{i} - 2\boldsymbol{j} - 2\boldsymbol{k}$.
15. Prove $(\boldsymbol{u} \times \boldsymbol{v}) \times \boldsymbol{w} = \boldsymbol{u} \times (\boldsymbol{v} \times \boldsymbol{w})$ if and only if $(\boldsymbol{u} \times \boldsymbol{w}) \times \boldsymbol{v} = 0$.
16. Prove $(\boldsymbol{u} \times \boldsymbol{v}) \times \boldsymbol{w} + (\boldsymbol{v} \times \boldsymbol{w}) \times \boldsymbol{u} + (\boldsymbol{w} \times \boldsymbol{u}) \times \boldsymbol{v} = 0$ (the **Jacobi identity**).
17. Prove

$$(\boldsymbol{u} \times \boldsymbol{v}) \cdot (\boldsymbol{u}' \times \boldsymbol{v}') = (\boldsymbol{u} \cdot \boldsymbol{u}')(\boldsymbol{v} \cdot \boldsymbol{v}') - (\boldsymbol{u} \cdot \boldsymbol{v}')(\boldsymbol{u}' \cdot \boldsymbol{v}) = \begin{vmatrix} \boldsymbol{u} \cdot \boldsymbol{u}' & \boldsymbol{u} \cdot \boldsymbol{v}' \\ \boldsymbol{u}' \cdot \boldsymbol{v} & \boldsymbol{u} \cdot \boldsymbol{v}' \end{vmatrix}$$

18. Prove, without recourse to geometry, that

$$\boldsymbol{u} \cdot (\boldsymbol{v} \times \boldsymbol{w}) = \boldsymbol{v} \cdot (\boldsymbol{w} \times \boldsymbol{u}) = \boldsymbol{w} \cdot (\boldsymbol{u} \times \boldsymbol{v}) = -\boldsymbol{u} \cdot (\boldsymbol{w} \times \boldsymbol{v})$$
$$= -\boldsymbol{w} \cdot (\boldsymbol{v} \times \boldsymbol{u}) = -\boldsymbol{v} \cdot (\boldsymbol{u} \times \boldsymbol{w})$$

4 PATHS

A function whose values are vectors is called a **vector-valued function**. The domain of a vector-valued function may be either a set of real numbers or a set of vectors. Here and in §13.5 we study only functions of the former type—that is, functions which send real numbers to vectors (called **vector-valued functions of a real variable**). For example, the rule $t \mapsto (\cos t)\boldsymbol{i} + (\sin t)\boldsymbol{j} + t\boldsymbol{k}$ defines a function whose domain is $\boldsymbol{R}$ and whose values are vectors in $\boldsymbol{R}^3$. We usually denote such functions by boldface Greek letters $\boldsymbol{\sigma}$, $\boldsymbol{\rho}$, $\boldsymbol{\tau}$, ...; the value of the function $\boldsymbol{\sigma}$ at the real number t is then denoted $\boldsymbol{\sigma}(t)$.

Vector-valued functions of a real variable are often used to describe the motion of point or particle moving in space. In this type of example, we interpret $\boldsymbol{\sigma}(t)$ to be the position vector of the moving point at time t.

Example 1 The function $\boldsymbol{\rho}: t \mapsto \boldsymbol{i} + t\boldsymbol{j}$ describes a point in the xy plane moving along a straight line parallel to the y axis. As t increases, the point moves in the direction of the vector $\boldsymbol{j}$.

Since vectors can be represented as ordered triples of real numbers, vector-valued functions can be represented in the form $\boldsymbol{\sigma}: t \mapsto (x(t), y(t), z(t))$. Here we can interpret $(x(t), y(t), z(t))$ as the position of a moving point at time t.

Example 2 The function $\boldsymbol{\sigma}: t \mapsto (t - \sin t, 1 - \cos t, 0)$ is the position function of a point on a rolling circle (see Example 9, §4.2). The circle lies in the xy plane and rolls along the x axis at constant speed; that is, the midpoint of the circle is moving to the right along the line $y = 1$, $z = 0$ at a constant speed of 1 radian per unit of time. The motion of the point $\boldsymbol{\sigma}(t)$ is more complicated; its locus is known as the **cycloid** (see Figure 4-1). We analyze the motion of $\boldsymbol{\sigma}(t)$ further on in Example 7.

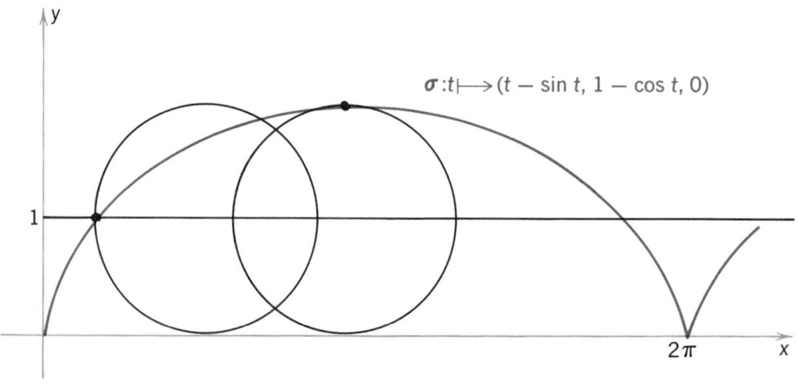

FIGURE 4-1

A vector-valued function $\boldsymbol{\sigma}:t \mapsto \boldsymbol{\sigma}(t)$ can be decomposed into a triple of real-valued functions, namely, the functions which give the coordinates of $\boldsymbol{\sigma}(t)$. It is convenient to denote these functions by σ_1, σ_2, and σ_3 and to write $\boldsymbol{\sigma} = (\sigma_1, \sigma_2, \sigma_3)$. For example, if $\boldsymbol{\sigma}:t \mapsto (\cos t, \sin t, t)$, then $\sigma_1 : t \mapsto \cos t$, $\sigma_2 : t \mapsto \sin t$, and $\sigma_3 : t \mapsto t$.

To indicate that the domain of $\boldsymbol{\sigma}$ is the interval I and that its rule is $t \mapsto \boldsymbol{\sigma}(t)$, we shall write $\boldsymbol{\sigma}: I \to \boldsymbol{R}^3 : t \mapsto \boldsymbol{\sigma}(t)$. A vector-valued function whose domain is a closed interval is called a **path**. If $\boldsymbol{\sigma}: I \to \boldsymbol{R}^3 : t \mapsto (\sigma_1(t), \sigma_2(t), \sigma_3(t))$ is a path, then the set of points in $\boldsymbol{R}^3$ of the form $(\sigma_1(t), \sigma_2(t), \sigma_3(t))$ for some $t \in I$ is called the **image** of $\boldsymbol{\sigma}$ or the **curve** traced out by $\boldsymbol{\sigma}$. (Thus the image of $\boldsymbol{\sigma}$ coincides with the range of $\boldsymbol{\sigma}$.)

Example 3 If $\boldsymbol{\sigma}:[0,2\pi] \to \boldsymbol{R}^3 : t \mapsto (\cos t, \sin t, 0)$, then the image of $\boldsymbol{\sigma}$ is the circle of radius 1 lying in the xy plane with center at the origin. If $\boldsymbol{\rho}$ is the function $\boldsymbol{\rho}:[0,4\pi] \to \boldsymbol{R}^3 : t \mapsto (\cos t, \sin t, 0)$, then the image of $\boldsymbol{\rho}$ coincides with the image of $\boldsymbol{\sigma}$, although $\boldsymbol{\rho}$ and $\boldsymbol{\sigma}$ are not the same function since they have different domains. The path $\boldsymbol{\sigma}$ describes a point $\boldsymbol{\sigma}(t)$ which moves once around the circle; the path $\boldsymbol{\rho}$ describes a point $\boldsymbol{\rho}(t)$ which moves twice around the circle.

In this example we have two functions $\boldsymbol{\sigma}$ and $\boldsymbol{\rho}$ with different domains but which trace out the same image or curve in $\boldsymbol{R}^3$. Thus a given curve in $\boldsymbol{R}^3$ may be the image of more than one path. It is also possible to have a function $\boldsymbol{\eta}$ with the same domain as $\boldsymbol{\sigma}$ such that a point $\boldsymbol{\eta}(t)$ moves around the circle twice. We simply define $\boldsymbol{\eta}:[0, 2\pi] \to \boldsymbol{R}^3 : t \mapsto (\cos 2t, \sin 2t, 0)$.

For vector-value functions, limits are defined in terms of the component functions. We say that the **limit** of $\boldsymbol{\sigma}(t)$ as t approaches t_0 is $\boldsymbol{u}$ if the limit as t approaches t_0 of $\sigma_1(t)$, $\sigma_2(t)$, and $\sigma_3(t)$ are u_1, u_2, and u_3, respectively. In symbols, we write

$$\lim_{t \to t_0} \boldsymbol{\sigma}(t) = \boldsymbol{u} \Leftrightarrow \lim_{t \to t_0} \sigma_1(t) = u_1, \ \lim_{t \to t_0} \sigma_2(t) = u_2 \text{ and } \lim_{t \to t_0} \sigma_3(t) = u_3$$

We say that $\boldsymbol{\sigma}$ is **continuous** (at t_0) if and only if σ_1, σ_2, and σ_3 are all continuous (at t_0). A vector-valued function $\boldsymbol{\sigma}$ is said to be **differentiable** (at t_0) if and only if each of its component functions is differentiable (at t_0). The derivative of $\boldsymbol{\sigma}$ at t_0 is denoted $\boldsymbol{\sigma}'(t_0)$ and is defined by

$$\boldsymbol{\sigma}'(t_0) = \lim_{\Delta t \to 0} \frac{\boldsymbol{\sigma}(t_0 + \Delta t) - \boldsymbol{\sigma}(t_0)}{\Delta t}$$

For a point $\boldsymbol{\sigma}(t)$ moving along a given path in space, it is clear from the definition that

$$\boldsymbol{\sigma}'(t_0) = (\sigma_1'(t_0), \sigma_2'(t_0), \sigma_3'(t_0)) = \sigma_1'(t_0)\boldsymbol{i} + \sigma_2'(t_0)\boldsymbol{j} + \sigma_3'(t_0)\boldsymbol{k}$$

Example 4 The vector-valued functions of Examples 1–3 are all continuous and differentiable. The derivative of $\boldsymbol{\sigma}:t \mapsto (\cos t, \sin t, t)$ is the function $\boldsymbol{\sigma}':t \mapsto (-\sin t, \cos t, 1)$; the derivative of $\boldsymbol{\sigma}$ at $t = 0$ is therefore the vector $(0,1,1)$.

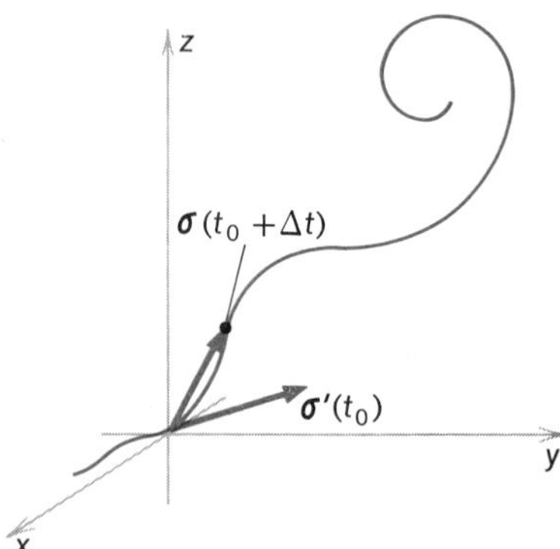

FIGURE 4-2

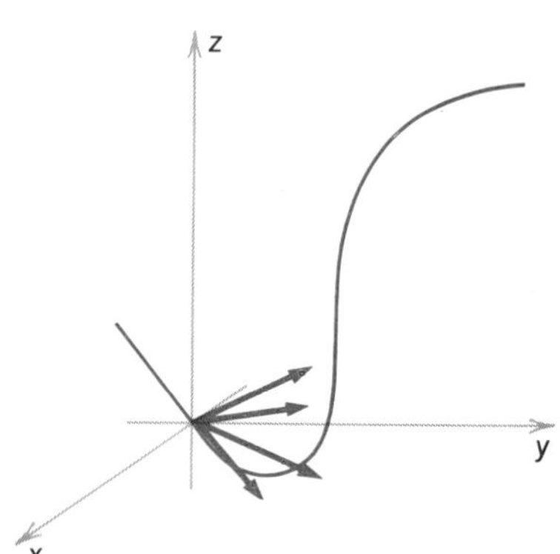

FIGURE 4-3

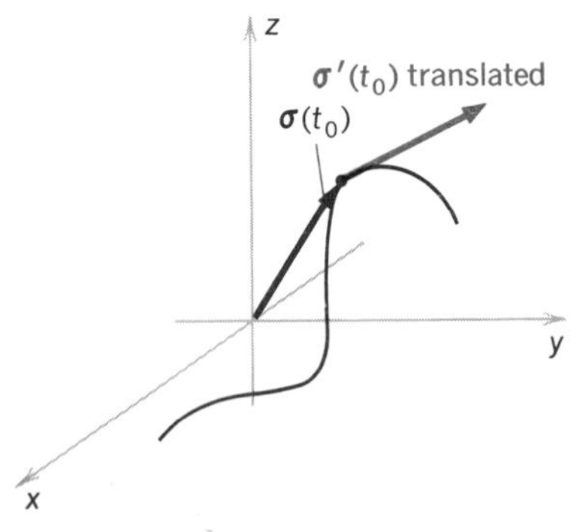

FIGURE 4-4

For real-valued functions $t \mapsto f(t)$, the derivative $f'(t_0)$ gives the slope of the tangent line to the graph of f at $(t_0, f(t_0))$. For vector-valued functions $\boldsymbol{\sigma}: t \mapsto \boldsymbol{\sigma}(t)$, the derivative $\boldsymbol{\sigma}'(t_0)$ is a vector which is parallel to the tangent line to the image of $\boldsymbol{\sigma}$ at $\boldsymbol{\sigma}(t_0)$. This follows from the fact that the vector $\boldsymbol{\sigma}'(t_0)$ can be obtained as a limit of secant vectors. Let us work this out in some detail. To simplify notation, we shall suppose that $\boldsymbol{\sigma}(t_0) = 0$. As we see in Figure 4-2, the vector $\boldsymbol{\sigma}(t_0 + \Delta t) - \boldsymbol{\sigma}(t_0)$ lies on the secant line to the curve. Since the vector $[\boldsymbol{\sigma}(t_0 + \Delta t) - \boldsymbol{\sigma}(t_0)]/\Delta t$ is just a scalar multiple of $\boldsymbol{\sigma}(t_0 + \Delta t) - \boldsymbol{\sigma}(t_0)$, it too lies on the secant line to the curve. In fact, when $0 < \Delta t < 1$ the vector $[\boldsymbol{\sigma}(t_0 + \Delta t) - \boldsymbol{\sigma}(t_0)]/\Delta t$ is an extension of the secant vector and cuts across the curve (Figure 4-3). As $\Delta t \to 0$, the vector $\boldsymbol{\sigma}(t_0 + \Delta t) - \boldsymbol{\sigma}(t_0)$ tends to $\mathbf{0}$ since $t \mapsto \boldsymbol{\sigma}(t)$ is continuous. However, if $\boldsymbol{\sigma}'(t_0) \neq \mathbf{0}$, the vector $[\boldsymbol{\sigma}(t_0 + \Delta t) - \boldsymbol{\sigma}(t_0)]/\Delta t$ will not vanish as $\Delta t \to 0$ and at the limit will swing over onto the tangent line to the curve (Figure 4-3). Therefore, the limit vector

$$\lim_{\Delta t \to 0} \frac{\boldsymbol{\sigma}(t_0 + \Delta t) - \boldsymbol{\sigma}(t_0)}{\Delta t} = \boldsymbol{\sigma}'(t_0)$$

will be tangent to the curve at $\boldsymbol{\sigma}(t_0)$. If $\boldsymbol{\sigma}'(t_0) = \mathbf{0}$, there may be no tangent line to the curve. This situation is discussed briefly at the end of Example 7. When $\boldsymbol{\sigma}(t_0) \neq 0$, the arrow obtained by translating $\boldsymbol{\sigma}'(t_0)$ to the point $\boldsymbol{\sigma}(t_0)$ will be tangent to the curve (Figure 4-4).

However, the vector $\boldsymbol{\sigma}'(t_0)$ is indicative of more than just the tangent line to the graph of $\boldsymbol{\sigma}$ at $\boldsymbol{\sigma}(t_0)$. We may also obtain the following information:

(1) The vector $\boldsymbol{\sigma}'(t_0)$ points in the direction of motion of the point $\boldsymbol{\sigma}(t)$; that is, $\boldsymbol{\sigma}'(t_0)$ not only determines the tangent but also indicates a direction or orientation for the tangent line. Intuitively, this means that we can tell which direction the point $\boldsymbol{\sigma}(t)$ would take along the tangent line at $\boldsymbol{\sigma}(t_0)$, if all restraining forces were suddenly removed and the point with position $\boldsymbol{\sigma}(t)$ were left to "fly off on a tangent."

(2) Since $\boldsymbol{\sigma}'(t_0)$ is a vector, it has magnitude as well as direction. In the discussion below, we show that the length of $\boldsymbol{\sigma}'(t_0)$ is, in fact, the **instantaneous speed** of the moving point $\boldsymbol{\sigma}(t)$ at $t = t_0$.

Because of properties (1) and (2), the vector $\boldsymbol{\sigma}'(t)$ can be thought of as the **velocity vector** and it is often denoted $\boldsymbol{v}(t)$.

Example 5 If $\boldsymbol{\sigma}: t \mapsto (\cos t, \sin t, t)$ then the velocity vector is $\boldsymbol{v}(t) = \boldsymbol{\sigma}'(t) = (-\sin t, \cos t, 1)$. The speed of a point is the magnitude of the velocity:

$$\text{speed} = |\boldsymbol{v}(t)| = (\sin^2 t + \cos^2 t + 1)^{1/2} = \sqrt{2}$$

Thus the point moves at constant speed although its velocity is not constant since it continually changes direction. The trajectory of the point whose motion is given by $\boldsymbol{\sigma}$ is called a (right-circular) **helix** (see Figure 4-5).

In order to justify (2) above, let us derive a formula for the distance traveled by the point $\boldsymbol{\sigma}(t)$ for $t \in I$, where I is a closed interval. This distance is called the **arc length** of the path $\boldsymbol{\sigma}: I \to \boldsymbol{R}^3 : t \mapsto \boldsymbol{\sigma}(t)$. For the arc length to be defined, we shall require that the component functions σ_1, σ_2, and σ_3 be continuously differentiable on I. When $\boldsymbol{\sigma}$ is a path with continuously differentiable component functions, we say that $\boldsymbol{\sigma}$ is $\boldsymbol{C}^1$.

To determine the arc length of the C^1 path, $\boldsymbol{\sigma}: I \to \boldsymbol{R}^3$, we shall use the technique of polygonal approximation, as we did for plane curves in §7.3. We first partition the interval I by means of a regular partition of order N:

$$a = t_0 < t_1 < \cdots < t_N = b$$

where a and b are the endpoints of I. We then consider the polygonal line obtained by joining the successive pairs of points $\boldsymbol{\sigma}(t_i)$, $\boldsymbol{\sigma}(t_{i+1})$ for $0 \leqslant i \leqslant N-1$. This yields a polygonal approximation to $\boldsymbol{\sigma}$ as in Figure 4-6. By the distance formula, it follows that the line segment from $\boldsymbol{\sigma}(t_i)$ to $\boldsymbol{\sigma}(t_{i+1})$ has length

$$|\boldsymbol{\sigma}(t_{i+1}) - \boldsymbol{\sigma}(t_i)|$$
$$= \sqrt{[\sigma_1(t_{i+1}) - \sigma_1(t_i)]^2 + [\sigma_2(t_{i+1}) - \sigma_2(t_i)]^2 + [\sigma_3(t_{i+1}) - \sigma_3(t_i)]^2}$$

Applying the Mean Value Theorem to σ_1, σ_2, and σ_3 on $[t_i, t_{i+1}]$ we obtain three points t_i^*, t_i^{**}, and t_i^{***} such that

$$\sigma_1(t_{i+1}) - \sigma_1(t_i) = \sigma_1'(t_i^*)(t_{i+1} - t_i)$$
$$\sigma_2(t_{i+1}) - \sigma_2(t_i) = \sigma_2'(t_i^{**})(t_{i+1} - t_i)$$
$$\sigma_3(t_{i+1}) - \sigma_3(t_i) = \sigma_3'(t_i^{***})(t_{i+1} - t_i)$$

Thus the line segment from $\boldsymbol{\sigma}(t_i)$ to $\boldsymbol{\sigma}(t_{i+1})$ has length

$$\sqrt{[\sigma_1'(t_i^*)]^2 + [\sigma_2'(t_i^{**})]^2 + [\sigma_3'(t_i^{***})]^2}\,(t_{i+1} - t_i)$$

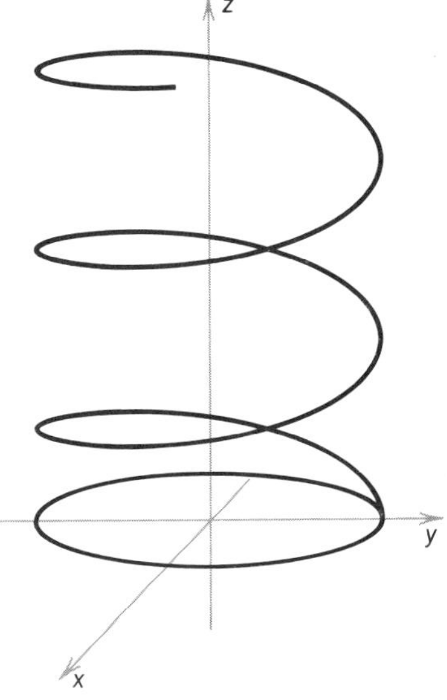

FIGURE 4-5

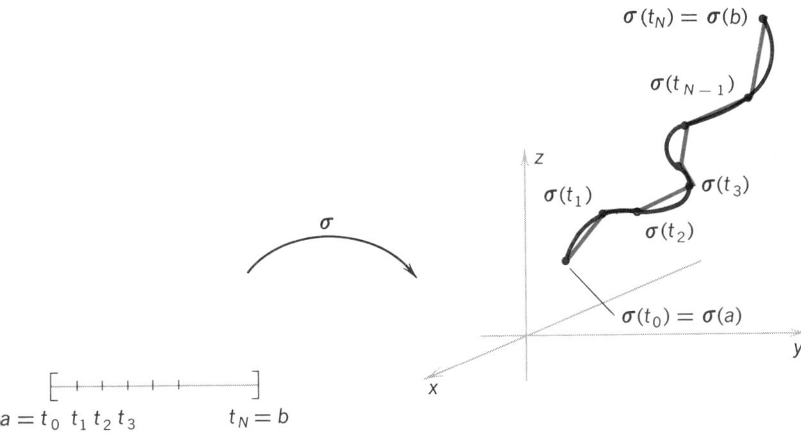

FIGURE 4-6

Therefore the length of our approximating polygonal line is

$$S_N = \sum_{i=0}^{N-1} \sqrt{[\sigma_1'(t_i^*)]^2 + [\sigma_2'(t_i^{**})]^2 + [\sigma_3'(t_i^{***})]^2}\,(t_{i+1}-t_i)$$

As $N \to \infty$, this polygonal line will approximate the image of $\boldsymbol{\sigma}$ more closely. Therefore, we take the limit of $\{S_N\}$ as $N \to \infty$, if it exists, as our definition of the arc length of $\boldsymbol{\sigma}$. Since the derivatives σ_1', σ_2', and σ_3' are all continuous on I, we can conclude (see §7.2 and §7.4) that, in fact, the limit does exist and

$$\lim_{N\to\infty} S_N = \int_a^b \sqrt{[\sigma_1'(t)]^2 + [\sigma_2'(t)]^2 + [\sigma_3'(t)]^2}\,dt$$

We define the **arc length** of $\boldsymbol{\sigma}$, denoted $l(\boldsymbol{\sigma})$, to be this real number. Since $\sqrt{[\sigma_1'(t)]^2+[\sigma_2'(t)]^2+[\sigma_3'(t)]^2} = |\boldsymbol{\sigma}'(t)|$, we have the formula

$$l(\boldsymbol{\sigma}) = \int_a^b |\boldsymbol{\sigma}'(t)|\,dt$$

When the endpoints of I are not specified we write this as

$$l(\boldsymbol{\sigma}) = \int_I |\boldsymbol{\sigma}'(t)|\,dt$$

Example 6 Find the arc length of the helix $\boldsymbol{\rho}:[0,4\pi]\to \boldsymbol{R}^3 : t \mapsto (\cos 2t, \sin 2t, \sqrt{5}\,t)$. The tangent vector is $\boldsymbol{\rho}'(t) = (-2\sin 2t, 2\cos 2t, \sqrt{5})$ which has magnitude

$$|\boldsymbol{\rho}'(t)| = [4(\sin 2t)^2 + 4(\cos 2t)^2 + 5]^{1/2} = (9)^{1/2} = 3$$

The arc length of $\boldsymbol{\rho}$ is therefore

$$l(\boldsymbol{\rho}) = \int_0^{4\pi} |\boldsymbol{\rho}'(t)|\,dt = \int_0^{4\pi} 3\,dt = 12\pi$$

It is important to realize that although two different paths $\boldsymbol{\sigma}$ and $\boldsymbol{\eta}$ may have the same image their arc lengths can be different; for example, $\boldsymbol{\eta}$ may trace out the curve twice, while $\boldsymbol{\sigma}$ may trace it out only once. Consider the two paths

$$\boldsymbol{\eta}: t \mapsto (\cos 2t, \sin 2t, 0) \qquad 0 \leqslant t \leqslant 2\pi$$

$$\boldsymbol{\sigma}: t \mapsto (\cos t, \sin t, 0) \qquad 0 \leqslant t \leqslant 2\pi$$

discussed earlier. Then image $(\boldsymbol{\sigma})$ = image $(\boldsymbol{\eta})$ = a unit circle centered at the origin and lying in $\boldsymbol{R}^2 \subset \boldsymbol{R}^3$. Now

$$l(\boldsymbol{\sigma}) = \int_0^{2\pi} |\boldsymbol{\sigma}'(t)|\,dt$$

$$= \int_0^{2\pi} \sqrt{\sin^2 t + \cos^2 t}\,dt = \int_0^{2\pi} dt = 2\pi$$

Similarly,

$$l(\boldsymbol{\eta}) = \int_0^{2\pi} |\boldsymbol{\eta}'(t)|\, dt$$

$$= \int_0^{2\pi} \sqrt{4\sin^2 2t + 4\cos^2 2t}\, dt$$

$$= 2\int_0^{2\pi} dt = 4\pi$$

which is twice the result for $\boldsymbol{\sigma}$. Thus $l(\boldsymbol{\eta}) = 2l(\boldsymbol{\sigma})$.

To see that the length $|\boldsymbol{\sigma}'(t)|$ of the tangent vector gives the speed of $\boldsymbol{\sigma}(t)$, we employ the formula

$$\text{average speed} = \frac{\text{change in distance}}{\text{change in time}}$$

and obtain the instantaneous speed by using a limiting process on the interval $[t_0, t_0+\Delta t]$. On $[t_0, t_0+\Delta t]$ the path $\boldsymbol{\sigma}(t)$ has length

$$\Delta s = \int_{t_0}^{t_0+\Delta t} |\boldsymbol{\sigma}'(t)|\, dt$$

The average speed of the point $\boldsymbol{\sigma}(t)$ is then $\Delta s/\Delta t$. We now take the limit of this ratio as Δt goes to zero. To evaluate this limit, we apply the Mean Value Theorem of Integral Calculus; this yields

$$\lim_{\Delta t\to 0} \frac{\Delta s}{\Delta t} = \lim_{\Delta t\to 0} \frac{1}{\Delta t}\int_{t_0}^{t_0+\Delta t} |\boldsymbol{\sigma}'(t)|\, dt = |\boldsymbol{\sigma}'(t_0)|$$

Hence, by using the formula for arc length we have justified our assertion in (2) above that $|\boldsymbol{\sigma}'(t)|$ is the speed of the moving point $\boldsymbol{\sigma}(t)$.

There is another notation for arc length which is sometimes used. If $\boldsymbol{\sigma}: I = [a, b] \to \boldsymbol{R}$ is a path, we define

$$s(t) = \int_a^t |\boldsymbol{\sigma}'(u)|\, du$$

Thus $s(b) = l(\boldsymbol{\sigma})$ and by the Fundamental Theorem

$$\frac{ds}{dt} = s'(t) = |\boldsymbol{\sigma}'(t)|$$

Using this notation we can write

$$l(\boldsymbol{\sigma}) = \int_I \frac{ds}{dt}\, dt = \int_I ds$$

where $ds = |\boldsymbol{\sigma}'(t)|\, dt$.

Example 7 Consider the point with position function $\boldsymbol{\sigma}: t \mapsto (t - \sin t, 1 - \cos t, 0)$ discussed in Example 2. The velocity is $\boldsymbol{\sigma}'(t) = (1 - \cos t, \sin t, 0)$ and the speed of the point $\boldsymbol{\sigma}(t)$ is

$$|\boldsymbol{\sigma}'(t)| = \sqrt{(1-\cos t)^2 + \sin^2 t} = \sqrt{2 - 2\cos t}$$

Hence, $\boldsymbol{\sigma}(t)$ moves at variable speed although, as we discovered earlier, the circle rolls at constant speed. Furthermore, the speed of $\boldsymbol{\sigma}(t)$ is zero when t is an integral multiple of 2π. At these values of t both the y and z coordinates of the point $\boldsymbol{\sigma}(t)$ are zero and so the point lies on the x axis.

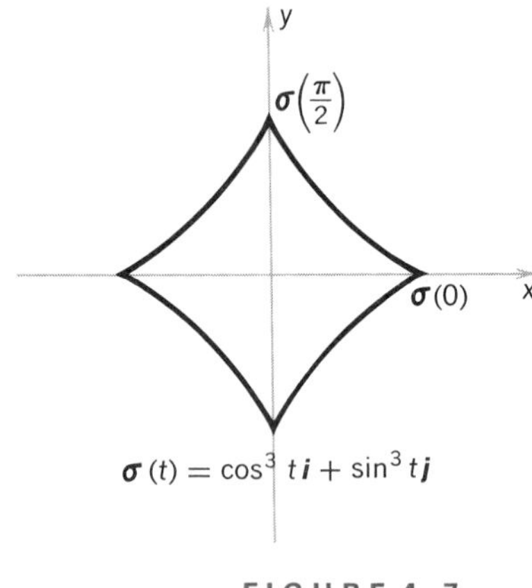

FIGURE 4-7

The image of a C^1 path is not necessarily "very smooth," since it may have sharp bends or changes of direction. For instance, the cycloid in the above example has cusps at all points where $\boldsymbol{\sigma}(t)$ touches the x axis (that is, where $t = 2\pi n$, $n = 0, \pm 1, \ldots$). Another example is the **hypocycloid of four cusps**, $\boldsymbol{\sigma}: [0, 2\pi] \to \boldsymbol{R}^2 : t \mapsto (\cos^3 t, \sin^3 t)$, which has cusps at four points (Figure 4-7). However, at all such points $\boldsymbol{\sigma}'(t) = 0$ so the tangent vector is not defined and the speed of the point $\boldsymbol{\sigma}(t)$ is zero. We see that the direction of $\boldsymbol{\sigma}(t)$ may change abruptly at points where it slows to rest.

Many books denote a C^1 path $\boldsymbol{\sigma}: t \mapsto (x(t), y(t), z(t))$ by

$$x = x(t) \qquad y = y(t) \qquad z = z(t) \qquad t \in I$$

Then the formula for arc length becomes

$$\text{arc length} = \int_I \sqrt{\left(\frac{dx}{dt}\right)^2 + \left(\frac{dy}{dt}\right)^2 + \left(\frac{dz}{dt}\right)^2}\, dt$$

When C^1 paths are given in this form, the usual notation for the tangent vector is

$$\frac{d\boldsymbol{s}}{dt} = \frac{dx}{dt}\boldsymbol{i} + \frac{dy}{dt}\boldsymbol{j} + \frac{dz}{dt}\boldsymbol{k}$$

where $\boldsymbol{s}(t) = (x(t)\, y(t), z(t))$. Then we have

$$\frac{ds}{dt} = \left|\frac{d\boldsymbol{s}}{dt}\right| = \sqrt{\left(\frac{dx}{dt}\right)^2 + \left(\frac{dy}{dt}\right)^2 + \left(\frac{dz}{dt}\right)^2} = |\boldsymbol{\sigma}'(t)|$$

The arc length is therefore equal to

$$\int_I ds = \int_I \left|\frac{d\boldsymbol{s}}{dt}\right| dt = \int_I |\boldsymbol{\sigma}'(t)| dt$$

This is just a restatement of the results above, but with slightly different notation.

Example 8 A particle moves along the hypocycloid according to the equations

$$x = \cos^3 t \qquad y = \sin^3 t \qquad z = 0 \qquad a \leqslant t \leqslant b$$

The velocity vector of the particle is

$$\frac{d\boldsymbol{s}}{dt} = \frac{dx}{dt}\boldsymbol{i} + \frac{dy}{dt}\boldsymbol{j} + \frac{dz}{dt}\boldsymbol{k} = -3\sin t\cos^2 t\boldsymbol{i} + 3\cos t\sin^2 t\boldsymbol{j}$$

and its speed is

$$\left|\frac{ds}{dt}\right| = (9\sin^2 t\cos^4 t + 9\cos^2 t\sin^4 t)^{1/2}$$

$$= 3\,|\sin t|\,|\cos t|$$

If we differentiate the velocity vector of a moving particle,

$$\boldsymbol{v}(t) = \frac{ds}{dt} = \frac{dx}{dt}\boldsymbol{i} + \frac{dy}{dt}\boldsymbol{j} + \frac{dz}{dt}\boldsymbol{k}$$

we obtain its **acceleration vector**

$$\frac{d\boldsymbol{v}}{dt} = \boldsymbol{a}(t) = \frac{d^2x}{dt^2}\boldsymbol{i} + \frac{d^2y}{dt^2}\boldsymbol{j} + \frac{d^2z}{dt^2}\boldsymbol{k}$$

The acceleration vector describes the change in the velocity of the moving point. Newton's Second Law of Motion says that the acceleration of a particle moving in space is proportional to the force on the particle

$$\boldsymbol{F} = m\boldsymbol{a}$$

where m is the mass of the particle. The direction of the acceleration vector can, therefore, be interpreted as the direction of the force applied to the particle.

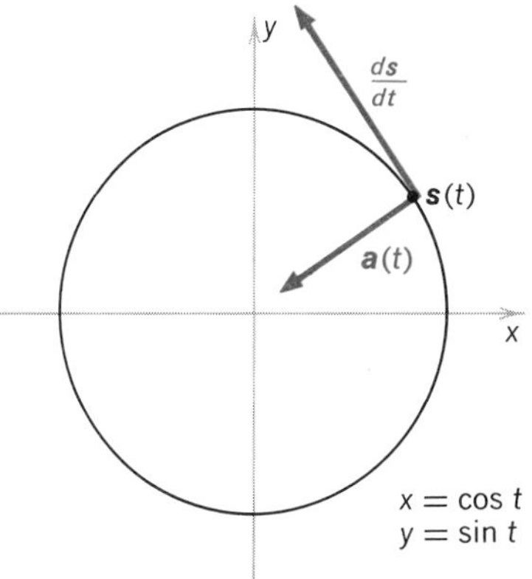

FIGURE 4-8

Example 9 A particle moves in a circle in the plane so that its position is given by

$$x = \cos t \qquad y = \sin t$$

Show that (a) the speed of the particle is constant, (b) its velocity vector is perpendicular to its position vector, (c) its acceleration is always directed towards the origin (see Figure 4-8).

We have $ds/dt = -\sin t\boldsymbol{i} + \cos t\boldsymbol{j}$ so the speed of the particle is $|ds/dt| = (\sin^2 t + \cos^2 t)^{1/2} = 1$; this proves (a). The position of the particle is $\boldsymbol{P}(t) = \cos t\boldsymbol{i} + \sin t\boldsymbol{j}$ so

$$\boldsymbol{P}(t)\cdot\frac{ds}{dt} = (\cos t)(-\sin t) + (\sin t)(\cos t) = 0$$

Hence, we have established (b). Finally, the acceleration is

$$\boldsymbol{a}(t) = -\cos t\boldsymbol{i} - \sin t\boldsymbol{j} = -\boldsymbol{P}(t)$$

The equation $\boldsymbol{a}(t) = -\boldsymbol{P}(t)$ says that the particle's acceleration is directed back along the position vector to the origin. This is an example of the phenomenon of **centripetal force**.

Note that even though the speed of the particle is constant, its acceleration is not zero because the direction of velocity is changing.

The definition of arc length can be extended to include paths which are not C^1 but which are formed by piecing together a finite number of C^1 paths. A path $\boldsymbol{\sigma}:[a,b]\to \boldsymbol{R}^3: t\to(x(t), y(t), z(t))$ is called **piecewise** C^1 if there is a partition of $[a,b]$

$$a = t_0 < t_1 < \cdots < t_N = b$$

such that the function $\boldsymbol{\sigma}$ restricted to each interval $[t_i, t_{i+1}]$, $0 \leqslant i \leqslant N-1$, is continuously differentiable. (The important point here is that the function need not be differentiable at the points t_i; only the right- and left-handed derivatives must exist at each of these points.) In the case of a path which is piecewise C^1, we define the arc length of the path to be the sum of the arc lengths of the C^1 paths which make it up. That is, if the partition

$$a = t_0 < t_1 < \cdots < t_N = b$$

satisfies the above conditions, we define

$$\begin{aligned}\text{arc length of } \boldsymbol{\sigma} &= \sum_{i=0}^{N-1} \int_{t_i}^{t_{i+1}} \left|\frac{ds}{dt}\right| dt \\ &= \sum_{i=0}^{N-1} \text{arc length of } \boldsymbol{\sigma} \text{ from } t_i \text{ to } t_{i+1}\end{aligned}$$

Example 10 The path $\boldsymbol{\tau}:[-1,1]\to \boldsymbol{R}^3: t\mapsto (|t|, |t|-|\frac{1}{2}|, 0)$ is not C^1 because $\tau_1: t\mapsto |t|$ is not differentiable at 0, nor is $\tau_2: t\mapsto |t-\frac{1}{2}|$ differentiable at $\frac{1}{2}$. However, if we take the partition

$$-1 = t_0 < 0 = t_1 < \tfrac{1}{2} = t_2 < 1 = t_3$$

we see that each of the τ_i is continuously differentiable on each of the intervals $[-1,0]$, $[0,\frac{1}{2}]$, and $[\frac{1}{2},1]$ and therefore, $\boldsymbol{\tau}$ is continuously differentiable on each interval. Thus, $\boldsymbol{\tau}$ is piecewise C^1.

Now, on $[-1,0]$, we have $x(t) = -t$, $y(t) = -t+\frac{1}{2}$, $z(t) = 0$, $|ds/dt| = \sqrt{2}$; hence, the arc length of $\boldsymbol{\tau}$ between -1 and 0 is $\int_{-1}^{0} \sqrt{2}\, dt = \sqrt{2}$. Similarly, on $[0,\frac{1}{2}]$, $x(t) = t$, $y(t) = -t+\frac{1}{2}$, $z(t) = 0$, and again $|ds/dt| = \sqrt{2}$, arc length of $\boldsymbol{\tau}$ between 0 and $\frac{1}{2}$ is $\frac{1}{2}\sqrt{2}$. Finally, on $[\frac{1}{2}, 1]$ we have $x(t) = t$, $y(t) = t-\frac{1}{2}$, $z(t) = 0$, and arc length of $\boldsymbol{\tau}$ between $\frac{1}{2}$ and 1 is $\frac{1}{2}\sqrt{2}$. Then the total arc length of $\boldsymbol{\tau}$ is $2\sqrt{2}$.

EXERCISES

In exercises 1–4 find $\boldsymbol{\sigma}'(t)$ and $\boldsymbol{\sigma}'(0)$ where $\boldsymbol{\sigma}: \boldsymbol{R}\to\boldsymbol{R}^3$ is the given path.

1. $\boldsymbol{\sigma}(t) = (\sin 2\pi t, \cos 2\pi t, 2t - t^2)$
2. $\boldsymbol{\sigma}(t) = (e^t, \cos t, \sin t)$
3. $\boldsymbol{\sigma}(t) = (t^2, t^3 - 4t, 0)$
4. $\boldsymbol{\sigma}(t) = (\sin 2t, \ln(1 + t), t)$

In exercises 5–8 find ds/dt, and $\ln(|ds/dt|)$.

5. $\mathbf{s} = e^t\mathbf{i} + e^{-t}\mathbf{j} + t\mathbf{k}$
6. $\mathbf{s} = (2\sin t)\mathbf{i} + (3\cos t)\mathbf{k}$
7. $\mathbf{s} = t\mathbf{i} + t^2\mathbf{j} + t^3\mathbf{k}$
8. $\mathbf{s} = (\cos^3 t)\mathbf{i} + (\sin^3 t)\mathbf{j} + t\mathbf{k}$

For each of the curves in exercises 9–15, determine the tangent and acceleration vectors, and the equation of the tangent line at the specified value of t.

9. $\mathbf{s} = 6t\mathbf{i} + 3t^2\mathbf{j} + t^3\mathbf{k},\ t = 0$
10. $\boldsymbol{\sigma}(t) = (\sin 3t, \cos 3t, 2t^{3/2}),\ t = 0$
11. The path in exercise 4, $t = 0$
12. The path in exercise 8, $t = \pi/4$
13. $\boldsymbol{\sigma}(t) = (t\sin t, t\cos t, \sqrt{3}t),\ t = 0$
14. $\mathbf{s} = \sqrt{2}t\mathbf{i} + e^t\mathbf{j} + e^{-t}\mathbf{k},\ t = 0$
15. $\mathbf{s} = t\mathbf{i} + t\mathbf{j} + \frac{2}{3}t^{3/2}\mathbf{k},\ t = 9$

In exercises 16–22 calculate the arc length of the given curve on the specified interval.

16. The path in exercise 9, $[0, 1]$
17. The path in exercise 10, $[0, 1]$
18. $\mathbf{s} = t\mathbf{i} + t\sin t\mathbf{j} + t\cos t\mathbf{k},\ [0, \pi]$
19. $\mathbf{s} = 2t\mathbf{i} + t\mathbf{j} + t^2\mathbf{k},\ [0, 2]$
20. The path in exercise 13, $[0, 1]$
21. The path in exercise 14, $[-1, 1]$
22. The path in exercise 15, $[t_0, t_1]$

5 THE PATH INTEGRAL

In this section we introduce the concept of a **path integral**; this is one of the ways in which integrals of functions of one variable can be generalized to functions of several variables (there are others). Path integrals have many applications in geometry and physics and they are also important tools in the study of higher mathematics.

Suppose we are given a function $F: \mathbf{R}^3 \to \mathbf{R}$ which sends points in $\mathbf{R}^3$ to real numbers. The value assigned to (x, y, z) by F is denoted $F(x, y, z)$. Such functions arise naturally in mathematical and physical situations. For example, $F(x, y, z)$ might measure distance from some fixed point or axis, or in a physical context, it might denote the temperature at (x, y, z). (We deal with these functions at greater length in Chapter 14.) It turns out that it is extremely useful to be able to integrate the function F along a path $\boldsymbol{\sigma}: I \to \mathbf{R}^3$. To illustrate this notion, suppose that the image of $\boldsymbol{\sigma}$ represents a wire. If $F(x, y, z)$ denotes the mass density at (x, y, z), we might ask what is the total amount of mass in the wire. If $F(x, y, z)$ indicates temperature, we might ask what is the average temperature along the wire. Both types of problems require integrating $F(x,y,z)$ over $\boldsymbol{\sigma}$.

The **path integral** or the **integral of $F(x,y,z)$ along the path σ** is defined when $\boldsymbol{\sigma}: I \to \boldsymbol{R}^3$ is piecewise C^1 and when the composite function $t \mapsto F(\sigma_1(t), \sigma_2(t), \sigma_3(t))$ is continuous on I. We define this integral by the equation

$$\int_\sigma F(x,y,z)\,ds = \int_I F(x,y,z)\frac{ds}{dt}\,dt = \int_I F(\sigma_1(t),\sigma_2(t),\sigma_3(t))\,|\sigma'(t)|\,dt$$

Example 1 Let $\boldsymbol{\sigma}$ be the helix $\boldsymbol{\sigma}:[0, 2\pi] \to \boldsymbol{R}^3: t \mapsto (\cos t, \sin t, t)$ and let $F(x,y,z) = x^2 + y^2 + z^2$. To evaluate the integral $\int_\sigma F(x,y,z)\,ds$, we first find

$$\frac{ds}{dt} = \sqrt{\left[\frac{d(\cos t)}{dt}\right]^2 + \left[\frac{d(\sin t)}{dt}\right]^2 + \left[\left(\frac{dt}{dt}\right)\right]^2}$$
$$= \sqrt{\sin^2 t + \cos^2 t + 1} = \sqrt{2}$$

We substitute for x, y, and z to obtain

$$F(x, y, z) = x^2 + y^2 + z^2 = \sin^2 t + \cos^2 t + t^2 = 1 + t^2$$

along $\boldsymbol{\sigma}$. This yields

$$\int_\sigma F(x,y,z)\,ds = \int_0^{2\pi} (1+t^2)\sqrt{2}\,dt = \sqrt{2}\left[t + \frac{t^3}{3}\right]_0^{2\pi}$$
$$= \frac{2\sqrt{2}\pi}{3}(3+4\pi^2)$$

If we think of the helix as an electric wire and $F(x,y,z) = x^2+y^2+z^2$ as the mass density, then the total mass of the wire is $(2\sqrt{2}\pi)/3[(3+4\pi^2)]$.

To motivate the definition of this integral we shall consider Riemann-"like" sums S_N in the same general way as we have done before to define arc length. For simplicity, let $\boldsymbol{\sigma}$ be C^1 on I. Next, subdivide the interval $I = [a,b]$ by means of a regular partition

$$a = t_0 < t_1 < \cdots < t_{N-1} = b$$

of order N. This leads to a decomposition of $\boldsymbol{\sigma}$ into paths (Figure 5-1) $\boldsymbol{\sigma}_i$

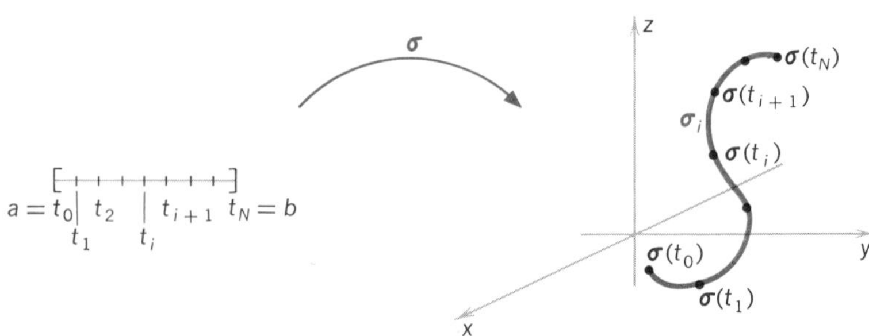

FIGURE 5-1

defined on $[t_i, t_{i+1}]$, $0 \leqslant i \leqslant N-1$. Denote the arc length of $\boldsymbol{\sigma}_i$ by Δs_i, thus

$$\Delta s_i = \int_{t_i}^{t_{i+1}} |\boldsymbol{\sigma}'(t)|\, dt$$

When N is large, the arc length Δs_i is small and $F(x,y,z)$ is approximately constant for points on $\boldsymbol{\sigma}_i$. We consider the sums

$$S_N = \sum_{i=0}^{N-1} F(x_i, y_i, z_i)\, \Delta s_i$$

where $(x_i, y_i, z_i) = \boldsymbol{\sigma}(t)$ for some $t \in [t_i, t_{i+1}]$. Although the proof is beyond the scope of this course, it can be shown, using the fact that σ is C^1, that

$$\lim_{N\to\infty} S_N = \int_I F(\sigma_1(t), \sigma_2(t), \sigma_3(t))\, |\sigma'(t)|\, dt = \int_{\boldsymbol{\sigma}} F(x,y,z)\, ds$$

An interesting application of the path integral is its use in defining the average value of a scalar function along a path. We define the number

$$\frac{\int_{\boldsymbol{\sigma}} F(x,y,z)\, ds}{l(\boldsymbol{\sigma})}$$

as the **average value** of F along $\boldsymbol{\sigma}$. We can justify our definition of average value by considering Riemann sums: The quotient

$$Q_N = \frac{\sum_{i=0}^{N-1} F(x_i, y_i, z_i)\, \Delta s_i}{\sum_{i=0}^{N-1} \Delta s_i} = \frac{\sum_{i=0}^{N-1} F(x_i, y_i, z_i)\, \Delta s_i}{l(\boldsymbol{\sigma})}$$

is the approximate average one obtains by considering $F(x,y,z)$ to be constant along each of the arcs $\boldsymbol{\sigma}_i$. The limit as $N \to \infty$ of the sequence $\{Q_N\}$ is the average value of $F(x,y,z)$ along $\boldsymbol{\sigma}$.

If we think of the helix in Example 1 as a heated wire and $F(x,y,z) = x^2 + y^2 + z^2$ as the temperature, then the average temperature along the wire is

$$\frac{\int_{\boldsymbol{\sigma}} F(x,y,z)\, ds}{l(\boldsymbol{\sigma})} = \frac{2\sqrt{2}\pi(3+4\pi^2)}{(3)2\sqrt{2}\pi} = \tfrac{1}{3}(3+4\pi^2) \text{ degrees}$$

Example 2 Find the average y coordinate of the points on the semicircle parametrized by $\boldsymbol{\rho}: [0,\pi] \to \boldsymbol{R}^3 : \theta \mapsto (0, a\sin\theta, a\cos\theta)$. In this example we have $F(x,y,z) = y$. We compute

$$\int_{\boldsymbol{\rho}} F(x,y,z)\, ds = \int_0^{\pi} a\sin\theta\,(a^2\cos^2\theta + a^2\sin^2\theta)^{1/2}\, d\theta = 2a^2$$

Since $l(\boldsymbol{\rho}) = \pi a$, the average y coordinate of points on the semicircle is $2a/\pi$.

The average x coordinate is clearly zero and the average z coordinate is easily computed also to be zero. If we let the semicircle represent a wire of uniform density, then the point $(\bar{x}, \bar{y}, \bar{z}) = (0, 2a/\pi, 0)$, whose coordinates are the average x, y, and z coordinates of points on the wire, is the center of gravity of the wire. We consider this topic at some length in Chapter 16.

An important special case of the path integral occurs when the path $\boldsymbol{\sigma}$ describes a plane curve. Let us examine this case in some detail. Suppose then that all points $\boldsymbol{\sigma}(t)$ lie in the xy plane. Let F be a real-valued function of two variables, $F:(x,y) \mapsto F(x,y)$. The path integral of F along $\boldsymbol{\sigma}$ is written

$$\int_{\boldsymbol{\sigma}} F(x,y)\, ds = \int_I F(x,y) \frac{ds}{dt}\, dt$$

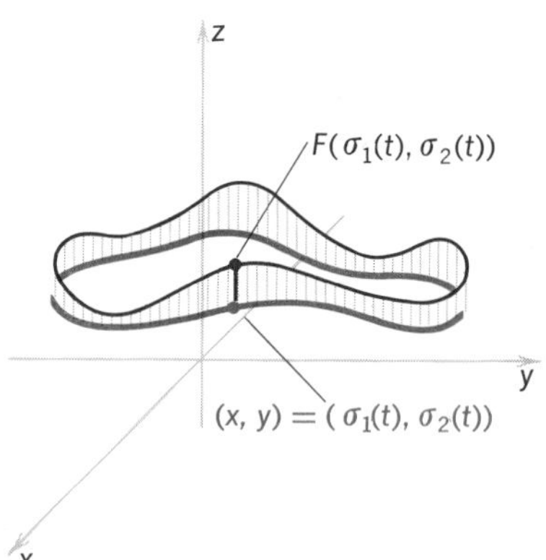

FIGURE 5-2

When $F(x,y) \geqslant 0$, this integral has a natural geometric interpretation as the "area of a fence." We can construct a "fence" with base the image of $\boldsymbol{\sigma}$ and with height $F(x,y)$ at (x,y) (Figure 5-2). If $\boldsymbol{\sigma}$ winds only once around the image of $\boldsymbol{\sigma}$ (see exercise 4.3), the integral $\int_{\boldsymbol{\sigma}} F(x,y)\, ds$ represents the area of a side of this fence. The reader should justify this interpretation for himself. One way to do this is to consider the appropriate area function and to compute its derivative. Alternatively, one may employ the technique of Riemann sums.

Example 3 Tom Sawyer's aunt has asked him to whitewash both sides of the old fence shown in Figure 5-3. Tom estimates that for each 25 square feet of whitewashing he lets someone do for him, the willing victim will pay 5 cents. How much can Tom hope to earn, assuming his aunt will provide whitewash free of charge?

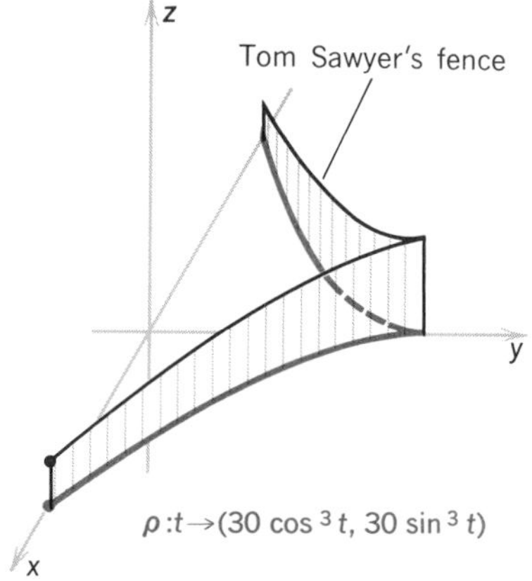

FIGURE 5-3

As indicated in the figure, the base of the fence is defined by the path $\boldsymbol{\rho}:[0, \pi/2] \rightarrow \boldsymbol{R}^2 : t \mapsto (30 \cos^3 t, 30 \sin^3 t)$ and the height of the fence at (x,y) is $F(x,y) = 1 + y/3$. The area of one side of the fence is then equal to the path integral $\int_{\boldsymbol{\rho}} F(x, y)\, ds = \int_{\boldsymbol{\rho}} (1 + y/3)\, ds$. Since $\boldsymbol{\rho}'(t) = (-90 \cos^2 t \sin t, 90 \sin^2 t \cos t)$, we have $|\boldsymbol{\rho}'(t)| = ds/dt = 90 \sin t \cos t$. So the integral is

$$\begin{aligned}\int_{\boldsymbol{\rho}} \left(1 + \frac{y}{3}\right) ds &= \int_0^{\pi/2} \left(1 + \frac{30 \sin^3 t}{3}\right) 90 \sin t \cos t\, dt \\ &= 90 \int_0^{\pi/2} (\sin t + 10 \sin^4 t) \cos t\, dt \\ &= 90 \left[\frac{\sin^2 t}{2} + 2 \sin^5 t\right]_0^{\pi/2} \\ &= 90(2 + \tfrac{1}{2}) = 225\end{aligned}$$

Hence, the area of one side of the fence is 225 square feet. Since both sides are to be whitewashed we must multiply by 2 to find the total area which is 450 square feet. Dividing by 25 and then multiplying by 5, we find that Tom could realize as much as 90 cents for the job.

Example 4 Since the area of the fence in the above example is 225 square feet on each side and the length of the base 45 feet, its average height is 225 feet/45 feet = 5 feet (Verify!).

We shall see many further applications of the path integral in Chapters 16 and 17.

EXERCISES

1. Show that when $F=1$, $\int_\sigma F(x,y,z)\,ds = l(\boldsymbol{\sigma})$.
2. Evaluate the following path integrals $\int_\sigma F(x,y,z)\,ds$ where
   (a) $F=x+y+z$ and $\boldsymbol{\sigma}:t\mapsto(\sin t,\cos t,t)$, $t\in[0,2\pi]$;
   (b) $F=\cos z$, $\boldsymbol{\sigma}$ as in (a);
   (c) $F=x\cos z$, $\boldsymbol{\sigma}:t\mapsto t\boldsymbol{i}+t^2\boldsymbol{j}$, $t\in[0,1]$.
3. Let $F:\boldsymbol{R}^3-\{\boldsymbol{0}\}\to\boldsymbol{R}$ be defined by $F(x,y,z)=1/y^3$. Evaluate $\int_\sigma F(x,y,z)\,ds$ where $\boldsymbol{\sigma}:[1,e]\to\boldsymbol{R}^3$ is given by $\boldsymbol{\sigma}(t)=\log t\boldsymbol{i}+t\boldsymbol{j}+2\boldsymbol{k}$.
4. In exercises 2(a), 2(b) above find the average value of F over the given curves.
5. Find the average y coordinate for
   (a) the path $\boldsymbol{\sigma}:t\mapsto(t^2,t,3)$; $t\in[0,1]$.
   (b) the graph of the function $f:x\mapsto x^3$ for $x\in[0,1]$.
6. Find the circumference of a circle of radius a using path integrals.
7. (a) Show that the path integral of F along a path given in polar coordinates by $r=r(\theta)$, $\theta_1\leqslant\theta\leqslant\theta_2$ is
   $$\int_{\theta_1}^{\theta_2} F(r,\theta)\sqrt{r^2+\left(\frac{dr}{d\theta}\right)^2}\,d\theta$$
   (b) Compute the arc length of $r=1+\cos\theta$, $0\leqslant\theta\leqslant 2\pi$.
8. Let f and g be C^1 functions on the path $\boldsymbol{\sigma}(t)=(x(t),y(t))\in\boldsymbol{R}^2$. Evaluate (notations as in the text, $t_i\leqslant t_i^*\leqslant t_{i+1}$)
   $$\lim_{N\to\infty}\sum_{i=1}^{N-1} f(\boldsymbol{\sigma}(t_i^*))\left|g(\boldsymbol{\sigma}(t_{i+1}))-g(\boldsymbol{\sigma}(t_i))\right|$$
9. Show that (in a sense you should make precise) the graph of $y=x\sin(1/x)$, $0<x\leqslant 1$ has infinite length whereas $y=x^2\sin(1/x)$, $0<x\leqslant 1$, has finite length. Sketch.
10. Let $F(x,y)=x-y$, $x=t^4$, $y=t^4$, $-1\leqslant t\leqslant 1$.
    (a) Compute the integral of F along this path and interpret the answer geometrically. Sketch.
    (b) Evaluate the arc length function $s(t)$ and redo (a) in terms of s.

REVIEW EXERCISES FOR CHAPTER 13

1. Find the length of each of the following paths:
   (a) $\boldsymbol{\sigma}(t)=(e^t\cos t,\,e^t\sin t,\,3)$, $\quad 0\leqslant t\leqslant 2\pi$
   (b) $\boldsymbol{\sigma}(t)=(\cos t,\,\sin t,\,t)$, $\quad 0\leqslant t\leqslant 2\pi$
   (c) $\boldsymbol{\sigma}(t)=\frac{3}{2}t^2\boldsymbol{i}+2t^2\boldsymbol{j}+t\boldsymbol{k}$, $\quad 0\leqslant t\leqslant 1$
   (d) $\boldsymbol{\sigma}(t)=t\boldsymbol{i}+\dfrac{1}{\sqrt{2}}t^2\boldsymbol{j}+\dfrac{1}{3}t^3\boldsymbol{k}$, $\quad 0\leqslant t\leqslant 1$

2. Let $f:[a,b]\to \boldsymbol{R}$. We define the **length of the graph** of f on $[a,b]$ to be the length of the path $t\mapsto (t,f(t))$, $a\leqslant t\leqslant b$. Show that the arc length of the graph of a function $y=f(x)$, $a\leqslant x\leqslant b$, is

$$\int_a^b \sqrt{1+[f'(x)]^2}\,dx$$

3. Find the length of the graph of $y=\ln x$ from $x=1$ to $x=2$.

4. Let $\boldsymbol{v}=3\boldsymbol{i}+4\boldsymbol{j}+5\boldsymbol{k}$ and $\boldsymbol{w}=\boldsymbol{i}-\boldsymbol{j}+\boldsymbol{k}$. Compute $\boldsymbol{v}+\boldsymbol{w}$, $3\boldsymbol{v}$, $6\boldsymbol{v}+8\boldsymbol{w}$, $-2\boldsymbol{v}$, $\boldsymbol{v}\cdot\boldsymbol{w}$, $\boldsymbol{v}\times\boldsymbol{w}$. Interpret each operation geometrically by graphing the vectors.

5. (a) Find the equation of the line through $(0,1,0)$ in the direction of $3\boldsymbol{i}+\boldsymbol{k}$.
   (b) Find the equation of the line passing through $(0,1,1)$ and $(0,1,0)$.

6. Use vector notation to describe the triangle in space whose vertices are the origin and the endpoints of vectors $\boldsymbol{a}$ and $\boldsymbol{b}$.

7. Show that three vectors $\boldsymbol{a},\boldsymbol{b},\boldsymbol{c}$ lie in a plane if and only if there are three scalars α,β,γ not all zero, such that $\alpha\boldsymbol{a}+\beta\boldsymbol{b}+\gamma\boldsymbol{c}=0$.

8. For real numbers a_1,a_2,a_3,b_1,b_2,b_3 show that

$$(a_1b_1+a_2b_2+a_3b_3)^2\leqslant(a_1^2+a_2^2+a_3^2)(b_1^2+b_2^2+b_3^2)$$

9. Let $\boldsymbol{u},\boldsymbol{v},\boldsymbol{w}$ be unit vectors which are orthogonal to each other. If $\boldsymbol{a}=\alpha\boldsymbol{u}+\beta\boldsymbol{v}+\gamma\boldsymbol{w}$, show that

$$\alpha=\boldsymbol{a}\cdot\boldsymbol{u},\qquad \beta=\boldsymbol{a}\cdot\boldsymbol{v},\qquad \gamma=\boldsymbol{a}\cdot\boldsymbol{w}$$

   Interpret the results geometrically.

10. Let $\boldsymbol{a},\boldsymbol{b}$ be two vectors in the plane, $\boldsymbol{a}=(a_1,a_2)$, $\boldsymbol{b}=(b_1,b_2)$.
    (a) Show $\begin{vmatrix} a_1 & b_1 \\ a_2 & b_2 \end{vmatrix}$ is the area of the parallelogram determined by $\boldsymbol{a},\boldsymbol{b}$.
    (b) Show that the area of the parallelogram determined by $\boldsymbol{a}$ and $\boldsymbol{b}+\lambda\boldsymbol{a}$ is the same as that in (a). Sketch and interpret as a property of determinants.

11. Find the volume of the parallelepiped determined by the vertices $(0,1,0)$, $(1,1,1)$, $(0,2,0)$, $(3,1,2)$.

12. Find the speed and velocity of the path described by $\boldsymbol{\sigma}(t)=t^3\boldsymbol{i}+\sin t\boldsymbol{j}+\cos(\sin t)\boldsymbol{k}$.

13. Find the length of the following paths:
    (a) $\boldsymbol{\sigma}(t)=t\boldsymbol{i}+3t\boldsymbol{j}+4t\boldsymbol{k},\quad 0\leqslant t\leqslant 1$
    (b) $\boldsymbol{\sigma}(t)=\sin t\boldsymbol{i}+\cos t\boldsymbol{j}+5t\boldsymbol{k},\quad 0\leqslant t\leqslant 2\pi$

14. (a) Integrate $F(x,y,z)=xyz$ along the paths (a) and (b) described in exercise 1.
    (b) Find the average value of F along the curves.

15. Show that the vector

$$\boldsymbol{v}=\frac{|\boldsymbol{a}|\,\boldsymbol{b}+|\boldsymbol{b}|\,\boldsymbol{a}}{|\boldsymbol{a}|+|\boldsymbol{b}|}$$

    bisects the angle between $\boldsymbol{a}$ and $\boldsymbol{b}$.

16. Use vector methods to prove that the distance from the point (x_1, y_1) to the line $ax+by=c$ is

$$\frac{|ax_1+by_1-c|}{\sqrt{a^2+b^2}}$$

17. Show that the direction of $\boldsymbol{b}\times\boldsymbol{c}$ is given by the right-hand rule.

18. Suppose $\boldsymbol{a}\cdot\boldsymbol{b}=\boldsymbol{a}'\cdot\boldsymbol{b}$ for all $\boldsymbol{b}$.
 (a) Show that $\boldsymbol{a}=\boldsymbol{a}'$.
 (b) Suppose $\boldsymbol{a}\times\boldsymbol{b}=\boldsymbol{a}'\times\boldsymbol{b}$ for all $\boldsymbol{b}$. Is it true that $\boldsymbol{a}=\boldsymbol{a}'$?

19. Find the distance (using vector methods) between the line l_1 determined by the points $(-1,-1,1)$, $(0,0,0)$ and the line l_2 determined by the points $(0,-2,0)$ and $(2,0,5)$.

20. Show that two planes given by the equations $Ax+By+Cz+D_1=0$ and $Ax+By+Cz+D_2=0$ are parallel and that the distance between two such planes is

$$\frac{|D_1-D_2|}{\sqrt{A^2+B^2+C^2}}$$

*21. In Example 7 of §4 we saw that $\boldsymbol{\sigma}: t\mapsto(t-\sin t,\ 1-\cos t, 0)$ had no tangent line at $\boldsymbol{\sigma}(0)$ since $\boldsymbol{\sigma}'(0)=\mathbf{0}$. Give an example of a $\boldsymbol{\sigma}$ with $\boldsymbol{\sigma}'(t_0)=\mathbf{0}$ for some t_0 and yet $\boldsymbol{\sigma}$ "has" a tangent line at $\boldsymbol{\sigma}(t_0)$.

22. Let $\boldsymbol{\sigma}: I\to\boldsymbol{R}^3$ represent the path of a moving particle as a function of time, with $\boldsymbol{\sigma}'(t)\neq\mathbf{0}$ for all t. Define the unit tangent vector $\boldsymbol{T}: I\to\boldsymbol{R}^3$ by

$$\boldsymbol{T}(t)=\frac{\boldsymbol{\sigma}'(t)}{|\boldsymbol{\sigma}'(t)|}$$

 (a) Show that the velocity vector

$$\boldsymbol{v}=\boldsymbol{T}\,\frac{ds}{dt}$$

 (b) Show that the acceleration of the particle is given by

$$\boldsymbol{a}=\frac{d\boldsymbol{v}}{dt}=\boldsymbol{T}\frac{d^2s}{dt}+\frac{ds}{dt}\frac{d\boldsymbol{T}}{dt}$$

23. Let $\boldsymbol{\sigma}$ and $\boldsymbol{\eta}$ be paths. Prove

$$\frac{d(\boldsymbol{\sigma}\cdot\boldsymbol{\eta})}{dt}=\frac{d\boldsymbol{\sigma}}{dt}\cdot\boldsymbol{\eta}(t)+\boldsymbol{\sigma}(t)\cdot\frac{d\boldsymbol{\eta}}{dt}=\boldsymbol{\sigma}'(t)\cdot\boldsymbol{\eta}(t)+\boldsymbol{\sigma}(t)\cdot\boldsymbol{\eta}'(t)$$

24. (a) In exercise 22 show that $\boldsymbol{T}$ is perpendicular to $d\boldsymbol{T}/dt$.
 (b) Using (a), show that the absolute value or norm of the acceleration vector is

$$|\boldsymbol{a}|=\sqrt{\left|\frac{d^2s}{dt^2}\right|+\left|\frac{ds}{dt}\right|^2\left|\frac{d\boldsymbol{T}}{dt}\right|^2}$$

14

DIFFERENTIATION

In this chapter we study functions from $\boldsymbol{R}^m$ to $\boldsymbol{R}^n$ where m and n are 1, 2, or 3. We shall examine the meaning of the notion of continuity and differentiability for these functions and consider some applications in geometry and physics.

1 FUNCTIONS FROM $\boldsymbol{R}^2$ to $\boldsymbol{R}$

Thus far our study of the calculus has concentrated on real functions (that is, those functions whose domain and range are both subsets of $\boldsymbol{R}$) and on paths (or, more precisely, functions from $\boldsymbol{R}$ to $\boldsymbol{R}^2$ or $\boldsymbol{R}^3$). In this section we turn our attention to functions whose domain is a subset of $\boldsymbol{R}^2$ and whose range is a subset of $\boldsymbol{R}$. We shall interpret the graphs of differentiable functions from $\boldsymbol{R}^2$ to $\boldsymbol{R}$ as surfaces in $\boldsymbol{R}^3$.

Let f be a function from a subset of $\boldsymbol{R}^2$ to $\boldsymbol{R}$. If (x_0, y_0) is a point in the domain of f, we denote the value of f at (x_0, y_0) by $f(x_0, y_0)$. Since (x, y) is a pair of variables, f is called a **real-valued function of two real variables.** Often, we shall shorten this expression and use "a function of two variables" in our discussions.

An example of such a function is $(x, y) \mapsto x^2 + y^2$. Sometimes it is preferable to define a function of two real variables by a single equation of the form

$$z = f(x, y)$$

Hence, for the function above, $z = x^2 + y^2$. The notation $z = f(x, y)$ is classical and is used in most textbooks. We shall see shortly that this notation is the most convenient for describing a real-valued function of two variables by means of its graph.

Two important examples of functions of two variables are

$$(x, y) \mapsto x, \qquad (x, y) \mapsto y$$

which are called the **projection functions** (on the first and second coordinate, respectively).

If $f: \boldsymbol{R}^2 \to \boldsymbol{R}$, then as (x_0, y_0) varies through the points in $\boldsymbol{R}^2$, the points $(x_0, y_0, f(x_0, y_0))$ describe or trace out a subset of $\boldsymbol{R}^3$; this subset is called the **graph** of f. As with functions of one real variable, it is often easier to understand the behavior of a function by examining its graph; in fact, the graph provides a complete picture of the global behavior of a function. Geometrically, we think of the graph of a function of two variables as a surface in $\boldsymbol{R}^3$. For most well-behaved functions, such as the ones given in the examples below, these graphs agree with our intuitive idea of a surface. We give a more sophisticated definition of a surface in §14.7. Before we describe some techniques of graphing functions of two variables, let us consider some examples.

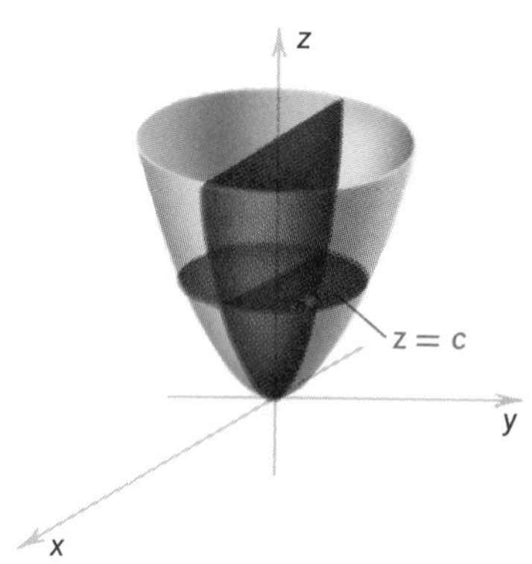

FIGURE 1-1

Example 1 If f is a constant function then for all (x, y), $z = f(x, y) = c$, where c is fixed. In this case the graph of f is a plane parallel to the xy plane at a distance c units above (if $c > 0$) or below (if $c < 0$) the xy plane.

Example 2 If $z = f(x, y) = x^2 + y^2$, the domain of f is all of $\boldsymbol{R}^2$. The graph of f is then a **paraboloid** (see Figure 1-1). To sketch this surface, we use the following procedure. First, from the given equation for f, we determine the equations of the curves in which the surface intersects the coordinate planes, and sketch these curves. Second, using the equation for f again, we decide what type of sections are made when the graph is cut by planes parallel to the coordinate planes. Hence, let us consider the intersection of the graph of f with the xz plane, then with the yz plane, and finally, with a plane $z = c$, for $c > 0$. To determine the intersection of the graph with the xz plane, we set $y = 0$ in the given equation $z = x^2 + y^2$; this yields $z = x^2 + 0^2 = x^2$. Thus, the intersection with the plane $y = 0$ (the xz plane) is the parabola described by the equation $z = x^2$. Similarly, the intersection with the plane $x = 0$ (the yz plane) is the parabola given by the equation $z = y^2$.

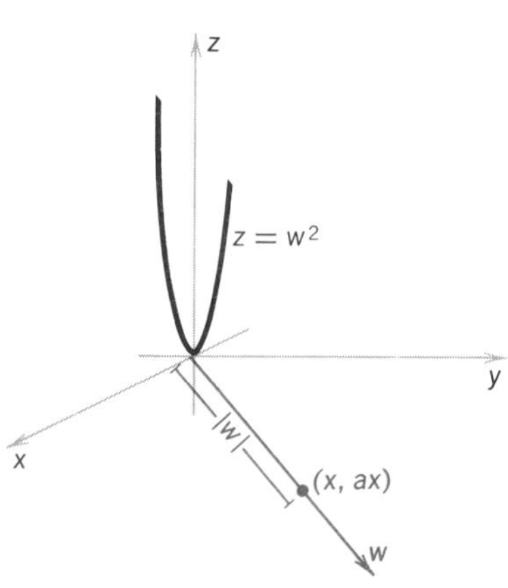

FIGURE 1-2

Now set $z = c$, $c > 0$. Then the equation of the curve traced out in this plane is $x^2 + y^2 = c$, which is the equation of a circle of radius $\sqrt{c}$. From this information we can make a good guess that the surface $z = x^2 + y^2$ is a parabola (such as $z = x^2$) rotated about the z axis. Note that this is not a proof—just a guess! To justify our intuition we introduce a new axis, called the w axis, where w is the line in $\boldsymbol{R}^2$ through (0, 0) with equation $y = ax$ (Figure 1-2). We would like to associate a real number to each point of the w axis in such a way that the equation of the curve obtained by intersecting the surface $z = x^2 + y^2$ with the zw plane is $z = w^2$. This would indeed show that the given surface is a paraboloid of revolution.

To each point (x, ax) on the w axis we associate the real number $w = \sqrt{1 + a^2}\, x$. The absolute value $|w|$ is equal to $\sqrt{1 + a^2}\, |x|$, the distance from (x, ax) to $(0, 0)$. Let us determine the intersections of the graph of f

with the zw plane (the plane perpendicular to the xy plane and passing through the line w). Substituting $y = ax$ into the equation for f we obtain $z = x^2 + (ax)^2 = (1 + a^2)x^2 = w^2$ or $z = w^2$. Thus the equation is of the form $z = w^2$—a parabola! Thus the graph of $z = x^2 + y^2$ does, in fact, describe a paraboloid of revolution.

Example 3 Consider the function $(x, y) \mapsto xy$ or $z = xy$. This function is defined on all of $\boldsymbol{R}^2$. Let us investigate the appearance of its graph.

In this case looking at the intersections of the graph with the xz plane and the yz plane gives us little information; if we set $x = 0$ we get $z = 0$ which is just the y axis (since y may take any value here). Similarly, when $y = 0$ we find that $z = 0$ or the graph is just the x axis. Thus the intersection of the graph with the zx and zy planes is the union of the axes, that is, the x axis and the y axis.

For this function, it is much more useful to consider intersections with the planes $z = c$; for $c > 0$ the intersections are hyperbolas of the form $xy = c$ and for $c < 0$, we also obtain hyperbolas. Therefore, as c varies from $-\infty$ to ∞ these hyperbolas lying in the plane $z = c$ trace out the graph of $z = f(x, y) = xy$ (see Figure 1-3).

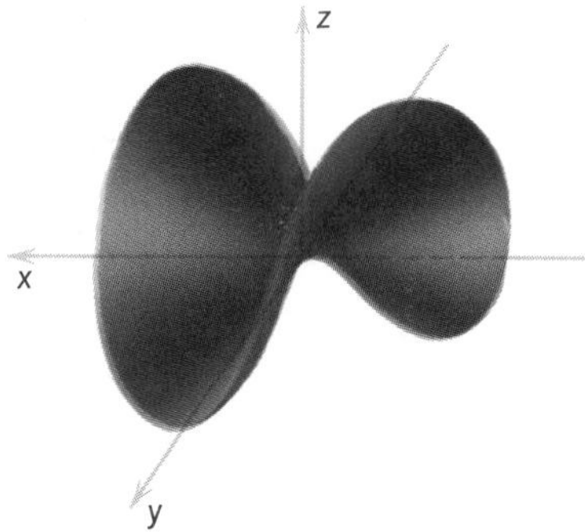

FIGURE 1-3

Example 4 Let us graph the function $z = x^2 - y^2$. We begin by finding the curves which result from intersecting the surface with the xz and yz planes. When $y = 0$, $z = x^2$; when $x = 0$, $z = -y^2$. Thus the intersections with these planes are $z = x^2$ and $z = -y^2$, respectively [see Figure 1-4 (a)].

Next we determine the intersection of the surface with the planes $z = c$, first for $c > 0$ and then for $c < 0$.

For $c > 0$, $z = c = x^2 - y^2$ and the set of (x, y) such that $x^2 - y^2 = c$ is a hyperbola [Figure 1-4 (b)]. As c varies between 0 and ∞ these intersections make up a family of hyperbolas whose vertices border on the curve $z = x^2$ in the xz plane. For $c < 0$ the intersections are the set of points (x, y, c) where $x^2 - y^2 = c$ or $y^2 - x^2 = -c > 0$. The set of (x, y) which satisfy this relation is a hyperbola [Figure 1-4 (c)] whose vertices are on the parabola $z = -y^2$. If one imagines (with some effort, of course) the hyperbolas on the plane $z = c$ varying from $-\infty$ to ∞, one sees that the surface traced out by the hyperbolas is the **saddle** surface shown in Figure 1-4 (d).

In the examples above it was very helpful to consider the intersections of the graph of the given function $z = f(x, y)$ with the plane $z = c$. If we project this set of points onto the xy plane we obtain what is called the **level set** or **level curve** of f at the level c. More explicitly, for any number c in the range of f we can form the set of all points in the plane which f sends to c. This set is nonempty because we assumed that c was in the range of f. Therefore, each level curve of f consists of the set of points (x, y) such that $f(x, y) = c$. (We are abusing, in some sense, the use of the term curve because it is not clear that the set of points we obtain will, in fact, be a curve in $\boldsymbol{R}^2$. There are conditions on f which guarantee that this will be the case but we shall not discuss them here.)

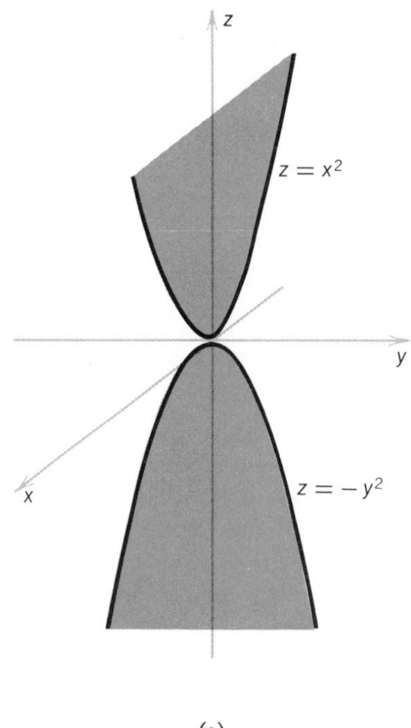

(a)

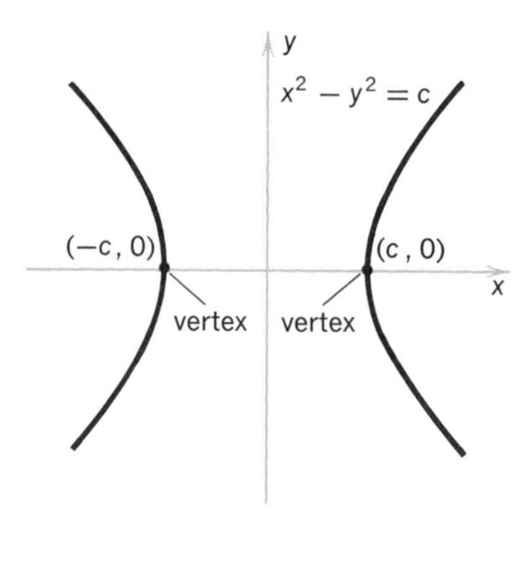

(b)

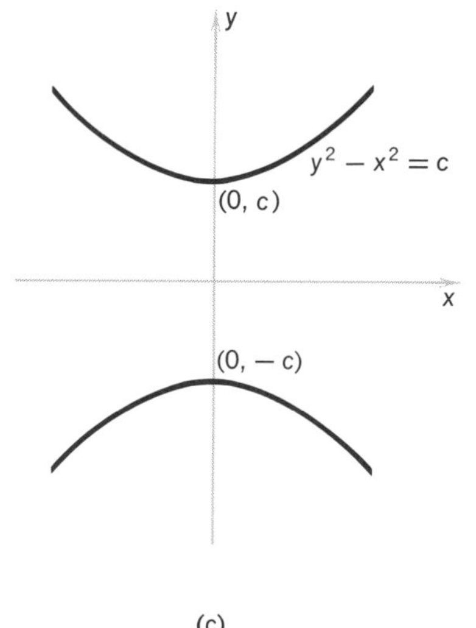

(c)

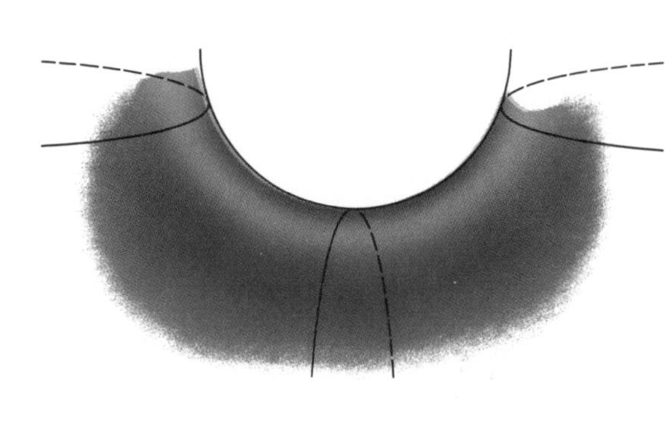

(d)

FIGURE 1-4

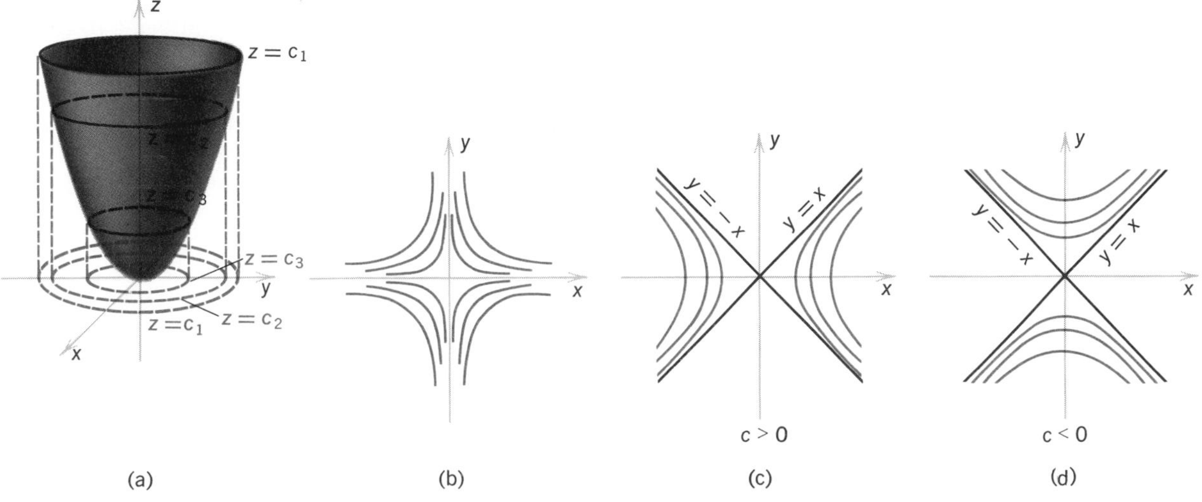

FIGURE 1-5

Level curves provide another method for representing f. One picks several constants c and draws the level curves for each one, obtaining a configuration of level curves in the xy plane. For instance, in Example 2, the level curves are circles of radius $\sqrt{c}$ (see Figure 1-5); in Example 3, they are the hyperbolas $xy = c$, $-\infty < c < \infty$; and in Example 4 they are the hyperbolas $x^2 - y^2 = c$, $-\infty < c < \infty$.

The concept of level curve has several physical applications. Cartographers draw **contour lines** on their maps; these lines are the level curves of the function h, where h is the height above sea level (Figure 1-6). In addition, meteorologists draw **isobars** on weather maps; these are just curves indicating constant barometric pressure (that is, the level curves of the barometric pressure function).

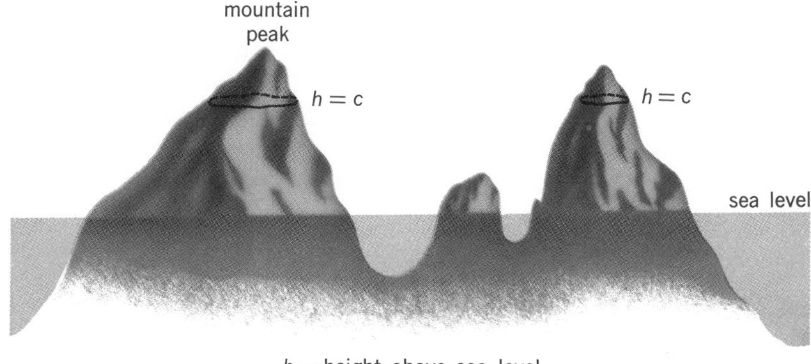

FIGURE 1-6

Just as the notion of an unbroken graph for a function of one variable led to the concept of continuity of the function, any attempt to define precisely the notion of an unbroken graph for a function of two variables will lead us to the concept of continuity of such functions.

The question here is how to describe mathematically the fact that a surface (or graph) is unbroken. For example, the surface

$$z = \begin{cases} 3, & x \geqslant 0 \\ 0, & x < 0 \end{cases}$$

(Figure 1-7) is a broken surface.

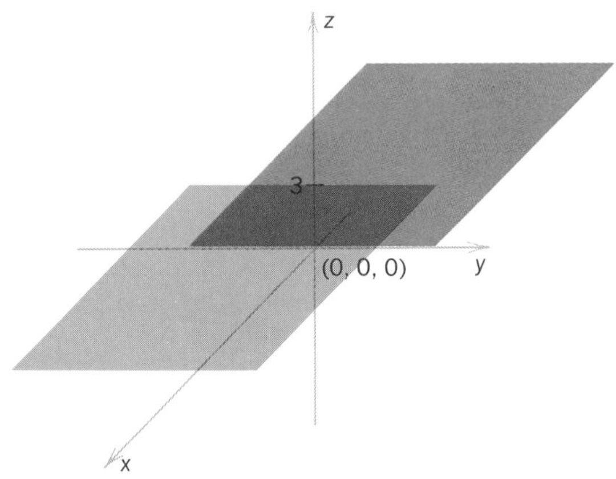

FIGURE 1-7

In order to define continuity of a function of two variables, we must first explain what we mean by continuity at a point $v_0 \in \boldsymbol{R}^2$. Intuitively, the function f is continuous if for v "close to" v_0, $f(v)$ is close to $f(v_0)$. The difficulty is how to say this in a precise mathematical way. We solve this problem by introducing the concept of a neighborhood in $\boldsymbol{R}^2$.

An **open disc** in $\boldsymbol{R}^2$ with center v_0 and radius r is defined to be the set of points v satisfying $|v - v_0| < r$. This is just the usual disc of radius r centered at v_0 with the bounding circle $|v - v_0| = r$ removed.

A **neighborhood** U of a point $v_0 \in \boldsymbol{R}^2$ is an open disc of some radius r with center v_0; that is,

$$|v - v_0| < r$$

A typical neighborhood U is shown in Figure 1-8.

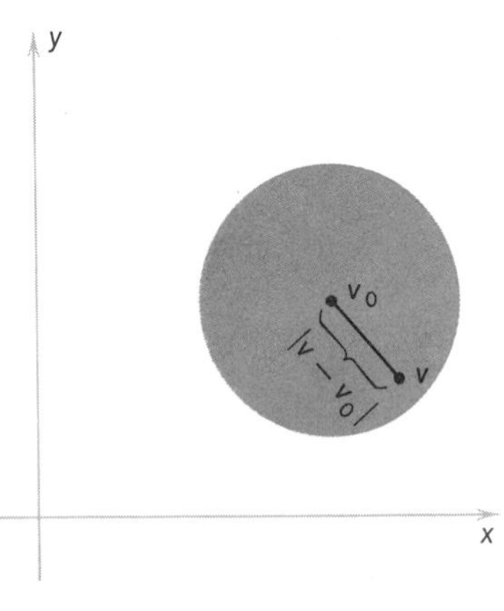

FIGURE 1-8

Therefore, when we say "v is close to v_0" we mean that v is in some neighborhood of v_0. Next, let us formalize the statement "$f(v)$ is close to $f(v_0)$ when v is close to v_0."

For $f(v)$ to be close to $f(v_0)$ means that $f(v)$ is in some neighborhood of $f(v_0)$. Hence, we say that f is continuous at v_0 if given any neighborhood N of $f(v_0)$, we can find a neighborhood U of v_0 such that $v \in U$ implies that $f(v) \in N$. The standard way of stating this is

> **f is continuous at v_0 if given any $\varepsilon > 0$ there is some $\delta > 0$ such that $|f(v) - f(v_0)| < \varepsilon$ if $|v - v_0| < \delta$**

This definition is equivalent to the one stated immediately above it.

A function $f: \boldsymbol{R}^2 \to \boldsymbol{R}$ is **continuous** if it is continuous at each point $v \in \boldsymbol{R}^2$.

Let us consider the example given earlier

$$f(x, y) = z = \begin{cases} 3, & x \geqslant 0 \\ 0, & x < 0 \end{cases}$$

This function is *not* continuous at any point $(0, y)$ for any $y \in \boldsymbol{R}$. To see this, let y be fixed and note that $f(0, y) = 3$ but $f(x, y) = 0$ if $x < 0$. Thus we can take x as close as we want to 0 [implying that (x, y) will be as close as we like to $(0, y)$] but the value $f(x, y)$ will *not* be close to the value $f(0, y)$ because one value is 0 and the other is 3.

Example 5 Show that $f(x, y) = x^2 - y^2$ is continuous. Let $v_0 = (x_0, y_0) \in \boldsymbol{R}^2$ and let $\varepsilon > 0$ be arbitrary. We must find a $\delta > 0$ such that if $|v - v_0| < \delta$ then $|f(v) - f(v_0)| < \varepsilon$. This means that if v belongs to the disc of radius δ with center at v_0, then $f(v)$ belongs to the interval length 2ε with center at $f(v_0)$ (see Figure 1-9). Now, if $v = (x, y)$,

$$\begin{aligned} |f(v) - f(v_0)| &= |x^2 + y^2 - x_0^2 - y_0^2| \leqslant |x^2 - x_0^2| + |y^2 - y_0^2| \\ &= |x - x_0|\,|x + x_0| + |y + y_0|\,|y - y_0| \\ &\leqslant (|x| + |x_0|)(|x - x_0|) + (|y| + |y_0|)(|y - y_0|) \qquad (1) \end{aligned}$$

If $|v - v_0| < 1$ then $|x - x_0| < 1$ and $|y - y_0| < 1$. So $|x| \leqslant |x_0| + 1$ and $|y| \leqslant |y_0| + 1$. Therefore, we can continue inequality (1) to obtain

$$\begin{aligned} |f(v) - f(v_0)| &\leqslant (2|x_0| + 1)(|x - x_0|) + (2|y_0| + 1)(|y - y_0|) \\ &\leqslant K(|x - x_0| + |y - y_0|) \end{aligned}$$

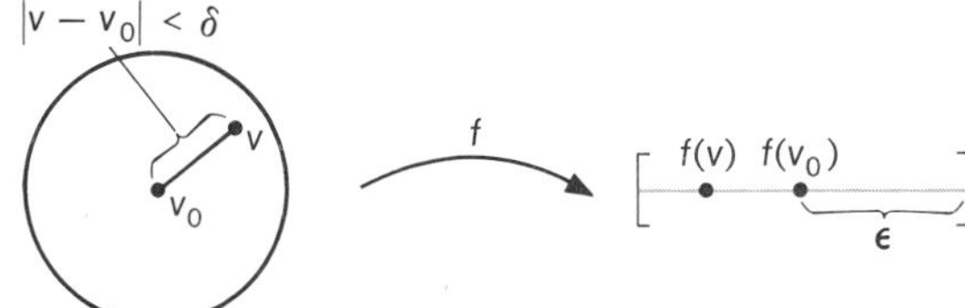

FIGURE 1-9

where $K = \max\{(2|x_0|+1), (2|y_0|+1)\}$. If we select $\delta = \min(1, \varepsilon/2K)$ then if $|v-v_0| < \delta$, $|x-x_0| < \delta$, and $|y-y_0| < \delta$ so

$$|f(v)-f(v_0)| \leqslant K\{(|x-x_0| + |y-y_0|)\} \leqslant K(\varepsilon/2K+\varepsilon/2K) = \varepsilon$$

This proves that f is continuous at any point $v_0 \in \boldsymbol{R}^2$ and, therefore, f is continuous; thus the graph of f which is the surface of Example 4 is unbroken.

If f and g have the same domain, and each is continuous, then the product function fg and the sum function $f+g$ are also continuous. In addition, the quotient f/g is continuous at all points v_0 where $g(v_0) \neq 0$.

For example, to show that the function

$$f(x,y) = \frac{3x^3y + 4x^2 + 2}{x^4 + y^2 + 1}$$

is continuous, all we need know is that the projection functions $p_1(x,y) = x$ and $p_2(x,y) = y$ are continuous (which we prove below), because

$$f(x,y) = \frac{3p_1^3(x,y) \cdot p_2(x,y) + 4 \cdot p_1^2(x,y) + 2}{p_1^4(x,y) + p_2^2(x,y) + 1}$$

where $p_i^k(x,y) = [p_i(x,y)]^k$. If the p_i, $i = 1,2$, are continuous then the numerator and the denominator of f are the sum of products of the p_i, thus they are both continuous. Moreover, the denominator of f is never 0 and so the quotient of the numerator and denominator is continuous.

Finally, to show that the projection function p_1 is indeed continuous (the proof for p_2 is analogous), let $v_0 \in \boldsymbol{R}^2$, $v_0 = (x_0, y_0)$ and let $\varepsilon > 0$ be arbitrary. Since

$$|p_1(v)-p_1(v_0)| = |x-x_0| \leqslant |v-v_0| = \sqrt{|x-x_0|^2 + |y-y_0|^2}$$

it follows that if $|v-v_0| < \varepsilon$ then $|p_1(v)-p_1(v_0)| < \varepsilon$. Therefore, picking $\delta = \varepsilon$ we can see that $|v-v_0| < \delta$ implies $|p_1(v)-p_1(v_0)| < \varepsilon$ and the proof is complete.

Our earlier results about continuity (§2.4) all have important analogues for functions of two variables. For example, if $h: \boldsymbol{R} \to \boldsymbol{R}$ is a continuous function then the composition $(h \circ f)(v) = h(f(v))$ is continuous. This result allows us to conclude that $\phi(x,y) = \sin xy$ is continuous because $\phi = (\sin \circ p_1 p_2)(x,y)$ and since $p_1 p_2$ is continuous, the composition is also continuous. (The proofs of other results are left as exercises.)

The fact that $f(v)$ approaches $f(v_0)$ if f is continuous at v_0 is often written

$$\lim_{v \to v_0} f(v) = f(v_0)$$

and read "the limit of $f(v)$ as v approaches v_0 equals $f(v_0)$." On the other hand, it may be that f is not continuous at v_0 but that $f(v)$ still gets close to some number A as v approaches v_0. Hence, we want to define what is meant by the more general statement

$$\lim_{v \to v_0} f(v) = A$$

As an example, consider the function

$$f(x, y) = \frac{\sin(x^2+y^2)}{x^2+y^2}$$

Even though f is not defined at $(0,0)$ we can ask whether $f(x,y)$ approaches some number as (x,y) approaches $(0,0)$. From one-variable calculus we know that

$$\lim_{\alpha \to 0} \frac{\sin \alpha}{\alpha} = 1$$

Thus a reasonable guess here is that

$$\lim_{v \to (0,0)} f(v) = \lim_{v \to (0,0)} \frac{\sin(|v|^2)}{|v|^2} = 1$$

Before we can prove this, we must define the limit notion in this context.

Paraphrasing one of our earlier definitions we say that

> $\lim_{v \to v_0} f(v) = A$ if given $\varepsilon > 0$ there is a $\delta > 0$ such that $0 < |v - v_0| < \delta$ implies $|f(v) - A| < \varepsilon$

We require $0 < |v - v_0|$ because we are interested only in the values of f near (but not necessarily at) v_0. The condition $0 < |v - v_0| < \delta$ defines a disk centered at v_0 with the point v_0 removed; we call such a punctured disk a **deleted neighborhood** of v_0. The above definition of limit can therefore be reworded to read

> $\lim_{v \to v_0} f(v) = A$ if and only if for every open interval neighborhood N of A there is a deleted neighborhood $D(v_0)$ of v_0 such that $v \in D(v_0) \Rightarrow f(v) \in N$

The proofs of the theorems in §2.3 on sums and products of limits now can be seen to carry over at once to limits of real-valued functions of two variables. We can also restate the definition of continuity in terms of limits; we have

> $f: \boldsymbol{R}^2 \to \boldsymbol{R}$ is continuous at $v_0 \in \boldsymbol{R}^2$ if $\lim_{v \to v_0} f(v) = f(v_0)$

[see exercise 3].

We are now ready to justify our guess. Since $\lim_{\alpha \to 0} (\sin \alpha / \alpha) = 1$, given $\varepsilon > 0$ we can find a $\delta > 0$, with $1 > \delta > 0$, such that $|\alpha| < \delta$ implies that $|(\sin \alpha/\alpha) - 1| < \varepsilon$. If $|v| < \delta$, then $|v|^2 < \delta^2 < \delta$, hence

$$|f(v) - 1| = \left| \frac{\sin(|v|^2)}{|v|^2} - 1 \right| < \varepsilon$$

Thus $\lim_{v \to (0,0)} f(v) = 1$.

EXERCISES

1. Show that if $f: \boldsymbol{R}^2 \to \boldsymbol{R}$ and $\lim_{v \to v_0} f(v)$ exists, then the limit is unique; that is, if there exist numbers b_1 and b_2 such that for every $\varepsilon > 0$ there is a $\delta_1 > 0$ and a $\delta_2 > 0$ such that $0 < |v - v_0| < \delta_i$ implies $|f(v) - b_i| < \varepsilon$ for $i = 1, 2$, then $b_1 = b_2$.

2. Prove the following limit theorems for functions $f, g: \boldsymbol{R}^2 \to \boldsymbol{R}$.

   (a) $\lim\limits_{v \to v_0} [f(v) + g(v)] = \lim\limits_{v \to v_0} f(v) + \lim\limits_{v \to v_0} g(v)$

   (b) $\lim\limits_{v \to v_0} f(v) g(v) = \lim\limits_{v \to v_0} f(v) \lim\limits_{v \to v_0} g(v)$

   (c) If $\lim\limits_{v \to v_0} g(v) \neq 0$ then $\lim\limits_{v \to v_0} [f(v)/g(v)] = \lim\limits_{v \to v_0} f(v) / \lim\limits_{v \to v_0} g(v)$

3. Use exercise 2 to show that the sum, product, and quotient (if the denominator is nonzero) of continuous functions is continuous.

4. (a) Suppose that $f: \boldsymbol{R}^2 \to \boldsymbol{R}$ and $h: \mathbf{R} \to \mathbf{R}$ are continuous. Prove that $h \circ f: \boldsymbol{R}^2 \to \boldsymbol{R}$ is continuous.

   (b) Suppose that h and f are not necessarily continuous. Can $h \circ f$ still be continuous?

In exercises 5–10 draw the level curves (in the xy plane) for the given function f and specified values of c. Sketch the graph of $z = f(x, y)$.

5. $f(x, y) = (100 - x^2 - y^2)^{1/2}$, $c = 0, 2, 4, 6, 8, 10$
6. $f(x, y) = (x^2 + y^2)^{1/2}$, $c = 0, 1, 2, 3, 4, 5$
7. $f(x, y) = x^2 + y^2$, $c = 0, 1, 2, 3, 4, 5$
8. $f(x, y) = 3x - 7y$, $c = 0, 1, 2, 3, -1, -2, -3$
9. $f(x, y) = x^2 + xy$, $c = 0, 1, 2, 3, -1, -2, -3$
10. $f(x, y) = x/y$

In exercises 11–13, find the indicated limits.

11. $\lim\limits_{(x,y) \to (0,0)} \dfrac{e^{xy} - 1}{x}$

*12. $\lim\limits_{(x,y) \to (0,0)} xy(x^2 + y^2)^{-1/2}$

*13. $\lim\limits_{(x,y) \to (0,0)} xy(x^2 - y^2)/(x^2 + y^2)$

14. For $(x, y) \neq (0, 0)$ define $f(x, y) = x^2 y / (x^4 + y^2)$. Show that

$$\lim_{(x,y) \to (0,0)} f(x, y)$$

does not exist as follows: First, compute $\lim_{x \to 0} g(x)$, where $g(x) = f(x, mx)$ [that is, the limit of $f(x, y)$ along the line $y = mx$]. Next, compute $\lim_{x \to 0} h(x)$, where $h(x) = f(x, x^2)$ [that is, the limit of $f(x, y)$ along the parabola $y = x^2$]. Now deduce a contradiction.

*15. Define $f: \boldsymbol{R}^2 \to \boldsymbol{R}$ by

$$f(x,y) = \begin{cases} \dfrac{xy}{(x^2+y^2)^{1/2}}, & (x,y) \neq 0 \\ 0, & (x,y) = 0 \end{cases}$$

Show that f is continuous.

2 DIFFERENTIABILITY

In this section we begin the study of the differential calculus of functions from $\boldsymbol{R}^2$ to $\boldsymbol{R}$. The primary question is what does it mean for a function of two variables $z = f(x, y)$ to be differentiable? This simple question is filled with pitfalls, and its answer is not as simple as it would seem. However, our discussion will clarify the concept of differentiability and provide sufficient conditions for determining when functions are differentiable.

In the past people tied themselves too strongly to the notion of the derivative of a function of one variable, and thus a truly intrinsic understanding of the derivative of a function of more than one real variable has developed only recently. Since we cannot present these modern ideas fully without deeper results from other areas of mathematics, we have chosen an approach which falls between the modern and classical.

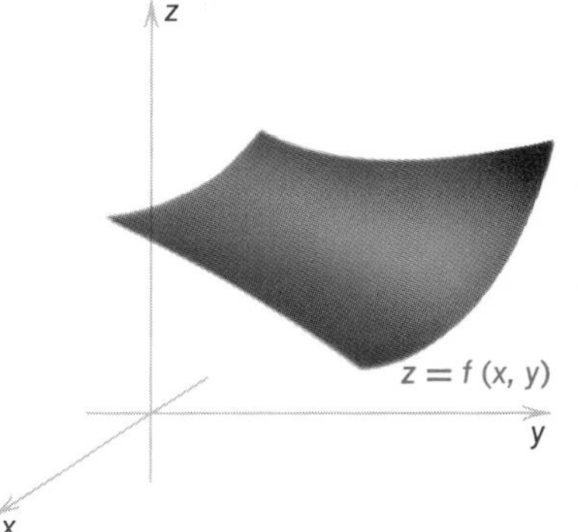

FIGURE 2-1

We know that if a function of one variable is differentiable, the geometric meaning is that its graph has a tangent line at each point. In fact, this notion leads to the definition of derivative of such functions. In the two-variable case, we think of the graph of a function $f: \boldsymbol{R}^2 \to \boldsymbol{R}$ as a surface $S \subset \boldsymbol{R}^3$ (Figure 2-1). For one variable, we found that differentiability implied continuity of the function and also a "smoothness" for the graph; that is, the function did not change direction too rapidly and, hence, the graph was free of cusps or corners. Analogously, for two variables, differentiability means not only that there are no breaks, but also that the surface is "smooth" in some sense; that is, S has no corners, edges, or peaks such as in Figure 2-2.

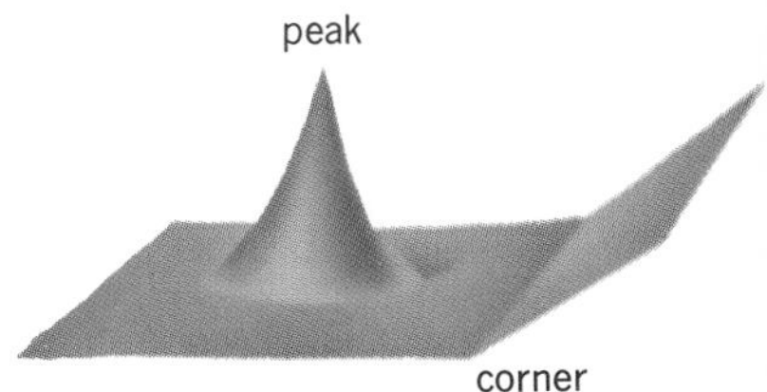

FIGURE 2-2

How can we make these intuitive ideas precise? Our first attempt might be to mimic the definition of differentiability for a function of one variable. Suppose we let $v_0 = (x_0, y_0) \in \boldsymbol{R}^2$ and consider

$$\lim_{\Delta v \to 0} \frac{f(v_0 + \Delta v) - f(v_0)}{\Delta v}$$

where

$$\Delta v = (\Delta x, \Delta y)$$

But wait! We cannot do this. Why? What does it mean to divide by a vector Δv? We do not know, and, in fact, the operation does not seem to have any meaning. So we must try again. Since we have a workable definition for one-variable calculus, let us continue trying to generalize it. We can make another observation: If we are trying to determine what it means for $f:\boldsymbol{R}^2 \to \boldsymbol{R}$ to be differentiable at a point $(x_0, y_0) \in \boldsymbol{R}^2$ suppose we "fix" one variable at a time to obtain functions of a single variable. Consider the two functions

$$x \mapsto f(x, y_0) \qquad \text{and} \qquad y \mapsto f(x_0, y) \tag{1}$$

These are real-valued functions of one real variable and it makes sense to require that they be differentiable if f is to be differentiable. Therefore, we now define the **partial derivatives** $(\partial f/\partial x)(x_0, y_0)$, $(\partial f/\partial y)(x_0, y_0)$; they are the derivatives of the functions in (1) above, respectively. Equivalently, these partial derivatives can be defined as follows:

$$\frac{\partial f}{\partial x}(x_0, y_0) = \lim_{\Delta x \to 0} \frac{f(x_0 + \Delta x, y_0) - f(x_0, y_0)}{\Delta x}$$

$$\frac{\partial f}{\partial y}(x_0, y_0) = \lim_{\Delta y \to 0} \frac{f(x_0, y_0 + \Delta y) - f(x_0, y_0)}{\Delta y}$$

provided that these limits exist. Note that $(\partial f/\partial x)(x_0, y_0)$, reads "the partial derivative of f with respect to x at (x_0, y_0)," and $(\partial f/\partial y)(x_0, y_0)$ are both real numbers. Now it is very tempting to say that f is differentiable at (x_0, y_0) if both limits exist. However, we must emphasize that although this sounds reasonable, there are pitfalls. Before we examine just what the problems are, let us give a geometric interpretation of the partial derivatives.

Let $(x_0, y_0) \in \boldsymbol{R}^2$ and suppose we pass a plane through the point (x_0, y_0) perpendicular to the xy plane and parallel to the x axis (see Figure 2-3). In other words, we hold y constant at y_0. If $f:\boldsymbol{R}^2 \to \boldsymbol{R}$, then, intuitively, the intersection of this plane with the graph of f is some curve C_1 in this plane. Since C_1 is given by $C_1 : x \mapsto f(x, y_0)$, we only need to deal with a function of one variable.

The existence of $(\partial f/\partial x)(x_0, y_0)$ means that the curve C_1 is differentiable at (x_0, y_0) and also that $(\partial f/\partial x)(x_0, y_0)$ is the slope of the tangent line T_1 to C_1 in this plane. Similar remarks hold for the curve $C_2 : y \mapsto f(x_0, y)$ in the plane passing through (x_0, y_0), perpendicular to the xy plane, and parallel to the y axis.

Now let us return to the problem of defining differentiability for functions of two variables. Consider the function $f(x, y) = \sin x - \sin y + x^{1/3} y^{1/3}$. We shall compute $(\partial f/\partial x)(0, 0)$ and $(\partial f/\partial y)(0, 0)$. By definition

$$\begin{aligned} \frac{\partial f}{\partial x}(0, 0) &= \lim_{\Delta x \to 0} \frac{f(0 + \Delta x, 0) - f(0, 0)}{\Delta x} \\ &= \lim_{\Delta x \to 0} \frac{\sin \Delta x}{\Delta x} = \sin'(0) = 1 \end{aligned}$$

and analogously we find that

$$\frac{\partial f}{\partial y}(0,0) = -1$$

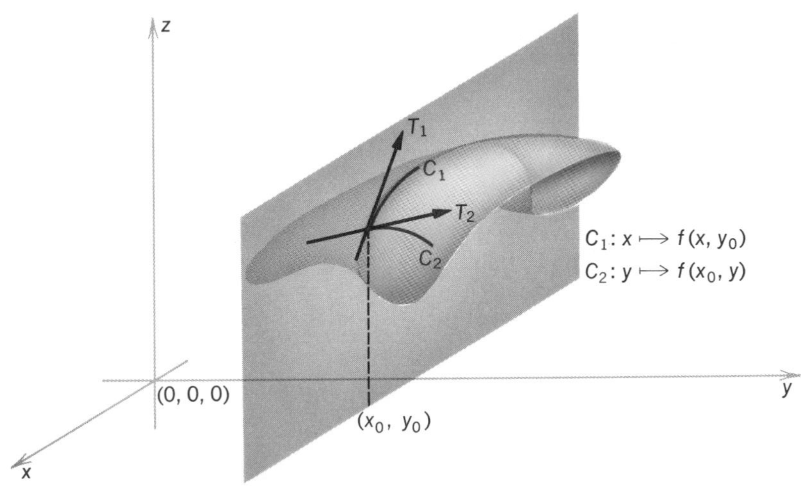

FIGURE 2-3

Now suppose that instead of approaching the origin along the x axis or the y axis, we approach along the line $y = x$; that is, we consider the curve C obtained when the plane perpendicular to the xy plane and passing through the line $y = x$ intersects the graph of the function (see Figure 2-4).

On the line $y = x$, the given function becomes

$$x \mapsto f(x,x) = \sin x - \sin x + x^{2/3} = x^{2/3} = h(x)$$

This function h has no derivative at $(0,0)$. To see this, note that

$$\frac{h(0+\Delta x) - h(0)}{\Delta x}$$

$$= \frac{(\Delta x)^{2/3}}{\Delta x} = \frac{1}{(\Delta x)^{1/3}}$$

which tends to ∞ as Δx tends toward 0. Therefore, even if $(\partial f/\partial x)(x_0, y_0)$ and $(\partial f/\partial y)(x_0, y_0)$ exist it may be that not all curves obtained by intersecting planes are differentiable at (x_0, y_0). Thus, if we defined "differentiable at (x_0, y_0)" to mean only the existence of $(\partial f/\partial x)(x_0, y_0)$ and $(\partial f/\partial y)(x_0, y_0)$, it would not be satisfactory.

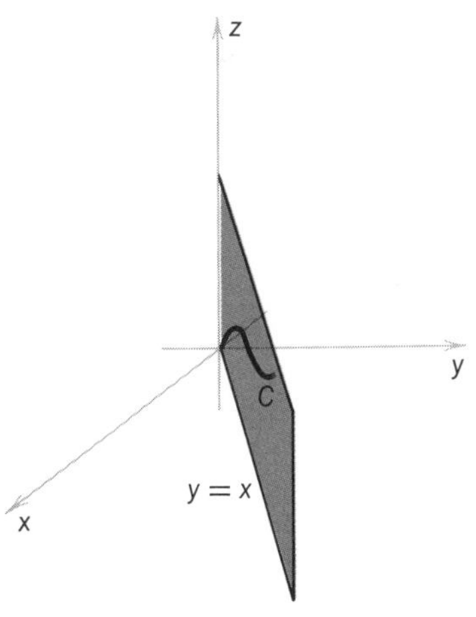

FIGURE 2-4

We have almost arrived at our final definition of differentiability for a function of two variables; however, we must again look at the one-variable case. To generalize the concept of differentiability to functions from $\boldsymbol{R}^2$ to $\boldsymbol{R}$, we use **differential approximation** as our starting point. If $f: \boldsymbol{R} \to \boldsymbol{R}$ is differentiable at x_0, we have

$$\lim_{\Delta x \to 0} \frac{f(x_0+\Delta x) - f(x_0)}{\Delta x} = f'(x_0)$$

This can be rewritten as

$$\lim_{\Delta x \to 0} \frac{f(x_0+\Delta x) - f(x_0) - f'(x_0)\,\Delta x}{\Delta x} = 0$$

Let $Q_{x_0}(\Delta x) = f(x_0+\Delta x) - f(x_0) - f'(x_0)\,\Delta x$; then from the equation immediately above it follows that

$$f(x_0+\Delta x) - f(x_0) = f'(x_0)\,\Delta x + Q_{x_0}(\Delta x)$$

where

$$\lim_{\Delta x \to 0} \frac{Q_{x_0}(\Delta x)}{\Delta x} = 0$$

Hence, $Q_{x_0}(\Delta x)$ is the "error term" in the approximation of the difference $f(x_0+\Delta x) - f(x_0)$ by $f'(x_0)\,\Delta x$. Since we have not only $\lim_{\Delta x \to 0} Q_{x_0}(\Delta x) = 0$, but also $\lim_{\Delta x \to 0} [Q_{x_0}(\Delta x)/\Delta x] = 0$, we say that $f'(x_0)\,\Delta x$ approximates the change in the function up to **second order.**

To extend this analysis to functions $f: \boldsymbol{R}^2 \to \boldsymbol{R}$ we consider a vector $u_0 = (x_0, y_0)$ and a vector $\Delta u = (\Delta x, \Delta y)$ which gives the change in u. Then we say that f is differentiable at u_0 if there are real numbers α and β (whose values we ignore for now) such that $\alpha\Delta x + \beta\Delta y$ approximates $f(u_0+\Delta u) - f(u_0)$ up to second order. Hence we have the following definition.

Definition The function $f: \boldsymbol{R}^2 \to \boldsymbol{R}$ is **differentiable** at a point $u_0 \in \boldsymbol{R}^2$ if there is a function $Q_{u_0}: \boldsymbol{R}^2 \to \boldsymbol{R}$ and numbers α, β such that

(i) $\displaystyle\lim_{\Delta u \to 0} \frac{Q_{u_0}(\Delta u)}{|\Delta u|} = 0$

(ii) $f(u_0+\Delta u) - f(u_0) = \alpha\Delta x + \beta\Delta y + Q_{u_0}(\Delta u)$

Notice that in condition (i) it would make no sense to divide by the vector Δu; instead we divide by the norm $|\Delta u|$. In terms of coordinates, if $\Delta u = (\Delta x, \Delta y)$ we have

$$\begin{aligned} Q_{u_0}(\Delta u) &= f(u_0+\Delta u) - f(u) - \alpha\Delta x - \beta\Delta y \\ &= f(x_0+\Delta x, y_0+\Delta y) - f(x_0, y_0) - \alpha\Delta x - \beta\Delta y \end{aligned}$$

Thus the differentiability of f at (x_0, y_0) may be expressed by the condition

$$\lim_{(\Delta x, \Delta y) \to (0,0)} \frac{f(x_0+\Delta x, y_0+\Delta y) - f(x_0, y_0) - \alpha\Delta x - \beta\Delta y}{|(\Delta x, \Delta y)|} = 0$$

Let us now determine exactly what the numbers α and β are. Our guess is that $\alpha = (\partial f/\partial x)(x_0, y_0)$ and $\beta = (\partial f/\partial y)(x_0, y_0)$. To show that our guess is indeed correct we fix y_0 and let $\Delta y = 0$. Then if f is differentiable at (x_0, y_0)

$$\lim_{\Delta x \to 0} \frac{f(x_0+\Delta x, y_0) - f(x_0, y_0) - \alpha\Delta x}{\Delta x} = 0$$

so

$$\lim_{\Delta x \to 0} \frac{f(x_0+\Delta x, y_0) - f(x_0, y_0) - \alpha\Delta x}{\Delta x} = 0$$

or

$$\lim_{\Delta x \to 0} \frac{f(x_0+\Delta x, y_0) - f(x_0, y_0)}{\Delta x} - \alpha = 0$$

But the limit of the first term is, by definition, $(\partial f/\partial x)(x_0, y_0)$. Therefore

$$\alpha = \lim_{\Delta x \to 0} \frac{f(x_0+\Delta x, y_0) - f(x_0, y_0)}{\Delta x} = \frac{\partial f}{\partial x}(x_0, y_0)$$

In a similar fashion it follows that $(\partial f/\partial y)(x_0, y_0) = \beta$.

Thus if f is differentiable at (x_0, y_0) the partial derivatives of f exist. Moreover, as we shall see in the section on directional derivatives, f will be differentiable along any line in $\boldsymbol{R}^2$ passing through (x_0, y_0); that is, the curve obtained from the intersection of the graph of f with *any* plane passing through the point (x_0, y_0) and perpendicular to the xy plane will be differentiable at (x_0, y_0).

Computation of partial derivatives plays a considerable role in the material which follows, so let us work out some examples.

Example 1 If $z = \cos xy + x\cos y = f(x, y)$, find $(\partial z/\partial x)(x_0, y_0)$ and $(\partial z/\partial y)(x_0, y_0)$. (Note we are using the notation $\partial z/\partial x$, $\partial z/\partial y$ for $\partial f/\partial x$, $\partial f/\partial y$.) First we fix y_0 and differentiate with respect to x. So

$$\begin{aligned}\frac{\partial z}{\partial x}(x_0, y_0) &= \left.\frac{d(\cos xy_0 + x\cos y_0)}{dx}\right|_{x=x_0} \\ &= -y_0 \sin xy_0 + \cos y_0\big|_{x=x_0} \\ &= -y_0 \sin x_0 y_0 + \cos y_0\end{aligned}$$

Similarly, we fix x_0 and differentiate with respect to y to obtain

$$\begin{aligned}\frac{\partial z}{\partial y}(x_0, y_0) &= \left.\frac{d(\cos x_0 y + x_0\cos y)}{dy}\right|_{y=y_0} \\ &= -x_0 \sin x_0 y - x_0 \sin y\big|_{y=y_0} \\ &= -x_0 \sin x_0 y_0 - x_0 \sin y_0\end{aligned}$$

Quite often the notation $\partial f/\partial x$, $\partial f/\partial y$ is used in place of $(\partial f/\partial x)(x_0, y_0)$ and $(\partial f/\partial y)(x_0, y_0)$. If f is differentiable at each point in $\boldsymbol{R}^2$ then for each (x_0, y_0) we have numbers $(\partial f/\partial x)(x_0, y_0)$ and $(\partial f/\partial y)(x_0, y_0)$. Then $(x, y) \mapsto (\partial f/\partial x)(x, y)$ and $(x, y) \mapsto (\partial f/\partial y)(x, y)$ are real-valued functions of two variables. They are denoted by $\partial f/\partial x$ and $\partial f/\partial y$ or $\partial z/\partial x$ and $\partial z/\partial y$, respectively.

Example 2 If $f(x, y) = x^2 y + y^3$, find $\partial f/\partial x$ and $\partial f/\partial y$. To find $\partial f/\partial x$ we hold y constant and differentiate only with respect to x; this yields

$$\frac{\partial f}{\partial x} = \frac{d(x^2 y + y^3)}{dx} = 2xy$$

Also,

$$\frac{\partial f}{\partial y} = \frac{d(x^2 y + y^3)}{dx} = x^2 + 3y^2$$

Example 3 Find $\partial f/\partial x$ if $f(x, y) = xy/\sqrt{x^2 + y^2}$. We have

$$\frac{\partial f}{\partial x} = \frac{y\sqrt{x^2+y^2} - xy(x/\sqrt{x^2+y^2})}{x^2 + y^2}$$

$$= \frac{y(x^2+y^2) - xy(x)}{(x^2+y^2)^{3/2}} = \frac{y^3}{(x^2+y^2)^{3/2}}$$

If f is differentiable everywhere it is defined and $\partial f/\partial x$ and $\partial f/\partial y$ are continuous, we say that f is **continuously differentiable** or C^1. The next example shows that there are functions which are differentiable but not continuously differentiable.

Example 4 Let

$$f(x, y) = \begin{cases} x^2 \sin(1/x), & x \neq 0 \\ 0 & x = 0 \end{cases}$$

Then if $x \neq 0$

$$\frac{\partial f}{\partial x}(x, y) = 2x \sin\frac{1}{x} + x^2 \cos\frac{1}{x}\left(\frac{-1}{x^2}\right)$$

$$= 2x \sin\frac{1}{x} - \cos\frac{1}{x}$$

For $x = 0$, we compute $(\partial f/\partial x)(0, y)$ directly from the definition:

$$\frac{\partial f}{\partial x}(0, y) = \lim_{\Delta x \to 0} \frac{f(\Delta x, y) - f(0, y)}{\Delta x} = \lim_{\Delta x \to 0} \frac{(\Delta x)^2 \sin\left(\dfrac{1}{\Delta x}\right)}{\Delta x}$$

$$= \lim_{\Delta x \to 0} \Delta x \cdot \sin\frac{1}{\Delta x}$$

But $|\sin h| \leqslant 1$ for all h, so in particular

$$0 \leqslant \left|\Delta x \cdot \sin \frac{1}{\Delta x}\right| \leqslant |\Delta x|$$

Therefore

$$0 \leqslant \lim_{\Delta x \to 0} \left|\Delta x \cdot \sin \frac{1}{\Delta x}\right| \leqslant \lim_{\Delta x \to 0} |\Delta x| = 0$$

from which we can conclude that

$$\lim_{\Delta x \to 0} \Delta x \cdot \sin \frac{1}{\Delta x} = 0$$

and thus

$$\frac{\partial f}{\partial x}(0, y) = 0$$

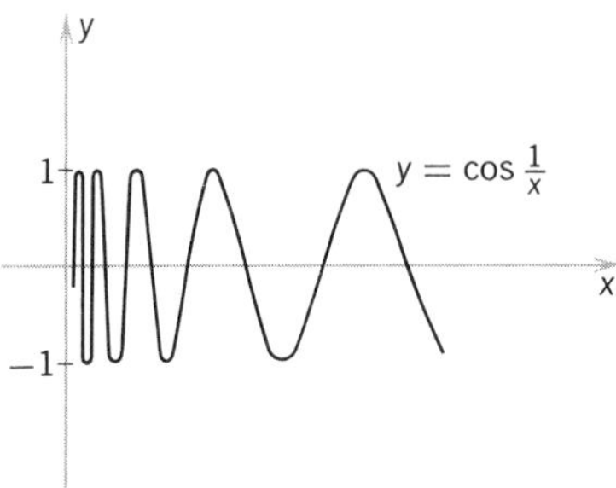

FIGURE 2-5

Hence $\partial f/\partial x$ exists everywhere. Also, $\partial f/\partial y = 0$ for all $(x, y) \in \boldsymbol{R}^2$ because y does not appear in the expression for f. Thus f has partial derivatives everywhere but $\partial f/\partial x$ is not a continuous function. Let us show why this is so.

Look at $(\partial f/\partial x)(1/n\pi, y)$. As $n \to \infty$, $x = 1/n\pi \to 0$. If x is very small, the term $2x \sin(1/x)$ is small because $|2x \sin(1/x)| \leqslant |2x|$. But the value of $\cos(1/x) = \cos n\pi$ for $x = 1/n\pi$ oscillates between ± 1 depending on the value of n (see Figure 2-5). Thus $(\partial f/\partial x)(1/n\pi, y)$ oscillates between values close to $+1$ and -1 as $n \to \infty$. Therefore, $\partial f/\partial x$ is not continuous at $(0, y)$ for any $y \in \boldsymbol{R}$. In exercise 10 the reader is asked to demonstrate the differentiability of the function f.

In order to show that a given $f: \boldsymbol{R}^2 \to \boldsymbol{R}$ is differentiable at $v_0 \in \boldsymbol{R}^2$ it is necessary to show that the quotient $Q_{v_0}(\Delta v)/|\Delta v| \to 0$ as $|\Delta v| \to 0$. This demonstration can often be complicated and time consuming. It would be useful to have at hand some alternative criterion for judging whether f is differentiable. We know from previous examples that the existence of $\partial f/\partial x$ and $\partial f/\partial y$ is not enough (that is, their existence is a necessary, but not a sufficient, condition). However, it turns out that if $\partial f/\partial x$ and $\partial f/\partial y$ exist and are continuous then f is differentiable. We state this formally as a theorem.

Theorem 1 Suppose $f: \boldsymbol{R}^2 \to \boldsymbol{R}$ is such that $\partial f/\partial x$ and $\partial f/\partial y$ exist and are continuous. Then f is differentiable. The proof of Theorem 1 is outlined in exercise 13.

Since we saw that $\partial f/\partial x$ is discontinuous for the function of Example 4, it follows that Theorem 1 provides a sufficient, but not a necessary condition for differentiability. In advanced courses one shows that, indeed, much weaker hypotheses will suffice.

The standard results concerning differentiable functions of one variable hold for real-valued differentiable functions of two variables; that is, if f and g are differentiable functions then so are $f+g$, fg, and f/g [at any point v_0 where $g(v_0) \neq 0$].

These statements are related to (but not a direct consequence of) the following result. Suppose $f: \boldsymbol{R}^2 \to \boldsymbol{R}$ is C^1 and let $g: \boldsymbol{R} \to \boldsymbol{R}$ be a continuously differentiable function of one variable. Then $g \circ f: \boldsymbol{R}^2 \to \boldsymbol{R}$ defined by $g \circ f(x, y) = g(f(x, y))$ is C^1. We shall leave the proof of the theorem and the verification that the above statements follow from it to the reader (see exercises 11 and 12).

We have already noted that if f is differentiable we have the partial derivatives $\partial f/\partial x$ and $\partial f/\partial y$ which are functions of two variables. If these functions are differentiable as functions of two variables and their derivatives are continuous, we say that f is **twice continuously differentiable** or C^2. Thus we may take $\partial f/\partial x$ and $\partial f/\partial y$ and form their partial derivative with respect to x or y to obtain higher-order partial derivatives defined and denoted by

$$\frac{\partial^2 f}{\partial x^2} = \frac{\partial}{\partial x}\left(\frac{\partial f}{\partial x}\right)$$

$$\frac{\partial^2 f}{\partial x\, \partial y} = \frac{\partial}{\partial x}\left(\frac{\partial f}{\partial y}\right)$$

$$\frac{\partial^2 f}{\partial y\, \partial x} = \frac{\partial}{\partial y}\left(\frac{\partial f}{\partial x}\right)$$

$$\frac{\partial^2 f}{\partial y^2} = \frac{\partial}{\partial y}\left(\frac{\partial f}{\partial y}\right)$$

These derivatives are again functions of two variables and if f is sufficiently differentiable we can continue taking derivatives *ad infinitum*. However, our primary concern for now is second derivatives. In the following examples remember that to differentiate with respect to one variable, you treat the other variable as if it were a constant.

Example 5 Let $f(x, y) = xy + (x+2y)^2$. Then

$$\frac{\partial f}{\partial x} = y + 2(x+2y), \qquad \frac{\partial f}{\partial y} = x + 4(x+2y)$$

$$\frac{\partial^2 f}{\partial x^2} = 2, \qquad \frac{\partial^2 f}{\partial y^2} = 8$$

$$\frac{\partial^2 f}{\partial x\, \partial y} = 5, \qquad \frac{\partial^2 f}{\partial y\, \partial x} = 5$$

Example 6 If $f(x, y) = \sin x \sin^2 y$, then

$$\frac{\partial f}{\partial x} = \cos x \sin^2 y, \qquad \frac{\partial f}{\partial y} = 2 \sin x \sin y \cos y = \sin x \sin 2y$$

$$\frac{\partial^2 f}{\partial x^2} = -\sin x \sin^2 y, \qquad \frac{\partial^2 f}{\partial y^2} = 2 \sin x \cos 2y$$

$$\frac{\partial^2 f}{\partial x\, \partial y} = \cos x \sin 2y, \qquad \frac{\partial^2 f}{\partial y\, \partial x} = 2 \cos x \sin y \cos y = \cos x \sin 2y$$

Notice that in the previous two examples $\partial^2 f/\partial x\, \partial y = \partial^2 y/\partial y\, \partial x$. Is this just luck or is it a general principle? The next theorem provides the answer.

Theorem 2 If f is C^2 (twice continuously differentiable) then the **mixed partials** are equal; that is,

$$\frac{\partial^2 f}{\partial x\, \partial y} = \frac{\partial^2 f}{\partial y\, \partial x}$$

Hence, when f is C^2, it does not matter whether we differentiate first with respect to x and then with respect to y, or in reverse.

Proof Consider the expression

$$f(x_0+\Delta x, y_0+\Delta y) - f(x_0+\Delta x, y_0) - f(x_0, y_0+\Delta y) + f(x_0, y_0) \quad (2)$$

We fix $y_0, \Delta y$, and introduce the function

$$g(x) = f(x, y_0+\Delta y) - f(x, y_0)$$

so that (2) equals $g(x_0+\Delta x) - g(x_0)$. By the Mean Value Theorem for functions of one variable this equals $g'(\bar{x})\Delta x$ for $\bar{x}$ between x_0 and $x_0+\Delta x$. Hence (2) equals

$$\left[\frac{\partial f}{\partial x}(\bar{x}, y_0+\Delta y) - \frac{\partial f}{\partial x}(\bar{x}, y_0)\right]\Delta x$$

Applying the Mean Value Theorem again we get for (2)

$$\frac{\partial^2 f}{\partial y\, \partial x}(\bar{x}, \bar{y})\, \Delta x\, \Delta y$$

Since $\partial^2 f/\partial y\, \partial x$ is continuous, we have proven that

$$\frac{\partial^2 f}{\partial y\, \partial x}(x_0, y_0) = \lim_{(\Delta x, \Delta y)\to(0,0)} \frac{1}{\Delta x\, \Delta y}[f(x_0+\Delta x, y_0+\Delta y)$$
$$-f(x_0+\Delta x, y_0) - f(x_0, y_0+\Delta y) + f(x_0, y_0)] \quad (3)$$

The right-hand side of (3) is symmetric in x and y and in this derivation we could have reversed the roles of x and y. In other words, in the same manner, one proves that $\partial^2 f/\partial x\, \partial y$ is given by the same limit, so we obtain the desired result. ∎

We conclude this section with a result which combines integral and differential calculus. If $G: \mathbf{R} \to \mathbf{R}$ is a function defined by an integral, for example,

$$G(t) = \int_a^b F(x, t)\, dx$$

where F and $\partial F/\partial t$ are continuous funtions, then, by more advanced methods, it can be shown that

$$G'(t) = \int_a^b \frac{\partial F}{\partial t}(x, t)\, dx \quad (4)$$

In other words, it is legitimate to differentiate under the integral sign. The idea of the proof is that

$$G'(t) = \lim_{\Delta t \to 0} \frac{1}{\Delta t} \int_a^b [F(x, t+\Delta t) - F(x, t)]\, dx$$

and this can be shown to equal

$$\int_a^b \lim_{\Delta t \to 0} \left[\frac{F(x, t+\Delta t) - F(x, t)}{\Delta t} \right] dx = \int_a^b \frac{\partial F}{\partial t}(x, t)\, dx$$

The justification for moving the limit under the integral sign is based on a property known as uniform continuity, but a complete discussion is beyond the scope of this course.

Example 7 Let $G(t) = \int_1^2 xe^{-tx}\, dx$. Then we may use the result of equation (4) to write $G'(t) = -\int_1^2 x^2 e^{-tx}\, dx$.

Example 8 Let $G(x, y) = \int_a^x f(t, y)\, dt$, where $f: \boldsymbol{R}^2 \to \boldsymbol{R}$ is C^2. Then we have

$$\frac{\partial G}{\partial x}(x, y) = f\ (x,\ y) \qquad \text{(by the Fundamental Theorem of Calculus)}$$

$$\frac{\partial G}{\partial y}(x, y) = \int_a^x \frac{\partial f}{\partial y}(t, y)\, dt \qquad \text{[by (4) above]}$$

Since f is C^2 we see that G is C^2. Moreover,

$$\frac{\partial^2 G}{\partial x\, \partial y}(x, y) = \frac{\partial f}{\partial y}(x, y) \qquad \text{(by the Fundamental Theorem)}$$

$$\frac{\partial^2 G}{\partial y\, \partial x}(x, y) = \frac{\partial f}{\partial y}(x, y)$$

and the mixed partials agree.

EXERCISES

1. Find $\partial f/\partial x, \partial f/\partial y$ if
   (a) $f(x, y) = xy$
   (b) $f(x, y) = e^{xy}$
   (c) $f(x, y) = x \cos x \cos y$
   (d) $f(x, y) = (x^2 + y^2) \log (x^2 + y^2)$
2. Evaluate the partial derivatives $\partial z/\partial x, \partial z/\partial y$ for the given function at the indicated points.
   (a) $z = \sqrt{a^2 - x^2 - y^2}$, $(0, 0)$, $(a/2, a/2)$
   (b) $z = \log \sqrt{1 + xy}$, $(1, 2)$, $(0, 0)$
   (c) $z = e^{ax} \cos (bx + y)$, $(2\pi/b, 0)$

3. In each case find the partial derivatives $\partial w/\partial x, \partial w/\partial y$.

   (a) $w = xe^{x^2+y^2}$ (b) $w = \dfrac{x^2+y^2}{x^2-y^2}$

   (c) $w = e^{xy}\log(x^2+y^2)$ (d) $w = x/y$

   (e) $w = \cos(ye^{xy})\sin x$

4. Let

$$f(x,y) = \begin{cases} xy(x^2-y^2)/(x^2+y^2), & (x,y) \neq (0,0) \\ 0, & (x,y) = (0,0) \end{cases}$$

   (a) If $(x,y) \neq (0,0)$ calculate $\partial f/\partial x$ and $\partial f/\partial y$.

   (b) Show that $\dfrac{\partial f}{\partial x}(0,0) = 0 = \dfrac{\partial f}{\partial y}(0,0)$

   (c) Show that $\dfrac{\partial^2 f}{\partial x \partial y}(0,0) = 1, \dfrac{\partial^2 f}{\partial y \partial x}(0,0) = -1$

   What went wrong? Why are the mixed partials not equal?

5. Show that each of the following functions is differentiable at each point in its domain. Decide which of the functions are C^1.

   (a) $f(x,y) = \dfrac{2xy}{(x^2+y^2)^2}$ (b) $f(x,y) = \dfrac{x}{y} + \dfrac{y}{x}$

   (c) $f(r,\theta) = \frac{1}{2}r\sin 2\theta,\ r \geqslant 0$ (d) $f(x,y) = \dfrac{xy}{\sqrt{x^2+y^2}}$

   (e) $f(x,y) = \dfrac{x^2y}{x^4+y^2}$

6. Differentiate each of the following functions.

   (a) $G(t) = \int_1^a \cos xt\, e^{-xt^2}\,dx$

   (b) $G(t) = \int_1^t \cos xt\, e^{-xt^2}\,dx$ (this one is tricky)

   (c) $G(t) = \int_0^1 \sqrt{1+\sin^2 xt}\,dx$

7. Show that

   (a) if $G(t) = \int_{-1}^1 \log(1+te^x)\,dx$ then $G'(t) = \dfrac{1}{t}\log\left(\dfrac{1+et}{1+t/e}\right)$

   (b) if $G(t) = \int_0^\pi \dfrac{e^{tx}\sin x}{x}\,dx$ then $G'(t) = \dfrac{e^{\pi t}+1}{t^2+1}$

8. Use part (a) of exercise 7 to show

$$\int_{-1}^1 \log(1+e^x)\,dx = \int_0^1 \frac{1}{t}\log\left(\frac{1+et}{1+t/e}\right)$$

9. Let $g: \boldsymbol{R} \to \boldsymbol{R}$ and $h: \boldsymbol{R} \to \boldsymbol{R}$ be C^2 and let $\alpha \in \boldsymbol{R}$ be a constant. Define $f: \boldsymbol{R}^2 \to \boldsymbol{R}$ by

$$f(x,y) = h(x+\alpha y) + g(x-\alpha y)$$

   Show that

$$\frac{\partial^2 f}{\partial y^2} = \alpha^2 \frac{\partial^2 f}{\partial x^2}$$

10. Show that the function

$$f(x,y) = \begin{cases} x^2 \sin\frac{1}{x}, & x \neq 0 \\ 0, & x = 0 \end{cases}$$

is differentiable.

*11. Prove that if $f:\mathbf{R}^2 \to \mathbf{R}$ and $g:\mathbf{R} \to \mathbf{R}$ are both C^1 then $g \circ f$ is C^1.

12. Show that the sum, product, and quotient (if the denominator is not zero) of C^1 functions is C^1.

*13. Prove Theorem 1. [Hint: To verify that

$$\frac{f(x_0+\Delta x, y_0+\Delta y) - f(x_0,y_0) - \frac{\partial f}{\partial x}(x_0,y_0)\Delta x - \frac{\partial f}{\partial y}(x_0,y_0)\Delta y}{\sqrt{(\Delta x)^2 + (\Delta y)^2}} \to 0$$

as $(\Delta x, \Delta y) \to (0,0)$, write

$$\begin{aligned} f(x_0+\Delta x, y_0+\Delta y) - f(x_0,y_0) &= [f(x_0+\Delta x, y_0+\Delta y) - f(x_0, y_0+\Delta y)] \\ &\quad + [f(x_0, y_0+\Delta y) - f(x_0,y_0)] \end{aligned}$$

and use the Mean Value Theorem from one-variable calculus.]

3 DIFFERENTIAL CALCULUS AND GEOMETRY

The last section was devoted mainly to the mechanical aspects of partial differentiation. Now we wish to use this machinery to introduce the gradient of a function and interpret this notion physically and geometrically in terms of level curves.

Let us begin our discussion by considering one of the most interesting and practical mathematical concepts—that of a vector field. Basically, a **vector field** $\boldsymbol{F}$ on $\mathbf{R}^2$ is a function from $\mathbf{R}^2$ (or a subset thereof) to $\mathbf{R}^2$.

Since we can identify points in $\mathbf{R}^2$ with the vectors ending at those points, we may think of a vector field as a function which assigns to each point $(x,y) \in \mathbf{R}^2$ a vector in $\mathbf{R}^2$ (Figure 3-1). Vector fields are important

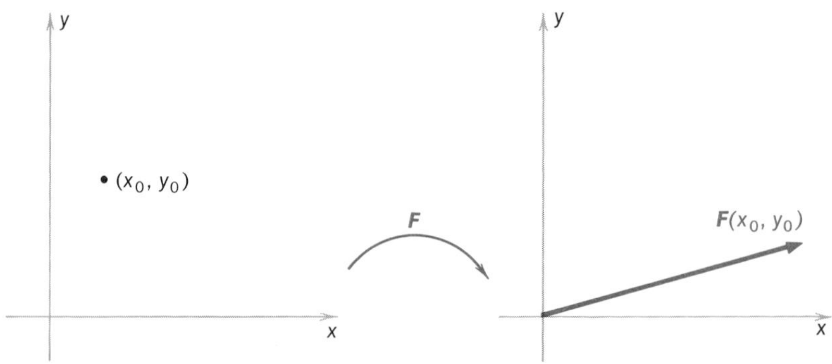

FIGURE 3-1

when dealing with quantities which involve both magnitude and direction. For example, $\boldsymbol{F}$ can be a gravitational force field or an electric force field.

If $\boldsymbol{F}:\boldsymbol{R}^2 \to \boldsymbol{R}^2$ is a vector field and we identify points in $\boldsymbol{R}^2$ with vectors, then $\boldsymbol{F}$ can be written in the form

$$\boldsymbol{F}(x,y) = f_1(x,y)\boldsymbol{i} + f_2(x,y)\boldsymbol{j}$$

where $\boldsymbol{i}$ and $\boldsymbol{j}$ are the unit basis vectors of $\boldsymbol{R}^2$. For example, $\boldsymbol{F}(x,y) = (x^2+yx)\boldsymbol{i}+(\sin x+2)\boldsymbol{j}$ is a vector field with $f_1(x,y)=x^2+yx$ and $f_2(x,y)=\sin x+2$. The functions f_1 and f_2 are called the **component** functions of $\boldsymbol{F}$. We say that $\boldsymbol{F}$ is continuous if and only if its component functions f_1 and f_2 are continuous; $\boldsymbol{F}$ is C^1 if and only if f_1 and f_2 are C^1.

For functions from $\boldsymbol{R}$ to $\boldsymbol{R}$ or from $\boldsymbol{R}^2$ to $\boldsymbol{R}$, we can get a geometric understanding of their behavior by drawing their graphs. The graphs would be subsets of $\boldsymbol{R}^2$ and $\boldsymbol{R}^3$, respectively. However, for vector fields we cannot picture their graphs (which would be subsets of $\boldsymbol{R}^4$, four-dimensional space), but we can represent them geometrically in another way.

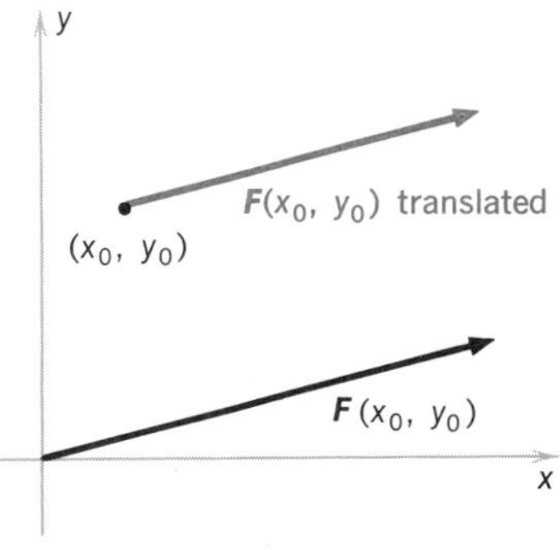

FIGURE 3-2

We said that a vector field associates to each point $(x_0, y_0) \in \boldsymbol{R}^2$ a vector $\boldsymbol{F}(x_0, y_0)$. Now, when we draw the graph of a function f, we show the association between p and values $f(p)$. In other words, both p and $f(p)$ are represented at the same time. To indicate the correspondence between (x, y) and $\boldsymbol{F}(x, y)$ in one picture, we can represent $\boldsymbol{F}(x_0, y_0)$ as a vector emanating from the point (x_0, y_0) instead of from the origin. That is, we parallel translate (Figure 3-2) $\boldsymbol{F}(x_0, y_0)$ so that its initial point is (x_0, y_0). In this manner both (x_0, y_0) and $\boldsymbol{F}(x_0, y_0)$ are represented in the same diagram. This way of viewing vector fields proved to be useful both mathematically and scientifically.

Example 1 For a point of mass m in the plane located at $(0,0)$, the gravitational force field $\boldsymbol{F}$ is of the form

$$\boldsymbol{F}(x,y) = -\alpha(x,y)\left[\frac{x}{\sqrt{x^2+y^2}}\boldsymbol{i} + \frac{y}{\sqrt{x^2+y^2}}\boldsymbol{j}\right] = -\alpha(x,y)\,\boldsymbol{n}$$

where $\alpha:\boldsymbol{R}^2 \to \boldsymbol{R}$, $\alpha(x,y) \geqslant 0$, and

$$\boldsymbol{n} = \frac{x}{\sqrt{x^2+y^2}}\boldsymbol{i} + \frac{y}{\sqrt{x^2+y^2}}\boldsymbol{j}$$

is the unit vector directed from $(0,0)$ to the point (x, y). The force field $\boldsymbol{F}$ is a vector field which at all points is directed "radially" inward (inward because there is a minus sign in front of the α) toward the origin where m is located. Therefore, $\boldsymbol{F}(x,y)$ is the force exerted on a unit mass at the point (x, y). The function $\alpha(x,y)$ is determined by Newton's law that two bodies attract each other by a force inversely proportional to the square of their distance and so

$$\boldsymbol{F}(x,y) = -\frac{Gm}{x^2+y^2}\boldsymbol{n}$$

where G is the gravitational constant.

Example 2 The electric force field induced by a "dipole" located at the origin in $\boldsymbol{R}^2$ is also a vector field. A dipole is a pair of equal and opposite charges in juxtaposition (see Figure 3-3). The lines in Figure 3-3 are the so-called lines of force or **flow lines.** The flow lines give a global picture of the vector field. A flow line is a curve $\boldsymbol{\sigma}:\boldsymbol{R}\to\boldsymbol{R}^2$ which is a solution to the differential equation $\boldsymbol{\sigma}'(t)=\boldsymbol{F}(\boldsymbol{\sigma}(t))$, that is, the tangent vector $\boldsymbol{\sigma}'(t)$ to the curve $t\mapsto\boldsymbol{\sigma}(t)$ is the vector $\boldsymbol{F}(\boldsymbol{\sigma}(t))$ at the point $\boldsymbol{\sigma}(t)$. Given some initial point p, the flow line through p describes the subsequent motion of a particle starting at p and only under the influence of the given force field. We discuss this further in Example 5.

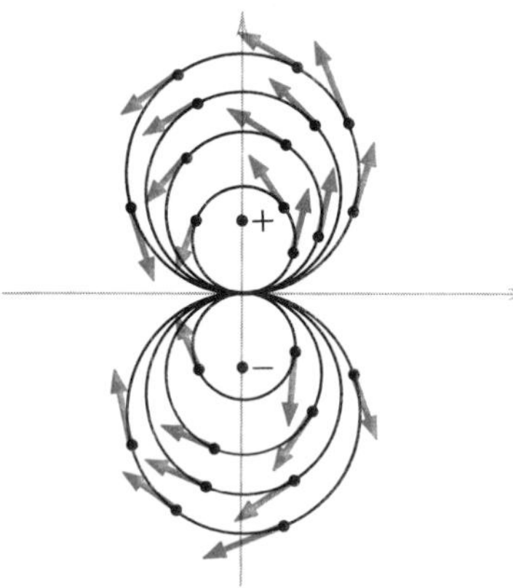

FIGURE 3-3

Example 3 The magnetic field of the earth is much like the dipole field of Example 2, with lines of force appearing as in Figure 3-4.

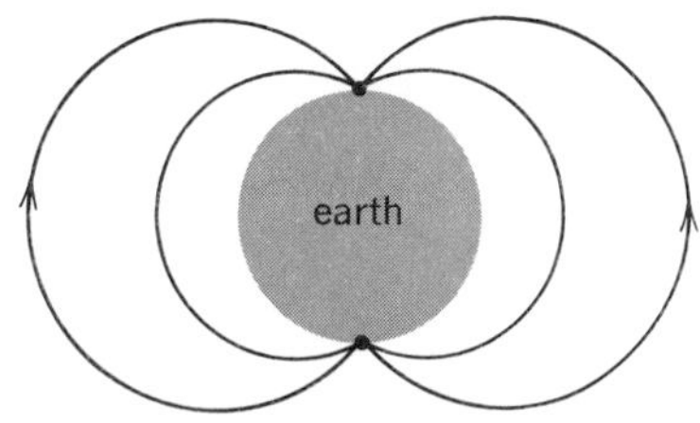

FIGURE 3-4

Example 4 Suppose a fluid is flowing over a flat river bed (see Figure 3-5). If we view the bed as a portion D of the xy plane, then we can define the vector field $\boldsymbol{V}$ by $\boldsymbol{V}:(x,y)\mapsto\boldsymbol{V}(x,y)$ where $\boldsymbol{V}(x,y)$ is the velocity vector of the fluid at each point $(x,y)\in D$. (We are assuming in this case that the velocity of the fluid at a point is independent of time.) In this example the meaning of the flow lines becomes somewhat clearer. To paraphrase it again, given a particle at point p the solution curve or flow line through p is simply the path a particle will travel if it is placed at p and then allowed to flow with the current.

Some of the most important vector fields arise in a very special way—from differentiable functions $f:\boldsymbol{R}^2\to\boldsymbol{R}$. For such differentiable mappings we define a vector field, called the **gradient** of f or **grad** f, as

$$\textbf{grad}\, f(x_0,y_0)=\frac{\partial f}{\partial x}(x_0,y_0)\boldsymbol{i}+\frac{\partial f}{\partial y}(x_0,y_0)\boldsymbol{j}$$

The vector grad $f(x_0,y_0)$ is often denoted

$$\nabla f(x_0,y_0)$$

and we shall adopt this notation. The symbol ∇ is usually read "del." Thus

$$\nabla f=\frac{\partial f}{\partial x}\boldsymbol{i}+\frac{\partial f}{\partial y}\boldsymbol{j}$$

is a vector field defined at all points where f is differentiable.

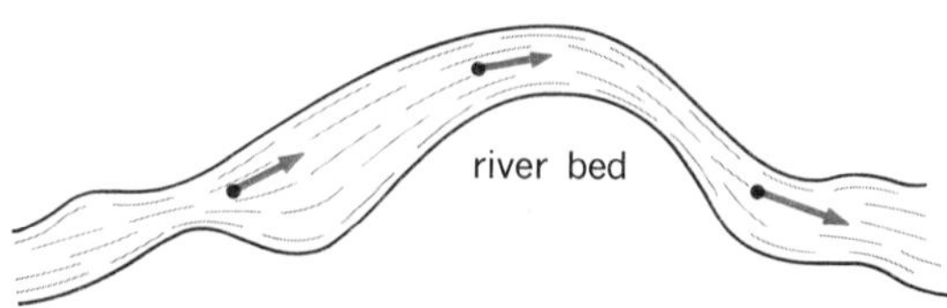

FIGURE 3-5

The vector fields in Examples 1–3 above are, in fact, gradient vector fields. The gravitational force field of Example 1

$$\boldsymbol{F}(x,y) = -\frac{Gm}{x^2+y^2}\boldsymbol{n}$$

can be written as $\boldsymbol{F} = \nabla(-U) = -\nabla U$, where $U(x,y) = -Gm/\sqrt{x^2+y^2}$. To see this note that

$$\frac{\partial U}{\partial x} = \frac{Gmx}{(x^2+y^2)^{3/2}}, \qquad \frac{\partial U}{\partial y} = \frac{Gmy}{(x^2+y^2)^{3/2}}$$

Thus

$$\begin{aligned} -\nabla U &= -\frac{Gmx}{(x^2+y^2)^{3/2}}\boldsymbol{i} - \frac{Gmy}{(x^2+y^2)^{3/2}}\boldsymbol{j} \\ &= -\frac{Gm}{(x^2+y^2)}\left(\frac{x}{\sqrt{x^2+y^2}}\boldsymbol{i} + \frac{y}{\sqrt{x^2+y^2}}\boldsymbol{j}\right) \\ &= -\frac{Gm}{x^2+y^2}\boldsymbol{n} \end{aligned}$$

The function U is called the **gravitational potential energy** and can be viewed as the "work done" bringing a particle of unit mass from the point (x, y) out to ∞. In Example 2 the vector field of a dipole is, in fact, the gradient of the function $U(x,y) = ax/(x^2+y^2)^{3/2}$. The vector field which describes the magnetic field of the earth (Example 3) is very similar to the vector field of a dipole.

There are many more sophisticated physical examples of vector fields, but for now let us return to the underlying mathematics.

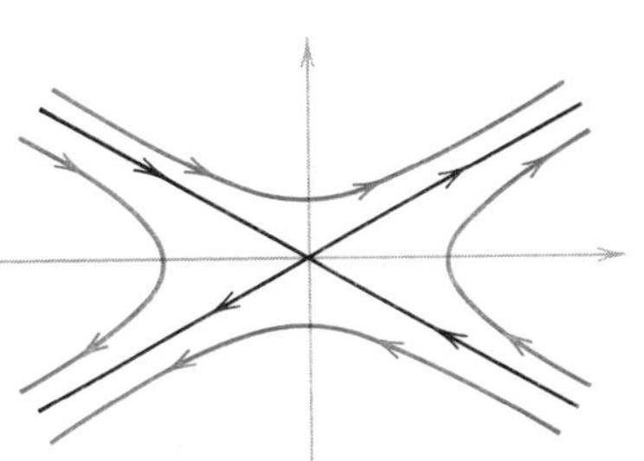

FIGURE 3-6

Example 5 The function $f(x,y) = xy$ has $\nabla f = y\boldsymbol{i} + x\boldsymbol{j}$ (see Figure 3-6). Hence, at the point $(2,0)$, $\nabla f = 2\boldsymbol{j}$, at the point $(-5,1)$, $\nabla f = \boldsymbol{i} - 5\boldsymbol{j}$, and so on. Notice that the vectors in Figure 3-6 seem to form a pattern of hyperbolas. These are the flow lines of this vector field. Since a flow line is a curve $t \mapsto \boldsymbol{\sigma}(t)$ whose tangent vector $\boldsymbol{\sigma}'(t)$ at each point $\boldsymbol{\sigma}(t)$ is the vector assigned to that point by the vector field, we can prove that for our vector field $\nabla f(x,y) = y\boldsymbol{i} + x\boldsymbol{j}$ the flow lines are actually hyperbolas. In fact, if the flow line (or line of force) is given by $\boldsymbol{\sigma}(t) = (\sigma_1(t), \sigma_2(t))$, then the tangent vector at $\boldsymbol{\sigma}(t)$ is

$$\sigma_1'(t)\boldsymbol{i} + \sigma_2'(t)\boldsymbol{j}$$

But the vector field ∇f must assign the same vector to the point $\boldsymbol{\sigma}(t) \in \boldsymbol{R}^2$ so we have

$$\sigma_1'(t)\boldsymbol{i} + \sigma_2'(t)\boldsymbol{j} = \sigma_2(t)\boldsymbol{i} + \sigma_1(t)\boldsymbol{j}$$

or $\sigma_1' = \sigma_2$ and $\sigma_2' = \sigma_1$. Differentiating again, we obtain the system of differential equations

$$\sigma_1''(t) = \sigma_1(t), \qquad \sigma_2''(t) = \sigma_2(t)$$

A solution of this system is $\sigma_1(t) = \sinh(t)$, $\sigma_2(t) = \cosh t$ which is a parametric representation of a hyperbola. Notice that in this example the gradient vanishes (equals the zero vector) at $(0,0)$ and also that the direction of the field varies continuously except at $(0,0)$. In general, a point where ∇f vanishes is called a **critical point** of f.

Before we discuss some interesting geometric examples, let us consider other important properties of the gradient.

Suppose $f: \boldsymbol{R}^2 \to \boldsymbol{R}$ and $g: \boldsymbol{R} \to \boldsymbol{R}^2$ are differentiable functions. Then we can form the composite function $f \circ g: \boldsymbol{R} \to \boldsymbol{R}$ defined by $(f \circ g)(t) = f(g(t))$. For example, if $f(x, y) = x^2 + \cos y$ and $g(t) = (t, t^2)$, then $(f \circ g)(t) = t^2 + \cos t^2$. The derivative of the composite function is $t \mapsto 2t - 2t \sin t^2$. Let us now compute $\nabla f(g(t))$. First $\partial f/\partial x = 2x$ so $(\partial f/\partial x)(g(t)) = 2t$. Second, $\partial f/\partial y = -\sin y$ so $(\partial f/\partial y)(g(t)) = -\sin t^2$; therefore, $\nabla f(g(t)) = 2t\boldsymbol{i} - \sin t^2 \boldsymbol{j}$. Also, $g'(t) = \boldsymbol{i} + 2t\boldsymbol{j}$. Now if we take the dot product $\nabla f(g(t)) \cdot g'(t)$ we get $2t - 2t \sin t^2$ which is exactly the derivative of the composition obtained above. This is not a coincidence but rather part of a generalization of the Chain Rule from one-variable calculus.

Theorem 1 **Chain Rule** Let $g: \boldsymbol{R} \to \boldsymbol{R}^2$ be differentiable at $t_0 \in \boldsymbol{R}$ and let $f: \boldsymbol{R}^2 \to \boldsymbol{R}$ be differentiable at $g(t_0)$. Then the composite function $f \circ g: \boldsymbol{R} \to \boldsymbol{R}$ is differentiable at t_0 and

$$(f \circ g)'(t_0) = \nabla f(g(t_0)) \cdot g'(t_0)$$

***Proof** According to the formulation of differentiability in §14.2 we must show that

$$f \circ g(t_0 + \Delta t) - f \circ g(t_0) = \nabla f(g(t_0)) \cdot g'(t_0) + Q_{t_0}(\Delta t)$$

where

$$\lim_{\Delta t \to 0} \frac{Q_{t_0}(\Delta t)}{|\Delta t|} = 0$$

Let $g(t) = (g_1(t), g_2(t))$. Since f is differentiable at $g(t_0)$

$$\begin{aligned} f \circ g(t_0 + \Delta t) - f \circ g(t_0) &= f(g(t_0) + [g(t_0 + \Delta t) - g(t_0)]) - f(g(t_0)) \\ &= f(g(t_0) + \Delta g(t_0)) - f(g(t_0)) \end{aligned}$$

$$\begin{aligned} \text{[where } \Delta g(t_0) &= g(t_0 + \Delta t) - g(t_0) \\ &= (g_1(t_0 + \Delta t) - g_1(t_0), g_2(t_0 + \Delta t) - g_2(t_0)) \\ &= (\Delta g_1(t_0), \Delta g_2(t_0))] \end{aligned}$$

so $f \circ g(t_0 + \Delta t) - f \circ g(t_0)$ can be written as

$$\frac{\partial f}{\partial x}(g(t_0)) \cdot \Delta g_1(t_0) + \frac{\partial f}{\partial y}(g(t_0)) \cdot \Delta g_2(t_0) + \tilde{Q}_{g(t_0)}(\Delta g(t_0)) \qquad (1)$$

and

$$\lim_{\Delta g(t_0)\to 0} \frac{\tilde{Q}_{g(t_0)}(\Delta g(t_0))}{|\Delta g(t_0)|} = 0$$

The fact that g is differentiable at t_0 implies

$$\Delta g_i(t_0) = g_i'(t_0)\cdot \Delta t + R_{t_0}^i(\Delta t), \qquad i = 1, 2$$

where

$$\lim_{\Delta t\to 0} \frac{R_{t_0}^i(\Delta t)}{|\Delta t|} = 0$$

Substituting this result into (1) we obtain

$$\frac{\partial f}{\partial x}(g(t_0))\cdot[g_1'(t_0)\cdot\Delta t + R_{t_0}^1(\Delta t)] + \frac{\partial f}{\partial y}(g(t_0))\cdot[g_2'(t_0)\cdot\Delta t + R_{t_0}^2(\Delta t)]$$

$$+ \tilde{Q}_{g(t_0)}(\Delta g(t_0)) = \left\{\frac{\partial f}{\partial x}(g(t_0))[g_1'(t_0)] + \frac{\partial f}{\partial y}(g(t_0))[g_2'(t_0)]\right\}\cdot\Delta t$$

$$+ \frac{\partial f}{\partial x}(g(t_0))\cdot R_{t_0}^1(\Delta t) + \frac{\partial f}{\partial y}(g(t_0))\cdot R_{t_0}^2(\Delta t) + \tilde{Q}_{g(t_0)}(\Delta g(t_0)) \qquad (2)$$

Denoting the last three terms of (2) by $Q_{t_0}(\Delta t)$, we can reduce this expression further to

$$\nabla f(g(t_0))\cdot g'(t_0) + Q_{t_0}(\Delta t)$$

Thus it only remains to show that

$$\lim_{\Delta t\to 0} \frac{Q_{t_0}(\Delta t)}{|\Delta t|} = 0$$

Now

$$\frac{Q_{t_0}(\Delta t)}{|\Delta t|} = \frac{\partial f}{\partial x}(g(t_0))\cdot\frac{R_{t_0}^1(\Delta t)}{|\Delta t|} + \frac{\partial f}{\partial y}(g(t_0))\cdot\frac{R_{t_0}^2(\Delta t)}{|\Delta t|} + \frac{\tilde{Q}_{g(t_0)}(\Delta g(t_0))}{|\Delta t|} \qquad (3)$$

Since $(\partial f/\partial x)(g(t_0))$ and $(\partial f/\partial y)(g(t_0))$ are fixed numbers and

$$\lim_{\Delta t\to 0} \frac{R_{t_0}^i(\Delta t)}{|\Delta t|} = 0, \qquad i = 1, 2$$

we find that the limit as $\Delta t \to 0$ of the first two terms in (3) is 0. The last term

$$\frac{\tilde{Q}_{g(t_0)}(\Delta g(t_0))}{|\Delta t|} = \frac{\tilde{Q}_{g(t_0)}(\Delta g(t_0))}{|\Delta g(t_0)|}\cdot\left|\frac{\Delta g(t_0)}{\Delta t}\right|$$

if $\Delta g(t_0) \neq 0$ and

$$\frac{\tilde{Q}_{g(t_0)}(\Delta g(t_0))}{|\Delta t|} = \frac{\tilde{Q}_{g(t_0)}(0)}{|\Delta t|} = 0$$

if $\Delta g(t_0)=0$ since $\tilde{Q}(0)=0$. Since we are trying to show that $\tilde{Q}_{g(t_0)}(\Delta g(t_0))/|\Delta t|$ is small if $|\Delta t|$ is small, we can assume that $\Delta g(t_0) \neq 0$ for if $\Delta g(t_0)=0$, $\tilde{Q}=0$. The assumption that f is differentiable at $g(t_0)$ implies

$$\lim_{\Delta g(t_0)\to 0} \frac{\tilde{Q}_{g(t_0)}(\Delta g(t_0))}{|\Delta g(t_0)|} = 0$$

Since g is differentiable at t_0, $\Delta g(t_0)\to 0$ as $\Delta t \to 0$, and

$$\left|\frac{\Delta g(t_0)}{\Delta t}\right| \to |g'(t_0)| \qquad \text{as} \qquad \Delta t \to 0$$

Therefore

$$\lim_{\Delta t\to 0} \frac{\tilde{Q}_{g(t_0)}(\Delta g(t_0))}{|\Delta t|} = 0 \cdot |g'(t_0)| = 0$$

which proves the Chain Rule. ∎

There is another way to express the Chain Rule which avoids the gradient; namely, if $z=f(x,y)$ is a function of two variables and $g: \boldsymbol{R}\to\boldsymbol{R}^2$ is given by $g(t)=(g_1(t), g_2(t))$ then

$$\frac{dz}{dt} = \frac{\partial f}{\partial x} g_1'(t) + \frac{\partial f}{\partial y} g_2'(t) = \frac{\partial f}{\partial x}\frac{dx}{dt} + \frac{\partial f}{\partial y}\frac{dy}{dt} \tag{4}$$

where the components of g are written as

$$x = g_1(t), \qquad y = g_2(t)$$

Example 6 If $f: \boldsymbol{R}^2 \to \boldsymbol{R}$ is given by $(x,y)\mapsto e^{xy}+\sin xy$ and $g: \boldsymbol{R}\to\boldsymbol{R}^2$ is given by $t\mapsto (t^3+1, t)$, then

$$\begin{aligned}\nabla f(x,y) &= (ye^{xy}+y\cos xy)\boldsymbol{i} + (xe^{xy}+x\cos xy)\boldsymbol{j}\\ &= (e^{xy}+\cos xy)(y\boldsymbol{i}+x\boldsymbol{j})\end{aligned}$$

and

$$g'(t) = 3t^2\boldsymbol{i}+\boldsymbol{j}$$

So

$$\begin{aligned}(f\circ g)'(t) &= \nabla f(g(t))\cdot g'(t)\\ &= (e^{(t^3+1)t} + \cos(t^3+1)t)(t\boldsymbol{i} + (t^3+1)\boldsymbol{j})\cdot(3t^2\boldsymbol{i}+\boldsymbol{j})\\ &= (e^{(t^3+1)t} + \cos(t^3+1)t)(4t^3+1)\end{aligned}$$

The reader can verify that the result is the same if (4) is used.

EXERCISES

In exercises 1–4, compute ∇f and g', and find $(f \circ g)'(1)$ directly and by evaluating $\nabla f(g(1)) \cdot g'(1)$.

1. $f(x,y) = xy$, $g(t) = (e^t, \cos t)$
2. $f(x,y) = e^{xy}$, $g(t) = (3t^2, t^3)$
3. $f(x,y) = (x^2 + y^2)\log\sqrt{x^2+y^2}$, $g(t) = (e^t, e^{-t})$
4. $f(x,y) = xe^{x^2+y^2}$, $g(t) = (t, -t)$

*5. This exercise demonstrates that the Chain Rule is not applicable if f is not differentiable. Consider the function

$$f(x,y) = \begin{cases} \dfrac{xy^2}{x^2+y^2}, & (x,y) \neq (0,0) \\ 0, & (x,y) = (0,0) \end{cases}$$

Show that

(a) $\dfrac{\partial f}{\partial x}(0,0)$ and $\dfrac{\partial f}{\partial y}(0,0)$ exist;

(b) if $g(t) = (at, bt)$ then $f \circ g$ is differentiable and $(f \circ g)'(0) = ab^2/(a^2+b^2)$, but $\nabla f(0,0) \cdot g'(0) = 0$.

*6. Try to find an example, similar to the one in exercise 5, in which the Chain Rule fails when f is differentiable but g is not C^1, or show that this cannot happen.

7. If $\phi: \boldsymbol{R}^2 \to \boldsymbol{R}$ and $f: \boldsymbol{R}^2 \to \boldsymbol{R}$ are C^1 we denote by $\phi\nabla f$ the vector field

$$(\phi\nabla f)(x,y) = \phi(x,y)\frac{\partial f}{\partial x}(x,y)\boldsymbol{i} + \phi(x,y)\frac{\partial f}{\partial y}(x,y)\boldsymbol{j}$$

Show that

$$\nabla(fg) = g\nabla f + f\nabla g$$

8. Let $\boldsymbol{F}$ and $\boldsymbol{G}$ be two C^1 vector fields. If $\boldsymbol{F} = F_1\boldsymbol{i} + F_2\boldsymbol{j}$ we denote by $\partial\boldsymbol{F}/\partial x$ the vector field

$$(x,y) \mapsto \frac{\partial F_1}{\partial x}(x,y)\boldsymbol{i} + \frac{\partial F_2}{\partial x}(x,y)\boldsymbol{j}$$

Using this notation, derive a formula for $\nabla(\boldsymbol{F} \cdot \boldsymbol{G})$.

9. Let $\boldsymbol{T}: \boldsymbol{R}^2 \to \boldsymbol{R}^2$ be a C^1 vector field and $f: \boldsymbol{R}^2 \to \boldsymbol{R}$ be C^1. Show that

$$\partial(f \circ \boldsymbol{T})/\partial x = \nabla f \cdot \partial\boldsymbol{T}/\partial x$$

Another way to present this problem is to denote $\boldsymbol{T}$ by $\boldsymbol{T}(x,y) = (u(x,y), v(x,y))$. Then the formula above says that

$$\frac{\partial}{\partial x} f(u(x,y),\, v(x,y)) = \frac{\partial f}{\partial u}\frac{\partial u}{\partial x} + \frac{\partial f}{\partial v}\frac{\partial v}{\partial x}$$

10. Let $\boldsymbol{T}(x,y) = (x^2 - y^2, 2xy)$ and $f(u,v) = u^2 + v^2$. Compute $(\partial/\partial x)(f \circ \boldsymbol{T})$ directly, using the result of exercise 9.

11. Find $(\partial/\partial s)(f \circ \boldsymbol{T})(0,0)$, where $f(u,v) = \cos u \sin v$ and $\boldsymbol{T}(s,t) = (\cos t^2 s, \log\sqrt{1+s^2})$.

4 THE DIRECTIONAL DERIVATIVE

In §14.2 we studied partial derivative $\partial f/\partial x$, $\partial f/\partial y$ of a function $f: \boldsymbol{R}^2 \to \boldsymbol{R}$. Recall that

$$\frac{\partial f}{\partial x}(x_0, y_0) = \lim_{\Delta x \to 0} \frac{f(x_0 + \Delta x, y_0) - f(x_0, y_0)}{\Delta x}$$

Thus $(\partial f/\partial x)(x_0, y_0) = d[f(x, y_0)]/dx$. If y_0 is fixed, then (x, y_0) belongs to a line L_x in the xy plane through (x_0, y_0) and parallel to the x axis (see Figure 4-1). Thus, $(\partial f/\partial x)(x_0, y_0)$ is obtained by taking the derivative of f when f is "restricted" to the line L_x—that is, for points $(x, y_0) \in L_x$ we consider the function $(x, y_0) \mapsto f(x, y_0)$. Similarly $(\partial f/\partial y)(x_0, y_0)$ is really the derivative of f along the line L_y (Figure 4-1) through (x_0, y_0) and parallel to the y axis. A natural question is must we restrict our attention to such special lines through (x_0, y_0)? What if we take some arbitrary line, say L, through (x_0, y_0). Let us compute the "derivative of f along the line L." To begin, we let $\boldsymbol{n} = \alpha\boldsymbol{i} + \beta\boldsymbol{j}$ be a unit vector ($\alpha^2 + \beta^2 = 1$) parallel to L. Then the line L is parametrized by

FIGURE 4-1

$$\lambda \mapsto (x_0 + \lambda\alpha, y_0 + \lambda\beta)$$

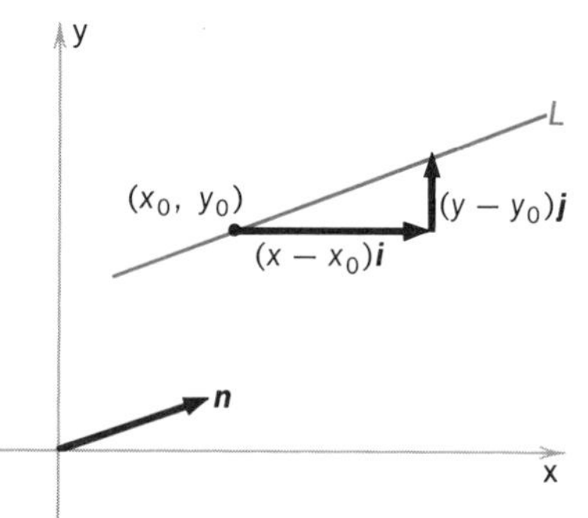

FIGURE 4-2

where $\lambda \in \boldsymbol{R}$ (Figure 4-2). As λ varies from $-\infty$ to $+\infty$, $(x_0 + \lambda\alpha, y_0 + \lambda\beta)$ varies through all the points on line L. Therefore, the function $\lambda \mapsto f(x_0 + \lambda\alpha, y_0 + \lambda\beta)$ is a function of one variable and it is, in fact, the "restriction" of the function f to the line L. Notice that if $\beta = 0$, L is the line L_x and if $\alpha = 0$, L is the line L_y. If f is differentiable at the point (x_0, y_0) the function $\lambda \mapsto f(x_0 + \lambda\alpha, y_0 + \lambda\beta)$ is differentiable at 0 and its derivative at $\lambda = 0$

$$\left.\frac{d[f(x_0 + \lambda\alpha, y_0 + \lambda\beta)]}{d\lambda}\right|_{\lambda=0} \tag{1}$$

is called the **directional derivative** of f in the direction $\boldsymbol{n} = \alpha\boldsymbol{i} + \beta\boldsymbol{j}$ or in the direction of the line L. This may be defined even if f is not differentiable.

Applying the Chain Rule to $\lambda \mapsto f \circ g(\lambda)$, where $g(\lambda) = (x_0 + \lambda\alpha, y_0 + \lambda\beta)$ we find that (1) is equal to

$$\nabla f(g(0)) \cdot g'(0) = \nabla f \cdot \boldsymbol{n} = \alpha\frac{\partial f}{\partial x}(x_0, y_0) + \beta\frac{\partial f}{\partial y}(x_0, y_0) \tag{2}$$

The number $\nabla f \cdot \boldsymbol{n}$ is the rate of change of f in the direction $\boldsymbol{n}$ or along L. If we had chosen some vector $\mu\boldsymbol{n} = \mu\alpha\boldsymbol{i} + \mu\beta\boldsymbol{j}$, then the directional derivative would be $\mu\alpha(\partial f/\partial x)(x_0, y_0) + \mu\beta(\partial f/\partial y)(x_0, y_0)$, clearly a different answer. Actually, the name directional derivative comes from the fact that unit vectors have the same length so they can differ only in their direction. If $\boldsymbol{n}$ is $\boldsymbol{i}$ or $\boldsymbol{j}$ then $\nabla f \cdot \boldsymbol{n}$ is $\partial f/\partial x$ or $\partial f/\partial y$, respectively.

Example 1 If $f: \mathbf{R}^2 \to \mathbf{R}$ is given by $f(x, y) = x^2 y + y^3$, find the directional derivative of f at (1, 1) in the direction

$$\frac{1}{\sqrt{5}}\boldsymbol{i} + \frac{2}{\sqrt{5}}\boldsymbol{j}$$

We have $\partial f/\partial x = 2xy$, $\partial f/\partial y = x^2 + 3y^2$ so

$$\nabla f(1, 1) = \frac{\partial f}{\partial x}(1, 1)\boldsymbol{i} + \frac{\partial f}{\partial y}(1, 1)\boldsymbol{j} = 2\boldsymbol{i} + 4\boldsymbol{j}$$

Thus our answer is

$$\nabla f(1, 1) \cdot \left(\frac{1}{\sqrt{5}}\boldsymbol{i} + \frac{2}{\sqrt{5}}\boldsymbol{j}\right) = \frac{2}{\sqrt{5}} + \frac{8}{\sqrt{5}} = 2\sqrt{5}$$

Given a differentiable function $f: \mathbf{R}^2 \to \mathbf{R}$ and a point $v_0 \in \mathbf{R}^2$, we can look at ***all*** the directional derivatives of f at v_0 and ask if there is a direction which maximizes the directional derivative. This, then, will be the direction in which f is changing fastest. Using formula (2) and results from Chapter 13, we can indeed find such a direction. If $\boldsymbol{r}$ is a unit vector denoting direction, then (by the Schwarz inequality) we can write

$$|\nabla f(v_0) \cdot \boldsymbol{r}| \leqslant |\nabla f(v_0)| \cdot |\boldsymbol{r}| = |\nabla f(v_0)|$$

and equality holds if and only if $\boldsymbol{r} = \mu \nabla f(v_0)$ for some scalar μ; i.e., some scalar multiple of the gradient. Since $\boldsymbol{r}$ is a unit vector we can set

$$\boldsymbol{r} = \pm \frac{\nabla f(v_0)}{|\nabla f(v_0)|}$$

These are the two unit vectors which are scalar multiples of $\nabla f(v_0)$. If we choose the positive sign we find

$$\nabla f(v_0) \cdot \frac{\nabla f(v_0)}{|\nabla f(v_0)|} = \frac{|\nabla f(v_0)|^2}{|\nabla f(v_0)|} = |\nabla f(v_0)|$$

which is positive; if we choose the negative sign,

$$\nabla f(v_0) \cdot \frac{-\nabla f(v_0)}{|\nabla f(v_0)|} = -|\nabla f(v_0)|$$

which is negative. Thus, the directional derivative is maximized when $\boldsymbol{r} = \nabla f(v_0)/|\nabla f(v_0)|$ and the maximum value of the derivative is $|\nabla f(v_0)|$.

There is one difficulty with what we have just done, namely, we have divided by $|\nabla f(v_0)|$. If $\nabla f(v_0) = \boldsymbol{0}$ then v_0 is a critical point of f and all directional derivatives are zero (Why?); in that case there is no maximizing direction.

In §4.1 we studied level curves of the form $f(x, y) - c = 0$ and used the method of implicit differentiation to compute tangent lines to these curves. This method is, in fact, justified by the geometric significance of the gradient

∇f which is that $\nabla f(x_0, y_0)$ is perpendicular to the level curve $f(x, y) = c$ at (x_0, y_0) (see Figure 4-3). This is intuitively plausible for on the curve $f(x, y) = c$ the function f is constant and therefore not changing at all; hence, the perpendicular direction from the level curve at $v_0 = (x_0, y_0)$ should be the direction in which $f(x, y)$ actually changes fastest, thus should be the direction of the gradient.

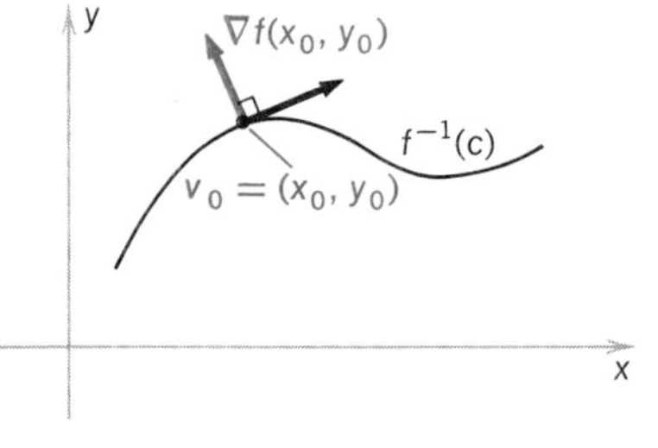

FIGURE 4-3

More precisely, if f is C^1 and $\nabla f(x_0, y_0) \neq 0$, then there exists a disk D centered at (x_0, y_0) and a C^1 path $\boldsymbol{\sigma}: [-1, 1] \to \boldsymbol{R}^2$ such that $\boldsymbol{\sigma}$ parametrizes the level curve $f(x, y) = c$ in the disk D with $\boldsymbol{\sigma}(0) = (x_0, y_0)$, and $\boldsymbol{\sigma}'(0) \neq 0$. [This assertion, which establishes that the level curve $f(x, y) = c$ is indeed a curve at (x_0, y_0) and has a tangent there, is quite difficult to prove and we shall not attempt to do so here.] Now the composition $f \circ \boldsymbol{\sigma}: t \mapsto f(\boldsymbol{\sigma}(t)) = c$ is a constant real function and so has zero derivative. By the Chain Rule,

$$0 = \left.\frac{d(f(\boldsymbol{\sigma}(t)))}{dt}\right|_{t=0} = \nabla f(\boldsymbol{\sigma}(0)) \cdot \boldsymbol{\sigma}'(0)$$

Therefore $\nabla f(x_0, y_0)$ is perpendicular to $\boldsymbol{\sigma}'(0)$ which is the tangent to $f(x, y) = c$ at (x_0, y_0). So the tangent line to $f(x, y) = c$ at (x_0, y_0) has equation $\nabla f(x_0, y_0) \cdot (x - x_0, y - y_0) = 0$. In the *dxdy* coordinate system which is centered at (x_0, y_0) this equation becomes

$$\nabla f(x_0, y_0) \cdot (dx, dy) = 0$$

This last equation defines the differential relation which we introduced informally in §4.1; that is, we define the differential relation $df(x, y, dx, dy) = 0$ by the equation

$$df(x, y, dx, dy) = \nabla f(x, y) \cdot (dx, dy) = 0$$

or

$$df(x, y, dx, dy) = \frac{\partial f}{\partial x}(x, y)\, dx + \frac{\partial f}{\partial y}(x, y)\, dy = 0$$

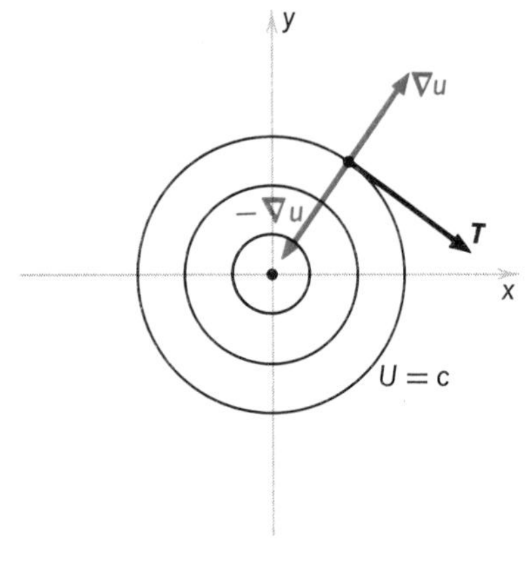

FIGURE 4-4

Example 2 Let us, in the light of our further knowledge, reexamine the gravitational force field $-\nabla U$, where

$$U = \frac{-Gm}{\sqrt{x^2 + y^2}}$$

The level curves of $U = c$ are all points (x, y) such that $U(x, y) = c$. If $c \neq 0$, we can set $U = c$ and solve the equation $x^2 + y^2 = G^2 m^2/c^2$. Thus, the required set of points is a circle with radius Gm/c. If $c = 0$, the set of points is empty; that is, $U \neq 0$ for any (x, y). Thus the level curves are all circles (Figure 4-4). These curves indicate levels of constant potential energy (recall that U is potential energy). Now $-\nabla U$ at any point $v_0 = (x_0, y_0)$ should be a vector orthogonal to the level curve through v_0; it follows from the remarks above that the level curve through v_0 is a circle of radius $|v_0|$. A unit tangent vector at v_0 to the circle of radius $|v_0|$ is

$$\frac{-y_0}{\sqrt{x_0^2+y_0^2}}\boldsymbol{i}+\frac{x_0}{\sqrt{x_0^2+y_0^2}}\boldsymbol{j}=\boldsymbol{T}$$

Since

$$\nabla U(v_0)=\frac{Gm}{\sqrt{x_0^2+y_0^2}}\left(\frac{x_0}{\sqrt{x_0^2+y_0^2}}\boldsymbol{i}+\frac{y_0}{\sqrt{x_0^2+y_0^2}}\boldsymbol{j}\right)$$

a short calculation shows that $\nabla U \cdot T = 0$. The gravitational force field $\boldsymbol{F}$ is given by $\boldsymbol{F} = -\nabla U$ and this explains why a particle in $\boldsymbol{R}^2$ would tend to move radially (perpendicular to the level curves) in toward our fixed mass. This principle is illustrated by the next example which does not exactly fit our mathematical theory but is illuminating in any case.

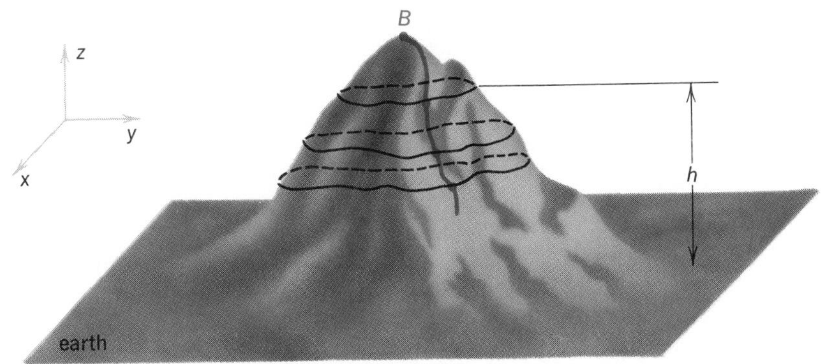

FIGURE 4-5

Example 3 Figure 4-5 shows a ball B placed at the top of a smooth surface S with equation $z=f(x,y)$. If the ball is not touched, it will not move (remember this is an abstraction and *not* reality). If the ball is given a slight push in a certain direction it will start to roll down the surface; the path of B will always be perpendicular to the level curves $h=c$, where $h:S\rightarrow \boldsymbol{R}$ is the height above the earth.

Example 4 This example is quite similar to the one above. Imagine a contour map of a hill with $f(x,y)$ the height of the hill and let M_c denote the level curves of f (Figure 4-6).

The directional derivatives $\nabla f \cdot \boldsymbol{n}$ in a direction $\boldsymbol{n}$ represents the slope or steepness of the hill in that direction. Clearly, if we walk along a level curve the slope is zero, and if we walk perpendicularly to a level curve, the slope is the greatest.

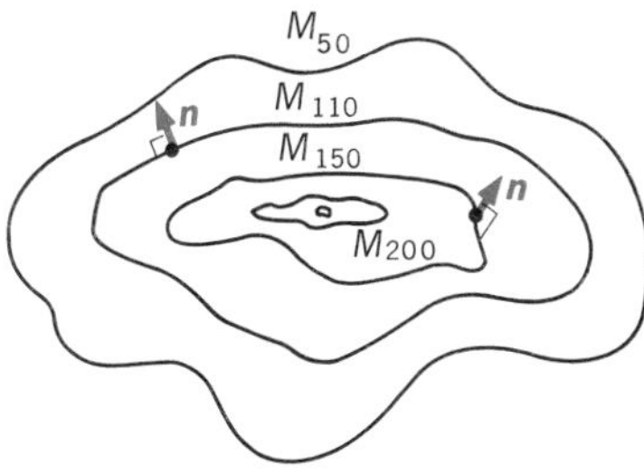

contour map of hill 250 feet high

FIGURE 4-6

EXERCISES

In exercises 1–5, compute the directional derivative of f at the given point p in the direction r.

1. $f(x,y) = x + 2xy - 3y^2, \quad p = (1,2), \quad \boldsymbol{r} = \frac{3}{5}\boldsymbol{i} + \frac{4}{5}\boldsymbol{j}$
2. $f(x,y) = \log\sqrt{x^2 + y^2}, \quad p = (1,0), \quad \boldsymbol{r} = \dfrac{1}{\sqrt{5}}(2\boldsymbol{i}+\boldsymbol{k})$
3. $f(x,y) = e^x \cos \pi y, \quad p = (0,-1), \quad \boldsymbol{r} = -\dfrac{1}{\sqrt{5}}\boldsymbol{i} + \dfrac{2}{\sqrt{5}}\boldsymbol{j}$
4. $f(x,y) = xy^2 + x^3 y, \quad p = (4,-2), \quad \boldsymbol{r} = \dfrac{1}{\sqrt{10}}\boldsymbol{i} + \dfrac{3}{\sqrt{10}}\boldsymbol{j}$
5. $f(x,y) = x^y, \quad p = (e,e), \quad \boldsymbol{r} = \frac{5}{13}\boldsymbol{i} + \frac{12}{13}\boldsymbol{j}$
6. Define the function $f\colon \boldsymbol{R}^2 \to \boldsymbol{R}$ by $f(x,y) = x + y - 3|xy|^{1/2}$. Show that
   (a) $(\partial f/\partial x)(0,0)$ and $(\partial f/\partial y)(0,0)$ both exist and equal 1, and hence $\nabla f(0,0)$ is defined and equals $\boldsymbol{i}+\boldsymbol{j}$;
   (b) $f(x,x) = -x$ if $x \geqslant 0$ and $f(x,x) = 5x$ if $x \leqslant 0$;
   (c) $\nabla P(0,0) \cdot \dfrac{\nabla f(0,0)}{|\nabla f(0,0)|} = \dfrac{2}{\sqrt{2}}$.

   Does the directional derivative of f exist at $(0,0)$ according to Definition 1? Is f differentiable?
7. For $\alpha \in \boldsymbol{R}$, let P_α be a family of parabolas defined by $y = x^2 + \alpha$.
   (a) Show that each point $(x_0, y_0) \in \boldsymbol{R}^2$ belongs to a unique P_α for some α.
   (b) Let $f\colon \boldsymbol{R}^2 \to \boldsymbol{R}$ be C^1 and suppose f is changing fastest at (x_0, y_0) in the direction of the unit vector perpendicular to the unit tangent vector to P_β at (x_0, y_0), where $(x_0, y_0) \in P_\beta$. If $f(0,0) = 0$ and $|\nabla f(x,y)| = \sqrt{1 + 4x^2}$, find f.
   (c) What are the P_α in relation to f?

5 FUNCTIONS FROM $\boldsymbol{R}^3$ to $\boldsymbol{R}$

In the last few sections we have been studying differentiable functions from $\boldsymbol{R}^2$ to $\boldsymbol{R}$. However, many topics in mathematics and physics require the ability to deal with functions of more than two variables—for example, the potential energy U of a particle in space relative to the earth is a real-valued function of three coordinates. If we all lived in a plane, perhaps the previous sections would suffice, but since we do not we must generalize our results to at least three dimensions.

Since most of the geometric intuition for differential calculus has already been achieved, we can proceed rather rapidly in this section to show that the theory is similar for functions from $\boldsymbol{R}^3$ to $\boldsymbol{R}$.

We can represent functions $f\colon \boldsymbol{R}^3 \to \boldsymbol{R}$ by $w = f(x,y,z)$. Unfortunately, the graphs of such functions cannot be pictured as a subset of $\boldsymbol{R}^3$ (this is why we deferred their study until now) and so we cannot gain much intuition from associated geometry. However, we shall depend upon the intuition we have developed for functions from $\boldsymbol{R}^2$ to $\boldsymbol{R}$ to help us here.

Example 1 As examples of functions of three variables we have

$$\text{(a) } f(x,y,z) = xyz$$

$$\text{(b) } f(x,y,z) = x^2 + y^2 + z^2$$

Given a function $f:\boldsymbol{R}^3 \to \boldsymbol{R}$ and a point $v_0 \in \boldsymbol{R}^3$, $v_0 = (x_0, y_0, z_0)$, we can ask whether the partial derivatives $(\partial f/\partial x)(v_0)$, $(\partial f/\partial y)(v_0)$, and $(\partial f/\partial z)(v_0)$ exist; that is, whether the following three limits exist:

$$\lim_{\Delta x \to 0} \frac{f(x_0+\Delta x, y_0, z_0) - f(x_0, y_0, z_0)}{\Delta x}$$

$$\lim_{\Delta y \to 0} \frac{f(x_0, y_0+\Delta y, z_0) + f(x_0, y_0, z_0)}{\Delta y}$$

$$\lim_{\Delta z \to 0} \frac{f(x_0, y_0, z_0+\Delta z) - f(x_0, y_0, z_0)}{\Delta z}$$

These derivatives are obtained just as in the two-variable case; here, we hold two variables fixed and differentiate with respect to the remaining variable. For example, if $v_0 = (x_0, y_0, z_0)$,

$$\frac{\partial f}{\partial z}(v_0) = \left.\frac{d(f(x_0, y_0, z))}{dz}\right|_{z=z_0}$$

If $f:\boldsymbol{R}^3 \to \boldsymbol{R}$ has partial derivatives at each point $v_0 \in \boldsymbol{R}^3$, then by varying v_0 we obtain a set of three real-valued functions $\partial f/\partial x$, $\partial f/\partial y$, $\partial f/\partial z$ of three variables.

Example 2 Compute $\partial f/\partial z$ for $f(x,y,z) = x^2yz + \cos zx + e^{xz^3}$.
We have

$$\frac{\partial f}{\partial z} = \frac{d[x^2yz + \cos zx + e^{xz^3}]}{dz} = x^2y - x\sin zx + 3z^2xe^{xz^3}$$

Now let us generalize the definition of differentiability.

Definition A function $f:\boldsymbol{R}^3 \to \boldsymbol{R}$ is **differentiable** at a point $v_0 \in \boldsymbol{R}^3$ if the partial derivatives $(\partial f/\partial x)(v_0)$, $(\partial f/\partial y)(v_0)$, and $(\partial f/\partial z)(v_0)$ exist and for $\Delta v = (\Delta x, \Delta y, \Delta z)$ we can express $f(v_0+\Delta v)$ by

$$f(v_0+\Delta v) = f(v_0) + \frac{\partial f}{\partial x}(v_0)\,\Delta x + \frac{\partial f}{\partial y}(v_0)\,\Delta y + \frac{\partial f}{\partial z}(v_0)\,\Delta z + Q_{v_0}(\Delta v)$$

where

$$\lim_{\Delta v \to 0} \frac{Q_{v_0}(\Delta v)}{|\Delta v|} = 0$$

There are some notions which need modification in order to apply to functions of three variables. Recall that for $f: \boldsymbol{R}^2 \to \boldsymbol{R}$ we defined the set of points (x, y) such that $f(x, y) = c$ is the level curve of f at level c. We can mimic the definition for functions of three variables, but in this case the set of points (x, y, z) such that $f(x, y, z) = c$ will not, in general, be a curve in $\boldsymbol{R}^3$ but a surface (but not always a surface, as we see below). Therefore, we call these sets **level surfaces** of f at the level c.

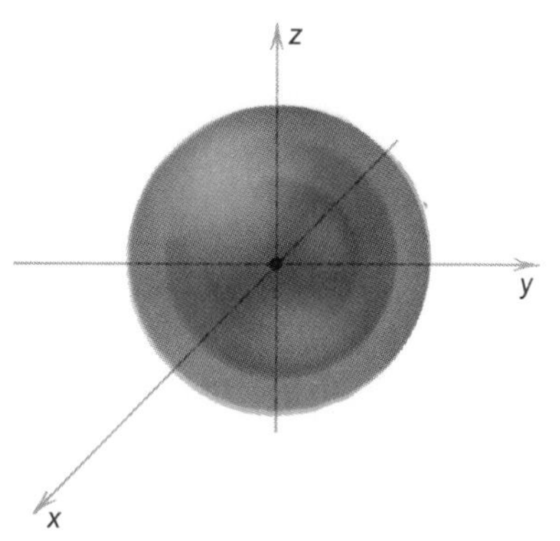

FIGURE 5-1

Example 3 Let $f(x, y, z) = x^2 + y^2 + z^2$. Then for $c > 0$ the level surface $x^2 + y^2 + z^2 = c$ is a sphere of radius $\sqrt{c}$ in $\boldsymbol{R}^3$. The case $c = 0$ is degenerate because the set of points for which $f(v) = 0$ has only one member, namely, $v = 0$. Clearly, we do not have a surface. (This phenomenon arises because, as we shall see later, 0 is a critical point for f.) Now if $c < 0$, there is no v for which $f(v) = c$ because f is always nonnegative. Thus we see that the level surfaces of f consist of concentric spheres and the origin in $\boldsymbol{R}^3$ (see Figure 5-1).

As in the two-variable case, we have the following definition.

Definition The function $f: \boldsymbol{R}^3 \to \boldsymbol{R}$ is **continuously differentiable** or C^1 if and only if it is differentiable and its first partial derivatives are continuous.

From now on, unless explicitly stated otherwise, all real-valued functions are assumed to be C^1.

We define a vector field on $\boldsymbol{R}^3$ to be a function from $\boldsymbol{R}^3$ to $\boldsymbol{R}^3$. In analogy to what we did in §14.3 we can identify $\boldsymbol{R}^3$ with vectors and thus we can say that a vector field $\boldsymbol{F}$ on $\boldsymbol{R}^3$ is a function which assigns to each point $v \in \boldsymbol{R}^3$ a directed line segment $\boldsymbol{F}(v)$ emanating from the origin. Therefore we can write

$$\boldsymbol{F}(v) = f_1(v)\boldsymbol{i} + f_2(v)\boldsymbol{j} + f_3(v)\boldsymbol{k}$$

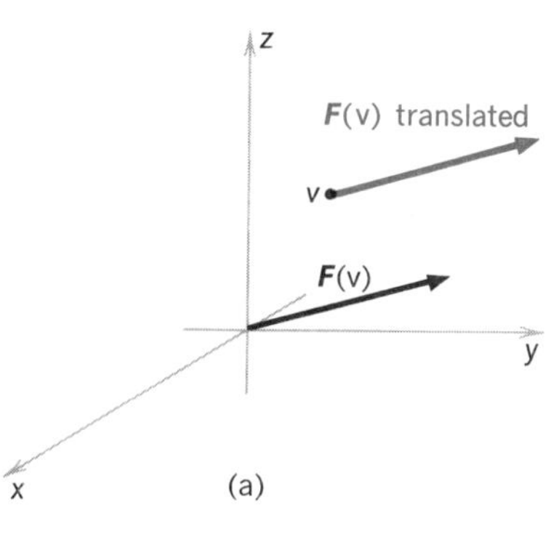

(a)

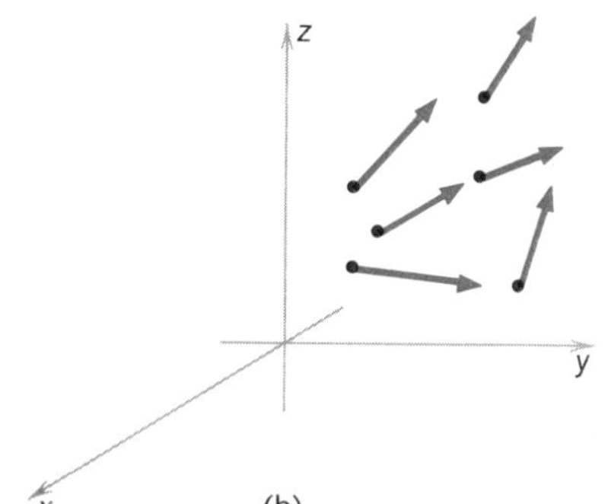

(b)

FIGURE 5-2

where each $f_i : \boldsymbol{R}^3 \to \boldsymbol{R}$. We say that $\boldsymbol{F}$ is C^1 if each f_i is C^1.

We can visualize or "graph" a vector field $\boldsymbol{F}$ by thinking of $\boldsymbol{F}$ as a function which assigns to each $v \in \boldsymbol{R}^3$ a directed line segment which is $\boldsymbol{F}(v)$ parallel translated to v [Figure 5-2(a)]. This method of "graphing" vector fields on $\boldsymbol{R}^2$ was discussed in §14.2. In Figure 5-2(b) we give an example of a vector field on $\boldsymbol{R}^3$. Although vector fields on $\boldsymbol{R}^3$ shall not be dealt with extensively in this chapter we shall study the calculus of these vector fields in great detail in Chapter 17.

For functions $f: \boldsymbol{R}^3 \to \boldsymbol{R}$ the associated gradient vector field $\nabla f: \boldsymbol{R}^3 \to \boldsymbol{R}^3$ is defined by

$$\nabla f(v) = \frac{\partial f}{\partial x}(v)\boldsymbol{i} + \frac{\partial f}{\partial y}(v)\boldsymbol{j} + \frac{\partial f}{\partial z}(v)\boldsymbol{k}$$

If $v_0 \in \boldsymbol{R}^3$ and $\nabla f(v_0) = 0$, then v_0 is a **critical point** of f. Given a unit vector $\boldsymbol{n} \in \boldsymbol{R}^3$, we define the **directional derivative** of f at v_0 in the direction $\boldsymbol{n}$ by

$$\nabla f(v_0) \cdot \boldsymbol{n}$$

We leave it as an exercise for the student to show that the directional derivative is maximized in the direction of the gradient.

Suppose now that v_0 belongs to the level surface $f(v_0) = c$. Then $\nabla f(v_0)$ will be perpendicular (or normal) to this level surface at the point v_0 (if the level surface is indeed a surface and not a point, as in Example 3). By perpendicular we mean that $\nabla f(v_0)$ is perpendicular to every curve on the surface $f(v) = c$ passing through v_0 (see Figure 5-3); that is, if $\boldsymbol{\sigma}:]a, b[\to \boldsymbol{R}^3$ is a C^1 curve on the surface with $\boldsymbol{\sigma}(t_0) = v_0$ then $\nabla f(v_0) \cdot \boldsymbol{\sigma}'(t_0) = 0$ for all curves. In the next section we use this fact to derive a formula for the equation of the tangent plane to a surface given as the graph of a function of two variables.

At this point, we state without proof a condition which guarantees that the level set of a function is, in fact, a surface: If the level set determined by $f(v) = c$ contains no critical points then it is either empty (has no points) or is a surface.

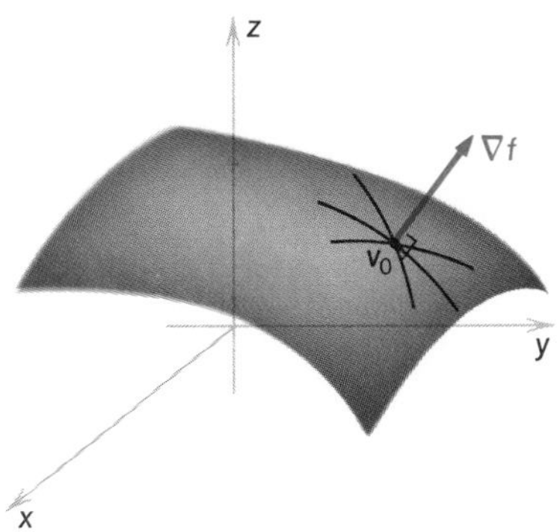

FIGURE 5-3

Example 4 Find a unit normal vector to the surface S given by $z = x^2 y^2 + y + 1$ at the point $(0, 0, 1)$.

Let $f(x, y, z) = x^2 y^2 + y + 1 - z$, and consider the surface defined by $f(x, y, z) = 0$. Since this is the set of points (x, y, z) with $f(x, y, z) = 0$ or $z = x^2 y^2 + y + 1$ we see that this is the surface S. Now

$$\begin{aligned}\nabla f(x, y, z) &= \frac{\partial f}{\partial x}\boldsymbol{i} + \frac{\partial f}{\partial y}\boldsymbol{j} + \frac{\partial f}{\partial z}\boldsymbol{k} \\ &= 2xy^2\boldsymbol{i} + (2x^2 y + 1)\boldsymbol{j} - \boldsymbol{k}\end{aligned}$$

and

$$\nabla f(0, 0, 1) = \boldsymbol{j} - \boldsymbol{k}$$

This vector is perpendicular to S at $(0, 0, 1)$, so to find a unit normal $\boldsymbol{n}$ we divide this vector by its length to obtain

$$\boldsymbol{n} = \frac{\nabla f(0, 0, 1)}{|f(0, 0, 1)|} = \frac{1}{\sqrt{2}}(\boldsymbol{j} - \boldsymbol{k})$$

We conclude this section with a discussion of some basic operations on vector fields—operations which are essential to our study of vector fields and surfaces in Chapter 17.

The del operator ∇ is often used to denote the vector expression

$$\nabla = \frac{\partial}{\partial x}\boldsymbol{i} + \frac{\partial}{\partial y}\boldsymbol{j} + \frac{\partial}{\partial z}\boldsymbol{k}$$

This operator is similar to the familiar differentiation operator d/dx; the del operator is meaningful only when applied to a function $f: \boldsymbol{R}^3 \to \boldsymbol{R}$, in which case it produces a vector. For example, when del operates on the differentiable function f, we obtain the gradient

$$\nabla f = \left(\frac{\partial}{\partial x}\boldsymbol{i} + \frac{\partial}{\partial y}\boldsymbol{j} + \frac{\partial}{\partial z}\boldsymbol{k}\right) f$$

$$= \frac{\partial f}{\partial x}\boldsymbol{i} + \frac{\partial f}{\partial y}\boldsymbol{j} + \frac{\partial f}{\partial z}\boldsymbol{k}$$

We shall see that ∇ is extremely useful in the study of other vector field operations.

The **curl** of a vector field $\boldsymbol{F} = F_1\boldsymbol{i} + F_2\boldsymbol{j} + F_3\boldsymbol{k}$ is defined as

$$\text{curl } \boldsymbol{F} = \left(\frac{\partial F_3}{\partial y} - \frac{\partial F_2}{\partial z}\right)\boldsymbol{i} + \left(\frac{\partial F_1}{\partial z} - \frac{\partial F_3}{\partial x}\right)\boldsymbol{j} + \left(\frac{\partial F_2}{\partial x} - \frac{\partial F_1}{\partial y}\right)\boldsymbol{k}$$

This lengthy formula is much easier to remember if we rewrite it using operator notation. If we view ∇ as a vector with components $\partial/\partial x$, $\partial/\partial y$, $\partial/\partial z$, then we can take the cross product

$$\nabla \times \boldsymbol{F} = \begin{vmatrix} \boldsymbol{i} & \boldsymbol{j} & \boldsymbol{k} \\ \dfrac{\partial}{\partial x} & \dfrac{\partial}{\partial y} & \dfrac{\partial}{\partial z} \\ F_1 & F_2 & F_3 \end{vmatrix}$$

$$= \left(\frac{\partial F_3}{\partial y} - \frac{\partial F_2}{\partial z}\right)\boldsymbol{i} + \left(\frac{\partial F_1}{\partial z} - \frac{\partial F_3}{\partial x}\right)\boldsymbol{j} + \left(\frac{\partial F_2}{\partial x} - \frac{\partial F_1}{\partial y}\right)\boldsymbol{k}$$

$$= \text{curl } \boldsymbol{F}$$

Thus, curl $\boldsymbol{F} = \nabla \times \boldsymbol{F}$ and we shall often use the latter expression in our work.

Example 5 Let $F(x, y, z) = x\boldsymbol{i} + xy\boldsymbol{j} + \boldsymbol{k}$. Find $\nabla \times \boldsymbol{F}$.

We have

$$\nabla \times \boldsymbol{F} = \begin{vmatrix} \boldsymbol{i} & \boldsymbol{j} & \boldsymbol{k} \\ \dfrac{\partial}{\partial x} & \dfrac{\partial}{\partial y} & \dfrac{\partial}{\partial z} \\ x & xy & 1 \end{vmatrix} = (0-0)\boldsymbol{i} - (0-0)\boldsymbol{j} + (y-0)\boldsymbol{k}$$

Thus $\nabla \times \boldsymbol{F} = y\boldsymbol{k}$.

The following theorem gives a basic property of the curl. The result should be compared with the fact that for any vector $\boldsymbol{v}$, we have $\boldsymbol{v} \times \boldsymbol{v} = 0$.

Theorem 1 For any C^1 function f we have

$$\nabla \times (\nabla f) = 0$$

that is, the curl of any gradient is zero.

Proof Let us write out the components: Since $\nabla f = (\partial f/\partial x, \partial f/\partial y, \partial f/\partial z)$ we have, by definition,

$$\nabla \times \nabla f = \begin{vmatrix} \boldsymbol{i} & \boldsymbol{j} & \boldsymbol{k} \\ \dfrac{\partial}{\partial x} & \dfrac{\partial}{\partial y} & \dfrac{\partial}{\partial z} \\ \dfrac{\partial f}{\partial x} & \dfrac{\partial f}{\partial y} & \dfrac{\partial f}{\partial z} \end{vmatrix} = \left(\frac{\partial^2 f}{\partial y\,\partial z} - \frac{\partial^2 f}{\partial z\,\partial y}\right)\boldsymbol{i} + \left(\frac{\partial^2 f}{\partial z\,\partial x} - \frac{\partial^2 f}{\partial x\,\partial z}\right)\boldsymbol{j} + \left(\frac{\partial^2 f}{\partial x\,\partial y} - \frac{\partial^2 f}{\partial y\,\partial x}\right)\boldsymbol{k}$$

Each component is zero because of the symmetry property of mixed partial derivatives, hence, the desired result follows. ∎

Although the physical significance of the curl will be brought out when we study Stoke's Theorem in Chapter 17, let us consider a simple situation which shows why the curl is associated with rotations.

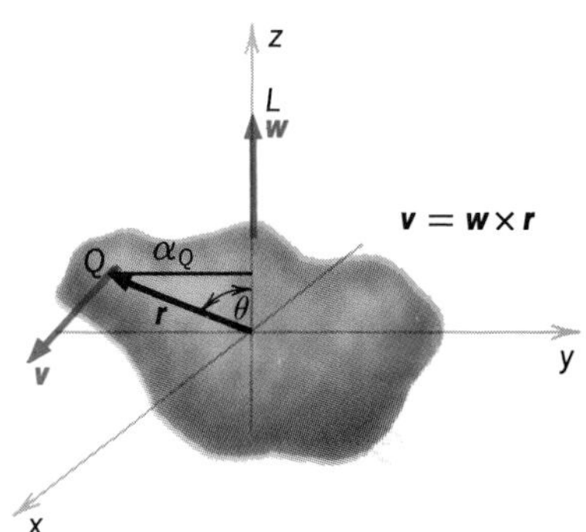

FIGURE 5-4

Example 6 Suppose we have a rigid body B rotating about an axis L. The rotational motion of the body can be completely described by a vector $\boldsymbol{w}$ along the axis of rotation, the direction being chosen so that the body rotates counterclockwise when $\boldsymbol{w}$ points out. Moreover, we take the length $\omega = |\boldsymbol{w}|$ to be the angular speed of the body B—that is, the tangential speed of any point of B divided by its distance to the axis L of rotation.

We also assume that we have selected a coordinate system so that L is the z axis. Let Q be any point of B and let α_Q be the distance from Q to L (Figure 5-4). Clearly,

$$\alpha_Q = |\boldsymbol{r}|\sin\theta$$

where $\boldsymbol{r}$ is the vector whose initial point is the origin and whose terminal point is Q. Then the tangential velocity $\boldsymbol{v}$ of Q is directed counterclockwise along the tangent to a circle parallel to the xy plane and with radius α_Q. The magnitude of this velocity is

$$|\boldsymbol{v}| = \omega\alpha_Q = \omega|\boldsymbol{r}|\sin\theta = |\boldsymbol{w}|\,|\boldsymbol{r}|\sin\theta$$

We have seen (page 618) that the direction and magnitude of $\boldsymbol{v}$ imply that

$$\boldsymbol{v} = \boldsymbol{w} \times \boldsymbol{r}$$

Because of our choice of axes, we can write $w = \omega\boldsymbol{k}$, $\boldsymbol{r} = x\boldsymbol{i} + y\boldsymbol{j} + z\boldsymbol{k}$. Therefore a small computation shows that

$$\boldsymbol{v} = \boldsymbol{w} \times \boldsymbol{r} = -\omega y\boldsymbol{i} + \omega x\boldsymbol{j}$$

and moreover

$$\operatorname{curl}\boldsymbol{v} = \begin{vmatrix} \boldsymbol{i} & \boldsymbol{j} & \boldsymbol{k} \\ \dfrac{\partial}{\partial x} & \dfrac{\partial}{\partial y} & \dfrac{\partial}{\partial z} \\ -\omega y & \omega x & 0 \end{vmatrix} = 2\omega\boldsymbol{k} = 2\boldsymbol{w}$$

Hence, for the rotation of a rigid body, the curl of the velocity vector field is a vector field directed along the axis of rotation with magnitude twice the angular speed.

The next basic operation is the **divergence** defined by

$$\textbf{div}\, \boldsymbol{F} = \nabla \cdot \boldsymbol{F} = \frac{\partial F_1}{\partial x} + \frac{\partial F_2}{\partial y} + \frac{\partial F_3}{\partial z}$$

In operator notation, div $\boldsymbol{F}$ is just the dot product of ∇ and $\boldsymbol{F}$. Note that $\nabla \times \boldsymbol{F}$ is a vector field whereas $\nabla \cdot \boldsymbol{F}: \boldsymbol{R}^3 \to \boldsymbol{R}$. We read $\nabla \cdot \boldsymbol{F}$ as "the divergence of $\boldsymbol{F}$."

The full significance of the divergence becomes apparent when we discuss Gauss' Theorem, but we can give some physical meaning here. If we imagine $\boldsymbol{F}$ as the velocity field of a gas (or a fluid), then div $\boldsymbol{F}$ represents the expansion per unit volume of the gas (or fluid). For example, if $\boldsymbol{F}(x, y, z) = x\boldsymbol{i} + y\boldsymbol{j} + z\boldsymbol{k}$, div $\boldsymbol{F} = 3$ and this means the gas is expanding at the rate of 3 cubic units per unit of volume. This is reasonable because $\boldsymbol{F}$ is an outward radial vector and as the gas moves outward along the flow lines it is expanding (see §14.3 for a discussion of flow lines).

The next theorem is the analog of Theorem 1.

Theorem 2 For any vector field $\boldsymbol{F}$,

$$\text{div}\,\text{curl}\, \boldsymbol{F} = \nabla \cdot (\nabla \times \boldsymbol{F}) = 0$$

that is, the divergence of any curl is zero.

The proof is similar to Theorem 1 and again rests on the equality of the mixed partial derivatives. The student should write out the details.

We have seen that $\nabla \times \boldsymbol{F}$ is related to rotations and $\nabla \cdot \boldsymbol{F}$ is related to compressions. This leads to the following common terminology. If $\nabla \times \boldsymbol{F} = 0$ we say $\boldsymbol{F}$ is **irrotational** and if $\nabla \cdot \boldsymbol{F} = 0$ we say $\boldsymbol{F}$ is **incompressible.**

The final operator we mention is the **Laplace operator** ∇^2 which operates on functions f as follows:

$$\nabla^2 f = \nabla \cdot (\nabla f) = \frac{\partial^2 f}{\partial x^2} + \frac{\partial^2 f}{\partial y^2} + \frac{\partial^2 f}{\partial z^2}$$

If $\boldsymbol{F} = F_1\boldsymbol{i} + F_2\boldsymbol{j} + F_3\boldsymbol{k}$ is a C^2 vector field we can also define $\nabla^2 \boldsymbol{F}$ in terms of components

$$\nabla^2 \boldsymbol{F} = \nabla^2 F_1 \boldsymbol{i} + \nabla^2 F_2 \boldsymbol{j} + \nabla^2 F_3 \boldsymbol{k}$$

We use this operator in formula 9 of Table 5-1 opposite. The operator ∇^2 is important in physical problems involving electric potential and heat conduction. These ideas are discussed more thoroughly in Chapter 17.

Table 5-1 contains some identities which are useful in dealing with the calculus of vector fields. Sometimes the results parallel the usual calculus formulas, but occasionally the identity is not so obvious. In the exercises, we ask the student to supply many of the proofs.

TABLE 5-1

Common Formulas of Vector Analysis

Here $f, g, \ldots$ denote real-valued functions; $\boldsymbol{F}, \boldsymbol{G}, \ldots$ denote vector fields.
1. $\nabla(\alpha f + \beta g) = \alpha\nabla f + \beta\nabla g$
2. $\nabla(fg) = f\nabla g + g\nabla f$
3. $\nabla(f/g) = (g\nabla f - f\nabla g)/g^2$, at points where $g(x) \neq 0$
4. $\operatorname{div}(\alpha \boldsymbol{F} + \beta \boldsymbol{G}) = \alpha \operatorname{div} \boldsymbol{F} + \beta \operatorname{div} \boldsymbol{G}$
5. $\operatorname{curl}(\alpha \boldsymbol{F} + \beta \boldsymbol{G}) = \alpha \operatorname{curl} \boldsymbol{F} + \beta \operatorname{curl} \boldsymbol{G}$
6. $\operatorname{div}(f\boldsymbol{F}) = f \operatorname{div} \boldsymbol{F} + \boldsymbol{F} \cdot \nabla f$
7. $\operatorname{div}(\boldsymbol{F} \times \boldsymbol{G}) = \boldsymbol{G} \cdot \operatorname{curl} \boldsymbol{F} - \boldsymbol{F} \cdot \operatorname{curl} \boldsymbol{G}$
8. $\operatorname{div} \operatorname{curl} \boldsymbol{F} = 0$
9. $\operatorname{curl} \operatorname{curl} \boldsymbol{F} = \operatorname{grad} \operatorname{div} \boldsymbol{F} - \nabla^2 \boldsymbol{F}$
10. $\operatorname{curl} \nabla f = 0$
11. $\operatorname{div}(\nabla f \times \nabla g) = 0$
12. $\operatorname{curl}(f\boldsymbol{F}) = f \operatorname{curl} \boldsymbol{F} + \nabla f \times \boldsymbol{F}$

Example 7 Prove formula 6 in Table 5-1.

Since $f\boldsymbol{F}$ has components fF_i,

$$\operatorname{div}(f\boldsymbol{F}) = \frac{\partial}{\partial x}(fF_1) + \frac{\partial}{\partial y}(fF_2) + \frac{\partial}{\partial z}(fF_3)$$

Now $(\partial/\partial x)(fF_1) = f(\partial F_1/\partial x) + F_1(\partial f/\partial x)$, with similar expressions for the other terms. We get

$$f\left(\frac{\partial F_1}{\partial x} + \frac{\partial F_2}{\partial y} + \frac{\partial F_3}{\partial z}\right) + F_1\frac{\partial f}{\partial x} + F_2\frac{\partial f}{\partial y} + F_3\frac{\partial f}{\partial z} = f\nabla\cdot F + F\cdot\nabla f$$

Example 8 Let $\boldsymbol{r}$ be the vector field $\boldsymbol{r}(x, y, z) = (x, y, z)$ (the position vector), and let $r = |\boldsymbol{r}|$. Compute ∇r and $\nabla \cdot (r\boldsymbol{r})$.

We have

$$\nabla r = \left(\frac{\partial r}{\partial x}, \frac{\partial r}{\partial y}, \frac{\partial r}{\partial z}\right)$$

Now $r(x, y, z) = \sqrt{x^2+y^2+z^2}$ so for $r \neq 0$ we get

$$\frac{\partial r}{\partial x} = \frac{x}{\sqrt{x^2+y^2+z^2}} = \frac{x}{r}$$

Thus

$$\nabla r = \left(\frac{x}{r}, \frac{y}{r}, \frac{z}{r}\right) = \frac{\boldsymbol{r}}{r}$$

For the second part, use formula 6 to write

$$\nabla \cdot (r\boldsymbol{r}) = r\nabla \cdot \boldsymbol{r} + \boldsymbol{r} \cdot \nabla r$$

Now $\nabla \cdot \boldsymbol{r} = \partial x/\partial x + \partial y/\partial y + \partial z/\partial z = 3$ and we just computed $\nabla r = \boldsymbol{r}/r$ above. Also, $\boldsymbol{r} \cdot \boldsymbol{r} = r^2$ so

$$\nabla \cdot (r\boldsymbol{r}) = 3r + \frac{r^2}{r} = 4r$$

EXERCISES

1. Compute the gradient ∇f for each of the following functions.
   (a) $f(x, y, z) = \sqrt{x^2 + y^2 + z^2}$
   (b) $f(x, y, z) = xy + yz + xz$
   (c) $f(x, y, z) = \dfrac{1}{x^2 + y^2 + z^2}$

For the following functions $f: \boldsymbol{R}^3 \to \boldsymbol{R}$ and $g: \boldsymbol{R} \to \boldsymbol{R}^3$ find ∇f and g' and evaluate $(f \circ g)'(1)$ in exercises 2–4.

2. $f(x, y, z) = xz + yz + xy, \quad g(t) = (e^t, \cos t, \sin t)$
3. $f(x, y, z) = e^{xyz}, \quad g(t) = (6t, 3t^2, t^3)$
4. $f(x, y, z) = (x^2 + y^2 + z^2) \log \sqrt{x^2 + y^2 + z^2}, \quad g(t) = (e^t, e^{-t}, t)$

In exercises 5 and 6, compute the directional derivative of f in the given direction at the given point.

5. $f(x, y, z) = xy^2 + y^2z^3 + z^3x$, $p = (4, -2, -1)$, $\boldsymbol{r} = \dfrac{1}{\sqrt{14}}(\boldsymbol{i} + 3\boldsymbol{j} + 2\boldsymbol{k})$
6. $f(x, y, z) = x^{y^z}$, $p = (e, e, 0)$, $\boldsymbol{r} = \dfrac{12}{13}\boldsymbol{i} + \dfrac{3}{13}\boldsymbol{j} + \dfrac{4}{13}\boldsymbol{k}$
7. Compute the divergence $\nabla \cdot \boldsymbol{F}$ for each of the following vector fields.
   (a) $\boldsymbol{F}(x, y, z) = x\boldsymbol{i} + y\boldsymbol{j} + z\boldsymbol{k}$
   (b) $\boldsymbol{F}(x, y, z) = yz\boldsymbol{i} + xz\boldsymbol{j} + xy\boldsymbol{k}$
   (c) $\boldsymbol{F}(x, y, z) = (x^2 + y^2 + z^2)(3\boldsymbol{i} + 4\boldsymbol{j} + 5\boldsymbol{k})$
8. Compute the curl $\nabla \times \boldsymbol{F}$ for each of (a)–(c) in exercise 7.
9. Suppose $\nabla \cdot \boldsymbol{F} = 0$ and $\nabla \cdot \boldsymbol{G} = 0$. Which of the following necessarily have zero divergence?
   (a) $\boldsymbol{F} + \boldsymbol{G}$
   (b) $\boldsymbol{F} \times \boldsymbol{G}$
   (c) $\boldsymbol{F}(\boldsymbol{F} \cdot \boldsymbol{G})$
10. Prove formulas 1–6 of Table 5-1.
11. Prove formulas 7–12 of Table 5-1.
12. Let $f(x, y, z) = x^2y^2 + y^2z^2$. Directly verify that $\nabla \times \nabla f = \boldsymbol{0}$.

13. Let $f(x,y,z) = 3x^2y\boldsymbol{i} + (x^3+y^3)\boldsymbol{j}$.
    (a) Verify that $\operatorname{curl}\boldsymbol{F} = \mathbf{0}$.
    (b) Find a function f such that $\boldsymbol{F} = \nabla f$.
    (c) Verify that for a general $\boldsymbol{F}$ such a function f can exist only if $\operatorname{curl}\boldsymbol{F} = \mathbf{0}$.
14. Prove the identity $\nabla\cdot(\nabla f\times\nabla g) = 0$.
15. Find a unit normal to the surface $x^3y+xz = 1$ at the point $(1,2,-1)$.
16. Let $\boldsymbol{r}(x,y,z) = (x,y,z)$ and $r = \sqrt{x^2+y^2+z^2} = |\boldsymbol{r}|$. Prove the following identities (make use of Table 5-1 as much as possible).

    (a) $\nabla\left(\dfrac{1}{r}\right) = \dfrac{-\boldsymbol{r}}{r^3}$, $r \neq 0$, and, in general, $\nabla(r^n) = nr^{n-2}\boldsymbol{r}$

    (b) $\nabla^2\left(\dfrac{1}{r}\right) = 0$, $r \neq 0$

    (c) $\nabla\cdot\left(\dfrac{\boldsymbol{r}}{r^3}\right) = 0$

    (d) $\nabla\times\boldsymbol{r} = \mathbf{0}$

6 THE TANGENT PLANE

Suppose $f:\boldsymbol{R}^2\to\boldsymbol{R}$ is differentiable. The graph of the function is a surface $z=f(x,y)$ and, intuitively, we expect that if the surface is sufficiently smooth near a point $z_0=f(x_0,y_0)$, then there will be a plane tangent to the surface of z_0.

Our first difficulty is to define the tangent plane and the second is to find its equation. To define the tangent plane let us use some geometric reasoning. Let $(x_0,y_0)\in\boldsymbol{R}^2$ and let $C:\boldsymbol{R}\to\boldsymbol{R}^2$ be any differentiable curve $C(t)=(C_1(t),C_2(t))$ which passes through (x_0,y_0) [that is, $C(t_0)=(x_0,y_0)$ for some $t_0\in\boldsymbol{R}$]. Then the function $t\mapsto(C_1(t),C_2(t),f(C_1(t),C_2(t)))$ describes a curve $\tilde{C}$ in $\boldsymbol{R}^3$ which lies on S. The derivative of this function at t_0 is, by results in Chapter 13, a tangent vector $\boldsymbol{v}_{\tilde{C}}(t_0)$ to $\tilde{C}$ at the point (see Figure 6-1):

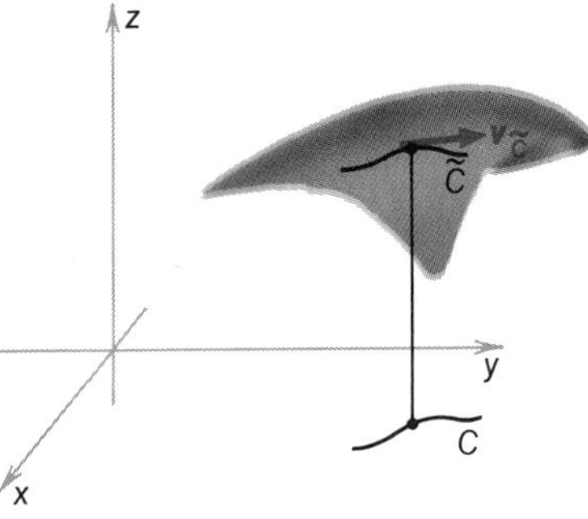

FIGURE 6-1

$$\begin{aligned}(x_0,y_0,f(x_0,y_0)) &= (C_1(t_0),C_2(t_0),f(C_1(t_0),C_2(t_0)))\\ &= C_1(t_0)\boldsymbol{i} + C_2(t_0)\boldsymbol{j} + f(C_1(t_0),C_2(t_0))\boldsymbol{k} \qquad (1)\end{aligned}$$

We differentiate (1) using the Chain Rule to obtain

$$\begin{aligned}\boldsymbol{v}_{\tilde{C}}(t_0) &= C_1'(t_0)\boldsymbol{i} + C_2'(t_0)\boldsymbol{j} + \left.\frac{d[f(C_1(t),C_2(t))]}{dt}\right|_{t=t_0}\boldsymbol{k}\\ &= C_1'(t_0)\boldsymbol{i} + C_2'(t_0)\boldsymbol{j} + [\nabla f(x_0,y_0)\cdot(C_1'(t_0)\boldsymbol{i} + C_2'(t_0)\boldsymbol{j})]\boldsymbol{k}\\ &= C_1'(t_0)\boldsymbol{i} + C_2'(t_0)\boldsymbol{j} + \left[\frac{\partial f}{\partial x}(x_0,y_0)\,C_1'(t_0) + \frac{\partial f}{\partial y}(x_0,y_0)\,C_2'(t_0)\right]\boldsymbol{k}\end{aligned}$$

If we can show that the tangent vectors $\boldsymbol{v}_{\tilde{C}}(t_0)$ for any curve $\tilde{C}$ on S obtained in this way all lie in the same plane, then this plane should be taken as the tangent plane to S at $(x_0, y_0, f(x_0, y_0))$.

Theorem 1 Let $\tilde{C}: t \mapsto (C_1(t), C_2(t), f(C_1(t), C_2(t)))$ be any differentiable curve on S obtained as above and suppose $\tilde{C}$ passes through $(x_0, y_0, f(x_0, y_0))$. Then the vector $\boldsymbol{v}_{\tilde{C}}$ lies in a plane P which is independent of the curve $\tilde{C}$.

Proof To prove this theorem we will find a normal vector $\boldsymbol{n}$ which does not depend on $\tilde{C}$ such that $\boldsymbol{n}\cdot\boldsymbol{v}_{\tilde{C}} = 0$ for any $\tilde{C}$. This will imply that all vectors $\boldsymbol{v}_{\tilde{C}}$ lie in the same plane (the plane perpendicular to $\boldsymbol{n}$).

The vector $\boldsymbol{n}$ is not difficult to find. We start by defining a new function $F: \boldsymbol{R}^3 \to \boldsymbol{R}$ by $F(x, y, z) = z - f(x, y)$. Then the set S of (x, y, z) such that $F(x, y, z) = 0$ is the level surface of F corresponding to the value 0. But this set is precisely the surface $z = f(x, y)$. We already know from §14.5 that $\nabla F(x_0, y_0, z_0)$ will be normal to S at $(x_0, y_0, z_0) = (x_0, y_0, f(x_0, y_0)) \in S$, and thus perpendicular to every curve $\tilde{C}$. We choose

$$\begin{aligned} \boldsymbol{n} &= \nabla F(x_0, y_0, z_0) = -\nabla f(x_0, y_0) + \boldsymbol{k} \\ &= -\frac{\partial f}{\partial y}(x_0, y_0)\,\boldsymbol{i} - \frac{\partial f}{\partial y}(x_0, y_0)\,\boldsymbol{j} + \boldsymbol{k} \end{aligned}$$

Thus $\boldsymbol{n}\cdot\boldsymbol{v}_{\tilde{C}} = 0$. Without using the result from §14.5, we can verify directly that $\boldsymbol{n}\cdot\boldsymbol{v}_{\tilde{C}} = 0$. We have

$$\begin{aligned} \boldsymbol{n}\cdot\boldsymbol{v}_{\tilde{C}} &= \left[-\frac{\partial f}{\partial x}(x_0, y_0)\,\boldsymbol{i} - \frac{\partial f}{\partial y}(x_0, y_0)\,\boldsymbol{j} + \boldsymbol{k}\right] \\ &\quad \cdot \left[C_1'(t_0)\,\boldsymbol{i} + C_2'(t_0)\,\boldsymbol{j} + \left(\frac{\partial f}{\partial x}(x_0, y_0)\,C_1'(t_0) + \frac{\partial f}{\partial y}(x_0, y_0)\,C_2'(t_0)\right)\boldsymbol{k}\right] \\ &= -\frac{\partial f}{\partial x}(x_0, y_0)\,C_1'(t_0) - \frac{\partial f}{\partial y}(x_0, y_0)\,C_2'(t_0) \\ &\quad + \frac{\partial f}{\partial x}(x_0, y_0)\,C_1'(t_0) + \frac{\partial f}{\partial y}(x_0, y_0)\,C_2'(t_0) = 0 \end{aligned}$$

Thus $\boldsymbol{n}$ is normal to $\boldsymbol{v}_{\tilde{C}}$ for any $\tilde{C}$; hence, all vectors $\boldsymbol{v}_{\tilde{C}}$ lie in the same plane. ∎

We define the **tangent plane** to a surface at $(x_0, y_0, f(x_0, y_0))$ to be the plane which contains all the tangent vectors $\boldsymbol{v}_{\tilde{C}}$. The key to finding an explicit formula for this plane is in the proof of Theorem 1—namely, once we know a normal vector to a plane and a point in the plane we can find the plane's equation. If $\boldsymbol{n} = A\boldsymbol{i} + B\boldsymbol{j} + C\boldsymbol{k}$ is a normal vector to a plane P and $(x_0, y_0, z_0) \in P$, then $A(x - x_0) + B(y - y_0) + C(z - z_0) = 0$ is the equation of P (page 621). Consequently if $z_0 = f(x_0, y_0)$, then

$$-\frac{\partial f}{\partial x}(x_0, y_0)[x-x_0] - \frac{\partial f}{\partial y}(x_0, y_0)[y-y_0] + [z-z_0] = 0$$

is the equation of the tangent plane; rewriting this we obtain

$$z = \frac{\partial f}{\partial x}(x_0, y_0)[x-x_0] + \frac{\partial f}{\partial y}(x_0, y_0)[y-y_0] + f(x_0, y_0)$$

Example 1 Find the equation of the plane tangent to the surface $z = x^2 + y^2 = f(x, y)$ at the point $(-1, 1, 2)$.

First we evaluate $(\partial f/\partial x)(-1, 1) = 2(-1) = -2$ and $(\partial f/\partial y)(-1, 1) = 2$. By the formula above, the equation of the tangent plane is

$$z = -2(x-(-1)) + 2(y-1) + 2 = -2(x+1) + 2(y-1) + 2$$

Example 2 Let S be the surface obtained as the graph of $z = \log(1 + x^2y^2)$. Find the equation of the tangent plane to S at the point $(0, 0, 0)$.

We have

$$\frac{\partial z}{\partial x} = \frac{2xy^2}{1 + x^2y^2}, \qquad \frac{\partial z}{\partial y} = \frac{2yx^2}{1 + x^2y^2}$$

Therefore, $(\partial z/\partial x)(0, 0) = 0 = (\partial z/\partial y)(0, 0)$; the equation of the tangent plane is

$$z = 0 \cdot (x - x_0) + 0 \cdot (y - y_0) + 0$$

or simply $z = 0$.

EXERCISES

1. In exercises 1–6, find the equation of the tangent plane at p to the surface described by $z = f(x, y)$.

1. $z = x^3 + y^3 - 6xy, \quad p = (1, 2)$
2. $z = \cos x \cos y, \quad p = (0, \pi/2)$
3. $z = \cos x \sin y, \quad p = (0, \pi/2)$
4. $z = \log(x+y) + x\cos y + \arctan(x+y), \quad p = (1, 0)$
5. $z = \sqrt{x^2 + y^2}, \quad p = (1, 1)$
6. $z = xy, \quad p = (2, 1)$
7. Let S be the surface described by $z = f(x, y)$, and assume f is C^1. Find all S such that at the point $(x, y, z) \in S$, $\boldsymbol{n}(x, y, z) = y\boldsymbol{i} + x\boldsymbol{j} - \boldsymbol{k}$ is the normal $\nabla f - \boldsymbol{k}$ to S.
8. Let $f: \boldsymbol{R}^2 \to \boldsymbol{R}$ and $g: \boldsymbol{R} \to \boldsymbol{R}$ be differentiable with $S = \text{graph } f$ and $\tilde{S} = \text{graph}(g \circ f)$. Suppose the tangent plane to $\tilde{S}$ at $(x_0, y_0, g(f(x_0, y_0)))$ is parallel to the tangent plane to S at $(x_0, y_0, f(x_0, y_0))$. Compute the possible values of $g'(f(x_0, y_0))$.

7 PARAMETRIZED SURFACES

It is limiting to think of (or describe) a smooth (without corners) surface as the graph of a C^1 function $f: \mathbf{R}^2 \to \mathbf{R}$ or $f: D \subset \mathbf{R}^2 \to \mathbf{R}$ For example suppose surface S is the set of points (x, y, z) where $x - z + z^3 = 0$. Here S is a sheet which (relative to the xy plane) doubles back on itself (see Figure 7-1). Obviously, we want to call S a surface, since it is just a plane with a wrinkle. However, S is **not** the graph of some function $z = f(x, y)$ because this would mean that for each $(x_0, y_0) \in \mathbf{R}^2$ there is only one z_0 with $(x_0, y_0, z_0) \in S$. As Figure 7-1 illustrates, this condition is violated.

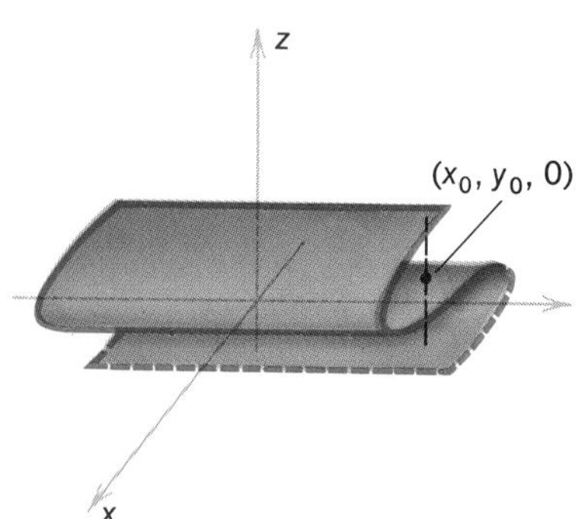

FIGURE 7-1

Another example we may use here is the torus T or surface of a doughnut which is depicted in Figure 7-2. Anyone would call T a surface and yet, by the same reasoning as above, T cannot be the graph of a differentiable function of two variables. These observations encourage us to extend our definition of a surface. The motivation for the definition which follows is partly that a surface can be thought of as being obtained from the plane by "rolling," "bending," and "pushing." For example, to get a torus you take a portion of the plane and roll it (like a cigarette—see Figure 7-3), and then take the two "ends" and bring them together until they meet (Figure 7-4).

FIGURE 7-2

Definition A **parametrization of a surface** S is a function $f: \mathbf{R}^2 \to \mathbf{R}^3$ or a function $f: D \subset \mathbf{R}^2 \to \mathbf{R}^3$, where S is the image of the parametrization f. Such a function f can be written

$$f(u, v) = (f_1(u, v), f_2(u, v), f_3(u, v))$$

where each $f_i: D \to \mathbf{R}$ is a real-valued function. The surface is differentiable if each $f_i: D \to \mathbf{R}$ is differentiable.

If $f: \mathbf{R}^2 \to \mathbf{R}^3$, we are deforming the whole plane; if $f: D \to \mathbf{R}^3$, where $D \subset \mathbf{R}^2$, then we are working only with part of the plane. We use the variables u and v because we think of the function f as deforming one plane (the uv plane) into the surface $S \subset \mathbf{R}^3$, where S has coordinates (x, y, z). Hence, we can write the function f as three equations

$$x = f_1(u, v), \qquad y = f_2(u, v), \qquad z = f_3(u, v)$$

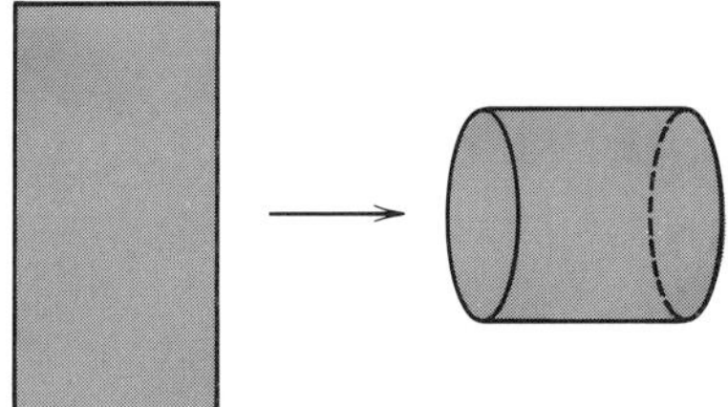

FIGURE 7-3

FIGURE 7-4

Suppose that f is differentiable at $(u_0, v_0) \in \boldsymbol{R}^2$. If we fix u, we get a map $\boldsymbol{R} \to \boldsymbol{R}^3$ given by $t \mapsto f(u_0, t)$ and the image of this map is a curve on the surface (Figure 7-5). From Chapter 13 we know that a tangent vector to this curve is given by

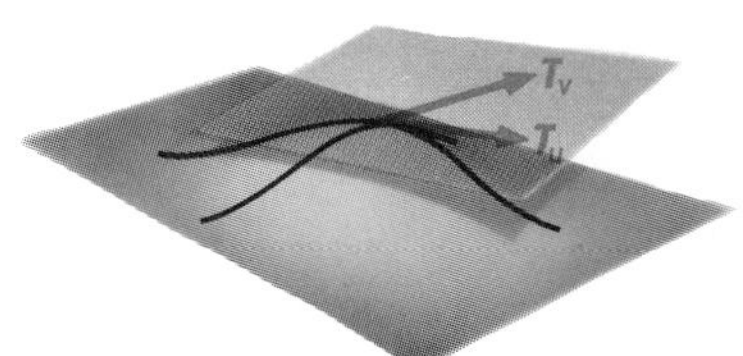

FIGURE 7-5

$$\boldsymbol{T}_v = \frac{\partial x}{\partial v}(u_0, v_0)\,\boldsymbol{i} + \frac{\partial y}{\partial v}(u_0, v_0)\,\boldsymbol{j} + \frac{\partial z}{\partial v}(u_0, v_0)\,\boldsymbol{k}$$

Similarly, if we fix v and consider the curve $t \mapsto f(t, v_0)$ we obtain another tangent vector

$$\boldsymbol{T}_u = \frac{\partial x}{\partial u}(u_0, v_0)\,\boldsymbol{i} + \frac{\partial y}{\partial u}(u_0, v_0)\,\boldsymbol{j} + \frac{\partial z}{\partial u}(u_0, v_0)\,\boldsymbol{k}$$

Since these vectors $\boldsymbol{T}_u$ and $\boldsymbol{T}_v$ are tangent to two curves on the surface at $f(u_0, v_0)$, they lie in the tangent plane to the surface at this point. (Note that we have not yet defined tangent plane to a parametrized surface. We can use the same geometric definition of a tangent plane for parametrized surfaces as we did for surfaces obtained as graphs. Below we give a slightly different definition which in fact turns out to be the "geometrically defined" tangent plane.) We say that the surface S is **smooth** at $f(u_0, v_0)$ if $\boldsymbol{T}_u \times \boldsymbol{T}_v \neq \boldsymbol{0}$ at (u_0, v_0). The surface is smooth if it is smooth for all points $f(u_0, v_0)$. The nonzero vector $\boldsymbol{T}_u \times \boldsymbol{T}_v$ will be a **normal** vector (recall the vector product of $\boldsymbol{T}_u$ and $\boldsymbol{T}_v$ is perpendicular to the plane spanned by $\boldsymbol{T}_u$ and $\boldsymbol{T}_v$) to the surface; the fact that the cross product is nonzero ensures that there will be a tangent plane.

As an example, consider the surface given by the equations

$$x = u\cos v, \quad y = u\sin v, \quad z = u \qquad u \geqslant 0$$

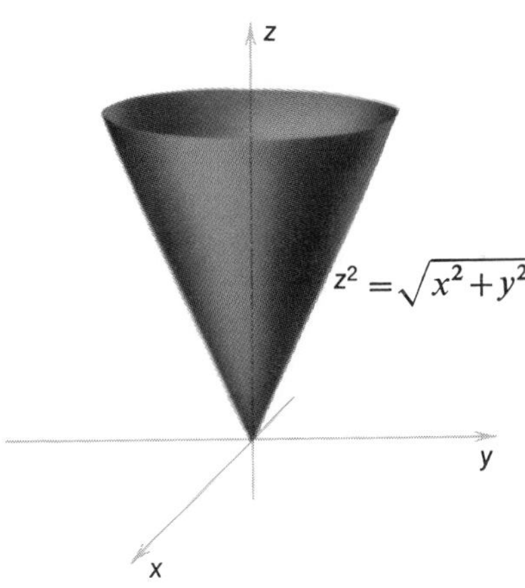

FIGURE 7-6

These equations describe the surface $z^2 = \sqrt{x^2+y^2}$ (just square x, y, z above to check this) which is shown in Figure 7-6. This surface is a cone with a "point" at $(0,0,0)$; it is a differentiable surface because each component function is differentiable. However, the surface is not smooth at $(0,0,0)$. To see this, compute $\boldsymbol{T}_u$ and $\boldsymbol{T}_v$ at $(0,0) \in \boldsymbol{R}^2$:

$$\begin{aligned}\boldsymbol{T}_u &= \frac{\partial x}{\partial u}(0,0)\,\boldsymbol{i} + \frac{\partial y}{\partial u}(0,0)\,\boldsymbol{j} + \frac{\partial z}{\partial u}(0,0)\,\boldsymbol{k} \\ &= \cos 0\,\boldsymbol{i} + \sin 0\,\boldsymbol{j} + \boldsymbol{k} = \boldsymbol{i} + \boldsymbol{k}\end{aligned}$$

and similarly

$$T_v = 0 \cdot -\sin 0 \boldsymbol{i} + 0 \cdot \cos 0 \boldsymbol{j} + 0 \cdot \boldsymbol{k} = \boldsymbol{0}$$

Thus $T_u \times T_v = \boldsymbol{0}$ and, by definition, the surface is not smooth at $(0,0,0)$.

If a parametrized surface $f: D \subset \boldsymbol{R}^2 \to \boldsymbol{R}^3$ is smooth at $f(u_0, v_0)$ we define the **tangent plane** to the surface at $f(u_0, v_0)$ to be the plane determined by the vectors T_u and T_v.

Example 1 Let $f: \boldsymbol{R}^2 \to \boldsymbol{R}^3$ be given by

$$x = u\cos v, \quad y = u \sin v, \quad z = u^2 + v^2$$

Then

$$T_u = \cos v \boldsymbol{i} + \sin v \boldsymbol{j} + 2u \boldsymbol{k}$$

$$T_v = -u \sin v \boldsymbol{i} + u \cos v \boldsymbol{j} + 2v \boldsymbol{k}$$

and the tangent plane at (u,v) is the set of vectors through (u,v) perpendicular to

$$T_u \times T_v = (-2u^2 \cos v + 2v \sin v, -2u^2 \sin v - 2v \cos v, u)$$

if this vector is nonzero. Since $T_u \times T_v$ is equal to zero at $(u,v) = (0,0)$, there is no tangent plane at $f(0,0) = (0,0,0)$. However, we can find the equation of a tangent plane at all the other points where $T_u \times T_v \neq 0$. For instance, let us find the equation of the tangent plane to the surface under consideration at the point $f(1,0) = (1,0,1)$. At this point

$$\boldsymbol{n} = T_u \times T_v = (-2,0,1) = -2\boldsymbol{i} + \boldsymbol{k}$$

Since we have the normal $\boldsymbol{n}$ to the surface and a point $(1,0,1)$ on the surface, we use the standard formula to obtain the equation of the tangent plane

$$-2(x-1) + (z-1) = 0$$

or

$$z = 1 + 2(x-1)$$

Example 2 Suppose a surface S is the graph of a differentiable function $g: \boldsymbol{R}^2 \to \boldsymbol{R}$. Then the surface will be smooth at all points $(u_0, v_0, g(u_0, v_0)) \in \boldsymbol{R}^2$. To show this, we write S in parametric form as follows:

$$x = u, \qquad y = v, \qquad z = g(u,v)$$

which is the same as $z = g(x,y)$. Then

$$T_u = \boldsymbol{i} + \frac{\partial g}{\partial u}(u_0, v_0)\boldsymbol{k}$$

$$T_v = \boldsymbol{j} + \frac{\partial g}{\partial v}(u_0, v_0)\boldsymbol{k}$$

and

$$T_u \times T_v = -\frac{\partial g}{\partial u}(u_0, v_0)\boldsymbol{i} - \frac{\partial g}{\partial v}(u_0, v_0)\boldsymbol{j} + \boldsymbol{k} \neq 0 \qquad (u_0, v_0) \in \boldsymbol{R} \qquad (1)$$

since the coefficient of $\boldsymbol{k}$ is 1.

Notice this example also shows that the definition of tangent plane for parametrized surfaces agrees with the one for surfaces obtained as graphs, since (1) is the same formula we obtained earlier for the normal to S at $(x_0, y_0, z_0) \in S$.

EXERCISES

In exercises 1–3, find the equation for the tangent plane to the given surface at the specified point.

1. $x = 2u, \quad y = u^2 + v, \quad z = v^2$ at $(0, 1, 1)$
2. $x = u^2 - v^2, \quad y = u + v, \quad z = u^2 + 4v$ at $(-\frac{1}{4}, \frac{1}{2}, 2)$
3. $x = u^2, \quad y = u \sin e^v, \quad z = \frac{1}{3} u \cos e^v$ at $(13, -2, 1)$

4. Find an expression for a unit normal vector to the surface
$$x = \cos v \sin u, \quad y = \sin v \sin u, \quad z = \cos u$$
for $u \in [0, 2\pi], \quad v \in [0, 2\pi]$. Identify this surface.

*5. Repeat exercise 4 for the surface
$$x = (2 - \cos v) \cos u, \quad y = (2 - \cos v) \sin u, \quad z = \sin v$$
for $-\pi \leqslant u \leqslant \pi, \quad -\pi \leqslant v \leqslant \pi$.

8 CRITICAL POINTS, MAXIMA, AND MINIMA

We return to the case of surfaces S which are represented as graphs of differentiable functions $z = g(x, y), g: D \subset \boldsymbol{R}^2 \to \boldsymbol{R}$. It would be useful to have a method of finding local maxima or local minima for these surfaces (see Figure 8-1).

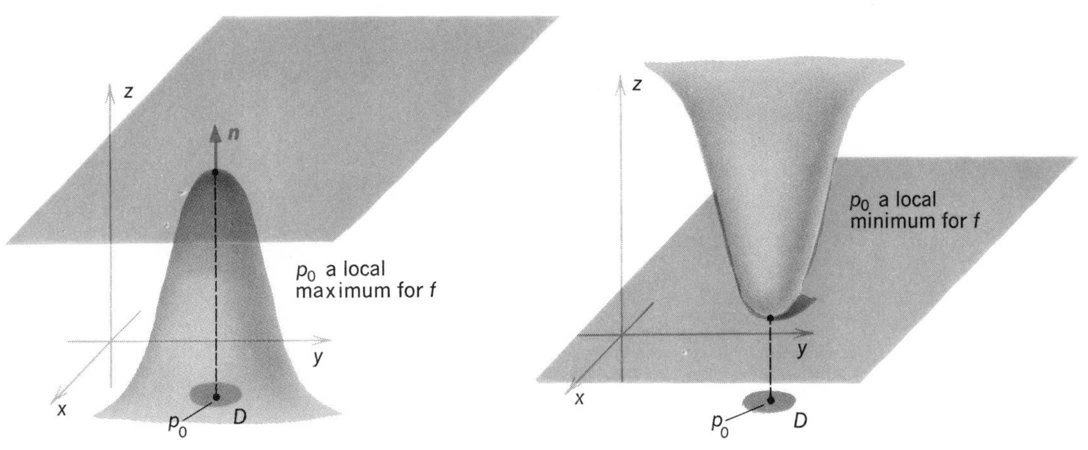

FIGURE 8-1

Definition A point p_0 is a **local maximum** of the surface which is the graph of $z = g(x, y)$ if there exists some disc D with center p_0 such that $g(p_0) \geqslant g(p)$ for all $p \in D$; the point p_0 is a **local minimum** if $g(p_0) \leqslant g(p)$ for all $p \in D$. Maximum and minimum points are called **extrema**.

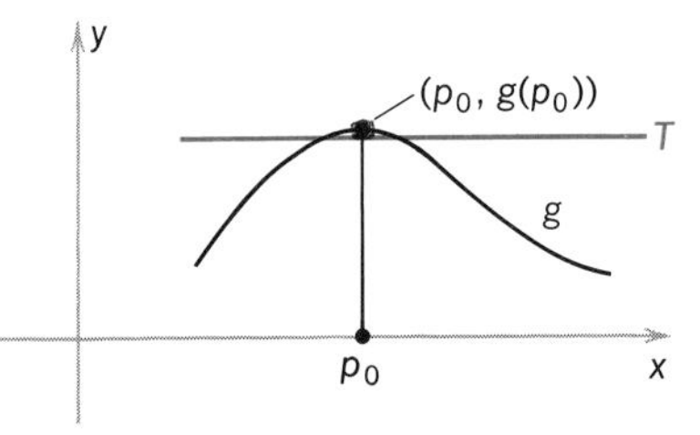

FIGURE 8-2

We know that for a function of one variable $g: \boldsymbol{R} \to \boldsymbol{R}$, the necessary condition for a point p_0 to be a local extremum is $g'(p_0) = 0$. Geometrically this condition means that the tangent T to the graph of g at $(p_0, g(p_0))$ is parallel to the x axis (see Figure 8-2). Remember that the requirement $g'(p_0) = 0$ is necessary but not sufficient. For example, $y = x^3$ is a function whose derivative is zero at $x = 0$ but the point $(0, 0)$ is neither a local maximum nor a local minimum.

If we carry over this geometrical reasoning to two dimensions, it seems clear that the tangent plane at a local maximum p_0 or a local minimum p_0 will be parallel to the xy plane. If this is true then the normal vector

$$\boldsymbol{n} = \frac{\partial g}{\partial x}(p_0)\boldsymbol{i} + \frac{\partial g}{\partial y}(p_0)\boldsymbol{j} - \boldsymbol{k}$$

will be parallel to the vector $\boldsymbol{k}$. This means that $\boldsymbol{n} = \lambda \boldsymbol{k}$ for some scalar λ which implies that

$$\frac{\partial g}{\partial x}(p_0) = \frac{\partial g}{\partial y}(p_0) = 0$$

or, equivalently, that

$$\nabla g(p_0) = 0$$

If $\nabla g(p_0) = 0$, then p_0 is a critical point of g. Hence, the first step in locating extrema is to set $\partial g/\partial x = 0$ and $\partial g/\partial y = 0$, and solve these two equations simultaneously (that is, find critical points of g). This is, generally, the easy part.

Example 1 Find all the critical points of $z = x^2 y + y^2 x$. Differentiating, we obtain

$$\frac{\partial z}{\partial x} = 2xy + y^2, \qquad \frac{\partial z}{\partial y} = 2xy + x^2$$

Equating the partials to zero yields

$$2xy + y^2 = 0, \qquad 2xy + x^2 = 0$$

Subtracting we obtain $x^2 = y^2$. Thus $x = \pm y$. Substituting $x = +y$ in the first equation above, we find that

$$2y^2 + y^2 = 3y^2 = 0$$

so $y = 0$ and thus $x = 0$. If $x = -y$ then

$$-2y^2 + y^2 = -y^2 = 0$$

so $y = 0$ and therefore $x = 0$.

Hence the only critical point is $(0,0)$. However, we do not know (without actually graphing the function) whether this point is a local maximum, a local minimum, or neither.

Example 2 Consider the surface in $\boldsymbol{R}^3$ which is the the graph $g(x,y) = x^3$; this surface is shown in Figure 8-3. In this case every point on the y axis is a critical point. To see this note that $(\partial z/\partial x)(0,y) = 0$ and $(\partial z/\partial y)(0,y) = 0$. So $\nabla g(0,y) = 0$ for all y. However, even though every point on the y axis is a critical point for g, it is clear from the figure that none of these points is a local maximum or minimum.

Do Examples 1 and 2 exhaust the possibilities of what can happen to a function at a critical point? The answer is no, as the next example shows.

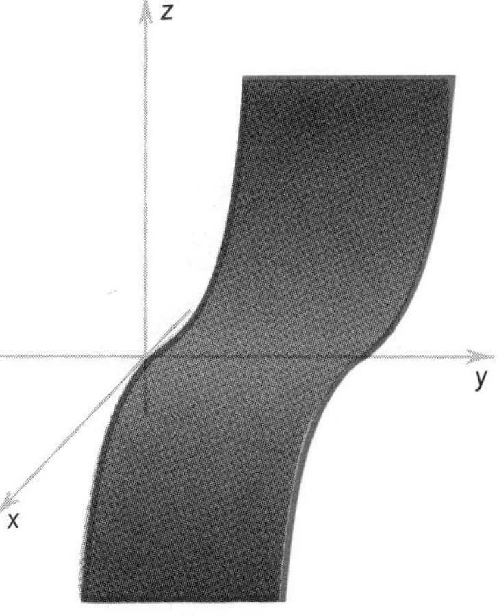

FIGURE 8-3

Example 3 Consider the function

$$z = g(x,y) = x^2 - y^2$$

which has

$$\nabla g(x,y) = 2x\boldsymbol{i} - 2y\boldsymbol{j}$$

It is easy to see that $(0,0)$ is the only critical point of g. However, the graph in Figure 8-4 shows that $(0,0)$ is neither a maximum nor a minimum. The point $(0,0)$ is what is called a **saddle point** [we graphed $z = g(x,y)$ in example 4, §14.1].

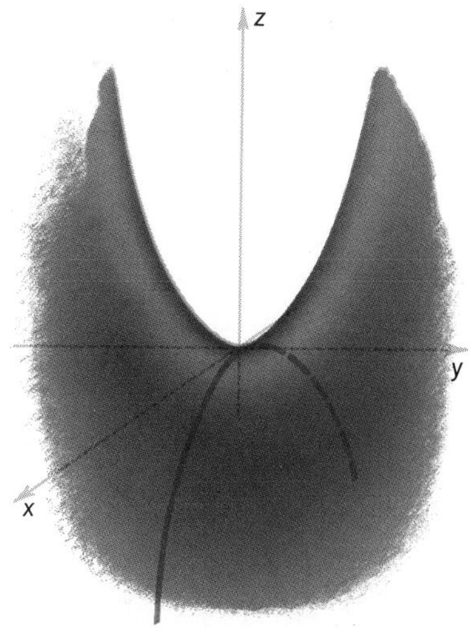

FIGURE 8-4

These examples show that we must establish some criteria for determining whether a critical point is

(1) a local maximum
(2) a local minimum
(3) a saddle point
(4) none of the above

Let us first consider the functions

$$f(x,y) = Ax^2 + Cy^2 + E, \qquad A \neq 0, C \neq 0$$

hence

$$\nabla f(x,y) = 2Ax\boldsymbol{i} + 2Cy\boldsymbol{j}$$

and

$$\nabla f(x,y) = 0 \Rightarrow (x,y) = (0,0)$$

and so $(0,0)$ is the only critical point.

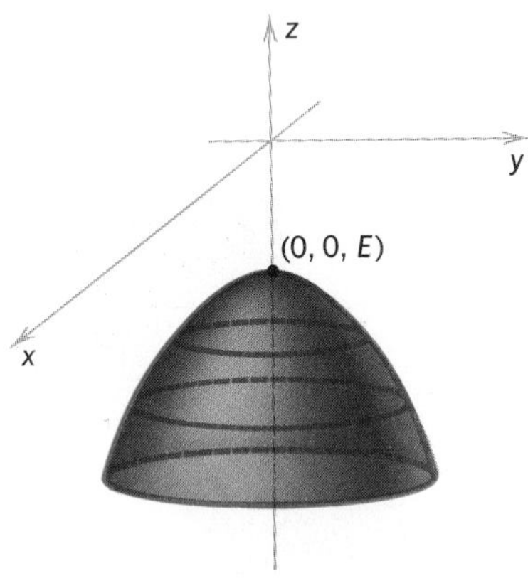

FIGURE 8-5

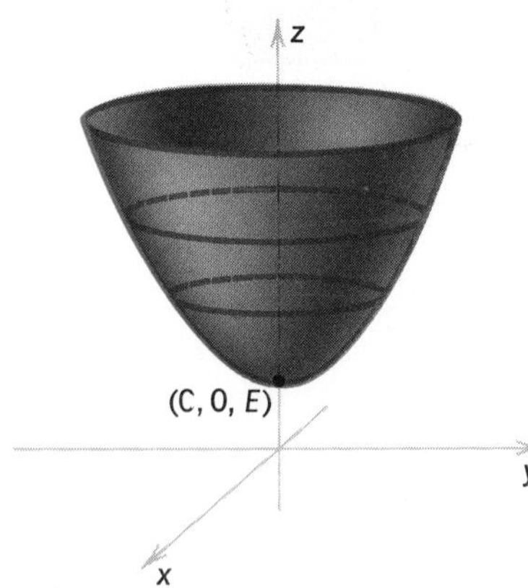

FIGURE 8-6

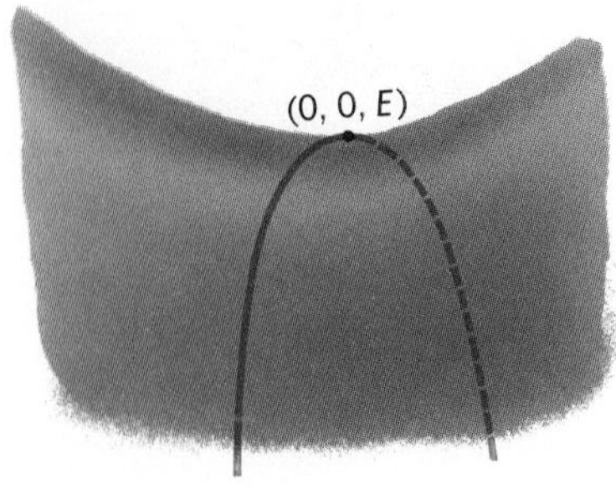

FIGURE 8-7

Case 1. If $A<0$ and $C<0$, the graph of f is like that in Figure 8-5 and so $(0,0)$ is a local maximum.
Case 2. If $A>0$ and $C>0$, the graph of f is like that in Figure 8-6 and therefore $(0,0)$ is a local minimum for f.
Case 3. If $A<0$ and $C>0$, then $(0,0)$ is a saddle point (see Figure 8-7).
Case 4. If $A>0$ and $C<0$, $(0,0)$ is still a saddle point.

The case where $A=0$ or $C=0$ is considered in exercise 13.

Let us now consider the more complicated case $z=f(x,y)=Ax^2+Bxy+Cy^2+E$. To find the critical points we set

$$\frac{\partial z}{\partial x} = 2Ax + By = 0$$

$$\frac{\partial z}{\partial y} = Bx + 2Cy = 0$$

Eliminating first the y term and then the x term, we obtain the equations

$$(4AC-B^2)\,x = 0$$

$$(4AC-B^2)\,y = 0$$

Thus if $4AC-B^2 \neq 0$ the only critical point is $(0,0)$. The case $4AC-B^2=0$ is more involved (see exercise 14). We shall assume that $4AC-B^2\neq 0$, $A\neq 0, C\neq 0$ and proceed to determine whether $(0,0)$ is a saddle point, a local maximum, or a local minimum. We do this by completing the square for $f(x,y)$ and then applying the four cases considered earlier.

First, notice that in the function $z=f(x,y)=Ax^2+Bxy+Cy^2+E$, the constant term E has no effect on the nature of the critical points—it merely shifts the graph of f up or down and this, of course, does not change the shape of the graph. Therefore, we may ignore E and write

$$\begin{aligned} z &= Ax^2 + Bxy + Cy^2 \\ &= A\left(x^2 + \frac{B}{A}xy\right) + Cy^2 = A\left(x + \frac{B}{2A}y\right)^2 + \left(C - \frac{B^2}{4A}\right)y^2 \\ &= A\left(x + \frac{B}{2A}y\right)^2 + \left(\frac{4AC-B^2}{4A}\right)y^2 = Ax_1^2 + C_1y^2 \end{aligned}$$

where

$$x_1 = x + \frac{B}{2A}y, \qquad C_1 = \frac{4AC-B^2}{4A}$$

Case 1′. If $4AC>B^2$ and $A<0$, then $C_1<0$. Therefore, Case 1 applies and $(0,0)$ is a local maximum.
Case 2′. If $4AC>B^2$ and $A>0$, then $C_1>0$. Therefore Case 2 applies and $(0,0)$ is a local minimum.
Case 3′. If $4AC<B^2$ and $A<0$, then $C_1>0$. Therefore Case 3 applies and $(0,0)$ is a saddle point.

Case 4′. If $4AC < B^2$ and $A > 0$, then $C_1 < 0$. Therefore Case 4 applies and $(0,0)$ is still a saddle point.

There is one important case we did not consider above, namely, if $4AC - B^2 < 0$ and $4AC = 0$; thus either A or C is zero. In this case $(0,0)$ is still a saddle point (see exercise 15).

To motivate the general condition which may be used to determine if a critical point p_0 of a map $f: \boldsymbol{R}^2 \to \boldsymbol{R}$ is either a local maximum, local minimum, or saddle point we must extend the Taylor Theorem with remainder to functions of two variables. We state the result without proof.

Theorem 1 Taylor's Theorem Let $f: \boldsymbol{R}^2 \to \boldsymbol{R}$ have continuous second-order partial derivatives in a disc D centered at (a,b). Then for any point $(x,y) \in D$

$$f(x,y) = f(a,b) + (x-a)\frac{\partial f}{\partial x}(a,b) + (y-b)\frac{\partial f}{\partial y}(a,b)$$

$$+\frac{1}{2}\left[(x-a)^2\frac{\partial^2 f}{\partial x^2}(a',b') + 2(x-a)(y-b)\frac{\partial^2 f}{\partial x\,\partial y}(a',b')\right.$$

$$\left.+(y-b)^2\frac{\partial^2 f}{\partial y^2}(a',b')\right]$$

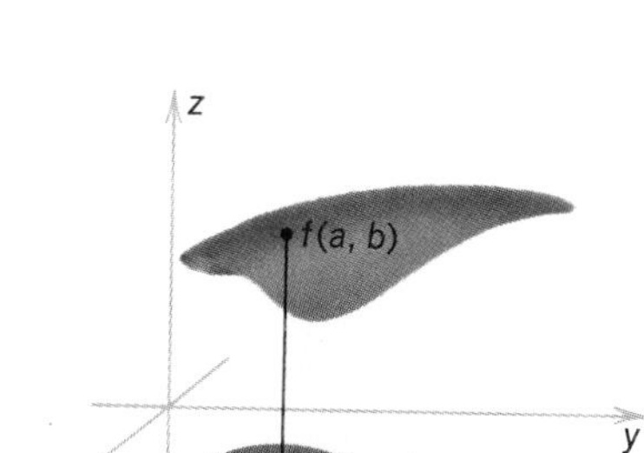

FIGURE 8-8

where (a',b') is a point which lies on the line segment from (a,b) to (x,y) (see Figure 8-8). If (a,b) is a critical point then the first derivatives $(\partial f/\partial x)(a,b)$, $(\partial f/\partial y)(a,b)$ are zero and f can be written as

$$f(x,y) = f(a,b) + \frac{1}{2}\frac{\partial^2 f}{\partial x^2}(a',b')x_1^2 + \frac{\partial^2 f}{\partial x\,\partial y}(a',b')x_1y_1 + \frac{1}{2}\frac{\partial^2 f}{\partial y^2}(a',b')y_1^2$$

where $x_1 = x - a$, $y_1 = y - b$.

Taylor's Theorem says that f is "almost" a quadratic function and hence we may apply the results obtained above with

$$A = \frac{1}{2}\frac{\partial^2 f}{\partial x^2}(a',b'), \qquad B = \frac{\partial^2 f}{\partial x\,\partial y}(a',b'), \qquad C = \frac{1}{2}\frac{\partial^2 f}{\partial y^2}(a',b')$$

The only difficulty is that Taylor's formula does not tell us what the point (a',b') is, but it turns out that this does not matter. If we instead set

$$A = \frac{1}{2}\frac{\partial^2 f}{\partial x^2}(a,b), \qquad B = \frac{\partial^2 f}{\partial x\,\partial y}(a,b), \qquad C = \frac{1}{2}\frac{\partial^2 f}{\partial y^2}(a,b)$$

then the relationship between A, B, and C provides enough information for us to decide what kind of critical point (a,b) is.

Case 1″. If $4AC - B^2 > 0$, $A < 0$ or, equivalently, if

$$D = \left[\frac{\partial^2 f}{\partial x^2}(a,b)\right]\left[\frac{\partial^2 f}{\partial y^2}(a,b)\right] - \left[\frac{\partial^2 f}{\partial x\,\partial y}(a,b)\right]^2 > 0$$

and $(\partial^2 f/\partial^2 x)(a,b) < 0$, then (a,b) is a local maximum. We call D the **characteristic** of f at the point (a,b).

Case 2″. If $4AC - B^2 > 0$, $A > 0$ or, equivalently, if

$$D = \left[\frac{\partial^2 f}{\partial x^2}(a,b)\right]\left[\frac{\partial^2 f}{\partial y^2}(a,b)\right] - \left[\frac{\partial^2 f}{\partial x\,\partial y}(a,b)\right]^2 > 0$$

and $(\partial^2 f/\partial x^2)(a,b) > 0$, then (a,b) is a local minimum.

Case 3″. If $D < 0$ then (a,b) is a saddle point.

If neither Cases 1″, 2″, or 3″ hold, these tests do not tell us anything and the behavior of the function at the critical points may (or may not) be much more complicated than these simple types (see exercises 3, 8, 9, 13, 14).

Example 4 Locate the maximum, minimum, and saddle points of the function

$$f(x,y) = \log(x^2 + y^2 + 1)$$

We must first locate the critical points of this function, so we calculate

$$\nabla f(x,y) = \frac{2x}{x^2+y^2+1}\boldsymbol{i} + \frac{2y}{x^2+y^2+1}\boldsymbol{j}$$

Thus $\nabla f(x,y) = 0$ if and only if $(x,y) = (0,0)$, and the only critical point of f is $(0,0)$. Now we must determine whether this is a maximum or minimum. The second partials are

$$\frac{\partial^2 f}{\partial x^2} = \frac{2(x^2+y^2+1) - (2x)(2x)}{(x^2+y^2+1)^2}$$

$$\frac{\partial^2 f}{\partial y^2} = \frac{2(x^2+y^2+1) - (2y)(2y)}{(x^2+y^2+1)^2}$$

and

$$\frac{\partial^2 f}{\partial x\,\partial y} = \frac{-2x(2y)}{(x^2+y^2+1)^2}$$

So

$$\frac{\partial^2 f}{\partial x^2}(0,0) = 2 = \frac{\partial^2 f}{\partial y^2}(0,0)$$

and

$$\frac{\partial^2 f}{\partial x\,\partial y}(0,0) = 0$$

which yields

$$D = 2 \cdot 2 = 4 > 0$$

Since $(\partial^2 f/\partial x^2)(0,0) > 0$ we conclude (by Case 2″) that $(0,0)$ is a local minimum.

Example 5 The graph of the function $f(x, y) = 1/xy$ is a surface S in $\boldsymbol{R}^3$. Find the points on S which are closest to the origin $(0, 0, 0)$.

The distance from (x, y, z) to $(0, 0, 0)$ is given by the formula

$$d(x, yz) = \sqrt{x^2 + y^2 + z^2}$$

If $(x, y, z) \in S$ then d can be expressed as a function $\tilde{d}(x, y) = d(x, y, 1/xy)$ of two variables:

$$\tilde{d}(x, y) = \sqrt{x^2 + y^2 + \frac{1}{x^2 y^2}}$$

Note that the minimum (if it exists) cannot occur "too near" the x axis or y axis because $\tilde{d}$ is not defined if either $x = 0$ or $y = 0$, and $\tilde{d}$ gets very large as x or y approaches the x axis or the y axis.

Since $\tilde{d} > 0$ it will be minimized when $\tilde{d}^2(x, y) = x^2 + y^2 + (1/x^2 y^2) = f(x, y)$ is minimized. (This function is much easier to deal with.) Now

$$\nabla f(x, y) = \nabla \tilde{d}^2(x, y) = \left(2x - \frac{2}{x^3 y^2}\right)\boldsymbol{i} + \left(2y - \frac{2}{y^3 x^2}\right)\boldsymbol{j} = 0$$

if and only if

$$\left(2x - \frac{2}{x^3 y^2}\right) = 0 = \left(2y - \frac{2}{y^3 x^2}\right)$$

or $x^4 y^2 - 1 = 0$ and $x^2 y^4 - 1 = 0$. From the first equation we get $y^2 = 1/x^4$ and substituting this into the second equation we obtain

$$\frac{x^2}{x^8} = 1 = \frac{1}{x^6}$$

Thus $x = \pm 1$ and $y = \pm 1$, and it therefore follows that $\tilde{d}^2$ has four critical points, namely, $(1, 1)$, $(1, -1)$, $(-1, 1)$, and $(-1, -1)$. To determine whether these are local minima, local maxima, or saddle points we proceed to apply our test:

$$\frac{\partial^2 f}{\partial x^2} = 2 + \frac{6}{x^4 y^2}, \qquad \frac{\partial^2 f}{\partial y^2} = 2 + \frac{6}{x^2 y^4}, \qquad \frac{\partial^2 f}{\partial y\, \partial x} = \frac{4}{x^3 y^3}$$

so

$$\frac{\partial^2 f}{\partial x^2}(a, b) = \frac{\partial^2 f}{\partial x^2}(a, b) = 8$$

where (a, b) is any one of the above four critical points, and $(\partial^2 f/\partial x\, \partial y)(a, b) = \pm 4$.

We see that in any of the above cases $D = 64 - 16 = 48 > 0$ and $(\partial^2 f/\partial x^2)(a, b) > 0$, so each critical point is a local minimum, and these are also all the minima for f.

Finally, note that $\tilde{d}^2(a, b) = 3$ for all these critical points and so the points on the surface that are closest to $(0,0,0)$ are $(1,1,1), (1,-1,-1), (-1,1,-1)$, and $(-1,-1,1)$ with $\tilde{d} = \sqrt{3}$ at these points. Thus $\tilde{d} \geqslant \sqrt{3}$ and is equal to $\sqrt{3}$ when $(x, y) = (a, b)$, where (a, b) is one of the four points above.

Example 6 Analyze the behavior of $z = x^5 y + xy^5 + xy$ at its critical points.

The first partial derivatives are

$$\frac{\partial z}{\partial x} = 5x^4 y + y^5 + y = y(5x^4 + y^4 + 1)$$

and

$$\frac{\partial z}{\partial y} = x(5y^4 + x^4 + 1)$$

The terms $5x^4 + y^4 + 1$ and $5y^4 + x^4 + 1$ are always greater than or equal to 1 so it follows that the only critical point is $(0, 0)$.

The second partials are

$$\frac{\partial^2 z}{\partial x^2} = 20x^3 y, \qquad \frac{\partial^2 z}{\partial y^2} = 20xy^3$$

and

$$\frac{\partial^2 z}{\partial x \, \partial y} = 5x^4 + 5y^4 + 1$$

Thus at $(0, 0)$, $D = -1$ and so $(0, 0)$ is a saddle point, by case 3″.

EXERCISES

In exercises 1–10, find the critical points of the given function and then determine whether they are local maxima, local minima, saddle points, or neither.

1. $f(x, y) = x^2 - y^2 + xy$
2. $f(x, y) = x^2 + y^2 - xy$
3. $f(x, y) = x^2 + y^2 + 2xy$
4. $f(x, y) = x^2 + y^2 + 3xy$
5. $f(x, y) = e^{1 + x^2 - y^2}$
6. $f(x, y) = x^2 - 3xy + 5x - 2y + 6y^2 + 8$
7. $f(x, y) = 3x^2 + 2xy + 2x + y^2 + y + 4$

*8. $f(x, y) = \sin(x^2 + y^2) \qquad -\frac{1}{2}\sqrt{\pi} \leqslant x \leqslant \frac{1}{2}\sqrt{\pi}, \quad -\frac{1}{2}\sqrt{\pi} \leqslant y \leqslant \frac{1}{2}\sqrt{\pi}$

*9. $f(x, y) = \cos(x^2 + y^2) \qquad -\frac{1}{2}\sqrt{\pi} \leqslant x \leqslant \frac{1}{2}\sqrt{\pi}, \quad -\frac{1}{2}\sqrt{\pi} \leqslant y \leqslant \frac{1}{2}\sqrt{\pi}$

10. $f(x, y) = \log(2 + \sin xy), \qquad 0 \leqslant x \leqslant \sqrt{\frac{\pi}{2}}, \quad 0 \leqslant y \leqslant \sqrt{\frac{\pi}{2}}$

*11. Prove that if (x_0, y_0) is a local minimum (or maximum) of $f: \boldsymbol{R}^2 \to \boldsymbol{R}$ then the partial derivatives of f at (x_0, y_0) are zero.

12. An examination of the function $f: \boldsymbol{R}^2 \to \boldsymbol{R}$; $(x, y) \to (y - 3x^2)(y - x^2)$ will give an idea of the difficulty of deducing conditions which guarantee that a critical point is a relative extremum. Show that
    (a) the origin is a critical point of f;
    (b) f has a relative minimum at $(0, 0)$ on every straight line through $(0, 0)$, that is, if $g(t) = (at, bt)$, then $f \circ g: \boldsymbol{R} \to \boldsymbol{R}$ has a relative minimum at 0, for every choice of a and b;
    (c) the origin is not a relative minimum for f.

13. Let $f(x, y) = Ax^2 + E$. What are the critical points of f? Are they local maxima or local minima?

14. Let $f(x, y) = x^2 - 2xy + y^2 = (x - y)^2$. Here $4AC - B^2 = D = 0$. Are the critical points local minima, local maxima, or saddle points?

*15. Let $f(x, y) = Ax^2 + Bxy + Cy^2 + E$. Assume that $4AC - B^2 < 0$ and $4AC = 0$. Show that $(0, 0)$ is still a saddle point. [Hint: Make a change of variables (coordinates).]

9 THE GEOMETRY OF MAPS FROM $\boldsymbol{R}^2$ to $\boldsymbol{R}^2$

In §14.3 we discussed vector fields on $\boldsymbol{R}^2$ and in the previous section, vector fields on $\boldsymbol{R}^3$. Now we wish to investigate what maps from $\boldsymbol{R}^2$ to $\boldsymbol{R}^2$ and $\boldsymbol{R}^3$ to $\boldsymbol{R}^3$ do to subsets of these spaces. This geometric understanding will be useful when we discuss the change of variables formula for multiple integrals in the next chapter.

Let $D^* \subset \boldsymbol{R}^2$ be a subset and suppose we apply a C^1 function or vector field $\boldsymbol{F}: \boldsymbol{R}^2 \to \boldsymbol{R}^2$ to D^*; that is, we look at $\boldsymbol{F}(x^*, y^*)$ for $(x^*, y^*) \in D^*$. Then $\boldsymbol{F}$ takes points in D^* to points in $\boldsymbol{R}^2$. We denote this image set of points by D or $\boldsymbol{F}(D^*)$; hence, $D = \boldsymbol{F}(D^*)$ is the set of all points (x, y) such that

$$(x, y) = \boldsymbol{F}(x^*, y^*) \text{ for some } (x^*, y^*) \in D^*$$

One way to understand the geometry of the map $\boldsymbol{F}$ is to see how it **deforms** or changes D^*. For example, Figure 9-1 illustrates a map $\boldsymbol{F}$ which

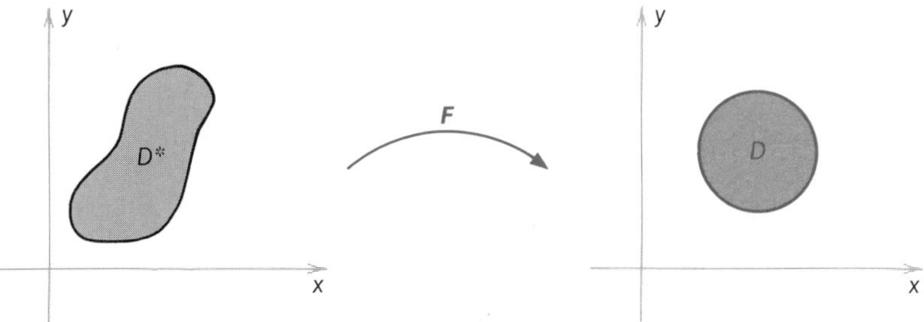

FIGURE 9-1

takes a slightly twisted region into a disc. We can make an analogous study of C^1 functions $\boldsymbol{F}: \boldsymbol{R}^3 \to \boldsymbol{R}^3$ and subsets $D^* \subset \boldsymbol{R}^3$.

Example 1 Let $D^* \subset \boldsymbol{R}^2$ be the rectangle $D^* = [0, 2\pi] \times [0, 1]$. Then all points in D^* are of the form (θ, r) where $0 \leqslant \theta \leqslant 2\pi, 0 \leqslant r \leqslant 1$. Let $\boldsymbol{F}$ be defined by $\boldsymbol{F}(\theta, r) = (r\cos\theta, r\sin\theta)$. Find the image set D.

We set $(x, y) = (r\cos\theta, r\sin\theta)$. Since $x^2 + y^2 = r^2\cos^2\theta + r^2\sin^2\theta = r^2 \leqslant 1$, the set of points $(x, y) \in \boldsymbol{R}^2$ such that $(x, y) \in D$ has the property that $x^2 + y^2 \leqslant 1$ and so $D \subset$ the unit disc. In addition, any point (x, y) in the unit disc can be written as $(r\cos\theta, r\sin\theta)$ for $0 \leqslant r \leqslant 1$ and $0 \leqslant \theta \leqslant 2\pi$. Thus D is the unit disc.

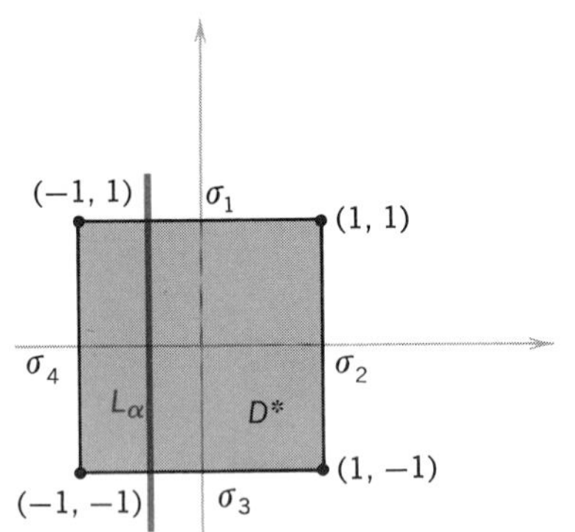

FIGURE 9-2

Example 2 Let $\boldsymbol{F}$ be defined by $\boldsymbol{F}(x, y) = ((x+y)/2, (x-y)/2)$. Let $D^* = [-1, 1] \times [-1, 1] \subset \boldsymbol{R}^2$ be a square with side of length 2 centered at the origin. Determine the image D obtained from D^* by applying $\boldsymbol{F}$.

Let us first determine the effect of $\boldsymbol{F}$ on the line $\boldsymbol{\sigma}_1(t) = (t, 1)$, where $-1 \leqslant t \leqslant 1$; then $\boldsymbol{F}(\boldsymbol{\sigma}_1(t)) = ((t+1)/2, (t-1)/2)$. The map $t \mapsto \boldsymbol{F}(\boldsymbol{\sigma}_1(t))$ is just a parameterization of the line $y = x - 1, 0 \leqslant x \leqslant 1$, since $(t-1)/2 = (t+1)/2 - 1$ (see Figure 9-2). Let

$$\boldsymbol{\sigma}_2(t) = (1, t), \qquad -1 \leqslant t \leqslant 1$$

$$\boldsymbol{\sigma}_3(t) = (t, -1), \qquad -1 \leqslant t \leqslant 1$$

$$\boldsymbol{\sigma}_4(t) = (-1, t), \qquad -1 \leqslant t \leqslant 1$$

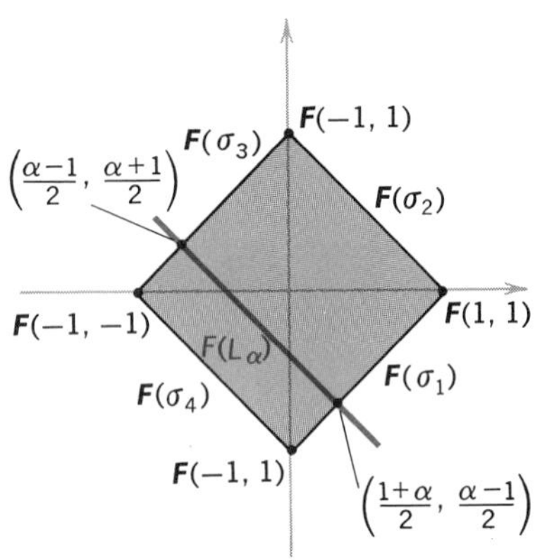

FIGURE 9-3

be parametrizations of the other edges of the square D^*. Using the same argument as above it is not hard to see that $\boldsymbol{F} \circ \boldsymbol{\sigma}_2$ is a parametrization of the line $y = 1 - x, 0 \leqslant x \leqslant 1$, $\boldsymbol{F} \circ \boldsymbol{\sigma}_3$ the line $y = x + 1$, $-1 \leqslant x \leqslant 0$, and $\boldsymbol{F} \circ \boldsymbol{\sigma}_4$ the line $y = -x - 1$, $-1 \leqslant x \leqslant 0$. By this time it seems reasonable to guess that $\boldsymbol{F}$ "flips" the square D^* around and takes it to the square D whose vertices are $(1, 0)$, $(0, 1)$, $(-1, 0)$, $(0, -1)$ (Figure 9-3). To prove that this is indeed the case let L_α (Figure 9-2) be a fixed line parametrized by $\boldsymbol{\sigma}(t) = (\alpha, t)$, $-1 \leqslant t \leqslant 1$ and $-1 \leqslant \alpha \leqslant 1$; then $\boldsymbol{F}(\boldsymbol{\sigma}(t)) = ((\alpha+t)/2, (\alpha-t)/2)$ which is a parametrization of the line $y = -x + \alpha$, $(\alpha-1)/2 \leqslant x \leqslant (\alpha+1)/2$. This line begins when $x = (\alpha-1)/2$ or at the point $((\alpha-1)/2, (1+\alpha)/2)$ and ends up at the point $((1+\alpha)/2, (\alpha-1)/2)$ and, as is easily checked, these points lie on the lines $\boldsymbol{F} \circ \boldsymbol{\sigma}_3$ and $\boldsymbol{F} \circ \boldsymbol{\sigma}_1$, respectively. Thus as α varies between -1 and 1, L_α sweeps the square D^* while $\boldsymbol{F}(L_\alpha)$ sweeps the square D determined by the vertices $(-1, 0)$, $(0, 1)$, $(1, 0)$, and $(0, -1)$.

Although we cannot visualize the graph of a function $\boldsymbol{F}: \boldsymbol{R}^2 \to \boldsymbol{R}^2$ it does help to consider how the function deforms subsets. However, simply looking at these deformations does not give us a complete picture of the behavior of $\boldsymbol{F}$. We may characterize $\boldsymbol{F}$ further using the notion of one to-one.

Definition The function $\boldsymbol{F}$ is **one-to-one** on D^* if $\boldsymbol{F}(u, v) = \boldsymbol{F}(u', v')$ for (u, v) and $(u', v') \in D^*$ implies that $u = u'$ and $v = v'$.

Geometrically, this statement means that two different points of D^* do not get sent into the same point of D by $\boldsymbol{F}$. For example, the function $\boldsymbol{F}(x, y) = (x^2 + y^2, y^4)$ is not one-to-one because $\boldsymbol{F}(1, -1) = (2, 1) = \boldsymbol{F}(1, 1)$ yet $(1, -1) \neq (1, 1)$. If a function is one-to-one it does not collapse points together.

Example 3 Consider the function $\boldsymbol{F}: \boldsymbol{R}^2 \to \boldsymbol{R}^2$ of Example 1, $\boldsymbol{F}(\theta, r) = (r\cos\theta, r\sin\theta)$. Show that $\boldsymbol{F}$ is not one-to-one.

If $\theta_1 \neq \theta_2$, then $\boldsymbol{F}(0, \theta_1) = \boldsymbol{F}(0, \theta_2)$ and so $\boldsymbol{F}$ cannot be one-to-one. This equality implies that if L is the side of the rectangle $D^* = [0, 2\pi] \times [0, 1]$ (Figure 9-4), where $0 \leqslant \theta \leqslant 2\pi$ and $r = 0$, then $\boldsymbol{F}$ maps all of L into a single point—the center of the unit disc D.

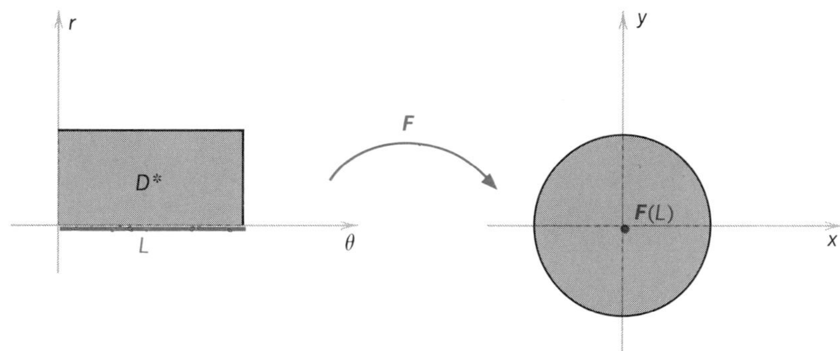

FIGURE 9-4

However, if we consider the set $S^* =]0, 2\pi] \times]0, 1]$ then $F: S^* \to S$ is one-to-one (see exercise 1).

Example 4 Show that the function $\boldsymbol{F}: \boldsymbol{R}^2 \to \boldsymbol{R}^2$ of Example 2 is one-to-one.

Suppose $\boldsymbol{F}(x, y) = \boldsymbol{F}(x', y')$; then $((x+y)/2, (x-y)/2) = ((x'+y')/2, (x'-y')/2)$ and we have

$$x + y = x' + y'$$

$$x - y = x' - y'$$

or

$$2x = 2x'$$

Thus $x = x'$ and $y = y'$, which shows that $\boldsymbol{F}$ is one-to-one.

EXERCISES

1. Let $S^* =]0, 2\pi] \times]0, 1]$ and define $\boldsymbol{F}(\theta, r) = (r\cos\theta, r\sin\theta)$. Determine the image set S. Show that $\boldsymbol{F}$ is one-to-one on S^*.

2. Define $\boldsymbol{F}(x, y) = \left(\dfrac{x-y}{\sqrt{2}}, \dfrac{x+y}{\sqrt{2}}\right)$. Show that $\boldsymbol{F}$ rotates the unit square, $D^* = [0, 1] \times [0, 1]$.

3. Let $D^* = [0, 1] \times [0, 1]$ and define $\boldsymbol{F}$ by $\boldsymbol{F}(u, v) = (u^2 + 4u, v)$. Find D. Is $\boldsymbol{F}$ one-to-one?

4. Let D^* be the parallelogram bounded by the lines $y = 3x - 4$, $y = 3x$, $y = \frac{1}{2}x$, and $y = \frac{1}{2}(x+4)$. Let $D = [0, 1] \times [0, 1]$. Find an $\boldsymbol{F}$ such that D is the image of D^* under $\boldsymbol{F}$.

5. Let $\boldsymbol{F}: \boldsymbol{R}^3 \to \boldsymbol{R}^3$ be defined by $\boldsymbol{F}: (\rho, \phi, \theta) \mapsto (x, y, z)$, where
$$x = \rho\sin\phi\cos\theta, \quad y = \rho\sin\phi\sin\theta, \quad z = \rho\cos\phi$$
Let D^* be the set of points (ρ, ϕ, θ) such that $\phi \in [0, \pi]$, $\theta \in [0, 2\pi]$, $\rho \in [0, 1]$. Find $D = \boldsymbol{F}(D^*)$. Is $\boldsymbol{F}$ one-to-one? If not, can we eliminate some subset of D^* as in exercise 1 so that on the remainder $\boldsymbol{F}$ will be one-to-one?

REVIEW EXERCISES FOR CHAPTER 14

1. Draw some level curves for the following functions.
   (a) $f(x, y) = 1/xy$
   (b) $f(x, y) = x^2 - xy - y^2$

2. Consider a temperature function $T(x, y) = x\sin y$. Plot a few level curves. Compute ∇T and explain its meaning.

3. Find the following limits if they exist.
   (a) $\displaystyle\lim_{(x,y)\to(0,0)} \frac{\cos xy - 1}{x}$
   (b) $\displaystyle\lim_{(x,y)\to(0,0)} \sqrt{\left|\frac{x+y}{x-y}\right|}, \; x \neq y$

4. Prove that any nth-degree polynomial function of two variables
$$P_n(x, y) = \sum_{j=0}^{n}\sum_{k=0}^{n} \alpha_{jk} x^j y^k$$
is continuous.

5. Compute the partial derivatives of the following functions.
   (a) $f(x, y, z) = xe^z + y\cos x$
   (b) $f(x, y, z) = (x + y + z)^{10}$
   (c) $f(x, y, z) = \dfrac{x^2 + y}{z}$

6. Compute the gradient of each function in exercise 5.

7. Evaluate all first and second partial derivatives of the following functions.
   (a) $f(x,y) = x \arctan(x/y)$
   (b) $f(x,y) = \cos\sqrt{x^2+y^2}$
   (c) $f(x,y) = e^{-x^2-y^2}$

8. Find the directional derivatives of the functions in exercise 7 at the point $(1,1)$ in the direction $(1\sqrt{2}, 1\sqrt{2})$.

9. Find $G'(t)$ if
   (a) $G(t) = \int_0^{t^2} \log(1+x^2)\,dx$;
   (b) $G(t) = \int_0^t e^{-t^2x}\cos(t+x)\,dx$.

10. Let $\boldsymbol{F} = F_1(x,y)\boldsymbol{i} + F_2(x,y)\boldsymbol{j}$ be a C^1 vector field. Show that if $\boldsymbol{F} = \nabla f$ for some f then $\partial F_1/\partial y = \partial F_2/\partial x$.

11. Show that $\boldsymbol{F} = y\cos x\boldsymbol{i} + x\sin y\boldsymbol{j}$ is not a gradient vector field.

12. Without looking in the book define
    (a) an irrotational vector field;
    (b) an incompressible vector field.
    Can a vector field be both irrotational and incompressible?

13. Compute the divergence of the following vector fields at the point indicated.
    (a) $\boldsymbol{F}(x,y,z) = x\boldsymbol{i} + 3xy\boldsymbol{j} + z\boldsymbol{k}$, $(0,1,0)$
    (b) $\boldsymbol{F}(x,y,z) = y\boldsymbol{i} + z\boldsymbol{j} + x\boldsymbol{k}$, $(1,1,1)$
    (c) $\boldsymbol{F}(x,y,z) = (x+y)^3\boldsymbol{i} + \sin(xy)\boldsymbol{j} + \cos(xyz)\boldsymbol{k}$, $(2,0,1)$

14. Compute the curl of each vector field in exercise 13 at the given point.

15. (a) Let $f(x,y,z) = xyz^2$; compute ∇f.
    (b) Let $\boldsymbol{F}(x,y,z) = xy\boldsymbol{i} + yz\boldsymbol{j} + zy\boldsymbol{k}$; compute $\nabla \times \boldsymbol{F}$.
    (c) Compute $\nabla \times (f\boldsymbol{F})$ using formula 12 of Table 5-1. Compare with a direct computation.

16. Compute $\nabla \cdot \boldsymbol{F}$ and $\nabla \times \boldsymbol{F}$ for the following vector fields.
    (a) $\boldsymbol{F} = 2x\boldsymbol{i} + 3y\boldsymbol{j} + 4z\boldsymbol{k}$
    (b) $\boldsymbol{F} = x^2\boldsymbol{i} + y^2\boldsymbol{j} + z^2\boldsymbol{k}$
    (c) $\boldsymbol{F} = (x+y)\boldsymbol{i} + (y+z)\boldsymbol{j} + (z+x)\boldsymbol{k}$

17. Find the equation of the plane tangent to the surfaces at the indicated points.
    (a) $z = x^2 + y^2$, $(0,0)$
    (b) $z = x^2 - y^2 + x$, $(1,0)$
    (c) $z = (x+y)^2$, $(3,2)$

18. Analyze the behavior of the following functions at the indicated points.
    (a) $z = x^2 - y^2 + 3xy$, $(0,0)$
    (b) $z = Ax^2 + By^2 + Cxy$, $(0,0)$

19. Find the equation of the tangent plane to the surface S given by the graph of
    (a) $f(x,y) = \sqrt{x^2+y^2} + (x^2+y^2)$ at $(1,0,2)$;
    (b) $f(x,y) = \sqrt{x^2+2xy-y^2+1}$ at $(1,1,\sqrt{3})$.

20. The temperature T around the world (assume the world is the unit sphere) is given by the function

$$T(x, y, z) = xyz$$

Find the points where the temperature is minimized and maximized.

*21. Suppose $f: \boldsymbol{R}^2 \to \boldsymbol{R}$ is C^1 and has the property that $\nabla f(v) \cdot \boldsymbol{n} = 0$ for some fixed unit vector $\boldsymbol{n}$ and all $v \in \boldsymbol{R}^2$. What can be said about f? [Hint: First consider the simple cases $\boldsymbol{n} = \boldsymbol{i}$, and $\boldsymbol{n} = \boldsymbol{j}$.]

22. Let $f(x, y) = \log\sqrt{x^2+y^2}$. Show that $\nabla \cdot (\nabla f) = 0$.

23. Repeat exercise 22 for $f(x, y, z) = 1/\sqrt{x^2+y^2+z^2}$

24. A cereal manufacturer wants to introduce a new kind of box with the top made out of special cardboard which costs twice as much as the material for the rest of the box. If the volume of the box is fixed at V in^3, what are the dimensions of the box which minimizes cost?

25. Describe how the following vector fields F transform the given region D^*. Is F one-to-one?

(a) $\boldsymbol{F}: (x, y) \mapsto (x^2+1, \cos y), \quad D^* = [-1, 1] \times [0, \pi]$

(b) $\boldsymbol{F}: (x, y) \mapsto \left(\dfrac{x}{\sqrt{x^2+y^2}}, \dfrac{y}{\sqrt{x^2+y^2}}\right), \quad D^* = \boldsymbol{R}^2 - \{0\}$

(c) $\boldsymbol{F}: (x, y) \mapsto (x^3, 3y+2), \quad D^* = [0, 1] \times [0, 1]$

(*d*) $\boldsymbol{F}: (x, y) \mapsto \left(\dfrac{x-1}{2}, \dfrac{y+3}{4}\right), \quad D^* =$ unit disc = the set of (x, y) with $x^2+y^2 \leqslant 1$

26. Find the minimum distance from the origin to the surface $f(x, y) = \sqrt{x^2-1}$.

27. Repeat exercise 26 for the surface $f(x, y) = 6xy+7$.

28. Let $f: \boldsymbol{R}^3 \to \boldsymbol{R}$ be C^1 and let $z: \boldsymbol{R}^2 \to \boldsymbol{R}$ be C^1. Consider the composite function

$$h(x, y) = f(x, y, z(x, y))$$

Show (using the formula in exercise 9, §14.3) that

$$\frac{\partial h}{\partial x} = \frac{\partial f}{\partial x} + \frac{\partial f}{\partial z}\frac{\partial z}{\partial x}$$

and

$$\frac{\partial h}{\partial y} = \frac{\partial f}{\partial y} + \frac{\partial f}{\partial z}\frac{\partial z}{\partial y}$$

15

MULTIPLE INTEGRATION

In this chapter we study the integration of real-valued functions of several variables; we are especially interested in integrals of continuous functions of two variables or **double integrals** as they are called. The double integral has a basic geometric interpretation as volume, and also can be defined rigorously as a limit of approximating sums. We present several techniques for evaluating double integrals, consider some applications, and then discuss improper integrals. Finally, we introduce the concept of the **triple integral.**

1 INTRODUCTION

In this introductory section, we briefly discuss some of the geometric aspects of the double integral, deferring the more rigorous discussion in terms of Riemann sums until §15.2.

Let us consider, then, a continuous function of two variables $f: R \subset \boldsymbol{R}^2 \to \boldsymbol{R}$ whose domain R is a rectangle with sides parallel to the coordinate axes. The rectangle R can be described in terms of two closed intervals $[a, b]$ and $[c, d]$ as

$$(x, y) \in R \Leftrightarrow x \in [a, b] \quad \text{and} \quad y \in [c, d]$$

In this case, we say that R is the **Cartesian product** of $[a, b]$ and $[c, d]$ and write $R = [a, b] \times [c, d]$.

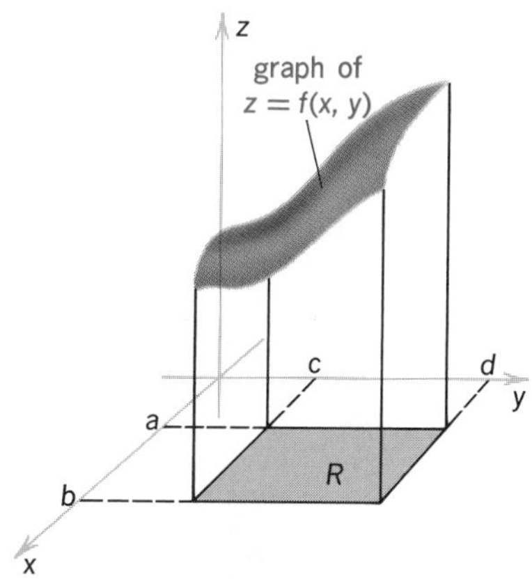

FIGURE 1-1

We assume too that $f(x, y) \geqslant 0$ on R. The graph of $z = f(x, y)$ is then a surface lying above the rectangle R. This surface, the rectangle R, and the four planes $x = a$, $x = b$, $y = c$, and $y = d$ form the boundary of a region in space (see Figure 1-1). The volume of this region is called the **(double)**

integral of f over R and is denoted by

$$\int_R f, \quad \int_R f(x,y)\,dA, \quad \int_R f(x,y)\,dx\,dy, \quad \iint_R f(x,y)\,dx\,dy$$

Example 1 (a) If $f(x,y)=k$, where k is a positive constant, then $\int_R f(x,y)\,dA = k(b-a)(d-c)$ since the integral is equal to the volume of a rectangular box with base R and height k. (b) If $f(x,y)=1-x$ and $R=[0,1]\times[0,1]$, then $\int_R f(x,y)\,dA=\frac{1}{2}$ since the integral is equal to the volume of the triangular solid shown in Figure 1-2.

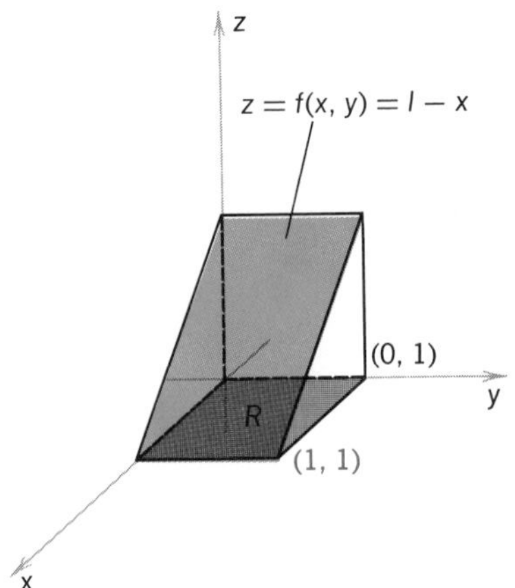

FIGURE 1-2

Example 2 Suppose $z=f(x,y)=x^2+y^2$ and $R=[-1,1]\times[0,1]$. Then the integral $\int_R f=\int_R (x^2+y^2)\,dx\,dy$ is equal to the volume of the solid sketched in Figure 1-3. We shall compute this integral in Example 3.

By Cavalieri's Principle (§6.4), it follows that the volume of a solid S (Figure 1-4) can be computed by integrating the cross-sectional area function $x \mapsto A(x)$ from a to b. In §15.2 we use Fubini's Theorem to establish this fact rigorously for the type of solid under consideration here. For now, let us see how this result provides a method for evaluating double integrals.

If we are considering the region under a graph $z=f(x,y)$, there are two natural cross-sectional area functions: the one obtained using cutting planes perpendicular to the x axis, and the one obtained using cutting planes perpendicular to the y axis. The cross section determined by a cutting plane $x=x_0$ of the first sort is the plane region under the graph of $z=f(x_0,y)$ from $y=c$ to $y=d$ (Figure 1-5). When we fix $x=x_0$, we have the real function $y\mapsto f(x_0,y)$ which is continuous on $[c,d]$. The cross-sectional area $A(x_0)$ is, therefore, equal to the integral $\int_c^d f(x_0,y)\,dy$. Thus the cross-sectional area function A has domain $[a,b]$ and rule $A: x\mapsto \int_c^d f(x,y)\,dy$. By Cavalieri's Principle, the volume V of the region under $z=f(x,y)$ must be equal to

$$V = \int_a^b A(x)\,dx = \int_a^b \left[\int_c^d f(x,y)\,dy\right] dx$$

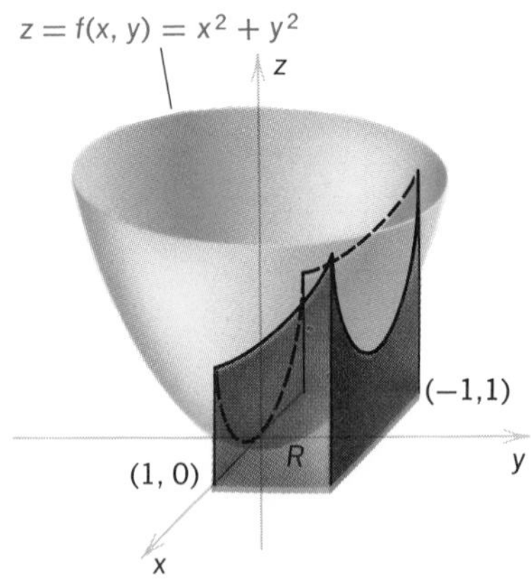

FIGURE 1-3

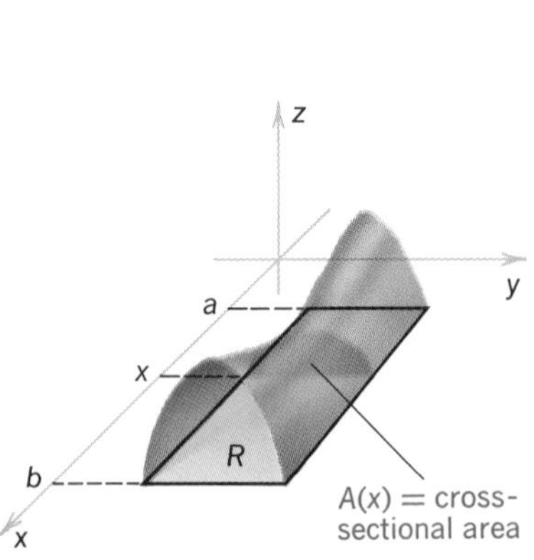

FIGURE 1-4

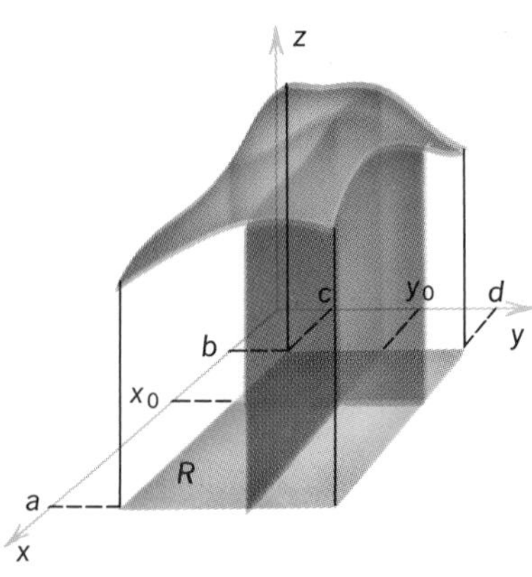

FIGURE 1-5

The integral $\int_a^b \left[\int_c^d f(x, y)\,dy\right] dx$ is known as an **iterated integral**, because it is obtained by integrating first with respect to y and then integrating the result with respect to x. Since $\int_R f(x, y)\,dA$ is equal to the volume V, it follows that

$$\int_R f(x,y)\,dA = \int_a^b \left[\int_c^d f(x,y)\,dy\right] dx \tag{1}$$

If we reverse the roles of x and y in the above discussion and use cutting planes perpendicular to the y axis, we obtain

$$\int_R f(x,y)\,dA = \int_c^d \left[\int_a^b f(x,y)\,dx\right] dy \tag{2}$$

Thus the iterated integrals (1) and (2) are equal. The expression on the right of (2) is the iterated integral obtained by integrating first with respect to x and then integrating the result with respect to y. As the following examples will illustrate, the notion of the iterated integral and equations (1) and (2) provide a powerful method for computing the double integral of a function of two variables.

Example 3 Let $z = f(x, y) = x^2 + y^2$ and let $R = [-1, 1] \times [0, 1]$. Let us evaluate the integral $\int_R (x^2 + y^2)\,dx\,dy$. By equation (2), we have

$$\int_R (x^2+y^2)\,dx\,dy = \int_0^1 \left[\int_{-1}^1 (x^2+y^2)\,dx\right] dy$$

To find $\int_{-1}^1 (x^2 + y^2)\,dx$, we treat y as a constant and integrate with respect to x. Since $x \mapsto x^3/3 + y^2 x$ is an antiderivative of $x \mapsto x^2 + y^2$, we apply the Fundamental Theorem of Calculus to obtain

$$\int_{-1}^1 (x^2+y^2)\,dx = \left[\frac{x^3}{3} + y^2 x\right]_{-1}^1 = \frac{2}{3} + 2y^2$$

Next we integrate $y \mapsto \frac{2}{3} + 2y^2$ with respect to y from 0 to 1:

$$\int_0^1 \left(\frac{2}{3} + 2y^2\right) dy = \left[\frac{2}{3}y + \frac{2}{3}y^3\right]_0^1 = \frac{4}{3} = \int_R (x^2+y^2)\,dx\,dy$$

Hence the volume of the solid in Figure 1-3 is $\frac{4}{3}$. For completeness, let us evalute $\int (x^2 + y^2)\,dx\,dy$ using (1)—that is, integrating first with respect to y and then with respect to x. We have

$$\int_R (x^2+y^2)\,dx\,dy = \int_{-1}^1 \left[\int_0^1 (x^2+y^2)\,dy\right] dx$$

Treating x as constant, we find

$$\int_0^1 (x^2+y^2)\,dy = \left[x^2 y + \frac{y^3}{3}\right]_0^1 = x^2 + \frac{1}{3}$$

Next we evaluate $\int_{-1}^1 (x^2 + \frac{1}{3})\,dx$ to obtain

$$\int_R (x^2+y^2)\,dx\,dy = \int_{-1}^1 \left(x^2 + \frac{1}{3}\right) dx = \left[\frac{x^3}{3} + \frac{x}{3}\right]_{-1}^1 = \frac{4}{3}$$

Example 4 Compute $\int_S \cos x \sin y \, dx \, dy$ where S is the square $[0, \pi/2] \times [0, \pi/2]$ (see Figure 1-6). We have by equation (2)

$$\int_S \cos x \sin y \, dx \, dy = \int_0^{\pi/2} \left[\int_0^{\pi/2} \cos x \, dx \right] \sin y \, dy = \int_0^{\pi/2} \sin y \, dy = 1$$

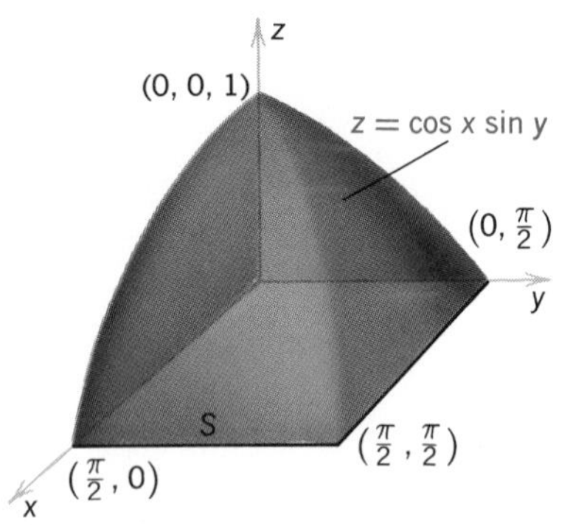

FIGURE 1-6

In the next section, we use Riemann sums to define the double integral for a wide class of functions of two variables. In §15.3, we treat double integrals over more general regions than rectangles. Although we shall drop the requirement that $f(x, y) \geqslant 0$, equations (1) and (2) remain valid. Therefore, the iterated integral will again provide the key to computing the double integral.

EXERCISES

1. Evaluate the following interated integrals.

   (a) $\int_{-1}^{1} \int_0^1 (x^3 y + y^2) \, dy \, dx$ (b) $\int_0^{\pi/2} \int_{-1}^{1} (y \cos x + 2) \, dy \, dx$

   (c) $\int_0^1 \int_0^1 (xye^{x+y}) \, dy \, dx$ (d) $\int_{-1}^{0} \int_1^2 (x \log y) \, dy \, dx$

2. Evaluate the integrals in (1) by first integrating with respect to x and then with respect to y.

2 THE DOUBLE INTEGRAL OVER A RECTANGLE

We are now ready to give a rigorous definition of the double integral as the limit of sequences of sums. In addition, we discuss some of the fundamental algebraic properties of the double integral and prove Fubini's Theorem which states that the double integral can be calculated as an iterated integral. To begin, let us establish some notation for partitions and sums.

Consider a rectangle $R \subset \boldsymbol{R}^2$, which is the Cartesian product $R = [a, b] \times [c, d]$. By the **regular partition** of order n of R we mean the pair of regular partitions of $[a, b]$ and of $[c, d]$ of order n, that is, the two collections of $n+1$ points $\{x_j^n\}_{j=0}^n$ and $\{y_k^n\}_{k=0}^n$ with

$$a = x_0^n < x_1^n < \cdots < x_n^n = b, \qquad c = y_0^n < y_1^n < \cdots < y_n^n = d$$

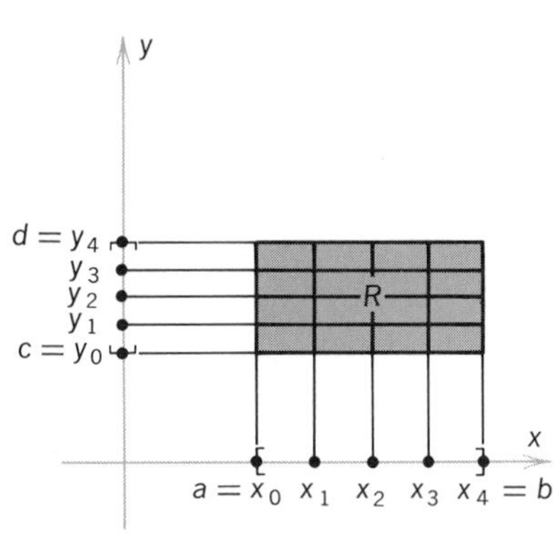

FIGURE 2-1

and

$$x_{j+1}^n - x_j^n = \frac{b-a}{n}, \qquad y_{k+1}^n - y_k^n = \frac{d-c}{n}$$

(see Figure 2-1).

Let R_{jk} be the rectangle $[x_j^n, x_{j+1}^n] \times [y_k^n, y_{k+1}^n]$, and let c_{jk} be any point in R_{jk}. Suppose $f: R \to \boldsymbol{R}$ is a continuous real-valued function. Form the sum

$$S_n = \sum_{k=0}^{n-1} \sum_{j=0}^{n-1} f(c_{jk}) \Delta x \, \Delta y = \sum_{k=0}^{n-1} \sum_{j=0}^{n-1} f(c_{jk}) \, \Delta A \tag{1}$$

where

$$\Delta x = \frac{b-a}{n} = x_{j+1}^n - x_j^n, \qquad \Delta y = \frac{d-c}{n} = y_{k+1}^n - y_k^n$$

and

$$\Delta A = \Delta x \, \Delta y$$

In a manner strictly analogous to the proof for the one-variable case (§7.2), it can be shown that when f is continuous, $\lim_{n\to\infty} S_n$ exists and is independent of the choice of points $c_{jk} \in R_{jk}$. This common limit of the sequences $\{S_n\}$ is called the **double integral** of f over R and is denoted by

$$\int_R f(x,y)\, dA$$

or alternatively, by $\int_R f(x,y)\,dx\,dy$ or $\int_R \int f(x,y)\,dx\,dy$. In general, whether or not f is continuous, if the Riemann sums converge then f is said to be **integrable** over R and the integral is defined as above. We state the important fact that continuous functions are integrable as a theorem.

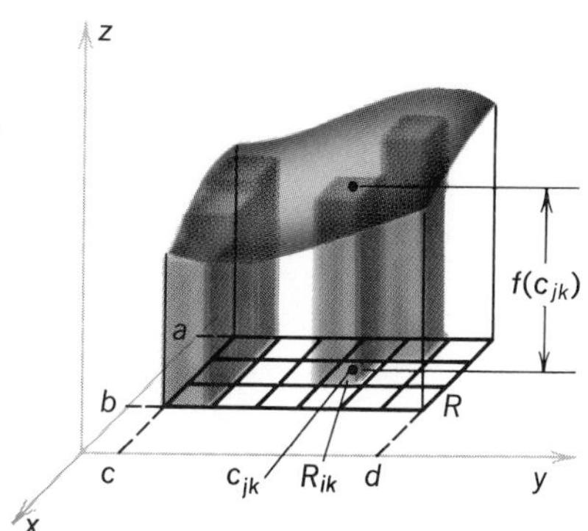

FIGURE 2-2

Theorem 1 In the notation introduced above, if f is continuous on the rectangle R, then the sequence

$$S_n = \sum_{j=0}^{n-1} \sum_{k=0}^{n-1} f(c_{jk})\, \Delta x \, \Delta y$$

converges as $n \to \infty$ to a limit which is the same for all choices of points $c_{jk} \in R_{jk}$. This common limit is called the integral of f over R.

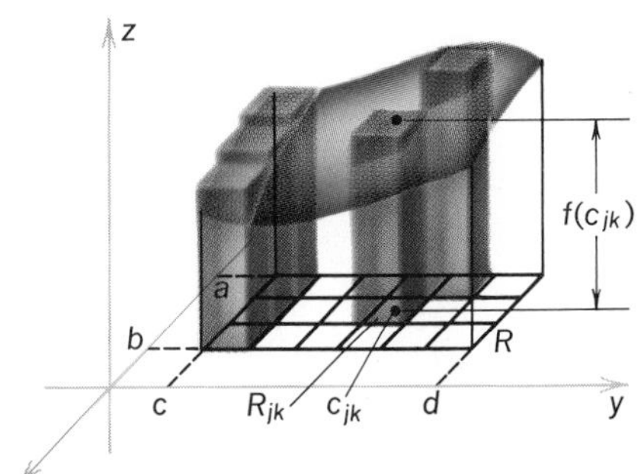

FIGURE 2-3

If $f(x,y) \geqslant 0$, the existence of $\lim_{n\to\infty} S_n$ has a straightforward geometric meaning. Consider the graph of $z = f(x,y)$ as the top of a solid whose base is the rectangle R. If we take each c_{jk} to be the point where $f(x,y)$ has its minimum value on R_{jk} (such c_{jk} exists by virtue of the continuity of f on R, but we shall not prove this), then $f(c_{jk})\,\Delta x\,\Delta y$ represents the volume of a rectangular box with base R_{jk}. The sum $\sum_{j=0}^{n-1}\sum_{k=0}^{n-1} f(c_{jk})\,\Delta x\,\Delta y$ represents the volume of an inscribed solid as in Figure 2-2. Similarly, if we take each c_{jk} to be the point where $f(x,y)$ has its maximum on R_{jk}, then the sum $\sum_{j=0}^{n-1}\sum_{k=0}^{n-1} f(c_{jk})\,\Delta x\,\Delta y$ is equal to the volume of a circumscribed solid as in Figure 2-3. Therefore, if $\lim_{n\to\infty} S_n$ exists and is independent of $c_{jk} \in R_{jk}$, we have established the Principle of Exhaustion for solids bounded by the graph of the continuous function $f: \boldsymbol{R}^2 \to \boldsymbol{R}$.

We would like to extend the double integral to a larger class of functions than the continuous ones. However, first we need some preliminary results.

Let $B \subset R$ be a subset of R (see Figure 2-4). In the nth regular partition of R, let b_n be the sum of the areas of those rectangles in the partition which intersect B; that is, those rectangles which contain at least one point of B. Then B is said to have **area zero** if $\lim_{n\to\infty} b_n = 0$.

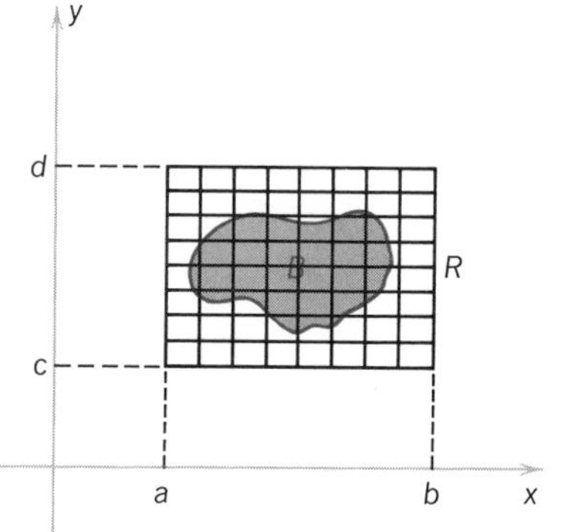

FIGURE 2-4

Example 1 Let B be a subset of R consisting of a finite number of points. Show that B has area zero.

Suppose B consists of m points $u_1, u_2, \ldots, u_m$. Then each point $u_j, j = 1, \ldots, m$ can be in at most four rectangles of the subdivision (see Figure 2-5). Since the area of each rectangle in the nth regular subdivision is $(b-a)(d-c)/n^2$, we have

$$0 \leqslant b_n \leqslant \frac{4m}{n^2}(b-a)(d-c)$$

Thus $\lim_{n\to\infty} b_n = 0$ and so B has area zero.

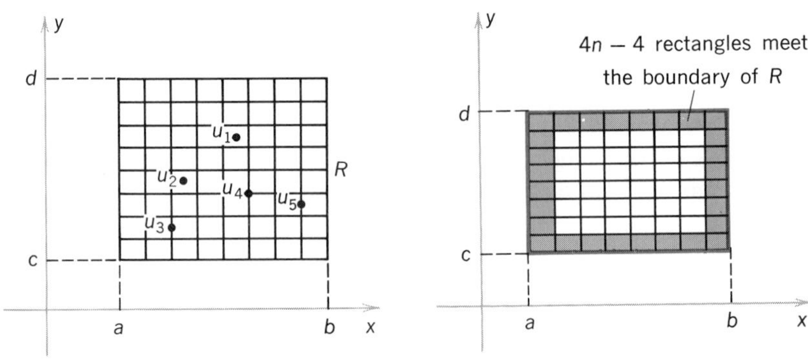

FIGURE 2-5

FIGURE 2-6

Example 2 Let B be the boundary of R (the set consisting of the four edges of R). Show that B has area zero.

The total area b_n of those rectangles in the nth regular subdivision containing points of the boundary can be given explicitly by

$$b_n = \left(\frac{b-a}{n}\right)\left(\frac{d-c}{n}\right)(4n-4) = \frac{4(n-1)}{n}\frac{(b-a)(d-c)}{n}$$

(see Figure 2-6). Thus

$$0 \leqslant b_n \leqslant \frac{4(b-a)(d-c)}{n} \quad \text{since} \quad \frac{n-1}{n} < 1$$

and hence

$$\lim_{n\to\infty} b_n = 0$$

The next theorem, which we state without proof, extends the results of the above two examples to an important class of curves in the plane, namely, the graphs of continuous real functions $y = \phi(x)$ and $x = \psi(y)$.

Theorem 2 The graph of a continuous real function $\phi:[a,b] \to R$ has area zero. In fact, any set composed of a finite number of such graphs has area zero.

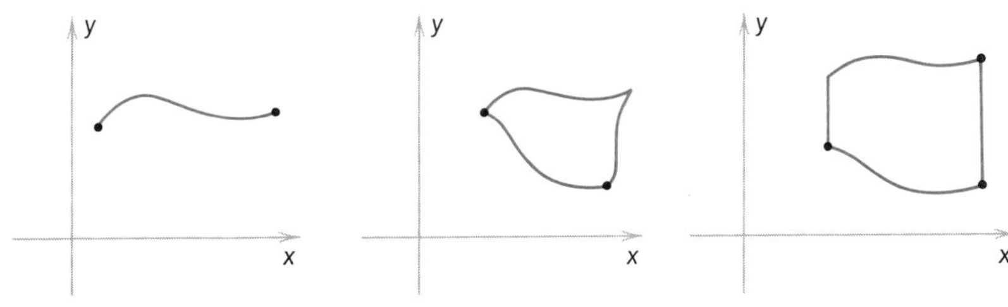

FIGURE 2-7

Thus the curves in Figure 2-7 are all sets of area zero. We use this notion of zero area in the next section when we define the integral over regions bounded by such curves. The proof of Theorem 2 is basically the same as the one given for Theorem 2 of §7.2, and it requires the uniform continuity of the function ϕ (some hints are given in exercise 5).

The following result, also stated without proof, provides an important criterion for determining if a function is integrable.

Theorem 3 Let $f: R \to \mathbf{R}$ be a bounded real-valued function on the rectangle R, and suppose that the set of points where f is discontinuous has area zero. Then f is integrable over R.

If we combine Theorem 2 and Theorem 3, we see that the functions sketched in Figure 2-8 are all integrable over R, since these functions are continuous except on sets of area zero.

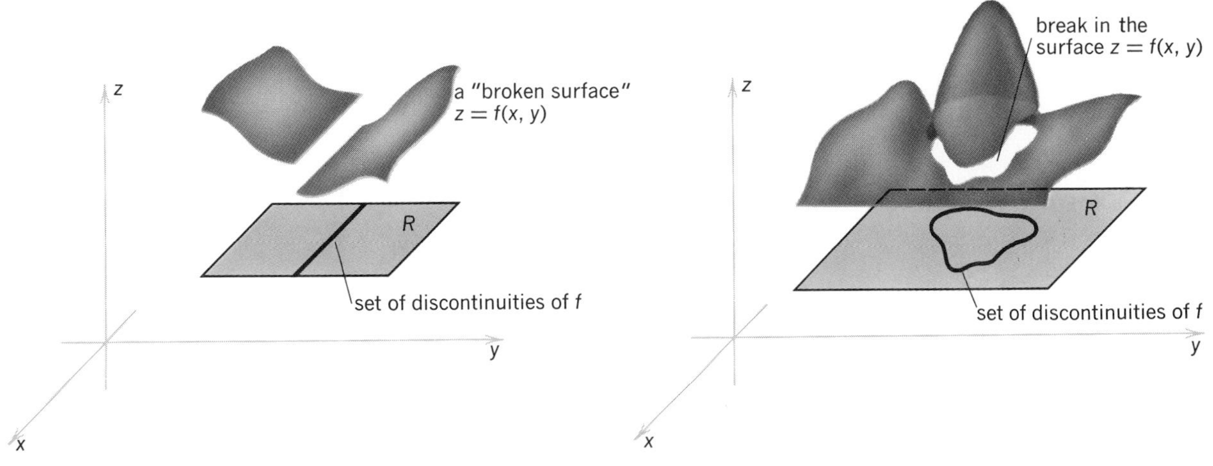

FIGURE 2-8

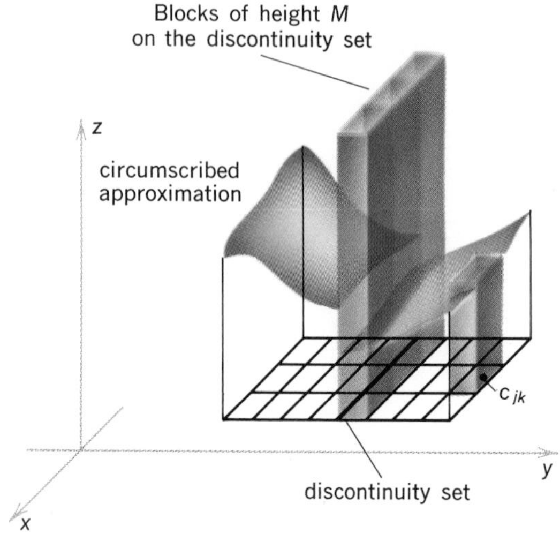

FIGURE 2-9

Geometrically, the results above mean that the Principle of Exhaustion can be demonstrated for solids of the kind shown in Figure 2-9 which are bounded by the graphs of integrable nonnegative functions, which are not necessarily continuous.

From the definition of the integral as a limit of sums, we can quickly deduce some fundamental properties of the integral $\int_R f(x,y)\,dA$; these properties are essentially the same as for the integral of a real-valued function of a single variable.

(1) **Linearity** $\int_R [f(x, y)+g(x, y)]\, dA = \int_R f(x, y)\, dA + \int_R g(x, y)\, dA$
(2) **Homogeneity** $\int_R cf(x,y)\, dA = c\int_R f(x,y)\, dA$, c a constant
(3) **Monotonicity** If $f(x,y) \geqslant g(x,y)$, then $\int_R f(x,y)\, dA \geqslant \int_R g(x,y)\, dA$
(4) **Additivity** If $R_i, i = 1, \ldots, m$ are pairwise disjoint rectangles, $R = \bigcup_{i=1}^m R_i$ is a rectangle, and f is integrable over each R_i, then $f: R \to \mathbf{R}$ is integrable over R and

$$\int_R f(x,y)\, dA = \sum_{i=1}^{m} \int_{R_i} f(x,y)\, dA$$

Properties (1) and (2) are a consequence of the definition of the integral as a limit of a sum and those facts for convergent sequences $\{S_n\}$, $\{T_n\}$ are

$$\lim_{n\to\infty} (T_n + S_n) = \lim_{n\to\infty} T_n + \lim_{n\to\infty} S_n$$

$$\lim_{n\to\infty} (cS_n) = c \lim_{n\to\infty} S_n$$

To demonstrate monotonicity we observe that if $f(x, y) \geqslant 0$, and S_n is a sequence which converges to $\int_R f(x, y)\, dA$, then $S_n \geqslant 0$ for all n so $\int_R f(x, y)\, dA = \lim_{n\to\infty} S_n \geqslant 0$. If $f(x, y) \geqslant g(x, y)$ for all $(x, y) \in R$, then $(f-g)(x, y) \geqslant 0$ for all (x, y) and using (1) and (2), we have

$$\int_R f(x, y)\, dA - \int_R g(x, y)\, dA = \int_R [f(x, y) - g(x, y)]\, dA \geqslant 0$$

This proves (3). The proof of (4) is left as an exercise.

Another important result which is somewhat more difficult to show is that if f is integrable then so is $|f|$. Moreover

$$\left| \int_R f \right| \leqslant \int_R |f| \tag{2}$$

This inequality is straightforward and we leave it for the exercises (see exercise 10).

Although we have noted the integrability of a variety of functions, we have not established rigorously a general method of computing the integral. In the case of one variable we avoid computing $\int_a^b g(x)\, dx$ from its definition as a limit of a sum by using the Fundamental Theorem of Integral Calculus. Although the same technique would not work in the two-variable situation (f is not the derivative of any real-valued function of two variables), we can formalize the method of iterated integrals of §15.1 by using the Fundamental Theorem in the one-variable case in conjunction with another powerful tool, Fubini's Theorem.

Theorem 4 Fubini's Theorem Let f be a bounded function with domain a rectangle $R = [a, b] \times [c, d]$ and suppose the discontinuities of f form a set of area zero. Then if

$$\int_c^d f(x, y)\, dy \quad \text{exists for each } x \in [a, b] \tag{3}$$

and

$$\int_a^b \left[\int_c^d f(x, y)\, dy \right] dx \quad \text{exists} \tag{4}$$

then

$$\int_a^b \int_c^d f(x, y)\, dy\, dx = \int_R f(x, y)\, dA$$

Similarly, if

$$\int_a^b f(x, y)\, dx \quad \text{exists for each } \quad y \in [c, d] \tag{3'}$$

and

$$\int_c^d \left[\int_a^b f(x, y)\, dx \right] dy \quad \text{exists} \tag{4'}$$

then

$$\int_c^d \int_a^b f(x, y)\, dx\, dy = \int_R f(x, y)\, dA$$

Thus if all these conditions hold simultaneously, then

$$\int_a^b \int_c^d f(x,y)\,dy\,dx = \int_c^d \int_a^b f(x,y)\,dx\,dy = \int_R f(x,y)\,dA$$

Proof We prove Fubini's Theorem first for f bounded and continuous on R and then, using this result, we prove the theorem for f whose discontinuities form a set of area zero.

Assume that (3) and (4) hold and f is continuous on R. We must show that

$$\int_a^b \int_c^d f(x,y)\,dy\,dx = \int_R f(x,y)\,dA$$

Let $c = y_0^n < y_1^n < \cdots < y_n^n = d$ be a partition of $[c,d]$ into n equal parts. Define $F(x) = \int_c^d f(x,y)\,dy$; then

$$F(x) = \sum_{i=0}^{n-1} \int_{y_i^n}^{y_{i+1}^n} f(x,y)\,dy$$

Using the integral version of the Mean Value Theorem, for each fixed x and for each i we have

$$\int_{y_i^n}^{y_{i+1}^n} f(x,y)\,dy = f(x, Y_i^n(x))(y_{i+1}^n - y_i^n)$$

where the point $Y_i^n(x)$ may depend on x (this requires continuity of f) and so

$$F(x) = \sum_{i=0}^{n-1} f(x, Y_i^n(x))(y_{i+1}^n - y_i^n)$$

Now by definition,

$$\int_a^b F(x)\,dx = \int_a^b \left[\int_c^d f(x,y)\,dy\right] dx$$

$$= \lim_{n\to\infty} \sum_{j=0}^{n-1} F(p_j^n)(x_{j+1}^n - x_j^n)$$

where $a = x_0^n < x_1^n < \cdots < x_n^n = b$ is a partition of the interval $[a,b]$ into n equal parts and p_j^n is any point in $[x_j^n, x_{j+1}^n]$. Therefore, letting p_j^n be any point in $[x_i^n, x_{i+1}^n]$ and setting $c_{ji} = (p_j^n, Y_i^n(p_j^n)) \in R$, we have (substituting p_j^n for x above)

$$F(p_j^n) = \sum_{i=0}^{n-1} f(c_{ji})(y_{i+1}^n - y_i^n)$$

and

$$\int_a^b \int_c^d f(x,y)\,dy\,dx = \int_a^b F(x)\,dx = \lim_{n\to\infty} \sum_{j=0}^{n-1} F(p_j^n)(x_{j+1}^n - x_j^n)$$

$$= \lim_{n\to\infty} \sum_{j=0}^{n-1} \sum_{i=0}^{n-1} f(c_{ji})(y_{i+1}^n - y_i^n)(x_{j+1}^n - x_j^n)$$

$$= \int_R f(x,y)\,dA$$

Thus

$$\int_a^b \int_c^d f(x, y)\, dy\, dx = \int_R f(x, y)\, dA$$

The case where (3) and (4) are replaced by (3′) and (4′) is analogous; therefore the proof of Fubini's theorem for f continuous is complete.

Now we assume that f is bounded, and continuous everywhere on R except for a set of area zero. Let D denote the set of discontinuities of f. Since f is bounded, there is an $M > 0$ such that $|f(x, y)| \leqslant M$ for all $(x, y) \in R$. Furthermore, because D has area zero, it follows that given $\varepsilon > 0$ there is a partition of R of order N such that the total area of the subrectangles in the partition which intersect D is less than $\varepsilon/2M$. Let U be the union of all those rectangles in this partition which intersect D. For simplicity, assume that U consists of only one rectangle (see Figure 2-10). The general case is entirely similar. Now

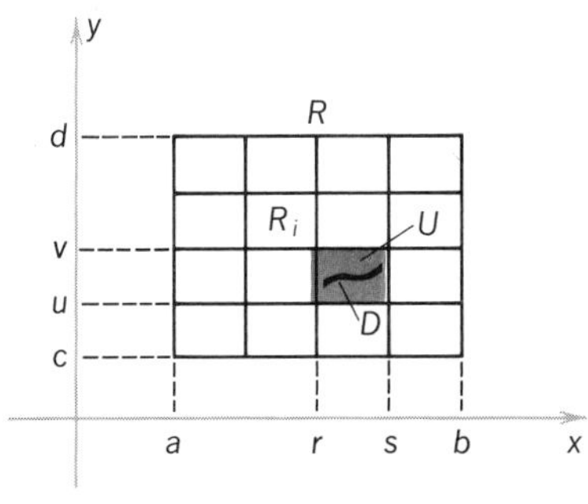

FIGURE 2-10

$$\text{Area } (U) \leqslant \varepsilon/2M$$

and

$$\int_R f(x, y)\, dA = \int_{R\text{-}U} f(x, y)\, dA + \int_U f(x, y)\, dA \tag{5}$$

But by inequality (2)

$$\left| \int_U f(x, y)\, dA \right| \leqslant \int_U |f(x, y)|\, dA \leqslant M \text{ Area } (U) = \varepsilon/2 \tag{6}$$

The set $R - U$ (that is, the set of points in R which are not in U) is the union of rectangles $\{R_i\}_{i=1}^k$ on each of which f is continuous and, therefore, on which Fubini's Theorem holds. If $U = [r, s] \times [u, v]$, then by additivity in the one-variable case we have

$$\begin{aligned}
\int_a^b \int_c^d f(x, y)\, dy\, dx &= \int_a^b \int_c^u f(x, y)\, dy\, dx + \int_a^b \int_v^d f(x, y)\, dy\, dx \\
&\quad + \int_a^b \int_u^v f(x, y)\, dy\, dx \\
&= \int_a^r \int_c^u f(x, y)\, dy\, dx + \int_s^b \int_c^u f(x, y)\, dy\, dx \\
&\quad + \int_a^r \int_v^d f(x, y)\, dy\, dx + \int_s^b \int_v^d f(x, y)\, dy\, dx \\
&\quad + \int_r^s \int_c^u f(x, y)\, dy\, dx + \int_r^s \int_v^d f(x, y)\, dy\, dx \\
&\quad + \int_a^r \int_u^v f(x, y)\, dy\, dx + \int_s^b \int_u^v f(x, y)\, dy\, dx \\
&\quad + \int_s^r \int_u^v f(x, y)\, dy\, dx
\end{aligned}$$

By Fubini's Theorem each of the first eight terms above is equal to

$$\sum_{i=1}^{k} \int_{R_i} f(x, y)\, dA = \int_{R-U} f(x, y)\, dA$$

for some set of subrectangles of R. Thus we have shown

$$\int_a^b \int_c^d f(x,y)\,dy\,dx = \int_{R-U} f(x,y)\,dA + \int_r^s \int_u^v f(x,y)\,dy\,dx \tag{7}$$

Also observe that

$$\left|\int_r^s \int_u^v f(x,y)\,dy\,dx\right| \leqslant \int_r^s \left|\int_u^v f(x,y)\,dy\right|\,dx$$
$$\leqslant \int_r^s \int_u^v |f(x,y)|\,dy\,dx \leqslant M\frac{\varepsilon}{2M} = \frac{\varepsilon}{2}$$

By equations (5)–(7) we conclude that

$$\left|\int_R f(x,y)\,dA - \int_a^b \int_c^d f(x,y)\,dy\,dx\right|$$
$$= \left|\int_R f(x,y)\,dA - \int_{R-U} f(x,y)\,dA + \int_{R-U} f(x,y)\,dA - \int_a^b \int_c^d f(x,y)\,dy\,dx\right|$$
$$\leqslant \left|\int_R f(x,y)\,dA - \int_{R-U} f(x,y)\,dA\right| + \left|\int_{R-U} f(x,y)\,dA - \int_a^b \int_c^d f(x,y)\,dy\,dx\right|$$
$$\leqslant \left|\int_U f(x,y)\,dA\right| + \left|\int_r^s \int_u^v f(x,y)\,dy\,dx\right| \leqslant \frac{\varepsilon}{2} + \frac{\varepsilon}{2} = \varepsilon$$

Since $\varepsilon > 0$ was arbitrary

$$\int_R f(x,y)\,dA = \int_a^b \int_c^d f(x,y)\,dy\,dx \quad \blacksquare$$

Example 3 Compute $\int_R (x^2+y)\,dA$ where R is the square $[0,1]\times[0,1]$. By Fubini's Theorem

$$\int_R (x^2+y)\,dA = \int_0^1 \int_0^1 (x^2+y)\,dx\,dy = \int_0^1 \left[\int_0^1 (x^2+y)\,dx\right] dy$$

By the Fundamental Theorem of Integral Calculus

$$\int_0^1 (x^2+y)\,dx = \left[\frac{x^3}{3} + yx\right]_0^1$$
$$= \tfrac{1}{3} + y$$

Thus

$$\int_R (x^2+y)\,dA = \int_0^1 (\tfrac{1}{3}+y)\,dy = \left[\frac{1}{3}y + \frac{y^2}{2}\right]_0^1 = \frac{5}{6}$$

What we have done is first hold y fixed, integrate with respect to x, and then evaluate the result between the given limits for the x variable. Next we integrate the remaining function (of y alone) with respect to y to obtain the final answer.

A consequence of Fubini's Theorem is that interchanging the order of integration in the iterated integrals does not change the answer. Let us verify this for the above example. We have

$$\int_0^1 \int_0^1 (x^2+y)\,dy\,dx = \int_0^1 \left[x^2y+\frac{y^2}{2}\right]_0^1 dx = \int_0^1 (x^2+\tfrac{1}{2})\,dx = \left[\frac{x^3}{3}+\frac{x}{2}\right]_0^1 = \frac{5}{6}$$

When $f(x,y) \geqslant 0$ on $R=[a,b]\times[c,d]$, we have seen that the integral $\int_R f(x,y)\,dA$ can be interpreted as a volume. If the function also takes on negative values, then the double integral can be thought of as the sum of all volumes lying between the surface $z=f(x,y)$, the plane $z=0$, and bounded by the planes $x=a$, $x=b$, $y=c$, $y=d$; here the volumes above $z=0$ are counted as positive and those below as negative. However, Fubini's Theorem remains valid in the case that $f(x,y)$ is negative or changes sign on R. This is expected since there is no restriction on the sign of f in the hypotheses of the theorem.

Example 4 Let R be the rectangle $[-2,1]\times[0,1]$ and let $f(x,y)=y(x^3-12x)$. Clearly, $f(x,y)$ takes on both positive and negative values on R. Let us evaluate the integral $\int_R f(x,y)\,dx\,dy = \int_R y(x^3-12x)\,dx\,dy$. By Fubini's Theorem, we may write

$$\int_R y(x^3-12x)\,dx\,dy = \int_0^1 \left[\int_{-2}^1 y(x^3-12x)\,dx\right] dy = \tfrac{57}{4}\int_0^1 y\,dy = 7\tfrac{1}{8}$$

Alternatively, integrating first with respect to y, we find

$$\int_R y(x^3-12x)\,dy\,dx = \int_{-2}^1 (x^3-12x)\left[\int_0^1 y\,dy\right] dx$$

$$= \frac{1}{2}\int_{-2}^1 (x^3-12x)\,dx = \frac{1}{2}\left[\frac{x^4}{4}-6x^2\right]_{-2}^1 = 7\tfrac{1}{8}$$

EXERCISES

1. Let A, B be two subsets of a rectangle R and suppose each of A and B have area zero. Prove that $A \cup B$ and $A \cap B$ have area zero.
2. Let f be continuous, $f \geqslant 0$, on the rectangle R. If $\int_R f\,dA = 0$ prove that $f=0$ on R.
3. Evaluate each of the following integrals if $R = [0,1]\times[0,1]$:
   (a) $\int_R (x^2+y^2)\,dA$
   (b) $\int_R ye^{xy}\,dA$
   (c) $\int_R (xy)^2\cos x^3\,dA$
4. Compute the volume of the solid bounded by the xz plane, the yz plane, the xy plane, the planes $x=1$ and $y=1$, and the surface $z=x^2+y^4$.

5. Let $\phi:[a,b]\to \mathbf{R}$ be continuous and let Γ be the graph of ϕ. Suppose Γ lies in a rectangle R. Show Γ has zero area. [Hint: Assume the projection of $R=[a,b]\times[c,d]$ on the x axis is $[a,b]$. Use the uniform continuity of ϕ (§7.2) to show that for every $\varepsilon>0$, there is an N such that $n\geqslant N\Rightarrow b_n\leqslant \varepsilon(b-a)$.]

*6. Let $f:[0,1]\times[0,1]\to \mathbf{R}$ be defined by

$$f(x,y)=\begin{cases}1, & x \text{ rational}\\ 2y, & x \text{ irrational}\end{cases}$$

Show that the iterated integral $\int_0^1(\int_0^1 f(x,y)\,dy)\,dx$ exists but the double integral does not; that is, show f is not integrable.

7. Let f be continuous on $R=[s,b]\times[c,d]$ and for $a<x<b$, $c<y<d$; define

$$F(x,y)=\int_a^x\int_c^y f(u,v)\,du\,dv$$

Show that $\dfrac{\partial^2 F}{\partial x\partial y}=\dfrac{\partial^2 F}{\partial y\partial x}=f(x,y)$

8. Let B be a subset of a rectangle R and define

$$f(x,y)=\begin{cases}0, & (x,y)\notin B\\ 1, & (x,y)\in B\end{cases}$$

(a) Indicate why $\int_R f\,dA$ should be interpreted as the area of B.
(b) Prove that if B has area zero then $\int_R f\,dA=0$.

9. Prove property (4) of integrals. [Hint: Consult the proof of Theorem 4, §7.2. First, give the proof for continuous f.]

10. Let f be integrable over D. Then we stated in the text that $|f|$ is integrable. Assuming this fact, prove inequality (2), namely,

$$\left|\int_R f\right|\leqslant\int_R|f|$$

3 THE DOUBLE INTEGRAL OVER MORE GENERAL REGIONS

Our goal in this section is twofold; first we wish to define the integral $\int_D f(x,y)\,dA$ for regions more general than simple rectangles and second, we want to develop a technique for evaluating this type of integral. To accomplish this, we shall define three special types of subsets of the xy plane and then extend the notion of the double integral to these sets.

Suppose we are given two continuous real-valued functions $\phi_1:[a,b]\to\mathbf{R}$, $\phi_2:[a,b]\to\mathbf{R}$ which satisfy $\phi_2(t)\leqslant\phi_1(t)$ for all $t\in[a,b]$. Let D be the set of all points (x,y) such that

$$x\in[a,b],\quad \phi_2(x)\leqslant y\leqslant\phi_1(x)$$

Then the region D is said to be of **type 1**. Figure 3-1 shows some regions of type 1. The curves and straight line segments which bound the region taken together constitute the **boundary** of D, denoted ∂D.

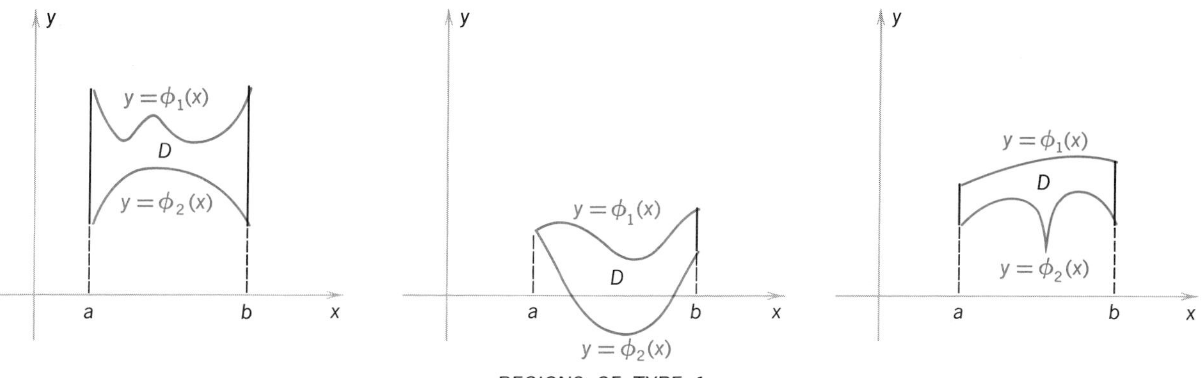

REGIONS OF TYPE 1

FIGURE 3-1

We say that a region D is of **type 2** if there are continuous functions $\phi_1, \phi_2:[c,d] \to \boldsymbol{R}$ such that D is the set of points (x, y) satisfying

$$y \in [c,d], \quad \phi_2(y) \leqslant x \leqslant \phi_1(y)$$

where $\phi_1(t) \leqslant \phi_2(t)$, $t \in [c,d]$. Again, the curves which bound the region D constitute its boundary ∂D. Some examples of type 2 regions are shown in Figure 3-2.

Finally, region of **type 3** is one which is both type 1 and type 2; an example of a type 3 region is the unit disc (Figure 3-3).

Sometimes we shall refer to regions of types 1, 2, and 3 as **elementary regions.** Note that by Theorem 2 of §15.2, the boundary ∂D of an elementary region has area zero.

If D is an elementary region in the plane, we can find a rectangle $R \supset D$. Assume we have chosen such an R; then given $f: D \to \boldsymbol{R}$, where f is continuous, we would like to define $\int_D f(x,y)\,dA$. To do this, we "extend" f to a function $\hat{f}$ defined on all of R by

$$\hat{f}(x,y) = \begin{cases} f(x,y), & (x,y) \in D \\ 0, & (x,y) \notin D \quad \text{and} \quad (x,y) \in R \end{cases}$$

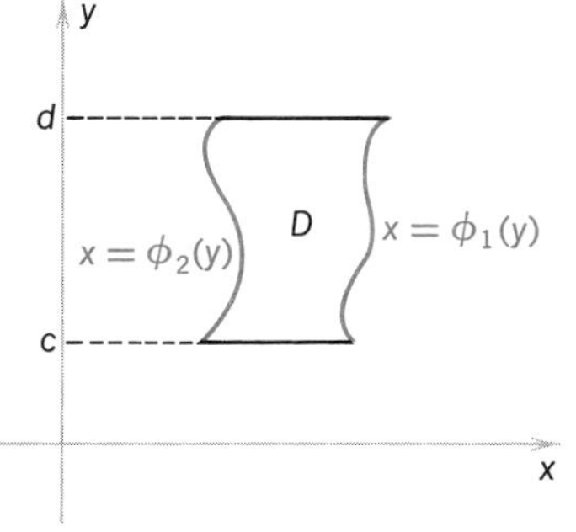

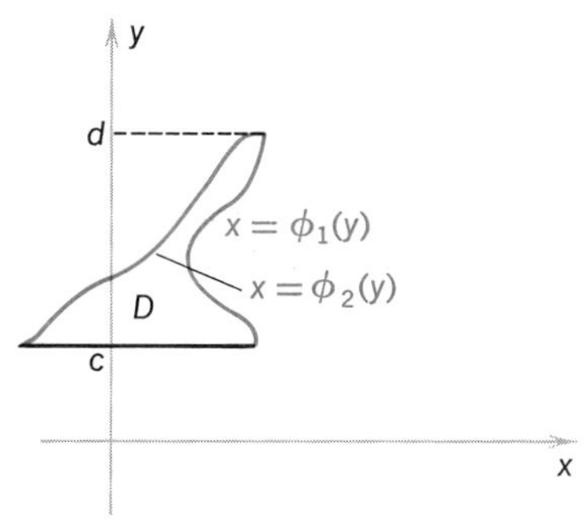

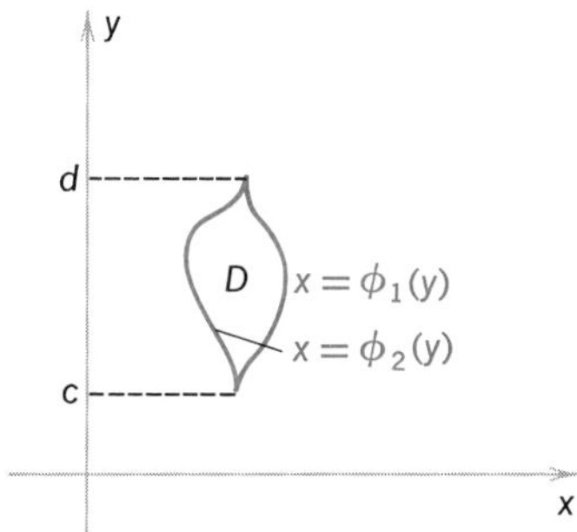

FIGURE 3-2

Now, $\hat{f}$ is defined on all of R and, moreover, $\hat{f}$ is bounded since f is, and $\hat{f}$ is continuous except perhaps on the boundary of D. The boundary of D has area zero and so $\hat{f}$ is integrable over R. Therefore, we define

$$\int_D f(x,y)\,dA = \int_R \hat{f}(x,y)\,dA$$

When $f(x,y) \geqslant 0$ on D, we can interpret the integral $\int_D f(x,y)\,dA$ as the volume of the three-dimensional region under the graph of f and over D (see Figure 3-4). We have defined $\int_D f(x,y)\,dx\,dy$ by choosing a rectangle $R \supset D$. It should be clear that the value of $\int_D f(x,y)\,dx\,dy$ does not depend on the particular R we select and we shall demonstrate this fact below.

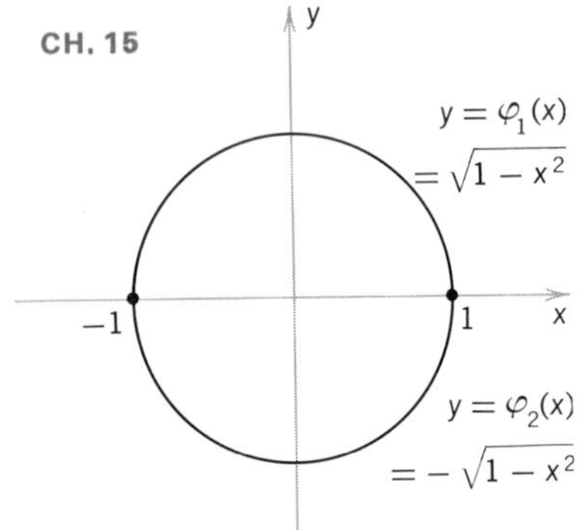

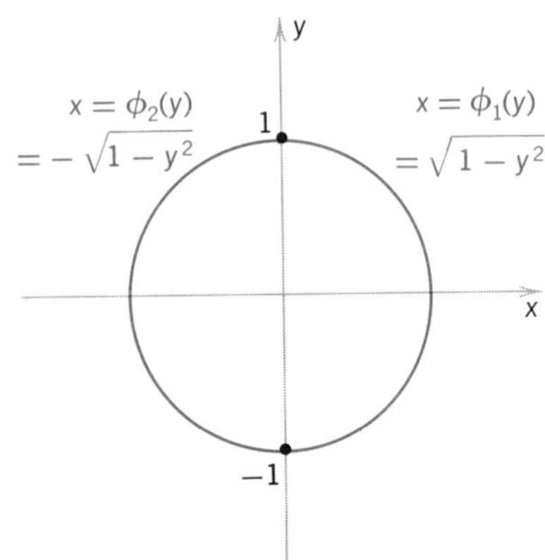

FIGURE 3-3

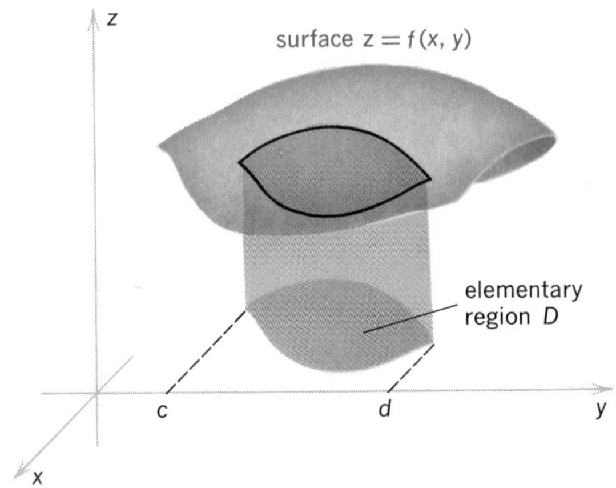

FIGURE 3-4

If $R = [a,b] \times [c,d]$ is a rectangle containing D, we can use the results on iterated integrals in §15.2 to obtain

$$\int_D f(x,y)\,dA = \int_R \hat{f}(x,y)\,dA = \int_a^b \int_c^d \hat{f}(x,y)\,dy\,dx = \int_c^d \int_a^b \hat{f}(x,y)\,dx\,dy$$

where $\hat{f}$ equals zero outside D. Assume D is a region of type 1 determined by functions $\phi_1 : [a,b] \to \boldsymbol{R}$, and $\phi_2 : [a,b] \to \boldsymbol{R}$. Consider the iterated integral

$$\int_a^b \int_c^d \hat{f}(x,y)\,dy\,dx$$

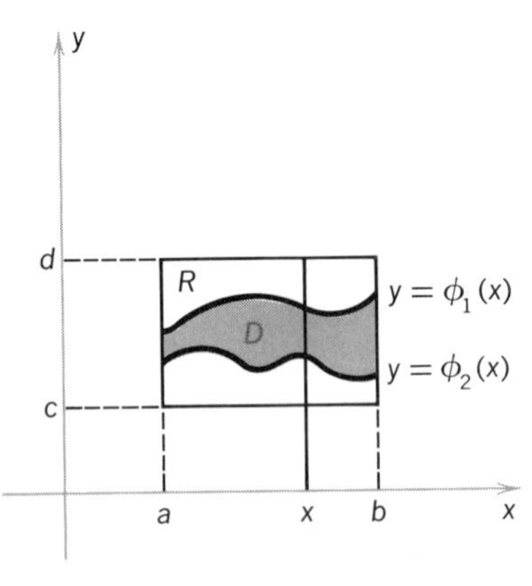

FIGURE 3-5

and, in particular, the inner integral $\int_c^d \hat{f}(x,y)\,dy$ for some fixed x (Figure 3-5). Since by definition, $\hat{f}(x,y) = 0$ if $y > \phi_1(x)$ or $y < \phi_2(x)$ we obtain

$$\int_c^d \hat{f}(x,y)\,dy = \int_{\phi_2(x)}^{\phi_1(x)} \hat{f}(x,y)\,dy = \int_{\phi_2(x)}^{\phi_1(x)} f(x,y)\,dy$$

Thus if D is a region of type 1, $\int_D f(x,y)\,dA$ is given by the iterated integral

$$\int_a^b \int_{\phi_2(x)}^{\phi_1(x)} f(x,y)\,dy\,dx \tag{1}$$

Furthermore, since we are assuming that for each fixed x, $f(x,y)$ is a continuous function of y for $\phi_2(x) \leqslant y \leqslant \phi_1(x)$, we can (in most cases) evaluate the iterated integrals in (1) using the Fundamental Theorem of Integral Calculus.

In the case $f(x,y) = 1$ for all $(x,y) \in D$ the value of (1) is

$$\int_a^b \int_{\phi_2(x)}^{\phi_1(x)} f(x,y)\,dy\,dx = \int_a^b [\phi_1(x) - \phi_2(x)]\,dx = A(D)$$

the area of D.

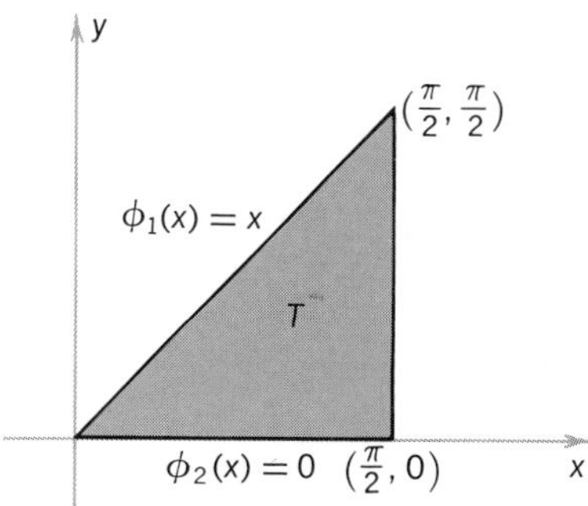

FIGURE 3-6

Example 1 Find $\int_T (x^3y + \cos x)\,dA$, where T is the triangle consisting of all points (x,y) such that $0 \leqslant x \leqslant \pi/2$, $0 \leqslant y \leqslant x$. Referring to Figure 3-6 and formula (1), we have

$$\begin{aligned}
\int_T (x^3y + \cos x)\,dA &= \int_0^{\pi/2} \int_0^x (x^3y + \cos x)\,dy\,dx \\
&= \int_0^{\pi/2} \left[\frac{x^3y^2}{2} + y\cos x\right]_0^x dx \\
&= \int_0^{\pi/2} \left(\frac{x^5}{2} + x\cos x\right) dx \\
&= \left[\frac{x^6}{12}\right]_0^{\pi/2} + \int_0^{\pi/2} (x\cos x)\,dx \\
&= \frac{\pi^6}{(12)(64)} + [x\sin x + \cos x]_0^{\pi/2} \\
&= \frac{\pi^6}{768} + \frac{\pi}{2} - 1
\end{aligned}$$

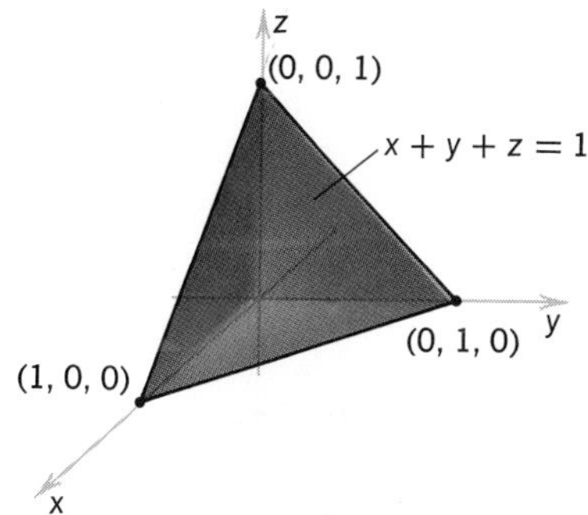

FIGURE 3-7

In the next example we apply the integral to find the volume of a solid whose base is a nonrectangular region D.

Example 2 Find the volume of the tetrahedron bounded by the planes $y = 0$, $z = 0$, $x = 0$, and the plane $z + x + y = 1$ (Figure 3-7).

We first note that the given tetrahedron has a triangular base D whose points (x,y) satisfy $0 \leqslant x \leqslant 1$, $0 \leqslant y \leqslant 1 - x$; hence D is a region of type 1 (in fact, D is type 3) (see Figure 3-8).

Now, for any point $(x,y) \in D$, the height of the surface $z = 1 - x - y$ above (x,y) is $1 - x - y$. Thus, the volume we seek is given by the integral

$$\int_D (1 - x - y)\,dA$$

Using the iterated integral formula (1) with $\phi_1(x) = 1 - x$ and $\phi_2(x) = 0$

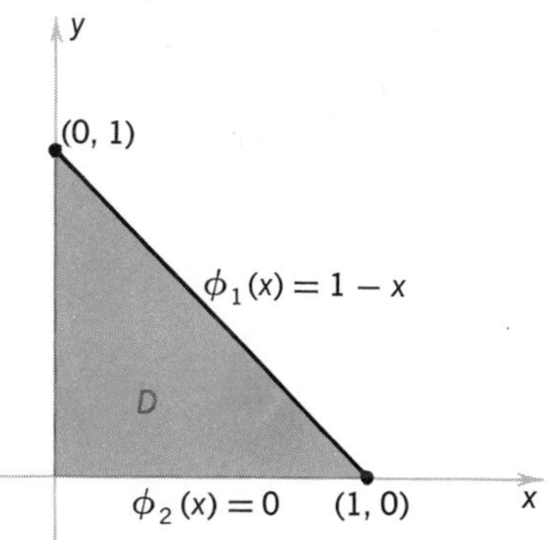

FIGURE 3-8

we have

$$\int_D (1-x-y)\,dA = \int_0^1 \int_0^{1-x} (1-x-y)\,dy\,dx$$

$$= \int_0^1 \left[(1-x)y - \frac{y^2}{2}\right]_0^{1-x} dx = \int_0^1 \left[\frac{(1-x)^2}{2}\right] dx$$

$$= -\frac{1}{2}\left[\frac{(1-x)^3}{3}\right]_0^1 = \frac{1}{6}$$

The methods for treating regions of type 2 are entirely analogous. Specifically, if D is the set of points (x, y) such that $y \in [c, d]$, $\phi_2(y) \leqslant x \leqslant \phi_1(y)$, then for f continuous, we have

$$\int_D f\,dA = \int_c^d \left[\int_{\phi_2(y)}^{\phi_1(y)} f(x, y)\,dx\right] dy \tag{2}$$

To find the area of D we substitute $f = 1$ in formula (2); this yields

$$\int_D dA = \int_c^d (\phi_1(y) - \phi_2(y))\,dy$$

Note that this result for area (and the corresponding formula for type 1 regions) agrees with our conclusions in single-variable calculus for the area between two curves.

Either the method for type 1 or type 2 regions can be used for regions of type 3.

It also follows from formulas (1) and (2) that $\int_D f$ is independent of the choice of the rectangle $R \supset D$ used in the definition of $\int_D f$. To see this let us consider the case when D is of type 1. Then formula (1) holds; moreover, on the right side of this formula R does not appear, and thus $\int_D f$ is independent of R.

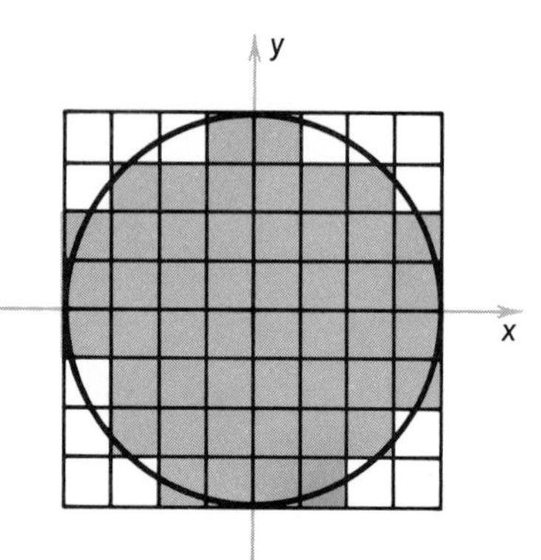

FIGURE 3-9

EXERCISES

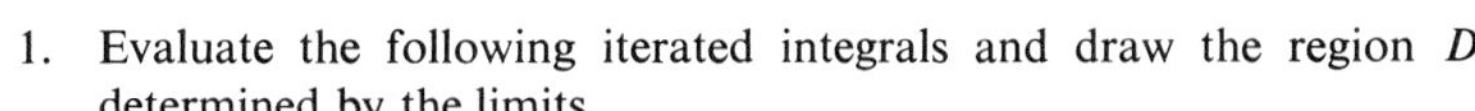

1. Evaluate the following iterated integrals and draw the region D determined by the limits.

   (a) $\int_0^1 \int_0^{x^2} dy\,dx$ (b) $\int_1^2 \int_{2x}^{3x+1} dy\,dx$

   (c) $\int_0^1 \int_1^{e^x} (x+y)\,dy\,dx$ (d) $\int_0^1 \int_{x^3}^{x^2} y\,dy\,dx$

*2. Form the Riemann sums needed to compute the area of the unit circle and show that the sum for a regular partition of order n is the area of rectangles of the type shaded in Figure 3-9 (whether to include rectangles meeting the boundary depends on the choice of the c_{ij}). To convince yourself further of the power of calculus try to show directly that the Riemann sums converge to some number.

3. Use double integrals to compute the area of a circle of radius r.

4. Determine the area of an ellipse with semiaxes of length a and b using double integrals.

5. What is the volume of a barn which has a rectangular base 20 ft by 40 ft and vertical walls 30 ft high at the front (which we assume is on the 20-ft side of the barn) and 40 ft high at the rear? The barn has a flat roof.

6. Let D be the region bounded by the positive x and y axes and the line $3x+4y=10$. Compute
$$\int_D (x^2+y^2)\,dA$$

7. Let D be the region bounded by the y axis and the parabola $x=-4y^2+3$. Compute
$$\int_D x^3 y\,dx\,dy$$

8. Evaluate $\int_0^1 \int_0^{x^2} (x^2+xy-y^2)\,dy\,dx$. Describe this iterated integral as a double integral over a certain region D.

9. Evaluate $\int_0^1 \int_0^{\sin x} xy\,dy\,dx$ and describe the region.

10. Find the volume of the region inside of the surface $z=x^2+y^2$ and between $z=0$ and $z=10$.

11. Find the area enclosed by one period of the sine function $\sin x$, for $0 \leqslant x \leqslant 2\pi$ and the x axis.

12. Compute the volume of a cone of base radius r and height h.

13. Evaluate $\int_D y\,dA$ where D is the set of points (x,y) such that $0 \leqslant x \leqslant y, y \leqslant \sqrt{x}$.

14. Prove that $\int_a^b \int_c^d f(x)g(y)\,dy\,dx = (\int_a^b f(x)\,dx)(\int_c^d g(y)\,dy)$. Is this result true if we integrate $f(x)g(y)$ over a general region D (for example, a region of type 1)?

15. Is there an analog of integration by parts which is valid for double integrals?

16. Use the methods of this section to show that the area of the parallelogram D determined by vectors $\boldsymbol{a}$ and $\boldsymbol{b}$ is $|a_1b_2-a_2b_1|$, where $\boldsymbol{a} = a_1\boldsymbol{i}+a_2\boldsymbol{j}$, $b = b_1\boldsymbol{i}+b_2\boldsymbol{j}$.

4 CHANGING THE ORDER OF INTEGRATION

Suppose that D is a region of type 3 given as the set of points (x,y) such that
$$a \leqslant x \leqslant b, \quad \phi_2(x) \leqslant y \leqslant \phi_1(x)$$
and also as the set of points (x,y) such that
$$c \leqslant y \leqslant d, \quad \psi_2(y) \leqslant x \leqslant \psi_1(y)$$
Hence we have the formula
$$\int_D f(x,y)\,dA = \int_a^b \int_{\phi_1(x)}^{\phi_2(x)} f(x,y)\,dy\,dx$$
$$= \int_c^d \int_{\psi_1(y)}^{\psi_2(y)} f(x,y)\,dx\,dy$$

If we are asked to compute one of the iterated integrals above, we may do so by evaluating the other iterated integral; this technique is called **changing the order of integration.** It is often useful to make such a change when evaluating iterated integrals, since one of the iterated integrals may be more difficult to compute than the other.

Example 1 By changing the order of integration, evaluate

$$\int_{-a}^{a}\int_{-\sqrt{a^2-x^2}}^{\sqrt{a^2-x^2}} x\sqrt{a^2-x^2-y^2}\,dy\,dx$$

Note that x varies between $-a$ and a, and for fixed x, $-\sqrt{a^2-x^2}\leqslant y\leqslant\sqrt{a^2-x^2}$. Thus the iterated integral is equivalent to the double integral

$$\int_D x\sqrt{a^2-x^2-y^2}\,dy\,dx$$

where D is the set of points (x, y) such that

$$-a\leqslant x\leqslant a,\quad -\sqrt{a^2-x^2}\leqslant y\leqslant\sqrt{a^2-x^2}$$

But this is the representation of the disc of radius a, hence D can also be described as the set of points (x, y) where

$$-a\leqslant y\leqslant a,\quad -\sqrt{a^2-y^2}\leqslant x\leqslant\sqrt{a^2-y^2}$$

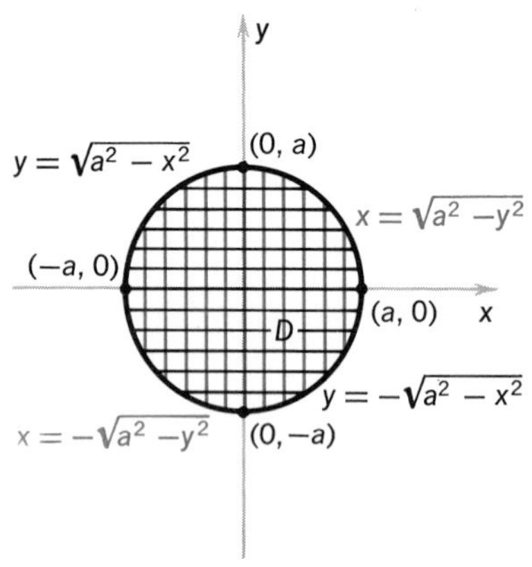

FIGURE 4-1

(see Figure 4-1). Thus

$$\begin{aligned}\int_{-a}^{a}\int_{-\sqrt{a^2-x^2}}^{\sqrt{a^2-x^2}} x\sqrt{a^2-x^2-y^2}\,dy\,dx &= \int_{-a}^{a}\int_{-\sqrt{a^2-y^2}}^{\sqrt{a^2-y^2}} x\sqrt{a^2-x^2-y^2}\,dx\,dy\\ &= -\tfrac{1}{2}\int_{-a}^{a}\int_{-\sqrt{a^2-y^2}}^{\sqrt{a^2-y^2}} -2x\sqrt{a^2-x^2-y^2}\,dx\,dy\\ &= -\tfrac{1}{3}\int_{-a}^{a}\Big[(a^2-x^2-y^2)^{3/2}\Big]_{-\sqrt{a^2-y^2}}^{\sqrt{a^2-y^2}}\,dy\\ &= -\tfrac{1}{3}\int_{-a}^{a}(0)\,dy = 0\end{aligned}$$

Could you have guessed that the answer would be zero earlier by a symmetry argument?

Example 2 Evaluate

$$\int_1^2\int_0^{\log x}(x-1)\sqrt{1+e^{2y}}\,dy\,dx$$

First note that we could not easily compute this integral in the order given by using the Fundamental Theorem. However, the integral is equal to $\int_D (x-1)\sqrt{1+e^{2y}}\,dA$, where D is the set of (x, y) such that

$$1\leqslant x\leqslant 2,\quad 0\leqslant y\leqslant \log x$$

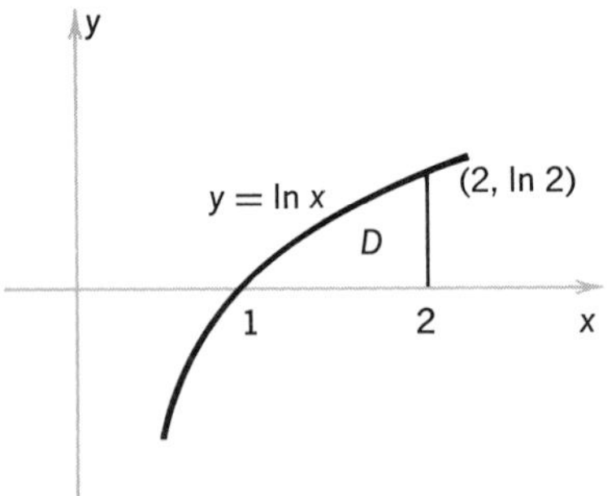

FIGURE 4-2

Now D is a region of type 3 (see Figure 4-2) and so it can be described by

$$0\leqslant y\leqslant \log 2,\quad e^y\leqslant x\leqslant 2$$

Thus the given iterated integral is equivalent to

$$\begin{aligned}
\int_0^{\log 2}\int_{e^y}^{2}(x-1)\sqrt{1+e^{2y}}\,dx\,dy
&= \int_0^{\log 2}\sqrt{1-e^{2y}}\left[\int_{e^y}^{2}(x-1)\,dx\right]dy\\
&= \int_0^{\log 2}\sqrt{1+e^{2y}}\left[\frac{x^2}{2}-x\right]_{e^y}^{2}dx\\
&= -\int_0^{\log 2}\left(\frac{e^{2y}}{2}-e^y\right)\sqrt{1+e^{2y}}\,dy\\
&= -\tfrac{1}{2}\int_0^{\log 2}e^{2y}\sqrt{1+e^{2y}}\,dy+\int_0^{\log 2}e^y\sqrt{1+e^{2y}}\,dy \qquad (1)
\end{aligned}$$

In the first integral immediately above we substitute $u=e^{2y}$ and in the second, $v=e^y$. Hence we obtain

$$-\tfrac{1}{4}\int_1^4\sqrt{1+u}\,du+\int_1^2\sqrt{1+v^2}\,dv \qquad (2)$$

Both integrals in (2) are easily found using techniques of first-year calculus. For the first integral we have

$$\begin{aligned}
\tfrac{1}{4}\int_1^4\sqrt{1+u}\,du &= \left[\tfrac{1}{6}(1+u)^{3/2}\right]_1^4 = \tfrac{1}{6}[(1+4)^{3/2}-2^{3/2}]\\
&= \tfrac{1}{6}[5^{3/2}-2^{3/2}] \qquad (3)
\end{aligned}$$

In the second integral, we make the further substitution $v=\tan\theta$; this yields

$$\begin{aligned}
\int_1^2\sqrt{1+v^2}\,dv &= \tfrac{1}{2}\left[v\sqrt{1+v^2}+\ln\left|\sqrt{1+v^2}+v\right|\right]_1^2\\
&= \tfrac{1}{2}[2\sqrt{5}+\ln|\sqrt{5}+2|]\\
&\quad -\tfrac{1}{2}[\sqrt{2}+\ln|\sqrt{2}+1|] \qquad (4)
\end{aligned}$$

Finally, we subtract (3) from (4) to obtain the final result

$$\frac{1}{2}\left(2\sqrt{5}-\sqrt{2}+\ln\frac{\sqrt{5}+2}{\sqrt{2}+1}\right)-\tfrac{1}{6}[5^{3/2}-2^{3/2}]$$

To conclude this section we mention an important analog for double integrals of the Mean Value Theorem of Integral Calculus.

Theorem 1 Mean Value Theorem for Double Integrals Suppose $f: D\to \mathbf{R}$ is continuous and D is an elementary region. Then for some point (x_0, y_0) in D we have

$$\int_D f(x,y)\,dA = f(x_0,y_0)\cdot A(D)$$

where $A(D)$ denotes the area of D.

The proof of this theorem is analogous to the proof in the one-variable case.

EXERCISES

1. In (a)–(d), change the order of integration, sketch the corresponding regions, and evaluate the integral both ways.
   (a) $\int_0^1 \int_x^1 xy\, dx\, dy$
   (b) $\int_0^{\pi/2} \int_0^{\cos\theta} \cos\theta\, dr\, d\theta$
   (c) $\int_0^1 \int_{2-y}^1 (x+y)^2\, dx\, dy$
   (d) $\int_a^b \int_a^y f(x,y)\, dx\, dy$
2. Find
   (a) $\int_{-1}^1 \int_{|y|}^1 (x+y)^2\, dx\, dy$
   (b) $\int_{-3}^3 \int_{-\sqrt{9-y^2}}^{\sqrt{9-y^2}} x^2\, dx\, dy$
   (c) $\int_0^1 \int_0^{e^y} \log x\, dx\, dy$
3. Prove
$$2\int_a^b \int_x^b f(x)f(y)\, dy\, dx = \left(\int_a^b f(x)\, dx\right)^2$$
4. If $m \leqslant f(x,y) \leqslant M$ prove that
$$m(\text{area } D) \leqslant \int_D f(x)\, dx\, dy \leqslant M(\text{area } D)$$
5. Compute $\int_D f(x,y)\, dA$ where $f(x,y) = \sqrt{x}\,y^2$ and D is the set of (x,y) where $x > 0, y > x^2, y < 10 - x^2$.
6. Compute the volume of an ellipsoid with semiaxes a, b, and c. [Hint: Use symmetry and first find the volume of one-half of the ellipsoid.]
7. Find the volume determined by $x^2+y^2+z^2 \leqslant 10, z \geqslant 2$.
8. State analog of the Extreme and Intermediate Value Theorems of §4.6 for a continuous function of two variables $f: D \to \boldsymbol{R}$. Outline proofs for the theorems (use the results of the one-variable case). Deduce Theorem 1 above.

5 IMPROPER INTEGRALS

In the previous sections we defined the notion of the integral for functions of two variables and stated theorems which guaranteed that f was indeed integrable over a set D. Recall that one of the hypotheses of Theorem 3, §15.2 was that f be bounded. The following example shows that the boundedness of f is necessary if the sums S_n are to converge.

Let S be the unit square $[0,1]\times[0,1]$ and let $f: S \to \boldsymbol{R}$ be defined by
$$f(x,y) = \begin{cases} \dfrac{1}{\sqrt{x}}, & x \neq 0 \\ 0, & x = 0 \end{cases}$$

Clearly, f is not bounded on S since, as x gets close to zero, f gets arbitrarily large. Let S_{ij} be a regular partition of S and form the sum (1) of §15.2:

$$S_n = \sum_{i=0}^{n-1} \sum_{j=0}^{n-1} f(c_{ij})\,\Delta x\,\Delta y$$

Let S_{11} be the subrectangle which contains $(0, 0)$ (see Figure 5-1) and choose some $c_{11} \in S_{11}$. For a fixed n, we can make S_n as large as we please by picking c_{11} closer and closer to $(0, 0)$—hence $\lim_{n\to\infty} S_n$ cannot exist independent of the choice of the c_{ij}.

However, let us formally evaluate the iterated integral of f following the rules for integrating a function of a single variable. We have

$$\int_0^1 \int_0^1 f(x, y)\,dx\,dy = \int_0^1 \int_0^1 \frac{dx}{\sqrt{x}}\,dy = \int_0^1 \left[(2\sqrt{x})\right]_0^1 dy$$

$$= \int_0^1 2\,dy = 2$$

Moreover, if we reverse the order of integration we also obtain

$$\int_0^1 \int_0^1 \frac{dy}{\sqrt{x}}\,dx = 2$$

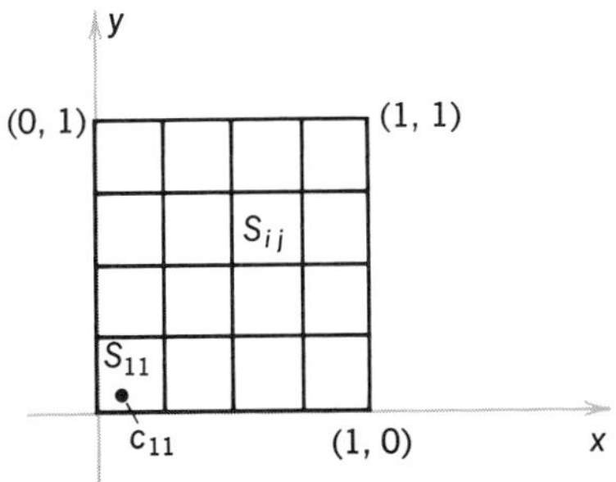

FIGURE 5-1

So in some sense this function is integrable. The question is, in what sense?

Recall that $\int_0^1 dx/\sqrt{x}$ is an improper integral; that is, $1/\sqrt{x}$ is unbounded on the interval $]0, 1]$ yet $\lim_{\delta\to 0}\int_\delta^1 (dx/\sqrt{x}) = 2$ and we define $\int_0^1 (dx/\sqrt{x})$ to be this limit. Similarly, for the two-variable case, we will allow the function to be unbounded at certain points on the boundary of its domain and thus broaden our notion of integrability.

More specifically, suppose the region D is of type 1 and $f: D \to \mathbf{R}$ is continuous and bounded except at certain points on the boundary. Assume that for all $(x, y) \in D$ we have $a \leqslant x \leqslant b, \phi_2(x) \leqslant y \leqslant \phi_1(x)$. Choose numbers $\delta, \eta > 0$ such that $D_{\eta,\delta}$ is the subset of D consisting of points (x, y) with $a+\eta \leqslant x \leqslant b-\eta, \phi_2(x)+\delta \leqslant y \leqslant \phi_1(x)-\delta$ (Figure 5-2), where S is chosen small enough so that $D_{\eta,\delta} \in D$. If either $\phi_2(a) = \phi_1(a)$ or $\phi_2(b) = \phi_1(b)$ we must modify this slightly because in this case $D_{\eta,\delta}$ may not be a subset of D (see Example 2). Since f is continuous and bounded on $D_{\eta,\delta}$ the integral $\int_{D_{\eta,\delta}} f$ exists. We can now ask what happens when the region $D_{\eta,\delta}$ expands to fill the region D, that is, as $(\eta, \delta) \to (0, 0)$.

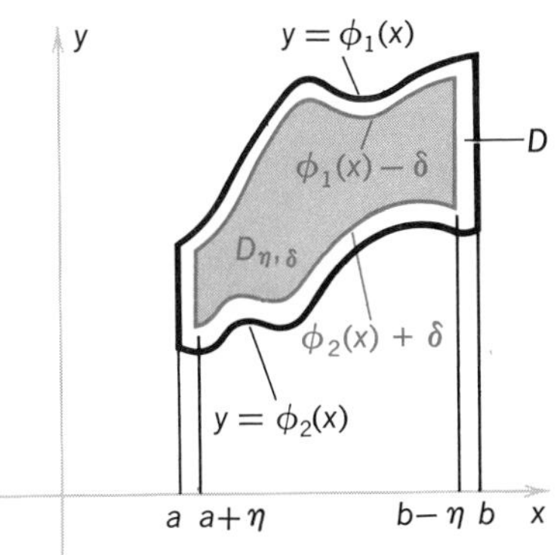

FIGURE 5-2

If

$$\lim_{(\eta,\delta)\to(0,0)} \int_{D_{\eta,\delta}} f$$

exists we define $\int_D f$ to be equal to this limit and say that it is the **improper integral** of f over D. This definition is exactly analogous to the definition of improper integral for a function of one variable.

Since f is integrable over $D_{\eta,\delta}$ we can apply Fubini's Theorem to obtain that

$$\int_{D_{\eta,\delta}} f = \int_{a+\eta}^{b-\eta} \int_{\phi_2(x)+\delta}^{\phi_1(x)-\delta} f(x, y)\,dy\,dx \tag{1}$$

So if f is integrable over D,

$$\int_D f = \lim_{(\eta,\delta)\to(0,0)} \int_{a+\eta}^{b-\eta} \int_{\phi_2(x)+\delta}^{\phi_1(x)-\delta} f(x,y)\, dy\, dx$$

Unfortunately, evaluating this limit is usually troublesome. However, we can simplify our work by considering

$$\lim_{\eta\to 0} \int_{a+\eta}^{b-\eta} \left[\lim_{\delta\to 0} \int_{\phi_2(x)+\delta}^{\phi_1(x)-\delta} f(x,y)\, dy\right] dx \tag{2}$$

if these limits exist. If the limits do exist, then we denote (2) by

$$\int_a^b \int_{\phi_2(x)}^{\phi_1(x)} f(x,y)\, dy\, dx$$

and call it the **iterated improper integral** of f over D. Using more advanced techniques it is possible to show that if f is a sufficiently well-behaved function (the kind we use in this text) then if the iterated improper integral exists it equals $\int_D f$.

The definition is analogous if D is a region of type 2. Finally, let us consider the case where D is a region of type 3 and f is unbounded at points on ∂D. For example, suppose D is the set of points (x,y) with

$$a \leqslant x \leqslant b, \qquad \phi_2(x) \leqslant y \leqslant \phi_1(x)$$

and

$$c \leqslant y \leqslant d, \qquad \psi_2(y) \leqslant x \leqslant \psi_1(y)$$

Then if f is well behaved and

$$\int_c^d \int_{\psi_2(y)}^{\psi_1(y)} f(x,y)\, dy\, dx, \qquad \int_a^b \int_{\phi_2(x)}^{\phi_1(x)} f(x,y)\, dy\, dx$$

exist, then both iterated integrals are equal and their common value is $\int_D f$. This is Fubini's Theorem for improper integrals.

Example 1 Evaluate $\int_D f(x,y)\, dy\, dx$ where $f(x,y) = 1/(\sqrt{1-x^2-y^2})$ and D is the unit disc $x^2+y^2 \leqslant 1$.

We can describe D as the set of points $(x.y)$ with $-1 \leqslant x \leqslant 1$, $-\sqrt{1-x^2} \leqslant y \leqslant \sqrt{1-x^2}$. Now since ∂D is the set of points (x,y) with $x^2+y^2=1$, f is unbounded at every point on ∂D since at such points the denominator of f is 0. We calculate the iterated improper integrals for one variable and obtain

$$\begin{aligned}
\int_{-1}^1 \int_{-\sqrt{1-x^2}}^{\sqrt{1-x^2}} \frac{dy\, dx}{\sqrt{1-x^2-y^2}}
&= \int_{-1}^1 \left\{\left[\sin^{-1}\left(\frac{y}{\sqrt{1-x^2}}\right)\right]_{-\sqrt{1-x^2}}^{\sqrt{1-x^2}}\right\} dx \\
&= \int_{-1}^1 [\sin^{-1}(1) - \sin^{-1}(-1)]\, dx \\
&= \pi \int_{-1}^1 dx = 2\pi
\end{aligned}$$

In this example we assumed that the iterated improper integral computed above is equal to the improper integral of $1/(\sqrt{1-x^2-y^2})$ over the unit disc. This is indeed the case as the interested student can verify for himself.

Example 2 Let $f(x,y)=1/(x-y)$ and let D be the set of (x,y) with $0\leqslant x\leqslant 1$ and $0\leqslant y\leqslant x$ (Figure 5-3). Show that f is not integrable over D.

Since the denominator of f is zero on the line $y=x$, f is unbounded on part of the boundary of D. Let $0<\eta<1$ and $0<\delta<\eta$ and let $D_{\eta,\delta}$ be the set of (x,y) with $0<\eta\leqslant x\leqslant 1-\eta$ and $\delta\leqslant y\leqslant x-\delta$ (Figure 5-3). We choose $\delta<\eta$ to guarantee that $D_{\eta,\delta}$ is contained in D. Consider

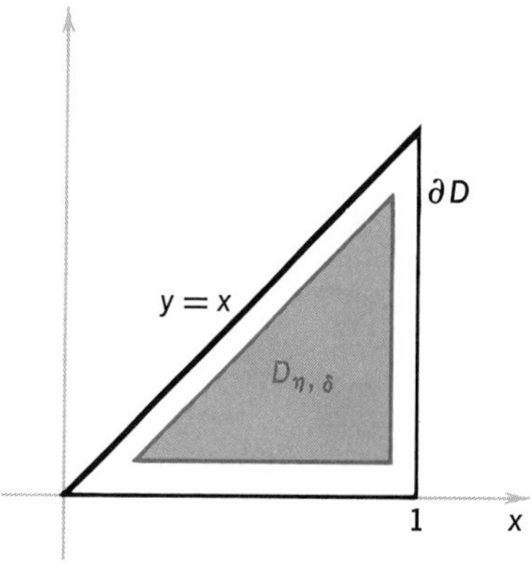

FIGURE 5-3

$$\begin{aligned}\int_{D_{\eta,\delta}} f &= \int_{0+\eta}^{1-\eta}\int_{\delta}^{x-\delta}\frac{1}{x-y}\,dy\,dx\\ &= \int_{\eta}^{1-\eta}\Big[-\log(x-y)\Big]_{\delta}^{x-\delta}\,dx\\ &= \int_{\eta}^{1-\eta}[-\log(\delta)+\log(x-\delta)]\,dx\\ &= [-\log\delta]\int_{\eta}^{1-\eta}dx+\int_{\eta}^{1-\eta}\log(x-\delta)\,dx\\ &= -(1-2\eta)\log\delta+[(x-\delta)\log(x-\delta)-(x-\delta)]_{\eta}^{1-n}\end{aligned}$$

In the last step we used the fact that $\int u\log u\,du=u\log u-u$. Continuing the above set of equalities, we have

$$-(1-2\eta)\log\delta+(1-\eta-\delta)\log(1-\eta-\delta)-(1-\eta-\delta)\\ -(\eta-\delta)\log(\eta-\delta)+(\eta-\delta)$$

As $(\eta,\delta)\to(0,0)$ the second term converges to 1 log 1 or 0, the third and fifth terms converge to 1 and 0, respectively. Let $v=\eta-\delta$. Since $v\log v\to 0$ as $v\to 0$ (use L'Hôpital's rule) we see that the fourth term goes to zero as $(\eta,\delta)\to(0,0)$. It is the first term which will give us trouble. Now

$$-(1-2\eta)\log\delta = -\log\delta+2\eta\log\delta \tag{3}$$

and it is not hard to see that this does not converge as $(\eta,\delta)\to(0,0)$. For example, let $\eta=2\delta$; then (3) becomes

$$-\log\delta+4\delta\log\delta$$

As before $-4\delta\log\delta\to 0$ as $\delta\to 0$ but $-\log\delta\to+\infty$ as $\delta\to 0$ which shows that (3) does not converge. Hence $\lim_{(\eta,\delta)\to(0,0)}\int_{D_{\eta,\delta}} f$ does not exist and f is not integrable.

EXERCISES

1. Evaluate the following integrals if they exist:

(a) $\int_D \frac{1}{\sqrt{xy}}\,dA$, $D=[0,1]\times[0,1]$

(b) $\int_D \frac{1}{\sqrt{|x-y|}}\,dx\,dy$, $D = [0,1] \times [0,1]$

[Hint: Divide D into two pieces.]

(c) $\int_D \frac{y}{x}\,dx\,dy$, D bounded by $x = 1$, $x = y$, and $x = 2y$

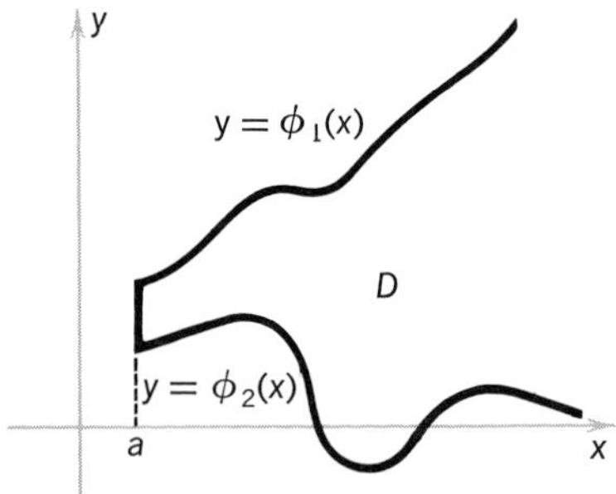

FIGURE 5-4

2. (a) Discuss how you would define $\int_D f\,dA$ if D is an unbounded region, for example, the set of (x, y) such that $a \leqslant x < \infty$, $\phi_2(x) \leqslant y \leqslant \phi_1(x)$ where $\phi_2 \leqslant \phi_1$ are given (Figure 5-4).
   (b) Evaluate $\int_D xye^{-(x^2+y^2)}\,dx\,dy$ if $x \geqslant 0, 0 \leqslant y \leqslant 1$.

3. Integrate e^{-xy} for $x \geqslant 0, 1 \leqslant y \leqslant 2$ in two ways (assume Fubini's Theorem) to show

$$\int_0^\infty \frac{e^x - e^{-2x}}{x}\,dx = \log 2$$

6 THE TRIPLE INTEGRAL

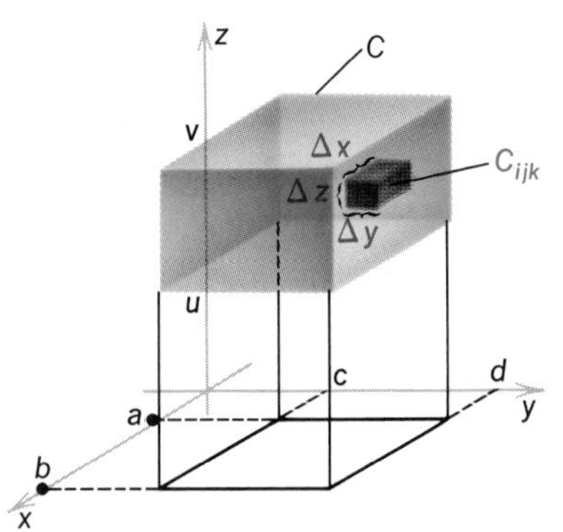

FIGURE 6-1

Given a continuous function $f: C \to \boldsymbol{R}$, where C is some rectangular parallelepiped in $\boldsymbol{R}^3$, we can define the integral of f over C as a limit of a sum, just as we did for a function of two variables. Briefly, we partition the sides of C into n equal parts and form the sum

$$S_n = \sum_{i=0}^{n-1} \sum_{j=0}^{n-1} \sum_{k=0}^{n-1} f(c_{ijk})\,\Delta V$$

where $c_{ijk} \in C_{ijk}$, the ijkth rectangular parallelepiped in the partition of C, and where ΔV is the volume of C_{ijk} (see Figure 6-1).

For continuous functions one can show $\lim_{n\to\infty} S_n$ exists and is independent of the points c_{ijk}; thus, it follows that all continuous functions are integrable. We call the limit of S_n the **triple integral** (or simply the integral) of f over C and denote it by

$$\int_C f, \int_C f(x, y, z)\,dV, \int_C f(x, y, z)\,dx\,dy\,dz$$

Suppose the rectangular parallelepiped C is the Cartesian product $[a, b] \times [c, d] \times [u, v]$. Then in analogy with functions of two variables we have six iterated integrals

$$\int_u^v \int_c^d \int_a^b f(x, y, z)\,dx\,dy\,dz, \quad \int_u^v \int_a^b \int_c^d f(x, y, z)\,dy\,dx\,dz,$$

$$\int_a^b \int_u^v \int_c^d f(x, y, z)\,dy\,dz\,dx, \ \ldots$$

The order of dx, dy, and dz indicates how the integration is carried out—for example, the first integral above stands for

$$\int_u^v \left[\int_c^d \left(\int_a^b f(x, y, z)\,dx\right) dy\right] dz$$

As in the two-variable case, if f is continuous then the iterated integrals are all equal. To complete the analogy with the double integral, we consider the problem of evaluating triple integrals over general regions. For bounded sets $W \subset R^3$ (that is, those sets which can be contained in some rectangular parallelepiped) such that ∂W has "volume zero" (we shall leave the definition of this to the student; see exercise 12), every continuous function $f: W \to \boldsymbol{R}$ is integrable via the same type of construction used in the two-dimensional case—that is, extend f to a function $\hat{f}$ which agrees with f on W and is zero outside of W. If B is a rectangular parallelepiped containing W we define

$$\int_W f(x, y, z)\, dV = \int_B \hat{f}(x, y, z)\, dV$$

As in the two-dimensional case this integral is independent of the choice of B. Fubini's Theorem is valid (by the same proof) and it reduces the computation of a triple integral to a threefold iterated integral.

In general, we shall restrict our attention to three-dimensional regions of types 1, 2, 3, and 4, defined as follows. A region W is of type 1 if it can be described as the set of all (x, y, z) such that

$$a \leqslant x \leqslant b, \quad \phi_2(x) \leqslant y \leqslant \phi_1(x), \quad \gamma_2(x, y) \leqslant z \leqslant \gamma_1(x, y) \qquad (1)$$

and where D is a two-dimensional region of type 1, $\gamma_i: D \to \boldsymbol{R}$, $i = 1, 2$, are continuous functions, and $\gamma_1(x, y) = \gamma_2(x, y)$ only for $(x, y) \in \partial D$. This last condition means that the surfaces $z = \gamma_1(x, y)$, $z = \gamma_2(x, y)$ intersect only for $(x, y) \in \partial D$.

A three-dimensional region is also of type 1 if it can be expressed as the set of all (x, y, z) such that

$$c \leqslant y \leqslant d, \quad \psi_2(y) \leqslant x \leqslant \psi_1(y), \quad \gamma_1(x, y) \leqslant z \leqslant \gamma_1(x, y) \qquad (2)$$

where D is a two-dimensional region of type 2 and $\gamma_i: D \to \boldsymbol{R}$ are as above. Figure 6-2 shows some examples of regions of type 1.

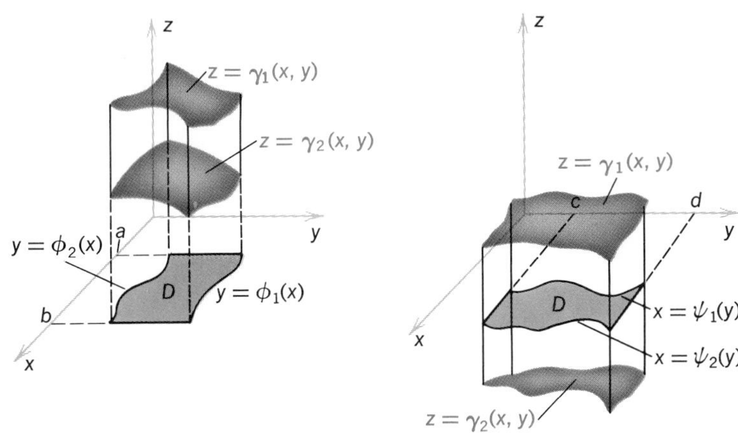

REGIONS OF TYPE 1

FIGURE 6-2

A region W is of type 2 if it can be expressed in the form (1) or (2) with the roles of x and z interchanged, and W is of type 3 if it can be expressed in the form (1) or (2) with y and z interchanged. A region W which is of type 1, 2, and 3 is said to be of type 4 (Figure 6-3). An example of a type 4 region is the ball of radius r, $x^2+y^2+z^2 \leqslant r^2$.

Suppose W is of type 1, then either

$$\int_W f(x,y,z)\,dV = \int_a^b \int_{\phi_2(x)}^{\phi_1(x)} \int_{\gamma_2(x,y)}^{\gamma_1(x,y)} f(x,y,z)\,dz\,dy\,dx \tag{3}$$

or

$$\int_W f(x,y,z)\,dV = \int_c^d \int_{\psi_2(y)}^{\psi_1(y)} \int_{\gamma_2(x,y)}^{\gamma_1(x,y)} f(x,y,z)\,dz\,dx\,dy \tag{4}$$

according to how W is defined. The proofs of (3) and (4) are the same as for the two-dimensional case and use Fubini's Theorem.

The formulas for regions of type 2 and 3 are similar and the reader should write them out in detail.

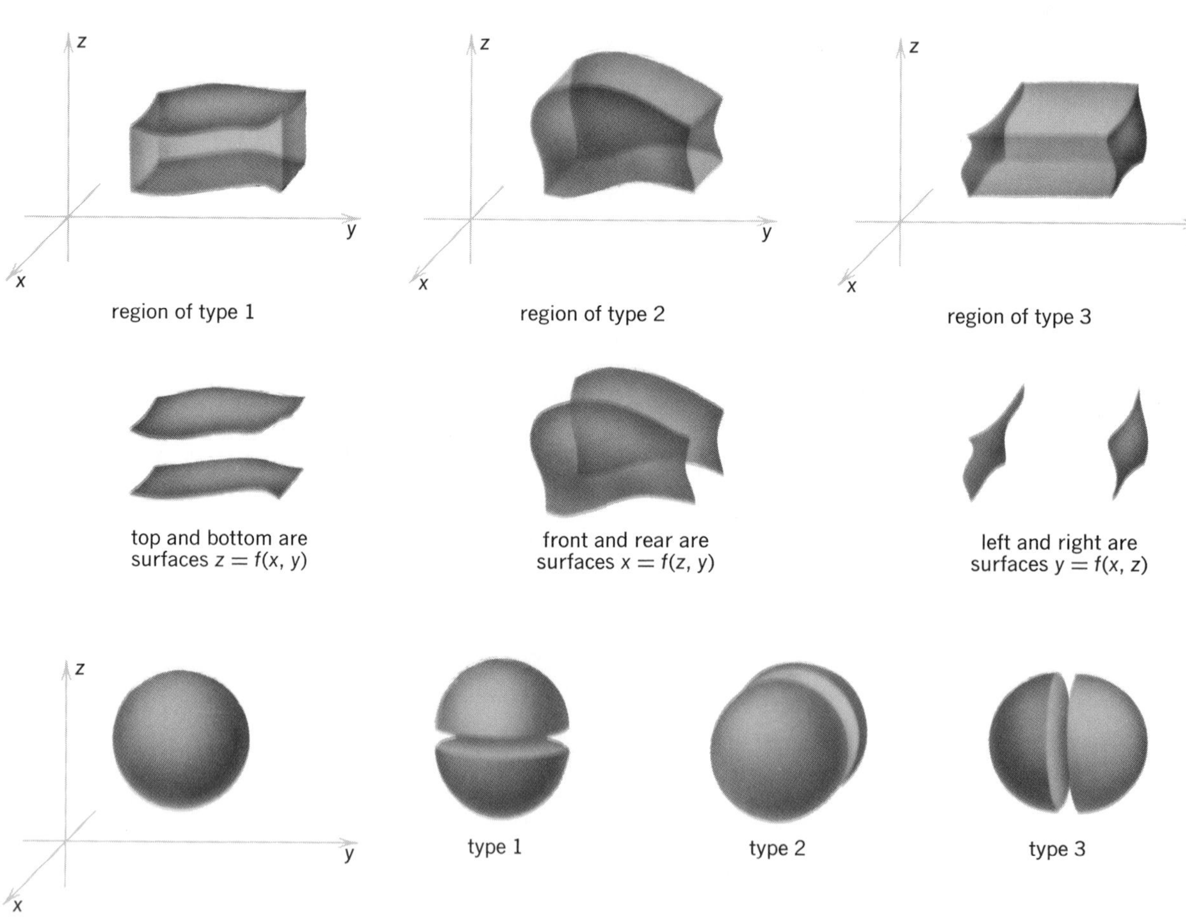

FIGURE 6-3

If $f(x,y,z)=1$ for all $(x,y,z)\in W$, then we obtain

$$\int_W dV = \int_W 1\, dV = \text{volume}\,(W)$$

In case W is of type 1 and formula (3) is applicable, we get

$$\begin{aligned}\text{volume}\,(W) &= \int_a^b \int_{\phi_2(x)}^{\phi_1(x)} \int_{\gamma_2(x,y)}^{\gamma_1(x,y)} dz\,dy\,dx \\ &= \int_a^b \int_{\phi_2(x)}^{\phi_1(x)} [\gamma_1(x,y)-\gamma_2(x,y)]\,dy\,dx\end{aligned}$$

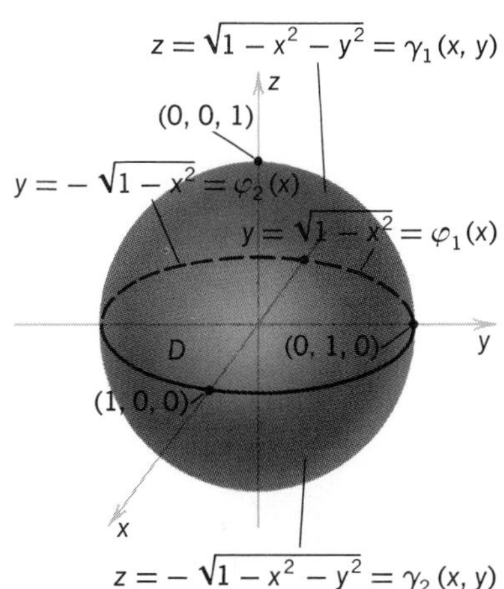

FIGURE 6-4

Example 1 Find $\int_W (1+y+z)\,dV$ where W is the unit ball $x^2+y^2+z^2 \leqslant 1$. The region W is of type 1 and we can describe it by the set of (x,y,z) with

$$-1 \leqslant x \leqslant 1, \quad -\sqrt{1-x^2} \leqslant y \leqslant \sqrt{1-x^2},$$
$$-\sqrt{1-x^2-y^2} \leqslant z \leqslant \sqrt{1-x^2-y^2}$$

(see Figure 6-4). Describing W is often the most difficult step, for once we have W expressed appropriately, it is relatively simple to evaluate the given triple integral using an equivalent iterated integral. In this case, we apply (3) to obtain

$$\int_W (1+y+z)\,dV = \int_{-1}^{1}\int_{-\sqrt{1-x^2}}^{\sqrt{1-x^2}}\int_{-\sqrt{1-x^2-y^2}}^{\sqrt{1-x^2-y^2}} (1+y+z)\,dz\,dy\,dx$$

First we hold y and x fixed and integrate with respect to z; this yields

$$\begin{aligned}&\int_{-1}^{1}\int_{-\sqrt{1-x^2}}^{\sqrt{1-x^2}} \left[z+yz+\frac{z^2}{2}\right]_{-\sqrt{1-x^2-y^2}}^{\sqrt{1-x^2-y^2}} dy\,dx \\ &= 2\int_{-1}^{1}\int_{-\sqrt{1-x^2}}^{\sqrt{1-x^2}} (\sqrt{1-x^2-y^2} + y\sqrt{1-x^2-y^2})\,dy\,dx \\ &= 2\int_{-1}^{1}\int_{-\sqrt{1-x^2}}^{\sqrt{1-x^2}} \sqrt{1-x^2-y^2}\,dy\,dx + 2\int_{-1}^{1}\int_{-\sqrt{1-x^2}}^{\sqrt{1-x^2}} y\sqrt{1-x^2-y^2}\,dy\,dx \qquad (5)\end{aligned}$$

Now the first integral in (5) is of the form $\int_{-a}^{a}\sqrt{a^2-y^2}\,dy$ since x is fixed. This integral represents the area of a semicircular region of radius a and so

$$\int_{-a}^{a}\sqrt{a^2-y^2}\,dy = \frac{a^2}{2}\pi$$

(Of course, we could have evaluated the integral directly, but this trick saves quite a bit of effort.) Thus

$$\int_{-\sqrt{1-x^2}}^{\sqrt{1-x^2}} \sqrt{1-x^2-y^2}\,dy = \frac{1-x^2}{2}\pi$$

and so

$$\begin{aligned}2\int_{-1}^{1}\int_{-\sqrt{1-x^2}}^{\sqrt{1-x^2}} \sqrt{1-x^2-y^2}\,dy\,dx &= 2\int_{-1}^{1} \pi\frac{1-x^2}{2}\,dx = \pi\int_{-1}^{1}(1-x^2)\,dx \\ &= \pi\left[x-\frac{x^3}{3}\right]_{-1}^{1} = \frac{4}{3}\pi\end{aligned}$$

For the second half of (5) we get

$$2\int_{-1}^{1}\int_{-\sqrt{1-x^2}}^{\sqrt{1-x^2}} y\sqrt{1-x^2-y^2}\,dy\,dx = -\int_{-1}^{1}\int_{-\sqrt{1-x^2}}^{\sqrt{1-x^2}} \sqrt{1-x^2-y^2}(-2y\,dy)\,dx$$

$$= -\frac{2}{3}\int_{-1}^{1}\left[(1-x^2-y^2)^{3/2}\right]_{-\sqrt{1-x^2}}^{\sqrt{1-x^2}}$$

$$= -\frac{2}{3}\int_{-1}^{1}(0)\,dx = 0$$

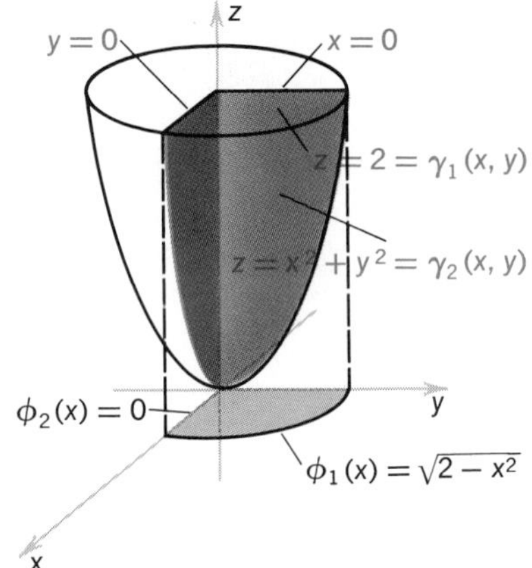

FIGURE 6-5

Combining both results, we have that

$$\int_W (1+y+z)\,dV = \tfrac{4}{3}\pi$$

Could you have guessed this directly (that is, observe $\int_W y\,dV = 0$ and $\int_W z\,dV = 0$ by symmetry, which leaves the volume of the sphere)?

Example 2 Let W be the region bounded by the planes $x=0$, $y=0$, $z=2$ and the surface $z=x^2+y^2$. Compute $\int_W x\,dx\,dy\,dz$.

The region W is sketched in Figure 6-5. To write this as a region of type 1 let $\gamma_1(x,y)=2$, $\gamma_2(x,y)=x^2+y^2$, $a=0$, $b=1$, $\phi_1(x)=\sqrt{2-x^2}$, and $\phi_2(x)=0$. Thus

$$\int_W x\,dx\,dy\,dz = \int_0^1\left[\int_0^{\sqrt{2-x^2}}\left(\int_{x^2+y^2}^{2} x\,dz\right)dy\right]dx$$

$$= \int_0^1\int_0^{\sqrt{2-x^2}} x(2-x^2+y^2)\,dy\,dx$$

$$= \int_0^1 x\left[(2-x^2)\sqrt{2-x^2} + \frac{(2-x^2)^{3/2}}{3}\right]dx$$

Notice that at this point the integral is somewhat tedious to compute. (It would require a trigonometric substitution which the reader should carry out.) The whole computation would be easier if it was done as a type 2 integral. We get

$$\int_W x\,dx\,dy\,dz = \int_0^2\left[\int_0^{\sqrt{z}}\left(\int_0^{\sqrt{z-y^2}} x\,dx\right)dy\right]dz$$

$$= \int_0^2\int_0^{\sqrt{z}}\left(\frac{z-y^2}{2}\right)dy\,dz$$

$$= \frac{1}{2}\int_0^2\left(z^{3/2} - \frac{z^{3/2}}{3}\right)dz$$

$$= \frac{1}{2}\int_0^2 \frac{2}{3}z^{3/2}\,dz$$

$$= \left[\frac{2}{15}z^{5/2}\right]_0^2 = \frac{2}{15}\cdot 2^{5/2} = \frac{8\sqrt{2}}{15}$$

EXERCISES

1. Evaluate $\int_W x^2\, dV$ where $W = [0,1] \times [0,1] \times [0,1]$.
2. Evaluate $\int_W e^{-xy} y\, dV$ where $W = [0,1] \times [0,1] \times [0,1]$.
3. Evaluate $\int_W x^2 \cos z\, dV$ where W is the region bounded by the planes $z = 0$, $z = \pi$, $y = 0$, $y = \pi$, $x = 0$, and $x + y = 1$.
4. Find the volume of the region bounded by $z = x^2 + 3y^2$ and $z = 9 - x^2$.
5. Evaluate $\int_0^1 \int_0^{2x} \int_{x+y}^{x^2+y^2} dz\, dy\, dx$ and sketch the region of integration.
6. Find the volume of the solid bounded by the surfaces $x^2 + 2y^2 = 2$, $z = 0$, and $x + y + 2z = 2$.
7. Change the order of integration in
$$\int_0^1 \int_0^x \int_0^y f(x,y,z)\, dz\, dy\, dx$$
to obtain five other forms of the answer. Sketch the region.
8. Find the volume of the region bounded by the surfaces $z^2 = x^2 + y^2$, $x^2 + y^2 + z^2 = 1$.
9. Let f be continuous and let B_ε be the ball of radius ε about a fixed point (x_0, y_0, z_0). Let $|B_\varepsilon|$ be the volume of B_ε. Prove
$$\lim_{\varepsilon \to 0} \frac{1}{|B_\varepsilon|} \int_{B_\varepsilon} f(x,y,z)\, dV = f(x_0, y_0, z_0)$$
10. Find the volume of the region bounded by the surfaces $z = x^2 + y^2$ and $z = 10 - x^2 - 2y^2$. Sketch.
11. Let W be symmetric in the xy plane: $(x,y,z) \in W$ implies $(x,y,-z) \in W$. Suppose $f(x,y,z) = -f(x,y,-z)$. Prove
$$\int_W f(x,y,z)\, dV = 0$$
12. Define volume zero by following the definition for area zero. Show that if A and B have volume zero, then $A \cup B$ and $A \cap B$ have volume zero.

7 CHANGING VARIABLES IN THE DOUBLE INTEGRAL

Given two regions D and D^* in $\boldsymbol{R}^2$ of type 1 or 2, a differentiable map T from D^* onto D [that is, $T(D^*) = D$—see §14.6], and any real-valued integrable function $f: D \to \boldsymbol{R}$, we would like to express $\int_D f(x,y)\, dA$ as an integral over D^* of the composite function $f \circ T$. In this section we shall see how to do this.

Assume the region D^* is a subset of $\boldsymbol{R}^2$ of type 1 with the coordinate variables designated by (u,v). Furthermore, assume that D is a subset of type 1 of the xy plane. The map T is given by two coordinate functions

$$T(u,v) = (x(u,v), y(u,v)) \quad \text{for} \quad (u,v) \in D^*$$

Now, as a first guess one might conjecture that

$$\int_D f(x,y)\,dx\,dy \stackrel{?}{=} \int_{D^*} f(x(u,v),y(u,v))\,du\,dv \tag{1}$$

where $f\circ T(u,v)=f(x(u,v),y(u,v))$ is the composite function defined on D^*. However, if we consider the function $f\colon D\to \boldsymbol{R}^2$ where $f(x,y)=1$, then equation (1) would imply

$$A(D) = \int_D dx\,dy \stackrel{?}{=} \int_{D^*} du\,dv = A(D^*) \tag{2}$$

It is easy to see that (2) will hold only for a few special cases and not for a general map T. For example, define T by $T(u,v)=(-u^2+4u,v)$. (This example was considered in the exercises of §14.9.) Restrict T to the unit square $D^*=[0,1]\times[0,1]$ in the uv plane (see Figure 7-1). Then T takes D^* onto $D=[0,3]\times[0,1]$. Clearly $A(D)\neq A(D^*)$ and so formula (2) is not valid.

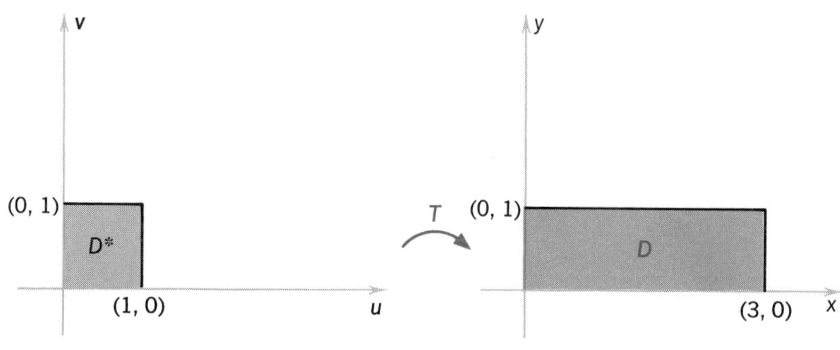

FIGURE 7-1

What is needed here is a measure of how a transformation $T\colon\boldsymbol{R}^2\mapsto\boldsymbol{R}^2$ distorts the area of a region. This is given by the **Jacobian** determinant.

Definition Let $T\colon D^*\subset\boldsymbol{R}^2\to\boldsymbol{R}^2$ be a C^1 transformation given by $x=x(u,v)$ and $y=y(u,v)$. The **Jacobian** of T, written $\partial(x,y)/\partial(u,v)$, is the determinant

$$\begin{vmatrix} \dfrac{\partial x}{\partial u} & \dfrac{\partial x}{\partial v} \\[2ex] \dfrac{\partial y}{\partial u} & \dfrac{\partial y}{\partial v} \end{vmatrix}$$

Example 1 The function $\boldsymbol{R}^2\to\boldsymbol{R}^2$ which transforms polar coordinates into Cartesian coordinates is given by

$$x = r\cos\theta, \qquad y = r\sin\theta$$

and its Jacobian is

$$\frac{\partial(x,y)}{\partial(r,\theta)} = \begin{vmatrix} \cos\theta & -r\sin\theta \\ \sin\theta & r\cos\theta \end{vmatrix} = r(\cos^2\theta+\sin^2\theta) = r$$

If we make suitable restrictions on the function T, we can show that the area of $D = T(D^*)$ is obtained by integrating the absolute value of the Jacobian $\partial(x,y)/\partial(u,v)$ over D^*; that is, we have the equations

$$A(D) = \int_D dx\,dy = \int_{D^*} \left|\frac{\partial(x,y)}{\partial(u,v)}\right| du\,dv \tag{3}$$

It is not easy to prove rigorously the assertion (3) that the Jacobian determinant is a measure of how a transformation distorts area, however, we do remark here that the Jacobian $\partial(x,y)/\partial(u,v)$ is equal in absolute value to the area of the parallelogram with adjacent sides the vectors

$$\boldsymbol{T}_u = \frac{\partial x}{\partial u}\boldsymbol{i} + \frac{\partial y}{\partial u}\boldsymbol{j}, \qquad \boldsymbol{T}_v = \frac{\partial x}{\partial v}\boldsymbol{i} + \frac{\partial y}{\partial v}\boldsymbol{j}$$

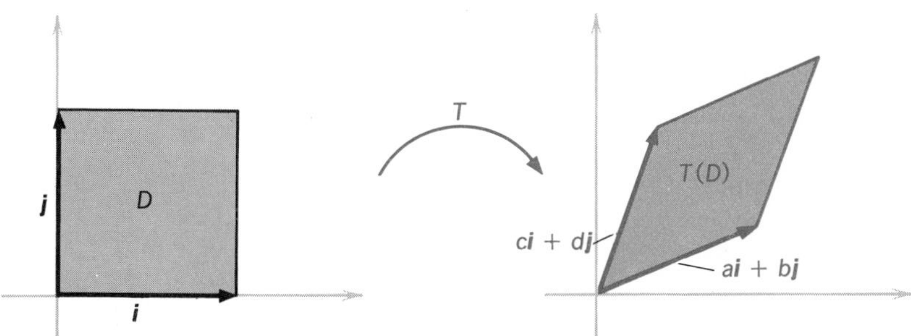

FIGURE 7-2

This follows from the results in §13.3 that the area of the parallelogram spanned by $\boldsymbol{ai}+\boldsymbol{bj}$ and $\boldsymbol{ci}+\boldsymbol{dj}$ is the absolute value of the determinant

$$\begin{vmatrix} a & b \\ c & d \end{vmatrix}$$

If T is given by $T(u,v) = (au+cv, bu+dv)$ where a, b, c, d are constants, then

$$\boldsymbol{T}_u = \boldsymbol{ai} + \boldsymbol{bj}, \qquad \boldsymbol{T}_v = \boldsymbol{ci} + \boldsymbol{dj}$$

Now T takes the unit square D spanned by $\boldsymbol{i}$ and $\boldsymbol{j}$ into the parallelogram $T(D)$ spanned by $\boldsymbol{ai}+\boldsymbol{bj}$ and $\boldsymbol{ci}+\boldsymbol{dj}$ (see Figure 7-2). Therefore

$$\text{Area } T(D) = \begin{vmatrix} a & b \\ c & d \end{vmatrix}$$

Let us consider an example.

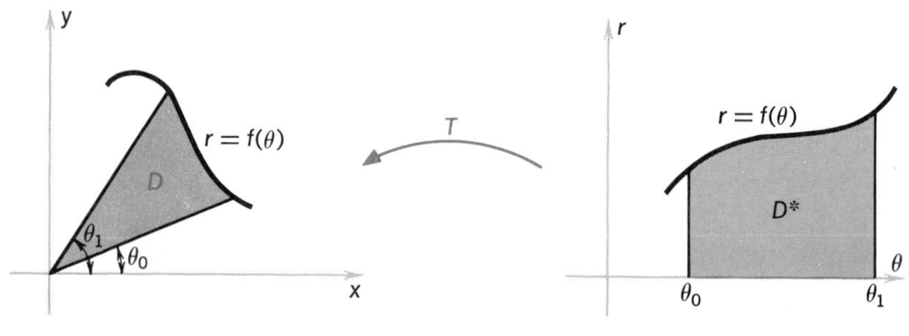

FIGURE 7-3

Example 2 Let the elementary region D in the xy plane be bounded by the graph of a polar equation $r = f(\theta)$ where $\theta_0 \leqslant \theta \leqslant \theta_1$ and $f(\theta) \geqslant 0$ (see Figure 7-3). In the θr plane we consider the type 1 region D^* where $\theta_0 \leqslant \theta \leqslant \theta_1$ and $0 \leqslant r \leqslant f(\theta)$. Under the transformation $x = r\cos\theta$, $y = r\sin\theta$, the region D^* is carried onto the region D. We have

$$\int_{D^*} \left|\frac{\partial(x,y)}{\partial(\theta,r)}\right| dr\,d\theta = \int_{D^*} r\,dr\,d\theta = \int_{\theta_0}^{\theta_1} \left[\int_0^{f(\theta)} r\,dr\right] d\theta$$

$$= \int_{\theta_0}^{\theta_1} \left[\frac{r^2}{2}\right]_0^{f(\theta)} d\theta = \int_{\theta_0}^{\theta_1} \frac{[f(\theta)^2]}{2}\,d\theta$$

But the last integral is equal to the area of D (see §8.8). Thus, we have verified that

$$\int_{D^*} \left|\frac{\partial(x,y)}{\partial(r,\theta)}\right| dr\,d\theta = A(D)$$

At this point, it would be useful for the reader to review the derivation of the area formula in §8.8. There, only the θ axis is partitioned. The partitioning of both r and θ axes leads to a grid as shown in Figure 7-4.

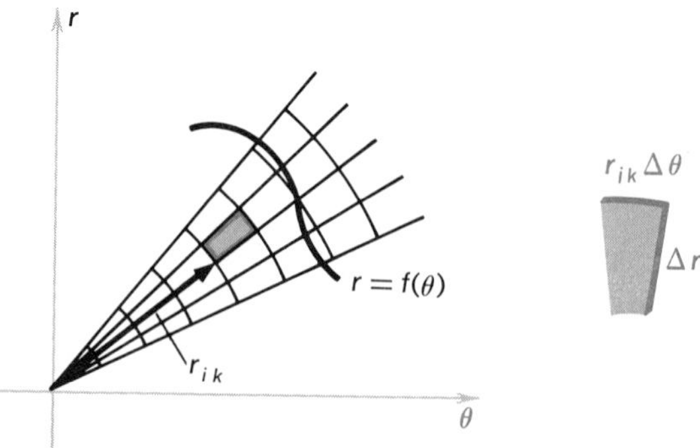

FIGURE 7-4

The jkth "polar rectangle" in the grid has area approximately equal to $r_{jk}\,\Delta r\,\Delta\theta$. (For n large, the jkth polar rectangle will look like a rectangle with sides of length $r_{jk}\,\Delta\theta$ and Δr.) This provides some insight into why we say the "area element $\Delta x\,\Delta y$" is transformed into the "area element $r\,\Delta r\,\Delta\theta$."

Now we are ready to state the Change of Variables Theorem for the double integral. First, however, let us recall the result for one-variable calculus; we have

$$\int_a^b f(x(u))\frac{dx}{du}\,du = \int_{x(a)}^{x(b)} f(x)\,dx$$

where f is continuous and $u \mapsto x(u)$ is continuously differentiable on $[a, b]$. It is not this formula which generalizes to multiple integrals but a slightly restricted version of it.

Suppose that the C^1 function $u \mapsto x(u)$ is one-to-one on $[a, b]$. Thus, we must have either $dx/du \geqslant 0$ on $[a, b]$ or $dx/du \leqslant 0$ on $[a, b]$ (for if dx/du is positive and then negative, the function $x = x(u)$ rises and then falls and thus is not one-to-one; a similar statement applies if dx/du is negative then positive). Let I^* denote the interval $[a, b]$, and let I denote the closed interval with endpoints $x(a)$ and $x(b)$. {Thus, $I = [x(a), x(b)]$ if $u \mapsto x(u)$ is increasing and $I = [x(b), x(a)]$ if $u \mapsto x(u)$ is decreasing.} With these notations we have

$$\int_{I^*} f(x(u))\left|\frac{dx}{du}\right| du = \int_I f(x)\,dx$$

This is the formula which generalizes to double integrals—I^* becomes D^*, I becomes D, and $|dx/du|$ is replaced by $|\partial(x, y)/\partial(u, v)|$. Let us state the result formally.

Theorem 1 Change of Variables for Double Integrals Let $T\colon D^* \to D$ be C^1 and suppose that T is one-to-one on D^*. Furthermore, suppose that $D = T(D^*)$. Then for any integrable function $f\colon D \to \mathbf{R}$, we have

$$\int_D f(x, y)\,dx\,dy = \int_{D^*} f(x(u, v), y(u, v))\left|\frac{\partial(x, y)}{\partial(u, v)}\right| du\,dv$$

Example 3 Let P be the parallelogram bounded by $y = 2x$, $y = 2x - 2$, $y = x$, and $y = x + 1$ (see Figure 7-5). Evaluate $\int_P xy\,dx\,dy$ by making the change of variables

$$x = u - v, \qquad y = 2u - v$$

that is, $T(u, v) = (u - v, 2u - v)$.

The transformation T takes the rectangle P^* bounded by $v = 0$, $v = -2$, $u = 0$, $u = 1$ onto P. Moreover,

$$\left|\frac{\partial(x, y)}{\partial(u, v)}\right| = 1\begin{vmatrix} 1 & -1 \\ 2 & -1 \end{vmatrix} = 1$$

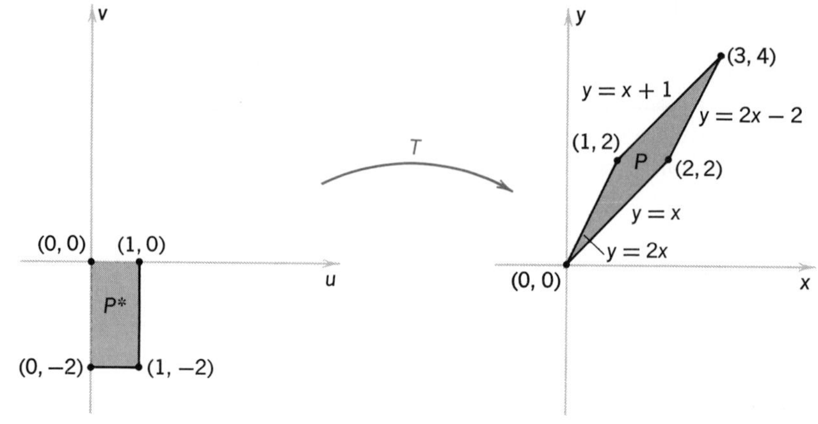

FIGURE 7-5

Therefore

$$\begin{aligned}\int_P xy\,dx\,dy &= \int_{P^*}(u-v)(2u-v)\,du\,dv = \int_{-2}^{0}\int_0^1 (2u^2-3vu+v^2)\,du\,dv\\ &= \int_{-2}^{0}\left[\frac{2}{3}u^3 - \frac{3u^2v}{2} + v^2u\right]_0^1 dv = \int_{-2}^{0}\left[\frac{2}{3} - \frac{3}{2}v + v^2\right] dv\\ &= \left[\frac{2}{3}v - \frac{3}{4}v^2 + \frac{v^3}{3}\right]_{-2}^{0} = -\left[\frac{2}{3}(-2) - 3 - \frac{8}{3}\right]\\ &= -\left[-\frac{12}{3} - 3\right] = 7\end{aligned}$$

Example 4 Evaluate $\int_D \ln(x^2+y^2)\,dx\,dy$, where D is the region in the first quadrant lying between the arcs of the circles

$$x^2 + y^2 = a^2, \qquad x^2 + y^2 = b^2 \qquad (0 < a < b)$$

These circles have equations $r = a$ and $r = b$ in polar coordinates. Therefore the transformation

$$x = r\cos\theta, \qquad y = r\sin\theta$$

sends the rectangle D^* given by $a \leqslant r \leqslant b$, $0 \leqslant \theta \leqslant \frac{1}{2}\pi$ onto the region D. This transformation is one-to-one on D^* and so, by Theorem 1, we have

$$\int_D \ln(x^2+y^2)\,dx\,dy = \int_{D^*} \ln r^2 \left|\frac{\partial(x,y)}{\partial(r,\theta)}\right| dr\,d\theta$$

Now $\partial(x,y)/\partial(r,\theta) = r$ as we have seen before, hence the right-hand integral becomes

$$\int_a^b \int_0^{\pi/2} r\ln r^2\,d\theta\,dr = \frac{\pi}{2}\int_a^b r\ln r^2\,dr = \frac{\pi}{2}\int_a^b 2r\ln r\,dr$$

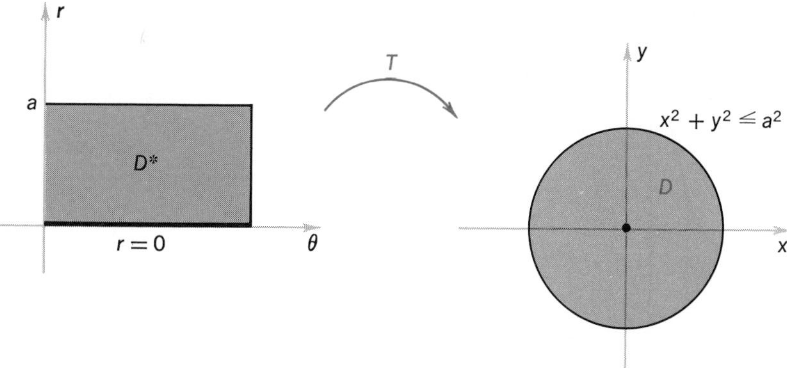

FIGURE 7-6

Applying integration by parts, we obtain the result

$$\frac{\pi}{2}\int_a^b 2r\ln r\,dr = \frac{\pi}{2}\left(a^2\ln a - b^2\ln b - \frac{1}{2}(a^2-b^2)\right)$$

Suppose we consider the rectangle D^* for $0 \leqslant \theta \leqslant 2\pi$, $0 \leqslant r \leqslant a$ in the θr plane. Then the transformation T given by $T(\theta, r) = (r\cos\theta, r\sin\theta)$ takes D^* onto the disk D with equation $x^2+y^2 \leqslant a^2$ in the xy plane. This transformation represents the change from Cartesian coordinates to polar coordinates. However, T does not satisfy the requirements of the Change of Variables Theorem since it is not one-to-one on D^*: T sends all points with $r=0$ to $(0,0)$ (Figure 7-6). However, the Change of Variables Theorem is still valid in this case. Basically, the reason for this is that the set of points in $\boldsymbol{R}$ for which T is not one-to-one is an "edge" of D^*; this set is of area zero and therefore can be "neglected." In summary, the formula

$$\int_D f(x,y)\,dx\,dy = \int_{D^*} f(r\cos\theta, r\sin\theta)\,r\,dr\,d\theta \tag{4}$$

is valid when T sends D^* onto D in a one-to-one fashion except possibly for points on the edges of D^*. Example 2 provides a simple example of this for the case where $f(x,y)$ is constantly 1. Below we consider a more challenging example.

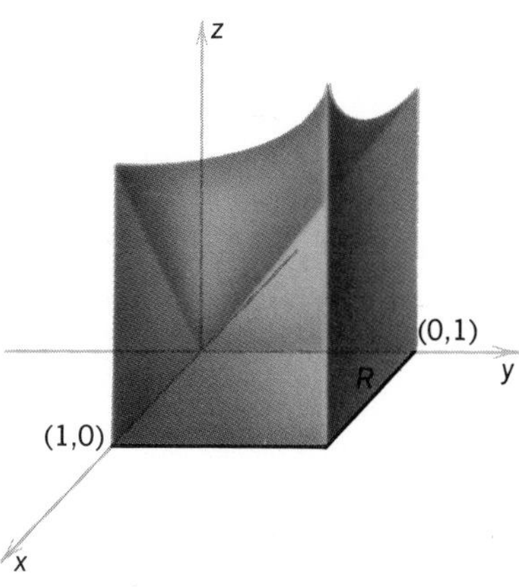

FIGURE 7-7

Example 5 Evaluate $\int_R \sqrt{x^2+y^2}\,dx\,dy$ where $R = [0,1]\times[0,1]$. This double integral is equal to the volume of the three-dimensional region in Figure 7-7. To apply Theorem 1 with polar coordinates, we refer to Figure 7-8. We see that R is the image under $T(\theta,r) = (r\cos\theta, r\sin\theta)$ of the region $D^* = D_1^* \cup D_2^*$ where for D_1^* we have $0 \leqslant \theta \leqslant \frac{1}{4}\pi$, $0 \leqslant r \leqslant \sec\theta$ and for D_2^* we have $\frac{1}{4}\pi \leqslant \theta \leqslant \frac{1}{2}\pi$, $0 \leqslant r \leqslant \csc\theta$. The transformation T sends D_1^* onto a triangle T_1 and D_2^* onto a triangle T_2. The transformation T is one-to-one except when $r=0$, so we can apply Theorem 1. By symmetry,

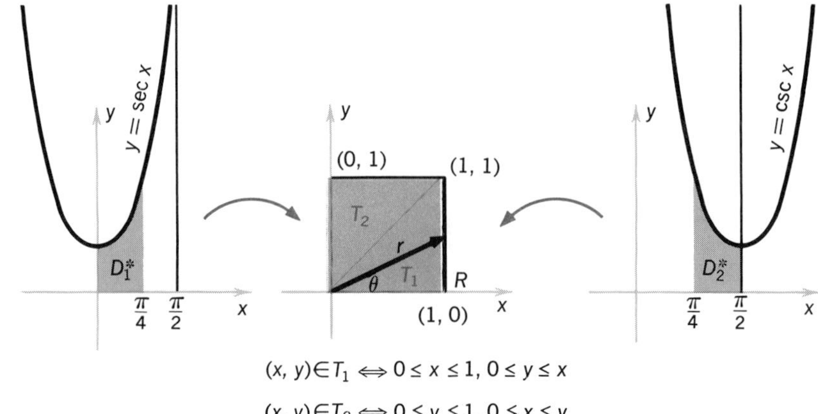

FIGURE 7-8

we see that

$$\int_R \sqrt{x^2+y^2}\,dx\,dy = 2\int_{T_1} \sqrt{x^2+y^2}\,dx\,dy$$

Applying formula (4), we obtain

$$\int_{T_1} \sqrt{x^2+y^2}\,dx\,dy = \int_{D_1^*} \sqrt{r^2}\,r\,dr\,d\theta = \int_{D_1^*} r^2\,dr\,d\theta$$

Next we use Fubini's Theorem

$$\int_{D_1^*} r^2\,dr\,d\theta = \int_0^{\pi/4}\left[\int_0^{\sec\theta} r^2\,dr\right]d\theta = \tfrac{1}{3}\int_0^{\pi/4} \sec^3\theta\,d\theta$$

We consult a table of integrals (see Appendix) to find $\int \sec^n x\,dx$; with $n = 3$, we have

$$\int_0^{\pi/4} \sec^3\,d\theta = \left[\frac{\sec\theta\tan\theta}{2}\right]_0^{\pi/4} + \frac{1}{2}\int_0^{\pi/4} \sec\theta\,d\theta$$
$$= \frac{\sqrt{2}}{2} + \frac{1}{2}\int_0^{\pi/4} \sec\theta\,d\theta$$

Consulting the table again for $\int \sec x\,dx$, we find

$$\tfrac{1}{2}\int_0^{\pi/4} \sec\theta\,d\theta = \tfrac{1}{2}\Big[\ln|\sec\theta+\tan\theta|\Big]_0^{\pi/4} = \tfrac{1}{2}\ln(1+\sqrt{2})$$

Combining these results and recalling the factor $\frac{1}{3}$, we obtain

$$\int_{D_1^*} r^2\,dr\,d\theta = \frac{1}{3}\left(\frac{\sqrt{2}}{2} + \frac{1}{2}\ln(1+\sqrt{2})\right) = \frac{1}{6}(\sqrt{2} + \ln(1+\sqrt{2}))$$

Multiplying by 2, we obtain the answer

$$\int_R \sqrt{x^2+y^2}\,dx\,dy = \tfrac{1}{3}(\sqrt{2} + \ln(1+\sqrt{2}))$$

EXERCISES

1. Let D be the unit circle. Evaluate
$$\int_D \exp(x^2+y^2)\,dx\,dy$$
by making the appropriate change of variables. [Hint: Use polar coordinates.]
2. Let D be the region $0 \leqslant y \leqslant x$ and $0 \leqslant x \leqslant 1$. Evaluate
$$\int_D (x+y)\,dx\,dy$$
by making the change of variables $x = u+v$, $y = u-v$. Check your answer by evaluating the iterated integral.
3. Let D^* be a region of type 1 in the uv plane bounded by
$$v = g(u), \qquad v = h(u) \leqslant g(u)$$
for $a \leqslant u \leqslant b$. Let $T: \mathbf{R}^2 \to \mathbf{R}^2$ be the transformation given by
$$x = u, \qquad y = \psi(u,v)$$
Assume $T(D^*) = D$ is a region of type 1; show that if $f: D^* \to \mathbf{R}$ is continuous then
$$\int_D f(x,y)\,dx\,dy = \int_{D^*} f(u,\psi(u,v)) \left|\frac{\partial\psi}{\partial v}\right| du\,dv$$
4. Find the area inside the curve $r = 1+\sin\theta$.
5. (a) Express $\int_0^1 \int_0^{x^2} xy\,dy\,dx$ as an integral over the triangle D^* which is the set of (u,v) where $0 \leqslant u \leqslant 1$, $0 \leqslant v \leqslant u$. [Hint: Find a one-to-one mapping T of D^* onto the given region of integration.]
   (b) Evaluate this integral directly and over D^*.

*6. Let D be the region bounded by $x^{3/2}+y^{3/2} = a^{3/2}$, for $x \geqslant 0$, $y \geqslant 0$, and the coordinate axes $x = 0$, $y = 0$. Express $\int_D f(x,y)\,dx\,dy$ as an integral over the triangle D^* which is the set of points $0 \leqslant u \leqslant a$, $0 \leqslant v \leqslant a-u$.

7. Let D be the unit disc. Express $\int_D (1+x^2+y^2)^{3/2}\,dx\,dy$ as an integral over the rectangle $[0, 2\pi] \times [0, 1]$ and evaluate.

8 CYLINDRICAL AND SPHERICAL COORDINATES

We have seen that both the Cartesian and polar coordinate systems are useful in working out integration problems. In many instances, a change of coordinates can greatly simplify the form of a problem. With this in mind, we introduce two new coordinate systems for $\mathbf{R}^3$ which are particularly well suited for certain types of problems.

The standard way to represent a point in $\mathbf{R}^3$ is by coordinates (x, y, z). We can also represent this point using **cylindrical coordinates** (r, θ, z) where

$$x = r\cos\theta, \qquad y = r\sin\theta, \qquad z = z \tag{1}$$

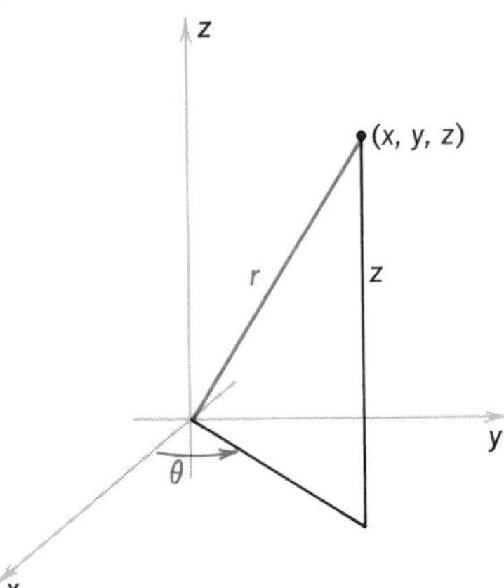

FIGURE 8-1

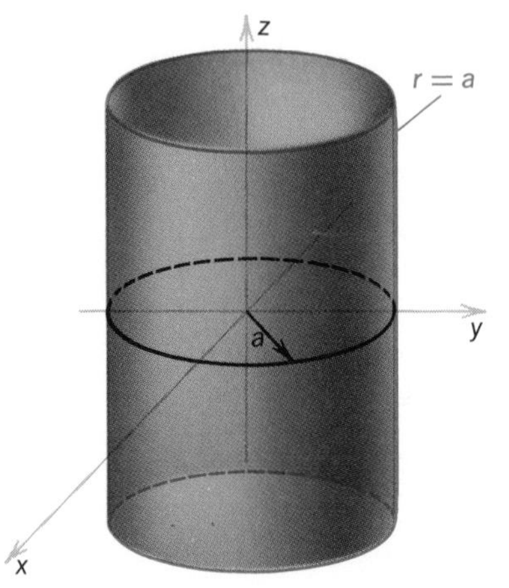

FIGURE 8-2

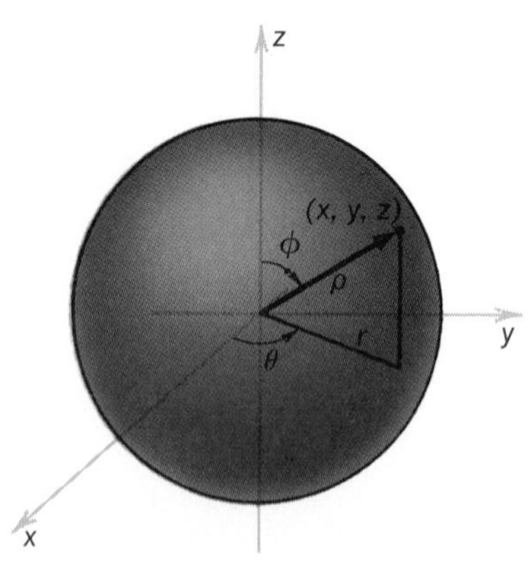

FIGURE 8-3

and

$$r^2 = x^2 + y^2, \qquad \theta = \tan^{-1} y/x$$

(see Figure 8-1).

In other words, for any point (x, y, z) we represent its first and second coordinates in terms of polar coordinates and leave the third coordinate unchanged. The formula (1) shows that given (r, θ, z) the coordinate (x, y, z) is completely determined and vice versa if we restrict $\theta \in [0, 2\pi[$ (sometimes the range $]-\pi, \pi]$ is convenient), and require that $r \geqslant 0$.

To see why we use the term cylindrical coordinates, note that if $0 \leqslant \theta \leqslant 2\pi$, $-\infty < z < \infty$, and $r = a$, some constant, then the locus of these points is a cylinder of radius a (see Figure 8-2).

Cylindrical coordinates are not the only possible generalization of polar coordinates to three dimensions. Recall that in two dimensions the magnitude of the vector $x\boldsymbol{i}+y\boldsymbol{j}$ ($\sqrt{x^2+y^2}$) is the r in the polar coordinate system. For cylindrical coordinates, the length of the vector $x\boldsymbol{i}+y\boldsymbol{j}+z\boldsymbol{k}$,

$$\rho = \sqrt{x^2+y^2+z^2}$$

is not one of the coordinates of the system—we use only the magnitude $r = \sqrt{x^2+y^2}$, the angle θ, and the "height" z.

We now rectify this by introducing the **spherical coordinate** system, which contains ρ as a coordinate. Spherical coordinates are often useful for problems which possess spherical symmetry (symmetry about a point) while cylindrical coordinates can be applied when cylindrical symmetry (symmetry about a line) is involved.

Given a point $(x, y, z) \in \boldsymbol{R}^3$, let

$$\rho = \sqrt{x^2+y^2+z^2}$$

and represent x and y by polar coordinates in the xy plane

$$x = r\cos\theta, \qquad y = r\sin\theta \tag{2}$$

where $r = \sqrt{x^2+y^2}$, $\theta = \tan^{-1} y/x$. The coordinate z is given by

$$z = \rho\cos\phi$$

where ϕ is the angle the radius vector $\boldsymbol{v} = x\boldsymbol{i}+y\boldsymbol{j}+z\boldsymbol{k}$ makes with the z axis, in the plane containing the vector $\boldsymbol{v}$ and the z axis (see Figure 8-3). Using the dot product we can express ϕ by

$$\cos\phi = \frac{\boldsymbol{v}\cdot\boldsymbol{k}}{|\boldsymbol{v}|}$$

or

$$\phi = \cos^{-1}\frac{\boldsymbol{v}\cdot\boldsymbol{k}}{|\boldsymbol{v}|}$$

We take as our coordinates the quantities ρ, θ, ϕ. Since

$$r = \rho\sin\phi$$

we can use (2) to find x, y, z in terms of the spherical coordinates ρ, θ, ϕ; we have the spherical change of coordinates s: $(\rho, \theta, \phi) \mapsto (x, y, z)$;

$$x = \rho\sin\phi\cos\theta, \qquad y = \rho\sin\phi\sin\theta, \qquad z = \rho\cos\phi \tag{3}$$

Note that in spherical coordinates the equation of the sphere of radius a takes on the particularly simple form

$$\rho = a$$

Example 1 Express (a) the surface $xz = 1$, (b) the surface $x^2 + y^2 - z^2 = 1$ in spherical coordinates.

We have from (3), $x = \rho \sin\phi \cos\theta$, $z = \rho\cos\phi$, so the surface (a) is

$$\rho^2 \sin\phi \cos\theta \cos\phi = 1$$

For part (b) we can write

$$x^2 + y^2 - z^2 = x^2 + y^2 + z^2 - 2z^2 = \rho^2 - 2\rho^2 \cos^2\phi$$

so the surface is $\rho^2(1 - 2\cos^2\phi) = 1 = \rho^2\cos(2\phi)$.

Spherical and cylindrical coordinates are very useful in conjunction with the Change of Variables formula for triple integrals which we state below. First we must define the Jacobian of a transformation from $\boldsymbol{R}^3$ to $\boldsymbol{R}^3$—it is a simple extension of the two-variable case.

Definition Let $T: W \subset \boldsymbol{R}^3 \to \boldsymbol{R}^3$ be a C^1 function defined by $x = x(u, v, w)$, $y = y(u, v, w)$, $z = z(u, v, w)$. Then the **Jacobian** of T, written $\partial(x, y, z)/\partial(u, v, w)$, is the determinant

$$\begin{vmatrix} \dfrac{\partial x}{\partial u} & \dfrac{\partial x}{\partial v} & \dfrac{\partial x}{\partial w} \\ \dfrac{\partial y}{\partial u} & \dfrac{\partial y}{\partial v} & \dfrac{\partial y}{\partial w} \\ \dfrac{\partial z}{\partial u} & \dfrac{\partial z}{\partial v} & \dfrac{\partial z}{\partial w} \end{vmatrix}$$

The above determinant is equal in absolute value to the volume of the parallelepiped determined by the vectors

$$\boldsymbol{T}_u = \frac{\partial x}{\partial u}\boldsymbol{i} + \frac{\partial y}{\partial u}\boldsymbol{j} + \frac{\partial z}{\partial u}\boldsymbol{k}$$

$$\boldsymbol{T}_v = \frac{\partial x}{\partial v}\boldsymbol{i} + \frac{\partial y}{\partial v}\boldsymbol{j} + \frac{\partial z}{\partial v}\boldsymbol{k}$$

$$\boldsymbol{T}_w = \frac{\partial x}{\partial w}\boldsymbol{i} + \frac{\partial y}{\partial w}\boldsymbol{j} + \frac{\partial z}{\partial w}\boldsymbol{k}$$

In this case the Jacobian measures how the transformation T distorts volume. Hence for volume (triple) integrals, we have the Change of Variables formula

$$\begin{aligned} &\int_D f(x, y, z)\, dx\, dy\, dz \\ &= \int_{D^*} f(x(u, v, w), y(u, v, w), z(u, v, w)) \left|\frac{\partial(x, y, z)}{\partial(u, v, w)}\right| du\, dv\, dw \qquad (4) \end{aligned}$$

where D^* is the region in uvw space corresponding to D in xyz space under a coordinate change $T:(u,v,w)\to(x(u,v,w),\, y(u,v,w),\, z(u,v,w))$. Again, if formula (4) is to hold, T should be C^1 and one-to-one except on a set of volume zero.

Let us apply formula (4) to cylindrical and spherical coordinates. First we compute the Jacobian for the map defining the change to cylindrical coordinates. Since

$$x = r\cos\theta, \qquad y = r\sin\theta, \qquad z = z$$

we have

$$\frac{\partial(x,y,z)}{\partial(r,\theta,z)} = \begin{vmatrix} \cos\theta & -r\sin\theta & 0 \\ \sin\theta & r\sin\theta & 0 \\ 0 & 0 & 1 \end{vmatrix} = r$$

Thus, we obtain the formula

$$\int_D f(x,y,z)\,dx\,dy\,dz = \int_{D^*} f(r,\theta,z)\,r\,dr\,d\theta\,dz \tag{5}$$

Next let us consider the spherical coordinate system. Since

$$x = \rho\sin\phi\cos\theta, \qquad y = \rho\sin\phi\sin\theta, \qquad z = \rho\cos\phi$$

we have

$$\begin{aligned}
\frac{\partial(x,y,z)}{\partial(\rho,\theta,\phi)} &= \begin{vmatrix} \sin\phi\cos\theta & -\rho\sin\phi\sin\theta & \rho\cos\phi\cos\theta \\ \sin\phi\sin\theta & \rho\sin\phi\cos\theta & \rho\cos\phi\sin\theta \\ \cos\phi & 0 & -\rho\sin\phi \end{vmatrix} \\
&= \sin\phi\cos\theta \begin{vmatrix} \rho\sin\phi\cos\theta & \rho\cos\phi\sin\theta \\ 0 & -\rho\sin\phi \end{vmatrix} \\
&\qquad + \rho\sin\phi\sin\theta \begin{vmatrix} \sin\phi\sin\theta & \rho\cos\phi\sin\theta \\ \cos\phi & -\rho\sin\phi \end{vmatrix} \\
&\qquad + \rho\cos\phi\cos\theta \begin{vmatrix} \sin\phi\sin\theta & \rho\sin\phi\cos\theta \\ \cos\phi & 0 \end{vmatrix} \\
&= \sin\phi\cos\theta(-\rho^2\sin^2\phi\cos\theta) \\
&\qquad + \rho\sin\phi\sin\theta(-\rho\sin^2\phi\sin\theta - \rho\cos^2\phi\sin\theta) \\
&\qquad + \rho\cos\phi\cos\theta(-\rho\sin\phi\cos\phi\cos\theta) \\
&= -\rho^2\sin^3\phi\cos^2\theta - \rho^2\sin^3\phi\sin^2\theta - \rho^2\cos^2\phi\sin\phi\sin^2\theta \\
&\qquad - \rho^2\sin\phi\cos^2\phi\cos^2\theta \\
&= -\rho^2\sin^3\phi(\cos^2\theta+\sin^2\theta) - \rho^2\sin\phi\cos^2\phi(\sin^2\theta+\cos^2\theta) \\
&= -\rho^2\sin^3\phi - \rho^2\sin\phi\cos^2\phi = -\rho^2\sin\phi(\sin^2\theta+\cos^2\phi) \\
&= -\rho^2\sin\phi
\end{aligned}$$

Thus we arrive at the formula

$$\int_D f(x,y,z)\,dx\,dy\,dz = \int_{D^*} f(\rho,\theta,\phi)\,\rho^2 \sin\phi\; d\rho\, d\theta\, d\phi \qquad (6)$$

In order to prove the validity of formula (6), we must show that the transformation S on the set D is one-to-one except on a set of volume zero. We do this in the next example.

Example 2 Evaluate $\int_D \exp(x^2+y^2+z^2)^{3/2}\,dV$ where D is the unit ball in $\boldsymbol{R}^3$.

First note that we cannot integrate this function using Fubini's Theorem and iterated integrals (try it!). Hence, let us try a change of variables. The transformation S into spherical coordinates seems appropriate since then the entire quantity $x^2+y^2+z^2$ can be replaced by one variable, namely, ρ^2. If D^* is the region such that

$$0 \leqslant \rho \leqslant 1, \quad 0 \leqslant \theta \leqslant 2\pi, \quad 0 \leqslant \phi \leqslant \pi$$

then $S: D^* \to D$. Assume for the moment that S is one-to-one except on a set $A \subset D^*$ of volume zero. Consequently we may apply formula (6) and write

$$\int_D \exp(x^2+y^2+z^2)^{3/2}\,dV = \int_{D^*} \rho^2 e^{\rho^3} \sin\phi\; dV$$

Using Fubini's Theorem, we see that this integral equals the iterated integral

$$\begin{aligned} &\int_0^1 \int_0^\pi \int_0^{2\pi} e^{\rho^3} \rho^2 \sin\phi\; d\theta\, d\phi\, d\rho \\ &= 2\pi \int_0^1 \int_0^\pi e^{\rho^3} \rho^2 \sin\phi\; d\phi\, d\rho \\ &= -2\pi \int_0^1 \rho^2 e^{\rho^3} \Big[\cos\phi\Big]_0^\pi d\rho = 4\pi \int_0^1 e^{\rho^3} \rho^2\, d\rho \\ &= \tfrac{4}{3}\pi \int_0^1 e^{\rho^3}(3\rho^2)\, d\rho = \Big[\tfrac{4}{3}\pi e^{\rho^3}\Big]_0^1 = \tfrac{4}{3}\pi(e-1) \end{aligned}$$

We now verify that S is one-to-one except on a set of volume zero. (The reader may omit this discussion on first reading.) Let $A_k, k = 1, \ldots, 4$ be subsets of D^* defined by $(\rho,\theta,\phi) \in A_1 \Rightarrow \phi = \pi$, $(p,\theta,\phi) \in A_2 \Rightarrow \phi = 0$, $(\rho,\theta,\phi) \in A_3 \Rightarrow \rho = 0$, and $(\rho,\theta,\phi) \in A_4 \Rightarrow \theta = 2\pi$; then if $A = A_1 \cup A_2 \cup A_3 \cup A_4$, the set A has area zero and

$$S:(D^* - A) \to D$$

is one-to-one. That A has volume zero follows from the fact that each A_i has volume zero (verify!). To see that $S:(D^*-A) \to D$ is one-to-one assume that $S(p_1) = S(p_2)$, where $p_i = (\rho_i, \theta_i, \phi_i)$, $i = 1, 2$. If $S(p_1) = 0$ then

$$\rho_1 = \rho_2 = |S(p_i)| = \sqrt{\rho_i^2 \sin^2\phi_i \cos^2\theta_i + \rho_i^2 \sin^2\phi_i \sin^2\theta_i + \rho_i^2 \cos^2\phi_i} = 0$$

which implies that $p_i \in A$. So for $p_i \in D^* - A$ we may assume that $S(p_i) \neq 0$. Let (x_i, y_i, z_i) denote the point $S(p_i)$. Thus

$$\rho_1 = |S(p_i)| = |S(p_2)| = \rho_2 \neq 0$$

Since $z_1 = z_2$, $\rho_1 \cos\phi_1 = \rho_2 \cos\phi_2$, whence $\cos\phi_1 = \cos\phi_2$. But $\phi_i \in]0, \pi[$ which implies that $\phi_1 = \phi_2$ and $\sin\phi_i \neq 0$. Using the fact that

$$\rho_1 \sin\theta_1 \sin\phi_1 = y_1 = y_2 = \rho_2 \sin\theta_2 \sin\phi_2$$

$$\rho_1 \cos\theta_1 \sin\phi_1 = x_1 = x_2 = \rho_2 \cos\theta_2 \sin\phi_2$$

we find that $\cos\theta_1 = \cos\theta_2$, $\sin\theta_1 = \sin\theta_2$. The latter equality implies that either both $\theta_i \in [\pi, 2\pi[$ or both $\theta_i \in [0, \pi]$. But for $\theta_i \in [0, \pi]$, and $\cos\theta_1 = \cos\theta_2$ we must have $\theta_1 = \theta_2$ and similarly for $\theta_i \in [\pi, 2\pi[$. Gathering the information, we have proved that $p_1 = p_2$ and hence $S:(D^* - A) \to D$ is one-to-one.

EXERCISES

1. (a) The following points are given in cylindrical coordinates; express each in retangular coordinates, and spherical coordinates: $(1, 45°, 1)$ $(2, \pi/2, -4)$, $(0, 45°, 10)$, $(3, \pi/6, 4)$.
   (b) Change each of the following points from rectangular coordinates to spherical coordinates; to cylindrical coordinates: $(2, 1, -2)$, $(0, 3, 4)$, $(\sqrt{2}, 1, 1)$, $(-2\sqrt{3}, -2, 3)$.
2. Describe the geometric meaning of the following mappings in cylindrical coordinates:
   (a) $(r, \theta, z) \mapsto (r, \theta, -z)$
   (b) $(r, \theta, z) \mapsto (r, \theta + \pi, -z)$
3. Describe the geometric meaning of the following mappings in spherical coordinates:
   (a) $(\rho, \theta, \phi) \mapsto (\rho, \theta + \pi, \phi)$
   (b) $(\rho, \theta, \phi) \mapsto (\rho, \theta, \phi + \pi/2)$
4. (a) Describe the surfaces $r = \text{constant}$, $\theta = \text{constant}$, $z = \text{constant}$ in the cylindrical coordinate system.
   (b) Describe the surfaces $\rho = \text{constant}$, $\theta = \text{constant}$, $\phi = \text{constant}$ in the spherical coordinate system.
5. Show that in order to represent each point in $\mathbf{R}^3$ by spherical coordinates it is only necessary to take values of θ between 0 and 2π, values of ϕ between 0 and π, and values of $\rho \geqslant 0$. Are coordinates unique if we allow $\rho \leqslant 0$?
6. Let $S: \mathbf{R}^3 \to \mathbf{R}^3$ be defined by

   $$S(u, v, w) = (u \cos v \cos w, u \sin v \cos w, u \sin w)$$

   Show that S is onto but not one-to-one.
7. Determine the equations of the following curves and surfaces in spherical and cylindrical coordinates:

   (a) $\dfrac{x^2}{a^2} + \dfrac{y^2}{b^2} + \dfrac{z^2}{a^2} = 1$

   (b) $z^2 = x^2 + y^2$
   (c) the line $y = x = z$

   (d) $z = \tan^{-1}\dfrac{y}{x}$, $x^2 + y^2 = 1$

8. Describe the surface $\sqrt{x^2+y^2}=\rho=\theta$. What is the Cartesian representation of this surface?
9. Describe the surface $r = z\cos\theta$ in rectangular coordinates.
10. Let D be the unit ball. Evaluate

$$\int_D \frac{dx\,dy\,dz}{\sqrt{2+x^2+y^2+z^2}}$$

by making the appropriate change of variables.
11. Let D be the first octant of the unit ball $x^2+y^2+z^2 \leqslant a^2$, where $x \geqslant 0$, $y \geqslant 0$, $z \geqslant 0$. Evaluate

$$\int_D \sqrt{\frac{(x^2+y^2+z^2)^{1/2}}{z+(x^2+y^2+z^2)^2}}\,dx\,dy\,dz$$

by changing variables.

REVIEW EXERCISES FOR CHAPTER 15

1. Evaluate each of the following integrals and describe the region determined by the limits.

   (a) $\int_0^3 \int_{-x^2+1}^{x^2+1} xy\,dy\,dx$ (b) $\int_0^1 \int_{\sqrt{x}}^{1} (x+y)^2\,dy\,dx$

   (c) $\int_0^1 \int_{e^x}^{e^{2x}} x\ln y\,dy\,dx$
2. Reverse the order of integration of the integrals in exercise 1 and evaluate [(*c*) is a little tricky].
3. Find the volume enclosed by the cone $x^2+y^2=z^2$ and the plane $z-y-1=0$.
4. Find the volume between the surfaces $x^2+y^2=z$ and $x^2+y^2+z^2=2$.
5. In exercise 2, §5 we discussed integrals over unbounded regions. Use the polar change of coordinates to show $\int_0^\infty e^{-x^2}\,dx=\sqrt{\pi}/2$. [Hint: Observe that

$$\left(\int_0^\infty e^{-x^2}\,dx\right)^2 = \int_0^\infty \int_0^\infty e^{-x^2-y^2}\,dx\,dy$$

   Assume Fubini's Theorem and the Change of Variables Theorem hold in the unbounded case.]
6. Use the ideas in exercise 5 to evaluate $\int_{\mathbf{R}^2} f(x,y)\,dx\,dy$ where $f(x,y)=1/(1+x^2+y^2)^{3/2}$.
7. Find $\int_{\mathbf{R}^3} f(x,y,z)\,dx\,dy\,dz$ where $f(x,y,z)=\exp[-(x^2+y^2+z^2)^{3/2}]$.
8. Find $\int_{\mathbf{R}^3} f(x,y,z)\,dx\,dy\,dz$ where

$$f(x,y,z)=\frac{1}{[1+(x^2+y^2+z^2)^{3/2}]^{3/2}}$$

9. A cylindrical hole of diameter 1 is bored through a sphere of radius 2. Assuming that the axis of the cylinder is the same as the axis of the sphere, find the volume of the sphere that remains.

10. Let C_1 and C_2 be two cylinders of infinite extent, of diameter 2, and with axes on the x and y axes respectively. Find the volume of $C_1 \cap C_2$.

11. Write the iterated integral $\int_0^1 \int_1^{1-x} \int_x^1 f(x,y,z)\,dz\,dy\,dx$ as an integral over a region in $\boldsymbol{R}^3$ and then rewrite it in five other possible orders of integration.

*12. Suppose D is the "unbounded" region on $\boldsymbol{R}^2$ given by the set of (x,y) with $0 \leqslant x < \infty, 0 \leqslant y \leqslant x$. Let $f(x,y) = x^{-3/2} e^{y-x}$. Does the improper integral $\int_D f$ exist?

13. Evaluate each of the following iterated integrals.

    (a) $\int_0^\infty \int_0^y xe^{-y^3}\,dx\,dy$

    (b) $\int_0^1 \int_y^{y^3} e^{x/y}\,dx\,dy$

    (c) $\int_0^{\pi/2} \int_0^{(\arcsin y)/y} y \cos xy\,dx\,dy$

14. Evaluate each of the following iterated integrals.

    (a) $\int_0^1 \int_0^z \int_0^y xy^2 z^3\,dx\,dy\,dz$

    (b) $\int_0^1 \int_0^y \int_0^{\sqrt{z}} \dfrac{x}{x^2+z^2}\,dx\,dz\,dy$

    (c) $\int_1^2 \int_1^z \int_{1/y}^2 yz^2\,dx\,dy\,dz$

15. Find the volume bounded by $x/a + y/b + z/c = 1$ and the coordinate planes.

16. Find the volume bounded by $z = 6 - x^2 - y^2$ and $z = x^2 + y^2$.

17. In (a)–(d) below, make the indicated change of variables. (Do not evaluate.)

    (a) $\int_0^1 \int_{-1}^1 \int_{-\sqrt{1-y^2}}^{\sqrt{1-y^2}} \sqrt{x^2+y^2}\,dx\,dy\,dz$, cylindrical coordinates

    (b) $\int_{-1}^1 \int_{-\sqrt{1-y^2}}^{\sqrt{1-y^2}} \int_{-\sqrt{4-x^2-y^2}}^{\sqrt{4-x^2-y^2}} xyz\,dz\,dx\,dy$, cylindrical coordinates

    (c) $\int_{-\sqrt{2}}^{\sqrt{2}} \int_{-\sqrt{2-y^2}}^{\sqrt{2-y^2}} \int_{\sqrt{x^2+y^2}}^{\sqrt{4-x^2-y^2}} z^2\,dz\,dx\,dy$, spherical coordinates.

    (d) $\int_0^1 \int_0^{\pi/4} \int_0^{2\pi} \rho^3 \sin 2\phi\,d\theta\,d\phi\,d\rho$, rectangular coordinates

16

FURTHER RESULTS OF MULTIPLE INTEGRATION

In this chapter we continue our study of multiple integration. The first two sections treat integrals over surfaces—a concept which is a generalization of the double integral. This material is basic to an understanding of Chapter 17 where we establish relations between surface and volume integrals and consider several physical applications. In §16.3 and §16.4 we use double integrals to calculate the center of mass of a given body. This classical application of multiple integration is important in theoretical mechanics.

1 AREA OF A SURFACE

In §14.7 we defined a surface S as the image of a C^1 function $\Phi: D \subset \boldsymbol{R}^2 \to \boldsymbol{R}^3$, written as $\Phi(u, v) = (h(u, v), g(u, v), f(u, v))$, or $x = h(u, v)$, $y = g(u, v), z = f(u, v)$. The map Φ was called the parametrization of S. Then Φ was said to be smooth at $(u_0, v_0) \in D$ if $\boldsymbol{T}_{u_0} \times \boldsymbol{T}_{v_0} \neq 0$, where

$$\boldsymbol{T}_{u_0} = \frac{\partial h}{\partial u}(u_0, v_0)\boldsymbol{i} + \frac{\partial g}{\partial u}(u_0, v_0)\boldsymbol{j} + \frac{\partial f}{\partial u}(u_0, v_0)\boldsymbol{k}$$

and

$$\boldsymbol{T}_{v_0} = \frac{\partial h}{\partial v}(u_0, v_0)\boldsymbol{i} + \frac{\partial g}{\partial v}(u_0, v_0)\boldsymbol{j} + \frac{\partial f}{\partial v}(u_0, v_0)\boldsymbol{k}$$

Recall that a smooth surface (loosely speaking) is one which has no corners or breaks.

Our objective here is to derive a formula for calculating the area of a surface, and there are various ways of getting at such formulas. In order to avoid certain difficulties involved in defining surface area by a direct limiting process, we take a simpler route and define surface area in terms of a double integral. Then we present a plausible argument to justify this definition; however, we admit that this argument is not based on an explicit definition of the area of a surface, but rather on our intuition of what the surface area ought to be.

In this chapter and the next we shall consider only surfaces which are smooth except perhaps at a finite number of points, and parametrizations $\Phi: D \to \boldsymbol{R}^3$ which are one-to-one (except perhaps on ∂D, the boundary of D) where D is an elementary region in $\boldsymbol{R}^2$.

We define the surface area $A(S)$ of S by

$$A(S) = \int_D |\boldsymbol{T}_u \times \boldsymbol{T}_v| \, du \, dv \tag{1}$$

where $|\boldsymbol{T}_u \times \boldsymbol{T}_v|$ is the norm of $\boldsymbol{T}_u \times \boldsymbol{T}_v$. As the reader can easily verify (see exercise 8) we have

$$|\boldsymbol{T}_u \times \boldsymbol{T}_v| = \sqrt{\left|\frac{\partial(x,y)}{\partial(u,v)}\right|^2 + \left|\frac{\partial(y,z)}{\partial(u,v)}\right|^2 + \left|\frac{\partial(x,z)}{\partial(u,v)}\right|^2} \tag{2}$$

where

$$\frac{\partial(x,y)}{\partial(u,v)} = \begin{vmatrix} \dfrac{\partial x}{\partial u} & \dfrac{\partial x}{\partial v} \\ \dfrac{\partial y}{\partial u} & \dfrac{\partial y}{\partial v} \end{vmatrix}$$

and so on. Thus formula (1) becomes

$$A(S) = \int_D \sqrt{\left[\frac{\partial(x,y)}{\partial(u,v)}\right]^2 + \left[\frac{\partial(y,z)}{\partial(u,v)}\right]^2 + \left[\frac{\partial(x,z)}{\partial(u,v)}\right]^2} \, du \, dv \tag{3}$$

We can justify this definition by analyzing the integral $\int_D |\boldsymbol{T}_u \times \boldsymbol{T}_v| \, du \, dv$ in terms of Riemann sums. For simplicity, suppose D is a rectangle; consider the nth regular partition of D, and let R_{ij} be the ijth rectangle in the partition with vertices (u_i, v_j), $0 \leqslant i \leqslant n-1$, $0 \leqslant j \leqslant n-1$. Denote the values of $\boldsymbol{T}_u$ and $\boldsymbol{T}_v$ at these points by $\boldsymbol{T}_{u_i}$ and $\boldsymbol{T}_{v_j}$. We can think of the vectors $\Delta u \boldsymbol{T}_{u_i}$ and $\Delta v \boldsymbol{T}_{v_j}$ as tangent to the surface at $\Phi(u_i, v_j) = (x_{ij}, y_{ij}, z_{ij})$ where $\Delta u = u_{i+1} - u_i$, $\Delta v = v_{j+1} - v_j$. Then these vectors form a parallelogram P_{ij} which lies in the tangent plane to the surface at (x_{ij}, y_{ij}, z_{ij}) (see Figure 1-1). We thus have a "patchwork cover" of the surface by the P_{ij}. For n large, the area of P_{ij} is a good approximation to the area of $\Phi(R_{ij})$. Since

$$A(P_{ij}) = |\Delta u \boldsymbol{T}_{u_i} \times \Delta v \boldsymbol{T}_{v_j}| = |\boldsymbol{T}_{u_i} \times \boldsymbol{T}_{v_j}| \, \Delta u \, \Delta v$$

the area of the cover is

$$\begin{aligned} R_n &= \sum_{i=0}^{n-1} \sum_{j=0}^{n-1} |\boldsymbol{T}_{u_i} \times \boldsymbol{T}_{v_j}| \, \Delta u \, \Delta v = \sum_{i=0}^{n-1} \sum_{j=0}^{n-1} A(P_{ij}) \\ &\approx \sum_{i=0}^{n-1} \sum_{j=0}^{n-1} A(\Phi(R_{ij})) = A(S) \end{aligned}$$

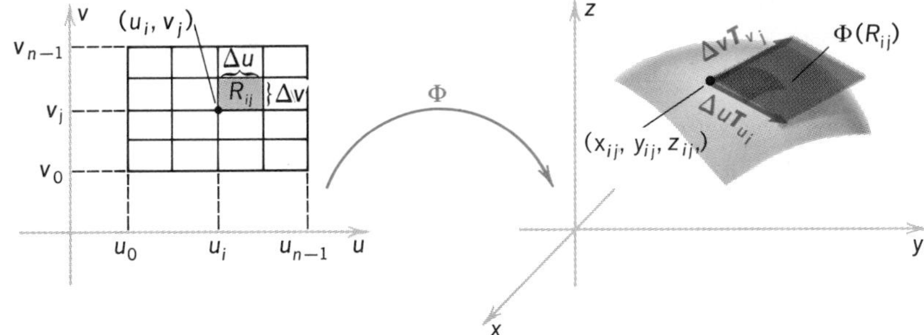

FIGURE 1-1

As $n \to \infty$, these sums converge to the integral

$$\int_D |\boldsymbol{T}_u \times \boldsymbol{T}_v| \, \Delta u \, \Delta v$$

and also to $A(S)$.

Just as a given curve in $\boldsymbol{R}^3$ could be parametrized in several ways, a given surface may also have different parametrizations. Although the formula for the area of S depends on the particular parametrization, we would expect that $A(S)$ is independent of this parametrization. This is, in fact, true, and will be discussed later (see Theorem 1, §17.3).

Example 1 Let D be the region determined by $0 \leqslant \theta \leqslant 2\pi$, $0 \leqslant r \leqslant 1$ and suppose the function $\Phi: D \to \boldsymbol{R}^3$ where

$$x = r\cos\theta, \quad y = r\sin\theta, \quad z = r$$

is a parametrization of a cone S. We compute

$$\frac{\partial(x, y)}{\partial(r, \theta)} = \begin{vmatrix} \cos\theta & -r\sin\theta \\ \sin\theta & r\cos\theta \end{vmatrix} = r$$

$$\frac{\partial(y, z)}{\partial(r, \theta)} = \begin{vmatrix} \sin\theta & r\cos\theta \\ 1 & 0 \end{vmatrix} = -r\cos\theta$$

$$\frac{\partial(x, z)}{\partial(r, \theta)} = \begin{vmatrix} \cos\theta & -r\sin\theta \\ 1 & 0 \end{vmatrix} = r\sin\theta$$

so the area integrand is

$$\begin{aligned} |\boldsymbol{T}_r \times \boldsymbol{T}_\theta| &= \sqrt{r^2 + r^2\cos^2\theta + r^2\sin^2\theta} \\ &= r\sqrt{2} \end{aligned}$$

Clearly, $|\boldsymbol{T}_r \times \boldsymbol{T}_\theta|$ vanishes for $r = 0$, but $\Phi(0, \theta) = (0, 0, 0)$ for any θ. Thus $(0, 0, 0)$ is the only point where the surface is not smooth (see §14.8). We have

$$\begin{aligned} \int_D |\boldsymbol{T}_r \times \boldsymbol{T}_\theta| \, dr \, d\theta &= \int_0^{2\pi} \int_0^1 \sqrt{2} r \, dr \, d\theta \\ &= \int_0^{2\pi} \tfrac{1}{2}\sqrt{2} \, d\theta \\ &= \sqrt{2}\pi \end{aligned}$$

To complete the example, we must verify that Φ is one-to-one (at least off the boundary of D). Let D^0 be the set of (r, θ) with $0<r<1$ and $0<\theta<2\pi$. Hence, D^0 is D without its boundary. Now, to see that $\Phi: D^0 \to \boldsymbol{R}^3$ is one-to-one, assume that $\Phi(r,\theta)=\Phi(r',\theta')$ for $(r,\theta) \in D^0$ $(r',\theta') \in D^0$. Then

$$r\cos\theta = r'\cos\theta', \quad r\sin\theta = r'\sin\theta', \quad r = r'$$

From these equations it follows that $\cos\theta = \cos\theta'$. Thus either $\theta=\theta'$, $\theta=\theta'+2\pi$ or $\theta'=\theta+2\pi$. But these last two cases are impossible since both θ and θ' belong to the open interval $]0, 2\pi[$ and thus cannot be 2π radians apart. This proves that off the boundary Φ is one-to-one. (Is $\Phi: D \to \boldsymbol{R}^3$ one-to-one?)

In future examples we shall usually omit the verification that the parametrization is one-to-one.

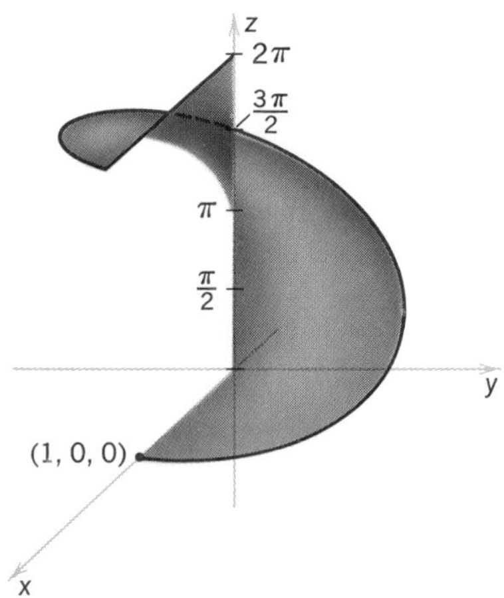

FIGURE 1-2

Example 2 A helicoid is defined by $\Phi: D \to \boldsymbol{R}^3$ where

$$x = r\cos\theta, \quad y = r\sin\theta, \quad z = \theta$$

and D is the region such that $0 \leqslant \theta \leqslant 2\pi$, $0 \leqslant r \leqslant 1$ (Figure 1-2). We compute $\partial(x,y)/\partial(r,\theta) = r$ as before, and

$$\frac{\partial(y,z)}{\partial(r,\theta)} = \begin{vmatrix} \sin\theta & r\cos\theta \\ 0 & 1 \end{vmatrix} = \sin\theta$$

$$\frac{\partial(x,z)}{\partial(r,\theta)} = \begin{vmatrix} \cos\theta & -r\sin\theta \\ 1 & 0 \end{vmatrix} = \cos\theta$$

The area integrand is, therefore, $\sqrt{r^2+1}$, which never vanishes so the surface is smooth. The area of the helicoid is

$$\begin{aligned} \int_D |\boldsymbol{T}_r \times \boldsymbol{T}_\theta| \, dr \, d\theta &= \int_0^{2\pi} \int_0^1 \sqrt{r^2+1} \, dr \, d\theta \\ &= \pi\left(\sqrt{2} + \ln(1+\sqrt{2})\right) \end{aligned}$$

A surface S given in the form $z = f(x,y)$, where $(x,y) \in D$ admits the parametrization

$$x = u, \quad y = v. \quad z = f(u,v)$$

for $(u,v) \in D$. When f is C^1, this parametrization is smooth, and the formula for surface area reduces to

$$A(S) = \int_D \sqrt{\left(\frac{\partial f}{\partial x}\right)^2 + \left(\frac{\partial f}{\partial y}\right)^2 + 1} \, dA \tag{4}$$

(Verify!)

In formula (4), we assume that $\partial f/\partial x$ and $\partial f/\partial y$ are continuous (and hence, bounded) functions on D. However, it is important, especially in the last sections of this chapter, to consider areas of surfaces for which either $(\partial f/\partial x)(x_0, y_0)$ or $(\partial f/\partial y)(x_0, y_0)$ get arbitrarily large as (x_0, y_0) approaches the boundary of D. For example, consider the hemisphere

$$z = \sqrt{1-x^2-y^2}$$

where D is the region $x^2+y^2 \leqslant 1$ (see Figure 1-3). We have

$$\frac{\partial f}{\partial x} = \frac{-x}{\sqrt{1-x^2-y^2}}, \qquad \frac{\partial f}{\partial y} = \frac{-y}{\sqrt{1-x^2-y^2}} \tag{5}$$

The boundary of D is the unit circle $x^2+y^2=1$, so as (x, y) gets close to ∂D, the value of x^2+y^2 approaches 1. Hence, the denominators in (5) go to zero.

We want to include cases such as these, so we define the area $A(S)$ of a surface S described by $z=f(x, y)$ over a region D, where f is differentiable with possible discontinuities of $\partial f/\partial x$ and $\partial f/\partial y$ on ∂D, as

$$A(S) = \int_D \sqrt{\left(\frac{\partial f}{\partial x}\right)^2 + \left(\frac{\partial f}{\partial y}\right)^2 + 1}\, dA$$

whenever $\sqrt{(\partial f/\partial x)^2+(\partial f/\partial y)^2+1}$ is integrable over D, even if the integral is improper.

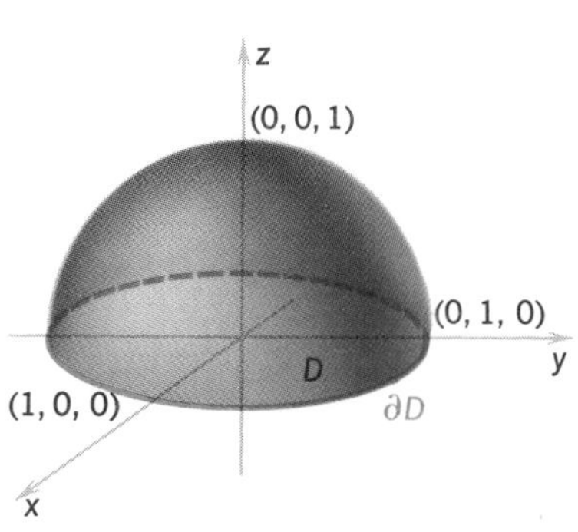

FIGURE 1-3

Example 3 Compute the area of the surface of the sphere S described by $x^2+y^2+z^2=1$.

We compute the area of the upper hemisphere S^+ where

$$x^2+y^2+z^2 = 1, \qquad z \geqslant 0$$

and then multiply our result by two. Hence we have

$$z = \sqrt{1-x^2-y^2}, \qquad x^2+y^2 \leqslant 1$$

Let D be the region $x^2+y^2 \leqslant 1$. Then

$$\begin{aligned}
A(S^+) &= \int_D \sqrt{\left(\frac{\partial z}{\partial x}\right)^2 + \left(\frac{\partial z}{\partial y}\right)^2 + 1}\, dA \\
&= \int_D \sqrt{\frac{x^2}{1-x^2-y^2} + \frac{y^2}{1-x^2-y^2} + 1}\, dA \\
&= \int_{-1}^{1} \int_{-\sqrt{1-x^2}}^{\sqrt{1-x^2}} \frac{1}{\sqrt{1-x^2-y^2}}\, dy\, dx \\
&= \int_{-1}^{1} \left[\sin^{-1} \frac{y}{\sqrt{1-x^2}}\right]_{-\sqrt{1-x^2}}^{\sqrt{1-x^2}} dx \\
&= \int_{-1}^{1} \left[\frac{\pi}{2} + \frac{\pi}{2}\right] dx = \int_{-1}^{1} \pi\, dx = 2\pi
\end{aligned}$$

Thus, the area of the entire sphere is 4π.

In Chapter 7 we found that the lateral surface area generated by revolving the graph of a function $y=f(x)$ about the x axis is given by

$$A_1 = 2\pi \int_a^b |f(x)| \sqrt{1+|f'(x)|^2}\, dx \tag{6}$$

If the graph is revolved about the y axis, we have

$$A_2 = 2\pi \int_a^b |x| \sqrt{1+|f'(x)|^2}\, dx \tag{7}$$

Now we will rederive (6) by using the methods developed above; one can obtain (7) in a similar fashion (exercise 5).

To derive formula (6) from formula (2), we must give a parametrization of S. Define the parametrization by

$$x = u, \quad y = f(u)\cos v, \quad z = f(u)\sin v$$

over the region D given by

$$a \leqslant u \leqslant b, \qquad 0 \leqslant v \leqslant 2\pi$$

This is indeed a parametrization of S because for fixed u,

$$(u, f(u)\cos v, f(u)\sin v)$$

traces out a circle of radius $|f(u)|$ with the center $(u, 0, 0)$. Now

$$\frac{\partial(x,y)}{\partial(u,v)} = f(u)\sin v$$

$$\frac{\partial(y,z)}{\partial(u,v)} = f(u)f'(u)$$

$$\frac{\partial(z,x)}{\partial(u,v)} = f(u)\cos v$$

So by formula (3),

$$\begin{aligned} A(S) &= \int_D \sqrt{\left|\frac{\partial(x,y)}{\partial(u,v)}\right|^2 + \left|\frac{\partial(y,z)}{\partial(u,v)}\right|^2 + \left|\frac{\partial(z,x)}{\partial(u,v)}\right|^2}\, du\, dv \\ &= \int_D \sqrt{|f(u)|^2\sin^2 v + |f(u)|^2|f'(u)|^2 + |f(u)|^2\cos^2 v}\, du\, dv \\ &= \int_D |f(u)| \sqrt{1+[f'(u)]^2}\, du\, dv \\ &= \int_a^b \int_0^{2\pi} |f(u)| \sqrt{1+[f'(u)]^2}\, dv\, du \\ &= 2\pi \int_a^b |f(u)| \sqrt{1+[f'(u)]^2}\, du \end{aligned}$$

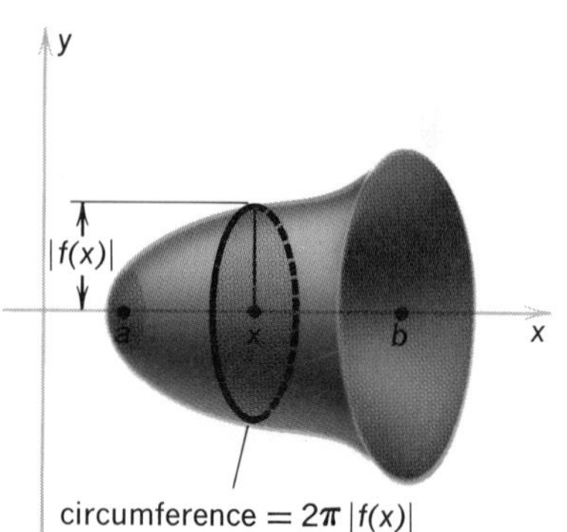

FIGURE 1-4

which is formula (6).

If S is the surface of revolution then $2\pi|f(x)|$ is the circumference of the vertical cross section to S at the point x (Figure 1-4). We remark that we can write

$$2\pi \int_a^b |f(x)| \sqrt{1+[f'(x)]^2}\, dx = \int_\sigma 2\pi |f(x)|\, ds$$

where the expression on the right is the integral of $2\pi|f(x)|$ along the path $\boldsymbol{\sigma}: [a,b] \to \boldsymbol{R}^2 : t \to (t, f(t))$. Therefore, the lateral surface area of a solid of revolution is obtained by integrating the cross-sectional circumference along the path.

EXERCISES

1. Find the surface area of the unit sphere S represented parametrically by $\Phi: D \to S \subset \boldsymbol{R}^3$ where D is the rectangle $0 \leqslant \theta \leqslant 2\pi$, $0 \leqslant \phi \leqslant \pi$ and Φ is given by the equations

$$x = \cos\theta \sin\phi, \quad y = \sin\theta \cos\phi, \quad z = \cos\phi$$

Note that we can represent the entire sphere parametrically but we cannot represent it in the form $z = f(x, y)$.

2. In exercise 1, what happens if we allow ϕ to vary from $-\pi/2$ to $\pi/2$; from 0 to 2π? Why do we obtain different answers?

3. The torus T can be represented parametrically by the function $\Phi: D \to \boldsymbol{R}^3$, where

$$x = (R - \sin\phi)\cos\theta, \quad y = (R - \sin\phi)\sin\theta, \quad z = \cos\phi$$

D is the region $\pi \leqslant \theta \leqslant \pi$, $0 \leqslant \phi \leqslant 2\pi$, and $R > 1$ is fixed. Show that $A(T) = (2\pi)^2 R$ first using formula (2) and then using formula (7).

4. Represent the ellipsoid

$$\mathscr{E}: \frac{x^2}{a^2} + \frac{y^2}{b^2} + \frac{z^2}{c^2} = 1$$

parametrically and write out the integral for its surface area $A(\mathscr{E})$ (do not evaluate the integral).

5. Let the curve $y = f(x)$, $a \leqslant x \leqslant b$ be rotated about the y axis. Show that the area of the surface swept out is

$$A = 2\pi \int_a^b \sqrt{1 + [f'(x)]^2}\, |x|\, dx$$

6. Use formula (7) to compute the surface area of the cone in example 1.

7. Find the area of the surface obtained by rotating the curve $y = x^2$, $0 \leqslant x \leqslant 1$, about the y axis.

8. Show that for the vectors $\boldsymbol{T}_u$ and $\boldsymbol{T}_v$ we have the formula

$$|\boldsymbol{T}_u \times \boldsymbol{T}_v| = \sqrt{\left[\frac{\partial(x, y)}{\partial(u, v)}\right]^2 + \left[\frac{\partial(y, z)}{\partial(u, v)}\right]^2 + \left[\frac{\partial(x, z)}{\partial(u, v)}\right]^2}$$

9. Find the area of the surface defined by $x + y + z = 1$, $x^2 + 2y^2 \leqslant 1$.

10. Compute the area of the surface given by

$$x = r\cos\theta, \quad y = 2r\cos\theta, \quad z = \theta, \quad 0 \leqslant r \leqslant 1, \quad 0 \leqslant \theta \leqslant 2\pi$$

Sketch.

11. Prove **Pappus' Theorem**: Let $\boldsymbol{\sigma}: [a, b] \to \boldsymbol{R}^2$ be a C^1 path. The lateral surface generated by rotating the graph of $\boldsymbol{\sigma}$ about the y axis is equal to $2\pi \bar{x} l(\boldsymbol{\sigma})$ where $\bar{x}$ is the average value of x coordinates of points on $\boldsymbol{\sigma}$.

2 INTEGRALS OVER SURFACES

The integral of a function over a surface is a natural generalization of the double integral. Later we shall use this notion in a variety of physical and geometric situations.

Let us start with a surface S parametrized by $\Phi: D \to S \subset \mathbf{R}^3$, $\Phi(u,v) = (h(u,v), g(u,v), f(u,v))$ or $x = h(u,v), y = g(u,v), z = f(u,v)$ as in §16.1.

If $F(x,y,z)$ is a real-valued continuous function defined on S, we define the **integral of F over S** to be

$$\int_S F(x,y,z)\, dS = \int_S F\, dS = \int_D F(\Phi(u,v))\, |\mathbf{T}_u \times \mathbf{T}_v|\, du\, dv \tag{1}$$

where $\mathbf{T}_u$ and $\mathbf{T}_v$ have the same meaning as in §16.1. Written out, equation (1) becomes

$$\int_S F\, dS = \int_D F(x(u,v), y(u,v), z(u,v)) \times \sqrt{\left[\frac{\partial(x,y)}{\partial(u,v)}\right]^2 + \left[\frac{\partial(y,z)}{\partial(u,v)}\right]^2 + \left[\frac{\partial(x,z)}{\partial(u,v)}\right]^2}\, du\, dv \tag{2}$$

Thus, if F is the function which is identically one, we recover the area formula (3) of §16.1. As with surface area, the independence of parametrization will be discussed later in Chapter 17 (see Theorem 1, §17.3).

We can obtain some intuition for this integral by considering it as a limit of sums. To do this, let D be a rectangle and partition D into n^2 rectangles R_{ij}. Let $(u_i v_j) \in R_{ij}$ and let $S_{ij} = \Phi(R_{ij})$ be the portion of the surface corresponding to R_{ij} (Figure 2-1). Again let $A(S_{ij})$ be the area of this portion of surface.

For large n, we consider F approximately constant on S_{ij} and we form the sum

$$\sum_{i=0}^{n-1} \sum_{j=0}^{n-1} F(\Phi(u_i, v_j))\, A(S_{ij})$$

But from §16.1, we have a formula for $A(S_{ij})$,

$$A(S_{ij}) = \int_{R_{ij}} |\mathbf{T}_u \times \mathbf{T}_v|\, du\, dv$$

which, by the Mean Value Theorem for integrals, equals $|\mathbf{T}_{u_i^*} \times \mathbf{T}_{v_j^*}|\, \Delta u\, \Delta v$ for some point (u_i^*, v_j^*) in R_{ij}. Hence our sum becomes

$$\sum_{i=0}^{n-1} \sum_{j=0}^{n-1} F(\Phi(u_i, v_j))\, |\mathbf{T}_{u_i^*} \times \mathbf{T}_{v_j^*}|\, \Delta u\, \Delta v$$

which is an approximating sum for the integral in (1).

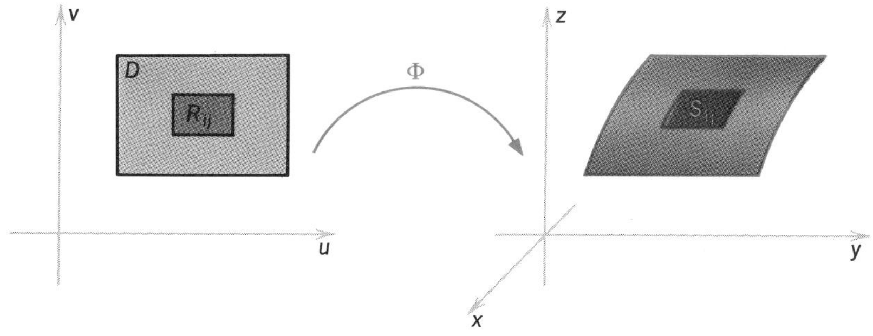

FIGURE 2-1

Example 1 Suppose a helicoid is described as in Example 2, §16.1, and let F be given by $F(x, y, z) = \sqrt{x^2+y^2+1}$. As before

$$\frac{\partial(x,y)}{\partial(r,\theta)} = r, \quad \frac{\partial(y,z)}{\partial(r,\theta)} = \sin\theta, \quad \frac{\partial(x,z)}{\partial(r,\theta)} = \cos\theta$$

Also, $F(r\cos\theta, r\sin\theta, \theta) = \sqrt{r^2+1}$. Therefore

$$\begin{aligned}\int_S F(x,y,z)\,dS &= \int_D F(\Phi(r,\theta))\,|\mathbf{T}_r \times \mathbf{T}_\theta|\,dr\,d\theta \\ &= \int_0^{2\pi}\int_0^1 \sqrt{r^2+1}\sqrt{r^2+1}\,dr\,d\theta \\ &= \int_0^{2\pi} \tfrac{4}{3}\,d\theta \\ &= \tfrac{8}{3}\pi\end{aligned}$$

Suppose S is the graph of a C^1 function $z=f(x, y)$. Then we can parametrize S by

$$x = u, \quad y = v, \quad z = f(u,v)$$

In this case

$$|\mathbf{T}_u \times \mathbf{T}_v| = \sqrt{1 + \left(\frac{\partial f}{\partial u}\right)^2 + \left(\frac{\partial f}{\partial v}\right)^2}$$

and so

$$\int_S F(x,y,z)\,dS = \int_D F(x,y,f(x,y))\sqrt{1 + \left(\frac{\partial f}{\partial x}\right)^2 + \left(\frac{\partial f}{\partial y}\right)^2}\,dx\,dy \qquad (3)$$

Example 2 Let S be the surface defined by $z=x^2+y$, where D is the region $0 \leqslant x \leqslant 1$, $-1 \leqslant y \leqslant 1$. Evaluate $\int_S x\,dS$.

By definition

$$\int_S x\,dS = \int_D x\sqrt{1+\left(\frac{\partial z}{\partial x}\right)^2+\left(\frac{\partial z}{\partial y}\right)^2}\,dx\,dy$$

$$= \int_{-1}^{1}\int_0^1 x\sqrt{1+4x^2+1}\,dx\,dy$$

$$= \tfrac{1}{8}\int_{-1}^{1}\left[\int_0^1 [2+4x^2]^{1/2}(8x\,dx)\right]dy$$

$$= \tfrac{2}{3}\cdot\tfrac{1}{8}\int_{-1}^{1}[(2+4x^2)^{3/2}]_0^1\,dy$$

$$= \tfrac{1}{12}\int_{-1}^{1}[6^{3/2}-2^{3/2}]\,dy = \tfrac{1}{6}[6^{3/2}-2^{3/2}]$$

$$= \sqrt{6}-\frac{\sqrt{2}}{3} = \sqrt{2}\left(\sqrt{3}-\frac{1}{3}\right)$$

Example 3 Evaluate $\int_S z^2\,dS$ where S is the unit sphere $x^2+y^2+z^2=1$.

For this problem it is convenient to represent the sphere parametrically by the equations $x=\cos\theta\cos\psi$, $y=\sin\theta\cos\psi$, $z=\cos\psi$, over the region D in the $\theta\psi$ plane given by $0\leqslant\psi\leqslant\pi$, $0\leqslant\theta\leqslant 2\pi$. From equation (1) we get

$$\int_S z^2\,dS = \int_D(\cos\psi)^2\,|\boldsymbol{T}_\theta\times\boldsymbol{T}_\psi|\,d\theta\,d\psi$$

Now a little computation [use formula (2) of §16.1] shows that

$$|\boldsymbol{T}_\theta\times\boldsymbol{T}_\psi| = |\sin\psi|$$

so

$$\int z^2\,dS = \int_0^{2\pi}\int_0^{\pi}\cos^2\psi\,|\sin\psi|\,d\psi\,d\theta$$

$$= \int_0^{2\pi}\int_0^{\pi}\cos^2\psi\sin\psi\,d\psi\,d\theta = \tfrac{1}{3}\int_0^{2\pi}[-\cos^3\psi]_0^{\pi}\,d\theta$$

$$= \frac{2}{3}\int_0^{2\pi}d\theta = \frac{4\pi}{3}$$

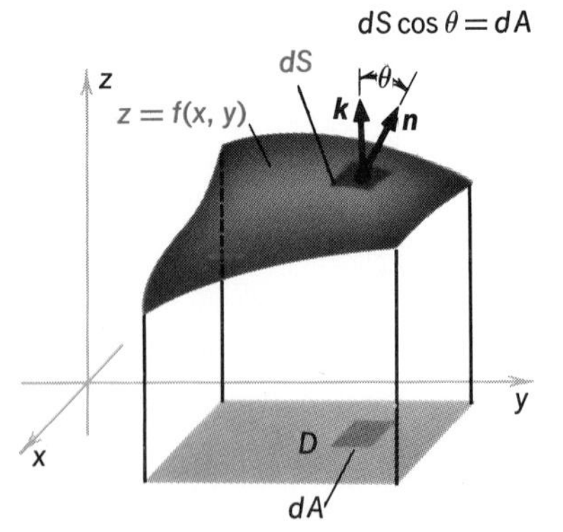

FIGURE 2-2

Once again, let S be the graph $z=f(x,y)$ and consider formula (3); we wish to interpret this result geometrically. We claim that

$$\int_S F(x,y,z)\,dS = \int_D\frac{F(x,y,f(x,y))}{\cos\theta}\,dx\,dy \tag{4}$$

where θ is the angle the normal to the surface makes with the unit vector $\boldsymbol{k}$ at the point $(x,y,f(x,y))$ (see Figure 2-2).

Let us prove this assertion. Since $z-f(x,y)=0$, a normal vector is $\boldsymbol{n}=-(\partial f/\partial x)\boldsymbol{i}-(\partial f/\partial y)\boldsymbol{j}+\boldsymbol{k}$ [recall that the normal to a surface $\phi(x,y,z)=$

constant is $\nabla\phi$]. Thus

$$\cos\theta = \frac{\boldsymbol{n}\cdot\boldsymbol{k}}{|\boldsymbol{n}|} = 1\bigg/\sqrt{\left(\frac{\partial f}{\partial x}\right)^2 + \left(\frac{\partial f}{\partial y}\right)^2 + 1}$$

The result is, in fact, obvious geometrically, for if a small rectangle in the xy plane has area dA then the area of the portion above it on the surface is $dS = dA/\cos\theta$ (Figure 2-2).

This approach helps us to remember formula (2) and to apply it in problems. We illustrate with an example.

Example 4 Compute $\int_S x\, dS$ where S is the triangle with vertices $(1,0,0)$, $(0,1,0)$, $(0,0,1)$.

We refer to Figure 2-3. This surface is a plane described by the equation $x+y+z=1$. Since the surface is a plane, the angle θ is constant and a unit normal vector is $\boldsymbol{n} = (1/\sqrt{3}, 1/\sqrt{3}, 1/\sqrt{3})$. Thus, $\cos\theta = 1/\sqrt{3}$ and

$$\int_S x\, dS = \sqrt{3}\int_D x\, dx\, dy$$

where D is the domain in the xy plane. But

$$\sqrt{3}\int_D x\, dx\, dy = \sqrt{3}\int_0^1\int_0^{1-x} x\, dy\, dx = \sqrt{3}\int_0^1 x(1-x)\, dx = \frac{\sqrt{3}}{6}$$

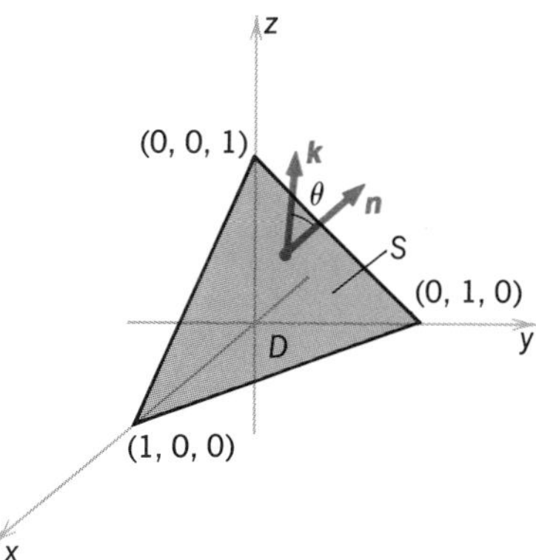

FIGURE 2-3

We can define the average value of $F(x,y,z)$ over S by the formula

$$\text{average value} = \frac{\int_S F(x,y,z)\, dS}{A(S)} = \frac{\int_S F(x,y,z)\, dS}{\int_S dS} \tag{5}$$

This definition is perfectly analogous to the definition of the average value of $F(x,y,z)$ along a path $\boldsymbol{\sigma}$ as $\int_{\boldsymbol{\sigma}} F(x,y,z)\, ds/\int_{\boldsymbol{\sigma}} ds$. We can justify equation (5) in the terms of Riemann sums as follows. The ratio

$$R_n = \frac{\sum_{i=0}^{n-1}\sum_{j=0}^{n-1} F(x_{ij}, y_{ij}, z_{ij})\, A(S_{ij})}{\sum_{i=0}^{n-1}\sum_{j=0}^{n-1} A(S_{ij})} = \frac{\sum_{i=0}^{n-1}\sum_{j=0}^{n-1} F(x_{ij}, y_{ij}, z_{ij})\, A(S_{ij})}{A(S)}$$

is the approximate average value of F on S obtained by considering F as constant on S_{ij}. The limit of the ratios R_n as $n\to\infty$ is the average value as given in equation (5).

Thus in Example 3 above, if $F(x,y,z) = z^2$ represents the temperature at (x,y,z), then the average temperature on the surface of the sphere is $\frac{1}{3}(4\pi)/4\pi = 1/3$ degree. In Example 4, we computed $\int_S x\, dS$ for the triangle whose vertices are the tips of the vectors $\boldsymbol{i}, \boldsymbol{j}$, and $\boldsymbol{k}$. The area of this triangle is

$$\int_S dS = \int_D \frac{1}{\cos\theta}\, dx\, dy = \int_0^1\int_0^{1-x}\sqrt{3}\, dy\, dx = \frac{\sqrt{3}}{2}$$

Hence, the average x coordinate of points on the triangle is the ratio $(\sqrt{3}/6)/(\sqrt{3}/2) = 1/3$.

In §16.3, we shall see that the average values of the x, y, and z coordinates determine the center of mass of a surface of uniform density.

EXERCISES

1. Compute $\int_S xy\,dS$ where S is the surface of the tetrahedron with sides $z=0$, $y=0$, $x+z=1$, and $x=y$.

2. Let $\Phi: D \subset \boldsymbol{R}^2 \to \boldsymbol{R}^3$ be a parametrization of a surface S defined by
$$x = h(u,v), \quad y = g(u,v), \quad z = f(u,v)$$
   (a) Let $\dfrac{\partial \Phi}{\partial u} = \left(\dfrac{\partial h}{\partial u}, \dfrac{\partial g}{\partial u}, \dfrac{\partial f}{\partial u}\right)$, $\dfrac{\partial \Phi}{\partial v} = \left(\dfrac{\partial h}{\partial v}, \dfrac{\partial g}{\partial v}, \dfrac{\partial f}{\partial v}\right)$ and set
$$E = \left|\frac{\partial \Phi}{\partial u}\right|^2, \quad F = \frac{\partial \Phi}{\partial u} \cdot \frac{\partial \Phi}{\partial v}, \quad G = \left|\frac{\partial \Phi}{\partial v}\right|^2$$
   Show that the surface area of S is $\int_D \sqrt{EG - F^2}\,du\,dv$. In this notation, how can we express $\int_S f\,dS$ for a general function f?
   (b) What does the formula become if the vectors $\partial\Phi/\partial u$ and $\partial\Phi/\partial v$ are orthogonal?
   (c) Use parts (a) and (b) to compute the surface area of a sphere of radius a.

3. Find an expression for the surface area between the planes $x = 1$ and $x = 2$. (Do not compute.)

   Find the surface area between the planes $x = 1$ and $x = 2$.

4. Compute $\int_S (x^2y^2 + y^2z^2 + z^2x^2)\,dS$, where S is the portion of the cone $z^2 = x^2 + y^2$ bounded by the planes $z = 0$, $z = 1$.

5. Evaluate $\int_S xyz\,dS$, where S is the triangle with vertices $(1,0,0)$, $(0,2,0)$, $(0,1,1)$.

6. Let a surface S be defined by $f(x,y,z) = z^3 + g(x,y) = 0$ for x, y in a domain D of $\boldsymbol{R}^2$. Show that the area of S is
$$\int_D \frac{\sqrt{\left(\dfrac{\partial f}{\partial x}\right)^2 + \left(\dfrac{\partial f}{\partial y}\right)^2 + \left(\dfrac{\partial f}{\partial z}\right)^2}}{\left|\dfrac{\partial f}{\partial z}\right|}\,dx\,dy$$
   Interpret the answer geometrically.

7. Evaluate $\int_S z\,dS$, where S is the surface $z = x^2 + y^2$, $x^2 + y^2 \leqslant 1$.

8. Evaluate $\int_S x^2y^2z^2\,dS$, where S is the surface $x = 2y^2 + 2z^2$, $2y^2 + 2z^2 \int \leqslant 1$.

3 CENTER OF MASS

We now apply the integral calculus of several variables to the problem of finding the **center of mass** or, as it is sometimes called, the **center of gravity** of a given body.

Suppose we have a system of n "point" masses $m_1, m_2, \ldots, m_n$, respectively. For simplicity, we wish to regard the system as a single point with total mass $M = \sum_{i=1}^{n} m_i$ concentrated at a point called the **center of mass** of the system. To be more precise, assume the coordinates of the n masses are given by the vectors $\boldsymbol{r}_i, i = 1, 2, \ldots, n$ in $\boldsymbol{R}^3$ (Figure 3-1). Let $\boldsymbol{R}$ be the position vector $(\bar{x}, \bar{y}, \bar{z})$ called the center of mass and defined by the relation

$$\boldsymbol{R} = \frac{1}{M}\sum_{i=1}^{n} m_i \boldsymbol{r}_i \tag{1}$$

or

$$M\boldsymbol{R} = \sum_{i=1}^{n} m_i \boldsymbol{r}_i \tag{2}$$

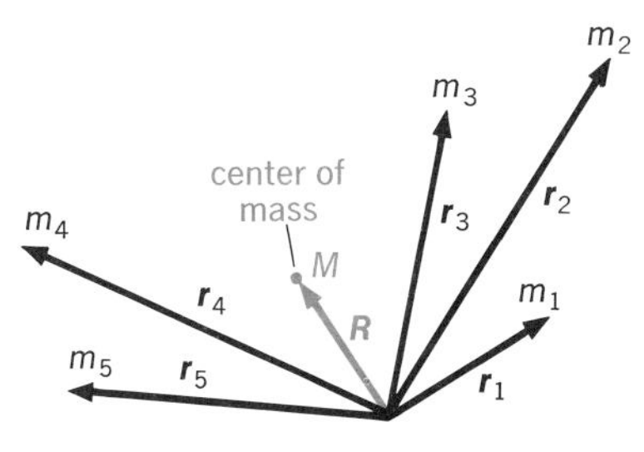

FIGURE 3-1

If $\boldsymbol{r}_i = (x_i, y_i, z_i)$, then, in terms of coordinates, $\bar{x}, \bar{y}, \bar{z}$ are given by

$$M\bar{x} = \sum_{i=1}^{n} m_i x_i, \quad M\bar{y} = \sum_{i=1}^{n} m_i y_i, \quad M\bar{z} = \sum_{i=1}^{n} m_i z_i$$

How can we justify our assertion that the system can be thought of as having all its mass concentrated at a point? Assume that each mass m_i is subjected to a force $\boldsymbol{F}_i$. The resulting acceleration of m_i is given by Newton's second law $\boldsymbol{F}_i = m_i \boldsymbol{a}_i$. The net force on the system is the vector sum

$$\boldsymbol{F} = \sum_{i=1}^{n} \boldsymbol{F}_i$$

Let $\boldsymbol{a} = d^2\boldsymbol{R}/dt^2$ denote the acceleration of the center of mass. Since the position vector $\boldsymbol{R}$ of the center of mass is given by equation (1), we have that

$$\boldsymbol{a} = \frac{d^2\boldsymbol{R}}{dt^2} = \frac{1}{M}\sum_{i=1}^{n} m_i \frac{d^2\boldsymbol{r}_i}{dt^2} = \frac{1}{M}\sum_{i} m_i \boldsymbol{a}_i$$

or

$$M\boldsymbol{a} = \sum_{i=1}^{n} m_i \boldsymbol{a}_i = \sum_{i=1}^{n} \boldsymbol{F}_i = \boldsymbol{F}$$

Thus we have the result

$$\boldsymbol{F} = M\boldsymbol{a} \tag{3}$$

where $\boldsymbol{F}$ is the net force on the system, M is the total mass, and $\boldsymbol{a}$ is the acceleration of the center of mass.

Equation (3) is precisely Newton's second law for a point of mass M whose position vector is $\boldsymbol{R}$. Therefore, the center of mass behaves as if all the mass were concentrated at this point. For example, suppose a shell fired from the earth's surface follows a parabolic path (neglecting air resistance, etc.). If the shell explodes in flight, the net force on it does not change and so the center of mass continues on a parabolic trajectory although the fragments do not.

Next, we want to extend the notion of center of mass from a finite system of point masses to a rigid body—that is, a body whose parts have a fixed location with respect to one another when the body is subjected to external forces.

Consider a body B and a fixed coordinate system (Figure 3-2). We can imagine that B is enclosed in some rectangular parallelepiped P. Partition P into n^3 rectangular parallelepipeds P_{ijk} for $i, j, k = 1, \ldots, n$, and consider the resulting subdivision of B. Select those indices i, j, k for which P_{ijk} intersects B; denote by Δm_{ijk} the mass enclosed in such a P_{ijk}. If we view each P_{ijk} which intersects B as a point mass located at $\boldsymbol{r}_{ijk}$, the center of mass $\boldsymbol{R}_n$ is given by the formula

$$\boldsymbol{R}_n = \frac{\sum_{k=1}^{n} \sum_{j=1}^{n} \sum_{i=1}^{n} \boldsymbol{r}_{ijk} \Delta m_{ijk}}{\sum_{k=1}^{n} \sum_{j=1}^{n} \sum_{i=1}^{n} \Delta m_{ijk}} \tag{4}$$

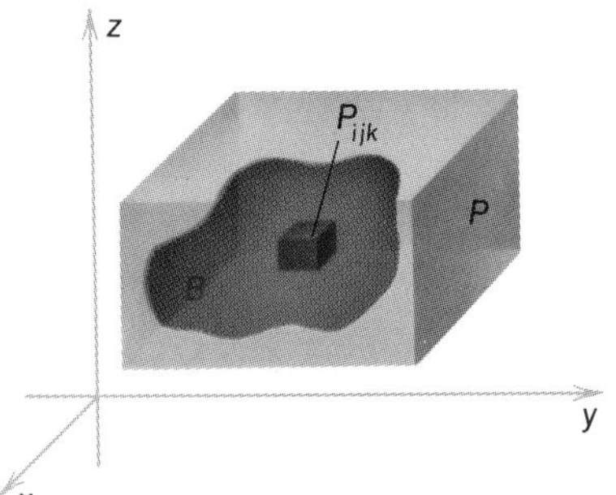

FIGURE 3-2

As $n \to \infty$ (or as the partition becomes finer and finer) we expect $\boldsymbol{R}_n$ to give us a better and better approximation to the actual center of mass. To see exactly what limit $\boldsymbol{R}_n$ approaches, we need some preliminary analysis of the increment Δm_{ijk} in our body B. Around any point $p_0 = (x_0, y_0, z_0)$ in B, we can construct a small rectangular parallelepiped of mass Δm and volume ΔV (Figure 3-3).

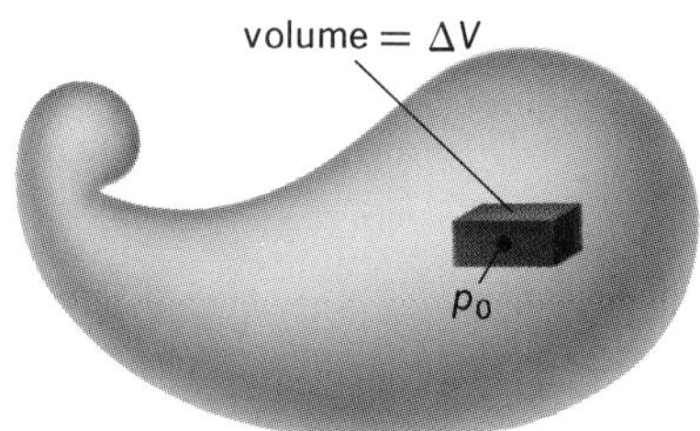

FIGURE 3-3

We shall assume that as ΔV tends to zero, the quotient $\Delta m/\Delta V$ tends to a limit $\delta(p_0) = \delta(x_0, y_0, z_0)$ (the limit depends on p_0). If we compute this limit for each point, we obtain a real-valued function δ with domain B which measures the **mass density** of the body at each point. For many bodies, δ is a continuous function and for our work we shall assume that this is always the case; however, there are simple examples for which it is not (consider a body composed of a nail driven into a block of wood or a metal skillet with a glass top).

Now we shall derive a formula in terms of the density function which will enable us to compute the center of mass of a body. Replace Δm_{ijk} in (4) by $\delta(\boldsymbol{r}_{ijk}) \times \Delta V_{ijk}$, where ΔV_{ijk} is the volume of the rectangular parallelepiped R_{ijk}. Passing to the limit in (4), we obtain our new formula

$$\boldsymbol{R} = \frac{\int_B \boldsymbol{r}\delta \, dV}{\int_B \delta \, dV} \tag{5}$$

where

$$\int_B \boldsymbol{r}\delta \, dV = \left(\int_B r_1 \delta \, dV\right)\boldsymbol{i} + \left(\int_B r_2 \delta \, dV\right)\boldsymbol{j} + \left(\int_B r_3 \delta \, dV\right)\boldsymbol{k}$$

If δ is a constant function, the body B is said to be **homogeneous.** In this case we may remove δ from under the integral sign in (5) to get

$$\boldsymbol{R} = \frac{\delta \int_B \boldsymbol{r} \, dV}{\delta \int_B dV} = \frac{\int_B \boldsymbol{r} \, dV}{\int_B dV} = \frac{1}{V} \int_B \boldsymbol{r} \, dV$$

In terms of components this is

$$\bar{x} = \frac{1}{V}\int_B x\,dV, \quad \bar{y} = \frac{1}{V}\int_B y\,dV, \quad \bar{z} = \frac{1}{V}\int_B z\,dV \tag{6}$$

Here, $\bar{x}$, $\bar{y}$, and $\bar{z}$ are the average values of the x, y, z coordinates of points in B. In connection with formula (6) some terminology is useful. The quantity $\int_B x\,dV$ is called the **moment** of B about the yz plane. If B were a thin slab (or a point mass) at a constant distance from the yz plane, the moment would be simply

$$(\text{volume}) \times (\text{distance from } yz \text{ plane})$$

It is conventional to define the center of mass of a geometric object—for example, a ball or hemisphere—to be the center of mass of the corresponding homogeneous body (in this case, a homogeneous sphere or hemisphere). Hence, we may use formula (6) for such problems.

Example 1 Find the center of mass of a hemisphere H described by $x^2+y^2+z^2 \leqslant a^2, z \geqslant 0$. Using equation (6) we have that the x component is

$$\bar{x} = \frac{1}{V}\int_H x\,dV$$

To evaulate this integral, we represent H by the equations

$$-a \leqslant y \leqslant a, \quad -\sqrt{a^2-y^2} \leqslant x \leqslant \sqrt{a^2-y^2}, \quad 0 \leqslant z \leqslant \sqrt{a^2-x^2-y^2}$$

so that

$$\bar{x} = \frac{1}{V}\int_{-a}^{a}\left\{\int_{-\sqrt{a^2-y^2}}^{\sqrt{a^2-y^2}}\left[\int_0^{\sqrt{a^2-x^2-y^2}} x\,dz\right]dx\right\}dy$$

We need not compute the volume of the hemisphere, since we already know that $V = \frac{2}{3}\pi a^3$. Then

$$\frac{1}{V}\int_H x\,dV = \frac{3}{2\pi a^3}\int_{-a}^{a}\left[\int_{-\sqrt{a^2-y^2}}^{\sqrt{a^2-y^2}} x\sqrt{a^2-x^2-y^2}\,dx\right]dy$$

$$\begin{aligned}\int_H x\,dV &= -\tfrac{1}{2}\int_{-a}^{a}\left[\int_{-\sqrt{a^2-y^2}}^{\sqrt{a^2-y^2}} -2x\sqrt{a^2-x^2-y^2}\,dx\right]dy \\ &= -\tfrac{1}{3}\int_{-a}^{a}\left[(a^2-x^2-y^2)^{3/2}\right]_{-\sqrt{a^2-y^2}}^{\sqrt{a^2-y^2}}dy \\ &= -\tfrac{1}{3}\int_{-a}^{a} 0\,dy = 0\end{aligned}$$

So $\bar{x}=0$. By a similar computation, it follows that $\bar{y}=0$. (These results also follow from general symmetry considerations.)

Now the z component of the center of mass is

$$\bar{z} = \frac{3}{2\pi a^3}\int_{-a}^{a}\int_{-\sqrt{a^2-y^2}}^{\sqrt{a^2-y^2}}\int_0^{\sqrt{a^2-x^2-y^2}} z\,dz\,dx\,dy$$

For the moment, let us forget about the factor $3/2\pi a^3$. Then

$$\int_{-a}^{a}\int_{-\sqrt{a^2-y^2}}^{\sqrt{a^2-y^2}}\int_{0}^{\sqrt{a^2-x^2-y^2}} z\,dz\,dx\,dy$$

$$= \int_{-a}^{a}\int_{-\sqrt{a^2-y^2}}^{\sqrt{a^2-y^2}}\left[\frac{z^2}{2}\right]_0^{\sqrt{a^2-x^2-y^2}} dx\,dy$$

$$= \tfrac{1}{2}\int_{-a}^{a}\int_{-\sqrt{a^2-y^2}}^{\sqrt{a^2-y^2}}(a^2-x^2-y^2)\,dx\,dy$$

$$= \tfrac{1}{2}\int_{-a}^{a}\left[a^2x-\frac{x^3}{3}-y^2x\right]_{-\sqrt{a^2-y^2}}^{\sqrt{a^2-y^2}} dy$$

$$= \int_{-a}^{a}\left[a^2\sqrt{a^2-y^2}-\frac{(a^2-y^2)^{3/2}}{3}-y^2\sqrt{a^2-y^2}\right]dy$$

$$= \int_{-a}^{a}\left[(a^2-y^2)\sqrt{a^2-y^2}-\frac{(a^2-y^2)^{3/2}}{3}\right]dy$$

$$= \tfrac{2}{3}\int_{-a}^{a}(a^2-y^2)^{3/2}\,dy$$

To evaluate this last integral, we substitute $y = a\sin\theta, dy = a\cos\theta\,d\theta$ and obtain

$$\tfrac{2}{3}\int_{-a}^{a}(a^2-y^2)^{3/2}\,dy = \tfrac{2}{3}\cdot\tfrac{3}{8}\pi a^4 = \tfrac{1}{4}\pi a^4$$

Thus, recalling the factor $3/2\pi a^3$, our final answer is

$$\bar{z} = \frac{3}{2\pi a^3}\cdot\frac{\pi a^4}{4} = \tfrac{3}{8}a$$

FIGURE 3-4

Therefore, the center of mass is given by the position vector $\boldsymbol{R} = (0, 0, \tfrac{3}{8}a)$ (Figure 3-4).

It is not unusual to encounter center of mass problems immediately following the development of the integral of a real-valued function of one variable. The reason for this is that there are many such problems which can be easily handled using a slicing technique and the corresponding integral of a function of one variable. However, we shall see that the theoretical justification of this technique depends on the ideas underlying the multiple integral. As an example, let us employ the slice technique to determine the center of mass of the hemisphere in Example 1. By symmetry we conclude $\bar{x} = 0,\ \ \bar{y} = 0$ as before (Figure 3-5).

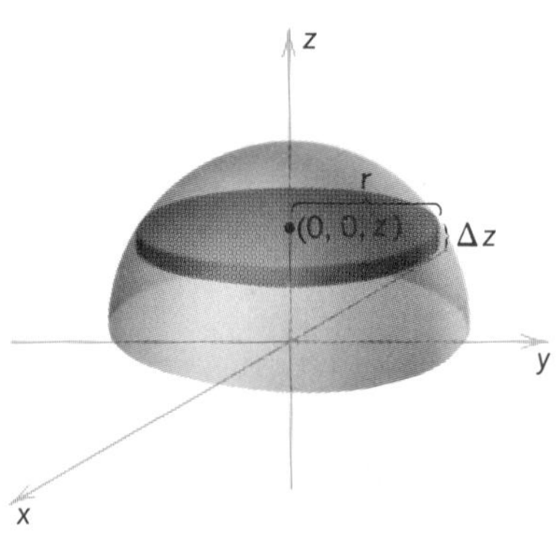

FIGURE 3-5

Next, consider a thin slab of radius r at height z above the xy plane and "thickness' Δz. The (**first**) moment of the slab about the xy plane is $M_{xy} = z(\pi r^2\,\Delta z)$—that is, the height z multiplied by the volume of the slab. Since $x^2+y^2+z^2 = a^2$ and $x^2+y^2 = r^2$, we have

$$r^2 = a^2 - z^2$$

The total moment M_{xy} of the hemisphere about the xy plane is given by the integral

$$M_{xy} = \int_0^a z\,dV = \pi \int_0^a z(a^2 - z^2)\,dz$$
$$= \pi\left[\left(\frac{a^2 z^2}{2} - \frac{z^4}{4}\right)\right]_0^a = \frac{\pi a^4}{4}$$

Thus, we find again that

$$\bar{z} = \frac{3}{2\pi a^3} \cdot M_{xy}$$
$$= \left(\frac{3}{2\pi a^3}\right)\frac{\pi a^4}{4} = \frac{3}{8}a$$

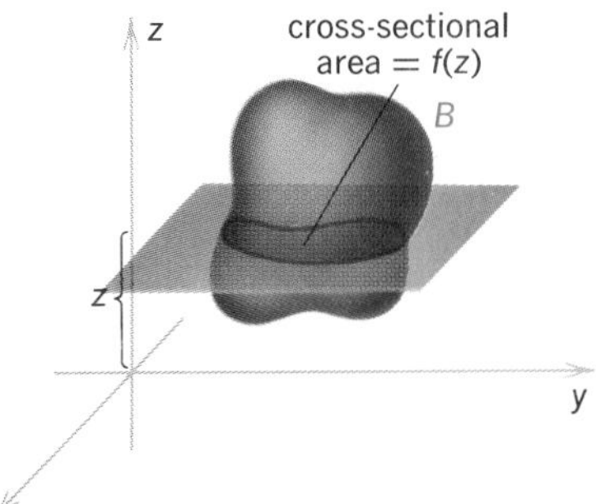

FIGURE 3-6

We can summarize this technique as follows. Suppose we have a region B in $\boldsymbol{R}^3$ with the property that the cross-sectional area of B determined by a plane parallel to the xy plane at height z is given by a continuous function $f(z)$ (Figure 3-6.) Then the formula for $\bar{z}$ is

$$\bar{z} = \frac{1}{V(B)} \int_r^s z f(z)\,dz \tag{7}$$

where r and s are the z coordinates of the upper and lower extremities of the body B. Similarly, if $g(x)$ and $h(y)$ represent the cross-sectional areas of B with respect to planes parallel to the yz plane and xz plane, respectively, we have

$$\bar{x} = \frac{1}{V(B)} \int_a^b x g(x)\,dx \tag{8}$$

$$\bar{y} = \frac{1}{V(B)} \int_c^d y h(y)\,dy \tag{9}$$

The proof follows from formula (6) by expressing the triple integral as an iterated integral, and will be given below.

If the given body is not homogeneous, this technique generally will not work; however, when the density function $\delta(z)$ depends only on the z coordinate, we can compute the $\bar{z}$ by the formula

$$\bar{z} = \frac{\int_r^s z f(z)\delta(z)\,dz}{\int_r^s f(z)\delta(z)\,dz} \tag{10}$$

Example 2 Find the $\bar{z}$ for a hemisphere of radius a, whose density at a point p varies directly with the distance of p to the base. We use formula (10) with $\delta(z) = kz$ for some constant k. Thus

$$\bar{z} = \frac{\int_0^a \pi z(a^2 - z^2)\,kz\,dz}{\int_0^a \pi(a^2 - z^2)\,kz\,dz} = \frac{\int_0^a z^2(a^2 - z^2)\,dz}{\int_0^a z(a^2 - z^2)\,dz}$$

For the numerator, we obtain

$$\int_0^a (a^2 z^2 - z^4)\, dz = \left[a^2 \cdot \frac{z^3}{3} - \frac{z^5}{5} \right]_0^a = \frac{2a^5}{15}$$

and for the denominator,

$$\int_0^a (a^2 z - z^3)\, dz = \left[a^2 \cdot \frac{z^2}{2} - \frac{z^4}{4} \right]_0^a = \frac{a^4}{4}$$

Therefore

$$\bar{z} = \frac{\frac{2}{15}a^5}{\frac{1}{4}a^4} = \frac{8}{15}a$$

By symmetry, we see that $\bar{x} = \bar{y} = 0$. Hence, the center of mass is $(0, 0, \frac{8}{15}a)$.

We conclude this section with a discussion of how formulas (7)–(9) are proved. We shall demonstrate (8) in the case where B is a three-dimensional region of type 1; that is, B is the set of (x, y, z) with

$$\gamma_2(x, y) \leqslant z \leqslant \gamma_1(x, y), \quad \psi_2(x) \leqslant y \leqslant \psi_1(x), \quad a \leqslant x \leqslant b$$

If $x \in [a, b]$ is fixed, then the cross section determined by a plane parallel to the yz plane passing through x can be projected onto the yz plane. Let us call this region R_x (see Figure 3-7). Then R_x is an elementary region in the yz plane and it can be described as the set of points (y, z) with $\psi_2(x) \leqslant y \leqslant \psi_1(x)$ (recall x is fixed) and $\gamma_2(x, y) \leqslant z \leqslant \gamma_1(x, y)$. Thus the area $A(R_x)$ is given by

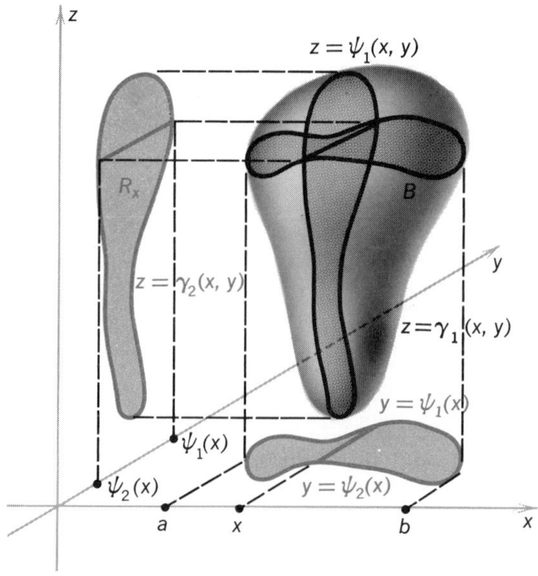

FIGURE 3-7

the iterated integral

$$A(R_x) = \int_{\psi_2(x)}^{\psi_1(x)} \int_{\gamma_2(x,y)}^{\gamma_1(x,y)} dz\, dy$$

But $g(x) = A(R_x)$. Substituting this in formula (8) we obtain

$$\bar{x} = \frac{1}{V(B)} \int_a^b x g(x)\, dx = \frac{1}{V(B)} \int_a^b x \left[\int_{\psi_2(x)}^{\psi_1(x)} \int_{\gamma_2(x,y)}^{\gamma_1(x,y)} dz\, dy \right] dx$$

Since x is a constant of integration in the first two integrals, this last expression is equal to

$$\frac{1}{V(B)} \int_a^b \int_{\psi_2(x)}^{\psi_1(x)} \int_{\gamma_2(x,y)}^{\gamma_1(x,y)} x\, dz\, dy\, dx$$

By Fubini's Theorem, the value of the integral above is

$$\frac{1}{V(B)} \int_B x\, dV$$

This is the definition of $\bar{x}$ and so we have proved (8).

EXERCISES

1. Find the center of mass of the system consisting of 10 g at $(0,0,1)$, 20 g at $(0,0,2)$, and 30 g at $(1,0,0)$.
2. Let B be the union of k regions, $B = B_1 \cup B_2 \cup \cdots \cup B_k$. Suppose the center of mass of B_i is $(\bar{x}_i, \bar{y}_i, \bar{z}_i)$ and m_i is the mass of B_i. Show that the center of mass of B is the same as a system with m_i concentrated at $(\bar{x}_i, \bar{y}_i, \bar{z}_i)$.
3. Find the center of mass of a homogeneous hemisphere of radius a plus a unit mass at $(r,0,0)$. [Hint: Use exercise 2.]
4. Find the center of mass of the homogeneous solid inside the cylinder $x^2+y^2=1$ and between $z=0$ and $z=x+y$.
5. Find the center of mass of the region inside $x^2+y^2=1$ and between $z=0$, $z=1$ if the density is $\delta(x,y,z)=z$.
6. Find the center of mass of the unit cube in the first octant with density $\delta(x,y,z)=x^2$.
7. Let l be a line in $\boldsymbol{R}^3$ and m a mass at the point (x,y,z). The **moment of inertia** of the mass about the line l is defined as $I_l = md^2$, where d is the distance from (x,y,z) to l. Show that the moment of inertia of a body B with mass density δ about the x axis is

$$I_x = \int_B (y^2+z^2)\, \delta(x,y,z)\, dx\, dy\, dz$$

with similar formulas for I_y, I_z. Derive a general formula for the moment of inertia about a line l. (The moment of inertia plays the role for angular motion that mass does for Newton's law. See any physics text for details.)

8. Prove that the moment of inertia of a body B about a line l equals the moment of inertia about l' (where l and l' are parallel and l' passes through the center of mass) plus Md^2, where M is the total mass and d is the distance of the mass center from l.
9. For a homogeneous sphere of radius a and total mass M, show that $I_x = I_y = I_z = \frac{2}{5} Ma^2$.
10. For a homogeneous ellipsoid with semiaxes a, b, c along the x, y, and z axes, respectively, and total mass M, show that $I_x = \frac{1}{5}M(b^2+c^2)$, $I_y = \frac{1}{5}M(a^2+c^2)$, $I_z = \frac{1}{5}M(a^2+b^2)$.

4 CENTERS OF MASS FOR PLANE REGIONS, CURVES, AND SURFACES

In this section we consider the problem of finding the center of mass of a plane region. This is relatively simple because of our earlier work, and the reader will find that the derivation of the formulas proceeds in much the same manner as in §16.3.

FIGURE 4-1

Suppose we have a region D in $\boldsymbol{R}^2$ and a mass density $\delta(x, y)$. (As a physical example, consider the metal plate in Figure 4-1.) Therefore, the mass of a small region of area $\Delta x \Delta y$ is approximately $\delta(x, y) \Delta x \Delta y$. The total mass is

$$M = \int_D \delta(x, y)\, dx\, dy$$

If we apply the same reasoning as we used for the three-dimensional case, we find that the center of mass $(\bar{x}, \bar{y})$ is

$$\begin{aligned} \bar{x} &= \frac{1}{M} \int_D x\, \delta(x, y)\, dA \\ \bar{y} &= \frac{1}{M} \int_D y\, \delta(x, y)\, dA \end{aligned} \tag{1}$$

If the region is homogeneous (δ is constant) then formulas (1) become

$$\begin{aligned} \bar{x} &= \frac{1}{A(D)} \int_D x\, dA \\ \bar{y} &= \frac{1}{A(D)} \int_D y\, dA \end{aligned} \tag{2}$$

where $A(D) =$ area of D. In this case, $\bar{x}$ and $\bar{y}$ are just the average x and y coordinates of points in D. (If a density is not mentioned we assume the region is homogeneous.)

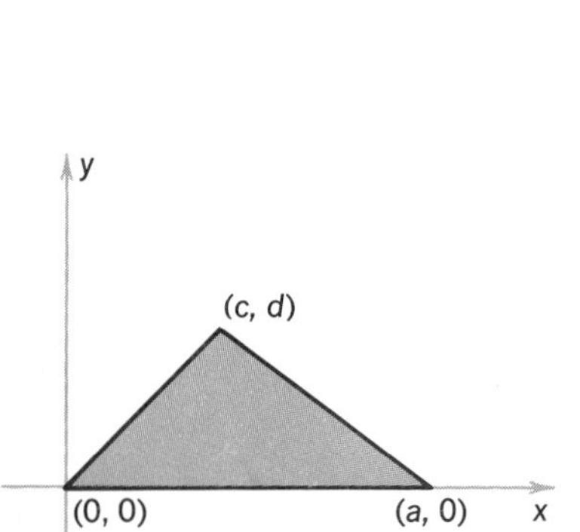

FIGURE 4-2

Example 1 Find the coordinates of the center of mass for the triangle T with vertices at $(0, 0)$, $(0, a)$, and (c, d) (see Figure 4-2).

The area of the triangle T is $ad/2$. Thus, by formulas (2)

$$\bar{x} = \frac{2}{ad}\int_T x\,dA$$

$$\bar{y} = \frac{2}{ad}\int_T y\,dA$$

We shall compute the value of $\bar{x}$; the computation for $\bar{y}$ is similar. The triangle T is a region of type 3; hence, in setting up the iterated integral for $\int_T x\,dA$ we may integrate with respect to x first. (Why is this better?) The equations of the three legs of the triangle are $y=0$, $x=(c/d)y$, $x=[(c-a)/d]y+a$. Therefore

$$\begin{aligned}\int_T x\,dA &= \int_0^d \int_{(c/d)y}^{[(c-a)/d]y+a} x\,dx\,dy = \tfrac{1}{2}\int_0^d [x^2]_{(c/d)y}^{[(c-a)/d]y+a}\,dy\\
&= \frac{1}{2}\int_0^d \left[\left(\frac{c-a}{d}y+a\right)^2 - \frac{c^2}{d^2}y^2\right]dy\\
&= \frac{1}{2}\int_0^d \left\{\left[\left(\frac{c-a}{d}\right)^2 - \frac{c^2}{d^2}\right]y^2 + \frac{2a}{d}(c-a)y+a^2\right\}dx\\
&= \frac{1}{2}\int_0^d \left[\left(\frac{2c-a}{d}\right)\left(\frac{-a}{d}\right)y^2 + \frac{2a}{d}(c-a)y+a^2\right]dy\\
&= \frac{1}{2}\left(\frac{a-2c}{d}\right)\left(\frac{a}{d}\right)\int_0^d y^2\,dy + \frac{a}{d}(c-a)\int_0^d y\,dy + \tfrac{1}{2}a^2\int_0^d dy\\
&= \frac{1}{2}\left(\frac{a-2c}{d}\right)\left(\frac{a}{d}\right)\frac{d^3}{3} + \frac{a}{d}(c-a)\frac{d^2}{2} + \tfrac{1}{2}a^2 d\\
&= \frac{ad}{6}(a-2c) + \tfrac{1}{2}ad(c-a) + \tfrac{1}{2}a^2 d\\
&= \frac{a^2 d}{6} - \frac{cad}{3} + \tfrac{1}{2}acd - \tfrac{1}{2}a^2 d + \tfrac{1}{2}a^2 d = \tfrac{1}{6}ad(c+a)\end{aligned}$$

Hence

$$\bar{x} = \frac{2}{ad}\int_T x\,dA = \tfrac{1}{3}(c+a)$$

A similar computation shows that

$$\bar{y} = \tfrac{1}{3}d$$

Therefore, $(\bar{x},\bar{y}) = (\frac{1}{3}(c+a), \frac{1}{3}d)$ [this point is the intersection of the medians of the triangle].

Essentially the same reasoning that led to the definition of center of mass for planar regions will lead to the corresponding notion of center of mass for paths and surfaces.

It would be useful to know how to compute the center of mass of a curve if, for example, we were interested in finding the center of mass of a homogeneous wire with a fixed cross-sectional diameter.

Therefore, let us suppose we have a curve C traced out by the coordinate functions

$$\boldsymbol{\sigma}(t) = (x(t), y(t), z(t)), \qquad t \in [a, b]$$

where $\boldsymbol{\sigma}'(t) \neq 0$, and $\boldsymbol{\sigma}(t_1) \neq \boldsymbol{\sigma}(t_2)$ for $t_1 \neq t_2$ (Figure 4-3). Then we define the center of mass $(\bar{x}, \bar{y}, \bar{z})$ of the curve by

$$\begin{aligned} \bar{x} &= \frac{1}{L(C)} \int_C x \, ds \\ \bar{y} &= \frac{1}{L(C)} \int_C y \, ds \\ \bar{z} &= \frac{1}{L(C)} \int_C z \, ds \end{aligned} \tag{3}$$

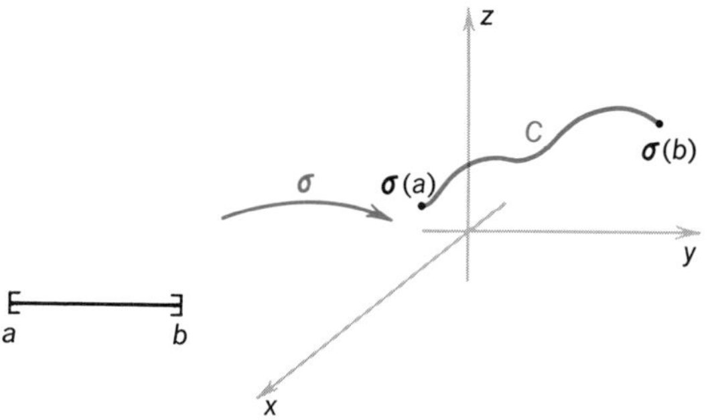

FIGURE 4-3

where

$$L(C) = \int_a^b \sqrt{\left(\frac{dx}{dt}\right)^2 + \left(\frac{dy}{dt}\right)^2 + \left(\frac{dz}{dt}\right)^2} \, dt$$

is the length of C, and as in §13.4 and §13.5, the path integral is defined by

$$\int_C x \, ds = \int_c^b x(t) \sqrt{\left(\frac{dx}{dt}\right)^2 + \left(\frac{dy}{dt}\right)^2 + \left(\frac{dz}{dt}\right)^2} \, dt$$

with similar expressions for $\int_C y \, ds$ and $\int_C z \, ds$. It can be shown that the center of mass does not depend on the particular $\boldsymbol{\sigma}$ chosen, but only on the range of $\boldsymbol{\sigma}$. In (3) $\bar{x}$, $\bar{y}$, and $\bar{z}$ are the average values of the x, y, and z coordinates on C. If the curve C represents a nonhomogeneous wire with mass density

$\rho(t)$, the formulas for center of mass become

$$\bar{x} = \frac{1}{M}\int_a^b x(t)\,\rho(t)\frac{ds}{dt}\,dt$$

$$\bar{y} = \frac{1}{M}\int_a^b y(t)\,\rho(t)\frac{ds}{dt}\,dt \qquad (4)$$

$$\bar{z} = \frac{1}{M}\int_a^b z(t)\,\rho(t)\frac{ds}{dt}\,dt$$

where

$$\frac{ds}{dt} = \sqrt{\left(\frac{dx}{dt}\right)^2 + \left(\frac{dy}{dt}\right)^2 + \left(\frac{dz}{dt}\right)^2}$$

and

$$M = \int_a^b \rho(t)\frac{ds}{dt}\,dt$$

is the mass of the wire.

Next, suppose S is a homogeneous surface given parametrically by $\Phi: D \to \boldsymbol{R}^3$ with $\Phi(u,v) = (h(u,v), g(u,v), f(u,v))$. Then the coordinates of the center of mass $(\bar{x}, \bar{y}, \bar{z})$ are given by

$$\bar{x} = \frac{1}{A(S)}\int_S x\,dS$$

$$\bar{y} = \frac{1}{A(S)}\int_S y\,dS \qquad (5)$$

$$\bar{z} = \frac{1}{A(S)}\int_S z\,dS$$

No doubt the student has already noticed the very strong similarity among the various formulas for center of mass. In general, to compute $\bar{x}$ (homogeneous case) we integrate the function x over the appropriate object (that is, solid, path, surface, plane region) and then divide by the total content of the object (that is, volume, length, surface area). Moreover, we perform similar calculations to obtain $\bar{y}$ and $\bar{z}$.

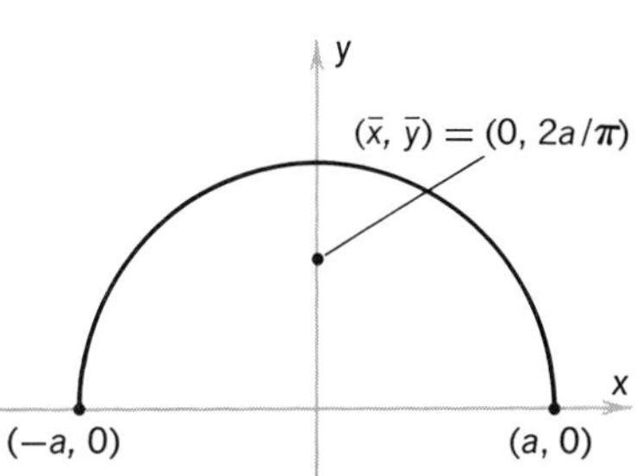

FIGURE 4-4

Example 2 Find $(\bar{x}, \bar{y})$ for the semicircular curve C traced out by $\boldsymbol{\sigma}$: $[0, \pi] \to \boldsymbol{R}^2 : \theta \mapsto (a\cos\theta, a\sin\theta)$ (see Figure 4-4). The parametric representation of $\boldsymbol{\sigma}$ is

$$x = a\cos,\theta, \qquad y = a\sin\theta$$

Therefore, we may apply formulas (3) and write

$$\bar{x} = \frac{1}{L(C)}\int_0^\pi x\sqrt{a^2\sin^2\theta + a^2\cos^2\theta}\,d\theta$$

$$= \frac{1}{L(C)}\int_0^\pi a^2\cos\theta\,d\theta = \frac{a^2}{L(C)}\int_0^\pi \cos\theta\,d\theta = 0$$

where

$$L(C) = \int_0^\pi \sqrt{a^2 \sin^2 \theta + a^2 \cos^2 \theta}\, d\theta$$

$$= a\int_0^\pi d\theta = \pi a$$

Furthermore,

$$\bar{y} = \frac{1}{L(C)}\int_0^\pi y\sqrt{a^2 \sin^2 \theta + a^2 \cos^2 \theta}\, d\theta$$

$$= \frac{1}{\pi a}\int_0^\pi a^2 \sin\theta\, d\theta = \frac{a}{\pi}\int_0^\pi \sin\theta\, d\theta$$

$$= \frac{a}{\pi}\Big[-\cos\theta\Big]_0^\pi = \frac{2a}{\pi}$$

Thus $(\bar{x}, \bar{y}) = (0, 2a/\pi)$.

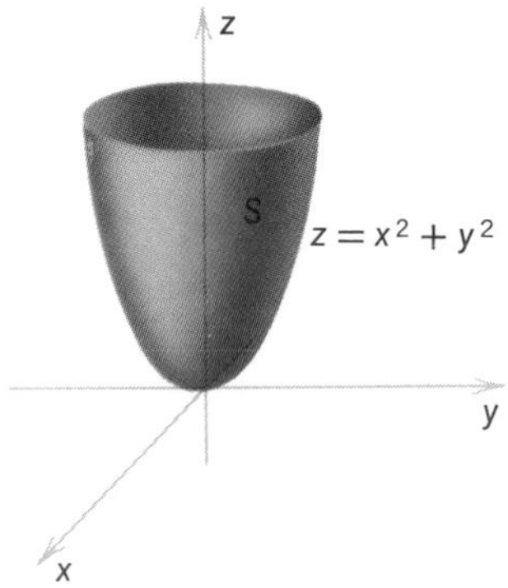

FIGURE 4-5

Example 3 Find the z coordinate of the center of mass of the surface S described by the function $\Phi: D \to R^3 : (u, v) \mapsto (u\cos v, u\sin v, u^2)$ where D is the set $0 \leqslant v \leqslant 2\pi, 0 \leqslant u \leqslant 1$. Notice that this S can also be described by the equation $z = x^2 + y^2$ over the unit disc $x^2 + y^2 \leqslant 1$ (see exercise 5 and Figure 4-5).

Using formulas (5) and the results in §16.2, we have

$$\bar{z} = \frac{1}{A(S)}\int_S z\, dS$$

where

$$A(S) = \int_D \sqrt{\left[\frac{\partial(x,y)}{\partial(u,v)}\right]^2 + \left[\frac{\partial(y,z)}{\partial(u,v)}\right]^2 + \left[\frac{\partial(z,x)}{\partial(u,v)}\right]^2}\, dA$$

$$= \int_D |\boldsymbol{T}_u \times \boldsymbol{T}_v|\, dA$$

Computing the appropriate determinants, we find

$$\frac{\partial(x,y)}{\partial(u,v)} = u, \qquad \frac{\partial(x,z)}{\partial(u,v)} = 2u^2 \sin v, \qquad \frac{\partial(y,z)}{\partial(u,v)} = -2u^2 \cos v$$

Thus the integrand above is $|\boldsymbol{T}_u \times \boldsymbol{T}_v| = \sqrt{u^2 + 4u^4 \sin^2 v + 4u^4 \cos^2 v} = \sqrt{u^2 + 4u^4} = u\sqrt{1 + 4u^2}$ and so

$$A(S) = \int_D u\sqrt{1+4u^2}\, dA = \int_0^{2\pi}\int_0^1 u\sqrt{1+4u^2}\, du\, dv$$

$$= \frac{1}{8}\int_0^{2\pi}\int_0^1 \sqrt{1+4u^2}\,(8u\, du)\, dv$$

$$= \frac{2}{3}\cdot\frac{1}{8}\int_0^{2\pi}\Big[(1+4u^2)^{3/2}\Big]_0^1 dv$$

$$= \frac{1}{12}\int_0^{2\pi}\Big[5^{3/2} - 1\Big]\, dv = \frac{\pi}{6}\Big[5^{3/2} - 1\Big]$$

Now

$$\int_S z\, dA = \int_D z\,|\boldsymbol{T}_u \times \boldsymbol{T}_v|\, dA = \int_D u^3\sqrt{1+4u^2} = \int_0^{2\pi}\int_0^1 u^3\sqrt{1+4u^2}\, du\, dv$$

$$= 2\pi \int_0^1 u^3\sqrt{1+4u^2}\, du$$

If $r^2 = 1+4u^2$, $u^2 = \frac{1}{4}(r^2-1)$, then $du = \frac{1}{4}(r/u)\, dr$. Thus

$$\int_0^1 u^3\sqrt{1+4u^2}\, du = \int_1^{\sqrt{5}} u^3 \cdot r\left(\frac{1}{4}\cdot\frac{r}{u}\right) dr = \frac{1}{4}\int_1^{\sqrt{5}} u^2 r^2\, dr$$

$$= \frac{1}{16}\int_1^{\sqrt{5}} (r^2-1)\, r^2\, dr = \frac{1}{16}\left[\frac{r^5}{5} - \frac{r^3}{3}\right]_1^{\sqrt{5}}$$

$$= \frac{1}{16}\left[5^{3/2} - \frac{5^{3/2}}{3} - \frac{1}{5} + \frac{1}{3}\right] = \frac{1}{16}\left[5^{3/2} - \frac{5^{3/2}}{3} + \frac{2}{15}\right]$$

Thus

$$\int_S z\, dS = \frac{2\pi}{16}\left[5^{3/2} - \frac{5^{3/2}}{3} + \frac{2}{15}\right]$$

and consequently

$$\bar{z} = \frac{\dfrac{\pi}{8}\left[5^{3/2} - \dfrac{5^{3/2}}{3} + \dfrac{2}{15}\right]}{\dfrac{\pi}{6}[5^{3/2}-1]} = \frac{1}{4}\cdot\frac{\left[3\cdot 5^{3/2} - 5^{3/2} + \dfrac{2}{5}\right]}{[5^{3/2}-1]}$$

EXERCISES

1. Prove the analog of exercise 2, §16.3, for plane regions.
2. Use exercise 1 to find the center of mass of the homogeneous plates shown in Figure 4-6.

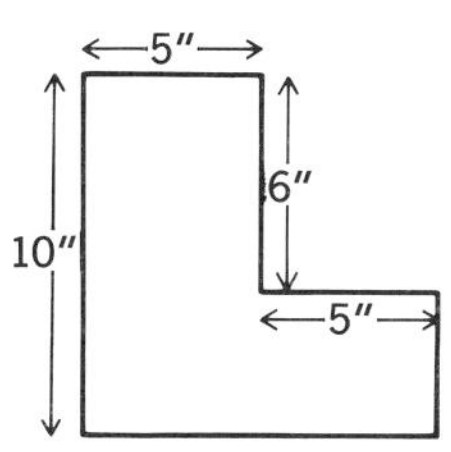

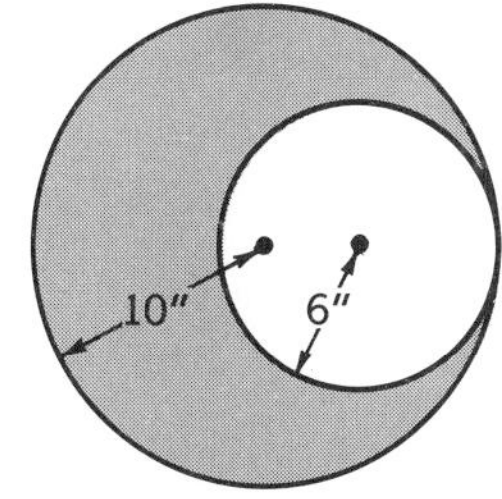

FIGURE 4-6

3. Let D be a type 1 region in $\boldsymbol{R}^2$ and let $W = D \times [c, d]$, that is, $(x, y, z) \in D \Rightarrow (x, y) \in D$, $c \leqslant z \leqslant d$ (see Figure 4-7). Let $\delta(x, y)$ be a mass density on D and consider the corresponding mass density $\delta(x, y, z) = \delta(x, y)$ on W. Show that the center of mass of W is

$$\left(\bar{x}, \bar{y}, \frac{c+d}{2}\right)$$

where $(\bar{x}, \bar{y})$ is the center of mass of D.

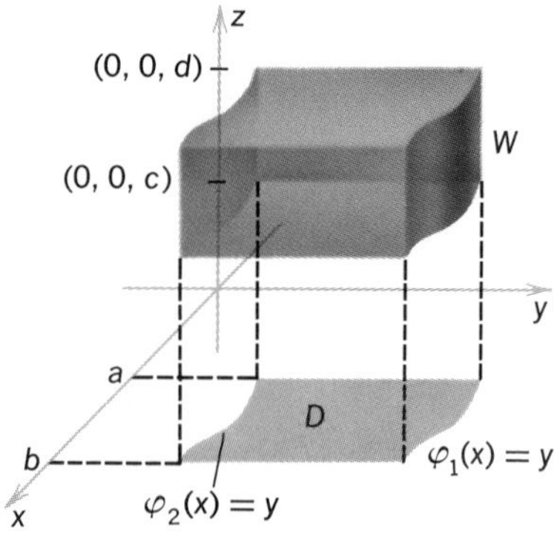

FIGURE 4-7

4. Justify equations (3) and (5) using an argument based on Riemann sums. Generalize equations (5) in case the surface has a mass density ρ which need not be constant.
5. Work Example 3 using the equation $z = x^2 + y^2$ to describe the same surface.
6. Find the center of mass of the region bounded by $y = x^3$, the x axis, and $x = 1$ with density $\delta(x, y) = x$.
7. Find the center of mass of the wire $y = \frac{2}{3}x^{3/2}$, $0 \leqslant x \leqslant 1$, with density $\rho(x) = \sqrt{x}$.
8. Find the moment of inertia of the wire $y = x^2$, $0 \leqslant x \leqslant 1$, about the x axis (see exercise 7, §16.3).
9. Prove **Pappus' Theorem**: Consider a curve rotated about an axis [for example, $y = f(x)$, $a \leqslant x \leqslant b$, rotated about the x axis]. Show that the area of the surface of revolution is

$$A = 2\pi h l$$

where l is the length of the curve and h is the distance of the center of gravity of the curve from the axis (see §16.1, exercise 11).

10. Formulate an analog to Pappus' Theorem for volumes of revolution.
11. Use exercise 10 to compute the volume of the region bounded by $x^2 + y^2 + z^2 = 1$, $x^2 + y^2 = \frac{1}{4}(z-1)^2$ and $x^2 + y^2 = 1$.
12. Find the center of mass of the surface of a hemisphere of radius a.

REVIEW EXERCISES FOR CHAPTER 16

1. Find the area of the surface defined by $\Phi: (u, v) \mapsto (x, y, z)$ where

$$x = h(u, v) = u + v, \quad y = g(u, v) = u, \quad z = f(u, v) = v$$

$0 \leqslant u \leqslant 1$, $0 \leqslant v \leqslant 1$. Sketch.

2. Write a formula for the surface area of $\Phi: (r, \theta) \mapsto (x, y, z)$ where

$$x = r\cos\theta, \quad y = 2r\sin\theta, \quad z = r$$

$0 \leqslant r \leqslant 1$, $0 \leqslant \theta \leqslant 2\pi$. Sketch.

3. Compute the integral of $f(x, y, z) = x^2 + y^2 + z^2$ over the surface in exercise 1.

4. Consider the curve $x(t)=t$, $y(t)=t^2$, $0 \leqslant t \leqslant 1$, and the surface determined by the set
$$x(t) = t, \quad y(t) = t^2, \quad 0 \leqslant z(t) \leqslant x(t)^2 + y(t)^2$$
Find its area and sketch.

5. Compute the center of mass of a wire bent into a circle $x(t)=\cos t$, $y(t)=\sin t$, $0 \leqslant t \leqslant 2\pi$, if the density is $\delta(t)=1+\cos t$.

6. Find the moment of inertia of a square plate, with side a, about an axis through the center of a plate and the center of a side.

7. Compute the center of mass of a sphere $x^2+y^2+z^2 \leqslant 1$ with density $\delta(x,y,z)=2+y$.

8. Repeat exercise 7 for the circle $x^2+y^2 \leqslant 1$ with $\delta(x,y) = 2+y$.

9. Find the center of mass of the region determined by $x^2+y^2 \leqslant 1$, $0 \leqslant z \leqslant 1-x$.

10. Find the moment of inertia about the z axis of the region in exercise 9.

11. Show that the mass center of the surface of a cubical box (with no lid) is $\frac{3}{5}$ the way down from the top on the axis of symmetry.

12. The curve $x(t)=t^2$, $y(t)=t$, $0 \leqslant t \leqslant \pi$, is rotated about the x axis. Compute the resulting surface area.

13. Consider a cone obtained by revolving the curve $y=ax+b$, $b>0$, $a<0$, $0 \leqslant x \leqslant -b/a$, about the y axis. Suppose the mass density is $\delta(x,y,z)=\sqrt{x^2+z^2}$. Compute the total mass and the center of mass. [Hint: Exploit the symmetry and consider a thin cylindrical shell of width Δx in the cone.]

14. Consider the solid B formed by rotating the region $0 \leqslant y \leqslant f(x)$, $0 \leqslant x \leqslant b$, about the y axis. Let the mass density $\delta(r)$ be a function of the radial distance from the y axis only.
    (a) Show that the total mass is
    $$M = 2\pi \int_0^b \delta(x)\, x f(x)\, dx$$
    (b) Find the formula for the center of mass.
    (c) Find a formula for the moment of inertia about the y axis.

*15. Supply a rigorous proof for exercise 14(a).

16. The circle $x=2-\cos\theta$, $y=\sin\theta$, $0 \leqslant \theta \leqslant 2\pi$, is rotated about the y axis to form a torus. The mass δ is a function only of the distance to the circle running around the center of the torus. Find a formula for the total mass. What is the center of gravity?

17. Compute the integral of e^{x+y+z} over the square with vertices $(1,0,0)$, $(2,0,0)$, $(1,1,1)$, and $(2,1,1)$.

18. Compute the integral of $x+y$ over the surface of the unit sphere.

19. Compute the integral of x over the rectangle with vertices $(1,1,1)$, $(2,1,1)$, and $(0,0,3)$.

17

VECTOR ANALYSIS

In Chapters 13 and 14 we studied vectors and vector functions in some detail. The emphasis was on the algebraic and geometric aspects in Chapter 13 and on the differential aspects in Chapter 14. Now our aim is to combine these earlier results with the material on integration from Chapters 15 and 16.

In the case of one variable, the key result for linking the differential and integral calculus is the Fundamental Theorem. For vector functions the basic idea is the same and the analogous result is contained in the three theorems of Green, Gauss, and Stokes. Hence, the goal of this chapter is to present these theorems and then consider some of their significant physical applications.

1 LINE INTEGRALS

If $\boldsymbol{F}$ is a force field in space, then any test particle (for example, a charge in an electric force field or a mass in a gravitational field) will experience the force $\boldsymbol{F}$. Suppose the particle moves along a path $\boldsymbol{\sigma}$ and is acted upon by $\boldsymbol{F}$. One of the very fundamental concepts in physics is the **work done** by $\boldsymbol{F}$ on the particle as it traces out the path $\boldsymbol{\sigma}$. If $\boldsymbol{\sigma}$ is a straight-line displacement given by the vector $\boldsymbol{d}$ and $\boldsymbol{F}$ is a constant force, then the work done by $\boldsymbol{F}$ in moving the particle along the path is $\boldsymbol{F}\cdot\boldsymbol{d}$. This basic physical notion leads to the equally basic mathematical concept of a line integral.

Let $\boldsymbol{F}$ be a vector field on $\boldsymbol{R}^3$ and let $\boldsymbol{\sigma}:[a,b]\to\boldsymbol{R}^3$ be a C^1 path. We define $\int_{\boldsymbol{\sigma}} \boldsymbol{F}$, the **line integral** of $\boldsymbol{F}$ along $\boldsymbol{\sigma}$, by the formula

$$\int_{\boldsymbol{\sigma}} \boldsymbol{F} = \int_a^b \boldsymbol{F}(\boldsymbol{\sigma}(t)) \cdot \boldsymbol{\sigma}'(t)\, dt$$

that is, we integrate the dot product of $\boldsymbol{F}$ with $\boldsymbol{\sigma}'$ over the interval $[a, b]$.

Sometimes we denote the path by $\boldsymbol{s}$ and use $d\boldsymbol{s}/dt$ for the tangent vector $\boldsymbol{\sigma}'(t)$. Then the line integral becomes

$$\int \boldsymbol{F}(\boldsymbol{s}(t)) \cdot \frac{d\boldsymbol{s}}{dt}\,dt$$

Recall that in Chapter 13 we defined the integral of a real-valued function f along the path $\boldsymbol{\sigma}$ as

$$\int_{\boldsymbol{\sigma}} f = \int_a^b f(\boldsymbol{\sigma}(t))|\boldsymbol{\sigma}'(t)|\,dt$$

where $|\boldsymbol{\sigma}'(t)|$ is the length of $\boldsymbol{\sigma}'(t)$. Now the line integral of $\boldsymbol{F}$ is actually an integral of a particular real-valued function f along $\boldsymbol{\sigma}$. To see this, choose

$$f(\boldsymbol{\sigma}(t)) = \boldsymbol{F}(\boldsymbol{\sigma}(t)) \cdot \boldsymbol{T}(t)$$

where

$$\boldsymbol{T}(t) = \boldsymbol{\sigma}'(t)/|\boldsymbol{\sigma}'(t)|$$

is the unit tangent vector [assume $\boldsymbol{\sigma}'(t) \neq 0$ so this is defined]. But $\boldsymbol{F}\cdot\boldsymbol{T}$ is exactly the component of $\boldsymbol{F}$ tangent to the curve (since $\boldsymbol{T}$ is a unit vector).

Consequently the line integral of a vector field $\boldsymbol{F}$ along a path $\boldsymbol{\sigma}'(t)$ is equal to the path integral of the real-valued function which is the tangential component of $\boldsymbol{F}$, that is,

$$\int_{\boldsymbol{\sigma}} \boldsymbol{F} = \int_a^b \left[\boldsymbol{F}(\boldsymbol{\sigma}(t)) \cdot \frac{\boldsymbol{\sigma}(t)}{|\boldsymbol{\sigma}'(t)|}\right] |\boldsymbol{\sigma}'(t)|\,dt = \int_a^b f(\boldsymbol{\sigma}(t))\,|\boldsymbol{\sigma}'(t)|\,dt$$

To compute a line integral in any particular case, one can use either the original definition or else integrate the tangential component, whichever is easier or more appropriate.

Example 1 Let $\boldsymbol{\sigma}(t) = (\sin t, \cos t, t)$, with $0 \leqslant t \leqslant 2\pi$. Let $\boldsymbol{F}(x, y, z) = x\boldsymbol{i} + y\boldsymbol{j} + z\boldsymbol{k}$. Then $\boldsymbol{F}(\boldsymbol{\sigma}(t)) = \boldsymbol{F}(\sin t, \cos t, t) = \sin t\boldsymbol{i} + \cos t\boldsymbol{j} + t\boldsymbol{k}$, and $\boldsymbol{\sigma}'(t) = \cos t\boldsymbol{i} - \sin t\boldsymbol{j} + \boldsymbol{k}$. Therefore,

$$\boldsymbol{F}(\boldsymbol{\sigma}(t)) \cdot \boldsymbol{\sigma}'(t) = \sin t \cos t - \cos t \sin t + t = t$$

and so

$$\int_{\boldsymbol{\sigma}} \boldsymbol{F} = \int_0^{2\pi} t\,dt = 2\pi^2$$

There is another way of writing line integrals. If we set $d\boldsymbol{s} = dx\boldsymbol{i} + dy\boldsymbol{j} + dz\boldsymbol{k}$, we can write

$$\int_{\boldsymbol{\sigma}} \boldsymbol{F} = \int_{\boldsymbol{\sigma}} \boldsymbol{F} \cdot d\boldsymbol{s} = \int_{\boldsymbol{\sigma}} F_1\,dx + F_2\,dy + F_3\,dz$$

where F_1, F_2, and F_3 are the components of the vector field $\boldsymbol{F}$. (We shall use both notations $\int_{\boldsymbol{\sigma}} \boldsymbol{F}$ and $\int_{\boldsymbol{\sigma}} \boldsymbol{F} \cdot d\boldsymbol{s}$ for the line integral of $\boldsymbol{F}$ along $\boldsymbol{\sigma}$.) We call the expression $F_1\,dx + F_2\,dy + F_3\,dz$ a **differential form.** By definition the integral of a form is

$$\int_\sigma F_1\,dx + F_2\,dy + F_3\,dz = \int_a^b \left(F_1\frac{dx}{dt} + F_2\frac{dy}{dt} + F_3\frac{dz}{dt}\right)dt$$

Forms provide an elegant notation that is very useful when working with line integrals.

Example 2 Evaluate $\int_\sigma x^2\,dx + xy\,dy + dz$, where $\boldsymbol{\sigma}:[0,1]\to\boldsymbol{R}^3$ is given by $\boldsymbol{\sigma}(t) = (t, t^2, 1) = (x(t), y(t), z(t))$

We compute $dx/dt = 1, dy/dt = 2t, dz/dt = 0$; therefore,

$$\int_\sigma x^2\,dx + xy\,dy + dz = \int_0^1 \left([x(t)]^2\frac{dx}{dt} + [x(t)y(t)]\frac{dy}{dt}\right)dt = \int_0^1 (t^2 + 2t^4)\,dt$$

$$= \left[\frac{1}{3}t^3 + \frac{2}{5}t^5\right]_0^1 = \frac{11}{15}$$

Example 3 Evaluate $\int_\sigma \cos z\,dx + e^x\,dy + e^y\,dz$, where $\boldsymbol{\sigma}(t) = (1, t, e^t)$ and $0 \leqslant t \leqslant 2$.

We compute $dx/dt = 0, dy/dt = 1, dz/dt = e^t$, and so

$$\int_\sigma \cos z\,dx + e^x\,dy + e^y\,dz = \int_0^2 (0 + e + e^{2t})\,dt$$

$$= \left[et + \tfrac{1}{2}e^{2t}\right]_0^2 = 2e + \tfrac{1}{2}e^4 - \tfrac{1}{2}$$

Example 4 Let $\boldsymbol{\sigma}$ be the path

$$x = \cos^3\theta, \quad y = \sin^3\theta, \quad z = \theta, \quad 0 \leqslant \theta \leqslant \frac{7\pi}{2}$$

(see Figure 1-1). Evaluate the integral $\int_\sigma \sin z\,dx + \cos z\,dy + (xy)^{1/3}\,dz$.

In this case we have

$$\frac{dx}{d\theta} = -3\cos^2\theta\sin\theta, \quad \frac{dy}{d\theta} = 3\sin^2\theta\cos\theta, \quad \frac{dz}{d\theta} = 1$$

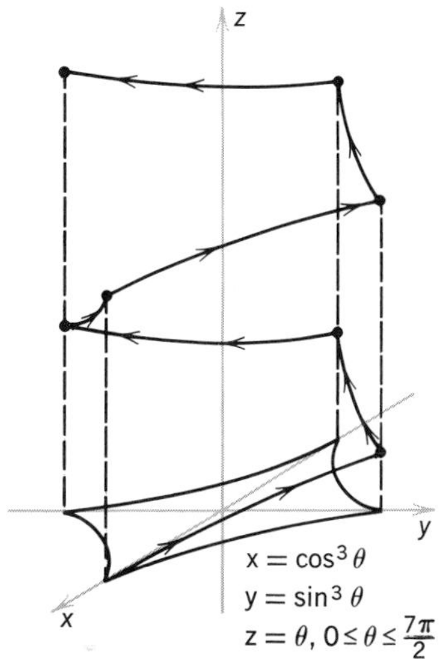

FIGURE 1-1

and so the integral is

$$\int_\sigma \sin z\,dx + \cos z\,dy + (xy)^{1/3}\,dz$$

$$= \int_0^{7\pi/2} (-3\cos^2\theta\sin^2\theta + 3\sin^2\theta\cos^2\theta + \cos\theta\sin\theta)\,d\theta$$

$$= \int_0^{7\pi/2} \cos\theta\sin\theta\,d\theta = [\tfrac{1}{2}\sin^2\theta]_0^{7\pi/2} = -\tfrac{1}{2}$$

We have already mentioned that the definition of line integral is motivated by the physical idea that if a particle is acted on by a constant force $\boldsymbol{F}$ and is moved through a displacement given by a vector $\boldsymbol{d}$, then the work done is

$$\boldsymbol{F}\cdot\boldsymbol{d} = (\text{force}) \times (\text{displacement in direction of force})$$

Suppose now that our particle traverses a curved path described by the function $\boldsymbol{\sigma}:[a,b]\to\boldsymbol{R}^3$. If t ranges over a small interval t to $t+\Delta t$, the particle moves from $\boldsymbol{\sigma}(t)$ to $\boldsymbol{\sigma}(t+\Delta t)$, a vector displacement $\boldsymbol{\Delta}=\boldsymbol{\sigma}(t+\Delta t)-\boldsymbol{\sigma}(t)$. Now, by the Mean Value Theorem we get, approximately, $\boldsymbol{\Delta}=\boldsymbol{\sigma}'(t)\,\Delta t$. The work done in going from $\boldsymbol{\sigma}(t)$ to $\boldsymbol{\sigma}(t+\Delta t)$ is therefore approximately $\boldsymbol{F}(\boldsymbol{\sigma}(t))\cdot\boldsymbol{\sigma}'(t)\,\Delta t$. We sum over all Δt and find that the total work is $\int_a^b \boldsymbol{F}(\boldsymbol{\sigma}(t))\cdot\boldsymbol{\sigma}'(t)\,dt$. Hence we may interpret $\int_{\boldsymbol{\sigma}}\boldsymbol{F}$ as the work done by $\boldsymbol{F}$ in moving a particle along $\boldsymbol{\sigma}$.

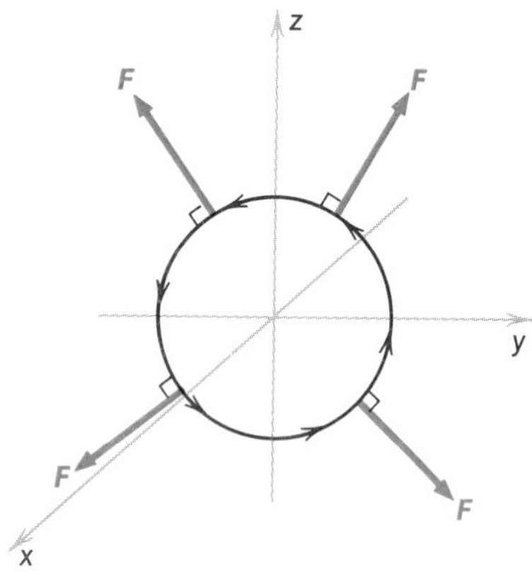

FIGURE 1-2

Example 5 Suppose $\boldsymbol{F}$ is the field $\boldsymbol{F}(x,y,z)=x^3\,\boldsymbol{i}+y\boldsymbol{j}+z\boldsymbol{k}$. We parametrize a circle σ of radius a in the yz plane by

$$x=0,\quad y=a\cos\theta,\quad z=a\sin\theta,\quad 0\leqslant\theta\leqslant 2\pi$$

Since the force field $\boldsymbol{F}$ is normal to the circle at every point on the circle, $\boldsymbol{F}$ will not contribute any work to a particle moving along the circle (Figure 1-2). The work done by $\boldsymbol{F}$ must, therefore, be 0. We can verify this by a direct computation:

$$\begin{aligned} W &= \int_{\sigma}\boldsymbol{F} = \int_{\sigma} x^3\,dx+y\,dy+z\,dz \\ &= \int_0^{2\pi}(0-a^2\cos\theta\sin\theta+a^2\cos\theta\sin\theta)\,d\theta \\ &= 0 \end{aligned}$$

as predicted.

Example 6 If we consider the field and curve of Example 4, we see that the work done by the field is $-\frac{1}{2}$, a negative quantity. This means that the field impedes movement along that path.

We have seen that the line integral $\int_{\boldsymbol{\sigma}}\boldsymbol{F}$ depends not only on the field $\boldsymbol{F}$ but also on the path $\boldsymbol{\sigma}:[a,b]\to\boldsymbol{R}^3$. In general, if $\boldsymbol{\sigma}$ and $\boldsymbol{\rho}$ are two different paths in $\boldsymbol{R}^3$, $\int_{\boldsymbol{\sigma}}\boldsymbol{F}\neq\int_{\boldsymbol{\rho}}\boldsymbol{F}$. On the other hand, we shall see that it is true that $\int_{\boldsymbol{\sigma}}\boldsymbol{F}=\pm\int_{\boldsymbol{\rho}}\boldsymbol{F}$ for every vector field $\boldsymbol{F}$ if $\boldsymbol{\rho}$ is what we call a **reparametrization of $\boldsymbol{\sigma}$.**

Definition Let $h:I\to I_1$ be a C^1 real-valued function which is a one-to-one map of an interval $I=[a,b]$ onto another interval $I_1=[a_1,b_1]$. Let $\boldsymbol{\sigma}:I_1\to\boldsymbol{R}^3$ be a piecewise C^1 path. Then we call the composition

$$\boldsymbol{\rho}=\boldsymbol{\sigma}\circ h:I\to\boldsymbol{R}^3$$

a **reparametrization** of $\boldsymbol{\sigma}$.

Hence $\boldsymbol{\rho}(t)=\boldsymbol{\sigma}(h(t))$, so h changes the variable; alternatively, one can think of h as changing the speed at which a point moves along the path. Indeed, observe $\boldsymbol{\rho}'(t)=\boldsymbol{\sigma}'(h(t))\,h'(t)$ so the length of the velocity vector for $\boldsymbol{\sigma}$ is multiplied by the scalar factor $h'(t)$.

It is implicit in the definition that h must carry endpoints to endpoints; that is, either $h(a)=a_1$ and $h(b)=b_1$, or $h(a)=b_1$ and $h(b)=a_1$. We thus distinguish two types of reparametrizations. If $\boldsymbol{\sigma}\circ h$ is a reparametrization of $\boldsymbol{\sigma}$ then either

$$\boldsymbol{\sigma}\circ h(a)=\boldsymbol{\sigma}(a_1) \quad \text{and} \quad \boldsymbol{\sigma}\circ h(b)=\boldsymbol{\sigma}(b_1)$$

or

$$\boldsymbol{\sigma}\circ h(a)=\boldsymbol{\sigma}(b_1) \quad \text{and} \quad \boldsymbol{\sigma}\circ h(b)=\boldsymbol{\sigma}(a_1)$$

In the first case, the reparametrization is called **orientation preserving** and a particle tracing the path $\boldsymbol{\sigma}\circ h$ moves in the same direction as a particle tracing $\boldsymbol{\sigma}$. In the second case, the reparametrization is called **orientation reversing**, and a particle tracing the path $\boldsymbol{\sigma}\circ h$ moves in the opposite direction to that of a particle tracing $\boldsymbol{\sigma}$ (Figure 1-3).

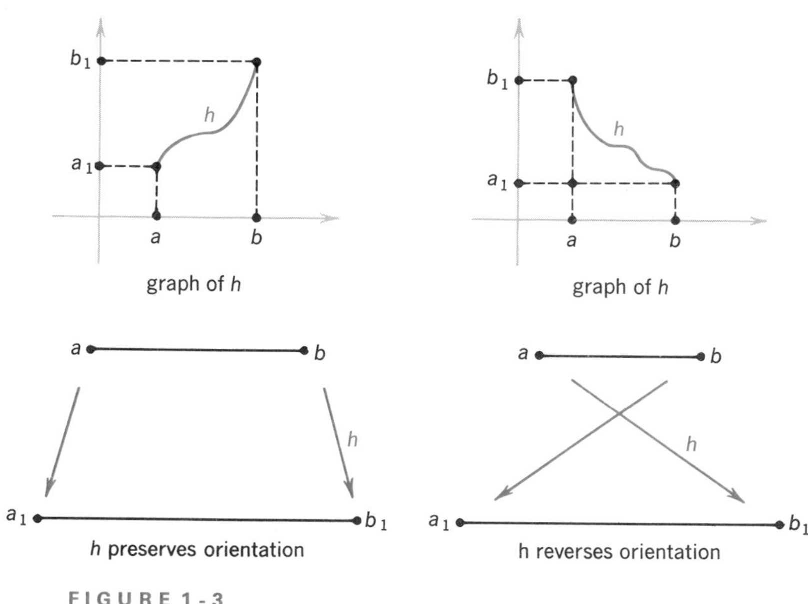

FIGURE 1-3

Example 7 Let $\boldsymbol{\sigma}:[a,b]\to \boldsymbol{R}^3$ be any piecewise C^1 path. Then:

(a) The path $\boldsymbol{\sigma}_{\text{op}}:[a,b]\to \boldsymbol{R}^3:t\mapsto\boldsymbol{\sigma}(a+b-t)$ is a reparametrization of $\boldsymbol{\sigma}$ corresponding to the map $h:[a,b]\to[a,b]:t\mapsto a+b-t$ (which is non-constant affine, and therefore clearly one-to-one, onto, and C^1); we call $\boldsymbol{\sigma}_{\text{op}}$ the **opposite path to $\boldsymbol{\sigma}$**. This reparametrization is orientation reversing.

(b) The path $\boldsymbol{\rho}:[0,1]\to \boldsymbol{R}^3:t\mapsto\boldsymbol{\sigma}(a+(b-a)t)$ is an orientation preserving reparametrization of $\boldsymbol{\sigma}$, again corresponding to a change of coordinates $h:[0,1]\to[a,b]:u\mapsto a+(b-a)u$.

Theorem 1 Let $\boldsymbol{F}$ be a vector field, continuous on the C^1 path $\boldsymbol{\sigma}:[a_1, b_1] \to \boldsymbol{R}^3$, and let $\boldsymbol{\rho}:[a, b] \to \boldsymbol{R}^3$ be a reparametrization of $\boldsymbol{\sigma}$. If $\boldsymbol{\rho}$ is orientation preserving, then

$$\int_{\rho} \boldsymbol{F} = \int_{\sigma} \boldsymbol{F}$$

while if $\boldsymbol{\rho}$ is orientation reversing, then

$$\int_{\rho} \boldsymbol{F} = -\int_{\sigma} \boldsymbol{F}$$

Proof We set $t = h(u)$, where h is the map such that $\boldsymbol{\rho} = \boldsymbol{\sigma} \circ h$. Then, since

$$\frac{dx}{du} = \frac{dx}{dt}\frac{dt}{du}, \quad \frac{dy}{du} = \frac{dy}{dt}\frac{dt}{du}, \quad \frac{dz}{du} = \frac{dz}{dt}\frac{dt}{du}$$

we have

$$\int_{\rho} \boldsymbol{F} = \int_a^b F_1 \frac{dx}{dt}\frac{dt}{du}\,du + F_2 \frac{dy}{dt}\frac{dt}{du}\,du + F_3 \frac{dz}{dt}\frac{dt}{du}\,du$$

which by ordinary Change of Variables becomes

$$\int_{h(a)}^{h(b)} F_1 \frac{dx}{dt}\,dt + F_2 \frac{dy}{dt}\,dt + F_3 \frac{dz}{dt}\,dt$$

$$= \begin{cases} \int_{\sigma} F_1\,dx + F_2\,dy + F_3\,dz & \text{if } \boldsymbol{\rho} \text{ is orientation preserving} \\ -\int_{\sigma} F_1\,dx + F_2\,dy + F_3\,dz & \text{if } \boldsymbol{\rho} \text{ is orientation reversing} \end{cases}$$ ∎

Thus, if it convenient to reparametrize a path when evaluating an integral, Theorem 1 assures us that the value of the integral will not be affected.

Example 8 Let $\boldsymbol{F}:(x, y, z) \to yz\boldsymbol{i} + xz\boldsymbol{j} + xy\boldsymbol{k}$ and $\boldsymbol{\sigma}:[-5, 10] \to \boldsymbol{R}^3 : t \to (t, t^2, t^3)$. Evaluate $\int_{\sigma} \boldsymbol{F}$ and $\int_{\sigma_{op}} \boldsymbol{F}$.

For $\boldsymbol{\sigma}$ we have $dx/dt = 1$, $dy/dt = 2t$, $dz/dt = 3t^2$, and $\boldsymbol{F}(\boldsymbol{\sigma}(t)) = t^5\,\boldsymbol{i} + t^4\,\boldsymbol{j} + t^3\,\boldsymbol{k}$, therefore,

$$\int_{\sigma} \boldsymbol{F} = \int_{-5}^{10} F_1 \frac{dx}{dt} + F_2 \frac{dy}{dt} + F_3 \frac{dz}{dt}\,dt$$

$$= \int_{-5}^{10} (t^5 + 2t^5 + 3t^5)\,dt = \left[t^6\right]_{-5}^{10} = 984\,375$$

On the other hand, for

$$\boldsymbol{\sigma}_{\text{op}}:[-5, 10] \to \boldsymbol{R}^3 : t \mapsto \sigma(5-t) = (5-t, (5-t)^2, (5-t)^3)$$

we have $dx/dt = -1$, $dy/dt = -10+2t = -2(5-t)$, $dz/dt = -75+30t-3t^2 = -3(5-t)^2$, and $\boldsymbol{F}(\boldsymbol{\sigma}_{op}(t)) = (5-t)^5\boldsymbol{i}+(5-t)^4\boldsymbol{j}+(5-t)^3\boldsymbol{k}$, therefore

$$\int_{\boldsymbol{\sigma}_{\text{op}}} \boldsymbol{F} = \int_{-5}^{10} (-(5-t)^5 - 2(5-t)^5 - 3(5-t)^5)\, dt$$

$$= \Big[(5-t)^6\Big]_{-5}^{10} = -984\,375$$

We are interested in reparametrizations because if the image of a particular $\boldsymbol{\sigma}$ can be represented in many ways, we want to be sure that our integrals do not depend on the particular parametrization. For example, for some problems the unit circle may be more conveniently represented by the map $\boldsymbol{\rho}$ given by

$$x(t) = \cos 2t, \qquad y(t) = \sin 2t, \qquad 0 \leqslant t \leqslant \pi$$

Theorem 1 guarantees that any integral computed for this representation will be the same as when we represent the circle by the map $\boldsymbol{\sigma}$ given by

$$x(t) = \cos t, \qquad y(t) = \sin t, \qquad 0 \leqslant t \leqslant 2\pi$$

since $\boldsymbol{\sigma} = \boldsymbol{\rho}\circ h$, where $h(t) = 2t$, and thus $\boldsymbol{\sigma}$ is a reparametrization of $\boldsymbol{\rho}$. However, notice that the map γ given by

$$x(t) = \cos t, \qquad y(t) = \sin t, \qquad 0 \leqslant t \leqslant 4\pi$$

is not a reparametrization of $\boldsymbol{\rho}$. Although it traces out the same image (the circle), it does so twice. (Why does this imply that γ is not a reparametrization of $\boldsymbol{\rho}$?)

The line integral $\int_{\boldsymbol{\sigma}} \boldsymbol{F}$ differs from the path integral $\int_{\boldsymbol{\sigma}} f$. We have seen that

$$\int_{\boldsymbol{\sigma}} \boldsymbol{F} = \int_{\boldsymbol{\sigma}} f \tag{1}$$

where

$$f(\boldsymbol{\sigma}(t)) = \boldsymbol{F}(\boldsymbol{\sigma}(t)) \cdot \frac{\boldsymbol{\sigma}'(t)}{|\boldsymbol{\sigma}'(t)|}$$

Although equation (1) establishes a relationship between line integrals and path integrals, there is a distinct difference between them, other than the obvious fact that the former is an integral of a vector field and the latter of a real-valued function.

The line integral is an **oriented integral,** in that a change of sign occurs (as we have seen in Theorem 1) if the orientation of the curve is reversed. The path integral does not have this property. This follows from the fact that changing t to $-t$ (reversing orientation) just changes the sign of $\boldsymbol{\sigma}'(t)$ and not its length.

The same technique used in Theorem 1 can be applied to prove the next result (see exercise 15).

Theorem 2 Let $\boldsymbol{\sigma}$ be piecewise C^1, f a continuous (real-valued) function on the image of $\boldsymbol{\sigma}$, and let $\boldsymbol{\rho}$ be any reparametrization of $\boldsymbol{\sigma}$. Then

$$\int_{\boldsymbol{\sigma}} f(x, y, z)\, ds = \int_{\boldsymbol{\rho}} f(x, y, z)\, ds$$

We conclude this section with a simple but often very useful technique for evaluating line integrals. A vector field $\boldsymbol{F}$ is a **gradient vector field** if $\boldsymbol{F} = \nabla f$ for some real-valued function f. Thus

$$\boldsymbol{F} = \frac{\partial f}{\partial x}\boldsymbol{i} + \frac{\partial f}{\partial y}\boldsymbol{j} + \frac{\partial f}{\partial z}\boldsymbol{k}$$

Suppose $G, g\colon [a, b] \to \boldsymbol{R}$ are real-valued functions with $G' = g$. Then by the Fundamental Theorem of Calculus $\int_a^b g = G(b) - G(a)$. Thus the value of the integral of g depends only on the value of G at the endpoints of the interval $[a, b]$. Since ∇f, in some sense, represents the derivative of f, one can ask whether $\int_{\boldsymbol{\sigma}} \nabla f$ is completely determined by the value of f at $\boldsymbol{\sigma}(a)$ and $\boldsymbol{\sigma}(b)$. The answer is contained in the following generalization of the Fundamental Theorem.

Theorem 3 Suppose $f\colon \boldsymbol{R}^3 \to \boldsymbol{R}$ and that $\boldsymbol{\sigma}\colon [a, b] \to \boldsymbol{R}^3$ is a (piecewise C^1) map. Then

$$\int_{\boldsymbol{\sigma}} \nabla f = f(\boldsymbol{\sigma}(b)) - f(\boldsymbol{\sigma}(a))$$

Proof We apply the Chain Rule to the composite function

$$F\colon t \mapsto f(\boldsymbol{\sigma}(t))$$

to obtain

$$F'(t) = (f \circ \boldsymbol{\sigma})'(t) = \nabla f(\boldsymbol{\sigma}(t)) \cdot \boldsymbol{\sigma}'(t)$$

The function F is a real function of the variable t, and so by the Fundamental Theorem, we have

$$\int_a^b F' = F(b) - F(a) = f(\boldsymbol{\sigma}(b)) - f(\boldsymbol{\sigma}(a))$$

Therefore,

$$\int_{\boldsymbol{\sigma}} \nabla f = \int_a^b \nabla f(\boldsymbol{\sigma}(t)) \cdot \boldsymbol{\sigma}'(t)\, dt = \int_a^b F'(t)\, dt = F(b) - F(a)$$
$$= f(\boldsymbol{\sigma}(b)) - f(\boldsymbol{\sigma}(a)) \quad \blacksquare$$

Example 9 Let $\boldsymbol{\sigma}$ be the path $\boldsymbol{\sigma}(t) = (t^4/4, \sin^3(t\pi/2)), 0 \leqslant t \leqslant 1$, in $\boldsymbol{R}^2$. Evaluate

$$\int_{\boldsymbol{\sigma}} y\, dx + x\, dy$$

We recognize $y\,dx + x\,dy$, or, equivalently, the vector field $y\boldsymbol{i} + x\boldsymbol{j}$ as the gradient of the function $f(x, y) = xy$. Thus

$$\int_{\boldsymbol{\sigma}} y\,dx + x\,dy = f(\boldsymbol{\sigma}(1)) - f(\boldsymbol{\sigma}(0))$$

$$= \tfrac{1}{4} \cdot 1 - 0 = \tfrac{1}{4}$$

Obviously if one can recognize the integrand as a gradient, then evaluation of the integral becomes much easier. In one-variable calculus, every integral is, in principle, obtainable by finding an antiderivative. For vector fields, however, this is not always true, because a vector field need not always be a gradient. This point will be examined in detail in §17.4.

EXERCISES

1. Let $\boldsymbol{F}(x, y, z) = x\boldsymbol{i} + y\boldsymbol{j} + z\boldsymbol{k}$. Evaluate the integral of $\boldsymbol{F}$ along each of the following paths:
   (a) $\boldsymbol{\sigma}(t) = (t, t, t), \quad 0 \leqslant t \leqslant 1$
   (b) $\boldsymbol{\sigma}(t) = (\cos t, \sin t, 0), \quad 0 \leqslant t \leqslant 2\pi$

2. Evaluate each of the following integrals:
   (a) $\int_{\boldsymbol{\sigma}} x\,dy - y\,dx, \quad \boldsymbol{\sigma}(t) = (\cos t, \sin t), \quad 0 \leqslant t \leqslant 2\pi$
   (b) $\int_{\boldsymbol{\sigma}} x\,dx + y\,dy, \quad \boldsymbol{\sigma}(t) = (\cos \pi t, \sin \pi t), \quad 0 \leqslant t \leqslant 2$
   (c) $\int_{\boldsymbol{\sigma}} yz\,dx + xz\,dy + xy\,dz$, where $\boldsymbol{\sigma}$ consists of straight line segments joining $(1, 0, 0)$ to $(0, 1, 0)$ to $(0, 0, 1)$

3. Consider the force $\boldsymbol{F}(x, y, z) = x\boldsymbol{i} + y\boldsymbol{j} + z\boldsymbol{k}$. Compute the work done in moving a particle along the parabola $y = x^2$, $z = 0$ from $x = -1$ to $x = 2$.

4. Let $\boldsymbol{\sigma}$ be a smooth path.
   (a) Suppose $\boldsymbol{F}$ is perpendicular to $\boldsymbol{\sigma}'(t)$ at $\boldsymbol{\sigma}(t)$. Show that
   $$\int_{\boldsymbol{\sigma}} \boldsymbol{F} = 0$$
   (b) If $\boldsymbol{F}$ is parallel to $\boldsymbol{\sigma}'(t)$ at $\boldsymbol{\sigma}(t)$, show that
   $$\int_{\boldsymbol{\sigma}} \boldsymbol{F} = \pm \int_{\boldsymbol{\sigma}} |\boldsymbol{F}|$$

5. Suppose $\boldsymbol{\sigma}(t)$ has length l, and $|\boldsymbol{F}| \leqslant M$. Then prove
   $$\left|\int_{\boldsymbol{\sigma}} \boldsymbol{F}\right| \leqslant Ml$$

6. Evaluate $\int_{\boldsymbol{\sigma}} \boldsymbol{F}$ where $\boldsymbol{F}(x, y, z) = y\boldsymbol{i} + 2x\boldsymbol{j} + y\boldsymbol{k}$ and $\boldsymbol{\sigma}(t) = t\boldsymbol{i} + t^2\boldsymbol{j} + t^3\boldsymbol{k}$, $0 \leqslant t \leqslant 1$.

7. Evaluate $\int_{\boldsymbol{\sigma}} y\,dx + (3y^3 - x)\,dy + z\,dz$ for each of the paths $\boldsymbol{\sigma}(t) = (t, t^n, 0)$, $0 \leqslant t \leqslant 1$, $n = 1, 2, 3, \ldots$.

8. The image of $t \to (\cos^3 t, \sin^3 t)$ in the plane is sketched in Figure 1–4. Evaluate the integral of the vector field $\boldsymbol{F}(x, y) = x\boldsymbol{i} + y\boldsymbol{j}$ around this curve.

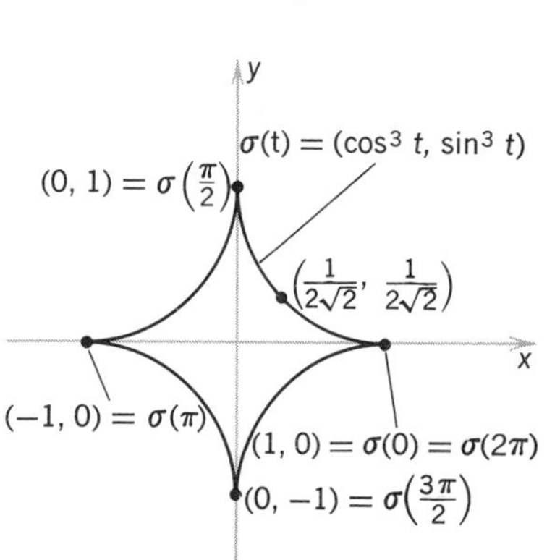

FIGURE 1-4

9. Suppose $\boldsymbol{\sigma}$, $\boldsymbol{\psi}$ are two paths with the same endpoints, and $\boldsymbol{F}$ is a vector field. Show that $\int_{\boldsymbol{\sigma}} \boldsymbol{F} = \int_{\boldsymbol{\psi}} \boldsymbol{F}$ is equivalent to $\int_c \boldsymbol{F} = 0$, where c is the closed curve obtained by first moving around $\boldsymbol{\sigma}$ and then moving around $\boldsymbol{\psi}$ in the opposite direction.

10. Let $\boldsymbol{\sigma}(t)$ be a path and $\boldsymbol{T}$ the unit tangent vector. What is $\int_{\boldsymbol{\sigma}} \boldsymbol{T}$?

11. Let $\boldsymbol{F} = (z^3 + 2xy)\boldsymbol{i} + x^2\boldsymbol{j} + 3xz^2\boldsymbol{k}$. Show that the integral of $\boldsymbol{F}$ around the circumference of the unit square is zero.

12. Using the path in exercise 8, argue that a C^1 map $\boldsymbol{\sigma}:[a,b] \to \boldsymbol{R}^3$ can have an image which does not "look smooth." Do you think this could happen if $\boldsymbol{\sigma}'(t)$ were always nonzero?

13. What is the value of the integral of any gradient field around a closed curve c, that is, the image of $\boldsymbol{\sigma}:[a,b] \to \boldsymbol{R}^3$, where $\boldsymbol{\sigma}(a) = \boldsymbol{\sigma}(b)$.

14. Evaluate $\int_{\boldsymbol{\sigma}} 2xyz\,dx + x^2z\,dy + x^2y\,dz$, where $\boldsymbol{\sigma}$ is a path joining $(1,1,1)$ to $(1,2,4)$.

15. Prove Theorem 2.

2 GREEN'S THEOREM

Green's Theorem is a result which relates line integrals along closed curves C in $\boldsymbol{R}^2$ to integrals over associated regions in $\boldsymbol{R}^2$ (the region "inside" the curve C). It is a fundamental theorem with many important applications; related results in $\boldsymbol{R}^3$ are given in the subsequent sections.

Before we start our analysis of Green's Theorem, we need the notions of simple curves, closed curves, and oriented curves.

We define a **simple curve** to be the image of a (piecewise C^1) map $\boldsymbol{\sigma}: I \to \boldsymbol{R}^3$ which is one-to-one on an interval I. Thus a simple curve is one which does not intersect itself (one-to-one). If $I = [a,b]$, we call $\boldsymbol{\sigma}(a)$ and $\boldsymbol{\sigma}(b)$ the endpoints of the curve. Each simple curve C has two orientations or directions associated with it. If P and Q are the endpoints of the curve, then we can consider C either as directed from P to Q or from Q to P. The simple curve C together with a sense of direction is called an **oriented simple curve** or **directed simple curve** (Figure 2-1).

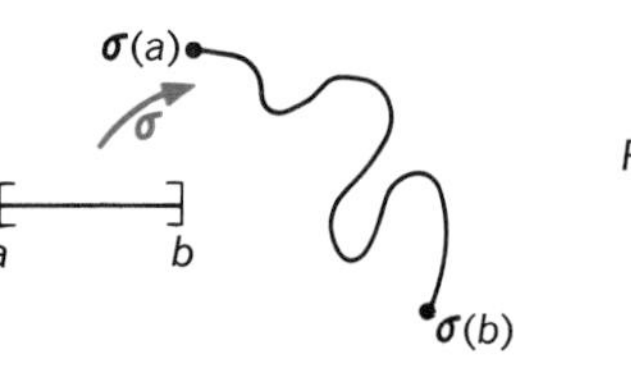

simple curve

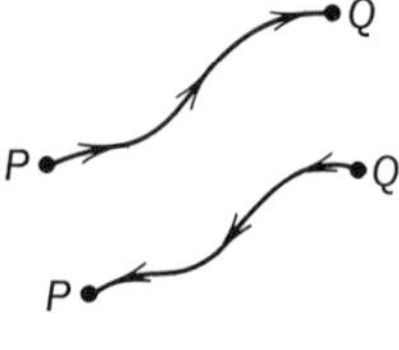

opposite orientations

FIGURE 2-1

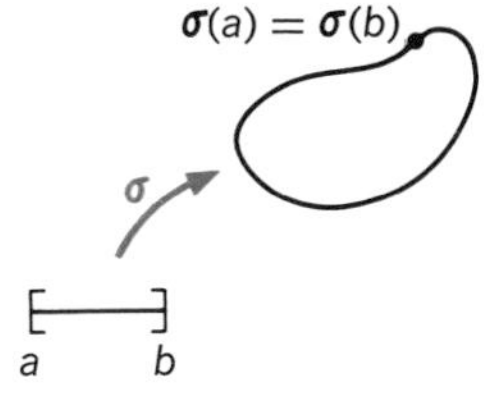

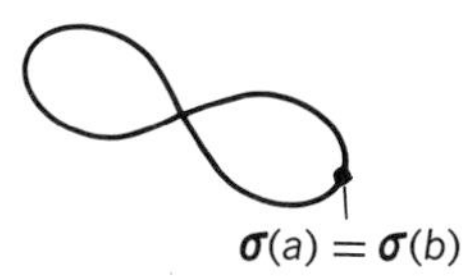

a simple closed curve

a closed curve which is not simple

FIGURE 2-2

If C is a simple curve and $\boldsymbol{\sigma}:[a,b]\rightarrow C\subset\boldsymbol{R}^3$ is a one-to-one (piecewise C^1) path with image C, we call $\boldsymbol{\sigma}$ a parametrization of C. Each parametrization $\boldsymbol{\sigma}$ of C determines an orientation for C. Thus $\boldsymbol{\sigma}$ traces C in one of the two possible directions. For example, the parametrizations $\boldsymbol{\sigma}$ and $\boldsymbol{\sigma}_{\text{op}}$ determine opposite orientations of C.

By a **simple closed curve** we mean the image of a (piecewise C^1) map $\boldsymbol{\sigma}:[a,b]\rightarrow\boldsymbol{R}^3$ which is one-to-one on $[a,b[$ and satisfies $\boldsymbol{\sigma}(a)=\boldsymbol{\sigma}(b)$ (Figure 2-2). If $\boldsymbol{\sigma}$ satisfies the condition $\boldsymbol{\sigma}(a)=\boldsymbol{\sigma}(b)$ but is not necessarily one-to-one on $[a,b[$, we call its image a **closed curve.** Simple closed curves also have two orientations, corresponding to the two possible directions of motion along the curve (Figure 2-3).

As above, if C is a simple closed curve, we say a map $\boldsymbol{\rho}:[a,b]\rightarrow C\subset\boldsymbol{R}^3$ which is one-to-one on $[a,b[$ and such that $\boldsymbol{\rho}(a)=\boldsymbol{\rho}(b)$ is a parametrization of the simple closed curve C; each parametrization determines one of the two orientations. In fact, as t increases from a to b, $\boldsymbol{\rho}(t)$ moves along the curve in some fixed direction until it comes back to the point $\boldsymbol{\rho}(a)=\boldsymbol{\rho}(b)$ (see Figure 2-4). The tangent vector $\boldsymbol{\rho}'(t)$ points in this direction at each point.

Notice that the requirements for parametrizations of simple closed curves are stronger than the ones we placed on parametrizations in the sense used earlier in §17.1. In the present case the map must be one-to-one on a half-open interval. This observation will be important in Theorem 1 below.

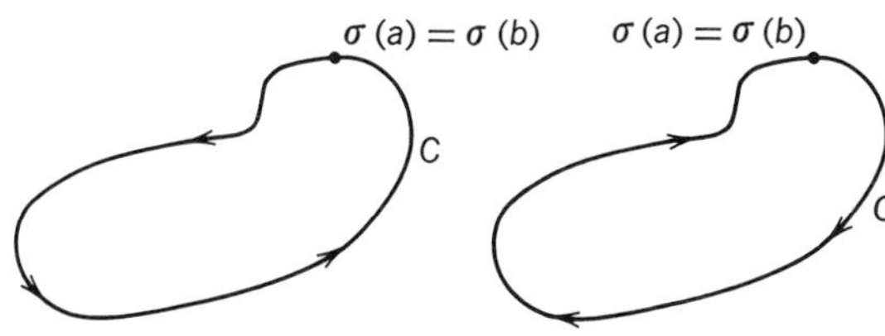

FIGURE 2-3

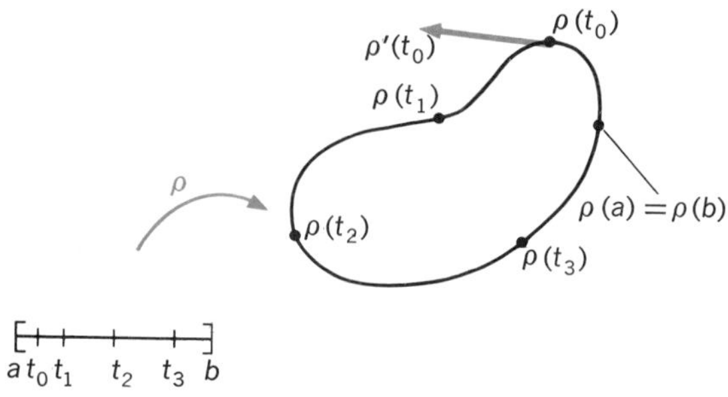

FIGURE 2-4

In previous sections we considered the notions of path integral and line integral along a path. We saw that $\int_\sigma f$ and $\int_\sigma \boldsymbol{F}$ depended on the mapping $\boldsymbol{\sigma}$. That is, it was possible to have two mappings $\boldsymbol{\sigma}$ and $\boldsymbol{\eta}$ with the same image and inducing the same orientation on the image, such that

$$\int_\sigma \boldsymbol{F} \neq \int_\eta \boldsymbol{F}$$

Now the image of a mapping $\boldsymbol{\eta}:[a, b] \to \boldsymbol{R}^3$ is a curve in space—thus it is a geometric object and not a function or path. The three central results of this chapter, namely, the theorems of Green, Stokes, and Gauss, are geometric theorems. To state them in the language of paths would be awkward. Therefore we would like to be able to speak about $\int_C f$ and $\int_C \boldsymbol{F}$, where C is an oriented simple curve or oriented simple closed curve. The next theorem allows us to do this.

Theorem 1 Let C be an oriented simple curve or an oriented simple closed curve and let $\boldsymbol{\sigma}$, $\boldsymbol{\eta}$ be two orientation-preserving parametrizations [that is, $\boldsymbol{\sigma}(t)$ and $\boldsymbol{\eta}(t)$ trace C in its preassigned direction]. Then if $\boldsymbol{F}$ is any continuous vector field on $\boldsymbol{R}^3$ and f is any continuous real-valued function

$$\int_\sigma \boldsymbol{F} = \int_\eta \boldsymbol{F} \qquad \text{and} \qquad \int_\sigma f = \int_\eta f$$

If $\boldsymbol{\sigma}$ is orientation preserving and $\boldsymbol{\eta}$ is orientation reversing then

$$\int_\sigma \boldsymbol{F} = -\int_\eta \boldsymbol{F}$$

Because of Theorems 1 and 2 of §17.1, we see that the above result is very plausible. The only difficulty is that $\boldsymbol{\sigma}$ and $\boldsymbol{\eta}$ need not be reparametrizations of one another. One can get around this point with some further arguments, which we shall omit here.

If C is an oriented simple curve or an oriented simple closed curve we define

$$\int_C \boldsymbol{F} = \int_\sigma \boldsymbol{F} \qquad \text{and} \qquad \int_C f = \int_\sigma f$$

where $\boldsymbol{\sigma}$ is any orientation-preserving parametrization of C. By Theorem 1, these integrals do not depend on the choice of $\boldsymbol{\sigma}$. If $\boldsymbol{F} = P\boldsymbol{i} + Q\boldsymbol{j} + R\boldsymbol{k}$ is a vector field, then in differential form notation we write

$$\int_C \boldsymbol{F} = \int_C P\,dx + Q\,dy + R\,dz$$

Example 1 Consider C, the perimeter of the unit square in $\boldsymbol{R}^2$, oriented in the counterclockwise sense (see Figure 2-5). Evaluate the line integral $\int_C x^2\,dx + xy\,dy$.

We may evaluate the integral using any parametrization of C which induces the given orientation. It is convenient to use the following piecewise C^1 parametrization:

$$\boldsymbol{\sigma}:[0,4] \to \boldsymbol{R}^2 : t \mapsto \begin{cases} (t,0), & 0 \leqslant t \leqslant 1 \\ (1,t-1), & 1 \leqslant t \leqslant 2 \\ (3-t,1), & 2 \leqslant t \leqslant 3 \\ (0,4-t), & 3 \leqslant t \leqslant 4 \end{cases}$$

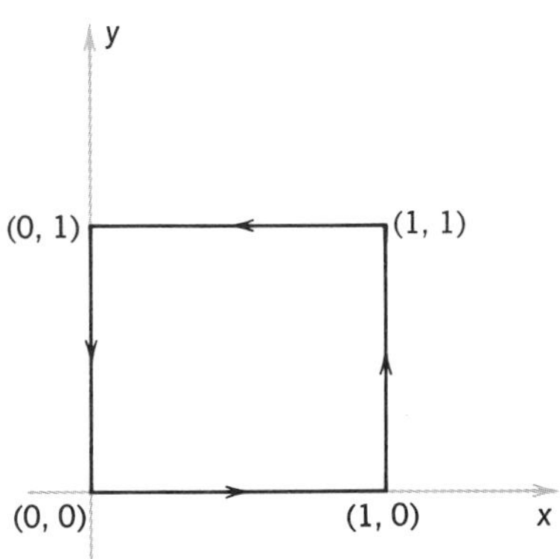

FIGURE 2-5

Then

$$\int_C x^2\,dx + xy\,dy = \int_0^1 (t^2+0)\,dt + \int_1^2 (0+(t-1))\,dt$$
$$+ \int_2^3 (-(3-t)^2+0)\,dt + \int_3^4 (0+0)\,dt = \tfrac{1}{3} + \tfrac{1}{2} + (-\tfrac{1}{3}) + 0 = \tfrac{1}{2}$$

Example 2 If $I = [a,b]$ is a closed interval on the real line, then I has two orientations: one corresponding to motion from a to b (left to right) and the other corresponding to motion from b to a (right to left). If f is a real function continuous on I, then denoting I with the first orientation by I^+, and I with the second orientation by I^-, we have

$$\int_{I^+} f(x)\,dx = \int_a^b f(x)\,dx = -\int_b^a f(x)\,dx = -\int_{I^-} f(x)\,dx$$

Let us now consider line integrals along simple closed curves in $\boldsymbol{R}^2$. We shall restrict our attention to curves which are the boundaries of elementary regions of type 1, 2, or 3 (see §15.3).

A simple closed curved C which is the boundary of a region of type 1, 2, or 3 has two orientations—counterclockwise or positive, and clockwise or negative. In analogy with Example 2 above we denote C with the counterclockwise orientation as C^+ and with clockwise orientation as C^- (Figure 2-6).

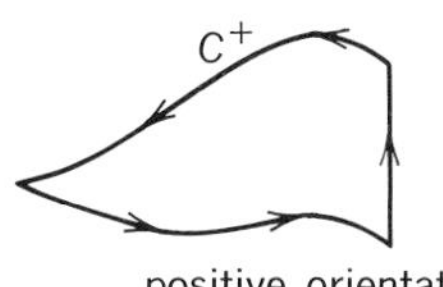

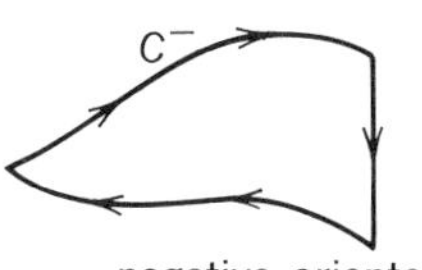

FIGURE 2-6

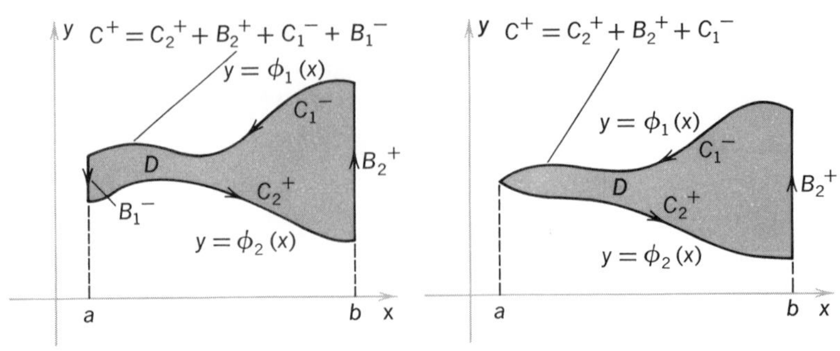

FIGURE 2-7

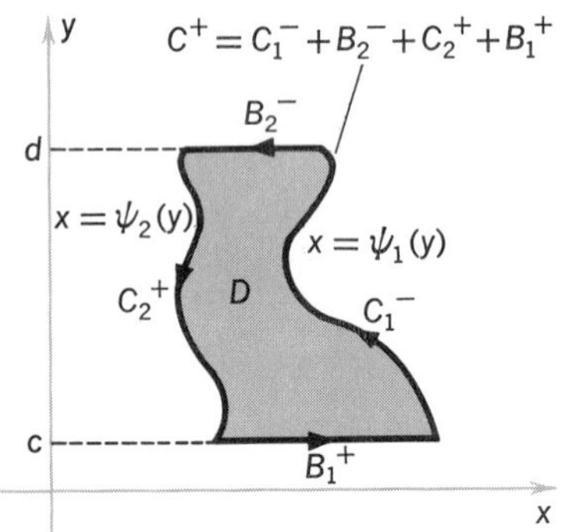

FIGURE 2-8

The boundary C of a region of type 1 can be decomposed into bottom and top portions, C_1 and C_2, and (if applicable) left and right vertical portions, B_1 and B_2. Then, following Figure 2-7, the integral is

$$\int_{C^+} \boldsymbol{F} = \int_{C_2^+} \boldsymbol{F} + \int_{B_2^+} \boldsymbol{F} + \int_{C_1^-} \boldsymbol{F} + \int_{B_1^-} \boldsymbol{F}$$

A similar remark applies to the decomposition of the boundary of a region of type 2 into left and right portions, and upper and lower horizontal portions (if applicable) (Figure 2-8).

The boundary of a region of type 3 has two decompositions—one into upper and lower halves, the other into left and right halves.

We shall now prove two lemmas in preparation for Green's theorem.

Lemma 1 Let D be a region of type 1 and let C be its boundary. Suppose $P: D \to R$ is C^1. Then

$$\int_{C^+} P\,dx = -\int_D \frac{\partial P}{\partial y}\,dx\,dy$$

Proof Suppose D is described by

$$a \leqslant x \leqslant b, \qquad \phi_1(x) \leqslant y \leqslant \phi_2(x)$$

We decompose C^+ by writing $C^+ = C_2^+ + B_2^+ + C_1^- + B_1^-$ (see Figure 2-7). By Fubini's Theorem, we may evaluate the double integral as an iterated integral:

$$\begin{aligned}\int_D \frac{\partial P}{\partial y}(x, y)\,dx\,dy &= \int_a^b \int_{\phi_2(x)}^{\phi_1(x)} \frac{\partial P}{\partial y}(x, y)\,dy\,dx \\ &= \int_a^b [P(x, \phi_1(x)) - P(x, \phi_2(x))]\,dx\end{aligned}$$

However, since C_2^+ can be parametrized by $x \mapsto (x, \phi_2(x))$ and C_1^+ can be parametrized by $x \mapsto (x, \phi_1(x))$ we have

$$\int_a^b P(x, \phi_2(x))\,dx = \int_{C_2^+} P(x, y)\,dx$$

and

$$\int_a^b P(x, \phi_1(x))\, dx = \int_{C_1^+} P(x, y)\, dx$$

Thus by Theorem 1

$$-\int_a^b P(x, \phi_1(x))\, dx = \int_{C_1^-} P(x, y)\, dx$$

Since x is constant on B_2^+ and B_1^- we have

$$\int_{B_2^+} P\, dx = 0 = \int_{B_1^-} P\, dx$$

and so

$$\int_D \frac{\partial P}{\partial y}\, dx\, dy = -\int_{C_2^+} P(x, y)\, dx - \int_{C_1^-} P(x, y)\, dx = -\int_{C^+} P\, dx \quad ∎$$

We now prove the analogous lemma with the roles of x and y interchanged.

Lemma 2 Let D be a region of type 2 with boundary C. Then if $Q: D \to R$ is C^1,

$$\int_{C^+} Q\, dy = \int_D \frac{\partial Q}{\partial x}\, dx\, dy$$

Proof The negative sign does not occur here because reversing the role of x and y corresponds to a change of orientation for the plane. To see this, suppose D is given by

$$\psi_2(y) \leqslant x \leqslant \psi_1(y), \qquad c \leqslant y \leqslant d$$

Using the notation of Figure 2-8 we have

$$\int_{C^+} Q\, dy = \int_{C_1^+ + B_1^+ + C_2^- + B_2^-} Q\, dy = \int_{C_1^+} Q\, dy + \int_{C_2^-} Q\, dy$$

where C_1^+ is the curve parametrized by $y \mapsto (y, \psi_1(y))$, $c \leqslant y \leqslant d$ and C_2^- is the curve $y \mapsto (y, \psi_2(y))$, $c \leqslant y \leqslant d$. Applying Fubini's Theorem, we obtain

$$\begin{aligned} \int_D \frac{\partial Q}{\partial x}\, dx\, dy &= \int_c^d \int_{\psi_2(y)}^{\psi_1(y)} \frac{\partial Q}{\partial x}\, dx\, dy \\ &= \int_c^d [Q(y, \psi_1(y)) - Q(y, \psi_2(y))]\, dy \\ &= \int_{C_1^+} Q\, dy - \int_{C_2^+} Q\, dy = \int_{C_1^+} Q\, dy + \int_{C_2^-} Q\, dy = \int_{C^+} Q\, dy \end{aligned}$$

∎

Combining Lemmas 1 and 2 yields the following important result.

Theorem 2 **Green's Theorem** Let D be a region of type 3 and let C be its boundary. Suppose $P: D \to \mathbf{R}$ and $Q: D \to \mathbf{R}$ are C^1. Then

$$\int_{C^+} P\,dx + Q\,dy = \int_D \left(\frac{\partial Q}{\partial x} - \frac{\partial P}{\partial y} \right) dx\,dy$$

Green's Theorem actually applies to any "decent" region in $\mathbf{R}^3$, but such a result is too deep to be included here. However, a somewhat generalized version is considered in exercise 4.

Let us use the notation ∂D for the oriented curve C^+. Then we have

$$\int_{\partial D} P\,dx + Q\,dy = \int_D \left(\frac{\partial Q}{\partial x} - \frac{\partial P}{\partial y} \right) dx\,dy$$

Henceforth, we shall use ∂D to denote the boundary curve of D oriented in the counterclockwise direction.

Green's Theorem is very useful because it relates a line integral around the boundary of a region to an area integral over the interior of the region, and in many cases it may be easier to evaluate the line integral than the area integral. For example, if we know that P vanishes on the boundary, we can immediately conclude that $\int_D (\partial P/\partial y)\,dx\,dy = 0$ even though $\partial P/\partial y$ need not vanish on the interior. (Can you construct such a P on the unit square?)

We can apply Green's Theorem to obtain a formula for the area of a region bounded by a simple closed curve.

Theorem 3 If C is a simple closed curve that bounds a region to which Green's Theorem applies, then the area of the region D bounded by C is

$$A = \frac{1}{2} \int_{\partial D} x\,dy - y\,dx$$

Proof Let $P(x,y) = -y$, $Q(x,y) = x$; then by Green's Theorem we have

$$\frac{1}{2} \int_{\partial D} x\,dy - y\,dx = \frac{1}{2} \int_D \left(\frac{\partial x}{\partial x} - \frac{\partial(-y)}{\partial y} \right) dx\,dy$$

$$= \int_D dx\,dy \quad \blacksquare$$

Example 3 The area of the region enclosed by the hypercycloid $x^{2/3} + y^{2/3} = a^{2/3}$ can be computed using the parametrization

$$x = a\cos^3\theta, \quad y = a\sin^3\theta, \quad 0 \leqslant \theta \leqslant 2\pi$$

(see Figure 2-9). Thus,

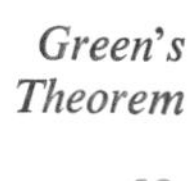

$$A = \frac{1}{2}\int_{\partial D} x\,dy - y\,dx = \frac{1}{2}\int_0^{2\pi} [(a\cos^3\theta)(3a\sin^2\theta\cos\theta)$$

$$- (a\sin^3\theta)(-3a\cos^2\sin\theta)]\,d\theta$$

$$= \frac{3}{2}a^2\int_0^{2\pi}[\sin^2\theta\cos^4\theta + \cos^2\theta\sin^4\theta]\,d\theta$$

$$= \frac{3}{2}a^2\int_0^{2\pi}\sin^2\theta\cos^2\theta(\sin^2\theta+\cos^2\theta)\,d\theta$$

$$= \frac{3}{2}a^2\int_0^{2\pi}\sin^2\theta\cos^2\theta\,d\theta$$

$$= \frac{3}{8}a^2\int_0^{2\pi}\sin^2 2\theta\,d\theta$$

$$= \frac{3}{8}a^2\int_0^{2\pi}\left[\frac{1-\cos 4\theta}{2}\right]d\theta$$

$$= \frac{3}{16}a^2\int_0^{2\pi}d\theta - \frac{3}{16}\int_0^{2\pi}\cos 4\theta\,d\theta$$

$$= \frac{3}{8}\pi a^2$$

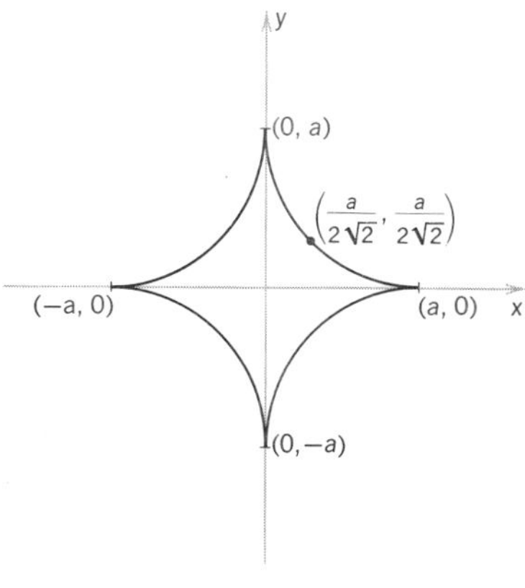

FIGURE 2-9

The statement of Green's Theorem contained in Theorem 2 is not the form which we generalize in the next section. We can formulate an alternate version in the language of vector fields.

Theorem 2′ **Green's Theorem** Let $D \subset \boldsymbol{R}^2$ be a region of type 3 and let ∂D be its boundary. Let $\boldsymbol{F} = P\boldsymbol{i} + Q\boldsymbol{j}$ be a C^1 vector field on D. Then

$$\int_{\partial D}\boldsymbol{F} = \int_D (\text{curl}\,\boldsymbol{F})\cdot\boldsymbol{k}\,dA = \int_D (\nabla\times\boldsymbol{F})\cdot\boldsymbol{k}\,dA$$

(see Figure 2-10).

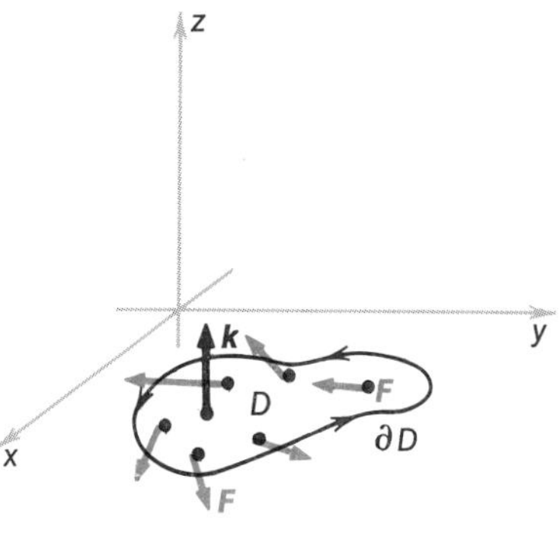

FIGURE 2-10

This result follows easily from Theorem 2 after we interpret the different symbols. We ask the reader to supply the details in exercise 11.

EXERCISES

1. Find the area bounded by one arc of the cycloid $x = a(\theta - \sin\theta)$, $y = a(1-\cos\theta)$, $a > 0$, $0 \leqslant \theta \leqslant 2\pi$, and the x axis. Use Green's Theorem.

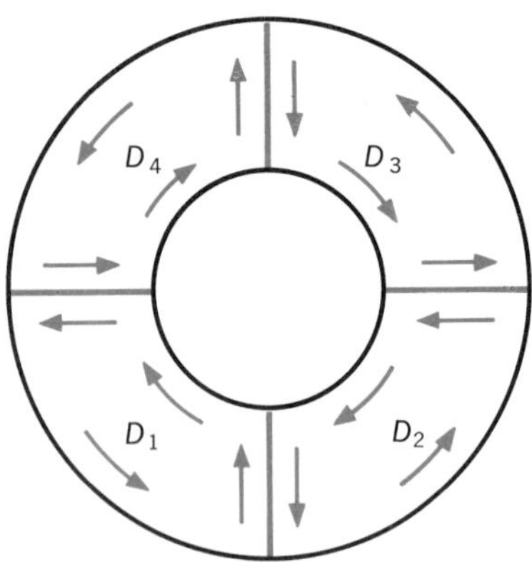

$D = D_1 \cup D_2 \cup D_3 \cup D_4$

FIGURE 2-11

2. Under the conditions of Green's Theorem, prove that

(a) $\int_{\partial D} PQ\,dx + PQ\,dy = \int_D \left[Q\left(\frac{\partial P}{\partial x} - \frac{\partial P}{\partial y}\right) + P\left(\frac{\partial Q}{\partial x} - \frac{\partial Q}{\partial y}\right)\right] dx\,dy$

(b) $\int_{\partial D} \left(Q\frac{\partial P}{\partial x} - P\frac{\partial Q}{\partial x}\right) dx + \left(P\frac{\partial Q}{\partial y} - Q\frac{\partial P}{\partial y}\right) dy$

$$= \int_D \left(P\frac{\partial^2 Q}{\partial x \partial y} - Q\frac{\partial^2 P}{\partial x \partial y}\right) dx\,dy$$

3. Evaluate $\int_{\sigma} (2x^3 - y^3)\,dx + (x^3 + y^3)\,dy$, where σ is the unit circle and verify Green's Theorem in this case.

4. Prove the following generalization of Green's Theorem: Let D be a region in the xy plane with boundary a simple closed curve C. Suppose that by means of a finite number of line segments parallel to the coordinate axes, D can be decomposed into a finite number of regions of type 3 (see Figure 2–11). Then if P and Q are C^1 on D we have

$$\int_D \left(\frac{\partial Q}{\partial x} - \frac{\partial Q}{\partial y}\right) dx\,dy = \int_{C^+} P\,dx + Q\,dy$$

[Hint: Let D_i, $i = 1, \ldots, k$ be the regions composing D. Apply Green's Theorem to each D_i.]

5. (a) Verify Green's Theorem using the integrand of exercise 3 ($P = 2x^3 - y^3$, $Q = x^3 + y^3$) and the annular region D described by $a \leqslant x^2 + y^2 \leqslant b$.

   (b) Does Green's Theorem as stated apply to this region? Discuss.

6. Let D be a region for which Green's Theorem holds. Suppose f is **harmonic**; that is,

$$\frac{\partial^2 f}{\partial x^2} + \frac{\partial^2 f}{\partial y^2} = 0$$

on D. Prove that

$$\int_{\partial D} \frac{\partial f}{\partial y}\,dx - \frac{\partial f}{\partial x}\,dy = 0$$

7. (a) Let D be a region for which Green's Theorem holds and let C^+ be a bounding curve. Suppose $\boldsymbol{\sigma}(t) = (x(t), y(t))$ is a parametrization of C^+ such that $|\boldsymbol{\sigma}'(t)| = 1$. Let

$$\boldsymbol{n}(t) = (y'(t), -x'(t))$$

Verify that $\boldsymbol{n}$ is a unit normal vector to C^+.

   (b) Let $\boldsymbol{F}$ be a vector field. Prove

$$\int_D \operatorname{div} \boldsymbol{F} = \int_{C^+} \boldsymbol{F} \cdot \boldsymbol{n}$$

where $\operatorname{div} \boldsymbol{F} = \nabla \cdot \boldsymbol{F} = \frac{\partial F_1}{\partial x} + \frac{\partial F_2}{\partial y}$.

8. Let $P(x,y) = -y/(x^2 + y^2)$, $Q(x,y) = x/(x^2 + y^2)$. Assuming D is the unit disc, investigate why Green's Theorem fails for this P and Q.

9. Use Green's Theorem to evaluate $\int_{C^+} (y^2 + x^3)\,dx + x^4\,dy$ where C^+ is the perimeter of the unit square, in the counterclockwise direction.

10. Use Theorem 3 to compute the area of the ellipse $x^2/a^2+y^2/b^2=1$.

11. Verify Theorem 2′.

12. Evaluate $\int_\sigma (x^5-2xy^3)\,dx-3x^2y^2\,dy$, where σ is the path $\sigma(t)=(t^8,t^{10})$, $0\leqslant t\leqslant 1$.

*13. Use Green's Theorem to prove the Change of Variables formula in the case

$$\int_D dx\,dy = \int_{D^*} \left|\frac{\partial(x,y)}{\partial(u,v)}\right| du\,dv$$

for a transformation $(u,v)\mapsto(x(u,v),y(u,v))$. Formulate the necessary hypotheses on the functions $x=x(u,v)$ and $y=y(u,v)$ and on $\partial(x,y)/\partial(u,v)$ for your proof.

*14. Use exercise 13 to prove the more general result

$$\int_D f(x,y)\,dx\,dy = \int_{D^*} f(x(u,v),\ y(u,v)) \left|\frac{\partial(x,y)}{\partial(u,v)}\right| du\,dv$$

for integrable f.

15. Show that for polar coordinates (r,θ),

$$x\,dy - y\,dx = r^2\,d\theta$$

Use this to find the area of one loop of the four-leafed rose $r=3\sin 2\theta$.

3 STOKES' THEOREM

Stokes' Theorem relates the integral of a vector field over a surface to an integral around the boundary curve of the surface. This result is a generalization of Green's Theorem and is quite fundamental in physical applications.

Let us begin by studying surface integrals of vector fields. We will consider surfaces S defined by $\Phi: D\to \boldsymbol{R}^3$, where Φ is given by the equations $x=g(u,v), y=h(u,v), z=f(u,v)$ and where, as in Chapter 16,

(i) D is an elementary region in the plane,
(ii) Φ is C^1 and one-to-one except possibly on the boundary of D,
(iii) $\boldsymbol{T}_u\times\boldsymbol{T}_v\neq 0$ except possibly for finitely many $\Phi(u,v)\in S$, where

$$\boldsymbol{T}_u = \frac{\partial g}{\partial u}\boldsymbol{i}+\frac{\partial h}{\partial u}\boldsymbol{j}+\frac{\partial f}{\partial u}\boldsymbol{k}$$

and

$$\boldsymbol{T}_v = \frac{\partial g}{\partial v}\boldsymbol{i}+\frac{\partial h}{\partial v}\boldsymbol{j}+\frac{\partial f}{\partial v}\boldsymbol{k}$$

We assume, as in Chapter 16, that all parametrizations are of the above type. Then if $\Phi: D\to\boldsymbol{R}^3$ is a parametrization of a surface S and $\boldsymbol{F}$ is a vector

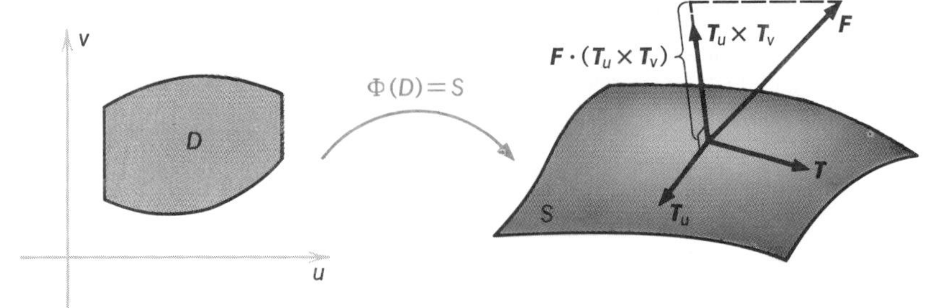

FIGURE 3-1

field defined on S, we define the **surface integral** of $\boldsymbol{F}$, $\int_\Phi \boldsymbol{F}$, by the equation

$$\int_\Phi \boldsymbol{F} = \int_D \boldsymbol{F} \cdot (\boldsymbol{T}_u \times \boldsymbol{T}_v)\, du\, dv$$

(see Figure 3-1).

Example 1 Let D be the rectangle in the $\theta\rho$ plane defined by

$$0 \leqslant \theta \leqslant 2\pi, \qquad -\tfrac{1}{2}\pi \leqslant \rho \leqslant \tfrac{1}{2}\pi$$

and let the surface S be defined by the parametrization $\Phi : D \to S \subset \boldsymbol{R}^3$ given by

$$x = \cos\theta\cos\rho, \quad y = \sin\theta\cos\rho, \quad z = \sin\rho$$

(Here ρ differs from the usual polar angle by $\frac{1}{2}\pi$.) Then S is the unit sphere parametrized by Φ. Also, let $\boldsymbol{r}$ be the position vector $\boldsymbol{r}(x, y, z) = x\boldsymbol{i} + y\boldsymbol{j} + z\boldsymbol{k}$. We compute $\int_\Phi \boldsymbol{r}$ as follows. First we find

$$\boldsymbol{T}_\theta = -\cos\rho\sin\theta\boldsymbol{i} + \cos\rho\cos\theta\boldsymbol{j}$$

$$\boldsymbol{T}_\rho = -\cos\theta\sin\rho\boldsymbol{i} - \sin\theta\sin\rho\boldsymbol{j} + \cos\rho\boldsymbol{k}$$

and hence

$$\boldsymbol{T}_\theta \times \boldsymbol{T}_\rho = \cos^2\rho\cos\theta\boldsymbol{i} + \cos^2\rho\sin\theta\boldsymbol{j} + \cos\rho\sin\rho\boldsymbol{k}$$

Then we evaluate

$$\begin{aligned}
\boldsymbol{r} \cdot (\boldsymbol{T}_\theta \times \boldsymbol{T}_\rho) &= (x\boldsymbol{i} + y\boldsymbol{j} + z\boldsymbol{k}) \cdot (\boldsymbol{T}_\theta \times \boldsymbol{T}_\rho) \\
&= (\cos\theta\cos\rho\boldsymbol{i} + \sin\theta\cos\rho\boldsymbol{j} + \sin\rho\boldsymbol{k}) \\
&\qquad \cdot \cos\rho(\cos\rho\cos\theta\boldsymbol{i} + \cos\rho\sin\theta\boldsymbol{j} + \sin\rho\boldsymbol{k}) \\
&= \cos\rho(\cos^2\rho\cos^2\theta + \cos^2\rho\sin^2\theta + \sin^2\rho) \\
&= \cos\rho
\end{aligned}$$

Then

$$\int_\Phi \boldsymbol{F} = \int_D \cos\rho\, d\rho\, d\theta = \int_0^{2\pi} 2\, d\theta = 4\pi$$

A very close analogy can be drawn between the surface integral $\int_\Phi \boldsymbol{F}$, and the line integral $\int_\sigma \boldsymbol{F}$. Recall that the line integral is an oriented integral. We needed the notion of orientation of a curve to extend the definition of $\int_\sigma \boldsymbol{F}$ to line integrals $\int_C \boldsymbol{F}$ over oriented curves. We would like to extend the definition of $\int_\Phi \boldsymbol{F}$ to surfaces in a similar fashion; that is, given a surface S parametrized by a mapping Φ, we want to define $\int_S \boldsymbol{F} = \int_\Phi \boldsymbol{F}$. In order to accomplish this, we need the notion of orientation of a surface.

Suppose a surface S can be described as having two sides (see Figure 3-2). The precise meaning of the term "side" will be given below in terms of the normal to the surface.

An **oriented surface** S is a two-sided surface along with a specific choice of one side of the surface. We call this special side of the surface the **outside** or **positive side,** and we call the other side the **inside** or **negative side.** At each point $(x, y, z) \in S$ there are two unit normal vectors $\boldsymbol{n}_1$ and $\boldsymbol{n}_2$, where $\boldsymbol{n}_1 = -\boldsymbol{n}_2$ (Figure 3-2). These normals determine the two sides of the surface. Thus to specify a side of a surface S, we choose a unit normal vector $\boldsymbol{n}$ which points to the positive side of S at each point.

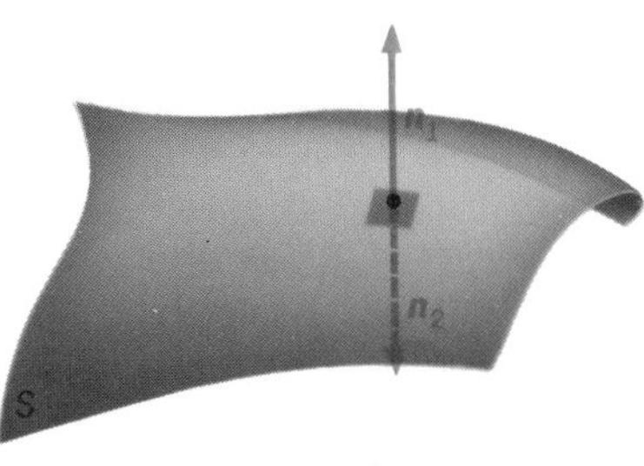

FIGURE 3-2

Let $\Phi: D \to \boldsymbol{R}^3$ be a parametrization of an oriented surface S and suppose S is smooth at $\Phi(u_0, v_0)$, $(u_0, v_0) \in D$; that is, the unit normal vector $(\boldsymbol{T}_{u_0} \times \boldsymbol{T}_{v_0})/|\boldsymbol{T}_{u_0} \times \boldsymbol{T}_{v_0}|$ is defined. If $\boldsymbol{n}(\Phi(u_0, v_0))$ denotes the unit normal to S at $\Phi(u_0, v_0)$ which points to the positive side of S at each point, it follows that $(\boldsymbol{T}_{u_0} \times \boldsymbol{T}_{v_0})/|\boldsymbol{T}_{u_0} \times \boldsymbol{T}_{v_0}| = \pm\boldsymbol{n}(\Phi(u_0, v_0))$. The parametrization Φ is said to be **orientation preserving** if $(\boldsymbol{T}_u \times \boldsymbol{T}_v)/|\boldsymbol{T}_u \times \boldsymbol{T}_v| = \boldsymbol{n}(\Phi(u, v))$ at all $(u, v) \in D$ for which S is smooth at $\Phi(u, v)$. In other words, Φ is orientation preserving if the vector $\boldsymbol{T}_u \times \boldsymbol{T}_v$ points to the outside of the surface. If $\boldsymbol{T}_u \times \boldsymbol{T}_v$ points to the inside of the surface at all $(u, v) \in D$ for which S is smooth at $\Phi(u, v)$, then Φ is said to be **orientation reversing**. Using the above notation, this condition corresponds to $(\boldsymbol{T}_u \times \boldsymbol{T}_v)/|\boldsymbol{T}_u \times \boldsymbol{T}_v| = -\boldsymbol{n}(\Phi(u, v))$.

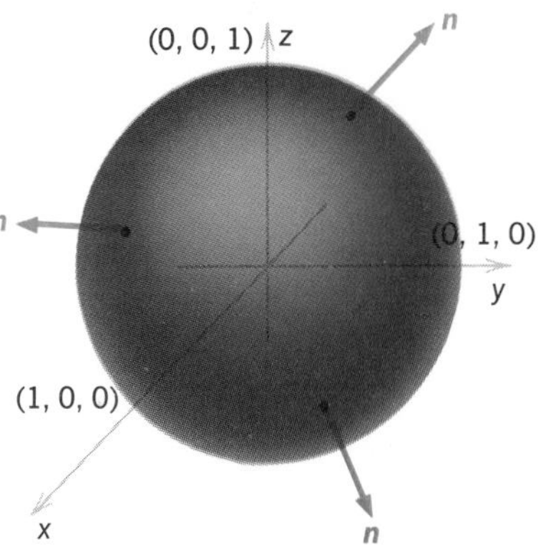

FIGURE 3-3

Example 2 We can give the unit sphere $x^2 + y^2 + z^2 = 1$ in $\boldsymbol{R}^3$ (Figure 3-3) an orientation by selecting the unit vector $\boldsymbol{n}(x, y, z) = \boldsymbol{r}/|\boldsymbol{r}|$, where $\boldsymbol{r} = x\boldsymbol{i} + y\boldsymbol{j} + z\boldsymbol{k}$ to point to the outside of the surface. This choice corresponds to our intuitive notion of outside for the sphere.

Now that the sphere S is an oriented surface, consider the parametrization Φ of S given in Example 1. The cross product of the tangent vectors $\boldsymbol{T}_\theta$ and $\boldsymbol{T}_\rho$—that is, the normal—is given by $\cos\rho(\cos\theta\cos\rho\boldsymbol{i} + \sin\theta\cos\rho\boldsymbol{j} + \sin\rho\boldsymbol{k}) = \boldsymbol{r}\cos\rho$. Since $\cos\rho > 0$ for $-\frac{1}{2}\pi < \rho < \frac{1}{2}\pi$, the normal vector points outward from the sphere. Thus the given parametrization Φ is orientation preserving.

Example 3 Let S be a surface described by $z = f(x, y)$. There are two unit normal directions to S at $(x_0, y_0, f(x_0, y_0))$, namely,

$$\boldsymbol{n} = \frac{-\dfrac{\partial f}{\partial x}(x_0, y_0)\boldsymbol{i} - \dfrac{\partial f}{\partial y}(x_0, y_0)\boldsymbol{j} + \boldsymbol{k}}{\sqrt{\left(\dfrac{\partial f}{\partial x}(x_0, y_0)\right)^2 + \left(\dfrac{\partial f}{\partial y}(x_0, y_0)\right)^2 + 1}}$$

and $-\boldsymbol{n}$. We can orient all such surfaces by taking the positive side of S to be the side to which $\boldsymbol{n}$ points (Figure 3-4). Thus the positive side of such a surface is determined by the unit normal $\boldsymbol{n}$ with positive $\boldsymbol{k}$ component. If we parametrize this surface by $\Phi(u, v) = (u, v, f(u, v))$ then Φ will be orientation preserving.

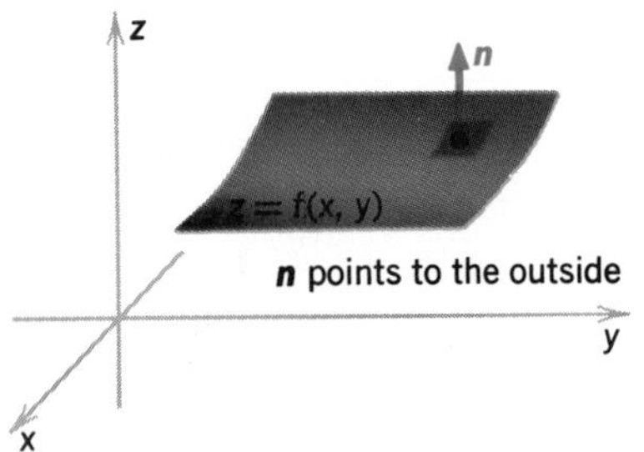

FIGURE 3-4

We now state without proof a general result which allows us to define the **integral of a vector field $\boldsymbol{F}$ over an oriented surface.**

Theorem 1 Let S be an oriented surface and let Φ_1 and Φ_2 be two orientation-preserving parametrizations, with $\boldsymbol{F}$ a continuous vector field defined on S. Then

$$\int_{\Phi_1} \boldsymbol{F} = \int_{\Phi_2} \boldsymbol{F}$$

If Φ_1 is orientation preserving and Φ_2 orientation reversing

$$\int_{\Phi_1} \boldsymbol{F} = -\int_{\Phi_2} \boldsymbol{F}$$

Moreover, if f is a real-valued continuous function defined on S, and if Φ_1 and Φ_2 are parametrizations then

$$\int_{\Phi_1} f\,dS = \int_{\Phi_2} f\,dS$$

If $f = 1$ we obtain

$$A(S) = \int_{\Phi_i} dS, \qquad i = 1, 2$$

thus showing that area is independent of parametrization.

We define the **surface integral** $\int_S \boldsymbol{F}$ over an oriented surface S by

$$\int_S \boldsymbol{F} = \int_{\Phi} \boldsymbol{F}$$

where Φ is an orientation-preserving parametrization. Theorem 1 guarantees that the value of the integral does not depend on the selection of Φ.

We mention that Theorem 1 is analogous to Theorem 1 of §17.2. The heart of the proof is contained in the Change of Variables formula—this time for double integrals.

The notation $\int_S \boldsymbol{F} \cdot d\boldsymbol{S}$ is sometimes used in place of $\int_S \boldsymbol{F}$. We can also express $\int_S \boldsymbol{F}$ as an integral of a real-valued function f over the surface S (see §15.2). This is analogous to the case where we expressed $\int_{\boldsymbol{\sigma}} \boldsymbol{F}$ as a path integral $\int_{\boldsymbol{\sigma}} f$ for a particular real-valued function f. Let S be an oriented smooth surface and let $\Phi: D \to \boldsymbol{R}^3$ be an orientation-preserving parametrization. Then $\boldsymbol{n} = (\boldsymbol{T}_u \times \boldsymbol{T}_v)/|\boldsymbol{T}_u \times \boldsymbol{T}_v|$ is a unit normal pointing to the outside of S, and

$$\int_S \boldsymbol{F} = \int_\Phi \boldsymbol{F} = \int_D \boldsymbol{F}\cdot(\boldsymbol{T}_u \times \boldsymbol{T}_v)\, du\, dv = \int_D \boldsymbol{F}\cdot\left(\frac{\boldsymbol{T}_u \times \boldsymbol{T}_v}{|\boldsymbol{T}_u \times \boldsymbol{T}_v|}\right)|\boldsymbol{T}_u \times \boldsymbol{T}_v|\, du\, dv$$

$$= \int_D (\boldsymbol{F}\cdot\boldsymbol{n})\,|\boldsymbol{T}_u \times \boldsymbol{T}_v|\, du\, dv = \int_S (\boldsymbol{F}\cdot\boldsymbol{n})\, dS = \int_S f\, dS$$

where $f = \boldsymbol{F}\cdot\boldsymbol{n}$.

Thus $\int_S \boldsymbol{F}$, the surface integral of $\boldsymbol{F}$ over S, is equal to the integral of the normal component of $\boldsymbol{F}$ over the surface.

The geometric and physical significance of this surface integral can be understood by expressing it as a limit of Riemann sums. For simplicity, we assume D is a rectangle. Fix a parametrization Φ of S which preserves orientation and partition the region D into n^2 pieces D_{ij}, $0 \leqslant i \leqslant n-1$, $0 \leqslant j \leqslant n-1$. We let Δu denote the length of the horizontal side of D_{ij} and Δv denote the length of the vertical side of D_{ij}. Let (u_i, v_j) be a point in D_{ij} and $(x_{ij}, y_{ij}, z_{ij}) = (x, y, z)$, the corresponding point on the surface. We consider the parallelogram with sides $\Delta u \boldsymbol{T}_u$ and $\Delta v \boldsymbol{T}_v$ lying in the tangent plane at (x_{ij}, y_{ij}, z_{ij}) and the parallelepiped formed by $\boldsymbol{F}$, $\Delta u \boldsymbol{T}_u$, and $\Delta v \boldsymbol{T}_v$. The volume of the parallelepiped is the absolute value of the triple product

$$\boldsymbol{F}\cdot(\Delta u \boldsymbol{T}_u \times \Delta v \boldsymbol{T}_v) = \boldsymbol{F}\cdot(\boldsymbol{T}_u \times \boldsymbol{T}_v)\,\Delta u \Delta v$$

The vector $\boldsymbol{T}_u \times \boldsymbol{T}_v$ is normal to the surface at (x_{ij}, y_{ij}, z_{ij}) and points to the outside of the surface. Thus the number $\boldsymbol{F}\cdot(\boldsymbol{T}_u \times \boldsymbol{T}_v)$ is positive when the parallelepiped lies on the outside of the surface (Figure 3-5).

$\mathbf{F}\cdot\mathbf{T}_u \times \mathbf{T}_v < 0$, (**F** points inward)

FIGURE 3-5

On the other hand, the parallelepiped lies on that side of the surface to which $\boldsymbol{F}$ is pointing. If we think of $\boldsymbol{F}$ as the velocity field of a fluid, $\boldsymbol{F}(x, y, z)$ is pointing in the direction in which fluid is moving across the surface near (x, y, z). Moreover, the number $|\boldsymbol{F}\cdot\boldsymbol{T}_u \Delta u \times \boldsymbol{T}_v \Delta v|$ measures the amount of fluid that passes through the tangent parallelogram per unit time

(Figure 3-6). Finally, the sign of $\boldsymbol{F}\cdot\Delta u\boldsymbol{T}_u\times\Delta v\boldsymbol{T}_v$ is positive if the vector $\boldsymbol{F}$ is pointing outward at (x_{ij}, y_{ij}, z_{ij}) and negative if $\boldsymbol{F}$ is pointing inward. Therefore, the sum $\sum_{i,j}\boldsymbol{F}\cdot\boldsymbol{T}_u\times\boldsymbol{T}_v\,\Delta u\Delta v$ is an approximate measure of the net quantity of fluid to flow outward across the surface per unit time. Hence, the integral $\int_S \boldsymbol{F}$ is the net quantity of fluid to flow across the surface per unit time, that is, the rate of fluid flow. Therefore this integral is also called the **flux** of $\boldsymbol{F}$ across the surface.

In case $\boldsymbol{F}$ represents electric or magnetic fields, $\int_S \boldsymbol{F}$ is also commonly known as the flux. The reader may be familiar with physical laws (such as Faraday's law) which relate flux of a vector field to a circulation (or current) in a bounding loop. This is the historical and physical basis of Stokes' Theorem. The corresponding principle in fluid mechanics is related to Kelvin's Circulation Theorem.

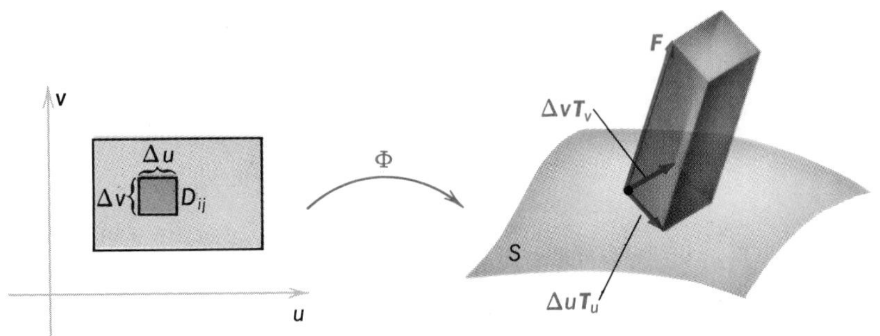

FIGURE 3-6

In preparation for Stokes' Theorem we shall now restrict our attention to surfaces S described by $z=f(x,y), (x,y)\in D$, where S is oriented so that

$$\boldsymbol{n}=\frac{-\dfrac{\partial f}{\partial x}\boldsymbol{i}-\dfrac{\partial f}{\partial y}\boldsymbol{j}+\boldsymbol{k}}{\sqrt{\left(\dfrac{\partial f}{\partial x}\right)^2+\left(\dfrac{\partial f}{\partial y}\right)^2+1}}$$

points outward. We have seen that we can parametrize S by $\Phi: D\to\boldsymbol{R}^3$ given by $\Phi(x,y)=(x,y,f(x,y))$. In this case, $\int_S \boldsymbol{F}$ can be written in a particularly simple form. We have

$$\boldsymbol{T}_x=\boldsymbol{i}+\frac{\partial z}{\partial x}\boldsymbol{k}$$

$$\boldsymbol{T}_y=\boldsymbol{j}+\frac{\partial z}{\partial y}\boldsymbol{k}$$

Thus $T_x \times T_y = -(\partial z/\partial x)\boldsymbol{i} - (\partial z/\partial y)\boldsymbol{j} + \boldsymbol{k}$. If $\boldsymbol{F} = F_1\boldsymbol{i} + F_2\boldsymbol{j} + F_3\boldsymbol{k}$ is a continuous vector field then we get the formula

$$\int_S \boldsymbol{F} = \int_D \boldsymbol{F} \cdot (\boldsymbol{T}_x \times \boldsymbol{T}_y)\, dx\, dy = \int_D \left[F_1\left(-\frac{\partial z}{\partial x}\right) + F_2\left(-\frac{\partial z}{\partial y}\right) + F_3 \right] dx\, dy \qquad (1)$$

Example 4 The equations

$$z = 12, \qquad x^2 + y^2 \leqslant 25$$

describe a disk of radius 5 lying in the plane $z = 12$. Suppose $\boldsymbol{r}$ is the field

$$\boldsymbol{r}(x, y, z) = x\boldsymbol{i} + y\boldsymbol{j} + z\boldsymbol{k}$$

Then $\int_S \boldsymbol{r}$ is easily computed; we have $\partial z/\partial x = \partial z/\partial y = 0$ since $z = 12$ is constant on the disk, so

$$\begin{aligned} \boldsymbol{r}(x, y, z) \cdot \boldsymbol{T}_x \times \boldsymbol{T}_y &= \boldsymbol{r}(x, y, z) \cdot (\boldsymbol{i} \times \boldsymbol{j}) \\ &= \boldsymbol{r}(x, y, z) \cdot \boldsymbol{k} = z \end{aligned}$$

and the integral becomes

$$\int_S \boldsymbol{F} = \int_D z\, dx\, dy = \int_D 12\, dx\, dy = 12\,(\text{area of } D) = 300\,\pi$$

On the other hand, since the disc is parallel to the xy plane, the outward unit normal is $\boldsymbol{k}$. So $\boldsymbol{n}(x, y, z) = \boldsymbol{k}$ and $\boldsymbol{F} \cdot \boldsymbol{n} = z$. However, $|\boldsymbol{T}_x \times \boldsymbol{T}_y| = |\boldsymbol{k}| = 1$ and so we know from the discussion on the top of page 801 that

$$\int_S \boldsymbol{F} = \int_S \boldsymbol{F} \cdot \boldsymbol{n}\, dS = \int_S z\, dS = \int_D z\, dx\, dy = 300\,\pi$$

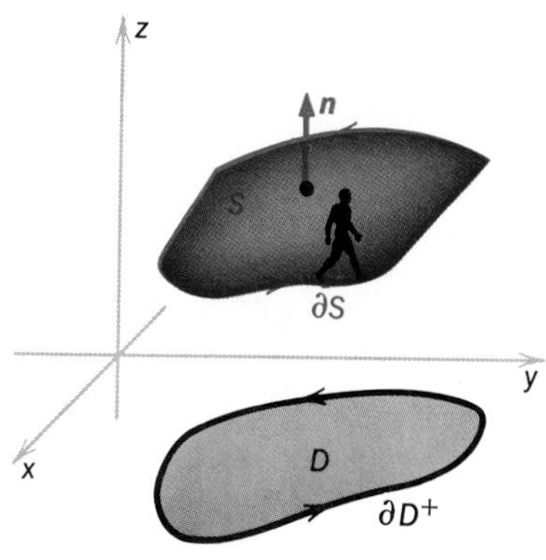

FIGURE 3-7

Once again, let S be the oriented surface described by $z = f(x, y)$, $(x, y) \in D$. Suppose that $\boldsymbol{\sigma}: [a, b] \to \boldsymbol{R}^3$, $\boldsymbol{\sigma}(t) = (\sigma_1(t), \sigma_2(t))$, is a parametrization of ∂D in the counterclockwise direction. Then we define the **boundary curve** ∂S to be the oriented simple closed curve which is the image of the mapping $\boldsymbol{\eta}: t \mapsto (\sigma_1(t), \sigma_2(t), f(\sigma_1(t), \sigma_2(t)))$ with the orientation induced by $\boldsymbol{\sigma}$ (Figure 3-7).

To remember the orientation or positive direction on ∂S, note that an "observer" walking along the boundary of the surface with the normal being his upright direction will be moving in the positive direction if the surface is on his left. The orientation on ∂S is often called the **orientation induced by an outward normal $\boldsymbol{n}$.**

We are now ready to state and prove the primary result of this section.

Theorem 2 **Stokes' Theorem** Let S be the oriented surface defined by $z = f(x, y)$, $(x, y) \in D$ and let $\boldsymbol{F}$ be a C^1 vector field on S. Then if ∂S denotes the oriented boundary curve of S as defined above we have

$$\int_S \operatorname{curl} \boldsymbol{F} = \int_S \nabla \times \boldsymbol{F} = \int_{\partial S} \boldsymbol{F}$$

Proof If $\boldsymbol{F} = F_1\boldsymbol{i} + F_2\boldsymbol{j} + F_3\boldsymbol{k}$ then

$$\operatorname{curl} \boldsymbol{F} = \left(\frac{\partial F_3}{\partial y} - \frac{\partial F_2}{\partial z}\right)\boldsymbol{i} + \left(\frac{\partial F_1}{\partial z} - \frac{\partial F_3}{\partial x}\right)\boldsymbol{j} + \left(\frac{\partial F_2}{\partial x} - \frac{\partial F_1}{\partial y}\right)\boldsymbol{k}$$

Therefore we use formula (1) to write

$$\int_S \operatorname{curl} \boldsymbol{F} = \int_D \left[\left(\frac{\partial F_3}{\partial y} - \frac{\partial F_2}{\partial z}\right)\left(-\frac{\partial z}{\partial x}\right) + \left(\frac{\partial F_1}{\partial z} - \frac{\partial F_3}{\partial x}\right)\left(-\frac{\partial z}{\partial y}\right) + \left(\frac{\partial F_2}{\partial x} - \frac{\partial F_1}{\partial y}\right)\right] dA \qquad (2)$$

On the other hand,

$$\int_{\partial S} \boldsymbol{F} = \int_{\boldsymbol{\eta}} \boldsymbol{F} = \int_{\boldsymbol{\eta}} F_1\,dx + F_2\,dy + F_3\,dz$$

where $\boldsymbol{\eta}:[a,b] \to \boldsymbol{R}^3$, $\boldsymbol{\eta}(t) = (\sigma_1(t), \sigma_2(t), f(\sigma_1(t), \sigma_2(t)))$ is the orientation-preserving parametrization of the oriented simple closed curve ∂S, discussed above. Thus

$$\int_{\partial S} \boldsymbol{F} = \int_a^b \left(F_1\frac{dx}{dt} + F_2\frac{dy}{dt} + F_3\frac{dz}{dt}\right) dt \qquad (3)$$

But by the Chain Rule

$$\frac{dz}{dt} = \frac{\partial z}{dx}\frac{dx}{dt} + \frac{\partial z}{\partial y}\frac{dy}{dt}$$

Substituting this expression into (3) we obtain

$$\begin{aligned}
\int_{\partial S} \boldsymbol{F} &= \int_a^b \left[\left(F_1 + F_3\frac{\partial z}{\partial x}\right)\frac{dx}{dt} + \left(F_2 + F_3\frac{\partial z}{\partial y}\right)\frac{dy}{dt}\right] dt \\
&= \int_{\boldsymbol{\sigma}} \left(F_1 + F_3\frac{\partial z}{\partial x}\right) dx + \left(F_2 + F_3\frac{\partial z}{\partial y}\right) dy \\
&= \int_{\partial D} \left(F_1 + F_3\frac{\partial z}{\partial x}\right) dx + \left(F_2 + F_3\frac{\partial z}{\partial y}\right) dy \qquad (4)
\end{aligned}$$

Applying Green's Theorem to (4) yields

$$\int_D \left[\frac{\partial(F_2 + F_3\,\partial z/\partial y)}{\partial x} - \frac{\partial(F_1 + F_3\,\partial z/\partial x)}{\partial y}\right] dA$$

which, using the formula of review exercise 28 in Chapter 14, is equal to

$$\int_D \left[\left(\frac{\partial F_2}{\partial x} + \frac{\partial F_2}{\partial z}\cdot\frac{\partial z}{\partial x} + \frac{\partial F_3}{\partial x}\frac{\partial z}{\partial y} + \frac{\partial F_3}{\partial z}\left(\frac{\partial z}{\partial x}\right)\left(\frac{\partial z}{\partial y}\right) + F_3\cdot\frac{\partial^2 z}{\partial x\,\partial y}\right) - \left(\frac{\partial F_1}{\partial y} + \frac{\partial F_1}{\partial z}\cdot\frac{\partial z}{\partial y} + \frac{\partial F_3}{\partial y}\cdot\frac{\partial z}{\partial x} + \frac{\partial F_3}{\partial z}\left(\frac{\partial z}{\partial y}\right)\left(\frac{\partial z}{\partial x}\right) + F_3\frac{\partial^2 z}{\partial y\,\partial x}\right)\right] dA$$

The last two terms in each parenthesis cancel each other, and then we can rearrange terms to obtain the integral (2) which completes the proof. ∎

Example 5 Use Stokes' Theorem to evaluate the line integral

$$\int_C -y^3\,dx + x^3\,dy - z^3\,dz$$

where C is the intersection of the cylinder $x^2+y^2=1$ and the plane $x+y+z=1$ and the orientation on C corresponds to counterclockwise motion in the xy plane.

The curve C bounds the surface S defined by $z=1-x-y=f(x,y)$ for $x^2+y^2\leqslant 1$ (Figure 3-8). We set $\boldsymbol{F}=-y^3\boldsymbol{i}+x^3\boldsymbol{j}-z^3\boldsymbol{k}$ which has curl $\nabla\times\boldsymbol{F}=(3x^2+3y^2)\boldsymbol{k}$. Then by Stokes' Theorem the line integral is equal to the surface integral

$$\int_S \nabla\times\boldsymbol{F}$$

But $\nabla\times\boldsymbol{F}$ has only a $\boldsymbol{k}$ component. Thus by formula (1) we have

$$\int_S \nabla\times\boldsymbol{F} = \int_D (3x^2+3y^2)\,dx\,dy$$

This integral can be evaluated by changing to polar coordinates. Doing this we get

$$3\int_D (x^2+y^2)\,dx\,dy = 3\int_0^1\int_0^{2\pi} r^2\cdot r\,dr\,d\theta = 6\pi\int_0^1 r^3\,dr = \frac{6\pi}{4} = \frac{3\pi}{2}$$

Let us verify this result by directly evaluating the line integral

$$\int_C -y^3\,dx + x^3\,dy - z^3\,dz$$

We can parametrize the curve ∂D by the equations

$$x=\cos t,\quad y=\sin t,\quad z=0,\quad 0\leqslant t\leqslant 2\pi$$

The curve C is therefore parametrized by the equations

$$x=\cos t,\quad y=\sin t,\quad z=1-\sin t-\cos t,\quad 0\leqslant t\leqslant 2\pi$$

Thus

$$\begin{aligned}\int_C & -y^3\,dx + x^3\,dy - z^3\,dz \\ &= \int_0^{2\pi} (-\sin^3 t)(-\sin t)\,dt + (\cos^3 t)(\cos t)\,dt \\ &\qquad + (1-\sin t-\cos t)^3(-\cos t+\sin t)\,dt \\ &= \int_0^{2\pi} (\cos^4 t+\sin^4 t)\,dt + \int_0^{2\pi} (1-\sin t-\cos t)^3(-\cos t+\sin t)\,dt\end{aligned}$$

The second integrand is of the form $u^n\,du$ and thus the integral is equal to

$$\tfrac{1}{4}[(1-\sin t-\cos t)^4]_0^{2\pi} = 0$$

Hence we are left with

$$\int_0^{2\pi} (\cos^4 t+\sin^4 t)\,dt$$

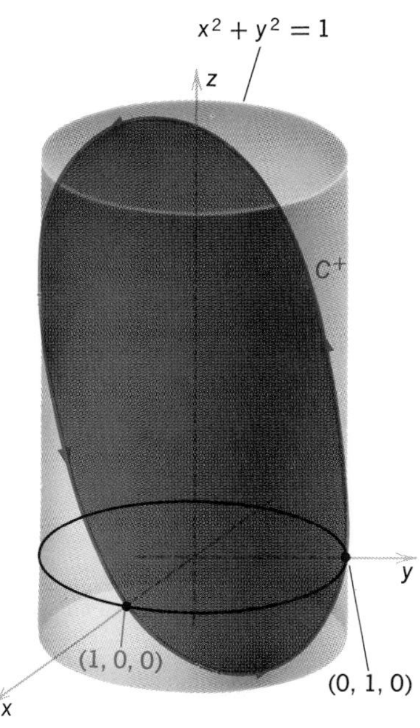

FIGURE 3-8

Since

$$\sin^2 t = \frac{1-\cos 2t}{2}, \qquad \cos^2 t = \frac{1+\cos 2t}{2}$$

we reduce the above integral to

$$\tfrac{1}{2}\int_0^{2\pi} (1+\cos^2 2t)\, dt = \pi + \tfrac{1}{2}\int_0^{2\pi} \cos^2 2t\, dt$$

Again using the fact that

$$\cos^2 2t = \frac{1+\cos 4t}{2}$$

we find

$$\begin{aligned}\pi + \tfrac{1}{4}\int_0^{2\pi} &(1+\cos 4t)\, dt \\ &= \pi + \tfrac{1}{4}\int_0^{2\pi} dt + \tfrac{1}{4}\int_0^{2\pi} \cos 4t\, dt \\ &= \pi + \frac{\pi}{2} + 0 = \frac{3\pi}{2}\end{aligned}$$

Stokes' Theorem can be proved for surfaces which are much more general than those we considered above. Suppose S is an oriented smooth surface with oriented boundary curve ∂S and $\boldsymbol{F}$ is a continuous vector field; then we still have

$$\int_S \operatorname{curl} \boldsymbol{F} = \int_{\partial S} \boldsymbol{F}$$

From our discussion on page 801 it follows that

$$\int_S (\operatorname{curl} \boldsymbol{F})\cdot \boldsymbol{n}\, dS = \int_S \operatorname{curl} \boldsymbol{F} = \int_{\partial S} \boldsymbol{F}$$

Thus Stokes' Theorem says that the integral of the normal component of the curl of a vector field over an oriented surface S is equal to the line integral of $\boldsymbol{F}$ along ∂S.

We conclude this section with some important physical applications of Stokes' Theorem. The first of these is the interpretation commonly given in physics of the curl of a vector field.

Suppose $\boldsymbol{V}$ represents the velocity vector field of a fluid. Consider a point P and a unit vector $\boldsymbol{n}$. Let S_ρ denote the disc of radius ρ and center P, which is perpendicular to $\boldsymbol{n}$, where $\boldsymbol{n}$ denotes the outward direction. By Stokes' Theorem

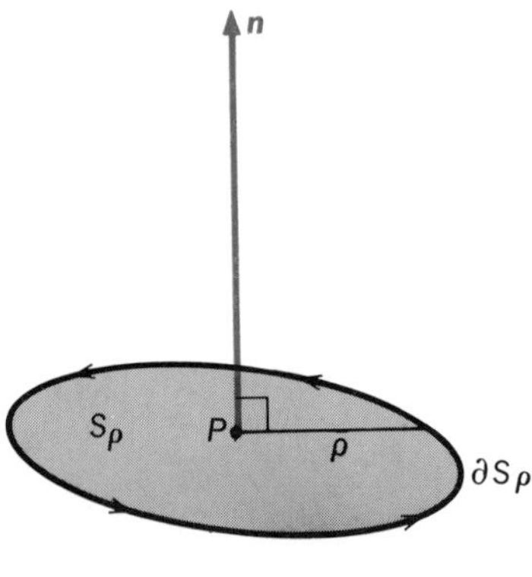

FIGURE 3-9

$$\int_{S_\rho} \operatorname{curl} \boldsymbol{V} = \int_{S_\rho} \operatorname{curl} \boldsymbol{V}\cdot \boldsymbol{n}\, dS = \int_{\partial S_\rho} \boldsymbol{V}$$

where ∂S_ρ has the orientation induced by $\boldsymbol{n}$ (see Figure 3-9). It is not difficult to show (see exercise 7) that there is a point $Q \in S_\rho$ such that

$$\int_{S_\rho} \operatorname{curl} \boldsymbol{V}\cdot \boldsymbol{n}\, dS = [\operatorname{curl} \boldsymbol{V}(Q)\cdot \boldsymbol{n}]\, A(S_\rho)$$

where $A(S_\rho) = \pi\rho^2$ is the area of S_ρ and curl $\boldsymbol{V}(Q)$ is the value of curl $\boldsymbol{V}$ at Q. Thus

$$\text{curl } \boldsymbol{V}(P)\cdot\boldsymbol{n} = \lim_{\rho\to 0} \frac{1}{A(S_\rho)} \int_{\partial S_\rho} \boldsymbol{V}\cdot d\boldsymbol{s} \tag{5}$$

This result allows us to see just what curl $\boldsymbol{V}$ means for the motion of a fluid. The **circulation** $\int_{\partial S_\rho} \boldsymbol{V}$ is the net velocity of the fluid around ∂S_ρ so curl $\boldsymbol{V}\cdot\boldsymbol{n}$ represents the turning or rotating effect of the fluid around the axis $\boldsymbol{n}$. More precisely, the above formula states that curl $\boldsymbol{V}(P)\cdot\boldsymbol{n}$ is the circulation of $\boldsymbol{V}$ per unit area on a surface perpendicular to $\boldsymbol{n}$. Observe that the magnitude of curl $\boldsymbol{V}\cdot\boldsymbol{n}$ is maximized when $\boldsymbol{n} = \text{curl } \boldsymbol{V}/|\text{curl } \boldsymbol{V}|$. Therefore, the rotating effect at P is greatest about the axis parallel to curl $\boldsymbol{V}/|\text{curl } \boldsymbol{V}|$. Thus curl $\boldsymbol{V}$ is aptly called the **vorticity vector.**

In the next section we prove that $\nabla\times\boldsymbol{F} = \boldsymbol{0}$ if and only if $\boldsymbol{F} = -\nabla f$ for some real-valued function f called the potential (or "voltage" if $\boldsymbol{F}$ is the electric field). The negative sign is used to conform to certain physical intuitions. However, for electric fields $\boldsymbol{E}$, we do not generally have $\nabla\times\boldsymbol{E} = \boldsymbol{0}$ unless the situation is static. On the other hand, a basic law of electromagnetic theory is that if $\boldsymbol{E}(t, x, y, z)$ and $\boldsymbol{H}(t, x, y, z)$ represent the electric and magnetic fields at time t, then $\nabla\times\boldsymbol{E} = -\partial\boldsymbol{H}/\partial t$.

Let us, using Stokes' Theorem, determine what this means physically. Assume S is a surface to which Stokes' Theorem applies. Then

$$\int_{\partial S} \boldsymbol{E} = \int_S \nabla\times\boldsymbol{E} = -\int_S \frac{\partial\boldsymbol{H}}{\partial t}$$

$$= -\frac{\partial}{\partial t}\int_S \boldsymbol{H}$$

(The last equality may be justified if $\boldsymbol{H}$ is C^1.) Thus we obtain

$$\int_{\partial S} \boldsymbol{E} = -\frac{\partial}{\partial t}\int_S \boldsymbol{H}$$

This equality is known as **Faraday's law.** The quantity $\int_{\partial S} \boldsymbol{E}$ represents the "voltage" around ∂S and if ∂S were a wire, a current would flow in proportion to this "voltage." Also $\int_S \boldsymbol{H}$ is the flux of $\boldsymbol{H}$, or the magnetic flux. Thus, Faraday's law says that the voltage around a loop equals the negative of the rate of change of magnetic flux through the loop.

EXERCISES

1. Evaluate $\int_S \nabla\times\boldsymbol{F}$, where S is the surface $x^2+y^2+3z^2 = 1$, $z \leqslant 0$, and $\boldsymbol{F} = y\boldsymbol{i} - x\boldsymbol{j} + zx^3y^2\boldsymbol{k}$.
2. Evaluate $\int_S \nabla\times\boldsymbol{F}$ where $\boldsymbol{F} = (x^2+y-4)\boldsymbol{i} + 3xy\boldsymbol{j} + (2xz+z^2)\boldsymbol{k}$ and S is the surface $x^2+y^2+z^2 = 16$, $z \geqslant 0$.

3. Let $\boldsymbol{\sigma}$ consist of straight lines joining $(1,0,0)$ to $(0,1,0)$ then to $(0,0,1)$ and let S be the triangle with these vertices. Verify Stokes' Theorem directly with $\boldsymbol{F} = yz\boldsymbol{i} + xz\boldsymbol{j} + xy\boldsymbol{k}$.

4. Prove that Faraday's law implies $\nabla \times \boldsymbol{E} = -\partial \boldsymbol{H}/\partial t$.

5. Let S be the surface of the unit sphere. Let $\boldsymbol{F}$ be a vector field and F_r its radial component. Prove

$$\int_S \boldsymbol{F} = \int_{\theta=0}^{2\pi} \int_{\phi=0}^{\pi} F_r \sin\phi \, d\phi \, d\theta$$

What is the formula for real-valued functions f?

6. Work out a formula like that in exercise 5 for integration over the surface of a cylinder.

*7. Prove the Mean Value Theorem for surface integrals:

$$\int_S \boldsymbol{F} \cdot \boldsymbol{n} \, dS = \boldsymbol{F}(Q) \cdot \boldsymbol{n}(Q) \, A(S)$$

for some $Q \in S$, where $A(S)$ is the area of S. [Hint: Prove it for real functions first, by reducing the problem to one of a double integral: show that if $g \geqslant 0$, then

$$\int_D fg = f(Q) \int_D g$$

for some $Q \in D$ (do it by considering $\int_D fg / \int_D g$ and using the Intermediate Value Theorem).]

8. Let S be a surface and let $\boldsymbol{F}$ be perpendicular to the tangent to the boundary of S. Show that

$$\int_S \nabla \times \boldsymbol{F} = 0$$

What does this mean physically?

9. Consider two surfaces S_1, S_2 with the same boundary ∂S. Describe how S_1 and S_2 must be oriented to ensure that

$$\int_{S_1} \nabla \times \boldsymbol{F} = \int_{S_2} \nabla \times \boldsymbol{F}$$

10. For a surface S and a fixed vector $\boldsymbol{v}$ prove

$$2 \int_S \boldsymbol{v} \cdot \boldsymbol{n} \, dS = \int_{\partial S} \boldsymbol{v} \times \boldsymbol{r}$$

where $\boldsymbol{r}(x, y, z) = (x, y, z)$.

11. Argue informally that if S is a closed surface then

$$\int_S \nabla \times \boldsymbol{F} = 0$$

(see exercise 9). [A **closed surface** is one that forms the boundary of a region in space; thus, for example, a sphere is a closed surface.]

12. If C is a closed curve which is the boundary of a surface S, show that

(a) $\int_C f \nabla g = \int_S \nabla f \times \nabla g$

(b) $\int_C (f \nabla g + g \nabla f) = 0$

13. If C is a closed curve which is the boundary of a surface S, show that

$$\int_C \boldsymbol{v} \cdot d\boldsymbol{s} = 0$$

for a constant vector $\boldsymbol{v}$.

14. Let $F = x^2 i + (2xy + x) j + zk$. Let C be the circle $x^2 + y^2 = 1$ and S the disc $x^2 + y^2 \leqslant 1$. Determine
    (a) the flux of F out of S.
    (b) the circulation of F around C.
    (c) Show that the flux of $\nabla \times F = 0$. Find an f with $F = \nabla f$. Verify Stokes' Theorem directly in this case.

4 CONSERVATIVE FIELDS

We saw in §17.1 that in the case of a gradient force field $F = \nabla f$, line integrals of F were evaluated as follows:

$$\int_\sigma F = f(\sigma(b)) - f(\sigma(a))$$

The value of the integral depends only on the endpoints $\sigma(b)$ and $\sigma(a)$ of the path. In other words, if we used another path with the same endpoints, we would still get the same answer. This leads us to say that the integral is **path independent.**

Gradient fields are very important in physical problems. Usually $-f = V$ represents a potential energy (gravitational, electrical, and so on) and F represents a force. Consider the example of a particle of mass m in the field of the earth; in this case one takes f to be GmM/r or $V = -GmM/r$, where G is the gravitational constant, M is the mass of the earth, and r is the distance from the center of the earth. Then the corresponding force is $F = (GmM/r^3)\,r = (GmM/r^2)\,n$, where n is the unit radial vector. (We discuss this case further below.) Note that F fails to be defined at the one point $r = 0$.

We now wish to characterize those vector fields which can be written as a gradient. Our job will be simplified considerably by Stokes' Theorem.

Theorem 1 Let F be a vector field defined and C^1 on R^3 except possibly for one point. The following conditions on F are all equivalent:

(i) For any closed curve C, $\int_C F = 0$.
(ii) For any two simple curves C_1, C_2 with the same endpoints, $\int_{C_1} F = \int_{C_2} F$.
(iii) F is the gradient of some function f; that is, $F = \nabla f$ (and if F has an exceptional point where it fails to be defined, so will f).
(iv) $\nabla \times F = 0$.

A vector field satisfying one (and hence, all) of these conditions is called a **conservative** vector field.

Proof We shall establish the following chain of implications which will prove the theorem:

$$\text{(i)} \Rightarrow \text{(ii)} \Rightarrow \text{(iii)} \Rightarrow \text{(iv)} \Rightarrow \text{(i)}$$

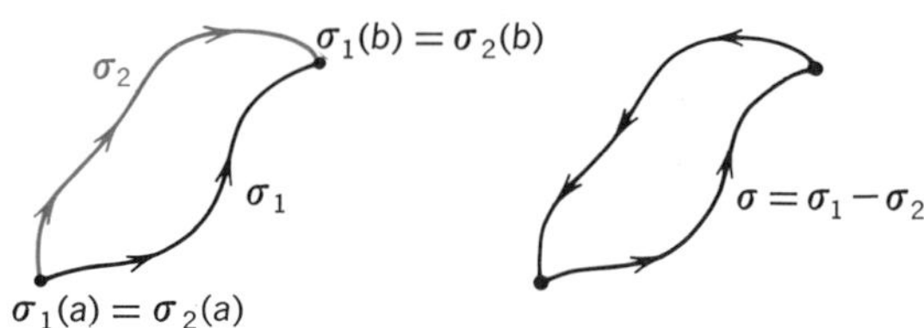

FIGURE 4-1

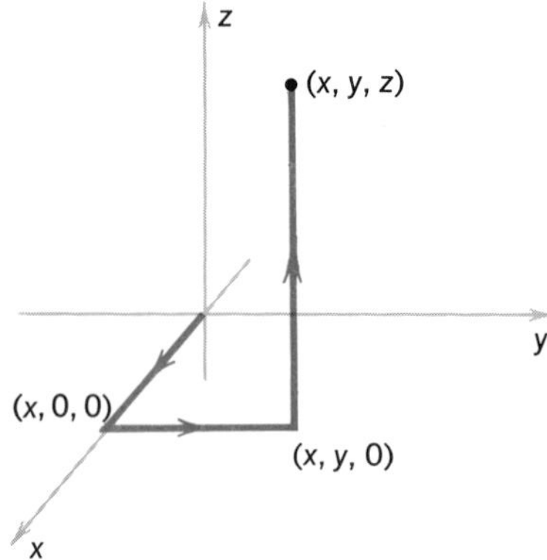

FIGURE 4-2

First we show (i)⇒(ii). In (ii) we are given parametrizations $\boldsymbol{\sigma}_1, \boldsymbol{\sigma}_2$ representing C_1 and C_2, with the same endpoints. Construct the closed curve $\boldsymbol{\sigma}$ obtained by first traversing $\boldsymbol{\sigma}_1$, then $\boldsymbol{\sigma}_2$ (Figure 4-1), or symbolically $\boldsymbol{\sigma} = \boldsymbol{\sigma}_1 - \boldsymbol{\sigma}_2$. But by (i)

$$\int_{\sigma} \boldsymbol{F} = \int_{\sigma_1} \boldsymbol{F} - \int_{\sigma_2} \boldsymbol{F} = 0$$

so (ii) holds.

Next we prove (ii)⇒(iii). Let C be any curve joining a point such as $(0, 0, 0)$ to $(x\,y, z)$, and suppose C is represented by the parametrization $\boldsymbol{\sigma}$ [if $(0, 0, 0)$ is the exceptional point of $\boldsymbol{F}$, we can choose a different starting point for $\boldsymbol{\sigma}$ without affecting the argument]. Define f to be $\int_{\sigma} \boldsymbol{F}$. By hypothesis (ii) f is independent of $\boldsymbol{\sigma}$. We will show that $\boldsymbol{F} = \operatorname{grad} f$. Indeed, choose $\boldsymbol{\sigma}$ to be the path shown in Figure 4-2, so that

$$f(x, y, z) = \int_0^x F_1(t, 0, 0)\, dt + \int_0^y F_2(x, t, 0)\, dt + \int_0^z F_3(x, y, t)\, dt$$

where $\boldsymbol{F} = (F_1, F_2, F_3)$. It then follows immediately that $\partial f/\partial z = F_3$. Using condition (ii) we can show that (see exercise 17) $\partial f/\partial x = F_1$, $\partial f/\partial y = F_2$, or $\nabla f = \boldsymbol{F}$. Thirdly, (iii)⇒(iv) because we proved in §14.5 that

$$\nabla \times \nabla f = \mathbf{0}$$

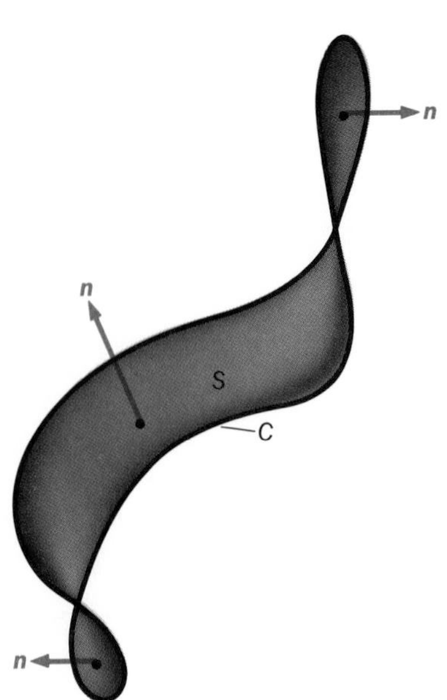

FIGURE 4-3

Finally, we demonstrate (iv)⇒(i). Let $\boldsymbol{\sigma}$ represent a closed curve C and let S be any surface whose boundary is $\boldsymbol{\sigma}$ (if $\boldsymbol{F}$ has an exceptional point, choose S to avoid it). Figure 4-3 shows that we can always find such a surface. By Stokes' Theorem

$$\int_C \boldsymbol{F} = \int_{\sigma} \boldsymbol{F} = \int_S (\nabla \times \boldsymbol{F}) \cdot \boldsymbol{n}\, dS = \int_S (\operatorname{curl} \boldsymbol{F}) \cdot \boldsymbol{n}\, dS$$

Since $\nabla \times \boldsymbol{F} = \mathbf{0}$, this integral vanishes. ∎

There are several useful physical interpretations of $\int_C \boldsymbol{F}$. We have already seen that one is the work done by $\boldsymbol{F}$ in moving a particle along C. A second interpretation is the notion of circulation which we encountered at the end of the last section. In this case we think of $\boldsymbol{F}$ as the velocity field of a fluid; that is, to each point P in space, $\boldsymbol{F}$ assigns the velocity vector of the fluid at P. Take C to be a closed curve, and let Δs be a small directed chord of C. Then $\boldsymbol{F}\cdot\Delta\boldsymbol{s}$ is approximately the tangential component of $\boldsymbol{F}$ times $|\Delta s|$. The integral $\int_C \boldsymbol{F}$ is the net tangential component around C. This means that if a small paddle wheel were placed in the fluid, it would rotate if the circulation of the fluid were nonzero, $\int_C \boldsymbol{F} \neq 0$ (see Figure 4-4). Thus we often speak of the line integral

$$\int_C \boldsymbol{F}$$

as being the circulation of $\boldsymbol{F}$ around C.

There is a similar interpretation in electromagnetic theory. If $\boldsymbol{F}$ represents an electric field, then a current will flow around a loop C if $\int_C \boldsymbol{F} \neq 0$.

By Theorem 1, a field $\boldsymbol{F}$ has no circulation if and only if curl $\boldsymbol{F} = \nabla \times \boldsymbol{F} = \boldsymbol{0}$. Hence, a vector field $\boldsymbol{F}$ with curl $\boldsymbol{F} = \boldsymbol{0}$ is called **irrotational**. We have therefore proved that a vector field in $\boldsymbol{R}^3$ is irrotational if and only if it is a gradient field for some function $\boldsymbol{F} = \nabla f$. The function f is called a **potential** for $\boldsymbol{F}$.

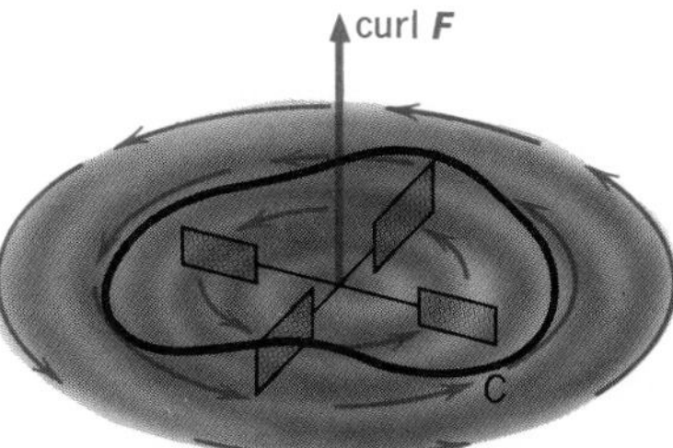

FIGURE 4-4

Example 1 Consider the vector field $\boldsymbol{F}$ on $\boldsymbol{R}^3$ defined by

$$\boldsymbol{F}(x, y, z) = y\boldsymbol{i} + (z\cos yz + x)\boldsymbol{j} + y\cos yz\,\boldsymbol{k}$$

Show that $\boldsymbol{F}$ is irrotational and find a scalar potential for $\boldsymbol{F}$.

We compute $\nabla \times \boldsymbol{F}$:

$$\begin{aligned}\nabla \times \boldsymbol{F} &= \left(\frac{\partial F_3}{\partial y} - \frac{\partial F_2}{\partial z}\right)\boldsymbol{i} + \left(\frac{\partial F_1}{\partial z} - \frac{\partial F_3}{\partial x}\right)\boldsymbol{j} + \left(\frac{\partial F_2}{\partial x} - \frac{\partial F_1}{\partial y}\right)\boldsymbol{k} \\ &= (\cos yz - yz\sin yz - \cos yz + yz\sin yz)\boldsymbol{i} \\ &\qquad + (0-0)\boldsymbol{j} + (1-1)\boldsymbol{k} \\ &= 0\boldsymbol{i} + 0\boldsymbol{j} + 0\boldsymbol{k} = \boldsymbol{0}\end{aligned}$$

so $\boldsymbol{F}$ is irrotational. To find a scalar potential we can proceed in several ways.

Method 1. By the technique used to prove (ii)$\Rightarrow$(iii) in Theorem 1, we may set

$$\begin{aligned}f(x, y, z) &= \int_0^x F_1(t, 0, 0)\,dt + \int_0^y F_2(x, t, 0)\,dt + \int_0^z F_3(x, y, t)\,dt \\ &= \int_0^x 0\,dt + \int_0^y x\,dt + \int_0^z y\cos yt\,dt \\ &= 0 + xy + \sin yz = xy + \sin yz\end{aligned}$$

One easily verifies that $\nabla f = \boldsymbol{F}$ as required:

$$\nabla f = \frac{\partial f}{\partial x}\boldsymbol{i} + \frac{\partial f}{\partial y}\boldsymbol{j} + \frac{\partial f}{\partial z}\boldsymbol{k} = y\boldsymbol{i} + (x + z\cos yz)\boldsymbol{j} + y\cos yz\,\boldsymbol{k}$$

Method 2. Because we know that f exists, we can solve the equations

$$\frac{\partial f}{\partial x} = y$$

$$\frac{\partial f}{\partial y} = x + z\cos yz$$

$$\frac{\partial f}{\partial z} = y\cos yz$$

These are equivalent to the simultaneous equations

$$f = xy + h_1(y,z)$$

$$f = \sin yz + xy + h_2(x,z)$$

$$f = \sin yz + h_3(x,y)$$

for functions h_1, h_2, h_3 independent of x, y, and z (respectively). When $h_1(y,z) = \sin yz$, $h_2 = 0$, and $h_3(x,y) = xy$, the three equations agree, and yield a potential for $\boldsymbol{F}$.

Example 2 Recall that if a mass M is at the origin of our coordinate system in $\boldsymbol{R}^3$, it exerts a force on a unit mass located at $\boldsymbol{r} = (x,y,z)$ with magnitude GM/r^2 and directed toward the origin. Here G is the gravitational constant which depends on the units of measurement, and $r = |\boldsymbol{r}| = \sqrt{x^2+y^2+z^2}$. If we remember that $-\boldsymbol{r}/r$ is a unit vector directed towards the origin, then we can write our force field as

$$\boldsymbol{F}(\boldsymbol{r}) = -\frac{GM\boldsymbol{r}}{r^3}$$

We show that $\boldsymbol{F}$ is irrotational and will find scalar potential for $\boldsymbol{F}$. Notice that $\boldsymbol{F}$ is not defined at the origin, but Theorem 1 still applies since it allows an exceptional point.

First let us verify that $\nabla \times \boldsymbol{F} = \boldsymbol{0}$. Referring to formula 12 of the table in §14.5 we get

$$\nabla \times \boldsymbol{F} = -GM\left\{\nabla\left(\frac{1}{r^3}\right) \times \boldsymbol{r} + \frac{1}{r^3}\nabla \times \boldsymbol{r}\right\}$$

But $\nabla\ (1/r^3) = -\boldsymbol{r}/r^5$ (see exercise 16, §14.5) so the first term vanishes as $\boldsymbol{r} \times \boldsymbol{r} = \boldsymbol{0}$. The second term vanishes because

$$\nabla \times \boldsymbol{r} = \begin{vmatrix} \boldsymbol{i} & \boldsymbol{j} & \boldsymbol{k} \\ \dfrac{\partial}{\partial x} & \dfrac{\partial}{\partial y} & \dfrac{\partial}{\partial z} \\ x & y & z \end{vmatrix} = \left(\frac{\partial z}{\partial y} - \frac{\partial y}{\partial z}\right)\boldsymbol{i} + \left(\frac{\partial x}{\partial z} - \frac{\partial z}{\partial x}\right)\boldsymbol{j} + \left(\frac{\partial y}{\partial x} - \frac{\partial x}{\partial y}\right)\boldsymbol{k} = \boldsymbol{0}$$

Hence $\nabla \times \boldsymbol{F} = \boldsymbol{0}$ (for $\boldsymbol{r} \neq \boldsymbol{0}$).

If we recall the formula $\nabla(r^n) = nr^{n-2}\boldsymbol{r}$ (exercise 16, §14.5) then we can read off a scalar potential for $\boldsymbol{F}$ by inspection. We have $\boldsymbol{F} = -\nabla\phi$, where $\phi(\boldsymbol{r}) = -GM/r$ is called, the **gravitational potential energy.** We discussed the two variable case in §14.3.

By Theorem 3 of §17.1, the work done by $\boldsymbol{F}$ in moving a unit mass particle from a point P_1 to P_2 is given by

$$\phi(P_1) - \phi(P_2) = GM\left(\frac{1}{r_2} - \frac{1}{r_1}\right)$$

where r_1 is the radial distance of P_1 from the origin, with r_2 similarly defined.

Theorem 1 is also true for smooth vector fields $\boldsymbol{F}$ on $\boldsymbol{R}^2$, and in fact the theorem takes a particularly simple form. However, for Theorem 1 to hold on $\boldsymbol{R}^2$, $\boldsymbol{F}$ cannot have any exceptional points; that is, $\boldsymbol{F}$ must be smooth everywhere (see exercise 12).

If $\boldsymbol{F} = P\boldsymbol{i} + Q\boldsymbol{j}$ then

$$\nabla \times \boldsymbol{F} = \left(\frac{\partial Q}{\partial x} - \frac{\partial P}{\partial y}\right)\boldsymbol{k}$$

and so the condition $\nabla \times \boldsymbol{F} = \boldsymbol{0}$ reduces to

$$\frac{\partial P}{\partial y} = \frac{\partial Q}{\partial x}$$

Therefore, if $\boldsymbol{F}$ is a C^1 vector field on $\boldsymbol{R}^2$ of the form $P\boldsymbol{i} + Q\boldsymbol{j}$ with $\partial P/\partial y = \partial Q/\partial x$, then by Theorem 1 applied to $\boldsymbol{R}^2$, $\boldsymbol{F} = \nabla f$ for some f in $\boldsymbol{R}^2$

Example 3 (a) Determine whether the vector field

$$\boldsymbol{F} = e^{xy}\boldsymbol{i} + e^{x+y}\boldsymbol{j}$$

is a gradient field. Here $P(x, y) = e^{xy}$ and $Q(x, y) = e^{x+y}$. So we compute

$$\frac{\partial P}{\partial y} = xe^y, \qquad \frac{\partial Q}{\partial x} = e^{x+y}$$

These are not equal so $\boldsymbol{F}$ cannot have a potential function.

(b) Repeat part (a) for

$$\boldsymbol{F} = 2x\cos y\,\boldsymbol{i} - x^2\sin y\boldsymbol{j}$$

In this case, we find

$$\frac{\partial P}{\partial y} = -2x\sin y = \frac{\partial Q}{\partial x}$$

and so $\boldsymbol{F}$ has a potential function f. To compute f we have the equations

$$\frac{\partial f}{\partial x} = 2x\cos y, \qquad \frac{\partial f}{\partial y} = -x^2\sin y$$

Thus

$$f(x,y) = x^2 \cos y + h_1(y)$$

and

$$f(x,y) = x^2 \cos y + h_2(y)$$

If $h_1 = h_2 = 0$ both equations are satisfied and we find that $f(x,y) = x^2 \cos y$, is a potential for $\boldsymbol{F}$.

Example 4 Let $\boldsymbol{\sigma}:[1,2] \to \boldsymbol{R}^2$ be given by

$$x = e^{t-1}, \qquad y = \sin(\pi/t)$$

Compute the integral

$$\int_{\sigma} \boldsymbol{F} = \int_{\sigma} 2x \cos y \, dx - x^2 \sin y \, dy$$

where $\boldsymbol{F} = 2x \cos y \boldsymbol{i} - x^2 \sin y \boldsymbol{j}$. We have $\boldsymbol{\sigma}(1) = (1,0)$ and $\boldsymbol{\sigma}(2) = (e,1)$. Since $\partial(2x \cos y)/\partial y = \partial(-x^2 \sin y)/\partial x$, $\boldsymbol{F}$ is irrotational and hence a gradient vector field (as we saw in Example 3). Thus by Theorem 1 we can replace $\boldsymbol{\sigma}$ by any piecewise C^1 curve having the same endpoints, in particular the polygonal path from $(1,0)$ to $(e,0)$ to $(e,1)$. So the line integral must be equal to

$$\begin{aligned}\int_{\sigma} \boldsymbol{F} &= \int_1^e 2x \cos 0 \, dx + \int_0^1 -e^2 \sin y \, dy \\ &= (e^2 - 1) + e^2(\cos 1 - 1) = e^2 \cos 1 - 1\end{aligned}$$

Alternatively, using Theorem 3 of §17.1 we have

$$\int_{\sigma} 2x \cos y \, dx - x^2 \sin y \, dy = \int_{\sigma} \nabla f = f(\sigma(2)) - f(\sigma(1)) = e^2 \cos 1 - 1$$

since $f(x,y) = x^2 \cos y$ is a potential function for $\boldsymbol{F}$.

The reader can see that this technique is simpler than trying to compute the integral directly.

We conclude this section with a theorem which is quite similar in spirit to Theorem 1. Theorem 1 was partly motivated as a converse to the result that curl $\nabla f = \boldsymbol{0}$ for any C^1 function $f:\boldsymbol{R}^3 \to \boldsymbol{R}$—or, if curl $\boldsymbol{F} = \boldsymbol{0}$ then $\boldsymbol{F} = \nabla f$. We also know (formula 8 in the table in §14.5) that div(curl $\boldsymbol{G}$) $= 0$ for any C^1 vector field $\boldsymbol{G}$. We can formulate the converse statement; if div $\boldsymbol{F} = 0$, is $\boldsymbol{F}$ the curl of a vector field $\boldsymbol{G}$? The affirmative answer is contained in the next result.

Theorem 2 If $\boldsymbol{F}$ is a C^1 vector field on $\boldsymbol{R}^3$ with div $\boldsymbol{F} = 0$ then there exists a C^1 vector field $\boldsymbol{G}$ with $\boldsymbol{F} = \text{curl } \boldsymbol{G}$.

The proof requires some care and we consider it in exercise 14.

EXERCISES

1. Show that any two potential functions for a vector field differ by at most a constant.

2. (a) Let $\boldsymbol{F}(x,y) = (xy, y^2)$ and let $\boldsymbol{\sigma}$ be the path $y = 2x^2$ joining $(0,0)$ to $(1,2)$ in $\boldsymbol{R}^2$. Evaluate $\int_{\sigma} \boldsymbol{F}$.
   (b) Does the integral in (a) depend on the path joining $(0,0)$ to $(1,2)$?

3. Let $\boldsymbol{F}(x,y,z) = (2xyz + \sin x)\boldsymbol{i} + x^2 z\boldsymbol{j} + x^2 y\boldsymbol{k}$. Find a function f such that $\boldsymbol{F} = \nabla f$.

4. Evaluate $\int_{\sigma} \boldsymbol{F}$, where $\boldsymbol{\sigma}(t) = (\cos^5 t, \sin^3 t, t^4)$, $0 \leqslant t \leqslant \pi$, and $\boldsymbol{F}$ is as in exercise 3.

5. What is the work required to move a particle from a point $r_0 \in \boldsymbol{R}^3$ "to ∞" under the force $\boldsymbol{F} = -\boldsymbol{r}/|\boldsymbol{r}|^3$, where $\boldsymbol{r}\,(x,y,z) = (x,y,z)$?

6. In exercise 5, show $\boldsymbol{F} = \nabla(1/r), r \neq 0,\ r = |\boldsymbol{r}|$. In what sense is the integral independent of path? Formulate a general theorem.

7. Let $\boldsymbol{F}(x,y,z) = xy\,\boldsymbol{i} + y\boldsymbol{j} + z\boldsymbol{k}$. Can there exist a function f such that $\boldsymbol{F} = \nabla f$?

8. Let $\boldsymbol{F} = F_1\boldsymbol{i} + F_2\boldsymbol{j} + F_3\boldsymbol{k}$ and suppose each F_1 satisfies the homogeneity condition

$$F_k(tx, ty, tz) = t\,F_k(x,y,z), \qquad k = 1,2,3$$

   Suppose also $\nabla \times \boldsymbol{F} = 0$. Prove $\boldsymbol{F} = \nabla f$ where

$$2f(x,y,z) = xF_1(x,y,z) + yF_2(x,y,z) + zF_3(x,y,z)$$

9. Are the following vector fields the curl of some vector fields? If so, find the vector field.

   (a) $\boldsymbol{F} = x\boldsymbol{i} + y\boldsymbol{i} + z\boldsymbol{k}$

   (b) $\boldsymbol{F} = (x^2+1)\boldsymbol{i} + (z - 2xy)\boldsymbol{j} + y\boldsymbol{k}$

   Then there is a function f such that $\boldsymbol{F} = \nabla f$.

10. Let a fluid have the velocity field $\boldsymbol{F}(x,y,z) = xy\boldsymbol{i} + yz\boldsymbol{j} + xz\boldsymbol{k}$. What is the circulation around the unit circle? Interpret your answer.

11. Let $\boldsymbol{F}(x,y,z) = (e^x \sin y)\boldsymbol{i} + (e^x \cos y)\boldsymbol{j} + z^2\boldsymbol{k}$. Evaluate the integral $\int_{\sigma}\boldsymbol{F}$, where $\boldsymbol{\sigma}(t) = (\sqrt{t}, t^3, e^{\sqrt{t}})$, $0 \leqslant t \leqslant 1$.

12. (a) Show that $\int_C (x\,dy - y\,dx)/(x^2+y^2) = 2\pi$, where C is the unit circle.
    (b) Conclude that the associated vector field $(x/(x^2+y^2))\boldsymbol{i} - (y/(x^2+y^2))\boldsymbol{j}$ is not a potential field.
    (c) Show, however, that $\partial P/\partial y = \partial Q/\partial x$. Does this contradict Theorem 1? If not, why not?

13. The mass of the earth is approximately 6×10^{27} g and that of the sun is 330000 times as much. The gravitational constant (in units of grams, seconds, and centimeters) is 6.7×10^{-8}. The distance of the earth from the sun is about 1.5×10^{12} cm. Compute, approximately, the work necessary to increase the distance of the earth from the sun by 1 cm.

14. Prove Theorem 2. [Hint: Define $\boldsymbol{G} = G_1\boldsymbol{i} + G_2\boldsymbol{j} + G_3\boldsymbol{k}$ by

$$G_1(x, y, z) = \int_0^z F_2(x, y, t)\,dt - \int_0^y F_3(x, t, 0)\,dt$$

$$G_2(x, y, z) = -\int_0^z F_1(x, y, t)\,dt$$

and $G_3(x, y, z) = 0$.]

15. Let $\boldsymbol{F} = xz\boldsymbol{i} - yz\boldsymbol{j} + y\boldsymbol{k}$. Verify that $\nabla \cdot \boldsymbol{F} = 0$. Find a $\boldsymbol{G}$ such that $\boldsymbol{F} = \nabla \times \boldsymbol{G}$.
16. Let $\boldsymbol{F} = x\cos y\boldsymbol{i} - x\sin y\boldsymbol{j} + \sin x\boldsymbol{k}$. Find a $\boldsymbol{G}$ such that $\boldsymbol{F} = \nabla \times \boldsymbol{G}$.
17. Show that the function f defined in Theorem 1 for (ii)$\Rightarrow$(iii) satisfies $\partial f/\partial x = F_1$, $\partial f/\partial y = F_2$.

5 GAUSS' THEOREM

Gauss' Theorem states that the flux of a vector field out of a closed surface equals the integral of the divergence of that vector field over the volume enclosed by the surface. The result is parallel to Stokes' Theorem and Green's Theorem in that it relates an integral over a closed geometrical object (curve or surface) to an integral over a contained region (surface or volume).

We begin by asking the reader to review the various regions in space that were introduced when we considered the volume integral; these regions were illustrated in Figure 6-3 of Chapter 15. As that figure indicates, the boundary of a region of type 1, 2, or 3 is a surface which is made up of a finite number (at most six, at least two) of surfaces given as graphs of functions from $\boldsymbol{R}^2$ to $\boldsymbol{R}$. This type of surface is called a **closed surface**; it has no (geometric) boundary. The surfaces S_i $(1 \leqslant i \leqslant N, N \leqslant 6)$ composing such a closed surface are called its **faces**. In this section we consider only those closed surfaces whose faces are surfaces of the type considered in §3.

Example 1 The cube in Figure 5-1 is a region of type 4 with six squares composing its boundary. The sphere is the boundary of a solid ball, which is a region of type 4.

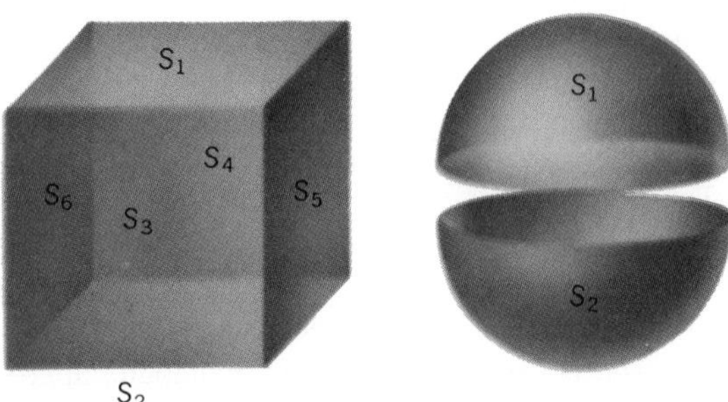

FIGURE 5-1

Closed surfaces can be oriented in two ways. The outward orientation makes the normal point outward into space, and the inward orientation makes the normal point into the bounded region (Figure 5-2).

Suppose S is a closed surface oriented in one of these two ways and $\boldsymbol{F}$ is a vector field on S. Then we define

$$\int_S \boldsymbol{F} = \sum_i \int_{S_i} \boldsymbol{F}$$

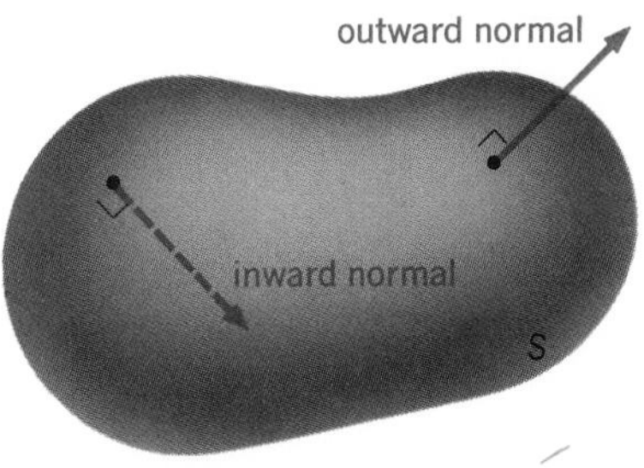

FIGURE 5-2

If S is given the outward orientation, the integral $\int_S \boldsymbol{F}$ measures the total flux of $\boldsymbol{F}$ outwards across S. That is, if we think of $\boldsymbol{F}$ as the velocity field of a fluid, $\int_S \boldsymbol{F}$ indicates the amount of fluid leaving the region bounded by S per unit time. If S is given the inward orientation, the integral $\int_S \boldsymbol{F}$ measures the total flux of $\boldsymbol{F}$ inwards across S.

There is another common way of writing these surface integrals, a way in which the specific orientation of S is mentioned. Let the orientation of S be given by a unit normal vector $\boldsymbol{n}(x, y, z)$ at each point of S. Then we have the oriented integral

$$\int_S \boldsymbol{F} = \int_S (\boldsymbol{F}\cdot\boldsymbol{n})\, dS$$

that is, the integral of the normal component of $\boldsymbol{F}$ over S. In the remainder of this section, if S is a closed surface which encloses a region Ω, we adopt the convention that $S = \partial\Omega$ is given the outward orientation with outward unit normal $\boldsymbol{n}(x, y, z)$ at each point $(x, y, z) \in S$. Furthermore, we denote the surface with the opposite (inward) orientation by $\partial\Omega_{\text{op}}$. Then the associated unit normal direction for this orientation is $-\boldsymbol{n}$. Thus

$$\int_{\partial\Omega} \boldsymbol{F} = \int_S (\boldsymbol{F}\cdot\boldsymbol{n})\, dS = -\int_S (\boldsymbol{F}\cdot(-\boldsymbol{n}))\, dS = -\int_{\partial\Omega_{\text{op}}} \boldsymbol{F}$$

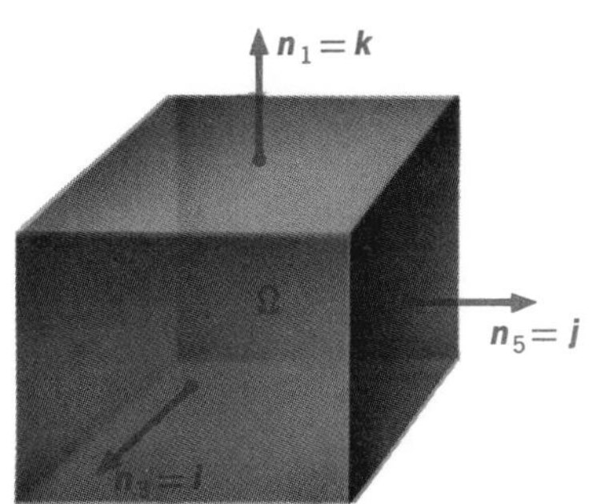

FIGURE 5-3

Example 2 The unit cube Ω given by

$$0 \leqslant x \leqslant 1, \qquad 0 \leqslant y \leqslant 1, \qquad 0 \leqslant z \leqslant 1$$

is a region in space of type 4 (Figure 5-3). We write the faces as

$$\begin{array}{lll} S_1: z = 1, & 0 \leqslant x \leqslant 1, & 0 \leqslant y \leqslant 1 \\ S_2: z = 0, & 0 \leqslant x \leqslant 1, & 0 \leqslant y \leqslant 1 \\ S_3: x = 1, & 0 \leqslant y \leqslant 1, & 0 \leqslant z \leqslant 1 \\ S_4: x = 0, & 0 \leqslant y \leqslant 1, & 0 \leqslant z \leqslant 1 \\ S_5: y = 1, & 0 \leqslant x \leqslant 1, & 0 \leqslant z \leqslant 1 \\ S_6: y = 0, & 0 \leqslant x \leqslant 1, & 0 \leqslant z \leqslant 1 \end{array}$$

From Figure 5-3 we see that

$$\boldsymbol{n}_1 = \boldsymbol{k} = -\boldsymbol{n}_2$$

$$\boldsymbol{n}_3 = \boldsymbol{i} = -\boldsymbol{n}_4$$

$$\boldsymbol{n}_5 = \boldsymbol{j} = -\boldsymbol{n}_6$$

So for a continuous field $\boldsymbol{F} = F_1\boldsymbol{i} + F_2\boldsymbol{j} + F_3\boldsymbol{k}$,

$$\begin{aligned}\int_{\partial\Omega} \boldsymbol{F} = \int_S \boldsymbol{F}\cdot\boldsymbol{n}\, dS &= \int_{S_1} F_3\, dS - \int_{S_2} F_3\, dS \\ &\quad + \int_{S_3} F_1\, dS - \int_{S_4} F_1\, dS \\ &\quad + \int_{S_5} F_2\, dS - \int_{S_6} F_2\, dS\end{aligned}$$

We now come to the last of the three central theorems of this chapter. This theorem relates surface integrals with volume integrals; in words, the theorem states that if Ω is a region in $\boldsymbol{R}^3$, then the flux of a field $\boldsymbol{F}$ outward across the closed surface $\partial\Omega$ is equal to the integral of $\operatorname{div}\boldsymbol{F}$ over Ω.

Theorem 1 Gauss' Divergence Theorem Let Ω be a region in space of type 4. Denote by $\partial\Omega$ the closed surface which bounds Ω. Let $\boldsymbol{F}$ be a smooth vector field defined on Ω. Then

$$\int_\Omega \nabla\cdot\boldsymbol{F} = \int_\Omega \operatorname{div}\boldsymbol{F} = \int_{\partial\Omega} \boldsymbol{F}$$

or using alternative notation

$$\int_\Omega \operatorname{div}\boldsymbol{F}\, dV = \int_{\partial\Omega} \boldsymbol{F} = \int_{\partial\Omega} (\boldsymbol{F}\cdot\boldsymbol{n})\, dS$$

Proof If $\boldsymbol{F} = P\boldsymbol{i} + Q\boldsymbol{j} + R\boldsymbol{k}$, then by definition, $\operatorname{div}\boldsymbol{F} = \partial P/\partial x + \partial Q/\partial y + \partial R/\partial z$, so we can write (using additivity of the volume integral)

$$\int_\Omega \operatorname{div}\boldsymbol{F}\, dV = \int_\Omega \frac{\partial P}{\partial x}\, dV + \int_\Omega \frac{\partial Q}{\partial y}\, dV + \int_\Omega \frac{\partial R}{\partial z}\, dV$$

On the other hand, the surface integral in question is

$$\begin{aligned}\int_{\partial\Omega} \boldsymbol{F} = \int_{\partial\Omega} \boldsymbol{F}\cdot\boldsymbol{n}\, dS &= \int_{\partial\Omega} (P\boldsymbol{i} + Q\boldsymbol{j} + R\boldsymbol{k})\cdot\boldsymbol{n}\, dS \\ &= \int_{\partial\Omega} P\boldsymbol{i}\cdot\boldsymbol{n}\, dS + \int_{\partial\Omega} Q\boldsymbol{j}\cdot\boldsymbol{n}\, dS + \int_{\partial\Omega} R\boldsymbol{k}\cdot\boldsymbol{n}\, dS\end{aligned}$$

The theorem will follow if we establish the equalities

$$\int_{\partial\Omega} P\boldsymbol{i}\cdot\boldsymbol{n}\, dS = \int_\Omega \frac{\partial P}{\partial x}\, dV \tag{1}$$

$$\int_{\partial\Omega} Q\boldsymbol{j}\cdot\boldsymbol{n}\, dS = \int_\Omega \frac{\partial Q}{\partial y}\, dV \tag{2}$$

$$\int_{\partial\Omega} R\boldsymbol{k}\cdot\boldsymbol{n}\, dS = \int_\Omega \frac{\partial R}{\partial z}\, dV \tag{3}$$

We shall prove (3); the other two equalities are proved in a perfectly analogous fashion.

Since Ω is a region of type 1 (as well as of types 2 and 3), there exists a pair of functions

$$z = f_1(x, y), \qquad z = f_2(x, y)$$

with common domain an elementary region D in the xy plane, such that Ω is the set of all points (x, y, z) satisfying

$$f_2(x, y) \leqslant z \leqslant f_1(x, y), \qquad (x, y) \in D$$

By Fubini's Theorem we have

$$\int_\Omega \frac{\partial R}{\partial z}\, dV = \int_D \left(\int_{z=f_2(x,y)}^{z=f_1(x,y)} \frac{\partial R}{\partial z}\, dz \right) dx\, dy$$

and so

$$\int_\Omega \frac{\partial R}{\partial z}\, dV = \int_D [R(x, y, f_1(x, y)) - R(x, y, f_2(x, y))]\, dx\, dy \tag{4}$$

The boundary $\partial\Omega$ of Ω is a closed surface whose top S_1 is the graph of $z = f_1(x, y), (x, y) \in D$, and whose bottom S_2 is the graph of $z = f_2(x, y)$, $(x, y) \in D$. The four other sides of $\partial\Omega$ consist of surfaces S_3, S_4, S_5, and S_6 whose normals are always perpendicular to the z axis. (See Figure 6-3 of Chapter 15; of course, some of the other four sides might vanish—for instance, if Ω is a solid ball, and $\partial\Omega$ is a sphere—but this will not affect the argument.) Now, by definition,

$$\int_{\partial\Omega} R\boldsymbol{k} \cdot \boldsymbol{n}\, dS = \int_{S_1} R\boldsymbol{k} \cdot \boldsymbol{n}_1\, dS + \int_{S_2} R\boldsymbol{k} \cdot \boldsymbol{n}_2\, dS + \sum_{i=3}^{6} \int_{S_i} R\boldsymbol{k} \cdot \boldsymbol{n}_i\, dS$$

Since on each of S_3, S_4, S_5, and S_6 the normal $\boldsymbol{n}_i$ is perpendicular to $\boldsymbol{k}$, we have $\boldsymbol{k} \cdot \boldsymbol{n}_i = 0$ along these faces, and so the integral reduces to

$$\int_{\partial\Omega} R\boldsymbol{k} \cdot \boldsymbol{n}\, dS = \int_{S_1} R\boldsymbol{k} \cdot \boldsymbol{n}_1\, dS + \int_{S_2} R\boldsymbol{k} \cdot \boldsymbol{n}_2\, dS \tag{5}$$

The surface S_2 is defined by $z = f_2(x, y)$, so

$$\boldsymbol{n}_2 = \frac{\dfrac{\partial f_2}{\partial x}\boldsymbol{i} + \dfrac{\partial f_2}{\partial y}\boldsymbol{j} - \boldsymbol{k}}{\sqrt{\left(\dfrac{\partial f_2}{\partial x}\right)^2 + \left(\dfrac{\partial f_2}{\partial y}\right)^2 + 1}}$$

(since S_2 is the bottom portion of Ω, for $\boldsymbol{n}_2$ to point outward it must have a negative $\boldsymbol{k}$ component; see Example 2). Thus

$$\boldsymbol{n}_2 \cdot \boldsymbol{k} = \frac{-1}{\sqrt{\left(\dfrac{\partial f_2}{\partial x}\right)^2 + \left(\dfrac{\partial f_2}{\partial y}\right)^2 + 1}}$$

and

$$\int_{S_2} R(\boldsymbol{k}\cdot\boldsymbol{n}_2)\,dS = \int_D R(x,y,f_2(x,y))\left(\frac{-1}{\sqrt{\left(\frac{\partial f_2}{\partial x}\right)^2+\left(\frac{\partial f_2}{\partial y}\right)^2+1}}\right)\sqrt{\left(\frac{\partial f_2}{\partial x}\right)^2+\left(\frac{\partial f_2}{\partial y}\right)^2+1}\,dA$$

$$= -\int_D R(x,y,f_2(x,y))\,dx\,dy \tag{6}$$

This equation also follows from formula (3), §16.2.

Similarly, on the top face S_1 we have

$$\boldsymbol{k}\cdot\boldsymbol{n} = \frac{1}{\sqrt{\left(\frac{\partial f_1}{\partial x}\right)^2+\left(\frac{\partial f_1}{\partial y}\right)^2+1}}$$

and

$$\int_{S_1} R(\boldsymbol{k}\cdot\boldsymbol{n}_1)\,dS = \int_D R(x,y,f_1(x,y))\,dx\,dy \tag{7}$$

Substituting (6) and (7) into equation (5) and then comparing with (4) we obtain

$$\int_\Omega \frac{\partial R}{\partial z}\,dV = \int_{\partial\Omega} R(\boldsymbol{k}\cdot\boldsymbol{n})\,dS$$

The remaining equalities (1) and (2) can be established in exactly the same way to complete the proof. ∎

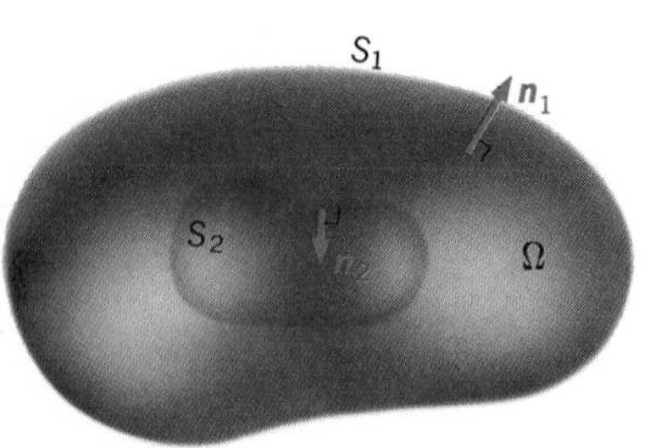

FIGURE 5-4

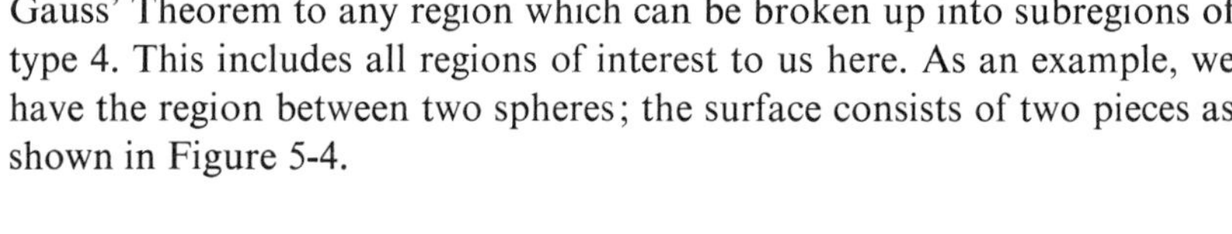

The reader should note that the proof is similar to that of Green's Theorem. By the same procedure used in exercise 15 of §17.2, we can extend Gauss' Theorem to any region which can be broken up into subregions of type 4. This includes all regions of interest to us here. As an example, we have the region between two spheres; the surface consists of two pieces as shown in Figure 5-4.

Example 3 Consider $\boldsymbol{F} = 2x\boldsymbol{i} + y^2\boldsymbol{j} + z^2\boldsymbol{k}$. Let S be the unit sphere $x^2+y^2+z^2=1$. Evaluate $\int_S \boldsymbol{F}\cdot\boldsymbol{n}\,dS$.

By Gauss' Theorem,

$$\int_\Omega \operatorname{div}\boldsymbol{F} = \int_S \boldsymbol{F}\cdot\boldsymbol{n}\,dS$$

where Ω is the ball bounded by the sphere. The integral on the left is $2\int_\Omega (1+y+z)\,dV = 8\pi/3$ (this volume integral was computed in Example 1, §15.6). The reader can convince himself that direct computation of $\int_S \boldsymbol{F}\cdot\boldsymbol{n}\,dS$ proves unwieldy.

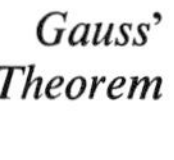

Example 4 Use the Divergence Theorem to evaluate

$$\int_{\partial W} (x^2+y+z)\, dS$$

where W is the solid ball $x^2+y^2+z^2 \leqslant 1$.

In order to apply Gauss' Divergence Theorem we must find some vector field

$$\boldsymbol{F} = F_1\boldsymbol{i} + F_2\boldsymbol{j} + F_3\boldsymbol{k}$$

on W with

$$\boldsymbol{F}\cdot\boldsymbol{n} = x^2+y+z$$

At any point $(x_0, y_0, z_0) \in \partial W$ (Figure 5-5) the unit normal $\boldsymbol{n}$ to ∂W at (x_0, y_0, z_0) is given by the vector

$$\boldsymbol{n} = x_0\boldsymbol{i} + y_0\boldsymbol{j} + z_0\boldsymbol{k}$$

since on ∂W, $x^2+y^2+z^2 = 1$, and the radius vector $\boldsymbol{r} = x\boldsymbol{i}+y\boldsymbol{j}+z\boldsymbol{k}$ is normal to the sphere ∂W. Therefore, if $\boldsymbol{F}$ is the desired vector field,

$$\boldsymbol{F}\cdot\boldsymbol{n} = F_1x + F_2y + F_3z$$

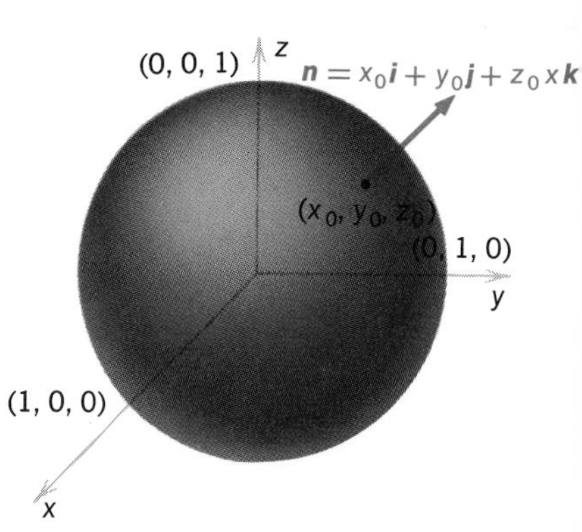

FIGURE 5-5

We set

$$F_1x = x^2, \qquad F_2y = y, \qquad F_3z = z$$

and solve for F_1, F_2, and F_3 to find that

$$\boldsymbol{F} = x\boldsymbol{i}+\boldsymbol{j}+\boldsymbol{k}$$

Computing $\operatorname{div}\boldsymbol{F}$ we get

$$\operatorname{div}\boldsymbol{F} = 1+0+0 = 1$$

Thus by Gauss' Divergence Theorem

$$\int_{\partial W} (x^2+y+z)\, dS = \int_W dV = \text{volume}(W) = \tfrac{4}{3}\pi$$

The physical meaning of divergence is that at a point P, $\operatorname{div} F(P)$ is the rate of net outward flux at P per unit volume. This follows from Gauss' Theorem: If Ω_ρ is a ball in $\boldsymbol{R}^3$ of radius ρ centered at P, then there is a point $Q \in \Omega$: such that

$$\int_{\partial\Omega} \boldsymbol{F}\cdot\boldsymbol{n}\, dS = \int_{\Omega_\rho} \operatorname{div}\boldsymbol{F}\, dV = \operatorname{div}\boldsymbol{F}(Q)\cdot\text{volume}(\Omega_\rho)$$

or

$$\operatorname{div}\boldsymbol{F}(P) = \lim_{\rho\to 0} \operatorname{div}\boldsymbol{F}(Q) = \lim_{\rho\to 0} \frac{1}{V(\Omega_\rho)} \int_{\partial\Omega_\rho} \boldsymbol{F}\cdot\boldsymbol{n}\, dS$$

This is analogous to the limit formulation of curl given at the end of §17.3. Thus, if $\operatorname{div} \boldsymbol{F}(P) > 0$, we consider P to be a **source**, for there is a net outwards flow near P. If $\operatorname{div} \boldsymbol{F}(P) < 0$, P is called a **sink** for $\boldsymbol{F}$.

A C^1 vector field $\boldsymbol{F}$ defined on $\boldsymbol{R}^3$ is called **divergence free** if $\operatorname{div} \boldsymbol{F} = 0$. If $\boldsymbol{F}$ is divergence free, we have $\int_S \boldsymbol{F} = 0$ for all closed surfaces S. The converse can also be demonstrated: If $\int_S \boldsymbol{F} = 0$ for all closed surfaces S, then $\boldsymbol{F}$ is divergence free. From formula 8 in the table of §14.5 and Theorem 2 of §17.4, it follows that $\boldsymbol{F}$ is divergence free if and only if $\boldsymbol{F} = \operatorname{curl} \boldsymbol{G}$ for some $\boldsymbol{G}$. If $\boldsymbol{F}$ is divergence free, we see that the flux of $\boldsymbol{F}$ across any closed surface S is 0, so if $\boldsymbol{F}$ is the velocity field of a fluid, the net amount of fluid to flow out of any region will be 0. Thus, exactly as much fluid must flow into the region as flows out (in unit time). A fluid with this property is **incompressible**.

Example 5 Evaluate $\int_S \boldsymbol{F}$, where $\boldsymbol{F}(x, y, z) = xy^2\,\boldsymbol{i} + x^2 y\boldsymbol{j} + y\boldsymbol{k}$ and S is the surface of the cylinder $x^2 + y^2 = 1$, $-1 < z < 1$ and $x^2 + y^2 \leqslant 1$ when $z = \pm 1$. Interpret physically.

One can compute this integral directly but it is usually easier to use the Divergence Theorem if it applies.

Now S is the boundary of the region Ω given by $x^2 + y^2 \leqslant 1$, $-1 \leqslant z \leqslant 1$. Thus $\int_S \boldsymbol{F} = \int_\Omega \operatorname{div} \boldsymbol{F}$. Moreover, $\int_\Omega \operatorname{div} \boldsymbol{F} = \int_\Omega (x^2 + y^2)\,dx\,dy\,dz = \int_{-1}^{1} (\int_{x^2+y^2 \leqslant 1} (x^2 + y^2)\,dx\,dy)\,dz = 2(\int_{x^2+y^2\leqslant 1} (x^2 + y^2)\,dx\,dy)$.

Before evaluating the double integral, we note that $\int_{\partial\Omega} \boldsymbol{F} \cdot \boldsymbol{n}\,dS = 2\int_{x^2+y^2\leqslant 1} (x^2 + y^2)\,dx\,dy > 0$. This means that $\int_{\partial\Omega} \boldsymbol{F}$, the net flux of $\boldsymbol{F}$ out of the cylinder, is positive, which agrees with the fact that $0 \leqslant \operatorname{div} \boldsymbol{F} = x^2 + y^2$ inside the cylinder.

We make the following change of variables to evaluate the double integral:

$$x = \rho\cos\theta, \qquad y = \rho\sin\theta, \qquad 0 \leqslant \rho \leqslant 1, \quad 0 \leqslant \theta \leqslant 2\pi$$

Hence, we have $\partial(x, y)/\partial(\rho, \theta) = \rho$ and $x^2 + y^2 = \rho^2$. Thus

$$\int_{x^2+y^2\leqslant 1} (x^2 + y^2)\,dx\,dy = \int_0^{2\pi} \left(\int_0^1 \rho^3\,d\rho\right) d\theta = \tfrac{1}{2}\pi$$

Therefore, $\int_\Omega \operatorname{div} \boldsymbol{F}\,dV = \pi$.

As we remarked above, Gauss' Divergence Theorem can be applied to region in space more general than those which are type 4. To conclude the section, we apply this observation to prove an important result.

Example 6 Prove **Gauss' law:** Let M be a region in $\boldsymbol{R}^3$ of type 4. Then if $(0, 0, 0) \notin \partial M$ we have

$$\int_{\partial M} \frac{\boldsymbol{r}\cdot\boldsymbol{n}}{r^3}\,dS = \begin{cases} 4\pi, & (0,0,0) \in M \\ 0, & (0,0,0) \notin M \end{cases}$$

where $\boldsymbol{r}(x, y, z) = x\boldsymbol{i} + y\boldsymbol{j} + z\boldsymbol{k}$ and $r(x, y, z) = |\boldsymbol{r}(x, y, z)| = \sqrt{x^2 + y^2 + z^2}$.

First suppose $(0, 0, 0) \notin M$. Then $\boldsymbol{r}/r^3$ is a C^1 vector field on M and ∂M (Why?) and by the Divergence Theorem,

$$\int_{\partial M} \frac{\boldsymbol{r}\cdot\boldsymbol{n}}{r^3}\,dS = \int_M \nabla \cdot \left(\frac{\boldsymbol{r}}{r^3}\right) dV$$

But $\nabla \cdot (\boldsymbol{r}/r^3) = 0$ for $r \neq 0$ as the reader can easily verify (see exercise 16, §14.5). Thus

$$\int_{\partial M} \frac{\boldsymbol{r} \cdot \boldsymbol{n}}{r^3}\, dS = 0$$

in this case.

Now let us suppose $(0,0,0) \in M$. We can no longer use the above method, because $\boldsymbol{r}/r^3$ is not smooth on M due to the singularity at $\boldsymbol{r} = (0,0,0)$. Since $(0,0,0) \in M$, there is an $\varepsilon > 0$ such that the ball N of radius ε centered at $(0,0,0)$ is contained completely inside M. Now let Ω be the region between M and N. Then Ω has boundary $\partial N \cup \partial M = S$. But the orientation on ∂N induced by the outward normal on Ω is the opposite of that obtained from N (see Figure 5-6). Now $\nabla \cdot (\boldsymbol{r}/r^3) = 0$ on S, so

$$\int_S \frac{\boldsymbol{r} \cdot \boldsymbol{n}}{r^3} dS = \int_\Omega \nabla \cdot \left(\frac{\boldsymbol{r}}{r^3}\right) = 0$$

FIGURE 5-6

Since

$$\int_S \frac{\boldsymbol{r} \cdot \boldsymbol{n}}{r^3}\, dS = \int_{\partial M} \frac{\boldsymbol{r} \cdot \boldsymbol{n}}{r^3}\, dS + \int_{\partial N} \frac{\boldsymbol{r} \cdot (-\boldsymbol{n})}{r^3}\, dS$$

where $-\boldsymbol{n}$ in the third integral represents the outward normal to S on ∂N and $\boldsymbol{n}$ in the first integral is the outward normal to ∂M. This means

$$\int_{\partial M} \frac{\boldsymbol{r} \cdot \boldsymbol{n}}{r^3}\, dS = \int_{\partial N} \frac{\boldsymbol{r} \cdot \boldsymbol{n}}{r^3}\, dS$$

where $\boldsymbol{n}$ is the outward normal on ∂M and ∂N.

Now on ∂N, $\boldsymbol{n} = \boldsymbol{r}/r$ and $r = \varepsilon$ (∂N is a sphere of radius ε), so

$$\int_{\partial N} \frac{\boldsymbol{r} \cdot \boldsymbol{n}}{r^3}\, dS = \int_{\partial N} \frac{\varepsilon^2}{\varepsilon^4}\, dS = \frac{1}{\varepsilon} \int_{\partial N} dS$$

But $\int_{\partial N} dS = 4\pi\varepsilon$, the area of the sphere of radius ε. This proves the result.

Gauss' law has the following physical interpretation. The potential due to a point charge Q at $(0,0,0)$ is given by

$$\phi(x, y, z) = \frac{Q}{4\pi r} = \frac{Q}{4\pi\sqrt{x^2 + y^2 + z^2}}$$

and the corresponding electric field is

$$\boldsymbol{E} = -\nabla\phi = \frac{Q}{4\pi}\left(\frac{\boldsymbol{r}}{r^3}\right)$$

Thus the result of Example 6 states that the total electric flux $\int_{\partial M} \boldsymbol{E}$ (that is, the flux of $\boldsymbol{E}$) out of a closed surface ∂M equals Q if the charge lies inside M and the flux is zero otherwise. (A generalization is given in exercise 10.) Note that even if $(0,0,0) \notin M$, $\boldsymbol{E}$ will still be nonzero on M.

For a continuous charge distribution described by a charge density ρ, the field $\boldsymbol{E}$ is related to the density ρ by

$$\operatorname{div} \boldsymbol{E} = \nabla \cdot \boldsymbol{E} = \rho$$

Thus by Gauss' Theorem,

$$\int_S \boldsymbol{E} = \int_\Omega \rho = Q$$

or the flux out of a surface is the total charge inside.

EXERCISES

1. Let S be a closed surface. Then use Gauss' Theorem to show $\int_S \nabla \times \boldsymbol{F} = 0$. (Compare with exercise 11 of §17.3.)
2. Let $\boldsymbol{F} = x^3\boldsymbol{i} + y^3\boldsymbol{j} + z^3\boldsymbol{k}$. Evaluate the surface integral of $\boldsymbol{F}$ over the unit sphere.
3. Evaluate $\int_{\partial\Omega} \boldsymbol{F}$, where $\boldsymbol{F} = x\boldsymbol{i} + y\boldsymbol{j} - z\boldsymbol{k}$ and Ω is the unit cube (in the first octant). Perform the calculation directly and check by using the Divergence Theorem.
4. Repeat exercise 3 for (a) $\boldsymbol{F} = \boldsymbol{i} + \boldsymbol{j} + \boldsymbol{k}$) (b) $\boldsymbol{F} = x^2\boldsymbol{i} + y^2\boldsymbol{j} + z^2\boldsymbol{k}$.
5. Let $\boldsymbol{F} = y\boldsymbol{i} + z\boldsymbol{j} + xz\boldsymbol{k}$. Evaluate $\int_{\partial\Omega} \boldsymbol{F}$, where Ω is the set (x, y, z) with $x^2 + y^2 \leqslant z \leqslant 1$.
6. Repeat exercise 5 for $\boldsymbol{F} = (x-y)\boldsymbol{i} + (y-z)\boldsymbol{j} + (z-x)\boldsymbol{k}$.
7. Let S be the surface of a region Ω. Show

$$\int_S \boldsymbol{r} \cdot \boldsymbol{n}\, dS = 3 \text{ volume}(\Omega)$$

Attempt to explain this geometrically. [Hint: Assume $(0, 0, 0) \in \Omega$ and consider the skew cone with its vertex at $(0,0,0)$ with base ΔS and altitude $|\boldsymbol{r}|$. Its volume is $\frac{1}{3}(\Delta S) \cdot (\boldsymbol{r} \cdot \boldsymbol{n})$.]

8. Evaluate $\int_S \boldsymbol{F} \cdot \boldsymbol{n}\, dS$, where $\boldsymbol{F} = 3xy^2\boldsymbol{i} + 3x^2y\boldsymbol{j} + z^3\boldsymbol{k}$ and S is the surface of the unit sphere.
9. Show $\int_\Omega \frac{1}{r^2} dx\, dy\, dz = \int_{\partial\Omega} \frac{\boldsymbol{r} \cdot \boldsymbol{n}}{r^2} dS$ where $\boldsymbol{r} = x\boldsymbol{i} + y\boldsymbol{j} + z\boldsymbol{k}$.
10. Fix vectors $\boldsymbol{v}_1, \ldots, \boldsymbol{v}_k \in \boldsymbol{R}^3$ and numbers (the "charges") $q_1, \ldots, q_k$. Set $\phi(x, y, z) = \sum_{i=1}^{k} q_i/(4\pi|\boldsymbol{r} - \boldsymbol{v}_k|)$, where $\boldsymbol{r} = (x, y, z)$. Show that for a closed surface S, and $\boldsymbol{E} = -\nabla\phi$,

$$\int_S \boldsymbol{E} = Q$$

where Q is the total charge inside S. (Assume Gauss' law from Example 6 and that none of the charges are on S.)

11. Prove **Green's identities**

$$\int_{\partial\Omega} f\nabla g \cdot \boldsymbol{n}\, dS = \int_\Omega [f\nabla^2 g + \nabla f \cdot \nabla g]\, dV$$

$$\int_{\partial\Omega} (f\nabla g - g\nabla f) \cdot \boldsymbol{n}\, dS = \int_\Omega (f\nabla^2 g - g\nabla^2 f)\, dV$$

12. Suppose $\boldsymbol{F}$ satisfies $\operatorname{div} \boldsymbol{F} = 0$ and $\operatorname{curl} \boldsymbol{F} = 0$. Then show we can write $\boldsymbol{F} = \nabla f$, where $\nabla^2 f = 0$.

13. Let ρ be a given (continuous) function. The potential of ρ is defined by

$$\phi(x) = \int_{\Omega} \frac{\rho(y)}{4\pi|x-y|} dy$$

Argue that

$$\int_{\partial\Omega} \nabla\phi \cdot \boldsymbol{n}\, dS = \int_{\Omega} \rho$$

for any region Ω and hence ϕ satisfies **Poisson's equation**

$$\nabla^2 \phi = -\rho$$

14. Suppose $\boldsymbol{F}$ is tangent to the closed surface S of a region Ω. Then prove

$$\int_{\Omega} \operatorname{div} \boldsymbol{F} = 0$$

15. Use Gauss' law and symmetry to prove that the electric field due to a charge Q evenly spread over the surface of a sphere is the same outside the surface as a point charge Q located at the center of the sphere. What is field inside the sphere?

16. Reformulate exercise 15 in terms of gravitational fields.

6 APPLICATIONS TO PHYSICS

We can apply the concepts developed in this chapter to the formulation of some physical theories. Let us first discuss an important equation which is referred to as a **conservation** equation. For fluids, it expresses the conservation of mass and for electromagnetic theory, the conservation of charge. We shall apply the equation to heat conduction and to electromagnetism.

Let $V(t, x, y, z)$ be a C^1 vector field for each t and let $\rho(t, x, y, z)$ be a C^1 real valued function. By the law of conservation of mass for V and ρ, we shall mean the condition

$$\frac{d}{dt}\int_{\Omega} \rho = -\int_{\partial\Omega} \boldsymbol{J}\cdot\boldsymbol{n}\, dS$$

where $\boldsymbol{J} = \rho V$ and Ω is an arbitrary region in $\boldsymbol{R}^3$

If we think of ρ as a mass density (ρ could also be charge density), that is, the mass per unit volume, and V as the velocity field of a fluid, the condition just says that the rate of change of total mass in Ω equals the rate at which mass flows into Ω. Recall that $\int_{\partial\Omega} \boldsymbol{J}\cdot\boldsymbol{n}\, dS$ is the flux of $\boldsymbol{J}$. We need the following result.

Theorem 1 For V and ρ defined on $\boldsymbol{R}^3$, the law of conservation of mass for V, ρ is equivalent to the condition

$$\operatorname{div} \boldsymbol{J} + \frac{\partial\rho}{\partial t} = 0$$

that is,

$$\rho \operatorname{div} V + V\cdot\nabla\rho + \frac{\partial\rho}{\partial t} = 0$$

Proof First, observe that $(d/dt)\int \rho\, dx\, dy\, dz = \int (\partial\rho/dt)\, dx\, dy\, dz$, and

$$\int_{\partial\Omega} \boldsymbol{J}\cdot\boldsymbol{n}\, dS = \int_{\Omega} \operatorname{div}\boldsymbol{J}$$

by the Divergence Theorem. Thus conservation of mass is equivalent to the condition

$$\int_{\Omega}\left(\operatorname{div}\boldsymbol{J} + \frac{\partial\rho}{dt}\right) dx\, dy\, dz = 0$$

Since this is to hold for all regions Ω this is equivalent to $\operatorname{div}\boldsymbol{J} + \partial\rho/\partial t = 0$. ∎

The equation $\operatorname{div}\boldsymbol{J} + \partial\rho/\partial t = 0$ is called the **equation of continuity.** This is not the only equation governing fluid motion and it does not determine the motion of the fluid, but is just one equation which must hold.

Notice that the fluids which this equation governs can be compressible. Even if $\operatorname{div} \boldsymbol{V} = 0$ (incompressible case, see discussion following Theorem 2 of §14.5), the equation is not automatic, because ρ can depend on (x, y, z) and t.

We shall now turn our attention to the heat equation which is one of the most important equations of applied mathematics. It has been, and remains, one of the prime motivations for the development of partial differential equations.

Let us argue intuitively. If $T(t, x, y, z)$ denotes the temperature in a body at time t, then ∇T represents the temperature gradient and heat "flows" with the vector field $-\nabla T = \boldsymbol{F}$. Now the energy density, that is, the energy per unit volume, is $c\rho_0 T$, where c is a constant (specific heat) and ρ_0 is the mass density, assumed constant. (We accept these assertions from elementary physics.) The **energy flux vector** is $\boldsymbol{J} = \kappa\boldsymbol{F}$, where κ is a constant called the **conductivity.**

We propose that energy be conserved. Formally, this means that $\boldsymbol{J}$ and $\rho = c\rho_0 T$ should obey the law of conservation of mass with ρ playing the role of "mass"; that is,

$$\frac{d}{dt}\int_{\Omega}\rho = -\int_{\partial\Omega}\boldsymbol{J}\cdot\boldsymbol{n}\, dS$$

By Theorem 1 this is equivalent to

$$\operatorname{div}\boldsymbol{J} + \frac{\partial\rho}{\partial t} = 0 \tag{1}$$

But $\operatorname{div}\boldsymbol{J} = \operatorname{div}(-\kappa\nabla T) = -\kappa\nabla^2 T$. (Recall that $\nabla^2 T = \partial^2 T/\partial x^2 + \partial^2 T/\partial y^2 + \partial^2 T/\partial z^2$ and ∇^2 is the Laplace operator.) Continuing, we have $\partial\rho/\partial t = \partial(c\rho_0 T)/\partial t = c\rho_0(\partial T/\partial t)$. Thus equation (1) becomes

$$\frac{\partial T}{\partial t} = \frac{\kappa}{c\rho_0}\nabla^2 T = k\nabla^2 T \tag{2}$$

where $k = \kappa/c\rho_0$ is called the **diffusivity.** Equation (2) is the important **heat equation.**

Unlike fluids, this equation does completely govern the conduction of heat in the following sense. If $T(0, x, y, z)$ is a given initial temperature distribution then there is a unique $T(t, x, y, z)$ determined satisfying equation (2). In other words, the initial condition at $t = 0$ gives us the result for $t > 0$.

Notice that if T does not change with time (steady-state case) then we must have $\nabla^2 T = 0$ (Laplace's equation).

To conclude this section we shall discuss Maxwell's equations and electromagnetic fields. To simplify computations we use the notation $\nabla \cdot \boldsymbol{F}$ for div $\boldsymbol{F}$ and $\nabla \times \boldsymbol{F}$ for curl $\boldsymbol{F}$.

The form of Maxwell's equations depends on the physical units one is employing, and changing units introduces factors like 4π, the velocity of light, and so on. We choose the system in which Maxwell's equations are simplest.

Let $\boldsymbol{E}(t, x, y, z)$ and $\boldsymbol{H}(t, x, y, z)$ be vector fields for each t. They satisfy (by definition) **Maxwell's equations** with charge density $\rho(x, y, z)$ and current density $\boldsymbol{J}(x, y, z)$ if

$$\nabla \cdot \boldsymbol{E} = \rho \qquad \text{[Gauss' law]} \tag{3}$$

$$\nabla \cdot \boldsymbol{H} = 0 \qquad \text{[no magnetic sources]} \tag{4}$$

$$\nabla \times \boldsymbol{E} + \frac{\partial \boldsymbol{H}}{\partial t} = \boldsymbol{0} \qquad \text{[Faraday's law]} \tag{5}$$

and

$$\nabla \times \boldsymbol{H} - \frac{\partial \boldsymbol{E}}{\partial t} = \boldsymbol{J} \qquad \text{[Ampere's law]} \tag{6}$$

Of these laws, (3) and (5) were discussed earlier in §17.5 and §17.3 in integral form; historically, they arose in these forms as physically observed laws. The origin of (4) and (6) is similar.

Physically, one interprets $\boldsymbol{E}$ as the **electric field** and $\boldsymbol{H}$ as the **magnetic field.** As time progresses, these fields interact with one another and charges and currents are produced according to the above equations. For example, electromagnetic waves (light) arise this way.

Since $\nabla \cdot \boldsymbol{H} = 0$ we can apply Theorem 2 of §17.4 and conclude that $\boldsymbol{H} = \nabla \times \boldsymbol{A}$ for some vector field $\boldsymbol{A}$. (We are assuming that $\boldsymbol{H}$ is defined on all of $\boldsymbol{R}^3$ for each time t.) This vector field $\boldsymbol{A}$ is not unique, and we can equally well use $\boldsymbol{A}' = \boldsymbol{A} + \nabla f$ for any function $f(t, x, y, z)$, since $\nabla \times \nabla f = \boldsymbol{0}$. For any such choice of $\boldsymbol{A}$, we have by (5)

$$\boldsymbol{0} = \nabla \times \boldsymbol{E} + \frac{\partial \boldsymbol{H}}{\partial t} = \nabla \times \boldsymbol{E} + \frac{\partial}{\partial t} \nabla \times \boldsymbol{A}$$

$$= \nabla \times \boldsymbol{E} + \nabla \times \frac{\partial \boldsymbol{A}}{\partial t} = \nabla \times \left(\boldsymbol{E} + \frac{\partial \boldsymbol{A}}{\partial t} \right) = \boldsymbol{0}$$

Hence applying Theorem 1, §17.4, there is a real-valued function ϕ on $\boldsymbol{R}^3$ such that

$$\boldsymbol{E} + \frac{\partial \boldsymbol{A}}{\partial t} = -\boldsymbol{\nabla}\phi$$

Substituting this equation, and $\boldsymbol{H} = \boldsymbol{\nabla} \times \boldsymbol{A}$ into equation (6) and using the identity

$$\boldsymbol{\nabla} \times (\boldsymbol{\nabla} \times \boldsymbol{A}) = \boldsymbol{\nabla}(\boldsymbol{\nabla} \cdot \boldsymbol{A}) - \boldsymbol{\nabla}^2 \boldsymbol{A}$$

we get

$$\boldsymbol{J} = \boldsymbol{\nabla} \times \boldsymbol{H} - \frac{\partial \boldsymbol{E}}{\partial t} = \boldsymbol{\nabla} \times (\boldsymbol{\nabla} \times \boldsymbol{A}) - \frac{\partial}{\partial t}\left(-\frac{\partial \boldsymbol{A}}{\partial t} - \boldsymbol{\nabla}\phi\right)$$

$$= \boldsymbol{\nabla}(\boldsymbol{\nabla} \cdot \boldsymbol{A}) - \boldsymbol{\nabla}^2 \boldsymbol{A} + \frac{\partial^2 \boldsymbol{A}}{\partial t^2} + \frac{\partial}{\partial t}(\boldsymbol{\nabla}\phi)$$

Thus

$$\boldsymbol{\nabla}^2 \boldsymbol{A} - \frac{\partial^2 \boldsymbol{A}}{\partial t^2} = -\boldsymbol{J} + \boldsymbol{\nabla}(\boldsymbol{\nabla} \cdot \boldsymbol{A}) + \frac{\partial}{\partial t}(\boldsymbol{\nabla}\phi)$$

or

$$\boldsymbol{\nabla}^2 \boldsymbol{A} - \frac{\partial^2 \boldsymbol{A}}{\partial t^2} = -\boldsymbol{J} + \boldsymbol{\nabla}\left(\boldsymbol{\nabla} \cdot \boldsymbol{A} + \frac{\partial \phi}{\partial t}\right) \tag{7}$$

Again using the equation $\boldsymbol{E} + \partial \boldsymbol{A}/\partial t = -\boldsymbol{\nabla}\phi$ and the equation $\boldsymbol{\nabla} \cdot \boldsymbol{E} = \rho$, we obtain

$$\rho = \boldsymbol{\nabla} \cdot \boldsymbol{E} = \boldsymbol{\nabla} \cdot \left(-\boldsymbol{\nabla}\phi - \frac{\partial \boldsymbol{A}}{\partial t}\right) = -\boldsymbol{\nabla}^2 \phi - \frac{\partial(\boldsymbol{\nabla} \cdot \boldsymbol{A})}{\partial t}$$

or

$$\boldsymbol{\nabla}^2 \phi = -\rho - \frac{\partial(\boldsymbol{\nabla} \cdot \boldsymbol{A})}{\partial t} \tag{8}$$

Now let us exploit the freedom in our choice of $\boldsymbol{A}$. We impose the "condition"

$$\boldsymbol{\nabla} \cdot \boldsymbol{A} + \frac{\partial \phi}{\partial t} = 0 \tag{9}$$

We must be sure we can do this. Suppose we have a given $\boldsymbol{A}_0$, and a corresponding ϕ_0, can we choose a new $\boldsymbol{A} = \boldsymbol{A}_0 + \boldsymbol{\nabla} f$ and then a new ϕ such that $\boldsymbol{\nabla} \cdot \boldsymbol{A} + \partial \phi/\partial t = 0$? We see that with this new $\boldsymbol{A}$, the new ϕ is $\phi = \phi_0 - \partial f/\partial t$. We leave the verification as an exercise for the reader. The condition (9) on f then becomes

$$0 = \boldsymbol{\nabla}(\boldsymbol{A}_0 + \boldsymbol{\nabla} f) + \frac{\partial(\phi_0 - \partial f/\partial t)}{\partial t} = \boldsymbol{\nabla} \cdot \boldsymbol{A}_0 + \boldsymbol{\nabla}^2 f + \frac{\partial \phi_0}{\partial t} - \frac{\partial^2 f}{\partial t^2}$$

or

$$\nabla^2 f - \frac{\partial^2 f}{\partial t^2} = -\left(\nabla \cdot A_0 + \frac{\partial \phi_0}{\partial t}\right) \tag{10}$$

Thus to be able to choose A, ϕ satisfying $\nabla \cdot A + \partial\phi/\partial t = 0$, we must be able to solve the above equation (10) for f. One can indeed do this under general conditions, although we cannot prove it here. Equation (10) is called the **inhomogeneous wave equation.**

If we accept that A, ϕ can be chosen to satisfy $\nabla \cdot A + \partial\phi/\partial t = 0$, then the equations (7) and (8) for A, ϕ become

$$\nabla^2 A - \frac{\partial^2 A}{\partial t^2} = -J \tag{7'}$$

$$\nabla^2 \phi - \frac{\partial^2 \phi}{\partial t^2} = -\rho \tag{8'}$$

(8′) follows from (8) after substituting $-\partial\phi/\partial t$ for $\nabla \cdot A$. Thus the wave equation appears again.

Conversely, if A and ϕ satisfy the equations $\nabla \cdot A + \partial\phi/\partial t = 0$, $\nabla^2 \phi + \partial^2 \phi/\partial t^2 = -\rho$, and $\nabla^2 A - \partial^2 A/\partial t^2 = -J$, then $E = -\nabla\phi - \partial A/\partial t$, $H = \nabla \times A$ satisfy Maxwell's equations. **This procedure then reduces Maxwell's equations to a study of the wave equation.**

REVIEW EXERCISES FOR CHAPTER 17

1. Use Green's Theorem to find the area of the loop of the curve $x = a \sin\theta\cos\theta$, $y = a\sin^2\theta$, for $a > 0$ and $0 \leqslant \theta \leqslant \pi$.
2. Let $r(x, y, z) = (x, y, z)$, $r = |r|$. Show $\nabla^2(\log r) = 1/r^2$ and $\nabla^2(r^n) = n(n+1)\, r^{n-2}$.
3. If $\nabla \times F = 0$ and $\nabla \times G = 0$, prove $\nabla \cdot (F \times G) = 0$.
4. Let m be a constant vector field and let

$$\phi = \frac{m \cdot r}{r^3}, \qquad F = \frac{m \times r}{r^3}$$

Compute $\nabla\phi$, $\nabla^2\phi$, $\nabla \cdot F$, and $\nabla \times F$.

5. Prove the identity

$$(F \cdot \nabla)F = \tfrac{1}{2}\nabla(F \cdot F) + (\nabla \times F) \times F$$

6. Prove that if $\nabla \times F = \nabla \times G$ then $F = G + \nabla f$ for some f.
7. Let $F = 2yz\,i + (-x + 3y + 2)j + (x^2 + z)k$. Evaluate $\int_S \nabla \times F$, where S is the cylinder $x^2 + y^2 = a^2$, $0 \leqslant z \leqslant 1$ (without the top and bottom). What if the top and bottom are included?

8. Let $\boldsymbol{F} = x^2y\boldsymbol{i} + z^8\boldsymbol{j} - 2xyz\boldsymbol{k}$. Evaluate the integral of $\boldsymbol{F}$ over the surface of the unit cube.

9. Let Ω be a region in $\boldsymbol{R}^3$ with boundary $\partial\Omega$. Prove the identity
$$\int_{\partial\Omega} \boldsymbol{F} \times (\boldsymbol{\nabla} \times \boldsymbol{G}) = \int_{\Omega} (\boldsymbol{\nabla} \times \boldsymbol{F}) \cdot (\boldsymbol{\nabla} \times \boldsymbol{G}) - \int_{\Omega} \boldsymbol{F} \cdot (\boldsymbol{\nabla} \times \boldsymbol{\nabla} \times \boldsymbol{G})$$

10. Verify Green's Theorem for the line integral
$$\int_C x^2y\,dx + y\,dy$$
when C is the boundary of the region between the curves $y = x$ and $y = x^3$, $0 \leqslant x \leqslant 1$.

11. Can you derive Green's Theorem in the plane from Gauss' Theorem?

12. (a) Show that $\boldsymbol{F} = (x^3 - 2xy^3)\boldsymbol{i} - 3x^2y^2\boldsymbol{j}$ is a gradient vector field.
    (b) Evaluate the integral of this form along the path $x = \cos^3\theta$, $y = \sin^3\theta$, $0 \leqslant \theta \leqslant \pi/2$.

13. (a) Show that $\boldsymbol{F} = 6xy\cos z\boldsymbol{i} + 3x^2\cos z\boldsymbol{j} - 3x^2y\sin z\boldsymbol{k}$ is conservative. (see §17.4).
    (b) Find f such that $\boldsymbol{F} = \boldsymbol{\nabla} f$.
    (c) Evaluate the integral of $\boldsymbol{F}$ along the curve $x = \cos^3\theta$, $y = \sin^3\theta$, $z = 0$, $0 \leqslant \theta \leqslant \pi/2$.

14. Let the velocity of a fluid be described by $\boldsymbol{F} = 6xz\boldsymbol{i} + x^2y\boldsymbol{j} + yz\boldsymbol{k}$. Compute the rate at which fluid is leaving the unit cube.

15. Suppose $\boldsymbol{\nabla} \cdot \boldsymbol{F}(x_0, y_0, z_0) > 0$. Show that for a sufficiently small sphere S centered at (x_0, y_0, z_0), the flux of $\boldsymbol{F}$ out of S is positive.

16. Let $\boldsymbol{F} = x^2\boldsymbol{i} + (x^2y - 2xy)\boldsymbol{j} - (x^2z)\boldsymbol{k}$. Does there exist a $\boldsymbol{G}$ such that $\boldsymbol{F} = \boldsymbol{\nabla} \times \boldsymbol{G}$?

17. Let $\boldsymbol{a}$ be a constant vector and $\boldsymbol{F} = \boldsymbol{a} \times \boldsymbol{r}$ [as usual, $\boldsymbol{r}(x, y, z) = (x, y, z)$]. Is F conservative? If so, find a potential for it.

18. Consider the case of incompressible fluid flow with velocity field $\boldsymbol{F}$ and density ρ.
    (a) If ρ is constant for each fixed t then ρ is constant in t as well.
    (b) If ρ is constant in t then $\boldsymbol{F} \cdot \boldsymbol{\nabla}\rho = 0$.

APPENDIX

MATHEMATICAL TABLES

1 *Trigonometric Functions*
2 *Common Logarithms*
3 *Exponentials*
4 *Derivatives*
5 *Integrals*

1 TRIGONOMETRIC FUNCTIONS

Angle				
Degree	*Radian*	*Sine*	*Cosine*	*Tangent*
0°	0.000	0.000	1.000	0.000
1°	.018	.018	1.000	.018
2°	.035	.035	0.999	.035
3°	.052	.052	.999	.052
4°	.070	.070	.998	.070
5°	.087	.087	.996	.088
6°	.105	.105	.995	.105
7°	.122	.122	.993	.123
8°	.140	.139	.990	.141
9°	.157	.156	.988	.158
10°	.175	.174	.985	.176
11°	.192	.191	.982	.194
12°	.209	.208	.978	.213
13°	.227	.225	.974	.231
14°	.244	.242	.970	.249
15°	.262	.259	.966	.268
16°	.279	.276	.961	.287
17°	.297	.292	.956	.306
18°	.314	.309	.951	.325
19°	.332	.326	.946	.344
20°	.349	.342	.940	.364
21°	.367	.358	.934	.384
22°	.384	.375	.927	.404
23°	.401	.391	.921	.425
24°	.419	.407	.914	.445
25°	.436	.423	.906	.466
26°	.454	.438	.899	.488
27°	.471	.454	.891	.510
28°	.489	.470	.883	.532
29°	.506	.485	.875	.554
30°	.524	.500	.866	.577
31°	.541	.515	.857	.601
32°	.559	.530	.848	.625
33°	.576	.545	.839	.649
34°	.593	.559	.829	.675
35°	.611	.574	.819	.700
36°	.628	.588	.809	.727
37°	.646	.602	.799	.754
38°	.663	.616	.788	.781
39°	.681	.629	.777	.810
40°	.698	.643	.766	.839
41°	.716	.656	.755	.869
42°	.733	.669	.743	.900
43°	.751	.682	.731	.933
44°	.768	.695	.719	.966
45°	.785	.707	.707	1.000
46°	0.803	0.719	0.695	1.036
47°	.820	.731	.682	1.072
48°	.838	.743	.669	1.111
49°	.855	.755	.656	1.150
50°	.873	.766	.643	1.192
51°	.890	.777	.629	1.235
52°	.908	.788	.616	1.280
53°	.925	.799	.602	1.327
54°	.942	.809	.588	1.376
55°	.960	.819	.574	1.428
56°	.977	.829	.559	1.483
57°	.995	.839	.545	1.540
58°	1.012	.848	.530	1.600
59°	1.030	.857	.515	1.664
60°	1.047	.866	.500	1.732
61°	1.065	.875	.485	1.804
62°	1.082	.883	.470	1.881
63°	1.100	.891	.454	1.963
64°	1.117	.899	.538	2.050
65°	1.134	.906	.423	2.145
66°	1.152	.914	.407	2.246
67°	1.169	.921	.391	2.356
68°	1.187	.927	.375	2.475
69°	1.204	.934	.358	2.605
70°	1.222	.940	.342	2.747
71°	1.239	.946	.326	2.904
72°	1.257	.951	.309	3.078
73°	1.274	.956	.292	3.271
74°	1.292	.961	.276	3.487
75°	1.309	.966	.259	3.732
76°	1.326	.970	.242	4.011
77°	1.344	.974	.225	4.331
78°	1.361	.978	.208	4.705
79°	1.379	.982	.191	5.145
80°	1.396	.985	.174	5.671
81°	1.414	.988	.156	6.314
82°	1.431	.990	.139	7.115
83°	1.449	.993	.122	8.144
84°	1.466	.995	.105	9.514
85°	1.484	.996	.087	11.43
86°	1.501	.998	.070	14.30
87°	1.518	.999	.052	19.08
88°	1.536	.999	.035	28.64
89°	1.553	1.000	.018	57.29
90°	1.571	1.000	.000	∞

2 COMMON LOGARITHMS

N	0	1	2	3	4	5	6	7	8	9
1.0	.0000	.0043	.0086	.0128	.0170	.0212	.0253	.0294	.0334	.0374
1.1	.0414	.0453	.0492	.0531	.0569	.0607	.0645	.0682	.0719	.0756
1.2	.0792	.0828	.0864	.0899	.0934	.0969	.1004	.1038	.1072	.1106
1.3	.1139	.1173	.1206	.1239	.1271	.1303	.1335	.1367	.1399	.1430
1.4	.1461	.1492	.1523	.1553	.1584	.1614	.1644	.1673	.1703	.1732
1.5	.1761	.1790	.1818	.1847	.1875	.1903	.1931	.1959	.1987	.2014
1.6	.2041	.2068	.2095	.2122	.2148	.2175	.2201	.2227	.2253	.2279
1.7	.2305	.2330	.2355	.2381	.2406	.2430	.2455	.2480	.2504	.2529
1.8	.2553	.2577	.2601	.2625	.2648	.2672	.2695	.2718	.2742	.2765
1.9	.2788	.2810	.2833	.2856	.2878	.2900	.2923	.2945	.2967	.2989
2.0	.3010	.3032	.3054	.3075	.3096	.3118	.3139	.3160	.3181	.3202
2.1	.3222	.3243	.3263	.3284	.3304	.3324	.3345	.3365	.3385	.3404
2.2	.3424	.3444	.3464	.3483	.3503	.3522	.3541	.3560	.3579	.3598
2.3	.3617	.3636	.3655	.3674	.3692	.3711	.3729	.3748	.3766	.3784
2.4	.3802	.3820	.3838	.3856	.3874	.3892	.3909	.3927	.3945	.3962
2.5	.3979	.3997	.4014	.4031	.4048	.4065	.4082	.4099	.4116	.4133
2.6	.4150	.4166	.4183	.4200	.4216	.4233	.4249	.4265	.4281	.4298
2.7	.4314	.4330	.4344	.4362	.4378	.4393	.4409	.4425	.4440	.4456
2.8	.4472	.4487	.4503	.4518	.4533	.4548	.4564	.4579	.4594	.4609
2.9	.4624	.4639	.4654	.4669	.4684	.4698	.4713	.4728	.4742	.4757
3.0	.4771	.4786	.4800	.4814	.4829	.4843	.4857	.4871	.4886	.4900
3.1	.4914	.4928	.4942	.4955	.4969	.4983	.4997	.5011	.5024	.5038
3.2	.5052	.5065	.5079	.5092	.5106	.5119	.5132	.5146	.5159	.5172
3.3	.5185	.5198	.5211	.5224	.5238	.5250	.5263	.5276	.5289	.5302
3.4	.5315	.5328	.5340	.5353	.5366	.5378	.5391	.5403	.5416	.5428
3.5	.5441	.5453	.5465	.5478	.5490	.5502	.5515	.5527	.5539	.5551
3.6	.5563	.5575	.5587	.5599	.5611	.5623	.5635	.5647	.5659	.5670
3.7	.5682	.5694	.5705	.5717	.5729	.5740	.5752	.5763	.5775	.5786
3.8	.5798	.5809	.5821	.5832	.5843	.5855	.5866	.5877	.5888	.5900
3.9	.5911	.5922	.5933	.5944	.5955	.5966	.5977	.5988	.5999	.6010
4.0	.6021	.6031	.6042	.6053	.6064	.6075	.6085	.6096	.6107	.6117
4.1	.6128	.6138	.6149	.6160	.6170	.6181	.6191	.6201	.6212	.6222
4.2	.6233	.6243	.6253	.6263	.6274	.6284	.6294	.6304	.6314	.6325
4.3	.6335	.6345	.6355	.6365	.6375	.6385	.6395	.6405	.6415	.6425
4.4	.6434	.6444	.6454	.6464	.6474	.6484	.6493	.6503	.6513	.6523
4.5	.6532	.6542	.6551	.6561	.6571	.6580	.6590	.6599	.6609	.6618
4.6	.6628	.6637	.6646	.6656	.6665	.6675	.6684	.6693	.6703	.6712
4.7	.6721	.6730	.6740	.6749	.6758	.6767	.6776	.6785	.6794	.6803
4.8	.6812	.6822	.6831	.6840	.6849	.6857	.6866	.6875	.6884	.6893
4.9	.6902	.6911	.6920	.6929	.6937	.6946	.6955	.6964	.6972	.6981
5.0	.6990	.6998	.7007	.7016	.7024	.7033	.7042	.7050	.7059	.7067
5.1	.7076	.7084	.7093	.7101	.7110	.7118	.7127	.7135	.7143	.7152
5.2	.7160	.7168	.7177	.7185	.7193	.7202	.7210	.7218	.7226	.7235
5.3	.7243	.7251	.7259	.7267	.7275	.7284	.7292	.7300	.7309	.7316
5.4	.7324	.7332	.7340	.7348	.7356	.7364	.7370	.7380	.7388	.7396
N	0	1	2	3	4	5	6	7	8	9

N	0	1	2	3	4	5	6	7	8	9
5.5	.7404	.7412	.7419	.7427	.7435	.7443	.7451	.7459	.7466	.7474
5.6	.7482	.7490	.7497	.7505	.7513	.7521	.7528	.7536	.7544	.7551
5.7	.7559	.7566	.7574	.7582	.7589	.7597	.7604	.7612	.7619	.7627
5.8	.7634	.7642	.7649	.7657	.7664	.7672	.7679	.7686	.7694	.7701
5.9	.7709	.7716	.7723	.7731	.7738	.7745	.7753	.7760	.7767	.7774
6.0	.7782	.7789	.7796	.7803	.7810	.7818	.7825	.7832	.7839	.7846
6.1	.7853	.7860	.7868	.7875	.7882	.7889	.7896	.7903	.7910	.7917
6.2	.7924	.7931	.7938	.7945	.7952	.7959	.7966	.7973	.7980	.7987
6.3	.7993	.8000	.8007	.8014	.8021	.8028	.8035	.8041	.8048	.8055
6.4	.8062	.8069	.8075	.8082	.8089	.8096	.8102	.8109	.8116	.8122
6.5	.8129	.8136	.8143	.8149	.8156	.8162	.8169	.8176	.8182	.8189
6.6	.8195	.8202	.8209	.8215	.8222	.8228	.8235	.8241	.8248	.8254
6.7	.8261	.8267	.8274	.8280	.8287	.8293	.8300	.8306	.8312	.8319
6.8	.8325	.8332	.8338	.8344	.8351	.8357	.8363	.8370	.8376	.8382
6.9	.8389	.8395	.8401	.8407	.8414	.8420	.8426	.8432	.8439	.8445
7.0	.8451	.8457	.8463	.8470	.8476	.8482	.8488	.8494	.8500	.8507
7.1	.8513	.8519	.8525	.8531	.8537	.8543	.8549	.8555	.8561	.8567
7.2	.8573	.8579	.8585	.8591	.8597	.8603	.8609	.8615	.8621	.8627
7.3	.8633	.8639	.8645	.8651	.8657	.8663	.8667	.8675	.8681	.8686
7.4	.8692	.8698	.8704	.8710	.8716	.8722	.8727	.8733	.8739	.8745
7.5	.8751	.8756	.8762	.8768	.8774	.8780	.8785	.8791	.8797	.8802
7.6	.8808	.8814	.8820	.8825	.8831	.8837	.8842	.8848	.8854	.8859
7.7	.8865	.8871	.8876	.8882	.8887	.8893	.8899	.8904	.8910	.8915
7.8	.8921	.8927	.8932	.8938	.8943	.8949	.8954	.8960	.8965	.8971
7.9	.8976	.8982	.8987	.8993	.8998	.9004	.9009	.9015	.9020	.9026
8.0	.9031	.9036	.9042	.9047	.9053	.9058	.9063	.9069	.9074	.9080
8.1	.9085	.9090	.9096	.9101	.9106	.9112	.9117	.9122	.9128	.9133
8.2	.9138	.9143	.9149	.9154	.9159	.9165	.9170	.9175	.9180	.9186
8.3	.9191	.9196	.9201	.9207	.9212	.9217	.9222	.9227	.9232	.9238
8.4	.9243	.9248	.9253	.9258	.9263	.9267	.9274	.9279	.9284	.9289
8.5	.9294	.9299	.9304	.9310	.9315	.9320	.9325	.9330	.9335	.9340
8.6	.9345	.9350	.9355	.9360	.9365	.9370	.9375	.9380	.9385	.9390
8.7	.9395	.9400	.9405	.9410	.9415	.9420	.9425	.9430	.9435	.9440
8.8	.9445	.9450	.9455	.9460	.9465	.9470	.9474	.9479	.9484	.9489
8.9	.9494	.9499	.9504	.9509	.9513	.9518	.9523	.9528	.9533	.9538
9.0	.9542	.9547	.9552	.9557	.9562	.9567	.9571	.9576	.9581	.9586
9.1	.9590	.9595	.9600	.9605	.9610	.9614	.9619	.9624	.9628	.9633
9.2	.9638	.9643	.9647	.9652	.9657	.9661	.9666	.9671	.9676	.9680
9.3	.9685	.9690	.9694	.9699	.9704	.9708	.9713	.9717	.9722	.9727
9.4	.9731	.9736	.9741	.9745	.9750	.9754	.9759	.9764	.9768	.9773
9.5	.9777	.9782	.9786	.9791	.9796	.9800	.9805	.9809	.9814	.9818
9.6	.9823	.9827	.9832	.9836	.9841	.9845	.9850	.9854	.9859	.9863
9.7	.9868	.9872	.9877	.9881	.9886	.9890	.9895	.9900	.9903	.9908
9.8	.9912	.9917	.9921	.9926	.9930	.9934	.9939	.9943	.9948	.9952
9.9	.9956	.9961	.9965	.9970	.9974	.9978	.9983	.9987	.9991	.9996
N	0	1	2	3	4	5	6	7	8	9

3 EXPONENTIALS

x	e^x	e^{-x}
0.00	1.0000	1.0000
0.05	1.0513	0.9512
0.10	1.1052	0.9048
0.15	1.1618	0.8607
0.20	1.2214	0.8187
0.25	1.2840	0.7788
0.30	1.3499	0.7408
0.35	1.4191	0.7047
0.40	1.4918	0.6703
0.45	1.5683	0.6376
0.50	1.6487	0.6065
0.55	1.7333	0.5769
0.60	1.8221	0.5488
0.65	1.9155	0.5220
0.70	2.0138	0.4966
0.75	2.1170	0.4724
0.80	2.2255	0.4493
0.85	2.3396	0.4274
0.90	2.4596	0.4066
0.95	2.5857	0.3867
1.0	2.7183	0.3679
1.1	3.0042	0.3329
1.2	3.3201	0.3012
1.3	3.6693	0.2725
1.4	4.0552	0.2466
1.5	4.4817	0.2231
1.6	4.9530	0.2019
1.7	5.4739	0.1827
1.8	6.0496	0.1653
1.9	6.6859	0.1496
2.0	7.3891	0.1353
2.1	8.1662	0.1225
2.2	9.0250	0.1108
2.3	9.9742	0.1003
2.4	11.023	0.0907

x	e^x	e^{-x}
2.5	12.182	0.0821
2.6	13.464	0.0743
2.7	14.880	0.0672
2.8	16.445	0.0608
2.9	18.174	0.0550
3.0	20.086	0.0498
3.1	22.198	0.0450
3.2	24.533	0.0408
3.3	27.113	0.0369
3.4	29.964	0.0334
3.5	33.115	0.0302
3.6	36.598	0.0273
3.7	40.447	0.0247
3.8	44.701	0.0224
3.9	49.402	0.0202
4.0	54.598	0.0183
4.1	60.340	0.0166
4.2	66.686	0.0150
4.3	73.700	0.0136
4.4	81.451	0.0123
4.5	90.017	0.0111
4.6	99.484	0.0101
4.7	109.95	0.0091
4.8	121.51	0.0082
4.9	134.29	0.0074
5	148.41	0.0067
6	403.43	0.0025
7	1096.6	0.0009
8	2981.0	0.0003
9	8103.1	0.0001
10	22026	0.00005

4 DERIVATIVES

$$\frac{dau}{dx} = a\frac{du}{dx}$$

$$\frac{d(u+v-w)}{dx} = \frac{du}{dx} + \frac{dv}{dx} - \frac{dw}{dx}$$

$$\frac{d(uv)}{dx} = u\frac{dv}{dx} + v\frac{du}{dx}$$

$$\frac{d(u/v)}{dx} = \frac{v(du/dx) - u(dv/dx)}{v^2}$$

$$\frac{d(u^n)}{dx} = nu^{n-1}\frac{du}{dx}$$

$$\frac{d(u^v)}{dx} = vu^{v-1}\frac{du}{dx} + u^v(\ln u)\frac{dv}{dx}$$

$$\frac{d(e^u)}{dx} = e^u\frac{du}{dx}$$

$$\frac{d(e^{au})}{dx} = ae^{au}\frac{du}{dx}$$

$$\frac{da^u}{dx} = a^u(\ln a)\frac{du}{dx}$$

$$\frac{d(\ln u)}{dx} = \frac{1}{u}\frac{du}{dx}$$

$$\frac{d(\log_a u)}{dx} = \frac{1}{u(\ln a)}\frac{du}{dx}$$

$$\frac{d\sin u}{dx} = \cos u\frac{du}{dx}$$

$$\frac{d\cos u}{dx} = -\sin u\frac{du}{dx}$$

$$\frac{d\tan u}{dx} = \sec^2 u\frac{du}{dx}$$

$$\frac{d\cot u}{dx} = -\csc^2 u\frac{du}{dx}$$

$$\frac{d\sec u}{dx} = \tan u\sec u\frac{du}{dx}$$

$$\frac{d\csc u}{dx} = -(\cot u)(\csc u)\frac{du}{dx}$$

$$\frac{d\arcsin u}{dx} = \frac{1}{\sqrt{1-u^2}}\frac{du}{dx}$$

$$\frac{d\arccos u}{dx} = \frac{-1}{\sqrt{1-u^2}}\frac{du}{dx}$$

$$\frac{d\arctan u}{dx} = \frac{1}{1+u^2}\frac{du}{dx}$$

$$\frac{d\,\mathrm{arc\,cot}\,u}{dx} = \frac{-1}{1+u^2}\frac{du}{dx}$$

$$\frac{d\,\mathrm{arc\,sec}\,u}{dx} = \frac{-1}{u\sqrt{u^2-1}}\frac{du}{dx}$$

$$\frac{d\sinh u}{dx} = \cosh u\frac{du}{dx}$$

$$\frac{d\cosh u}{dx} = \sinh u\frac{du}{dx}$$

$$\frac{d\tanh u}{dx} = \mathrm{sech}^2\, u\frac{du}{dx}$$

$$\frac{d\coth u}{dx} = -(\mathrm{csch}^2\, u)\frac{du}{dx}$$

$$\frac{d\,\mathrm{sech}\, u}{dx} = -(\mathrm{sech}\, u)(\tanh u)\frac{du}{dx}$$

$$\frac{d\,\mathrm{csch}\, u}{dx} = -(\mathrm{csch}\, u)(\coth u)\frac{du}{dx}$$

$$\frac{d\sinh^{-1} u}{dx} = \frac{1}{\sqrt{1+u^2}}\frac{du}{dx}$$

$$\frac{d\cosh^{-1} u}{dx} = \frac{1}{\sqrt{u^2-1}}\frac{du}{dx}$$

$$\frac{d\tanh^{-1} u}{dx} = \frac{1}{1-u^2}\frac{du}{dx}$$

$$\frac{d\coth^{-1} u}{dx} = \frac{1}{1-u^2}\frac{du}{dx}$$

$$\frac{d\,\mathrm{sech}^{-1} u}{dx} = \frac{-1}{u\sqrt{1-u^2}}\frac{du}{dx}$$

$$\frac{d\,\mathrm{csch}^{-1} u}{dx} = \frac{-1}{|u|\sqrt{1+u^2}}\frac{du}{dx}$$

5 INTEGRALS

$$\int x^n\,dx = \frac{1}{n+1}x^{n+1} \qquad (n \neq -1)$$

$$\int \frac{1}{x}\,dx = \ln|x| + C$$

$$\int e^x\,dx = e^x + C$$

$$\int a^x\,dx = \frac{a^x}{\ln a} + C$$

$$\int \sin x\,dx = -\cos x + C$$

$$\int \cos x\,dx = \sin x + C$$

$$\int \tan x\,dx = \ln|\cos x| + C$$

$$\int \cot x\,dx = \ln|\sin x| + C$$

$$\int \sec x\,dx = \ln|\sec x + \tan x| + C$$
$$= \ln|\tan \tfrac{1}{2}x + \tfrac{1}{4}\pi| + C$$

$$\int \csc x\,dx = \ln|\csc x - \cot x| + C$$
$$= \ln|\tan \tfrac{1}{2}x| + C$$

$$\int \arcsin\frac{x}{a}\,dx = x\arcsin\frac{x}{a} + \sqrt{a^2 - x^2} + C \qquad (a > 0)$$

$$\int \arccos\frac{x}{a}\,dx = x\arccos\frac{x}{a} - \sqrt{a^2 - x^2} + C \qquad (a > 0)$$

$$\int \arctan\frac{x}{a}\,dx = x\arctan\frac{x}{a} - \frac{a}{2}\ln(a^2 + x^2) + C \qquad (a > 0)$$

$$\int \sin^2 mx\,dx = \frac{1}{2m}(mx - \sin mx\cos mx) + C$$

$$\int \cos^2 mx\,dx = \frac{1}{2m}(mx + \sin mx\cos mx) + C$$

$$\int \sec^2 x\,dx = \tan x + C$$

$$\int \csc^2 x\,dx = -\cot x + C$$

$$\int \sin^n x\,dx = -\frac{\sin^{n-1} x \cos x}{n} + \frac{n-1}{n}\int \sin^{n-2} x\,dx$$

$$\int \cos^n x\,dx = \frac{\cos^{n-1} x \sin x}{n} + \frac{n-1}{n}\int \cos^{n-2} x\,dx$$

$$\int \tan^n x\,dx = \frac{\tan^{n-1} x}{n-1} - \int \tan^{n-2} x\,dx$$

$$\int \cot^n x\,dx = -\frac{\cot^{n-1} x}{n-1} - \int \cot^{n-2} x\,dx$$

$$\int \sec^n x\,dx = \frac{\tan x \sec^{n-2} x}{n-1} + \frac{n-2}{n-1}\int \sec^{n-2} x\,dx$$

$$\int \csc^n x\,dx = -\frac{\cot x \csc^{n-2} x}{n-1} + \frac{n-2}{n-1}\int \csc^{n-2} x\,dx$$

$$\int \sinh x\,dx = \cosh x + C$$

$$\int \cosh x\,dx = \sinh x + C$$

$$\int \tanh x\,dx = \ln(\cosh x) + C$$

$$\int \coth x\,dx = \ln|\sinh x| + C$$

$$\int \operatorname{sech} x\,dx = \arctan(\sinh x) + C = 2\arctan(e^x) + C$$

$$\int \operatorname{csch} x\,dx = \ln\left|\tanh\frac{x}{2}\right| + C = -\frac{1}{2}\ln\frac{\cosh x + 1}{\cosh x - 1} + C$$

$$\int \sinh^2 x\,dx = \frac{1}{4}\sinh 2x - \frac{1}{2}x + C$$

$$\int \cosh^2 x\,dx = \frac{1}{4}\sinh 2x + \frac{1}{2}x + C$$

$$\int \operatorname{sech}^2 x\,dx = \tanh x + C$$

$$\int \sinh^{-1}\frac{x}{a}\,dx = x\sinh^{-1}\frac{x}{a} - \sqrt{x^2+a^2} \qquad (a > 0)$$

$$\int \cosh^{-1}\frac{x}{a}\,dx = \begin{cases} x\cosh^{-1}\dfrac{x}{a} - \sqrt{x^2-a^2} + C & \left[\cosh^{-1}\left(\dfrac{x}{a}\right) > 0\right] \\[2ex] x\cosh^{-1}\dfrac{x}{a} + \sqrt{x^2-a^2} + C & \left[\cosh^{-1}\left(\dfrac{x}{a}\right) < 0\right] \end{cases}$$

$$\int \tanh^{-1}\frac{x}{a}\,dx = x\tanh^{-1}\frac{x}{a} + \frac{a}{2}\ln|a^2 - x^2| + C$$

$$\int \sqrt{a^2+x^2}\,dx = \int \sqrt{x^2 \pm a^2}\,dx$$

$$\int \frac{1}{\sqrt{a^2 + x^2}}\,dx = \ln\,(x + \sqrt{a^2 + x^2}) + C = \sinh^{-1}\frac{x}{a} + C \qquad (a > 0)$$

$$\int \frac{1}{a^2 + x^2}\,dx = \frac{1}{a}\arctan\frac{x}{a} + C \qquad (a > 0)$$

$$\int \sqrt{a^2 - x^2}\,dx = \frac{x}{2}\sqrt{a^2 - x^2} + \frac{a^2}{2}\arcsin\frac{x}{a} + C \qquad (a > 0)$$

$$\int (a^2 - x^2)^{3/2}\,dx = \frac{x}{8}(5a^2 - 2x^2)\sqrt{a^2 - x^2} + \frac{3a^4}{8}\arcsin\frac{x}{a} + C$$

$$\int \frac{1}{\sqrt{a^2 - x^2}}\,dx = \arcsin\frac{x}{a} + C \qquad (a > 0)$$

$$\int \frac{1}{a^2 - x^2}\,dx = \frac{1}{2a}\ln\left|\frac{a + x}{a - x}\right| + C \qquad (a > 0)$$

$$\int \frac{1}{(a^2 - x^2)^{3/2}}\,dx = \frac{x}{a^2\sqrt{a^2 - x^2}} + C$$

$$\int \sqrt{x^2 \pm a^2}\,dx = \frac{x}{2}\sqrt{x^2 \pm a^2} \pm \frac{a^2}{2}\ln|x + \sqrt{x^2 \pm a^2}| + C \qquad (a > 0)$$

$$\int \frac{1}{\sqrt{x^2 - a^2}}\,dx = \ln|x + \sqrt{x^2 - a^2}| + C = \cosh^{-1}\frac{x}{a} + C \qquad (a > 0)$$

$$\int \frac{1}{x(a + bx)}\,dx = \frac{1}{a}\ln\left|\frac{x}{a + bx}\right| + C$$

$$\int x\sqrt{a + bx}\,dx = \frac{2(3bx - 2a)(a + bx)^{3/2}}{15b^2} + C$$

$$\int \frac{\sqrt{a + bx}}{x}\,dx = 2\sqrt{a + bx} + a\int \frac{1}{x\sqrt{a + bx}}\,dx$$

$$\int \frac{x}{\sqrt{a + bx}}\,dx = \frac{2(bx - 2a)\sqrt{a + bx}}{3b^2} + C$$

$$\int \frac{1}{x\sqrt{a + bx}}\,dx = \frac{1}{\sqrt{a}}\ln\left|\frac{\sqrt{a + bx} - \sqrt{a}}{\sqrt{a + bx} + \sqrt{a}}\right| + C \qquad (a > 0)$$

$$= \frac{2}{\sqrt{-a}}\arctan\sqrt{\frac{a + bx}{-a}} + C \qquad (a < 0)$$

$$\int \frac{\sqrt{a^2 - x^2}}{x}\,dx = \sqrt{a^2 - x^2} - a\ln\frac{a + \sqrt{a^2 - x^2}}{|x|} + C$$

$$\int x\sqrt{a^2 - x^2}\,dx = -\tfrac{1}{3}(a^2 - x^2)^{3/2} + C$$

$$\int x^2\sqrt{a^2 - x^2}\,dx = \frac{x}{8}(2x^2 - a^2)\sqrt{a^2 - x^2} + \frac{a^4}{8}\arcsin\frac{x}{a} + C$$

$$\int \frac{1}{x\sqrt{a^2 - x^2}}\,dx = \frac{1}{a}\ln\frac{a - \sqrt{a^2 - x^2}}{|x|} + C$$

$$\int \frac{x}{\sqrt{a^2 - x^2}}\,dx = -\sqrt{a^2 - x^2} + C$$

$$\int \frac{x^2}{\sqrt{a^2 - x^2}}\,dx = -\frac{x}{2}\sqrt{a^2 - x^2} + \frac{a^2}{2}\arcsin\frac{x}{a} + C$$

$$\int \frac{\sqrt{x^2 + a^2}}{x}\,dx = \sqrt{x^2 + a^2} - a\ln\frac{a + \sqrt{x^2 + a^2}}{|x|} + C$$

$$\int \frac{\sqrt{x^2 - a^2}}{x}\,dx = \sqrt{x^2 - a^2} - a\arccos\frac{a}{|x|} + C$$

$$\int x\sqrt{x^2 \pm a^2}\,dx = \tfrac{1}{3}(x^2 \pm a^2)^{2/3} + C$$

$$\int \frac{1}{x\sqrt{x^2 + a^2}}\,dx = \frac{1}{a}\ln\frac{|x|}{a + \sqrt{x^2 + a^2}} + C$$

$$\int \frac{1}{x\sqrt{x^2 - a^2}}\,dx = \frac{1}{a}\arccos\frac{a}{|x|} + C$$

$$\int \frac{1}{x^2\sqrt{x^2 \pm a^2}}\,dx = \mp\frac{\sqrt{x^2 \pm a^2}}{a^2 x} + C$$

$$\int \frac{x}{\sqrt{x^2 \pm a^2}}\,dx = \sqrt{x^2 \pm a^2} + C$$

$$\int \frac{1}{ax^2 + bx + c}\,dx = \frac{1}{\sqrt{b^2 - 4ac}}\ln\frac{2ax + b - \sqrt{b^2 - 4ac}}{2ax + b + \sqrt{b^2 - 4ac}} \qquad (b^2 > 4ac)$$

$$= \frac{2}{\sqrt{4ac - b^2}}\arctan\frac{2ax + b}{\sqrt{4ac - b^2}} \qquad (b^2 < 4ac)$$

$$\int \frac{x}{ax^2 + bx + c}\,dx = \frac{1}{2a}\ln(ax^2 + bx + c) - \frac{b}{2a}\int\frac{1}{ax^2 + bx + c}\,dx$$

$$\int \frac{1}{\sqrt{ax^2 + bx + c}}\,dx = \frac{1}{\sqrt{a}}\ln\left(2ax + b + 2\sqrt{a}\sqrt{ax^2 + bx + c}\right) \qquad (a > 0)$$

$$= \frac{1}{\sqrt{-a}}\arcsin\frac{-2ax - b}{\sqrt{b^2 - 4ac}} \qquad (a < 0)$$

$$\int \sqrt{ax^2 + bx + c}\,dx = \frac{2ax + b}{4a}\sqrt{ax^2 + bx + c} + \frac{4ac - b^2}{8a}\int\frac{1}{\sqrt{ax^2 + bx + c}}\,dx$$

$$\int \frac{x}{\sqrt{ax^2 + bx + c}}\,dx = \frac{\sqrt{ax^2 + bx + c}}{a} - \frac{b}{2a}\int\frac{1}{\sqrt{ax^2 + bx + c}}\,dx$$

$$\int \frac{1}{x\sqrt{ax^2 + bx + c}}\,dx = \frac{1}{\sqrt{c}}\ln\left(\frac{\sqrt{ax^2 + bx + c} + \sqrt{c}}{x} + \frac{b}{2\sqrt{c}}\right) \qquad c > 0$$

$$= \frac{1}{\sqrt{-c}}\arcsin\frac{bx + 2c}{x\sqrt{b^2 - 4ac}} \qquad c < 0$$

$$\int \sin ax \sin bx\,dx = \frac{\sin(a - b)x}{2(a - b)} - \frac{\sin(a + b)x}{2(a + b)} + C \qquad (a^2 \neq b^2)$$

$$\int \sin ax \cos bx\,dx = -\frac{\cos(a - b)x}{2(a - b)} - \frac{\cos(a + b)x}{2(a + b)} + C \qquad (a^2 \neq b^2)$$

$$\int \cos ax \cos bx \, dx = \frac{\sin (a-b)x}{2(a-b)} + \frac{\sin (a+b)x}{2(a+b)} + C \qquad (a^2 \neq b^2)$$

$$\int \sec x \tan x \, dx = \sec x + C$$

$$\int \csc x \cot x \, dx = -\csc x + C$$

$$\int \cos^m x \sin^n x \, dx = \frac{\cos^{m-1} x \sin^{n+1} x}{m+n} + \frac{m-1}{m+n}\int \cos^{m-2} x \sin^n x \, dx$$

$$= -\frac{\sin^{n-1} x \cos^{m+1} x}{m+n} + \frac{n-1}{m+n}\int \cos^m x \sin^{n-2} x \, dx$$

$$= -\frac{\sin^{n+1} x \cos^{m+1} x}{m+1} + \frac{m+n+2}{m+1}\int \cos^{m+2} x \sin^n x \, dx$$

$$= \frac{\sin^{n+1} x \cos^{m+1} x}{n+1} + \frac{m+n+2}{n+1}\int \cos^m x \sin^{n+2} x \, dx$$

$$\int x^n \sin ax \, dx = -\frac{1}{a} x^n \cos ax + \frac{n}{a}\int x^{n-1} \cos ax \, dx$$

$$\int x^n \cos ax \, dx = \frac{1}{a} x^n \sin ax - \frac{n}{a}\int x^{n-1} \sin ax \, dx$$

$$\int x^n e^{ax} \, dx = \frac{x^n e^{ax}}{a} - \frac{n}{a}\int x^{n-1} e^{ax} \, dx$$

$$\int x^n \ln ax \, dx = x^{n+1}\left[\frac{\ln ax}{n+1} - \frac{1}{(n+1)^2}\right] + C$$

$$\int x^n (\ln ax)^m \, dx = \frac{x^{n+1}}{n+1} (\ln ax)^m - \frac{m a^{n+1}}{n+1}\int x^n (\ln ax)^{m-1} \, dx$$

$$\int e^{ax} \sin bx \, dx = \frac{e^{ax}(a \sin bx - b \cos bx)}{a^2 + b^2} + C$$

$$\int e^{ax} \cos bx \, dx = \frac{e^{ax}(b \sin bx + a \cos bx)}{a^2 + b^2} + C$$

$$\int \operatorname{sech} x \tanh x \, dx = -\operatorname{sech} x + C$$

$$\int \operatorname{csch} x \coth x \, dx = -\operatorname{csch} x + C$$

ANSWERS TO SELECTED EXERCISES

1.1

1. $1\frac{1}{2}$, $7/4$, $1+1/n$ for $n \geqslant 2$
5. (a) $]-\infty, +\infty[$
   (b) $]-\infty, +\infty[$
   (c) all real numbers except those of the form $(n+\frac{1}{2})\pi$ with n an integer
   (d) all real numbers except 2 and 1
   (e) $[0, +\infty[$ (f) $]-\infty, +\infty[$
6. (a) $]-\infty, +\infty[$
   (b) $[-1, 1]$
   (c) $[0, \infty[$
   (d) $]0, 1]$
   (e) the set whose only element is 6
7. (a) 0, 2, π
   (b) 0, $1/\sqrt{2}$, 0
   (c) 0, 1, 0
   (d) -1, $3/2$, $\dfrac{\pi^3-2}{\pi^3-3\pi+2}$
   (e) 0, 2
   (f) -1, 0, $3\sqrt[3]{3}$
   (g) 6, 6
8. Rule (b) does not define a function since it assigns two values to all negative real numbers; rule (c) assigns two values to zero; rule (e) assigns two values to π; rule (f) assigns two values to $|h|$ when $h \neq 0$; the remaining rules all define functions.

1.2

3. (a) $lwd = 96$; $2(lw+wd+ld) = 154$; $2(w+d) = 16$
   (b) $T = kP/V$
   (c) $E = \frac{1}{2}mv^2$; $p = mv$; $E = \dfrac{p^2}{2m}$
   (d) $s = \sqrt{2500t^2 + (5+50t)^2}$
   (e) $A = 4l + \dfrac{80l}{l-3}$

1.3

1. (a) $x \mapsto 16x^2 + 32x + 16$
   (c) $\theta \mapsto \dfrac{16\sec^4\theta - 1}{16\sec^4\theta - 4}$
   (e) The composite is not defined since $\operatorname{sgn} 0 = 0$.
   (g) $t \mapsto \sqrt{1-t^2}$
   (i) $r \mapsto \dfrac{7\pi}{1+49\pi^2}$
   (k) $c \mapsto 4c^2 + 4$
   (m) $k \mapsto -63k^2 - 84k - 32$
   (n) $z \mapsto \begin{cases} 1, & z > 1/\sqrt{2} \text{ or } 0 > z > -1/\sqrt{2} \\ 0, & z = \pm 1/\sqrt{2} \\ -1, & z < -1/\sqrt{2} \text{ or } 0 < z < 1/\sqrt{2} \end{cases}$

(o) $\beta \mapsto \pi - \beta$
(q) $z \mapsto 0$
(s) $x \mapsto x$

2. No. Yes.
3. (a) $y \mapsto (y+2)^2$
(c) $y \mapsto 6 - 4y$
(e) $x \mapsto \begin{cases} 0, & x \geqslant 0 \\ -2x, & x < 0 \end{cases}$
4. (a) $y \mapsto 4y^3 + 4y^2$
(c) $y \mapsto -21y^2 - 2y + 8$
(e) $x \mapsto \begin{cases} -x^2, & x \geqslant 0 \\ x^2, & x < 0 \end{cases}$
5. (a) $y \mapsto y^2 - 4y - 4$
(c) $y \mapsto 10y - 2$
(e) $x \mapsto \begin{cases} 2x, & x \geqslant 0 \\ 0, & x < 0 \end{cases}$
6. (a) $\theta \mapsto \sin\theta$
(c) $t \mapsto -\operatorname{sgn} t \quad (t \neq 0)$
7. (a) No.
(c) Yes and monotone increasing.
(e) No.
(g) No.
(i) Yes and monotone decreasing.
(k) No, when $a \neq 0$; yes, when $a = 0$ and $b \neq 0$.
(m) Yes and monotone increasing.
9. (a) Monotone increasing.
(c) Monotone increasing.
(e) Neither.
(g) Neither.
(i) Neither.
10. (c) $v \mapsto \begin{cases} v-1, & v > 0 \\ 0, & v = 0 \\ v+1, & v < 0 \end{cases}$
(f) $s \mapsto s^3$
(h) $y \mapsto \sqrt[3]{y}$
(i) $t \mapsto t^2 \quad (t > 0)$
(j) $\theta \mapsto (y-b)/m$
(m) $t \mapsto t$
(n) $y \mapsto \begin{cases} \sqrt{y}, & y \geqslant 0 \\ -\sqrt{|y|}, & y \leqslant 0 \end{cases}$
The function of Example 10 is its own inverse.

1.4

2. (a) $y - 1 = \frac{2}{3}(x-4)$
(c) $y + 3 = \frac{5}{2}(x-1)$
(e) $y + 1 = -3(x-4)$
(g) $y + 4 = x - 3$
(i) $x = -3$
3. (e) -3; $(0, 11)$
(f) 10; $(0, -5)$
(g) 1; $(0, -7)$
(h) $\frac{4}{7}$; $(0, -\frac{20}{7})$
(i) This line has no slope and no y intercept.
4. (a) $4x - y = 7$; $x + 4y = 23$
(c) $3x = 4y + 6$; $4x + 3y + 17 = 0$
(e) $x + 2 = 0$; $y = 0$
(g) $x + y + 5 = 0$; $x - y + 5 = 0$
7. (a) $3\sqrt{5}$
(c) $\frac{3}{4}\sqrt{10}$
8. (a) $3\sqrt{2}$
(c) $\frac{3}{5}\sqrt{17}$
(e) $\dfrac{15}{\sqrt{17}}$
9. $x^2 + y^2 + 2xy - 4ax + 2a^2 = 0$
11. $y + 6 = -\frac{9}{7}(x-9)$; $y + 6 = -\frac{3}{4}(x-9)$
16. (a) $\theta = \arcsin\dfrac{2}{\sqrt{13}} = \arccos\dfrac{3}{\sqrt{13}}$
(c) $\theta = \arcsin\dfrac{5}{\sqrt{29}} = \arccos\dfrac{2}{\sqrt{29}}$
(e) $\theta = \arcsin\dfrac{3}{\sqrt{10}} = \arccos\dfrac{-1}{\sqrt{10}}$
(g) $\theta = \frac{1}{4}\pi$
(i) $\theta = \frac{1}{2}\pi$
17. $m = -x_0/y_0$; $b = 1/y_0$
18. $y + 1 = (x-4)/\sqrt{3}$

1.5

2. (a) $(x+5)^2 + (y-3)^2 = 2$
(c) $(x-a)^2 + (y-2a)^2 = 4a^2$
(e) $(x+1)^2 + (y-2)^2 = 1$
(g) $(x-4)^2 + (y+1)^2 = 5$
(i) $(x-3)^2 + (y+\frac{5}{4})^2 = \frac{145}{16}$
(k) $(x+\frac{1}{2}a)^2 + (y-\frac{1}{2}a)^2 = \frac{1}{2}a^2$
(m) $(x+2)^2 + (y-9)^2 = 100$ or
$(x-12)^2 + (y+5)^2 = 100$

3. $7y + 20x = 0$ or $7y + 4x = 0$

4. (a) $\dfrac{(x-2)^2}{4} + \dfrac{(y-1)^2}{3} = 1$;
   axes on $x = 2$ and $y = 1$; foci at $(3,1)$ and $(1,1)$
   (c) $\dfrac{x^2}{10} + \dfrac{(y+2)^2}{9} = 1$;
   axes on $x = 0$ and $y = -2$; foci at $(\pm 1, -2)$
   (e) $x^2/5 + y^2/3 = 1$; axes on x and y axes; foci at $(\pm\sqrt{2}, 0)$

5. (a) $8(y+1) + x^2 = 0$; $p = -2$; vertex at $(0, -1)$; focus at $(0, -3)$; axis the y axis
   (c) $4(y+1) - (x-2)^2 = 0$; $p = 1$; vertex at $(2, -1)$; focus at $(2, 0)$; axis the line $x = 2$
   (e) $\frac{1}{2}(y+4) + (x+1)^2 = 0$; $p = -\frac{1}{8}$; vertex $(-1, -4)$; focus $(-1, -4\frac{1}{8})$; axis the line $x = -1$
   (g) $\frac{3}{2}(y+\frac{239}{24}) + (x+\frac{9}{4})^2 = 0$; $p = -\frac{3}{8}$; vertex at $(-\frac{9}{4}, -\frac{239}{24})$; focus at $(-\frac{9}{4}, -10\frac{1}{3})$; axis the line $x = -\frac{9}{4}$

6. (a) center $(-\frac{1}{2}a, 0)$; foci $(\frac{1}{2}(-a \pm 3a), 0)$; axes $y = 0$, $x = -\frac{1}{2}a$; vertices $(0,0)$, $(-a, 0)$
   (c) center $(0, -a)$; foci $(0, -a \pm \frac{1}{2}\sqrt{5}a)$; axes $x = 0$, $y = -a$; vertices $(0,0)$, $(0, -2a)$
   (e) center $(-\frac{1}{2}, -1)$; foci $(-\frac{1}{2}, 0)$, $(-\frac{1}{2}, -2)$; axes $x = -\frac{1}{2}$, $y = -1$; vertices $(-\frac{1}{2}, -1 \pm \frac{1}{2}\sqrt{2})$

1.6

1. (a) $9(x'+1)^2 + 4(y'-2)^2 = 25$
   (c) $x'^2 - 4y'^2 = 84$

2. (a) $y'^2 - \sqrt{3}x'y' + x' + \sqrt{3}y' = 10$
   (c) $x'^2 - 3y'^2 = 24$
   (e) $y'^2 - x'^2 = 25$

3. (a) $\dfrac{(x-\frac{1}{8})^2}{\frac{3}{8}} - \dfrac{(y-\frac{1}{2})^2}{6} = 1$
   (c) $\dfrac{x'^2}{\frac{100}{103}} + \dfrac{y'^2}{\frac{100}{47}} = 1$; $\tan\theta = \frac{1}{3}$
   (e) $(x'-\sqrt{\frac{3}{2}})^2 - (y'-\sqrt{\frac{1}{2}})^2 = 1$; $\theta = \frac{1}{4}\pi$
   (g) $x'^2 + \dfrac{y'^2}{6} = 1$; $\tan\theta = \frac{1}{2}$
   (i) $(x'-1)^2 - y'^2 = 1$; $\theta = \frac{1}{4}\pi$
   (k) $\dfrac{x'^2}{\frac{5}{4}} + \dfrac{y'^2}{\frac{5}{13}} = 1$; $\tan\theta = 3$
   (l) No locus.

   (n) $x^2 + \dfrac{(y-1)^2}{2} = 1$
   [This is an ellipse with vertical major axis.]
   (o) $(x-1) = 4(y-1)^2$ [This is a parabola with directrix parallel to the y axis.]
   (p) $x^2 + \dfrac{y^2}{10000} = 1$
   [This is a "skinny" ellipse with vertical major axis.]

REVIEW EXERCISES FOR CHAPTER 1

12. The tangent function on $]-\frac{1}{2}\pi, \frac{1}{2}\pi[$ is an example.

*19. $x^2 - a^2 = 0$

21. The domain of f is $]-\infty, +\infty[$, and its range is $[0, +\infty[$; f is neither monotone nor one-one. g has domain consisting of all real numbers except 1 and its range is the same as its domain; g is not monotone but is one-one and is its own inverse. We have

$$g \circ f: x \mapsto \frac{x^2+1}{x^2-1}$$

$$f + g: x \mapsto x^2 + \frac{x+1}{x-1}$$

$$f - g: x \mapsto x^2 - \frac{x+1}{x-1}$$

$$fg: x \mapsto \frac{x^3+x^2}{x-1}$$

$$f/g: x \mapsto \frac{x^3-x^2}{x+1}$$

23. The domain of f is $]-\infty, +\infty[$ and its range is $]-\infty, 6\frac{1}{4}]$; f is neither monotone nor one-one. The domain of g is $]-\infty, +\infty[$ and its range is $[-4, \infty[$;

g is neither monotone nor one-one. We have $g \circ f: x \mapsto x^4 - 2x^3 - 9x^2 + 10x + 21$; $fg: x \mapsto -x^4 - x^3 + 11x^2 + 9x - 18$;

$$f/g: x \mapsto \frac{3-x}{6+x-x^2}$$

25. The domain of f is $]-\infty, +\infty[$ and its range is $[-\sqrt{2}, +\sqrt{2}]$; f is neither monotone nor one-one. The domain of g is $]-\infty, +\infty[$ and its range is
$$\left[\frac{1}{2} - \frac{3\sqrt{5}}{10}, \frac{1}{2} + \frac{3\sqrt{5}}{10}\right];$$
g is neither monotone nor one-one. The functions $g \circ f, f+g, f-g$, and fg all have domain $]-\infty, +\infty[$; the domain of f/g consists of all real numbers except -2.
27. The domain of f is $]-\infty, +\infty[$ and its range is $[-1, +1]$; f is neither monotone nor one-one. The domain of g is $]-\infty, +\infty[$ and its range is $[-2, 2]$; g is neither monotone nor one-one. The functions $g \circ f, f+g, f-g$, and fg all have domain $]-\infty, +\infty[$; we have $f/g: x \mapsto \cos x$ if $x \neq n\pi$, n an integer.
39. $y = -(b/a)x + b$
41. $x + y = 1$
43. $\dfrac{x^2}{25} + \dfrac{(y-1)^2}{9} = 1$
45. $\dfrac{(x' - \frac{15}{7})^2}{\frac{147}{950}} + \dfrac{(y' - \frac{5}{3})^2}{\frac{63}{950}} = 1$; $\tan\theta = 3$
47. $\dfrac{(x' - 2\sqrt{2})^2}{156} - \dfrac{(y' - 6\sqrt{2})^2}{156} = 1$; $\theta = \frac{1}{4}\pi$
49. $(x - \frac{3}{4})^2 + (y + \frac{2}{3})^2 = \frac{25}{16}$
51. $\frac{1}{4}x^2 - \frac{1}{5}y^2 = 1$

2.1

1. -9
3. $\frac{1}{2}$
5. -13
7. -6
9. $\frac{1}{8}(\sqrt{2} - 1)$
11. $\frac{1}{2}$
13. $\frac{1}{2}\sqrt{2}$
15. The limit does not exist.
17. $-\frac{1}{25}$
19. $\frac{1}{4}$
21. (a) $(b, 0)$
    (b) $\frac{1}{2}$
    (c) 1
23. E approaches $\left(2, \dfrac{1 - 2x_Q}{\sqrt{1 - x_Q{}^2}}\right)$.

2.2

1. Yes.
3. Yes.
5. (a) No; (b) No.
7. (a) Yes. (b) No. (c) Yes. (d) Yes.
9. No.
11. No, unless $N =]A, \infty[$.
13. (a) Yes. (b) Yes. (c) Yes.

2.3

1. (a) 8
   (c) 4
   (e) 3
   (g) $\frac{3}{4}$
   (i) 0
   (k) 1
2. (a) $\sqrt{2}$
   (c) $\frac{1}{2}\sqrt{x}$
   (e) $\frac{37}{25}$
   (g) $\frac{3}{2}$
9. (a) $(0, \frac{1}{2})$
   (b) $(0, 1)$
11. The position D.

2.4

2. (a) 1
   (c) n
   (e) $\frac{5}{2}$
   (g) 2

3. Equations (a), (b), (d), (f), (g), (h) and (j) must hold.
5. $b = 0$; m can have any value.
10. $\left(\frac{n}{n+1}, 0\right)$
12. (a) $(b, 0)$
    (b) -1
    (c) -2

REVIEW EXERCISES FOR CHAPTER 2

3. 2
5. $2\sqrt{3}$
7. $-\frac{1}{9}$
9. 0
11. $a - b$
13. 3
15. $-\frac{3}{4}$
17. $\sqrt{2}$
25. For (a) the solution is $(b, 0)$; however, for parts (b) and (c) neither C nor D approaches a limit value.

3.1

1. (a) $f' : x \mapsto 2x$
   (c) $f' : x \mapsto 1$
   (e) $f' : x \mapsto 2$
2. (a) $\frac{dy}{dx} = 2x - 1$
   (c) $\frac{dy}{dx} = -\frac{2}{x^3}$
   (e) $\frac{dy}{dx} = 4x - 1$
   (g) $\frac{dy}{dx} = 2ax + b$
   (i) $\frac{dy}{dx} = \cos x(1 - 2\sin x)$
3. .(a) Monotone increasing on $[\frac{1}{2}, +\infty[$; monotone decreasing on $]-\infty, \frac{1}{2}]$; critical point at $\frac{1}{2}$.
   (c) Monotone increasing on $]-\infty, -2]$ and on $[2, +\infty[$; monotone decreasing on $[-2, 2]$; critical points at ± 2.
   (e) Monotone increasing on $[1, +\infty[$; monotone decreasing on $]-\infty, 1]$; critical point at 1.

3.2

1. All real numbers except -1; $f'(x) = 1/(x+1)^2$.
3. All real numbers except zero; $f'(x) = 0$.
5. Continuous but not differentiable.
7. Continuous but not differentiable.
9. Limits (a), (c) and (d) must exist; (b) might not.
11. $y - \frac{1}{2}\sqrt{2} = \frac{1}{2}\sqrt{2}(x - \frac{1}{4}\pi)$
13. $y = mx + b$
15. The indicated point is not in the domain of the function.

3.3

1. $ds/dt = 2t - 4$
3. $ds/dt = gt + v_0$
5. $ds/dt = 8t + 12$
7. $dy/dx = 15(x^2 - x^4)$
9. $dy/dx = x^3 - x^2 + x - 1$
11. $dy/dx = 5x^4 - 2x$
13. $dy/dx = 12x + 13$
15. $dy/dx = -19/(3x-2)^2$
17. $\frac{dr}{dq} = 2q(q+1)^{-1} - q^2(q+1)^{-2} = \frac{q^2+2q}{(q+1)^2}$
19. $\frac{d(\cot x)}{dx} = -\csc^2 x$, $\frac{d(\sec x)}{dx} = \tan x \sec x$, $\frac{d(\csc x)}{dx} = -\cot x \csc x$
21. (1) Monotone increasing on $[2, +\infty[$; monotone decreasing on $]-\infty, 2]$; critical point at 2.

(3) Monotone increasing on $[-v_0/g, +\infty[$; monotone decreasing on $]-\infty, -v_0/g]$; critical point at $-v_0/g$.

(5) Monotone increasing on $[-\frac{3}{2}, +\infty[$; monotone decreasing on $]-\infty, -\frac{3}{2}]$; critical point at $-\frac{3}{2}$.

(7) Monotone increasing on $[-1, 1]$; monotone decreasing on $]-\infty, -1]$ and on $[1, +\infty[$; critical points at ± 1.

(9) Monotone increasing on $[1, +\infty[$; monotone decreasing on $]-\infty, 1]$; critical point at 1.

(13) Monotone increasing on $[-\frac{13}{12}, +\infty[$; monotone decreasing on $]-\infty, -\frac{13}{12}]$; critical point at $-\frac{13}{12}$.

(15) Monotone decreasing on $]-\infty, \frac{2}{3}[$ and on $]\frac{2}{3}, +\infty[$; no critical points.

3.4

1. $d^2s/dt^2 = 10$; $d^3s/dt^3 = 0$
3. $d^2s/dt^2 = g$; $d^3s/dt^3 = 0$
5. $d^2y/dx^2 = 30x - 60x^3$; $d^3y/dx^3 = 30 - 180x^2$
7. $d^2y/dx^2 = 3x^2 - 2x + 1$; $d^3y/dx^3 = 6x - 2$
9. $d^2y/dx^2 = 20x^3 - 2$; $d^3y/dx^3 = 60x^2$
11. $d^2r/dq^2 = 2 + 6q^{-4}$; $d^3r/dq^3 = -24q^{-5}$
13. $d^2r/d\theta^2 = -\cos\theta$; $d^3r/d\theta^3 = \sin\theta$
15. $d^3r/d\theta^3 = \sec^3\theta + \sec\theta\tan^2\theta$;
    $d^3r/d\theta^3 = 5\sec^3\theta\tan\theta + \sec\theta\tan^3\theta$
17. $f'' : x \mapsto \dfrac{n(n+1)}{x^{n+2}}$; $f''' : x \mapsto \dfrac{-n(n+1)(n+2)}{x^{n+3}}$

3.5

1. 5000 cu in per minute; 4500 cu in per minute
2. $4\pi r^2$, which is equal to the surface area of the sphere
3. (a) 44 ft/sec
   (b) $3\frac{3}{8}$ secs; $182\frac{1}{4}$ ft
   (c) 108 ft/sec
   (d) -32 ft/sec^2
4. (a) -3; $\frac{3}{2}$
   (b) -7; $1 \pm \dfrac{\sqrt{80}}{3}$
   (c) -2; $\dfrac{2 \pm \sqrt{7}}{3}$
   (d) -62; $\frac{1}{16}$
5. $v = 2t - 4$; $a = 2$
7. $v = \dfrac{4 - 2t - t^2}{(4+t^2)^2}$; $a = \dfrac{2t^3 + 10t^2 - 24t + 8}{(4+t^2)^3}$
9. $v = 3(2t\cos t - (t^2+7)\sin t)$;
   $a = 3(-5\cos t - 4t\sin t - t^2\cos t)$
11. $A = 16\pi t^2$; 1600 sq ft/sec.

*12. $V = \pi x^2(18 - \frac{9}{4}x)$; monotone increasing on $[0, \frac{16}{3}]$; monotone decreasing on $[\frac{16}{3}, 8]$; maximum volume $512\pi/3$ at $x = \frac{16}{3}$.

*13. The acceleration is $-\frac{2}{5}$ ft/sec^2 and the deceleration is $\frac{2}{5}$ ft/sec^2.

3.6

1. $8x + 10$
3. $6x(x^2+1)^2$
5. $-6\cos(3\cos 2x)\sin 2x$
7. $\sin x/\cos^2 x$
8. $(h \circ g)'(x) \begin{cases} 2, & x > -2 \\ -2, & x < -2 \end{cases} = 2\,\mathrm{sgn}(x+2)$ if $x \neq -2$
9. $30(3x+5)^9$
11. $-3(x+5)^{-4}$
13. $2(x^2+1)(x^3-2x)[(x^3-2x)2x + (3x^2-2)(x^2+1)]$
15. $(2x + x^{-2})(x^3+2x-6)^4$
    $+ 4(3x^2+2)(x^2 - x^{-1} + 1)(x^3+2x-6)^3$
17. $8(2x-6)^3(x^2+5)^{-2} + 4x(2x-6)^4(x^2+5)^{-3}$
19. $2(\tan^3 x + 3\cos x - 7)(3\tan^2 x\sec^2 x - 3\sin x)$
21. $-anx^{n-1}\cos(ax^n)$
23. $\dfrac{x\cos x - \sin x}{x^2}$
25. $\tan^2 x$

REVIEW EXERCISES FOR CHAPTER 3

2. $10x^9 - 32x^7 - 96x^5 + 256x^3$;
   $90x^8 - 224x^6 - 480x^4 + 768x^2$;
   $0, 2048$; $y = 0$

4. $-6x^{-4}+6x^{-3}-36$; $24x^{-5}-18x^{-4}$; $-36, 6$; $y=-36x+19$

5. $-9x^{-4}-4x^{-3}+5x^{-2}$; $36x^{-5}+12x^{-4}-10x^{-3}$; the derivatives are not defined at zero.

7. $\dfrac{1-\cos 4ax}{8}$; $\frac{1}{2}a\sin 4ax$; $0,0$

9. $\sin^{m-1}x(m\cos mx+\cos x\sin mx)$;
$\sin^{m-2}x((m-1)(\sin x\cos x)(m\cos mx+\cos x\sin mx) - (m^2+\sin x)\sin mx + m\cos x\cos mx)$;
$(\frac{1}{2}\sqrt{2})^m(m\pm\frac{1}{2}\sqrt{2})$ if $m=8n\pm 1$,
$\pm(\frac{1}{2}\sqrt{2})^m$ if $m=8n\pm 2$,
$(\frac{1}{2}\sqrt{2})^m(-m\pm\frac{1}{2}\sqrt{2})$ if $m=8n\pm 3$,
$m(\frac{1}{2}\sqrt{2})^{m-1}$ if $m=8n$,
$-m(\frac{1}{2}\sqrt{2})^{m-1}$ if $m=8n+4$;
$\frac{1}{2}(\frac{1}{2}\sqrt{2})^{m-2}(m^2+(\sqrt{2}-1)m)$ if $m=8n$,
$\mp\frac{1}{4}(\frac{1}{2}\sqrt{2})^{m-2}(\sqrt{2}m^2-(1-\sqrt{2})m+3)$ if $m=8n\pm 1$,
$\pm(\frac{1}{2}\sqrt{2})^{m-2}(-m^2+\frac{1}{4}\sqrt{2}m-\frac{3}{4}\sqrt{2})$ if $m=8n\pm 2$,
$\mp(\frac{1}{2}\sqrt{2})^{m-2}(3\sqrt{2}+(1+\sqrt{2})m+3)$ if $m=8n\pm 3$,
$-\frac{1}{2}(\frac{1}{2}\sqrt{2})^{m-2}(m^2+(\sqrt{2}-1)m)$ if $m=8n+4$

11. $\dfrac{12x^2\sin(\frac{1}{2}x)-4x^3\cos(\frac{1}{2}x)}{\sin^3(\frac{1}{2}x)}$,
$\dfrac{6x^3-12x^2\sin x+24x\sin^2(\frac{1}{2}x)}{\sin^4(\frac{1}{2}x)}$;
$6\pi^2-\pi^3$, $3\pi^3-12\pi^2+24$

13. $15x^4+57x^2+6$; $60x^3+114x$; $3734, -672$

15. Monotone increasing on $]-\infty,0]$; monotone decreasing on $[0,+\infty[$; critical point at 0.

16. If $a>0$, the function is monotone increasing on
$$\left]-\infty, a\frac{1-\sqrt{5}}{2}\right],$$
on
$$[0,a[, \text{ and on } \left]a, a\frac{1+\sqrt{5}}{2}\right],$$
monotone decreasing on
$$\left[a\frac{1-\sqrt{5}}{2}, 0\right] \text{ and on } \left[a\frac{1+\sqrt{5}}{2}, +\infty\right[;$$
if $a<0$, the function is monotone increasing on
$$\left]-\infty, a\frac{1+\sqrt{5}}{2}\right], \text{ and on } \left[0, a\frac{1-\sqrt{5}}{2}\right],$$
monotone decreasing on
$$\left[a\frac{1+\sqrt{5}}{2}, a\right[,$$
on
$$]a,0[\text{ and on } \left[a\frac{1-\sqrt{5}}{2}, +\infty\right[;$$
critical points at 0 and
$$a\frac{1\pm\sqrt{5}}{2}.$$

17. Monotone increasing on $]-\infty,2]$ and on $[6,+\infty[$; monotone decreasing on $[2,4[$ and on $]4,6]$; critical points at $2,6$.

19. Monotone increasing on $[0,\infty[$; monotone decreasing on $]-\infty,0]$; critical point at 0.

21. Monotone increasing on $[0,\frac{1}{2}[$ and on $[1,+\infty[$; monotone decreasing on $]-\infty,0]$ and on $]\frac{1}{2},1]$; critical points at 0 and 1.

23. Monotone increasing on $[(2n-\frac{1}{2})\pi, (2n+\frac{1}{2})\pi[$; monotone decreasing on $](2n+\frac{1}{2})\pi, (2n+\frac{3}{2})\pi]$; critical points at $(2n-\frac{1}{2})\pi$.

25. $v=2t-1$; $a=2$

27. $v=3t^2-5$; $a=6t$

29. $v=-t^3/3+t^2+t+5$;
$s=-t^4/12+t^3/3+t^2/2+5t$

31. $v=96-32t$; $a=-32$

33. $v=\frac{1}{2}t^2$; $s=t^3/6$

35. $v=12$; $s=12t-17$

4.1

1. $(1-3x^2)\,dx$

3. $4\sin\theta\cos\theta\,d\theta$

5. $\dfrac{a^2-\frac{3}{2}ax}{(a-x)^{3/2}}dx$

7. $\left(\frac{1}{2}(2ax-x^2)^{1/2}-\dfrac{(x-a)^2}{4(2ax-x^2)^{1/2}}\right)dx$

8. $2y\,dy-2ax\,dx=0$; $dy/dx=ax/y$;
$2a\,dy+3a^2\,dx=0$; $2a(y-a)+3a^2(x+\frac{3}{2}a)=0$

10. $3y\,dx+(3x-4y)\,dy=0$; $dy/dx=3y/(4y-3x)$;
$\frac{15}{2}dx-3\,dy=0$; $15(x-\frac{7}{3})-6(y-\frac{5}{2})=0$

12. $(4x^3-3y^3-1)\,dx+(2-9xy^2)\,dy=0$;
$\dfrac{dy}{dx}=\dfrac{1+3y^3-4x^3}{2-9xy^2}$;

$(3+2\sqrt{\tfrac{2}{3}})\,dx - 4\,dy = 0;$
$(3+2\sqrt{\tfrac{2}{3}})(x-1) = 4(y+\sqrt{\tfrac{2}{3}})$

13. Differentiating $\sqrt{x/y^3}-2\sqrt{x^3/y}=0$ yields $(1-6x)\,dx + (2x^2/y-3xy^4)\,dy=0$; and
$$\frac{dy}{dx}=\frac{6x-1}{(2x^2/y-3xy^4)};$$
simplifying the relation to $xy-\frac{1}{2}=0$ yields $y\,dx+x\,dy=0$ and $dy/dx=-y/x$. At $(\frac{1}{2},1)$ both differential relations give the tangent line $dy=-2\,dx$ or $y=-2x+2$.

14. Differentiating $x^2+(x-y)/(x+y)=0$ yields $2y+2x(x+y)^2\,dx-2x\,dy=0$ and $dy/dx=y/x+(x+y)^2$; simplifying the relation to $x^2(x+y)+x-y=0$ yields $(3x^2+2yx+1)\,dx+(x^2-1)\,dy=0$ and
$$\frac{dy}{dx}=\frac{3x^2+2yx+1}{1-x^2}.$$
At $(2,\frac{10}{3})$ both differential relations give the tangent line $9\,dy=dx$ or $9(y+\frac{10}{3})=x-2$.

16. $-\dfrac{6}{5}(1+t)^{-2}\left(\dfrac{1-t}{1+t}\right)^{-2/5}$

18. $(1-x^2)^{1/3}-\dfrac{2x^2}{(1-x^2)^{2/3}}$

19. $dr/dh = 28/129$

21. 126/169 ft/sec

4.2

1. $dy/dx = \frac{4}{3}t;\ 3y = 8x - 16$
3. $dy/dx = -1;\ x + y = r$
5. $\dfrac{dy}{dx} = \dfrac{\sin\phi}{1-\cos\phi};\ y = 2$
7. $-600/\sqrt{481}$ mph
9. $\frac{1}{50}\pi$ ft/sec
11. $25\pi/\sqrt{2}$ ft/min, $25\pi/\sqrt{2}$ ft/min
13. $1325/\sqrt{815}$ ft/sec
15. $\frac{1}{120}\pi(\sqrt{381}) \approx \frac{1}{6}\pi$ in/min

*17. $x = \phi R - r\sin\phi;\ y = R - r\cos\phi$

4.3

1. (a) $(4,5),\ (2,-1),\ 8$
   (c) $(-6,6),\ (8,-8),\ -14$
   (e) $8\boldsymbol{i}+4\boldsymbol{j},\ 2\boldsymbol{i},\ 19$
   (g) $\left(\dfrac{\sqrt{2}+\sqrt{3}}{2}\right)\boldsymbol{i}+\left(\dfrac{\sqrt{2}+1}{2}\right)\boldsymbol{j},$
   $\left(\dfrac{\sqrt{2}-\sqrt{3}}{2}\right)\boldsymbol{i}+\left(\dfrac{\sqrt{2}-1}{2}\right)\boldsymbol{j},\quad \dfrac{\sqrt{6}+\sqrt{2}}{4}$

4. (a) $\sqrt{13},\ (-9,-6),\ \left(\dfrac{-3}{\sqrt{13}},\dfrac{-2}{\sqrt{13}}\right),\ \left(\dfrac{-2}{\sqrt{13}},\dfrac{3}{\sqrt{13}}\right)$
   (c) $5\sqrt{2},\ (-\sqrt{5},7\sqrt{5}),\ \left(\dfrac{-1}{5\sqrt{2}},\dfrac{7}{5\sqrt{2}}\right),\ \left(\dfrac{7}{5\sqrt{2}},\dfrac{1}{5\sqrt{2}}\right)$
   (e) $\sqrt{41},\ 10\boldsymbol{i}+8\boldsymbol{j},\ \dfrac{5}{\sqrt{41}}\boldsymbol{i}+\dfrac{4}{\sqrt{41}}\boldsymbol{j},\ \dfrac{4}{\sqrt{41}}\boldsymbol{i}-\dfrac{5}{\sqrt{41}}\boldsymbol{j}$
   (g) $\sqrt{641},\ \dfrac{-25}{2}\boldsymbol{i}+2\boldsymbol{j},$
   $\dfrac{25}{\sqrt{641}}\boldsymbol{i}-\dfrac{4}{\sqrt{641}}\boldsymbol{j},\ \dfrac{4}{\sqrt{641}}\boldsymbol{i}+\dfrac{25}{\sqrt{641}}\boldsymbol{j}$

4.4

1. (a) $f_1: t \mapsto 8t,\ f_2: t \mapsto 96-16t^2;$
   $\boldsymbol{f}(5) = (40,-304),\ \boldsymbol{f}(\sqrt{3}) = (8\sqrt{3},48)$
   (c) $h_1: p \mapsto p-\sin p,\ h_2: p \mapsto 1-\cos p;$
   $\boldsymbol{h}(0) = (0,0),$
   $$\boldsymbol{h}\left(\frac{-3}{4}\pi\right) = \left(\frac{-3\pi+2\sqrt{2}}{4},\frac{2+\sqrt{2}}{2}\right)$$
   (e) $g_1: t \mapsto t,\ g_2: t \mapsto t^{3/5};\ \boldsymbol{g}(-1) = (-1,1);$
   $\boldsymbol{g}(-32) = (-32,-8)$

2. (a) $\boldsymbol{f}': t \mapsto (8,96-32t);$
   $\boldsymbol{f}'(5) = (8,16),\ \boldsymbol{f}'(\sqrt{3}) = (8,96-32\sqrt{3})$
   (c) $\boldsymbol{h}': p \mapsto (1-\cos p,\sin p);$
   $\boldsymbol{h}'(0) = (0,0),\ \boldsymbol{h}'(-\frac{3}{4}\pi) = \left(\dfrac{2+\sqrt{2}}{2},\dfrac{-\sqrt{2}}{2}\right)$
   (e) $\boldsymbol{g}': t \mapsto (1,\frac{3}{5}t^{-2/5}),\ \boldsymbol{g}^1(-1) = (1,\frac{3}{5}),$
   $\boldsymbol{g}'(-32) = (1,\frac{3}{20})$

3. $\left(\dfrac{-5\sqrt{7}}{21},\dfrac{10\sqrt{14}}{7}\right)$

4. $(3,-4),\ (-4,-3)$

4.5

1. (a) 1
   (c) -1
   (e) $\frac{1}{4}\pi$
   (g) 1
   (i) $+\infty$
4. An example is $G(x) = (|x|)^{3/2}$.
5. (a) 1
   (b) 1
   (c) 0
   (d) 0
6. Functions (a), (c), (d) and (e) have infinite derivative $+\infty$ at zero; (b) and (f) have infinite right-hand derivative $+\infty$ at zero and infinite left-hand derivative $-\infty$ at zero.
13. The function is discontinuous at every integer.

4.6

1. (a) max $= 5$ at $x = -2$; min $= -1$ at $x = -1$
   (c) max $= -\frac{1}{6}$ at $x = -6$; min $= -\frac{1}{2}$ at $x = -2$
   (e) max $= 1$ at $x = \pi/2$; min $= -1$ at $3\pi/2$
7. (a) $8, 9, 10$; $\pi, \pi+1, \pi+2$; $b, b+1, b+2$; $b, b+1, b+2$; $q, q+1, q+2$; $1, 2, 3$; $0, 1, 2$
   (c) $]-\infty, q[$; $1, 0, a, a, \frac{1}{2}, 0$

4.7

1. (a) max $= 5$ at $x = -2$; min $= -20$ at $x = 3$
   (c) max $= \frac{23}{12} + \frac{4}{3}\sqrt{2}$ at $x = 1 + \sqrt{2}$; min $= -162$ at $x = 6$
   (e) max $= 1$ at $\operatorname{arc\,cos} \frac{1}{2}\pi$; min $= -1$ at $x = \operatorname{arc\,cos}(-\frac{1}{2}\pi)$
3. (a) $1 - 1/\sqrt{3}$
   (b) $\frac{27}{8}$
   (c) 1
   (d) 2
4. (b) -2 and -1 or -1 and 0
   (d) -2 and -1 or 3 and 4

4.8

5. (a) 1
   (b) $\frac{1}{3}\sqrt{10}$
   (c) 0
7. (a) $f\colon x \mapsto (1-x^{2/3})^{3/2}$; $f'(\frac{1}{8}) = -\sqrt{3}$
   (c) $f\colon x \mapsto \dfrac{x + \sqrt{216-23x^2}}{6}$; $f'(3) = \dfrac{-15}{2}$

REVIEW EXERCISES FOR CHAPTER 4

5. 0 7. 2 9. -1
15. $+\infty$ 17. $+\infty$ 19. 1
25. $\dfrac{-3(1-x)^{1/2}}{(1+x)^{5/2}}$; $\dfrac{9-6x}{(1-x)^{1/2}(1+x)^{7/2}}$; 0, not twice differentiable at $x = 1$
27. $x^{-2/3} + x^{-5/4} + x^{4/3}$; $-\frac{2}{3}x^{-5/3} - \frac{5}{4}x^{-9/4} + \frac{4}{3}x^{1/3}$; $3, -\frac{7}{12}$
29. (The authors apologize for this one.)
$$\tfrac{1}{2}\csc^2(\tfrac{1}{2}x) - \frac{\cot^2(\frac{1}{2}x) - \cot(\frac{1}{2}x) - 1}{(1-\cot^2(\frac{1}{2}x))^{3/2}};$$
$$-\tfrac{1}{2}\csc^2(\tfrac{1}{2}x)\cot(\tfrac{1}{2}x)\frac{\cot^2(\frac{1}{2}x) - \cot(\frac{1}{2}x) - 1}{(1-\cot^2(\frac{1}{2}x))^{3/2}} +$$
$$+ \tfrac{1}{2}\csc^2(\tfrac{1}{2}x)\frac{(1-\cot^2(\frac{1}{2}x))(\frac{3}{2} - \cot(\frac{1}{2}x)\csc^2(\frac{1}{2}x) + \csc^2(\frac{1}{2}x)) + \frac{3}{2}\cot(\frac{1}{2}x)}{(1-\cot^2(\frac{1}{2}x))^{5/2}}$$
the derivatives are not defined at π
31. $dy/dx = 1/y$; vertical at $(\frac{1}{2}, 0)$; no horizontal tangents
33. $\dfrac{dy}{dx} = \dfrac{2xy+3y}{8y-x^2-3x}$; horizontal at $\left(\dfrac{-3}{2}, \dfrac{-9 \pm \sqrt{337}}{32}\right)$; no vertical tangents
35. $|y_0|^{1/2}\,dx + |x_0|^{1/2}\,dy = 0$; $|y_0|^{1/2}(x-x_0) + |x_0|^{1/2}(y-y_0) = 0$
37. $dy + (9x_0^{-2/3} - 1)^{1/2}\,dx = 0$; $y - y_0 = -(9x_0^{-2/3}-1)^{1/2}(x-x_0)$
39. $dy + (\sec^2 x_0 \sin(\tan x_0))\,dx = 0$; $y = y_0 - (\sec^2 x_0 \sin(\tan x_0))(x-x_0)$

41. $\cos t_0\,dy = -\cos(\cos t_0)\sin t_0\,dx$ or
$(1-x_0^2)^{1/2}\,dy = -\cos((1-x_0^2)^{1/2})\,x_0\,dx$;
$(1-x_0^2)^{1/2}(y-y_0) = -\cos((1-x_0^2)^{1/2})\,x_0(x-x_0)$

43. $2t_0\,dy = (3t_0^2-3)\,dx$ or
$2(x_0+2)^{1/2}\,dy = 3(x_0+1)\,dx$ if $t_0 \geqslant 0$;
$-2(x_0+2)^{1/2}\,dy = 3(x_0+1)\,dx$ if $t_0 \leqslant 0$;
$2(x_0+2)^{1/2}(y-y_0) = 3(x_0+1)(x-x_0)$ if $t_0 \geqslant 0$,
$-2(x_0+2)^{1/2}(y-y_0) = 3(x_0+1)(x-x_0)$ if $t_0 \leqslant 0$

45. $(t_0-1)\,dy = (t_0+\frac{1}{2})\,dx$ or
$(x_0-2)^{1/2}\,dy = (\frac{3}{2}+(x_0-2)^{1/2})\,dx$ if $t_0 \geqslant 1$,
$(x_0-2)^{1/2}\,dy = ((x_0-2)^{1/2}-\frac{3}{2})\,dx$ if $t_0 \leqslant 1$;
$(x_0-2)^{1/2}(y-y_0) = (\frac{3}{2}+(x_0-2)^{1/2})(x-x_0)$ if $t_0 \geqslant 1$,
$(x_0-2)^{1/2}(y-y_0) = ((x_0-2)^{1/2}-\frac{3}{2})(x-x_0)$ if $t_0 \leqslant 1$

46. $\sin^2\theta_0\cos\theta_0\,dx + \cos^2\theta_0\sin\theta_0\,dy = 0$ or
$y_0^{2/3}x_0^{1/3}\,dx + x_0^{2/3}y_0^{1/3}\,dy = 0$;
$y_0^{2/3}x_0^{1/3}(x-x_0) + x_0^{2/3}y_0^{1/3}(y-y_0) = 0$

47. $dy = \frac{3}{8}s_0\,dx$ or $dy = \frac{3}{4}(y_0+1)^{1/3}\,dx$;
$y - y_0 = \frac{3}{4}(y_0+1)^{1/3}(x-x_0)$

49. 18

51. 64

55. The locus has a cusp at the origin and so no tangent line is defined there; hence $dx/dt = dy/dt = 0$ when the particle passes through the origin.

57. 0 ft/sec

5.1

8. Local maximum at $(\frac{1}{3},\frac{31}{27})$; local minimum at $(1,1)$; point of inflection at $(\frac{2}{3},\frac{29}{27})$.

10. Discontinuity at $x=0$; asymptote $y=1$; local maxima at $x=(1/2n)\pi$; local minima at $x=1/(2n+1)\pi$; points of inflection where $x=-\tan(1/x)$.

12. The function is not differentiable at $x=\frac{1}{2}n\pi$ where there are points of inflection; local maxima at $((n+\frac{1}{4})\pi, 1)$; local minima at $((n+\frac{3}{4}\pi), -1)$.

14. Critical points at $x = n\pi \pm \pi/4$; points of inflection at $x = n(\pi/2)$.

16. Discontinuities at $x = \pm 1$; asymptotes $x = \pm 1$, $y = 3x+1$; local maxima at $-3 \pm \sqrt{10}$; no inflection points. (This problem is simplified if you first divide out the fraction.)

18. The function is not defined for $x \in]-2,0[$; asymptotes $y=x-2$, $y=-x-4$.

20. Critical points at $x=(2n+1)\pi$; points of inflection at $x=n\pi$. (Notice that there are no local extrema since every critical point is a point of inflection.)

22. The function is not defined for $x<0$; no critical points or points of inflection.

24. Critical point and point of inflection at $x=0$.

26. Discontinuities at $x=(n+\frac{1}{2})\pi$, where there are vertical asymptotes; local maxima at $(2n-\frac{1}{2})\pi$; local minima at $(2n+\frac{1}{2})\pi$.

28. Discontinuity at $x=0$ where there is a vertical asymptote; the function is not defined for $x<0$; local minimum at $(9,\frac{2}{3})$; point of inflection at $(27,\frac{4}{9}\sqrt{3})$.

30. Local extrema at points x where $x=\tan(-x)$; points of inflection at points x where $\cot x = x/2$.

5.2

1. Square of side 5/2.

3. $r = \sqrt[3]{V/5\pi},\ h = \sqrt[3]{V/25\pi}$

5. 10 in

7. At the point 9 ft from the 12-ft pole and 6 ft from the 8-ft pole.

11. $\frac{1}{4}w$

13. $2ab$

15. $\sqrt{\dfrac{a^3}{b+a}}$

16. $(\frac{1}{3}a+\frac{1}{3}b, \frac{1}{3}c)$

18. $10\sqrt{117}$ ft

19. 50 mph

21. 20 ft.

5.3

1. 10.003333…; 10.003332…; $E \approx 10^{-6}$

3. 31.875; $\approx$31.875; $E \approx 10^{-4}$

5. 3.935; 3.920…; $E \approx 10^{-2}$

6. $1 + 10^{-3}n$; $1 + 10^{-3}n + 10^{-6}\dfrac{n(n-1)}{2} + \cdots$; $E \approx 10^{-6}(\frac{1}{2}n^2)$

9. $1\frac{25}{96}$ cu ft

5.4

1. (a) tangent $17x - y = 12$; normal $x + 17y = 376$
   (c) tangent $3x - y = 4a^2$; normal $x + 3y = 28a^2$
   (e) tangent $x + 4y = 2$; normal $y - 4x = 9$
2. (a) $9y - 2x = \frac{39}{4}\sqrt{2}$ at $(\frac{3}{2}\sqrt{2}, -\frac{3}{4}\sqrt{2})$ and $9y - 2x = \frac{117}{4}\sqrt{2}$ at $(-\frac{3}{2}\sqrt{2}, \frac{3}{4}\sqrt{2})$
   (c) $x + 2y = 10$
   (e) $x = 1$
3. (a) $K(x) = \dfrac{-2}{(17-4x)^{3/2}}$, $R(x) = \frac{1}{2}(17-4x)^{3/2}$;
   $K(0) = \dfrac{-2}{17\sqrt{17}}$, $R(0) = \frac{1}{2}17\sqrt{17}$
   (c) $K(\theta) = \dfrac{-\cos\theta}{(2-\cos^2\theta)^{3/2}}$; $R(\theta) = \dfrac{(2-\cos^2\theta)^{3/2}}{|\cos\theta|}$;
   $K(0) = -1$, $R(0) = 1$
   (d) $K(x) = \dfrac{6x}{(1+9x^4)^{3/2}}$, $R(x) = \dfrac{(1+9x^4)^{3/2}}{|6x|}$;
   $K(0) = 0$, $R(0)$ is not defined
4. (a) $K'(x)$ is never zero.
   (c) $K'(\theta) = 0$ at $\theta = n\pi$.
   (d) $K'(x) = 0$ at $x = \pm 1/(3\sqrt{5})^{1/2}$.
   (f) $K'(x) = 0$ for all x.

5.5

1. 0
3. $\frac{1}{2}$
5. 0
7. $-\frac{1}{9}$
9. $-\frac{1}{2}$
11. $\pm\infty\,(n > m)$, $a_n/b_m\,(n = m)$, $0\,(n < m)$

5.6

1. (a) $x \mapsto x^5/5$
   (c) $y \mapsto \frac{1}{10}(2y+1)^5$
   (e) $v \mapsto -1/2v^2$
   (g) $z \mapsto \frac{3}{4}z^4 - 2z^3 + \frac{3}{2}z^2$
2. (a) $y = \sin x + \frac{1}{2}$
   (c) $y = 2x^2 - 4x + 3$
   (e) $y = x^2 - 4x + 5$
3. 3.5 sec; $183\frac{3}{4}$ ft; $s = 35t - 5t^2$
4. $\sqrt{(8000-\frac{1}{2})g}$; $\sqrt{8000g} \approx 10\sqrt{\frac{16}{33}}$ mi/sec
5. (a) $y^2 - x^2 + C = 0$
   (c) $y^{2/3} - x^{2/3} + C = 0$
   (e) $y^{1/2} + \frac{1}{3}x^{3/2} + C = 0$

REVIEW EXERCISES FOR CHAPTER 5

1. (b) No relative extrema.
   (c) $x = \frac{8}{3}$
   (e) No asymptotes.
   *(f) $K = \dfrac{-16}{(4097)^{3/2}}$; normal line $y - 8 = -\frac{1}{64}x$
3. (b) Relative maximum at $x = 0$.
   (c) $x = \pm\dfrac{4\sqrt{5}}{3}$
   (e) The x axis.
   *(f) $K = -\frac{3}{40}$; normal line $x = 0$
5. (b) No relative extrema.
   (c) $x = 0$
   (e) $x = \pm 1$
   *(f) $K = 0$; normal line $y = -x$
7. (b) No relative extrema.
   (c) No points of inflection.
   (e) $x = 1$, $y = x$
   *(f) $K = \dfrac{2}{5\sqrt{10}}$; normal line $y - 2 = -\frac{1}{3}x$
9. (b) Relative minimum at $x = \frac{2}{9}$.
   (c) No points of inflection.
   (e) No asymptotes.
   *(f) $K = \dfrac{9}{5\sqrt{10}}$; normal line $x = 3y + 12$

13. 5000 square meters
17. 6 square feet
19. A sphere.
21. $x \mapsto \frac{1}{4}x^4 - \frac{8}{3}x^3 + 32x^2 + 8x$
23. $x \mapsto -80/3x$
25. $x \mapsto \frac{2}{9}(1+3x)^{3/2}$
27. 1
29. $-\frac{5}{3}$
31. $-1/2n$

6.1

1. (a) 3
   (b) $\frac{25}{2}$
   (c) 6
3. 0
5. (a) 9
   (b) $\frac{3}{2}$; $\frac{3}{2}$; $\frac{5}{6}$
7. $G'(t) = |t|$

6.2

2. 34,641
5. 10; $8\frac{1}{4}$; 10
7. 44; $43/\sqrt{2}$; $44/\sqrt{2}$
9. 4; 2; 2
11. 1; $\sqrt{2}/2$; $2/3\pi$
13. 51/2
14. 2
16. $\sqrt{\frac{2}{3}}$

6.3

1. (a) $\frac{1}{2}$
   (c) 20
   (e) $53\frac{1}{3}$
   (g) 4
3. $\frac{56}{3}$
4. (a) $\frac{5}{6}$
   (c) $\dfrac{512\sqrt{5}}{15}$
   (e) $\dfrac{512\sqrt{2}}{15}$
5. 9

6.4

1. $\frac{1}{3}\pi r^2$
3. $\frac{2}{3}\pi rx$
7. 12π
9. 64π
11. $31\pi/160$
13. $212\pi/5$
15. $2\pi/3(26\sqrt{6} - 19\sqrt{19})$
17. π
19. $\frac{1}{30}\sqrt{15}$; $\frac{1}{10}$; $dx/dt = 1/\sqrt{100-x}$

6.5

1. $x \mapsto x^3$
3. $x \mapsto -x$
5. $x \mapsto \sin(\cos x)$
*7. $x \mapsto 3x\cos x^3 - \dfrac{\sin x^3}{x^2}$
9. $]-1,1[$; $x \mapsto \dfrac{1}{1-\sqrt{x}}$
11. $]-\infty, +\infty[$; $x^2 - x + 1$
13. 0
15. -52
17. $\frac{7}{48}$
19. $\frac{1}{3}$

6.6

1. $\frac{3}{2}x^2 + C$
3. $-\frac{10}{3}x^3 + x + C$
5. $\frac{1}{3}(x+10)^3 + C$

7. $\frac{125}{7}x^7 + \frac{75}{6}x^6 + 18x^5 + \frac{31}{4}x^4 + 6x^3 + \frac{3}{2}x^2 + x + C$
9. $-x^4/100 + C$
11. $\frac{1}{10}(x+10)^{10} + C$
13. $-\frac{1}{3}\cos(3x+4) + C$
15. π^2
17. π
19. 2π
21. πab

REVIEW EXERCISES FOR CHAPTER 6

2. $\frac{1}{3}x^3 + x^2 - x + C$
3. 0
5. $2x^5$
7. $\cos a + \sin a - (\cos b + \sin b)$
9. $\frac{2}{3}(x^3 - x + 2)^{3/2} + C$
11. $\frac{1}{15}\sin^5(3x) + C$
13. 0
15. 0
17. $\frac{1}{8}$
19. No value, since the integrand is not defined on the interval determined by the limits of integration.
21. $\frac{1}{3}(b^3 - a^3)$
23. $]-\sqrt{\frac{1}{2}\pi}, \sqrt{\frac{1}{2}\pi}[$; $]-\sqrt{\frac{1}{2}\pi}, \sqrt{\frac{1}{2}\pi}[$
25. $]-\infty, +\infty[$; everywhere except at zero
27. $\frac{4}{3}$
29. 12π
31. $\frac{5}{14}\pi$
33. $\pi/54$
35. $\frac{512}{15}\pi$
37. $16\pi/15$
39. $277\pi/7$

7.1

3. (a) $7, \frac{7}{2}, \frac{7}{3}, \frac{7}{4}, \frac{7}{5}$; the limit is 0.
   (c) $0, \frac{3}{5}, \frac{4}{5}, \frac{15}{17}, \frac{12}{13}$; the limit is 1.
   (e) $\frac{1}{4}, \frac{4}{9}, \frac{9}{16}, \frac{16}{25}, \frac{25}{36}$; the limit is 1.
   (g) r, r^2, r^3, r^4, r^5; the limit is 0.
4. (a) $\frac{1}{2}$
   (c) 1
   (e) 1
   (g) $\sqrt{2}/2$
5. (a) Diverges.
   (c) Converges to 1.
   (e) Converges to 1.
10. (a) 1
    (c) 0

7.2

3. $\frac{1}{22}(11^{11} - 10^{11})$
4. $\frac{1}{2}(1 - \cos 4)$
5. $\sqrt{2} - 1$

7.3

1. (a) $19\frac{13}{20}$
   (c) $\frac{2}{3}\left(\left(\frac{\pi+4}{4}\right)^{3/2} - 1\right)$
   (e) $\pi^2/8$

7.4

2. (a) $\frac{1}{6}\pi(17\sqrt{17} - 1)$
   (b) $135\pi/2$
3. (a) $\frac{1}{27}\pi(10\sqrt{10} - 1)$
   (b) $4m\pi(1 + m^2)^{1/2}$
   (c) $2\pi a^2$
4. $m \leqslant \sqrt{3}$

7.5

1. 197/12
3. (a) 40 in lbs
   (b) 120 in lbs

5. $\frac{1}{2}mgR$
7. 210,000 (62.4)
9. $3k/10$, where k is the proportionality constant.

7.6

1. $\frac{1}{3}$
3. $\frac{1}{18}$
5. $\frac{1}{2}\pi$
7. $\frac{3}{2}(\sqrt[3]{9}-1)$
9. Converges.
11. Converges.

REVIEW EXERCISES FOR CHAPTER 7

1. $\frac{46}{3}$
3. $\frac{32}{3}$
5. $\frac{10}{3}$
7. $\frac{1}{2}$
9. $\frac{125}{6}$
11. $\frac{1}{3}$
13. $\frac{49}{12}$
15. $\frac{27}{20}$
17. $\frac{9}{2}(\sqrt[3]{4}) - \frac{9}{2}$
19. $\dfrac{1179\pi}{256}$
21. $\frac{1}{9}\pi(82\sqrt{82}-1)$
23. $\frac{1}{2}\pi^2$
25. 2500π
27. $\frac{8}{3}\pi$
29. $\frac{1}{6}\pi$
31. $16a^3/3$
33. $\frac{1}{15}\pi(88\sqrt{2}+107)$
35. $112\pi a^3/15$
37. 800 mi lbs
39. 187,500 ft lbs

8.1

4. $\frac{1}{6}\pi$, $\frac{2}{3}\pi$, π, $\frac{1}{2}\pi$, 0.6, 0.96, $\frac{1}{3}\pi$
7. (a) $x \mapsto \dfrac{1}{\sqrt{4-x^2}}$
   (c) $x \mapsto \sec^{-1}\left(\dfrac{2}{x}\right) - \dfrac{x}{\sqrt{4-x^2}} - \dfrac{8x}{\sqrt{1+4x^2}}$
   (e) $x \mapsto 0$
8. (a) $\pi/3$
   (c) $-\frac{1}{12}\pi$

8.2

4. (a) $1/x$
   (c) x
5. (a) $4b$
   (c) $2b - 3a$
   (e) $\frac{3}{2}b$
   (g) $2a - b$
6. (a) $\dfrac{2x+2}{x^2+2x}$; $\dfrac{-2(x^2+2x+2)}{(x^2+2x)^2}$
   (c) $\cot x$; $-\csc^2 x$
   (e) $\dfrac{1}{x^4+x}$; $-\dfrac{4x^3+1}{(x^4+x)^2}$
   (g) $\dfrac{-\sin(\ln x)}{x}$; $\dfrac{-\cos(\ln x) + \sin(\ln x)}{x^2}$
   (i) $\ln(a^2+x^2)$; $\dfrac{2x}{a^2+x^2}$
   (k) $e^{2x}(13\cos 3x - 3\sin 3x)$;
   $e^{2x}(17\cos 3x - 45\sin 3x)$
   (m) $e^x\cos(e^x) - e^{2x}\sin(e^x)$;
   $(e^x - e^{3x})\cos(e^x) - 3e^{2x}\sin(e^x)$
   (o) $\frac{1}{2}(e^{-x}-1)^{-1/2}$;
   $\frac{1}{4}e^x(e^x-1)^{-3/2}$
7. (a) $\frac{1}{2}e^{2x} + C$
   (c) $-\frac{4}{3}e^{-3x} + C$
   (e) $e^{\sin x} + C$
   (g) $-\ln(\cos x) + C$
   (i) $2\ln(1+\sqrt{x}) + C$

8.3

1. (a) 0
   (c) $\dfrac{1}{\log_2 3+1}$
   (e) No solution exists.
6. (a) $x \mapsto (\ln 7)\,7^x$
   (c) $x \mapsto (1+\ln 2)\,e^{x(1+\ln 2)}$
   (e) $x \mapsto (k\ln z+m\ln w)\,z^{kx}\,w^{mx}$
   (g) $x \mapsto -e^{x-e^x}$
   (i) $x \mapsto 2x^{(\ln x-1)}\ln x$
7. (a) $8/\ln 3$
   (c) $4/27\ln 3$
11. (a) 2
   (c) 2
   (e) $+\infty\,(a>b),\ 1(a=b),\ 0(a<b)$
   (g) $1/e$
   (i) e^2

8.4

5.

	$\sinh x$	$\cosh x$	$\tanh x$	$\coth x$	$\operatorname{sech} x$	$\operatorname{csch} x$
(a)	$\frac{5}{12}$	$\frac{13}{12}$	$\frac{5}{13}$	$\frac{13}{5}$	$\frac{12}{13}$	$\frac{12}{5}$
(c)	$-\frac{3}{4}$	$\frac{5}{4}$	$-\frac{3}{5}$	$-\frac{5}{3}$	$\frac{4}{5}$	$-\frac{4}{3}$
(e)	$\frac{25}{7}$	$\frac{\sqrt{674}}{7}$	$\frac{25}{\sqrt{674}}$	$\frac{\sqrt{674}}{25}$	$\frac{7}{\sqrt{674}}$	$\frac{7}{25}$

8. (a) $x \mapsto -\sec^2 x\,\operatorname{csch}^2(\tan x)$
   (c) $x \mapsto \dfrac{\sinh x}{2(\cosh x-1)^{1/2}}$
   (e) $x \mapsto (2x^2-2x)^{-1}$
   (g) $x \mapsto -\operatorname{csch} x$
9. (a) $-\frac{1}{3}\cosh^{-3}x + C$
   (c) $2\cosh\sqrt{x} + C$

8.5

1. 3 minutes; $\frac{5}{16}$ milligrams
2. $\approx 21\%$
5. ≈ 17 sec; ≈ 19.7 sec; ≈ 19.9 sec
7. $\approx 40.8°$F
9. $x = 3\cos\frac{1}{2}t - 4\sin\frac{1}{2}t$; 5; $t \approx 1.292$ rad

8.6

2. (a) $(0,2)$
   (c) $(-\frac{1}{2}, -\frac{1}{2}\sqrt{3})$
   (e) $(-\sqrt{2}, -\sqrt{2})$
3. (a) $(-3, \frac{1}{4}\pi+2n\pi)$, $(3, -\frac{3}{4}\pi+2n\pi)$, n an integer
   (c) $(-1, -\frac{1}{4}\pi+2n\pi)$, $(1, \frac{3}{4}\pi+2n\pi)$, n an integer
4. (a) $(\sqrt{2}, -\frac{1}{4}\pi)$
   (c) $(2,0)$
   (e) $(2, -\frac{3}{4}\pi)$
13. (a) $r\cos\theta = 3$
   (c) $r^2\sin 2\theta = 8$
   (e) $r = 2a\sin\theta$
14. (a) $x^2 + y^2 = 4x$
   (c) $x^2 = 2y$
   (e) $y^2 = 4 - 4x$
18. $r = \dfrac{2C}{\cos\theta+1}$ if $C > 0$; $r = \dfrac{2C}{\cos\theta-1}$ if $C < 0$
20. (a) $(0,0)$, $(2, \pm\frac{1}{2}\pi)$, $(1, \pm\frac{2}{3}\pi)$, $(0.44, \pm 141.3°)$

8.7

1. (a) $\frac{1}{4}\pi$
   (c) $\frac{5}{6}\pi$
4. (a) $\frac{1}{3}\pi$
6. $16\sqrt{2} - 5\sqrt{5}$
9. $\sqrt{2} + \ln(\sqrt{2}+1)$
11. $\frac{\pi^3}{48}$
13. $\frac{1}{2}(1-\frac{1}{4}\pi)$
15. $\frac{1}{2}(1+\frac{1}{2}\pi)$
17. $\frac{3}{8}\pi$
19. 2
21. $15\pi/2$
23. $8\pi - 16$

REVIEW EXERCISES FOR CHAPTER 8

9. $\dfrac{2}{\pi(1+x^2)}$; $-\dfrac{4x}{\pi(1+x^2)^2}$
11. $\ln x + 1$; $1/x$
13. $(x^2+1)^{-1/2}$; $-x(x^2+1)^{-3/2}$
15. $1; 0$
17. $\frac{1}{2}\pi \operatorname{sech}^2 x \sec^2(\frac{1}{2}\pi \tanh x)$;
$\frac{1}{2}\pi \operatorname{sech}^2 x \sec^2(\frac{1}{2}\pi \tanh x)$
$\times (\frac{1}{2}\pi \operatorname{sech}^2 x \tan(\frac{1}{2}\pi \tanh x) - \tanh x)$
19. $(1+x^2-x\sqrt{1-x^2}\sin^{-1} x)/(1+x^2)\sqrt{1-x^4}$
21. $\dfrac{12\cos x+15}{5\cos x+4+9\sin^2 x}$
23. $-\csc x$
25. $(\sec^2 x)e^{\tan x}$
27. $(\sin x)^x(\ln(\sin x) + x\cot x)$
29. $(\ln x)^x(\ln(\ln x) + 1/\ln x)$
31. $e^{-x}((1-x)\cosh^{-1}(1-x) - x(x^2-2x)^{-1/2})$
35. $1/\pi$
37. (a) $-\frac{1}{5}\ln(3-5x) + C$
(c) $\frac{1}{3}\ln(x^3-2) + C$
(e) $(2x^2-4)^{-1} + C$
(g) $-e^{-\theta} + C$
39. ≈ 5577 years
41. ≈ 4.6 hours
43. 34 miles
45. (a) $(a, \frac{1}{6}\pi)$, $(a, \frac{5}{6}\pi)$
(b) $r = 0$
(c) $(2, \pm\frac{1}{3}\pi)$, $(2, \pi \pm \frac{1}{3}\pi)$
49. $16a/3$
51. (a) $\sqrt{3} - \frac{1}{3}\pi$
(b) $4a^2$
(c) a^2

9.1

1. $\frac{2}{9}(3x+2)^{3/2} + C$
3. $-\dfrac{1}{2(x-1)^2} + C$
5. $-\frac{1}{3}(16-x^2)^{3/2} + C$
7. $\frac{4}{9}(x^3+2)^{3/4} + C$
9. $\ln|\sin x| + C$
11. $\dfrac{\sec 3x}{3} + C$
13. $\dfrac{(\ln x)^2}{2} + C$
15. $2\ln(1+\sqrt{x}) + C$
17. $-\dfrac{e^{3\cos 2x}}{6} + C$
19. $\frac{1}{4}\ln^2(1+x^2) + C$
21. $\frac{2}{3}$
23. $\frac{8}{3}$
25. $\frac{1}{3}$

9.2

1. $\frac{1}{2}\sin x - \frac{1}{10}\sin 5x + C$
3. $\frac{1}{4}\sin 2x + \frac{1}{12}\sin 6x + C$
5. $-\frac{1}{10}\cos^5 2x + \frac{1}{14}\cos^7 2x + C$
7. $\frac{2}{11}\cos^{11/2} x - \frac{2}{7}\cos^{7/2} x + C$
9. $\frac{1}{2}x - \frac{1}{8}\sin 4x + C$
11. $\frac{3}{8}x - \frac{1}{4}\sin 2x + \frac{1}{32}\sin 4x + C$
13. $-\frac{1}{144}\sin^3 6x + \frac{1}{12}\sin 6x + \frac{5}{16}x + \frac{1}{64}\sin 12x + C$
15. $\frac{1}{12}\tan^4 3x + \frac{1}{18}\tan^6 3x + C$
17. $-\frac{1}{6}\cot^2 3x - \frac{1}{12}\cot^4 3x + C$
19. $\frac{2}{11}\tan^{11/2} x + \frac{4}{7}\tan^{7/2} x + \frac{2}{3}\tan^{3/2} x + C$
23. $\dfrac{32 - 19\sqrt{2}}{20}$

9.3

3. $x\ln x - x + C$
5. $x(\ln x)^2 - 2x\ln x + 2x + C$
7. $-(x^2+2x+2)e^{-x} + C$
9. $\frac{1}{2}x[\sin(\ln x) - \cos(\ln x)] + C$

11. $\frac{1}{3}x^3 \ln x - \frac{1}{9}x^3 + C$
13. $x \tan x - \ln|\sec x| + C$
15. $(\frac{1}{2}x^2 - \frac{1}{4})\sin^{-1} x + \frac{1}{4}x\sqrt{1-x^2} + C$
17. $x \cos^{-1} 2x - \frac{1}{2}\sqrt{1-4x^2} + C$
19. $x \ln(x^2+2) - 2x + 2\sqrt{2}\tan^{-1}\left(\frac{x\sqrt{2}}{2}\right) + C$
23. (a) $-\cos\theta + \frac{1}{3}\cos^3\theta + C$
(b) $-\frac{\sin^2 x \cos x}{3} - \frac{2}{3}\cos\theta + C$
25. $-\frac{1}{4}\left(\frac{\sqrt{3}}{2} + \frac{2\sqrt{3}}{9}\pi^2 + \frac{\pi}{3}\right)$
27. -4π

9.4

3. $\frac{1}{\sqrt{5}}\sin^{-1}(x\sqrt{\tfrac{5}{2}}) + C$
5. $\sqrt{4+x^2} + C$
7. $\sqrt{x^2-9} - 3\sec^{-1}\left(\frac{x}{3}\right) + C$
9. $\frac{1}{3}\ln\left|\frac{\sqrt{x^2-9}}{x+3}\right| + C$
11. $3\ln\left|\frac{3-\sqrt{9-4x^2}}{x}\right| + \sqrt{9-4x^2} + C$
13. $\frac{1}{2}\tan^{-1}\left(\frac{x+1}{2}\right) + C$
15. $\sqrt{x^2-2x+5} + \ln|x-1+\sqrt{x^2-2x+5}| + C$
17. $\frac{x+1}{4(x^2+2x+2)^2} + \frac{3x+3}{8(x^2+2x+2)} + \frac{3}{8}\tan^{-1}(x+1) + C$
19. $\ln(1+\sqrt{2})$
21. $\pi/8$
23. 8π
25. $\frac{1}{\sqrt{3}} - \frac{\pi}{6}$

9.5

1. $\frac{1}{4}\ln\left|\frac{x-2}{x+2}\right| + C$
3. $\frac{x^2}{2} + \ln\frac{|x+1|}{(2x+1)^{3/2}} + C$
5. $\frac{1}{6}\ln|(x+5)^5(x-1)| + C$
7. $-\ln|x+1| + \ln|x+2| - \frac{2}{x+2} + C$
9. $\tan^{-1} x + \frac{1}{2}\ln(x^2+2) + C$
11. $\ln[(x+1)^4(x^2+4)] - 3\tan^{-1}\frac{1}{2}x + C$
13. $\frac{5}{2}\tan^{-1} x + \frac{x}{2(x^2+1)} + C$
15. $2\ln|2x-1| + \frac{1}{x^2+1} - \ln(x^2+1) - \tan^{-1} x + C$

9.6

1. $\sec x - \tan x + x + C$
3. $\ln\left|\frac{\tan\frac{1}{2}x}{1+\tan\frac{1}{2}x}\right| + C$
5. $\frac{2}{3}\tan^{-1}\left(\frac{5\tan\frac{1}{2}x + 4}{3}\right) + C$
7. $2\sqrt{x} + 4\sqrt[4]{x} + \ln(\sqrt[4]{x}-1)^4 + C$
9. $-\frac{2}{45}(1-x^3)^{3/2}(2+3x^3) + C$
11. $2\sqrt{x+2} - 6\ln(3+\sqrt{x+2}) + C$
13. $\frac{\pi}{3\sqrt{3}}$
15. $\frac{11}{2} + \frac{3}{8}\ln\frac{3}{5}$

REVIEW EXERCISES FOR CHAPTER 9

1. $\sqrt{x^2+2x-4} + C$
3. $\sin(\ln x) + C$
5. $5\ln\left|\frac{5-\sqrt{25-x^2}}{x}\right| + \sqrt{25-x^2} + C$
7. $x + \ln|(x+2)(x-4)^4| + C$
9. $-\frac{2}{105}(1-x)^{3/2}(15x^2+12x+8) + C$

11. $\ln|x-2| - \dfrac{2}{x-2} + C$
13. $\ln(e^{2x}+3)^{2/3} - \frac{1}{3}x + C$
15. $\frac{5}{16}x + \frac{1}{2}\sin x + \frac{3}{32}\sin 2x - \frac{1}{24}\sin^3 x + C$
17. $\dfrac{2}{\sqrt{3}}\tan^{-1}\left(\dfrac{2\tan\frac{1}{2}x+1}{\sqrt{3}}\right) + C$
19. $\ln\sqrt{x^2+3} + \tan^{-1}x + C$
21. $\frac{1}{48}\cos^3 2x - \frac{1}{16}\cos 2x + C$
23. $\frac{1}{13}e^{2x}(3\sin 3x + 2\cos 3x) + C$
25. $\dfrac{1}{3e^x} + \dfrac{1}{9}\ln\left|\dfrac{e^x-3}{e^x}\right| + C$
27. $\frac{1}{2}x^2\sin^{-1}x^2 + \frac{1}{2}\sqrt{1-x^4} + C$
29. $\frac{1}{12}(\sin x^2 + \cos x^2)(4+\sin 2x^2) + C$
31. $\ln(x-2+\sqrt{x^2-4x+13}) + C$
33. $\frac{1}{7}\tan^7 x + \frac{1}{5}\tan^5 x + C$
35. $-\frac{1}{3}\ln\left|\dfrac{3+\sqrt{x^2+9}}{x}\right| + C$
37. $x\ln(x^2+1) - 2x + 2\tan^{-1}x + C$
39. $2\sqrt{\sec x} + C$
41. $\frac{1}{8}\ln\left|\dfrac{x-2}{x+2}\right| + \frac{1}{4}\tan^{-1}\frac{1}{2}x + C$
43. $\dfrac{\sqrt{2}}{4}\ln\left|\dfrac{\tan^2\frac{1}{2}x+3-2\sqrt{2}}{\tan^2\frac{1}{2}x+3+2\sqrt{2}}\right| + C$
45. $\frac{1}{3}\tan^3 x + C$
47. $2x^3\sin^{-1}2x + \frac{1}{4}\sqrt{1-4x^2} - \frac{1}{12}(1-4x^2)^{3/2} + C$
49. $\frac{2}{9}(\tan x)^{9/2} + \frac{2}{5}(\tan x)^{5/2} + C$
51. $-\frac{2}{5}(x-1)^{-1} - \frac{4}{25}\ln|x-1| + \frac{2}{25}\ln(x^2+4)$
    $+ \frac{19}{50}\tan^{-1}\frac{1}{2}x + C$
53. $-\sin x - \csc x + C$
55. 2
57. $\dfrac{1}{2e}$
59. π
61. $\dfrac{1}{\ln 2}$
63. 4π
65. $\frac{1}{4}\pi$

10.1

1. (a) 35/6
   (c) Divergent.
   (e) Divergent.
2. (a) 68/111
   (c) 10/3

10.2

1. Divergent.
3. Convergent.
5. Convergent.
7. Divergent.
9. Convergent.
11. Convergent.
13. Convergent.
15. Divergent.
17. Divergent.

10.3

1. (a) Convergent.
   (c) Convergent.
   (e) Divergent.
   (g) Divergent.
   (i) Convergent.
2. (a) Absolutely convergent.
   (c) Conditionally convergent.
   (e) Absolutely convergent.
   (g) Absolutely convergent.
   (i) Absolutely convergent.
   (k) Conditionally convergent.

10.4

1. (a) 1
   (c) 1
   (e) ∞
2. (a) $]-1,1]$
   (c) $]-1,1[$
3. (a) $[2,4]$
   (c) $[-2,0[$
   (e) $[0,4[$
   (g) $]-6,10[$

4. (a) $[0,4[$

   (b) $\sum_{n=1}^{\infty}\frac{(x-2)^{n-1}}{2^n}$; $]0,4[$

8. (a) $\sum_{n=0}^{\infty}(n+1)(-1)^n x^n$

   (b) $\sum_{n=0}^{\infty}\frac{(-1)^n x^{2n+1}}{2n+1}, |x|<1$

10.5

2. (a) $\sum_{n=0}^{\infty}(n+1)x^n$

   (c) $\sum_{n=1}^{\infty}\frac{(-1)^{n-1}(x-1)^n}{n}$

   (e) $\sum_{n=1}^{\infty}\frac{(-1)^{n-1}x^{2n-1}}{2n-1}$

3. (a) $T_4(x) = 1 - x^2 + x^4$

   (c) $T_3(x) = 2 + 2\sqrt{3}(x-\frac{1}{3}\pi) + 7(x-\frac{1}{3}\pi)^2 + \frac{23\sqrt{3}}{3}(x-\frac{1}{3}\pi)^3$

6. $y = 2 + x^2 + \frac{x^4}{12} + \cdots + \frac{2x^{2n}}{(2n)!} + \cdots$

8. (a) $1 + bx + \frac{(bx)^2}{2!} + \frac{(bx)^3}{3!} + \cdots;\ ]-\infty,+\infty[$

   (c) $\ln(b+x) = \ln b + \frac{x}{b} - \frac{x^2}{2b^2} + \frac{x^3}{3b^3} - \cdots + \frac{(-1)^{n-1}x^n}{nb^n} + \cdots;\ ]-b,b]$

   (e) $\sin 3x = 3x - \frac{(3x)^3}{3!} + \frac{(3x)^5}{5!} - \frac{(3x)^7}{7!} + \cdots;$ $]-\infty,+\infty[$

REVIEW EXERCISES FOR CHAPTER 10

11. 1
13. $\frac{1}{4}$
19. Divergent.
21. Convergent.
23. Convergent.
25. Divergent.
27. Convergent.
29. Convergent.
31. Absolutely convergent.
33. $]-5,5]$
35. $[-1,1[$
37. $[-1,1]$
39. $]-1,1]$
41. $\ln 3 + \sum_{n=1}^{\infty}\frac{(-1)^{n-1}(x-3)^n}{3^n\cdot n}$
43. $1 + x - \frac{x^3}{3} - \frac{x^4}{6} + \cdots$
45. $e\left[1 + \frac{1}{2}(x-2) + \frac{1}{4}\frac{(x-2)^2}{2!} + \cdots + \frac{1}{2^{n-1}}\frac{(x-2)^{n-1}}{(n-1)!} + \cdots\right];\ ]-\infty,+\infty[$
47. $x + \frac{1}{2}\frac{x^3}{3} + \frac{1\cdot 3}{2\cdot 4}\frac{x^5}{5} + \frac{1\cdot 3\cdot 5}{2\cdot 4\cdot 6}\frac{x^7}{7} + \cdots$

11.1

1. $s(t) = \frac{1}{3}\sqrt{3}\sin\sqrt{3}t$
3. $s(t) = e^t - 1$
5. $q(t) = \mathscr{E}C(1-e^{-t/RC})$
7. $s(t) = e^{-(25/2)t}\left(\cos\frac{\sqrt{75}}{2}t + 2\sqrt{\frac{25}{3}}\sin\frac{\sqrt{75}}{2}t\right)$

11.2

1. $y = e^{-x}$
3. $y = -\frac{1}{24}x^4 + x$
4. $y = e^{-\frac{1}{2}x}(3\cos\frac{1}{2}\sqrt{3}x + 3\sin\frac{1}{2}\sqrt{3}x)$

11.3

1. $y = \frac{1}{2}\left(C+\frac{B}{x^2}\right)$, B a constant
3. $e^y = C + (3x+4)e^{-x}$
5. $r = C(\cos 2\theta)^{1/2}$
7. $2y^2(C+\sin^{-1}x) = 1$
9. $y = C_1 \ln|x| + C_2$
11. $y = C_2 + C_1 \ln(x+\sqrt{1+x^2})$
13. $y = Cx$
15. $y = (1-x^2)^{1/2} - \ln\left(\frac{1-\sqrt{1-x^2}}{x}\right)$
16. (a) $2\tan^{-1}(y/x) + \ln(x^2+y^2) = C$
    (c) $\ln|x| + e^{-y/x} = C$

11.4

1. $y = x^2(2+\ln|x|) - x$
3. $y = \frac{1}{3}x^2(x^2+1) - \frac{2}{15}(x^2+1)^2 + \frac{17}{15}(x^2+1)^{-1/2}$
5. $y\cosh x = C - e^x$
7. $y = xe^{\sin x}$
8. (b) 1 . $1\frac{2}{3}\%$
9. 0.29%

11.5

1. $y = C_1 + C_2 e^{-2x}$
3. $y = C_1 \sin wx + C_2 \cos wx$
5. $y = e^{3x}(C_1 + C_2 x)$
7. $y = e^{-x-\frac{1}{4}\pi}(\cos x - \sin x)$
9. $y = -\cos x \ln|\sec x + \tan x| + C_1 \cos x + C_2 \sin x$
11. $y = e^{-2x}(C_1 \cos x + C_2 \sin x) + \frac{1}{5}x + \frac{6}{25}$
13. $y = e^{-2x}(C_1 + C_2 x) + x^2 e^{-2x} - \frac{1}{8}\cos 2x$
15. $x = C_1 \cos at + C_2 \sin at + \frac{E}{a^2-\rho^2}\cos \rho t$

11.6

1. $$y = C_1\left[1 + \sum_{k=1}^{\infty} \frac{1\cdot 4\cdot\dots\cdot(3k-2) \times (-2)1\cdot\dots\cdot(3k-5)}{(3k)!}x^{3k}\right] + C_2\left[x + \sum_{k=1}^{\infty}\frac{2\cdot 5\cdot\dots\cdot(3k-1) \times(-1)\cdot 2\cdot\dots\cdot(3k-4)}{(3k+1)!}x^{3k+1}\right]$$
2. $$y = C_1\left[1 + \sum_{k=1}^{\infty}\frac{(-1)^{k+1}x^{3k}}{12^k(k!)(3k-1)}\right] + C_2 x$$
3. $$y = C_1(1+\tfrac{1}{2}x^2) + C_2\left(x + \sum_{k=1}^{\infty}\frac{[(-1)\cdot 1\cdot 3\cdot\dots\cdot(2k-3)]^2}{4^k(2k+1)!}x^{2k+1}\right)$$
5. $$y = C_1\left[1 - \frac{\rho(\rho+1)x^2}{2!} + \frac{\rho(\rho+1)(\rho-2)(\rho+3)x^4}{4!} - \dots + \dots\right] + C_2\left[x - \frac{(\rho-1)(\rho+2)}{3!}x^3 + \frac{(\rho-1)(\rho+2)(\rho-3)(\rho+4)}{5!}x^5 - \dots + \dots\right]$$
7. (a) $]-\infty, +\infty[$; e^{x^2}
   (b) $]-\infty, +\infty[$; $e^x - x - 1$

REVIEW EXERCISES FOR CHAPTER 11

1. $\ln(y^2-1) + x^2 = C$
3. $y = C_1 e^{-2x} + C_2 e^x + \frac{1}{4}e^{2x}$
5. $y^2 = \sin x - \cos x + Ce^{-x}$
7. $y^3 = x - 2 + Ce^{-x}$
11. $1/y = 2 - x^2 + Ce^{-\frac{1}{2}x^2}$
13. (a) $y = x^{\frac{1}{2}(1-a)}[C_1 u_1(\ln x) + C_2 u_2(\ln x)]$, where $u_1(x)$ and $u_2(x)$ are determined by the sign of $4b-(a-1)^2$.
    (b) $4b < a^2 - 2a + 1$ where a is an odd integer.
14. The kth cup contains $Q(1-e_k/e)$ cubic inches of wine where $e_k = \sum_{n=1}^{k} 1/n!$
16. (b) $y = x - \frac{e^{\frac{1}{2}x^4}}{C + \int_0^x t^2 e^{\frac{1}{2}t^4}\,dt}$

12.1

1. $p(x) = x^4 + x$
2. $p(x) = x^3 + 2x^2 + x + 1$
5. $h(x) = x^3 + 3x^2 + x + 1$
6. $h(x) = x^5 + x$
7. Four decimal places.

12.2

2. 5.67
4. 3.13
6. 1.1
8. $\dfrac{8011}{10200}$
18. 0.53; 0.5111; 0.510955; 0.510828
19. 1.5; 1.22; 1.21; 1.205

12.3

1. ≈ 0.510822 ($n = 3$); ≈ 0.51082555 ($n = 4$)
3. 0.32087
4. ≈ 1.26 ($n = 2$); ≈ 1.258 ($n = 5$)

12.4

3. (a) ≈ 0.85211
   (b) ≈ 0.98398
   (c) ≈ 0.93910
7. 0.0032; 0.0044; 0.0071; 0.0135; 0.0827
10. 2π or six years, one hundred days.

12.5

1. $y_4^* = 0.6359,\ y_4 = 0.66255$
2. $y_4^* = 0.6862,\ y_4 = 0.6667$
5. The exact solution is $y = 2x^3 - y^3$
6. The exact solution is $y = e^{-x}$.
7. The exact solution is $x = \frac{1}{2}y + 1/2y$.
8. The exact solution is $x = -y(1+y^2)^{-1}$.

REVIEW EXERCISES FOR CHAPTER 12

1. $p(x) = x^3 + x^2 + 10$
2. $p(x) = x^6 + x$
3. $p(x) = x^2 + x + 1$
4. $h(x) = x^3 + x^2 + 10$
5. $h(x) = x^5$
6. $h(x) = x^3 + x^2 + x + 1$
7. $f(3.927) \approx 0.75449$
11. (d) 1.95289
12. $n = 0$
13. 0.8358
15. 0.4298
17. ≈ 0.32080505; ≈ 0.32080559
18. $y = x^3 + x^2 \ln x$
19. $y = e^{-x} - e^{-2x}$
20. $r = (\cos 2u)^{1/2}$

13.1

1. $x = 0$ and $z = 0$; $x = 0$ and $y = 0$; $y = 0$; $x = 0$
5. $(-21, 23) - (4, 6) = (-25, 17)$
7. $(8a, -2b, 13c) = (52, 12, 11) + \frac{1}{2}(16a - 104, -4b - 24, 26c - 22)$
9. $24\boldsymbol{i}$
11. $(2\alpha, 7\alpha + 2\beta, 7\beta)$, where $\alpha, \beta \in \boldsymbol{R}$
13. $(-1+2t)\boldsymbol{i} - \boldsymbol{j} + (-1+4t)\boldsymbol{k}$

15. There are four possible parallelograms: $(\boldsymbol{v}_0+\lambda\boldsymbol{a})+(\boldsymbol{v}_0+\beta\boldsymbol{b})$ where $\boldsymbol{v}_0=(x_0,y_0,z_0)$:
 (i) $0\leqslant\lambda\leqslant 1$ and $0\leqslant\beta\leqslant 1$
 (ii) $-1\leqslant\lambda\leqslant 0$ and $-1\leqslant\beta\leqslant 0$
 (iii) $-1\leqslant\lambda\leqslant 0$ and $0\leqslant\beta\leqslant 1$
 (iv) $0\leqslant\lambda\leqslant 1$ and $-1\leqslant\beta\leqslant 0$.

13.2

3. $9°$
5. 75
7. $|\boldsymbol{u}|=\sqrt{5},\quad |\boldsymbol{v}|=\sqrt{2},\quad \boldsymbol{u}\cdot\boldsymbol{v}=-3$
9. $|\boldsymbol{u}|=\sqrt{11},\quad |\boldsymbol{v}|=\sqrt{62},\quad \boldsymbol{u}\cdot\boldsymbol{v}=-14$
11. $|\boldsymbol{u}|=\sqrt{14},\quad |\boldsymbol{v}|=\sqrt{26},\quad \boldsymbol{u}\cdot\boldsymbol{v}=-17$

13.3

3. $-3\boldsymbol{i}+\boldsymbol{j}+5\boldsymbol{k}$
5. $\sqrt{35}$
7. $\dfrac{1}{\sqrt{1+t^2}}(\boldsymbol{i}-\boldsymbol{j}+t\boldsymbol{k})$ where t is a real number.
9. $\pm\dfrac{1}{\sqrt{23567}}(113\boldsymbol{i}+17\boldsymbol{j}-103\boldsymbol{k})$
 $\approx\pm\dfrac{1}{153}(113\boldsymbol{i}+17\boldsymbol{j}-103\boldsymbol{k})$; these are the only ones.
11. $\dfrac{\sqrt{2}}{13}$
13. $\boldsymbol{u}+\boldsymbol{v}=3\boldsymbol{i}-3\boldsymbol{j}+3\boldsymbol{k},\quad \boldsymbol{u}\cdot\boldsymbol{v}=6,\quad |\boldsymbol{u}|=\sqrt{6},$
 $|\boldsymbol{v}|=3,\quad \boldsymbol{u}\times\boldsymbol{v}=-3\boldsymbol{i}+\boldsymbol{k}$

13.4

1. $\boldsymbol{\sigma}'(t)=(2\pi\cos 2\pi t,-2\pi\sin 2\pi t,2-2t)$;
 $\boldsymbol{\sigma}'(0)=(2\pi,0,2)$
3. $\boldsymbol{\sigma}'(t)=(2t,35^2-4,0)$; $\boldsymbol{\sigma}'(0)=(0,-4,0)$
5. $d\boldsymbol{s}/dt=e^t\boldsymbol{i}-e^{-t}\boldsymbol{j}+\boldsymbol{k}$;
 $\ln(|d\boldsymbol{s}/dt|)=\frac{1}{2}\ln(e^{2t}+e^{-2t}+1)$
7. $d\boldsymbol{s}/dt=\boldsymbol{i}+2t\boldsymbol{j}+3t^2\boldsymbol{k}$;
 $\ln(|d\boldsymbol{s}/dt|)=\frac{1}{2}\ln(1+4t^2+9t^4)$
9. $d\boldsymbol{s}/dt=6\boldsymbol{i}+6t\boldsymbol{j}+3t^2\boldsymbol{k}$; $\boldsymbol{a}(t)=6\boldsymbol{j}+6t\boldsymbol{k}$;
 $d\boldsymbol{s}/dt(0)=6\boldsymbol{i}$
11. $d\boldsymbol{\sigma}/dt=\left(2\cos 2t,\dfrac{1}{1+t},1\right)$;
 $\boldsymbol{a}(t)=\left(-4\sin 2t,\dfrac{-1}{(1+t)^2},0\right)$; $d\boldsymbol{\sigma}/dt(0)=(1,1,1)$
13. $d\boldsymbol{\sigma}/dt=(t\cos t+\sin t,\cos t-t\sin t,\sqrt{3})$;
 $d\boldsymbol{\sigma}/dt(0)=(0,1,\sqrt{3})$;
 $\boldsymbol{a}(t)=(2\cos t-t\sin t,-2\sin t-t\cos t,0)$
15. $d\boldsymbol{s}/dt=(1,1,t^{1/2})$; $\boldsymbol{a}(t)=(0,0,\frac{1}{2}t^{-1/2})$;
 $d\boldsymbol{s}/dt(9)=(1,1,3)$
17. $4\sqrt{2}-2$
19. $\sqrt{21}+\frac{5}{4}\sinh^{-1}(4/\sqrt{5})$
21. $2\left(e-\dfrac{1}{e}\right)$

13.5

1. $\int_{\boldsymbol{\sigma}}F(x,y,z)\,ds=\int_I F(\sigma_1(t),\sigma_2(t),\sigma_3(t))\cdot|\boldsymbol{\sigma}'(t)|\,ds$,
 so if $F=1$ then
 $F(\sigma_1(t),\sigma_2(t),\sigma_3(t))=1$. Hence,
 $\int_{\boldsymbol{\sigma}}F(x,y,z)\,ds=\int_I|\boldsymbol{\sigma}'(t)|\,ds=l(\boldsymbol{\sigma})$.
3. $-\dfrac{1}{3}\left(1+\dfrac{1}{e^2}\right)^{3/2}+\dfrac{1}{3}(2)^{3/2}$
5. (a) $\dfrac{5\sqrt{5}-1}{6\sqrt{5}+6\log(\sqrt{5}+2)}$
 (b) 1/4

REVIEW EXERCISES FOR CHAPTER 13

1. (a) $\sqrt{2}(e^{2\pi}-1)$
 (b) 2π
 (c) $\dfrac{5}{2}\sqrt{\dfrac{1}{26}}+\dfrac{1}{50}\sinh^{-1}(5)$
 (d) 2

3. $\log 2$
5. (a) $35\boldsymbol{i}+\boldsymbol{j}+t\boldsymbol{k}$
   (b) $\boldsymbol{j}+t\boldsymbol{k}$
9. $$\begin{aligned}\boldsymbol{a}\cdot\boldsymbol{u} &= (\alpha\boldsymbol{u}+\beta\boldsymbol{v}+\gamma\boldsymbol{w})\cdot\boldsymbol{u}\\ &= \alpha(\boldsymbol{u}\cdot\boldsymbol{u})+\beta(\boldsymbol{v}\cdot\boldsymbol{u})+\gamma(\boldsymbol{w}\cdot\boldsymbol{u})\\ &= \alpha\cdot 1+0+0=\alpha\end{aligned}$$
   $$\begin{aligned}\boldsymbol{a}\cdot\boldsymbol{v} &= (\alpha\boldsymbol{u}+\beta\boldsymbol{v}+\gamma\boldsymbol{w})\cdot\boldsymbol{v}\\ &= \alpha(\boldsymbol{u}\cdot\boldsymbol{v})+\beta(\boldsymbol{v}\cdot\boldsymbol{v})+\gamma(\boldsymbol{w}\cdot\boldsymbol{v})\\ &= 0+\beta\cdot 1+0=\beta\end{aligned}$$
   $$\begin{aligned}\boldsymbol{a}\cdot\boldsymbol{w} &= (\alpha\boldsymbol{u}+\beta\boldsymbol{v}+\gamma\boldsymbol{w})\cdot\boldsymbol{w}\\ &= \alpha(\boldsymbol{u}\cdot\boldsymbol{w})+\beta(\boldsymbol{v}\cdot\boldsymbol{w})+\gamma(\boldsymbol{w}\cdot\boldsymbol{w})\\ &= 0+0+1\cdot\gamma=\gamma\end{aligned}$$
   Geometrically, $\boldsymbol{a}=\boldsymbol{v}(\boldsymbol{a}\ \boldsymbol{u},\boldsymbol{a}\ \boldsymbol{v},\boldsymbol{a}\ \boldsymbol{w})$, i.e., the point where $\boldsymbol{a}$ terminates is just $(\boldsymbol{a}\ \boldsymbol{u},\boldsymbol{a}\ \boldsymbol{v},\boldsymbol{a}\ \boldsymbol{w})$.
11. 1
13. (a) $\sqrt{26}$
    (b) $2\pi\sqrt{26}$
19. $14/\sqrt{66}$
21. $\boldsymbol{\sigma}(t)=(t^2,t^4)$, $\boldsymbol{\sigma}'(0)=(0,0)$, and $\boldsymbol{\sigma}$ represents the parabola $y=x^2$ which has the line $y=0$ as its tangent at $(0,0)$.

14.1

1. If $b_1\neq b_2$ let $\varepsilon=|b_1-b_2|$. Let δ_i be such that $0<|v-v_0|<\delta_i$ implies that $|f(v)-b_i|<\varepsilon/3$. Then
   $$\begin{aligned}|b_1-b_2| &= |f(v)-b_2+b_1-f(v)|\\ &\leqslant |f(v)-b_2|+|f(v)-b_1|\end{aligned}$$
   If $|v-v_0|<\min(\delta_1,\delta_2)$ then continuing the inequality we get
   $$|b_1-b_2|\leqslant\varepsilon/3+\varepsilon/3=2\varepsilon/3=\tfrac{2}{3}|b_1-b_2|$$
   Therefore, we have $1\leqslant 2/3$, a contradiction.
11. 0

*13. 0

14.2

1. (a) $\dfrac{\partial f}{\partial x}=y,\quad \dfrac{\partial f}{\partial y}=x$
   (b) $\dfrac{\partial f}{\partial x}=ye^{xy},\quad \dfrac{\partial f}{\partial y}=xe^{xy}$
   (c) $\dfrac{\partial f}{\partial x}=\cos x\cos y-x\sin x\cos y,\quad \dfrac{\partial f}{\partial y}=-x\cos x$
   $\sin y$
   (d) $\dfrac{\partial f}{\partial x}=2x\log(x^2+y^2)+2x=2x\log e(x^2+y^2),$
   $\dfrac{\partial f}{\partial y}=2y\log e(x^2+y^2)$
3. (a) $\dfrac{\partial w}{\partial x}=\exp(x^2+y^2)+2x^2\exp(x^2+y^2),$
   $\dfrac{\partial w}{\partial y}=2xy\exp(x^2+y^2)$
   (b) $\dfrac{\partial w}{\partial x}=-\dfrac{4xy^2}{(x^2-y^2)^2},\quad \dfrac{\partial w}{\partial y}=\dfrac{4x^2y}{(x^2-y^2)^2}$
   (c) $\dfrac{\partial w}{\partial x}=ye^{xy}\log(x^2+y^2)+\dfrac{2xe^{xy}}{x^2+y^2},$
   $\dfrac{\partial w}{\partial y}=xe^{xy}\log(x^2+y^2)+\dfrac{2ye^{xy}}{x^2+y^2}$
   (d) $\dfrac{\partial w}{\partial x}=\dfrac{1}{y},\quad \dfrac{\partial w}{\partial y}=-\dfrac{x}{y^2}$
   (e) $\dfrac{\partial w}{\partial x}=-y^2e^{xy}\sin(ye^{xy})\sin x+\cos(ye^{xy})\cos x$
   $\dfrac{\partial w}{\partial y}=-[e^{xy}+yxe^{xy}]\sin(ye^{xy})\sin x$
5. (a) f is C^1 on its domain $\boldsymbol{R}^2-(0,0)$.
   (b) f is C^1 on its domain $\boldsymbol{R}^2-$coordinate axes.
   (c) f is C^1 on all of $\boldsymbol{R}^2$.
   (d) f is C^1 on $\boldsymbol{R}^2-(0,0)$.
   (e) f is C^1 on $\boldsymbol{R}^2-(0,0)$.

14.3

1. $e(\cos 1-\sin 1)$
3. $(e^{2t}-e^{-2t})\log(e^{2t}-e^{-2t})+(e^{2t}+e^{-2t})$
5. (a) $\dfrac{\partial f}{\partial x}(0,0)=\lim\limits_{\Delta x\to 0}\dfrac{f(\Delta x,0)-f(0,0)}{\Delta x}=0$
   $\dfrac{\partial f}{\partial y}(0,0)=\lim\limits_{\Delta y\to 0}\dfrac{f(0,\Delta y)-f(0,0)}{\Delta y}=0$
11. 0

14.4

1. -5
3. $e/\sqrt{5}$
5. $(17/13)e^e$
7. (a) The unique α is given by $\alpha = y_0 - x_0^2$.
   (c) The P_α are the level curves of f.

14.5

1. (a) $\nabla f(x,y,z) = \dfrac{1}{\sqrt{x^2+y^2+z^2}}(x\boldsymbol{i}+y\boldsymbol{j}+z\boldsymbol{k})$
   (b) $\nabla f(x,y,z) = (y+z)\boldsymbol{i} + (x+z)\boldsymbol{j} + (x+y)\boldsymbol{k}$
   (c) $\nabla f(x,y,z) = \dfrac{-2}{(x^2+y^2+z^2)^2}(x\boldsymbol{i}+y\boldsymbol{j}+z\boldsymbol{k})$
3. $108e^{18}$
5. $15/\sqrt{14}$
7. (a) $\nabla \cdot \boldsymbol{F} = 3$
   (b) $\nabla \cdot \boldsymbol{F} = 0$
   (c) $\nabla \cdot \boldsymbol{F} = 6x + 8y + 10z$
9. Only (a) need have zero divergence.
13. (a)
$$\operatorname{curl} \boldsymbol{F} = \begin{vmatrix} \boldsymbol{i} & \boldsymbol{j} & \boldsymbol{k} \\ \dfrac{\partial}{\partial x} & \dfrac{\partial}{\partial y} & \dfrac{\partial}{\partial z} \\ 3x^2y & (x^3+y^3) & 0 \end{vmatrix} = \boldsymbol{0}$$
   (b) $f(x,y) = x^3y + y^4/4$
   (c) If $\boldsymbol{F} = \nabla f$, then $\operatorname{curl} f = \operatorname{curl} \nabla f = \boldsymbol{0}$ by formula 10, Table 5–1.
15. $\dfrac{1}{3\sqrt{3}}(5\boldsymbol{i}+\boldsymbol{j}+\boldsymbol{k})$

14.6

1. $z = -9(x-1) + 6(y-2) - 3$
3. $z = 1$
5. $z = \dfrac{1}{\sqrt{2}}(x-1) + \dfrac{1}{\sqrt{2}}(y-1) + \sqrt{2}$
7. $f(x,y) = xy + c$, c a constant

14.7

1. $z = 2(y-1) + 1$
3. $6(z-1) + 4(y+2) - (x-13) = 0$

*5. $\cos u \cos v\,\boldsymbol{i} + \sin u \cos v\,\boldsymbol{j} - \sin v\,\boldsymbol{k}$

14.8

1. $(0,0)$ is a saddle point.
3. The critical points are the line $y = -x$ and they are local minima because $f \geqslant 0$ everywhere but equals zero on this line.
5. $(0,0)$ is a saddle point.
7. $(-\frac{1}{4}, -\frac{1}{4})$ is a local minimum.
9. $(0,0)$ is a local maximum but the test fails because all partial derivatives are zero at $(0,0)$. However, for all (x,y) close to $(0,0)$, $x^2+y^2 \geqslant 0$; therefore, $\cos(x^2+y^2) \leqslant 1 = \cos 0$ for all (x,y) close to $(0,0)$ and it follows that $(0,0)$ is a local maximum.
13. If $A = 0$, then the critical points of f are all of $\boldsymbol{R}^2$. If $A \neq 0$, the critical points consist of the y axis. They are local minima if $A > 0$ and local maxima if $A < 0$.

14.9

1. $S =$ unit disc minus its center.
3. $D = [0,5] \times [0,1]$
5. D is the set of (x,y,z) with $x^2+y^2+z^2 \leqslant 1$ (the unit ball). $\boldsymbol{F}$ is not one-to-one but is one-to-one on $]0,1] \times]0,\pi[\times]0,2\pi]$.

REVIEW EXERCISES FOR CHAPTER 14

3. (a) 0
   (b) The limit does not exist.
5. (a) $\dfrac{\partial f}{\partial x} = e^z - y\sin x$, $\dfrac{\partial f}{\partial y} = \cos x$, $\dfrac{\partial f}{\partial z} = xe^z$

(b) $\dfrac{\partial f}{\partial x} = \dfrac{\partial f}{\partial y} = \dfrac{\partial f}{\partial z} = 10(x+y+z)^9$

(c) $\dfrac{\partial f}{\partial x} = \dfrac{2x}{z}, \quad \dfrac{\partial f}{\partial y} = \dfrac{1}{z}, \quad \dfrac{\partial f}{\partial z} = -\dfrac{(x^2+y)}{z^2}$

7. (a) $\dfrac{\partial f}{\partial x} = \arctan\left(\dfrac{x}{y}\right) + \dfrac{xy}{x^2+y^2}, \quad \dfrac{\partial f}{\partial y} = \dfrac{-x}{x^2+y^2}$

(b) $\dfrac{\partial f}{\partial x} = \dfrac{-x}{\sqrt{x^2+y^2}} \sin\sqrt{x^2+y^2},$

$\dfrac{\partial f}{\partial y} = \dfrac{-y}{\sqrt{x^2+y^2}} \sin\sqrt{x^2+y^2}$

(c) $\dfrac{\partial f}{\partial x} = -2xe^{-x^2-y^2}, \quad \dfrac{\partial f}{\partial y} = -2ye^{-x^2-y^2}$

9. (a) $G'(t) = 2t\log(1+t^4)$

(b) $G'(t) = e^{-t^3}\cos 2t + \int_0^t -2txe^{-t^2x}\cos(t+x)\,dx$
$- \int_0^t e^{-t^2x}\sin(1+x)\,dx$

13. (a) 2
(b) 0
(c) 12

15. (a) $\nabla f(x,y,z) = yz^2\boldsymbol{i} + xz^2\boldsymbol{j} + 2xyz\,\boldsymbol{k}$
(b) $\nabla \times \boldsymbol{F} = (z-y)\boldsymbol{i} - x\boldsymbol{k}$
(c) $\nabla \times (f\boldsymbol{F}) = (-x^2z^2+xyz^3-xy^2z^2)\boldsymbol{i}$
$+ (-3xyz^2-2xy^2z)\boldsymbol{j}$
$+ (xyz^2-xz^3-x^2z^2y)\boldsymbol{k}$

17. (a) $z = 0$
(b) $z = 3(x-1)+2$
(c) $z = 10(x-3)+10(y-2)+25$

19. (a) $z = 3(x-1)+2$
(b) $z = \dfrac{2}{\sqrt{3}}(x-1) + \sqrt{3}$

15.1

1. (a) $\frac{2}{3}$
(b) 2π
(c) 1
(d) $\frac{1}{2} - \ln 2$

15.2

3. (a) $\frac{2}{3}$
(b) $e-2$
(c) $\frac{1}{9}\sin 1$

4. $\frac{8}{15}$

15.3

1. (a) $\frac{1}{3}$
(b) $\frac{5}{2}$
(c) $\dfrac{e^2-1}{4}$
(d) $\frac{1}{35}$

5. 28000 cu ft

6. 30625/144

8. 33/140; D is the region $0 \leqslant x \leqslant 1$, $0 \leqslant y \leqslant x^2$

10. 50π

14. No.

15.4

1. (a) $\frac{1}{8}$
(b) $\dfrac{\pi}{4}$
(c) $-\frac{17}{12}$
(d) $G(b) - G(a)$ where $\partial G/\partial y = F(y,y) - F(a,y)$ and $\partial F/\partial x = f(x,y)$

2. (a) $\frac{4}{3}$
(c) $3-e$

6. $\frac{4}{3}\pi abc$

8. To prove the Extreme Value Theorem for continuous $f: D \mapsto \boldsymbol{R}$, proceed one variable at a time. Suppose D is of type 1; then consider $x \mapsto M(x)$, where $M(x_0)$ is the maximum of $y \mapsto f(x_0, y)$ on $[\phi_1(x_0), \phi_2(x_0)]$. Show $M: x \mapsto M(x)$ is continuous on $[a,b]$ and take its maximum. For the Intermediate Value Theorem, note that any two points in D can be connected by a continuous curve lying in D.

15.5

1. (a) 4
(c) $\frac{3}{16}$

2. (b) $\dfrac{1}{4}\left(1 - \dfrac{1}{e}\right)$

15.6

1. $\frac{1}{3}$
5. $-\frac{1}{6}$
8. $\frac{4\pi}{3}(1-\sqrt{2})$

15.7

1. $2\pi(e^{1/2}-1)$
5. (a) $T(u,v) = (1-u, v^2)$
   (b) $\frac{1}{60}$
7. $\frac{2\pi}{5}$

15.8

4. (a) A cylinder, a plane through the origin perpendicular to the xy plane, a plane parallel to the xy plane.
   (b) A sphere, a plane through the origin perpendicular to the xy plane, a cone.
8. $x\tan(x^2+y^2+z^2) = y$
10. The integral reduces to $2\pi\int_0^1 \rho^2(2+\rho^2)^{1/2}\,d\rho$.
11. $\frac{2}{9}\pi(1-(a^3)^{3/2}+(1+a^3)^{3/2})$

REVIEW EXERCISES FOR CHAPTER 15

1. (c) $\frac{1}{2}e^2-1$
4. $\frac{\pi}{3}(4\sqrt{2}-\frac{1}{2})$
6. 2π

*12. The integral $\int_D f$ does not exist. Note that if we require $0 < a \leqslant x$, the integral converges.

16.1

2. For $-\pi/2 \leqslant \phi \leqslant \pi/2$, we are computing twice the area of the upper hemisphere; for $0 \leqslant \phi \leqslant 2\pi$, we are computing twice the area of the sphere.
9. $\frac{\pi}{2}\sqrt{6}$

16.2

5. $\frac{\sqrt{14}}{6}$

16.3

1. $(\frac{1}{2}, 0, \frac{5}{6})$
3. $\frac{1}{\frac{2}{3}\pi a^3+1}(r, 0, \frac{1}{4}\pi a^4)$
5. $\bar{x} = \bar{y} = 0, \quad \bar{z} = \frac{2}{3}$

16.4

10. In the notation of exercise 9 with $f(x) \geqslant 0$, $V = 2\pi\bar{y}A$ where A is the area under $y = f(x)$ and $\bar{y}$ is the center of mass of the region.
11. 2π

REVIEW EXERCISES FOR CHAPTER 16

1. $\sqrt{3}$
2. $2\sqrt{3}$
5. 2π
16. $M = 8\pi^2\int_0^1 \delta(r)\,dr$

17.1

1. (a) $\frac{3}{2}$
   (b) 0
3. 9
6. 34/15
7. $\frac{3}{4} - (n-1)/(n+1)$
10. The length of $\boldsymbol{\sigma}$.
13. 0
14. 7

17.2

1. $2\pi a^2$
3. $3\pi/2$ (Evaluate the double integral using polar coordinates.)
5. (a) $3\pi(a^4 - b^4)/4$
9. $-\frac{1}{4}$
12. $-\frac{5}{6}$
15. (b) $9\pi/4$

17.3

1. 2π
3. 0
6. For a cylinder with radius R,
   $\int_S F = R\int_{z=a}^{b}\int_{\theta=0}^{2\pi} F_r \, d\theta \, dz$
9. The orientations on $\partial S_1 = \partial S_2$ must agree.
14. (a) 0
    (b) π

17.4

2. (a) 37/10
   (b) Yes.
3. $x^2yz - \cos x$
5. $1/|r_0|$
7. No.
11. $e \sin 1 + e^3/3 - \frac{1}{3}$
13. 59 ergs
15. $\boldsymbol{G} = (-yz - z, -xz, 0)$

17.5

2. 3π
3. 1
5. 0
6. $3\pi/2$
8. 3π
15. The field inside is zero.

REVIEW EXERCISES FOR CHAPTER 17

1. $\pi a^2/2$
4. (a) $\nabla\phi = \boldsymbol{m}/r^3 - (\boldsymbol{m}\cdot\boldsymbol{r})\boldsymbol{r}/r^5$
   (b) $\nabla^2\phi = 0$
   (c) $\nabla\cdot\boldsymbol{F} = 0$
   (d) $\nabla\times\boldsymbol{F} = 0$
7. (a) 4π
   (b) 0
8. 0
10. $\frac{1}{6}$
11. Yes.
12. (a) $f = x^4/4 - x^2y^3$
    (b) $-\frac{1}{4}$
13. (b) $f = 3x^2y\cos z$
    (c) 0
14. 23/6
17. Yes; $f = a_1yz - a_2xz + a_3xy$

INDEX

D

E

F

J

K

L

M

T

U

V

W

2
B 3
C 4
D 5
E 6
F 7
G 8
H 9
I 0
J 1